FE

Rapid Preparation for the Chemical Fundamentals of Engineering Exam

CHEMICAL
REVIEW MANUAL

Pass The Exam—Guaranteed

Michael R. Lindeburg, PE

The Power to Pass®
www.ppi2pass.com

Professional Publications, Inc. • Belmont, California

Report Errors and View Corrections for This Book

Everyone benefits when you report typos and other errata, comment on existing content, and suggest new content and other improvements. You will receive a response to your submission; other engineers will be able to update their books; and, PPI will be able to correct and reprint the content for future readers.

PPI provides two easy ways for you to make a contribution. If you are reviewing content on **feprep.com**, click on the "Report a Content Error" button in the Help and Support area of the footer. If you are using a printed copy of this book, go to **ppi2pass.com/errata**. To view confirmed errata, go to **ppi2pass.com/errata**.

FE Chemical Review Manual

Current printing of this edition: 1

Printing History

date	edition number	printing number	update
May 2016	1	1	New book.

© 2016 Professional Publications, Inc. All rights reserved.

All content is copyrighted by Professional Publications, Inc. (PPI). No part, either text or image, may be used for any purpose other than personal use. Reproduction, modification, storage in a retrieval system or retransmission, in any form or by any means, electronic, mechanical, or otherwise, for reasons other than personal use, without prior written permission from the publisher is strictly prohibited. For written permission, contact PPI at permissions@ppi2pass.com.

Printed in the United States of America.

PPI
1250 Fifth Avenue
Belmont, CA 94002
(650) 593-9119
ppi2pass.com

ISBN: 978-1-59126-445-3

Library of Congress Control Number: 2016932944

F E D C B A

PPI's Guarantee

This *FE Chemical Review Manual* is your best choice to prepare for the Chemical Fundamentals of Engineering (FE) examination. It is the only review manual that

- covers every Chemical FE exam knowledge area
- is based on the NCEES *FE Reference Handbook* (*NCEES Handbook*)
- provides example questions in true exam format
- provides instructional material for essentially every relevant equation, figure, and table in the *NCEES Handbook*
- can be accessed online at **feprep.com**

PPI is confident that if you use this book conscientiously to prepare for the Chemical FE exam, following the guidelines described in the "How to Use This Book" section, you'll pass the exam. Otherwise, regardless of where you purchased this book, with no questions asked, we will refund the purchase price of your printed book (up to PPI's published website price).

To request a refund, you must provide the following items within three months of taking the exam:

1. A summary letter listing your name, email address, and mailing address
2. Your original packing slip, store sales receipt, or online order acknowledgment showing the price you paid
3. A dated email, notification letter, or printout of your MyNCEES webpage showing that you did not pass the FE exam
4. Your book

Mail all items to:

PPI
FE Chemical Review Manual Refund
1250 Fifth Avenue
Belmont, CA 94002

This guarantee does not extend to other products, packages, or bundled products. Guarantees do not apply to web access books. To be eligible for a refund, the price of this book must be individually listed on your receipt. Packages and bundles may be covered by their own guarantees.

The Power to Pass®
www.ppi2pass.com

Topics

Mathematics

Chem. Reaction Engineering

Probability/ Statistics

Mass Transfer

Fluid Mechanics/ Dynamics

Unit Processes

Thermodynamics

Process Control

Chemistry

Process Design/ Economics

Heat Transfer

Safety/Health/ Environment

Engineering Sciences

Computational Tools

Materials Science

Ethics/ Prof. Practice

Table of Contents

Preface

The purpose of this book is to prepare you for the National Council of Examiners for Engineering and Surveying (NCEES) Fundamentals of Engineering (FE) exam.

In 2014, the NCEES adopted revised specifications for the exam. The council also transitioned from a paper-based version of the exam to a computer-based testing (CBT) version. The FE exam now requires you to sit in front of a monitor, solve problems served up by the CBT system, access an electronic reference document, and perform your scratch calculations on a reusable note-pad. You may also use an on-screen calculator with which you will likely be unfamiliar. The experience of taking the FE exam will probably be unlike anything you have ever, or will ever again, experience in your career. Similarly, preparing for the exam will be unlike preparing for any other exam.

The CBT FE exam presented three new challenges to me when I began preparing instructional material for it. (1) The subjects in the testable body of knowledge are oddly limited and do not represent a complete cross section of the traditional engineering fundamentals subjects. (2) The NCEES *FE Reference Handbook* (*NCEES Handbook*) is poorly organized, awkwardly formatted, inconsistent in presentation, and idiomatic in convention. (3) Traditional studying, doing homework while working toward a degree, and working at your own desk as a career engineer are poor preparations for the CBT exam experience.

No existing exam review book overcomes all of these challenges. But I wanted you to have something that does. So, in order to prepare you for the CBT FE exam, this book was designed and written from the ground up. In many ways, this book is as unconventional as the exam.

This book covers all of the knowledge areas listed in the NCEES Chemical FE exam specifications. For all practical purposes, this book contains the equivalent of all of the equations, tables, and figures presented in the *NCEES Handbook*, ninth edition (June 2015 revision) that you will need for the Chemical FE exam. And, with the exceptions listed in the "Variables" section, for better or worse, this book duplicates the terms, variables, and formatting of the *NCEES Handbook* equations.

NCEES has selected what it believes to be all of the engineering fundamentals important to an early-career, minimally-qualified engineer, and has distilled them into its single reference, the *NCEES Handbook*. Personally, I cannot accept the premise that engineers learn and use so little engineering while getting their degrees and during their first few career years. However, regardless of whether you accept the NCEES subset of engineering fundamentals, one thing is certain: In serving as your sole source of formulas, theory, methods, and data during the exam, the *NCEES Handbook* severely limits the types of problems that can be included in the FE exam.

The obsolete paper-based exam required very little knowledge outside of what was presented in the previous editions of the *NCEES Handbook*. That *NCEES Handbook* supported a plug-and-chug examinee performance within a constrained body of knowledge. Based on the current FE exam specifications and the *NCEES Handbook*, the CBT FE exam is even more limited than the old paper-based exam. The number (breadth) of knowledge areas, the coverage (depth) of knowledge areas, the number of problems, and the duration of the exam are all significantly reduced. If you are only concerned about passing and/or "getting it over with" before graduation, these reductions are all in your favor. Your only deterrents will be the cost of the exam and the inconvenience of finding a time and place to take it.

Accepting that "it is what it is," I designed this book to guide you through the exam's body of knowledge.

I have several admissions to make: (1) This book contains nothing magical or illicit. (2) This book, by itself, is only one part of a complete preparation. (3) This book stops well short of being perfect. What do I mean by those admissions?

First, this book does not contain anything magical. It's called a "review" manual, and you might even learn something new from it. It will save you time in assembling review material and questions. However, it won't learn the material for you. Merely owning it is not enough. You will have to put in the time to use it.

Similarly, there is nothing clandestine or unethical about this book. It does not contain any actual exam problems. It was written in a vacuum, based entirely on the NCEES Chemical FE exam specifications. This book is not based on feedback from actual examinees.

Truthfully, I expect that many exam problems will be similar to the problems I have used, because NCEES and I developed content with the same set of constraints. (If anything, NCEES is even more constrained when it comes to fringe, outlier, eccentric, or original topics.)

There are a finite number of ways that problems involving Ohm's law ($V = IR$) and Newton's second law of motion ($F = ma$) can be structured. Any similarity between problems in this book and problems in the exam is easily attributed to the limited number of engineering formulas and concepts, the shallowness of the coverage, and the need to keep the entire solution process (reading, researching, calculating, and responding) to less than three minutes for each problem.

Let me give an example to put some flesh on the bones. As any competent engineer can attest, in order to calculate the pressure drop in a pipe network, you would normally have to (1) determine fluid density and viscosity based on the temperature, (2) convert the mass flow rate to a volumetric flow rate, (3) determine the pipe diameter from the pipe size designation (e.g., pipe schedule), (4) calculate the internal pipe area, (5) calculate the flow velocity, (6) determine the specific roughness from the conduit material, (7) calculate the relative roughness, (8) calculate the Reynolds number, (9) calculate or determine the friction factor graphically, (10) determine the equivalent length of fittings and other minor losses, (11) calculate the head loss, and finally, (12) convert the head loss to pressure drop. Length, flow quantity, and fluid property conversions typically add even more complexity. (SSU viscosity? Diameter in inches? Flow rate in SCFM?) As reasonable and conventional as that solution process is, a problem of such complexity is beyond the upper time limit for an FE exam problem.

To make it possible to be solved in the time allowed, any exam problem you see is likely to be more limited. In fact, most or all of the information you need to answer a problem will be given to you in its problem statement. If only the real world were so kind!

Second, by itself, this book is inadequate. It was never intended to define the entirety of your preparation activity. While it introduces essentially all of the exam knowledge areas and content in the *NCEES Handbook*, an introduction is only an introduction. To be a thorough review, this book needs augmentation.

By design, this book has three significant inadequacies.

1. This book has a limited number of pages, so it cannot contain enough of everything for everyone. The number of example problems that can fit in it is also limited. The number of problems needed by you, personally, to come up to speed in a particular subject may be inadequate. For example, how many problems will you have to review in order to feel comfortable about divergence, curl, differential equations, and linear algebra? (Answer: Probably more than are in all the books you will ever own!) So, additional exposure is inevitable if you want to be adequately prepared in every subject.

2. This book does not contain the *NCEES Handbook*, per se. This book is limited in helping you become familiar with the idiosyncratic sequencing, formatting, variables, omissions, and presentation of topics in the *NCEES Handbook*. The only way to remedy this is to obtain your own copy of the *NCEES Handbook* (available in printed format from PPI and as a free download from the NCEES website) and use it in conjunction with your review.

3. This book does not contain a practice examination (mock exam, sample exam, etc.). With the advent of the CBT format, any sample exam in printed format is little more than another collection of practice problems. The actual FE exam is taken sitting in front of a computer using an online reference book, so the only way to practice is to sit in front of a computer while you answer problems. Using an online reference is very different from the work environment experienced by most engineers, and it will take some getting used to.

Third, and finally, I reluctantly admit that I have never figured out how to write or publish a completely flawless first (or even subsequent) edition. The PPI staff comes pretty close to perfection in the areas of design, editing, typography, and illustrating. Subject matter experts help immensely with calculation checking, and beta testing before you see a book helps smooth out wrinkles. However, I still manage to muck up the content. So, I hope you will "let me have it" when you find my mistakes. PPI has established an easy way for you to report an error, as well as to review changes that resulted from errors that others have submitted. Just go to **ppi2pass.com/errata**. When you submit something, I'll receive it via email. When I answer it, you'll receive a response. We'll both benefit.

Best wishes in your examination experience. Stay in touch!

Michael R. Lindeburg, PE

Acknowledgments

Developing a book specific to the computerized Chemical FE exam has been a monumental project. It involved the usual (from an author's and publisher's standpoint) activities of updating and repurposing existing content and writing new content. However, the project was made extraordinarily more difficult by two factors: (1) a new book design, and (2) the publication schedule.

PPI staff members have had a lot of things to say about this book during its development. In reference to you and other examinees being unaware of what PPI staff did, one of the often-heard statements was, "They will never know."

However, I want you to know, so I'm going to tell you.

Editorial project managers Dennis Fitzgerald, Matt Schneiderman, and Sam Webster and associate project managers Thomas Bliss, Sierra Cirimelli-Low, Nicole Evans, and Tracy Katz managed the gargantuan operation, with considerable support from Sarah Hubbard, associate editorial director. Production services manager Cathy Schrott kept the process moving smoothly and swiftly, despite technical difficulties that seemed determined to stall the process at every opportunity. Serena Cavanaugh, acquisitions editor, arranged for all of the outside subject matter experts who were involved with this book. All of the content was eventually reviewed for consistency, PPI style, and accuracy by Jennifer Lindeburg King, associate editor-in-chief.

Though everyone in PD&I has a specialty, this project pulled everyone from his or her comfort zone. The entire staff worked on "building" the chapters of this book from scratch, piecing together existing content with new content. Everyone learned (with amazing speed) how to grapple with the complexities of XML and MathML while wrestling misbehaving computer code into submission. Tom Bergstrom, technical illustrator, and Kate Hayes, production associate (retired), updated existing illustrations and created new ones. They also paginated and made corrections. Copy editors David Chu, Hilary Flood, Brian Gonzalez, Tyler Hayes, Richard Iriye, Julia Lopez, Scott Marley, Ellen Nordman, Ceridwen Quattrin, Heather Turbeville, and Ian A. Walker copy edited, proofread, corrected, and paginated.

Paying customers (such as you) shouldn't have to be test pilots. So, close to the end of the process, when content was starting to coalesce out of the shapelessness of the PPI content management system, several subject matter experts became crash car dummies "for the good of engineering." They pretended to be examinees and worked through all of the content, looking for calculation errors, references that went nowhere, and logic that was incomprehensible. These engineers and their knowledge area contributions are: Prajesh Gongal (Chemical Reaction Engineering; Mass Transfer; Process Design and Economics), Eric C. Huang (Fluid Mechanics/Dynamics; Chemical Reaction Engineering), Gennaro Maffia (Chemical Reaction Engineering; Mass Transfer; Process Design and Economics), N. S. Nandagopal, PE (Materials Science; Process Control; Process Design and Economics; Safety, Health, and Environment; Ethics and Professional Practice), Dr. John R. Richards, PE (Mathematics; Fluid Mechanics/Dynamics; Chemistry; Heat Transfer; Chemical Reaction Engineering; Mass Transfer; Unit Processes; Process Control; Process Design and Economics; Safety, Health, and Environment), Kodi Jean Verhalen, PE (Mass Transfer; Process Control; Safety, Health, and Environment; Ethics and Professional Practice), and L. Adam Williamson, PE (Chemical Reaction Engineering).

In addition to their own contributions to the validity of the chemical engineering chapters, N. S. Nandagopal, PE and Dr. John R. Richards, PE served as "super reviewers." They consolidated the comments received from various PhD reviewers and industry subject matter experts, adjudicating the differences and validating criticisms and suggestions. The result is that the material in this book is not only correct, it is complete, logical, and easier to understand.

In addition to the subject matter experts who reviewed the content remotely, several subject matter experts came to PPI's office to perform secondary, in-house technical reviews of the new chemical engineering content. These engineers and their knowledge area contributions are: Suraj Bhaskar (Mathematics; Probability and Statistics; Chemical Reaction Engineering; Mass Transfer; Unit Processes; Process Control; Process Design and Economics), Nathan N. Knapp (Fluid Mechanics/Dynamics; Materials Science; Chemical Reaction Engineering; Mass Transfer; Unit Processes; Process Control; Process Design and Economics), Bryan Li (Heat Transfer; Chemical Reaction Engineering; Mass Transfer), Richard Luna (Mathematics; Probability and Statistics; Fluid Mechanics/Dynamics; Chemistry; Heat Transfer; Materials Science; Chemical Reaction Engineering; Mass Transfer; Process Control; Process Design and Economics; Safety, Health, and Environment), and Qudus Omotayo Lawal (Process Design and Economics).

Consistent with the past 38 years, I continue to thank my wife, Elizabeth, for accepting and participating in a writer's life that is full to overflowing. Even though our children have been out on their own for a long time, we seem to have even less time than we had before. As a corollary to Aristotle's "Nature abhors a vacuum," I propose: "Work expands to fill the void."

To my granddaughter, Sydney, who had to share her Grumpus with his writing, I say, "I only worked when you were taking your naps. And besides, you hog the bed!"

I also appreciate the grant of permission to reproduce materials from several other publishers. In each case, attribution is provided where the material has been included. Neither PPI nor the publishers of the reproduced material make any representations or warranties as to the accuracy of the material, nor are they liable for any damages resulting from its use.

Thank you, everyone! I'm really proud of what you've accomplished. Your efforts will be pleasing to examinees and effective in preparing them for the Chemical FE exam.

Michael R. Lindeburg, PE

Codes and References Used to Prepare This Book

This book is based on the NCEES *FE Reference Handbook* (*NCEES Handbook*), ninth edition (June 2015 revision). The other documents, codes, and standards that were used to prepare this book were the most current available at the time.

NCEES does not specifically tie the FE exam to any edition (version) of any code or standard. Rather than make the FE exam subject to the vagaries of such codes and standards as are published by the American Chemical Society (ACS), the American Concrete Institute (ACI), the American Institute of Chemical Engineers (AIChE), the American Institute of Steel Construction (AISC), the American National Standards Institute (ANSI), the American Society of Civil Engineers (ASCE), the American Society of Heating, Refrigerating and Air-Conditioning Engineers (ASHRAE), the American Society of Mechanical Engineers (ASME), ASTM International (ASTM), the International Code Council (ICC), the Institute of Electrical and Electronic Engineers (IEEE), the National Fire Protection Association (NFPA), and so on, NCEES effectively writes its own "code," the *NCEES Handbook*.

Most surely, every standard- or code-dependent concept (e.g., flammability) in the *NCEES Handbook* can be traced back to some section of some edition of a standard or code (e.g., 29CFR). So, it would be logical to conclude that you need to be familiar with everything (the limitations, surrounding sections, and commentary) in the code related to that concept. However, that does not seem to be the case. The *NCEES Handbook* is a code unto itself, and you won't need to study the parent documents. Nor will you need to know anything pertaining to related, adjacent, similar, or parallel code concepts. For example, although square concrete columns are covered in the *NCEES Handbook*, round columns are not.

Therefore, although methods and content in the *NCEES Handbook* can be ultimately traced back to some edition (version) of a relevant code, you don't need to know which. You don't need to know whether that content is current, limited in intended application, or relevant. You only need to use the content.

Introduction

PART 1: ABOUT THIS BOOK

This book is intended to guide you through the Chemical Fundamentals of Engineering (FE) examination body of knowledge and the idiosyncrasies of the National Council of Examiners for Engineers and Surveyors (NCEES) *FE Reference Handbook* (*NCEES Handbook*). This book is not intended as a reference book, because you cannot use it while taking the FE examination. The only reference you may use is the *NCEES Handbook*. However, the *NCEES Handbook* is not intended as a teaching tool, nor is it an easy document to use. The *NCEES Handbook* was never intended to be something you study or learn from, or to have value as anything other than an exam-day compilation. Many of its features may distract you because they differ from what you were expecting, were exposed to, or what you currently use.

To effectively use the *NCEES Handbook*, you must become familiar with its features, no matter how odd they may seem. *FE Chemical Review Manual* will help you become familiar with the format, layout, organization, and odd conventions of the *NCEES Handbook*. This book, which displays the *NCEES Handbook* material in blue for easy identification, satisfies two important needs: it is (1) something to learn from, and (2) something to help you become familiar with the *NCEES Handbook*.

Organization

This book is organized into topics (e.g., "Unit Processes") that correspond to the knowledge areas listed by NCEES in its Chemical FE exam specifications. However, unlike the *NCEES Handbook*, this book arranges subtopics into chapters (e.g., "Solid-Liquid Processes") that build logically on one another. Each chapter contains sections (e.g., "Reflux Ratio") organized around *NCEES Handbook* equations, but again, the arrangement of those equations is based on logical development, not the *NCEES Handbook*. Equations that are presented together in this book may actually be many pages apart in the *NCEES Handbook*.

The presentation of each subtopic or related group of equations uses similar components and follows a specific sequence. The components of a typical subtopic are:

- general section title
- background and developmental content
- equation name (or description) and equation number

- equation with *NCEES Handbook* formatting
- any relevant variations of the equation
- any values typically associated with the equation
- additional explanation and development
- worked quantitative example using the *NCEES Handbook* equation
- footnotes

Not all sections contain all of these features. Some features may be omitted if they are not needed. For example, "$g = 9.81$ m/s^2" would be a typical value associated with the equation $W = mg$. There would be no typical values associated with the equation $F = ma$.

Much of the information in this book and in the *NCEES Handbook* is relevant to more than one knowledge area or subtopic. For example, equations related to the Fluid Mechanics and Dynamics of Liquids knowledge area also pertain to Fluid Mechanics and Dynamics of Gases. Many Strength of Materials concepts correlate with Statics subtopics. The index will help you locate all information related to any of the topics or subtopics you wish to review.

Content

This book presents equations, figures, tables, and other data equivalent to those given in the *NCEES Handbook*. For example, the *NCEES Handbook* includes tables for conversion factors, material properties, and areas and centroids of geometric shapes, so this book provides equivalent tables. Occasionally, a redundant element of the *NCEES Handbook*, or some item having no value to examinees, has been omitted.

Some elements, primarily figures and tables, that were originally published by authoritative third parties (and for whom reproduction permission has been granted) have been reprinted exactly as they appear in the *NCEES Handbook*. Other elements have been editorially and artistically reformulated, but they remain equivalent in utility to the originals.

Colors

Due to the selective nature of topics included in the *NCEES Handbook*, coverage of some topics in the *NCEES Handbook* may be incomplete. This book aims to offer more comprehensive coverage, and so, it contains material that is not covered in the *NCEES Handbook*.

This book uses color to differentiate between what is available to you during the exam, and what is supplementary content that makes a topic more interesting or easier to understand. Anything that closely parallels or duplicates the *NCEES Handbook* is printed in blue. Headings that introduce content related to *NCEES Handbook* equations are printed in blue. Titles of figures and tables that are essentially the same as in the *NCEES Handbook* are similarly printed in blue. Headings that introduce sections, equations, figures, and tables that are NOT in the *NCEES Handbook* are printed in **black**. The **black** content is background, preliminary and supporting material, explanations, extensions to theory, and application rules that are generally missing from the *NCEES Handbook*.

Numbering

The equations, figures, and tables in the *NCEES Handbook* are unnumbered. All equations, figures, and tables in this book include unique numbers provided to help you navigate through the content.

You will find many equations in this book that have no numbers and are printed in **black**, not blue. These equations represent instructional materials which are often missing pieces or interim results not presented in the *NCEES Handbook*. In some cases, the material was present in the eighth edition of the *NCEES Handbook*, but is absent in the ninth edition. In some cases, I included instruction in deleted content. (This book does not contain all of the deleted eighth edition *NCEES Handbook* content, however.)

Equation and Variable Names

This book generally uses the *NCEES Handbook* terminology and naming conventions, giving standard, normal, and customary alternatives within parentheses or footnotes. For example, the *NCEES Handbook* refers to what is commonly known as the Bernoulli equation as the "energy equation." This book acknowledges the *NCEES Handbook* terminology when introducing the equation, but uses the term "Bernoulli equation" thereafter.

Variables

This book makes every effort to include the *NCEES Handbook* equations exactly as they appear in the *NCEES Handbook*. While any symbol can be defined to represent any quantity, in many cases, the *NCEES Handbook*'s choice of variables will be dissimilar to what most engineers are accustomed to. For example, although there is no concept of weight in the SI system, the *NCEES Handbook* defines W as the symbol for weight with units of newtons. While engineers are comfortable with E, E_k, KE, and U representing kinetic energy, after introducing KE in its introductory pages, the *NCEES Handbook* uses T (which is used sparingly by some scientists) for kinetic energy. The *NCEES Handbook* designates power as $\dot{W}$ instead of P. Because you have to be familiar with them, this book reluctantly follows all of those conventions.

This book generally follows the *NCEES Handbook* convention regarding use of italic fonts, even when doing so results in ambiguity. For example, as used by the *NCEES Handbook*, aspect ratio, AR, is indistinguishable from $A \times R$, area times radius. Occasionally, the *NCEES Handbook* is inconsistent in how it represents a particular variable, or in some sections, it drops the italic font entirely and presents all of its variables in roman font. This book maintains the publishing convention of showing all variables as italic.

There are a few important differences between the ways the *NCEES Handbook* and this book present content. These differences are intentional for the purpose of maintaining clarity and following PPI's publication policies.

- *pressure:* The *NCEES Handbook* primarily uses P for pressure, an atypical engineering convention. This book always uses p so as to differentiate it from P, which is reserved for power, momentum, and axial loading in related chapters.

- *velocity:* The *NCEES Handbook* uses v and occasionally Greek nu, ν, for velocity. This book always uses v to differentiate it from Greek upsilon, υ, which represents specific volume in some topics (e.g., thermodynamics), and Greek nu, ν, which represents absolute viscosity and Poisson's ratio.

- *specific volume:* The *NCEES Handbook* uses v for specific volume. This book always uses Greek upsilon, υ, a convention that most engineers will be familiar with.

- *units:* The *NCEES Handbook* and the FE exam generally do not emphasize the difference between pounds-mass and pounds-force. "Pounds" ("lb") can mean either force or mass. This book always distinguishes between pounds-force (lbf) and pounds-mass (lbm).

Distinction Between Mass and Weight

The *NCEES Handbook* specifies the unit weight of water, γ_w, as 9.810 kN/m^3. This book follows that convention but takes every opportunity to point out that there is no concept of weight in the SI system.

Equation Formatting

The *NCEES Handbook* writes out many multilevel equations as an awkward string of characters on a single line, using a plethora of parentheses and square and curly brackets to indicate the precedence of mathematical operations. So, this book does also. However, in examples using the equations, this book reverts to normal publication style after presenting the base equation styled as it is in the *NCEES Handbook*. The change in style will show you the equations as the *NCEES Handbook* presents them, while presenting the calculations in a normal and customary typographic manner.

Footnotes

I have tried to anticipate the kinds of questions about this book and the *NCEES Handbook* that an instructor would be asked in class. Footnotes are used in this book as the preferred method of answering those questions and of drawing your attention to features in the *NCEES Handbook* that may confuse, confound, and infuriate you. Basically, *NCEES Handbook* conventions are used within the body of this book, and any inconsistencies, oddities and unconventionalities, and occasionally, even errors, are pointed out in the footnotes.

If you know the NCEES knowledge areas backward as well as forward, many of the issues pointed out in the footnotes will seem obvious. However, if you have only a superficial knowledge of the knowledge areas, the footnotes will answer many of your questions. The footnotes are intended to be factual and helpful.

Indexed Terms

The print version of this book contains an index with thousands of terms. The index will help you quickly find just what you are looking for, as well as identify related concepts and content.

PART 2: HOW YOU CAN USE THIS BOOK

IF YOU ARE A STUDENT

In reference to Isaac Asimov's *Foundation and Empire* trilogy, you'll soon experience a Seldon crisis. Given all the factors (the exam you're taking, what you learned as a student, how much time you have before the exam, and your own personality), the behaviors (strategies made evident through action) required of you will be self-evident.

Here are some of those strategies.

Get the NCEES *FE Reference Handbook*

Get a copy of the *NCEES Handbook*. Use it as you read through this book. You will want to know the sequence of the sections, what data is included, and the approximate locations of important figures and tables in the *NCEES Handbook*. You should also know the terminology (words and phrases) used in the *NCEES Handbook* to describe equations or subjects, because those are the terms you will have to look up during the exam.

The *NCEES Handbook* is available both in printed and PDF format. The index of the print version may help you locate an equation or other information you are looking for, but few terms are indexed thoroughly. The PDF version includes search functionality that is similar to what you'll have available when taking the computer-based exam. In order to find something using the PDF search function, your search term will have to match the content exactly (including punctuation).

Diagnose Yourself

Use the diagnostic exams in this book to determine how much you should study in the various knowledge areas. You can use diagnostic exams (and other assessments) in two ways: take them before you begin studying to determine which subjects you should emphasize, or take them after you finish studying to determine if you are ready to move on.

Make a Schedule

In order to complete your review of all examination subjects, you must develop and adhere to a review schedule. If you are not taking a live review course (where the order of your preparation is determined by the lectures), you'll want to prepare your own schedule. If you want to pencil out a schedule on paper, a blank study schedule template is provided at the end of this Introduction.

The amount of material in each chapter of this book was designed to fit into a practical schedule. You should be able to review one chapter of the book each day. There are 58 chapters and 16 diagnostic exams in this book. So, you need at least 74 study days. This requires you to treat every day the same and work through weekends.

If you'd rather take all the weekends off and otherwise stick with the one-chapter-per-study-day concept, you will have to begin approximately 128 days before the exam. Use the off days to rest, review, and study problems from other books. If you are pressed for time or get behind schedule, you don't have to take the days off. That will be your choice.

Near the exam date, give yourself a week to take a realistic practice exam, to remedy any weaknesses it exposes, and to recover from the whole ordeal.

Work Through Everything

NCEES has greatly reduced the number of subjects about which you are expected to be knowledgeable and has made nothing optional. Skipping your weakest subjects is no longer a viable preparation strategy. You should study all examination knowledge areas, not just your specialty areas. That means you study every chapter in this book and skip nothing. Do not limit the number of chapters you study in hopes of finding enough problems in your areas of expertise to pass the exam.

Be Thorough

Being thorough means really doing the work. Read the material, don't skim it. Solve each numerical example using your calculator. Read through the solution, and refer back to the equations, figures, and tables it references.

Don't jump into answering problems without first reviewing the instructional text in this book. Unlike reference books that you skim or merely refer to when needed, this book requires you to read everything. That reading is going to be your only review. Reading the instructional text is a "high value" activity. There isn't much text to read in the first place, so the value per word is high. There aren't any derivations or proofs, so the text is useful. Everything in blue titled sections is in the *NCEES Handbook*, so it has a high probability of showing up on the exam.

Work Problems

You have less than an average of three minutes to answer each problem on the exam. You must be able to quickly recall solution procedures, formulas, and important data. You will not have time to derive solution methods—you must know them instinctively. The best way to develop fast recall is to work as many practice problems as you can find.

Finish Strong

There will be physical demands on your body during the examination. It is very difficult to remain alert, focused, and attentive for six hours or more. Unfortunately, the more time you study, the less time you have to maintain your physical condition. Thus, most examinees arrive at the examination site in high mental condition but in deteriorated physical condition. While preparing for the FE exam is not the only good reason for embarking on a physical conditioning program, it can serve as a good incentive to get in shape.

Claim Your Reward

As Hari Seldon often said in Isaac Asimov's *Foundation and Empire* trilogy, the outcome of your actions will be inevitable.

IF YOU ARE AN INSTRUCTOR

CBT Challenges

The computer-based testing (CBT) FE exam format, content, and frequent administration present several challenges to teaching a live review course. Some of the challenges are insurmountable to almost all review courses. Live review courses cannot be offered year round, a different curriculum is required for each engineering discipline, and a hard-copy, in-class mock exam taken at the end of the course no longer prepares examinees for the CBT experience. The best that instructors can do is to be honest about the limitations of their courses, and to refer examinees to any other compatible resources.

Many of the standard, tried-and-true features of live FE review courses are functionally obsolete. These obsolete features include general lectures that cover "everything," complex numerical examples with more than two or three simple steps, instructor-prepared handouts containing notes and lists of reference materials, and a hard-copy mock exam. As beneficial as those features were in the past, they are no longer best commercial practice for the CBT FE exam. However, they may still be used and provide value to examinees.

This book parallels the content of the *NCEES Handbook* and, with the exceptions listed in this Introduction, uses the same terminology and nomenclature. The figures and tables are equivalent to those in the *NCEES Handbook*. You can feel confident that I had your students and the success of your course in mind when I designed this book.

Instruction for Multiple Exams

Since this book is intended to be used by those studying for the Chemical FE exam, there is no easy way to use it as the basis for more than a Chemical FE exam review course.

Historically, most commercial review courses (taken primarily by engineers who already have their degrees) prepared examinees for the Other Disciplines FE examination. That is probably the only logical (practical, sustainable, etc.) course of action, even now. Few commercial review course providers have the large customer base and diverse instructors needed to offer simultaneous courses for every discipline.

University review courses frequently combine students from multiple disciplines, focusing the review course content on the core overlapping concepts and the topics covered by the Other Disciplines FE exam. The change in the FE exam scope has made it more challenging than ever to adequately prepare a diverse student group.

If you are tasked with teaching a course to examinees who are taking more than one exam, contact PPI at **feprep.com/instruct** for guidelines and suggestions. For information about all of the individual FE exams, the most recent information is posted and updated at **feprep.com/faqs**.

Lectures

Your lectures should duplicate what the examinees would be doing in a self-directed review program. That means walking through each chapter in this book in its entirety. You're basically guiding a tour through the book. By covering everything in this book, you'll cover everything on the exam.

Handouts

Everything you do in a lecture should be tied back to the *NCEES Handbook*. You will be doing your students a great disservice if you get them accustomed to using your course handouts or notes to solve problems. They can't use your notes in the exam, so train them to use the only reference they are allowed to use.

NCEES allows that the exam may require broader knowledge than the *NCEES Handbook* contains. However, there are very few areas that require formulas not present in the *NCEES Handbook*. Therefore, you shouldn't deviate too much from the subject matter of each chapter.

Homework

Students like to see and work a lot of problems. They experience great reassurance in working exam-like problems and finding out how easy the problems are. Repetition and reinforcement should come from working additional problems, not from more lecture.

It is unlikely that your students will be working to capacity if their work is limited to what is in this book. You will have to provide or direct your students to more problems in order to help them effectively master the concepts you will be teaching.

Schedule

I have found that a 15-week format works best for a live FE exam review course that covers everything and is intended for working engineers who already have their degrees. This schedule allows for one 2- to $2^1/_2$-hour lecture per week, with a 10-minute break each hour.

Table 1 outlines a typical format for a live commercial Chemical FE review course. To some degree, the lectures build upon one another. However, a credible decision can be made to present the knowledge areas in the order they appear in the *NCEES Handbook*.

However, a 15-week course is too long for junior and senior engineering majors still working toward a degree. College students and professors don't have that much time. And, students don't need as thorough of a review as do working engineers who have forgotten more of the fundamentals. College students can get by with the most cursory of reviews in some knowledge areas, such as mathematics, fluid mechanics, and chemistry.

For college students, an 8-week course consisting of six weeks of lectures followed by two weeks of open questions seems appropriate. If possible, two 1-hour lectures per week are more likely to get students to attend than a single 2- or 3-hour lecture per week. The course consists of a comprehensive march through all knowledge areas except mathematics, with the major emphasis being on problem-solving rather than lecture. For current engineering majors, the main goals are to keep the students focused and to wake up their latent memories, not to teach the subjects.

Table 1 Recommended 15-Week Chemical FE Exam Review Course Format for Commercial Review Courses

week	*FE Chemical Review Manual* chapter titles	*FE Chemical Review Manual* chapter numbers
1	Units; Algebra; Vectors; Analytic Geometry; Trigonometry; Linear Algebra; Calculus; Differential Equations; Numerical Methods	1–9
2	Probabililty and Statistics	10
3	Computer Software; Professional Practice; Ethics; Licensure	55–58
4	Fluid Properties; Fluid Statics; Fluid Dynamics; Fluid Measurement and Similitude; Compressible Fluid Dynamics; Fluid Machines	11–16
5	Properties of Substances; Laws of Thermodynamics; Power Cycles and Entropy; Mixtures of Gases, Vapors, and Liquids	17–20
6	Inorganic Chemistry; Combustion; Organic Chemistry; Biochemistry	21–24
7	Conduction; Convection; Radiation; Transport Phenomena	25–28
8	Trusses; Pulleys, Cables, and Friction; Centroids and Moments of Inertia; Indeterminate Statics; Kinematics; Kinetics; Kinetics of Rotational Motion; Energy and Work; Electrostatics; Direct-Current Circuits; Alternating-Current Circuits; Rotating Machines	29–40
9	Material Properties and Testing; Engineering Materials	41–42
10	Reaction Kinetics	43
11	Diffusion	44
12	Vapor-Liquid and Gas-Liquid Separation Processes; Solid-Liquid Processes	45–46
13	Control Loops and Control Systems; Instrumentation and Control System Hardware	47–48
14	Plant and Process Design; Engineering Economics	49–50
15	Toxicology; Industrial Hygiene; Discharge Water Quality; Air Quality	51–54

Table 2 outlines a typical format for a live university review course. The sequence of the lectures is less important for a university review course than for a commercial course, because students will have recent experience in the subjects. Some may actually be enrolled in some of the related courses while you are conducting the review.

I strongly believe in the benefits of exposing all review course participants to a realistic sample examination. Unless you have made arrangements with **feprep.com** for your students to take an online exam, you probably cannot provide them with an experience equivalent to the actual exam. A written take-home exam is better than nothing, but since it will not mimic the exam experience, it must be presented as little more than additional problems to solve.

I no longer recommend an in-class group final exam. It seems inhumane to make students sit for hours into the late evening for the final exam when they could be learning from it in the comfort of their own homes. So, if you are going to use a written mock exam, I recommend distributing it at the first meeting of the review course and assigning it as a take-home exercise.

Table 2 *Recommended 8-Week Chemical FE Exam Review Course Format for University Courses*

class	*FE Chemical Review Manual* chapter titles	*FE Chemical Review Manual* chapter numbers
1	Computer Software; Professional Practice; Ethics; Licensure	55–58
2	Fluid Properties; Fluid Statics; Fluid Dynamics; Fluid Measurement and Similitude; Compressible Fluid Dynamics; Fluid Machines; Properties of Substances; Laws of Thermodynamics; Power Cycles and Entropy; Mixtures of Gases, Vapors, and Liquids	11–20
3	Inorganic Chemistry; Combustion; Organic Chemistry; Biochemistry; Conduction; Convection; Radiation; Transport Phenomena	21–28
4	Trusses; Pulleys, Cables, and Friction; Centroids and Moments of Inertia; Indeterminate Statics; Kinematics; Kinetics; Kinetics of Rotational Motion; Energy and Work; Electrostatics; Direct-Current Circuits; Alternating-Current Circuits; Rotating Machines	29–40
5	Material Properties and Testing; Engineering Materials; Reaction Kinetics; Diffusion; Vapor-Liquid and Gas-Liquid Separation Processes; Solid-Liquid Processes	41–46
6	Control Loops and Control Systems; Instrumentation and Control System Hardware; Plant and Process Design; Engineering Economics; Toxicology; Industrial Hygiene; Discharge Water Quality; Air Quality	47–54
7	open questions	–
8	open questions	–

PART 3: ABOUT THE EXAM

EXAM STRUCTURE

The FE exam is a computer-based test that contains 110 multiple-choice problems given over two consecutive sessions (sections, parts, etc.). Each session contains approximately 55 multiple-choice problems that are grouped together by knowledge area (subject, topic, etc.). The subjects are not explicitly labeled, and the beginning and ending of the subjects are not noted. No subject spans the two exam sessions. That is, if a subject appears in the first session of the exam, it will not appear in the second.

Each problem has four possible answer options, labeled (A), (B), (C), and (D). Only one problem and its answer options are given on-screen at a time. The exam is not adaptive (i.e., your response to one problem has no bearing on the next problem you are given). Even if you answer the first five mathematics problems correctly, you'll still have to answer the sixth problem.

In essence, the FE exam is two separate, partial exams given in sequence. During either session, you cannot view or respond to problems in the other session.

Your exam will include a limited (unknown) number of problems (known as "pretest items") that will not be scored and will not have an impact on your results. NCEES does this to determine the viability of new problems for future exams. You won't know which problems are pretest items. They are not identifiable and are randomly distributed throughout the exam.

EXAM DURATION

The exam is six hours long and includes an 8-minute tutorial, a 25-minute break, and a brief survey at the conclusion of the exam. The total time you'll have to actually answer the exam problems is 5 hours and 20 minutes. The problem-solving pace works out to slightly less than 3 minutes per problem. However, the exam does not pace you. You may spend as much time as you like on each problem. Although the on-screen navigational interface is slightly awkward, you may work through the problems (in that session) in any sequence. If you want to go back and check your answers before you submit a session for grading, you may. However, once you submit a section you are not able to go back and review it.

You can divide your time between the two sessions any way you'd like. That is, if you want to spend 4 hours on the first section, and 1 hour and 20 minutes on the second section, you could do so. Or, if you want to spend 2 hours and 10 minutes on the first section, and 3 hours and 10 minutes on the second section, you could do that instead. Between sessions, you can take a 25-minute break. (You can take less, if you would like.) You cannot work through the break, and the break time cannot be added to the time permitted for either session. Once each session begins, you can leave your seat for personal reasons, but the "clock" does not stop for your absence. Unanswered problems are scored the same as problems answered incorrectly, so you should use the last few minutes of each session to guess at all unanswered problems.

THE NCEES NONDISCLOSURE AGREEMENT

At the beginning of your CBT experience, a nondisclosure agreement will appear on the screen. In order to begin the exam, you must accept the agreement within two minutes. If you do not accept within two minutes, your CBT experience will end, and you will forfeit your appointment and exam fees. The CBT nondisclosure agreement is discussed in the section entitled "Subversion After the Exam." The nondisclosure agreement, as stated in the *NCEES Examinee Guide*, is as follows.

> This exam is confidential and secure, owned and copyrighted by NCEES and protected by the laws of the United States and elsewhere. It is made available to you, the examinee, solely for valid assessment and licensing purposes. In order to take this exam, you must agree not to disclose, publish, reproduce, or transmit this exam, in whole or in part, in any form or by any means, oral or written, electronic or mechanical, for any purpose, without the prior express written permission of NCEES. This includes agreeing not to post or disclose any test questions or answers from this exam, in whole or in part, on any websites, online forums, or chat rooms, or in any other electronic transmissions, at any time.

YOUR EXAM IS UNIQUE

The exam that you take will not be the exam taken by the person sitting next to you. Differences between exams go beyond mere sequencing differences. NCEES says that the CBT system will randomly select different, but equivalent, problems from its database for each examinee using a linear-on-the-fly (LOFT) algorithm. Each examinee will have a unique exam of equivalent difficulty. That translates into each examinee having a slightly different minimum passing score.

So, you may conclude either that many problems are static clones of others, or that NCEES has an immense database of trusted problems with supporting econometric data.[1,2] However, there is no way to determine exactly how NCEES ensures that each examinee is given an equivalent exam. All that can be said is that looking at your neighbor's monitor would be a waste of time.

THE EXAM INTERFACE

The on-screen exam interface contains only minimal navigational tools. On-screen navigation is limited to selecting an answer, advancing to the next problem, going back to the previous problem, and flagging the current problem for later review. The interface also includes a timer, the current problem number (e.g., 45 of 110), a pop-up scientific calculator, and access to an on-screen version of the *NCEES Handbook*.

During the exam, you can advance sequentially through the problems, but you cannot jump to any specific problem, whether or not it has been flagged. After you have completed the last problem in a session, however, the navigation capabilities change, and you are permitted to review problems in any sequence and navigate to flagged problems.

THE *NCEES HANDBOOK* INTERFACE

Examinees are provided with a 24-inch computer monitor that will simultaneously display both the exam problems and a searchable PDF of the *NCEES Handbook*. The PDF's table of contents consists of live links. The search function is capable of finding anything in the *NCEES Handbook*, down to and including individual variables. However, the search function finds only precise search terms (e.g., "Hazenwilliams" will not locate "Hazen-Williams"). Like the printed version of the *NCEES Handbook*, the PDF also contains an index, but its terms and phrases are fairly limited and likely to be of little use.

WHAT IS THE REQUIRED PASSING SCORE?

Scores are based on the total number of problems answered correctly, with no deductions made for problems answered incorrectly. Raw scores may be adjusted slightly, and the adjusted scores are then scaled.

Since each problem has four answer options, the lower bound for a minimum required passing score is the performance generated by random selection, 25%. While it is inevitable that some examinees can score less than 25%, it is more likely that most examinees can score

---

[1]The FE exam draws upon a simple database of finished problems. The CBT system does not construct each examinee's problems from a set of "master" problems using randomly generated values for each problem parameter constrained to predetermined ranges.
[2]Problems used in the now-obsolete paper-and-pencil exam were either 2-minute or 4-minute problems, based on the number of problems and time available in morning and afternoon sessions. Since all of the CBT exam problems are 3-minute problems, a logical conclusion is that 100% of the problems are brand new, or (more likely) that morning and afternoon problems are comingled within each subject.

slightly more than 25% simply with judicious guessing and elimination of obvious incorrect options. So, the goal of all examinees should be to increase their scores from 25% to the minimum required passing score.

NCEES does not post minimum required passing scores for the CBT FE exam because the required passing score varies depending on the difficulty of the exam. While all exams have approximately equivalent difficulty, each exam has different problems, and so, minor differences in difficulty exist between the exam you take and the exam your neighbor will be taking. To account for these differences in difficulty, the exam's raw score is turned into a scaled score, and the scaled score is used to determine the passing rate.

For the CBT examination, each examinee will have a unique exam of approximately equivalent difficulty. This translates into a different minimum passing score for each examination. NCEES "accumulates" the passing score by summing each problem's "required performance value" (RPV).[3] The RPV represents the fraction of minimally qualified examinees that it thinks will solve the problem correctly. In the past, RPVs for new problems were dependent on the opinions of experts that it polled with the question, "What fraction of minimally qualified examinees do you think should be able to solve this problem correctly?" For problems that have appeared in past exams, including the "pre-test" items that are used on the CBT exam, NCEES actually knows the fraction. Basically, out of all of the examinees who passed the FE exam (the "minimally qualified" part), NCEES knows how many answered a pre-test problem correctly (the "fraction of examinees" part). A particularly easy problem on Ohm's law might have an RPV of 0.88, while a more difficult problem on Bayes' theorem might have an RPV of 0.37. Add up all of the RPVs, and bingo, you have the basis for a passing score. What could be simpler?[4]

WHAT IS THE AVERAGE PASSING RATE?

For July through November 2015, approximately 75% of first-time CBT test takers passed the written discipline-specific Chemical FE exam. The average failure rate was, accordingly, 25%. Some of those who failed the first time retook the FE exam, although the percentage of successful examinees declined precipitously with each subsequent attempt.

---

[3]NCEES does not actually use the term "required performance value," although it does use the method described.

[4]The flaw in this logic, of course, is that water seeks its own level. Deficient educational background and dependency on automation results in lower RPVs, which the NCEES process translates into a lower minimum passing score requirement. In the past, an "equating subtest" (a small number of problems in the exam that were associated with the gold standard of econometric data) was used to adjust the sum of RPVs based on the performance of the candidate pool. Though unmentioned in NCEES literature, that feature may still exist in the CBT exam process. However, the adjustment would still be based on the performance (good or bad) of the examinees.

WHAT REFERENCE MATERIAL CAN I BRING TO THE EXAM?

Since October 1993, the FE exam has been what NCEES calls a "limited-reference exam." This means that nothing except what is supplied by NCEES may be used during the exam. Therefore, the FE exam is really an "NCEES-publication only" exam. NCEES provides its own searchable, electronic version of the *NCEES Handbook* for use during the exam. Computer screens are 24 inches wide so there is enough room to display the exam problems and the *NCEES Handbook* side-by-side. No printed books from any publisher may be used.

WILL THE *NCEES HANDBOOK* HAVE EVERYTHING I NEED DURING THE EXAM?

In addition to not allowing examinees to be responsible for their own references, NCEES also takes no responsibility for the adequacy of coverage of its own reference. Nor does it offer any guidance or provide examples as to what else you should know, study, or memorize. The following warning statement comes from the *NCEES Handbook* preface.

> The *FE Reference Handbook* does not contain all the information required to answer every question on the exam. Basic theories, conversions, formulas, and definitions examinees are expected to know have not been included.

As open-ended as that warning statement sounds, the exam does not actually expect much knowledge outside of what is covered in the *NCEES Handbook*. For all practical purposes, the *NCEES Handbook* will have everything that you need. For example, if the *NCEES Handbook* covers only copper resistivity, you won't be asked to demonstrate a knowledge of aluminum resistivity. If the *NCEES Handbook* covers only common-emitter circuits, you won't be expected to know about common-base or common-collector circuits.

That makes it pretty simple to predict the kinds of problems that will appear on the exam. If you take your preparation seriously, the *NCEES Handbook* is pretty much a guarantee that you won't waste any time learning subjects that are not on the FE exam.

WILL THE *NCEES HANDBOOK* HAVE EVERYTHING I NEED TO STUDY FROM?

Saying that you won't need to work outside of the content published in the *NCEES Handbook* is not the same as saying the *NCEES Handbook* is adequate to study from.

From several viewpoints, the *NCEES Handbook* is marginally adequate in organization, presentation, and consistency as an examination reference. The *NCEES Handbook* was never intended to be something you study or learn from, so it is most definitely inadequate for that purpose. Background, preliminary and supporting material, explanations, extensions to the theory, and application rules are all missing from the *NCEES Handbook*. Many subtopics (e.g., contract law) listed in the exam specifications are not represented in the *NCEES Handbook*.

That is why you will notice many equations, figures, and tables in this book that are not **blue**. You may, for example, read several paragraphs in this book containing various **black** equations before you come across a **blue** equation section. While the **black** material may be less likely to appear on the exam than the **blue** material, it provides background information that is essential to understanding the **blue** material. Although memorization of the **black** material is not generally required, this material should at least make sense to you.

CHEMICAL FE EXAM KNOWLEDGE AREAS AND PROBLEM DISTRIBUTION

The following Chemical FE exam specifications have been published by NCEES. Some of the topics listed are not covered in any meaningful manner (or at all) by the *NCEES Handbook*. The only conclusion that can be drawn is that the required knowledge of these subjects is shallow, qualitative, and/or nonexistent.

1. **mathematics (8–12 problems):** analytic geometry; roots of equations; calculus; differential equations

2. **probability and statistics (4–6 problems):** probability distributions; expected value in decision making; hypothesis testing; measures of central tendencies and dispersions; estimation for a single mean; regression and curve fitting

3. **engineering sciences (4–6 problems):** applications of vector analysis; basic dynamics; work, energy, and power; electricity and current and voltage laws

4. **computational tools (4–6 problems):** numerical methods and concepts; spreadsheets for chemical engineering calculations; simulators

5. **materials science (4–6 problems):** chemical, electrical, mechanical, and physical properties; material types and compatibility; corrosion mechanisms and control

6. **chemistry (8–12 problems):** inorganic chemistry; organic chemistry

7. **fluid mechanics/dynamics (8–12 problems):** fluid properties; dimensionless numbers; mechanical energy balance; Bernoulli equation; laminar and turbulent flow; flow measurement; pumps, turbines, and compressors; compressible flow and non-Newtonian fluids

8. **thermodynamics (8–12 problems):** thermodynamic properties; properties data and phase diagrams; thermodynamics laws; thermodynamic processes; cyclic processes and efficiency; phase equilibrium; chemical equilibrium; heats of reaction and mixing

9. **material/energy balances (8–12 problems):** mass balance; energy balance; recycle/bypass processes; reactive systems

10. **heat transfer (8–12 problems):** conductive heat transfer; convective heat transfer; radiation heat transfer; heat transfer coefficients; heat transfer equipment, operation, and design

11. **mass transfer and separation (8–12 problems):** molecular diffusion; convective mass transfer; separated systems; equilibrium stage methods; continuous contact methods; humidification and drying

12. **chemical reaction engineering (8–12 problems):** reaction rates and order; rate constant; conversion, yield, and selectivity; types of reactions; reactor types

13. **process design and economics (8–12 problems):** process flow diagrams and piping and instrumentation diagrams; equipment selection; cost estimation; comparison of economic alternatives; process design and optimization

14. **process control (5–8 problems):** dynamics; control strategies; control loop design and hardware

15. **safety, health, and environment (5–8 problems):** hazardous properties of materials; industrial hygiene; process safety and hazard analysis; overpressure and underpressure protection; waste minimization, waste treatment, and regulation

16. **ethics and professional practice (2–3 problems):** codes of ethics; agreements and contracts; ethical and legal considerations; professional liability; public protection issues

DOES THE EXAM REQUIRE LOOKING UP VALUES IN TABLES?

For some problems, you might have to look up a value, but in those cases, you must use the value in the *NCEES Handbook*. For example, you might know that the density of atmospheric air is $0.075 \ \mathrm{lbm/ft^3}$ for all comfortable conditions. If you needed the density of air for a particle settling problem, you would find the official *NCEES Handbook* value is "$0.0734 \ \mathrm{lbm/ft^3}$ at 80°F." Whether or not using $0.075 \ \mathrm{lbm/ft^3}$ will result in an (approximate) correct answer or an incorrect answer depends on whether the problem writer wants to reward you for knowing something or punish you for not using the *NCEES Handbook*.

However, in order to reduce the time required to solve problems, and to reduce the variability of answers caused by examinees using different starting values, problems generally provide all required information. Unless the problem is specifically determining whether you can read a table or figure, all relevant values (resistivity, permittivity, permeability, density, modulus of elasticity, viscosity, enthalpy, yield strength, etc.) needed to solve the problem are often included in the problem statement. NCEES does not want the consequences of using correct methods with ambiguous data.

DO PROBLEM STATEMENTS INCLUDE SUPERFLUOUS INFORMATION?

Particularly since all relevant information is provided in the problem statements, some problems end up being pretty straightforward. In order to obfuscate the solution method, some irrelevant, superfluous information will be provided in the problem statement. For example, when finding the capacitance from a given plate area and separation (i.e., $C = \varepsilon A/d$), the temperature and permeability of the surrounding air might be given. However, if you understand the concept, this practice will be transparent to you.

Problems in this book typically do not include superfluous information. The purpose of this book is to teach you, not confuse you.

REGISTERING FOR THE EXAM

The CBT exams are administered at approved Pearson VUE testing centers. Registration is open year-round and can be completed online through your MyNCEES account.[5] Registration fees may be paid online. Once you receive notification from NCEES that you are eligible to schedule your exam, you can do so online through your MyNCEES account. Select the location where you would like to take your exam, and select from the list of available dates. You will receive a letter from Pearson VUE (via email) confirming your exam location and date.

Whether or not applying for and taking the exam is the same as applying for an FE certificate from your state depends on the state. In most cases, you might take the exam without your state board ever knowing about it. In fact, as part of the NCEES online exam application process, you will have to agree to the following statement:

> Passage of the FE exam alone does not ensure certification as an engineer intern or engineer-in-training in any U.S. state or territory. To obtain certification, you must file an application with an engineering licensing board and meet that board's requirements.

After graduation, when you are ready to obtain your FE (EIT, IE, etc.) suitable-for-wall-hanging certificate, you can apply and pay an additional fee to your state. In some cases, you will be required to take an additional nontechnical exam related to professional practice in your state. Actual procedures will vary from state to state.

WHEN YOU CAN TAKE THE EXAM

The FE exam is administered throughout all 12 months of the year.

WHAT TO BRING TO THE EXAM

You do not need to bring much with you to the exam. For admission, you must bring a current, signed, government-issued photographic identification. This is typically a driver's license or passport. A student ID card is not acceptable for admittance. The first and last name on the photographic ID must match the name on your appointment confirmation letter. NCEES recommends that you bring a copy of your appointment confirmation letter in order to speed up the check-in process. Pearson VUE will email this to you once you create a MyNCEES account and register for the exam.

Earplugs, noise-cancelling headphones, and tissues are provided at the testing center. Additionally, all examinees are provided with a reusable, erasable notepad and compatible writing instrument to use for scratchwork during the exam.

Pearson VUE staff may visually examine any approved item without touching you or the item. In addition to the items provided at the testing center, the following items are permitted during the FE exam.[6]

- your ID (same one used for admittance to the exam)
- key to your test center locker
- NCEES-approved calculator without a case
- inhalers
- cough drops and prescription and nonprescription pills, including headache remedies, all unwrapped and not bottled, unless the packaging states they must remain in the packaging
- bandages, braces (for your neck, back, wrist, leg, or ankle), casts, and slings
- eyeglasses (without cases); eye patches; handheld, nonelectric magnifying glasses (without cases); and eyedrops[7]
- hearing aids
- medical/surgical face masks, medical devices attached to your body (e.g., insulin pumps and spinal cord stimulators), and medical alert bracelets (including those with USB ports)
- pillows and cushions
- light sweaters or jackets
- canes, crutches, motorized scooters and chairs, walkers, and wheelchairs

---

[5]PPI is not associated with NCEES. Your MyNCEES account is not your PPI account.

[6]All items are subject to revision and reinterpretation at any time.
[7]Eyedrops can remain in their original bottle.

WHAT ELSE TO BRING TO THE EXAM

Depending on your situation, any of the following items may prove useful but should be left in your test center locker.

- calculator batteries

- contact lens wetting solution

- spare calculator

- spare reading glasses

- loose shoes or slippers

- extra set of car keys

- eyeglass repair kit, including a small screwdriver for fixing glasses (or removing batteries from your calculator)

WHAT NOT TO BRING TO THE EXAM

Leave all of these items in your car or at home: pens and pencils, erasers, scratch paper, clocks and timers, unapproved calculators, cell phones, pagers, communication devices, computers, tablets, cameras, audio recorders, and video recorders.

WHAT CALCULATORS ARE PERMITTED?

To prevent unauthorized transcription and distribution of the exam problems, calculators with communicating and text editing capabilities have been banned by NCEES. You may love the reverse Polish notation of your HP 48GX, but you'll have to get used to one of the calculators NCEES has approved. If you start using one of these approved calculators at the beginning of your review, you should be familiar enough with it by the time of the exam. Calculators permitted by NCEES are listed at **ppi2pass.com/calculators**. All of the listed calculators have sufficient engineering/scientific functionality for the exam.

At the beginning of your review program, you should purchase or borrow a spare calculator. It is preferable, but not essential, that your primary and spare calculators be identical. If your spare calculator is not identical to your primary calculator, spend some time familiarizing yourself with its functions.

Examinees found using a calculator that is not approved by NCEES will be discharged from the testing center and charged with exam subversion by their states. (See the section "Exam Subversion.")

WHAT UNITS ARE USED ON THE EXAM?

You will need to learn the SI system if you are not already familiar with it. Contrary to engineering practice in the United States, the FE exam primarily uses SI units.

The *NCEES Handbook* generally presents only dimensionally consistent equations. (For example, $F = ma$ is consistent with units of newtons, kilograms, meters, and seconds. However, it is not consistent for units of pounds-force, pounds-mass, feet, and seconds.) Although pound-based data is provided parallel to the SI data in most tables, many equations cannot use the pound-based data without including the gravitational constant. After being mentioned in the first few pages, the gravitational constant ($g_c = 32.2$ ft-lbm/lbf-sec^2), which is necessary to use for equations with inconsistent U.S. units, is barely mentioned in the *NCEES Handbook* and does not appear in most equations.

Outside of the table of conversions and introductory material at its beginning, the *NCEES Handbook* does not consistently differentiate between pounds-mass and pounds-force. The labels "pound" and "lb" are used to represent both force and mass. Densities are listed in tables with units of lb/in^3.

Kips are always units of force that can be incorporated into ft-kips, units for moment, and ksi, units of stress or strength.

IS THE EXAM HARD AND/OR TRICKY?

Whether or not the exam is hard or tricky depends on who you talk to. Other than providing superfluous data (so as not to lead you too quickly to the correct formula) and anticipating common mistakes, the FE exam is not a tricky exam. The exam does not overtly try to get you to fail. The problems are difficult in their own right. NCEES does not need to provide you misleading or vague statements. Examinees manage to fail on a regular basis with perfectly straightforward problems.

Commonly made mistakes are routinely incorporated into the available answer choices. Thus, the alternative answers (known as distractors) will seem logical to many examinees. For example, if you forget to convert the pipe diameter from millimeters to meters, you'll find an answer option that is off by a factor of 1000. Perhaps that meets your definition of "tricky."

Problems are generally practical, dealing with common and plausible situations that you might encounter on the job. In order to avoid the complications of being too practical, the ideal or perfect case is often explicitly called for in the problem statement (e.g., "Assume an ideal gas."; "Disregard the effects of air friction."; or "The steam expansion is isentropic.").

You won't have to draw on any experiential knowledge or make reasonable assumptions. If a motor efficiency is required, it will be given to you. You won't have to assume a reasonable value. If a wire is to be sized to limit current density, the limit will be explicitly given to you. If a temperature increase requires a factor of safety, the factor of safety will be given to you.

IS THE EXAM SOPHISTICATED?

Considering the features available with computerized testing, the sophistication of the FE testing algorithm is relatively low. All of the problems are fixed and pre-defined; new problems are not generated from generic stubs. You will get the same number of problems in each knowledge area, regardless of how well or poorly you do on previous problems in that knowledge area; adaptive testing is not used. The testing software randomly selects problems from a limited database; it is possible to see some of the same problems if you take the exam a second time.

Only two levels of categorization are used in the database: discipline and knowledge area. For example, a problem would be categorized as "Electrical and Computer Discipline" and "Electronics." With problems randomly selected from the database, the variation (breadth) of coverage follows the variation of the database. Within the limitations imposed by the need for an equivalent exam, it is statistically possible for the testing program to present you with ten bipolar junction transistor problems or seventeen differential equation problems.

Although the overall difficulty level of the exam is intended to be equivalent for all examinees, the difficulty level within a particular knowledge area can vary significantly. For example, within the Probability and Statistics knowledge area, you might have to solve nine Bayes' theorem problems, while your friend may get nine coin flip problems. In order to keep the overall difficulty level the same, after calculating all of those conditional probabilities, you may be rewarded with nine simple $F = ma$ and $v = Q/A$ type problems, while your friend gets to work problems involving organic chemistry, entropy, and three-dimensional tripods.

GOOD-FAITH EFFORT

Let's be honest. Some examinees take the FE exam because they have to, not because they want to. This situation is usually associated with university degree programs that require taking the exam as a condition of graduation. In most cases, such programs require only that students take the exam, not pass it. Accordingly, some short-sighted students consider the exam to be a formality, and they give it only token attention.

NCEES uses several methods to determine if you have made a "good-faith effort" on the exam. Some of the criteria for determining that you haven't include marking all of the answer choices the same (all "A," all "B," etc.), using a repeating sequence of responses (e.g., "A, B, C, D" over and over), leaving the exam site significantly early, and achieving a raw score of less than 30%. These criteria may be used by themselves or together.

The test results of examinees who are deemed not to have given a "good-faith effort" are separated statistically from other test results. Releasing to the universities the names of specific examinees whose test results

are in that category is at the discretion of NCEES, which has not yet formalized its policy.

WHAT DOES "MOST NEARLY" REALLY MEAN?

One of the more disquieting aspects of exam problems is that answer choices generally have only two or three significant digits, and the answer choices are seldom exact. An exam problem may prompt you to complete the sentence, "The value is most nearly...", or may ask "Which answer choice is closest to the correct value?" A lot of self-confidence is required to move on to the next problem when you don't find an exact match for the answer you calculated, or if you have had to split the difference because no available answer choice is close.

At one time, NCEES provided this statement regarding the use of "most nearly."

> Many of the questions on NCEES exams require calculations to arrive at a numerical answer. Depending on the method of calculation used, it is very possible that examinees working correctly will arrive at a range of answers. The phrase "most nearly" is used to accommodate all these answers that have been derived correctly but which may be slightly different from the correct answer choice given on the exam. You should use good engineering judgment when selecting your choice of answer. For example, if the question asks you to calculate an electrical current or determine the load on a beam, you should literally select the answer option that is most nearly what you calculated, regardless of whether it is more or less than your calculated value. However, if the question asks you to select a fuse or circuit breaker to protect against a calculated current or to size a beam to carry a load, you should select an answer option that will safely carry the current or load. Typically, this requires selecting a value that is closest to but larger than the current or load.

The difference is significant. Suppose you were asked to calculate "most nearly" the diameter of a wire needed to limit the current density to 2.34 A/mm^2. Suppose, also, that you calculated 8.25 mm. If the answer options were (A) 7 mm, (B) 8 mm, (C) 9 mm, and (D) 10 mm, you would go with answer option (B), because it is most nearly what you calculated. If, however, you were asked to select the minimum wire diameter to limit the current density to that value, you would have to go with option (C). Got it? If not, stop reading until you understand the distinction.

WHEN DO I FIND OUT IF I PASSED?

You will receive an email notification that your exam results are ready for viewing through your MyNCEES account 7–10 days after the exam. That email will also include instructions that you can use to proceed with your state licensing board. If you fail, you will be shown your percentage performance in each knowledge area. The diagnostic report may help you figure out what to study before taking the exam again. Because each examinee answers different problems in each knowledge area, the diagnostic report probably should not be used to compare the performance of two examinees, to determine how much smarter than another examinee you are, to rate employees, or to calculate raises and bonuses.

If you fail the exam, you may take it again. NCEES's policy is that examinees may take the exam once per testing window, up to three times per 12-month period. However, you should check with your state board to see whether it imposes any restrictions on the number and frequency of retakes.

SUBVERSION DURING THE EXAM

With the CBT exam, you can no longer get kicked out of the exam room for not closing your booklet or putting down your pencil in time. However, there are still plenty of ways for you to run afoul of the rules imposed on you by NCEES, your state board, and Pearson VUE. For example, since communication devices are prohibited in the exam, occurrences as innocent as your cell phone ringing during the exam can result in the immediate invalidation of your exam.

The *NCEES Examinee Guide* gives the following statement regarding fraudulent and/or unprofessional behavior. Somewhere along the way, you will probably have to read and accept it, or something similar, before you can take the FE exam.

Fraud, deceit, dishonesty, unprofessional behavior, and other irregular behavior in connection with taking any NCEES exam are strictly prohibited. Irregular behavior includes but is not limited to the following: failing to work independently; impersonating another individual or permitting such impersonation (surrogate testing); possessing prohibited items; communicating with other examinees or any outside parties by way of cell phone, personal computer, the Internet, or any other means during an exam; disrupting other examinees; creating safety concerns; and possessing, reproducing, or disclosing nonpublic exam questions, answers, or other information regarding the content of the exam before, during, or after the exam administration. Evidence of an exam irregularity may be based on your performance on the exam, a report from an administrator or a third party, or other information.

The chief proctor is authorized to take appropriate action to investigate, stop, or correct any observed or suspected irregular behavior, including discharging you from the test center and confiscating any prohibited devices or materials. You must cooperate fully in any investigation of a suspected irregularity. NCEES reserves the right to pursue all available remedies for exam irregularities, including canceling scores and pursuing administrative, civil, and/or criminal remedies.

If you are involved in an exam irregularity, the following may occur: invalidation of results, notification to your licensing board, forfeiture of exam fees, and restrictions on future testing. Some violations may incur additional consequences, to be pursued at the discretion of NCEES.

Based on the grounds for dismissal used with previous exam administrations, you can expect harsh treatment for

- having a cell phone in your possession

- having a device with copying, recording, or communication capabilities in your possession. These include but are not limited to cameras, pagers, personal digital assistants (PDAs), radios, headsets, tape players, calculator watches, electronic dictionaries, electronic translators, transmitting devices, digital media players (e.g., iPods), and tablets (e.g., iPads, Kindles, or Nooks)

- having papers, books, or notes

- having a calculator that is not on the NCEES-approved list

- appearing to copy or actually copying someone else's work

- talking to another examinee during the exam

- taking notes or writing on anything other than your NCEES-provided reusable, erasable notepad

- removing anything from the exam area

- leaving the exam area without authorization

- violating any other restrictions that are cause for dismissal or exam invalidation (e.g., whistling while you work, chewing gum, or being intoxicated)

If you are found to be in possession of a prohibited item (e.g., a cell phone) after the exam begins, that item will be confiscated and sent to NCEES. While you will probably eventually get your cell phone back, you won't get a refund of your exam fees.

Cheating and what is described as "subversion" are dealt with quite harshly. Proctors who observe you giving or receiving assistance, compromising the integrity of the exam, or participating in any other form of cheating during an exam will require you to surrender all exam materials and leave the test center. You won't be permitted to continue with the exam. It will be a summary execution, carried out without due process and mercy.

Of course, if you arrive with a miniature camera disguised as a pen or eyeglasses, your goose will be cooked. Talk to an adjacent examinee, and your goose will be cooked. Use a mirror to look around the room while putting on your lipstick or combing your hair, and your goose will be cooked. Bring in the wrong calculator, and your goose will be cooked. Loan your calculator to someone whose batteries have died, and your goose will be cooked. Though you get the idea, many of the ways that you might inadvertently get kicked out of the CBT exam are probably (and, unfortunately) yet to be discovered. Based on this fact, you shouldn't plan on being the first person to bring a peppermint candy in a crackly cellophane wrapper.

And, as if being escorted with your personal items out of the exam room wasn't embarrassing enough, your ordeal still won't be over. NCEES and your state will bar you from taking any exam for one or more years. Any application for licensure pending an approval for exam will be automatically rejected. You will have to reapply and pay your fees again later. By that time, you probably will have decided that the establishment's response to a minor infraction was so out of proportion that licensure as a professional engineer isn't even in the cards.

SUBVERSION AFTER THE EXAM

The NCEES testing (and financial) model is based on reusing all of its problems forever. To facilitate such reuse, the FE (and PE) exams are protected by non-disclosure agreements and a history of aggressive pursuit of actual and perceived offenses. In order to be allowed to take its exams, NCEES requires examinees to agree to its terms.

Copyright protection extends to only the exact words, phrases, and sentences, and sequences thereof, used in problems. However, the intent of the NCEES nondisclosure agreement is to grant NCEES protection beyond what is normally available through copyright protection—to prevent you from even discussing a problem in general terms (e.g., "There was a problem on structural bolts that stumped me. Did anyone else think the problem was unsolvable?").

Most past transgressions have been fairly egregious.[8] In several prominent instances, NCEES has incurred substantial losses and expenses. In those cases, offenders have gotten what they deserved. But, even innocent public disclosures of the nature of "Hey, did anyone else

---

[8]A candidate in Puerto Rico during the October 2006 Civil PE exam administration was found with scanning and transmitting equipment during the exam. She had recorded the entire exam, as well as the 2005 FE exam. The candidate pled guilty to two counts of fourth-degree aggravated fraud and was sentenced to six months' probation. All of the problems in both exams were compromised. NCEES obtained a civil judgment of over $1,000,000 against her.

have trouble solving that vertical crest curve problem?" have been aggressively pursued.

A restriction against saying anything at all to anybody about any aspect of a problem is probably too broad to be legally enforceable. Unfortunately, most examinees don't have the time, financial resources, or sophistication to resist what NCEES throws at them. Their only course of action is to accept whatever punishment is meted out to them by their state boards and by NCEES.

In the past, NCEES has used the U.S courts and aggressively pursued financial redress for loss of its intellectual property and violation of its copyright. It has administratively established a standard (accounting) value of thousands of dollars for each disclosed or compromised problem. You can calculate your own *pro forma* invoice from NCEES by multiplying this amount by the number of problems you discuss with others.

DOING YOUR PART, NCEES STYLE

NCEES has established a security tip line so that you can help it police the behavior of other examinees. Before, during, or after the exam, if you see any of your fellow examinees acting suspiciously, NCEES wants you to report them by phone or through the NCEES website. You'll have to identify yourself, but NCEES promises that the information you provide will be strictly confidential, and that your personal contact information will not be shared outside the NCEES compliance and security staff. Unless required by statute, rules of discovery, or a judge, of course.

PART 4: STRATEGIES FOR PASSING THE EXAM

A FEW DAYS BEFORE THE EXAM

There are a few things you should do a week or so before the examination date. For example, visit the exam site in order to find the testing center building, parking areas, examination room, and restrooms. You should also make arrangements for childcare and transportation. Since your examination may not start or end exactly at the designated times, make sure that your childcare and transportation arrangements can allow for some flexibility.

Second in importance to your scholastic preparation is the preparation of your two examination kits. (See "What to Bring to the Exam" and "What Else to Bring to the Exam" in this Introduction.) The first kit includes items that can be left in your assigned locker (e.g., your admittance letter, photo ID, and extra calculator batteries). The second kit includes items that should be left in your car in case you need them (e.g., copy of your application, warm sweater, and extra snacks or beverages).

THE DAY BEFORE THE EXAM

If possible, take the day before the examination off from work to relax. Do not cram the last night. A good prior night's sleep is the best way to start the examination. If you live far from the examination site, consider getting a hotel room in which to spend the night.

Make sure your exam kits are packed and ready to go.

THE DAY OF THE EXAM

You should arrive at least 30 minutes before your scheduled start time. This will allow time for finding a convenient parking place, bringing your items to the testing center, and checking in.

DURING THE EXAM

Once the examination has started, observe the following suggestions.

Do not spend more than four minutes working a problem. (The average time available per problem is slightly less than three minutes.) If you have not finished a problem in that time, flag it for later review if you have time, and continue on.

Don't ask your proctors technical questions. Proctors are pure administrators. They don't know anything about the exam or its subjects.

Even if you do not discover them, errors in the exam (and in the *NCEES Handbook*) do occur. Rest assured that errors are almost always discovered during the scoring process, and that you will receive the performance credit for all flawed items.

However, NCEES has a form for reporting errors, and the test center should be able to provide it to you. If you encounter a problem with (a) missing information, (b) conflicting information, (c) no correct response from the four answer choices, or (d) more than one correct answer, use your provided reusable, erasable notepad to record the problem identification numbers. It is not necessary to tell your proctor during the exam. Wait until after the exam to ask your proctor about the procedure for reporting errors on the exam.

AFTER YOU PASS

[ ]   Celebrate. Take someone out to dinner. Go off your diet. Get dessert.

[ ]   Thank your family members and anyone who had to put up with your grouchiness before the exam.

[ ]   Thank your old professors.

[ ]   Tell everyone at the office.

[ ]   Ask your employer for new business cards and a raise.

[ ]   Tell your review course provider and instructors.

[ ]   Tell the folks at PPI who were rootin' for you all along.

[ ]   Start thinking about the PE exam.

Sample Study Schedule (for Individuals)

Time required to complete study schedule (excluding taking a practice exam):
    74 days for a "crash course," going straight through, with no rest and review days and no weekends
    92 days going straight through, taking off rest and review days, but no weekends
    128 days using only the five-day work week, taking off rest and review days and weekends

Your examination date: _____

Number of days: _____

Latest day you can start: _____

day no.	date	chap. no.	knowledge area	subject
1	_____	Introduction	Mathematics	Introduction; Diagnostic Exam
2	_____	1		Units
3	_____	2		Algebra
4	_____	3		Vectors
5	_____	none		**rest; review**
6	_____	4		Analytic Geometry
7	_____	5		Trigonometry
8	_____	6		Linear Algebra
9	_____	7		Calculus
10	_____	none		**rest; review**
11	_____	8		Differential Equations
12	_____	9		Numerical Methods
13	_____	II	Probability and Statistics	Diagnostic Exam
14	_____	10		Probability and Statistics
15	_____	none		**rest; review**
16	_____	III	Fluid Mechanics/Dynamics	Diagnostic Exam
17	_____	11		Fluid Properties
18	_____	12		Fluid Statics
19	_____	13		Fluid Dynamics
20	_____	none		**rest; review**
21	_____	14		Fluid Measurement and Similitude
22	_____	15		Compressible Fluid Dynamics
23	_____	16		Fluid Machines
24	_____	none		**rest; review**
25	_____	IV	Thermodynamics	Diagnostic Exam
26	_____	17		Properties of Substances
27	_____	18		Laws of Thermodynamics
28	_____	19		Power Cycles and Entropy
29	_____	20		Mixtures of Gases, Vapors, and Liquids
30	_____	none		**rest; review**
31	_____	V	Chemistry	Diagnostic Exam
32	_____	21		Inorganic Chemistry
33	_____	22		Combustion
34	_____	23		Organic Chemistry
35	_____	24		Biochemistry
36	_____	none		**rest; review**
37	_____	VI	Heat Transfer	Diagnostic Exam
38	_____	25		Conduction
39	_____	26		Convection
40	_____	27		Radiation
41	_____	28		Transport Phenomena
42	_____	none		**rest; review**

day no.	date	chap. no.	knowledge area	subject
43		VII	Engineering Sciences	Diagnostic Exam
44		29		Trusses
45		30		Pulleys, Cables, and Friction
46		31		Centroids and Moments of Inertia
47		32		Indeterminate Statics
48		none		**rest; review**
49		33		Kinematics
50		34		Kinetics
51		35		Kinetics of Rotational Motion
52		36		Energy and Work
53		none		**rest; review**
54		37		Electrostatics
55		38		Direct-Current Circuits
56		39		Alternating-Current Circuits
57		40		Rotating Machines
58		none		**rest; review**
59		VIII	Materials Science	Diagnostic Exam
60		41		Material Properties and Testing
61		42		Engineering Materials
62		none		**rest; review**
63		IX	Chemical Reaction Engineering	Diagnostic Exam
64		43		Reaction Kinetics
65		X	Mass Transfer	Diagnostic Exam
66		44		Diffusion
67		none		**rest; review**
68		XI	Unit Processes	Diagnostic Exam
69		45		Vapor-Liquid and Gas-Liquid Separation Processes
70		46		Solid-Liquid Processes
71		none		**rest; review**
72		XII	Process Control	Diagnostic Exam
73		47		Control Loops and Control Systems
74		48		Instrumentation and Control System Hardware
75		none		**rest; review**
76		XIII	Process Design and Economics	Diagnostic Exam
77		49		Plant and Process Design
78		50		Engineering Economics
79		none		**rest; review**
80		XIV	Safety, Health, and Environment	Diagnostic Exam
81		51		Toxicology
82		52		Industrial Hygiene
83		53		Discharge Water Quality
84		54		Air Quality
85		none		**rest; review**
86		XV	Computational Tools	Diagnostic Exam
87		55		Computer Software
88		XVI	Ethics and Professional Practice	Diagnostic Exam
89		56		Professional Practice
90		57		Ethics
91		58		Licensure
92		none		**rest; review**
93-97		none	none	Practice Exam
98		none		FE Examination

Diagnostic Exam

Topic I: Mathematics

1. Which of the following equations correctly describes the shaded area of the x-y plane?

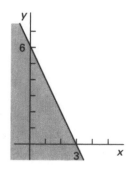

(A) $2x - y \le 6$

(B) $2x + y \le 6$

(C) $2x - y \ge 6$

(D) $x + 2y \ge 6$

2. What is most nearly the slope of the line tangent to the parabola $y = 12x^2 + 3$ at a point where $x = 5$?

(A) 24

(B) 120

(C) 140

(D) 300

3. What is most nearly the interior angle, θ, of a regular polygon with seven sides?

(A) 51°

(B) 64°

(C) 120°

(D) 130°

4. Which statement about a quadratic equation of the form $ax^2 + bx + c = 0$ is true?

(A) It has two different roots.

(B) If one of its roots is real, the other root can be imaginary.

(C) The curve defined by the equation will pass through the y-axis.

(D) The curve defined by the equation will pass through the x-axis.

5. Four vector fields are shown.

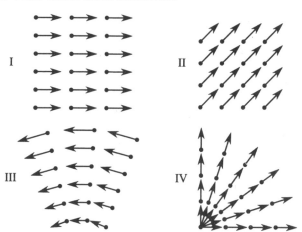

Which field has a positive vector curl?

(A) I

(B) II

(C) III

(D) IV

6. What are the coordinates of the point of intersection of the following lines?

$$y_1 = x + 2$$
$$y_2 = x^2 + 5x + 6$$

(A) $(-3, 0)$

(B) $(-2, 0)$

(C) $(-1, 1)$

(D) $(2, 0)$

7. $x = -2$ is one of the roots of the equation $x^3 + x^2 - 22x - 40 = 0$. What are the other two roots?

(A) -5 and 5

(B) -5 and 2

(C) -4 and 5

(D) -4 and 2

8. A vector originates at point $(2, 3, 11)$ and terminates at point $(10, 15, 20)$. What is the magnitude of the vector?

(A) 12

(B) 14

(C) 15

(D) 17

9. What is the unit vector of $\mathbf{R} = 12\mathbf{i} - 20\mathbf{j} - 9\mathbf{k}$?

(A) $0.25\mathbf{i} - 0.80\mathbf{j} - 0.54\mathbf{k}$

(B) $0.48\mathbf{i} - 0.80\mathbf{j} - 0.36\mathbf{k}$

(C) $0.54\mathbf{i} - 0.77\mathbf{j} - 0.35\mathbf{k}$

(D) $0.64\mathbf{i} - 0.64\mathbf{j} - 0.42\mathbf{k}$

10. Which of these vector identities is INCORRECT?

(A) $\mathbf{A} \cdot \mathbf{A} = 0$

(B) $\mathbf{A} \times \mathbf{A} = 0$

(C) $\mathbf{A} \cdot \mathbf{B} = \mathbf{B} \cdot \mathbf{A}$

(D) $\mathbf{A} \times \mathbf{B} = -\mathbf{B} \times \mathbf{A}$

SOLUTIONS

1. $y = 6 - 2x$ is the equation of the line. $2x + y \leq 6$ describes the shaded area.

The answer is (B).

2. The slope of the line is

$$m = \frac{dy}{dx}\bigg|_{x=5} = 24x$$
$$= (24)(5)$$
$$= 120$$

The answer is (B).

3. The interior angle is

$$\theta = \left[\frac{\pi(n-2)}{n}\right] = \pi\left(1 - \frac{2}{n}\right) = (180°)\left(1 - \frac{2}{7}\right)$$
$$= 128.6° \quad (130°)$$

The answer is (D).

4. The curve defined by a quadratic equation of the form $ax^2 + bx + c = 0$ is always a parabola that opens vertically. The parabola will always pass through the y-axis at some value of x, so option C is true. When the minimum or maximum point of the parabola is on the x-axis, the two roots are the same (a double root), so option A is false. The two roots must be both real or both imaginary, so option B is false. The parabola may open upward and be entirely above the x-axis, or open downward and be entirely below it, so option D is false.

The answer is (C).

5. Curl is a vector operator that describes a vector field's vorticity (rotation) at a point. Illustrations I and II are linear vector fields without rotation or accumulation (divergence). Illustration IV has divergence, but no rotation. Only Illustration III has rotation.

The answer is (C).

6. At the point where the two lines cross, the x- and y-values satisfy both equations.

$$y_1 = y_2$$
$$x + 2 = x^2 + 5x + 6$$
$$x^2 + 4x + 4 = 0$$

The roots of this quadratic equation are

$$x = \frac{-b \pm \sqrt{b^2 - 4ac}}{2a} = \frac{-4 \pm \sqrt{(4)^2 - (4)(1)(4)}}{(2)(1)}$$
$$= -2$$

The discriminant (the portion under the radical sign) is equal to zero, so the quadratic equation has one double root of -2. Inserting $x = -2$ into the original equations gives

$$y_1 = (-2) + 2 = 0$$
$$y_2 = (-2)^2 + 5(-2) + 6 = 0$$

The lines cross at $(x, y) = (-2, 0)$.

The answer is (B).

7. One root of the equation is given as -2. Divide both sides of the equation by $x + 2$ to get $x^2 - x - 20 = 0$, then use the quadratic formula to find the remaining two roots.

$$
\begin{array}{r}
x^2 - x - 20 \\
x + 2 \overline{\big)\, x^3 + x^2 - 22x - 40} \\
\underline{x^3 + 2x^2} \\
- x^2 - 22x \\
\underline{- x^2 - 2x} \\
- 20x - 40 \\
\underline{- 20x - 40} \\
0
\end{array}
$$

$$x = \frac{-b \pm \sqrt{b^2 - 4ac}}{2a} = \frac{-(-1) \pm \sqrt{(-1)^2 - (4)(1)(-20)}}{(2)(1)}$$

$$= -4 \text{ and } 5$$

The answer is (C).

8. In three-dimensional space, the distance between two points is

$$d = \sqrt{(x_2 - x_1)^2 + (y_2 - y_1)^2 + (z_2 - z_1)^2}$$
$$= \sqrt{(10 - 2)^2 + (15 - 3)^2 + (20 - 11)^2}$$
$$= 17$$

The answer is (D).

9. The magnitude of $\mathbf{R} = a\mathbf{i} + b\mathbf{j} + c\mathbf{k}$ is

$$|\mathbf{R}| = \sqrt{a^2 + b^2 + c^2} = \sqrt{(12)^2 + (-20)^2 + (-9)^2}$$
$$= 25$$

Divide vector $\mathbf{R}$ by its magnitude to find the unit vector that is parallel with $\mathbf{R}$.

$$\frac{\mathbf{R}}{|\mathbf{R}|} = \frac{12\mathbf{i} - 20\mathbf{j} - 9\mathbf{k}}{25} = 0.48\mathbf{i} - 0.80\mathbf{j} - 0.36\mathbf{k}$$

The answer is (B).

10. If the dot product of two vectors is zero, either one or both of the vectors is zero or the two vectors are perpendicular. The equation $\mathbf{A} \cdot \mathbf{A} = 0$ is, therefore, true only when $\mathbf{A} = 0$, and it is not an identity. The other three options are identities.

The answer is (A).

1 Units

Figure 1.1 *Common Units of Mass*

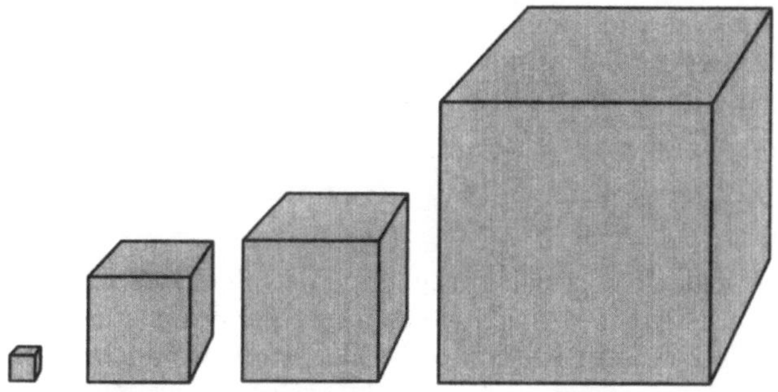

1 gram 1 pound   1 kilogram   1 slug
  (1)   (454)   (1000)   (14 594)

1. INTRODUCTION

The purpose of this chapter is to eliminate some of the confusion regarding the many units available for each engineering variable. In particular, an effort has been made to clarify the use of the so-called English systems, which for years have used the *pound* unit both for force and mass—a practice that has resulted in confusion for even those familiar with it.

It is expected that most engineering problems will be stated and solved in either English engineering or SI units. Therefore, a discussion of these two systems occupies the majority of this chapter.

2. COMMON UNITS OF MASS

The choice of a mass unit is the major factor in determining which system of units will be used in solving a problem. Obviously, you will not easily end up with a force in pounds if the rest of the problem is stated in meters and kilograms. Actually, the choice of a mass unit determines more than whether a conversion factor will be necessary to convert from one system to another (e.g., between the SI and English systems). An inappropriate choice of a mass unit may actually require a conversion factor *within* the system of units.

The common units of mass are the gram, pound, kilogram, and slug. There is nothing mysterious about these units. All represent different quantities of matter, as Fig. 1.1 illustrates. In particular, note that the pound and slug do not represent the same quantity of matter. One slug is equal to 32.1740 pounds-mass.

3. MASS AND WEIGHT

The SI system uses kilograms for mass and newtons for weight (force). The units are different, and there is no confusion between the variables. However, for years, the term *pound* has been used for both mass and weight. This usage has obscured the distinction between the two: mass is a constant property of an object; weight varies with the gravitational field. Even the conventional use of the abbreviations *lbm* and *lbf* (to distinguish between pounds-mass and pounds-force) has not helped eliminate the confusion.

An object with a mass of one pound will have an earthly weight of one pound, but this is true only on the earth. The weight of the same object will be much less on the moon. Therefore, care must be taken when working with mass and force in the same problem.

The relationship that converts mass to weight is familiar to every engineering student.

$$W = mg$$

This equation illustrates that an object's weight will depend on the local acceleration of gravity as well as the object's mass. The mass will be constant, but gravity will depend on location. Mass and weight are not the same.

4. ACCELERATION OF GRAVITY

Gravitational acceleration on the earth's surface is usually taken as 32.2 ft/sec² or 9.81 m/s². These values are rounded from the more exact standard values of

32.1740 ft/sec² and 9.8066 m/s². However, the need for greater accuracy must be evaluated on a problem-by-problem basis. Usually, three significant digits are adequate, since gravitational acceleration is not constant anyway, but is affected by location (primarily latitude and altitude) and major geographical features.

5. CONSISTENT SYSTEMS OF UNITS

A set of units used in a calculation is said to be *consistent* if no conversion factors are needed. (The terms *homogeneous* and *coherent* are also used to describe a consistent set of units.) For example, a moment is calculated as the product of a force and a lever arm length.

$$M = dF$$

A calculation using the previous equation would be consistent if M was in newton-meters, F was in newtons, and d was in meters. The calculation would be inconsistent if M was in ft-kips, F was in kips, and d was in inches (because a conversion factor of 1/12 would be required).

The concept of a consistent calculation can be extended to a system of units. A *consistent system of units* is one in which no conversion factors are needed for any calculation. For example, Newton's second law of motion can be written without conversion factors. Newton's second law for an object with a constant mass simply states that the force required to accelerate the object is proportional to the acceleration of the object. The constant of proportionality is the object's mass.

$$F = ma$$

Notice that this relationship is $F = ma$, not $F = Wa/g$ or $F = ma/g_c$. $F = ma$ is consistent: It requires no conversion factors. This means that in a consistent system where conversion factors are not used, once the units of m and a have been selected, the units of F are fixed. This has the effect of establishing units of work and energy, power, fluid properties, and so on.

The decision to work with a consistent set of units is desirable but unnecessary, depending often on tradition and environment. Problems in fluid flow and thermodynamics are routinely solved in the United States with inconsistent units. This causes no more of a problem than working with inches and feet when calculating moments. It is necessary only to use the proper conversion factors.

6. THE ENGLISH ENGINEERING SYSTEM

Through common and widespread use, pounds-mass (lbm) and pounds-force (lbf) have become the standard units for mass and force in the *English Engineering System*.

There are subjects in the United States where the practice of using pounds for mass is firmly entrenched. For example, most thermodynamics, fluid flow, and heat transfer problems have traditionally been solved using the units of lbm/ft³ for density, Btu/lbm for enthalpy, and Btu/lbm-°F for specific heat. Unfortunately, some equations contain both lbm-related and lbf-related variables, as does the steady flow conservation of energy equation, which combines enthalpy in Btu/lbm with pressure in lbf/ft².

The units of pounds-mass and pounds-force are as different as the units of gallons and feet, and they cannot be canceled. A mass conversion factor, g_c, is needed to make the equations containing lbf and lbm dimensionally consistent. This factor is known as the *gravitational constant* and has a value of 32.1740 lbm-ft/lbf-sec². The numerical value is the same as the standard acceleration of gravity, but g_c is not the local gravitational acceleration, g. (It is acceptable, and recommended, that g_c be rounded to the same number of significant digits as g. Therefore, a value of 32.2 for g_c would typically be used.) g_c is a conversion constant, just as 12.0 is the conversion factor between feet and inches.

Equation 1.1: Force in Pounds-Force

$$F = ma/g_c \qquad 1.1$$

Variation

$$F \text{ in lbf} = \frac{(m \text{ in lbm})\left(a \text{ in } \frac{\text{ft}}{\text{sec}^2}\right)}{g_c \text{ in } \frac{\text{lbm-ft}}{\text{lbf-sec}^2}}$$

Description

The English Engineering System is an inconsistent system, as defined according to Newton's second law. $F = ma$ cannot be written if lbf, lbm, and ft/sec² are the units used. The g_c term must be included.

g_c does more than "fix the units." Since g_c has a numerical value of 32.1740, it actually changes the calculation numerically. A force of 1.0 pound will not accelerate a 1.0 pound-mass at the rate of 1.0 ft/sec².

In the English Engineering System, work and energy are typically measured in ft-lbf (mechanical systems) or in British thermal units, Btu (thermal and fluid systems). One Btu is equal to approximately 778 ft-lbf.

Example

What is most nearly the weight in lbf of a 1.00 lbm object in a gravitational field of 27.5 ft/sec²?

(A) 0.85 lbf

(B) 1.2 lbf

(C) 28 lbf

(D) 32 lbf

Solution

The weight is

$$F = ma/g_c$$

$$= \frac{(1.00 \text{ lbm})\left(27.5 \frac{\text{ft}}{\text{sec}^2}\right)}{32.2 \frac{\text{lbm-ft}}{\text{lbf-sec}^2}}$$

$$= 0.854 \text{ lbf} \quad (0.85 \text{ lbf})$$

The answer is (A).

7. OTHER FORMULAS AFFECTED BY INCONSISTENCY

It is not a significant burden to include g_c in a calculation, but it may be difficult to remember when g_c should be used. Knowing when to include the gravitational constant can be learned through repeated exposure to the formulas in which it is needed, but it is safer to carry the units along in every calculation.

Equation 1.2 Through Eq. 1.6: Equations Requiring g_c

$$KE = mv^2/2g_c \quad \text{[in ft-lbf]} \qquad 1.2$$
$$PE = mgh/g_c \quad \text{[in ft-lbf]} \qquad 1.3$$
$$p = \rho gh/g_c \quad \text{[in lbf/ft}^2] \qquad 1.4$$
$$SW = \rho g/g_c \quad \text{[in lbf/ft}^3] \qquad 1.5$$
$$\tau = (\mu/g_c)(dv/dy) \quad \text{[in lbf/ft}^2] \qquad 1.6$$

Description

Equation 1.2 through Eq. 1.6 are some representative equations that require the g_c term. Equation 1.2 calculates kinetic energy, Eq. 1.3 calculates potential energy, Eq. 1.4 calculates pressure, Eq. 1.5 calculates specific weight, and Eq. 1.6 calculates shear stress.[1] In all cases, it is assumed that the standard English Engineering System units will be used.

Example

A rocket that has a mass of 4000 lbm travels at 27,000 ft/sec. What is most nearly its kinetic energy?

(A) 1.4×10^9 ft-lbf
(B) 4.5×10^{10} ft-lbf
(C) 1.5×10^{12} ft-lbf
(D) 4.7×10^{13} ft-lbf

---

[1]The NCEES *FE Reference Handbook* (*NCEES Handbook*) is not consistent in the variables it uses in Eq. 1.2 through Eq. 1.6. For example, T is used for kinetic energy and U is used for potential energy in the Dynamics section, and γ is used for specific weight in the Fluids section.

Solution

From Eq. 1.2, the kinetic energy is

$$KE = mv^2/2g_c = \frac{(4000 \text{ lbm})\left(27,000 \frac{\text{ft}}{\text{sec}}\right)^2}{(2)\left(32.2 \frac{\text{lbm-ft}}{\text{lbf-sec}^2}\right)}$$

$$= 4.53 \times 10^{10} \text{ ft-lbf} \quad (4.5 \times 10^{10} \text{ ft-lbf})$$

The answer is (B).

8. WEIGHT AND SPECIFIC WEIGHT

Weight is a force exerted on an object due to its placement in a gravitational field. If a consistent set of units is used, $W = mg$ can be used to calculate the weight of a mass. In the English Engineering System, however, the following equation must be used.

$$W = \frac{mg}{g_c}$$

Both sides of this equation can be divided by the volume of an object to derive the *specific weight* (*unit weight, weight density*), γ, of the object. The following equation illustrates that the weight density (in lbf/ft^3) can also be calculated by multiplying the mass density (in lbm/ft^3) by g/g_c.

$$\frac{W}{V} = \left(\frac{m}{V}\right)\left(\frac{g}{g_c}\right)$$

Since g and g_c usually have the same numerical values, the only effect of the following equation is to change the units of density.

$$\gamma = \frac{W}{V} = \left(\frac{m}{V}\right)\left(\frac{g}{g_c}\right) = \frac{\rho g}{g_c}$$

Weight does not occupy volume; only mass has volume. The concept of weight density has evolved to simplify certain calculations, particularly fluid calculations. For example, pressure at a depth is calculated from

$$p = \gamma h$$

Compare this to the equation for pressure at a depth.

9. THE ENGLISH GRAVITATIONAL SYSTEM

Not all English systems are inconsistent. Pounds can still be used as the unit of force as long as pounds are not used as the unit of mass. Such is the case with the consistent *English Gravitational System.*

If acceleration is given in ft/sec², the units of mass for a consistent system of units can be determined from Newton's second law.

$$\text{units of } m = \frac{\text{units of } F}{\text{units of } a} = \frac{\text{lbf}}{\dfrac{\text{ft}}{\text{sec}^2}} = \frac{\text{lbf-sec}^2}{\text{ft}}$$

The combination of units in this equation is known as a *slug*. g_c is not needed since this system is consistent. It would be needed only to convert slugs to another mass unit.

Slugs and pounds-mass are not the same, as Fig. 1.1 illustrates. However, both are units for the same quantity: mass. The following equation will convert between slugs and pounds-mass.

$$\text{no. of slugs} = \frac{\text{no. of lbm}}{g_c}$$

The number of slugs is not derived by dividing the number of pounds-mass by the local gravity. g_c is used regardless of the local gravity. The conversion between feet and inches is not dependent on local gravity; neither is the conversion between slugs and pounds-mass.

Since the English Gravitational System is consistent, the following equation can be used to calculate weight. Notice that the local gravitational acceleration is used.

$$W \text{ in lbf} = (m \text{ in slugs})\left(g \text{ in } \frac{\text{ft}}{\text{sec}^2}\right)$$

10. METRIC SYSTEMS OF UNITS

Strictly speaking, a *metric system* is any system of units that is based on meters or parts of meters. This broad definition includes *mks systems* (based on meters, kilograms, and seconds) as well as *cgs systems* (based on centimeters, grams, and seconds).

Metric systems avoid the pounds-mass versus pounds-force ambiguity in two ways. First, matter is not measured in units of force. All quantities of matter are specified as mass. Second, force and mass units do not share a common name.

The term *metric system* is not explicit enough to define which units are to be used for any given variable. For example, within the cgs system there is variation in how certain electrical and magnetic quantities are represented (resulting in the ESU and EMU systems). Also, within the mks system, it is common engineering practice today to use kilocalories as the unit of thermal energy, while the SI system requires the use of joules. Thus, there is a lack of uniformity even within the metricated engineering community.

The "metric" parts of this book are based on the SI system, which is the most developed and codified of the so-called metric systems. It is expected that there will be occasional variances with local engineering custom, but it is difficult to anticipate such variances within a book that must be consistent.

11. SI UNITS (THE MKS SYSTEM)

SI units comprise an mks system (so named because it uses the meter, kilogram, and second as dimensional units). All other units are derived from the dimensional units, which are completely listed in Table 1.1. This system is fully consistent, and there is only one recognized unit for each physical quantity (variable).

Two types of units are used: base units and derived units. The *base units* (see Table 1.1) are dependent only on accepted standards or reproducible phenomena. The previously unclassified *supplementary units*, radian and steradian, have been classified as derived units. The *derived units* (see Table 1.2 and Table 1.3) are made up of combinations of base and supplementary units.

Table 1.1 SI Base Units

quantity	name	symbol
length	meter	m
mass	kilogram	kg
time	second	s
electric current	ampere	A
temperature	kelvin	K
amount of substance	mole	mol
luminous intensity	candela	cd

Table 1.2 Some SI Derived Units with Special Names

quantity	name	symbol	expressed in terms of other units
frequency	hertz	Hz	1/s
force	newton	N	kg·m/s²
pressure, stress	pascal	Pa	N/m²
energy, work, quantity of heat	joule	J	N·m
power, radiant flux	watt	W	J/s
quantity of electricity, electric charge	coulomb	C	
electric potential, potential difference, electromotive force	volt	V	W/A
electric capacitance	farad	F	C/V
electric resistance	ohm	Ω	V/A
electric conductance	siemen	S	A/V
magnetic flux	weber	Wb	V·s
magnetic flux density	tesla	T	Wb/m²
inductance	henry	H	Wb/A
luminous flux	lumen	lm	
illuminance	lux	lx	lm/m²
plane angle	radian	rad	
solid angle	steradian	sr	

In addition, there is a set of non-SI units that may be used. This concession is primarily due to the significance and widespread acceptance of these units. Use of the non-SI units listed in Table 1.4 will usually create an inconsistent expression requiring conversion factors.

Table 1.3 Some SI Derived Units

quantity	description	expressed in terms of other units
area	square meter	m^2
volume	cubic meter	m^3
speed		
linear	meter per second	m/s
angular	radian per second	rad/s
acceleration		
linear	meter per second squared	m/s^2
angular	radian per second squared	rad/s^2
density, mass density	kilogram per cubic meter	kg/m^3
concentration (of amount of substance)	mole per cubic meter	mol/m^3
specific volume	cubic meter per kilogram	m^3/kg
luminance	candela per square meter	cd/m^2
absolute viscosity	pascal second	Pa·s
kinematic viscosity	square meters per second	m^2/s
moment of force	newton meter	N·m
surface tension	newton per meter	N/m
heat flux density, irradiance	watt per square meter	W/m^2
heat capacity, entropy	joule per kelvin	J/K
specific heat capacity, specific entropy	joule per kilogram kelvin	J/kg·K
specific energy	joule per kilogram	J/kg
thermal conductivity	watt per meter kelvin	W/m·K
energy density	joule per cubic meter	J/m^3
electric field strength	volt per meter	V/m
electric charge density	coulomb per cubic meter	C/m^3
surface density of charge, flux density	coulomb per square meter	C/m^2
permittivity	farad per meter	F/m
current density	ampere per square meter	A/m^2
magnetic field strength	ampere per meter	A/m
permeability	henry per meter	H/m
molar energy	joule per mole	J/mol
molar entropy, molar heat capacity	joule per mole kelvin	J/mol·K
radiant intensity	watt per steradian	W/sr

The units of force can be derived from Newton's second law.

$$\text{units of force} = (m \text{ in } kg)\left(a \text{ in } \frac{m}{s^2}\right) = \frac{kg \cdot m}{s^2}$$

This combination of units for force is known as a *newton*. Figure 1.2 illustrates common force units.

Table 1.4 Acceptable Non-SI Units

quantity	unit name	symbol name	relationship to SI unit
area	hectare	ha	1 ha = 10 000 m^2
energy	kilowatt-hour	kW·h	1 kW·h = 3.6 MJ
mass	metric ton[a]	t	1 t = 1000 kg
plane angle	degree (of arc)	°	1° = 0.017453 rad
speed of rotation	revolution per minute	r/min	1 r/min = $2\pi/60$ rad/s
temperature interval	degree Celsius	°C	1°C = 1K ($\Delta T_{°C} = \Delta T_K$)
time	minute	min	1 min = 60 s
	hour	h	1 h = 3600 s
	day (mean solar)	d	1 d = 86 400 s
	year (calendar)	a	1 a = 31 536 000 s
velocity	kilometer per hour	km/h	1 km/h = 0.278 m/s
volume	liter[b]	L	1 L = 0.001 m^3

[a]The international name for metric ton is *tonne*. The metric ton is equal to the *megagram* (Mg).
[b]The international symbol for liter is the lowercase l, which can be easily confused with the numeral 1. Several English-speaking countries have adopted the script ℓ and uppercase L as a symbol for liter in order to avoid any misinterpretation.

Figure 1.2 Common Force Units and Relative Sizes

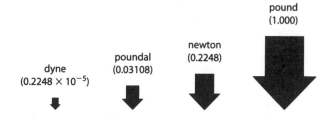

dyne
(0.2248×10^{-5})

poundal
(0.03108)

newton
(0.2248)

pound
(1.000)

Energy variables in the SI system have units of N·m, or equivalently, $kg \cdot m^2/s^2$. Both of these combinations are known as a *joule*. The units of power are joules per second, equivalent to a *watt*.

Example

A 10 kg block is raised vertically 3 m. What is most nearly the change in potential energy?

(A) 30 J

(B) 98 J

(C) 290 J

(D) 880 J

Solution

The change in potential energy is

$$\Delta PE = mg\Delta h$$
$$= (10 \text{ kg})\left(9.81 \frac{m}{s^2}\right)(3 \text{ m})$$
$$= 294 \text{ kg} \cdot m^2/s^2 \quad (290 \text{ J})$$

The answer is (C).

Mathematics

12. RULES FOR USING THE SI SYSTEM

In addition to having standardized units, the SI system also has rigid syntax rules for writing the units and combinations of units. Each unit is abbreviated with a specific symbol. The following rules for writing and combining these symbols should be adhered to.

- The expressions for derived units in symbolic form are obtained by using the mathematical signs of multiplication and division; for example, units of velocity are m/s, and units of torque are N·m (not N-m or Nm).

- Scaling of most units is done in multiples of 1000.

- The symbols are always printed in roman type, regardless of the type used in the rest of the text. The only exception to this is in the use of the symbol for liter, where the use of the lowercase el (l) may be confused with the numeral one (1). In this case, "liter" should be written out in full, or the script ℓ or L should be used.

- Symbols are not pluralized: 1 kg, 45 kg (not 45 kgs).

- A period after a symbol is not used, except when the symbol occurs at the end of a sentence.

- When symbols consist of letters, there is always a full space between the quantity and the symbols: 45 kg (not 45kg). However, when the first character of a symbol is not a letter, no space is left: 32°C (not 32° C or 32 °C); or 42°12′45″ (not 42° 12′ 45″).

- All symbols are written in lowercase, except when the unit is derived from a proper name: m for meter; s for second; A for ampere, Wb for weber, N for newton, W for watt.

- Prefixes are printed without spacing between the prefix and the unit symbol (e.g., km is the symbol for kilometer). (See Table 1.5 for a list of SI prefixes.)

- In text, symbols should be used when associated with a number. However, when no number is involved, the unit should be spelled out: The area of the carpet is 16 m^2, not 16 square meters. Carpet is sold by the square meter, not by the m^2.

- A practice in some countries is to use a comma as a decimal marker, while the practice in North America, the United Kingdom, and some other countries is to use a period (or dot) as the decimal marker. Furthermore, in some countries that use the decimal comma, a dot is frequently used to divide long numbers into groups of three. Because of these differing practices, spaces must be used instead of commas to separate long lines of digits into easily readable blocks of three digits with respect to the decimal marker: 32 453.246 072 5. A space (half-space preferred) is optional with a four-digit number: 1 234 or 1234.

- Where a decimal fraction of a unit is used, a zero should always be placed before the decimal marker: 0.45 kg (not .45 kg). This practice draws attention to the decimal marker and helps avoid errors of scale.

- Some confusion may arise with the word "tonne" (1000 kg). When this word occurs in French text of Canadian origin, the meaning may be a ton of 2000 pounds.

Table 1.5 SI Prefixes[*]

prefix	symbol	value
exa	E	10^{18}
peta	P	10^{15}
tera	T	10^{12}
giga	G	10^{9}
mega	M	10^{6}
kilo	k	10^{3}
hecto	h	10^{2}
deka	da	10^{1}
deci	d	10^{-1}
centi	c	10^{-2}
milli	m	10^{-3}
micro	μ	10^{-6}
nano	n	10^{-9}
pico	p	10^{-12}
femto	f	10^{-15}
atto	a	10^{-18}

[*]There is no "B" (billion) prefix. In fact, the word billion means 10^9 in the United States but 10^{12} in most other countries. This unfortunate ambiguity is handled by avoiding the use of the term billion.

13. CONVERSION FACTORS AND CONSTANTS

Commonly used equivalents are given in Table 1.6. Temperature conversions are given in Table 1.7. Table 1.8 gives commonly used constants in customary U.S. and SI units, respectively. Conversion factors are given in Table 1.9.

Table 1.6 Commonly Used Equivalents

1 gal of water weighs	8.34 lbf
1 ft^3 of water weighs	62.4 lbf
1 in^3 of mercury weighs	0.491 lbf
The mass of 1 m^3 of water is	1000 kg
1 mg/L is[*]	8.34 lbf/Mgal

[*]mg/L is a mass concentration, and its equivalent should also be a mass concentration. However, the *NCEES Handbook* lists a weight concentration.

Table 1.7 Temperature Conversions

$$°F = 1.8(°C) + 32°$$

$$°C = \frac{°F - 32°}{1.8}$$

$$°R = °F + 459.69°$$

$$K = °C + 273.15°$$

Table 1.8 Fundamental Constants

quantity	symbol	customary U.S.	SI
Charge			
electron	e		-1.6022×10^{-19} C
proton	p		$+1.6021 \times 10^{-19}$ C
Density			
air [STP, 32°F, (0°C)]		0.0805 lbm/ft^3	1.29 kg/m^3
air [70°F, (20°C), 1 atm]		0.0749 lbm/ft^3	1.20 kg/m^3
earth [mean]		345 lbm/ft^3	5520 kg/m^3
mercury		849 lbm/ft^3	1.360×10^4 kg/m^3
seawater		64.0 lbm/ft^3	1025 kg/m^3
water [mean]		62.4 lbm/ft^3	1000 kg/m^3
Distance [mean]			
earth radius		2.09×10^7 ft	6.370×10^6 m
earth-moon separation		1.26×10^9 ft	3.84×10^8 m
earth-sun separation		4.89×10^{11} ft	1.49×10^{11} m
moon radius		5.71×10^6 ft	1.74×10^6 m
sun radius		2.28×10^9 ft	6.96×10^8 m
first Bohr radius	a_0	1.736×10^{-10} ft	5.292×10^{-11} m
Gravitational Acceleration			
earth [mean]	g	$32.174 \ (32.2)$ ft/sec^2	$9.807 \ (9.81)$ m/s^2
moon [mean]		5.47 ft/sec^2	1.67 m/s^2
Mass			
atomic mass unit	u	3.66×10^{-27} lbm	1.6606×10^{-27} kg
earth		1.32×10^{25} lbm	6.00×10^{24} kg
electron [rest]	m_e	2.008×10^{-30} lbm	9.109×10^{-31} kg
moon		1.623×10^{23} lbm	7.36×10^{22} kg
neutron [rest]	m_n	3.693×10^{-27} lbm	1.675×10^{-27} kg
proton [rest]	m_p	3.688×10^{-27} lbm	1.673×10^{-27} kg
sun		4.387×10^{30} lbm	1.99×10^{30} kg
Pressure, atmospheric		$14.696 \ (14.7)$ lbf/in^2	1.0133×10^5 Pa
Temperature, standard		$32°F \ (492°R)$	$0°C \ (273K)$
Velocity			
earth escape (from surface, average)		3.67×10^4 ft/sec	1.12×10^4 m/s
light [vacuum]	c	9.84×10^8 ft/sec	$2.99792 \ (3.00) \times 10^8$ m/s
sound [air, STP]	a	1090 ft/sec	331 m/s
[air, 70°F (20°C)]		1130 ft/sec	344 m/s
Volume			
molar ideal gas [STP]	V_m	359 ft^3/lbmol	22.414 m^3/kmol
			$22\,414$ L/kmol
Fundamental Constants			
Avogadro's number	N_A		$6.0221 \ (6.022) \times 10^{23}$ mol^{-1}
Bohr magneton	μ_B		9.2732×10^{-24} J/T
Boltzmann constant	k	5.65×10^{-24} ft-lbf/°R	1.3807×10^{-23} J/K
Faraday constant	F		$96\,485$ C/mol
gravitational constant	g_c	$32.174 \ (32.2)$ lbm-ft/lbf-sec^2	
gravitational constant	G	3.44×10^{-8} ft^4/lbf-sec^4	6.673×10^{-11} N·m^2/kg^2 (m^3/kg·s^2)
nuclear magneton	μ_N		5.050×10^{-27} J/T
permeability of a vacuum	μ_0		1.2566×10^{-6} N/A^2 (H/m)
permittivity of a vacuum	ϵ_0		$8.854 \ (8.85) \times 10^{-12}$ C^2/N·m^2 (F/m)
Planck's constant	h		6.6256×10^{-34} J·s
Rydberg constant	R_∞		1.097×10^7 m^{-1}
specific gas constant, air	R	53.3 ft-lbf/lbm-°R	287 J/kg·K
Stefan-Boltzmann constant	σ	1.71×10^{-9} Btu/ft^2-hr-°R^4	5.67×10^{-8} W/m^2·K^4
triple point, water		$32.02°F, 0.0888$ psia	$0.01109°C, 0.6123$ kPa
universal gas constant	$\overline{R}$	1545 ft-lbf/lbmol-°R	8314 J/kmol·K
	$\overline{R}$	1.986 Btu/lbmol-°R	8.314 kPa·m^3/kmol·K
			0.08206 atm·L/mol·K

Mathematics

Table 1.9 *Conversion Factors*

multiply	by	to obtain
ac	43,560	ft^2
ampere-hr	3600	coulomb
angstrom	1×10^{-10}	m
atm	76.0	cm Hg
atm	29.92	in Hg
atm	14.70	lbf/in^2 (psia)
atm	33.90	ft water
atm	1.013×10^5	Pa
bar	1×10^5	Pa
bar	0.987	atm
barrels of oil	42	gallons of oil
Btu	1055	J
Btu	2.928×10^{-4}	kW·h
Btu	778	ft-lbf
Btu/hr	3.930×10^{-4}	hp
Btu/hr	0.293	W
Btu/hr	0.216	ft-lbf/sec
cal (g-cal)	3.968×10^{-3}	Btu
cal	1.560×10^{-6}	hp-hr
cal (g-cal)	4.186	J
cal/sec	4.184	W
cm	3.281×10^{-2}	ft
cm	0.394	in
cP	0.001	Pa·s
cP	1	g/m·s
cP	2.419	lbm/hr-ft
cSt	1×10^{-6}	m^2/s
cfs	0.646371	MGD
ft^3	7.481	gal
m^3	1000	L
eV	1.602×10^{-19}	J
ft	30.48	cm
ft	0.3048	m
ft-lbf	1.285×10^{-3}	Btu
ft-lbf	3.766×10^{-7}	kW·h
ft-lbf	0.324	g-cal
ft-lbf	1.35582	J
ft-lbf/sec	1.818×10^{-3}	hp
gal	3.785	L
gal	0.134	ft^3
gal water	8.3453	lbf water
gamma (γ, Γ)[*]	1×10^{-9}	T
gauss	1×10^{-4}	T
gram	2.205×10^{-3}	lbm
hectare	1×10^{-4}	m^2
hectare	2.47104	ac
hp	42.4	Btu/min
hp	745.7	W
hp	33,000	ft-lbf/min
hp	550	ft-lbf/sec
hp-hr	2545	Btu
hp-hr	1.98×10^{-4}	ft-lbf
hp-hr	2.68×10^{-4}	J
hp-hr	0.746	kW·h
in	2.54	cm
in of Hg	0.0334	atm
in of Hg	13.60	in of H$_2$O
in of H$_2$O	0.0361	lbf/in^2
in of H$_2$O	0.002458	atm
J	9.478×10^{-4}	Btu
J	0.7376	ft-lbf
J	1	N·m

multiply	by	to obtain
J/s	1	W
kg	2.205	lbm
kgf	9.8066	N
km	3281	ft
km/h	0.621	mi/hr
kPa	0.145	lbf/in^2
kW	1.341	hp
kW	737.6	ft-lbf/sec
kW	3413	Btu/hr
kW·h	3413	Btu
kW·h	1.341	hp-hr
kW·h	3.6×10^6	J
kip	1000	lbf
kip	4448	N
L	61.02	in^3
L	0.264	gal
L	10×10^{-3}	m^3
L/s	2.119	ft^3/min
L/s	15.85	gal/min
m	3.281	ft
m	1.094	yd
m	196.8	ft/min
mi	5280	ft
mi	1.609	km
mph	88.0	ft/min
mph	1.609	kph
mm of Hg	1.316×10^{-3}	atm
mm of H$_2$O	9.678×10^{-5}	atm
N	0.225	lbf
N	1	kg·m/s^2
N·m	0.7376	ft-lbf
N·m	1	J
Pa	9.869×10^{-6}	atm
Pa	1	N/m^2
Pa·s	10	P
lbm	0.454	kg
lbf	4.448	N
lbf-ft	1.356	N·m
lbf/in^2	0.068	atm
lbf/in^2	2.307	ft water
lbf/in^2	2.036	in Hg
lbf/in^2	6895	Pa
radian	180/π	deg
stokes	1×10^{-4}	m^2/s
therm	10^5	Btu
ton (metric)	1000	kg
ton (short)	2000	lbf
W	3.413	Btu/hr
W	1.341×10^{-3}	hp
W	1	J/s
Wb/m^2	10,000	gauss

(Atmospheres are standard; calories are gram-calories; gallons are U.S. liquid; miles are statute; pounds-mass are avoirdupois.)

[*]The *NCEES Handbook* is incorrect in assigning uppercase gamma (Γ) as the symbol for the magnetic field intensity unit gamma. Only the symbol γ is used.

2 Algebra

1. LOGARITHMS

Logarithms can be considered to be exponents. In the equation $b^c = x$, for example, the exponent c is the logarithm of x to the base b. The two equations $\log_b x = c$ and $b^c = x$ are equivalent.

Equation 2.1 Through Eq. 2.3: Common and Natural Logarithms

$$\log_b(x) = c \quad [b^c = x] \qquad 2.1$$
$$\ln x \quad [\text{base} = e] \qquad 2.2$$
$$\log x \quad [\text{base} = 10] \qquad 2.3$$

Description

Although any number may be used as a base for logarithms, two bases are most commonly used in engineering. The base for a *common logarithm* is 10. The notation used most often for common logarithms is *log*, although log_{10} is sometimes seen.

The base for a *natural logarithm* is 2.71828..., an irrational number that is given the symbol e. The most common notation for a natural logarithm is *ln*, but log_e is sometimes seen.

Example

What is the value of $\log_{10} 1000$?

(A) 2

(B) 3

(C) 8

(D) 10

Solution

$\log_{10} 1000$ is the power of 10 that produces 1000. Use Eq. 2.1.

$$\log_b(x) = c \quad [b^c = x]$$
$$\log_{10} 1000 = c$$
$$10^c = 1000$$
$$c = 3$$

The answer is (B).

Equation 2.4 Through Eq. 2.10: Logarithmic Identities

$$\log_b b^n = n \qquad 2.4$$
$$\log x^c = c \log x \qquad 2.5$$
$$x^c = \text{antilog}(c \log x) \qquad 2.6$$
$$\log xy = \log x + \log y \qquad 2.7$$
$$\log_b b = 1 \qquad 2.8$$
$$\log 1 = 0 \qquad 2.9$$
$$\log x/y = \log x - \log y \qquad 2.10$$

Description

Logarithmic identities are useful in simplifying expressions containing exponentials and other logarithms.

Example

Which of the following is equal to $(0.001)^{2/3}$?

(A) $\text{antilog}\left(\frac{3}{2} \log 0.001\right)$

(B) $\frac{2}{3} \text{antilog}(\log 0.001)$

(C) $\text{antilog}\left(\log \dfrac{0.001}{\frac{2}{3}}\right)$

(D) $\text{antilog}\left(\frac{2}{3} \log 0.001\right)$

Solution

Use Eq. 2.5 and Eq. 2.6.

$$\log x^c = c \log x$$
$$\log (0.001)^{2/3} = \frac{2}{3} \log 0.001$$
$$(0.001)^{2/3} = \text{antilog}\left(\frac{2}{3} \log 0.001\right)$$

The answer is (D).

Equation 2.11: Changing the Base

$$\log_b x = (\log_a x)/(\log_a b) \qquad 2.11$$

Variations

$$\log_{10} x = \ln x \log_{10} e$$

$$\ln x = \frac{\log_{10} x}{\log_{10} e}$$
$$\approx 2.302585 \log_{10} x$$

Description

Equation 2.11 is often useful for calculating a logarithm with any base quickly when the available resources produce only natural or common logarithms. Equation 2.11 can also be used to convert a logarithm to a different base, such as from a common logarithm to a natural logarithm.

Example

Given that $\log_{10} 5 = 0.6990$ and $\log_{10} 9 = 0.9542$, what is the value of $\log_5 9$?

(A) 0.2550

(B) 0.7330

(C) 1.127

(D) 1.365

Solution

Use Eq. 2.11.

$$\log_b x = (\log_a x)/(\log_a b)$$
$$\log_5 9 = \frac{\log_{10} 9}{\log_{10} 5} = \frac{0.9542}{0.6990}$$
$$= 1.365$$

The answer is (D).

2. COMPLEX NUMBERS

A *complex number* is the sum of a *real number* and an *imaginary number*. Real numbers include the *rational numbers* and the *irrational numbers*, while imaginary numbers represent the square roots of negative numbers. Every imaginary number can be expressed in the form ib, where i represents the square root of -1 and b is a real number. Another term for i is the *imaginary unit vector*.

$$i = \sqrt{-1}$$

j is commonly used to represent the imaginary unit vector in the fields of electrical engineering and control systems engineering to avoid confusion with the variable for current, i.[1]

$$j = \sqrt{-1}$$

When a complex number is expressed in the form $a + ib$, the complex number is said to be in *rectangular* or *trigonometric form*. In the expression $a + ib$, a is the real component (or real part), and b is the imaginary component (or imaginary part). (See Fig. 2.1.)

Figure 2.1 Graphical Representation of a Complex Number

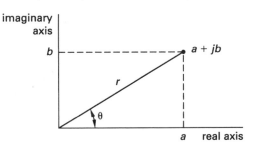

Most algebraic operations (addition, multiplication, exponentiation, etc.) work with complex numbers. When adding two complex numbers, real parts are added to real parts, and imaginary parts are added to imaginary parts.

$$(a + jb) + (c + jd) = (a + c) + j(b + d)$$

$$(a + jb) - (c + jd) = (a - c) + j(b - d)$$

Multiplication of two complex numbers in rectangular form uses the algebraic distributive law and the equivalency $j^2 = -1$.

$$(a + jb)(c + jd) = (ac - bd) + j(ad + bc)$$

Division of complex numbers in rectangular form requires use of the *complex conjugate*. The complex conjugate of a complex number $a + jb$ is $a - jb$. When both the numerator and the denominator are multiplied by the complex conjugate of the denominator, the denominator becomes the real number $a^2 + b^2$. This technique is known as *rationalizing the denominator*.

$$\frac{a + jb}{c + jd} = \frac{(a + jb)(c - jd)}{(c + jd)(c - jd)} = \frac{(ac + bd) + j(bc - ad)}{c^2 + d^2}$$

---

[1]The NCEES *FE Reference Handbook* (*NCEES Handbook*) uses only j to represent the imaginary unit vector. This book uses both j and i.

Example

Which of the following is most nearly equal to $(7 + 5.2j)/(3 + 4j)$?

   (A) $-0.3 + 1.8j$

   (B) $1.7 - 0.5j$

   (C) $2.3 - 1.2j$

   (D) $2.3 + 1.3j$

Solution

When the numerator and denominator are multiplied by the complex conjugate of the denominator, the denominator becomes a real number.

$$\frac{a + jb}{c + jd} = \frac{(a + jb)(c - jd)}{(c + jd)(c - jd)} = \frac{(ac + bd) + j(bc - ad)}{c^2 + d^2}$$

$$\frac{7 + 5.2j}{3 + 4j} = \frac{((7)(3) + (5.2)(4)) + j((5.2)(3) - (7)(4))}{(3)^2 + (4)^2}$$

$$= 1.672 - 0.496j \quad (1.7 - 0.5j)$$

The answer is (B).

3. POLAR COORDINATES

Equation 2.12: Polar Form of a Complex Number

$$x + jy = r(\cos\theta + j\sin\theta) = re^{j\theta} \qquad 2.12$$

Variations

$$z \equiv r(\cos\theta + i\sin\theta)$$

$$z \equiv r\operatorname{cis}\theta$$

$$z = r\angle\theta$$

Description

A complex number can be expressed in the *polar form* $r(\cos\theta + j\sin\theta)$, where θ is the angle from the x-axis and r is the distance from the origin. r and θ are the *polar coordinates* of the complex number. Another notation for the polar form of a complex number is $re^{j\theta}$.

Equation 2.13 and Eq. 2.14: Converting from Polar Form to Rectangular Form

$$x = r\cos\theta \qquad 2.13$$

$$y = r\sin\theta \qquad 2.14$$

Description

The rectangular form of a complex number, $x + jy$, can be determined from the complex number's polar coordinates r and θ using Eq. 2.13 and Eq. 2.14.

Equation 2.15 and Eq. 2.16: Converting from Rectangular Form to Polar Form

$$r = |x + jy| = \sqrt{x^2 + y^2} \qquad 2.15$$

$$\theta = \arctan(y/x) \qquad 2.16$$

Description

The polar form of a complex number, $r(\cos\theta + j\sin\theta)$, can be determined from the complex number's rectangular coordinates x and y using Eq. 2.15 and Eq. 2.16.

Example

The rectangular coordinates of a complex number are $(4, 6)$. What are the complex number's approximate polar coordinates?

   (A) $(4.0, 33°)$

   (B) $(4.0, 56°)$

   (C) $(7.2, 33°)$

   (D) $(7.2, 56°)$

Solution

The radius and angle of the polar form can be determined from the x- and y-coordinates using Eq. 2.15 and Eq. 2.16.

$$r = \sqrt{x^2 + y^2} = \sqrt{(4)^2 + (6)^2}$$

$$= 7.211 \quad (7.2)$$

$$\theta = \arctan(y/x) = \arctan\frac{6}{4}$$

$$= 56.3° \quad (56°)$$

The answer is (D).

Equation 2.17 and Eq. 2.18: Multiplication and Division with Polar Forms

$$[r_1(\cos\theta_1 + j\sin\theta_1)][r_2(\cos\theta_2 + j\sin\theta_2)]$$
$$= r_1 r_2 [\cos(\theta_1 + \theta_2) + j\sin(\theta_1 + \theta_2)] \qquad 2.17$$

$$\frac{r_1(\cos\theta_1 + j\sin\theta_1)}{r_2(\cos\theta_2 + j\sin\theta_2)} = \frac{r_1}{r_2}[\cos(\theta_1 - \theta_2) + j\sin(\theta_1 - \theta_2)]$$
$$\qquad 2.18$$

Mathematics

Mathematics

Variations

$$z_1 z_2 = (r_1 r_2) \angle (\theta_1 + \theta_2)$$

$$\frac{z_1}{z_2} = \frac{r_1}{r_2} \angle (\theta_1 - \theta_2)$$

Description

The multiplication and division rules defined for complex numbers expressed in rectangular form can be applied to complex numbers expressed in polar form. Using the trigonometric identities, these rules reduce to Eq. 2.17 and Eq. 2.18.

Equation 2.19: de Moivre's Formula

$$(x + jy)^n = [r(\cos \theta + j \sin \theta)]^n$$
$$= r^n (\cos n\theta + j \sin n\theta) \qquad \textbf{2.19}$$

Description

Equation 2.19 is *de Moivre's formula*. This equation is valid for any real number x and integer n.

Equation 2.20 Through Eq. 2.23: Euler's Equations

$$e^{j\theta} = \cos \theta + j \sin \theta \qquad \textbf{2.20}$$

$$e^{-j\theta} = \cos \theta - j \sin \theta \qquad \textbf{2.21}$$

$$\cos \theta = \frac{e^{j\theta} + e^{-j\theta}}{2} \qquad \textbf{2.22}$$

$$\sin \theta = \frac{e^{j\theta} - e^{-j\theta}}{2j} \qquad \textbf{2.23}$$

Description

Complex numbers can also be expressed in exponential form. The relationship of the exponential form to the trigonometric form is given by *Euler's equations*, also known as *Euler's identities*.

Example

If $j = \sqrt{-1}$, which of the following is equal to j^j?

(A) j^2

(B) e^{2j}

(C) -1

(D) $e^{-\frac{\pi}{2}}$

Solution

j is the imaginary unit vector, so $r = 1$ and $\theta = 90°(\pi/2)$ in Fig. 2.1. From Eq. 2.19,

$$(j)^n = (\cos \theta + j \sin \theta)^n$$

From Eq. 2.20,

$$e^{j\theta} = \cos \theta + j \sin \theta$$

Since $\theta = \pi/2$,

$$j^j = \left(e^{j\frac{\pi}{2}}\right)^j = e^{j^2\frac{\pi}{2}} = e^{-\frac{\pi}{2}}$$

The answer is (D).

4. POLYNOMIALS

A *polynomial* is a rational expression—usually the sum of several variable terms known as *monomials*—that does not involve division. The *degree of the polynomial* is the highest power to which a variable in the expression is raised. The following *standard polynomial forms* are useful when trying to find the roots of an equation.

$$(a + b)(a - b) = a^2 - b^2$$

$$(a \pm b)^2 = a^2 \pm 2ab + b^2$$

$$(a \pm b)^3 = a^3 \pm 3a^2 b + 3ab^2 \pm b^3$$

$$(a^3 \pm b^3) = (a \pm b)(a^2 \mp ab + b^2)$$

$$(a^n - b^n) = (a - b)\left(\begin{array}{c} a^{n-1} + a^{n-2}b + a^{n-3}b^2 \\ + \cdots + b^{n-1} \end{array}\right)$$

[n is any positive integer]

$$(a^n + b^n) = (a + b)\left(\begin{array}{c} a^{n-1} - a^{n-2}b + a^{n-3}b^2 \\ - \cdots + b^{n-1} \end{array}\right)$$

[n is any positive odd integer]

The *binomial theorem* defines a polynomial of the form $(a + b)^n$.

$$(a + b)^n = \underset{[i=0]}{a^n} + \underset{[i=1]}{na^{n-1}b} + \underset{[i=2]}{C_2 a^{n-2}b^2}$$
$$+ \cdots + C_i a^{n-i}b^i + \cdots + nab^{n-1} + b^n$$

$$C_i = \frac{n!}{i!(n-i)!} \qquad [i = 0, 1, 2, \ldots, n]$$

The coefficients of the expansion can be determined quickly from *Pascal's triangle*—each entry is the sum of the two entries directly above it. (See Fig. 2.2.)

The values $x_1, x_2, \ldots, x_n$ of the independent variable x that satisfy a polynomial equation $f(x) = 0$ are known as *roots* or *zeros* of the polynomial. A polynomial of degree n with real coefficients will have at most n real roots, although they need not all be distinctly different.

Figure 2.2 Pascal's Triangle

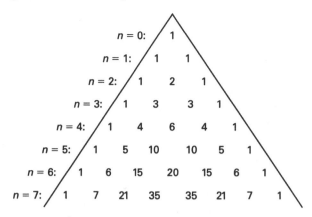

5. POLYNOMIAL FUNCTIONS

Equation 2.24 and Eq. 2.25: Quadratic Equations

$$ax^2 + bx + c = 0 \qquad 2.24$$

$$x = \frac{-b \pm \sqrt{b^2 - 4ac}}{2a} \qquad 2.25$$

Description

A *quadratic equation* is a second-degree polynomial equation with a single variable. A quadratic equation can be written in the form of Eq. 2.24, where x is the variable and a, b, and c are constants. (If a is zero, the equation is linear.)

The *roots*, x_1 and x_2, of a quadratic equation are the two values of x that satisfy the equation (i.e., make it true). These values can be found from the *quadratic formula*, Eq. 2.25.

The quantity under the radical in Eq. 2.25 is called the *discriminant*. By inspecting the discriminant, the types of roots of the equation can be determined.

- If $b^2 - 4ac > 0$, the roots are real and unequal.
- If $b^2 - 4ac = 0$, the roots are real and equal. This is known as a *double root*.
- If $b^2 - 4ac < 0$, the roots are complex and unequal.

Example

What are the roots of the quadratic equation $-7x + x^2 = -10$?

(A) -5 and 2

(B) -2 and 0.4

(C) 0.4 and 2

(D) 2 and 5

Solution

Rearrange the equation into the form of Eq. 2.24.

$$x^2 + (-7x) + 10 = 0$$

Use the quadratic formula, Eq. 2.25, with $a = 1$, $b = -7$, and $c = 10$.

$$x = \frac{-b \pm \sqrt{b^2 - 4ac}}{2a}$$

$$= \frac{-(-7) \pm \sqrt{(-7)^2 - (4)(1)(10)}}{(2)(1)}$$

$$= 2 \text{ and } 5$$

The answer is (D).

6. ROOTS

Equation 2.26: kth Roots of a Complex Number

$$w = \sqrt[k]{r}\left[\cos\left(\frac{\theta}{k} + n\frac{360°}{k}\right) + j\sin\left(\frac{\theta}{k} + n\frac{360°}{k}\right)\right] \qquad 2.26$$

Description

Use Eq. 2.26 to find the *kth root* of the complex number $z = r(\cos\theta + j\sin\theta)$. n can be any integer number.

Example

What is the cube root of the complex number $8e^{j60°}$?

(A) $2(\cos 60° + j\sin 60°)$

(B) $2(j\cos 20° + \sin 20°)$

(C) $2.7(\cos 20° + j\sin 20°)$

(D) $2(\cos(20° + 120°n) + j\sin(20° + 120°n))$

Solution

From Eq. 2.26, the *k*th root of a complex number is

$$w = \sqrt[k]{r}\left[\cos\left(\frac{\theta}{k} + n\frac{360°}{k}\right) + j\sin\left(\frac{\theta}{k} + n\frac{360°}{k}\right)\right]$$

$$= \sqrt[3]{8}\left(\begin{array}{c}\cos\left(\frac{60°}{3} + n\left(\frac{360°}{3}\right)\right) \\ + j\sin\left(\frac{60°}{3} + n\left(\frac{360°}{3}\right)\right)\end{array}\right)$$

$$= 2\left(\cos(20° + 120°n) + j\sin(20° + 120°n)\right)$$

$$[n = 0, 1, 2, \ldots]$$

The answer is (D).

7. PROGRESSIONS AND SERIES

A *progression* or *sequence*, $\{A\}$, is an ordered set of numbers a_i, such as 1, 4, 9, 16, 25, ... The *terms* in a sequence can be all positive, negative, or of alternating signs. l is the last term and is also known as the *general term* of the sequence.

$$\{A\} = a_1, a_2, a_3, \ldots, l$$

A sequence is said to *diverge* (i.e., be *divergent*) if the terms approach infinity, and it is said to *converge* (i.e., be *convergent*) if the terms approach any finite value (including zero).

A *series* is the sum of terms in a sequence. There are two types of series: A *finite series* has a finite number of terms. An *infinite series* has an infinite number of terms, but this does not imply that the sum is infinite. The main tasks associated with series are determining the sum of the terms and determining whether the series converges. A series is said to converge if the sum, S_n, of its terms exists. A finite series is always convergent. An infinite series may be convergent.

Equation 2.27 and Eq. 2.28: Arithmetic Progression

$$l = a + (n-1)d \qquad \text{2.27}$$

$$S = n(a+l)/2 = n[2a + (n-1)d]/2 \qquad \text{2.28}$$

Description

The *arithmetic progression* is a standard sequence that diverges. It has the form shown in Eq. 2.27.

In Eq. 2.27 and Eq. 2.28, a is the *first term*, d is a constant called the *common difference*, and n is the number of terms.

The difference of adjacent terms is constant in an arithmetic progression. The sum of terms in a finite arithmetic series is shown by Eq. 2.28.

Example

What is the sum of the following finite sequence of terms?

$$18, 25, 32, 39, \ldots, 67$$

(A) 181

(B) 213

(C) 234

(D) 340

Solution

Each term is 7 more than the previous term. This is an arithmetic sequence. The general mathematical representation for an arithmetic sequence is

$$l = a + (n-1)d$$

In this case, the difference term is $d = 7$. The first term is $a = 18$, and the last term is $l = 67$.

$$l = a + (n-1)d$$

$$n = \frac{l-a}{d} + 1$$

$$= \frac{67-18}{7} + 1$$

$$= 8$$

The sum of n terms is

$$S = n[2a + (n-1)d]/2$$

$$= \frac{(8)\big((2)(18) + (8-1)(7)\big)}{2}$$

$$= 340$$

The answer is (D).

Equation 2.29 Through Eq. 2.32: Geometric Progression

$$l = ar^{n-1} \qquad \text{2.29}$$

$$S = a(1-r^n)/(1-r) \quad [r \neq 1] \qquad \text{2.30}$$

$$S = (a-rl)/(1-r) \quad [r \neq 1] \qquad \text{2.31}$$

$$\lim_{n \to \infty} S_n = a/(1-r) \quad [r < 1] \qquad \text{2.32}$$

Variations

$$S_n = \sum_{i=1}^{n} ar^{i-1} = \frac{a-rl}{1-r} = \frac{a(1-r^n)}{1-r}$$

$$S_n = \sum_{i=1}^{\infty} ar^{i-1} = \frac{a}{1-r}$$

Description

The *geometric progression* is another standard sequence. The quotient of adjacent terms is constant in a geometric progression. It converges for $-1 < r < 1$ and diverges otherwise.

In Eq. 2.29 through Eq. 2.32, a is the first term, and r is known as the *common ratio*.

The sum of a finite geometric series is given by Eq. 2.30 and Eq. 2.31. The sum of an infinite geometric series is given by Eq. 2.32.

Example

What is the sum of the following geometric sequence?

$$32, 80, 200, \ldots, 19531.25$$

(A) 21,131.25

(B) 24,718.25

(C) 31,250.00

(D) 32,530.75

Solution

The common ratio is

$$r = \frac{80}{32} = \frac{200}{80} = 2.5$$

Since the ratio and both the initial and final terms are known, the sum can be found using Eq. 2.31.

$$S = (a - rl)/(1 - r)$$
$$= \frac{32 - (2.5)(19531.25)}{1 - 2.5}$$
$$= 32,530.75$$

The answer is (D).

Equation 2.33 Through Eq. 2.36: Properties of Series

$$\sum_{i=1}^{n} c = nc \qquad 2.33$$

$$\sum_{i=1}^{n} cx_i = c\sum_{i=1}^{n} x_i \qquad 2.34$$

$$\sum_{i=1}^{n}(x_i + y_i - z_i) = \sum_{i=1}^{n} x_i + \sum_{i=1}^{n} y_i - \sum_{i=1}^{n} z_i \qquad 2.35$$

$$\sum_{x=1}^{n} x = (n + n^2)/2 \qquad 2.36$$

Description

Equation 2.33 through Eq. 2.36 list some basic properties of series. The terms x_i, y_i, and z_i represent general terms in any series. Equation 2.33 describes the obvious result of n repeated additions of a constant, c. Equation 2.34 shows that the product of a constant, c, and a serial summation of series terms is distributive. Equation 2.35 shows that addition of series is associative. Equation 2.36 gives the sum of n consecutive integers. This is not really a property of series in general; it is the property of a special kind of arithmetic sequence. It is a useful identity for use with *sum-of-the-years' depreciation*.

Equation 2.37: Power Series

$$\sum_{i=0}^{\infty} a_i(x - a)^i \qquad 2.37$$

Variation

$$\sum_{i=1}^{n} a_i x^{i-1} = a_1 + a_2 x + a_3 x^2 + \cdots + a_n x^{n-1}$$

Description

A *power series* is a series of the form shown in Eq. 2.37. The *interval of convergence* of a power series consists of the values of x for which the series is convergent. Due to the exponentiation of terms, an infinite power series can only be convergent in the interval $-1 < x < 1$.

A power series may be used to represent a function that is continuous over the interval of convergence of the series. The *power series representation* may be used to find the derivative or integral of that function.

Power series behave similarly to polynomials: They may be added together, subtracted from each other, multiplied together, or divided term by term within the interval of convergence. They may also be differentiated and integrated within their interval of convergence. If $f(x) = \sum_{i=1}^{n} a_i x^i$, then over the interval of convergence,

$$f'(x) = \sum_{i=1}^{n} \frac{d(a_i x^i)}{dx}$$

$$\int f(x)\,dx = \sum_{i=1}^{n} \int a_i x^i\,dx$$

Equation 2.38: Taylor's Series

$$f(x) = f(a) + \frac{f'(a)}{1!}(x - a) + \frac{f''(a)}{2!}(x - a)^2$$
$$+ \cdots + \frac{f^{(n)}(a)}{n!}(x - a)^n + \cdots \qquad 2.38$$

Description

Taylor's series (*Taylor's formula*), Eq. 2.38, can be used to expand a function around a point (i.e., to approximate the function at one point based on the function's value at another point). The approximation consists of a series, each term composed of a derivative of the original function and a polynomial. Using Taylor's formula requires that the original function be continuous in the interval $[a, b]$. To expand a function, $f(x)$, around a point, a, in order to obtain $f(b)$, use Eq. 2.38.

If $a = 0$, Eq. 2.38 is known as a *Maclaurin series.*

To be a useful approximation, two requirements must be met: (1) Point a must be relatively close to point b, and (2) the function and its derivatives must be known or be easy to calculate.

Example

Taylor's series is used to expand the function $f(x)$ about $a = 0$ to obtain $f(b)$.

$$f(x) = \frac{1}{3x^3 + 4x + 8}$$

What are the first two terms of Taylor's series?

(A) $\dfrac{1}{16} + \dfrac{b}{8}$

(B) $\dfrac{1}{8} - \dfrac{b}{16}$

(C) $\dfrac{1}{8} + \dfrac{b}{16}$

(D) $\dfrac{1}{4} - \dfrac{b}{16}$

Solution

The first two coefficient terms of Taylor's series are

$$f(0) = \frac{1}{(3)(0)^3 + (4)(0) + 8}$$

$$= 1/8$$

$$f'(x) = \frac{-(9x^2 + 4)}{(3x^3 + 4x + 8)^2}$$

$$f'(0) = \frac{-((9)(0)^2 + 4)}{((3)(0)^3 + (4)(0) + 8)^2} = \frac{-4}{64} = -1/16$$

Using Eq. 2.38, find the first two complete terms of Taylor's series.

$$f(b) = f(a) + \frac{f'(a)}{1!}(b - a)$$

$$= \frac{1}{8} + \frac{\left(\frac{-1}{16}\right)(b - 0)}{1}$$

$$= \frac{1}{8} - \frac{b}{16}$$

The answer is (B).

3 Vectors

1. VECTORS

A physical property or quantity can be a scalar, vector, or tensor. A *scalar* has only magnitude. Knowing its value is sufficient to define a scalar. Mass, enthalpy, density, and speed are examples of scalars.

Force, momentum, displacement, and velocity are examples of *vectors*. A vector is a directed straight line with a specific magnitude. A vector is specified completely by its direction (consisting of the vector's *angular orientation* and its *sense*) and magnitude. A vector's *point of application* (*terminal point*) is not needed to define the vector. Two vectors with the same direction and magnitude are said to be equal vectors even though their *lines of action* may be different.

Unit vectors are vectors with unit magnitudes (i.e., magnitudes of one). They are represented in the same notation as other vectors. Although they can have any direction, the standard unit vectors (i.e., the *Cartesian unit vectors*, **i**, **j**, and **k**) have the directions of the *x*-, *y*-, and *z*-coordinate axes, respectively, and constitute the *Cartesian triad*.

A *tensor* has magnitude in a specific direction, but the direction is not unique. A tensor in three-dimensional space is defined by nine components, compared with the three that are required to define vectors. These components are written in matrix form. Stress, dielectric constant, and magnetic susceptibility are examples of tensors.

Equation 3.1: Components of a Vector

$$\mathbf{A} = a_x\mathbf{i} + a_y\mathbf{j} + a_z\mathbf{k} \qquad 3.1$$

Description

A vector **A** can be written in terms of unit vectors and its components. (See Fig. 3.1.)

If a vector is based (i.e., starts) at the origin $(0, 0, 0)$, its magnitude (length) can be calculated as

$$|\mathbf{A}| = L_{\mathbf{A}} = \sqrt{a_x^2 + a_y^2 + a_z^2}$$

Figure 3.1 *Components of a Vector*

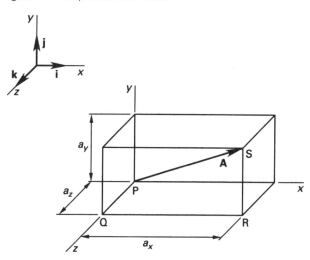

Example

Find the unit vector (i.e., the direction vector) associated with the origin-based vector $18\mathbf{i} + 3\mathbf{j} + 29\mathbf{k}$.

(A) $0.525\mathbf{i} + 0.088\mathbf{j} + 0.846\mathbf{k}$

(B) $0.892\mathbf{i} + 0.178\mathbf{j} + 0.416\mathbf{k}$

(C) $1.342\mathbf{i} + 0.868\mathbf{j} + 2.437\mathbf{k}$

(D) $6\mathbf{i} + \mathbf{j} + \frac{29}{3}\mathbf{k}$

Solution

The unit vector of a particular vector is the vector itself divided by its length.

$$\text{unit vector} = \frac{18\mathbf{i} + 3\mathbf{j} + 29\mathbf{k}}{\sqrt{(18)^2 + (3)^2 + (29)^2}}$$

$$= 0.525\mathbf{i} + 0.088\mathbf{j} + 0.846\mathbf{k}$$

The answer is (A).

Equation 3.2: Vector Addition

$$\mathbf{A} + \mathbf{B} = (a_x + b_x)\mathbf{i} + (a_y + b_y)\mathbf{j} + (a_z + b_z)\mathbf{k} \qquad 3.2$$

Description

Addition of two vectors by the *polygon method* is accomplished by placing the tail of the second vector at the

Mathematics

head (tip) of the first. The sum (i.e., the *resultant vector*) is a vector extending from the tail of the first vector to the head of the second (see Fig. 3.2). Alternatively, the two vectors can be considered as two adjacent sides of a parallelogram, while the sum represents the diagonal. This is known as addition by the *parallelogram method*. The components of the resultant vector are the sums of the components of the added vectors.

Figure 3.2 *Addition of Two Vectors*

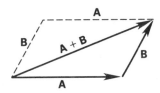

Example

What is the sum of the two vectors $5i + 3j - 7k$ and $10i - 12j + 5k$?

(A) $8i - 7j - k$

(B) $10i - 9j + 3k$

(C) $15i - 9j - 2k$

(D) $15i + 7j - 3k$

Solution

Use Eq. 3.2.

$$\mathbf{A} + \mathbf{B} = (a_x + b_x)\mathbf{i} + (a_y + b_y)\mathbf{j} + (a_z + b_z)\mathbf{k}$$
$$= (5 + 10)\mathbf{i} + (3 + (-12))\mathbf{j} + ((-7) + 5)\mathbf{k}$$
$$= 15\mathbf{i} - 9\mathbf{j} - 2\mathbf{k}$$

The answer is (C).

Equation 3.3: Vector Subtraction

$$\mathbf{A} - \mathbf{B} = (a_x - b_x)\mathbf{i} + (a_y - b_y)\mathbf{j} + (a_z - b_z)\mathbf{k} \qquad 3.3$$

Description

Vector subtraction is similar to vector addition, as shown by Eq. 3.3.

Equation 3.4 and Eq. 3.5: Vector Dot Product

$$\mathbf{A} \cdot \mathbf{B} = a_x b_x + a_y b_y + a_z b_z \qquad 3.4$$

$$\mathbf{A} \cdot \mathbf{B} = |\mathbf{A}||\mathbf{B}|\cos\theta = \mathbf{B} \cdot \mathbf{A} \qquad 3.5$$

Variation

$$\theta = \arccos\left(\frac{\mathbf{A} \cdot \mathbf{B}}{|\mathbf{A}||\mathbf{B}|}\right)$$
$$= \arccos\left(\frac{a_x b_x + a_y b_y + a_z b_z}{|\mathbf{A}||\mathbf{B}|}\right)$$

Description

The *dot product* (*scalar product*) of two vectors is a scalar that is proportional to the length of the projection of the first vector onto the second vector. (See Fig. 3.3.)

Use the variation to find the angle, θ, formed between two given vectors.

Figure 3.3 *Vector Dot Product*

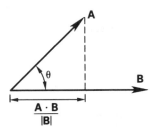

The dot product can be calculated in two ways, as Eq. 3.4 and Eq. 3.5 indicate. θ is limited to 180° and is the acute angle between the two vectors.

Example

What is the dot product, $\mathbf{A} \cdot \mathbf{B}$, of the vectors $\mathbf{A} = 2i + 4j + 8k$ and $\mathbf{B} = -2i + j - 4k$?

(A) $-4i + 4j - 32k$

(B) $-4i - 4j - 32k$

(C) -40

(D) -32

Solution

Use Eq. 3.4.

$$\mathbf{A} \cdot \mathbf{B} = a_x b_x + a_y b_y + a_z b_z$$
$$= (2)(-2) + (4)(1) + (8)(-4)$$
$$= -32$$

The answer is (D).

Equation 3.6 and Eq. 3.7: Vector Cross Product

$$\mathbf{A} \times \mathbf{B} = \begin{vmatrix} \mathbf{i} & \mathbf{j} & \mathbf{k} \\ a_x & a_y & a_z \\ b_x & b_y & b_z \end{vmatrix} = -\mathbf{B} \times \mathbf{A} \qquad 3.6$$

$$\mathbf{A} \times \mathbf{B} = |\mathbf{A}||\mathbf{B}|\mathbf{n} \sin \theta \qquad 3.7$$

Description

The *cross product (vector product)*, $\mathbf{A} \times \mathbf{B}$, of two vectors is a vector that is orthogonal (perpendicular) to the plane of the two vectors. (See Fig. 3.4.) The unit vector representation of the cross product can be calculated as a third-order determinant. $\mathbf{n}$ is the unit vector in the direction perpendicular to the plane containing $\mathbf{A}$ and $\mathbf{B}$.

Figure 3.4 Vector Cross Product

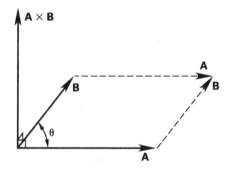

Example

What is the cross product, $\mathbf{A} \times \mathbf{B}$, of vectors $\mathbf{A}$ and $\mathbf{B}$?

$$\mathbf{A} = \mathbf{i} + 4\mathbf{j} + 6\mathbf{k}$$
$$\mathbf{B} = 2\mathbf{i} + 3\mathbf{j} + 5\mathbf{k}$$

(A) $\mathbf{i} - \mathbf{j} - \mathbf{k}$

(B) $-\mathbf{i} + \mathbf{j} + \mathbf{k}$

(C) $2\mathbf{i} + 7\mathbf{j} - 5\mathbf{k}$

(D) $2\mathbf{i} + 7\mathbf{j} + 5\mathbf{k}$

Solution

Use Eq. 3.6. The cross product of two vectors is the determinant of a third-order matrix as shown.

$$\mathbf{A} \times \mathbf{B} = \begin{bmatrix} \mathbf{i} & \mathbf{j} & \mathbf{k} \\ a_x & a_y & a_z \\ b_x & b_y & b_z \end{bmatrix} = \begin{bmatrix} \mathbf{i} & \mathbf{j} & \mathbf{k} \\ 1 & 4 & 6 \\ 2 & 3 & 5 \end{bmatrix}$$

$$= \mathbf{i}[(4)(5) - (6)(3)] - \mathbf{j}[(1)(5) - (6)(2)]$$
$$+ \mathbf{k}[(1)(3) - (4)(2)]$$

$$= 2\mathbf{i} + 7\mathbf{j} - 5\mathbf{k}$$

The answer is (C).

2. VECTOR IDENTITIES

Equation 3.8 Through Eq. 3.10: Dot Product Identities

$$\mathbf{A} \cdot \mathbf{B} = \mathbf{B} \cdot \mathbf{A} \qquad 3.8$$

$$\mathbf{A} \cdot (\mathbf{B} + \mathbf{C}) = \mathbf{A} \cdot \mathbf{B} + \mathbf{A} \cdot \mathbf{C} \qquad 3.9$$

$$\mathbf{A} \cdot \mathbf{A} = |\mathbf{A}|^2 \qquad 3.10$$

Description

The dot product for vectors is commutative and distributive, as shown by Eq. 3.8 and Eq. 3.9. Equation 3.10 gives the dot product of a vector with itself, the square of its magnitude.

Equation 3.11: Dot Product of Parallel Unit Vectors

$$\mathbf{i} \cdot \mathbf{i} = \mathbf{j} \cdot \mathbf{j} = \mathbf{k} \cdot \mathbf{k} = 1 \qquad 3.11$$

Description

As indicated in Eq. 3.11, the dot product of two parallel unit vectors is one.

Example

What is the dot product $\mathbf{A} \cdot \mathbf{B}$ of unit vectors $\mathbf{A} = 3\mathbf{i}$ and $\mathbf{B} = 2\mathbf{i}$?

(A) −6

(B) −5

(C) 5

(D) 6

Solution

Use Eq. 3.11.

$$\mathbf{A} \cdot \mathbf{B} = 3\mathbf{i} \cdot 2\mathbf{i} = (3 \cdot 2)\mathbf{i} \cdot \mathbf{i} = (6)(1)$$
$$= 6$$

The answer is (D).

Equation 3.12: Dot Product of Orthogonal Vectors

$$\mathbf{i} \cdot \mathbf{j} = \mathbf{j} \cdot \mathbf{k} = \mathbf{k} \cdot \mathbf{i} = 0 \qquad 3.12$$

Description

The dot product can be used to determine whether a vector is a unit vector and to show that two vectors are orthogonal (perpendicular). As indicated in Eq. 3.12, the dot product of two non-null (nonzero) orthogonal vectors is zero.

Equation 3.13 Through Eq. 3.15: Cross Product Identities

$$\mathbf{A} \times \mathbf{B} = -\mathbf{B} \times \mathbf{A} \qquad 3.13$$

$$\mathbf{A} \times (\mathbf{B} + \mathbf{C}) = (\mathbf{A} \times \mathbf{B}) + (\mathbf{A} \times \mathbf{C}) \qquad 3.14$$

$$(\mathbf{B} + \mathbf{C}) \times \mathbf{A} = (\mathbf{B} \times \mathbf{A}) + (\mathbf{C} \times \mathbf{A}) \qquad 3.15$$

Description

The vector cross product is distributive, as demonstrated in Eq. 3.14 and Eq. 3.15. However, as Eq. 3.13 shows, it is not commutative.

Equation 3.16: Cross Product of Parallel Unit Vectors

$$\mathbf{i} \times \mathbf{i} = \mathbf{j} \times \mathbf{j} = \mathbf{k} \times \mathbf{k} = 0 \qquad 3.16$$

Description

If two non-null vectors are parallel, their cross product will be zero.

Equation 3.17 and Eq. 3.18: Cross Product of Normal Unit Vectors

$$\mathbf{i} \times \mathbf{j} = \mathbf{k} = -\mathbf{j} \times \mathbf{i} \qquad 3.17$$

$$\mathbf{k} \times \mathbf{i} = \mathbf{j} = -\mathbf{i} \times \mathbf{k} \qquad 3.18$$

Description

If two non-null vectors are normal (perpendicular), their vector cross product will be perpendicular to both vectors.

4 Analytic Geometry

1. MENSURATION OF AREAS

The dimensions, perimeter, area, and other geometric properties constitute the *mensuration* (i.e., the measurements) of a geometric shape.

Equation 4.1 and Eq. 4.2: Parabolic Segments

$$A = 2bh/3 \qquad \textbf{4.1}$$

$$A = bh/3 \qquad \textbf{4.2}$$

Description

Equation 4.1 and Eq. 4.2 give the area of a parabolic segment. (See Fig. 4.1.) Equation 4.1 gives the area within the curve of the parabola, and Eq. 4.2 gives the area outside the curve of the parabola.

Figure 4.1 Parabolic Segments

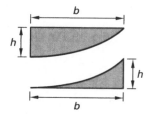

Equation 4.3 Through Eq. 4.6: Ellipses

$$A = \pi ab \qquad \textbf{4.3}$$

$$P_{\text{approx}} = 2\pi \sqrt{(a^2 + b^2)/2} \qquad \textbf{4.4}$$

$$P = \pi(a+b) \left| \begin{array}{l} 1 + (1/2)^2 \lambda^2 + (1/2 \times 1/4)^2 \lambda^4 \\ + (1/2 \times 1/4 \times 3/6)^2 \lambda^6 \\ + (1/2 \times 1/4 \times 3/6 \times 5/8)^2 \lambda^8 \\ + (1/2 \times 1/4 \times 3/6 \times 5/8 \times 7/10)^2 \lambda^{10} \\ + \cdots \end{array} \right| \qquad \textbf{4.5}$$

$$\lambda = (a-b)/(a+b) \qquad \textbf{4.6}$$

Description

Equation 4.3 gives the area of an ellipse. (See Fig. 4.2.) a and b are the semimajor and semiminor axes, respectively. Equation 4.4 gives an approximation of the perimeter of an ellipse. Equation 4.5 expresses the perimeter exactly, but one factor is the sum of an infinite series in which λ is defined as in Eq. 4.6.

Figure 4.2 Ellipse

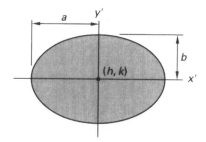

Example

An ellipse has a semimajor axis with length $a = 12$ and a semiminor axis with length $b = 3$. What is the approximate length of the perimeter of the ellipse?

(A) 24

(B) 47

(C) 55

(D) 180

Solution

Use Eq. 4.4.

$$P_{\text{approx}} = 2\pi \sqrt{(a^2 + b^2)/2} = 2\pi \sqrt{\frac{12^2 + 3^2}{2}}$$

$$= 54.96 \quad (55)$$

The answer is (C).

Equation 4.7 and Eq. 4.8: Circular Segments

$$A = [r^2(\phi - \sin\phi)]/2 \qquad 4.7$$

$$\phi = s/r = 2\{\arccos[(r-d)/r]\} \qquad 4.8$$

Description

A *circular segment* is a region bounded by a circular arc and a chord, as shown by the shaded portion in Fig. 4.3. The arc and chord are both limited by a central angle, ϕ. Use Eq. 4.7 to find the area of a circular segment when its central angle, ϕ, and the radius of the circle, r, are known; in Eq. 4.7, the central angle must be in radians. Use Eq. 4.8 to find the central angle when the radius of the circle and either the height of the circular segment, d, or the length of the arc, s, are known.

Figure 4.3 Circular Segment

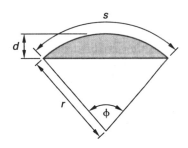

Example

Two 20 m diameter circles are placed so that the circumference of each just touches the center of the other.

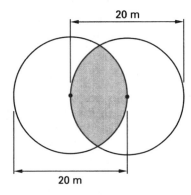

What is most nearly the area of the shared region?

(A) 62 m^2

(B) 110 m^2

(C) 120 m^2

(D) 170 m^2

Solution

The shared region can be thought of as two equal circular segments, each as shown in the illustration. The

radius of each circle is $r = 10$ m. The height of each circular segment is half the radius, so $d = 5$ m.

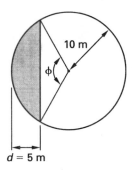

Use Eq. 4.8 to find the angle ϕ.

$$\phi = 2\{\arccos[(r-d)/r]\} = 2\arccos\left(\frac{10 \text{ m} - 5 \text{ m}}{10 \text{ m}}\right)$$
$$= 120°$$

Convert ϕ to radians.

$$\phi = (120°)\left(\frac{2\pi}{360°}\right) = 2.094 \text{ rad}$$

From Eq. 4.7, the area of a circular segment is

$$A = [r^2(\phi - \sin\phi)]/2$$
$$= \frac{(10 \text{ m})^2\Big(2.094 \text{ rad} - \sin(2.094 \text{ rad})\Big)}{2}$$
$$= 61.4 \text{ m}^2$$

The area of the shared region is twice this amount.

$$A_{\text{shared}} = 2A = (2)(61.4 \text{ m}^2)$$
$$= 122.8 \text{ m}^2 \quad (120 \text{ m}^2)$$

The answer is (C).

Equation 4.9 and Eq. 4.10: Circular Sectors

$$A = \phi r^2/2 = sr/2 \qquad 4.9$$

$$\phi = s/r \qquad 4.10$$

Description

A *circular sector* is a portion of a circle bounded by two radii and an arc, as shown in Fig. 4.4. Between the two radii is the central angle, ϕ. Use Eq. 4.9 to find the area of a circular sector when its radius, r, and either its central angle or the length of its arc, s, are known; the central angle must be in radians. Use Eq. 4.10 to find the central angle in radians when the arc length and radius are known.

Figure 4.4 Circular Sector

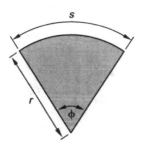

Figure 4.5 Parallelogram

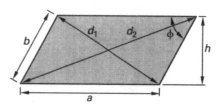

Example

A circular sector has an area of 3 m² and a central angle of 50°. What is most nearly the radius?

(A) 1.5 m

(B) 2.6 m

(C) 3.0 m

(D) 3.3 m

Solution

The central angle must be converted to radians.

$$\phi = (50°)\left(\frac{2\pi}{360°}\right) = 0.873 \text{ rad}$$

Use Eq. 4.9.

$$A = \phi r^2/2$$

$$r = \sqrt{\frac{2A}{\phi}} = \sqrt{\frac{(2)(3 \text{ m}^2)}{0.873 \text{ rad}}}$$

$$= 2.62 \text{ m} \quad (2.6 \text{ m})$$

The answer is (B).

Equation 4.11 Through Eq. 4.15: Parallelograms

$$P = 2(a+b) \qquad \text{4.11}$$

$$d_1 = \sqrt{a^2 + b^2 - 2ab(\cos \phi)} \qquad \text{4.12}$$

$$d_2 = \sqrt{a^2 + b^2 + 2ab(\cos \phi)} \qquad \text{4.13}$$

$$d_1^2 + d_2^2 = 2(a^2 + b^2) \qquad \text{4.14}$$

$$A = ah = ab(\sin \phi) \qquad \text{4.15}$$

Description

Equation 4.11 is the formula for the perimeter of a parallelogram. (See Fig. 4.5.) Equation 4.12 and Eq. 4.13 give the diagonals of the parallelogram when its sides and acute included angle are known. Equation 4.14 relates the sides and the diagonals, and Eq. 4.15 gives the parallelogram's area. A parallelogram with all sides of equal length is called a *rhombus*.

Equation 4.16 Through Eq. 4.20: Regular Polygons

$$\phi = 2\pi/n \qquad \text{4.16}$$

$$\theta = \left[\frac{\pi(n-2)}{n}\right] = \pi\left(1 - \frac{2}{n}\right) \qquad \text{4.17}$$

$$P = ns \qquad \text{4.18}$$

$$s = 2r[\tan(\phi/2)] \qquad \text{4.19}$$

$$A = (nsr)/2 \qquad \text{4.20}$$

Description

A *regular polygon* is a polygon with equal sides and equal angles. (See Fig. 4.6.) n is the number of sides. Equation 4.16 gives the central angle, ϕ, formed by two line segments drawn from the center to adjacent vertices. Equation 4.17 gives the measure of each interior angle, θ. Equation 4.18 gives the perimeter of the polygon (s is the length of one side).

Figure 4.6 Regular Polygon (n equal sides)

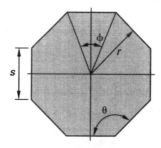

Equation 4.19 can be used to find the length of one side when the polygon's central angle and apothem, r, are known. The *apothem* is a line segment drawn from the center of the polygon to the midpoint of one side; this is also the radius of a circle inscribed within the polygon. Equation 4.20 is the formula for the area of the polygon.

Example

A regular polygon has six sides, each with a length of 25 cm. What is most nearly the length of the apothem, r?

(A) 10 cm

(B) 15 cm

(C) 20 cm

(D) 22 cm

Solution

Use Eq. 4.16 to find the central angle.

$$\phi = 2\pi/n = \frac{2\pi \text{ rad}}{6} = 1.047 \text{ rad}$$

Convert radians to degrees.

$$\phi = (1.047 \text{ rad})\left(\frac{360°}{2\pi}\right) = 60.0°$$

Use Eq. 4.19 to find the length of the apothem.

$$s = 2r[\tan(\phi/2)]$$

$$r = \frac{s}{2\tan\frac{\phi}{2}} = \frac{25 \text{ cm}}{2\tan\frac{60°}{2}}$$

$$= 21.65 \text{ cm} \quad (22 \text{ cm})$$

The answer is (D).

2. MENSURATION OF VOLUMES

Equation 4.21: Prismoids

$$V = (h/6)(A_1 + A_2 + 4A) \qquad \textbf{4.21}$$

Variation

$$h = \frac{6V}{A_1 + A_2 + 4A}$$

Description

A *polyhedron* is a three-dimensional solid whose faces are all flat and whose edges are all straight.

If all the vertices (corners) of the polyhedron are contained within two parallel planes, the solid is a *prismoid* (*prismatoid*). A simple example is a truncated pyramid whose top and bottom faces are parallel. Less obviously, a complete (not truncated) pyramid is also a prismoid; one plane contains the bottom of the pyramid, while the other is parallel to the bottom and contains the single vertex at the top. Figure 4.7 shows an irregular prismoid with all vertices contained in the top and bottom planes.

Use Eq. 4.21 to find the volume of a prismoid. h is the distance between the two parallel planes measured along a direction perpendicular to both. A_1 and A_2 are the areas of the faces contained within these two planes; if one of these planes contains only a single point (such as at the top vertex of a pyramid), the area is zero. A is the cross-sectional area of the solid halfway between the two parallel planes. Each vertex of this cross section is halfway between a vertex on the top face and another one on the bottom face, but A is not necessarily the average of A_1 and A_2.

Figure 4.7 *Prismoid*

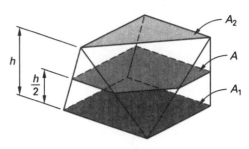

Example

A prismoid has a volume of 100 cm³. The area of the bottom face is 20 cm², the area of the top face is 5 cm², and the cross-sectional area halfway between the top and bottom faces is 10 cm². What is the approximate height of the prismoid?

(A) 8.0 cm

(B) 9.2 cm

(C) 11 cm

(D) 13 cm

Solution

Use Eq. 4.21, the formula for the volume of a prismoid.

$$V = (h/6)(A_1 + A_2 + 4A)$$

$$h = \frac{6V}{A_1 + A_2 + 4A} = \frac{(6)(100 \text{ cm}^3)}{20 \text{ cm}^2 + 5 \text{ cm}^2 + (4)(10 \text{ cm}^2)}$$

$$= 9.23 \text{ cm} \quad (9.2 \text{ cm})$$

The answer is (B).

Equation 4.22 and Eq. 4.23: Spheres

$$V = 4\pi r^3/3 = \pi d^3/6 \qquad \textbf{4.22}$$
$$A = 4\pi r^2 = \pi d^2 \qquad \textbf{4.23}$$

Description

Equation 4.22 and Eq. 4.23 are the formulas for the volume and surface area, respectively, of a sphere whose radius, r, or diameter, d, is known. (See Fig. 4.8.)

Figure 4.8 *Sphere*

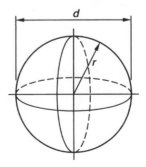

Example

A sphere has a radius of 10 cm. What is approximately the sphere's volume?

(A) 3600 cm^3

(B) 4000 cm^3

(C) 4200 cm^3

(D) 4800 cm^3

Solution

Use Eq. 4.22.

$$V = 4\pi r^3/3$$

$$= \frac{4\pi(10 \text{ cm})^3}{3}$$

$$= 4188.79 \text{ cm}^3 \quad (4200 \text{ cm}^3)$$

The answer is (C).

Equation 4.24 Through Eq. 4.26: Right Circular Cones

$$V = (\pi r^2 h)/3 \qquad 4.24$$

$$A = \text{side area} + \text{base area} = \pi r\left(r + \sqrt{r^2 + h^2}\right) \qquad 4.25$$

$$A_x:A_b = x^2:h^2 \qquad 4.26$$

Description

A *right circular cone* is a cone whose base is a circle and whose axis is perpendicular to the base. (See Fig. 4.9.) Equation 4.24 gives the volume of a right circular cone whose height, h, and base radius, r, are known. Equation 4.25 gives the cone's area. Equation 4.26 says that the cross-sectional area of the cone varies with the square of the distance from the apex.

Figure 4.9 Right Circular Cone

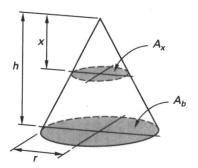

Example

A cone has a height of 100 cm. The cross section of the cone at a distance of 5 cm from the apex is a circle with an area of 20 cm^2. What is most nearly the area of the cone's base?

(A) 5000 cm^2

(B) 6000 cm^2

(C) 8000 cm^2

(D) 9000 cm^2

Solution

Use Eq. 4.26.

$$A_x:A_b = x^2:h^2$$

$$A_b = \frac{h^2 A_x}{x^2} = \frac{(100 \text{ cm})^2 (20 \text{ cm}^2)}{(5 \text{ cm})^2}$$

$$= 8000 \text{ cm}^2$$

The answer is (C).

Equation 4.27 and Eq. 4.28: Right Circular Cylinders

$$V = \pi r^2 h = \frac{\pi d^2 h}{4} \qquad 4.27$$

$$A = \text{side area} + \text{end areas} = 2\pi r(h + r) \qquad 4.28$$

Description

A *right circular cylinder* is a cylinder whose base is a circle and whose axis is perpendicular to the base. (See Fig. 4.10.) Equation 4.27 gives the volume of a right circular cylinder, and Eq. 4.28 gives the total surface area.

Figure 4.10 Right Circular Cylinder

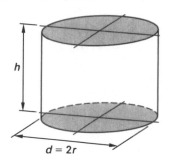

$d = 2r$

Equation 4.29: Paraboloids of Revolution

$$V = \frac{\pi d^2 h}{8} \qquad 4.29$$

Description

A *paraboloid of revolution* is the surface that is obtained by rotating a parabola around its axis. Equation 4.29 can be used to find the volume of a paraboloid of revolution if its height and diameter are known. (See Fig. 4.11.)

Figure 4.11 *Paraboloid of Revolution*

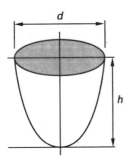

3. STRAIGHT LINES

Figure 4.12 is a straight line in two-dimensional space. The *slope* of the line is m, the y-intercept is b, and the x-intercept is a. A known point on the line is represented as (x_1, y_1).

Figure 4.12 *Straight Line*

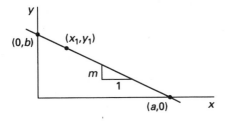

Equation 4.30: Slope

$$m = (y_2 - y_1)/(x_2 - x_1) \qquad \textbf{4.30}$$

Description

Given two points on a straight line, (x_1, y_1) and (x_2, y_2), Eq. 4.30 gives the slope of the line. The slopes of two parallel lines are equal.

Example

Find the slope of the line that passes through points $(-3, 2)$ and $(5, -2)$.

    (A) −2

    (B) −0.5

    (C) 0.5

    (D) 2

Solution

Use Eq. 4.30.

$$m = (y_2 - y_1)/(x_2 - x_1) = \frac{-2 - 2}{5 - (-3)} = -0.5$$

The answer is (B).

Equation 4.31: Slopes of Perpendicular Lines

$$m_1 = -1/m_2 \qquad \textbf{4.31}$$

Description

If two lines are perpendicular to each other, then their slopes, m_1 and m_2, are negative reciprocals of each other, as shown by Eq. 4.31. For example, if the slope of a line is 5, the slope of a line perpendicular to it is −1/5.

Example

A line goes through the point $(4, -6)$ and is perpendicular to the line $y = 4x + 10$. What is the equation of the line?

    (A) $y = -\frac{1}{4}x - 20$

    (B) $y = -\frac{1}{4}x - 5$

    (C) $y = \frac{1}{5}x + 5$

    (D) $y = \frac{1}{4}x + 5$

Solution

The slopes of two lines that are perpendicular are related by

$$m_1 = -1/m_2$$

The slope of the line perpendicular to the line with slope $m_1 = 4$ is

$$m_2 = -1/m_1 = -1/4$$

The equation of the line is in the form $y = mx + b$. $m = -1/4$, and a known point is $(x, y) = (4, -6)$.

$$-6 = \left(-\frac{1}{4}\right)(4) + b$$

$$b = -6 - \left(-\frac{1}{4}\right)(4)$$

$$= -5$$

The equation of the line is

$$y = -\frac{1}{4}x - 5$$

The answer is (B).

Equation 4.32: Standard Form of the Equation of a Line

$$y = mx + b \qquad 4.32$$

Description

The equation of a line can be represented in several forms. The procedure for finding the equation depends on the form chosen to represent the line. In general, the procedure involves substituting one or more known points on the line into the equation in order to determine the constants.

Equation 4.32 is the *standard form* of the equation of a line. This is also known as the *slope-intercept form* because the constants in the equation are the line's slope, m, and its y-intercept, b.

Example

What is the slope of the line defined by $y - x = 5$?

   (A) -1
   (B) $-1/5$
   (C) $1/4$
   (D) 1

Solution

The standard (or slope-intercept) form of the equation of a straight line is $y = mx + b$, where m is the slope and b is the y-intercept. Rearrange the given equation into standard form.

$$y - x = 5$$
$$y = x + 5$$

The slope, m, is the coefficient of x, which is 1.

The answer is (D).

Equation 4.33: General Form of the Equation of a Line

$$Ax + By + C = 0 \qquad 4.33$$

Description

Equation 4.33 is the *general form* of the equation of a line.

Example

What is the general form of the equation for a line whose x-intercept is 4 and y-intercept is -6?

   (A) $2x - 3y - 18 = 0$
   (B) $2x + 3y + 18 = 0$
   (C) $3x - 2y - 12 = 0$
   (D) $3x + 2y + 12 = 0$

Solution

Find the slope of the line.

$$m = \frac{y_2 - y_1}{x_2 - x_1}$$
$$= \frac{-6 - 0}{0 - 4}$$
$$= 3/2$$

Write the equation of the line in standard (or slope-intercept) form, then arrange it in the form of Eq. 4.33.

$$y = mx + b$$
$$mx - y + b = 0$$
$$\tfrac{3}{2}x - y + (-6) = 0$$
$$3x - 2y - 12 = 0$$

The answer is (C).

Equation 4.34: Point-Slope Form of the Equation of a Line

$$y - y_1 = m(x - x_1) \qquad 4.34$$

Description

Equation 4.34 is the *point-slope form* of the equation of a line. This equation defines the line in terms of its slope, m, and one known point, (x_1, y_1).

Example

A circle with a radius of 5 is centered at the origin.

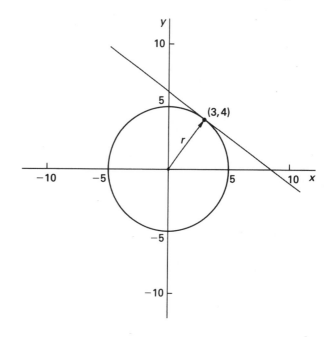

What is the standard form of the equation of the line tangent to this circle at the point $(3, 4)$?

(A) $x = \dfrac{-4}{3} y - \dfrac{25}{4}$

(B) $y = \dfrac{3}{4} x + \dfrac{25}{4}$

(C) $y = \dfrac{-3}{4} x + \dfrac{9}{4}$

(D) $y = \dfrac{-3}{4} x + \dfrac{25}{4}$

Solution

The slope of the radius line from point $(0, 0)$ to point $(3, 4)$ is $4/3$. Since the radius and tangent line are perpendicular, the slope of the tangent line is $-3/4$.

The point-slope form of a straight line with slope $m = -3/4$ and containing point $(x_1, y_1) = (3, 4)$ is

$$y - y_1 = m(x - x_1)$$

$$y - 4 = \left(\dfrac{-3}{4}\right)(x - 3)$$

Rearranging this into standard form gives

$$y = \dfrac{-3}{4} x + \left(\dfrac{9}{4} + 4\right)$$

$$= \dfrac{-3}{4} x + \dfrac{25}{4}$$

The answer is (D).

Equation 4.35: Angle Between Two Lines

$$\alpha = \arctan[(m_2 - m_1)/(1 + m_2 \cdot m_1)] \qquad \textit{4.35}$$

Description

Two intersecting lines in two-dimensional space are shown in Fig. 4.13.

Figure 4.13 *Two Lines Intersecting in Two-Dimensional Space*

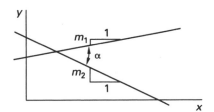

The slopes of the two lines are m_1 and m_2. The acute angle, α, between the lines is given by Eq. 4.35.[1]

Example

The angle between the line $y = -7x + 12$ and the line $y = 3x$ is most nearly

(A) $22°$

(B) $27°$

(C) $33°$

(D) $37°$

Solution

Use Eq. 4.35.

$$\alpha = \arctan[(m_2 - m_1)/(1 + m_2 \cdot m_1)]$$

$$= \arctan \dfrac{-7 - 3}{1 + (-7)(3)} = \arctan 0.5$$

$$= 26.57° \quad (27°)$$

The answer is (B).

4. DISTANCE BETWEEN POINTS IN SPACE

Equation 4.36: Distance Between Two Points in Space

$$d = \sqrt{(x_2 - x_1)^2 + (y_2 - y_1)^2 + (z_2 - z_1)^2} \qquad \textit{4.36}$$

Description

The distance between two points (x_1, y_1, z_1) and (x_2, y_2, z_2) in three-dimensional space can be found using Eq. 4.36.

Example

What is the distance between point P at $(1, -3, 5)$ and point Q at $(-3, 4, -2)$?

(A) $\sqrt{10}$

(B) $\sqrt{14}$

(C) 8

(D) $\sqrt{114}$

---

[1]Since the NCEES *FE Reference Handbook* (*NCEES Handbook*) does not specify an absolute value, Eq. 4.35 is sensitive to which slopes are designated m_1 and m_2. The slopes should be intuitively selected to avoid a negative arctan argument. Alternatively, since the tangent exhibits odd symmetry, if a negative angle is calculated, the negative sign can simply be disregarded.

Mathematics

Solution

The distance between points P and Q is

$$d_{PQ} = \sqrt{(x_2 - x_1)^2 + (y_2 - y_1)^2 + (z_2 - z_1)^2}$$

$$= \sqrt{(-3-1)^2 + \left(4 - (-3)\right)^2 + (-2-5)^2}$$

$$= \sqrt{114}$$

The answer is (D).

5. CONIC SECTIONS

A *conic section* is any of several kinds of curves that can be produced by passing a plane through a cone as shown in Fig. 4.14.

Figure 4.14 *Conic Sections Produced by Cutting Planes*

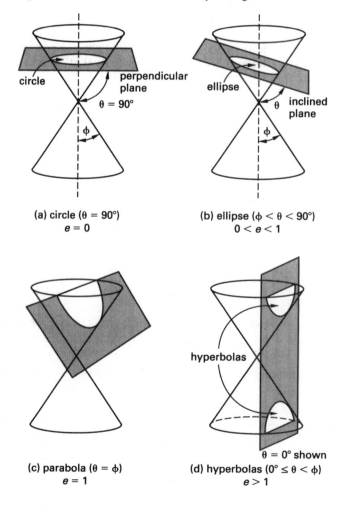

(a) circle ($\theta = 90°$)
$e = 0$

(b) ellipse ($\phi < \theta < 90°$)
$0 < e < 1$

(c) parabola ($\theta = \phi$)
$e = 1$

(d) hyperbolas ($0° \leq \theta < \phi$)
$e > 1$

Equation 4.37: Eccentricity of a Cutting Plane

$$e = \cos\theta / (\cos\phi) \qquad 4.37$$

Description

If θ is the angle between the vertical axis and the cutting plane and ϕ is the *cone-generating angle*, then the *eccentricity*, e, of the conic section is given by Eq. 4.37.

Equation 4.38 Through Eq. 4.41: General Form and Normal Form of the Conic Section Equation

$$Ax^2 + Bxy + Cy^2 + Dx + Ey + F = 0 \qquad 4.38$$

$$x^2 + y^2 + 2ax + 2by + c = 0 \qquad 4.39$$

$$h = -a; \quad k = -b \qquad 4.40$$

$$r = \sqrt{a^2 + b^2 - c} \qquad 4.41$$

Description

All conic sections are described by second-degree (quadratic) polynomials with two variables. The *general form* of the conic section equation is given by Eq. 4.38. x and y are variables, and A, B, C, D, E, and F are constants.

h and k are the coordinates (h, k) of the conic section's center. r is a size parameter, usually the radius of a circle or a sphere. If $r = 0$, then the conic section describes a point. If r is negative, the equation does not describe a conic section.

If $A = C$, then B must be zero for a conic section. If $A = C = 0$, the conic section is a *line*, and if $A = C \neq 0$, the conic section is a *circle*. If $A \neq C$, then if

- $B^2 - 4AC < 0$, the conic section is an *ellipse*

- $B^2 - 4AC > 0$, the conic section is a *hyperbola*

- $B^2 - 4AC = 0$, the conic section is a *parabola*

The general form of the conic section equation can be applied when the conic section is at any orientation relative to the coordinate axes. Equation 4.39 is the *normal form* of the conic section equation. It can be applied when one of the principal axes of the conic section is parallel to a coordinate axis, thereby eliminating certain terms of the general equation and reducing the number of constants needed to three: a, b, and c.

Example

What kind of conic section is described by the following equation?

$$4x^2 - y^2 + 8x + 4y = 15$$

(A) circle

(B) ellipse

(C) parabola

(D) hyperbola

Solution

The general form of a conic section is given by Eq. 4.38 as

$$Ax^2 + Bxy + Cy^2 + Dx + Ey + F = 0$$

In this case, $A = 4$, $B = 0$, and $C = -1$. Since $A \neq C$, the conic section is not a circle or line.

Calculate the discriminant.

$$B^2 - 4AC = (0)^2 - (4)(4)(-1) = 16$$

This is greater than zero, so the section is a hyperbola.

The answer is (D).

Equation 4.42: Standard Form of the Equation of a Horizontal Parabola

$$(y - k)^2 = 2p(x - h) \quad [\text{center at } (h, k)] \qquad \text{4.42}$$

Description

A *parabola* is the locus of points equidistant from the focus (point F in Fig. 4.15) and a line called the *directrix*. The distance between the focus and the directrix is p.[2] The directrix is defined by the equation $x = h - (p/2)$. When the vertex of the parabola is at the origin, $h = k = 0$, and Eq. 4.43 and Eq. 4.44 apply.

Figure 4.15 *Parabola*

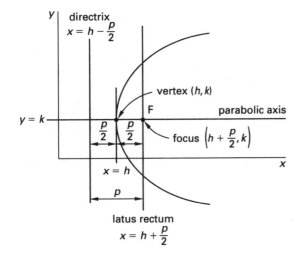

A parabola is symmetric with respect to its *parabolic axis*. The line normal to the parabolic axis and passing through the focus is known as the *latus rectum*. The eccentricity of a parabola is equal to 1.

Equation 4.42 is the *standard form* of the equation of a horizontal parabola. It can be applied when the principal axes of the parabola coincide with the coordinate axes.

The equation for a vertical parabola is similar.

$$(x - h)^2 = 2p(y - k)$$

If p is positive, the parabola opens upward. If p is negative, the parabola opens downward.

Equation 4.43 and Eq. 4.44: Parabola with Vertex at the Origin

$$\text{focus: } (p/2, 0) \qquad \qquad \text{4.43}$$
$$x = -p/2 \qquad \qquad \text{4.44}$$

Description

The definitions in Eq. 4.43 and Eq. 4.44 apply when the vertex of the parabola is at the origin—that is, when $(h, k) = (0, 0)$.

The parabola opens to the right (points to the left) if $p > 0$, and it opens to the left (points to the right) if $p < 0$.

Example

What is the equation of a parabola with a vertex at $(4, 8)$ and a directrix at $y = 5$?

    (A) $(x - 8)^2 = 12(y - 4)$

    (B) $(x - 4)^2 = 12(y - 8)$

    (C) $(x - 4)^2 = 6(y - 8)$

    (D) $(y - 8)^2 = 12(x - 4)$

Solution

The directrix, described by $y = 5$, is parallel to the x-axis, so this is a vertical parabola. The vertex (at $y = 8$) is above the directrix, so the parabola opens upward.

The distance from the vertex to the directrix is

$$\frac{p}{2} = 8 - 5 = 3$$
$$p = 6$$

The focus is located a distance $p/2$ from the vertex. The focus is at $(4, 8 + 3)$ or $(4, 11)$.

The standard form equation for a parabola with vertex at (h, k) and opening upward is

$$(x - h)^2 = 2p(y - k)$$
$$(x - h)^2 = (2)(6)(y - 8)$$
$$(x - 4)^2 = 12(y - 8)$$

The answer is (B).

---

[2]There are two conventions used to define the parameters of a parabola. One convention, as used in the *NCEES Handbook*, is to define p as the distance from the focus to the directrix. This results in the $2p$ term in Eq. 4.42. Another convention, arguably more prevalent, is to define p as the distance from the focus to the vertex (i.e., the distance from the focus to the directrix is $2p$), which would result in a corresponding term of $4p$ in Eq. 4.42.

Equation 4.45: Standard Form of the Equation of an Ellipse

$$\frac{(x-h)^2}{a^2}+\frac{(y-k)^2}{b^2}=1 \quad \text{[center at } (h,k)\text{]}$$ *4.45*

Description

An *ellipse* has two foci, F_1 and F_2, separated along the *major axis* by a distance $2c$. (See Fig. 4.16.) The line perpendicular to the major axis passing through the center of the ellipse is the *minor axis*. The lines perpendicular to the major axis passing through the foci are the *latera recta*. The distance between the two vertices is $2a$. The ellipse is the locus of points such that the sum of the distances from the two foci is $2a$. The eccentricity of the ellipse is always less than one. If the eccentricity is zero, the ellipse is a circle.

Figure 4.16 Ellipse (with horizontal major axis)

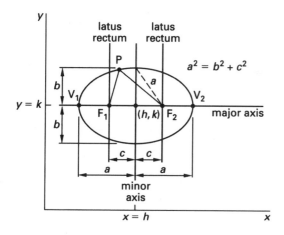

Equation 4.45 is the *standard form* of the equation of an ellipse with center at (h,k), *semimajor distance a*, and *semiminor distance b*. Equation 4.45 can be applied when the principal axes of the ellipse coincide with the coordinate axes and when the major axis is oriented vertically.

Example

What is the equation of the ellipse with center at $(0,0)$, with vertical major axis, and that passes through the points $(2,0)$, $(0,3)$, and $(-2,0)$?

(A) $\dfrac{x^2}{9}-\dfrac{y^2}{4}=1$

(B) $\dfrac{x^2}{4}-\dfrac{y^2}{9}=1$

(C) $\dfrac{x^2}{9}+\dfrac{y^2}{4}=1$

(D) $\dfrac{x^2}{4}+\dfrac{y^2}{9}=1$

Solution

An ellipse has the standard form

$$\frac{(x-h)^2}{a^2}+\frac{(y-k)^2}{b^2}=1$$

The center is at $(h,k)=(0,0)$.

$$\frac{(x-0)^2}{a^2}+\frac{(y-0)^2}{b^2}=1$$

Substitute the known values of (x,y) to determine a and b.

For $(x,y)=(2,0)$,

$$\frac{(2)^2}{a^2}+\frac{(0)^2}{b^2}=1$$
$$a^2=4$$
$$a=2$$

For $(x,y)=(0,3)$,

$$\frac{(0)^2}{a^2}+\frac{(3)^2}{b^2}=1$$
$$b^2=9$$
$$b=3$$

This ellipse is oriented vertically since $b>a$.

Check: For $(x,y)=(-2,0)$,

$$\frac{(-2)^2}{a^2}+\frac{(0)^2}{b^2}=1$$
$$a^2=4$$
$$a=2 \quad \begin{bmatrix}\text{This step is not necessary}\\\text{as } a \text{ is determined}\\\text{from the first point.}\end{bmatrix}$$

The equation of the ellipse is

$$\frac{x^2}{(2)^2}+\frac{y^2}{(3)^2}=1$$
$$\frac{x^2}{4}+\frac{y^2}{9}=1$$

The answer is (D).

Equation 4.46 Through Eq. 4.49: Ellipse with Center at the Origin

foci: $(\pm ae,0)$	*4.46*
$x=\pm a/e$	*4.47*
$e=\sqrt{1-(b^2/a^2)}=c/a$	*4.48*
$b=a\sqrt{1-e^2}$	*4.49*

Description

When the center of the ellipse is at the origin ($h = k = 0$), the foci are located at $(ae, 0)$ and $(-ae, 0)$, the directrices are located at $\pm x = a/e$, and the eccentricity and semiminor distance are given by Eq. 4.48 and Eq. 4.49, respectively. Each directrix is a vertical line located outside of the ellipse. The location of each directrix is such that the distance from a point on the ellipse to the nearest directrix is equal to the distance from that point on the ellipse to the nearest focus.

Equation 4.50: Standard Form of the Equation of a Hyperbola

$$\frac{(x-h)^2}{a^2} - \frac{(y-k)^2}{b^2} = 1 \quad [\text{center at } (h,k)] \qquad \textit{4.50}$$

Description

As shown in Fig. 4.17, a *hyperbola* has two foci separated along the *transverse axis* by a distance $2c$. The two lines perpendicular to the transverse axis that pass through the foci are the *conjugate axes*. As the distance from the center increases, the hyperbola approaches two straight lines, called the *asymptotes*, that intersect at the hyperbola's center.

The distance from the center to either vertex is a. The distance from either vertex to either asymptote in a direction perpendicular to the transverse axis is b. The hyperbola is the locus of points such that the distances from any point to the two foci differ by $2a$. The distance from the center to either focus is c.

Equation 4.50 is the *standard form* of the equation of a hyperbola with center at (h, k) and opening horizontally.

Figure 4.17 *Hyperbola*

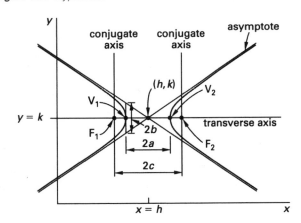

Equation 4.51 Through Eq. 4.54: Hyperbola with Center at the Origin

$$\text{foci: } (\pm ae, 0) \qquad \textit{4.51}$$
$$x = \pm a/e \qquad \textit{4.52}$$
$$e = \sqrt{1 + (b^2/a^2)} = c/a \qquad \textit{4.53}$$
$$b = a\sqrt{e^2 - 1} \qquad \textit{4.54}$$

Description

When the hyperbola is centered at the origin ($h = k = 0$), the foci are located at $(ae, 0)$ and $(-ae, 0)$, the directrices are located at $x = a/e$ and $x = -a/e$, and the eccentricity, e, and distance b are given by Eq. 4.53 and Eq. 4.54, respectively.

Example

What is most nearly the eccentricity of the hyperbola shown?

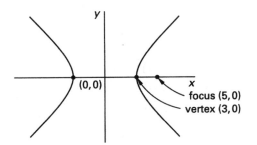

(A) 1.33

(B) 1.67

(C) 2.00

(D) 3.00

Solution

Use Eq. 4.53. a is the distance from the center to either vertex, and c is the distance from the center to either focus. The eccentricity is

$$e = c/a = \frac{5}{3} = 1.67$$

The answer is (B).

Equation 4.55 and Eq. 4.56: Standard Form of the Equation of a Circle

$$(x - h)^2 + (y - k)^2 = r^2 \qquad \textit{4.55}$$

$$r = \sqrt{(x - h)^2 + (y - k)^2} \qquad \textit{4.56}$$

Description

Equation 4.55 is the *standard form* (also called the *center-radius form*) of the equation of a circle with center at (h, k) and radius r. (See Fig. 4.18.) The radius is given by Eq. 4.56.

Figure 4.18 Circle

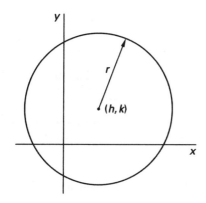

Example

What is the equation of the circle passing through the points $(0, 0)$, $(0, 4)$, and $(-4, 0)$?

   (A) $(x - 2)^2 + (y - 2)^2 = \sqrt{8}$

   (B) $(x - 2)^2 + (y - 2)^2 = 8$

   (C) $(x + 2)^2 + (y - 2)^2 = 8$

   (D) $(x + 2)^2 + (y + 2)^2 = \sqrt{8}$

Solution

From Eq. 4.55, the center-radius form of the equation of a circle is

$$(x - h)^2 + (y - k)^2 = r^2$$

Substitute the first two points, $(0, 0)$ and $(0, 4)$.

$$(0 - h)^2 + (0 - k)^2 = r^2$$
$$(0 - h)^2 + (4 - k)^2 = r^2$$

Since both are equal to the unknown r^2, set the left-hand sides equal. Simplify and solve for k.

$$h^2 + k^2 = h^2 + (4 - k)^2$$
$$k^2 = (4 - k)^2$$
$$k = 2$$

Substitute the third point, $(-4, 0)$, into the center-radius form.

$$(-4 - h)^2 + (0 - k)^2 = r^2$$

Set this third equation equal to the first equation. Simplify and solve for h.

$$(-4 - h)^2 + k^2 = h^2 + k^2$$
$$(-4 - h)^2 = h^2$$
$$h = -2$$

Now that h and k are known, substitute them into the first equation to determine r^2.

$$h^2 + k^2 = r^2$$
$$(-2)^2 + (2)^2 = 8$$

Substitute the known values of h, k, and r^2 into the center-radius form.

$$(x + 2)^2 + (y - 2)^2 = 8$$

The answer is (C).

Equation 4.57: Distance Between Two Points on a Plane

$$d = \sqrt{(y_2 - y_1)^2 + (x_2 - x_1)^2} \qquad \textbf{4.57}$$

Description

The distance, d, between two points (x_1, y_1) and (x_2, y_2) is given by Eq. 4.57.

Equation 4.58: Length of Tangent to Circle from a Point

$$t^2 = (x' - h)^2 + (y' - k)^2 - r^2 \qquad \textbf{4.58}$$

Description

The length, t, of a *tangent* to a circle from a point (x', y') in two-dimensional space is illustrated in Fig. 4.19 and can be found from Eq. 4.58.

Figure 4.19 Tangent to a Circle from a Point

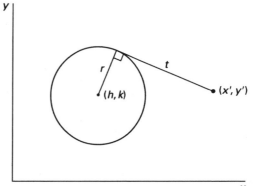

Example

What is the length of the line tangent from point $(7,1)$ to the circle shown?

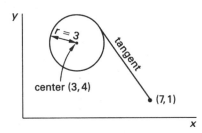

(A) 3

(B) 4

(C) 5

(D) 7

Solution

Use Eq. 4.58.

$$t^2 = (x' - h)^2 + (y' - k)^2 - r^2$$
$$= (7 - 3)^2 + (1 - 4)^2 - 3^2$$
$$= 16$$
$$t = 4$$

The answer is (B).

6. QUADRIC SURFACE (SPHERE)

Equation 4.59: Standard Form of the Equation of a Sphere

$$(x - h)^2 + (y - k)^2 + (z - m)^2 = r^2 \qquad 4.59$$

Description

Equation 4.59 is the *standard form* of the equation of a sphere centered at (h, k, m) with radius r.

Example

Most nearly, what is the radius of a sphere with a center at the origin and that passes through the point $(8, 1, 6)$?

(A) 9.2

(B) $\sqrt{101}$

(C) 65

(D) 100

Solution

Use Eq. 4.59.

$$r^2 = (x - h)^2 + (y - k)^2 + (z - m)^2$$
$$r = \sqrt{(8 - 0)^2 + (1 - 0)^2 + (6 - 0)^2}$$
$$= \sqrt{101}$$

The answer is (B).

7. SOLID ANGLES

Equation 4.60: Solid Angle

$$\omega = \frac{\text{surface area}}{r^2} \qquad 4.60$$

Description

A *solid angle*, ω, is a measure of the angle subtended at the vertex of a cone, as shown in Fig. 4.20. The solid angle has units of *steradians* (abbreviated *sr*). A steradian is the solid angle subtended at the center of a unit sphere (i.e., a sphere with a radius of one) by a unit area on its surface. Since the surface area of a sphere of radius r is r^2 times the surface area of a unit sphere, the solid angle is equal to the area cut out by the cone divided by r^2. (See Eq. 4.60.)

Figure 4.20 *Solid Angle*

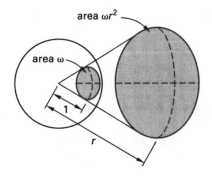

5 Trigonometry

1. DEGREES AND RADIANS

Degrees and *radians* are two units for measuring angles. One complete circle is divided into 360 degrees (written 360°) or 2π radians (abbreviated *rad*).[1] The conversions between degrees and radians are

multiply	by	to obtain
radians	$\dfrac{180}{\pi}$	degrees
degrees	$\dfrac{\pi}{180}$	radians

The number of radians in an angle, θ, corresponds to twice the area within a circular sector with arc length θ and a radius of one, as shown in Fig. 5.1. Alternatively, the area of a sector with central angle θ radians is $\theta/2$ for a *unit circle* (i.e., a circle with a radius of one unit).

Figure 5.1 *Radians and Area of Unit Circle*

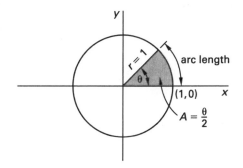

2. PLANE ANGLES

A *plane angle* (usually referred to as just an *angle*) consists of two intersecting lines and an intersection point known as the *vertex*. The angle can be referred to by a capital letter representing the vertex (e.g., B in Fig. 5.2), a letter representing the angular measure (e.g., B or β), or by three capital letters, where the middle letter is the vertex and the other two letters are two points on different lines, and either the symbol $\angle$ or $\sphericalangle$ (e.g., $\sphericalangle$ ABC).

Figure 5.2 *Angle*

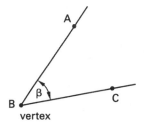

The angle between two intersecting lines generally is understood to be the smaller angle created.[2] Angles have been classified as follows.

- *acute angle:* an angle less than 90° ($\pi/2$ rad)

- *obtuse angle:* an angle more than 90° ($\pi/2$ rad) but less than 180° (π rad)

- *reflex angle:* an angle more than 180° (π rad) but less than 360° (2π rad)

- *related angle:* an angle that differs from another by some multiple of 90° ($\pi/2$ rad)

- *right angle:* an angle equal to 90° ($\pi/2$ rad)

- *straight angle:* an angle equal to 180° (π rad); that is, a straight line

Complementary angles are two angles whose sum is 90° ($\pi/2$ rad). *Supplementary angles* are two angles whose sum is 180° (π rad). *Adjacent angles* share a common vertex and one (the interior) side. Adjacent angles are supplementary if, and only if, their exterior sides form a straight line.

Vertical angles are the two angles with a common vertex and with sides made up by two intersecting straight lines, as shown in Fig. 5.3. Vertical angles are equal.

Angle of elevation and *angle of depression* are surveying terms referring to the angle above and below the horizontal plane of the observer, respectively.

---

[1]The abbreviation *rad* is also used to represent *radiation absorbed dose*, a measure of radiation exposure.

[2]In books on geometry, the term *ray* is used instead of *line*.

Figure 5.3 Vertical Angles

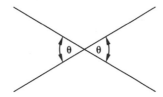

3. TRIANGLES

A *triangle* is a three-sided closed polygon with three angles whose sum is 180° (π rad). Triangles are identified by their vertices and the symbol Δ (e.g., ΔABC in Fig. 5.4). A side is designated by its two endpoints (e.g., AB in Fig. 5.4) or by a lowercase letter corresponding to the capital letter of the opposite vertex (e.g., c).

Figure 5.4 Similar Triangles

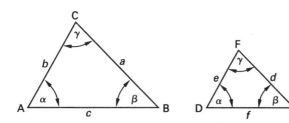

In *similar triangles*, the corresponding angles are equal and the corresponding sides are in proportion. (Since there are only two independent angles in a triangle, showing that two angles of one triangle are equal to two angles of the other triangle is sufficient to show similarity.) The symbol for similarity is $\sim$. In Fig. 5.4, $\Delta ABC \sim \Delta DEF$ (i.e., ΔABC is similar to ΔDEF).

4. RIGHT TRIANGLES

A *right triangle* is a triangle in which one of the angles is 90° ($\pi/2$ rad), as shown in Fig. 5.5. Choosing one of the acute angles as a reference, the sides of the triangle are called the *adjacent side*, x, the *opposite side*, y, and the *hypotenuse*, r.

Equation 5.1 Through Eq. 5.6: Trigonometric Functions

$\sin\theta = y/r$	5.1
$\cos\theta = x/r$	5.2
$\tan\theta = y/x$	5.3
$\csc\theta = r/y$	5.4
$\sec\theta = r/x$	5.5
$\cot\theta = x/y$	5.6

Figure 5.5 Right Triangle

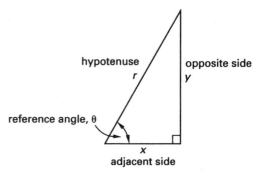

Description

The trigonometric functions given in Eq. 5.1 through Eq. 5.6 are calculated from the sides of the right triangle.

The trigonometric functions correspond to the lengths of various line segments in a right triangle in a unit circle. Figure 5.6 shows such a triangle inscribed in a unit circle.

Figure 5.6 Trigonometric Functions in a Unit Circle

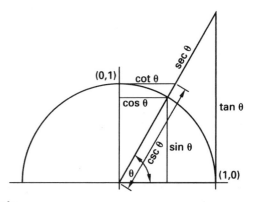

Example

The values of $\cos 45°$ and $\tan 45°$, respectively, are

(A) 1 and $\sqrt{2}/2$

(B) 1 and $\sqrt{2}$

(C) $\sqrt{2}/2$ and 1

(D) $\sqrt{2}$ and 1

Solution

For convenience, let the adjacent side of a 45° right triangle have a length of $x = 1$. Then the opposite side has a length of $y = 1$, and the hypotenuse has a length of $r = \sqrt{2}$.

Using Eq. 5.2 and Eq. 5.3,

$$\cos 45° = x/r = \frac{1}{\sqrt{2}} = \sqrt{2}/2$$

$$\tan 45° = y/x = \frac{1}{1} = 1$$

The answer is (C).

5. GENERAL TRIANGLES

The term *general triangle* refers to any triangle, including but not limited to right triangles. Figure 5.7 shows a general triangle.

Figure 5.7 *General Triangle*

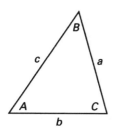

Equation 5.7: Law of Sines

$$\frac{a}{\sin A} = \frac{b}{\sin B} = \frac{c}{\sin C} \qquad 5.7$$

Description

For a general triangle, the *law of sines* relates the sines of the three angles A, B, and C and their opposite sides, a, b, and c, respectively.

Example

The vertical angle to the top of a flagpole from point A on the ground is observed to be $37°\,11'$. The observer walks 17 m directly away from the flagpole from point A to point B and finds the new angle to be $25°\,43'$.

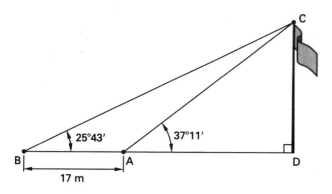

What is the approximate height of the flagpole?

(A) 10 m

(B) 22 m

(C) 82 m

(D) 300 m

Solution

The two observations lead to two triangles with a common leg, h.

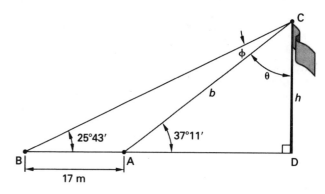

Find angle θ in triangle ADC.

$$37°\,11' + 90° + \theta = 180°$$

$$\theta = 52°\,49'$$

Find angle ϕ in triangle BAC.

$$25°\,43' + 90° + (52°\,49' + \phi) = 180°$$

$$\phi = 11°\,28'$$

Use the law of sines on triangle BAC to find side b.

$$\frac{\sin 11°\,28'}{17\text{ m}} = \frac{\sin 25°\,43'}{b}$$

$$b = 37.11\text{ m}$$

Find the flagpole height, h, using triangle ADC.

$$\sin 37°\,11' = \frac{h}{b}$$

$$h = b \sin 37°\,11'$$

$$= (37.11\text{ m})\sin 37°\,11'$$

$$= 22.43\text{ m} \quad (22\text{ m})$$

The answer is (B).

Equation 5.8 Through Eq. 5.10: Law of Cosines

$$a^2 = b^2 + c^2 - 2bc \cos A \qquad 5.8$$
$$b^2 = a^2 + c^2 - 2ac \cos B \qquad 5.9$$
$$c^2 = a^2 + b^2 - 2ab \cos C \qquad 5.10$$

Variations

$$\cos A = \frac{b^2 + c^2 - a^2}{2bc}$$

$$\cos B = \frac{a^2 + c^2 - b^2}{2ac}$$

$$\cos C = \frac{a^2 + b^2 - c^2}{2ab}$$

Description

For a general triangle, the *law of cosines* relates the cosines of the three angles A, B, and C and their opposite sides, a, b, and c, respectively.

Example

Three circles of radii 110 m, 140 m, and 220 m are tangent to one another. What are the interior angles of the triangle formed by joining the centers of the circles?

    (A) 34.2°, 69.2°, and 76.6°

    (B) 36.6°, 69.1°, and 74.3°

    (C) 42.2°, 62.5°, and 75.3°

    (D) 47.9°, 63.1°, and 69.0°

Solution

The three circles and the triangle are shown.

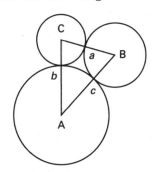

Calculate the length of each side of the triangle.

$$a = 110 \text{ m} + 140 \text{ m} = 250 \text{ m}$$

$$b = 110 \text{ m} + 220 \text{ m} = 330 \text{ m}$$

$$c = 140 \text{ m} + 220 \text{ m} = 360 \text{ m}$$

From Eq. 5.8,

$$a^2 = b^2 + c^2 - 2bc \cos A$$

$$\cos A = \frac{b^2 + c^2 - a^2}{2bc}$$

$$= \frac{(330 \text{ m})^2 + (360 \text{ m})^2 - (250 \text{ m})^2}{(2)(330 \text{ m})(360 \text{ m})}$$

$$= 0.7407$$

$$A = 42.2°$$

From Eq. 5.9,

$$b^2 = a^2 + c^2 - 2ac \cos B$$

$$\cos B = \frac{a^2 + c^2 - b^2}{2ac}$$

$$= \frac{(250 \text{ m})^2 + (360 \text{ m})^2 - (330 \text{ m})^2}{(2)(250 \text{ m})(360 \text{ m})}$$

$$= 0.4622$$

$$B = 62.5°$$

From Eq. 5.10,

$$c^2 = a^2 + b^2 - 2ab \cos C$$

$$\cos C = \frac{a^2 + b^2 - c^2}{2ab}$$

$$= \frac{(250 \text{ m})^2 + (330 \text{ m})^2 - (360 \text{ m})^2}{(2)(250 \text{ m})(330 \text{ m})}$$

$$= 0.2533$$

$$C = 75.3°$$

The answer is (C).

6. TRIGONOMETRIC IDENTITIES

Equation 5.11 through Eq. 5.42 are some of the most commonly used trigonometric identities.

Equation 5.11 Through Eq. 5.13: Reciprocal Functions

$\csc \theta = 1/\sin \theta$	*5.11*
$\sec \theta = 1/\cos \theta$	*5.12*
$\cot \theta = 1/\tan \theta$	*5.13*

Description

Three pairs of the trigonometric functions are reciprocals of each other. The prefix "co-" is not a good indicator of the reciprocal functions; while the tangent and cotangent functions are reciprocals of each other, two other pairs—the sine and cosine functions and the secant and cosecant functions—are not.

Example

Simplify the expression $\cos \theta \sec \theta / \tan \theta$.

    (A) 1

    (B) $\cot \theta$

    (C) $\csc \theta$

    (D) $\sin \theta$

Solution

Use the reciprocal functions given in Eq. 5.12 and Eq. 5.13.

$$\frac{\cos\theta\sec\theta}{\tan\theta} = \frac{\cos\theta\left(\dfrac{1}{\cos\theta}\right)}{\tan\theta}$$

$$= \frac{1}{\tan\theta}$$

$$= \cot\theta$$

The answer is (B).

Equation 5.14 Through Eq. 5.19: General Identities

$$\cos\theta = \sin(\theta + \pi/2) = -\sin(\theta - \pi/2) \qquad 5.14$$

$$\sin\theta = \cos(\theta - \pi/2) = -\cos(\theta + \pi/2) \qquad 5.15$$

$$\tan\theta = \sin\theta/\cos\theta \qquad 5.16$$

$$\sin^2\theta + \cos^2\theta = 1 \qquad 5.17$$

$$\tan^2\theta + 1 = \sec^2\theta \qquad 5.18$$

$$\cot^2\theta + 1 = \csc^2\theta \qquad 5.19$$

Description

Equation 5.14 through Eq. 5.19 give some general trigonometric identities.

Example

Which of the following expressions is equivalent to the expression $\csc\theta\cos^3\theta\tan\theta$?

- (A) $\sin\theta$
- (B) $\cos\theta$
- (C) $1 - \sin^2\theta$
- (D) $1 + \sin^2\theta$

Solution

Simplify the expression using the trigonometric identities given in Eq. 5.11, Eq. 5.16, and Eq. 5.17.

$$\csc\theta\cos^3\theta\tan\theta = \left(\frac{1}{\sin\theta}\right)\cos^3\theta\left(\frac{\sin\theta}{\cos\theta}\right)$$

$$= \cos^2\theta$$

$$= 1 - \sin^2\theta$$

The answer is (C).

Equation 5.20 Through Eq. 5.23: Double-Angle Identities

$$\sin 2\alpha = 2\sin\alpha\cos\alpha \qquad 5.20$$

$$\cos 2\alpha = \cos^2\alpha - \sin^2\alpha = 1 - 2\sin^2\alpha = 2\cos^2\alpha - 1 \qquad 5.21$$

$$\tan 2\alpha = (2\tan\alpha)/(1 - \tan^2\alpha) \qquad 5.22$$

$$\cot 2\alpha = (\cot^2\alpha - 1)/(2\cot\alpha) \qquad 5.23$$

Description

The identities given in Eq. 5.20 through Eq. 5.23 show equivalent expressions of trigonometric functions of double angles.

Example

What is an equivalent expression for $\sin 2\alpha$?

- (A) $-2\sin\alpha\cos\alpha$
- (B) $\frac{1}{2}\sin\alpha\cos\alpha$
- (C) $\dfrac{2\sin\alpha}{\sec\alpha}$
- (D) $2\sin\alpha\cos\dfrac{\alpha}{2}$

Solution

Use Eq. 5.20, the double-angle formula for the sine function.

$$\sin 2\alpha = 2\sin\alpha\cos\alpha = \frac{2\sin\alpha}{\sec\alpha}$$

The answer is (C).

Equation 5.24 Through Eq. 5.31: Two-Angle Identities

$$\sin(\alpha + \beta) = \sin\alpha\cos\beta + \cos\alpha\sin\beta \qquad 5.24$$

$$\cos(\alpha + \beta) = \cos\alpha\cos\beta - \sin\alpha\sin\beta \qquad 5.25$$

$$\tan(\alpha + \beta) = (\tan\alpha + \tan\beta)/(1 - \tan\alpha\tan\beta) \qquad 5.26$$

$$\cot(\alpha + \beta) = (\cot\alpha\cot\beta - 1)/(\cot\alpha + \cot\beta) \qquad 5.27$$

$$\sin(\alpha - \beta) = \sin\alpha\cos\beta - \cos\alpha\sin\beta \qquad 5.28$$

$$\cos(\alpha - \beta) = \cos\alpha\cos\beta + \sin\alpha\sin\beta \qquad 5.29$$

$$\tan(\alpha - \beta) = (\tan\alpha - \tan\beta)/(1 + \tan\alpha\tan\beta) \qquad 5.30$$

$$\cot(\alpha - \beta) = (\cot\alpha\cot\beta + 1)/(\cot\beta - \cot\alpha) \qquad 5.31$$

Description

The identities given in Eq. 5.24 through Eq. 5.31 show equivalent expressions of two-angle trigonometric functions.

Example

Simplify the following expression.

$$\frac{\cos(\alpha + \beta) + \cos(\alpha - \beta)}{\cos \beta}$$

(A) $\cos \alpha/2$

(B) $2 \cos \alpha$

(C) $\sin 2\alpha$

(D) $\sin^2 \alpha$

Solution

Use Eq. 5.25 and Eq. 5.29.

$$\frac{\cos(\alpha + \beta)}{+ \cos(\alpha - \beta)}{\cos \beta} = \frac{\left(\begin{array}{c}\cos \alpha \cos \beta \\ - \sin \alpha \sin \beta \end{array}\right) + \left(\begin{array}{c}\cos \alpha \cos \beta \\ + \sin \alpha \sin \beta \end{array}\right)}{\cos \beta}$$

$$= \frac{2 \cos \alpha \cos \beta}{\cos \beta}$$

$$= 2 \cos \alpha$$

The answer is (B).

Equation 5.32 Through Eq. 5.35: Half-Angle Identities

$$\sin(\alpha/2) = \pm \sqrt{(1 - \cos \alpha)/2} \quad \textbf{5.32}$$
$$\cos(\alpha/2) = \pm \sqrt{(1 + \cos \alpha)/2} \quad \textbf{5.33}$$
$$\tan(\alpha/2) = \pm \sqrt{(1 - \cos \alpha)/(1 + \cos \alpha)} \quad \textbf{5.34}$$
$$\cot(\alpha/2) = \pm \sqrt{(1 + \cos \alpha)/(1 - \cos \alpha)} \quad \textbf{5.35}$$

Description

The identities given in Eq. 5.32 through Eq. 5.35 show equivalent expressions of half-angle trigonometric functions.

Equation 5.36 Through Eq. 5.42: Miscellaneous Identities

$$\sin \alpha \sin \beta = (1/2)[\cos(\alpha - \beta) - \cos(\alpha + \beta)] \quad \textbf{5.36}$$
$$\cos \alpha \cos \beta = (1/2)[\cos(\alpha - \beta) + \cos(\alpha + \beta)] \quad \textbf{5.37}$$
$$\sin \alpha \cos \beta = (1/2)[\sin(\alpha + \beta) + \sin(\alpha - \beta)] \quad \textbf{5.38}$$
$$\sin \alpha + \sin \beta = 2 \sin[(1/2)(\alpha + \beta)]\cos[(1/2)(\alpha - \beta)] \quad \textbf{5.39}$$
$$\sin \alpha - \sin \beta = 2 \cos[(1/2)(\alpha + \beta)]\sin[(1/2)(\alpha - \beta)] \quad \textbf{5.40}$$
$$\cos \alpha + \cos \beta = 2 \cos[(1/2)(\alpha + \beta)]\cos[(1/2)(\alpha - \beta)] \quad \textbf{5.41}$$
$$\cos \alpha - \cos \beta = -2 \sin[(1/2)(\alpha + \beta)]\sin[(1/2)(\alpha - \beta)] \quad \textbf{5.42}$$

Description

The identities given in Eq. 5.36 through Eq. 5.42 show equivalent expressions of other trigonometric functions.

7. HYPERBOLIC FUNCTIONS AND IDENTITIES

Hyperbolic transcendental functions (normally referred to as *hyperbolic functions*) are specific equations containing combinations of the terms e^{θ} and $e^{-\theta}$. These combinations appear regularly in certain types of problems (e.g., analysis of cables and heat transfer through fins) and are given specific names and symbols to simplify presentation.[3]

- *hyperbolic sine*

$$\sinh \theta = \frac{e^{\theta} - e^{-\theta}}{2}$$

- *hyperbolic cosine*

$$\cosh \theta = \frac{e^{\theta} + e^{-\theta}}{2}$$

- *hyperbolic tangent*

$$\tanh \theta = \frac{e^{\theta} - e^{-\theta}}{e^{\theta} + e^{-\theta}} = \frac{\sinh \theta}{\cosh \theta}$$

- *hyperbolic cotangent*

$$\coth \theta = \frac{e^{\theta} + e^{-\theta}}{e^{\theta} - e^{-\theta}} = \frac{\cosh \theta}{\sinh \theta}$$

- *hyperbolic secant*

$$\operatorname{sech} \theta = \frac{2}{e^{\theta} + e^{-\theta}} = \frac{1}{\cosh \theta}$$

- *hyperbolic cosecant*

$$\operatorname{csch} \theta = \frac{2}{e^{\theta} - e^{-\theta}} = \frac{1}{\sinh \theta}$$

Hyperbolic functions cannot be related to a right triangle, but they are related to a rectangular (equilateral) hyperbola, as shown in Fig. 5.8. For a *unit hyperbola* $(a^2 = 1)$, the shaded area has a value of $\theta/2$ and is sometimes given the units of *hyperbolic radians*.

$$\sinh \theta = \frac{y}{a}$$

$$\cosh \theta = \frac{x}{a}$$

$$\tanh \theta = \frac{y}{x}$$

$$\coth \theta = \frac{x}{y}$$

$$\operatorname{sech} \theta = \frac{a}{x}$$

$$\operatorname{csch} \theta = \frac{a}{y}$$

---

[3]The hyperbolic sine and cosine functions are pronounced (by some) as "sinch" and "cosh," respectively.

Figure 5.8 Equilateral Hyperbola and Hyperbolic Functions

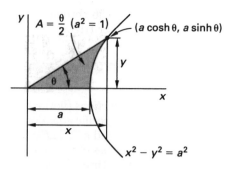

The hyperbolic identities are different from the standard trigonometric identities. Some of the most important identities are presented as follows.

$$\cosh^2 \theta - \sinh^2 \theta = 1$$

$$1 - \tanh^2 \theta = \operatorname{sech}^2 \theta$$

$$1 - \coth^2 \theta = -\operatorname{csch}^2 \theta$$

$$\cosh \theta + \sinh \theta = e^\theta$$

$$\cosh \theta - \sinh \theta = e^{-\theta}$$

$$\sinh(\theta \pm \phi) = \sinh \theta \cosh \phi \pm \cosh \theta \sinh \phi$$

$$\cosh(\theta \pm \phi) = \cosh \theta \cosh \phi \pm \sinh \theta \sinh \phi$$

$$\tanh(\theta \pm \phi) = \frac{\tanh \theta \pm \tanh \phi}{1 \pm \tanh \theta \tanh \phi}$$

8. FOURIER SERIES

Any periodic waveform can be written as the sum of an infinite number of sinusoidal terms (i.e., an infinite series), known as *harmonic terms*. Such a sum of sinusoidal terms is known as a *Fourier series*, and the process of finding the terms is *Fourier analysis*. Since most series converge rapidly, it is possible to obtain a good approximation to the original waveform with a limited number of sinusoidal terms.

Equation 5.43 and Eq. 5.44: Fourier's Theorem

$$f(t) = a_0 + \sum_{n=1}^{\infty} [a_n \cos(n\omega_0 t) + b_n \sin(n\omega_0 t)] \qquad 5.43$$

$$T = 2\pi/\omega_0 \qquad 5.44$$

Variation

$$\omega_0 = \frac{2\pi}{T} = 2\pi f$$

Description

Fourier's theorem is Eq. 5.43. The object of a Fourier analysis is to determine the *Fourier coefficients* a_n and b_n. The term a_0 can often be determined by inspection since it is the average value of the waveform.

ω_0 is the *natural (fundamental) frequency* of the waveform. It depends on the actual waveform *period*, T.

Equation 5.45 Through Eq. 5.47: Fourier Coefficients

$$a_0 = (1/T) \int_0^T f(t)\,dt \qquad 5.45$$

$$a_n = (2/T) \int_0^T f(t)\cos(n\omega_0 t)\,dt \quad [n = 1, 2, \ldots] \qquad 5.46$$

$$b_n = (2/T) \int_0^T f(t)\sin(n\omega_0 t)\,dt \quad [n = 1, 2, \ldots] \qquad 5.47$$

Description

The *Fourier coefficients* are found from the relationships shown in Eq. 5.45 through Eq. 5.47.

Example

What are the first terms in the Fourier series of the repeating function shown?

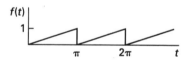

(A) $\dfrac{1}{2} - \cos 2t - \dfrac{1}{2}\cos 4t - \dfrac{1}{3}\cos 6t$

(B) $\dfrac{1}{2} - \dfrac{1}{\pi}\sin 2t - \dfrac{1}{2\pi}\sin 4t - \dfrac{1}{3\pi}\sin 6t$

(C) $\dfrac{1}{4} - \dfrac{1}{\pi}\left(\begin{array}{l} \cos 2t + \sin 2t + \cos 4t \\ \quad + \dfrac{1}{2}\sin 4t + \cos 6t + \dfrac{1}{3}\sin 6t \end{array} \right)$

(D) $\dfrac{1}{4} - \dfrac{1}{\pi}\left(\begin{array}{l} \dfrac{1}{\pi}\cos 2t + \sin 2t \\ \quad + \dfrac{1}{2\pi}\cos 4t + \dfrac{1}{2}\sin 4t \\ \quad + \dfrac{1}{3\pi}\cos 6t + \dfrac{1}{3}\sin 6t \end{array} \right)$

Solution

A Fourier series has the form given by Eq. 5.43.

$$f(t) = a_0 + \sum_{n=1}^{\infty} [a_n \cos(n\omega_0 t) + b_n \sin(n\omega_0 t)]$$

The constant term a_0 corresponds to the average of the function. The average is seen by observation to be $1/2$, so $a_0 = 1/2$.

In this problem, the triangular pulses are ramps, so $f(t)$ has the form of kt, where k is a scalar. A cycle is completed at $t = \pi$, so $T = \pi$, and $\omega_0 = 2\pi/T = 2$. Since $f(T) = 1$ (that is, $f(t) = 1$ at $t = \pi$), $f(t) = t/\pi$.

Calculate the general form of the a_n terms using Eq. 5.46.

$$a_n = (2/T) \int_0^T f(t) \cos(n\omega_0 t)\, dt$$

$$= \frac{2}{\pi^2} \int_0^{\pi} t \cos(2nt)\, dt$$

$$= \frac{1}{n\pi^2} \left(\cos(2nt) + 2nt \sin(2nt) \right) \Big|_0^{\pi}$$

$$= 0$$

There are no a_n terms in the series. From Eq. 5.43, there are no cosine terms in the expansion. There are only sine terms in the expansion.

Only choice (B) satisfies both of these requirements.

Alternatively, the values can be derived, though this would be a lengthy process.

The answer is (B).

Equation 5.48: Parseval Relation

$$F_N^2 = a_0^2 + (1/2) \sum_{n=1}^{N} (a_n^2 + b_n^2) \qquad \textbf{5.48}$$

Variation

$$F_{\text{rms}} = \sqrt{a_0^2 + \frac{a_1^2 + a_2^2 + \cdots a_N^2 + b_1^2 + b_2^2 + \cdots b_N^2}{2}}$$

Description

The *Parseval relation* (also known as *Parseval's equality*) calculates the root-mean-square (rms) value of a Fourier series that has been truncated after N terms. The rms value, F_{rms}, is the square root of Eq. 5.48.

6 Linear Algebra

Description

Addition and subtraction of two matrices are possible only if both matrices have the same number of rows and columns. They are accomplished by adding or subtracting the corresponding entries of the two matrices.

1. MATRICES

A *matrix* is an ordered set of *entries* (*elements*) arranged rectangularly and set off by brackets. The entries can be variables or numbers. A matrix by itself has no particular value; it is merely a convenient method of representing a set of numbers.

The size of a matrix is given by the number of rows and columns, and the nomenclature $m \times n$ is used for a matrix with m rows and n columns. For a square matrix, the numbers of rows and columns are the same and are equal to the *order of the matrix*.

Matrices are designated by bold uppercase letters. Matrix entries are designated by lowercase letters with subscripts, for example, a_{ij}. The term a_{23} would be the entry in the second row and third column of matrix $\mathbf{A}$. The matrix $\mathbf{C}$ can also be designated as (c_{ij}), meaning "the matrix made up of c_{ij} entries."

Equation 6.1: Addition of Matrices

$$\begin{bmatrix} A & B & C \\ D & E & F \end{bmatrix} + \begin{bmatrix} G & H & I \\ J & K & L \end{bmatrix}$$
$$= \begin{bmatrix} A+G & B+H & C+I \\ D+J & E+K & F+L \end{bmatrix} \quad 6.1$$

Variation

$$\mathbf{C} = \mathbf{A} + \mathbf{B} \equiv (c_{ij}) \equiv (a_{ij} + b_{ij})$$

Equation 6.2: Multiplication of Matrices

$$\mathbf{C} = \begin{bmatrix} A & B \\ C & D \\ E & F \end{bmatrix} \cdot \begin{bmatrix} H & I \\ J & K \end{bmatrix}$$
$$= \begin{bmatrix} (A{\cdot}H + B{\cdot}J) & (A{\cdot}I + B{\cdot}K) \\ (C{\cdot}H + D{\cdot}J) & (C{\cdot}I + D{\cdot}K) \\ (E{\cdot}H + F{\cdot}J) & (E{\cdot}I + F{\cdot}K) \end{bmatrix} \quad 6.2$$

Variations

$$\mathbf{C} = \mathbf{AB}$$

$$\mathbf{C} = \mathbf{A} \times \mathbf{B}$$

$$\mathbf{C} \equiv (c_{ij}) = \left(\sum_{l=1}^{n} a_{il} b_{lj} \right)$$

Description

A matrix can be multiplied by another matrix, but only if the left-hand matrix has the same number of columns as the right-hand matrix has rows. *Matrix multiplication* occurs by multiplying the elements in each left-hand matrix row by the entries in each corresponding right-hand matrix column, adding the products, and placing the sum at the intersection point of the participating row and column.

The commutative law does not apply to matrix multiplication. That is, $\mathbf{A} \times \mathbf{B}$ is not equivalent to $\mathbf{B} \times \mathbf{A}$.

Example

What is the matrix product $\mathbf{AB}$ of matrices $\mathbf{A}$ and $\mathbf{B}$?

$$\mathbf{A} = \begin{bmatrix} 2 & 1 \\ 1 & 0 \end{bmatrix} \qquad \mathbf{B} = \begin{bmatrix} 4 & 3 \\ 2 & 1 \end{bmatrix}$$

(A) $\begin{bmatrix} 10 & 4 \\ 7 & 3 \end{bmatrix}$

(B) $\begin{bmatrix} 11 & 4 \\ 5 & 2 \end{bmatrix}$

(C) $\begin{bmatrix} 8 & 3 \\ 2 & 0 \end{bmatrix}$

(D) $\begin{bmatrix} 10 & 7 \\ 4 & 3 \end{bmatrix}$

Solution

Use Eq. 6.2. Multiply the elements of each row in matrix $\mathbf{A}$ by the elements of the corresponding column in matrix $\mathbf{B}$.

$$\mathbf{C} = \begin{bmatrix} A & B \\ C & D \\ E & F \end{bmatrix} \cdot \begin{bmatrix} H & I \\ J & K \end{bmatrix}$$

$$= \begin{bmatrix} (A{\cdot}H + B{\cdot}J) & (A{\cdot}I + B{\cdot}K) \\ (C{\cdot}H + D{\cdot}J) & (C{\cdot}I + D{\cdot}K) \\ (E{\cdot}H + F{\cdot}J) & (E{\cdot}I + F{\cdot}K) \end{bmatrix}$$

$$= \begin{bmatrix} 2 \times 4 + 1 \times 2 & 2 \times 3 + 1 \times 1 \\ 1 \times 4 + 0 \times 2 & 1 \times 3 + 0 \times 1 \end{bmatrix}$$

$$= \begin{bmatrix} 10 & 7 \\ 4 & 3 \end{bmatrix}$$

The answer is (D).

Equation 6.3: Transposes of Matrices

$$\mathbf{A} = \begin{bmatrix} A & B & C \\ D & E & F \end{bmatrix} \qquad \mathbf{A}^T = \begin{bmatrix} A & D \\ B & E \\ C & F \end{bmatrix} \qquad 6.3$$

Variation

$$\mathbf{B} = \mathbf{A}^T$$

Description

The *transpose*, $\mathbf{A}^T$, of an $m \times n$ matrix $\mathbf{A}$ is an $n \times m$ matrix constructed by taking the ith row and making it the ith column.

Example

What is the transpose of matrix $\mathbf{A}$?

$$\mathbf{A} = \begin{bmatrix} 5 & 8 & 5 & 8 \\ 8 & 7 & 6 & 2 \end{bmatrix}$$

(A) $\begin{bmatrix} 8 & 7 & 6 & 2 \\ 5 & 8 & 5 & 8 \end{bmatrix}$

(B) $\begin{bmatrix} 2 & 6 & 7 & 8 \\ 8 & 5 & 8 & 5 \end{bmatrix}$

(C) $\begin{bmatrix} 8 & 5 \\ 7 & 8 \\ 6 & 5 \\ 2 & 8 \end{bmatrix}$

(D) $\begin{bmatrix} 5 & 8 \\ 8 & 7 \\ 5 & 6 \\ 8 & 2 \end{bmatrix}$

Solution

The transpose of a matrix is constructed by taking the ith row and making it the ith column.

The answer is (D).

Equation 6.4: Determinants of 2×2 Matrices

$$\begin{vmatrix} a_1 & a_2 \\ b_1 & b_2 \end{vmatrix} = a_1 b_2 - a_2 b_1 \qquad 6.4$$

Variation

$$|\mathbf{A}| = \begin{vmatrix} a_1 & a_2 \\ b_1 & b_2 \end{vmatrix} = a_1 b_2 - a_2 b_1$$

Description

A *determinant* is a scalar calculated from a square matrix. The determinant of matrix $\mathbf{A}$ can be represented as $D\{\mathbf{A}\}$, $Det(\mathbf{A})$, or $|\mathbf{A}|$. The following rules can be used to simplify the calculation of determinants.

- If $\mathbf{A}$ has a row or column of zeros, the determinant is zero.

- If $\mathbf{A}$ has two identical rows or columns, the determinant is zero.

- If $\mathbf{B}$ is obtained from $\mathbf{A}$ by adding a multiple of a row (column) to another row (column) in $\mathbf{A}$, then $|\mathbf{B}| = |\mathbf{A}|$.

- If $\mathbf{A}$ is *triangular* (a square matrix with zeros in all positions above or below the diagonal), the determinant is equal to the product of the diagonal entries.

- If $\mathbf{B}$ is obtained from $\mathbf{A}$ by multiplying one row or column in $\mathbf{A}$ by a scalar k, then $|\mathbf{B}| = k|\mathbf{A}|$.

- If $\mathbf{B}$ is obtained from the $n \times n$ matrix $\mathbf{A}$ by multiplying by the scalar matrix k, then $|\mathbf{B}| = |\mathbf{k} \times \mathbf{A}| = k^n|\mathbf{A}|$.

- If $\mathbf{B}$ is obtained from $\mathbf{A}$ by switching two rows or columns in $\mathbf{A}$, then $|\mathbf{B}| = -|\mathbf{A}|$.

Calculation of determinants is laborious for all but the smallest or simplest of matrices. For a 2×2 matrix, the formula used to calculate the determinant is easy to remember.

Example

What is the determinant of matrix $\mathbf{A}$?

$$\mathbf{A} = \begin{bmatrix} 3 & 6 \\ 2 & 4 \end{bmatrix}$$

(A) 0

(B) 15

(C) 14

(D) 26

Solution

From Eq. 6.4, for a square 2×2 matrix,

$$\begin{vmatrix} a_1 & a_2 \\ b_1 & b_2 \end{vmatrix} = a_1 b_2 - a_2 b_1$$

$$\begin{vmatrix} 3 & 6 \\ 2 & 4 \end{vmatrix} = 3 \times 4 - 6 \times 2$$

$$= 0$$

The answer is (A).

Equation 6.5: Determinants of 3×3 Matrices

$$\begin{vmatrix} a_1 & a_2 & a_3 \\ b_1 & b_2 & b_3 \\ c_1 & c_2 & c_3 \end{vmatrix} = a_1 b_2 c_3 + a_2 b_3 c_1 + a_3 b_1 c_2 - a_3 b_2 c_1 - a_2 b_1 c_3 - a_1 b_3 c_2$$

$$6.5$$

Variations

$$\mathbf{A} = \begin{bmatrix} a_1 & a_2 & a_3 \\ b_1 & b_2 & b_3 \\ c_1 & c_2 & c_3 \end{bmatrix}$$

$$|\mathbf{A}| = a_1 b_2 c_3 + a_2 b_3 c_1 + a_3 b_1 c_2 - a_3 b_2 c_1 - a_2 b_1 c_3 - a_1 b_3 c_2$$

Description

In addition to the formula-based method expressed as Eq. 6.5, two methods are commonly used for calculating the determinants of 3×3 matrices by hand. The first uses an augmented matrix constructed from the original matrix and the first two columns. The determinant is calculated as the sum of the products in the left-to-right downward diagonals less the sum of the products in the left-to-right upward diagonals.

$$\text{augmented } \mathbf{A} = \begin{bmatrix} a_1 & a_2 & a_3 & a_1 & a_2 \\ b_1 & b_2 & b_3 & b_1 & b_2 \\ c_1 & c_2 & c_3 & c_1 & c_2 \end{bmatrix}$$

The second method of calculating the determinant is somewhat slower than the first for a 3×3 matrix but illustrates the method that must be used to calculate determinants of 4×4 and larger matrices. This method is known as *expansion by cofactors* (cofactors are explained in the following section). One row (column) is selected as the base row (column). The selection is arbitrary, but the number of calculations required to obtain the determinant can be minimized by choosing the row (column) with the most zeros. The determinant is equal to the sum of the products of the entries in the base row (column) and their corresponding cofactors.

$$\mathbf{A} = \begin{bmatrix} a_1 & a_2 & a_3 \\ b_1 & b_2 & b_3 \\ c_1 & c_2 & c_3 \end{bmatrix} \quad \begin{bmatrix} \text{first column chosen} \\ \text{as base column} \end{bmatrix}$$

$$|\mathbf{A}| = a_1 \begin{vmatrix} b_2 & b_3 \\ c_2 & c_3 \end{vmatrix} - b_1 \begin{vmatrix} a_2 & a_3 \\ c_2 & c_3 \end{vmatrix} + c_1 \begin{vmatrix} a_2 & a_3 \\ b_2 & b_3 \end{vmatrix}$$

$$= a_1(b_2 c_3 - b_3 c_2) - b_1(a_2 c_3 - a_3 c_2)$$

$$+ c_1(a_2 b_3 - a_3 b_2)$$

$$= a_1 b_2 c_3 - a_1 b_3 c_2 - b_1 a_2 c_3 + b_1 a_3 c_2$$

$$+ c_1 a_2 b_3 - c_1 a_3 b_2$$

Example

For the following set of equations, what is the determinant of the coefficient matrix?

$$10x + 3y + 10z = 5$$
$$8x - 2y + 9z = 5$$
$$8x + y - 10z = 5$$

(A) 598

(B) 620

(C) 714

(D) 806

Solution

Calculate the determinant of the coefficient matrix.

$$|\mathbf{A}| = \begin{vmatrix} a_1 & a_2 & a_3 \\ b_1 & b_2 & b_3 \\ c_1 & c_2 & c_3 \end{vmatrix}$$

$$= a_1 b_2 c_3 + a_2 b_3 c_1 + a_3 b_1 c_2 - a_3 b_2 c_1$$
$$\quad - a_2 b_1 c_3 - a_1 b_3 c_2$$

$$= (10)(-2)(-10) + (3)(9)(8) + (10)(8)(1)$$
$$\quad - (8)(-2)(10) - (1)(9)(10)$$
$$\quad - (-10)(8)(3)$$

$$= 806$$

The answer is (D).

Inverse of a Matrix

The *inverse*, $\mathbf{A}^{-1}$, of an invertible matrix, $\mathbf{A}$, is a matrix such that the product $\mathbf{A}\mathbf{A}^{-1}$ produces a matrix with ones along its diagonal and zeros elsewhere (i.e., above and below the diagonal). Only square matrices have inverses, but not all square matrices are invertible (i.e., have inverses). The product of a matrix and its inverse produces an identity matrix. For 3×3 matrices,

$$\mathbf{A}\mathbf{A}^{-1} = \mathbf{I} = \begin{bmatrix} 1 & 0 & 0 \\ 0 & 1 & 0 \\ 0 & 0 & 1 \end{bmatrix}$$

The inverse of a 2×2 matrix is easily determined by the following formula.

$$\mathbf{A} = \begin{bmatrix} a_1 & a_2 \\ b_1 & b_2 \end{bmatrix}$$

$$\mathbf{A}^{-1} = \frac{\begin{bmatrix} b_2 & -a_2 \\ -b_1 & a_1 \end{bmatrix}}{|\mathbf{A}|}$$

Equation 6.6: Identity Matrix

$$[\mathbf{A}][\mathbf{A}]^{-1} = [\mathbf{A}]^{-1}[\mathbf{A}] = [\mathbf{I}] \qquad 6.6$$

Variation

$$\mathbf{A} \times \mathbf{A}^{-1} = \mathbf{A}^{-1} \times \mathbf{A} = \mathbf{I}$$

Description

The product of a matrix $\mathbf{A}$ and its *inverse*, $\mathbf{A}^{-1}$, is the *identity matrix*, $\mathbf{I}$. A matrix has an inverse if and only if it is *nonsingular* (i.e., its determinant is nonzero).

Example

Using the property that $|\mathbf{AB}| = |\mathbf{A}||\mathbf{B}|$ for two square matrices, what is $|\mathbf{A}^{-1}|$ in terms of $|\mathbf{A}|$ for any invertible square matrix $\mathbf{A}$?

(A) $\dfrac{1}{|\mathbf{A}|}$

(B) $\dfrac{1}{|\mathbf{A}^{-1}|}$

(C) $\dfrac{|\mathbf{A}|}{|\mathbf{A}^{-1}|}$

(D) $\dfrac{|\mathbf{A}^{-1}|}{|\mathbf{A}|}$

Solution

Since $|\mathbf{AB}| = |\mathbf{A}||\mathbf{B}|$,

$$|\mathbf{AA}^{-1}| = |\mathbf{A}||\mathbf{A}^{-1}|$$

Solving for $|\mathbf{A}^{-1}|$,

$$|\mathbf{A}^{-1}| = \frac{|\mathbf{AA}^{-1}|}{|\mathbf{A}|}$$

But $|\mathbf{AA}^{-1}| = |\mathbf{I}| = 1$. Therefore,

$$|\mathbf{A}^{-1}| = \frac{|\mathbf{AA}^{-1}|}{|\mathbf{A}|} = \frac{1}{|\mathbf{A}|}$$

The answer is (A).

Cofactors

Cofactors are determinants of submatrices associated with particular entries in the original square matrix. The *minor* of entry a_{ij} is the determinant of a submatrix resulting from the elimination of the single row i and the single column j. For example, the minor corresponding to entry a_{12} in a 3×3 matrix $\mathbf{A}$ is the determinant of the matrix created by eliminating row 1 and column 2.

$$\text{minor of } a_{12} = \begin{vmatrix} a_{21} & a_{23} \\ a_{31} & a_{33} \end{vmatrix}$$

The cofactor of entry a_{ij} is the minor of a_{ij} multiplied by either $+1$ or -1, depending on the position of the entry (i.e., the cofactor either exactly equals the minor or it differs only in sign). The sign of the cofactor of a_{ij} is positive if $(i+j)$ is even, and it is negative if $(i+j)$ is odd. For a 3×3 matrix, the multipliers in each position are

$$\begin{bmatrix} +1 & -1 & +1 \\ -1 & +1 & -1 \\ +1 & -1 & +1 \end{bmatrix}$$

For example, the cofactor of entry a_{12} in a 3×3 matrix $\mathbf{A}$ is

$$\text{cofactor of } a_{12} = -\begin{vmatrix} a_{21} & a_{23} \\ a_{31} & a_{33} \end{vmatrix}$$

Equation 6.7: Classical Adjoint

$$\mathbf{B} = \mathbf{A}^{-1} = \frac{\text{adj}(\mathbf{A})}{|\mathbf{A}|} \qquad \textit{6.7}$$

Description

The *classical adjoint*, or *adjugate*, is the transpose of the cofactor matrix. The resulting matrix can be designated as $\mathbf{A}_{adj}$, $\text{adj}\{\mathbf{A}\}$, or $\mathbf{A}^{adj}$.

For a 3×3 or larger matrix, the inverse is determined by dividing every entry in the classical adjoint by the determinant of the original matrix, as shown in Eq. 6.7.

Example

The cofactor matrix of matrix $\mathbf{A}$ is $\mathbf{C}$.

$$\mathbf{A} = \begin{bmatrix} 4 & 2 & 3 \\ 3 & 2 & 2 \\ 2 & 1 & 4 \end{bmatrix} \qquad \mathbf{C} = \begin{bmatrix} 6 & -8 & -1 \\ -5 & 10 & 0 \\ -2 & 1 & 2 \end{bmatrix}$$

What is the inverse of matrix $\mathbf{A}$?

(A) $\begin{bmatrix} 0.25 & 0 & 0 \\ 0 & 0.50 & 0 \\ 0 & 0 & 0.25 \end{bmatrix}$

(B) $\begin{bmatrix} 0.25 & 0.50 & 0.33 \\ 0.33 & 0.50 & 0.50 \\ 0.50 & 1.0 & 0.25 \end{bmatrix}$

(C) $\begin{bmatrix} 1.2 & -1.0 & -0.40 \\ -1.6 & 2.0 & 0.20 \\ -0.20 & 0 & 0.40 \end{bmatrix}$

(D) $\begin{bmatrix} 0.80 & 0.40 & -0.60 \\ 0.20 & -0.40 & 0.40 \\ -0.40 & 0.60 & 0.80 \end{bmatrix}$

Solution

The classical adjoint is the transpose of the cofactor matrix.

$$\text{adj}(\mathbf{A}) = \mathbf{C}^T = \begin{bmatrix} 6 & -5 & -2 \\ -8 & 10 & 1 \\ -1 & 0 & 2 \end{bmatrix}$$

Using Eq. 6.4, calculate the determinant of $\mathbf{A}$ by expanding along the top row.

$$\begin{aligned} |\mathbf{A}| &= (4)(8-2) - (2)(12-4) + (3)(3-4) \\ &= 24 - 16 - 3 \\ &= 5 \end{aligned}$$

Using Eq. 6.7, divide the classical adjoint by the determinant.

$$\begin{aligned} \mathbf{A}^{-1} &= \frac{\text{adj}(\mathbf{A})}{|\mathbf{A}|} \\ &= \frac{\begin{bmatrix} 6 & -5 & -2 \\ -8 & 10 & 1 \\ -1 & 0 & 2 \end{bmatrix}}{5} \\ &= \begin{bmatrix} 1.2 & -1.0 & -0.40 \\ -1.6 & 2.0 & 0.20 \\ -0.20 & 0 & 0.40 \end{bmatrix} \end{aligned}$$

The answer is (C).

2. WRITING SIMULTANEOUS LINEAR EQUATIONS IN MATRIX FORM

Matrices are used to simplify the presentation and solution of sets of simultaneous linear equations. For example, the following three methods of presenting simultaneous linear equations are equivalent:

$$a_{11}x_1 + a_{12}x_2 = b_1$$

$$a_{21}x_1 + a_{22}x_2 = b_2$$

$$\begin{bmatrix} a_{11} & a_{12} \\ a_{21} & a_{22} \end{bmatrix} \begin{bmatrix} x_1 \\ x_2 \end{bmatrix} = \begin{bmatrix} b_1 \\ b_2 \end{bmatrix}$$

$$\mathbf{AX} = \mathbf{B}$$

In the second and third representations, $\mathbf{A}$ is known as the *coefficient matrix*, $\mathbf{X}$ as the *variable matrix*, and $\mathbf{B}$ as the *constant matrix*.

Not all systems of simultaneous equations have solutions, and those that do may not have unique solutions. The existence of a solution can be determined by calculating the determinant of the coefficient matrix. Solution-existence rules are summarized in Table 6.1.

- If the system of linear equations is homogeneous (i.e., $\mathbf{B}$ is a zero matrix) and $|\mathbf{A}|$ is zero, there are an infinite number of solutions.

- If the system is homogeneous and $|\mathbf{A}|$ is nonzero, only the trivial solution exists.

- If the system of linear equations is nonhomogeneous (i.e., $\mathbf{B}$ is not a zero matrix) and $|\mathbf{A}|$ is nonzero, there is a unique solution to the set of simultaneous equations.

- If $|\mathbf{A}|$ is zero, a nonhomogeneous system of simultaneous equations may still have a solution. The requirement is that the determinants of all substitutional matrices (mentioned in Sec. 6.1) are zero, in which case there will be an infinite number of solutions. Otherwise, no solution exists.

Table 6.1 *Solution Existence Rules for Simultaneous Equations*

	$\mathbf{B} = 0$	$\mathbf{B} \neq 0$
$|\mathbf{A}| = 0$	infinite number of solutions (linearly dependent equations)	either an infinite number of solutions or no solution at all
$|\mathbf{A}| \neq 0$	trivial solution only ($x_i = 0$)	unique nonzero solution

3. SOLVING SIMULTANEOUS LINEAR EQUATIONS WITH CRAMER'S RULE

Gauss-Jordan elimination can be used to obtain the solution to a set of simultaneous linear equations. The coefficient matrix is augmented by the constant matrix.

Then, elementary row operations are used to reduce the coefficient matrix to canonical form. All of the operations performed on the coefficient matrix are performed on the constant matrix. The variable values that satisfy the simultaneous equations will be the entries in the constant matrix when the coefficient matrix is in canonical form.

Determinants are used to calculate the solution to linear simultaneous equations through a procedure known as *Cramer's rule*.

The procedure is to calculate determinants of the original coefficient matrix $\mathbf{A}$ and of the n matrices resulting from the systematic replacement of a column in $\mathbf{A}$ by the constant matrix $\mathbf{B}$ (i.e., the *substitutional matrices*). For a system of three equations in three unknowns, there are three substitutional matrices, $\mathbf{A}_1$, $\mathbf{A}_2$, and $\mathbf{A}_3$, as well as the original coefficient matrix, for a total of four matrices whose determinants must be calculated.

The values of the unknowns that simultaneously satisfy all of the linear equations are

$$x_1 = \frac{|\mathbf{A}_1|}{|\mathbf{A}|}$$

$$x_2 = \frac{|\mathbf{A}_2|}{|\mathbf{A}|}$$

$$x_3 = \frac{|\mathbf{A}_3|}{|\mathbf{A}|}$$

Example

Using Cramer's rule, what values of x, y, and z will satisfy the following system of simultaneous equations?

$$2x + 3y - 4z = 1$$

$$3x - y - 2z = 4$$

$$4x - 7y - 6z = -7$$

(A) $x = 1$, $y = -4$, $z = -1$

(B) $x = 1$, $y = 3$, $z = 1$

(C) $x = 3$, $y = -2$, $z = 4$

(D) $x = 3$, $y = 1$, $z = 2$

Solution

The determinant of the coefficient matrix is

$$|\mathbf{A}| = \begin{vmatrix} 2 & 3 & -4 \\ 3 & -1 & -2 \\ 4 & -7 & -6 \end{vmatrix} = 82$$

The determinants of the substitutional matrices are

$$|\mathbf{A}_1| = \begin{vmatrix} 1 & 3 & -4 \\ 4 & -1 & -2 \\ -7 & -7 & -6 \end{vmatrix} = 246$$

$$|\mathbf{A}_2| = \begin{vmatrix} 2 & 1 & -4 \\ 3 & 4 & -2 \\ 4 & -7 & -6 \end{vmatrix} = 82$$

$$|\mathbf{A}_3| = \begin{vmatrix} 2 & 3 & 1 \\ 3 & -1 & 4 \\ 4 & -7 & -7 \end{vmatrix} = 164$$

The values of x, y, and z that will satisfy the linear equations are

$$x = \frac{246}{82} = 3$$

$$y = \frac{82}{82} = 1$$

$$z = \frac{164}{82} = 2$$

The answer is (D).

7 Calculus

1. DERIVATIVES

In most cases, it is possible to transform a continuous function, $f(x_1, x_2, x_3, \ldots)$, of one or more independent variables into a derivative function. In simple two-dimensional cases, the *derivative* can be interpreted as the slope (tangent or rate of change) of the curve described by the original function.

Equation 7.1 Through Eq. 7.3: Definitions of the Derivative

$$y' = \underset{\Delta x \to 0}{\text{limit}}[(\Delta y)/(\Delta x)] \qquad \textit{7.1}$$

$$y' = \underset{\Delta x \to 0}{\text{limit}}\{[f(x + \Delta x) - f(x)]/(\Delta x)\} \qquad \textit{7.2}$$

$$y' = \text{the slope of the curve } f(x) \qquad \textit{7.3}$$

Variation

$$f'(x) = \lim_{\Delta x \to 0}\left(\frac{f(x + \Delta x) - f(x)}{\Delta x}\right)$$

Description

Since the slope of a curve depends on x, the derivative function will also depend on x. The derivative, $f'(x)$, of a function $f(x)$ is defined mathematically by the variation given here. However, limit theory is seldom needed to actually calculate derivatives.

Equation 7.4 and Eq. 7.5: First Derivative

$$y = f(x) \qquad \textit{7.4}$$

$$D_x y = dy/dx = y' \qquad \textit{7.5}$$

Variations

$$f'(x), \quad \frac{df(x)}{dx}, \quad \mathbf{D}f(x), \quad \mathbf{D}_x f(x)$$

Description

The derivative of a function $y = f(x)$, also known as the *first derivative*, is represented in various ways, as shown by the variations.

Example

What is the slope of the curve $y = 10x^2 - 3x - 1$ when it crosses the positive part of the x-axis?

(A) 3/20

(B) 1/5

(C) 1/3

(D) 7

Solution

The curve crosses the x-axis when $y = 0$. At this point,

$$10x^2 - 3x - 1 = 0$$

Use the quadratic equation or complete the square to determine the two values of x where the curve crosses the x-axis.

$$x^2 - 0.3x = 0.1$$
$$(x - 0.15)^2 = 0.1 + (0.15)^2$$
$$x = \pm 0.35 + 0.15$$
$$= -0.2, \ 0.5$$

Since x must be positive, $x = 0.5$. The slope of the function is the first derivative.

$$\frac{dy}{dx} = 20x - 3$$
$$= (20)(0.5) - 3$$
$$= 7$$

The answer is (D).

Mathematics

Equation 7.6 Through Eq. 7.32: Derivatives

$$dc/dx = 0 \qquad \text{7.6}$$

$$dx/dx = 1 \qquad \text{7.7}$$

$$d(cu)/dx = c\,du/dx \qquad \text{7.8}$$

$$d(u + v - w)/dx = du/dx + dv/dx - dw/dx \qquad \text{7.9}$$

$$d(uv)/dx = u\,dv/dx + v\,du/dx \qquad \text{7.10}$$

$$d(uvw)/dx = uv\,dw/dx + uw\,dv/dx + vw\,du/dx \qquad \text{7.11}$$

$$\frac{d(u/v)}{dx} = \frac{v\,du/dx - u\,dv/dx}{v^2} \qquad \text{7.12}$$

$$d(u^n)/dx = nu^{n-1}\,du/dx \qquad \text{7.13}$$

$$d[f(u)]/dx = \{d[f(u)]/du\}\,du/dx \qquad \text{7.14}$$

$$du/dx = 1/(dx/du) \qquad \text{7.15}$$

$$\frac{d(\log_a u)}{dx} = (\log_a e)\frac{1}{u}\frac{du}{dx} \qquad \text{7.16}$$

$$\frac{d(\ln u)}{dx} = \frac{1}{u}\frac{du}{dx} \qquad \text{7.17}$$

$$\frac{d(a^u)}{dx} = (\ln a)a^u\frac{du}{dx} \qquad \text{7.18}$$

$$d(e^u)/dx = e^u\,du/dx \qquad \text{7.19}$$

$$d(u^v)/dx = vu^{v-1}\,du/dx + (\ln u)u^v\,dv/dx \qquad \text{7.20}$$

$$d(\sin u)/dx = \cos u\,du/dx \qquad \text{7.21}$$

$$d(\cos u)/dx = -\sin u\,du/dx \qquad \text{7.22}$$

$$d(\tan u)/dx = \sec^2 u\,du/dx \qquad \text{7.23}$$

$$d(\cot u)/dx = -\csc^2 u\,du/dx \qquad \text{7.24}$$

$$d(\sec u)/dx = \sec u\tan u\,du/dx \qquad \text{7.25}$$

$$d(\csc u)/dx = -\csc u\cot u\,du/dx \qquad \text{7.26}$$

$$\frac{d(\sin^{-1}u)}{dx} = \frac{1}{\sqrt{1-u^2}}\frac{du}{dx} \quad \left[-\pi/2 \le \sin^{-1}u \le \pi/2\right] \qquad \text{7.27}$$

$$\frac{d(\cos^{-1}u)}{dx} = -\frac{1}{\sqrt{1-u^2}}\frac{du}{dx} \quad \left[0 \le \cos^{-1}u \le \pi\right] \qquad \text{7.28}$$

$$\frac{d(\tan^{-1}u)}{dx} = \frac{1}{1+u^2}\frac{du}{dx} \quad \left[-\pi/2 < \tan^{-1}u < \pi/2\right] \qquad \text{7.29}$$

$$\frac{d(\cot^{-1}u)}{dx} = -\frac{1}{1+u^2}\frac{du}{dx} \quad \left[0 < \cot^{-1}u < \pi\right] \qquad \text{7.30}$$

$$\frac{d(\sec^{-1}u)}{dx} = \frac{1}{u\sqrt{u^2-1}}\frac{du}{dx}$$
$$\left[0 < \sec^{-1}u < \pi/2 \text{ or } -\pi \le \sec^{-1}u < -\pi/2\right] \qquad \text{7.31}$$

$$\frac{d(\csc^{-1}u)}{dx} = -\frac{1}{u\sqrt{u^2-1}}\frac{du}{dx}$$
$$\left[0 < \csc^{-1}u \le \pi/2 \text{ or } -\pi < \csc^{-1}u \le -\pi/2\right] \qquad \text{7.32}$$

Description

Formulas for the derivatives of some common functional forms are listed in Eq. 7.6 through Eq. 7.32.

Example

Evaluate dy/dx for the following expression.

$$y = e^{-x}\sin 2x$$

(A) $e^{-x}(2\cos 2x - \sin 2x)$

(B) $-e^{-x}(2\sin 2x + \cos 2x)$

(C) $e^{-x}(2\sin 2x + \cos 2x)$

(D) $-e^{-x}(2\cos 2x - \sin 2x)$

Solution

Use the product rule (Eq. 7.10).

$$\frac{d}{dx}\left(e^{-x}\sin 2x\right) = e^{-x}\frac{d}{dx}\left(\sin 2x\right)$$
$$+ (\sin 2x)\frac{d}{dx}\left(e^{-x}\right)$$
$$= e^{-x}(\cos 2x)(2)$$
$$+ (\sin 2x)(e^{-x})(-1)$$
$$= e^{-x}(2\cos 2x - \sin 2x)$$

The answer is (A).

2. CRITICAL POINTS

Derivatives are used to locate the local *critical points*, that is, *extreme points* (also known as *maximum* and *minimum points*) as well as the *inflection points (points of contraflexure)* of functions of one variable. The plurals *extrema*, *maxima*, and *minima* are used without the word "points." These points are illustrated in Fig. 7.1. There is usually an inflection point between two adjacent local extrema.

Figure 7.1 Critical Points

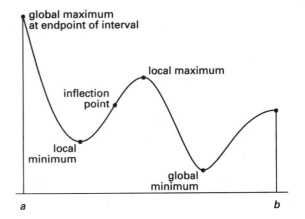

The first derivative, $f'(x)$, is calculated to determine where the critical points might be. The second derivative, $f''(x)$, is calculated to determine whether a located point is a maximum, minimum, or inflection point. With this method, no distinction is made between local and global extrema. The extrema should be compared to the function values at the endpoints of the interval.

Critical points are located where the first derivative is zero. This is a necessary, but not sufficient, requirement. That is, for a function $y = f(x)$, the point $x = a$ is a critical point if

$$f'(a) = 0$$

Equation 7.33 and Eq. 7.34: Test for a Maximum

$f'(a) = 0$	7.33
$f''(a) < 0$	7.34

Description

For a function $f(x)$ with an extreme point at $x = a$, if the point is a maximum, then the second derivative is negative.

Example

What is the maximum value of the function $f(x) = -x^2 - 8x + 1$?

(A) 1

(B) 4

(C) 8

(D) 17

Solution

Use Eq. 7.33 and Eq. 7.34.

$$f(x) = -x^2 - 8x + 1$$
$$f'(x) = -2x - 8$$
$$f''(x) = -2$$

$f'(x) = 0$ when x is equal to –4, and $f''(x)$ is less than zero, so $f(x)$ has its maximum value at $x = -4$.

$$f(x) = -x^2 - 8x + 1$$
$$= -(-4)^2 - (8)(-4) + 1$$
$$= 17$$

The answer is (D).

Equation 7.35 and Eq. 7.36: Test for a Minimum

$f'(a) = 0$	7.35
$f''(a) > 0$	7.36

Description

For a function $f(x)$ with a critical point at $x = a$, if the point is a minimum, then the second derivative is positive.

Example

What is the minimum value of the function $f(x) = 3x^2 + 3x - 5$?

(A) −12.0

(B) −8.0

(C) −5.75

(D) −5.00

Solution

Use Eq. 7.35 and Eq. 7.36.

$$f(x) = 3x^2 + 3x - 5$$
$$f'(x) = 6x + 3$$
$$f''(x) = 6$$

$f'(x) = 0$ when x is equal to –0.5, and $f''(x)$ is greater than zero, so $f(x)$ has its minimum value at $x = -0.5$.

$$f(x) = 3x^2 + 3x - 5$$
$$= (3)(-0.5)^2 + (3)(-0.5) - 5$$
$$= -5.75$$

The answer is (C).

Equation 7.37: Test for a Point of Inflection

$f''(a) = 0$	7.37

Description

For a function $f(x)$ with $f'(x) = 0$ at $x = a$, if the point is a point of inflection, then Eq. 7.37 is true.

3. PARTIAL DERIVATIVES

Derivatives can be taken with respect to only one independent variable at a time. For example, $f'(x)$ is the derivative of $f(x)$ and is taken with respect to the independent variable x. If a function, $f(x_1, x_2, x_3 \ldots)$, has more than one independent variable, a *partial derivative* can be found, but only with respect to one of the independent variables. All other variables are treated as constants.

Equation 7.38 and Eq. 7.39: Partial Derivative

$$z = f(x, y) \qquad 7.38$$

$$\frac{\partial z}{\partial x} = \frac{\partial f(x, y)}{\partial x} \qquad 7.39$$

Variations

Symbols for a partial derivative of $f(x, y)$ taken with respect to variable x are $\partial f / \partial x$ and $f_x(x, y)$.

Description

The geometric interpretation of a partial derivative $\partial f / \partial x$ is the slope of a line tangent to the surface (a sphere, an ellipsoid, etc.) described by the function when all variables except x are held constant. In three-dimensional space with a function described by Eq. 7.38, the partial derivative $\partial f / \partial x$ (equivalent to $\partial z / \partial x$) is the slope of the line tangent to the surface in a plane of constant y. Similarly, the partial derivative $\partial f / \partial y$ (equivalent to $\partial z / \partial y$) is the slope of the line tangent to the surface in a plane of constant x.

Example

What is the partial derivative with respect to x of the following function?

$$z = e^{xy}$$

(A) e^{xy}

(B) $\dfrac{e^{xy}}{x}$

(C) $\dfrac{e^{xy}}{y}$

(D) ye^{xy}

Solution

Use Eq. 7.19 and Eq. 7.39. The partial derivative is

$$d(e^u)/dx = e^u \, du/dx$$

$$\frac{\partial z}{\partial x} = \frac{\partial e^{xy}}{\partial x} = e^{xy} \frac{\partial(xy)}{\partial x}$$

$$= ye^{xy}$$

The answer is (D).

4. CURVATURE

The sharpness of a curve between two points on the curve can be defined as the rate of change of the inclination of the curve with respect to the distance traveled along the curve. As shown in Fig. 7.2, the rate of change of the inclination of the curve is the change in the angle formed by the tangents to the curve at each point and the x-axis. The distance, s, traveled along the curve is the arc length of the curve between points 1 and 2.

Figure 7.2 Curvature

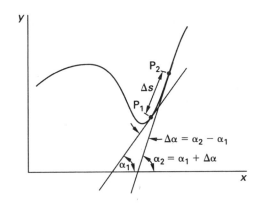

Equation 7.40: Curvature

$$K = \lim_{\Delta s \to 0} \frac{\Delta \alpha}{\Delta s} = \frac{d\alpha}{ds} \qquad 7.40$$

Description

On roadways, a "sharp" curve is one that changes direction quickly, corresponding to a small curve radius. The smaller the curve radius, the sharper the curve. Some roadway curves are circular, some are parabolic, and some are spiral. Not all curves are circular, but all curves described by polynomials have an instantaneous sharpness and radius of curvature. The sharpness, K, of a curve at a point is given by Eq. 7.40.

Equation 7.41 Through Eq. 7.43: Curvature in Rectangular Coordinates

$$K = \frac{y''}{[1 + (y')^2]^{3/2}} \qquad 7.41$$

$$x' = dx/dy \qquad 7.42$$

$$K = \frac{-x''}{[1 + (x')^2]^{3/2}} \qquad 7.43$$

Description

For an equation of a curve $f(x, y)$ given in rectangular coordinates, the curvature is defined by Eq. 7.41.

If the function $f(x, y)$ is easier to differentiate with respect to y instead of x, then Eq. 7.43 may be used.

Equation 7.44 and Eq. 7.45: Radius of Curvature

$$R = \frac{1}{|K|} \quad [K \neq 0] \qquad 7.44$$

$$R = \left| \frac{[1 + (y')^2]^{3/2}}{|y''|} \right| \quad [y'' \neq 0] \qquad 7.45$$

Description

The *radius of curvature*, R, of a curve describes the radius of a circle whose center lies on the concave side of the curve and whose tangent coincides with the tangent to the curve at that point. Radius of curvature is the absolute value of the reciprocal of the curvature.

Example

What is the approximate radius of curvature of the function $f(x)$ at the point $(x, y) = (8, 16)$?

$$f(x) = x^2 + 6x - 96$$

(A) 1.9×10^{-4}

(B) 9.8

(C) 96

(D) 5300

Solution

The first and second derivatives are

$$f'(x) = 2x + 6$$
$$f''(x) = 2$$

At $x = 8$,

$$f'(8) = (2)(8) + 6 = 22$$

From Eq. 7.45, the radius of curvature, R, is

$$R = \left| \frac{[1 + f'(x)^2]^{3/2}}{|f''(x)|} \right|$$
$$= \left| \frac{(1 + (22)^2)^{3/2}}{2} \right|$$
$$= 5340.5 \quad (5300)$$

The answer is (D).

5. GRADIENT, DIVERGENCE, AND CURL

The *vector del operator*, ∇, is defined as

$$\nabla = \frac{\partial}{\partial x}\mathbf{i} + \frac{\partial}{\partial y}\mathbf{j} + \frac{\partial}{\partial z}\mathbf{k}$$

Equation 7.46: Gradient of a Scalar Function

$$\nabla \phi = \left(\frac{\partial}{\partial x}\mathbf{i} + \frac{\partial}{\partial y}\mathbf{j} + \frac{\partial}{\partial z}\mathbf{k} \right) \phi \qquad 7.46$$

Variation

$$\nabla f(x, y, z) = \left(\frac{\partial f(x, y, z)}{\partial x} \right)\mathbf{i} + \left(\frac{\partial f(x, y, z)}{\partial y} \right)\mathbf{j} + \left(\frac{\partial f(x, y, z)}{\partial z} \right)\mathbf{k}$$

Description

A *scalar function* is a mathematical expression that returns a single numerical value (i.e., a *scalar*). The function may be of one or multiple variables (i.e., $f(x)$ or $f(x, y, z)$ or $f(x_1, x_2 \ldots x_n)$), but it must calculate a single number for each location. The *gradient vector field*, $\nabla \phi$, gives the maximum rate of change of the scalar function $\phi = \phi(x, y, z)$.

Equation 7.47: Divergence of a Vector Field

$$\nabla \cdot \mathbf{V} = \left(\frac{\partial}{\partial x}\mathbf{i} + \frac{\partial}{\partial y}\mathbf{j} + \frac{\partial}{\partial z}\mathbf{k} \right) \cdot (V_1\mathbf{i} + V_2\mathbf{j} + V_3\mathbf{k})$$
$$7.47$$

Description

In three dimensions, $\mathbf{V}$ is a vector field with components V_1, V_2, and V_3. V_1, V_2, and V_3 may be specified as functions of variables, such as $P(x, y, z)$, $Q(x, y, z)$, and $R(x, y, z)$. The *divergence* of a vector field $\mathbf{V}$ is the scalar function defined by Eq. 7.47, the dot product of the del operator and the vector (i.e., the divergence is a scalar). The divergence of $\mathbf{V}$ can be interpreted as the *accumulation* of flux (i.e., a flowing substance) in a small region (i.e., at a point).

If $\mathbf{V}$ represents a flow (e.g., air moving from hot to cool regions), then $\mathbf{V}$ is incompressible if $\nabla \cdot \mathbf{V} = 0$, since the substance is not accumulating.

Example

What is the divergence of the following vector field?

$$\mathbf{V} = 2x\mathbf{i} + 2y\mathbf{j}$$

(A) 0

(B) 2

(C) 3

(D) 4

Mathematics

Solution

Use Eq. 7.47.

$$\nabla\cdot\mathbf{V} = \left(\frac{\partial}{\partial x}\mathbf{i} + \frac{\partial}{\partial y}\mathbf{j} + \frac{\partial}{\partial z}\mathbf{k}\right)\cdot(V_1\mathbf{i} + V_2\mathbf{j} + V_3\mathbf{k})$$

$$= \left(\frac{\partial}{\partial x}\mathbf{i} + \frac{\partial}{\partial y}\mathbf{j} + \frac{\partial}{\partial z}\mathbf{k}\right)\cdot(2x\mathbf{i} + 2y\mathbf{j} + 0\mathbf{k})$$

$$= \frac{\partial(2x)}{\partial x} + \frac{\partial(2y)}{\partial y} + \frac{\partial(0)}{\partial z}$$

$$= 2 + 2 + 0$$

$$= 4$$

The answer is (D).

Equation 7.48: Curl of a Vector Field

$$\nabla\times\mathbf{V} = \left(\frac{\partial}{\partial x}\mathbf{i} + \frac{\partial}{\partial y}\mathbf{j} + \frac{\partial}{\partial z}\mathbf{k}\right)\times(V_1\mathbf{i} + V_2\mathbf{j} + V_3\mathbf{k})$$

$$7.48$$

Variation

$$\operatorname{curl}\mathbf{V} = \nabla\times\mathbf{V}$$

$$= \begin{vmatrix} \mathbf{i} & \mathbf{j} & \mathbf{k} \\ \dfrac{\partial}{\partial x} & \dfrac{\partial}{\partial y} & \dfrac{\partial}{\partial z} \\ P(x,y,z) & Q(x,y,z) & R(x,y,z) \end{vmatrix}$$

Description

The *curl*, $\nabla\times\mathbf{V}$, of a vector field $\mathbf{V}(x,y,z)$ is the vector field defined by Eq. 7.48, the cross (vector) product of the del operator and vector. For any location, the curl vector has both magnitude and direction. That is, the curl vector determines how fast the flux is rotating and in what direction the flux is going. The curl of a vector field can be interpreted as the *vorticity* per unit area of flux (i.e., a flowing substance) in a small region (i.e., at a point). One of the uses of the curl is to determine whether flow (represented in direction and magnitude by $\mathbf{V}$) is rotational. Flow is irrotational if curl $\nabla\times\mathbf{V} = 0$.

Example

Determine the curl of the vector function $\mathbf{V}(x,y,z)$.

$$\mathbf{V}(x,y,z) = 3x^2\mathbf{i} + 7e^x y\mathbf{j}$$

(A) $7e^x y$

(B) $7e^x y\mathbf{i}$

(C) $7e^x y\mathbf{j}$

(D) $7e^x y\mathbf{k}$

Solution

Using the variation of Eq. 7.48,

$$\operatorname{curl}\mathbf{V} = \begin{vmatrix} \mathbf{i} & \mathbf{j} & \mathbf{k} \\ \dfrac{\partial}{\partial x} & \dfrac{\partial}{\partial y} & \dfrac{\partial}{\partial z} \\ 3x^2 & 7e^x y & 0 \end{vmatrix}$$

Expand the determinant across the top row.

$$\left(\frac{\partial}{\partial y}0 - \frac{\partial}{\partial z}7e^x y\right)\mathbf{i} - \left(\frac{\partial}{\partial x}0 - \frac{\partial}{\partial z}3x^2\right)\mathbf{j}$$

$$+ \left(\frac{\partial}{\partial x}7e^x y - \frac{\partial}{\partial y}3x^2\right)\mathbf{k}$$

$$= (0-0)\mathbf{i} - (0-0)\mathbf{j} + (7e^x y - 0)\mathbf{k}$$

$$= 7e^x y\mathbf{k}$$

The answer is (D).

Equation 7.49 Through Eq. 7.52: Vector Identities

$$\nabla^2\phi = \nabla\cdot(\nabla\phi) = (\nabla\cdot\nabla)\phi \qquad 7.49$$

$$\nabla\times\nabla\phi = 0 \qquad 7.50$$

$$\nabla\cdot(\nabla\times\mathbf{A}) = 0 \qquad 7.51$$

$$\nabla\times(\nabla\times\mathbf{A}) = \nabla(\nabla\cdot\mathbf{A}) - \nabla^2\mathbf{A} \qquad 7.52$$

Description

Equation 7.49 through Eq. 7.52 are identities associated with gradient, divergence, and curl.

Equation 7.53: Laplacian of a Scalar Function

$$\nabla^2\phi = \frac{\partial^2\phi}{\partial x^2} + \frac{\partial^2\phi}{\partial y^2} + \frac{\partial^2\phi}{\partial z^2} \qquad 7.53$$

Description

The *Laplacian* of a scalar function, $\phi = \phi(x,y,z)$, is the divergence of the gradient function. (This is essentially the second derivative of a scalar function.) A function that satisfies Laplace's equation $\nabla^2 = 0$ is known as a *potential function*. Accordingly, the operator ∇^2 is commonly written as $\nabla\cdot\nabla$ or Δ. The potential function quantifies the attraction of the flux to move in a particular direction. It is used in electricity (voltage potential), mechanics (gravitational potential), mixing and diffusion (concentration gradient), hydraulics (pressure gradient), and heat transfer (thermal gradient). The term Laplacian almost always refers to three-dimensional functions, and usually functions in rectangular coordinates. The term *d'Alembertian* is used when working with four-dimensional functions. The symbol $\square^2$ (with four sides) is used in place of ∇^2. The d'Alembertian is encountered frequently when working with wave functions (including those involving relativity and quantum mechanics) of x, y, and z for location, and t for time.

Example

Determine the Laplacian of the scalar function $\frac{1}{3}x^3 - 9y + 5$ at the point $(3, 2, 7)$.

   (A) 0

   (B) 1

   (C) 6

   (D) 9

Solution

The Laplacian of the function is

$$\nabla^2 \phi = \frac{\partial^2 \phi}{\partial x^2} + \frac{\partial^2 \phi}{\partial y^2} + \frac{\partial^2 \phi}{\partial z^2}$$

$$\nabla^2 \left(\tfrac{1}{3}x^3 - 9y + 5\right) = \frac{\partial^2 \left(\tfrac{1}{3}x^3 - 9y + 5\right)}{\partial x^2}$$

$$+ \frac{\partial^2 \left(\tfrac{1}{3}x^3 - 9y + 5\right)}{\partial y^2}$$

$$+ \frac{\partial^2 \left(\tfrac{1}{3}x^3 - 9y + 5\right)}{\partial z^2}$$

$$= 2x + 0 + 0$$

$$= 2x$$

At $(3, 2, 7)$, $2x = (2)(3) = 6$.

The answer is (C).

6. LIMITS

A *limit* is the value a function approaches when an independent variable approaches a target value. For example, suppose the value of $y = x^2$ is desired as x approaches 5. This could be written as

$$y(5) = \lim_{x \to 5} x^2$$

The power of limit theory is wasted on simple calculations such as this one, but limit theory is appreciated when the function is undefined at the target value. The object of limit theory is to determine the limit without having to evaluate the function at the target. The

general case of a limit evaluated as x approaches the target value a is written as

$$\lim_{x \to a} f(x)$$

It is not necessary for the actual value, $f(a)$, to exist for the limit to be calculated. The function $f(x)$ may be undefined at point a. However, it is necessary that $f(x)$ be defined on both sides of point a for the limit to exist. If $f(x)$ is undefined on one side, or if $f(x)$ is discontinuous at $x = a$, as in Fig. 7.3(c) and Fig. 7.3(d), the limit does not exist at $x = a$.

Figure 7.3 *Existence of Limits*

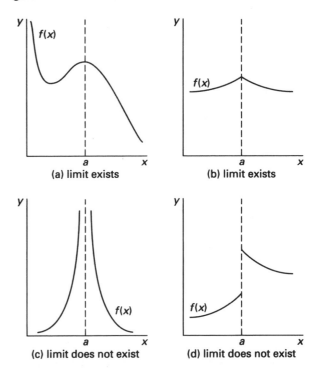

     (a) limit exists   (b) limit exists

   (c) limit does not exist   (d) limit does not exist

Equation 7.54: L'Hopital's Rule

$$\lim_{x \to \alpha} \frac{f'(x)}{g'(x)}, \quad \lim_{x \to \alpha} \frac{f''(x)}{g''(x)}, \quad \lim_{x \to \alpha} \frac{f'''(x)}{g'''(x)} \qquad 7.54$$

Variation

$$\lim_{x \to a} \frac{f(x)}{g(x)} = \lim_{x \to a} \frac{f^k(x)}{g^k(x)}$$

Description

L'Hôpital's rule may be used only when the numerator and denominator of the expression are both indeterminate (i.e., are both zero or are both infinite) at the limit point. $f^k(x)$ and $g^k(x)$ are the kth derivatives of the functions $f(x)$ and $g(x)$, respectively. L'Hôpital's rule can be applied repeatedly as required as long as the numerator and denominator are both indeterminate.

Example

Evaluate the following limit.

$$\lim_{x \to 0} \frac{1 - e^{3x}}{4x}$$

(A) $-\infty$

(B) $-3/4$

(C) 0

(D) $1/4$

Solution

This limit has the indeterminate form $0/0$, so use L'Hôpital's rule.

$$\lim_{x \to \alpha} \frac{f(x)}{g(x)} = \lim_{x \to \alpha} \frac{f'(x)}{g'(x)}$$

$$\lim_{x \to 0} \frac{1 - e^{3x}}{4x} = \lim_{x \to 0} \frac{-3e^{3x}}{4}$$

$$= -3/4$$

The answer is (B).

7. INTEGRALS

Equation 7.55 Through Eq. 7.77: Indefinite Integrals

$$\int df(x) = f(x) \qquad 7.55$$

$$\int dx = x \qquad 7.56$$

$$\int a f(x) \, dx = a \int f(x) \, dx \qquad 7.57$$

$$\int [u(x) \pm v(x)] \, dx = \int u(x) \, dx \pm \int v(x) \, dx \qquad 7.58$$

$$\int x^m \, dx = \frac{x^{m+1}}{m+1} \quad [m \neq -1] \qquad 7.59$$

$$\int u(x) \, dv(x) = u(x)v(x) - \int v(x) \, du(x) \qquad 7.60$$

$$\int \frac{dx}{ax + b} = \frac{1}{a} \ln|ax + b| \qquad 7.61$$

$$\int \frac{dx}{\sqrt{x}} = 2\sqrt{x} \qquad 7.62$$

$$\int a^x \, dx = \frac{a^x}{\ln a} \qquad 7.63$$

$$\int \sin x \, dx = -\cos x \qquad 7.64$$

$$\int \cos x \, dx = \sin x \qquad 7.65$$

$$\int \sin^2 x \, dx = \frac{x}{2} - \frac{\sin 2x}{4} \qquad 7.66$$

$$\int \cos^2 x \, dx = \frac{x}{2} + \frac{\sin 2x}{4} \qquad 7.67$$

$$\int x \sin x \, dx = \sin x - x \cos x \qquad 7.68$$

$$\int x \cos x \, dx = \cos x + x \sin x \qquad 7.69$$

$$\int \sin x \cos x \, dx = (\sin^2 x)/2 \qquad 7.70$$

$$\int \sin ax \cos bx \, dx = -\frac{\cos(a - b)x}{2(a - b)}$$
$$- \frac{\cos(a + b)x}{2(a + b)} \quad [a^2 \neq b^2] \qquad 7.71$$

$$\int \tan x \, dx = -\ln|\cos x| = \ln|\sec x| \qquad 7.72$$

$$\int \cot x \, dx = -\ln|\csc x| = \ln|\sin x| \qquad 7.73$$

$$\int \tan^2 x \, dx = \tan x - x \qquad 7.74$$

$$\int \cot^2 x \, dx = -\cot x - x \qquad 7.75$$

$$\int e^{ax} \, dx = (1/a)e^{ax} \qquad 7.76$$

$$\int x e^{ax} \, dx = (e^{ax}/a^2)(ax - 1) \qquad 7.77$$

Description

Integration is the inverse operation of differentiation. There are two types of integrals: *definite integrals*, which are restricted to a specific range of the independent variable, and *indefinite integrals*, which are unrestricted. Indefinite integrals are sometimes referred to as *antiderivatives*.

Equation 7.78: Fundamental Theorem of Integral Calculus

$$\lim_{n \to \infty} \sum_{i=1}^{n} f(x_i) \Delta x_i = \int_{a}^{b} f(x) \, dx \qquad 7.78$$

Description

The definition of a definite integral is given by the *fundamental theorem of integral calculus*. The right-hand side of Eq. 7.78 represents the area bounded by $f(x)$ above, $y = 0$ below, $x = a$ to the left, and $x = b$ to the right. This is commonly referred to as the "*area under the curve.*"

Example

What is the approximate total area bounded by $y = \sin x$ over the interval $0 \le x \le 2\pi$? (x is in radians.)

(A) 0

(B) $\pi/2$

(C) 2

(D) 4

Solution

The integral of $f(x)$ represents the area under the curve $f(x)$ between the limits of integration. However, since the value of $\sin x$ is negative in the range $\pi \le x \le 2\pi$, the total area would be calculated as zero if the integration was carried out in one step. The integral could be calculated over two ranges, but it is easier to exploit the symmetry of the sine curve.

$$A = \int_{x_1}^{x_2} f(x)\,dx = \int_0^{2\pi} |\sin x|\,dx$$

$$= 2\int_0^\pi \sin x\,dx$$

$$= -2\cos x\Big|_0^\pi$$

$$= (-2)(-1-1)$$

$$= 4$$

The answer is (D).

8. CENTROIDS AND MOMENTS OF INERTIA

Applications of integration include the determination of the *centroid of an area* and various moments of the area, including the *area moment of inertia*.

The integration method for determining centroids and moments of inertia is not necessary for basic shapes. Formulas for basic shapes can be found in tables.

Equation 7.79 Through Eq. 7.82: Centroid of an Area

$$x_c = \frac{\int x\,dA}{A} \quad\quad 7.79$$

$$y_c = \frac{\int y\,dA}{A} \quad\quad 7.80$$

$$A = \int f(x)\,dx \quad\quad 7.81$$

$$dA = f(x)\,dx = g(y)\,dy \quad\quad 7.82$$

Description

The centroid of an area is analogous to the *center of gravity* of a homogeneous body. The location, (x_c, y_c), of the centroid of the area bounded by the x- and y-axis and the mathematical function $y = f(x)$ can be found from Eq. 7.79 through Eq. 7.82.

Example

What is most nearly the x-coordinate of the centroid of the area bounded by $y = 0$, $f(x)$, $x = 0$, and $x = 20$?

$$f(x) = x^3 + 7x^2 - 5x + 6$$

(A) 7.6

(B) 9.4

(C) 14

(D) 16

Solution

Use Eq. 7.79 and Eq. 7.82.

$$\int xf(x)\,dx = \int_0^{20} (x^4 + 7x^3 - 5x^2 + 6x)\,dx$$

$$= \frac{x^5}{5} + \frac{7x^4}{4} - \frac{5x^3}{3} + \frac{6x^2}{2}\Big|_0^{20}$$

$$= 907{,}867$$

From Eq. 7.81, the area under the curve is

$$A = \int_a^b f(x)\,dx = \int_0^{20} (x^3 + 7x^2 - 5x + 6)\,dx$$

$$= \tfrac{1}{4}x^4 + \tfrac{7}{3}x^3 - \tfrac{5}{2}x^2 + 6x\Big|_0^{20}$$

$$= \left(\tfrac{1}{4}\right)(20)^4 + \left(\tfrac{7}{3}\right)(20)^3 - \left(\tfrac{5}{2}\right)(20)^2 + (6)(20)$$

$$= 57{,}786.67 \quad (57{,}787)$$

Use Eq. 7.79 to find the x-coordinate of the centroid.

$$x_c = \frac{\int x\,dA}{A}$$

$$= \frac{\int xf(x)\,dx}{A}$$

$$= \frac{907{,}867}{57{,}787}$$

$$= 15.71 \quad (16)$$

The answer is (D).

Equation 7.83 and Eq. 7.84: First Moment of the Area

$$M_y = \int x\,dA = x_c A \qquad 7.83$$

$$M_x = \int y\,dA = y_c A \qquad 7.84$$

Description

The quantity $\int x\,dA$ is known as the *first moment of the area* or *first area moment* with respect to the y-axis. Similarly, $\int y\,dA$ is known as the *first moment of the area* with respect to the x-axis. Equation 7.83 and Eq. 7.84 show that the first moment of the area can be calculated from the area and centroidal distance.

Equation 7.85 and Eq. 7.86: Moment of Inertia

$$I_y = \int x^2\,dA \qquad 7.85$$

$$I_x = \int y^2\,dA \qquad 7.86$$

Description

The *second moment of the area* or *moment of inertia, I,* of the area is needed in mechanics of materials problems. The symbol I_x is used to represent a moment of inertia with respect to the x-axis. Similarly, I_y is the moment of inertia with respect to the y-axis.

Example

What is most nearly the moment of inertia about the y-axis of the area bounded by $y = 0$, $f(x) = x^3 + 7x^2 - 5x + 6$, $x = 0$, and $x = 20$?

(A) 6.3×10^5

(B) 8.2×10^6

(C) 9.9×10^6

(D) 1.5×10^7

Solution

From Eq. 7.85, the moment of inertia about the y-axis is

$$I_y = \int x^2\,dA = \int x^2 f(x)\,dx$$

$$= \int_0^{20} (x^5 + 7x^4 - 5x^3 + 6x^2)\,dx$$

$$= \frac{x^6}{6} + \frac{7x^5}{5} - \frac{5x^4}{4} + \frac{6x^3}{3}\Big|_0^{20}$$

$$= 1.5 \times 10^7$$

The answer is (D).

Equation 7.87 and Eq. 7.88: Centroidal Moment of Inertia

$$I_{\text{parallel axis}} = I_c + Ad^2 \qquad 7.87$$

$$J = \int r^2\,dA = I_x + I_y \qquad 7.88$$

Description

Moments of inertia can be calculated with respect to any axis, not just the coordinate axes. The moment of inertia taken with respect to an axis passing through the area's centroid is known as the *centroidal moment of inertia, I_c.* The centroidal moment of inertia is the smallest possible moment of inertia for the area.

If the moment of inertia is known with respect to one axis, the moment of inertia with respect to another parallel axis can be calculated from the *parallel axis theorem,* also known as the *transfer axis theorem.* (See Eq. 7.87.) This theorem is also used to evaluate the moment of inertia of areas that are composed of two or more basic shapes. In Eq. 7.87, d is the distance between the centroidal axis and the second, parallel axis.

Example

The moment of inertia about the x'-axis of the cross section shown is $334\,000$ cm^4. The cross-sectional area is 86 cm^2, and the thicknesses of the web and the flanges are the same.

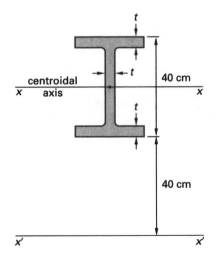

What is most nearly the moment of inertia about the centroidal axis?

(A) 2.4×10^4 cm^4

(B) 7.4×10^4 cm^4

(C) 2.0×10^5 cm^4

(D) 6.4×10^5 cm^4

Solution

Use Eq. 7.87. The moment of inertia around the centroidal axis is

$$I'_x = I_{x_c} + d_x^2 A$$

$$
\begin{aligned}
I_{x_c} &= I'_x - d_x^2 A \\
&= 334\,000 \text{ cm}^4 - (86 \text{ cm}^2)\left(40 \text{ cm} + \frac{40 \text{ cm}}{2}\right)^2 \\
&= 24\,400 \text{ cm}^4 \quad (2.4 \times 10^4 \text{ cm}^4)
\end{aligned}
$$

The answer is (A).

 **Differential Equations**

1. INTRODUCTION TO DIFFERENTIAL EQUATIONS

A *differential equation* is a mathematical expression combining a function (e.g., $y = f(x)$) and one or more of its derivatives. The *order* of a differential equation is the highest derivative in it. *First-order differential equations* contain only first derivatives of the function, *second-order differential equations* contain second derivatives (and may contain first derivatives as well), and so on.

The purpose of solving a differential equation is to derive an expression for the function in terms of the independent variable. The expression does not need to be explicit in the function, but there can be no derivatives in the expression. Since, in the simplest cases, solving a differential equation is equivalent to finding an indefinite integral, it is not surprising that *constants of integration* must be evaluated from knowledge of how the system behaves. Additional data are known as *initial values*, and any problem that includes them is known as an *initial value problem*.

Equation 8.1: Linear Differential Equation with Constant Coefficients

$$b_n \frac{d^n y(x)}{dx^n} + \cdots + b_1 \frac{dy(x)}{dx} + b_0 y(x) = f(x)$$
$$[b_n, \ldots, b_i, \ldots, b_1, \text{ and } b_0 \text{ are constants}] \qquad 8.1$$

Description

A *linear differential equation* can be written as a sum of multiples of the function $y(x)$ and its derivatives. If the multipliers are scalars, the differential equation is said to have *constant coefficients*. Equation 8.1 shows the general form of a linear differential equation with constant coefficients. $f(x)$ is known as the forcing function. If the forcing function is zero, the differential equation is said to be *homogeneous*.

If the function $y(x)$ or one of its derivatives is raised to some power (other than one) or is embedded in another function (e.g., y embedded in $\sin y$ or e^y), the equation is said to be *nonlinear*.

Example

Which of the following is NOT a linear differential equation?

(A) $5\dfrac{d^2 y}{dt^2} - 8\dfrac{dy}{dt} + 16y = 4te^{-7t}$

(B) $5\dfrac{d^2 y}{dt^2} - 8t^2 \dfrac{dy}{dt} + 16y = 0$

(C) $5\dfrac{d^2 y}{dt^2} - 8\dfrac{dy}{dt} + 16y = \dfrac{dy}{dy}$

(D) $5\left(\dfrac{dy}{dt}\right)^2 - 8\dfrac{dy}{dt} + 16y = 0$

Solution

A linear differential equation consists of multiples of a function, $y(t)$, and its derivatives, $d^n y/dt^n$. The multipliers may be scalar constants or functions, $g(t)$, of the independent variable, t. The forcing function, $f(t)$, (i.e., the right-hand side of the equation) may be 0, a constant, or any function of the independent variable, t. The multipliers cannot be higher powers of the function, $y(t)$.

The answer is (D).

2. LINEAR HOMOGENEOUS DIFFERENTIAL EQUATIONS WITH CONSTANT COEFFICIENTS

Each term of a *homogeneous differential equation* contains either the function or one of its derivatives. The forcing function is zero. That is, the sum of the function and its derivative terms is equal to zero.

$$b_n \frac{d^n y(x)}{dx^n} + \cdots + b_1 \frac{dy(x)}{dx} + b_0 y(x) = 0$$

Equation 8.2: Characteristic Equation

$$P(r) = b_n r^n + b_{n-1} r^{n-1} + \cdots + b_1 r + b_0 \qquad 8.2$$

Description

A *characteristic equation* can be written for a homogeneous linear differential equation with constant coefficients, regardless of order. This characteristic equation is simply the polynomial formed by replacing all derivatives with variables raised to the power of their respective derivatives. That is, all instances of $d^n y(x)/dx^n$ are replaced with r^n, resulting in an equation of the form of Eq. 8.2.

Equation 8.3: Solving Linear Differential Equations with Constant Coefficients

$$y_h(x) = C_1 e^{r_1 x} + C_2 e^{r_2 x} + \cdots + C_i e^{r_i x} + \cdots + C_n e^{r_n x}$$

$$\text{8.3}$$

Description

Homogeneous linear differential equations are most easily solved by finding the n roots of Eq. 8.2, the characteristic polynomial $P(r)$. If the roots of Eq. 8.2 are real and different, the solution is Eq. 8.3.

Equation 8.4 and Eq. 8.5: Homogeneous First-Order Linear Differential Equations

$$y' + ay = 0 \qquad \text{8.4}$$

$$y = Ce^{-at} \qquad \text{8.5}$$

Variations

$$\frac{dy}{dt} + ay = 0$$

$$f(t) = Ce^{-at}$$

Description

A homogeneous, first-order, linear differential equation with constant coefficients has the general form of Eq. 8.4.

The characteristic equation is $r + a = 0$ and has a root of $r = -a$. Equation 8.5 is the solution.

Example

Which of the following is the general solution to the differential equation and boundary conditions?

$$\frac{dy}{dt} - 5y = 0$$

$$y(0) = 3$$

(A) $-\frac{1}{3}e^{-5t}$

(B) $3e^{5t}$

(C) $5e^{-3t}$

(D) $\frac{1}{5}e^{-3t}$

Solution

This is a first-order, linear differential equation. The characteristic equation is $r - 5 = 0$. The root, r, is 5.

The solution is in the form of Eq. 8.5.

$$y = Ce^{5t}$$

The initial condition is used to find C.

$$y(0) = Ce^{(5)(0)} = 3$$

$$C = 3$$

$$y = 3e^{5t}$$

The answer is (B).

Equation 8.6 Through Eq. 8.8: Homogeneous Second-Order Linear Differential Equations with Constant Coefficients

$$y'' + ay' + by = 0 \qquad \text{8.6}$$

$$(r^2 + ar + b)Ce^{rx} = 0 \qquad \text{8.7}$$

$$r^2 + ar + b = 0 \qquad \text{8.8}$$

Description

A second-order, homogeneous, linear differential equation has the general form given by Eq. 8.6.

The characteristic equation is Eq. 8.8.

Depending on the form of the forcing function, the solutions to most second-order differential equations will contain sinusoidal terms (corresponding to oscillatory behavior) and exponential terms (corresponding to decaying or increasing unstable behavior). Behavior of real-world systems (electrical circuits, spring-mass-dashpot, fluid flow, heat transfer, etc.) depends on the amount of system *damping* (electrical resistance, mechanical friction, pressure drop, thermal insulation, etc.).

With *underdamping* (i.e., with "light" damping) without continued energy input (i.e., a free system without a forcing function), the transient behavior will gradually decay to the steady-state equilibrium condition. Behavior in underdamped free systems will be oscillatory with diminishing magnitude. The damping is known as underdamping because the amount of damping is less than the critical damping, and the *damping ratio*, ζ, is less than 1. The characteristic equation of underdamped systems has two complex roots.

With *overdamping* ("heavy" damping), damping is greater than critical, and the damping ratio is greater than 1. Transient behavior is a sluggish gradual decrease into the steady-state equilibrium condition without oscillations. The characteristic equation of overdamped systems has two distinct real roots (zeros).

With *critical damping*, the damping ratio is equal to 1. There is no overshoot, and the behavior reaches the steady-state equilibrium condition the fastest of the three cases, without oscillations. The characteristic equation of critically damped systems has two identical real roots (zeros).

Equation 8.9 Through Eq. 8.14: Roots of the Characteristic Equation

$$r_{1,2} = \frac{-a \pm \sqrt{a^2 - 4b}}{2} \qquad 8.9$$

$$y = C_1 e^{r_1 x} + C_2 e^{r_2 x} \qquad 8.10$$

$$y = (C_1 + C_2 x) e^{r_1 x} \qquad 8.11$$

$$y = e^{\alpha x}(C_1 \cos \beta x + C_2 \sin \beta x) \qquad 8.12$$

$$\alpha = -a/2 \qquad 8.13$$

$$\beta = \frac{\sqrt{4b - a^2}}{2} \qquad 8.14$$

Description

The roots of the characteristic equation are given by the quadratic equation, Eq. 8.9.

If $a^2 > 4b$, then the two roots are real and different, and the solution is overdamped, as shown in Eq. 8.10.

If $a^2 = 4b$, then the two roots are real and the same (i.e., are *double roots*), and the solution is critically damped, as shown in Eq. 8.11.

If $a^2 < 4b$, then the two roots are imaginary and of the form $(\alpha + i\beta)$ and $(\alpha - i\beta)$, and the solution is underdamped, as shown in Eq. 8.12.

Example

What is the general solution to the following homogeneous differential equation?

$$y'' - 8y' + 16y = 0$$

(A) $y = C_1 e^{4x}$

(B) $y = (C_1 + C_2 x) e^{4x}$

(C) $y = C_1 e^{-4x} + C_2 e^{4x}$

(D) $y = C_1 e^{2x} + C_2 e^{4x}$

Solution

Find the roots of the characteristic equation.

$$r^2 - 8r + 16 = 0$$

$$a = -8$$

$$b = 16$$

From Eq. 8.9,

$$\begin{aligned} r_{1,2} &= \frac{-a \pm \sqrt{a^2 - 4b}}{2} \\ &= \frac{-(-8) \pm 2\sqrt{(-8)^2 - (4)(16)}}{2} \\ &= 4, 4 \end{aligned}$$

Because $a^2 = 4b$, the characteristic equation has double roots. With $r = 4$, the solution takes the form

$$\begin{aligned} y &= (C_1 + C_2 x) e^{rx} \\ &= (C_1 + C_2 x) e^{4x} \end{aligned}$$

The answer is (B).

3. LINEAR NONHOMOGENEOUS DIFFERENTIAL EQUATIONS WITH CONSTANT COEFFICIENTS

In a nonhomogeneous differential equation, the sum of derivative terms is equal to a nonzero *forcing function* of the independent variable (i.e., $f(x)$ in Eq. 8.1 is nonzero). In order to solve a nonhomogeneous equation, it is often necessary to solve the homogeneous equation first. The homogeneous equation corresponding to a nonhomogeneous equation is known as the *reduced equation* or *complementary equation*.

Equation 8.15: Complete Solution to Nonhomogeneous Differential Equation

$$y(x) = y_h(x) + y_p(x) \qquad 8.15$$

Description

The complete solution to the nonhomogeneous differential equation is shown in Eq. 8.15. The term $y_h(x)$ is the *complementary solution*, which solves the complementary (i.e., homogeneous) case. The *particular solution*, $y_p(x)$, is any specific solution to the nonhomogeneous Eq. 8.1 that is known or can be found. Initial values are used to evaluate any unknown coefficients in the complementary solution after $y_h(x)$ and $y_p(x)$ have been combined. The particular solution will not have any unknown coefficients.

Table 8.1: Method of Undetermined Coefficients

Table 8.1 Method of Undetermined Coefficients

form of $f(x)$	form of $y_p(x)$
A	B
$A e^{\alpha x}$	$B e^{\alpha x}, \ a \neq r_n$
$A_1 \sin \omega x + A_2 \cos \omega x$	$B_1 \sin \omega x + B_2 \cos \omega x$

Description

Two methods are available for finding a particular solution. The *method of undetermined coefficients*, as presented here, can be used only when $f(x)$ in Eq. 8.1 takes on one of the forms given in Table 8.1. $f(x)$ is known as the *forcing function*.

The particular solution can be read from Table 8.1 if the forcing function is one of the forms given. Of course, the coefficients A_i and B_i are not known—these are the *undetermined coefficients*. The exponent s is the smallest non-negative number (and will be zero, one, or two, etc.), which ensures that no term in the particular solution is also a solution to the complementary equation. s must be determined prior to proceeding with the solution procedure.

Once $y_p(x)$ (including s) is known, it is differentiated to obtain $dy_p(x)/dx$, $d^2y_p(x)/dx^2$, and all subsequent derivatives. All of these derivatives are substituted into the original nonhomogeneous equation. The resulting equation is rearranged to match the forcing function, $f(x)$, and the unknown coefficients are determined, usually by solving simultaneous equations.

The presence of an exponential of the form e^{rx} in the solution indicates that *resonance* is present to some extent.

Equation 8.16 Through Eq. 8.20: First-Order Linear Nonhomogeneous Differential Equations with Constant Coefficients, with Step Input

$$\tau\frac{dy}{dt} + y = Kx(t) \qquad 8.16$$

$$x(t) = \begin{cases} A & t<0 \\ B & t>0 \end{cases} \qquad 8.17$$

$$y(0) = KA \qquad 8.18$$

$$y(t) = KA + (KB - KA)\left(1 - \exp\left(\frac{-t}{\tau}\right)\right) \qquad 8.19$$

$$\frac{t}{\tau} = \ln\left[\frac{KB - KA}{KB - y}\right] \qquad 8.20$$

Variation

$$b_1\frac{dy(t)}{dt} + b_0 y(t) = u(t) \qquad [u(t) = \text{unit step function}]$$

Description

As the variation equation for Eq. 8.16 implies, a first-order, linear, nonhomogeneous differential equation with constant coefficients is an extension of Eq. 8.1. Equation 8.16 builds on the differential equation of Eq. 8.1 in the context of a specific control system scenario. It also changes the independent variable from x to t and changes the notation for the forcing function used in Eq. 8.1.

The *time constant*, τ, is the amount of time a homogeneous system (i.e., one with a zero forcing function, $x(t)$) would take to reach $(e-1)/e$, or approximately 63.2% of its final value. This could also be described as the time required to grow to within 36.8% of the final value or as the time to decay to 36.8% of the initial value. The *system gain*, K, or *amplification ratio* is a scalar constant that gives the ratio of the output response to the input response at steady state.

Equation 8.16 describes a *step function*, a special case of a generic forcing function. The forcing function is some value, typically zero ($A = 0$) until $t = 0$, at which time the forcing function immediately jumps to a constant value. Equation 8.19 gives the *step response*, the solution to Eq. 8.16.

Example

A spring-mass-dashpot system starting from a motionless state is acted upon by a step function. The response is described by the differential equation in which time, t, is given in seconds measured from the application of the ramp function.

$$\frac{dy}{dt} + 2y = 2u(0) \quad [y(0) = 0]$$

How long will it take for the system to reach 63% of its final value?

(A) 0.25 s

(B) 0.50 s

(C) 1.0 s

(D) 2.0 s

Solution

To fit this problem into the format used by Eq. 8.16, the coefficient of y must be 1. Dividing by 2,

$$0.5\frac{dy}{dt} + y = tu(0)$$

$$\tau = 0.50 \text{ s}$$

The answer is (B).

 **Numerical Methods**

1. INTRODUCTION TO NUMERICAL METHODS

Although the roots of second-degree polynomials are easily found by a variety of methods (by factoring, completing the square, or using the quadratic equation), easy methods of solving cubic and higher-order equations exist only for specialized cases. However, cubic and higher-order equations occur frequently in engineering, and they are difficult to factor. Trial and error solutions, including graphing, are usually satisfactory for finding only the general region in which the root occurs.

Numerical analysis is a general subject that covers, among other things, iterative methods for evaluating roots to equations. The most efficient numerical methods are too complex to present and, in any case, work by hand. However, some of the simpler methods are presented here. Except in critical problems that must be solved in real time, a few extra calculator or computer iterations will make no difference.[1]

2. CONVERGENCE AND TOLERANCE

Each time an iterative method is repeated, it produces a value. If this series of values gets closer and closer to a single, finite limit value, the method is said to be *convergent*. A method that is convergent will approach a single, finite limit value as the solution.

The mathematical criterion for the convergence of a numerical method is written as

$$\lim_{n \to \infty} |A - A_n| = 0$$

Here, A is the single, finite limit value, and A_n is the numerical result after n iterations. If the method is convergent, then for any positive number, however small, the difference between $|A - A_n|$ and zero can be made still smaller by choosing a large enough value for n.

A method may be convergent even when $|A - A_n|$ does not actually reach zero after any finite number of iterations. A practical criterion for convergence is needed, then, that doesn't depend on an infinite number of iterations. Commonly, a method is said to be convergent when the difference between the values produced by two consecutive iterations is no greater than a given *tolerance*, ε.

$$|A_n - A_{n-1}| \le \varepsilon$$

It is crucial to choose an appropriate value for the tolerance. If ε is too large, this criterion may be met by methods that are not actually convergent. If ε is too small, the computation effort and time needed may be too great to be practical.

3. ROOT EXTRACTION

Equation 9.1: Newton's Method

$$a^{j+1} = a^j - \frac{f(x)}{\left. \dfrac{df(x)}{dx} \right|_{x=a^j}} \qquad 9.1$$

Description

Newton's method (also known as the *Newton-Raphson method*) is a particular form of *fixed-point iteration*. In this sense, "fixed point" is often used as a synonym for "root" or "zero."

All fixed-point techniques require a starting point. Preferably, the starting point will be close to the actual root.[2] And, while Newton's method converges quickly, it requires the function to be continuously differentiable.

At each iteration ($j = 0$, 1, 2, etc.), Eq. 9.1 estimates the root. The maximum error is determined by looking at how much the estimate changes after each iteration. If the change between the previous and current estimates (representing the magnitude of error in the estimate) is

---

[1]Most advanced handheld calculators have "root finder" functions that use numerical methods to iteratively solve equations.

[2]Theoretically, the only penalty for choosing a starting point too far away from the root will be a slower convergence to the root. However, for some nonlinear relations, the penalty will be nonconvergence.

too large, the current estimate is used as the independent variable for the subsequent iteration.[3]

Example

Newton's method is being used to find the roots of the equation $f(x) = (x-2)^2 - 1$. What is the third approximation of the root if $x = 9.33$ is chosen as the first approximation?

(A) 1.0

(B) 2.0

(C) 3.0

(D) 4.0

Solution

Perform two iterations of Newton's method with an initial guess of 9.33.

$$f(x) = (x-2)^2 - 1$$
$$f'(x) = (2)(x-2)$$
$$f(x_1) = (9.33 - 2)^2 - 1 = 52.73$$
$$f'(x_1) = (2)(9.33 - 2) = 14.66$$

From Eq. 9.1,

$$x^{j+1} = x^j - \frac{f(x)}{f'(x)}$$

$$x_2 = x_1 - \frac{f(x_1)}{f'(x_1)} = 9.33 - \frac{f(9.33)}{f'(9.33)} = 9.33 - \frac{52.73}{14.66}$$
$$= 5.73$$

$$f(x_2) = (5.73 - 2)^2 - 1 = 12.91$$
$$f'(x_2) = (2)(5.73 - 2) = 7.46$$

$$x_3 = x_2 - \frac{f(x_2)}{f'(x_2)} = 5.73 - \frac{f(5.73)}{f'(5.73)} = 5.73 - \frac{12.91}{7.46}$$
$$= 4.0$$

The answer is (D).

4. MINIMIZATION

Equation 9.2 Through Eq. 9.4: Newton's Method of Minimization

$$x_{k+1} = x_k - \left(\left. \frac{\partial^2 h}{\partial x^2} \right|_{x=x_k} \right)^{-1} \left. \frac{\partial h}{\partial x} \right|_{x=x_k} \qquad 9.2$$

---

[3]Actually, the theory defining the maximum error is more definite than this. For example, for a large enough value of j, the error decreases approximately linearly. The consecutive values of a^j converge linearly to the root as well.

$$\frac{\partial h}{\partial x} = \begin{bmatrix} \dfrac{\partial h}{\partial x_1} \\ \dfrac{\partial h}{\partial x_2} \\ \cdots \\ \cdots \\ \dfrac{\partial h}{\partial x_n} \end{bmatrix} \qquad 9.3$$

$$\frac{\partial^2 h}{\partial x^2} = \begin{bmatrix} \dfrac{\partial^2 h}{\partial x_1^2} & \dfrac{\partial^2 h}{\partial x_1 \partial x_2} & \cdots & \cdots & \dfrac{\partial^2 h}{\partial x_1 \partial x_n} \\ \dfrac{\partial^2 h}{\partial x_1 \partial x_2} & \dfrac{\partial^2 h}{\partial x_2^2} & \cdots & \cdots & \dfrac{\partial^2 h}{\partial x_2 \partial x_n} \\ \cdots & \cdots & \cdots & \cdots & \cdots \\ \cdots & \cdots & \cdots & \cdots & \cdots \\ \dfrac{\partial^2 h}{\partial x_1 \partial x_n} & \dfrac{\partial^2 h}{\partial x_2 \partial x_n} & \cdots & \cdots & \dfrac{\partial^2 h}{\partial x_n^2} \end{bmatrix} \qquad 9.4$$

Description

Newton's algorithm for minimization is given by Eq. 9.2. Equation 9.2 applies to a scalar value function, $h(\mathbf{x}) = h(x_1, x_2, \ldots, x_n)$, and is used to calculate a vector, $\mathbf{x}^* \in R_n$, where $h(\mathbf{x}^*) \leq h(\mathbf{x})$ for all $\mathbf{x}$ values.

5. NUMERICAL INTEGRATION

Equation 9.5: Euler's Rule

$$\int_a^b f(x)\, dx \approx \Delta x \sum_{k=0}^{n-1} f(a + k\Delta x) \qquad 9.5$$

Description

Equation 9.5 is known as *Euler's rule* or the *forward rectangular rule*.

Equation 9.6 and Eq. 9.7: Trapezoidal Rule

$$\int_a^b f(x)\, dx \approx \Delta x \left[\frac{f(a) + f(b)}{2} \right] \qquad [n=1] \qquad 9.6$$

$$\int_a^b f(x)\, dx \approx \frac{\Delta x}{2} \left[f(a) + 2\sum_{k=0}^{n-1} f(a + k\Delta x) + f(b) \right]$$
$$[n>1] \qquad 9.7$$

Variation

$$A = \frac{d}{2}\left(h_0 + 2\sum_{i=1}^{n-1} h_i + h_n \right)$$

Description

Areas of sections with irregular boundaries cannot be determined precisely, and approximation methods must be used. Figure 9.1 shows an example of an *irregular area*. If the irregular side can be divided into a series of n cells of equal width, and if the irregular side of each cell is fairly straight, the *trapezoidal rule* is appropriate. Equation 9.6 and Eq. 9.7 describe the trapezoidal rule for $n = 1$ and $n > 1$, respectively.

Figure 9.1 *Irregular Areas*

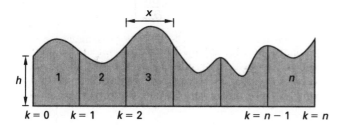

Example

For the irregular area under the curve shown, $a = 3$, and $b = 15$. The formula of the curve is

$$f(x) = (1 - x)(x - 30)$$

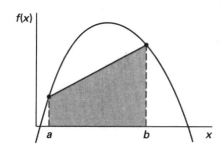

What is the approximate area using the trapezoidal rule?

(A) 1200

(B) 1300

(C) 1600

(D) 1900

Solution

Calculate Δx.

$$\Delta x = b - a = 15 - 3 = 12$$

Calculate $f(a)$ and $f(b)$.

$$f(a) = (1 - a)(a - 30) = (1 - 3)(3 - 30) = 54$$

$$f(b) = (1 - b)(b - 30) = (1 - 15)(15 - 30) = 210$$

This is a one-trapezoid integration (i.e., $n = 1$), so use Eq. 9.6. The area under the curve is

$$\text{area} = \Delta x \left[\frac{f(a) + f(b)}{2} \right] = (12) \left(\frac{54 + 210}{2} \right)$$

$$= 1584 \quad (1600)$$

The answer is (C).

Equation 9.8 Through Eq. 9.10: Simpson's Rule

$$\int_a^b f(x)\, dx \approx \left(\frac{b - a}{6} \right) \left[f(a) + 4f \left(\frac{a + b}{2} \right) + f(b) \right]$$

$$[n = 2] \quad 9.8$$

$$\int_a^b f(x)\, dx \approx \frac{\Delta x}{3} \left[f(a) + 2 \sum_{k=2,4,6,\ldots}^{n-2} f(a + k\Delta x) + 4 \sum_{k=1,3,5,\ldots}^{n-1} f(a + k\Delta x) + f(b) \right]$$

$$[n \geq 4] \quad 9.9$$

$$\Delta x = (b - a)/n \quad 9.10$$

Variation

$$A = \frac{d}{3} \left(h_0 + 2 \sum_{\substack{i \text{ even} \\ i=2}}^{n-2} h_i + 4 \sum_{\substack{i \text{ odd} \\ i=1}}^{n-1} h_i + h_n \right)$$

Description

If the irregular side of each cell is curved (parabolic), *Simpson's rule* (*parabolic rule*) should be used. n must be even to use Simpson's rule.

The Simpson's rule equations for $n = 2$ and $n \geq 4$ are given by Eq. 9.8 and Eq. 9.9, respectively.

Example

For the irregular area under the curve shown, $a = 3$, and $b = 15$. The formula of the curve is

$$f(x) = (1 - x)(x - 30)$$

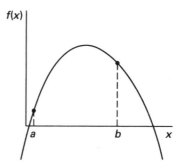

What is the approximate area using Simpson's rule?

(A) 1310

(B) 1870

(C) 1960

(D) 2000

Solution

Calculate $f(a)$, $f(b)$, and $f((a + b)/2)$.

$$f(a) = (1-a)(a-30) = (1-3)(3-30) = 54$$

$$f(b) = (1-b)(b-30) = (1-15)(15-30) = 210$$

$$\frac{a+b}{2} = \frac{3+15}{2} = 9$$

$$f\left(\frac{a+b}{2}\right) = f(9) = (1-9)(9-30) = 168$$

Calculate Δx, using the smallest even number of cells, $n = 2$, and Eq. 9.10.

$$\Delta x = (b-a)/n = \frac{15-3}{2} = 6$$

Using Eq. 9.8, the area is

$$\int_a^b f(x)\,dx \approx \left(\frac{b-a}{6}\right)\left[f(a) + 4f\left(\frac{a+b}{2}\right) + f(b)\right]$$

$$= \left(\frac{15-3}{6}\right)\left(54 + (4)(168) + 210\right)$$

$$= 1872 \quad (1870)$$

The answer is (B).

6. NUMERICAL SOLUTION OF ORDINARY DIFFERENTIAL EQUATIONS

Equation 9.11 Through Eq. 9.14: Euler's Approximation

$$x[(k+1)\Delta t] \cong x(k\Delta t) + \Delta t f[x(k\Delta t), k\Delta t] \qquad 9.11$$

$$x[(k+1)\Delta t] \cong x(k\Delta t) + \Delta t f[x(k\Delta t)] \qquad 9.12$$

$$x_{k+1} = x_k + \Delta t(dx_k/dt) \qquad 9.13$$

$$x_{k+1} = x + \Delta t[f(x(k), t(k))] \qquad 9.14$$

Description

Euler's approximation is a method for estimating the value of a function given the value and slope of the function at an adjacent location. The simplicity of the concept is illustrated by writing Euler's approximation in terms of the traditional two-dimensional x-y coordinate system.

$$y(x_2) = y(x_1) + (x_2 - x_1)y'(x_1)$$

As long as the derivative can be evaluated, Euler's method can be used to predict the value of any function whose values are limited to discrete, sequential points in time or space (e.g., a difference equation).

Euler's approximation applies to a differential equation of the form $f(x, t) = dx/dt$, where $x(0) = x_0$. Equation 9.11 applies to a general time $k\Delta t$. Equation 9.12 applies when $f(x) = dx/dt$ and can be expressed recursively as Eq. 9.13, or as Eq. 9.14.

The error associated with Euler's approximation is zero for linear systems. Euler's approximation can be used as a quick estimate for nonlinear systems as long as the presence of error is recognized. Figure 9.2 shows the geometric interpretation of Euler's approximation for a curvilinear function.

Figure 9.2 *Geometric Interpretation of Euler's Approximation*

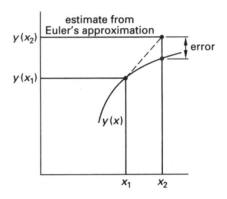

Diagnostic Exam

Topic II: Probability and Statistics

1. A fair coin is tossed three times. What is the approximate probability of heads appearing at least one time?

(A) 0.67

(B) 0.75

(C) 0.80

(D) 0.88

2. Samples of aluminum-alloy channels are tested for stiffness. Stiffness is normally distributed. The following frequency distribution is obtained.

stiffness	frequency
2480	23
2440	35
2400	40
2360	33
2320	21

What is the approximate probability that the stiffness of any given channel section is less than 2350?

(A) 0.08

(B) 0.16

(C) 0.23

(D) 0.36

3. Most nearly, what is the sample variance of the following data?

$$0.50, 0.80, 0.75, 0.52, 0.60$$

(A) 0.015

(B) 0.018

(C) 0.11

(D) 0.12

4. Two students are working independently on a problem. Their respective probabilities of solving the problem are 1/3 and 3/4. What is the probability that at least one of them will solve the problem?

(A) 1/2

(B) 5/8

(C) 2/3

(D) 5/6

5. Most nearly, what is the arithmetic mean of the following values?

$$9.5, 2.4, 3.6, 7.5, 8.2, 9.1, 6.6, 9.8$$

(A) 6.3

(B) 7.1

(C) 7.8

(D) 8.1

6. A normal distribution has a mean of 12 and a standard deviation of 3. If a sample is taken from the normal distribution, most nearly, what is the probability that the sample will be between 15 and 18?

(A) 0.091

(B) 0.12

(C) 0.14

(D) 0.16

7. A marksman can hit a bull's-eye from 100 m three times out of every four shots. What is the probability that he will hit a bull's-eye with at least one of his next three shots?

(A) 3/4

(B) 15/16

(C) 31/32

(D) 63/64

8. The final scores of students in a graduate course are distributed normally with a mean of 72 and a standard deviation of 10. Most nearly, what is the probability that a student's score will be between 65 and 78?

(A) 0.42

(B) 0.48

(C) 0.52

(D) 0.65

9. What is the sample standard deviation of the following 50 data points?

data value	frequency
1.5	3
2.5	8
3.5	18
4.5	12
5.5	9

(A) 1.12

(B) 1.13

(C) 1.26

(D) 1.28

10. 15% of a batch of mixed-color gumballs are green. Out of a random sample of 20 gumballs, what is the probability of getting two green balls?

(A) 0.12

(B) 0.17

(C) 0.23

(D) 0.46

SOLUTIONS

1. Calculate the probability of no heads, and then subtract that from 1 to get the probability of at least one head. If there are no heads, then all tosses must be tails.

$$P\left(\begin{array}{c}\text{three tails in}\\\text{three tosses}\end{array}\right) = P\left(\begin{array}{c}\text{one tail in}\\\text{one toss}\end{array}\right)^3 = \left(\frac{1}{2}\right)^3$$

$$= 0.125$$

$$P(E) = 1 - P(\text{not } E) = 1 - 0.125$$

$$= 0.875 \quad (0.88)$$

The answer is (D).

2. The probability can be found using the standard normal table. In order to use the standard normal table, the population mean and standard deviation must be found.

The arithmetic mean is an unbiased estimator of the population mean.

$$n = 23 + 35 + 40 + 33 + 21$$

$$= 152$$

$$\overline{X} = (1/n)\sum_{i=1}^{n} X_i$$

$$= \left(\frac{1}{152}\right)\left(\begin{array}{c}(2480)(23) + (2440)(35)\\+ (2400)(40) + (2360)(33)\\+ (2320)(21)\end{array}\right)$$

$$= 2402$$

The sample standard deviation is an unbiased estimator of the standard deviation.

$$s = \sqrt{[1/(n-1)]\sum_{i=1}^{n}(X_i - \overline{X})^2}$$

$$= \sqrt{\left(\frac{1}{152-1}\right)\left(\begin{array}{c}(23)(2480 - 2402)^2\\+ (35)(2440 - 2402)^2\\+ (40)(2400 - 2402)^2\\+ (33)(2360 - 2402)^2\\+ (21)(2320 - 2402)^2\end{array}\right)}$$

$$= 50.82$$

Find the standard normal variable corresponding to 2350.

$$Z = \frac{x - \mu}{\sigma} = \frac{2350 - 2402}{50.82}$$

$$= -1.0$$

Since the unit normal distribution is symmetrical about $x=0$, the probability of x being in the interval $[-\infty, -1]$ is the same as x being in the interval $[+1, +\infty]$. This corresponds to the value of $R(x)$ in Table 10.2.

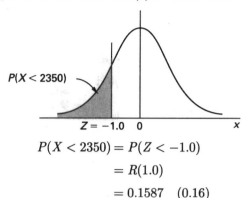

$$P(X < 2350) = P(Z < -1.0)$$
$$= R(1.0)$$
$$= 0.1587 \quad (0.16)$$

The answer is (B).

3. The arithmetic mean of the data is

$$\overline{X} = (1/n)\sum_{i=1}^{n} X_i$$
$$= \left(\frac{1}{5}\right)(0.50 + 0.80 + 0.75 + 0.52 + 0.60)$$
$$= 0.634$$

Use the mean to find the sample variance.

$$s^2 = [1/(n-1)]\sum_{i=1}^{n}(X_i - \overline{X})^2$$
$$= \left(\frac{1}{5-1}\right)\begin{pmatrix}(0.50 - 0.634)^2 + (0.80 - 0.634)^2 \\ + (0.75 - 0.634)^2 + (0.52 - 0.634)^2 \\ + (0.60 - 0.634)^2\end{pmatrix}$$
$$= 0.0183 \quad (0.018)$$

The answer is (B).

4. The probability that either or both of the students solve the problem is given by the laws of total and joint probability.

Since the two students are working independently, the joint probability of both students solving the problem is

$$P(A, B) = P(A)P(B)$$
$$= \left(\frac{1}{3}\right)\left(\frac{3}{4}\right)$$
$$= 1/4$$

The total probability is

$$P(A + B) = P(A) + P(B) - P(A, B)$$
$$= \frac{1}{3} + \frac{3}{4} - \frac{1}{4}$$
$$= 5/6$$

The answer is (D).

5. The arithmetic mean is the sum of the values divided by the total number of items.

$$\overline{X} = (1/n)\sum_{i=1}^{n} X_i$$
$$= \left(\frac{1}{8}\right)(9.5 + 2.4 + 3.6 + 7.5 + 8.2 + 9.1 + 6.6 + 9.8)$$
$$= 7.09 \quad (7.1)$$

The answer is (B).

6. Find the standard normal values for the minimum and maximum values.

$$Z_1 = \frac{x_1 - \mu}{\sigma} = \frac{15 - 12}{3} = 1$$
$$Z_2 = \frac{x_2 - \mu}{\sigma} = \frac{18 - 12}{3} = 2$$

Plot these values on a normal distribution curve.

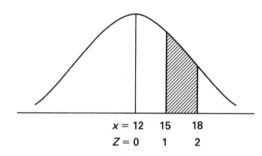

From the standard normal table, the probabilities are

$$P(Z < 1) = 0.8413$$
$$P(Z < 2) = 0.9772$$

The probability that the outcome will be between 15 and 18 is

$$P(15 < x < 18) = P(x < 18) - P(x < 15)$$
$$= P(Z < 2) - P(Z < 1)$$
$$= 0.9772 - 0.8413$$
$$= 0.1359 \quad (0.14)$$

The answer is (C).

7. Solving this problem requires calculating three probabilities.

$$P(\text{at least 1 hit in 3 shots}) = P(\text{1 hit in 3 shots})$$
$$+ P(\text{2 hits in 3 shots})$$
$$+ P(\text{3 hits in 3 shots})$$

An easier way to find the probability of making at least one hit is to solve for its complementary probability, that of making zero hits.

$$P(\text{miss}) = 1 - P(\text{hit}) = 1 - \frac{3}{4}$$

$$= 1/4$$

$$P(\text{at least one hit}) = 1 - P(\text{no hits})$$

$$= 1 - \left(\begin{array}{c} P(\text{miss}) \times P(\text{miss}) \\ \times P(\text{miss}) \end{array} \right)$$

$$= 1 - \left(\tfrac{1}{4}\right)\left(\tfrac{1}{4}\right)\left(\tfrac{1}{4}\right)$$

$$= 63/64$$

The answer is (D).

8. Calculate standard normal values for the points of interest, 65 and 78.

$$Z = \frac{x_0 - \mu}{\sigma}$$

$$Z_{65} = \frac{65 - 72}{10}$$

$$= -0.70$$

$$Z_{78} = \frac{78 - 72}{10}$$

$$= 0.60$$

The probability of a score falling between 65 and 78 is equal to the area under the unit normal curve between -0.70 and 0.60. Determine this area by subtracting $F(Z_{65})$ from $F(Z_{78})$. Although the $F(x)$ statistic is not tabulated for negative x values, the curve's symmetry allows the $R(x)$ statistic to be used instead.

$$F(-x) = R(x)$$

$$P(65 < X < 78) = F(0.60) - R(0.70)$$

$$= 0.7257 - 0.2420$$

$$= 0.4837 \quad (0.48)$$

The answer is (B).

9. The number of data points is given as 50. The arithmetic mean is

$$\overline{X} = (1/n)\sum_{i=1}^{n} X_i$$

$$= \left(\frac{1}{50}\right)\left(\begin{array}{c} (3)(1.5) + (8)(2.5) + (18)(3.5) \\ + (12)(4.5) + (9)(5.5) \end{array} \right)$$

$$= 3.82$$

The sample standard deviation is

$$s = \sqrt{[1/(n-1)]\sum_{i=1}^{n}(X_i - \overline{X})^2}$$

$$= \sqrt{ \left(\frac{1}{50-1}\right)\left(\begin{array}{c} (3)(1.5 - 3.82)^2 + (8)(2.5 - 3.82)^2 \\ + (18)(3.5 - 3.82)^2 \\ + (12)(4.5 - 3.82)^2 \\ + (9)(5.5 - 3.82)^2 \end{array} \right) }$$

$$= 1.133 \quad (1.13)$$

The answer is (B).

10. Use the binomial distribution.

$$p = 0.15$$

$$P_{20}(2) = \frac{n!}{x!(n-x)!} p^x q^{n-x}$$

$$= \left(\frac{20!}{(2!)(20-2)!}\right)(0.15)^2(1 - 0.15)^{20-2}$$

$$= 0.229 \quad (0.23)$$

The answer is (C).

10 Probability and Statistics

Figure 10.1 *Venn Diagrams*

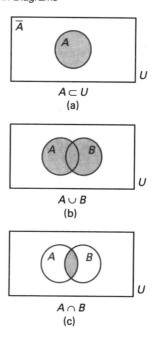

1. SET THEORY

A *set* (usually designated by a capital letter) is a population or collection of individual items known as *elements* or *members*. The *null set*, $\emptyset$, is empty (i.e., contains no members). If A and B are two sets, A is a *subset* of B if every member in A is also in B. A is a *proper subset* of B if B consists of more than the elements in A. These relationships are denoted as follows.[1]

$$A \subseteq B \quad \text{[subset]}$$

$$A \subset B \quad \text{[proper subset]}$$

The *universal set*, U, is one from which other sets draw their members. If A is a subset of U, then $\overline{A}$ (also designated as A', A^{-1}, $\tilde{A}$, and $-A$) is the *complement* of A and consists of all elements in U that are not in A. This is illustrated in a *Venn diagram* in Fig. 10.1(a).

The *union of two sets*, denoted by $A \cup B$ and shown in Fig. 10.1(b), is the set of all elements that are either in A or B or both. The *intersection of two sets*, denoted by $A \cap B$ and shown in Fig. 10.1(c), is the set of all elements that belong to both A and B. If $A \cap B = \emptyset$, A and B are said to be *disjoint sets*.

If A, B, and C are subsets of the universal set, the following laws apply.

Identity Laws

$$A \cup \emptyset = A$$

$$A \cup U = U$$

$$A \cap \emptyset = \emptyset$$

$$A \cap U = A$$

Idempotent Laws

$$A \cup A = A$$

$$A \cap A = A$$

---

[1]The *NCEES FE Reference Handbook* (*NCEES Handbook*) is inconsistent in its representation of sets, set members, matrices, matrix elements, and relations. Uppercase and lowercase, bold and not bold, italic, and not italic are all used interchangeably. For example, uppercase italic letters are used for set theory, while nonitalic letters are used in discrete math. In order to present these subjects in the same chapter, this book has adopted a consistent presentation style that may differ somewhat from the *NCEES Handbook*.

Complement Laws

$$A \cup \overline{A} = U$$
$$\overline{(\overline{A})} = A$$
$$A \cap \overline{A} = \varnothing$$
$$\overline{U} = \varnothing$$

Commutative Laws

$$A \cup B = B \cup A$$
$$A \cap B = B \cap A$$

Equation 10.1 and Eq. 10.2: Associative Laws

$$A \cup (B \cup C) = (A \cup B) \cup C \qquad 10.1$$

$$A \cap (B \cap C) = (A \cap B) \cap C \qquad 10.2$$

Equation 10.3 and Eq. 10.4: Distributive Laws

$$A \cup (B \cap C) = (A \cup B) \cap (A \cup C) \qquad 10.3$$

$$A \cap (B \cup C) = (A \cap B) \cup (A \cap C) \qquad 10.4$$

Equation 10.5 and Eq. 10.6: De Morgan's Laws

$$\overline{A \cup B} = \overline{A} \cap \overline{B} \qquad 10.5$$

$$\overline{A \cap B} = \overline{A} \cup \overline{B} \qquad 10.6$$

2. MATRIX OF RELATION

Elements (objects, items, etc.) are members of sets. For example, Sam is an element of the set of eighth grade boys at his school. Brenda is also an element, although she is an element of a different set—the set of eighth grade girls at the school. An element from one set may or may not interact (combine, respond, react, etc.) with an element from another set. For example, Sam and Brenda may or may not eventually get married. In terms of a "married" variable, $m_{\text{Sam-Brenda}} = 1$ if they get married, and $m_{\text{Sam-Brenda}} = 0$ if they do not. Similarly, "married" values, m_{ij}, can be assigned to all possible combinations of unique boy-girl pairs of eighth graders. A "relationship" club (set), R, could be formed that only accepts married couples as members. The (Sam, Brenda) ordered pair would be a member of the relationship club.

Equation 10.7: Matrix of Relation

$$m_{ij} = \begin{cases} 1 \text{ if } (a_i, b_j) \in R \\ 0 \text{ if } (a_i, b_j) \notin R \end{cases} \qquad 10.7$$

Description

Consider two finite sets, A and B, with m and n elements, respectively.[2]

$$A = \{a_1, a_2, a_3, ..., a_m\}$$
$$B = \{b_1, b_2, b_3, ..., b_n\}$$

The result, R, of associating a member from A with a member from B (through a function, a relation, or simple combining) is known as a *binary relation* "from A to B," represented as "aRb." The *composition* (i.e., combination of elements) (a_i, b_j) taken one at a time from each set is known as a *2-tuple* or *pair*. If the sequence of associated set members is always the same, the composition is known as an ordered pair. Ordered pairs can be summarized in table form.

	b_1	b_2	b_3	...	b_n
a_1	(a_1, b_1)	(a_1, b_2)	(a_1, b_3)	...	(a_1, b_n)
a_2	(a_2, b_1)	(a_2, b_2)	(a_2, b_3)	...	(a_2, b_n)
a_3	(a_3, b_1)	(a_3, b_2)	(a_3, b_3)	...	(a_3, b_n)
...	...	...	...	...	...
a_m	(a_m, b_1)	(a_m, b_2)	(a_m, b_3)	...	(a_m, b_n)

A composition relation variable, m_{ij}, can be used to specify whether or not an ordered pair (a_i, b_j) belongs to a third set, R. If an ordered pair (a_i, b_j) is a member of R, the value of the composition relation, m_{ij}, is 1.[3] If the pair is not a member of R, the value of the composition relation is 0. Equation 10.7 defines the composition relation values, m_{ij}. The *matrix of relation*, $\mathbf{M}_R$, is a *0-1 matrix* that contains all of the composition relation values, m_{ij}.[4] When the matrix describes connections between nodes, it is known as an *adjacency matrix*.

---

[2]The *NCEES Handbook* is inconsistent in how it defines an *n-tuple* (i.e., a set and its members): "$A = a_1, a_2, a_3...$," "$A = \{a_1, a_2, a_3...\}$," and "$A = (a_1, a_2, a_3...)$" are all used.

[3]Considering all of the available letters of the alphabet, it is unfortunate that the *NCEES Handbook* uses "m" to represent both the number of elements in set A and the values in the composition relation matrix.

[4](1) By convention, the rows and columns for a matrix representing a function are reversed from those of a matrix representing a relation. That is, a relation matrix is the transpose of its function matrix. The number of rows in the matrix of relation is equal to the number of members in the first (i.e., "from") set. (2) The *NCEES Handbook* designates the *matrix of relation* as $\mathbf{M}_R < [m_{ij}]$. This notation is not defined in the *NCEES Handbook*'s list of symbols. The symbol "$<$" is used (by some) in set theory to designate a proper subset (i.e., "$G < H$" means "G is a proper subset of H"). However, $\mathbf{M}_R$ is a matrix, not a set. The symbol "$<$" doesn't have any meaning associated with matrices.

$$\mathbf{M}_R = \begin{bmatrix} m_{11} & m_{12} & m_{13} & \cdots & m_{1n} \\ m_{21} & m_{22} & m_{23} & \cdots & m_{2n} \\ m_{31} & m_{32} & m_{33} & \cdots & m_{3n} \\ \cdots & \cdots & \cdots & \cdots & \cdots \\ m_{m1} & m_{m2} & m_{m3} & \cdots & m_{mn} \end{bmatrix}$$

Example

Two sets, A and B, are defined as $A = \{a_1, a_2, a_3, a_4, a_5\}$, and $B = \{b_1, b_2, b_3\}$. R is a relation from A to B defined by the following set of ordered pairs: $R = \{(a_1, b_1), (a_1, b_2), (a_1, b_3), (a_2, b_1), (a_3, b_1), (a_3, b_2), (a_3, b_3),$ and $(a_5, b_1)\}$. What is the matrix of the relation?

(A) $\begin{bmatrix} 1 & 1 & 1 & 0 & 1 \\ 1 & 0 & 1 & 0 & 0 \\ 1 & 0 & 1 & 0 & 0 \end{bmatrix}$

(B) $\begin{bmatrix} 1 & 0 & 0 \\ 0 & 0 & 0 \\ 1 & 1 & 1 \\ 0 & 0 & 0 \\ 1 & 0 & 0 \end{bmatrix}$

(C) $\begin{bmatrix} 1 & 1 & 1 \\ 1 & 0 & 0 \\ 1 & 1 & 1 \\ 0 & 0 & 0 \\ 1 & 0 & 0 \end{bmatrix}$

(D) $\begin{bmatrix} 1 & 1 & 1 \\ 0 & 0 & 1 \\ 1 & 1 & 1 \\ 0 & 0 & 0 \\ 0 & 0 & 1 \end{bmatrix}$

Solution

Since the relation is from A to B, and since set A has five members, the matrix of relation has five rows. Option A is the transpose of the correct matrix.

The answer is (C).

3. COMBINATIONS AND PERMUTATIONS

There are a finite number of ways in which n elements can be combined into distinctly different groups of r items. For example, suppose a farmer has a chicken, a rooster, a duck, and a cage that holds only two birds. The possible *combinations* of three birds taken two at a time are (chicken, rooster), (chicken, duck), and (rooster, duck). The birds in the cage will not remain stationary, so the combination (rooster, chicken) is not

distinctly different from (chicken, rooster). That is, the combinations are not *order conscious*.

Equation 10.8: Combinations

$$C(n, r) = \frac{P(n, r)}{r!} = \frac{n!}{r!(n-r)!} \qquad \textit{10.8}$$

Description

The number of *combinations* of n items taken r at a time is written $C(n, r)$, C_r^n, nC_r, $_nC_r$, or $\binom{n}{r}$ (pronounced "n choose r"). It is sometimes referred to as the *binomial coefficient* and is given by Eq. 10.8.

Example

Six design engineers are eligible for promotion to pay grade G8, but only four spots are available. How many different combinations of promoted engineers are possible?

(A) 4

(B) 6

(C) 15

(D) 20

Solution

The number of combinations of $n = 6$ items taken $r = 4$ items at a time is

$$C(6, 4) = \frac{n!}{r!(n-r)!} = \frac{6!}{4!(6-4)!}$$

$$= \frac{6 \times 5 \times 4 \times 3 \times 2 \times 1}{4 \times 3 \times 2 \times 1 \times 2 \times 1}$$

$$= 15$$

The answer is (C).

Equation 10.9: Permutations

$$P(n, r) = \frac{n!}{(n-r)!} \qquad \textit{10.9}$$

Description

An order-conscious subset of r items taken from a set of n items is the *permutation*, $P(n, r)$, also written P_r^n, nP_r, and $_nP_r$. A permutation is order conscious because the arrangement of two items (e.g., a_i and b_i) as $a_i b_i$ is different from the arrangement $b_i a_i$. The number of permutations is found from Eq. 10.9.

Example

An identification code begins with three letters. The possible letters are A, B, C, D, and E. If none of the letters are used more than once, how many different ways can the letters be arranged to make a code?

(A) 10

(B) 20

(C) 40

(D) 60

Solution

Since the order of the letters affects the identification code, determine the number of permutations of $n = 5$ items taken $r = 3$ items at a time using Eq. 10.9.

$$P(5, 3) = \frac{n!}{(n-r)!}$$

$$= \frac{5!}{(5-3)!}$$

$$= \frac{5 \times 4 \times 3 \times 2 \times 1}{2 \times 1}$$

$$= 60$$

The answer is (D).

Equation 10.10: Permutations of Different Object Types

$$P(n; n_1, n_2, \ldots, n_k) = \frac{n!}{n_1! n_2! \ldots n_k!} \qquad 10.10$$

Description

Suppose n_1 objects of one type (e.g., color, size, shape, etc.) are combined with n_2 objects of another type and n_3 objects of yet a third type, and so on, up to k types. The collection of $n = n_1 + n_2 + \cdots + n_k$ objects forms a population from which arrangements of n items can be formed. The number of permutations of n objects taken n at a time from a collection of k types of objects is given by Eq. 10.10.

Example

An urn contains 13 marbles total: 4 black marbles, 2 red marbles, and 7 yellow marbles. Arrangements of 13 marbles are made. Most nearly, how many unique ways can the 13 marbles be ordered (arranged)?

(A) 800

(B) 1200

(C) 14,000

(D) 26,000

Solution

The marble colors represent different types of objects. The number of permutations of the marbles taken 13 at a time is

$$P(13; 4, 2, 7) = \frac{n!}{n_1! n_2! \ldots n_k!} = \frac{13!}{4! 2! 7!}$$

$$= \frac{\begin{array}{c}13 \times 12 \times 11 \times 10 \times 9 \times 8 \times 7 \\ \times 6 \times 5 \times 4 \times 3 \times 2 \times 1\end{array}}{\begin{array}{c}4 \times 3 \times 2 \times 1 \times 2 \times 1 \\ \times 7 \times 6 \times 5 \times 4 \times 3 \times 2 \times 1\end{array}}$$

$$= 25,740 \quad (26,000)$$

The answer is (D).

4. LAWS OF PROBABILITY

Probability theory determines the relative likelihood that a particular event will occur. An *event*, E, is one of the possible outcomes of a *trial*. The *probability* of E occurring is denoted as $P(E)$.

Probabilities are real numbers in the range of zero to one. If an event E is certain to occur, then the probability $P(E)$ of the event is equal to one. If the event is certain *not* to occur, then the probability $P(E)$ of the event is equal to zero. The probability of any other event is between zero and one.

The probability of an event occurring is equal to one minus the probability of the event not occurring. This is known as a *complementary probability*.

$$P(E) = 1 - P(\text{not } E)$$

Complementary probability can be used to simplify some probability calculations. For example, calculation of the probability of numerical events being "greater than" or "less than" or quantities being "at least" a certain number can often be simplified by calculating the probability of the complementary event.

Probabilities of multiple events can be calculated from the probabilities of individual events using a variety of methods. When multiple events are considered, those events can either be independent or dependent. The probability of an *independent event* does not affect (and is not affected by) other events. The assumption of independence is appropriate when sampling from infinite or very large populations, when sampling from finite populations with replacement, or when sampling from different populations (universes). For example, the outcome of a second coin toss is generally not affected by the outcome of the first coin toss. The probability of a *dependent event* is affected by what has previously happened. For example, drawing a second card from a deck of cards without replacement is affected by what was drawn as the first card.

Events can be combined in two basic ways, according to the way the combination is described. Events can be connected by the words "and" and "or." For example, the question, "What is the probability of event A and event B occurring?" is different than the question, "What is the probability of event A or event B occurring?" The combinatorial "and" is designated in various ways: AB, $A \cdot B$, $A \times B$, $A \cap B$, and A, B, among others. In this book, the probability of A and B both occurring is designated as $P(A, B)$.

The combinatorial "or" is designated as: $A + B$ and $A \cup B$. In this book, the probability of A or B occurring is designated as $P(A + B)$.

Equation 10.11: Law of Total Probability

$$P(A + B) = P(A) + P(B) - P(A, B) \qquad \text{10.11}$$

Description

Equation 10.11 gives the probability that either event A or B will occur. $P(A, B)$ is the probability that both A and B will occur.

Example

A deck of ten children's cards contains three fish cards, two dog cards, and five cat cards. What is the probability of drawing either a cat card or a dog card from a full deck?

(A) 1/10

(B) 2/10

(C) 5/10

(D) 7/10

Solution

The two events are mutually exclusive, so the probability of both happening, $P(A, B)$, is zero. The total probability of drawing either a cat card or a dog card is

$$P(A + B) = P(A) + P(B) - P(A, B) = \frac{5}{10} + \frac{2}{10} - 0$$
$$= 7/10$$

The answer is (D).

Equation 10.12: Law of Compound (Joint) Probability

$$P(A, B) = P(A)P(B|A) = P(B)P(A|B) \qquad \text{10.12}$$

Variation

$$P(A, B) = P(A)P(B) \qquad \begin{bmatrix} \text{independent} \\ \text{events} \end{bmatrix}$$

Description

Equation 10.12, the *law of compound (joint) probability*, gives the probability that events A and B will both occur. $P(B|A)$ is the *conditional probability* that B will occur given that A has already occurred. Likewise, $P(A|B)$ is the conditional probability that A will occur given that B has already occurred. It is possible that the events come from different populations (universes, sample spaces, etc.), such as when one marble is drawn from one urn and another marble is drawn from a different urn. In that case, the events will be independent and won't affect each other. If the events are independent, then $P(B|A) = P(B)$ and $P(A|B) = P(A)$. Examples of dependent events for which the probability is conditional include drawing objects from a container or cards from a deck, without replacement.

Example

A bag contains seven orange balls, eight green balls, and two white balls. Two balls are drawn from the bag without replacing either of them. Most nearly, what is the probability that the first ball drawn is white and the second ball drawn is orange?

(A) 0.036

(B) 0.052

(C) 0.10

(D) 0.53

Solution

There is a total of 17 balls. There are 2 white balls. The probability of picking a white ball as the first ball is

$$P(A) = 2/17$$

After picking a white ball first, there are 16 balls remaining, 7 of which are orange. The probability of picking an orange ball second given that a white ball was chosen first is

$$P(B|A) = 7/16$$

The probability of picking a white ball first and an orange ball second is

$$P(A, B) = P(A)P(B|A) = \left(\frac{2}{17}\right)\left(\frac{7}{16}\right)$$
$$= 0.05147 \quad (0.052)$$

The answer is (B).

Probability
Statistics

Equation 10.13: Bayes' Theorem

$$P(B_j|A) = \frac{P(B_j)P(A|B_j)}{\sum\limits_{i=1}^{n} P(A|B_i)P(B_i)} \qquad 10.13$$

Variation

$$P(B_j|A) = \frac{P(B \text{ and } A)}{P(A)}$$

Description

Given two dependent sets of events, A and B, the probability that event B will occur given the fact that the dependent event A has already occurred is written as $P(B_j|A)$ and is given by *Bayes' theorem*, Eq. 10.13.

Example

A medical patient exhibits a symptom that occurs naturally 10% of the time in all people. The symptom is also exhibited by all patients who have a particular disease. The incidence of that particular disease among all people is 0.0002%. What is the probability of the patient having that particular disease?

(A) 0.002%

(B) 0.01%

(C) 0.3%

(D) 4%

Solution

This problem is asking for a conditional probability: the probability that a person has a disease, D, given that the person has a symptom, S. Use Bayes' theorem to calculate the probability. The probability that a person has the symptom S given that they have the disease D is $P(S|D)$ and is 100%. Multiply by 100% to get the answer as a percentage.

$$P(D|S) = \frac{P(D)P(S|D)}{P(S|D)P(D) + P(S|\text{not } D)P(\text{not } D)}$$

$$= \frac{(0.000002)(1.00)}{(1.00)(0.000002) + (0.10)(0.999998)}$$

$$= 0.00002 \quad (0.002\%)$$

The answer is (A).

5. MEASURES OF CENTRAL TENDENCY

It is often unnecessary to present experimental data in their entirety, either in tabular or graphic form. In such cases, the data and distribution can be represented by various parameters. One type of parameter is a measure of *central tendency*. The mode, median, and mean are measures of central tendency.

Mode

The *mode* is the observed value that occurs most frequently. The mode may vary greatly between series of observations; its main use is as a quick measure of the central value since little or no computation is required to find it. Beyond this, the usefulness of the mode is limited.

Median

The *median* is the point in the distribution that partitions the total set of observations into two parts containing equal numbers of observations. It is not influenced by the extremity of scores on either side of the distribution. The median is found by counting from either end through an ordered set of data until half of the observations have been accounted for. If the number of data points is odd, the median will be the exact middle value. If the number of data points is even, the median will be the average of the middle two values.

Equation 10.14: Arithmetic Mean

$$\overline{X} = (1/n)(X_1 + X_2 + \cdots + X_n) = (1/n)\sum_{i=1}^{n} X_i$$

$$10.14$$

Variation

$$\overline{X} = \frac{\sum f_i X_i}{\sum f_i} \quad \begin{bmatrix} f_i \text{ are frequencies} \\ \text{of occurrence of} \\ \text{events } i \end{bmatrix}$$

Description

The *arithmetic mean* is the arithmetic average of the observations. The *sample mean*, $\overline{X}$, can be used as an unbiased estimator of the *population mean*, μ. The term *unbiased estimator* means that on the average, the sample mean is equal to the population mean. The mean may be found without ordering the data (as was necessary to find the mode and median) from Eq. 10.14.

Example

100 random samples were taken from a large population. A particular numerical characteristic of sampled items was measured. The results of the measurements were as follows.

- 45 measurements were between 0.859 and 0.900.

- 0.901 was observed once.

- 0.902 was observed three times.

- 0.903 was observed twice.

- 0.904 was observed four times.

- 45 measurements were between 0.905 and 0.958.

The smallest value was 0.859, and the largest value was 0.958. The sum of all 100 measurements was 91.170. Except those noted, no measurements occurred more than twice.

What are the (a) mean, (b) mode, and (c) median of the measurements, respectively?

(A) 0.908; 0.902; 0.902

(B) 0.908; 0.904; 0.903

(C) 0.912; 0.902; 0.902

(D) 0.912; 0.904; 0.903

Solution

(a) From Eq. 10.14, the arithmetic mean is

$$\overline{X} = (1/n)\sum_{i=1}^{n} X_i = \left(\frac{1}{100}\right)(91.170) = 0.9117 \quad (0.912)$$

(b) The mode is the value that occurs most frequently. The value of 0.904 occurred four times, and no other measurements repeated more than four times. 0.904 is the mode.

(c) The median is the value at the midpoint of an ordered (sorted) set of measurements. There were 100 measurements, so the middle of the ordered set occurs between the 50th and 51st measurements. Since these measurements are both 0.903, the average of the two is 0.903.

The answer is (D).

Equation 10.15: Weighted Arithmetic Mean

$$\overline{X}_w = \frac{\sum w_i X_i}{\sum w_i} \qquad \textit{10.15}$$

Description

If some observations are considered to be more significant than others, a *weighted mean* can be calculated. Equation 10.15 defines a *weighted arithmetic mean*, $\overline{X}_w$, where w_i is the weight assigned to observation X_i.

Example

A course has four exams that comprise the entire grade for the course. Each exam is weighted. A student's scores on all four exams and the weight for each exam are as given.

exam	student score	weight
1	80%	1
2	95%	2
3	72%	2
4	95%	5

What is most nearly the student's final grade in the course?

(A) 82%

(B) 85%

(C) 87%

(D) 89%

Solution

The student's final grade is the weighted arithmetic mean of the individual exam scores.

$$\begin{aligned}
\overline{X}_w &= \frac{\sum w_i X_i}{\sum w_i} \\
&= \frac{(1)(80\%) + (2)(95\%) + (2)(72\%) + (5)(95\%)}{1+2+2+5} \\
&= 88.9\% \quad (89\%)
\end{aligned}$$

The answer is (D).

Equation 10.16: Geometric Mean

$$\text{sample geometric mean} = \sqrt[n]{X_1 X_2 X_3 \ldots X_n} \qquad \textit{10.16}$$

Description

The *geometric mean* of n nonnegative values is defined by Eq. 10.16. The geometric mean is the number that, when raised to the power of the sample size, produces the same result as the product of all samples. It is appropriate to use the geometric mean when the values being averaged are used as consecutive multipliers in other calculations. For example, the total revenue earned on an investment of C earning an effective interest rate of i_k in year k is calculated as $R = C(i_1 i_2 i_3 \ldots i_k)$. The interest rate, i, is a multiplicative element. If a \$100 investment earns 10% in year 1 (resulting in \$110 at the end of the year), then the \$110 earns 30% in year 2 (resulting in \$143), and the \$143 earns 50% in year 3 (resulting in \$215), the average interest earned each year would not be the arithmetic mean of $(10\% + 30\% + 50\%)/3 = 30\%$. The average would be calculated as a geometric mean (24.66%).

Example

What is most nearly the geometric mean of the following data set?

$$0.820, \ 1.96, \ 2.22, \ 0.190, \ 1.00$$

(A) 0.79

(B) 0.81

(C) 0.93

(D) 0.96

Solution

The geometric mean of the data set is

$$\text{sample geometric mean} = \sqrt[n]{X_1 X_2 X_3 \ldots X_n}$$
$$= \sqrt[5]{\begin{array}{c} (0.820)(1.96)(2.22) \\ \times (0.190)(1.00) \end{array}}$$
$$= 0.925 \quad (0.93)$$

The answer is (C).

Equation 10.17: Root-Mean-Square

$$\text{sample root-mean-square value} = \sqrt{(1/n)\sum X_i^2}$$
$$\text{10.17}$$

Description

The *root-mean-square* (rms) value of a series of observations is defined by Eq. 10.17. The variable X_{rms} is often used to represent the rms value.

Example

The water level on a tank in a chemical plant is measured every six hours. The tank has a depth of 6 m. The water levels on the tank on a certain day were found to be 2.5 m, 4.2 m, 5.6 m, and 3.3 m. What is most nearly the root-mean-square value of water level for that day?

    (A) 2.0 m

    (B) 3.3 m

    (C) 4.1 m

    (D) 5.8 m

Solution

Use Eq. 10.17 to find the root-mean-square value of water level for the day.

$$X_{\text{rms}} = \sqrt{(1/n)\sum X_i^2}$$
$$= \sqrt{\left(\frac{1}{4}\right)\left(\begin{array}{c} (2.5 \text{ m})^2 + (4.2 \text{ m})^2 \\ + (5.6 \text{ m})^2 + (3.3 \text{ m})^2 \end{array}\right)}$$
$$= 4.07 \text{ m} \quad (4.1 \text{ m})$$

The answer is (C).

6. MEASURES OF DISPERSION

Measures of dispersion describe the variability in observed data.

Equation 10.18 Through Eq. 10.22: Standard Deviation

$$\sigma_{\text{population}} = \sqrt{(1/N)\sum(X_i - \mu)^2} \qquad \text{10.18}$$

$$\sigma_{\text{sum}} = \sqrt{\sigma_1^2 + \sigma_2^2 + \cdots + \sigma_n^2} \qquad \text{10.19}$$

$$\sigma_{\text{series}} = \sigma\sqrt{n} \qquad \text{10.20}$$

$$\sigma_{\text{mean}} = \frac{\sigma}{\sqrt{n}} \qquad \text{10.21}$$

$$\sigma_{\text{product}} = \sqrt{A^2\sigma_b^2 + B^2\sigma_a^2} \qquad \text{10.22}$$

Variation

$$\sigma = \sqrt{\frac{\sum f_i(X_i - \mu)^2}{\sum f_i}}$$

Description

One measure of dispersion is the *standard deviation*, defined in Eq. 10.18. N is the total population size, not the sample size, n. This implies that the entire population is measured.

Equation 10.18 can be used to calculate the standard deviation only when the entire population can be included in the calculation. When only a small subset is available, as when a sample is taken (see Eq. 10.23), there are two obstacles to its use. First, the population mean, μ, is not known. This obstacle is overcome by using the sample average, $\overline{X}$, which is an unbiased estimator of the population mean. Second, Eq. 10.18 is inaccurate for small samples.

When combining two or more data sets for which the standard deviations are known, the standard deviation for the combined data is found using Eq. 10.19. This equation is used even if some of the data sets are subtracted; subtracting one data set from another increases the standard deviation of the result just as adding the two data sets does.

When a series of samples is taken from the same population, the sum of the standard deviations for the series is calculated from Eq. 10.20, where σ is the population standard deviation and n is the number of samples. The standard deviation of the mean values of these samples is called the *standard deviation* (or *standard error*) *of the mean* and is found with Eq. 10.21.

The standard deviation of the product of two random variables is given by Eq. 10.22. A and B are the expected values of the two variables, and σ_a^2 and σ_b^2 are the population variances for the two variables.

Example

A cat colony living in a small town has a total population of seven cats. The ages of the cats are as shown.

age	number
7 yr	1
8 yr	1
10 yr	2
12 yr	1
13 yr	2

What is most nearly the standard deviation of the age of the cat population?

(A) 1.7 yr

(B) 2.0 yr

(C) 2.2 yr

(D) 2.4 yr

Solution

Using Eq. 10.14, the arithmetic mean of the ages is the population mean, μ.

$$\mu = (1/n)\sum_{i=1}^{n} X_i$$

$$= \left(\frac{1}{7}\right)\left(\begin{array}{c}(1)(7 \text{ yr}) + (1)(8 \text{ yr}) + (2)(10 \text{ yr}) \\ + (1)(12 \text{ yr}) + (2)(13 \text{ yr})\end{array}\right)$$

$$= 10.4 \text{ yr}$$

From Eq. 10.18, the standard deviation of the ages is

$$\sigma_{\text{population}} = \sqrt{(1/N)\sum(X_i - \mu)^2}$$

$$= \sqrt{\left(\frac{1}{7}\right)\left(\begin{array}{c}(7 \text{ yr} - 10.4 \text{ yr})^2 \\ + (8 \text{ yr} - 10.4 \text{ yr})^2 \\ + (2)(10 \text{ yr} - 10.4 \text{ yr})^2 \\ + (12 \text{ yr} - 10.4 \text{ yr})^2 \\ + (2)(13 \text{ yr} - 10.4 \text{ yr})^2\end{array}\right)}$$

$$= 2.19 \text{ yr} \quad (2.2 \text{ yr})$$

The answer is (C).

Equation 10.23: Sample Standard Deviation

$$s = \sqrt{[1/(n-1)]\sum_{i=1}^{n}(X_i - \overline{X})^2} \qquad 10.23$$

Description

The *standard deviation of a sample* (particularly a small sample) of n items calculated from Eq. 10.18 is a *biased estimator* of (i.e., on the average, it is not equal to) the population standard deviation. A different measure of

dispersion called the *sample standard deviation, s* (not the same as the standard deviation of a sample), is an unbiased estimator of the population standard deviation. The sample standard deviation can be found using Eq. 10.23.

Example

Samples of aluminum-alloy channels were tested for stiffness. The following distribution of results was obtained.

stiffness	frequency
2480	23
2440	35
2400	40
2360	33
2320	21

If the mean of the samples is 2402, what is the approximate standard deviation of the population from which the samples are taken?

(A) 48.2

(B) 49.7

(C) 50.6

(D) 50.8

Solution

The number of samples is

$$n = 23 + 35 + 40 + 33 + 21 = 152$$

The sample standard deviation, s, is the unbiased estimator of the population standard deviation, σ.

$$s = \sqrt{[1/(n-1)]\sum_{i=1}^{n}(X_i - \overline{X})^2}$$

$$= \sqrt{\left(\frac{1}{152-1}\right)\left(\begin{array}{c}(23)(2480 - 2402)^2 \\ + (35)(2440 - 2402)^2 \\ + (40)(2400 - 2402)^2 \\ + (33)(2360 - 2402)^2 \\ + (21)(2320 - 2402)^2\end{array}\right)}$$

$$= 50.82 \quad (50.8)$$

The answer is (D).

Equation 10.24 Through Eq. 10.26: Variance and Sample Variance

$$\sigma^2 = (1/N)\left[(X_1 - \mu)^2 + (X_2 - \mu)^2 + \cdots + (X_N - \mu)^2\right] \qquad 10.24$$

$$\sigma^2 = (1/N) \sum_{i=1}^{N} (X_i - \mu)^2 \qquad \textbf{10.25}$$

$$s^2 = [1/(n-1)] \sum_{i=1}^{n} (X_i - \overline{X})^2 \qquad \textbf{10.26}$$

Description

The *variance* is the square of the standard deviation. Since there are two standard deviations, there are two variances. The *variance of the population* (i.e., the *population variance*) is σ^2, and the *sample variance* is s^2. The population variance can be found using either Eq. 10.24 or Eq. 10.25, both derived from Eq. 10.18, and the sample variance can be found using Eq. 10.26, derived from Eq. 10.23.

Example

Most nearly, what is the sample variance of the following data set?

$$2, 4, 6, 8, 10, 12, 14$$

(A) 4.3

(B) 5.2

(C) 8.0

(D) 19

Solution

Find the mean using Eq. 10.14.

$$\overline{X} = (1/n) \sum_{i=1}^{n} X_i = \left(\frac{1}{7}\right)(2 + 4 + 6 + 8 + 10 + 12 + 14)$$

$$= 8$$

From Eq. 10.26, the sample variance is

$$s^2 = [1/(n-1)] \sum_{i=1}^{n} (X_i - \overline{X})^2$$

$$= \left(\frac{1}{7-1}\right)\left(\begin{array}{c} (2-8)^2 + (4-8)^2 + (6-8)^2 \\ + (8-8)^2 + (10-8)^2 + (12-8)^2 \\ + (14-8)^2 \end{array}\right)$$

$$= 18.67 \quad (19)$$

The answer is (D).

Equation 10.27: Sample Coefficient of Variation

$$CV = s/\overline{X} \qquad \textbf{10.27}$$

Description

The *relative dispersion* is defined as a measure of dispersion divided by a measure of central tendency. The *sample coefficient of variation*, CV, is a relative dispersion calculated from the sample standard deviation and the mean.

Example

The following data were recorded from a laboratory experiment.

$$20, 25, 30, 32, 27, 22$$

The mean of the data is 26. What is most nearly the sample coefficient of variation of the data?

(A) 0.18

(B) 1.1

(C) 2.4

(D) 4.6

Solution

Find the sample standard deviation of the data using Eq. 10.23.

$$s = \sqrt{[1/(n-1)] \sum_{i=1}^{n} (X_i - \overline{X})^2}$$

$$= \sqrt{\left(\frac{1}{6-1}\right)\left(\begin{array}{c} (20-26)^2 + (25-26)^2 \\ + (30-26)^2 + (32-26)^2 \\ + (27-26)^2 + (22-26)^2 \end{array}\right)}$$

$$= 4.6$$

From Eq. 10.27, the sample coefficient of variation is

$$CV = s/\overline{X} = \frac{4.6}{26} = 0.177 \quad (0.18)$$

The answer is (A).

7. NUMERICAL EVENTS

A *discrete numerical event* is an occurrence that can be described by an integer. For example, 27 cars passing through a bridge toll booth in an hour is a discrete numerical event. Most numerical events are *continuously distributed* and are not constrained to discrete or integer values. For example, the resistance of a 10% 1 Ω resistor may be any value between 0.9 Ω and 1.1 Ω.

8. EXPECTED VALUES

Equation 10.28: Expected Value of a Discrete Variable

$$\mu = E[X] = \sum_{k=1}^{n} x_k f(x_k) \qquad \textbf{10.28}$$

Probability/ Statistics

Description

The *expected value*, E, of a discrete random variable, X, is given by Eq. 10.28. $f(x_k)$ is the probability mass function as defined in Eq. 10.35.

Example

The probability distribution of the number of calls, X, that a customer service agent receives each hour is shown.

x	$f(x)$
0	0.00
2	0.04
4	0.05
6	0.10
8	0.35
10	0.46

What is most nearly the average number of phone calls that a customer service agent expects to receive in an hour?

- (A) 5
- (B) 7
- (C) 8
- (D) 9

Solution

The expected number of received calls is

$$\mu = E[X] = \sum_{k=1}^{n} x_k f(x_k)$$

$$= (0)(0.00) + (2)(0.04) + (4)(0.05)$$
$$\quad + (6)(0.10) + (8)(0.35) + (10)(0.46)$$
$$= 8.28 \quad (8)$$

The answer is (C).

Equation 10.29: Variance of a Discrete Variable

$$\sigma^2 = V[X] = \sum_{k=1}^{n} (x_k - \mu)^2 f(x_k) \qquad \textit{10.29}$$

Description

Equation 10.29 gives the variance, σ^2, of a discrete function of variable X. To use Eq. 10.29, the population mean, μ, must be known, having been calculated from the total population of n values. The name "discrete" requires only that n be a finite number and all values of x be known. It does not limit the values of x to integers.

Equation 10.30 and Eq. 10.31: Expected Value (Mean) of a Continuous Variable

$$\mu = E[X] = \int_{-\infty}^{\infty} x f(x)\, dx \qquad \textit{10.30}$$

$$E[Y] = E[g(X)] = \int_{-\infty}^{\infty} g(x) f(x)\, dx \qquad \textit{10.31}$$

Description

Equation 10.30 calculates the population mean, μ, of a continuous variable, X, from the probability density function, $f(x)$. Equation 10.31 calculates the mean of any continuously distributed variable defined by $Y = g(x)$, whose values are observed according to the probabilities given by the probability density function (PDF) $f(x)$. Equation 10.31 is the general form of Eq. 10.30, where $g(x) = x$.

Equation 10.32: Variance of a Continuous Variable

$$\sigma^2 = V[X] = E[(X - \mu)^2] = \int_{-\infty}^{\infty} (x - \mu)^2 f(x)\, dx \qquad \textit{10.32}$$

Description

Equation 10.32 gives the variance of a continuous random variable, X. μ is the mean of X, and $f(x)$ is the density function of X.

Equation 10.33: Standard Deviation of a Continuous Variable

$$\sigma = \sqrt{V[X]} \qquad \textit{10.33}$$

Variation

$$\sigma = \sqrt{\sigma^2}$$

Description

The standard deviation is always the square root of the variance, as shown in the variation equation. Equation 10.33 gives the standard deviation for a continuous random variable, X.

Equation 10.34: Coefficient of Variation of a Continuous Variable

$$CV = \sigma/\mu \qquad \textit{10.34}$$

Description

The coefficient of variation of a continuous variable is calculated from Eq. 10.34.

9. PROBABILITY DENSITY FUNCTIONS (DISCRETE)

Equation 10.35: Probability Mass Function

$$f(x_k) = P(X = x_k) \quad [k = 1, 2, \ldots, n] \qquad 10.35$$

Description

A *discrete random variable*, X, can take on values from a set of discrete values, x_i. The set of values can be finite or infinite, as long as each value can be expressed as an integer. The *probability mass function*, defined by Eq. 10.35, gives the probability that a discrete random variable, X, is equal to each of the set's possible values, x_k. The probabilities of all possible outcomes add up to unity.

Equation 10.36: Probability Density Function

$$P(a \le X \le b) = \int_a^b f(x)\,dx \qquad 10.36$$

Description

A *density function* is a nonnegative function whose integral taken over the entire range of the independent variable is unity. A *probability density function* (PDF) is a mathematical formula that gives the probability of a numerical event.

Various mathematical models are used to describe probability density functions. Figure 10.2 shows a graph of a continuous probability density function. The area under the probability density function is the probability that the variable will assume a value between the limits of evaluation. The total probability, or the probability that the variable will assume any value over the interval, is 1.0. The probability of an exact numerical event is zero. That is, there is no chance that a numerical event will be exactly a. It is possible to determine only the probability that a numerical event will be less than a, greater than b, or between the values of a and b.

Figure 10.2 Probability Density Function

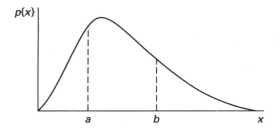

If a random variable, X, is continuous over an interval, then a nonnegative *probability density function* of that variable exists over the interval as defined by Eq. 10.36.

10. BINOMIAL DISTRIBUTION

Equation 10.37 and Eq. 10.38: Binomial Probability Density Function

$$b(x; n, p) = \binom{n}{x} p^x (1 - p)^{n-x} \qquad 10.37$$

$$\binom{n}{x} = \frac{n!}{x!(n-x)!} \qquad 10.38$$

Description

The *binomial probability density function* (*binomial distribution*) is used when all outcomes can be categorized as either successes or failures. The probability of success in a single trial is p, and the probability of failure is the complement, $1 - p$. The population is assumed to be infinite in size so that sampling does not change the values of p and $1 - p$. (The binomial distribution can also be used with finite populations when sampling with replacement.)

Equation 10.37 gives the probability of x successes in n independent *successive trials*. The quantity $\binom{n}{x}$ is the *binomial coefficient*, identical to the number of combinations of n items taken x at a time. Equation 10.37 is a discrete distribution, taking on values only for discrete integer values up to n.

Example

Five percent of a large batch of high-strength steel bolts purchased for bridge construction are defective. If seven bolts are randomly sampled, what is the probability that exactly three will be defective?

(A) 0.36%

(B) 0.48%

(C) 2.1%

(D) 43%

Solution

The bolts are either defective or not, so the binomial distribution can be used.

$$p = 0.05 \quad [\text{success} = \text{defective}]$$

$$q = 1 - 0.05 = 0.95 \quad [\text{failure} = \text{not defective}]$$

From Eq. 10.37,

$$p\{3\} = f(3) = \binom{n}{x} p^x q^{n-x} = \binom{7}{3}(0.05)^3(0.95)^{7-3}$$

$$= \left(\frac{7 \cdot 6 \cdot 5 \cdot 4 \cdot 3 \cdot 2 \cdot 1}{4 \cdot 3 \cdot 2 \cdot 1 \cdot 3 \cdot 2 \cdot 1}\right)(0.05)^3(0.95)^4$$

$$= 0.00356 \quad (0.36\%)$$

The answer is (A).

Equation 10.39 and Eq. 10.40: Mean and Variance of Binomial Probability Density Function: Discrete Distribution

$$\mu = np \qquad \text{10.39}$$
$$\sigma^2 = np(1 - p) \qquad \text{10.40}$$

Description

Equation 10.39 is the mean, μ, and Eq. 10.40 is the variance, σ^2, of the binomial distribution.

11. PROBABILITY DISTRIBUTION FUNCTIONS (CONTINUOUS)

A *cumulative probability distribution function*, $F(x)$, gives the probability that a numerical event will occur or the probability that the numerical event will be less than or equal to some value, x.

Equation 10.41: Cumulative Distribution Function: Discrete Random Variable

$$F(x_m) = \sum_{k=1}^{m} P(x_k) = P(X \le x_m) \quad [m = 1, 2, \ldots, n]$$
$$\text{10.41}$$

Description

For a *discrete random variable*, X, the probability distribution function is the sum of the individual probabilities of all possible events up to and including event x_m. The *cumulative distribution function* (CDF) is a function that calculates the cumulative sum of all values up to and including a particular end point. For discrete probability density functions (PDFs), $F(x_m)$, the CDF can be calculated as a summation, as shown in Eq. 10.41.

Because calculating cumulative probabilities can be cumbersome, tables of values are often used. Table 10.1 at the end of this chapter gives values for cumulative binomial probabilities, where n is the number of trials, P is the probability of success for a single trial, and x is the maximum number of successful trials.

Equation 10.42: Cumulative Distribution Function: Continuous Random Variable

$$F(x) = \int_{-\infty}^{x} f(t)\, dt \qquad \text{10.42}$$

Description

For continuous functions, the CDF is calculated as an integral of the PDF from minus infinity to the limit of integration, as in Eq. 10.42. This integral corresponds to the area under the curve up to the limit of integration

and represents the probability that the variable is less than or equal to the limit of integration. That is, $F(x) = P(x \le a)$. A CDF has a maximum value of 1.0, and for a continuous probability density function, $F(x)$ will approach 1.0 asymptotically.

Example

For the probability density function shown, what is the probability of the random variable x being less than 1/3?

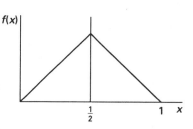

(A) 0.11

(B) 0.22

(C) 0.25

(D) 0.33

Solution

The total area under the probability density function is equal to 1. The area of two triangles is

$$A = (2)\left(\tfrac{1}{2}\right) bh = (2)\left(\tfrac{1}{2}\right)\left(\tfrac{1}{2}\right) h = 1$$

Therefore, the height of the curve at its peak is 2.

The equation of the line from $x = 0$ up to $x = \tfrac{1}{2}$ is

$$f(x) = 4x \quad \left[0 \le x \le \tfrac{1}{2}\right]$$

The probability that $x < \tfrac{1}{3}$ is equal to the area under the curve between 0 and $\tfrac{1}{3}$. From Eq. 10.42,

$$F\left(0 < x < \tfrac{1}{3}\right) = \int_{0}^{1/3} f(x)\, dx = \int_{0}^{1/3} 4x \, dx = 2x^2 \Big|_{0}^{1/3}$$

$$= (2)\left(\tfrac{1}{3}\right)^2 - 0$$

$$= 0.222 \quad (0.22)$$

The answer is (B).

12. PROBABILITY DISTRIBUTIONS

Equation 10.43 and Eq. 10.44: Binomial Distribution

$$P_n(x) = C(n, x) p^x q^{n-x} = \frac{n!}{x!(n-x)!} p^x q^{n-x} \qquad \text{10.43}$$
$$q = 1 - p \qquad \text{10.44}$$

Description

The *binomial probability function* is used when all outcomes are discrete and can be categorized as either successes or failures. The probability of success in a single trial is designated as p, and the probability of failure is the complement, q, calculated from Eq. 10.44.

Equation 10.43 gives the probability of x successes in n independent successive trials. The quantity $C(n, x)$ is the *binomial coefficient*, identical to the number of combinations of n items taken x at a time. (See Eq. 10.8.)

Example

A cat has a litter of seven kittens. If the probability that any given kitten will be female is 0.52, what is the probability that exactly two of the seven will be male?

(A) 0.07

(B) 0.18

(C) 0.23

(D) 0.29

Solution

Since the outcomes are "either-or" in nature, the outcomes (and combinations of outcomes) follow a binomial distribution. A male kitten is defined as a success. The probability of a success is

$$p = 1 - 0.52 = 0.48 = P(\text{male kitten})$$

$$q = 0.52 = P(\text{female kitten})$$

$$n = 7 \text{ trials}$$

$$x = 2 \text{ successes}$$

$$P_n(x) = \frac{n!}{x!(n-x)!} p^x q^{n-x}$$

$$P_7(2) = \left(\frac{7!}{2!(7-2)!}\right)(0.48)^2(0.52)^{7-2}$$

$$= 0.184 \quad (0.18)$$

The answer is (B).

Equation 10.45 Through Eq. 10.48: Normal Distribution

$$f(x) = \frac{1}{\sigma\sqrt{2\pi}} e^{-\frac{1}{2}\left(\frac{x-\mu}{\sigma}\right)^2} \quad [-\infty \le x \le \infty] \qquad 10.45$$

$$f(x) = \frac{1}{\sqrt{2\pi}} e^{-x^2/2} \quad [-\infty \le x \le \infty] \qquad 10.46$$

$$Z = \frac{x-\mu}{\sigma} \qquad 10.47$$

$$F(-x) = 1 - F(x) \qquad 10.48$$

Description

The *normal distribution* (*Gaussian distribution*) is a symmetrical continuous distribution, commonly referred to as the *bell-shaped curve*, which describes the distribution of outcomes of many real-world experiments, processes, and phenomena. (See Fig. 10.3.)

Figure 10.3 Normal Distribution

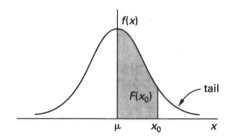

The probability density function for the normal distribution with population mean μ and population variance σ^2 is illustrated in Fig. 10.4 and is described by Eq. 10.45.

Figure 10.4 Normal Curve with Mean μ and Standard Deviation σ

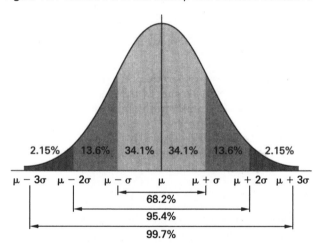

Since $f(x)$ is difficult to integrate (i.e., Eq. 10.45 is difficult to evaluate), Eq. 10.45 is seldom used directly, and a *unit normal table* is used instead. (See Table 10.2 at the end of this chapter.) The unit normal table (also called the *standard normal table*) is based on a normal distribution with a mean of zero and a standard deviation of one. The standard normal distribution is given by Eq. 10.46. In Table 10.2, $F(x)$ is the area under the curve from $-\infty$ to x, $R(x)$ is the area under the curve from x to ∞, and $W(x)$ is the area under the curve between $-x$ and x. The generic variable x used in Table 10.2 is the *standard normal variable*, Z, calculated in Eq. 10.47, not the actual measurement of the random variable, X. That is, the x used in Table 10.2 is not the x used in Eq. 10.47.

Since the range of values from an experiment or phenomenon will not generally correspond to the unit normal table, a value, x, must be converted to a standard normal variable, Z. In Eq. 10.47, μ and σ are the population mean and standard deviation, respectively, of the distribution from which x comes. The unbiased estimators for μ and σ are $\overline{X}$ and s, respectively, when a sample is used to estimate the population parameters. Both $\overline{X}$ and s approach the population values as the sample size, n, increases.

Example

The heights of several thousand fifth grade boys in Santa Clara County are measured. The mean of the heights is 1.20 m, and the variance is 25×10^{-4} m². Approximately what percentage of these boys is taller than 1.23 m?

(A) 27%

(B) 31%

(C) 69%

(D) 73%

Solution

To convert the normal distribution to unit normal distribution, the new variable, Z, is constructed from the height, x, mean μ, and standard deviation, σ. The mean is known; the standard deviation is found from the variance and a variation of Eq. 10.33.

$$\sigma = \sqrt{\sigma^2} = \sqrt{25 \times 10^{-4} \text{ m}^2}$$
$$= 0.05 \text{ m}$$

For a height less than or equal to 1.23 m, from Eq. 10.47,

$$Z = \frac{x - \mu}{\sigma} = \frac{1.23 \text{ m} - 1.20 \text{ m}}{0.05 \text{ m}}$$
$$= 0.6$$

From Table 10.2, the cumulative distribution function at $Z = 0.6$ is $F(Z) = 0.7257$. The percentage of boys having height greater than 1.23 m is

percentage taller than 1.23 m $= 100\% - (0.7257)(100\%)$
$$= 27.43\% \quad (27\%)$$

The answer is (A).

Equation 10.49 and Eq. 10.50: Central Limit Theorem

$$\mu_{\overline{y}} = \mu \qquad \text{10.49}$$
$$\sigma_{\overline{y}} = \frac{\sigma}{\sqrt{n}} \qquad \text{10.50}$$

Description

The *central limit theorem* states that the distribution of a significantly large number of sample means of n items where all items are drawn from the same (i.e., parent) population will be normal. According to the central limit theorem, the mean of sample means, $\mu_{\overline{y}}$, is equal to the population mean of the parent distribution, μ, as shown in Eq. 10.49. The standard deviation of the sample means, $\sigma_{\overline{y}}$, is equal to the standard deviation of the parent population divided by the square root of the sample size, as shown in Eq. 10.50.

Equation 10.51 and Eq. 10.52: t-Distribution

$$f(t) = \frac{\Gamma\left(\frac{\nu+1}{2}\right)}{\sqrt{\nu\pi}\Gamma\left(\frac{\nu}{2}\right)}\left(1+\frac{t^2}{\nu}\right)^{-\frac{\nu+1}{2}} \quad [-\infty \le t \le \infty] \qquad \text{10.51}$$

$$t = \frac{\overline{x}-\mu}{s/\sqrt{n}} \quad [-\infty \le t \le \infty] \qquad \text{10.52}$$

Description

For the *t-distribution* (commonly referred to as *Student's t-distribution*), the probability distribution function with ν degrees of freedom (sample size of $n+1$) is given by Eq. 10.51. The t-distribution is tabulated in Table 10.4 at the end of this chapter, with t as a function of ν and α. In Table 10.4, the column labeled "ν" lists the degrees of freedom, one less than the sample size. Degrees of freedom is sometimes given the symbol "df."

In Eq. 10.52, x is a unit normal variable, and r is the root-mean-squared value of $n+1$ other random variables (i.e., the sample size is $n+1$).

Equation 10.53: Gamma Function

$$\Gamma(n) = \int_0^\infty t^{n-1}e^{-t}dt \quad [n>0] \qquad \text{10.53}$$

Description

The *gamma function*, $\Gamma(n)$, is an extension of the factorial function and is used to determine values of the factorial for complex numbers greater than zero (i.e., positive integers).

Equation 10.54: Chi-Squared Distribution

$$\chi^2 = Z_1^2 + Z_2^2 + \cdots + Z_n^2 \qquad \text{10.54}$$

Description

The sum of the squares of n independent normal random variables will be distributed according to the *chi-squared distribution* and will have n degrees of freedom.

The chi-squared distribution is often used with hypothesis testing of variances. Chi-squared values, $\chi^2_{\alpha,n}$, for selected values of α and n can be found from Table 10.5 at the end of this chapter.

13. STUDENT'S *t*-TEST

Equation 10.55: Exceedance

$$\alpha = \int_{t_{\alpha,\nu}}^{\infty} f(t)\,dt \qquad 10.55$$

Description

The *t-test* is a method of comparing two variables, usually to test the significance of the difference between samples. For example, the *t*-test can be used to test whether the populations from which two samples are drawn have the same means.

The *exceedance* (i.e., the probability of being incorrect), α, is equal to the total area under the upper tail. (See Fig. 10.5.) For a *one-tail test*, $\alpha = 1 - C$. For a *two-tail test*, $\alpha = 1 - C/2$. Since the *t*-distribution is symmetric about zero, $t_{1-\alpha,n} = -t_{1-\alpha,n}$. As n increases, the *t*-distribution approaches the normal distribution.

α can be used to find the critical values of F, as listed in Table 10.6 at the end of this chapter.

Figure 10.5 *Exceedance*

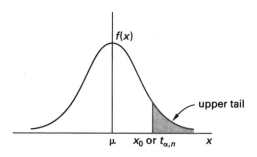

14. CONFIDENCE LEVELS

The results of experiments are seldom correct 100% of the time. Recognizing this, researchers accept a certain probability of being wrong. In order to minimize this probability, an experiment is repeated several times. The number of repetitions required depends on the level of confidence wanted in the results. For example, if the results have a 5% probability of being wrong, the *confidence level*, C, is 95% that the results are correct.

15. SUMS OF RANDOM VARIABLES

Equation 10.56: Sums of Random Variables

$$Y = a_1 X_1 + a_2 X_2 + \cdots + a_n X_n \qquad 10.56$$

Description

The sum of random variables, Y, is found from Eq. 10.56.

Equation 10.57: Expected Value of the Sum of Random Variables

$$\mu_y = E(Y) = a_1 E(X_1) + a_2 E(X_2) + \cdots + a_n E(X_n) \qquad 10.57$$

Description

The expected value of the sum of random variables, μ_y, is calculated using Eq. 10.57.

Equation 10.58 and Eq. 10.59: Variance of the Sum of Independent Random Variables

$$\sigma_y^2 = V(Y) = a_1^2 V(X_1) + a_2^2 V(X_2) + \cdots + a_n^2 V(X_n) \qquad 10.58$$

$$\sigma_y^2 = a_1^2 \sigma_1^2 + a_2^2 \sigma_2^2 + \cdots + a_n^2 \sigma_n^2 \qquad 10.59$$

Description

The variance of the sum of independent random variables can be calculated from Eq. 10.58 and Eq. 10.59.

Equation 10.60: Standard Deviation of the Sum of Independent Random Variables

$$\sigma_y = \sqrt{\sigma_y^2} \qquad 10.60$$

Description

The standard deviation of the sum of independent random variables (see Eq. 10.56) is found from Eq. 10.60.

16. SUMS AND DIFFERENCES OF MEANS

When two variables are sampled from two different standard normal variables (i.e., are independent), their sums will be distributed with mean $\mu_{\text{new}} = \mu_1 + \mu_2$ and variance $\sigma_{\text{new}}^2 = \sigma_1^2/n_1 + \sigma_2^2/n_2$. The sample sizes, n_1 and n_2, do not have to be the same. The relationships for confidence intervals and hypothesis testing can be used for a new variable, $x_{\text{new}} = x_1 + x_2$, if μ is replaced by μ_{new} and σ is replaced by σ_{new}.

For the difference in two standard normal variables, the mean is the difference in two population means, $\mu_{\text{new}} = \mu_1 - \mu_2$, but the variance is the sum, as it was for the sum of two standard normal variables.

17. CONFIDENCE INTERVALS

Population properties such as means and variances must usually be estimated from samples. The sample mean, $\overline{X}$, and sample standard deviation, s, are unbiased estimators, but they are not necessarily precisely equal to the true population properties. For estimated values, it is common to specify an interval expected to contain the true population properties. The interval is known as a confidence interval because a confidence level, C (e.g., 99%), is associated with it. (There is still a $1 - C$ chance that the true population property is outside of the interval.) The interval will be bounded below by its *lower confidence limit* (LCL) and above by its *upper confidence limit* (UCL).

As a consequence of the *central limit theorem*, means of samples of n items taken from a distribution that is normally distributed with mean μ and standard deviation σ will be normally distributed with mean μ and variance σ^2/n. Therefore, the probability that any given average, $\overline{X}$, exceeds some value, L, is

$$p\{\overline{X} > L\} = p\left\{ x > \left| \frac{L - \mu}{\frac{\sigma}{\sqrt{n}}} \right| \right\}$$

L is the *confidence limit* for the confidence level $1 - p\{\overline{X} > L\}$ (expressed as a percentage). Values of $p\{x\}$ are read directly from the unit normal table. (See Table 10.2.) As an example, $x = 1.645$ for a 95% confidence level since only 5% of the curve is above that x in the upper tail. This is known as a *one-tail confidence limit* because all of the exceedance probability is given to one side of the variation.

With *two-tail confidence limits*, the probability is split between the two sides of variation. There will be upper and lower confidence limits: UCL and LCL, respectively. This is appropriate when it is not specifically known that the calculated parameter is too high or too low. Table 10.3 at the end of this chapter lists standard normal variables and t values for two-tail confidence limits.

$$p\{\text{LCL} < \overline{X} < \text{UCL}\}$$

$$= p\left\{ \frac{\text{LCL} - \mu}{\frac{\sigma}{\sqrt{n}}} < x < \frac{\text{UCL} - \mu}{\frac{\sigma}{\sqrt{n}}} \right\}$$

Equation 10.61 and Eq. 10.62: Confidence Limits and Interval for Mean of a Normal Distribution

$$\overline{X} - Z_{\alpha/2} \frac{\sigma}{\sqrt{n}} \leq \mu \leq \overline{X} + Z_{\alpha/2} \frac{\sigma}{\sqrt{n}} \quad \text{[known } \sigma] \quad 10.61$$

$$\overline{X} - t_{\alpha/2} \frac{s}{\sqrt{n}} \leq \mu \leq \overline{X} + t_{\alpha/2} \frac{s}{\sqrt{n}} \quad \text{[unknown } \sigma] \quad 10.62$$

Variations

$$\text{LCL} = \overline{X} - t_{\alpha/2, n-1} \left(\frac{s}{\sqrt{n}} \right)$$

$$\text{UCL} = \overline{X} + t_{\alpha/2, n-1} \left(\frac{s}{\sqrt{n}} \right)$$

Description

The *confidence limits for the mean*, μ, of a normal distribution can be calculated from Eq. 10.61 when the standard deviation, σ, is known.

If the standard deviation, σ, of the underlying distribution is not known, the confidence limits must be estimated from the sample standard deviation, s, using Eq. 10.62. Accordingly, the standard normal variable is replaced by the t-distribution parameter, $t_{\alpha/2}$, with $n-1$ degrees of freedom, where n is the sample size. $\alpha = 1 - C$, and $\alpha/2$ is the t-distribution parameter since half of the exceedance is allocated to each confidence limit.

Equation 10.63 and Eq. 10.64: Confidence Limits for the Difference Between Two Means

$$\overline{X}_1 - \overline{X}_2 - Z_{\alpha/2} \sqrt{\frac{\sigma_1^2}{n_1} + \frac{\sigma_2^2}{n_2}}$$

$$\leq \mu_1 - \mu_2 \leq \overline{X}_1 - \overline{X}_2$$

$$+ Z_{\alpha/2} \sqrt{\frac{\sigma_1^2}{n_1} + \frac{\sigma_2^2}{n_2}} \quad \text{[known } \sigma_1 \text{ and } \sigma_2] \qquad 10.63$$

$$\overline{X}_1 - \overline{X}_2 - t_{\alpha/2} \sqrt{\frac{\left(\frac{1}{n_1} + \frac{1}{n_2}\right)\left[(n_1 - 1)S_1^2 + (n_2 - 1)S_2^2\right]}{n_1 + n_2 - 2}}$$

$$\leq \mu_1 - \mu_2 \leq \overline{X}_1 - \overline{X}_2$$

$$+ t_{\alpha/2} \sqrt{\frac{\left(\frac{1}{n_1} + \frac{1}{n_2}\right)\left[(n_1 - 1)S_1^2 + (n_2 - 1)S_2^2\right]}{n_1 + n_2 - 2}}$$

$$\text{[unknown } \sigma_1 \text{ and } \sigma_2]$$

$$10.64$$

Description

The difference in two standard normal variables will be distributed with mean $\mu_{\text{new}} = \mu_1 - \mu_2$. Use Eq. 10.63 to calculate the confidence interval for the difference between two means, μ_1 and μ_2, if the standard deviations σ_1 and σ_2 are known. If the standard deviations σ_1 and σ_2 are unknown, use Eq. 10.64. S_1 and S_2 are the sample standard deviations for the two sample sets. The t-distribution parameter, $t_{\alpha/2}$, has $n_1 + n_2 - 2$ degrees of freedom.

Example

100 resistors produced by company A and 150 resistors produced by company B are tested to find their limits before burning out. The test results show that the company A resistors have a mean rating of 2 W before burning out, with a standard deviation of 0.25 W; and the company B resistors have a 3 W mean rating before burning out, with a standard deviation of 0.30 W. What are the 95% confidence limits for the difference between the two means for the company A resistors and company B resistors (i.e., A − B)?

- (A) −1.1 W; −1.0 W
- (B) −1.1 W; −0.93 W
- (C) −1.1 W; −0.90 W
- (D) −1.0 W; −0.99 W

Solution

From Table 10.3, the value of the standard normal variable for a two-tail test with 95% confidence is 1.9600.

From Eq. 10.63, the confidence limits for the difference between the two means are

$$\mathrm{LCL}(\mu_1 - \mu_2) = \overline{X}_1 - \overline{X}_2 - Z_{\alpha/2}\sqrt{\frac{\sigma_1^2}{n_1} + \frac{\sigma_2^2}{n_2}}$$

$$= 2\ \mathrm{W} - 3\ \mathrm{W}$$

$$- 1.9600\sqrt{\frac{(0.25\ \mathrm{W})^2}{100} + \frac{(0.30\ \mathrm{W})^2}{150}}$$

$$= -1.0686\ \mathrm{W}\quad(-1.1\ \mathrm{W})$$

$$\mathrm{UCL}(\mu_1 - \mu_2) = \overline{X}_1 - \overline{X}_2 + Z_{\alpha/2}\sqrt{\frac{\sigma_1^2}{n_1} + \frac{\sigma_2^2}{n_2}}$$

$$= 2\ \mathrm{W} - 3\ \mathrm{W}$$

$$+ 1.9600\sqrt{\frac{(0.25\ \mathrm{W})^2}{100} + \frac{(0.30\ \mathrm{W})^2}{150}}$$

$$= -0.9314\ \mathrm{W}\quad(-0.93\ \mathrm{W})$$

The answer is (B).

Equation 10.65: Confidence Limits and Interval for the Variance of a Normal Distribution

$$\frac{(n-1)s^2}{\chi_{\alpha/2,n-1}^2} \le \sigma^2 \le \frac{(n-1)s^2}{\chi_{1-\alpha/2,n-1}^2} \qquad 10.65$$

Description

Equation 10.65 gives the limits of a confidence interval (confidence $C = 1 - \alpha$) for an estimate of the population variance calculated as the sample variance from Eq. 10.26 with a sample size of n drawn from a normal distribution. Since the variance is a squared variable, it will be distributed as a chi-squared distribution with $n - 1$ degrees of freedom. Therefore, the denominators are the χ^2 values taken from Table 10.5 at the end of this chapter. (The values in Table 10.5 are already squared and should not be squared again.) Since the chi-squared distribution is not symmetrical, the table values for $\alpha/2$ and for $1 - (\alpha/2)$ will be different for the two confidence limits.

18. HYPOTHESIS TESTING

A *hypothesis test* is a procedure that answers the question, "Did these data come from [a particular type of] distribution?" There are many types of tests, depending on the distribution and parameter being evaluated. The most simple hypothesis test determines whether an average value obtained from n repetitions of an experiment could have come from a population with known mean μ and standard deviation σ. A practical application of this question is whether a manufacturing process has changed from what it used to be or should be. Of course, the answer (i.e., yes or no) cannot be given with absolute certainty—there will be a confidence level associated with the answer.

The following procedure is used to determine whether the average of n measurements can be assumed (with a given confidence level) to have come from a known normal population, or to determine the sample size required to make the decision with the desired confidence level.

Equation 10.66 Through Eq. 10.71: Test on Mean of Normal Distribution, Population Mean and Variance Known

step 1: Assume random sampling from a normal population.

The *null hypothesis* is

$$H_0: \mu = \mu_0 \qquad 10.66$$

The *alternative hypothesis* is

$$H_1: \mu = \mu_1 \qquad 10.67$$

A *type I error* is rejecting H_0 when it is true. The probability of a type I error is the *level of significance*.

$$\alpha = \mathrm{probability(type\ I\ error)} \qquad 10.68$$

A *type II error* is accepting H_0 when it is false.

$$\beta = \text{probability}(\text{type II error}) \qquad \textit{10.69}$$

step 2: Choose the desired confidence level, C.

step 3: Decide on a one-tail or two-tail test. If the hypothesis being tested is that the average has or has not *increased* or has not *decreased*, use a one-tail test. If the hypothesis being tested is that the average has or has not *changed*, use a two-tail test.

step 4: Use Table 10.3 or the unit normal table to determine the x-value corresponding to the confidence level and number of tails.

$$Z = \frac{\overline{X} - \mu_0}{\sigma/\sqrt{n}} \qquad \textit{10.70}$$

$$n = \left[\frac{Z_{\alpha/2}\sigma}{\overline{x} - \mu}\right]^2 \qquad \textit{10.71}$$

step 5: Calculate the actual standard normal variable, Z, from Eq. 10.70. The relationship of the sample size, n, and the actual standard normal variable is illustrated in Eq. 10.71.

step 6: If $Z \geq x$, the average can be assumed (with confidence level C) to have come from a different distribution.

Equation 10.72 Through Eq. 10.79: Sample Size for Normal Distribution, α and β Known

$$H_0: \mu = \mu_0 \qquad \textit{10.72}$$

$$H_1: \mu \neq \mu_0 \qquad \textit{10.73}$$

$$\beta = \Phi\left(\frac{\mu_0 - \mu}{\sigma/\sqrt{n}} + Z_{\alpha/2}\right) - \Phi\left(\frac{\mu_0 - \mu}{\sigma/\sqrt{n}} - Z_{\alpha/2}\right) \qquad \textit{10.74}$$

$$n \simeq \frac{(Z_{\alpha/2} + Z_\beta)^2 \sigma^2}{(\mu_1 - \mu_0)^2} \qquad \textit{10.75}$$

$$H_0: \mu = \mu_0 \qquad \textit{10.76}$$

$$H_1: \mu > \mu_0 \qquad \textit{10.77}$$

$$\beta = \Phi\left(\frac{\mu_0 - \mu}{\sigma/\sqrt{n}} + Z_\alpha\right) \qquad \textit{10.78}$$

$$n = \frac{(Z_\alpha + Z_\beta)^2 \sigma^2}{(\mu_1 - \mu_0)^2} \qquad \textit{10.79}$$

Description

Equation 10.72 through Eq. 10.79 are used to determine the required sample size when the probabilities of type 1 and type 2 errors, α and β, respectively, are known. μ_1 is the assumed true mean. The notation $\Phi(z)$ designates the cumulative normal distribution function (i.e., the fraction of the normal curve from $-\infty$ up to z).[5] Equation 10.72 through Eq. 10.75 are used when the test is to determine if the sample mean is the same as the population mean, while Eq. 10.76 through Eq. 10.79 are used when the test is to determine if the sample mean is larger or smaller than the population mean.

Example

When it is operating properly, a chemical plant has a daily production rate that is normally distributed with a mean of 880 tons/day and a standard deviation of 21 tons/day. During an analysis period, the output is measured with random sampling on 50 consecutive days, and the mean output is found to be 871 tons/day. With a 95% confidence level, determine if the plant is operating properly.

(A) There is at least a 5% probability that the plant is operating properly.

(B) There is at least a 95% probability that the plant is operating properly.

(C) There is at least a 5% probability that the plant is not operating properly.

(D) There is at least a 95% probability that the plant is not operating properly.

Solution

Since a specific direction in the variation is not given (i.e., the example does not ask if the average has decreased), use a two-tail hypothesis test.

From Table 10.3, $x = 1.9600$.

Use Eq. 10.70 to calculate the actual standard normal variable.

$$Z = \frac{\overline{X} - \mu}{\sigma/\sqrt{n}} = \frac{871 - 880}{\dfrac{21}{\sqrt{50}}} = -3.03$$

Since $-3.03 < 1.9600$, the distributions are not the same. There is at least a 95% probability that the plant is not operating correctly.

The answer is (D).

19. LINEAR REGRESSION

Equation 10.80 Through Eq. 10.86: Method of Least Squares

If it is necessary to draw a straight line $(y = \hat{a} + \hat{b}x)$ through n two-dimensional data points (x_1, y_1),

---

[5]Not only is $\Phi(z)$ undefined in the *NCEES Handbook*, but it is the same as what the *NCEES Handbook* designated earlier as $F(x)$.

$(x_2, y_2), \ldots, (x_n, y_n)$, the following method based on the *method of least squares* can be used.

step 1: Calculate the following seven quantities.

$$\sum x_i \quad \sum x_i^2 \quad \left(\sum x_i\right)^2 \quad \sum x_i y_i$$
$$\sum y_i \quad \sum y_i^2 \quad \left(\sum y_i\right)^2$$

$$\bar{x} = (1/n)\left(\sum_{i=1}^{n} x_i\right) \qquad 10.80$$

$$\bar{y} = (1/n)\left(\sum_{i=1}^{n} y_i\right) \qquad 10.81$$

step 2: Calculate the slope, $\hat{b}$, of the line.

$$\hat{b} = S_{xy}/S_{xx} \qquad 10.82$$

$$S_{xy} = \sum_{i=1}^{n} x_i y_i - (1/n)\left(\sum_{i=1}^{n} x_i\right)\left(\sum_{i=1}^{n} y_i\right) \qquad 10.83$$

$$S_{xx} = \sum_{i=1}^{n} x_i^2 - (1/n)\left(\sum_{i=1}^{n} x_i\right)^2 \qquad 10.84$$

step 3: Calculate the y-intercept, $\hat{a}$.

$$\hat{a} = \bar{y} - \hat{b}\bar{x} \qquad 10.85$$

The equation of the straight line is

$$y = \hat{a} + \hat{b}x \qquad 10.86$$

Example

The least squares method is used to plot a straight line through the data points $(1,6)$, $(2,7)$, $(3,11)$, and $(5,13)$. The slope of the line is most nearly

(A) 0.87
(B) 1.7
(C) 1.9
(D) 2.0

Solution

First, calculate the following values.

$$\sum x_i = 1 + 2 + 3 + 5 = 11$$
$$\sum y_i = 6 + 7 + 11 + 13 = 37$$
$$\sum x_i^2 = (1)^2 + (2)^2 + (3)^2 + (5)^2 = 39$$
$$\sum x_i y_i = (1)(6) + (2)(7) + (3)(11) + (5)(13) = 118$$

Find the value of S_{xy} using Eq. 10.83.

$$S_{xy} = \sum_{i=1}^{n} x_i y_i - (1/n)\left(\sum_{i=1}^{n} x_i\right)\left(\sum_{i=1}^{n} y_i\right)$$
$$= 118 - \left(\frac{1}{4}\right)(11)(37)$$
$$= 16.25$$

Find the value of S_{xx} from Eq. 10.84.

$$S_{xx} = \sum_{i=1}^{n} x_i^2 - (1/n)\left(\sum_{i=1}^{n} x_i\right)^2$$
$$= 39 - \left(\frac{1}{4}\right)(11)^2$$
$$= 8.75$$

From Eq. 10.82, the slope is

$$\hat{b} = S_{xy}/S_{xx}$$
$$= \frac{16.25}{8.75}$$
$$= 1.857 \quad (1.9)$$

The answer is (C).

Equation 10.87 and Eq. 10.88: Standard Error of Estimate

$$S_e^2 = \frac{S_{xx}S_{yy} - S_{xy}^2}{S_{xx}(n-2)} = MSE \qquad 10.87$$

$$S_{yy} = \sum_{i=1}^{n} y_i^2 - (1/n)\left(\sum_{i=1}^{n} y_i\right)^2 \qquad 10.88$$

Description

Equation 10.87 gives the *mean squared error*, S_e^2 or *MSE*, which estimates the likelihood of a value being close to an observed value by averaging the square of the errors (i.e., the difference between the estimated value and observed value). Small *MSE* values are favorable, as they indicate a smaller likelihood of error.

Equation 10.89 and Eq. 10.90: Confidence Intervals for Slope and Intercept

$$\hat{b} \pm t_{\alpha/2, n-2}\sqrt{\frac{MSE}{S_{xx}}} \qquad 10.89$$

$$\hat{a} \pm t_{\alpha/2, n-2}\sqrt{\left(\frac{1}{n} + \frac{\bar{x}^2}{S_{xx}}\right)MSE} \qquad 10.90$$

Description

The confidence intervals for calculated slope and intercept are calculated from the mean square error using Eq. 10.89 and Eq. 10.90, respectively.

Equation 10.91 and Eq. 10.92: Sample Correlation Coefficient

$$R = \frac{S_{xy}}{\sqrt{S_{xx}S_{yy}}} \qquad 10.91$$

$$R^2 = \frac{S_{xy}^2}{S_{xx}S_{yy}} \qquad 10.92$$

Description

Once the slope of the line is calculated using the least squares method, the *goodness of fit* can be determined by calculating the *sample correlation coefficient, R*. The goodness of fit describes how well the calculated regression values, plotted as a line, match actual observed values, plotted as points.

If $\hat{b}$ is positive, R will be positive; if $\hat{b}$ is negative, R will be negative. As a general rule, if the absolute value of R exceeds 0.85, the fit is good; otherwise, the fit is poor. R equals 1.0 if the fit is a perfect straight line.

Example

The least squares method is used to plot a straight line through the data points $(5, -5)$, $(3, -2)$, $(2, 3)$, and $(-1, 7)$. The correlation coefficient is most nearly

(A) -0.97

(B) -0.92

(C) -0.88

(D) -0.80

Solution

First, calculate the following values.

$$\sum x_i = 5 + 3 + 2 + (-1) = 9$$

$$\sum y_i = (-5) + (-2) + 3 + 7 = 3$$

$$\sum x_i^2 = (5)^2 + (3)^2 + (2)^2 + (-1)^2 = 39$$

$$\sum y_i^2 = (-5)^2 + (-2)^2 + (3)^2 + (7)^2 = 87$$

$$\sum x_i y_i = (5)(-5) + (3)(-2) + (2)(3) + (-1)(7) = -32$$

From Eq. 10.91, and substituting Eq. 10.83, Eq. 10.84, and Eq. 10.88 for S_{xy}, S_{xx}, and S_{yy}, respectively, the correlation coefficient is

$$R = \frac{S_{xy}}{\sqrt{S_{xx}S_{yy}}}$$

$$= \frac{\sum x_i y_i - (1/n)\left(\sum x_i\right)\left(\sum y_i\right)}{\sqrt{\left(\sum x_i^2 - (1/n)\left(\sum x_i\right)^2\right)\left(\sum y_i^2 - (1/n)\left(\sum y_i\right)^2\right)}}$$

$$= \frac{-32 - \left(\frac{1}{4}\right)(9)(3)}{\sqrt{\left(39 - \left(\frac{1}{4}\right)(9)^2\right)\left(87 - \left(\frac{1}{4}\right)(3)^2\right)}}$$

$$= -0.972 \quad (-0.97)$$

The answer is (A).

20. NONLINEAR REGRESSION

Even if linear regression using the least squares or another method fails to find a valid linear model for the data set, there may still be a nonlinear relationship between the variables. The data may be described by a parabolic, logarithmic, or other nonlinear relationship between the variables.

Sometimes the relationship can be *linearized* (i.e., made linear) by a *variable transformation*; that is, by appropriately changing one or both variables. For example, one or both variables may be transformed by taking squares, square roots, cubes, logarithms, or reciprocals.

A plot of the untransformed data will often provide a clue to the type of variable transformation needed. For example, in Fig. 10.6(a), the plot of untransformed data suggests an exponential relationship in which $y = ae^{bx}$, where a and b are the unknown constants. This relationship can be linearized, as in Fig. 10.6(b), by taking the logarithm (either base-10 or natural) of both sides. Figure 10.7 illustrates some of the most common variable transformations.

Example

The experimental (x, y) data point pairs $(0, 0.5)$, $(0.2, 1.4)$, $(0.5, 6.1)$, and $(0.9, 45.0)$ are represented by an exponential curve of the form $y = ae^{bx}$. Using least-squares linear regression, what is most nearly the value of exponent b?

(A) 1.0

(B) 2.0

(C) 4.0

(D) 5.0

Solution

Since $\ln(e^x) = x$, the data can be easily linearized by taking the natural logarithm of both sides of the equation. The linear equation for the transformed data is $\ln y = \ln a + bx$. In this form, the exponent, b, is the slope of the straight line. Transform the experimental data points

Figure 10.6 Nonlinear Data Curves

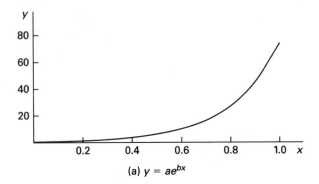

(a) $y = ae^{bx}$

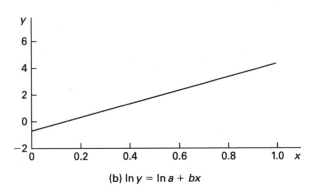

(b) $\ln y = \ln a + bx$

from (x, y) to $(x, \ln y)$ by taking the natural logarithm of the y values. For example, since $\ln(0.5) \approx -0.69$, the first transformed data pair becomes $(0, -0.69)$. The transformed data pair table is

x	y	$\ln y$
0	0.5	−0.69
0.2	1.4	0.34
0.5	6.1	1.81
0.9	45.0	3.81

Use the method of least squares to develop the equation of a straight line in the form of $y = bx + c$ describing the transformed values.

Calculate the following values for use in the least squares method.

$$\sum x_i = 0 + 0.2 + 0.5 + 0.9 = 1.6$$

$$\sum \ln y_i = -0.69 + 0.34 + 1.81 + 3.81 = 5.27$$
$$\sum x_i^2 = (0)^2 + (0.2)^2 + (0.5)^2 + (0.9)^2 = 1.1$$

$$\sum \ln y_i^2 = (-0.69)^2 + (0.34)^2 + (1.81)^2 + (3.81)^2$$
$$= 18.3839$$

$$\sum x_i \ln y_i = (0)(-0.69) + (0.2)(0.34)$$
$$+ (0.5)(1.81) + (0.9)(3.81)$$
$$= 4.402$$

Figure 10.7 Common Variable Transformations

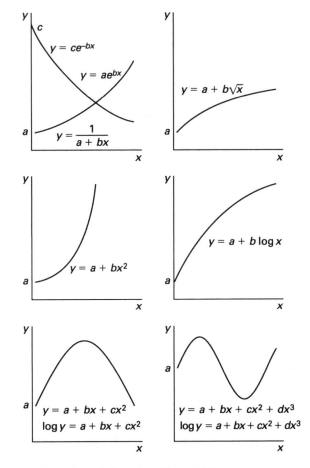

Find the value of S_{xy} using Eq. 10.83.

$$S_{xy} = \sum_{i=1}^{n} x_i \ln y_i - (1/n)\left(\sum_{i=1}^{n} x_i\right)\left(\sum_{i=1}^{n} \ln y_i\right)$$

$$= 4.402 - \left(\frac{1}{4}\right)(1.6)(5.27)$$

$$= 2.294$$

Find the value of S_{xx} using Eq. 10.84.

$$S_{xx} = \sum_{i=1}^{n} x_i^2 - (1/n)\left(\sum_{i=1}^{n} x_i\right)^2$$

$$= 1.1 - \left(\frac{1}{4}\right)(1.6)^2$$

$$= 0.46$$

The slope, b, of the equation $y = bx + c$ is found from Eq. 10.82. This value corresponds directly to the value of b in $y = ae^{bx}$.[6]

$$\hat{b} = S_{xy}/S_{xx} = \frac{2.294}{0.46} = 4.99 \quad (5.0)$$

The answer is (D).

---

[6]If base-10 logarithms had been used instead of natural logarithms, least-squares regression would produce a straight-line equation of $y = -0.295 + 2.162 \log_{10} x$. The slope of 2.162 corresponds to $b \log_{10} b / \ln b$, not b.

Table 10.1 Cumulative Binomial Probabilities $P(X \leq x)$

| | | \multicolumn{11}{c}{P} | | | | | | | | | |
n	x	0.1	0.2	0.3	0.4	0.5	0.6	0.7	0.8	0.9	0.95	0.99
1	0	0.9000	0.8000	0.7000	0.6000	0.5000	0.4000	0.3000	0.2000	0.1000	0.0500	0.0100
2	0	0.8100	0.6400	0.4900	0.3600	0.2500	0.1600	0.0900	0.0400	0.0100	0.0025	0.0001
	1	0.9900	0.9600	0.9100	0.8400	0.7500	0.6400	0.5100	0.3600	0.1900	0.0975	0.0199
3	0	0.7290	0.5120	0.3430	0.2160	0.1250	0.0640	0.0270	0.0080	0.0010	0.0001	0.0000
	1	0.9720	0.8960	0.7840	0.6480	0.5000	0.3520	0.2160	0.1040	0.0280	0.0073	0.0003
	2	0.9990	0.9920	0.9730	0.9360	0.8750	0.7840	0.6570	0.4880	0.2710	0.1426	0.0297
4	0	0.6561	0.4096	0.2401	0.1296	0.0625	0.0256	0.0081	0.0016	0.0001	0.0000	0.0000
	1	0.9477	0.8192	0.6517	0.4752	0.3125	0.1792	0.0837	0.0272	0.0037	0.0005	0.0000
	2	0.9963	0.9728	0.9163	0.8208	0.6875	0.5248	0.3483	0.1808	0.0523	0.0140	0.0006
	3	0.9999	0.9984	0.9919	0.9744	0.9375	0.8704	0.7599	0.5904	0.3439	0.1855	0.0394
5	0	0.5905	0.3277	0.1681	0.0778	0.0313	0.0102	0.0024	0.0003	0.0000	0.0000	0.0000
	1	0.9185	0.7373	0.5282	0.3370	0.1875	0.0870	0.0308	0.0067	0.0005	0.0000	0.0000
	2	0.9914	0.9421	0.8369	0.6826	0.5000	0.3174	0.1631	0.0579	0.0086	0.0012	0.0000
	3	0.9995	0.9933	0.9692	0.9130	0.8125	0.6630	0.4718	0.2627	0.0815	0.0226	0.0010
	4	1.0000	0.9997	0.9976	0.9898	0.6988	0.9222	0.8319	0.6723	0.4095	0.2262	0.0490
6	0	0.5314	0.2621	0.1176	0.0467	0.0156	0.0041	0.0007	0.0001	0.0000	0.0000	0.0000
	1	0.8857	0.6554	0.4202	0.2333	0.1094	0.0410	0.0109	0.0016	0.0001	0.0000	0.0000
	2	0.9842	0.9011	0.7443	0.5443	0.3438	0.1792	0.0705	0.0170	0.0013	0.0001	0.0000
	3	0.9987	0.9830	0.9295	0.8208	0.6563	0.4557	0.2557	0.0989	0.0159	0.0022	0.0000
	4	0.9999	0.9984	0.9891	0.9590	0.9806	0.7667	0.5798	0.3446	0.1143	0.0328	0.0015
	5	1.0000	0.9999	0.9993	0.9959	0.9844	0.9533	0.8824	0.7379	0.4686	0.2649	0.0585
7	0	0.4783	0.2097	0.0824	0.0280	0.0078	0.0106	0.0002	0.0000	0.0000	0.0000	0.0000
	1	0.8503	0.5767	0.3294	0.1586	0.0625	0.0188	0.0038	0.0004	0.0000	0.0000	0.0000
	2	0.9743	0.8520	0.6471	0.4199	0.2266	0.0963	0.0288	0.0047	0.0002	0.0000	0.0000
	3	0.9973	0.9667	0.8740	0.7102	0.5000	0.2898	0.1260	0.0333	0.0027	0.0002	0.0000
	4	0.9998	0.9953	0.9712	0.9037	0.7734	0.5801	0.3529	0.1480	0.0257	0.0038	0.0000
	5	1.0000	0.9996	0.9962	0.9812	0.9375	0.8414	0.6706	0.4233	0.1497	0.0444	0.0020
	6	1.0000	1.0000	0.9998	0.9984	0.9922	0.9720	0.9176	0.7903	0.5217	0.3017	0.0679
8	0	0.4305	0.1678	0.0576	0.0168	0.0039	0.0007	0.0001	0.0000	0.0000	0.0000	0.0000
	1	0.8131	0.5033	0.2553	0.1064	0.0352	0.0085	0.0013	0.0001	0.0000	0.0000	0.0000
	2	0.9619	0.7969	0.5518	0.3154	0.1445	0.0498	0.0113	0.0012	0.0000	0.0000	0.0000
	3	0.9950	0.9437	0.8059	0.5941	0.3633	0.1737	0.0580	0.0104	0.0004	0.0000	0.0000
	4	0.9996	0.9896	0.9420	0.8263	0.6367	0.4059	0.1941	0.0563	0.0050	0.0004	0.0000
	5	1.0000	0.9988	0.9887	0.9502	0.8555	0.6846	0.4482	0.2031	0.0381	0.0058	0.0001
	6	1.0000	0.9999	0.9987	0.9915	0.9648	0.8936	0.7447	0.4967	0.1869	0.0572	0.0027
	7	1.0000	1.0000	0.9999	0.9993	0.9961	0.9832	0.9424	0.8322	0.5695	0.3366	0.0773
9	0	0.3874	0.1342	0.0404	0.0101	0.0020	0.0003	0.0000	0.0000	0.0000	0.0000	0.0000
	1	0.7748	0.4362	0.1960	0.0705	0.0195	0.0038	0.0004	0.0000	0.0000	0.0000	0.0000
	2	0.9470	0.7382	0.4628	0.2318	0.0889	0.0250	0.0043	0.0003	0.0000	0.0000	0.0000
	3	0.9917	0.9144	0.7297	0.4826	0.2539	0.0994	0.0253	0.0031	0.0001	0.0000	0.0000
	4	0.9991	0.9804	0.9012	0.7334	0.5000	0.2666	0.0988	0.0196	0.0009	0.0000	0.0000
	5	0.9999	0.9969	0.9747	0.9006	0.7461	0.5174	0.2703	0.0856	0.0083	0.0006	0.0000
	6	1.0000	0.9997	0.9957	0.9750	0.9102	0.7682	0.5372	0.2618	0.0530	0.0084	0.0001
	7	1.0000	1.0000	0.9996	0.9962	0.9805	0.9295	0.8040	0.5638	0.2252	0.0712	0.0034
	8	1.0000	1.0000	1.0000	0.9997	0.9980	0.9899	0.9596	0.8658	0.6126	0.3698	0.0865
10	0	0.3487	0.1074	0.0282	0.0060	0.0010	0.0001	0.0000	0.0000	0.0000	0.0000	0.0000
	1	0.7361	0.3758	0.1493	0.0464	0.0107	0.0017	0.0001	0.0000	0.0000	0.0000	0.0000
	2	0.9298	0.6778	0.3828	0.1673	0.0547	0.0123	0.0016	0.0001	0.0000	0.0000	0.0000
	3	0.9872	0.8791	0.6496	0.3823	0.1719	0.0548	0.0106	0.0009	0.0000	0.0000	0.0000
	4	0.9984	0.9672	0.8497	0.6331	0.3770	0.1662	0.0473	0.0064	0.0001	0.0000	0.0000
	5	0.9999	0.9936	0.9527	0.8338	0.6230	0.3669	0.1503	0.0328	0.0016	0.0001	0.0000
	6	1.0000	0.9991	0.9894	0.9452	0.8281	0.6177	0.3504	0.1209	0.0128	0.0010	0.0000
	7	1.0000	0.9999	0.9984	0.9877	0.9453	0.8327	0.6172	0.3222	0.0702	0.0115	0.0001
	8	1.0000	1.0000	0.9999	0.9983	0.9893	0.9536	0.8507	0.6242	0.2639	0.0861	0.0043
	9	1.0000	1.0000	1.0000	0.9999	0.9990	0.9940	0.9718	0.8926	0.6513	0.4013	0.0956

Probability/ Statistics

Table 10.2 *Unit Normal Distribution*

x	$f(x)$	$F(x)$	$R(x)$	$2R(x)$	$W(x)$
0.0	0.3989	0.5000	0.5000	1.0000	0.0000
0.1	0.3970	0.5398	0.4602	0.9203	0.0797
0.2	0.3910	0.5793	0.4207	0.8415	0.1585
0.3	0.3814	0.6179	0.3821	0.7642	0.2358
0.4	0.3683	0.6554	0.3446	0.6892	0.3108
0.5	0.3521	0.6915	0.3085	0.6171	0.3829
0.6	0.3332	0.7257	0.2743	0.5485	0.4515
0.7	0.3123	0.7580	0.2420	0.4839	0.5161
0.8	0.2897	0.7881	0.2119	0.4237	0.5763
0.9	0.2661	0.8159	0.1841	0.3681	0.6319
1.0	0.2420	0.8413	0.1587	0.3173	0.6827
1.1	0.2179	0.8643	0.1357	0.2713	0.7287
1.2	0.1942	0.8849	0.1151	0.2301	0.7699
1.3	0.1714	0.9032	0.0968	0.1936	0.8064
1.4	0.1497	0.9192	0.0808	0.1615	0.8385
1.5	0.1295	0.9332	0.0668	0.1336	0.8664
1.6	0.1109	0.9452	0.0548	0.1096	0.8904
1.7	0.0940	0.9554	0.0446	0.0891	0.9109
1.8	0.0790	0.9641	0.0359	0.0719	0.9281
1.9	0.0656	0.9713	0.0287	0.0574	0.9426
2.0	0.0540	0.9772	0.0228	0.0455	0.9545
2.1	0.0440	0.9821	0.0179	0.0357	0.9643
2.2	0.0355	0.9861	0.0139	0.0278	0.9722
2.3	0.0283	0.9893	0.0107	0.0214	0.9786
2.4	0.0224	0.9918	0.0082	0.0164	0.9836
2.5	0.0175	0.9938	0.0062	0.0124	0.9876
2.6	0.0136	0.9953	0.0047	0.0093	0.9907
2.7	0.0104	0.9965	0.0035	0.0069	0.9931
2.8	0.0079	0.9974	0.0026	0.0051	0.9949
2.9	0.0060	0.9981	0.0019	0.0037	0.9963
3.0	0.0044	0.9987	0.0013	0.0027	0.9973
Fractiles					
1.2816	0.1755	0.9000	0.1000	0.2000	0.8000
1.6449	0.1031	0.9500	0.0500	0.1000	0.9000
1.9600	0.0584	0.9750	0.0250	0.0500	0.9500
2.0537	0.0484	0.9800	0.0200	0.0400	0.9600
2.3263	0.0267	0.9900	0.0100	0.0200	0.9800
2.5758	0.0145	0.9950	0.0050	0.0100	0.9900

Table 10.3 *Values of x for Various Two-Tail Confidence Intervals*

confidence interval level, C	two-tail limit, x $(Z_{\alpha/2})$
80%	1.2816
90%	1.6449
95%	1.9600
96%	2.0537
98%	2.3263
99%	2.5758

Table 10.4 Student's t-Distribution (values of t for ν degrees of freedom (sample size n + 1); 1 − α confidence level)

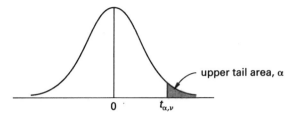

area under the upper tail

ν^*	$\alpha = 0.25$	$\alpha = 0.20$	$\alpha = 0.15$	$\alpha = 0.10$	$\alpha = 0.05$	$\alpha = 0.025$	$\alpha = 0.01$	$\alpha = 0.005$	ν^*
1	1.000	1.376	1.963	3.078	6.314	12.706	31.821	63.657	1
2	0.816	1.061	1.386	1.886	2.920	4.303	6.965	9.925	2
3	0.765	0.978	1.350	1.638	2.353	3.182	4.541	5.841	3
4	0.741	0.941	1.190	1.533	2.132	2.776	3.747	4.604	4
5	0.727	0.920	1.156	1.476	2.015	2.571	3.365	4.032	5
6	0.718	0.906	1.134	1.440	1.943	2.447	3.143	3.707	6
7	0.711	0.896	1.119	1.415	1.895	2.365	2.998	3.499	7
8	0.706	0.889	1.108	1.397	1.860	2.306	2.896	3.355	8
9	0.703	0.883	1.100	1.383	1.833	2.262	2.821	3.250	9
10	0.700	0.879	1.093	1.372	1.812	2.228	2.764	3.169	10
11	0.697	0.876	1.088	1.363	1.796	2.201	2.718	3.106	11
12	0.695	0.873	1.083	1.356	1.782	2.179	2.681	3.055	12
13	0.694	0.870	1.079	1.350	1.771	2.160	2.650	3.012	13
14	0.692	0.868	1.076	1.345	1.761	2.145	2.624	2.977	14
15	0.691	0.866	1.074	1.341	1.753	2.131	2.602	2.947	15
16	0.690	0.865	1.071	1.337	1.746	2.120	2.583	2.921	16
17	0.689	0.863	1.069	1.333	1.740	2.110	2.567	2.898	17
18	0.688	0.862	1.067	1.330	1.734	2.101	2.552	2.878	18
19	0.688	0.861	1.066	1.328	1.729	2.093	2.539	2.861	19
20	0.687	0.860	1.064	1.325	1.725	2.086	2.528	2.845	20
21	0.686	0.859	1.063	1.323	1.721	2.080	2.518	2.831	21
22	0.686	0.858	1.061	1.321	1.717	2.074	2.508	2.819	22
23	0.685	0.858	1.060	1.319	1.714	2.069	2.500	2.807	23
24	0.685	0.857	1.059	1.318	1.711	2.064	2.492	2.797	24
25	0.684	0.856	1.058	1.316	1.708	2.060	2.485	2.787	25
26	0.684	0.856	1.058	1.315	1.706	2.056	2.479	2.779	26
27	0.684	0.855	1.057	1.314	1.703	2.052	2.473	2.771	27
28	0.683	0.855	1.056	1.313	1.701	2.048	2.467	2.763	28
29	0.683	0.854	1.055	1.311	1.699	2.045	2.462	2.756	29
30	0.683	0.854	1.055	1.310	1.697	2.042	2.457	2.750	30
∞	0.674	0.842	1.036	1.282	1.645	1.960	2.326	2.576	∞

[*]The number of independent degrees of freedom, ν, is always one less than the sample size, n.

Table 10.5 *Critical Values of Chi-Squared Distribution*

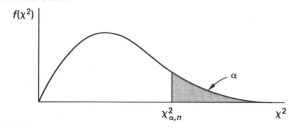

degrees of freedom, ν	$\chi^2_{0.995}$	$\chi^2_{0.990}$	$\chi^2_{0.975}$	$\chi^2_{0.950}$	$\chi^2_{0.900}$	$\chi^2_{0.100}$	$\chi^2_{0.050}$	$\chi^2_{0.025}$	$\chi^2_{0.010}$	$\chi^2_{0.005}$
1	0.0000393	0.0001571	0.0009821	0.0039321	0.0157908	2.70554	3.84146	5.02389	6.6349	7.87944
2	0.0100251	0.0201007	0.0506356	0.102587	0.21072	4.60517	5.99147	7.37776	9.21034	10.5966
3	0.0717212	0.114832	0.215795	0.351846	0.584375	6.25139	7.81473	9.3484	11.3449	12.8381
4	0.20699	0.29711	0.484419	0.710721	1.063623	7.77944	9.48773	11.1433	13.2767	14.8602
5	0.41174	0.5543	0.831211	1.145476	1.61031	9.23635	11.0705	12.8325	15.0863	16.7496
6	0.675727	0.872085	1.237347	1.63539	2.20413	10.6446	12.5916	14.4494	16.8119	18.5476
7	0.989265	1.239043	1.68987	2.16735	2.83311	12.017	14.0671	16.0128	18.4753	20.2777
8	1.344419	1.646482	2.17973	2.73264	3.48954	13.3616	15.5073	17.5346	20.0902	21.955
9	1.734926	2.087912	2.70039	3.32511	4.16816	14.6837	16.919	19.0228	21.666	23.5893
10	2.15585	2.55821	3.24697	3.9403	4.86518	15.9871	18.307	20.4831	23.2093	25.1882
11	2.60321	3.05347	3.81575	4.57481	5.57779	17.275	19.6751	21.92	24.725	26.7569
12	3.07382	3.57056	4.40379	5.22603	6.3038	18.5494	21.0261	23.3367	26.217	28.2995
13	3.56503	4.10691	5.00874	5.89186	7.0415	19.8119	22.3621	24.7356	27.6883	29.8194
14	4.07468	4.66043	5.62872	6.57063	7.78953	21.0642	23.6848	26.119	29.1413	31.3193
15	4.60094	5.22935	6.26214	7.26094	8.54675	22.3072	24.9958	27.4884	30.5779	32.8013
16	5.14224	5.81221	6.90766	7.96164	9.31223	23.5418	26.2962	28.8454	31.9999	34.2672
17	5.69724	6.40776	7.56418	8.67176	10.0852	24.769	27.5871	30.191	33.4087	35.7185
18	6.26481	7.01491	8.23075	9.39046	10.8649	25.9894	28.8693	31.5264	34.8053	37.1564
19	6.84398	7.63273	8.90655	10.117	11.6509	27.2036	30.1435	32.8523	36.1908	38.5822
20	7.43386	8.2604	9.59083	10.8508	12.4426	28.412	31.4104	34.1696	37.5662	39.9968
21	8.03366	8.8972	10.28293	11.5913	13.2396	29.6151	32.6705	35.4789	38.9321	41.401
22	8.64272	9.54249	10.9823	12.338	14.0415	30.8133	33.9244	36.7807	40.2894	42.7956
23	9.26042	10.19567	11.6885	13.0905	14.8479	32.0069	35.1725	38.0757	41.6384	44.1813
24	9.88623	10.8564	12.4011	13.8484	15.6587	33.1963	36.4151	39.3641	42.9798	45.5585
25	10.5197	11.524	13.1197	14.6114	16.4734	34.3816	37.6525	40.6465	44.3141	46.9278
26	11.1603	12.1981	13.8439	15.3791	17.2919	35.5631	38.8852	41.9232	45.6417	48.2899
27	11.8076	12.8786	14.5733	16.1513	18.1138	36.7412	40.1133	43.1944	46.963	49.6449
28	12.4613	13.5648	15.3079	16.9279	18.9392	37.9159	41.3372	44.4607	48.2782	50.9933
29	13.1211	14.2565	16.0471	17.7083	19.7677	39.0875	42.5569	45.7222	49.5879	52.3356
30	13.7867	14.9535	16.7908	18.4926	20.5992	40.256	43.7729	46.9792	50.8922	53.672
40	20.7065	22.1643	24.4331	26.5093	29.0505	51.805	55.7585	59.3417	63.6907	66.7659
50	27.9907	29.7067	32.3574	34.7642	37.6886	63.1671	67.5048	71.4202	76.1539	79.49
60	35.5346	37.4848	40.4817	43.1879	46.4589	74.397	79.0819	83.2976	88.3794	91.9517
70	43.2752	45.4418	48.7576	51.7393	55.329	85.5271	90.5312	95.0231	100.425	104.215
80	51.172	53.54	57.1532	60.3915	64.2778	96.5782	101.879	106.629	112.329	116.321
90	59.1963	61.7541	65.6466	69.126	73.2912	107.565	113.145	118.136	124.116	128.299
100	67.3276	70.0648	74.2219	77.9295	82.3581	118.498	124.342	129.561	135.807	140.169

Table 10.6 Critical Values of F

For a particular combination of numerator and denominator degrees of freedom, entry represents the critical values of F corresponding to a specified upper tail area (α).

df_2	1	2	3	4	5	6	7	8	9	10	12	15	20	24	30	40	60	120	∞
1	161.4	199.5	215.7	224.6	230.2	234.0	236.8	238.9	240.5	241.9	243.9	245.9	248.0	249.1	250.1	251.1	252.2	253.3	254.3
2	18.51	19.00	19.16	19.25	19.30	19.33	19.35	19.37	19.38	19.40	19.41	19.43	19.45	19.45	19.46	19.47	19.48	19.49	19.50
3	10.13	9.55	9.28	9.12	9.01	8.94	8.89	8.85	8.81	8.79	8.74	8.70	8.66	8.64	8.62	8.59	8.57	8.55	8.53
4	7.71	6.94	6.59	6.39	6.26	6.16	6.09	6.04	6.00	5.96	5.91	5.86	5.80	5.77	5.75	5.72	5.69	5.66	5.63
5	6.61	5.79	5.41	5.19	5.05	4.95	4.88	4.82	4.77	4.74	4.68	4.62	4.56	4.53	4.50	4.46	4.43	4.40	4.36
6	5.99	5.14	4.76	4.53	4.39	4.28	4.21	4.15	4.10	4.06	4.00	3.94	3.87	3.84	3.81	3.77	3.74	3.70	3.67
7	5.59	4.74	4.35	4.12	3.97	3.87	3.79	3.73	3.68	3.64	3.57	3.51	3.44	3.41	3.38	3.34	3.30	3.27	3.23
8	5.32	4.46	4.07	3.84	3.69	3.58	3.50	3.44	3.39	3.35	3.28	3.22	3.15	3.12	3.08	3.04	3.01	2.97	2.93
9	5.12	4.26	3.86	3.63	3.48	3.37	3.29	3.23	3.18	3.14	3.07	3.01	2.94	2.90	2.86	2.83	2.79	2.75	2.71
10	4.96	4.10	3.71	3.48	3.33	3.22	3.14	3.07	3.02	2.98	2.91	2.85	2.77	2.74	2.70	2.66	2.62	2.58	2.54
11	4.84	3.98	3.59	3.36	3.20	3.09	3.01	2.95	2.90	2.85	2.79	2.72	2.65	2.61	2.57	2.53	2.49	2.45	2.40
12	4.75	3.89	3.49	3.26	3.11	3.00	2.91	2.85	2.80	2.75	2.69	2.62	2.54	2.51	2.47	2.43	2.38	2.34	2.30
13	4.67	3.81	3.41	3.18	3.03	2.92	2.83	2.77	2.71	2.67	2.60	2.53	2.46	2.42	2.38	2.34	2.30	2.25	2.21
14	4.60	3.74	3.34	3.11	2.96	2.85	2.76	2.70	2.65	2.60	2.53	2.46	2.39	2.35	2.31	2.27	2.22	2.18	2.13
15	4.54	3.68	3.29	3.06	2.90	2.79	2.71	2.64	2.59	2.54	2.48	2.40	2.33	2.29	2.25	2.20	2.16	2.11	2.07
16	4.49	3.63	3.24	3.01	2.85	2.74	2.66	2.59	2.54	2.49	2.42	2.35	2.28	2.24	2.19	2.15	2.11	2.06	2.01
17	4.45	3.59	3.20	2.96	2.81	2.70	2.61	2.55	2.49	2.45	2.38	2.31	2.23	2.19	2.15	2.10	2.06	2.01	1.96
18	4.41	3.55	3.16	2.93	2.77	2.66	2.58	2.51	2.46	2.41	2.34	2.27	2.19	2.15	2.11	2.06	2.02	1.97	1.92
19	4.38	3.52	3.13	2.90	2.74	2.63	2.54	2.48	2.42	2.38	2.31	2.23	2.16	2.11	2.07	2.03	1.98	1.93	1.88
20	4.35	3.49	3.10	2.87	2.71	2.60	2.51	2.45	2.39	2.35	2.28	2.20	2.12	2.08	2.04	1.99	1.95	1.90	1.84
21	4.32	3.47	3.07	2.84	2.68	2.57	2.49	2.42	2.37	2.32	2.25	2.18	2.10	2.05	2.01	1.96	1.92	1.87	1.81
22	4.30	3.44	3.05	2.82	2.66	2.55	2.46	2.40	2.34	2.30	2.23	2.15	2.07	2.03	1.98	1.94	1.89	1.84	1.78
23	4.28	3.42	3.03	2.80	2.64	2.53	2.44	2.37	2.32	2.27	2.20	2.13	2.05	2.01	1.96	1.91	1.86	1.81	1.76
24	4.26	3.40	3.01	2.78	2.62	2.51	2.42	2.36	2.30	2.25	2.18	2.11	2.03	1.98	1.94	1.89	1.84	1.79	1.73
25	4.24	3.39	2.99	2.76	2.60	2.49	2.40	2.34	2.28	2.24	2.16	2.09	2.01	1.96	1.92	1.87	1.82	1.77	1.71
26	4.23	3.37	2.98	2.74	2.59	2.47	2.39	2.32	2.27	2.22	2.15	2.07	1.99	1.95	1.90	1.85	1.80	1.75	1.69
27	4.21	3.35	2.96	2.73	2.57	2.46	2.37	2.31	2.25	2.20	2.13	2.06	1.97	1.93	1.88	1.84	1.79	1.73	1.67
28	4.20	3.34	2.95	2.71	2.56	2.45	2.36	2.29	2.24	2.19	2.12	2.04	1.96	1.91	1.87	1.82	1.77	1.71	1.65
29	4.18	3.33	2.93	2.70	2.55	2.43	2.35	2.28	2.22	2.18	2.10	2.03	1.94	1.90	1.85	1.81	1.75	1.70	1.64
30	4.17	3.32	2.92	2.69	2.53	2.42	2.33	2.27	2.21	2.16	2.09	2.01	1.93	1.89	1.84	1.79	1.74	1.68	1.62
40	4.08	3.23	2.84	2.61	2.45	2.34	2.25	2.18	2.12	2.08	2.00	1.92	1.84	1.79	1.74	1.69	1.64	1.58	1.51
60	4.00	3.15	2.76	2.53	2.37	2.25	2.17	2.10	2.04	1.99	1.92	1.84	1.75	1.70	1.65	1.59	1.53	1.47	1.39
120	3.92	3.07	2.68	2.45	2.29	2.17	2.09	2.02	1.96	1.91	1.83	1.75	1.66	1.61	1.55	1.50	1.43	1.35	1.25
∞	3.84	3.00	2.60	2.37	2.21	2.10	2.01	1.94	1.88	1.83	1.75	1.67	1.57	1.52	1.46	1.39	1.32	1.22	1.00

denominator df_2 / numerator df_1

$F(\alpha, df_1, df_2)$ — $\alpha = 0.05$

Probability/Statistics

Diagnostic Exam

Topic III: Fluid Mechanics/Dynamics

1. Oil flows through a 0.12 m diameter pipe at a velocity of 1 m/s. The density and the dynamic viscosity of the oil are 870 kg/m^3 and 0.082 kg/s·m^2, respectively. If the pipe length is 100 m, the head loss due to friction is most nearly

(A) 1.2 m

(B) 1.5 m

(C) 1.8 m

(D) 2.1 m

2. A thin metal disc of mass 0.01 kg is kept balanced by a jet of air, as shown.

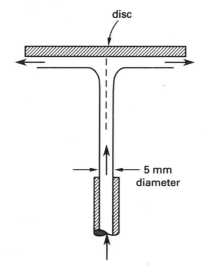

The diameter of the jet at the nozzle exit is 5 mm. Assuming atmospheric conditions at 101.3 kPa and 20°C, the velocity of the jet as it leaves the nozzle is most nearly

(A) 45 m/s

(B) 65 m/s

(C) 85 m/s

(D) 95 m/s

3. Water flows at 14 m^3/s in a 6 m wide rectangular open channel. The critical velocity is most nearly

(A) 0.82 m/s

(B) 1.8 m/s

(C) 2.8 m/s

(D) 14 m/s

4. To measure low flow rates of air, a laminar flow meter is used. It consists of a large number of small-diameter tubes in parallel. One design uses 4000 tubes, each with an inside diameter of 2 mm and a length of 25 cm. The pressure difference through the flow meter is 0.5 kPa, and the absolute viscosity of the air is 1.81×10^{-8} kPa·s. The flow rate of atmospheric air at 20°C is most nearly

(A) 0.1 m^3/s

(B) 0.2 m^3/s

(C) 0.4 m^3/s

(D) 0.5 m^3/s

5. Carbon tetrachloride has a specific gravity of 1.56. The height of a column of carbon tetracholoride that supports a pressure of 1 kPa is most nearly

(A) 0.0065 cm

(B) 6.5 cm

(C) 10 cm

(D) 64 cm

6. A model of a dam has been constructed so that the scale of dam to model is 15:1. The similarity is based on Froude numbers. At a certain point on the spillway of the model, the velocity is 5 m/s. At the corresponding point on the spillway of the actual dam, the velocity would most nearly be

(A) 6.7 m/s

(B) 7.5 m/s

(C) 15 m/s

(D) 19 m/s

7. A 10 cm diameter sphere floats half submerged in 20°C water. The density of water at 20°C is 998 kg/m^3. The mass of the sphere is most nearly

(A) 0.26 kg

(B) 0.52 kg

(C) 0.80 kg

(D) 2.6 kg

8. From the illustration shown, what is most nearly the pressure difference between tanks A and B?

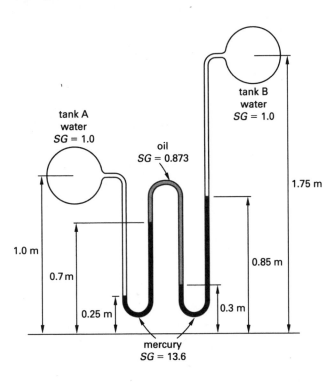

(A) 110 kPa

(B) 120 kPa

(C) 130 kPa

(D) 140 kPa

9. Water flows through a horizontal, frictionless pipe with an inside diameter of 20 cm as shown. A pitot-static meter measures the flow. The deflection of the mercury manometer attached to the pitot tube is 5 cm. The specific gravity of mercury is 13.6.

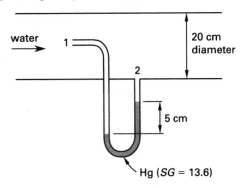

The flow rate in the pipe is most nearly

(A) 0.08 m³/s

(B) 0.1 m³/s

(C) 0.2 m³/s

(D) 0.3 m³/s

10. A horizontal pipe 10 cm in diameter carries 0.05 m³/s of water to a nozzle, through which the water exits to atmospheric pressure. The exit diameter of the nozzle is 4 cm. Losses through the nozzle are negligible. The pressure at the entrance to the nozzle is most nearly

(A) 420 kPa

(B) 560 kPa

(C) 680 kPa

(D) 770 kPa

SOLUTIONS

1. The Reynolds number is

$$\mathrm{Re} = \frac{\mathrm{v}D\rho}{\mu} = \frac{\left(1\ \frac{\mathrm{m}}{\mathrm{s}}\right)(0.12\ \mathrm{m})\left(870\ \frac{\mathrm{kg}}{\mathrm{m}^3}\right)}{0.082\ \frac{\mathrm{kg}}{\mathrm{s\cdot m}^2}}$$

$$= 1273$$

Since Re < 2300, the flow is laminar.

$$f = \frac{64}{\mathrm{Re}} = \frac{64}{1273}$$

$$= 0.05027$$

The head loss is

$$h_f = f\frac{L}{D}\frac{\mu_m^2}{2g} = (0.05027)\left(\frac{100\ \mathrm{m}}{0.12\ \mathrm{m}}\right)\left(\frac{\left(1\ \frac{\mathrm{m}}{\mathrm{s}}\right)^2}{(2)\left(9.81\ \frac{\mathrm{m}}{\mathrm{s}^2}\right)}\right)$$

$$= 2.135\ \mathrm{m} \quad (2.1\ \mathrm{m})$$

The answer is (D).

2. Applying the momentum equation in the vertical direction, the weight of the disc is equal to the rate of change of momentum of the air jet.

$$mg = \rho A\mathrm{v}^2$$

The specific gas constant for air is 0.2870 kJ/kg·K. The density of the air at the nozzle exit is

$$\rho = \frac{p}{RT}$$

$$= \frac{101.3\ \mathrm{kPa}}{\left(0.2870\ \frac{\mathrm{kJ}}{\mathrm{kg\cdot K}}\right)(20°\mathrm{C} + 273°)}$$

$$= 1.205\ \mathrm{kg/m}^3$$

The velocity of the air jet is

$$\mathrm{v} = \sqrt{\frac{mg}{\rho A}} = \sqrt{\frac{mg}{\rho\left(\frac{\pi D^2}{4}\right)}}$$

$$= \sqrt{\frac{(0.01\ \mathrm{kg})\left(9.81\ \frac{\mathrm{m}}{\mathrm{s}^2}\right)}{\left(1.205\ \frac{\mathrm{kg}}{\mathrm{m}^3}\right)\left(\frac{\pi(5\ \mathrm{mm})^2}{(4)\left(1000\ \frac{\mathrm{mm}}{\mathrm{m}}\right)^2}\right)}}$$

$$= 64.39\ \mathrm{m/s} \quad (65\ \mathrm{m/s})$$

The answer is (B).

3. Find the critical depth.

$$y_c = \left(\frac{q^2}{g}\right)^{1/3} = \sqrt[3]{\frac{Q^2}{gw^2}}$$

$$= \sqrt[3]{\frac{\left(14\ \frac{\mathrm{m}^3}{\mathrm{s}}\right)^2}{\left(9.81\ \frac{\mathrm{m}}{\mathrm{s}^2}\right)(6\ \mathrm{m})^2}}$$

$$= 0.822\ \mathrm{m}$$

The critical velocity is the velocity that makes the Froude number equal to one when the characteristic length, y_h, is equal to the critical depth.

$$\mathrm{Fr} = \frac{\mathrm{v}}{\sqrt{gy_h}} = \frac{\mathrm{v}}{\sqrt{gy_c}} = 1$$

$$\mathrm{v} = \sqrt{gy_c}$$

$$= \sqrt{\left(9.81\ \frac{\mathrm{m}}{\mathrm{s}^2}\right)(0.822\ \mathrm{m})}$$

$$= 2.84\ \mathrm{m/s} \quad (2.8\ \mathrm{m/s})$$

The answer is (C).

4. For laminar flow in a circular pipe, the flow rate can be calculated with the Hagen-Poiseuille equation. The flow in one tube is

$$Q = \frac{\pi D^4 \Delta p_f}{128\mu L}$$

The flow in N tubes, then, is

$$Q_N = \frac{\pi D^4 \Delta p_f N}{128\mu L}$$

At 20°C and 1 atm (101.3 kPa), the density of the air is

$$\rho = \frac{p}{RT} = \frac{101.3\ \mathrm{kPa}}{\left(0.2870\ \frac{\mathrm{kJ}}{\mathrm{kg\cdot K}}\right)(20°\mathrm{C} + 273°)}$$

$$= 1.205\ \mathrm{kg/m}^3$$

The flow rate of the 20°C atmospheric air is

$$Q_N = \frac{\pi D^4 \Delta p_f N}{128\mu L}$$

$$= \frac{\pi(2\ \mathrm{mm})^4(0.5\ \mathrm{kPa})(4000)\left(100\ \frac{\mathrm{cm}}{\mathrm{m}}\right)}{(128)(1.81 \times 10^{-8}\ \mathrm{kPa\cdot s})(25\ \mathrm{cm})\left(1000\ \frac{\mathrm{mm}}{\mathrm{m}}\right)^4}$$

$$= 0.174\ \mathrm{m}^3/\mathrm{s} \quad (0.2\ \mathrm{m}^3/\mathrm{s})$$

To confirm that the flow is laminar, calculate the Reynolds number. The velocity of the airflow is

$$v = \frac{Q}{A} = \frac{Q}{N\left(\frac{\pi D_{\text{tube}}^2}{4}\right)}$$

$$= \frac{0.174 \ \frac{m^3}{s}}{(4000)\left(\frac{\pi(2 \ mm)^2}{(4)\left(1000 \ \frac{mm}{m}\right)^2}\right)}$$

$$= 13.81 \ m/s$$

The Reynolds number is

$$\text{Re} = \frac{v D \rho}{\mu} = \frac{\left(13.81 \ \frac{m}{s}\right)(2 \ mm)\left(1.205 \ \frac{kg}{m^3}\right)}{(1.81 \times 10^{-5} \ \text{Pa·s})\left(1000 \ \frac{mm}{m}\right)}$$

$$= 1839$$

The Reynolds number is less than 2300, so the flow is laminar, and the device is suitable to measure the flow. The calculated airflow rate, 0.2 m³/s, is correct.

The answer is (B).

5. Use the relationship between pressure, density, and fluid depth, and solve for the column height.

$$p = \rho g h$$

$$h = \frac{p}{\rho g} = \frac{p}{SG\rho_{\text{water}} g} = \frac{\left(1000 \ \frac{N}{m^2}\right)\left(100 \ \frac{cm}{m}\right)}{(1.56)\left(1000 \ \frac{kg}{m^3}\right)\left(9.81 \ \frac{m}{s^2}\right)}$$

$$= 6.53 \ cm \quad (6.5 \ cm)$$

The answer is (B).

6. The Froude numbers must be equal.

$$\text{Fr}_{\text{dam}} = \text{Fr}_{\text{model}}$$

$$\frac{v_{\text{dam}}}{\sqrt{g y_{h,\text{dam}}}} = \frac{v_{\text{model}}}{\sqrt{g y_{h,\text{model}}}}$$

$$v_{\text{dam}} = v_{\text{model}}\sqrt{\frac{y_{h,\text{dam}}}{y_{h,\text{model}}}} = \left(5 \ \frac{m}{s}\right)\sqrt{\frac{15}{1}}$$

$$= 19.36 \ m/s \quad (19 \ m/s)$$

The answer is (D).

7. The volume of the sphere is

$$V_{\text{sphere}} = \frac{\pi D^3}{6} = \frac{\pi(10 \ cm)^3}{(6)\left(100 \ \frac{cm}{m}\right)^3} = 0.0005236 \ m^3$$

The buoyant force is equal to the weight of the entire sphere, and also equal to the weight of the displaced water.

$$F_b = W_{\text{sphere}} = W_{\text{displaced}}$$

$$m_{\text{sphere}} g = m_{\text{displaced}} g$$

Dividing both weights by g gives

$$m_{\text{sphere}} = m_{\text{displaced}}$$

$$= \rho_{\text{water}} V_{\text{displaced}}$$

The volume of the displaced water is equal to half the volume of the sphere.

$$V_{\text{displaced}} = \frac{V_{\text{sphere}}}{2}$$

$$m_{\text{sphere}} = \rho_{\text{water}}\left(\frac{V_{\text{sphere}}}{2}\right)$$

$$= \left(1000 \ \frac{kg}{m^3}\right)\left(\frac{0.0005236 \ m^3}{2}\right)$$

$$= 0.262 \ kg \quad (0.26 \ kg)$$

The answer is (A).

8. The pressures in tanks A and B are related by the equation

$$p_A + \gamma_w h_1 - \gamma_{\text{Hg}} h_2 + \gamma_{\text{oil}} h_3 - \gamma_{\text{Hg}} h_4 - \gamma_w h_5 = p_B$$

The heights are

$$h_1 = 1.0 \ m - 0.25 \ m = 0.75 \ m$$

$$h_2 = 0.7 \ m - 0.25 \ m = 0.45 \ m$$

$$h_3 = 0.7 \ m - 0.3 \ m = 0.4 \ m$$

$$h_4 = 0.85 \ m - 0.3 \ m = 0.55 \ m$$

$$h_5 = 1.75 \ m - 0.85 \ m = 0.90 \ m$$

Also, $\gamma_w = \rho_w g$. Rearrange to solve for the difference in pressures.

$$p_A - p_B = \rho_w g(h_5 - h_1) + \gamma_{\text{Hg}}(h_2 + h_4) - \gamma_{\text{oil}} h_3$$

$$= \left(1000 \ \frac{kg}{m^3}\right)\left(9.81 \ \frac{m}{s^2}\right)$$

$$\times \left(\begin{array}{c}(0.90 \ m - 0.75 \ m) \\ + (13.6)(0.45 \ m + 0.55 \ m) \\ - (0.873)(0.4 \ m)\end{array}\right)$$

$$= 1.315 \times 10^5 \ \text{Pa} \quad (130 \ \text{kPa})$$

The answer is (C).

9. As there is no head loss between location 1 and location 2,

$$\frac{p_2}{\rho_w} + \frac{v_2^2}{2} + z_2 g = \frac{p_1}{\rho_w} + \frac{v_1^2}{2} + z_1 g$$

The velocity at location 1 is zero, and the difference in elevations is negligible, so

$$
\begin{aligned}
v_2 &= \sqrt{\frac{2(p_1 - p_2)}{\rho_w}} = \sqrt{\frac{2(\rho_{Hg} - \rho_w)g\Delta h}{\rho_w}} \\
&= \sqrt{2(SG_{Hg} - SG_w)g\Delta h} \\
&= \sqrt{(2)(13.6 - 1)\left(9.81 \ \frac{m}{s^2}\right)\left(\frac{5 \ cm}{100 \ \frac{cm}{m}}\right)} \\
&= 3.516 \ m/s
\end{aligned}
$$

The flow rate is

$$
\begin{aligned}
Q &= Av = \left(\frac{\pi D^2}{4}\right)v \\
&= \left(\frac{\pi (20 \ cm)^2}{(4)\left(100 \ \frac{cm}{m}\right)^2}\right)\left(3.516 \ \frac{m}{s}\right) \\
&= 0.11 \ m^3/s \quad (0.1 \ m^3/s)
\end{aligned}
$$

The answer is (B).

10. The areas of the nozzle entrance (location 1) and the nozzle exit (location 2) are

$$A_1 = \frac{\pi D^2}{4} = \frac{\pi (10 \ cm)^2}{(4)\left(100 \ \frac{cm}{m}\right)^2} = 0.007854 \ m^2$$

$$A_2 = \frac{\pi D^2}{4} = \frac{\pi (4 \ cm)^2}{(4)\left(100 \ \frac{cm}{m}\right)^2} = 0.001257 \ m^2$$

The velocities of the water at the nozzle entrance and exit are

$$v_1 = \frac{Q}{A_1} = \frac{0.05 \ \frac{m^3}{s}}{0.007854 \ m^2} = 6.366 \ m/s$$

$$v_2 = \frac{Q}{A_2} = \frac{0.05 \ \frac{m^3}{s}}{0.001257 \ m^2} = 39.79 \ m/s$$

Use the Bernoulli equation, and solve for the pressure at the entrance to the nozzle. The nozzle is horizontal, and the elevations at locations 1 and 2 are the same, so these terms cancel.

$$
\begin{aligned}
\frac{p_2}{\rho_w} + \frac{v_2^2}{2} + z_2 g &= \frac{p_1}{\rho_w} + \frac{v_1^2}{2} + z_1 g \\
p_1 &= p_2 + \left(\frac{v_2^2 - v_1^2}{2}\right)\rho_w \\
&= 0 \ Pa + \left(\frac{\left(39.79 \ \frac{m}{s}\right)^2 - \left(6.366 \ \frac{m}{s}\right)^2}{2}\right) \\
&\quad \times \left(1000 \ \frac{kg}{m^3}\right) \\
&= 771\,308 \ Pa \quad (770 \ kPa)
\end{aligned}
$$

The answer is (D).

11 Fluid Properties

Nomenclature

A	area	m^2
d	diameter	m
F	force	N
g	gravitational acceleration, 9.81	m/s^2
h	height	m
K	power law consistency index	–
L	length	m
m	mass	kg
n	power law index	–
p	pressure	Pa
r	radius	m
SG	specific gravity	–
v	velocity	m/s
V	volume	m^3
W	weight	N

Symbols[1]

β	angle of contact	deg
γ	specific (unit) weight	N/m^3
δ	thickness of fluid	m
μ	absolute viscosity	Pa·s
ν	kinematic viscosity	m^2/s
ρ	density	kg/m^3
σ	surface tension	N/m
τ	stress	Pa
υ	specific volume	m^3/kg

Subscripts

n	normal
t	tangential (shear)
v	vapor
w	water

---

[1]The NCEES *FE Reference Handbook* (*NCEES Handbook*) uses the symbol τ for both normal and shear stress. τ is almost universally interpreted in engineering practice as the symbol for shear stress. The use of τ stems from Cauchy stress tensor theory and the desire to use the same symbol for all nine stress directions. However, the stress tensor concept is not developed in the *NCEES Handbook*, and σ is used as the symbol for normal stress elsewhere, so the use of τ_n for normal stress and τ_t for tangential (shear) stress may be confusing. (This usage does avoid a symbol conflict with surface tension, σ, which is used in the *NCEES Handbook* in contexts unrelated to stress.)

1. FLUIDS

A *fluid* is a substance in either the liquid or gas phase. Fluids cannot support shear, and they deform continuously to minimize applied shear forces.

In fluid mechanics, a fluid is modeled as a *continuum*—that is, a substance that can be divided into infinitesimally small volumes, with properties that are continuous functions over the entire volume. For the infinitesimally small volume ΔV, Δm is the infinitesimal mass, and ΔW is the infinitesimal weight.

Equation 11.1: Density

$$\rho = \lim_{\Delta V \to 0} \Delta m / \Delta V \qquad \textit{11.1}$$

Variation

$$\rho = \frac{m}{V}$$

Description

The *density*, ρ, also called *mass density*, of a fluid is its mass per unit volume. The density of a fluid in a liquid form is usually given, known in advance, or easily obtained from tables.

If ΔV is the volume of an infinitesimally small element, the density is given as Eq. 11.1. Density is typically measured in kg/m^3.

Specific Volume

Specific volume, υ, is the volume occupied by a unit mass of fluid.

$$\upsilon = \frac{1}{\rho}$$

Specific volume is the reciprocal of density and is typically measured in m^3/kg.

Equation 11.2 Through Eq. 11.4: Specific Weight

$$\gamma = \lim_{\Delta V \to 0} \Delta W / \Delta V \qquad \textit{11.2}$$

$$\gamma = \lim_{\Delta V \to 0} g \Delta m / \Delta V = \rho g \qquad \textit{11.3}$$

$$\gamma = \rho g \qquad \textit{11.4}$$

Fluid Mechanics/
Dynamics

Variation

$$\gamma = \frac{W}{V} = \frac{mg}{V}$$

Description

Specific weight, γ, also known as *unit weight*, is the weight of substance per unit volume.

The use of specific weight is most often encountered in civil engineering work in the United States, where it is commonly called *density*. The usual units of specific weight are N/m^3. Specific weight is not an absolute property of a substance since it depends on the local gravitational field.

Example

The density of a gas is 1.5 kg/m^3. The specific weight of the gas is most nearly

(A) 9.0 N/m^3

(B) 15 N/m^3

(C) 76 N/m^3

(D) 98 N/m^3

Solution

Use Eq. 11.4.

$$\gamma = \rho g = \left(1.5 \ \frac{kg}{m^3}\right)\left(9.81 \ \frac{m}{s^2}\right)$$

$$= 14.715 \ kg/s^2 \cdot m^2 \quad (15 \ N/m^3)$$

The answer is (B).

Equation 11.5: Specific Gravity

$$SG = \gamma/\gamma_w = \rho/\rho_w \qquad \textit{11.5}$$

Description

Specific gravity, SG, is the dimensionless ratio of a fluid's density to a standard reference density. For liquids and solids, the reference is the density of pure water, which is approximately 1000 kg/m^3 over the normal ambient temperature range. The temperature at which water density should be evaluated is not standardized, so some small variation in the reference density is possible. See Table 11.1 and Table 11.2 for the properties of water in SI and customary U.S. units, respectively.

Since the SI density of water is very nearly 1.000 g/cm^3 (1000 kg/m^3), the numerical values of density in g/cm^3 and specific gravity are the same.

Example

A fluid has a density of 860 kg/m^3. The specific gravity of the fluid is most nearly

(A) 0.63

(B) 0.82

(C) 0.86

(D) 0.95

Solution

Use Eq. 11.5. The specific gravity is

$$SG = \rho/\rho_w = \frac{860 \ \dfrac{kg}{m^3}}{1000 \ \dfrac{kg}{m^3}} = 0.86$$

The answer is (C).

2. PRESSURE

Fluid pressures are measured with respect to two pressure references: zero pressure and atmospheric pressure. Pressures measured with respect to a true zero pressure reference are known as *absolute pressures*. Pressures measured with respect to atmospheric pressure are known as *gage pressures*. To distinguish them, the word "gage" or "absolute" can be added to the measurement (e.g., 25.1 kPa absolute). Alternatively, the letter "g" can be added to the measurement for gage pressures (e.g., 15 kPag), and the pressure is assumed to be absolute otherwise.

Equation 11.6 and Eq. 11.7: Absolute Pressure

$$\text{absolute pressure} = \text{atmospheric pressure} \\ + \text{gage pressure reading}$$

$$\textit{11.6}$$

$$\text{absolute pressure} = \text{atmospheric pressure} \\ - \text{vacuum gage pressure} \\ \text{reading}$$

$$\textit{11.7}$$

Values

Standard atmospheric pressure is equal to 101.3 kPa or 29.921 inches of mercury.

Description

Absolute and gage pressures are related by Eq. 11.6. In this equation, "atmospheric pressure" is the actual atmospheric pressure that exists when the gage measurement is taken. It is not standard atmospheric pressure unless that pressure is implicitly or explicitly applicable. Also, since a barometer measures atmospheric pressure, *barometric pressure* is synonymous with atmospheric pressure.

A *vacuum* measurement is implicitly a pressure below atmospheric pressure (i.e., a negative gage pressure). It must be assumed that any measured quantity given as a vacuum is a quantity to be subtracted from the atmospheric pressure. (See Eq. 11.7.) When a condenser is operating with a vacuum of 4.0 inches of mercury, the absolute pressure is approximately $29.92 - 4.0 = 25.92$ inches of mercury (25.92 in Hg). Vacuums are always stated as positive numbers.

Table 11.1 Properties of Water (SI units)

temperature (°C)	specific weight, γ (kN/m³)	density, ρ (kg/m³)	viscosity, $\mu \times 10^3$ (Pa·s)	kinematic viscosity, $\nu \times 10^6$ (m²/s)	vapor pressure, p_v (kPa)
0	9.805	999.8	1.781	1.785	0.61
5	9.807	1000.0	1.518	1.518	0.87
10	9.804	999.7	1.307	1.306	1.23
15	9.798	999.1	1.139	1.139	1.70
20	9.789	998.2	1.002	1.003	2.34
25	9.777	997.0	0.890	0.893	3.17
30	9.764	995.7	0.798	0.800	4.24
40	9.730	992.2	0.653	0.658	7.38
50	9.689	988.0	0.547	0.553	12.33
60	9.642	983.2	0.466	0.474	19.92
70	9.589	977.8	0.404	0.413	31.16
80	9.530	971.8	0.354	0.364	47.34
90	9.466	965.3	0.315	0.326	70.10
100	9.399	958.4	0.282	0.294	101.33

Table 11.2 Properties of Water (customary U.S. units)

temperature (°F)	specific weight, γ (lbf/ft³)	density, ρ (lbm-sec²/ft⁴)	viscosity, $\mu \times 10^{-5}$ (lbf-sec/ft²)	kinematic viscosity, $\nu \times 10^{-5}$ (ft²/sec)	vapor pressure, p_v (lbf/ft²)
32	62.42	1.940	3.746	1.931	0.09
40	62.43	1.940	3.229	1.664	0.12
50	62.41	1.940	2.735	1.410	0.18
60	62.37	1.938	2.359	1.217	0.26
70	62.30	1.936	2.050	1.059	0.36
80	62.22	1.934	1.799	0.930	0.51
90	62.11	1.931	1.595	0.826	0.70
100	62.00	1.927	1.424	0.739	0.95
110	61.86	1.923	1.284	0.667	1.24
120	61.71	1.918	1.168	0.609	1.69
130	61.55	1.913	1.069	0.558	2.22
140	61.38	1.908	0.981	0.514	2.89
150	61.20	1.902	0.905	0.476	3.72
160	61.00	1.896	0.838	0.442	4.74
170	60.80	1.890	0.780	0.413	5.99
180	60.58	1.883	0.726	0.385	7.51
190	60.36	1.876	0.678	0.362	9.34
200	60.12	1.868	0.637	0.341	11.52
212	59.83	1.860	0.593	0.319	14.70

Fluid Mechanics/ Dynamics

Example

A vessel is initially connected to a reservoir open to the atmosphere. The connecting valve is then closed, and a vacuum of 65.5 kPa is applied to the vessel. Assume standard atmospheric pressure. What is most nearly the absolute pressure in the vessel?

(A) 36 kPa

(B) 66 kPa

(C) 86 kPa

(D) 110 kPa

Solution

From Eq. 11.7, for vacuum pressures,

$$\begin{aligned}
\text{absolute pressure} &= \text{atmospheric pressure} \\
&\quad - \text{vacuum gage pressure} \\
&\quad \text{reading} \\
&= 101.3 \text{ kPa} - 65.5 \text{ kPa} \\
&= 35.8 \text{ kPa} \quad (36 \text{ kPa})
\end{aligned}$$

The answer is (A).

3. STRESS

Stress, τ, is force per unit area. There are two primary types of stress, differing in the orientation of the loaded area: *normal stress* and *tangential* (or *shear*) *stress*. With *normal stress*, τ_n, the area is normal to the force carried. With *tangential* (or *shear*) *stress*, τ_t, the area is parallel to the force.

Ideal fluids that are inviscid and incompressible respond to normal stresses, but they cannot support shear, and they deform continuously to minimize applied shear forces.

Equation 11.8 and Eq. 11.9: Normal Stress[2]

$$\tau(1) = \lim_{\Delta A \to 0} \Delta F / \Delta A \qquad \textit{11.8}$$
$$\tau_n = -p \qquad \textit{11.9}$$

Description

At some arbitrary point 1, with an infinitesimal area, ΔA, subjected to a force, ΔF, the normal or shear stress is defined as in Eq. 11.8.

Normal stress is equal to the pressure of the fluid, as indicated by Eq. 11.9.

4. VISCOSITY

The *viscosity* of a fluid is a measure of that fluid's resistance to flow when acted upon by an external force, such as a pressure gradient or gravity.

The viscosity of a fluid can be determined with a *sliding plate viscometer* test. Consider two plates of area A separated by a fluid with thickness δ. The bottom plate is fixed, and the top plate is kept in motion at a constant velocity, v, by a force, F. (See Fig. 11.1.)

Figure 11.1 *Sliding Plate Viscometer*

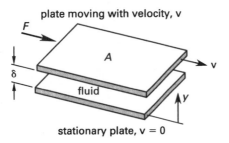

plate moving with velocity, v

stationary plate, v = 0

Experiments with many fluids have shown that the force, F, that is needed to maintain the velocity, v, is proportional to the velocity and the area but is inversely proportional to the separation of the plates. That is,

$$\frac{F}{A} \propto \frac{\text{v}}{\delta}$$

The constant of proportionality needed to make this an equality for a particular fluid is the fluid's *absolute viscosity*, μ, also known as the *absolute dynamic viscosity*. Typical units for absolute viscosity are Pa·s (N·s/m^2).

$$\frac{F}{A} = \mu \left(\frac{\text{v}}{\delta}\right)$$

F/A is the *fluid shear stress* (tangential stress), τ_t.

Equation 11.10 Through Eq. 11.12: Newton's Law of Viscosity

$$\text{v}(y) = \text{v}y/\delta \qquad \textit{11.10}$$
$$d\text{v}/dy = \text{v}/\delta \qquad \textit{11.11}$$
$$\tau_t = \mu(d\text{v}/dy) \quad \text{[one-dimensional]} \qquad \textit{11.12}$$

Variation

$$\tau_t = \frac{F}{A} = \mu \left(\frac{\text{v}}{\delta}\right)$$

Description

For a thin Newtonian fluid film, Eq. 11.10 and Eq. 11.11 describe the linear velocity profile. The quantity $d\text{v}/dy$ is known by various names, including *rate of strain, shear rate, velocity gradient*, and *rate of shear formation*.

Equation 11.12 is known as *Newton's law of viscosity*, from which Newtonian fluids get their name. (Not all fluids are Newtonian, although many are.) For a Newtonian fluid, strains are proportional to the applied shear stress (i.e., the stress versus strain curve is a straight line with slope μ). The straight line will be closer to the τ axis if the fluid is highly viscous. For low-viscosity fluids, the straight line will be closer to the $d\text{v}/dy$ axis. Equation 11.12 is applicable only to Newtonian fluids, for which the relationship is linear.

Equation 11.13: Power Law

$$\tau_t = K(d\text{v}/dy)^n \qquad \textit{11.13}$$

Values

fluid	power law index, n
Newtonian	1
non-Newtonian	
pseudoplastic	< 1
dilatant	> 1

---

[2]Equation 11.8, as given in the *NCEES Handbook*, is vague. The *NCEES Handbook* calls $\tau(1)$ the "surface stress at point 1." Point 1 is undefined, and the term "surface stress" is used without explanation, although it apparently refers to both normal and shear stress. The format of using parentheses to designate the location of a stress is not used elsewhere in the *NCEES Handbook*.

Description

Many fluids are not Newtonian (i.e., do not behave according to Eq. 11.12). Non-Newtonian fluids have viscosities that change with shear rate, dv/dt. For example, *pseudoplastic fluids* exhibit a decrease in viscosity the faster they are agitated. Such fluids present no serious pumping difficulties. On the other hand, pumps for *dilatant fluids* must be designed carefully, since dilatant fluids exhibit viscosities that increase the faster they are agitated. The fluid shear stress for most non-Newtonian fluids can be predicted by the *power law*, Eq. 11.13. In Eq. 11.13, the constant K is known as the *consistency index*. The consistency index, also known as the *flow consistency index*, is actually the average fluid viscosity across the range of viscosities being modeled. For *pseudoplastic non-Newtonian fluids*, $n < 1$; for *dilatant non-Newtonian fluids*, $n > 1$. For Newtonian fluids, $n = 1$.

Equation 11.14: Kinematic Viscosity

$$\nu = \mu/\rho \qquad 11.14$$

Description

Another quantity with the name viscosity is the ratio of absolute viscosity to mass density. This combination of variables, known as *kinematic viscosity*, ν, appears often in fluids and other problems and warrants its own symbol and name. Kinematic viscosity is merely the name given to a frequently occuring combination of variables. Typical units are m^2/s.

Example

$32°C$ water flows at 2 m/s through a pipe that has an inside diameter of 3 cm. The viscosity of the water is 769×10^{-6} $N \cdot s/m^2$, and the density of the water is 995 kg/m^3. The kinematic viscosity of the water is most nearly

(A) 0.71×10^{-6} m^2/s

(B) 0.77×10^{-6} m^2/s

(C) 0.84×10^{-6} m^2/s

(D) 0.92×10^{-6} m^2/s

Solution

The kinematic viscosity is

$$\nu = \mu/\rho = \dfrac{769 \times 10^{-6} \dfrac{N \cdot s}{m^2}}{995 \dfrac{kg}{m^3}}$$

$$= 0.773 \times 10^{-6} \ m^2/s \quad (0.77 \times 10^{-6} \ m^2/s)$$

The answer is (B).

5. SURFACE TENSION AND CAPILLARITY

Equation 11.15: Surface Tension

$$\sigma = F/L \qquad 11.15$$

Description

The membrane or "skin" that seems to form on the free surface of a fluid is caused by intermolecular cohesive forces and is known as *surface tension*, σ. Surface tension is the reason that insects are able to sit on a pond and a needle is able to float on the surface of a glass of water, even though both are denser than the water that supports them. Surface tension also causes bubbles and droplets to form in spheres, since any other shape would have more surface area per unit volume.

Surface tension can be interpreted as the tensile force between two points a unit distance apart on the surface, or as the amount of work required to form a new unit of surface area in an apparatus similar to that shown in Fig. 11.2. Typical units of surface tension are N/m, J/m^2, and dynes/cm. (Dynes/cm are equivalent to mN/m.)

Figure 11.2 *Wire Frame for Stretching a Film*

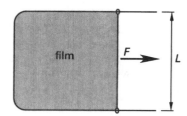

Surface tension is defined as a force, F, acting along a line of length L, as indicated by Eq. 11.15.

The apparatus shown in Fig. 11.2 consists of a wire frame with a sliding side that has been dipped in a liquid to form a film. Surface tension is determined by measuring the force necessary to keep the sliding side stationary against the surface tension pull of the film. However, since the film has two surfaces (i.e., two surface tensions), the surface tension is

$$\sigma = \dfrac{F}{2L} \quad \begin{bmatrix} \text{wire frame} \\ \text{apparatus} \end{bmatrix}$$

Surface tension can also be measured by measuring the force required to pull a *Du Nouy wire ring* out of a liquid, as shown in Fig. 11.3. Because the ring's inner and outer sides are both in contact with the liquid, the wetted perimeter is twice the circumference. The surface tension is therefore

$$\sigma = \dfrac{F}{4\pi r} \quad \begin{bmatrix} \text{Du Nouy ring} \\ \text{apparatus} \end{bmatrix}$$

Figure 11.3 *Du Nouy Ring Surface Tension Apparatus*

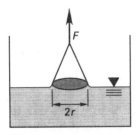

Equation 11.16: Capillary Rise or Depression

$$h = 4\sigma \cos \beta / \gamma d \qquad \textit{11.16}$$

Variation

$$h = \frac{4\sigma \cos \beta}{\rho g d_{\text{tube}}}$$

Description

Capillary action is the name given to the behavior of a liquid in a thin-bore tube. Capillary action is caused by surface tension between the liquid and a vertical solid surface. In water, the adhesive forces between the liquid molecules and the surface are greater than (i.e., dominate) the cohesive forces between the water molecules themselves. The adhesive forces cause the water to attach itself to and climb a solid vertical surface; the water rises above the general water surface level. (See Fig. 11.4.) This is called *capillary rise*, and the curved surface of the liquid within the tube is known as a *meniscus*.

Figure 11.4 *Capilarity of Liquids*

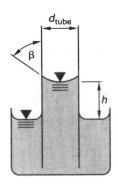 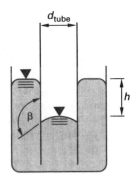

(a) adhesive force dominates   (b) cohesive force dominates

For a few liquids, such as mercury, the molecules have a strong affinity for each other (i.e., the cohesive forces dominate). These liquids avoid contact with the tube surface. In such liquids, the meniscus will be below the general surface level, a state called *capillary depression.*

The *angle of contact*, β, is an indication of whether adhesive or cohesive forces dominate. For contact angles less than 90°, adhesive forces dominate. For contact angles greater than 90°, cohesive forces dominate. For water in a glass tube, the contact angle is zero; for mercury in a glass tube, the contact angle is 140°.

Equation 11.16 can be used to predict the capillary rise (if the result is positive) or capillary depression (if the result is negative) in a small-bore tube. Surface tension is a material property of a fluid, and contact angles are specific to a particular fluid-solid interface. Both may be obtained from tables.

Example

An open glass tube with a diameter of 1 mm contains mercury at 20°C. At this temperature, mercury has a surface tension of 0.519 N/m and a density of 13 600 kg/m³. The contact angle for mercury in a glass tube is 140°. The capillary depression is most nearly

- (A) 6.1 mm
- (B) 8.6 mm
- (C) 12 mm
- (D) 17 mm

Solution

Use Eq. 11.4 and Eq. 11.16 to find the capillary depression (or negative rise).

$$h = 4\sigma \cos \beta / \gamma d = \frac{4\sigma \cos \beta}{\rho g d_{\text{tube}}}$$

$$= \frac{(4)\left(0.519 \, \frac{\text{N}}{\text{m}}\right)\cos 140° \left(1000 \, \frac{\text{mm}}{\text{m}}\right)}{\left(13\,600 \, \frac{\text{kg}}{\text{m}^3}\right)\left(9.81 \, \frac{\text{m}}{\text{s}^2}\right)(1 \text{ mm})}$$

$$= -0.0119 \text{ m} \quad (12 \text{ mm})$$

The answer is (C).

12 Fluid Statics

Nomenclature

A	area	m^2
F	force	N
g	gravitational acceleration, 9.81	m/s^2
h	vertical depth or difference in vertical depth	m
I	moment of inertia	m^4
p	pressure	Pa
R	resultant force	N
SG	specific gravity	–
V	volume	m^3
y	distance	m
z	elevation	m

Symbols

α	angle	deg
γ	specific (unit) weight	N/m^3
θ	angle	deg
ρ	density	kg/m^3

Subscripts

0	atmospheric
atm	atmospheric
B	barometer fluid
C	centroid
CP	center of pressure
f	fluid
m	manometer
R	resultant
v	vapor
x	horizontal

1. HYDROSTATIC PRESSURE

Hydrostatic pressure is the pressure a fluid exerts on an immersed object or on container walls. The term *hydrostatic* is used with all fluids, not only with water.

Pressure is equal to the force per unit area of surface.

$$p = \frac{F}{A}$$

Hydrostatic pressure in a stationary, incompressible fluid behaves according to the following characteristics.

- Pressure is a function of vertical depth and density only. If density is constant, then the pressure will be the same at two points with identical depths.

- Pressure varies linearly with vertical depth. The relationship between pressure and depth for an incompressible fluid is given by the equation

$$p = \rho g h = \gamma h$$

Since ρ and g are constants, this equation shows that p and h are linearly related. One determines the other.

- Pressure is independent of an object's area and size, and of the weight (or mass) of water above the object. Figure 12.1 illustrates the *hydrostatic paradox*. The pressures at depth h are the same in all four columns because pressure depends only on depth, not on volume.

Figure 12.1 Hydrostatic Paradox

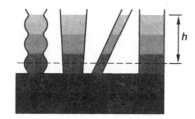

- Pressure at a point has the same magnitude in all directions (*Pascal's law*). Therefore, pressure is a scalar quantity.

- Pressure is always normal to a surface, regardless of the surface's shape or orientation. (This is a result of the fluid's inability to support shear stress.)

Equation 12.1: Pressure Difference in a Static Fluid[1]

$$p_2 - p_1 = -\gamma(z_2 - z_1) = -\gamma h = -\rho g h \qquad \text{12.1}$$

---

[1]Although the variable y is used elsewhere in the NCEES *FE Reference Handbook* (*NCEES Handbook*) to represent vertical direction, the *NCEES Handbook* uses z to measure some vertical dimensions within fluid bodies. As referenced to a Cartesian coordinate system (a practice that is not continued elsewhere in the *NCEES Handbook*), Eq. 12.1 is academically correct, but it is inconsistent with normal practice, which measures z from the fluid surface, synonymous with "depth." The *NCEES Handbook* reverts to common usage of h, y, and z in its subsequent discussion of forces on submerged surfaces.

Description

As pressure in a fluid varies linearly with depth, difference in pressure likewise varies linearly with difference in depth. This is expressed in Eq. 12.1. The variable z decreases with depth while the pressure increases with depth, so pressure and elevation have an inverse linear relationship, as indicated by the negative sign.

2. MANOMETERS

Manometers can be used to measure small pressure differences, and for this purpose, they provide good accuracy. A difference in manometer fluid surface heights indicates a pressure difference. When both ends of the manometer are connected to pressure sources, the name *differential manometer* is used. If one end of the manometer is open to the atmosphere, the name *open manometer* is used. An open manometer indicates gage pressure. It is theoretically possible, but impractical, to have a manometer indicate absolute pressure, since one end of the manometer would have to be exposed to a perfect vacuum.

Consider the simple manometer in Fig. 12.2. The pressure difference $p_2 - p_1$ causes the difference h_m in manometer fluid surface heights. Fluid column h_2 exerts a hydrostatic pressure on the manometer fluid, forcing the manometer fluid to the left. This increase must be subtracted out. Similarly, the column h_1 restricts the movement of the manometer fluid. The observed measurement must be increased to correct for this restriction. The typical way to solve for pressure differences in a manometer is to start with the pressure on one side, and then add or subtract changes in hydrostatic pressure at known points along the column until the pressure on the other side is reached.

$$p_2 = p_1 + \rho_1 g h_1 + \rho_m g h_m - \rho_2 g h_2$$
$$= p_1 + \gamma_1 h_1 + \gamma_m h_m - \gamma_2 h_2$$

Figure 12.2 Manometer Requiring Corrections

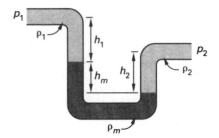

Equation 12.2 and Eq. 12.3: Pressure Difference in a Simple Manometer

$$p_0 = p_2 + \gamma_2 h_2 - \gamma_1 h_1 = p_2 + g(\rho_2 h_2 - \rho_1 h_1) \quad \textbf{12.2}$$

$$p_0 = p_2 + (\gamma_2 - \gamma_1)h = p_2 + (\rho_2 - \rho_1)gh \quad [h_1 = h_2 = h]$$
$$\textbf{12.3}$$

Description

Figure 12.3 illustrates an open manometer. Neglecting the air in the open end, the pressure difference is given by Eq. 12.2. $p_0 - p_2$ is the gage pressure in the vessel.

Figure 12.3 Open Manometer

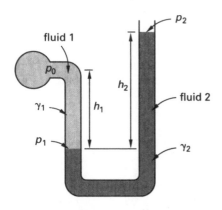

Equation 12.2 is a version of Eq. 12.1 as applied to an open manometer. Equation 12.3 is a simplified version that can be used only when h_1 is equal to h_2.[2]

Example

One leg of a mercury U-tube manometer is connected to a pipe containing water under a gage pressure of 100 kPa. The mercury in this leg stands 0.75 m below the water. The mercury in the other leg is open to the air. The density of the water is 1000 kg/m³, and the specific gravity of the mercury is 13.6.

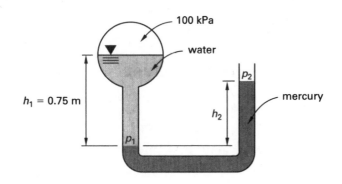

---

[2]h_1 and h_2 would be equal only in the most contrived situations.

The height of the mercury in the open leg is most nearly

(A) 0.05 m

(B) 0.5 m

(C) 0.8 m

(D) 1 m

Solution

Find the specific weights of water, γ_1, and mercury, γ_2.

$$\gamma_1 = \rho_{\text{water}}g = \left(1000 \ \frac{\text{kg}}{\text{m}^3}\right)\left(9.81 \ \frac{\text{m}}{\text{s}^2}\right)$$
$$= 9810 \ \text{N/m}^3$$
$$\gamma_2 = \rho_{\text{Hg}}g = (SG_{\text{Hg}})\rho_{\text{water}}g$$
$$= (13.6)\left(1000 \ \frac{\text{kg}}{\text{m}^3}\right)\left(9.81 \ \frac{\text{m}}{\text{s}^2}\right)$$
$$= 133\,416 \ \text{N/m}^3$$

Use Eq. 12.2 to find the height of the mercury in the open leg. Since the 100 kPa pressure is a gage pressure, the atmospheric pressure, p_2, is zero.

$$p_0 = p_2 + \gamma_2 h_2 - \gamma_1 h_1$$
$$h_2 = \frac{p_0 + \gamma_1 h_1 - p_2}{\gamma_2}$$
$$= \frac{(100 \ \text{kPa})\left(1000 \ \frac{\text{Pa}}{\text{kPa}}\right) + \left(9810 \ \frac{\text{N}}{\text{m}^3}\right)(0.75 \ \text{m}) - 0 \ \text{Pa}}{133\,416 \ \frac{\text{N}}{\text{m}^3}}$$
$$= 0.805 \ \text{m} \quad (0.8 \ \text{m})$$

The answer is (C).

3. BAROMETERS

The *barometer* is a common device for measuring the absolute pressure of the atmosphere. It is constructed by filling a long tube open at one end with mercury (alcohol or another liquid can also be used) and inverting the tube such that the open end is below the level of the mercury-filled container. The vapor pressure of the mercury in the tube is insignificant; if this is neglected, the fluid column is supported only by the atmospheric pressure transmitted through the container fluid at the lower, open end. (See Fig. 12.4.) In such a case, the atmospheric pressure is given by

$$p_{\text{atm}} = \rho g h = \gamma h$$

Figure 12.4 *Barometer*

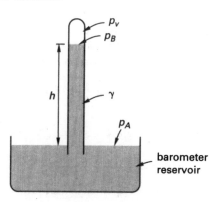

barometer reservoir

Equation 12.4: Vapor Pressure[3]

$$p_{\text{atm}} = p_A = p_v + \gamma h = p_B + \gamma h = p_B + \rho g h \qquad 12.4$$

Variation

$$p_{\text{atm}} = p_v + (SG)\rho_{\text{water}}gh$$

Description

When the vapor pressure of the barometer liquid is significant, as it is with alcohol or water, the vapor pressure effectively reduces the height of the fluid column, as Eq. 12.4 indicates.

Example

A fluid with a vapor pressure of 0.2 Pa and a specific gravity of 12 is used in a barometer. If the fluid's column height is 1 m, the atmospheric pressure is most nearly

(A) 9.8 kPa

(B) 12 kPa

(C) 98 kPa

(D) 120 kPa

Solution

From Eq. 12.4,

$$p_{\text{atm}} = p_B + \rho g h = p_v + (SG)\rho_{\text{water}}gh$$
$$= 0.2 \ \text{Pa} + (12)\left(1000 \ \frac{\text{kg}}{\text{m}^3}\right)\left(9.81 \ \frac{\text{m}}{\text{s}^2}\right)(1 \ \text{m})$$
$$= 117\,720 \ \text{Pa} \quad (120 \ \text{kPa})$$

The answer is (D).

---

[3]In Eq. 12.4, the *NCEES Handbook* uses A as a subscript to designate location A in Fig. 12.4, not to designate "atmosphere," which is inconsistently designated by the subscripts "atm" and "0." Figure 12.4 shows both p_v and p_B, although in fact, these two pressures are the same.

4. FORCES ON SUBMERGED PLANE SURFACES

The pressure on a horizontal plane surface is uniform over the surface because the depth of the fluid above is uniform. The resultant of the pressure distribution acts through the *center of pressure* of the surface, which corresponds to the centroid of the surface. (See Fig. 12.5.)

Figure 12.5 *Hydrostatic Pressure on a Horizontal Plane Surface*

The total vertical force on the horizontal plane of area A is given by the equation

$$R = pA$$

It is not always correct to calculate the vertical force on a submerged surface as the weight of the fluid above it. Such an approach works only when there is no change in the cross-sectional area of the fluid above the surface. This is a direct result of the *hydrostatic paradox*. (See Fig. 12.1.) The two containers in Fig. 12.6 have the same distribution of pressure (or force) over their bottom surfaces.

Figure 12.6 *Two Containers with the Same Pressure Distribution*

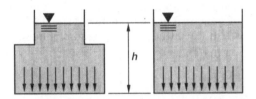

The pressure on a vertical rectangular plane surface increases linearly with depth. The pressure distribution will be triangular, as in Fig. 12.7(a), if the plane surface extends to the surface; otherwise, the distribution will be trapezoidal, as in Fig. 12.7(b).

Figure 12.7 *Hydrostatic Pressure on a Vertical Plane Surface*

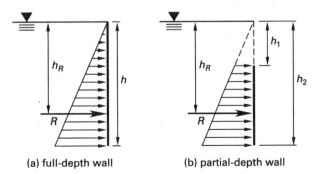

(a) full-depth wall   (b) partial-depth wall

The resultant force on a vertical rectangular plane surface is

$$R = \overline{p}A$$

$\overline{p}$ is the *average pressure*, which is also equal to the pressure at the centroid of the plane area. The average pressure is

$$\overline{p} = \tfrac{1}{2}(p_1 + p_2) = \tfrac{1}{2}\rho g(h_1 + h_2) = \tfrac{1}{2}\gamma(h_1 + h_2)$$

Although the resultant is calculated from the average depth, the resultant does not act at the average depth. The resultant of the pressure distribution passes through the centroid of the pressure distribution. For the triangular distribution of Fig. 12.7(a), the resultant is located at a depth of $h_R = \tfrac{2}{3}h$. For the more general case, the center of pressure can be calculated by the method described in Sec. 12.5.

The average pressure and resultant force on an inclined rectangular plane surface are calculated in the same fashion as for the vertical plane surface. (See Fig. 12.8.) The pressure varies linearly with depth. The resultant is calculated from the average pressure, which, in turn, depends on the average depth.

Figure 12.8 *Hydrostatic Pressure on an Inclined Rectangular Plane Surface*

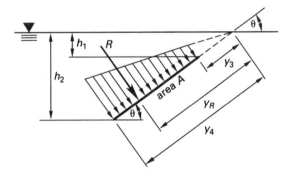

The resultant and average pressure on an inclined plane surface are given by the same equations as for a vertical plane surface.

$$R = \overline{p}A$$

$$\overline{p} = \tfrac{1}{2}(p_1 + p_2) = \tfrac{1}{2}\rho g(h_1 + h_2) = \tfrac{1}{2}\gamma(h_1 + h_2)$$

As with a vertical plane surface, the resultant acts at the centroid of the pressure distribution, not at the average depth.

5. CENTER OF PRESSURE

For the case of pressure on a general plane surface, the resultant force depends on the average pressure and acts through the *center of pressure*, CP. Figure 12.9 shows a nonrectangular plane surface of area A that may or may not extend to the liquid surface and that may or may

Figure 12.9 *Hydrostatic Pressure on a Submerged Plane Surface**

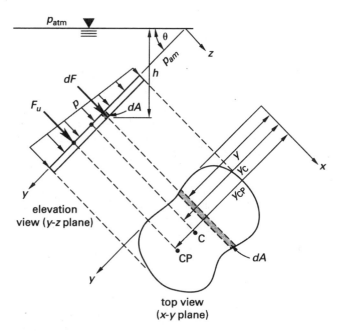

*The meaning of the inclined p_{am} in this figure as presented in the *NCEES Handbook* is unknown. Its location implies that it is not the same as p_{atm}.

not be inclined. The average pressure is calculated from the location of the plane surface's centroid, C, where y_C is measured parallel to the plane surface. That is, if the plane surface is inclined, y_C is an inclined distance.

The center of pressure is always at least as deep as the area's centroid. In most cases, it is deeper.

The pressure at the centroid is

$$p_C = \bar{p} = p_{atm} + \rho g y_C \sin\theta$$
$$= p_{atm} + \gamma y_C \sin\theta$$

....

Equation 12.5: Absolute Pressure on a Point[4]

$$p = p_0 + \rho g h \quad [h \geq 0] \qquad \textbf{12.5}$$

Description

Equation 12.5 gives the absolute pressure on a point at a vertical distance of h under the surface. Depth, h, must be greater than or equal to 0.[5]

---
[4]The *NCEES Handbook* is inconsistent in how it designates atmospheric pressure. Although p_{atm} is used in Eq. 12.4 and in Fig. 12.9, Eq. 12.5 uses p_0.
[5]Equation 12.5 calculates the *absolute pressure* because it includes the atmospheric pressure term is omitted, the *gauge pressure* (also known as *gage pressure*) will be calculated.

Example

A closed tank with the dimensions shown contains water. The air pressure in the tank is 700 kPa. Point P is located halfway up the inclined wall.

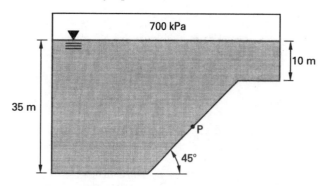

The pressure at point P is most nearly

(A) 920 kPa

(B) 1900 kPa

(C) 7200 kPa

(D) 8100 kPa

Solution

The tank and its geometry are shown.

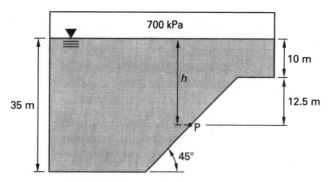

Point P is halfway up the inclined surface, so it is at a depth of

$$h = 10 \text{ m} + \frac{35 \text{ m} - 10 \text{ m}}{2} = 22.5 \text{ m}$$

The pressure at point P is

$$p = p_0 + \rho g h$$
$$= (700 \text{ kPa})\left(1000 \frac{\text{Pa}}{\text{kPa}}\right)$$
$$+ \left(1000 \frac{\text{kg}}{\text{m}^3}\right)\left(9.81 \frac{\text{m}}{\text{s}^2}\right)(22.5 \text{ m})$$
$$= 920\,725 \text{ Pa} \quad (920 \text{ kPa})$$

The answer is (A).

Equation 12.6 Through Eq. 12.8: Distance to Center of Pressure

$$y_{CP} = y_C + I_{xC}/y_C A \qquad 12.6$$

$$y_{CP} = y_C + \rho g \sin\theta I_{xC}/p_C A \qquad 12.7$$

$$y_C = h_C/\sin\alpha \qquad 12.8$$

Description

Equation 12.6 and Eq. 12.7 apply when the atmospheric pressure acts on the liquid surface and on the dry side of the submerged surface. The distance from the surface of the liquid to the center of pressure measured along the slanted surface, y_{CP}, is found from Eq. 12.6 and Eq. 12.7. Equation 12.7 is derived from Eq. 12.6 by using the pressure-height relationship $p_C = \rho g h_C = \rho g y_C \sin\theta$.[6] y_C is the distance from the surface of the liquid to the centroid of the area, C, calculated using Eq. 12.8. In Eq. 12.6 and Eq. 12.7, the subscript x refers to a horizontal (centroidal) axis parallel to the surface, which might not be obvious from Fig. 12.9, as presented in the *NCEES Handbook*.

Equation 12.9 and Eq. 12.10: Resultant Force

$$F_R = (p_0 + \rho g y_C \sin\theta)A \qquad 12.9$$

$$F_{R_{net}} = (\rho g y_C \sin\theta)A \qquad 12.10$$

Description

The resultant force, F_R, on the wetted side of the surface is found from Eq. 12.9. Equation 12.10 calculates the net resultant force when p_0 acts on both sides of the surface.[7]

6. BUOYANCY

Buoyant force is an upward force that acts on all objects that are partially or completely submerged in a fluid. The fluid can be a liquid or a gas. There is a buoyant force on all submerged objects, not only on those that are stationary or ascending. A buoyant force caused by displaced air also exists, although it may be insignificant. Examples include the buoyant force on a rock sitting at the bottom of a pond, the buoyant force on a rock sitting exposed on the ground (since the rock is "submerged" in air), and the buoyant force on partially exposed floating objects, such as icebergs.

Buoyant force always acts to cancel the object's weight (i.e., buoyancy acts against gravity). The magnitude of the buoyant force is predicted from *Archimedes' principle* (the *buoyancy theorem*), which states that the buoyant force on a submerged or floating object is equal to the weight of the displaced fluid. An equivalent statement of Archimedes' principle is that a floating object displaces liquid equal in weight to its own weight. In the situation of an object floating at the interface between two immiscible liquids of different densities, the buoyant force equals the sum of the weights of the two displaced fluids.

In the case of stationary (i.e., not moving vertically) floating or submerged objects, the buoyant force and object weight are in equilibrium. If the forces are not in equilibrium, the object will rise or fall until equilibrium is reached—that is, the object will sink until its remaining weight is supported by the bottom, or it will rise until the weight of liquid is reduced by breaking the surface.

The two forces acting on a stationary floating object are the *buoyant force* and the *object's weight*. The buoyant force acts upward through the centroid of the displaced volume (not the object's volume). This centroid is known as the *center of buoyancy*. The gravitational force on the object (i.e., the object's weight) acts downward through the entire object's center of gravity.

---

[6]p_C in Eq. 12.7 is defined as the "pressure at the centroid of the area," but it cannot be calculated from Eq. 12.5, because Eq. 12.5 calculates an absolute pressure. As used in Eq. 12.7, p_C is a gauge pressure, not an absolute pressure. $\sin\alpha$ in Eq. 12.8 is an error, and it should be $\sin\theta$.

[7]For all practical purposes, atmospheric pressure always acts on both sides of an object. It acts through the liquid on both sides of a submerged plate, it acts on both sides of a submerged gate, and it acts on both sides of a discharge gate/door. Except for objects in a vacuum, Eq. 12.9 is of purely academic interest.

13 Fluid Dynamics

Nomenclature

A	area	m^2
AR	aspect ratio	–
b	span	m
B	channel width	m
c	chord length	m
C	coefficient	–
$C_{D\infty}$	drag coefficient at zero lift	–
d	depth	m
d	diameter	m
D	diameter	m
E	specific energy	J/kg
f	friction factor	–
F	force	N
Fr	Froude number	–
g	gravitational acceleration, 9.81	m/s^2
h	head	m
h	height	m
I	impulse	N·s
k	conversion constant	–
k_1	constant of proportionality	–
K	constant	–
K	consistency index	–
L	length	m
m	mass	kg
$\dot{m}$	mass flow rate	kg/s
M	moment	N·m
n	Manning roughness coefficient	–
n	power law index	–

p	pressure	Pa
Δp_f	pressure drop due to friction	Pa
P	momentum	kg·m/s
q	unit discharge	$m^3/m \cdot s$
Q	flow rate	m^3/s
r	distance from centerline	m
R	radius	m
Re	Reynolds number (Newtonian fluid)	–
Re$'$	Reynolds number (non-Newtonian fluid)	–
S	slope of energy grade line	–
t	time	s
T	surface width	m
v	velocity	m/s
W	weight	N
$\dot{W}$	power	J/s
y	depth	m
y_h	hydraulic depth (characteristic length)	m
z	elevation	m

Symbols

α	angle	deg
α	geometric angle of attack	deg
α	kinetic energy correction factor	–
β	negative of angle of attack for zero lift	deg
γ	specific (unit) weight	N/m^3
ε	specific roughness	m
μ	absolute viscosity	Pa·s
ρ	density	kg/m^3
τ	shear stress	Pa
ν	kinematic viscosity	m^2/s

Subscripts

b	blade or bulk
c	critical
D	drag
f	friction
h	hydraulic
H	hydraulic
j	jet
L	lift or loss
max	maximum
M	moment
p	constant pressure, perpendicular, or plan
s	surface
t	total
v	constant volume or velocity
w	wall

Fluid Mechanics/Dynamics

1. INTRODUCTION

Fluid dynamics is the theoretical science of all fluids in motion. In a general sense, *hydraulics* is the study of the practical laws of incompressible fluid flow and resistance in pipes and open channels. Hydraulic formulas are often developed from experimentation, empirical factors, and curve fitting, without an attempt to justify why the fluid behaves the way it does.

2. CONSERVATION LAWS

Equation 13.1 Through Eq. 13.3: Continuity Equation

$$A_1 v_1 = A_2 v_2 \qquad 13.1$$

$$Q = Av \qquad 13.2$$

$$\dot{m} = \rho Q = \rho A v \qquad 13.3$$

Description

Fluid mass is always conserved in fluid systems, regardless of the pipeline complexity, orientation of the flow, and fluid. This single concept is often sufficient to solve simple fluid problems.

$$\dot{m}_1 = \dot{m}_2$$

When applied to fluid flow, the conservation of mass law is known as the *continuity equation*.

$$\rho_1 A_1 v_1 = \rho_2 A_2 v_2$$

If the fluid is incompressible, then $\rho_1 = \rho_2$. Equation 13.1 is the continuity equation for incompressible flow.

Volumetric flow rate, Q, is defined as the product of cross-sectional area and velocity, as shown in Eq. 13.2. From Eq. 13.1 and Eq. 13.2, it follows that

$$Q_1 = Q_2$$

Various units are used for volumetric flow rate. MGD (millions of gallons per day) and MGPCD (millions of gallons per capita day) are units commonly used in municipal water works problems. MMSCFD (millions of standard cubic feet per day) may be used to express gas flows.

Calculation of flow rates is often complicated by the interdependence between flow rate and friction loss. Each affects the other, so many pipe flow problems must be solved iteratively. Usually, a reasonable friction factor is assumed and used to calculate an initial flow rate. The flow rate establishes the flow velocity, from which a revised friction factor can be determined.

Example

An incompressible fluid flows through a pipe with an inner diameter of 10 cm at a velocity of 4 m/s. The pipe contracts to an inner diameter of 8 cm. What is most nearly the velocity of the fluid in the narrower section?

(A) 4.7 m/s

(B) 5.0 m/s

(C) 5.8 m/s

(D) 6.3 m/s

Solution

The cross-sectional areas of the two pipes are

$$A_1 = \frac{\pi D_1^2}{4} = \frac{\pi (10 \text{ cm})^2}{4} = 78.54 \text{ cm}^2$$

$$A_2 = \frac{\pi D_2^2}{4} = \frac{\pi (8 \text{ cm})^2}{4} = 50.27 \text{ cm}^2$$

Use Eq. 13.1.

$$A_1 v_1 = A_2 v_2$$

$$v_2 = \frac{A_1 v_1}{A_2} = \frac{(78.54 \text{ cm}^2)\left(4 \dfrac{\text{m}}{\text{s}}\right)}{50.27 \text{ cm}^2}$$

$$= 6.25 \text{ m/s} \quad (6.3 \text{ m/s})$$

The answer is (D).

Equation 13.4 and Eq. 13.5: Bernoulli Equation

$$\frac{p_2}{\gamma} + \frac{v_2^2}{2g} + z_2 = \frac{p_1}{\gamma} + \frac{v_1^2}{2g} + z_1 \qquad 13.4$$

$$\frac{p_2}{\rho} + \frac{v_2^2}{2} + z_2 g = \frac{p_1}{\rho} + \frac{v_1^2}{2} + z_1 g \qquad 13.5$$

Description

The *Bernoulli equation*, also known as the *field equation* or the *energy equation*, is an energy conservation equation that is valid for incompressible, frictionless flow. The Bernoulli equation states that the total energy of a fluid flowing without friction losses in a pipe is constant. The total energy possessed by the fluid is the sum of its pressure, kinetic, and potential energies. In other words, the Bernoulli equation states that the total head at any two points is the same.

Example

The diameter of a water pipe gradually changes from 5 cm at the entrance, point A, to 15 cm at the exit, point B. The exit is 5 m higher than the entrance. The pressure is 700 kPa at the entrance and 664 kPa at the exit. Friction between the water and the pipe walls is negligible. The water density is 1000 kg/m^3.

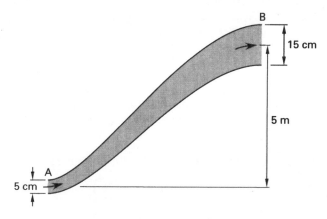

What is most nearly the rate of discharge at the exit?

(A) 0.0035 m³/s

(B) 0.0064 m³/s

(C) 0.010 m³/s

(D) 0.018 m³/s

Solution

Find the relationship between the entrance and exit velocities, v_1 and v_2, respectively. From Eq. 13.1 and substituting the pipe area equation,

$$A_1 v_1 = A_2 v_2$$

$$\left(\frac{\pi D_1^2}{4}\right) v_1 = \left(\frac{\pi D_2^2}{4}\right) v_2$$

$$v_1 = \left(\frac{D_2^2}{D_1^2}\right) v_2 = \left(\frac{(15 \text{ cm})^2 v_2}{(5 \text{ cm})^2}\right) = 9 v_2$$

Use the Bernoulli equation, Eq. 13.5, to find the velocity at the exit.

$$\frac{p_2}{\rho} + \frac{v_2^2}{2} + z_2 g = \frac{p_1}{\rho} + \frac{v_1^2}{2} + z_1 g$$

$$= \frac{p_1}{\rho} + \frac{(9 v_2)^2}{2} + z_1 g$$

$$\frac{p_2 - p_1}{\rho} + g(z_2 - z_1) = \frac{81 v_2^2}{2} - \frac{v_2^2}{2}$$

$$v_2 = \sqrt{\frac{p_2 - p_1}{40 \rho} + \frac{g(z_2 - z_1)}{40}}$$

$$= \sqrt{\frac{\begin{array}{c}(664 \text{ kPa} - 700 \text{ kPa}) \\ \times \left(1000 \frac{\text{Pa}}{\text{kPa}}\right)\end{array}}{(40)\left(1000 \frac{\text{kg}}{\text{m}^3}\right)} + \frac{\left(9.81 \frac{\text{m}}{\text{s}^2}\right)(5 \text{ m})}{40}}$$

$$= 0.571 \text{ m/s}$$

Multiply the velocity at the exit with the cross-sectional area to get the rate of flow.

$$Q = A_2 v_2$$

$$= \left(\frac{\pi D_2^2}{4}\right) v_2$$

$$= \left(\frac{\pi (15 \text{ cm})^2}{(4)\left(100 \frac{\text{cm}}{\text{m}}\right)^2}\right)\left(0.571 \frac{\text{m}}{\text{s}}\right)$$

$$= 0.0101 \text{ m}^3/\text{s} \quad (0.010 \text{ m}^3/\text{s})$$

The answer is (C).

3. REYNOLDS NUMBER

The *Reynolds number*, Re, is a dimensionless number interpreted as the ratio of inertial forces to viscous forces in the fluid.

The inertial forces are proportional to the flow diameter, velocity, and fluid density. (Increasing these variables will increase the momentum of the fluid in flow.) The viscous force is represented by the fluid's *absolute viscosity*, μ.

....

Equation 13.6: Reynolds Number, Newtonian Fluids

$$\text{Re} = v D \rho / \mu = v D / \nu \qquad \textbf{13.6}$$

Description

Since μ / ρ is the *kinematic viscosity*, ν, the equation can be simplified.

If all of the fluid particles move in paths parallel to the overall flow direction (i.e., in layers), the flow is said to be *laminar*. This occurs when the Reynolds number is less than approximately 2100. *Laminar flow* is typical when the flow channel is small, the velocity is low, and the fluid is viscous. Viscous forces are dominant in laminar flow.

Turbulent flow is characterized by a three-dimensional movement of the fluid particles superimposed on the overall direction of motion. A fluid is said to be in turbulent flow if the Reynolds number is greater than approximately 4000. (This is the most common case.)

The flow is said to be in the *critical zone* or *transition region* when the Reynolds number is between 2100 and 4000. These numbers are known as the lower and upper *critical Reynolds numbers*, respectively.

Example

The mean velocity of 40°C water in a 44.7 mm (inside diameter) tube is 1.5 m/s. The kinematic viscosity is $\nu = 6.58 \times 10^{-7}$ m^2/s. What is most nearly the Reynolds number?

    (A) 8.1×10^3

    (B) 8.5×10^3

    (C) 9.1×10^4

    (D) 1.0×10^5

Solution

From Eq. 13.6,

$$\mathrm{Re} = \mathrm{v}D\rho/\mu = \mathrm{v}D/\nu$$

$$= \frac{\left(1.5\ \dfrac{\mathrm{m}}{\mathrm{s}}\right)(44.7\ \mathrm{mm})}{\left(6.58 \times 10^{-7}\ \dfrac{\mathrm{m}^2}{\mathrm{s}}\right)\left(1000\ \dfrac{\mathrm{mm}}{\mathrm{m}}\right)}$$

$$= 1.02 \times 10^5 \quad (1.0 \times 10^5)$$

The answer is (D).

Equation 13.7: Reynolds Number, Non-Newtonian Fluids

$$\mathrm{Re}' = \frac{\mathrm{v}^{(2-n)} D^n \rho}{K\left(\dfrac{3n+1}{4n}\right)^n 8^{(n-1)}} \qquad \textit{13.7}$$

Description

Many fluids are not Newtonian (i.e., do not behave according to Eq. 11.12). Non-Newtonian fluids have viscosities that change with shear rate, $d\mathrm{v}/dt$. For example, *pseudoplastic fluids* exhibit a decrease in viscosity the faster they are agitated. Such fluids present no serious pumping difficulties. On the other hand, pumps for *dilatant fluids* must be designed carefully, since dilatant fluids exhibit viscosities that increase the faster they are agitated. For non-Newtonian fluids, *power law* parameters must be used when calculating the Reynolds number, Re'. In Eq. 13.7, the constant K is known as the *consistency index*. For *pseudoplastic non-Newtonian fluids*, $n < 1$; for *dilatant non-Newtonian fluids*, $n > 1$. For Newtonian fluids, $n = 1$, and Eq. 13.7 reduces to Eq. 13.6.

4. FLOW DISTRIBUTION

With laminar flow in a circular pipe or between two parallel plates, viscosity makes some fluid particles adhere to the wall. The closer to the wall, the greater the tendency will be for the fluid to adhere. In general, the fluid velocity will be zero at the wall and will follow a parabolic distribution away from the wall. The

average flow velocity (also known as the *bulk velocity*) is found from the flow rate and cross-sectional area.

$$\mathrm{v} = \frac{Q}{A} \quad [\text{average}]$$

Because of the parabolic distribution, velocity will be maximum at the centerline, midway between the two walls (i.e., at the center of a pipe). (See Fig. 13.1.)

Figure 13.1 *Laminar and Turbulent Velocity Distributions*

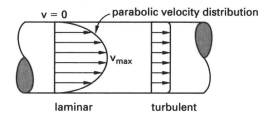

Equation 13.8 Through Eq. 13.11: Flow Velocity

$$\mathrm{v}(r) = \mathrm{v}_{\max}\left[1 - \left(\frac{r}{R}\right)^2\right] \qquad \textit{13.8}$$

$$\mathrm{v}_{\max} = 2\overline{\mathrm{v}} \quad [\text{laminar flow in circular pipe}] \qquad \textit{13.9}$$

$$\mathrm{v}_{\max} = 1.5\overline{\mathrm{v}} \quad [\text{laminar flow between plates}] \qquad \textit{13.10}$$

$$\mathrm{v}_{\max} = 1.18\overline{\mathrm{v}} \quad [\text{fully turbulent flow}] \qquad \textit{13.11}$$

Variation

$$\overline{\mathrm{v}} = \frac{Q}{A}$$

Description

For flow through a pipe with diameter $2R$ or between parallel plates with separation distance $2R$, the velocity at any point a distance r from the centerline is given by Eq. 13.8. The value of $\mathrm{v}_{\max}$ varies depending on the conditions of the flow. Equation 13.9 and Eq. 13.10 give the values of $\mathrm{v}_{\max}$ for laminar flow in circular pipes and between plates, respectively.

With turbulent flow, a distinction between velocities of particles near the pipe wall or centerline is usually not made. All the fluid particles are assumed to flow at the bulk velocity. In reality, no flow is completely turbulent, and there is a slight difference between the centerline velocity and the average velocity. For fully turbulent flow ($\mathrm{Re} > 10\,000$), a good approximation of the average velocity is approximately 85% of the maximum velocity, as stated by Eq. 13.11.

Equation 13.12: Shear Stress

$$\frac{\tau}{\tau_w} = \frac{r}{R} \qquad \textit{13.12}$$

Description

Like flow velocity, the shear stress created by the flow also varies with location. The shear stress, τ, at any point a distance r from the centerline can be found from the shear stress at the wall, τ_w, using the relationship in Eq. 13.12.

5. STEADY INCOMPRESSIBLE FLOW IN PIPES AND CONDUITS

Equation 13.13 and Eq. 13.14: Energy Conservation Equation

$$\frac{p_1}{\gamma} + z_1 + \frac{v_1^2}{2g} = \frac{p_2}{\gamma} + z_2 + \frac{v_2^2}{2g} + h_f \qquad 13.13$$

$$\frac{p_1}{\rho g} + z_1 + \frac{v_1^2}{2g} = \frac{p_2}{\rho g} + z_2 + \frac{v_2^2}{2g} + h_f \qquad 13.14$$

Description

The *energy conservation (extended field) equation*, also known as the *steady-flow energy equation*, for steady incompressible flow is shown in Eq. 13.13 and Eq. 13.14. Equation 13.13 and Eq. 13.14 do not include the effects of *shaft devices* (so named because they have rotating shafts adding or extracting power) such as pumps, compressors, fans, and turbines. (Effects of shaft devices would be represented by *shaft work* terms.)

If the cross-sectional area of the pipe is the same at points 1 and 2, then $v_1 = v_2$ and $v_1^2/2g = v_2^2/2g$. If the elevation of the pipe is the same at points 1 and 2, then $z_1 = z_2$. When analyzing discharge from reservoirs and large tanks, it is common to use gauge pressures, so that $p_1 = 0$ at the surface. In addition, since the surface elevation changes slowly (or not at all) when drawing from a large tank or reservoir, $v_1 = 0$.

The *head loss due to friction* is denoted by the symbol h_f.

Example

An open reservoir with a water surface level at an elevation of 200 m drains through a 1 m diameter pipe with the outlet at an elevation of 180 m. The pipe outlet discharges to atmospheric pressure. The total head losses in the pipe and fittings are 18 m. Assume steady incompressible flow. The flow rate from the outlet is most nearly

(A) 4.9 m³/s

(B) 6.3 m³/s

(C) 31 m³/s

(D) 39 m³/s

Solution

Use the energy conservation equation, Eq. 13.14. Take point 1 at the reservoir surface and point 2 at the pipe outlet.

$$\frac{p_1}{\rho g} + z_1 + \frac{v_1^2}{2g} = \frac{p_2}{\rho g} + z_2 + \frac{v_2^2}{2g} + h_f$$

The pressure is atmospheric at the reservoir and the outlet, so $p_1 = p_2$. The velocity at the reservoir surface is $v_1 \approx 0$ m/s, so the equation reduces to

$$z_1 = z_2 + \frac{v_2^2}{2g} + h_f$$

Solve for the velocity at the pipe outlet, v_2.

$$
\begin{aligned}
v_2 &= \sqrt{2g(z_1 - z_2 - h_f)} \\
&= \sqrt{(2)\left(9.81\ \frac{m}{s^2}\right)(200\ m - 180\ m - 18\ m)} \\
&= 6.26\ m/s
\end{aligned}
$$

The flow rate out of the pipe outlet is

$$
\begin{aligned}
Q &= v_2 A = v_2 \left(\frac{\pi D^2}{4}\right) \\
&= \left(6.26\ \frac{m}{s}\right)\left(\frac{\pi(1\ m)^2}{4}\right) \\
&= 4.92\ m^3/s \quad (4.9\ m^3/s)
\end{aligned}
$$

The answer is (A).

Equation 13.15: Pressure Drop

$$p_1 - p_2 = \gamma h_f = \rho g h_f \qquad 13.15$$

Description

For a pipe of constant cross-sectional area and constant elevation, the *pressure change (pressure drop)* from one point to another is given by Eq. 13.15.

6. FRICTION LOSS

Equation 13.16 Through Eq. 13.18: Darcy-Weisbach Equation[1]

$$h_f = f \frac{L}{D} \frac{v^2}{2g} \qquad 13.16$$

$$h_f = (4f_{\text{Fanning}}) \frac{Lv^2}{D2g} = \frac{2f_{\text{Fanning}} Lv^2}{Dg} \qquad 13.17$$

$$f_{\text{Fanning}} = \frac{f}{4} \qquad 13.18$$

---

[1](1) The Weisbach frictional head loss equation is commonly presented as $h_f = fLv^2/2Dg$, as shown in the variation equation. The NCEES *FE Reference Handbook* (*NCEES Handbook*) separates out three terms, f, L/d, and $v^2/2g$, and changes the sequence of the denominator to show that the head loss is a multiple of velocity head. (2) The *Fanning friction factor* is primarily of interest to chemical engineers. Civil and mechanical engineers rarely encounter the Fanning friction factor and, to them, "friction factor" always refers to the Darcy friction factor.

Variation

$$h_f = \frac{fLv^2}{2Dg}$$

Values

Table 13.1 Specific Roughness of Typical Materials

material	ε ft	ε mm
asphalted cast iron	0.0002–0.0006	0.06–0.2
blasted rock tunnel	1.0–2.0	300–600
cast iron	0.0006–0.003	0.2–0.9
commercial steel or wrought iron	0.0001–0.0003	0.03–0.09
concrete	0.001–0.01	0.3–3.0
corrugated metal pipe	0.1–0.2	30–60
galvanized iron	0.0002–0.0008	0.06–0.2
glass, drawn brass, copper, or lead	smooth	smooth
concrete- or steel-lined large tunnel	0.002–0.004	0.6–1.2
riveted steel	0.003–0.03	0.9–9.0

Description

The *Darcy-Weisbach equation* (*Darcy equation*) is one method for calculating the frictional energy loss for fluids. It can be used for both laminar and turbulent flow.

The *Darcy friction factor*, f, is one of the parameters that is used to calculate the friction loss. One of the advantages to using the Darcy equation is that the assumption of laminar or turbulent flow does not need to be confirmed if f is known. The friction factor is not constant, but decreases as the Reynolds number (fluid velocity) increases, up to a certain point, known as *fully turbulent flow*. Once the flow is fully turbulent, the friction factor remains constant and depends only on the relative roughness of the pipe surface and not on the Reynolds number. For very smooth pipes, fully turbulent flow is achieved only at very high Reynolds numbers.

The friction factor is not dependent on the material of the pipe, but affected by its roughness. For example, for a given Reynolds number, the friction factor will be the same for any smooth pipe material (glass, plastic, smooth brass, copper, etc.).

The friction factor is determined from the *relative roughness*, ε/D, and the Reynolds number, Re. The relative roughness is calculated from the *specific roughness* of the material, ε, given in tables, and the inside diameter of the pipe. (See Table 13.1.)[2] The *Moody friction factor chart* (also known as the *Stanton diagram*), Fig. 13.2, presents the friction factor graphically. There are different lines for selected discrete values of relative roughness. Because of the complexity of this graph, it is easy to incorrectly

---

[2]The information in Table 13.1 is presented as part of the Moody (Stanton) friction factor chart, Fig. 13.2.

locate the Reynolds number or use the wrong curve. Nevertheless, the Moody chart remains the most common method of obtaining the friction factor.

Example

Water at 10°C is pumped through 300 m of steel pipe at a velocity of 2.3 m/s. The pipe has an inside diameter of 84.45 mm and a friction factor of 0.0195. The friction loss is most nearly

(A) 2.0 m

(B) 8.6 m

(C) 19 m

(D) 24 m

Solution

Use Eq. 13.16, the Darcy-Weisbach equation.

$$h_f = f\frac{L}{D}\frac{v^2}{2g}$$

$$= (0.0195)\left(\frac{300 \text{ m}}{\dfrac{84.45 \text{ mm}}{1000 \frac{\text{mm}}{\text{m}}}}\right)\left(\frac{\left(2.3 \frac{\text{m}}{\text{s}}\right)^2}{(2)\left(9.81 \frac{\text{m}}{\text{s}^2}\right)}\right)$$

$$= 18.7 \text{ m} \quad (19 \text{ m})$$

The answer is (C).

Equation 13.19: Hagen-Poiseuille Equation

$$Q = \frac{\pi R^4 \Delta p_f}{8\mu L} = \frac{\pi D^4 \Delta p_f}{128\mu L} \qquad 13.19$$

Variation

$$v = \frac{D^2 \Delta p_f}{32\mu L}$$

Description

If the flow is laminar and in a circular pipe, then the *Hagen-Poiseuille equation* can be used to calculate the flow rate. In Eq. 13.19, the Hagen-Poiseuille equation is presented in the form of a pressure drop, $\Delta p_f = \gamma h_f$.

Example

Water flows at 1.2 m/s through a horizontal pipe with an inner diameter of 8 cm. The absolute viscosity of the water is 0.001002 Pa·s. What is most nearly the pressure drop after 60 m of pipe?

(A) 97 Pa

(B) 190 Pa

(C) 300 Pa

(D) 360 Pa

Figure 13.2 Moody Friction Factor Chart

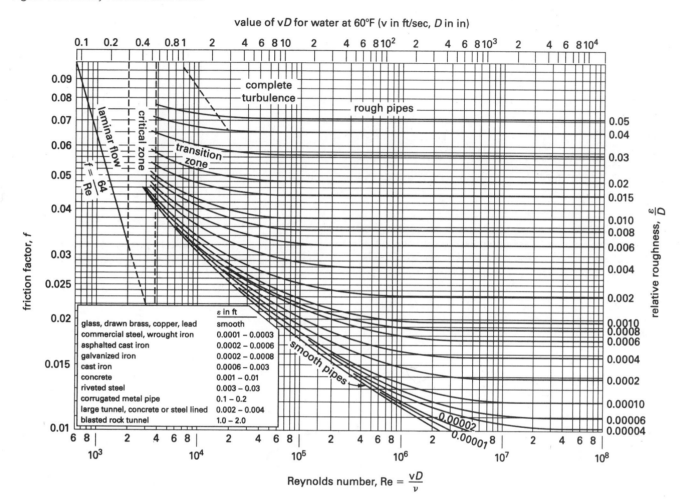

From *Friction Factor for Pipe Flow*, by L. F. Moody, copyright © 1944, by ASME. Reproduced with permission.

Solution

Use the Hagen-Poiseuille equation, Eq. 13.19.

$$Q = \frac{\pi D^4 \Delta p_f}{128 \mu L}$$

$$\Delta p_f = \frac{128 \mu L Q}{\pi D^4}$$

$$= \frac{32 \mu L v}{D^2}$$

$$= \frac{(32)(0.001002 \text{ Pa·s})(60 \text{ m})\left(1.2 \dfrac{\text{m}}{\text{s}}\right)\left(100 \dfrac{\text{cm}}{\text{m}}\right)^2}{(8 \text{ cm})^2}$$

$$= 361 \text{ Pa} \quad (360 \text{ Pa})$$

The answer is (D).

7. FLOW IN NONCIRCULAR CONDUITS

Equation 13.20: Hydraulic Radius

$$R_H = \frac{\text{cross-sectional area}}{\text{wetted perimeter}} = \frac{D_H}{4} \qquad \textit{13.20}$$

Description

The *hydraulic radius* is defined as the cross-sectional area in flow divided by the *wetted perimeter*.

The area in flow is the cross-sectional area of the fluid flowing. When a fluid is flowing under pressure in a pipe (i.e., *pressure flow*), the area in flow will be the internal area of the pipe. However, the fluid may not completely fill the pipe and may flow simply because of a sloped surface (i.e., *gravity flow* or *open channel flow*).

The wetted perimeter is the length of the line representing the interface between the fluid and the pipe or channel. It does not include the *free surface* length (i.e., the interface between fluid and atmosphere).

For a circular pipe flowing completely full, the area in flow is πR^2. The wetted perimeter is the entire circumference, $2\pi R$. The hydraulic radius in this case is half the radius of the pipe.

$$R_H = \frac{\pi R^2}{2\pi R} = \frac{R}{2} = \frac{D}{4}$$

The hydraulic radius of a pipe flowing half full is also $R/2$, since the flow area and wetted perimeter are both halved.

Many fluid, thermodynamic, and heat transfer processes are dependent on the physical length of an object. The general name for this controlling variable is *characteristic dimension*. The characteristic dimension in evaluating fluid flow is the *hydraulic diameter*, D_H. The hydraulic diameter for a full-flowing circular pipe is simply its inside diameter. If the hydraulic radius of a noncircular duct is known, it can be used to calculate the hydraulic diameter.

$$D_H = 4R_H = 4 \times \frac{\text{area in flow}}{\text{wetted perimeter}}$$

The frictional energy loss by a fluid flowing in a rectangular, annular, or other noncircular duct can be calculated from the Darcy equation by using the hydraulic diameter, D_H, in place of the diameter, D. The friction factor, f, is determined in any of the conventional manners.

Example

The 8 cm × 12 cm rectangular flume shown is filled to three-quarters of its height.

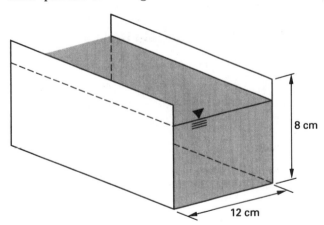

What is most nearly the hydraulic radius of the flow?

- (A) 1.5 cm
- (B) 2.5 cm
- (C) 3.0 cm
- (D) 5.0 cm

Solution

The hydraulic radius is

$$R_H = \frac{\text{cross-sectional area}}{\text{wetted perimeter}}$$

$$= \frac{(12 \text{ cm})\left(\frac{3}{4}\right)(8 \text{ cm})}{(2)\left(\frac{3}{4}\right)(8 \text{ cm}) + 12 \text{ cm}}$$

$$= 3.0 \text{ cm}$$

The answer is (C).

8. MINOR LOSSES IN PIPE FITTINGS, CONTRACTIONS, AND EXPANSIONS

In addition to the frictional energy lost due to viscous effects, friction losses also result from fittings in the line, changes in direction, and changes in flow area. These losses are known as *minor losses*, since they are usually much smaller in magnitude than the pipe wall frictional loss.

Equation 13.21 Through Eq. 13.25: Energy Conservation Equation[3]

$$\frac{p_1}{\gamma} + z_1 + \frac{v_1^2}{2g} = \frac{p_2}{\gamma} + z_2 + \frac{v_2^2}{2g} + h_f + h_{f,\text{fitting}} \qquad 13.21$$

$$\frac{p_1}{\rho g} + z_1 + \frac{v_1^2}{2g} = \frac{p_2}{\rho g} + z_2 + \frac{v_2^2}{2g} + h_f + h_{f,\text{fitting}} \qquad 13.22$$

$$h_{f,\text{fitting}} = C\frac{v^2}{2g} \qquad 13.23$$

$$\frac{v^2}{2g} = 1 \text{ velocity head} \qquad 13.24$$

$$h_{f,\text{fitting}} = 0.04v^2/2g \quad \begin{bmatrix} \text{gradual} \\ \text{contraction} \end{bmatrix} \qquad 13.25$$

Description

The energy conservation equation accounting for minor losses is Eq. 13.21 and Eq. 13.22.

The minor losses can be calculated using the *method of loss coefficients*. Each fitting has a *loss coefficient*, C, associated with it, which, when multiplied by the kinetic energy, gives the head loss. A loss coefficient is the minor head loss expressed in fractions (or multiples) of the velocity head.

---

[3](1) Although the *NCEES Handbook* uses C as the loss coefficient, it is far more common in engineering practice to use K (or sometimes k). When C is used, it is almost exclusively used for calculating flow through valves, in which case, C_v has very different values and, technically, has units. (2) Certainly, the numerical value of $v^2/2g$ is not always 1.0. Equation 13.24 in the *NCEES Handbook* is a simple definition intended to make the point that the method of loss coefficients, Eq. 13.23, calculates minor losses as multiples of velocity head.

Loss coefficients for specific fittings and valves must be known in order to be used. They cannot be derived theoretically.

Losses at pipe exits and entrances in tanks also fall under the category of minor losses. The values of C given in Fig. 13.3 account for minor losses in various exit and entrance conditions.

Figure 13.3 *C-Values for Head Loss*

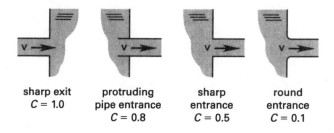

| sharp exit $C = 1.0$ | protruding pipe entrance $C = 0.8$ | sharp entrance $C = 0.5$ | round entrance $C = 0.1$ |

Some fittings have specific loss values, which are generally provided in the problem statement. The nominal value for head loss, C, in well-streamlined gradual contractions is given by Eq. 13.25.

Example

Water exits a tank through a nozzle with a diameter of 5 cm located 5 m below the surface of the water. The water level in the tank is kept constant. The loss coefficient for the nozzle is 0.5.

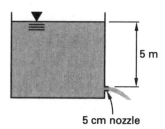

5 m

5 cm nozzle

The flow rate through the nozzle is most nearly

(A) 7.1 L/s

(B) 11 L/s

(C) 16 L/s

(D) 24 L/s

Solution

Use Eq. 13.22 and Eq. 13.23 to find the velocity through the nozzle, with location 1 at the surface of the water in the tank and location 2 at the nozzle exit.

$$\frac{p_1}{\rho g} + z_1 + \frac{v_1^2}{2g} = \frac{p_2}{\rho g} + z_2 + \frac{v_2^2}{2g} + h_f + h_{f,\text{fitting}}$$

$$= \frac{p_2}{\rho g} + z_2 + \frac{v_2^2}{2g} + h_f + C\frac{v_2^2}{2g}$$

The pressure is atmospheric at both locations, so the pressure terms cancel. The velocity at location 1 is zero. The head loss due to friction between locations 1 and 2 is zero. Rearrange to solve for v_2.

$$v_2 = \sqrt{\frac{2g(z_1 - z_2)}{C + 1}} = \sqrt{\frac{(2)\left(9.81 \ \frac{m}{s^2}\right)(5 \ m)}{0.5 + 1}}$$

$$= 8.087 \ m/s$$

The flow rate is

$$Q = Av_2 = \left(\frac{\pi D^2}{4}\right)v_2$$

$$= \left(\frac{\pi(5 \ cm)^2}{(4)\left(100 \ \frac{cm}{m}\right)^2}\right)\left(8.087 \ \frac{m}{s}\right)\left(1000 \ \frac{L}{m^3}\right)$$

$$= 15.88 \ L/s \quad (16 \ L/s)$$

The answer is (C).

9. MULTIPATH PIPELINES

A *pipe loop* is a set of two pipes placed in parallel, both originating and terminating at the same junction. (See Fig. 13.4.) Adding a second pipe in parallel with the first is a standard method of increasing the capacity of a line.

Figure 13.4 *Parallel Pipe Loop System*

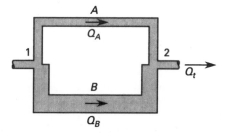

If the pipe diameters are known, the multipath flow equations can be solved simultaneously for the branch velocities. In such problems, it is common to neglect minor losses, the velocity head, and the variation in the friction factor, f, with velocity.

Equation 13.26 and Eq. 13.27: Head Loss and Flow Rate in Multipath Pipelines[4]

$$h_L = f_A \frac{L_A}{D_A} \frac{v_A^2}{2g} = f_B \frac{L_B}{D_B} \frac{v_B^2}{2g} \qquad 13.26$$

$$(\pi D^2/4)v = (\pi D_A^2/4)v_A + (\pi D_B^2/4)v_B \qquad 13.27$$

---

[4]The *NCEES Handbook* is inconsistent in the symbol used for friction head loss. Head loss, h_L, used in Eq. 13.26 is no different than head loss due to friction, h_f, used in Eq. 13.16.

Description

The relationships between flow, velocity, and pipe diameter and length are illustrated in Fig. 13.5.

Figure 13.5 *Multipath Pipeline*

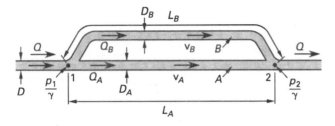

The flow divides in such a manner as to make the head loss in each branch the same.

$$h_{f,A} = h_{f,B}$$

The head loss between the two junctions is the same as the head loss in each branch.

$$h_{f,1-2} = h_{f,A} = h_{f,B}$$

The total flow rate is the sum of the flow rates in the two branches.

$$Q_t = Q_A + Q_B$$

Example

Water flows at 6 m³/s in a 1 m diameter pipeline, then divides into two branch lines that discharge to the atmosphere 1000 m from the junction. Branch A uses 0.75 m diameter pipe, and branch B uses 0.60 m diameter pipe. All branches are at the same elevation, and all pipes have the same friction factor of 0.0023. Friction losses from fittings are negligible. The flow rate in branch A is most nearly

(A) 3.5 m³/s

(B) 3.8 m³/s

(C) 4.1 m³/s

(D) 4.4 m³/s

Solution

The cross-sectional areas of the main and two branch pipes are

$$A_{main} = \frac{\pi D^2}{4} = \frac{\pi (1\ m)^2}{4} = 0.785\ m^2$$

$$A_A = \frac{\pi D_A^2}{4} = \frac{\pi (0.75\ m)^2}{4} = 0.442\ m^2$$

$$A_B = \frac{\pi D_B^2}{4} = \frac{\pi (0.60\ m)^2}{4} = 0.283\ m^2$$

The velocity in the 1 m main pipe is

$$v_{main} = \frac{Q}{A_{main}} = \frac{6\ \frac{m^3}{s}}{0.785\ m^2} = 7.64\ m/s$$

Find the relationship between v_A and v_B from Eq. 13.26.

$$f_A \frac{L_A}{D_A} \frac{v_A^2}{2g} = f_B \frac{L_B}{D_B} \frac{v_B^2}{2g}$$

$$(0.0023)\left(\frac{1000\ m}{0.75\ m}\right)\left(\frac{v_A^2}{2g}\right) = (0.0023)\left(\frac{1000\ m}{0.60\ m}\right)\left(\frac{v_B^2}{2g}\right)$$

$$v_B = v_A \sqrt{\frac{0.60\ m}{0.75\ m}} = 0.894 v_A$$

Find another relationship between v_A and v_B from Eq. 13.27, then solve for v_A.

$$Q = A_A v_A + A_B v_B$$

$$6\ \frac{m^3}{s} = (0.442\ m^2) v_A + (0.283\ m^2) v_B$$

$$\begin{aligned} v_A &= \frac{6\ \frac{m^3}{s}}{0.442\ m^2} - 0.64 v_B \\ &= 13.58\ \frac{m}{s} - (0.64)(0.894 v_A) \\ &= 8.637\ m/s \end{aligned}$$

$$\begin{aligned} v_B &= 0.894 v_A = (0.894)\left(8.637\ \frac{m}{s}\right) \\ &= 7.725\ m/s \end{aligned}$$

The flow rates in the two branches are

$$\begin{aligned} Q_A &= A_A v_A = (0.442\ m^2)\left(8.637\ \frac{m}{s}\right) \\ &= 3.816\ m^3/s \quad (3.8\ m^3/s) \end{aligned}$$

$$\begin{aligned} Q_B &= A_B v_B = (0.283\ m^2)\left(7.725\ \frac{m}{s}\right) \\ &= 2.184\ m^3/s \quad (2.2\ m^3/s) \end{aligned}$$

Calculate the total flow rate to check the calculation.

$$\begin{aligned} Q_t &= Q_A + Q_B = 3.8\ \frac{m^3}{s} + 2.2\ \frac{m^3}{s} \\ &= 6\ m^3/s \end{aligned}$$

The answer is (B).

10. OPEN CHANNEL AND PARTIAL-AREA PIPE FLOW

An *open channel* is a fluid passageway that allows part of the fluid to be exposed to the atmosphere. This type of channel includes natural waterways, canals, culverts, flumes, and pipes flowing under the influence of gravity (as opposed to pressure conduits, which always flow full). A *reach* is a straight section of open channel with uniform shape, depth, slope, and flow quantity.

Equation 13.28: Manning's Equation

$$v = (K/n)R_H^{2/3}S^{1/2} \qquad \textbf{13.28}$$

Values

SI units	$K = 1$
customary U.S. units	$K = 1.486$

Description

Manning's equation has typically been used to estimate the velocity of flow in any open channel. It depends on the hydraulic radius, R_H, the slope of the energy grade line, S, and a dimensionless *Manning's roughness coefficient*, n. A conversion constant, K, modifies the equation for use with SI or customary U.S. units. The slope of the energy grade line is the terrain grade (slope) for uniform open-channel flow. The Manning roughness coefficient is typically taken as 0.013 for concrete.

Example

Water flows through the open concrete channel shown. Assume a Manning roughness coefficient of 0.013 for concrete.

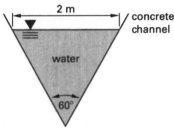

What is most nearly the minimum slope needed to maintain the flow at 3 m³/s?

(A) 0.00015

(B) 0.00052

(C) 0.0015

(D) 0.0052

Solution

The area of flow is that of an equilateral triangle 2 m on each side, so

$$A = \frac{\sqrt{3}a^2}{4} = \frac{\sqrt{3}(2 \text{ m})^2}{4} = 1.732 \text{ m}^2$$

The hydraulic radius is

$$R_H = \frac{\text{cross-sectional area}}{\text{wetted perimeter}} = \frac{1.732 \text{ m}^2}{2 \text{ m} + 2 \text{ m}} = 0.433 \text{ m}$$

The velocity needed is

$$v = \frac{Q}{A} = \frac{3 \frac{\text{m}^3}{\text{s}}}{1.732 \text{ m}^2} = 1.732 \text{ m/s}$$

Rearrange Manning's equation to solve for slope.

$$v = (K/n)R_H^{2/3}S^{1/2}$$

$$S = \left(\frac{vn}{KR_H^{2/3}} \right)^2$$

$$= \left(\frac{\left(1.732 \frac{\text{m}}{\text{s}}\right)(0.013)}{(1)(0.433 \text{ m})^{2/3}} \right)^2$$

$$= 0.001507 \quad (0.0015)$$

The answer is (C).

Equation 13.29: Hazen-Williams Equation

$$v = k_1 C R_H^{0.63} S^{0.54} \qquad \textbf{13.29}$$

Values

SI units	$k_1 = 0.849$
customary U.S. units	$k_1 = 1.318$

Description

Although Manning's equation can be used for circular pipes flowing less than full, the *Hazen-Williams equation* is used more often. A conversion constant, k_1, is used to modify the Hazen-Williams equation for use with SI or customary U.S. units.[5] The *Hazen-Williams roughness coefficient*, C, has a typical range of 100 to 130 for most materials as shown in Table 13.2, although very smooth materials can have higher values.

Table 13.2 Values of Hazen-Williams Coefficient, C

pipe material	C
ductile iron	140
concrete (regardless of age)	130
cast iron:	
new	130
5 yr old	120
20 yr old	100
welded steel, new	120
wood stave (regardless of age)	120
vitrified clay	110
riveted steel, new	110
brick sewers	100
asbestos-cement	140
plastic	150

---

[5]There is no significance to the subscript or uppercase and lowercase usage in the conversion factor. While the *NCEES Handbook* chose to use K in Eq. 13.28 and Eq. 13.29, it could just as easily have chosen to use K_1 and K_2, or k_1 and k_2.

Example

Stormwater flows through a square concrete pipe that has a hydraulic radius of 1 m and an energy grade line slope of 0.8. Using the Hazen-Williams equation, the velocity of the water is most nearly

(A) 43 m/s

(B) 67 m/s

(C) 98 m/s

(D) 150 m/s

Solution

Use the Hazen-Williams equation, Eq. 13.29, to calculate the velocity. From Table 13.2, the Hazen-Williams roughness coefficient for concrete is 130.

$$v = k_1 C R_H^{0.63} S^{0.54} = (0.849)(130)(1 \text{ m})^{0.63}(0.8)^{0.54}$$
$$= 97.8 \text{ m/s} \quad (98 \text{ m/s})$$

The answer is (C).

Equation 13.30 Through Eq. 13.32: Circular Pipe Head Loss[6]

$$h_f = \frac{4.73L}{C^{1.852}D^{4.87}} Q^{1.852} \quad \text{[U.S. only]} \qquad \textit{13.30}$$

$$p = \frac{4.52Q^{1.85}}{C^{1.85}D^{4.87}} \quad \text{[U.S. only]} \qquad \textit{13.31}$$

$$p = \frac{6.05Q^{1.85}}{C^{1.85}D^{4.87}} \times 10^5 \quad \text{[SI only]} \qquad \textit{13.32}$$

Description

Civil engineers commonly use the *Hazen-Williams equation* to calculate head loss. This method requires knowing the Hazen-Williams *roughness coefficient*, C. The advantage of using this equation is that C does not depend on the Reynolds number. The Hazen-Williams equation is empirical and is not dimensionally homogeneous.

Equation 13.30 gives the head loss expressed in feet. L is in feet, Q is in cubic feet per second, and D is in inches.[7]

Equation 13.31 and Eq. 13.32 give the head loss expressed as pressure in customary U.S. and SI units, respectively. In Eq. 13.31, p is in pounds per square inch per foot of pipe, Q is in gallons per minute, and D is in

inches.[8] In Eq. 13.32, p is in bars per meter of pipe, Q is in liters per minute, and D is in millimeters.[9]

Equation 13.33: Specific Energy

$$E = \alpha \frac{v^2}{2g} + y = \frac{\alpha Q^2}{2gA^2} + y \qquad \textit{13.33}$$

Description

Specific energy, E, is a term used primarily with open channel flow. It is the total head with respect to the channel bottom. Because the channel bottom is the reference elevation for potential energy, potential energy does not contribute to specific energy; only kinetic energy and pressure energy contribute. α is the kinetic energy correction factor, which is usually equal to 1.0. The pressure head at the channel bottom is equal to the depth of the channel, y.

In *uniform flow* (flow with constant width and depth), total head decreases due to the frictional effects (e.g., elevation increase), but specific energy is constant. In nonuniform flow, total head also decreases, but specific energy may increase or decrease.

Equation 13.34 Through Eq. 13.36: Critical Depth

$$y_c = \left(\frac{q^2}{g}\right)^{1/3} \qquad \textit{13.34}$$

$$q = Q/B \qquad \textit{13.35}$$

$$\frac{Q^2}{g} = \frac{A^3}{T} \qquad \textit{13.36}$$

Description

For any channel, there is some depth of flow that will minimize the energy of flow. (The depth is not minimized, however.) This depth is known as the *critical depth*, y_c. The critical depth depends on the shape of the channel, but it is independent of the channel slope.

For rectangular channels, the critical depth can be found with Eq. 13.34. q in this equation is *unit discharge*, the flow per unit width. q is defined in Eq. 13.35 as the ratio of the flow rate, Q, to the channel width, B.

For channels with nonrectangular shapes (including trapezoidal channels), the critical depth can be found by trial and error from Eq. 13.36. T is the surface width. To use this equation, assume trial values of the critical depth,

---

[6]The *NCEES Handbook* uses the exponents 1.852 and 1.85 interchangeably.

[7]In many expressions of the Hazen-Williams equation in engineering literature, including Eq. 13.31 in the *NCEES Handbook*, Q is in gallons per minute. Without an expressed statement, it is not possible to know which units are to be used.

[8]The *NCEES Handbook* is inconsistent in how it represents frictional pressure loss. p used in Eq. 13.31 is the same concept as Δp_f used in Eq. 13.19 except expressed on a per unit length basis. The units used in Eq. 13.31 are not the same as the units used in Eq. 13.30.

[9]Some of these units are traditional metric but not standard SI units. Without an expressed statement, it is not possible to know which units are to be used.

use them to calculate the dependent quantities in the equation, and then verify the equality.

Figure 13.6 illustrates how specific energy is affected by depth, and accordingly, how specific energy relates to critical depth.

Figure 13.6 Specific Energy Diagram

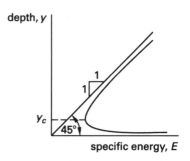

Equation 13.37 and Eq. 13.38: Froude Number

$$Fr = \frac{v}{\sqrt{gy_h}} \qquad 13.37$$

$$y_h = A/T \qquad 13.38$$

Description

Equation 13.37 is the formula for the dimensionless *Froude number*, Fr, a convenient index of the flow regime. v is velocity. y_h is the *characteristic length*, also referred to as the *characteristic (length) scale, hydraulic depth, mean hydraulic depth,* and others, depending on the channel configuration.[10] For a circular channel flowing half full, $y_h = \pi D/8$. For a rectangular channel, $y_h = d$, the depth corresponding to velocity v. For trapezoidal and semicircular channels, and in general, y_h is the area in flow divided by the top width.

The Froude number can be used to determine whether the flow is subcritical or supercritical. When the Froude number is less than one, the flow is subcritical. The depth of flow is greater than the critical depth, and the flow velocity is less than the critical velocity.

When the Froude number is greater than one, the flow is supercritical. The depth of flow is less than critical depth, and the flow velocity is greater than the critical velocity.

When the Froude number is equal to one, the flow is critical.[11]

Example

A 150 m long surface vessel with a speed of 40 km/h is modeled at a scale of 1:50. Similarity based on Froude

numbers is appropriate for the modeling of surface vessels, so the model should travel at a speed of most nearly

(A) 0.22 m/s

(B) 1.6 m/s

(C) 2.2 m/s

(D) 16 m/s

Solution

The surface vessel and the model should have the same Froude numbers. From Eq. 13.37,

$$Fr_{vessel} = Fr_{model}$$

$$\frac{v_{vessel}}{\sqrt{gy_{h,vessel}}} = \frac{v_{model}}{\sqrt{gy_{h,model}}}$$

$$v_{model} = v_{vessel}\sqrt{\frac{y_{h,model}}{y_{h,vessel}}}$$

$$= \frac{\left(40\ \frac{km}{h}\right)\left(1000\ \frac{m}{km}\right)}{3600\ \frac{s}{h}}\sqrt{\frac{1}{50}}$$

$$= 1.57\ m/s \quad (1.6\ m/s)$$

The answer is (B).

11. THE IMPULSE-MOMENTUM PRINCIPLE

The *momentum*, **P**, of a moving object is a vector quantity defined as the product of the object's mass and velocity.

$$\mathbf{P} = m\mathbf{v}$$

The *impulse*, **I**, of a constant force is calculated as the product of the force's magnitude and the length of time the force is applied.

$$\mathbf{I} = \mathbf{F}\Delta t$$

The *impulse-momentum principle* states that the impulse applied to a body is equal to the change in that body's momentum. This is one way of stating *Newton's second law.*

$$\mathbf{I} = \Delta \mathbf{P}$$

From this,

$$F\Delta t = m\Delta v$$

$$= m(v_2 - v_1)$$

For fluid flow, there is a mass flow rate, $\dot{m}$, but no mass per se. Since $\dot{m} = m/\Delta t$, the impulse-momentum equation can be rewritten as

$$F = \dot{m}\Delta v$$

---

[10]The *NCEES Handbook* uses both H and h as subscripts to designate "hydraulic."

[11]The similarity of the Froude number to the Mach number used to classify gas flows is more than coincidental. Both bodies of knowledge employ parallel concepts.

Substituting for the mass flow rate, $\dot{m} = \rho A v$. The quantity $Q \rho v$ is the *rate of momentum*.

$$F = \rho A v \Delta v$$
$$= Q \rho \Delta v$$

Equation 13.39: Control Volume[12]

$$\sum F = Q_2 \rho_2 v_2 - Q_1 \rho_1 v_1 \qquad \text{13.39}$$

Description

Equation 13.39 results from applying the impulse-momentum principle to a control volume. $\sum F$ is the resultant of all external forces acting on the control volume. $Q_1 \rho_1 v_1$ and $Q_2 \rho_2 v_2$ represent the rate of momentum of the fluid entering and leaving the control volume, respectively, in the same direction as the force.

12. PIPE BENDS, ENLARGEMENTS, AND CONTRACTIONS

The impulse-momentum principle illustrates that fluid momentum is not always conserved when the fluid is acted upon by an external force. Examples of external forces are gravity (considered zero for horizontal pipes), gage pressure, friction, and turning forces from walls and vanes. Only if these external forces are absent is fluid momentum conserved.

When a fluid enters a pipe fitting or bend, as illustrated in Fig. 13.7, momentum is changed. Since the fluid is confined, the forces due to static pressure must be included in the analysis. The effects of gravity and friction are neglected.

Equation 13.40 Through Eq. 13.43: Forces on Pipe Bends

$$p_1 A_1 - p_2 A_2 \cos \alpha - F_x = Q \rho (v_2 \cos \alpha - v_1) \qquad \text{13.40}$$

$$F_y - W - p_2 A_2 \sin \alpha = Q \rho (v_2 \sin \alpha - 0) \qquad \text{13.41}$$

$$F = \sqrt{F_x^2 + F_y^2} \qquad \text{13.42}$$

$$\theta = \tan^{-1} \frac{F_y}{F_x} \qquad \text{13.43}$$

[12]While any symbol can be used to represent any quantity, the use of bold letters is usually reserved for vector quantities. The *NCEES Handbook* uses bold F to designate force in its fluid dynamics section. While force can indeed be a vector quantity, the fluid equations in this section are not vector equations. F in text and $\mathbf{F}$ in illustrations should be interpreted as F, as it is presented in this book.

Figure 13.7 Forces on a Pipe Bend

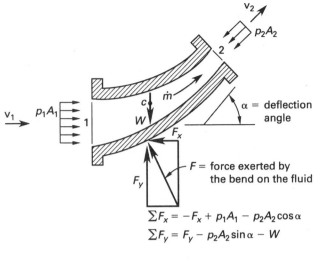

$$\sum F_x = -F_x + p_1 A_1 - p_2 A_2 \cos \alpha$$
$$\sum F_y = F_y - p_2 A_2 \sin \alpha - W$$

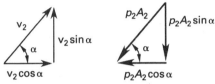

Variations

$$-F_x = p_2 A_2 \cos \alpha - p_1 A_1 + Q \rho (v_2 \cos \alpha - v_1)$$

$$F_y = (p_2 A_2 + Q \rho v_2) \sin \alpha + m_{\text{fluid}} g \quad \begin{bmatrix} \text{fluid weight} \\ \text{included} \end{bmatrix}$$

Description

Applying Eq. 13.39 to the fluid in the pipe bend in Fig. 13.7 gives these equations for the force of the bend on the fluid. m_{fluid} and $W_{\text{fluid}} = m_{\text{fluid}} g$ are the mass and weight, respectively, of the fluid in the bend (often neglected). Equation 13.40 is the thrust equation for a pipe bend. The resultant force is calculated from Eq. 13.42. The value of the angle θ is found from Eq. 13.43.

Example

A 1 m penstock is anchored by a thrust block at a point where the flow makes a 20° change in direction. The water flow rate is 5.25 m³/s. The water pressure is 140 kPa everywhere in the penstock. If the initial flow direction is parallel to the x-direction, the magnitude of the force on the thrust block in the x-direction is most nearly

(A) 6.8 kN
(B) 8.3 kN
(C) 8.7 kN
(D) 9.2 kN

Solution

The pipe area is

$$A = \frac{\pi D^2}{4} = \frac{\pi (1 \text{ m})^2}{4}$$
$$= 0.7854 \text{ m}^2$$

The velocity is

$$v = \frac{Q}{A} = \frac{5.25 \frac{\text{m}^3}{\text{s}}}{0.7854 \text{ m}^2}$$
$$= 6.68 \text{ m/s}$$

Use the thrust equation for a pipe bend, Eq. 13.40.

$$p_1 A_1 - p_2 A_2 \cos \alpha - F_x = Q\rho(v_2 \cos \alpha - v_1)$$
$$-F_x = p_2 A_2 \cos \alpha - p_1 A_1$$
$$+ Q\rho(v_2 \cos \alpha - v_1)$$
$$= (1.4 \times 10^5 \text{ Pa})(0.7854 \text{ m}^2)$$
$$\times (\cos 20° - 1)$$
$$+ \left(5.25 \frac{\text{m}^3}{\text{s}}\right)\left(1000 \frac{\text{kg}}{\text{m}^3}\right)$$
$$\times \left(6.68 \frac{\text{m}}{\text{s}}\right)$$
$$\times (\cos 20° - 1)$$
$$= -8748 \text{ N} \quad (8.7 \text{ kN})$$

The answer is (C).

13. JET PROPULSION[13]

A basic application of the impulse-momentum principle is *jet propulsion*. The velocity of a fluid jet issuing from an orifice in a tank can be determined by comparing the total energies at the free fluid surface and at the jet itself. (See Fig. 13.8.) At the fluid surface, $p_1 = 0$ (atmospheric) and $v_1 = 0$. The only energy the fluid has is potential energy. At the jet, $p_2 = 0$ and $z_2 = 0$. All of the potential energy difference has been converted to kinetic energy. The change in momentum of the fluid produces a force.

Impulse due to a constant force, F, acting over a duration Δt, is $F\Delta t$. The change in momentum of a constant mass is $m\Delta v$. The developed force can be derived from the impulse-momentum principle.

$$F\Delta t = m\Delta v$$
$$F = \frac{m}{\Delta t}\Delta v = \dot{m}v_2$$

---

[13]"Jet propulsion" is probably too generic a description since this is a special case of a *reaction force* due to a liquid discharging from an orifice.

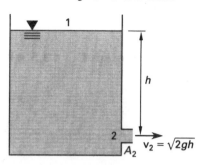

Figure 13.8 *Fluid Jet Issuing from a Tank Orifice*

Equation 13.44 Through Eq. 13.47: Discharge from an Orifice

$F = Q\rho(v_2 - 0)$	*13.44*
$F = 2\gamma h A_2$	*13.45*
$Q = A_2\sqrt{2gh}$	*13.46*
$v_2 = \sqrt{2gh}$	*13.47*

Description

The governing equations for reaction force due to discharge from an orifice are Eq. 13.44 and Eq. 13.45.

For the Bernoulli equation (see Eq. 13.4 and Eq. 13.5), it is easy to calculate the initial jet velocity (known as *Torricelli's speed of efflux*), Eq. 13.47.

14. DEFLECTORS AND BLADES

Equation 13.48 and Eq. 13.49: Fixed Blade

$-F_x = Q\rho(v_2 \cos \alpha - v_1)$	*13.48*
$F_y = Q\rho(v_2 \sin \alpha - 0)$	*13.49*

Description

Figure 13.9 illustrates a fluid jet being turned through an angle, α, by a *fixed blade* (also called a *fixed* or *stationary vane*). It is common to assume that $|v_2| = |v_1|$, although this will not be strictly true if friction between the blade and fluid is considered. Since the fluid is both retarded (in

Figure 13.9 *Open Jet on a Stationary Blade*

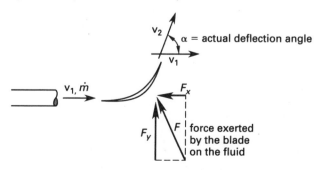

the x-direction) and accelerated (in the y-direction), there will be two components of blade force on the fluid.

Equation 13.50 Through Eq. 13.53: Moving Blade

$$-F_x = Q\rho(v_{2x} - v_{1x}) \qquad \textbf{13.50}$$

$$-F_x = -Q\rho(v_1 - v)(1 - \cos\alpha) \qquad \textbf{13.51}$$

$$F_y = Q\rho(v_{2y} - v_{1y}) \qquad \textbf{13.52}$$

$$F_y = +Q\rho(v_1 - v)\sin\alpha \qquad \textbf{13.53}$$

Variation

$$F_y = \frac{Q\rho(v_1 - v)\sin\alpha}{g_c}$$

Description

If a blade is moving away at velocity v from the source of the fluid jet, only the *relative velocity difference* between the jet and blade produces a momentum change. Furthermore, not all of the fluid jet overtakes the moving blade. (See Fig. 13.10.)

Figure 13.10 *Open Jet on a Moving Blade*

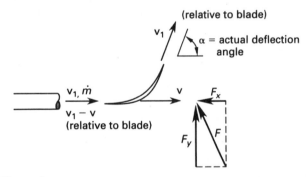

Example

A water jet impinges upon a retreating blade at a velocity of 10 m/s and mass rate of 1 kg/s, as shown.

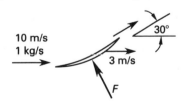

If the blade velocity is a constant 3 m/s, the magnitude of the force, F, that the blade imposes on the jet is most nearly

(A) 2.8 N

(B) 3.6 N

(C) 4.7 N

(D) 6.1 N

Solution

Using Eq. 13.51, find the x-component of the force.

$$\begin{aligned}
-F_x &= -Q\rho(v_1 - v)(1 - \cos\alpha) \\
&= -\dot{m}(v_1 - v)(1 - \cos\alpha) \\
&= -\left(1\,\frac{\text{kg}}{\text{s}}\right)\left(10\,\frac{\text{m}}{\text{s}} - 3\,\frac{\text{m}}{\text{s}}\right)(1 - \cos 30°) \\
&= -0.9378\ \text{N}
\end{aligned}$$

Find the y-component of the force from Eq. 13.53.

$$\begin{aligned}
F_y &= +Q\rho(v_1 - v)\sin\alpha \\
&= \dot{m}(v_1 - v)\sin\alpha \\
&= \left(1\,\frac{\text{kg}}{\text{s}}\right)\left(10\,\frac{\text{m}}{\text{s}} - 3\,\frac{\text{m}}{\text{s}}\right)\sin 30° \\
&= 3.5\ \text{N}
\end{aligned}$$

From Eq. 13.42, the magnitude of the force is

$$\begin{aligned}
F &= \sqrt{F_x^2 + F_y^2} \\
&= \sqrt{(-0.9378\ \text{N})^2 + (3.5\ \text{N})^2} \\
&= 3.62\ \text{N} \quad (3.6\ \text{N})
\end{aligned}$$

The answer is (B).

15. IMPULSE TURBINE

An *impulse turbine* consists of a series of blades (buckets or vanes) mounted around a wheel. (See Fig. 13.11.) The power transferred from a fluid jet to the blades of a turbine is calculated from the x-component of force on the blades. The y-component of force does no work. v is the tangential blade velocity.

Figure 13.11 *Impulse Turbine*

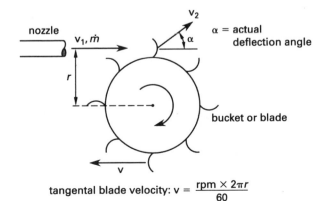

tangental blade velocity: $v = \dfrac{\text{rpm} \times 2\pi r}{60}$

Equation 13.54 Through Eq. 13.56: Power from an Impulse Turbine

$$\dot{W} = Q\rho(v_1 - v)(1 - \cos\alpha)v \qquad 13.54$$

$$\dot{W}_{\text{max}} = Q\rho(v_1^2/4)(1 - \cos\alpha) \qquad 13.55$$

$$\dot{W}_{\text{max}} = (Q\rho v_1^2)/2 = (Q\gamma v_1^2)/2g \quad [\alpha = 180°] \qquad 13.56$$

Description

The maximum theoretical tangential blade velocity is the velocity of the jet: $v = v_1$. This is known as the *runaway speed* and can only occur when the turbine is unloaded. If Eq. 13.54 is maximized with respect to v, however, the maximum power will be found to occur when the blade is traveling at half of the jet velocity: $v = v_1/2$. The power (force) is also affected by the deflection angle of the blade. Power is maximized when $\alpha = 180°$. Figure 13.12 illustrates the relationship between power and the variables α and v.

Figure 13.12 Turbine Power

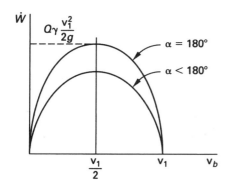

Equation 13.56 is a simplified version of Eq. 13.55, derived by substituting $\alpha = 180°$ and $v = v_1/2$.

Example

A 200 kg/s stream of water leaves a nozzle and strikes a bucket whose angle is 120° from the horizontal. The total head as the water leaves the nozzle is 15 m, and the 25 cm diameter turbine runner is turning at 500 rpm. The power produced is most nearly

(A) 9.3 kW

(B) 14 kW

(C) 21 kW

(D) 32 kW

Solution

The radius of the turbine runner is

$$r = \frac{D}{2} = \frac{25 \text{ cm}}{(2)\left(100 \frac{\text{cm}}{\text{m}}\right)} = 0.125 \text{ m}$$

The blade velocity is

$$v = \frac{\text{rpm} \times 2\pi r}{60}$$

$$= \frac{\left(500 \frac{\text{rev}}{\text{min}}\right)2\pi(0.125 \text{ m})}{60 \frac{\text{s}}{\text{min}}}$$

$$= 6.54 \text{ m/s}$$

From Eq. 13.47, the velocity of the stream is

$$v_1 = \sqrt{2gh}$$

$$= \sqrt{(2)\left(9.81 \frac{\text{m}}{\text{s}^2}\right)(15 \text{ m})}$$

$$= 17.16 \text{ m/s}$$

Use Eq. 13.54 to find the power produced.

$$\dot{W} = Q\rho(v_1 - v)(1 - \cos\alpha)v$$

$$= \dot{m}(v_1 - v)(1 - \cos\alpha)v$$

$$= \left(200 \frac{\text{kg}}{\text{s}}\right)\left(17.16 \frac{\text{m}}{\text{s}} - 6.54 \frac{\text{m}}{\text{s}}\right)$$

$$\times (1 - \cos 120°)\left(6.54 \frac{\text{m}}{\text{s}}\right)$$

$$= 20\,833 \text{ W} \quad (21 \text{ kW})$$

The answer is (C).

16. DRAG

Equation 13.57: Drag Force

$$F_D = \frac{C_D \rho v^2 A}{2} \qquad 13.57$$

Description

Drag is a frictional force that acts parallel but opposite to the direction of motion. Drag is made up of several components (e.g., skin friction and pressure drag), but the total drag force can be calculated from a dimensionless *drag coefficient*, C_D. Dimensional analysis shows that the drag coefficient depends only on the Reynolds number.

In most cases, the area, A, to be used is the projected area (i.e., the frontal area) normal to the stream. This is appropriate for spheres, disks, and vehicles. It is also appropriate for cylinders and ellipsoids that are oriented such that their longitudinal axes are perpendicular to the flow. In a few cases (e.g., airfoils and flat plates parallel to the flow), the area is the projection of the object onto a plane parallel to the stream.

Equation 13.58 and Eq. 13.59: Drag Coefficients for Flat Plates Parallel with the Flow

$$C_D = 1.33/\text{Re}^{0.5} \quad [10^4 < \text{Re} < 5 \times 10^5] \qquad \textbf{13.58}$$

$$C_D = 0.031/\text{Re}^{1/7} \quad [10^6 < \text{Re} < 10^9] \qquad \textbf{13.59}$$

Description

Drag coefficients vary considerably with Reynolds numbers, often showing regions of distinctly different behavior. For that reason, the drag coefficient is often plotted. (Figure 13.13 illustrates the drag coefficient for spheres and circular flat disks oriented perpendicular to the flow.) Semiempirical equations can be used to calculate drag coefficients as long as the applicable ranges of Reynolds numbers are stated. For example, for flat plates placed parallel to the flow, Eq. 13.58 and Eq. 13.59 can be used.

Equation 13.58 and Eq. 13.59 calculate the drag coefficient for one side of a plate. If the drag force for an entire plate is needed (the usual case), the drag force calculated from Eq. 13.57 would be doubled.

Equation 13.58 and Eq. 13.59 represent the average *skin friction coefficient* over an entire plate length, L, measured parallel to the flow moving with *far-field* (undisturbed) *velocity*, v_∞. The value of the local skin friction coefficient varies along the length of the plate.

Whether to use Eq. 13.58 or Eq. 13.59 depends on the length of the plate, which in turn, affects the Reynolds number. In skin friction calculations, the laminar flow regime extends up to a Reynolds number of approximately 5×10^5. Equation 13.58 is valid for laminar flow, while Eq. 13.59 is valid for turbulent flow. For long plates, both regimes are present simultaneously with a *transition region* in between, although Eq. 13.59 takes that into consideration in its averaging.

$$\text{Re} = \frac{\rho v_\infty L}{\mu} = \frac{v_\infty L}{\nu}$$

Equation 13.60 and Eq. 13.61: Drag Coefficient for Airfoils

$$C_D = C_{D\infty} + \frac{C_L^2}{\pi AR} \qquad \textbf{13.60}$$

$$AR = \frac{b^2}{A_p} = \frac{A_p}{c^2} \qquad \textbf{13.61}$$

Variations

$$AR = \frac{b}{c}$$

$$A_p = bc$$

Description

Drag force also acts on airfoils parallel but opposite to the direction of motion. The drag coefficient for airfoils is approximated by Eq. 13.60. $C_{D\infty}$ is the *drag coefficient at zero lift*, also known as the *zero lift drag coefficient* and *infinite span drag coefficient*. C_L is the lift coefficient.[14] This value is determined by the type of object in motion.

The dimensions of an airfoil or wing are frequently given in terms of chord length and aspect ratio. The *chord length*, c, is the front-to-back dimension of the airfoil. The *aspect ratio*, AR, is the ratio of the *span* (wing length) to chord length.[15] The area, A_p, in Eq. 13.61 is the airfoil's area projected onto the plane of the chord. For a rectangular airfoil, $A_p = $ chord $\times$ span.

17. LIFT

Lift is a force that is exerted on an object (flat plate, airfoil, rotating cylinder, etc.) as the object passes through a fluid. Lift combines with drag to form the resultant force on the object, as shown in Fig. 13.14.

The generation of lift from air flowing over an airfoil is predicted by the Bernoulli equation. Air molecules must travel a longer distance over the top surface of the airfoil than over the lower surface, and, therefore, they travel faster over the top surface. Since the total energy of the air is constant, the increase in kinetic energy comes at the expense of pressure energy. The static pressure on the top of the airfoil is reduced, and a net upward force is produced.

Within practical limits, the lift produced can be increased at lower speeds by increasing the curvature of the wing. This increased curvature is achieved by the use of *flaps*. (See Fig. 13.15.) When a plane is traveling slowly (e.g., during takeoff or landing), its flaps are extended to create the lift needed.

Equation 13.62: Lift Force

$$F_L = \frac{C_L \rho v^2 A_p}{2} \qquad \textbf{13.62}$$

Description

The lift produced, F_L, can be calculated from Eq. 13.62, whose use is not limited to airfoils.

---

[14]The drag coefficient at zero lift is traditionally (and, almost universally) represented by the symbol $C_{D,0}$ or similar. The *NCEES Handbook* calls this the "infinite span drag coefficient" and uses the symbol $C_{D\infty}$, both practices of which are unusual and confusing. For an infinite span, the aspect ratio will be infinitely large, and Eq. 13.60 will reduce to an "infinite span drag coefficient" by definition. Since Eq. 13.60 is for an airfoil, the exclusion of the *wing span efficiency factor* (*Oswald efficiency factor*) without identifying the wing as having an elliptical shape is misleading.

[15]The nomenclature used by the *NCEES Handbook* for aspect ratio makes it difficult to distinguish AR from $A \times R$, except through context.

Figure 13.13 Drag Coefficients for Spheres and Circular Flat Disks

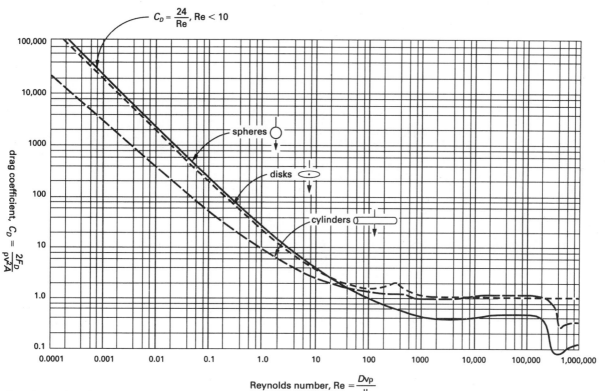

Note: Intermediate divisions are 2, 4, 6, and 8.

Figure 13.14 Lift and Drag on an Airfoil

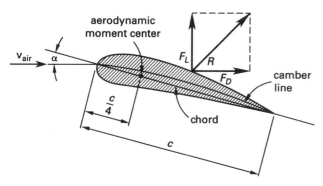

Figure 13.15 Use of Flaps in an Airfoil

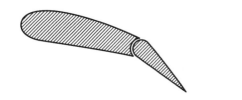

Equation 13.63: Coefficient of Lift (Flat Plate)

$$C_L = 2\pi k_1 \sin(\alpha + \beta) \qquad \textbf{13.63}$$

Description

The dimensionless *coefficient of lift*, C_L, is used to measure the effectiveness of the airfoil. The coefficient of lift depends on the shape of the airfoil and the Reynolds number. No simple relationship can be given for calculating the coefficient of lift for airfoils, but the theoretical coefficient of lift for a thin plate in two-dimensional flow at a low *angle of attack*, α, is given by Eq. 13.63.[16] Actual airfoils are able to achieve only 80–90% of this theoretical value. k_1 is a constant of proportionality, and β is the negative of the angle of attack for zero lift.[17]

For various reasons (primarily wingtip vortices), real airfoils of finite length generate a small amount of *downwash*. This downwash reduces the *geometric angle of attack* (or equivalently, reduces the coefficient of lift) slightly by the *induced (drag) angle of attack*, resulting in an *effective angle of attack*. Equation 13.63 conveys this reduction in angle of attack and is an important part of *Prandtl lifting-line theory*.[18]

$$\alpha_{\text{effective}} = \alpha_{\text{geometric}} - \alpha_{\text{induced}}$$

The coefficient of lift for an airfoil cannot be increased without limit merely by increasing α. Eventually, the *stall angle* is reached, at which point the coefficient of lift decreases dramatically. (See Fig. 13.16.)

---

[16]The angle of attack is the geometric angle between the relative wind and the straight chord line, as shown in Fig. 13.14.

[17]k_1 in Eq. 13.63 in the *NCEES Handbook* is not the same as k_1 in Eq. 13.29.

[18]Rather than simply showing a subtraction operation, the *NCEES Handbook* combines the induced angle of attack with the mathematical operation by referring to it as "the negative of the angle of attack for zero lift."

Figure 13.16 *Typical Plot of Lift Coefficient*

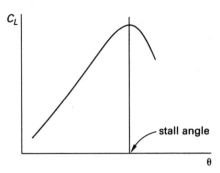

Equation 13.64: Aerodynamic Moment

$$M = \frac{C_M \rho \mathrm{v}^2 A_p c}{2} \qquad 13.64$$

Description

The *aerodynamic moment (pitching moment)*, M, is applied at the aerodynamic center of an airfoil. It is calculated at the quarter point using Eq. 13.64. C_M is the moment coefficient.

14 Fluid Measurement and Similitude

Nomenclature

A	area	m^2
C	coefficient	–
Ca	Cauchy number	–
d	depth	m
D	diameter	m
E	specific energy	J/kg
F	force	N
Fr	Froude number	–
g	gravitational acceleration, 9.81	m/s^2
h	head	m
h	head loss	m
h	height	m
l	characteristic length	m
p	pressure	Pa
Q	flow rate	m^3/s
Re	Reynolds number	–
v	velocity	m/s
We	Weber number	–
z	elevation	m

Symbols

γ	specific (unit) weight	N/m^3
μ	absolute viscosity	Pa·s
ρ	density	kg/m^3
σ	surface tension	N/m

Subscripts

0	stagnation (zero velocity)
c	contraction
E	elastic
G	gravitational
I	inertial
m	manometer fluid or model
p	constant pressure or prototype
s	static
T	surface tension
v	velocity
v	constant volume

1. PITOT TUBE

A *pitot tube* is simply a hollow tube that is placed longitudinally in the direction of fluid flow, allowing the flow to enter one end at the fluid's *velocity of approach*. (See Fig. 14.1.) A pitot tube is used to measure velocity of flow.

Figure 14.1 Pitot Tube

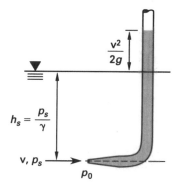

When the fluid enters the pitot tube, it is forced to come to a stop (at the *stagnation point*), and its kinetic energy is transformed into static pressure energy.

Equation 14.1: Fluid Velocity[1]

$$v = \sqrt{(2/\rho)(p_0 - p_s)} = \sqrt{2g(p_0 - p_s)/\gamma} \qquad 14.1$$

Description

The Bernoulli equation can be used to predict the static pressure at the stagnation point. Since the velocity of the fluid within the pitot tube is zero, the upstream velocity can be calculated if the *static*, p_s, and *stagnation*, p_0, *pressures* are known.

$$\frac{p_s}{\rho} + \frac{v^2}{2} = \frac{p_0}{\rho}$$

$$\frac{p_s}{\gamma} + \frac{v^2}{2g} = \frac{p_0}{\gamma}$$

---

[1]As used in the NCEES *FE Reference Handbook* (*NCEES Handbook*), there is no significance to the inconsistent placement of the density terms in the two forms of Eq. 14.1.

In reality, the fluid may be compressible. If the Mach number is less than approximately 0.3, Eq. 14.1 for incompressible fluids may be used.

Example

The density of air flowing in a duct is 1.15 kg/m³. A pitot tube is placed in the duct as shown. The static pressure in the duct is measured with a wall tap and pressure gage.

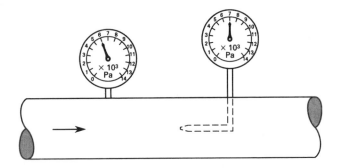

From the gage readings, the velocity of the air is most nearly

(A) 42 m/s

(B) 100 m/s

(C) 110 m/s

(D) 150 m/s

Solution

The static pressure is read from the first static pressure gage as 6000 Pa. The impact pressure is 7000 Pa. From Eq. 14.1,

$$v = \sqrt{(2/\rho)(p_0 - p_s)}$$

$$= \sqrt{\left(\dfrac{2}{1.15\,\frac{\text{kg}}{\text{m}^3}}\right)(7000\text{ Pa} - 6000\text{ Pa})}$$

$$= 41.7 \text{ m/s} \quad (42 \text{ m/s})$$

The answer is (A).

2. VENTURI METER

Figure 14.2 illustrates a simple *venturi meter*. This flow-measuring device can be inserted directly into a pipeline. Since the diameter changes are gradual, there is very little friction loss. Static pressure measurements are taken at the throat and upstream of the diameter change. The difference in these pressures is directly indicated by a *differential manometer*.

Figure 14.2 *Venturi Meter with Differential Manometer*

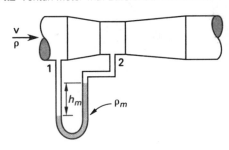

The pressure differential across the venturi meter shown can be calculated from the following equations.

$$p_1 - p_2 = (\rho_m - \rho)gh_m = (\gamma_m - \gamma)h_m$$

$$\frac{p_1 - p_2}{\rho} = \left(\frac{\rho_m}{\rho} - 1\right)gh_m$$

$$\frac{p_1 - p_2}{\gamma} = \left(\frac{\gamma_m}{\gamma} - 1\right)h_m$$

Equation 14.2: Flow Rate Through Venturi Meter

$$Q = \frac{C_v A_2}{\sqrt{1 - (A_2/A_1)^2}} \sqrt{2g\left(\frac{p_1}{\gamma} + z_1 - \frac{p_2}{\gamma} - z_2\right)} \quad 14.2$$

Variation

$$Q = \frac{C_v A_2}{\sqrt{1 - \left(\frac{A_2}{A_1}\right)^2}} \sqrt{2\left(\frac{p_1}{\rho} + z_1 - \frac{p_2}{\rho} - z_2\right)}$$

Values

The *coefficient of velocity*, C_v, accounts for the small effect of friction and is very close to 1.0, usually 0.98 or 0.99.

Description

The flow rate, Q, can be calculated from venturi measurements using Eq. 14.2. For a horizontal venturi meter, $z_1 = z_2$. The quotients, p/γ, in Eq. 14.2 represent the heads of the fluid flowing through a venturi meter. Therefore, the specific weight, γ, of the fluid should be used, not the specific weight of the manometer fluid.

Example

A venturi meter is installed horizontally to measure the flow of water in a pipe. The area ratio of the meter, A_2/A_1, is 0.5, the velocity through the throat of the meter is 3 m/s, and the coefficient of velocity is 0.98. The pressure differential across the venturi meter is most nearly

(A) 1.5 kPa

(B) 2.3 kPa

(C) 3.5 kPa

(D) 6.8 kPa

Solution

From Eq. 14.2, for a venturi meter,

$$Q = \frac{C_v A_2}{\sqrt{1 - (A_2/A_1)^2}} \sqrt{2g\left(\frac{p_1}{\gamma} + z_1 - \frac{p_2}{\gamma} - z_2\right)}$$

Dividing both sides by the area at the throat, A_2, gives

$$\frac{Q}{A_2} = v_2 = \frac{C_v}{\sqrt{1 - \left(\frac{A_2}{A_1}\right)^2}} \sqrt{2g\left(\frac{p_1}{\gamma} + z_1 - \frac{p_2}{\gamma} - z_2\right)}$$

Since the venturi meter is horizontal, $z_1 = z_2$. Reducing and solving for the pressure differential gives

$$\begin{aligned}
p_1 - p_2 &= \frac{v_2^2\left(1 - \left(\frac{A_2}{A_1}\right)^2\right)\gamma}{2gC_v^2} \\
&= \frac{\left(3\,\frac{m}{s}\right)^2\left(1 - (0.5)^2\right)\left(9.81\,\frac{kN}{m^3}\right)\left(1000\,\frac{N}{kN}\right)}{(2)\left(9.81\,\frac{m}{s^2}\right)(0.98)^2} \\
&= 3514\,Pa \quad (3.5\,kPa)
\end{aligned}$$

The answer is (C).

3. ORIFICE METER

The *orifice meter* (or *orifice plate*) is used more frequently than the venturi meter to measure flow rates in small pipes. It consists of a thin or sharp-edged plate with a central, round hole through which the fluid flows.

As with the venturi meter, pressure taps are used to obtain the static pressure upstream of the orifice plate and at the *vena contracta* (i.e., at the point of minimum area and minimum pressure). A differential manometer connected to the two taps conveniently indicates the difference in static pressures. The pressure differential equations, derived for the manometer in Fig. 14.2, are also valid for the manometer configuration of the orifice shown in Fig. 14.3.

Figure 14.3 Orifice Meter with Differential Manometer

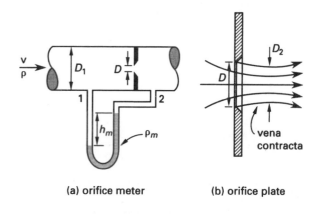

(a) orifice meter   (b) orifice plate

Equation 14.3: Orifice Area

$$A_2 = C_c A \qquad \text{14.3}$$

Description

The area of the orifice is A, and the area of the pipeline is A_1. The area at the vena contracta, A_2, can be calculated from the orifice area and the *coefficient of contraction*, C_c, using Eq. 14.3.

Equation 14.4: Coefficient of the Meter (Orifice Plate)[2]

$$C = \frac{C_v C_c}{\sqrt{1 - C_c^2(A_0/A_1)^2}} \qquad \text{14.4}$$

Description

The *coefficient of the meter*, C, combines the coefficients of velocity and contraction in a way that corrects the theoretical discharge of the meter for frictional flow and for contraction at the vena contracta. The coefficient of the meter is also known as the flow coefficient.[3] Approximate orifice coefficients are listed in Table 14.1.

---

[2]The *NCEES Handbook*'s use of the symbol C for coefficient of the meter is ambiguous. In literature describing orifice plate performance, when C_d is not used, C is frequently reserved for the coefficient of discharge. The symbols C_M, C_F (for coefficient of the meter and *flow coefficient*), K, and F are typically used to avoid ambiguity.

[3]The *NCEES Handbook* lists "orifice coefficient" as a synonym for the "coefficient of the meter." However, this ambiguous usage should be avoided, as four orifice coefficients are attributed to an orifice: coefficient of contraction, coefficient of velocity, coefficient of discharge, and coefficient of resistance.

Table 14.1 Approximate Orifice Coefficients for Turbulent Water

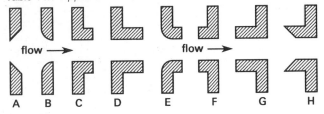

illustration	description	C	C_c	C_v
A	sharp-edged	0.61	0.62	0.98
B	round-edged	0.98	1.00	0.98
C	short tube (fluid separates from walls)	0.61	1.00	0.61
D	short tube (no separation)	0.80	1.00	0.80
E	short tube with rounded entrance	0.97	0.99	0.98
F	reentrant tube, length less than one-half of pipe diameter	0.54	0.55	0.99
G	reentrant tube, length 2–3 pipe diameters	0.72	1.00	0.72
H	Borda	0.51	0.52	0.98
(none)	smooth, well-tapered nozzle	0.98	0.99	0.99

Fluid Mechanics/ Dynamics

Equation 14.5: Flow Through Orifice Plate

$$Q = CA_0\sqrt{2g\left(\frac{p_1}{\gamma} + z_1 - \frac{p_2}{\gamma} - z_2\right)} \qquad 14.5$$

Variation

$$Q = CA_0\sqrt{2\left(\frac{p_1}{\rho} + z_1 - \frac{p_2}{\rho} - z_2\right)}$$

Description

The flow rate through the orifice meter is given by Eq. 14.5. Generally, z_1 and z_2 are equal.

4. SUBMERGED ORIFICE

The flow rate of a jet issuing from a *submerged orifice* in a tank can be determined by modifying Eq. 14.5 in terms of the potential energy difference, or head difference, on either side of the orifice. (See Fig. 14.4.)

Equation 14.6 Through Eq. 14.8: Flow Through Submerged Orifice[4]

$$Q = A_2 v_2 = C_c C_v A \sqrt{2g(h_1 - h_2)} \qquad 14.6$$

$$Q = CA\sqrt{2g(h_1 - h_2)} \qquad 14.7$$

$$C = C_c C_v \qquad 14.8$$

---

[4]The *NCEES Handbook*'s use of the symbol C for both coefficient of discharge (submerged orifice) and coefficient of the meter (see Eq. 14.4) makes it difficult to determine the meaning of C.

Figure 14.4 Submerged Orifice

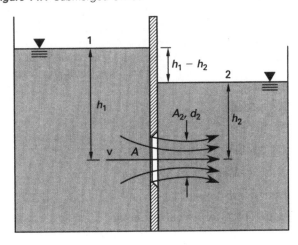

Description

The coefficients of velocity and contraction can be combined into the *coefficient of discharge*, C, calculated from Eq. 14.8.

5. ORIFICE DISCHARGING FREELY INTO ATMOSPHERE

If the orifice discharges from a tank into the atmosphere, Eq. 14.7 can be further simplified. (See Fig. 14.5.)

Figure 14.5 Orifice Discharging Freely into the Atmosphere

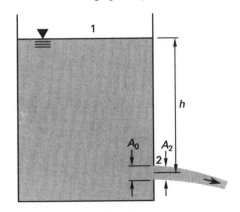

Equation 14.9: Orifice Flow with Free Discharge

$$Q = CA_0\sqrt{2gh} \qquad 14.9$$

Variation

$$v_2 = \frac{Q}{A_2} = C_v\sqrt{2gh}$$

Description

A_0 is the orifice area. A_2 is the area at the vena contracta (see Eq. 14.3 and Fig. 14.5).

Example

Water under an 18 m head discharges freely into the atmosphere through a 25 mm diameter orifice. The orifice is round-edged and has a coefficient of discharge of 0.98.

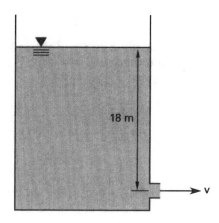

The velocity of the water as it passes through the orifice is most nearly

(A) 1.2 m/s

(B) 3.2 m/s

(C) 8.2 m/s

(D) 18 m/s

Solution

From Eq. 14.9, for an orifice discharging freely into the atmosphere,

$$Q = CA_0\sqrt{2gh}$$

Dividing both sides by A_0 gives

$$v = C\sqrt{2gh}$$
$$= 0.98\sqrt{(2)\left(9.81\ \frac{m}{s^2}\right)(18\ m)}$$
$$= 18.4\ m/s \quad (18\ m/s)$$

The answer is (D).

6. SIMILITUDE

Similarity considerations between a *model* (subscript *m*) and a full-size object (subscript *p*, for *prototype*) imply that the model can be used to predict the performance of the prototype. Such a model is said to be *mechanically similar* to the prototype.

Complete mechanical similarity requires geometric, kinematic, and dynamic similarity. *Geometric similarity*

means that the model is true to scale in length, area, and volume. *Kinematic similarity* requires that the flow regimes of the model and prototype be the same. *Dynamic similarity* means that the ratios of all types of forces are equal for the model and the prototype. These forces result from inertia, gravity, viscosity, elasticity (i.e., fluid compressibility), surface tension, and pressure.

For dynamic similarity, the number of possible ratios of forces is large. For example, the ratios of viscosity/inertia, inertia/gravity, and inertia/surface tension are only three of the ratios of forces that must match for every corresponding point on the model and prototype. Fortunately, some force ratios can be neglected because the forces are negligible or are self-canceling.

Equation 14.10 Through Eq. 14.14: Dynamic Similarity[5]

$$\left[\frac{F_I}{F_p}\right]_p = \left[\frac{F_I}{F_p}\right]_m = \left[\frac{\rho v^2}{p}\right]_p = \left[\frac{\rho v^2}{p}\right]_m \quad 14.10$$

$$\left[\frac{F_I}{F_V}\right]_p = \left[\frac{F_I}{F_V}\right]_m = \left[\frac{v l \rho}{\mu}\right]_p = \left[\frac{v l \rho}{\mu}\right]_m = [Re]_p = [Re]_m \quad 14.11$$

$$\left[\frac{F_I}{F_G}\right]_p = \left[\frac{F_I}{F_G}\right]_m = \left[\frac{v^2}{lg}\right]_p = \left[\frac{v^2}{lg}\right]_m = [Fr]_p = [Fr]_m \quad 14.12$$

$$\left[\frac{F_I}{F_E}\right]_p = \left[\frac{F_I}{F_E}\right]_m = \left[\frac{\rho v^2}{E_v}\right]_p = \left[\frac{\rho v^2}{E_v}\right]_m = [Ca]_p = [Ca]_m \quad 14.13$$

$$\left[\frac{F_I}{F_T}\right]_p = \left[\frac{F_I}{F_T}\right]_m = \left[\frac{\rho l v^2}{\sigma}\right]_p = \left[\frac{\rho l v^2}{\sigma}\right]_m = [We]_p = [We]_m \quad 14.14$$

Description

If Eq. 14.10 through Eq. 14.14 are satisfied for model and prototype, complete dynamic similarity will be achieved. In practice, it is rare to be able (or to even attempt) to demonstrate *complete similarity*. Usually, *partial similarity* is based on only one similarity law, and correlations, experience, and general rules of thumb are used to modify the results. For completely submerged objects (i.e., where there is no free surface), such as torpedoes in water and aircraft in the atmosphere, similarity is usually based on Reynolds numbers. For objects partially submerged and

[5]The *NCEES Handbook* is inconsistent in its presentation and definition of the Froude number. While some equations use y_h as the symbol for *characteristic length*, Eq. 14.12 uses l. This leads to some potentially confusing and misleading conflicts.

experiencing wave activity, such as surface ships, open channels, spillways, weirs, and hydraulic jumps, partial similarity is usually based on Froude numbers.

Example

A 200 m long submarine is being designed to travel underwater at 3 m/s. The corresponding underwater speed for a 6 m model is most nearly

(A) 0.5 m/s

(B) 2 m/s

(C) 100 m/s

(D) 200 m/s

Solution

The Reynolds numbers should be equal for model and prototype. From Eq. 14.11,

$$\left[\frac{F_I}{F_V}\right]_p = \left[\frac{F_I}{F_V}\right]_m = \left[\frac{\mathrm{v}l\rho}{\mu}\right]_p = \left[\frac{\mathrm{v}l\rho}{\mu}\right]_m = [\mathrm{Re}]_p = [\mathrm{Re}]_m$$

The density and absolute viscosity of the water will be the same for both prototype and model, so

$$\mathrm{v}_p l_p = \mathrm{v}_m l_m$$

$$\mathrm{v}_m = \frac{\mathrm{v}_p l_p}{l_m} = \frac{\left(3\,\dfrac{\mathrm{m}}{\mathrm{s}}\right)(200\text{ m})}{6\text{ m}} = 100\text{ m/s}$$

The answer is (C).

Fluid Mechanics/ Dynamics

15 Compressible Fluid Dynamics

Nomenclature

A	area	m^2
A^*	critical area	m^2
c	specific heat	J/kg·K
c	speed of sound	m/s
k	ratio of specific heats	–
m	mass	kg
Ma	Mach number	–
p	pressure	Pa
R	specific gas constant	J/kg·K
$\overline{R}$	universal gas constant, 8314	J/kmol·K
T	absolute temperature	K
v	velocity	m/s
V	volume	m^3

Symbols

ρ	density	kg/m^3
v	specific volume	m^3/kg

Subscripts

0	stagnation
p	constant pressure
v	constant volume

1. COMPRESSIBLE FLUID DYNAMICS

A *high-velocity gas* is defined as a gas moving with a velocity in excess of approximately 100 m/s. A high gas velocity is often achieved at the expense of internal energy. A drop in internal energy, u, is seen as a drop in enthalpy, h, since $h = u + pv$. Since the Bernoulli equation does not account for this conversion, it cannot be used to predict the thermodynamic properties of the gas. Furthermore, density changes and shock waves complicate the use of traditional evaluation tools such as energy and momentum conservation equations.

2. IDEAL GAS

Equation 15.1 and Eq. 15.2: Ideal Gas Law

$$pv = RT \qquad \text{15.1}$$
$$pV = mRT \qquad \text{15.2}$$

Variations

$$p = \left(\frac{m}{V}\right)RT = \rho RT$$

$$pv = p\left(\frac{V}{m}\right)$$

Description

The *ideal gas law* is an *equation of state* for ideal gases. An equation of state is a relationship that predicts the state (i.e., a property, such as pressure, temperature, volume, etc.) from a set of two other independent properties.

Equation 15.3: Specific Gas Constant

$$R = \frac{\overline{R}}{\text{mol. wt}} \qquad \text{15.3}$$

Values

	customary U.S.	SI
universal gas constant, $\overline{R}$	1545 ft-lbf/lbmol-°R	8314 J/kmol·K
		8.314 kPa·m^3/kmol·K
		0.08206 L·atm/mol·K
specific gas constant, R (dry air)	53.3 ft-lbf/lbm-°R	287 J/kg·K

Description

The *specific gas constant*, R, can be determined from the *molecular weight* of the substance, mol. wt, and the *universal gas constant*, $\overline{R}$.

The universal gas constant, $\overline{R}$, is "universal" (within a system of units), because the same value can be used for any gas. Its value depends on the units used for pressure, temperature, and volume, as well as on the units of mass.

Example

A vessel of air is kept at 97 kPa and 300K. The molecular weight of air is 29 kg/kmol. The density of the air in the vessel is most nearly

(A) 0.039 kg/m^3

(B) 0.53 kg/m^3

(C) 1.1 kg/m^3

(D) 3.2 kg/m^3

Solution

From Eq. 15.3, the specific gas constant for air is

$$R = \frac{\overline{R}}{\text{mol. wt}} = \frac{8314 \, \frac{\text{J}}{\text{kmol·K}}}{29 \, \frac{\text{kg}}{\text{kmol}}}$$

$$= 287 \, \text{J/kg·K}$$

Use the equation of state for an ideal gas, Eq. 15.2.

$$pV = mRT$$

$$\rho = \frac{m}{V} = \frac{p}{RT} = \frac{(97 \text{ kPa})\left(1000 \, \frac{\text{Pa}}{\text{kPa}}\right)}{\left(287 \, \frac{\text{J}}{\text{kg·K}}\right)(300\text{K})}$$

$$= 1.128 \text{ kg/m}^3 \quad (1.1 \text{ kg/m}^3)$$

The answer is (C).

Equation 15.4: Boyle's Law

$$p_1 v_1 / T_1 = p_2 v_2 / T_2 \qquad \text{15.4}$$

Variation

$$\frac{pv}{T} = \text{constant}$$

Description

The equation of state leads to another general relationship, as shown in Eq. 15.4. When temperature is held constant, Eq. 15.4 reduces to *Boyle's law*.

$$pv = \text{constant}$$

Equation 15.5: Ratio of Specific Heats

$$k = c_p / c_v \qquad \text{15.5}$$

Values

For air, the ratio of specific heats is $k = 1.40$.

Description

There is no heat loss in an *adiabatic process*. An *isentropic process* is an adiabatic process in which there is no change in system *entropy* (i.e., the process is reversible). For such a process, the following equation is valid.

$$pv^k = \text{constant}$$

For gases, the *ratio of specific heats*, k, is defined by Eq. 15.5, in which c_p is the *specific heat at constant pressure* and c_v is the *specific heat at constant volume*. An implicit assumption (requirement) for ideal gases is that the ratio of specific heats is constant throughout all processes.

3. COMPRESSIBLE FLOW

Equation 15.6: Speed of Sound

$$c = \sqrt{kRT} \qquad \text{15.6}$$

Values

Table 15.1 *Approximate Speeds of Sound (at one atmospheric pressure)*

material	speed of sound	
	(ft/sec)	(m/s)
air	1130 at 70°F	330 at 0°C
aluminum	16,400	4990
carbon dioxide	870 at 70°F	260 at 0°C
hydrogen	3310 at 70°F	1260 at 0°C
steel	16,900	5150
water	4880 at 70°F	1490 at 20°C

(Multiply ft/sec by 0.3048 to obtain m/s.)

Description

The *speed of sound*, c, in a fluid is a function of its bulk modulus, or equivalently, of its compressibility. Equation 15.6 gives the speed of sound in an ideal gas. The temperature, T, must be in degrees absolute (i.e., °R or K).

Approximate speeds of sound for various materials at 1 atm are given in Table 15.1.

Equation 15.7: Mach Number[1]

$$\text{Ma} \equiv \frac{\text{v}}{c} \qquad \text{15.7}$$

Description

The *Mach number*, Ma, of an object is the ratio of the object's speed to the speed of sound in the medium through which the object is traveling.

The term *subsonic travel* implies Ma < 1.[2] Similarly, *supersonic travel* implies Ma > 1, but usually Ma < 5. Travel above Ma = 5 is known as *hypersonic travel*. Travel in the transition region between subsonic and supersonic (i.e., 0.8 < Ma < 1.2) is known as *transonic travel*. A *sonic boom* (a shock-wave phenomenon) occurs when an object travels at supersonic speed.

---

[1]The symbol ≡ means "is defined as." It is not a mathematical operator and should not be used in mathematical equations.
[2]In the language of compressible fluid flow, this is known as the *subsonic flow regime*.

Example

Air at 300K flows at 1.0 m/s. The ratio of specific heats for air is 1.4, and the specific gas constant is 287 J/kg·K. The Mach number for the air is most nearly

(A) 0.003

(B) 0.008

(C) 0.02

(D) 0.07

Solution

Calculate the speed of sound in 300K air using Eq. 15.6.

$$c = \sqrt{kRT} = \sqrt{(1.4)\left(287 \; \frac{\text{J}}{\text{kg·K}}\right)(300\text{K})}$$
$$= 347 \text{ m/s}$$

From Eq. 15.7, the Mach number is

$$\text{Ma} = \frac{\text{v}}{c} = \frac{1.0 \; \frac{\text{m}}{\text{s}}}{347 \; \frac{\text{m}}{\text{s}}} = 0.00288 \quad (0.003)$$

The answer is (A).

4. ISENTROPIC FLOW RELATIONSHIPS

If the gas flow is adiabatic and frictionless (that is, reversible), the entropy change is zero, and the flow is known as *isentropic flow*. As a practical matter, completely isentropic flow does not exist. However, some high-velocity, steady-state flow processes proceed with little increase in entropy and are considered to be isentropic. The irreversible effects are accounted for by various correction factors, such as nozzle and discharge coefficients.

Equation 15.8: Isentropic Flow

$$\frac{p_2}{p_1} = \left(\frac{T_2}{T_1}\right)^{\frac{k}{k-1}} = \left(\frac{\rho_2}{\rho_1}\right)^k \qquad \textbf{15.8}$$

Description

Equation 15.8 gives the relationship between static properties for any two points in the flow. k is the ratio of specific heats, given by Eq. 15.5.

Example

Through an isentropic process, a piston compresses 2 kg of an ideal gas at 150 kPa and 35°C in a cylinder to a pressure of 300 kPa. The specific heat of the gas for constant pressure processes is 5 kJ/kg·K; for constant volume processes, the specific heat is 3 kJ/kg·K. The final temperature of the gas is most nearly

(A) 130°C

(B) 190°C

(C) 210°C

(D) 260°C

Solution

Find the ratio of specific heats using Eq. 15.5.

$$k = \frac{c_p}{c_v}$$
$$= \frac{5 \; \frac{\text{kJ}}{\text{kg·K}}}{3 \; \frac{\text{kJ}}{\text{kg·K}}}$$
$$= 1.67$$

Use Eq. 15.8, converting Celsius to absolute temperature, and solve for the final temperature.

$$\frac{p_2}{p_1} = \left(\frac{T_2}{T_1}\right)^{\frac{k}{k-1}}$$
$$T_2 = T_1 \left(\frac{p_2}{p_1}\right)^{(k-1)/k}$$
$$= (35°\text{C} + 273°)\left(\frac{300 \text{ kPa}}{150 \text{ kPa}}\right)^{(1.67-1)/1.67}$$
$$= 406.4\text{K}$$

Converting the result back to Celsius,

$$T_2 = 406.4\text{K} - 273° = 133.4°\text{C} \quad (130°\text{C})$$

The answer is (A).

Equation 15.9: Stagnation Temperature[3]

$$T_0 = T + \frac{\text{v}^2}{2 \cdot c_p} \qquad \textbf{15.9}$$

Description

In an ideal gas for an isentropic process, for a given point in the flow, *stagnation temperature* (also known as *total temperature*), T_0, is related to *static temperature*, T, as shown in Eq. 15.9.

---

[3]As used in the NCEES *FE Reference Handbook* (*NCEES Handbook*), there is no significance to the multiplication "dot" in Eq. 15.9 other than scalar multiplication. This is not a vector calculation.

Example

Air enters a straight duct at 300K and 300 kPa and with a velocity of 150 m/s. The specific heat of the air is 1004.6 J/kg·K. Assuming adiabatic flow and an ideal gas, the stagnation temperature is most nearly

(A) 310K

(B) 330K

(C) 350K

(D) 380K

Solution

The stagnation temperature is

$$T_0 = T + \frac{v^2}{2 \cdot c_p}$$

$$= 300\text{K} + \frac{\left(150 \, \frac{\text{m}}{\text{s}}\right)^2}{(2)\left(1004.6 \, \frac{\text{J}}{\text{kg·K}}\right)}$$

$$= 311.2\text{K} \quad (310\text{K})$$

The answer is (A).

Equation 15.10 Through Eq. 15.12: Isentropic Flow Factors[4]

$$\frac{T_0}{T} = 1 + \frac{k-1}{2} \cdot \text{Ma}^2 \qquad \textbf{\textit{15.10}}$$

$$\frac{p_0}{p} = \left(\frac{T_0}{T}\right)^{\frac{k}{k-1}} = \left(1 + \frac{k-1}{2} \cdot \text{Ma}^2\right)^{\frac{k}{k-1}} \qquad \textbf{\textit{15.11}}$$

$$\frac{\rho_0}{\rho} = \left(\frac{T_0}{T}\right)^{\frac{1}{k-1}} = \left(1 + \frac{k-1}{2} \cdot \text{Ma}^2\right)^{\frac{1}{k-1}} \qquad \textbf{\textit{15.12}}$$

Description

In isentropic flow, total pressure, total temperature, and total density remain constant, regardless of the flow area and velocity. The instantaneous properties, known as *static properties*, do change along the flow path, however.[5] Equation 15.10 through Eq. 15.12 predict these static properties as functions of the Mach number, Ma, and ratio of specific heats, k, for ideal gas flow. Therefore, the ratios can be easily tabulated. The numbers in such tables are known as *isentropic flow factors*.

---

[4]See Ftn. 3.
[5]The *static properties* are not the same as the *stagnation properties*.

Example

At the entry of a diverging section, the temperature of a gas is 27°C, and the Mach number is 1.5. The exit Mach number is 2.5. Assume isentropic flow of a perfect gas with a ratio of specific heats of 1.3. The temperature of the gas at exit is most nearly

(A) −100°C

(B) −85°C

(C) −66°C

(D) −31°C

Solution

Use Eq. 15.10 with the entrance conditions to find the stagnation temperature.

$$\frac{T_0}{T} = 1 + \frac{k-1}{2} \cdot \text{Ma}^2$$

$$T_0 = \left(1 + \left(\frac{k-1}{2}\right)(\text{Ma})^2\right) T$$

$$= \left(1 + \left(\frac{1.3-1}{2}\right)(1.5)^2\right)(27°\text{C} + 273°)$$

$$= 401.25\text{K}$$

Total temperature does not change. Use Eq. 15.10 again, this time with the stagnation temperature just found and the exit Mach number, to find the temperature at the exit.

$$\frac{T_0}{T} = 1 + \frac{k-1}{2} \cdot \text{Ma}^2$$

$$T = \frac{T_0}{1 + \left(\frac{k-1}{2}\right)(\text{Ma})^2}$$

$$= \frac{401.25\text{K}}{1 + \left(\frac{1.3-1}{2}\right)(2.5)^2} - 273°$$

$$= -65.90°\text{C} \quad (-66°\text{C})$$

The answer is (C).

5. CRITICAL AREA

In order to design a nozzle capable of expanding a gas to some given velocity or Mach number, it is sufficient to have an expression for the flow area versus Mach number. Since the reservoir cross-sectional area is an unrelated variable, it is not possible to use the stagnation area as a reference area and to develop the ratio $[A_0/A]$ as was done for temperature, pressure, and density. The usual choice for a reference area is the *critical area*—the area at which the gas velocity is (or could be) sonic. This area is designated as A^*.

Equation 15.13: Critical Area

$$\frac{A}{A^*} = \frac{1}{\text{Ma}} \left[\frac{1 + \frac{1}{2}(k-1)\text{Ma}^2}{\frac{1}{2}(k+1)} \right]^{\frac{k+1}{2(k-1)}} \qquad 15.13$$

Description

Figure 15.1 is a plot of Eq. 15.13 versus the Mach number. As long as the Mach number is less than 1.0, the area must decrease in order for the velocity to increase. However, if the Mach number is greater than 1.0, the area must increase in order for the velocity to increase.

Figure 15.1 *A/A* versus Ma*

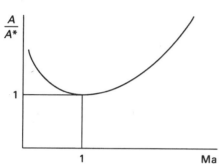

It is not possible to change the Mach number to any desired value over any arbitrary distance along the axis of a converging-diverging nozzle. If the rate of change, dA/dx, is too great, the assumptions of one-dimensional flow become invalid. Usually, the converging section has a steeper angle (known as the *convergent angle*) than the diverging section. If the diverging angle is too great, a normal shock wave may form in that part of the nozzle.

6. NORMAL SHOCK RELATIONSHIPS

A nozzle must be designed to the design pressure ratio in order to keep the flow supersonic in the diverging section of the nozzle. It is possible, though, to have supersonic velocity only in part of the diverging section. Once the flow is supersonic in a part of the diverging section, however, it cannot become subsonic by an isentropic process.

Therefore, the gas experiences a *shock wave* as the velocity drops from supersonic to subsonic. Shock waves are very thin (several molecules thick) and separate areas of radically different thermodynamic properties. Since the shock wave forms normal to the flow direction, it is known as a *normal shock wave*. The strength of a shock wave is measured by the change in Mach number across it. (See Fig. 15.2.)

The velocity always changes from supersonic to subsonic across a shock wave. Since there is no loss of heat energy, a shock wave is an adiabatic process, and total temperature is constant. However, the process is not isentropic, and total pressure decreases. Momentum is also conserved. Table 15.2 lists the property changes across a shock wave.

Figure 15.2 *Normal Shock Relationships*

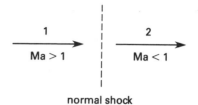

normal shock

Table 15.2 *Property Changes Across a Normal Shock Wave*

property	change
total temperature	is constant
total pressure	decreases
total density	decreases
velocity	decreases
Mach number	decreases
pressure	increases
density	increases
temperature	increases
entropy	increases
internal energy	increases
enthalpy	is constant
momentum	is constant

Equation 15.14 Through Eq. 15.18: Properties Across a Normal Shock Wave

$$\text{Ma}_2 = \sqrt{\frac{(k-1)\text{Ma}_1^2 + 2}{2k\text{Ma}_1^2 - (k-1)}} \qquad 15.14$$

$$\frac{T_2}{T_1} = \left[2 + (k-1)\text{Ma}_1^2 \right] \frac{2k\text{Ma}_1^2 - (k-1)}{(k+1)^2 \text{Ma}_1^2} \qquad 15.15$$

$$\frac{p_2}{p_1} = \frac{1}{k+1} \left[2k\text{Ma}_1^2 - (k-1) \right] \qquad 15.16$$

$$\frac{\rho_2}{\rho_1} = \frac{\text{v}_1}{\text{v}_2} = \frac{(k+1)\text{Ma}_1^2}{(k-1)\text{Ma}_1^2 + 2} \qquad 15.17$$

$$T_{01} = T_{02} \qquad 15.18$$

Description

Equation 15.14 through Eq. 15.18 give the relationships between downstream and upstream flow conditions for a normal shock wave.

16 Fluid Machines

Nomenclature

c	specific heat	J/kg·K
C	unit conversion constant	varies
D	diameter	m
e	efficiency	–
g	gravitational acceleration, 9.81	m/s^2
h	enthalpy	J/kg
h	head	m
H	head	m
k	ratio of specific heats	–
KE	kinetic energy	J/kg
$\dot{m}$	mass flow rate	kg/s
n	substitutional variable for $(k-1)/k$	–
N	rotational speed	rev/min
$NPSH$	net positive suction head	m
p	pressure	Pa or psi
p_1	inlet pressure	Pa or psi
p_2	exit pressure	Pa or psi
P_w	power	hp
Q	volumetric flow rate	L/s
R	specific gas constant	J/kg·K or ft-lbf/lbm-°R
T	absolute temperature	K
T_1	absolute inlet temperature	K or °R
v	velocity	m/s
w	work per unit mass	J/kg
W	weight flow of air being compressed[1]	N/s or lbf/sec
W	work	J
$\dot{W}$	power	W

Symbols

γ	specific (unit) weight	N/m^3
η	efficiency	–
ρ	density	kg/m^3

Subscripts

a	actual
A	available
c	compressor
comp	compressor
C	compressor
e	exit
es	exit after an isentropic process
f	fan or friction
H	hydraulic
i	inlet
p	constant pressure
R	required
s	isentropic or static
t	total
turb	turbine
T	turbine
v	constant volume

1. PUMPS AND COMPRESSORS

Pumps, *compressors*, *fans*, and *blowers* convert mechanical energy (i.e., work input) into fluid energy, increasing the total energy content of the fluid flowing through them. (See Fig. 16.1.) Pumps can be considered adiabatic devices because the fluid gains (or loses) very little heat during the short time it passes through them. If the inlet and outlet are the same size and at the same elevation, the kinetic and potential energy changes can be neglected.[2]

Figure 16.1 *Compressor*

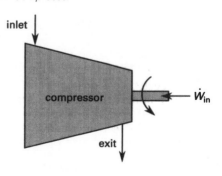

---

[1]If W is considered to be the mass of air being compressed, in lbm/sec, then the term g/g_c must be added where appropriate, such as in Eq. 16.25.

[2]Even if the pump inlet and outlet are different sizes and at different elevations, the kinetic and potential energy changes are small compared to the pressure energy increase.

2. PUMP POWER

Pumps convert mechanical enery into fluid energy, increasing the energy of the fluid. Pump power is known as *hydraulic power* or *water power*. Hydraulic power is the net power transferred to the fluid by the pump.

Horsepower is the unit of power used in the United States and other non-SI countries, which gives rise to the terms *hydraulic horsepower* and *water horsepower*. The unit of power in SI units is the watt.

Equation 16.1 and Eq. 16.2: Motor Input Power[3]

$$\dot{W} = Q\gamma h/\eta = Q\rho g h/\eta_t \qquad 16.1$$

$$\eta_t = \eta_{\text{pump}} \times \eta_{\text{motor}} \qquad 16.2$$

Description

A pump adds energy to a fluid flow. The increase in energy, ΔE, is manifested primarily by an increase in pressure, and to much lesser degrees, changes in velocity and, sometimes, elevation across the pump inlet and outlet. From the *work-energy principle*, the energy added is equal to the work, W, done on the fluid. The *rate of work done*, $\dot{W}$, is equal to the power. The power that is drawn in increasing the pressure is known as *hydraulic power* (*fluid power, hydraulic horsepower*, etc.). The hydraulic power is calculated from the head added to the fluid, h, and the fluid flow rate, Q.[4]

$$\dot{W}_H = Q\gamma h$$

A pump is a mechanical device, and some power delivered to it is lost due to friction between the fluid and the pump and friction in the pump bearings. This loss is accounted for by a *pump efficiency* term, η_{pump}. The power delivered to the pump is

$$\dot{W}_{\text{pump}} = \frac{\dot{W}_H}{\eta_{\text{pump}}}$$

A pump is usually driven by an electrical motor. The electrical power drawn from the power line is referred to as the *electrical power, consumed power,* or *purchased power*. Since the motor is an electromechanical device, some of the power delivered to it is lost due to frictional losses in the motor bearings, air windage, and electrical heating. This loss is accounted for by a *motor efficiency*

term, η_{motor}. The power delivered to the motor (purchased from the power line) is

$$\dot{W}_{\text{purchased}} = \frac{\dot{W}_{\text{pump}}}{\eta_{\text{motor}}}$$

Electrical motors are rated according to their output power. (So, a 2 horsepower motor would have a useful power of 2 horsepower.) Output power for any device is referred to as *brake power*.

Since the *total pump efficiency* (calculated from Eq. 16.2), η_t, is included, Eq. 16.1 calculates the power drawn by the motor (purchased). If the efficiency term were omitted, the power would be the hydraulic power. Since the hydraulic power is what is left over after losses, it is sometimes described as the *net power*.

Applying the continuity equation to Eq. 16.1,

$$\dot{W} = \frac{\dot{m}gh}{\eta}$$

Example

A pump with 70% efficiency moves water from ground level to a height of 5 m. The flow rate is 10 m^3/s. Most nearly, what is the power delivered to the pump?

- (A) 80 kW
- (B) 220 kW
- (C) 700 kW
- (D) 950 kW

Solution

From Eq. 16.1,

$$\dot{W} = Q\rho g h/\eta_{\text{pump}} = \frac{\left(10 \, \frac{m^3}{s}\right)\left(1000 \, \frac{kg}{m^3}\right)\left(9.81 \, \frac{m}{s^2}\right)(5 \, m)}{0.70}$$

$$= 700\,714 \text{ W} \quad (700 \text{ kW})$$

The answer is (C).

Equation 16.3 Through Eq. 16.5: Centrifugal Pump Power[5]

$$\dot{W}_{\text{fluid}} = \rho g H Q \qquad 16.3$$

$$\dot{W} = \frac{\rho g H Q}{\eta_{\text{pump}}} \qquad 16.4$$

$$\dot{W}_{\text{purchased}} = \frac{\dot{W}}{\eta_{\text{motor}}} \qquad 16.5$$

---

[3]As used in the NCEES *FE Reference Handbook* (*NCEES Handbook*) in Eq. 16.1, the symbols η and η_t represent the same quantity, the *total pump efficiency*.
[4]The *NCEES Handbook* uses lowercase h to represent head added by a pump. While any symbol can be used to represent any quantity, the symbols used for head added by a pump are most commonly H, h_A, TH (for total head), and TDH (for total dynamic head).
[5]The *NCEES Handbook* uses H for head added in Eq. 16.3 and Eq. 16.4, which is the same as h in Eq. 16.1.

Values[6]

efficiency type	efficiency range
pump, η_{pump}	0–1
motor, η_{motor}	0–1

Description

The hydraulic power delivered by a centrifugal pump is calculated from Eq. 16.3. Brake (motor) power drawn by the pump can be found from Eq. 16.4, and *purchased power* is determined using Eq. 16.5. Efficiency ranges for pumps and motors are given in the values section. H is the increase in head delivered by the pump.

Example

Water at 10°C is pumped through smooth steel pipes from tank 1 to tank 2 as shown. The discharge rate is 0.1 m³/min. The pump efficiency is 80%.

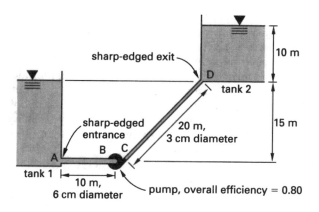

If the pump adds a total of 15 m of head, the pump motor's brake power is most nearly

(A) 19 W

(B) 32 W

(C) 200 W

(D) 310 W

Solution

From Eq. 16.4, the brake power is

$$\dot{W} = \frac{\rho g H Q}{\eta_{pump}} = \frac{\left(1000 \ \dfrac{kg}{m^3}\right)\left(9.81 \ \dfrac{m}{s^2}\right)(15 \ m)\left(0.1 \ \dfrac{m^3}{min}\right)}{(0.80)\left(60 \ \dfrac{s}{min}\right)}$$

$$= 306.6 \ W \quad (310 \ W)$$

The answer is (D).

---

[6]The values of efficiencies provided in the *NCEES Handbook* for pumps and motors are not particularly useful. All real devices fall into the range of 0–1. No efficiencies in the real universe are outside of those ranges.

Equation 16.6 Through Eq. 16.11: Pump and Compressor Power (Rate of Work)[7,8]

$$\dot{W}_{comp} = -\dot{m}(h_e - h_i) \quad \text{[adiabatic compressor]} \qquad 16.6$$

$$\dot{W}_{comp} = -\dot{m}c_p(T_e - T_i) \quad \begin{bmatrix} \text{ideal gas; constant} \\ \text{specific heat} \end{bmatrix}$$
$$16.7$$

$$w_{comp} = -c_p(T_e - T_i) \quad \text{[per unit mass]} \qquad 16.8$$

$$\dot{W}_{comp} = -\dot{m}\left(h_e - h_i + \frac{v_e^2 - v_i^2}{2}\right)$$
$$= -\dot{m}\left(c_p(T_e - T_i) + \frac{v_e^2 - v_i^2}{2}\right) \quad \begin{bmatrix} \Delta KE \\ \text{included} \end{bmatrix}$$
$$16.9$$

$$\dot{W}_{comp} = \frac{\dot{m}p_i k}{(k-1)\rho_i \eta_c}\left[\left(\frac{p_e}{p_i}\right)^{1-1/k} - 1\right] \quad \begin{bmatrix} \text{adiabatic} \\ \text{compression} \end{bmatrix}$$
$$16.10$$

$$\dot{W}_{comp} = \frac{\overline{R}T_i}{M\eta_c} \ln \frac{p_e}{p_i} \dot{m} \quad \text{[isothermal compression]}$$
$$16.11$$

Description

Equation 16.6 through Eq. 16.11 are used to calculate *compressor power*, referred to as a "rate of work," based on the fluid type and other appropriate assumptions.

The first form of Eq. 16.9 is the general equation, applicable to both vapors (such as steam) and real and ideal gases, while the second form of Eq. 16.9 is specifically only for ideal gases (for which specific heat is assumed to be constant). Equation 16.6 and Eq. 16.7 are simplifications of both forms of Eq. 16.9 that assume kinetic energy (velocity) changes are insignificant. Equation 16.8 is Eq. 16.7 restated on a per-mass basis. While Eq. 16.7 and the second form of Eq. 16.9 depend on temperature changes, Eq. 16.10 and Eq. 16.11 depend on the pressure changes. Equation 16.9 and Eq. 16.10 are strictly for ideal gases. Equation 16.10 calculates the input power, $\dot{W}_{in}$, shown in Fig. 16.1.

---

[7]Although Eq. 16.6 through Eq. 16.11 are presented together in the *NCEES Handbook* and purport to calculate "compressor power," Eq. 16.10 is not the same quantity as the other equations. Equation 16.10 includes the compressor efficiency, and therefore, is the power drawn by (into) the compressor from its prime mover, before compressor losses. All of the other equations are based on working fluid properties, and therefore, represent the net power delivered to the fluid, after compressor losses. Only Eq. 16.10 calculates the input power shown in Fig. 16.1.

[8]The *NCEES Handbook* contains some inconsistencies in nomenclature and sign convention. Although M is used for molecular weight in Eq. 16.11, "mol. wt" is used elsewhere. (M does not represent "mass.") The subscript c used in η_c is the same as "comp" (compressor) used in Eq. 16.6 through Eq. 16.11 (and C used in Eq. 16.12). In keeping with the "standard" thermodynamic convention for systems that work (energy) is negative when work is done on a system, Eq. 16.6 through Eq. 16.9 have negative signs. This convention is not followed in Eq. 16.10 and Eq. 16.11.

Fluid Mechanics/ Dynamics

Example

15 kg/s of air are compressed from 1 atm and 300K (enthalpy of 300.19 kJ/kg) to 2 atm and 900K (enthalpy of 732.93 kJ/kg). Heat transfer is negligible. The air compressor's power output is most nearly

(A) 5.6 MW

(B) 6.5 MW

(C) 7.6 MW

(D) 8.9 MW

Solution

Heat transfer is negligible, so the compression process may be considered adiabatic. Use Eq. 16.6 to find the rate of work.

$$\dot{W}_{comp} = -\dot{m}(h_e - h_i)$$
$$= -\left(15\,\frac{kg}{s}\right)\left(300.19\,\frac{kJ}{kg} - 732.93\,\frac{kJ}{kg}\right)$$
$$= 6491.1\,kJ/s \quad (6.5\,MW)$$

The answer is (B).

Equation 16.12: Compressor Isentropic Efficiency

$$\eta_C = \frac{w_s}{w_a} = \frac{T_{es} - T_i}{T_e - T_i} \qquad 16.12$$

Description

The *isentropic efficiency* (*adiabatic efficiency*), η_C, of a compressor is the ratio of ideal (isentropic) energy extraction to actual energy extraction. Actual isentropic efficiencies vary from approximately 65% for 1 MW unit to over 90% for 100 MW and larger units.

3. TURBINES

Turbines can generally be thought of as pumps operating in reverse. A turbine extracts energy from the fluid, converting fluid energy into mechanical energy. These devices can be considered to be adiabatic because the fluid gains (or loses) very little heat during the short time it passes through them. The kinetic and potential energy changes can be neglected. (See Fig. 16.2.)

Figure 16.2 *Turbine*

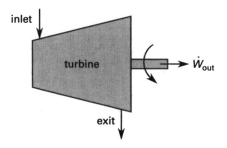

Equation 16.13 Through Eq. 16.16: Turbine Power (Rate of Work)[9]

$$\dot{W}_{turb} = \dot{m}(h_i - h_e) \quad \text{[adiabatic turbine]} \qquad 16.13$$

$$\dot{W}_{turb} = \dot{m}c_p(T_i - T_e) \quad \begin{bmatrix} \text{ideal gas; constant} \\ \text{specific heat} \end{bmatrix} \qquad 16.14$$

$$w_{turb} = c_p(T_i - T_e) \quad \text{[per unit mass]} \qquad 16.15$$

$$\dot{W}_{turb} = \dot{m}\left(h_e - h_i + \frac{v_e^2 - v_i^2}{2}\right)$$
$$= \dot{m}\left(c_p(T_e - T_i) + \frac{v_e^2 - v_i^2}{2}\right) \quad \begin{bmatrix} \Delta KE \\ \text{included} \end{bmatrix}$$
$$16.16$$

Description

Equation 16.13 through Eq. 16.16 give the power (rate of work) for turbines. Since they are all based on actual fluid exit properties, these equations determine the actual energy extracted from the fluid. The first form of Eq. 16.16 is the general equation, applicable to both vapors (such as steam) and real and ideal gases, while the second form of Eq. 16.16 is specifically only for ideal gases (for which specific heat is assumed to be constant). Equation 16.13 and Eq. 16.14 are simplifications of both forms of Eq. 16.16 that assume kinetic energy (velocity) changes are insignificant. Equation 16.15 is Eq. 16.14 restated on a per-mass basis.

Example

Steam flows at 34 kg/s through a steam turbine. The steam decreases in temperature from 150°C to 145°C, while the specific heat of the steam remains constant at 2.1 kJ/kg·°C. Assume a negligible change in kinetic energy. The ideal power generated by the turbine is most nearly

(A) 36 kW

(B) 43 kW

(C) 360 kW

(D) 430 kW

Solution

Use Eq. 16.14 to determine the power developed by the turbine.

$$\dot{W}_{turb} = \dot{m}c_p(T_i - T_e)$$
$$= \left(34\,\frac{kg}{s}\right)\left(2.1\,\frac{kJ}{kg\cdot°C}\right)(150°C - 145°C)$$
$$= 357\,kJ/s \quad (360\,kW)$$

The answer is (C).

---

[9]Equation 16.13 through Eq. 16.16 can be used to calculate the output power, $\dot{W}_{out}$, in Fig. 16.2 only if the mechanical efficiency of the turbine is 100%.

Equation 16.17: Turbine Isentropic Efficiency[10]

$$\eta_T = \frac{w_a}{w_s} = \frac{T_i - T_e}{T_i - T_{es}} \qquad 16.17$$

Variation

$$\eta_T = \frac{h_i - h_e}{h_i - h_{es}}$$

Description

The definition of *isentropic efficiency* (*adiabatic efficiency*), η_T, for a turbine is the inverse of what it is for a compressor. For a turbine, isentropic efficiency is the ratio of actual to ideal (isentropic) energy extraction. Actual isentropic efficiencies vary from approximately 65% for 1 MW unit to over 90% for 100 MW and larger units.

4. CAVITATION

Cavitation is a spontaneous vaporization of the fluid inside the pump, resulting in a degradation of pump performance. Wherever the fluid pressure is less than the vapor pressure, small pockets of vapor will form. These pockets usually form only within the pump itself, although cavitation slightly upstream within the suction line is also possible. As the vapor pockets reach the surface of the impeller, the local high fluid pressure collapses them. Noise, vibration, impeller pitting, and structural damage to the pump casing are manifestations of cavitation.

Cavitation can be caused by any of the following conditions.

- discharge head far below the pump head at peak efficiency
- high suction lift or low suction head
- excessive pump speed
- high liquid temperature (i.e., high vapor pressure)

5. NET POSITIVE SUCTION HEAD

The occurrence of cavitation is predictable. Cavitation will occur when the net pressure in the fluid drops below the vapor pressure. This criterion is commonly stated in terms of head: Cavitation occurs when the net positive suction head available is less than the net positive suction head required for satisfactory operation (e.g., $NPSH_A < NPSH_R$).

The minimum fluid energy required at the pump inlet for satisfactory operation (i.e., the required head) is known as the *net positive suction head required*, $NPSH_R$.[11] $NPSH_R$ is a function of the pump and will be given by the pump manufacturer as part of the pump performance data.[12] (See Fig. 16.3.) $NPSH_R$ is dependent on the flow rate.

Figure 16.3 Centrifugal Pump Characteristics

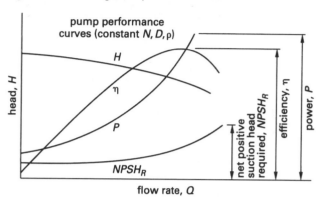

Equation 16.18: Net Positive Suction Head Available[13]

$$NPSH_A = \frac{p_{atm}}{\rho g} + H_s - H_f - \frac{v^2}{2g} - \frac{p_{vapor}}{\rho g} \qquad 16.18$$

Description

Net positive suction head available, $NPSH_A$, is the actual total fluid energy at the pump inlet. H_s is the static head at the pump inlet, and H_f is the friction loss between the fluid source and the pump inlet.

6. SIMILARITY (SCALING) LAWS

Equation 16.19 Through Eq. 16.23: Similarity (Scaling) Laws

$$\left(\frac{Q}{ND^3}\right)_2 = \left(\frac{Q}{ND^3}\right)_1 \qquad 16.19$$

$$\left(\frac{\dot{m}}{\rho ND^3}\right)_2 = \left(\frac{\dot{m}}{\rho ND^3}\right)_1 \qquad 16.20$$

---

[10]The subscript T in turbine isentropic efficiency, η_T, designates "turbine" and is the same as subscript "turb" used in Eq. 16.13 through Eq. 16.16.

[11]If $NPSH_R$ (a head term) is multiplied by the fluid specific weight, it is known as the *net inlet pressure required*, NIPR. Similarly, $NPSH_A$ can be converted to NIPA.

[12]It is also possible to calculate $NPSH_R$ from other information, such as suction specific speed. However, this still depends on information provided by the manufacturer.

[13]The *NCEES Handbook* is inconsistent in its nomenclature as it relates to $NPSH_A$. In Eq. 16.18, H_f is the same as h_f used elsewhere in the fluids sections; p_{vapor} corresponds to p_v used in Eq. 12.4; p_{atm} corresponds to p_0 in Eq. 12.5; H_s corresponds to h used in Eq. 12.1; and v corresponds to the inlet velocity, v_i, in Eq. 16.9.

$$\left(\frac{H}{N^2 D^2}\right)_2 = \left(\frac{H}{N^2 D^2}\right)_1 \qquad 16.21$$

$$\left(\frac{p}{\rho N^2 D^2}\right)_2 = \left(\frac{p}{\rho N^2 D^2}\right)_1 \qquad 16.22$$

$$\left(\frac{\dot{W}}{\rho N^3 D^5}\right)_2 = \left(\frac{\dot{W}}{\rho N^3 D^5}\right)_1 \qquad 16.23$$

Description

The performance of one pump can be used to predict the performance of a *dynamically similar* (*homologous*) pump. This can be done by using Eq. 16.19 through Eq. 16.23.

These *similarity laws* (also known as *scaling laws*) assume that both pumps

- operate in the turbulent region
- have the same pump efficiency
- operate at the same percentage of wide-open flow

These relationships assume that the efficiencies of the larger and smaller pumps are the same. In reality, larger pumps will be more efficient than smaller pumps. Therefore, extrapolations to much larger or much smaller sizes should be avoided.

Example

A 200 mm pump operating at 1500 rpm discharges 120 L/s of water. The capacity of a homologous 250 mm pump operating at the same speed is most nearly

(A) 150 L/s

(B) 180 L/s

(C) 200 L/s

(D) 230 L/s

Solution

Use Eq. 16.19, and solve for the flow rate of the second pump. The rotational speed, N, is unchanged.

$$\left(\frac{Q}{ND^3}\right)_2 = \left(\frac{Q}{ND^3}\right)_1$$

$$Q_2 = \left(\frac{Q_1}{D_1^3}\right) D_2^3$$

$$= \left(\frac{120 \, \frac{\text{L}}{\text{s}}}{(200 \, \text{mm})^3}\right)(250 \, \text{mm})^3$$

$$= 234.4 \, \text{L/s} \quad (230 \, \text{L/s})$$

The answer is (D).

7. FANS

There are two main types of fans: axial and centrifugal. Typical fan characteristics are given in Fig. 16.4.

Figure 16.4 *Fan Characteristics*

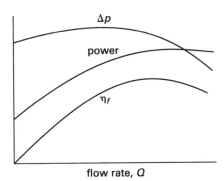

Axial-flow fans are essentially propellers mounted with small tip clearances in ducts. They develop static pressure by changing the airflow velocity. Axial flow fans are usually used when it is necessary to move large quantities of air (i.e., greater than 250 m^3/s) against low static pressures (i.e., less than 3 kPa), although the pressures and flow rates are much lower at most installations. An axial-flow fan may be followed by a *diffuser* (i.e., an *evase*) to convert some of the kinetic energy to static pressure.

Compared with centrifugal fans, axial flow fans are more compact and less expensive. However, they run faster than centrifugals, draw more power, are less efficient, and are noisier. Axial flow fans are capable of higher velocities than centrifugal fans.

Centrifugal fans are used in installations moving less than 500 m^3/s and pressures less than 15 kPa. Like centrifugal pumps, they develop static pressure by imparting a centrifugal force on the rotating air. Depending on the blade curvature, kinetic energy can be made greater (forward-curved blades) or less (backward-curved blades) than the tangential velocity of the impeller blades.

Forward-curved centrifugals (also called *squirrel cage fans*) are the most widely used centrifugals for general ventilation and packaged units. They operate at relatively low speeds, about half that of backward-curved fans. This makes them useful in high-temperature applications where stress due to rotation is a factor. Compared with backward-curved centrifugals, forward-curved blade fans have a greater capacity (due to their higher velocities) but require larger *scrolls*. However, since the fan blades are "cupped," they cannot be used when the air contains particles or contaminants. Efficiencies are the lowest of all centrifugals—between 70% and 75%.

Motors driving centrifugal fans with forward-curved blades can be overloaded if the duct losses are not calculated correctly. The power drawn increases rapidly with increases in the delivery rate. The motors are usually sized with some safety factor to compensate for the possibility that the actual system pressure will be less than the design pressure. For forward-curved blades, the maximum efficiency occurs near the point of maximum static pressure. Since their tip speeds are low, these fans are quiet. The fan noise is lowest at maximum pressure.

Radial fans (also called *straight-blade fans*, *paddle wheel fans*, and *shaving wheel fans*) have blades that are neither forward- nor backward-inclined. Radial fans are the workhorses of most industrial exhaust applications and can be used in material-handling and conveying systems where large amounts of bulk material pass through them. Such fans are low-volume, high-pressure (up to 15 kPa), high-noise, high-temperature, and low-efficiency (65–70%) units. *Radial tip fans* constitute a subcategory of radial fans. Their performance characteristics are between those of forward-curved and conventional radial fans.

Backward-curved centrifugals are quiet, medium- to high-volume and pressure, and high-efficiency units. They can be used in most applications with clean air below 540°C and up to about 10 kPa. They are available in three styles: flat, curved, and airfoil. Airfoil fans have the highest efficiency (up to 90%), while the other types have efficiencies between 80% and 90%. Because of these high efficiencies, power savings easily compensate for higher installation or replacement costs.

Equation 16.24: Backward-Curved Fan Power

$$\dot{W} = \frac{\Delta p Q}{\eta_f} \qquad \textit{16.24}$$

Description

Equation 16.24 is used to calculate the required power of a *fan motor* with backward-curved blades. η_f is the fan efficiency, and Δp is the rise in pressure. Power will be in watts if the pressure change is in pascals and the flow rate is in cubic meters per second. In heating, ventilating, and air conditioning work, pressure increase is often stated in centimeters of water. Heights of water must be converted to heights of air in order to determine the pressure rise.

$$\Delta p = \rho_{\text{air}} g h_{\text{air}} = \rho_{\text{air}} g \left(\frac{\rho_{\text{water}}}{\rho_{\text{air}}} \right) h_{\text{water}} = \rho_{\text{water}} g h_{\text{water}}$$

8. BLOWERS

The term *blower* is generally used to describe a device that supplies large volumes of low-pressure air to another process. For example, in wastewater plants, *sparging*[14] *blowers* supply air to nozzles submerged in aeration tanks and lagoons, while compressors are used to treat contaminated groundwater by injecting air into

the water table. Blowers are, for all practical purposes, compressors (fans, pumps, etc.) for which the pump equations are equally applicable.

Since blowers almost always move air, which is a compressible gas, it is possible to calculate the power required to move the air by calculating the thermodynamic work of compression. Isentropic (i.e., "adiabatic"), steady-flow equations are used, with the effects of non-ideal compression being incorporated into an isentropic efficiency.

Equation 16.25 and Eq. 16.26: Blower Power

$$P_w = \frac{WRT_1}{Cne} \left[\left(\frac{p_2}{p_1} \right)^{0.283} - 1 \right] \qquad \textit{16.25}$$

$$n = \frac{k-1}{k} \qquad \textit{16.26}$$

Values

$C = 550$ ft-lbf/hp-sec
$\quad = 29.7$ (for SI use)[15]

$n_{\text{air}} = 0.283$

$R_{\text{air}} = 53.3$ ft-lbf/lbm-°R

Description

Equation 16.25 calculates the power required to compress air[16] from p_1 to p_2 at the rate of W pounds per hour.[17] P_w is the power, in horsepower.[18] p_2 is the exit pressure,

---

[14]In chemistry, *sparging* is the act of bubbling a gas through a liquid (usually water) in order to transfer oxygen to the liquid or transfer some substance in the liquid to the air.

[15]The *NCEES Handbook* is entirely misleading in providing this value, as it cannot be used to obtain a blower power in horsepower, as the *Handbook* nomenclature implies. In fact, this unit cannot be used with customary U.S. units at all. In order to use the 29.7 conversion factor, the following units MUST be used: W in kg/h; R in kJ/kg·K; and T in K. If these units are used, then the power will be calculated in kilowatts (not horsepower).

[16]Some authorities refer to this as the "aeration" power. However, Eq. 16.25 does not include the power required to drive rotating arms, paddles, and other moving equipment.

[17](1) The *NCEES Handbook* presents Eq. 16.25 in terms of customary U.S. units, treating blowers as something different from compressors, which are covered on the same page. After presenting thermodynamic relations for compressors, the *NCEES Handbook* unnecessarily introduces an entirely new set of variables for almost every term for blowers. (2) The *NCEES Handbook* also fails to describe how to use Eq. 16.25 with values in SI units, although the compressor equations are applicable and the conversions can be determined from basic principles. (3) Although $\dot{m}$ is used for mass flow rate in the compressor section, W (without the dot) is used for "weight of flow of air" in the blower equation. (4) It is odd that the *NCEES Handbook* would use the unnecessary variable n in the denominator of Eq. 16.25, but then insert air's actual value of n for the exponent. (5) The value of $n = 0.283$ corresponds to a ratio of specific heats of $k = 1.395$, which in turn corresponds to an average process temperature of 250°F, accounting for the heat of compression.

[18]It is not clear what the "w" subscript refers to. The subscript is a vestige of the source from which the equation was taken, and the subscript does not have any connection with other content in the *NCEES Handbook*. The "w" is probably a reference to "water," since the equation was taken from a wastewater treatment book, although it could also be a reference to external work.

Fluid Mechanics/ Dynamics

which is some amount (i.e., a few psi) greater than the hydrostatic pressure at the exit depth (for blowers used to supply air to aeration tanks). W is the weight (in lbf) of air being compressed per second. R is the specific gas constant for air, in units of ft-lbf/lbm-°R. T_1 is the inlet temperature of the air in absolute temperature units. C is a constant needed to convert the power to either kilowatts or horsepower. n is a substitutional variable given by Eq. 16.26 used to simplify Eq. 16.25.[19] e is the *adiabatic efficiency (isentropic efficiency)*, which typically varies from 0.70 to 0.90. Figure 16.5 shows a typical electric air blower.

Figure 16.5 *A Typical Electric Air Blower*

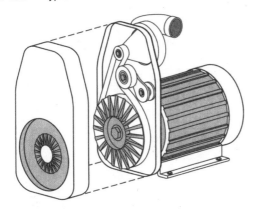

Example

100 small-bubble diffusers are fed by a blower. Each diffuser needs 3 SCFM (i.e., standard cubic feet per minute) of air. Standard conditions are defined as 68°F and 14.7 lbf/in² absolute (psia). The average air temperature is 60°F, and the plant is located at an altitude where the average atmospheric pressure is 14.5 psia. The average blower efficiency is 0.70. The inlet loss is 0.5 in water gage (w.g.). The air will be introduced in 50°F water at a depth of 8 ft, and the discharge pressure is the static pressure plus 1 psi. Most nearly, what blower power is needed?

(A) 6.6 hp

(B) 7.2 hp

(C) 7.6 hp

(D) 8.1 hp

Solution

The air requirement, Q, is

$$Q = (100 \text{ diffusers})\left(3 \ \frac{\text{SCFM}}{\text{diffuser}}\right) = 300 \text{ SCFM}$$

The volume is at standard conditions (i.e., standard temperature and pressure). At standard conditions, the density, ρ, is

$$\rho = \frac{p}{RT} = \frac{\left(14.7 \ \frac{\text{lbf}}{\text{in}^2}\right)\left(12 \ \frac{\text{in}}{\text{ft}}\right)^2}{\left(53.3 \ \frac{\text{ft-lbf}}{\text{lbm-°R}}\right)(68°F + 460°)}$$

$$= 0.07522 \text{ lbm/ft}^3$$

The mass flow rate, $\dot{m}$, is

$$\dot{m} = Q\rho = \frac{\left(300 \ \frac{\text{ft}^3}{\text{min}}\right)\left(0.07522 \ \frac{\text{lbm}}{\text{ft}^3}\right)}{60 \ \frac{\text{s}}{\text{min}}}$$

$$= 0.3761 \text{ lbm/s}$$

The density of 50°F water is 62.41 lbm/ft³.

The exit gage pressure (in excess of atmospheric pressure), Δp_2, at a depth of 8 ft is

$$\Delta p_2 = \frac{\rho g(h + h_{\text{inlet}})}{g_c} + \Delta p$$

$$= \frac{\left(62.41 \ \frac{\text{lbm}}{\text{ft}^3}\right)\left(32.2 \ \frac{\text{ft}}{\text{sec}^2}\right)\left(8 \text{ ft} + \frac{0.5 \text{ in}}{12 \ \frac{\text{in}}{\text{ft}}}\right)}{\left(32.2 \ \frac{\text{ft-lbm}}{\text{lbf-sec}^2}\right)\left(12 \ \frac{\text{in}}{\text{ft}}\right)^2} + 1 \ \frac{\text{lbf}}{\text{in}^2}$$

$$= 4.485 \text{ lbf/in}^2$$

The power requirement, P_w, is[20]

$$P_w = \frac{WRT_1}{Cne}\left(\left(\frac{p_2}{p_1}\right)^n - 1\right) \times \frac{g}{g_c}$$

$$= \left(\frac{\left(0.3761 \ \frac{\text{lbm}}{\text{sec}}\right)\left(53.3 \ \frac{\text{ft-lbf}}{\text{lbm-°R}}\right)(60°F + 460°)}{\left(550 \ \frac{\text{ft-lbf}}{\text{hp-sec}}\right)(0.283)(0.70)}\right)$$

$$\times \left(\left(\frac{14.5 \ \frac{\text{lbf}}{\text{in}^2} + 4.485 \ \frac{\text{lbf}}{\text{in}^2}}{14.5 \ \frac{\text{lbf}}{\text{in}^2}}\right)^{0.283} - 1\right)$$

$$\times \left(\frac{32.2 \ \frac{\text{ft}}{\text{sec}^2}}{32.2 \ \frac{\text{ft-lbm}}{\text{lbf-sec}^2}}\right)$$

$$= 7.581 \text{ hp} \quad (7.6 \text{ hp})$$

The answer is (C).

---

[19]Compare this equation to the equation for adiabatic work earlier on the same page of the *NCEES Handbook*. There is no need to use n, as it was not used with the compressor work equation.

[20]The last term (g/g_c) was added because the mass flow rate is used, rather than the "weight" flow rate.

Diagnostic Exam

Topic IV: Thermodynamics

1. 2 m^3 of an ideal gas is compressed from 100 kPa to 200 kPa. As a result of the process, the internal energy of the gas increases by 10 kJ, and 140 kJ of heat is lost to the surroundings. What is most nearly the work done by the gas during the process?

(A) −150 kJ

(B) −130 kJ

(C) −85 kJ

(D) −45 kJ

2. A cylinder fitted with a frictionless piston contains an ideal gas at temperature T and pressure p. The gas expands isothermally and reversibly until the pressure is $p/3$. Which statement is true regarding the work done by the gas during expansion?

(A) It is equal to the change in enthalpy of the gas.

(B) It is equal to the change in internal energy of the gas.

(C) It is equal to the heat absorbed by the gas.

(D) It is greater than the heat absorbed by the gas.

3. Consider the following balanced actual combustion reaction for propane.

$$C_3H_8 + 14.29(air) \rightarrow 4H_2O + 3CO + 11.29N_2$$

Assume air is 21% oxygen and 79% nitrogen by volume. What is most nearly the percent theoretical air?

(A) 50%

(B) 60%

(C) 68%

(D) 75%

4. In 1 hour, approximately how much black-body radiation escapes a 1 cm × 2 cm rectangular opening in a kiln whose internal temperature is 980°C?

(A) 20 kJ

(B) 100 kJ

(C) 130 kJ

(D) 150 kJ

5. Refrigerant HFC-134a at 0.8 MPa and 70°C is cooled and condensed at constant pressure in a steady-state process until it is a saturated liquid. Cooling water enters the condenser at 20°C and leaves at 30°C. If the mass flow rate of the refrigerant is 0.1 kg/s, the mass flow rate of the cooling water is most nearly

(A) 0.51 kg/s

(B) 0.65 kg/s

(C) 0.70 kg/s

(D) 0.75 kg/s

6. What is most nearly the melting temperature of sodium chloride, given that the latent heat of fusion is 30 kJ/mol, and the associated entropy change is 28 J/mol·K?

(A) 370K

(B) 880K

(C) 930K

(D) 1100K

7. Most nearly, what is the volume of 0.05 kg of refrigerant HFC-134a at 1.3 MPa with a quality of 37.5%?

(A) 9.6×10^{-5} m^3

(B) 1.7×10^{-4} m^3

(C) 3.1×10^{-4} m^3

(D) 2.2×10^{-3} m^3

8. Hot air at an average temperature of 100°C flows through a 3 m long tube with an inside diameter of 60 mm. The temperature of the tube is 20°C along its entire length. The average convective film coefficient is 20.1 W/m^2·K. What is most nearly the rate of convective heat transfer from the air to the tube?

(A) 520 W

(B) 850 W

(C) 910 W

(D) 1100 W

9. 80 kg of water is sealed in a 0.5 m^3 rigid container and is heated to 425°C. The critical temperature of steam is 647.1K, and the critical pressure of steam is 22.06 MPa. Treating steam as a real gas, what is most nearly the pressure in the container?

(A) 20 MPa

(B) 33 MPa

(C) 40 MPa

(D) 52 MPa

10. A refrigerant is saturated at 312K and 0.9334 MPa. Under these saturated conditions, the specific volume of the saturated liquid is 0.000795 m^3/kg, and the specific volume of the saturated gas is 0.01872 m^3/kg. The quality is 0.02538. The total mass of the refrigerant is 400 kg. What is most nearly the volume of the liquid refrigerant?

(A) 0.16 m^3

(B) 0.18 m^3

(C) 0.31 m^3

(D) 0.39 m^3

SOLUTIONS

1. Use the first law of thermodynamics.

$$W = Q - \Delta U$$

Using the standard sign convention, heat transferred to the surroundings is negative.

$$W = -140 \text{ kJ} - 10 \text{ kJ}$$
$$= -150 \text{ kJ}$$

The minus sign indicates that work was done on the gas.

The answer is (A).

2. Use the first law of thermodynamics.

$$Q = \Delta U + W$$

Since the internal energy remains constant during an isothermal process, $\Delta U = 0$.

$$Q = W$$

The work done by the gas is equal to the heat absorbed by the gas.

The answer is (C).

3. Balance the combustion reaction equation for complete combustion.

$$C_3H_8 + a(\text{air}) \rightarrow bH_2O + dN_2 + eCO_2$$

$$C_3H_8 + a(0.21O_2 + 0.79N_2) \rightarrow bH_2O + dN_2 + eCO_2$$

From the hydrogen balance, 4 moles of water are produced. From the carbon balance, 3 moles of carbon dioxide are produced. Therefore 5 moles of oxygen are required to react 1 mole of propane.

$$5 = a(0.21)$$
$$a = 23.81$$

The percent theoretical air is

$$\frac{N_{\text{actual}}}{N_{\text{complete combustion}}} \times 100\% = \frac{14.29 \text{ mol}}{23.81 \text{ mol}} \times 100\%$$
$$= 60\%$$

The answer is (B).

4. Calculate the area of the opening.

$$A = (1 \text{ cm})(2 \text{ cm})$$
$$= 2 \text{ cm}^2$$

The black-body radiation is

$$\dot{Q}_{black} = \varepsilon \sigma A T^4$$

$$= \frac{(1)\left(5.67 \times 10^{-8} \ \frac{W}{m^2 \cdot K^4}\right)(2 \ cm^2)}{\left(100 \ \frac{cm}{m}\right)^2} \times (980°C + 273°)^4$$

$$= 28 \ W$$

The energy emitted in an hour is

$$Q = \dot{Q}t$$

$$= \frac{(28 \ W)(1 \ h)\left(3600 \ \frac{s}{h}\right)}{1000 \ \frac{J}{kJ}}$$

$$= 100.6 \ kJ \quad (100 \ kJ)$$

The answer is (B).

5. The inlet and exit enthalpies are found from the HFC-134a pressure-enthalpy diagram. At 0.8 MPa and 70°C, the inlet enthalpy is 455.3 kJ/kg. At 0.8 MPa, the saturated exit enthalpy is 243.7 kJ/kg.

The inlet and exit enthalpies of water are found from steam tables. At 20°C, the saturated inlet enthalpy is 83.96 kJ/kg. At 30°C, the exit enthalpy is 125.79 kJ/kg.

The energy balance is

$$\dot{m}_{HFC\text{-}134a}(h_i - h_e)_{HFC\text{-}134a} = \dot{m}_{water}(h_e - h_i)_{water}$$

$$\left(0.1 \ \frac{kg}{s}\right)\left(455.3 \ \frac{kJ}{kg} - 243.7 \ \frac{kJ}{kg}\right)$$

$$= \dot{m}_{water}\left(125.79 \ \frac{kJ}{kg} - 83.96 \ \frac{kJ}{kg}\right)$$

$$\dot{m}_{water} = 0.506 \ kg/s \quad (0.51 \ kg/s)$$

The answer is (A).

6. The NaCl will be in an equilibrium state when all of the substance is melted. At equilibrium, the change in Gibbs energy is zero.

$$\Delta g = \Delta h - T_m \Delta s = 0$$

$$T_m = \frac{\Delta h}{\Delta s} = \frac{\left(30 \ \frac{kJ}{mol}\right)\left(1000 \ \frac{J}{kJ}\right)}{28 \ \frac{J}{mol \cdot K}}$$

$$= 1071K \quad (1100K)$$

The answer is (D).

7. Use the HFC-134a pressure-enthalpy diagram. At 1.3 MPa, the specific volume of the saturated liquid (from the almost vertical broken line) is approximately 0.00090 m³/kg. Similarly, the specific volume of the saturated vapor is approximately 0.015 m³/kg.

The volume is

$$V = mv = m(v_f + xv_{fg}) = m(v_f + x(v_g - v_f))$$

$$= (0.05 \ kg)\left(\begin{array}{c} 0.00090 \ \frac{m^3}{kg} \\ + (0.375)\left(0.015 \ \frac{m^3}{kg} - 0.00090 \ \frac{m^3}{kg}\right)\end{array}\right)$$

$$= 3.094 \times 10^{-4} \ m^3 \quad (3.1 \times 10^{-4} \ m^3)$$

The answer is (C).

8. The heat transfer area is

$$A = \pi dL = \frac{\pi(60 \ mm)(3 \ m)}{1000 \ \frac{mm}{m}}$$

$$= 0.565 \ m^2$$

Use Newton's law of convection. The temperature difference is the same for temperatures expressed in Celsius (°C) and kelvins (K).

$$\dot{Q} = hA(T_w - T_\infty)$$

$$= hA(T_{air} - T_{wall})$$

$$= \left(20.1 \ \frac{W}{m^2 \cdot K}\right)(0.565 \ m^2)(100°C - 20°C)$$

$$= 908.3 \ W \quad (910 \ W)$$

The answer is (C).

9. Use the ideal gas law to find the approximate pressure.

$$p \approx \frac{mRT}{V}$$

$$= \frac{(80 \ kg)\left(0.4615 \ \frac{kJ}{kg \cdot K}\right)(425°C + 273°)}{(0.5 \ m^3)\left(1000 \ \frac{kPa}{MPa}\right)}$$

$$= 51.54 \ MPa$$

The reduced temperature and pressure, respectively, are

$$T_r = \frac{T}{T_c} = \frac{425°C + 273°}{647.1K} = 1.08$$

$$p_r = \frac{p}{p_c} = \frac{51.54 \ MPa}{22.06 \ MPa} = 2.34$$

Thermodynamics

Use the generalized compressibility chart. At the reduced temperature and pressure, the compressibility factor is $Z \approx 0.38$. The pressure is

$$p' = Z\frac{mRT}{V} = Zp$$
$$= (0.38)(51.54 \ \text{MPa})$$
$$= 19.59 \ \text{MPa} \quad (20 \ \text{MPa})$$

The answer is (A).

10. The mass of the liquid refrigerant is

$$m_f = (1 - x_2)m_{\text{total}}$$
$$= (1 - 0.02538)(400 \ \text{kg})$$
$$= 389.8 \ \text{kg}$$

The volume of the liquid refrigerant is

$$V_f = m_f v_f$$
$$= (389.8 \ \text{kg})\left(0.000795 \ \frac{\text{m}^3}{\text{kg}}\right)$$
$$= 0.31 \ \text{m}^3$$

The answer is (C).

Thermodynamics

17 Properties of Substances

Nomenclature

a	constant	–	–
a	Helmholtz function	Btu/lbm	kJ/kg
b	constant	–	–
c	specific heat	Btu/lbm	kJ/kg·K
$\overline{c}$	mean heat capacity	Btu/lbm	kJ/kg·K
C	heat	Btu	kJ
C	number of components in the system	–	–
F	number of independent variables	–	–
g	Gibbs function	Btu/lbm	kJ/kg
h	specific enthalpy	Btu/lbm	kJ/kg
H	enthalpy	Btu	kJ
J	Joule's constant, 778	ft-lbf/Btu	n.a.
k	ratio of specific heats	–	–
m	mass	lbm	kg
M	molecular weight	–	kg/kmol
N	number	–	–
p	pressure[1]	lbf/in^2	Pa
P	number of phases existing simultaneously	–	–
R	specific gas constant	ft-lbf/lbm-°R	kJ/kg·K
$\overline{R}$	universal gas constant, 1545 (8314)	ft-lbf/lbmol-°R	J/kmol·K
s	specific entropy	Btu/lbm-°R	kJ/kg·K
S	entropy	Btu/°R	kJ/K
T	absolute temperature	°R	K
u	specific internal energy	Btu/lbm	kJ/kg
U	internal energy	Btu	kJ
V	volume	ft^3	m^3
x	quality	–	–
z	compressibility factor	–	–
Z	compressibility factor	–	–

Symbols

ρ	density	lbm/ft^3	kg/m^3
v	specific volume[2]	ft^3/lbm	m^3/kg
$\overline{v}$	molar specific volume	ft^3/lbmol	m^3/kmol

Subscripts

c	critical
f	fluid
fg	liquid-to-gas (vaporization)
g	gas
p	constant pressure
r	reduced
T	constant temperature
v	constant volume

1. PHASES OF A PURE SUBSTANCE

Thermodynamics is the study of a substance's energy-related properties. The properties of a substance and the procedures used to determine those properties depend on the state and the phase of the substance. The thermodynamic *state* of a substance is defined by two or more independent thermodynamic properties. For example, the temperature and pressure of a substance are two properties commonly used to define the state of a superheated vapor.

The common *phases* of a substance are solid, liquid, and gas. However, because substances behave according to different rules, it is convenient to categorize them into more than only these three phases.

Solid: A solid does not take on the shape or volume of its container.

Saturated liquid: A saturated liquid has absorbed as much heat energy as it can without vaporizing. Liquid water at standard atmospheric pressure and 212°F (100°C) is an example of a saturated liquid.

Subcooled liquid: If a liquid is not saturated (i.e., the liquid is not at its boiling point), it is said to be subcooled. Water at 1 atm and room temperature is subcooled, as it can absorb additional energy without vaporizing.

---

[1]With the exception of use as a subscript, the NCEES *FE Reference Handbook* (*NCEES Handbook*) uses uppercase P as the symbol for pressure. In contrast to the *NCEES Handbook*, this book generally uses lowercase p to represent pressure. The equations in this book involving pressure will differ slightly in appearance from the *NCEES Handbook*.

[2]The *NCEES Handbook* uses lowercase italic v for specific volume. This book uses Greek upsilon, v, to avoid confusion with the symbol for velocity in kinetic energy calculations. The equations in this book involving specific volume will differ slightly in appearance from those in the *NCEES Handbook*.

Liquid-vapor mixture: A liquid and vapor of the same substance can coexist at the same temperature and pressure. This is called a two-phase, liquid-vapor mixture.

Perfect gas: A perfect gas is an ideal gas whose specific heats (and hence ratio of specific heats) are constant.

Saturated vapor: A vapor (e.g., steam at standard atmospheric pressure and 212°F (100°C)) that is on the verge of condensing is said to be saturated.

Superheated vapor: A superheated vapor is one that has absorbed more energy than is needed merely to vaporize it. A superheated vapor will not condense when small amounts of energy are removed.

Ideal gas: A gas is a highly superheated vapor. If the gas behaves according to the ideal gas law, $pV = mRT$, it is called an ideal gas.

Real gas: A real gas does not behave according to the ideal gas laws.

Gas mixtures: Most gases mix together freely. Two or more pure gases together constitute a gas mixture.

Vapor-gas mixtures: Atmospheric air is an example of a mixture of several gases and water vapor.

It is theoretically possible to develop a three-dimensional surface that predicts a substance's phase based on the properties of pressure, temperature, and specific volume. Such a three-dimensional p-v-T diagram is illustrated in Fig. 17.1.

Figure 17.1 Three-Dimensional p-v-T Phase Diagram

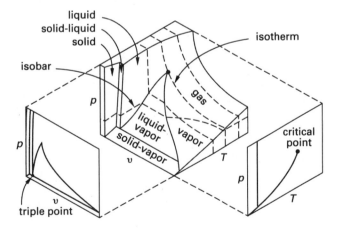

If one property is held constant during a process, a two-dimensional projection of the p-v-T diagram can be used. Figure 17.2 is an example of this projection, which is known as an *equilibrium diagram* or a *phase diagram*.

The most important part of a phase diagram is limited to the liquid-vapor region. A general phase diagram showing this region and the bell-shaped dividing line (known as the *vapor dome*) is shown in Fig. 17.3.

Figure 17.2 Pressure-Volume Phase Diagram

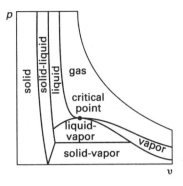

Figure 17.3 Vapor Dome with Isobars

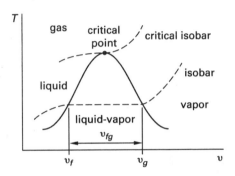

The vapor dome region can be drawn with many variables for the axes. For example, either temperature or pressure can be used for the vertical axis. Internal energy, enthalpy, specific volume, or entropy can be chosen for the horizontal axis. However, the principles presented here apply to all combinations.

The left-hand part of the vapor dome curve separates the liquid phase from the liquid-vapor phase. This part of the line is known as the *saturated liquid line*. Similarly, the right-hand part of the line separates the liquid-vapor phase from the vapor phase. This line is called the *saturated vapor line.*

Lines of constant pressure (*isobars*) can be superimposed on the vapor dome. Each isobar is horizontal as it passes through the two-phase region, verifying that both temperature and pressure remain unchanged as a liquid vaporizes.

There is no dividing line between liquid and vapor at the top of the vapor dome. Above the vapor dome, the phase is a gas.

The implicit dividing line between liquid and gas is the isobar that intersects the topmost part of the vapor dome. This is known as the *critical isobar*, and the highest point of the vapor dome is known as the *critical point.* This critical isobar also provides a way to distinguish between a vapor and a gas. A substance below the critical isobar (but to the right of the vapor dome) is a vapor. Above the critical isobar, it is a gas.

The *triple point* of a substance is a unique state at which solid, liquid, and gaseous phases can coexist. For instance, the triple point of water occurs at a pressure of 0.00592 atm and a temperature of 491.71°R (273.16K).

Figure 17.4 illustrates a vapor dome for which pressure has been chosen as the vertical axis and enthalpy has been chosen as the horizontal axis. The shape of the dome is essentially the same, but the lines of constant temperature (*isotherms*) have slopes of different signs than the isobars.

Figure 17.4 Vapor Dome with Isotherms

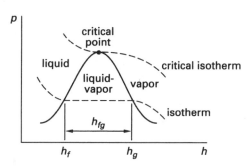

Figure 17.4 also illustrates the subscripting convention used to identify points on the saturation line. The subscript f (fluid) is used to indicate a saturated liquid. The subscript g (gas) is used to indicate a saturated vapor. The subscript fg is used to indicate the difference in saturation properties.

The vapor dome is a good tool for illustration, but it cannot be used to determine a substance's phase. Such a determination must be made based on the substance's pressure and temperature according to the following rules.

rule 1: A substance is a subcooled liquid if its temperature is less than the saturation temperature corresponding to its pressure.

rule 2: A substance is in the liquid-vapor region if its temperature is equal to the saturation temperature corresponding to its pressure.

rule 3: A substance is a superheated vapor if its temperature is greater than the saturation temperature corresponding to its pressure.

rule 4: A substance is a subcooled liquid if its pressure is greater than the saturation pressure corresponding to its temperature.

rule 5: A substance is in the liquid-vapor region if its pressure is equal to the saturation pressure corresponding to its temperature.

rule 6: A substance is a superheated vapor if its pressure is less than the saturation pressure corresponding to its temperature.

Equation 17.1: Gibbs Phase Rule (Non-Reacting Systems)

$$P + F = C + 2 \qquad \text{17.1}$$

Description

Gibbs phase rule defines the relationship between the number of phases and components in a mixture at equilibrium.

P is the number of phases existing simultaneously; F is the number of independent variables, known as *degrees of freedom*; and C is the number of components in the system. Composition, temperature, and pressure are examples of degrees of freedom that can be varied.

For example, if water is to be stored such that three phases (solid, liquid, gas) are present simultaneously, then $P = 3$, $C = 1$, and $F = 0$. That is, neither pressure nor temperature can be varied independently. This state is exemplified by water at its triple point.

Example

How many phases can exist in equilibrium for a fixed proportion water-alcohol mixture held at constant pressure?

(A) 0

(B) 1

(C) 2

(D) 3

Solution

There are two components, water and alcohol, so $C = 2$. Normally, composition, pressure, and temperature can be varied (i.e., three degrees of freedom), but with a specific composition and pressure held constant, $F = 1$. From Gibbs phase rule, Eq. 17.1, the number of phases, P, is

$$
\begin{aligned}
P + F &= C + 2 \\
P &= C + 2 - F \\
&= 2 + 2 - 1 \\
&= 3
\end{aligned}
$$

The answer is (D).

2. STATE FUNCTIONS (PROPERTIES)

The thermodynamic state or condition of a substance is determined by its properties. *Intensive properties* are independent of the amount of substance present. Temperature, pressure, and stress are examples of intensive properties. *Extensive properties* are dependent on (i.e., are proportional to) the amount of substance present. Examples are volume, strain, charge, and mass.

Thermodynamics

In most books on thermodynamics, both lowercase and uppercase forms of the same characters are used to represent property variables. The two forms are used to distinguish between the units of mass. For example, lowercase h represents specific enthalpy (usually called "enthalpy") in units of Btu/lbm or kJ/kg. Uppercase H is used to represent the molar enthalpy in units of Btu/lbmol or kJ/kmol.

Properties of gases in tabulated form are useful or necessary for solving many thermodynamic problems. The properties of saturated and superheated steam are tabulated in Table 17.1[3] and Table 17.2, respectively, at the end of this chapter. A pressure-enthalpy (p-h) diagram for refrigerant HFC-134a in SI units is presented in Fig. 17.5 at the end of this chapter.[4]

Mass

The mass, m, of a substance is a measure of its quantity. Mass is independent of location and gravitational field strength. In thermodynamics, the customary U.S. and SI units of mass are pound-mass (lbm) and kilogram (kg), respectively.

Pressure

Customary U.S. pressure units are pounds per square inch (lbf/in^2). Standard SI pressure units are kPa or MPa, although bars are also used in tabulations of thermodynamic data: (1 bar = 1 atm = 10^5 Pa.)

Most pressure gauges read atmospheric pressures, but in general, thermodynamic calculations will be performed using absolute pressures. The values of a standard atmosphere in various units are given in Table 17.3.

Table 17.3 Standard Atmospheric Pressure

1.000 atm	(atmosphere)
14.696 psia	(pounds per square inch absolute)
2116.2 psfa	(pounds per square foot absolute)
407.1 in w.g.	(inches of water; inches water gage)
33.93 ft w.g.	(feet of water; feet water gage)
29.921 in Hg	(inches of mercury)
760.0 mm Hg	(millimeters of mercury)
760.0 torr	
1.013 bars	
1013 millibars	
1.013×10^5 Pa	(pascals)
101.3 kPa	(kilopascals)

Temperature

Temperature is a thermodynamic property of a substance that depends on energy content. Heat energy entering a substance will increase the temperature of that substance. Normally, heat energy will flow only

from a hot object to a cold object. If two objects are in *thermal equilibrium* (are at the same temperature), no heat energy will flow between them.

If two systems are in thermal equilibrium, they must be at the same temperature. If both systems are in equilibrium with a third system, then all three systems are at the same temperature. This concept is known as the *zeroth law of thermodynamics*.

The *absolute temperature scale* defines temperature independently of the properties of any particular substance. This is unlike the Celsius and Fahrenheit scales, which are based on the freezing point of water. The absolute temperature scale should be used for all thermodynamic calculations.

In the customary U.S. system, the absolute temperature scale is the *Rankine scale*.

$$T_{\circ R} = T_{\circ F} + 459.67^\circ$$

$$\Delta T_{\circ R} = \Delta T_{\circ F}$$

The absolute temperature scale in SI is the *Kelvin scale*.

$$T_K = T_{\circ C} + 273.15^\circ$$

$$\Delta T_K = \Delta T_{\circ C}$$

The relationships between the four temperature scales are illustrated in Fig. 17.6, which also defines the approximate *boiling point*, *triple point*, *ice point*, and *absolute zero* temperatures.

Figure 17.6 Temperature Scales

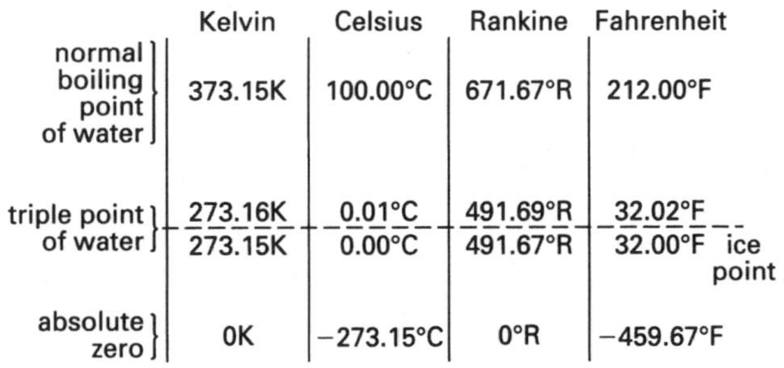

	Kelvin	Celsius	Rankine	Fahrenheit
normal boiling point of water	373.15K	100.00°C	671.67°R	212.00°F
triple point of water	273.16K	0.01°C	491.69°R	32.02°F
	273.15K	0.00°C	491.67°R	32.00°F ice point
absolute zero	0K	−273.15°C	0°R	−459.67°F

Equation 17.2: Specific Volume

$$v = V/m \qquad \qquad 17.2$$

Variation

$$v = \frac{1}{\rho}$$

Description

Specific volume, v, is the volume occupied by one unit mass of a substance. Customary U.S. units in tabulations of thermodynamic data are cubic feet per pound-mass (ft^3/lbm). Standard SI specific volume units are cubic meters per kilogram (m^3/kg). Molar specific

---

[3]The *NCEES Handbook* is not consistent with how it uses zero in Table 17.1. There is no difference between "0.00" in the first line and "0" in the last line.

[4]The *NCEES Handbook* labels this figure as a *P-h* diagram, consistent with its convention to use uppercase P as the symbol for pressure.

volume has units of ft³/lbmol (m³/kmol) and is seldom tabulated. Specific volume is the reciprocal of density.

Example

A 1 m³ volume of gas has a mass of 1.2 kg. What is most nearly the specific volume of the gas?

- (A) 730 cm³/g
- (B) 830 cm³/g
- (C) 890 cm³/g
- (D) 940 cm³/g

Solution

Use Eq. 17.2 to find the specific volume of the gas.

$$v = V/m = \frac{(1 \text{ m}^3)\left(100 \frac{\text{cm}}{\text{m}}\right)^3}{(1.2 \text{ kg})\left(1000 \frac{\text{g}}{\text{kg}}\right)}$$

$$= 833 \text{ cm}^3/\text{g} \quad (830 \text{ cm}^3/\text{g})$$

The answer is (B).

Equation 17.3: Specific Internal Energy

$$u = U/m \qquad 17.3$$

Description

Internal energy accounts for all of the energy of the substance excluding pressure, potential, and kinetic energy. The internal energy is a function of the state of a system. Examples of internal energy are the translational, rotational, and vibrational energies of the molecules and atoms in the substance. Since the movement of atoms and molecules increases with temperature, internal energy is a function of temperature. It does not depend on the process or path taken to reach a particular temperature.

In the United States, the *British thermal unit*, Btu, is used in thermodynamics to represent the various forms of energy. (One Btu is approximately the energy given off by burning one wooden match.) Standard units of specific internal energy, u, are Btu/lbm and kJ/kg. The units of molar internal energy are Btu/lbmol and kJ/kmol. Equation 17.3 gives the relationship between the specific and system internal energy.[5]

---

[5]Some authorities use uppercase letters to represent the molar properties of a substance (e.g., H for molar enthalpy) and lowercase letters to represent the specific (i.e., per unit mass) properties (e.g., h for specific enthalpy). The *NCEES Handbook* uses uppercase letters to represent the total properties of the system, regardless of the amount of system substance. For example, V is the total volume of the substance in the system, U is the total internal energy of the substance in the system, and H is the total enthalpy of the substance in the system. Except for use with equations of state, the molar properties are not represented in the *NCEES Handbook*.

Example

A system contains 1 kg of saturated steam vapor heated to 120°C. The specific internal energy of saturated steam vapor is 2529.3 kJ/kg. What is most nearly the total internal energy of the steam?

- (A) 2525 kJ
- (B) 2529 kJ
- (C) 3525 kJ
- (D) 3552 kJ

Solution

Use Eq. 17.3 to calculate the total internal energy.

$$u = U/m$$

$$U = mu = (1 \text{ kg})\left(2529.3 \frac{\text{kJ}}{\text{kg}}\right)$$

$$= 2529.3 \text{ kJ} \quad (2529 \text{ kJ})$$

The answer is (B).

Equation 17.4: Specific Enthalpy

$$h = u + pv = H/m \qquad 17.4$$

Description

Enthalpy represents the total useful energy of a substance. Useful energy consists of two parts: the specific internal energy, u, and the *flow energy* (also known as *flow work* and *p-V work*), pv. Therefore, enthalpy has the same units as internal energy.

Enthalpy is defined as useful energy because, ideally, all of it can be used to perform useful tasks. It takes energy to increase the temperature of a substance. If that internal energy is recovered, it can be used to heat something else (e.g., to vaporize water in a boiler). Also, it takes energy to increase pressure and volume (as in blowing up a balloon). If pressure and volume are decreased, useful energy is given up.

Strictly speaking, the customary U.S. units of Eq. 17.4 are not consistent, since flow work (as written) has units of ft-lbf/lbm, not Btu/lbm. (There is also a consistency problem if pressure is defined in lbf/ft² and given in lbf/in².) Equation 17.4 should be written as

$$h = u + \frac{pv}{J}$$

The conversion factor, J, in the above equation is known as *Joule's constant*. It has a value of approximately 778 ft-lbf/Btu. (In SI units, Joule's constant has a value of 1.0 N·m/J and is unnecessary.) As in Eq. 17.4, Joule's constant is often omitted from the statement of generic thermodynamic equations, but it is always needed with customary U.S. units for dimensional consistency.

Example

Steam at 416 Pa and 166K has a specific volume of 0.41 m³/kg and a specific enthalpy of 29.4 kJ/kg. What is most nearly the internal energy per kilogram of steam?

(A) 28.5 kJ/kg

(B) 29.2 kJ/kg

(C) 30.2 kJ/kg

(D) 30.4 kJ/kg

Solution

From Eq. 17.4,

$$h = u + pv$$
$$u = h - pv$$

$$= 29.4 \ \frac{kJ}{kg} - \frac{(416 \ \text{Pa}) \left(0.41 \ \frac{m^3}{kg} \right)}{1000 \ \frac{Pa}{kPa}}$$

$$= 29.2 \ \text{kJ/kg}$$

The answer is (B).

Equation 17.5: Specific Entropy

$$s = S/m \qquad \textbf{17.5}$$

Description

Entropy is a measure of the energy that is no longer available to perform useful work within the current environment. Other definitions (the "disorder of the system," the "randomness of the system," etc.) are frequently quoted. Although these alternate definitions cannot be used in calculations, they are consistent with the third law of thermodynamics (also known as the *Nernst theorem*). This law states that the absolute entropy of a perfect crystalline solid in thermodynamic equilibrium is (approaches) zero when the temperature is (approaches) absolute zero.

The units of specific entropy are Btu/lbm-°R and kJ/kg·K.

Equation 17.6: Gibbs Function

$$g = h - Ts \qquad \textbf{17.6}$$

Variation

$$g = u + pv - Ts$$

Description

The Gibbs function for a pure substance is defined by Eq. 17.6 and the variation equation. It is used in investigating latent heat changes and chemical reactions.

For a constant-temperature, constant-pressure, nonflow process that is approaching equilibrium, the Gibbs function approaches a minimum value.

$$(dg)_{T,p} < 0 \quad \text{[nonequilibrium]}$$

Once the minimum value is obtained, equilibrium is attained, and the Gibbs function is constant.

$$(dg)_{T,p} = 0 \quad \text{[equilibrium]}$$

The *Gibbs function of formation*, g^0, has been tabulated at the standard reference conditions of 25°C and 1 atm. A chemical reaction can occur spontaneously only if the change in Gibbs function is negative (i.e., the Gibbs function of formation for the products is less than the Gibbs function of formation for the reactants).

$$\sum_{\text{products}} Ng^0 < \sum_{\text{reactants}} Ng^0$$

Example

Water at 50°C and 1 atm has a specific enthalpy of 209.33 kJ/kg and a specific entropy of 0.7038 kJ/kg·K. What is most nearly the Gibbs function for the water?

(A) −18 kJ/kg

(B) −3.0 kJ/kg

(C) 300 kJ/kg

(D) 440 kJ/kg

Solution

Use Eq. 17.6. The Gibbs function is

$$g = h - Ts$$

$$= 209.33 \ \frac{kJ}{kg} - (50°C + 273°) \left(0.7038 \ \frac{kJ}{kg \cdot K} \right)$$

$$= -17.997 \ \text{kJ/kg} \quad (-18 \ \text{kJ/kg})$$

The answer is (A).

Equation 17.7: Helmholtz Function

$$a = u - Ts \qquad \textbf{17.7}$$

Variation

$$a = h - pv - Ts$$

Description

The Helmholtz function for a pure substance is defined by Eq. 17.7. Like the Gibbs function, the Helmholtz function is used in investigating equilibrium conditions. For a constant-temperature, constant-volume nonflow process approaching equilibrium, the Helmholtz function approaches its minimum value.

$$(dA)_{T,V} < 0 \quad \text{[nonequilibrium]}$$

Once the minimum value is obtained, equilibrium is attained, and the Helmholtz function will be constant.

$$(dA)_{T,V} = 0 \quad \text{[equilibrium]}$$

The Helmholtz function is sometimes known as the *free energy of the system* because its change in a reversible isothermal process equals the maximum energy that can be "freed" and converted to mechanical work. The same term has also been used for the Gibbs function under analogous conditions. For example, the difference in standard Gibbs functions of reactants and products has often been called the "free energy difference."

Since there is a possibility for confusion, it is better to refer to the Gibbs and Helmholtz functions by their actual names.

Example

A superheated vapor has a specific internal energy of 2733.7 kJ/kg and a specific entropy of 8.0333 kJ/kg·K. What is most nearly the Helmholtz free energy function of the superheated vapor at 250°C and 0.1 MPa?

(A) −2700 kJ/kg

(B) −1500 kJ/kg

(C) 8.0 kJ/kg

(D) 2700 kJ/kg

Solution

Use Eq. 17.7. The Helmholtz function is

$$a = u - Ts$$

$$= 2733.7 \ \frac{\text{kJ}}{\text{kg}} - (250°\text{C} + 273°)\left(8.0333 \ \frac{\text{kJ}}{\text{kg·K}}\right)$$

$$= -1468 \text{ kJ/kg} \quad (-1500 \text{ kJ/kg})$$

The answer is (B).

Equation 17.8 and Eq. 17.9: Specific Heat

$$c_p = \left(\frac{\partial h}{\partial T}\right)_p \quad \text{[at constant pressure]} \qquad 17.8$$

$$c_v = \left(\frac{\partial u}{\partial T}\right)_v \quad \text{[at constant volume]} \qquad 17.9$$

Variation

$$c = \frac{Q}{m \Delta T}$$

Values

Table 17.4 *Approximate Specific Heats of Selected Liquids and Solids (at room temperature)*

substance	c_p		density	
	$\frac{\text{kJ}}{\text{kg·K}}$	$\frac{\text{Btu}}{\text{lbm-°R}}$	$\frac{\text{kg}}{\text{m}^3}$	$\frac{\text{lbm}}{\text{ft}^3}$
liquids				
ammonia	4.80	1.146	602	38
mercury	0.139	0.033	13 560	847
water	4.18	1.000	997	62.4
solids				
aluminum	0.900	0.215	2700	170
copper	0.386	0.092	8900	555
ice (0°C; 32°F)	2.11	0.502	917	57.2
iron	0.450	0.107	7840	490
lead	0.128	0.030	11 310	705

Description

An increase in internal energy is needed to cause a rise in temperature. Different substances differ in the quantity of heat needed to produce a given temperature increase. The heat energy, Q, required to change the temperature of a mass, m, by an amount, ΔT, is called the *specific heat (heat capacity) of the substance*, c (see the variation equation). Because specific heats of solids and liquids are slightly temperature dependent, the mean specific heats are used for processes covering large temperature ranges.

For gases, the specific heat depends on the type of process during which the heat exchange occurs. Equation 17.8 defines the specific heats for constant-pressure processes, c_p, and Eq. 17.9 defines the specific heat for constant-volume processes, c_v.

c_p and c_v for solids and liquids are essentially the same and are given in Table 17.4. Approximate values of c_p and c_v for common gases are given in Table 17.5.

Example

The temperature-dependent molar heat capacity of nitrogen in units of kJ/kmol·K is

$$C_p = 39.06 - 512.79 T^{-1.5} + 1072.7 T^{-2} - 820.4 T^{-3}$$

What is most nearly the change in enthalpy per kg of nitrogen when it is heated at constant pressure from 1000K to 1500K?

(A) 600 kJ/kg

(B) 700 kJ/kg

(C) 800 kJ/kg

(D) 900 kJ/kg

Table 17.5 *Approximate Specific Heats of Selected Gases (at room temperature)*

gas	mol. wt	c_p kJ/kg·K	c_p Btu/lbm-°R	c_v kJ/kg·K	c_v Btu/lbm-°R	k	R kJ/kg·K
air	29	1.00	0.240	0.718	0.171	1.40	0.2870
argon	40	0.520	0.125	0.312	0.0756	1.67	0.2081
butane	58	1.72	0.415	1.57	0.381	1.09	0.1430
carbon dioxide	44	0.846	0.203	0.657	0.158	1.29	0.1889
carbon monoxide	28	1.04	0.249	0.744	0.178	1.40	0.2968
ethane	30	1.77	0.427	1.49	0.361	1.18	0.2765
helium	4	5.19	1.25	3.12	0.753	1.67	2.0769
hydrogen	2	14.3	3.43	10.2	2.44	1.40	4.1240
methane	16	2.25	0.532	1.74	0.403	1.30	0.5182
neon	20	1.03	0.246	0.618	0.148	1.67	0.4119
nitrogen	28	1.04	0.248	0.743	0.177	1.40	0.2968
octane vapor	114	1.71	0.409	1.64	0.392	1.04	0.0729
oxygen	32	0.918	0.219	0.658	0.157	1.40	0.2598
propane	44	1.68	0.407	1.49	0.362	1.12	0.1885
steam	18	1.87	0.445	1.41	0.335	1.33	0.4615

Solution

Use separation of variables with Eq. 17.8 to find the change in specific enthalpy. (Use uppercase letters to designate the molar quantities.)

$$C_p = \left(\frac{\partial H}{\partial T}\right)_p$$

$$\partial H = C_p\, \partial T$$

$$\Delta H = \int C_p\, dT$$

$$= \int_{1000\text{K}}^{1500\text{K}} \left(\begin{array}{c} 39.06 - 512.79\,T^{-1.5} \\ + 1072.7\,T^{-2} - 820.4\,T^{-3} \end{array} \right) dT$$

$$= 19\,524 \text{ kJ/kmol}$$

The molecular weight of nitrogen is 28 kg/kmol. The change in specific enthalpy is

$$\Delta h = \frac{\Delta H}{M}$$

$$= \frac{19\,524 \,\dfrac{\text{kJ}}{\text{kmol}}}{28 \,\dfrac{\text{kg}}{\text{kmol}}}$$

$$= 697.3 \text{ kJ/kg} \quad (700 \text{ kJ/kg})$$

The answer is (B).

3. TWO-PHASE SYSTEMS: LIQUID-VAPOR MIXTURES

Equation 17.10: Quality

$$x = m_g/(m_g + m_f) \qquad \text{17.10}$$

Description

Within the vapor dome, water is at its saturation pressure and temperature. When saturated, water can simultaneously exist in liquid and vapor phases in any proportion between 0 and 1. The *quality* is the fraction by weight of the total mass that is vapor.

Example

If the ratio of vapor mass to liquid mass in a mixture is 0.8, the quality of the mixture is most nearly

(A) 0.20

(B) 0.25

(C) 0.44

(D) 0.80

Solution

Write the mass of the vapor in terms of the mass of liquid.

$$\frac{m_g}{m_f} = 0.8$$

$$m_g = 0.8 m_f$$

From Eq. 17.10, the quality of the mixture is

$$x = m_g/(m_g + m_f)$$

$$= \frac{0.8 m_f}{0.8 m_f + m_f}$$

$$= 0.44$$

The answer is (C).

Equation 17.11 Through Eq. 17.18: Primary Thermodynamic Properties

$$v = xv_g + (1 - x)v_f \qquad \textit{17.11}$$

$$v = v_f + xv_{fg} \qquad \textit{17.12}$$

$$u = xu_g + (1 - x)u_f \qquad \textit{17.13}$$

$$u = u_f + xu_{fg} \qquad \textit{17.14}$$

$$h = xh_g + (1 - x)h_f \qquad \textit{17.15}$$

$$h = h_f + xh_{fg} \qquad \textit{17.16}$$

$$s = xs_g + (1 - x)s_f \qquad \textit{17.17}$$

$$s = s_f + xs_{fg} \qquad \textit{17.18}$$

Description

When the thermodynamic state of a substance is within the vapor dome, there is a one-to-one correspondence between the saturation temperature and saturation pressure. One determines the other. The thermodynamic state is uniquely defined by any two independent properties (temperature and quality, pressure and enthalpy, entropy and quality, etc.).

If the quality of a liquid-vapor mixture is known, it can be used to calculate all of the primary thermodynamic properties (specific volume, specific internal energy, specific enthalpy, and specific entropy). If a thermodynamic property has a value between the saturated liquid and saturated vapor values (e.g., h is between h_f and h_g), then Eq. 17.11 through Eq. 17.18 can be solved for the quality.

Each pair of equations (e.g., Eq. 17.11 and Eq. 17.12, Eq. 17.13 and Eq. 17.14, Eq. 17.15 and Eq. 17.16, and Eq. 17.17 and Eq. 17.18) is equivalent by the following relationships.

$$v_{fg} = v_g - v_f$$

$$u_{fg} = u_g - u_f$$

$$h_{fg} = h_g - h_f$$

$$s_{fg} = s_g - s_f$$

If temperatures are known, the quantities in the previous relationships can be obtained from *saturation tables* (or *steam tables* in the case of water). (See Table 17.1.)

Example

A saturated steam supply line is analyzed for specific internal energy. At 270°C, the internal energy of the steam is 2160.1 kJ/kg. The specific internal energy of the saturated liquid is 1178.1 kJ/kg. The specific internal energy of the saturated vapor is 2593.7 kJ/kg. What is most nearly the quality of the steam?

(A) 20%

(B) 50%

(C) 70%

(D) 80%

Solution

Solve Eq. 17.13 for the quality.

$$u = xu_g + (1 - x)u_f$$

$$x = \frac{u - u_f}{u_g - u_f} = \frac{2160.1 \; \frac{\mathrm{m}^3}{\mathrm{kg}} - 1178.1 \; \frac{\mathrm{m}^3}{\mathrm{kg}}}{2593.7 \; \frac{\mathrm{m}^3}{\mathrm{kg}} - 1178.1 \; \frac{\mathrm{m}^3}{\mathrm{kg}}}$$

$$= 0.69 \quad (70\%)$$

The answer is (C).

4. PHASE RELATIONS

Equation 17.19: Clapeyron Equation (for Phase Transitions)

$$\left(\frac{dp}{dT}\right)_{\mathrm{sat}} = \frac{h_{fg}}{Tv_{fg}} = \frac{s_{fg}}{v_{fg}} \qquad \textit{17.19}$$

Description

The change in enthalpy during a phase transition, although it cannot be measured directly, can be determined from the pressure, temperature, and specific volume changes through the *Clapeyron equation*, Eq. 17.19.[6] $(dp/dT)_{\mathrm{sat}}$ is the slope of the vapor-liquid saturation line.

Equation 17.20: Clausius-Clapeyron Equation

$$\ln\left(\frac{p_2}{p_1}\right) = \frac{h_{fg}}{R} \cdot \frac{T_2 - T_1}{T_1 T_2} \qquad \textit{17.20}$$

Description

Under the assumptions that $v_{fg} = v_g$, the system behaves ideally, and h_{fg} is independent of temperature, Eq. 17.19 reduces to the Clausius-Clapeyron equation, Eq. 17.20.[7] Although more accurate methods exist (e.g., the *Antoine equation*), the Clausius-Clapeyron equation can be used to estimate the vapor pressure of pure solids and pure liquids. For solids, the heat of vaporization, h_{fg}, is replaced by the heat of sublimation.

---

[6]The *NCEES Handbook* presents both forms of Eq. 17.19 as the Clapeyron equation, and both forms as equalities, which is not accurate. $(dp/dT)_{\mathrm{sat}} = \Delta s/\Delta v$ is the actual Clapeyron equation, and it is thermodynamically exact, needing no approximations in its derivation. $(dp/dT)_{\mathrm{sat}} \approx h_{fg}/Tv_{fg}$ is the differential form of the *two-point Clausius-Clapeyron equation* (presented separately as Eq. 17.20), and it is a thermodynamic approximation derived from $\Delta s \approx \Delta h/T$. In addition to being an approximation, Δs, Δh, and their ratio are not constant over the temperature range. Δh usually varies more slowly with temperature than Δs.

[7]In the *NCEES Handbook*, the natural logarithm is represented as $\ln_e$. The subscript e is a redundant notation, and so has been deleted from Eq. 17.20.

Example

At 100°C, the heat of vaporization of water is 40.7 kJ/mol. What is the vapor pressure at 110°C?

(A) 0.71 atm

(B) 0.80 atm

(C) 0.88 atm

(D) 1.4 atm

Solution

A substance vaporizes when the pressure on it is reduced to its saturation pressure. Water boils at 100°C, so the saturation pressure at that temperature is 1.0 atm. (The steam table could also be used to determine $p_{\text{sat},100°C}$.)

Use Eq. 17.20.

$$T_1 = 100°C + 273° = 373K$$

$$T_2 = 110°C + 273° = 383K$$

$$\ln\left(\frac{p_{v,383K}}{p_{v,373K}}\right) = \frac{h_{fg}}{\overline{R}} \cdot \frac{T_2 - T_1}{T_1 T_2}$$

$$p_{v,383K} = p_{v,373K} e^{\frac{h_{fg}(T_2-T_1)}{\overline{R}(T_1 T_2)}}$$

$$= (1.0 \text{ atm}) e^{\frac{\left(40.7 \frac{kJ}{mol}\right)\left(1000 \frac{J}{kJ}\right)(383K-373K)}{\left(8.314 \frac{J}{mol \cdot K}\right)(383K)(373K)}}$$

$$= 1.4 \text{ atm}$$

The answer is (D).

5. IDEAL GASES

A gas can be considered to behave ideally if its pressure is very low or the temperature is much higher than its critical temperature. (Otherwise, the substance is in vapor form.) Under these conditions, the molecule size is insignificant compared with the distance between molecules, and molecules do not interact. By definition, an ideal gas behaves according to the various ideal gas laws.

Equation 17.21: Specific Gas Constant

$$R = \frac{\overline{R}}{\text{mol. wt}} \qquad 17.21$$

Values

	customary U.S.	SI
universal gas constant, $\overline{R}$	1545 ft-lbf/lbmol-°R	8314 J/kmol·K
		0.08206 atm·L/mol·K
		287 J/kg·K

Description

R is the *specific gas constant*. It is specific because it is valid only for a gas with a particular molecular weight.[8]

$\overline{R}$, given in Eq. 17.21, is known as the *universal gas constant*. It is "universal" (within a system of units) because the same value can be used with any gas. Its value depends on the units used for pressure, temperature, and volume, as well as on the units of mass. Selected values of the universal gas constant in various units are given in the values section.

Example

Assume air to be an ideal gas with a molecular weight of 28.967 kg/kmol. What is most nearly the specific gas constant of air?

(A) 0.11 kJ/kg·K

(B) 0.29 kJ/kg·K

(C) 3.5 kJ/kg·K

(D) 8.3 kJ/kg·K

Solution

The universal gas constant is 8314 J/kmol·K. Use Eq. 17.21 to find the specific gas constant.

$$R = \frac{\overline{R}}{\text{mol. wt}} = \frac{8314 \frac{J}{kmol \cdot K}}{\left(28.967 \frac{kg}{kmol}\right)\left(1000 \frac{J}{kJ}\right)}$$

$$= 0.2870 \text{ kJ/kg·K} \quad (0.29 \text{ kJ/kg·K})$$

The answer is (B).

Equation 17.22 Through Eq. 17.24: Ideal Gas Law

$$pv = RT \qquad 17.22$$

$$pV = mRT \qquad 17.23$$

$$p_1 v_1 / T_1 = p_2 v_2 / T_2 \qquad 17.24$$

Description

An *equation of state* is a relationship that predicts the state (i.e., a property, such as pressure, temperature, volume) from a set of two other independent properties.

Avogadro's law states that equal volumes of different gases at the same temperature and pressure contain equal numbers of molecules. For one mole of any gas, Avogadro's law can be stated as the equation of state for

---

[8]The *NCEES Handbook* is inconsistent in the symbol it uses for molecular weight. In Eq. 17.21, "mol. wt" is used. In other thermodynamics equations, M is used.

ideal gases, equivalent formulations of which are given by Eq. 17.22 through Eq. 17.24. Temperature, T, in Eq. 17.22 through Eq. 17.24 must be in degrees absolute.

Since R is constant for any ideal gas of a particular molecular weight, it follows that the quantity pv/T is constant for an ideal gas undergoing any process, as shown by Eq. 17.24.

Example

0.5 m³ of superheated steam has a pressure of 400 kPa and temperature of 300°C. What is most nearly the mass of the steam?

   (A) 0.040 kg

   (B) 0.76 kg

   (C) 42 kg

   (D) 55 kg

Solution

Since the steam is superheated, consider it to be an ideal gas. The molecular weight of water is 18 kg/kmol. From Eq. 17.21, the specific gas constant is

$$R = \frac{\overline{R}}{\text{mol. wt}} = \frac{8314 \, \frac{\text{J}}{\text{kmol·K}}}{\left(18 \, \frac{\text{kg}}{\text{kmol}}\right)\left(1000 \, \frac{\text{J}}{\text{kJ}}\right)}$$

$$= 0.4619 \text{ kJ/kg·K}$$

Use the ideal gas law as given by Eq. 17.23 to find the mass of the steam.

$$pV = mRT$$

$$m = \frac{pV}{RT}$$

$$= \frac{(400 \text{ kPa})\left(1000 \, \frac{\text{Pa}}{\text{kPa}}\right)(0.5 \text{ m}^3)}{\left(0.4619 \, \frac{\text{kJ}}{\text{kg·K}}\right)\left(1000 \, \frac{\text{J}}{\text{kJ}}\right)(300°\text{C} + 273°)}$$

$$= 0.7557 \text{ kg} \quad (0.76 \text{ kg})$$

The answer is (B).

Equation 17.25 Through Eq. 17.27: Ideal Gas Criteria

$$c_p - c_v = R \qquad 17.25$$

$$\left(\frac{\partial h}{\partial p}\right)_T = 0 \qquad 17.26$$

$$\left(\frac{\partial u}{\partial v}\right)_T = 0 \qquad 17.27$$

Description

Specific enthalpy, and similarly specific internal energy, can be related to the equation of state for ideal gases (i.e., Eq. 17.22). Depending on the units chosen, a conversion factor may be needed.

$$h = u + pv = u + RT$$

$$u = h - pv = h - RT$$

Equation 17.25 through Eq. 17.27 can be derived from these relationships.

For an ideal gas, specific heats are related by the specific gas constant of the ideal gas, as shown by Eq. 17.25. Furthermore, some thermodynamic properties of ideal gases do not depend on other thermodynamic properties. In particular, the specific enthalpy of an ideal gas is independent of pressure for constant-temperature processes. That is, changes in pressure do not affect changes in specific enthalpy when temperature is constant, as shown by Eq. 17.26. Similarly, the specific internal energy of an ideal gas undergoing a constant-temperature process is independent of specific volume, as shown by Eq. 17.27.

Equation 17.28 Through Eq. 17.31: Changes in Thermodynamic Properties of Perfect Gases

$$\Delta u = c_v \Delta T \qquad 17.28$$

$$\Delta h = c_p \Delta T \qquad 17.29$$

$$\Delta s = c_p \ln(T_2/T_1) - R\ln(p_2/p_1) \qquad 17.30$$

$$\Delta s = c_v \ln(T_2/T_1) - R\ln(v_2/v_1) \qquad 17.31$$

Description

Some relations for determining property changes in an ideal gas do not depend on the type of process. This is particularly true for *perfect gases*, which are defined as ideal gases whose specific heats are constant.[9] For perfect gases, changes in enthalpy, internal energy, and entropy are independent of the process, as shown by Eq. 17.28 through Eq. 17.31. Equation 17.28 through Eq. 17.31 can be used for any process. Equation 17.28 does not require a constant-volume process. Similarly, Eq. 17.29 does not require a constant-pressure process.

---

[9]The *NCEES Handbook* introduces Eq. 17.28 through Eq. 17.31 with the statement, "...For cold air standard, heat capacities are assumed to be constant at their room temperature values. In that case, the following are true:" In truth, the four equations are valid for an ideal (perfect) gas at any temperature. The equations are not limited to cold air, nor are they limited to the simplified analysis of an air turbine or reciprocating internal combustion engine, which is where the *cold air standard*, or more commonly, just *air standard*, analysis is encountered.

Example

Nitrogen behaving as an ideal gas undergoes a temperature change from 260°C to 93°C. The specific heat at constant pressure is 1.04 kJ/kg·K. What is most nearly the change in enthalpy per kilogram of nitrogen gas?

(A) −200 kJ/kg

(B) −170 kJ/kg

(C) 110 kJ/kg

(D) 170 kJ/kg

Solution

Specific heats of ideal (perfect) gases are defined to be constant. Therefore, Eq. 17.29 can be used to find the change in enthalpy.

$$\Delta h = c_p \Delta T$$
$$= \left(1.04 \ \frac{kJ}{kg \cdot K}\right)\left((93°C + 273°) - (260°C + 273°)\right)$$
$$= -173.7 \ kJ/kg \quad (-170 \ kJ/kg)$$

The answer is (B).

Mean Heat Capacity

The *mean heat capacity*, $\overline{c}_p$ is to be used in Eq. 17.28 through Eq. 17.31 when heat capacities are temperature dependent (i.e., not constant). As the following equation shows, for most calculations, the average specific heat is either calculated as the average of the specific heats at the end-point temperatures, or is taken as the specific heat at the average temperature. A third approximation, used in linear processes (such as fluid flowing through a long pipe or heat exchanger), is to the specific heat at midpoint (mid-length) along the process. Similar equations can be used to calculate $\overline{c}_v$ for use with Eq. 17.28.

$$\overline{c}_p \approx \frac{c_{p,T_2} - c_{p,T_1}}{T_2 - T_1} \approx c_{p,\frac{1}{2}(T_1 + T_2)}$$

Equation 17.32: Ratio of Specific Heats

$$k = c_p / c_v \qquad 17.32$$

Values

For air, $k = 1.40$. For gases with monoatomic molecules (e.g., helium and argon) at standard conditions, $k \approx 1.67$. For diatomic gases (e.g., nitrogen and oxygen) at standard conditions, $k \approx 1.4$.

Description

Equation 17.32 is the *ratio of specific heats*. The value of that ratio of specific heats depends on the gas; and for maximum accuracy, the value also depends on other properties, such as temperature and pressure. The values primarily depend on the type of molecule formed by the gas.

6. REAL GASES

Real gases do not meet the basic assumptions defining an ideal gas. Specifically, the molecules of a real gas occupy a volume that is not negligible in comparison with the total volume of the gas. (This is especially true for gases at low temperatures.) Furthermore, real gases are subject to *van der Waals' forces*, which are attractive forces between gas molecules.

Equation 17.33 and Eq. 17.34: Theorem of Corresponding States

$$T_r = \frac{T}{T_c} \qquad 17.33$$

$$p_r = \frac{p}{p_c} \qquad 17.34$$

Description

The *theorem of corresponding states* says that the behavior (e.g., properties) of all liquid and gaseous substances can be correlated with "normalized" temperature and pressure. These so-called normalized characteristics are known as the *reduced temperature* (see Eq. 17.33) and *reduced pressure* (see Eq. 17.34), calculated from the *critical properties* of the substance. These are the properties at the critical point. A substance's thermodynamic *critical point* corresponds to the point at the top of the vapor dome in a plot of thermodynamic properties. (See Fig. 17.7 at the end of this chapter.) Above the critical point, liquid and vapor phases are indistinguishable, and the substance is a vapor that cannot be liquefied at any pressure. (However, the substance can be solidified at a high enough pressure.)

Equation 17.35 Through Eq. 17.37: Equations of State (Real Gas)[10]

$$p = \left(\frac{RT}{v}\right) Z \qquad 17.35$$

$$p = \left(\frac{RT}{v}\right)\left(1 + \frac{B}{v} + \frac{C}{v^2} + \cdots\right) \qquad 17.36$$

$$p = \frac{RT}{v - b} - \frac{a(T)}{(v + c_1 b)(v + c_2 b)} \qquad 17.37$$

[10](1) Although the specific volume, v, with typical units of m^3/kg, is typically used with Eq. 17.35 and the compressibility factor, almost every other real gas equation of state is correlated with the molar volume, with typical units of $m^3/kmol$. The *NCEES Handbook* uses molar volume with Eq. 17.38, but presents Eq. 17.36 and Eq. 17.37 in terms of the variable it uses for specific volume. (2) In some engineering literature, molar volume is commonly designated ν, V, V_m, or $\hat{V}$. The *NCEES Handbook* uses $\overline{\nu}$ for molar volume to distinguish it from the symbol used for specific volume.

Description

Equation 17.35 is the *generalized compressibility* equation of state, where Z is the *compressiblity factor*. Values of Z can be found using Fig. 17.7.[11] Equation 17.35 can be used with any substance (solid, liquid, or gas) for which the compressibility factor is known.

Equation 17.36 is the *virial equation of state*, where B, C, and so on are the *virial coefficients*. Equation 17.36 can only be used with gases.

Equation 17.37 is the *cubic equation of state*, where $a(T)$, c_1, and c_2 are species dependent.[12] Equation 17.37 is primarily used with gases, although it can also be used with liquids since the molecular spacing is explicitly incorporated. The van der Waals' equation of state, Eq. 17.38, is one form of the cubic equation of state.

Equation 17.38 Through Eq. 17.40: Van der Waals' Equation of State (Real Gas)

$$\left(p + \frac{a}{\bar{v}^2}\right)(\bar{v} - b) = \bar{R}T \qquad 17.38$$

$$a = \left(\frac{27}{64}\right)\left(\frac{\bar{R}^2 T_c^2}{p_c}\right) \qquad 17.39$$

$$b = \frac{\bar{R}T_c}{8p_c} \qquad 17.40$$

---

[11](1) The compressibility factor is typically represented by uppercase Z, as it is in Eq. 17.35. As presented in the *NCEES Handbook*, the use of lowercase z in Fig. 17.7 is inconsistent. (2) The label "$z_e = 0.27$," which appears within Fig. 17.7, is a reference to the *critical compressibility factor*, the value of z at $p_r = T_r = 1.00$ (i.e., at the critical point). Although any symbol can be used to represent any variable, z_e is probably a typographic deviation from z_c. (3) Real values can be used to "fine-tune" the results derived from generalized graphs. The "$z_e = 0.27$" label means that the graph has been "calibrated" (adjusted) and drawn such that the compressibility factor has a minimum value of 0.27 at $p_r = T_r = 1.00$. (4) Without having additional information, the graph is most likely improperly labeled, since the critical compressibility factor has a graphed minimum value of approximately 0.23. (5) For well-behaved real gases, minimum values of z_c may deviate slightly from 0.27. The value ranges from 0.26 to 0.29 for common hydrocarbon gases and vapors, and essentially all well-behaved gases fall within the range of 0.20 to 0.30. (6) For well-behaved gases, the error in the compressibility factor may still be as high as 6% near the minimum value. Some elements and compounds, such as hydrogen, helium, and neon, as well as polar, nonspherical, and chain molecules, do not follow the two-parameter model (i.e., are not well-behaved) and require different methods.

[12](1) The *NCEES Handbook* presents the numerator of the second term of Eq. 17.37 as $a(T)$. This is intended to indicate that the constant, a, is a function of temperature and temperature only. (Other models may include a dependency on the *acentric factor*, which accounts for molecules with nonspherical shapes.) The product $a \times T$ is not the intended meaning. (2) Although Eq. 17.37 is presented in its common form, the dependency is actually on the reduced temperature, $T_r = T/T_c$, not specifically on the absolute temperature, T. (3) Since all of the constants in the real gas equations of state depend on some combination of properties and their ranges, it is unnecessary (and confusing) to show the dependency in Eq. 17.37.

Description

One of the methods of accounting for real gas behavior is to modify the ideal gas equation of state with various empirical correction factors. Since the modifications are empirical, the resulting equations of state are known as *correlations*. One well-known correlation is *van der Waals' equation of state*, given by Eq. 17.38. In Eq. 17.39 and Eq. 17.40, T_c and p_c are the substance's *critical temperature* and *critical pressure*, respectively.

The van der Waals corrections are particularly accurate when a gas is above its critical temperature but is also useful when a low-pressure gas is below its critical temperature. For an ideal gas, the a and b terms are zero. When the spacing between molecules is close, as it would be at low temperatures, the molecules attract each other and reduce the pressure exerted by the gas. The pressure is then corrected by the $a/\bar{v}^2$ term, where $\bar{v}$ is the molar specific volume. b is a constant that accounts for the molecular volume in a dense state.

Example

Steam in a rigid 3 m³ vessel has a critical temperature of 647.1K and a critical pressure of 22.06 MPa. The steam is heated to 500°C and 10 MPa. Using the van der Waals' equation of state, what is most nearly the molar specific volume of the steam?

(A) 0.2 m³/kmol

(B) 0.6 m³/kmol

(C) 1 m³/kmol

(D) 10 m³/kmol

Solution

Calculate the van der Waals constants, a and b, from Eq. 17.39 and Eq. 17.40, respectively.

$$a = \left(\frac{27}{64}\right)\left(\frac{\bar{R}^2 T_c^2}{p_c}\right)$$

$$= \left(\frac{27}{64}\right)\left(\frac{\left(8314 \frac{J}{kmol \cdot K}\right)^2 (647.1K)^2}{(22.06 \text{ MPa})\left(1000 \frac{J}{kJ}\right)^2 \left(1000 \frac{kPa}{MPa}\right)}\right)$$

$$= 553.53 \text{ kPa·m}^6/\text{kmol}^2$$

$$b = \frac{\bar{R}T_c}{8p_c}$$

$$= \frac{\left(8314 \frac{J}{kmol \cdot K}\right)(647.1K)}{(8)(22.06 \text{ MPa})\left(1000 \frac{J}{kJ}\right)\left(1000 \frac{kPa}{MPa}\right)}$$

$$= 0.0305 \text{ m}^3/\text{kmol}$$

From Eq. 17.38, the molar specific volume is

$$\left(p + \frac{a}{\bar{v}^2}\right)(\bar{v} - b) = \overline{R}T$$

$$\begin{pmatrix} (10 \text{ MPa})\left(1000 \ \dfrac{\text{kPa}}{\text{MPa}}\right) \\[2mm] + \dfrac{553.53 \ \dfrac{\text{kPa·m}^6}{\text{kmol}^2}}{\bar{v}^2} \end{pmatrix}$$

$$\times \left(\bar{v} - 0.0305 \ \frac{\text{m}^3}{\text{kmol}}\right) = \frac{\left(8314 \ \dfrac{\text{J}}{\text{kmol·K}}\right) \times (500°\text{C} + 273°)}{1000 \ \dfrac{\text{J}}{\text{kJ}}}$$

$$10{,}000\bar{v}^3 - 6731.7\bar{v}^2 + 553.53\bar{v} - 16.88 = 0$$

This is a cubic equation.[13]

$$\bar{v} = 0.583 \text{ m}^3/\text{kmol} \quad (0.6 \text{ m}^3/\text{kmol})$$

The answer is (B).

---

[13]For the FE exam, the easiest way to solve cubic equations is to substitute the four answer choices to determine which one solves the cubic equation.

Thermodynamics

Table 17.1 Saturated Water—Temperature Table

temp., T (°C)	sat. press., p_{sat} (kPa)	specific volume (m³/kg)		internal energy (kJ/kg)			enthalpy (kJ/kg)			entropy (kJ/kg·K)		
		sat. liquid, v_f	sat. vapor, v_g	sat. liquid, u_f	evap., u_{fg}	sat. vapor, u_g	sat. liquid, h_f	evap., h_{fg}	sat. vapor, h_g	sat. liquid, s_f	evap., s_{fg}	sat. vapor, s_g
0.01	0.6113	0.001 000	206.14	0.00	2375.3	2375.3	0.01	2501.3	2501.4	0.0000	9.1562	9.1562
5	0.8721	0.001 000	147.12	20.97	2361.3	2382.3	20.98	2489.6	2501.6	0.0761	8.9496	9.0257
10	1.2276	0.001 000	106.38	42.00	2347.2	2389.2	42.01	2477.7	2519.8	0.1510	8.7498	8.9008
15	1.7051	0.001 001	77.93	62.99	2333.1	2396.1	62.99	2465.9	2528.9	0.2245	8.5569	8.7814
20	2.339	0.001 002	57.79	83.95	2319.0	2402.9	83.96	2454.1	2538.1	0.2966	8.3706	8.6672
25	3.169	0.001 003	43.36	104.88	2304.9	2409.8	104.89	2442.3	2547.2	0.3674	8.1905	8.5580
30	4.246	0.001 004	32.89	125.78	2290.8	2416.6	125.79	2430.5	2556.3	0.4369	8.0164	8.4533
35	5.628	0.001 006	25.22	146.67	2276.7	2423.4	146.68	2418.6	2565.3	0.5053	7.8478	8.3531
40	7.384	0.001 008	19.52	167.56	2262.6	2430.1	167.57	2406.7	2574.3	0.5725	7.6845	8.2570
45	9.593	0.001 010	15.26	188.44	2248.4	2436.8	188.45	2394.8	2583.2	0.6387	7.5261	8.1648
50	12.349	0.001 012	12.03	209.32	2234.2	2443.5	209.33	2382.7	2592.1	0.7038	7.3725	8.0763
55	15.758	0.001 015	9.568	230.21	2219.9	2450.1	230.23	2370.7	2600.9	0.7679	7.2234	7.9913
60	19.940	0.001 017	7.671	251.11	2205.5	2456.6	251.13	2358.5	2609.6	0.8312	7.0784	7.9096
65	25.03	0.001 020	6.197	272.02	2191.1	2463.1	272.06	2346.2	2618.3	0.8935	6.9375	7.8310
70	31.19	0.001 023	5.042	292.95	2176.6	2569.6	292.98	2333.8	2626.8	0.9549	6.8004	7.7553
75	38.58	0.001 026	4.131	313.90	2162.0	2475.9	313.93	2321.4	2635.3	1.0155	6.6669	7.6824
80	47.39	0.001 029	3.407	334.86	2147.4	2482.2	334.91	2308.8	2643.7	1.0753	6.5369	7.6122
85	57.83	0.001 033	2.828	355.84	2132.6	2488.4	355.90	2296.0	2651.9	1.1343	6.4102	7.5445
90	70.14	0.001 036	2.361	376.85	2117.7	2494.5	376.92	2283.2	2660.1	1.1925	6.2866	7.4791
95	84.55	0.001 040	1.982	397.88	2102.7	2500.6	397.96	2270.2	2668.1	1.2500	6.1659	7.4159
	MPa											
100	0.101 35	0.001 044	1.6729	418.94	2087.6	2506.5	419.04	2257.0	2676.1	1.3069	6.0480	7.3549
105	0.120 82	0.001 048	1.4194	440.02	2072.3	2512.4	440.15	2243.7	2683.8	1.3630	5.9328	7.2958
110	0.143 27	0.001 052	1.2102	461.14	2057.0	2518.1	461.30	2230.2	2691.5	1.4185	5.8202	7.2387
115	0.169 06	0.001 056	1.0366	482.30	2041.4	2523.7	482.48	2216.5	2699.0	1.4734	5.7100	7.1833
120	0.198 53	0.001 060	0.8919	503.50	2025.8	2529.3	503.71	2202.6	2706.3	1.5276	5.6020	7.1296
125	0.2321	0.001 065	0.7706	524.74	2009.9	2534.6	524.99	2188.5	2713.5	1.5813	5.4962	7.0775
130	0.2701	0.001 070	0.6685	546.02	1993.9	2539.9	546.31	2174.2	2720.5	1.6344	5.3925	7.0269
135	0.3130	0.001 075	0.5822	567.35	1977.7	2545.0	567.69	2159.6	2727.3	1.6870	5.2907	6.9777
140	0.3613	0.001 080	0.5089	588.74	1961.3	2550.0	589.13	2144.7	2733.9	1.7391	5.1908	6.9299
145	0.4154	0.001 085	0.4463	610.18	1944.7	2554.9	610.63	2129.6	2740.3	1.7907	5.0926	6.8833
150	0.4758	0.001 091	0.3928	631.68	1927.9	2559.5	632.20	2114.3	2746.5	1.8418	4.9960	6.8379
155	0.5431	0.001 096	0.3468	653.24	1910.8	2564.1	653.84	2098.6	2752.4	1.8925	4.9010	6.7935
160	0.6178	0.001 102	0.3071	674.87	1893.5	2568.4	675.55	2082.6	2758.1	1.9427	4.8075	6.7502
165	0.7005	0.001 108	0.2727	696.56	1876.0	2572.5	697.34	2066.2	2763.5	1.9925	4.7153	6.7078
170	0.7917	0.001 114	0.2428	718.33	1858.1	2576.5	719.21	2049.5	2768.7	2.0419	4.6244	6.6663
175	0.8920	0.001 121	0.2168	740.17	1840.0	2580.2	741.17	2032.4	2773.6	2.0909	4.5347	6.6256
180	1.0021	0.001 127	0.194 05	762.09	1821.6	2583.7	763.22	2015.0	2778.2	2.1396	4.4461	6.5857
185	1.1227	0.001 134	0.174 09	784.10	1802.9	2587.0	785.37	1997.1	2782.4	2.1879	4.3586	6.5465
190	1.2544	0.001 141	0.156 54	806.19	1783.8	2590.0	807.62	1978.8	2786.4	2.2359	4.2720	6.5079
195	1.3978	0.001 149	0.141 05	828.37	1764.4	2592.8	829.98	1960.0	2790.0	2.2835	4.1863	6.4698
200	1.5538	0.001 157	0.127 36	850.65	1744.7	2595.3	852.45	1940.7	2793.2	2.3309	4.1014	6.4323
205	1.7230	0.001 164	0.115 21	873.04	1724.5	2597.5	875.04	1921.0	2796.0	2.3780	4.0172	6.3952
210	1.9062	0.001 173	0.104 41	895.53	1703.9	2599.5	897.76	1900.7	2798.5	2.4248	3.9337	6.3585
215	2.104	0.001 181	0.094 79	918.14	1682.9	2601.1	920.62	1879.9	2800.5	2.4714	3.8507	6.3221
220	2.318	0.001 190	0.086 19	940.87	1661.5	2602.4	943.62	1858.5	2802.1	2.5178	3.7683	6.2861
225	2.548	0.001 199	0.078 49	963.73	1639.6	2603.3	966.78	1836.5	2803.3	2.5639	3.6863	6.2503
230	2.795	0.001 209	0.071 58	986.74	1617.2	2603.9	990.12	1813.8	2804.0	2.6099	3.6047	6.2146
235	3.060	0.001 219	0.065 37	1009.89	1594.2	2604.1	1013.62	1790.5	2804.2	2.6558	3.5233	6.1791
240	3.344	0.001 229	0.059 76	1033.21	1570.8	2604.0	1037.32	1766.5	2803.8	2.7015	3.4422	6.1437
245	3.648	0.001 240	0.054 71	1056.71	1546.7	2603.4	1061.23	1741.7	2803.0	2.7472	3.3612	6.1083
250	3.973	0.001 251	0.050 13	1080.39	1522.0	2602.4	1085.36	1716.2	2801.5	2.7927	3.2802	6.0730
255	4.319	0.001 263	0.045 98	1104.28	1596.7	2600.9	1109.73	1689.8	2799.5	2.8383	3.1992	6.0375
260	4.688	0.001 276	0.042 21	1128.39	1470.6	2599.0	1134.37	1662.5	2796.9	2.8838	3.1181	6.0019
265	5.081	0.001 289	0.038 77	1152.74	1443.9	2596.6	1159.28	1634.4	2793.6	2.9294	3.0368	5.9662
270	5.499	0.001 302	0.035 64	1177.36	1416.3	2593.7	1184.51	1605.2	2789.7	2.9751	2.9551	5.9301
275	5.942	0.001 317	0.032 79	1202.25	1387.9	2590.2	1210.07	1574.9	2785.0	3.0208	2.8730	5.8938
280	6.412	0.001 332	0.030 17	1227.46	1358.7	2586.1	1235.99	1543.6	2779.6	3.0668	2.7903	5.8571
285	6.909	0.001 348	0.027 77	1253.00	1328.4	2581.4	1262.31	1511.0	2773.3	3.1130	2.7070	5.8199
290	7.436	0.001 366	0.025 57	1278.92	1297.1	2576.0	1289.07	1477.1	2766.2	3.1594	2.6227	5.7821
295	7.993	0.001 384	0.023 54	1305.2	1264.7	2569.9	1316.3	1441.8	2758.1	3.2062	2.5375	5.7437
300	8.581	0.001 404	0.021 67	1332.0	1231.0	2563.0	1344.0	1404.9	2749.0	3.2534	2.4511	5.7045
305	9.202	0.001 425	0.019 948	1359.3	1195.9	2555.2	1372.4	1366.4	2738.7	3.3010	2.3633	5.6643
310	9.856	0.001 447	0.018 350	1387.1	1159.4	2546.4	1401.3	1326.0	2727.3	3.3493	2.2737	5.6230
315	10.547	0.001 472	0.016 867	1415.5	1121.1	2536.6	1431.0	1283.5	2714.5	3.3982	2.1821	5.5804
320	11.274	0.001 499	0.015 488	1444.6	1080.9	2525.5	1461.5	1238.6	2700.1	3.4480	2.0882	5.5362
330	12.845	0.001 561	0.012 996	1505.3	993.7	2498.9	1525.3	1140.6	2665.9	3.5507	1.8909	5.4417
340	14.586	0.001 638	0.010 797	1570.3	894.3	2464.6	1594.2	1027.9	2622.0	3.6594	1.6763	5.3357
350	16.513	0.001 740	0.008 813	1641.9	776.6	2418.4	1670.6	893.4	2563.9	3.7777	1.4335	5.2112
360	18.651	0.001 893	0.006 945	1725.2	626.3	2351.5	1760.5	720.3	2481.0	3.9147	1.1379	5.0526
370	21.03	0.002 213	0.004 925	1844.0	384.5	2228.5	1890.5	441.6	2332.1	4.1106	0.6865	4.7971
374.14	22.09	0.003 155	0.003 155	2029.6	0	2029.6	2099.3	0	2099.3	4.4298	0	4.4298

Thermodynamics

Table 17.2 *Superheated Water Tables*

temperature, T (°C)	specific volume, v (m³/kg)	internal energy, u (kJ/kg)	enthalpy, h (kJ/kg)	entropy, s (kJ/kg·K)	specific volume, v (m³/kg)	internal energy, u (kJ/kg)	enthalpy, h (kJ/kg)	entropy, s (kJ/kg·K)
	\multicolumn{4}{c} $p=0.01$ MPa (45.81°C)		$p=0.05$ MPa (81.33°C)			
sat.	14.674	2437.9	2584.7	8.1502	3.240	2483.9	2645.9	7.5939
50	14.869	2443.9	2592.6	8.1749				
100	17.196	2515.5	2687.5	8.4479	3.418	2511.6	2682.5	7.6947
150	19.512	2587.9	2783.0	8.6882	3.889	2585.6	2780.1	7.9401
200	**21.825**	**2661.3**	**2879.5**	**8.9038**	**4.356**	**2659.9**	**2877.7**	**8.1580**
250	24.136	2736.0	2977.3	9.1002	4.820	2735.0	2976.0	8.3556
300	26.445	2812.1	3076.5	9.2813	5.284	2811.3	3075.5	8.5373
400	31.063	2968.9	3279.6	9.6077	6.209	2968.5	3278.9	8.8642
500	35.679	3132.3	3489.1	9.8978	7.134	3132.0	3488.7	9.1546
600	**40.295**	**3302.5**	**3705.4**	**10.1608**	**8.057**	**3302.2**	**3705.1**	**9.4178**
700	44.911	3479.6	3928.7	10.4028	8.981	3479.4	3928.5	9.6599
800	49.526	3663.8	4159.0	10.6281	9.904	3663.6	4158.9	9.8852
900	54.141	3855.0	4396.4	10.8396	10.828	3854.9	4396.3	10.0967
1000	58.757	4053.0	4640.6	11.0393	11.751	4052.9	4640.5	10.2964
1100	**63.372**	**4257.5**	**4891.2**	**11.2287**	**12.674**	**4257.4**	**4891.1**	**10.4859**
1200	67.987	4467.8	5147.8	11.4091	13.597	4467.8	5147.7	10.6662
1300	72.602	4683.7	5409.7	11.5811	14.521	4683.6	5409.6	10.8382
	\multicolumn{4}{c} $p=0.10$ MPa (99.63°C)		$p=0.20$ MPa (120.23°C)			
sat.	1.6940	2506.1	2675.5	7.3594	0.8857	2529.5	2706.7	7.1272
100	1.6958	2506.7	2676.2	7.3614				
150	1.9364	2582.8	2776.4	7.6134	0.9596	2576.9	2768.8	7.2795
200	2.172	2658.1	2875.3	7.8343	1.0803	2654.4	2870.5	7.5066
250	**2.406**	**2733.7**	**2974.3**	**8.0333**	**1.1988**	**2731.2**	**2971.0**	**7.7086**
300	2.639	2810.4	3074.3	8.2158	1.3162	2808.6	3071.8	7.8926
400	3.103	2967.9	3278.2	8.5435	1.5493	2966.7	3276.6	8.2218
500	3.565	3131.6	3488.1	8.8342	1.7814	3130.8	3487.1	8.5133
600	4.028	3301.9	3704.4	9.0976	2.013	3301.4	3704.0	8.7770
700	**4.490**	**3479.2**	**3928.2**	**9.3398**	**2.244**	**3478.8**	**3927.6**	**9.0194**
800	4.952	3663.5	4158.6	9.5652	2.475	3663.1	4158.2	9.2449
900	5.414	3854.8	4396.1	9.7767	2.705	3854.5	4395.8	9.4566
1000	5.875	4052.8	4640.3	9.9764	2.937	4052.5	4640.0	9.6563
1100	6.337	4257.3	4891.0	10.1659	3.168	4257.0	4890.7	9.8458
1200	**6.799**	**4467.7**	**5147.6**	**10.3463**	**3.399**	**4467.5**	**5147.5**	**10.0262**
1300	7.260	4683.5	5409.5	10.5183	3.630	4683.2	5409.3	10.1982
	\multicolumn{4}{c} $p=0.40$ MPa (143.63°C)		$p=0.60$ MPa (158.85°C)			
sat.	0.4625	2553.6	2738.6	6.8959	0.3157	2567.4	2756.8	6.7600
150	0.4708	2564.5	2752.8	6.9299				
200	0.5342	2646.8	2860.5	7.1706	0.3520	2638.9	2850.1	6.9665
250	0.5951	2726.1	2964.2	7.3789	0.3938	2720.9	2957.2	7.1816
300	**0.6548**	**2804.8**	**3066.8**	**7.5662**	**0.4344**	**2801.0**	**3061.6**	**7.3724**
350	0.7137	2884.6	3170.1	7.7324	0.4742	2881.2	3165.7	7.5464
400	0.7726	2964.4	3273.4	7.8985	0.5137	2962.1	3270.3	7.7079
500	0.8893	3129.2	3484.9	8.1913	0.5920	3127.6	3482.8	8.0021
600	1.0055	3300.2	3702.4	8.4558	0.6697	3299.1	3700.9	8.2674
700	**1.1215**	**3477.9**	**3926.5**	**8.6987**	**0.7472**	**3477.0**	**3925.3**	**8.5107**
800	1.2372	3662.4	4157.3	8.9244	0.8245	3661.8	4156.5	8.7367
900	1.3529	3853.9	4395.1	9.1362	0.9017	3853.4	4394.4	8.9486
1000	1.4685	4052.0	4639.4	9.3360	0.9788	4051.5	4638.8	9.1485
1100	1.5840	4256.5	4890.2	9.5256	1.0559	4256.1	4889.6	9.3381
1200	**1.6996**	**4467.0**	**5146.8**	**9.7060**	**1.1330**	**4466.5**	**5146.3**	**9.5185**
1300	1.8151	4682.8	5408.8	9.8780	1.2101	4682.3	5408.3	9.6906
	\multicolumn{4}{c} $p=0.80$ MPa (170.43°C)		$p=1.00$ MPa (179.91°C)			
sat.	0.2404	2576.8	2769.1	6.6628	0.194 44	2583.6	2778.1	6.5865
200	0.2608	2630.6	2839.3	6.8158	0.2060	2621.9	2827.9	6.6940
250	0.2931	2715.5	2950.0	7.0384	0.2327	2709.9	2942.6	6.9247
300	0.3241	2797.2	3056.5	7.2328	0.2579	2793.2	3051.2	7.1229
350	**0.3544**	**2878.2**	**3161.7**	**7.4089**	**0.2825**	**2875.2**	**3157.7**	**7.3011**
400	0.3843	2959.7	3267.1	7.5716	0.3066	2957.3	3263.9	7.4651
500	0.4433	3126.0	3480.6	7.8673	0.3541	3124.4	3478.5	7.7622
600	0.5018	3297.9	3699.4	8.1333	0.4011	3296.8	3697.9	8.0290
700	0.5601	3476.2	3924.2	8.3770	0.4478	3475.3	3923.1	8.2731
800	**0.6181**	**3661.1**	**4155.6**	**8.6033**	**0.4943**	**3660.4**	**4154.7**	**8.4996**
900	0.6761	3852.8	4393.7	8.8153	0.5407	3852.2	4392.9	8.7118
1000	0.7340	4051.0	4638.2	9.0153	0.5871	4050.5	4637.6	8.9119
1100	0.7919	4255.6	4889.1	9.2050	0.6335	4255.1	4888.6	9.1017
1200	0.8497	4466.1	5145.9	9.3855	0.6798	4465.6	5145.4	9.2822
1300	**0.9076**	**4681.8**	**5407.9**	**9.5575**	**0.7261**	**4681.3**	**5407.4**	**9.4543**

Figure 17.5 *p-h Diagram for Refrigerant HFC-134a (SI units)*

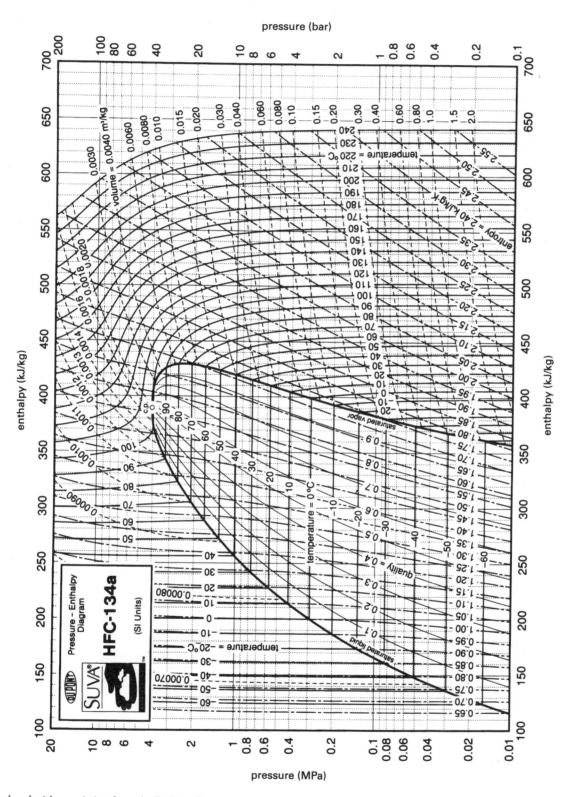

Reproduced with permission from the DuPont Company.

Thermodynamics

Figure 17.7 Compressibility Factors

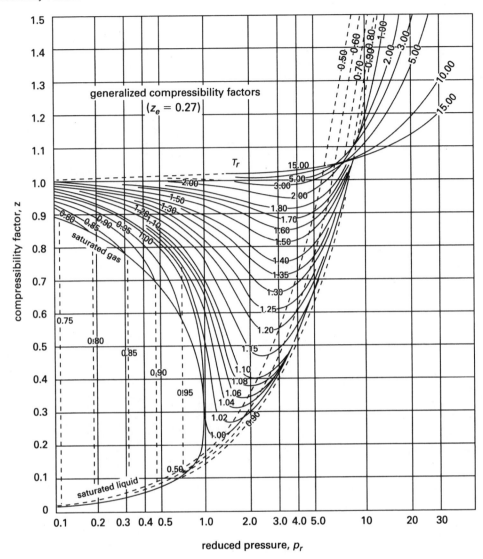

de Nevers, N. (2012) Appendix A: Useful Tables and Charts, in *Physical and Chemical Equilibrium for Chemical Engineers*, Second Edition, John Wiley & Sons, Inc., Hoboken, NJ.

18 Laws of Thermodynamics

Nomenclature

c	specific heat	kJ/kg·K
D	diameter	m
g	gravitational acceleration, 9.81	m/s^2
h	specific enthalpy	kJ/kg
H	total enthalpy	kJ
k	ratio of specific heats	–
KE	kinetic energy	kJ
L	length	m
m	mass	kg
$\dot{m}$	mass flow rate	kg/s
MW	molecular weight	kg/kmol
n	number of moles	–
n	polytropic exponent	–
p	pressure	Pa
PE	potential energy	kJ
q	heat energy	kJ/kg
Q	total heat energy	kJ
$\dot{Q}$	rate of heat transfer	kW
r	radius	m
R	specific gas constant	kJ/kg·K
s	specific entropy	kJ/kg·K
S	total entropy	kJ/K
T	absolute temperature	K
u	specific internal energy	kJ/kg
U	total internal energy	kJ
v	velocity	m/s
V	volume	m^3
w	specific work	kJ/kg
W	work	kJ
$\dot{W}$	rate of work (power)	kW
z	elevation[1]	m

Symbols

η	efficiency	–
υ	specific volume[2]	m^3/kg

Subscripts

b	boundary
C	cold
e	exit
es	ideal (isentropic) exit state
f	fluid (liquid)
fg	liquid-to-gas (vaporization)
g	gas (vapor)
H	high or hot
i	in or inlet (entrance)
L	low
p	constant pressure
rev	reversible
s	isentropic
v	constant volume

1. SYSTEMS

A *thermodynamic system* is defined as the matter enclosed within an arbitrary but precisely defined *control volume*. Everything external to the system is defined as the *surroundings*, *environment*, or *universe*. The environment and system are separated by the *system boundaries*. The surface of the control volume is known as the *control surface*. The control surface can be real (e.g., piston and cylinder walls) or imaginary.

If mass flows through the system across system boundaries, the system is an *open system*. Pumps, heat exchangers, and jet engines are examples of open systems. An important type of open system is the *steady-flow open system* in which matter enters and exits at the same rate. Pumps, turbines, heat exchangers, and boilers are all steady-flow open systems.

If no mass crosses the system boundaries, the system is said to be a *closed system*. The matter in a closed system may be referred to as a *control mass*. Closed systems can have variable volumes. The gas compressed by a piston in a cylinder is an example of a closed system with a variable control volume.

---

[1]The NCEES *FE Reference Handbook* (*NCEES Handbook*) is inconsistent in the variable used to represent elevation above the datum in energy equations. For fluids subjects and in the Bernoulli equation, the *NCEES Handbook* uses z; for thermodynamics subjects, the *NCEES Handbook* uses Z. Since lowercase z is the most common symbol used in engineering practice, this book uses that convention. The equations in this book involving elevation above a datum will differ slightly in appearance from the *NCEES Handbook*.

[2]The *NCEES Handbook* uses lowercase italic v for specific volume. This book uses Greek upsilon, υ, to avoid confusion with the symbol for velocity in kinetic energy calculations. The equations in this book involving specific volume will differ slightly in appearance from those in the *NCEES Handbook*.

In most cases, energy in the form of heat, work, or electrical energy can enter or exit any open or closed system. Systems closed to both matter and energy transfer are known as *isolated systems*.

2. TYPES OF PROCESSES

Changes in thermodynamic properties of a system often depend on the type of process experienced. This is particularly true for gaseous systems. The following list describes several common types of processes.

- *adiabatic process*—a process in which no energy crosses the system boundary. Adiabatic processes include isentropic and throttling processes.

- *isentropic process*—an adiabatic process in which there is no entropy production (i.e., it is reversible). Also known as a *constant entropy process*.

- *throttling process*—an adiabatic process in which there is no change in enthalpy, but for which there is a significant pressure drop.

- *constant pressure process*—also known as an *isobaric process*.

- *constant temperature process*—also known as an *isothermal process*.

- *constant volume process*—also known as an *isochoric* or *isometric process*.

- *polytropic process*—a process that obeys the polytropic equation of state (see Eq. 18.18). Gases always constitute the system in polytropic processes. *n* is the *polytropic exponent*, a property of the equipment, not of the gas.

A system that is in equilibrium at the start and finish of a process may or may not be in equilibrium during the process. A *quasistatic process* (*quasiequilibrium process*) is one that can be divided into a series of infinitesimal deviations (steps) from equilibrium. During each step, the property changes are small, and all intensive properties are uniform throughout the system. The interim equilibrium at each step is often called *quasiequilibrium*.

A *reversible process* is one that is performed in such a way that, at the conclusion of the process, both the system and the local surroundings can be restored to their initial states. Quasiequilibrium processes are assumed to be reversible processes.

3. STANDARD SIGN CONVENTION

A standard sign convention is used in calculating work, heat, and property changes in systems. This sign convention takes the system (not the environment) as the reference. For example, a net heat gain, Q, would mean the system gained energy and the environment lost energy.

Changes in enthalpy, entropy, and internal energy (ΔH, ΔS, and ΔU, respectively) are positive if these properties increase within the system. ΔU will be negative if the internal energy of the system decreases.

4. FIRST LAW OF THERMODYNAMICS

There is a basic principle that underlies all property changes as a system undergoes a process: All energy must be accounted for. Energy that enters a system must either leave the system or be stored in some manner, and energy cannot be created or destroyed. These statements are the primary manifestations of the *first law of thermodynamics*: The net energy crossing the system boundary is the change in energy inside the system.

The first law applies whether or not a process is reversible. So, the first law can also be stated as: The work done in an adiabatic process depends only on the system's endpoint conditions, not on the nature of the process.

Equation 18.1: Heat

$$q = Q/m \qquad \text{18.1}$$

Description

Heat, Q, transferred due to a temperature difference is positive if heat flows into the system. In accordance with the standard sign convention, Q will be negative if the net heat exchange is a loss of heat to the surroundings.

Equation 18.2: Specific Work

$$w = W/m \qquad \text{18.2}$$

Description

Work, W, is positive if the system does work on the surroundings. W will be negative if the surroundings do work on the system (e.g., a piston compressing gas in a cylinder).

5. CLOSED SYSTEMS

Equation 18.3: The First Law of Thermodynamics for Closed Systems

$$Q - W = \Delta U + \Delta KE + \Delta PE \qquad \text{18.3}$$

Variation

$$Q = \Delta U + W$$

Description

The first law of thermodynamics, as given by Eq. 18.3, states that the heat energy, Q, entering a closed system can either increase the temperature (increase U) or be used to perform work (increase W) on the surroundings. The Q term is understood to be the net heat entering the system, which is the heat energy entering the system less the heat energy lost to the surroundings. ΔKE is the change in kinetic energy, and ΔPE is the change in potential energy.[3]

In most cases, the changes in kinetic and potential energy can be disregarded and the first law of thermodynamics for closed systems can be written as in the variation equation.

Example

A half-cylinder tub is full of water at 30°C. The tub has a length of 1.5 m and a diameter of 0.8 m. The specific heat of water is 4.18 kJ/kg·K. What is most nearly the temperature of the water after 3 MJ of heat is added to the tub?

(A) 28°C

(B) 30°C

(C) 32°C

(D) 36°C

Solution

Find the volume of the tub.

$$V = \frac{\pi r^2 L}{2} = \frac{\pi D^2 L}{8} = \frac{\pi (0.8 \text{ m})^2 (1.5 \text{ m})}{8}$$
$$= 0.377 \text{ m}^3$$

Find the mass of the water in the tub. The density of water at 30°C is 995.7 kg/m³.

$$m = \rho V = \left(995.7 \ \frac{\text{kg}}{\text{m}^3}\right)(0.377 \text{ m}^3)$$
$$= 375.37 \text{ kg}$$

Since the only thing being added to the tub and water is heat, the tub and water constitute a closed system. Use Eq. 18.3 to find the final temperature. Because there is no work done by or on the system, the work, difference in potential energy, and difference in kinetic energy can be disregarded.

$$Q - W = \Delta U + \Delta KE + \Delta PE$$
$$Q = \Delta U$$

Solve for the temperature of the water after heat is added.

$$Q = \Delta U = mc_p \Delta T = mc_p(T_2 - T_1)$$
$$T_2 = \frac{Q + mc_p T_1}{mc_p}$$

$$= \frac{\begin{array}{c}(3 \text{ MJ})\left(1000 \ \dfrac{\text{kJ}}{\text{MJ}}\right) \\[6pt] + (375.37 \text{ kg})\left(4.18 \ \dfrac{\text{kJ}}{\text{kg·K}}\right)(30°\text{C})\end{array}}{(375.37 \text{ kg})\left(4.18 \ \dfrac{\text{kJ}}{\text{kg·K}}\right)}$$

$$= 31.9° \text{ C} \quad (32°\text{C})$$

The answer is (C).

Equation 18.4: Reversible Boundary Work

$$w_b = \int p \, dv \qquad \text{18.4}$$

Description

The work done by or on a closed system during a process, w_b, is calculated by the area under the curve in the p-V plane, and is called *reversible work*, *boundary work*, *p-V work*, or *flow work*.[4]

Boundary work gets its name from the fact that the boundary of the system changes, resulting in a volume change. This volume change can be used to separate boundary work from other energy/work due to electrical, thermal, and mechanical sources. For example, a propeller within a closed volume can circulate and increase the kinetic energy of a gas, but this "shaft work" is not boundary work.

Equation 18.4 is defined as "reversible boundary work." If the system volume changes gradually (i.e., the boundary moves slowly), the process will be "quasistatic," achieving a continuum of equilibrium throughout. In that case, the work will be reversible. This would seem to imply that Eq. 18.4 is valid only for reversible processes. In fact, Eq. 18.4 can be used with irreversible processes as long as the total work performed is recognized as the sum of reversible and irreversible parts. Consider a heated gas expanding within a vertical cylinder with a rusty, heavy piston. To move the piston, the gas pressure has to overcome the weight of the piston (which might represent a reversible compression in some cases), and it also has to overcome the friction (which is not reversible). Equation 18.4 can be used to calculate the total of the two parts. However, the total isn't all reversible.

---

[3]The *NCEES Handbook* is inconsistent in the variables it uses to represent kinetic and potential energy. In its dynamics section, the *NCEES Handbook* uses T and U for kinetic and potential energy, respectively; for thermodynamics subjects, the *NCEES Handbook* uses KE and PE (presented in this book as KE and PE).

[4]Usually, the thermodynamic interpretation of the term "boundary work" is the work done on the entire boundary (i.e., on the entire system). The form of Eq. 18.4 calculates the "specific" boundary work (i.e., the work done per unit mass of the system). This is not explicitly stated in the *NCEES Handbook*.

Values calculated from Eq. 18.4 will have signs that are derived from the definition of a definite integral. A negative boundary work will indicate that the volume has decreased. A negative sign is not related to the standard thermodynamic sign convention, which is arbitrary and independent of the definition of an integral. Equation 18.4 is defined as boundary work without specifying "on the system" or "by the system." Whether or not boundary work is positive or negative must be determined on a case-by-case basis, primarily depending on the context.[5]

Example

5 kmol of water vapor at 100°C and 1 atm are condensed from an initial volume of 153 L to a liquid state at 100°C. The molecular weight of water is 18.016 kg/kmol, and the specific volume of the liquid is 0.001044 m³/kg. What is most nearly the work done by the atmosphere on the water?

(A) 6.0 kJ

(B) 6.2 kJ

(C) 6.0 MJ

(D) 6.2 MJ

Solution

Calculate the initial and final volumes of the system.

$$V_1 = \frac{153 \text{ L}}{1000 \frac{\text{L}}{\text{m}^3}} = 0.153 \text{ m}^3$$

$$V_2 = n(MW)_{H_2O} v_f$$

$$= (5 \text{ kmol})\left(18.016 \frac{\text{kg}}{\text{kmol}}\right)\left(0.001044 \frac{\text{m}^3}{\text{kg}}\right)$$

$$= 0.094 \text{ m}^3$$

As the vapor condenses, the total volume of water changes. The constant atmospheric pressure (p; 101 325 Pa) acts on the system boundary as the volume changes. Use Eq. 18.4 to calculate the work done. Use uppercase letters to designate the work done on the entire system.

$$W_b = \int_{V_1}^{V_2} p \, dV$$

$$= p(V_2 - V_1)$$

$$= \frac{(101\,325 \text{ Pa})(0.094 \text{ m}^3 - 0.153 \text{ m}^3)}{1000 \frac{\text{J}}{\text{kJ}}}$$

$$= -5.97 \text{ kJ} \quad (6.0 \text{ kJ})$$

The answer is (A).

---

[5]Answers to the question, "How much work did you do today?" rarely include the word "negative." For that reason, boundary work is always positive in casual answers. The standard thermodynamic sign convention is relevant only when work is a component of an energy balance such as the first law.

6. SPECIAL CASE OF CLOSED SYSTEMS (FOR IDEAL GASES)

Equation 18.5 and Eq. 18.6: Constant Pressure Process (Charles' Law)

$$T/v = \text{constant} \qquad 18.5$$
$$w_b = p\Delta v \qquad 18.6$$

Description

A system whose pressure remains constant is known as an *isobaric system.*

The ideal gas law reduces to Eq. 18.5 and is known as *Charles' law.*[6] When pressure is constant in a closed system, Eq. 18.4 reduces to Eq. 18.6.

Example

A gas goes through the following thermodynamic processes.

A to B: constant-temperature compression
B to C: constant-volume cooling
C to A: constant-pressure expansion

The pressure and volume at state C are 140 kPa and 0.028 m³, respectively. The net work during the C-to-A process is 10.5 kJ.

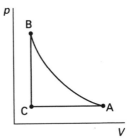

Most nearly, what is the volume at state A?

(A) 0.07 m³

(B) 0.10 m³

(C) 0.19 m³

(D) 0.24 m³

Solution

Since pressure is constant during the C-to-A process, use Eq. 18.6 to find the volume at A.

$$w_b = p\Delta v$$

$$w_{C-A} = p(v_A - v_C)$$

$$(10.5 \text{ kJ})\left(1000 \frac{\text{J}}{\text{kJ}}\right) = (140 \text{ kPa})\left(1000 \frac{\text{Pa}}{\text{kPa}}\right)$$
$$\times (v_A - 0.028 \text{ m}^3)$$

$$v_A = 0.103 \text{ m}^3 \quad (0.10 \text{ m}^3)$$

The answer is (B).

---

[6]The *NCEES Handbook* uses "= constant" to mean "is constant," not to mean "a constant." There is no single number that describes the ratio T/v for all isobaric processes.

Thermodynamics

Equation 18.7 and Eq. 18.8: Constant Volume Process (Guy-Lussac's Law)

$$T/p = \text{constant} \qquad 18.7$$
$$w_b = 0 \qquad 18.8$$

Description

A system whose volume remains constant is known as an *isochoric system*.

The ideal gas law reduces to Eq. 18.7 and is known as *Guy-Lussac's law*. When the volume of a closed system is held constant, Eq. 18.4 reduces to Eq. 18.8.

Equation 18.9 and Eq. 18.10: Constant Temperature Process (Boyle's Law)

$$pv = \text{constant} \qquad 18.9$$
$$w_b = RT \ln(v_2/v_1) = RT \ln(p_1/p_2) \qquad 18.10$$

Description

A system whose temperature remains constant is known as an *isothermal system*.

The ideal gas law reduces to Eq. 18.9 and is known as *Boyle's law*. When the temperature of a closed system is held constant, Eq. 18.4 reduces to Eq. 18.10.

Example

A gas goes through the following thermodynamic processes.

A to B: constant-temperature compression
B to C: constant-volume cooling
C to A: constant-pressure expansion

The pressure and volume at state C are 140 kPa and 0.028 m³, respectively. The net work during the C-to-A process is 10.5 kJ. The volume at state A is 0.103 m³.

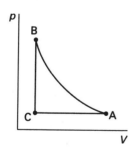

What is most nearly the work performed in the A-to-B process?

(A) 0 kJ
(B) 5.3 kJ
(C) 13 kJ
(D) 19 kJ

Solution

Solution

The pressure at state A is the same as the pressure at state C. By Eq. 18.10, the work in constant-temperature processes per unit mass is

$$w_{\text{A-B}} = RT_A \ln(v_B/v_A)$$

The mass of the gas (from the ideal gas law) is

$$m = \frac{p_A v_A}{RT_A}$$

The total work performed in the A-to-B process is

$$W_{\text{A-B}} = m w_{\text{A-B}} = \left(\frac{p_A v_A}{RT_A}\right)\left(RT_A \ln \frac{v_B}{v_A}\right)$$
$$= p_A v_A \ln \frac{v_B}{v_A}$$
$$= \frac{(140 \text{ kPa})\left(10^3 \frac{\text{Pa}}{\text{kPa}}\right)}{1000 \frac{\text{J}}{\text{kJ}}}$$
$$\times (0.103 \text{ m}^3) \ln \frac{0.028 \text{ m}^3}{0.103 \text{ m}^3}$$
$$= -18.8 \text{ kJ} \quad (19 \text{ kJ})$$

The answer is (D).

Equation 18.11 Through Eq. 18.13: Isentropic Process

$$pv^k = \text{constant} \qquad 18.11$$
$$w = (p_2 v_2 - p_1 v_1)/(1-k) \qquad 18.12$$
$$w = R(T_2 - T_1)/(1-k) \qquad 18.13$$

Description

A process where the heat transfer (i.e., heat loss or heat gain) is zero is known as an *adiabatic process*. Since $\Delta s = Q/T_0$, if $Q = 0$, then entropy is unchanged. A process where entropy remains constant is known as an *isentropic process*, also known as a *reversible adiabatic process*.

Equation 18.11 describes the behavior of a closed, isentropic system. k is the ratio of specific heats (see Eq. 18.17). Equation 18.12 and Eq. 18.13 are related by the ideal gas law.

Equation 18.14 Through Eq. 18.17: Isentropic Process for Ideal Gases

$$\frac{p_2}{p_1} = \left(\frac{v_1}{v_2}\right)^k \qquad 18.14$$

(side tab) **Thermodynamics**

$$\frac{T_2}{T_1} = \left(\frac{p_2}{p_1}\right)^{\frac{k-1}{k}} \qquad \textbf{18.15}$$

$$\frac{T_2}{T_1} = \left(\frac{v_1}{v_2}\right)^{k-1} \qquad \textbf{18.16}$$

$$k = c_p/c_v \qquad \textbf{18.17}$$

Description

Equation 18.14 through Eq. 18.16 are valid for ideal gases undergoing *isentropic processes* (i.e., entropy is constant). The *ratio of specific heats, k,* is given by Eq. 18.17. For closed, isentropic systems, Eq. 18.4 reduces to Eq. 18.12 and Eq. 18.13.

Example

Air is compressed isentropically in a piston-cylinder arrangement to 1/10 of its initial volume. The initial temperature is 35°C, and the ratio of specific heats is 1.4. What is most nearly the final temperature?

(A) 350K

(B) 360K

(C) 620K

(D) 770K

Solution

Since the process is isentropic, use Eq. 18.16 to calculate the final temperature.

$$\frac{T_2}{T_1} = \left(\frac{v_1}{v_2}\right)^{k-1}$$

$$T_2 = T_1\left(\frac{v_1}{v_2}\right)^{k-1} = (35°C + 273°)\left(\frac{10}{1}\right)^{1.4-1}$$

$$= 773.7\text{K} \quad (770\text{K})$$

The answer is (D).

Equation 18.18 and Eq. 18.19: Polytropic Process

$$pv^n = \text{constant} \qquad \textbf{18.18}$$

$$w = (p_2v_2 - p_1v_1)/(1-n) \quad [n \neq 1] \qquad \textbf{18.19}$$

Description

A process is polytropic if it satisfies Eq. 18.18 for some real number n, known as the *polytropic exponent*. Polytropic processes represent a larger class of processes. This is due to the fact that Eq. 18.18 reduces to other processes depending on the value of the polytropic exponent. Any process which is either constant-pressure,

constant-temperature, constant-volume, or isentropic is also polytropic.

$n = 0$ [constant pressure process]
$n = 1$ [constant temperature process]
$n = k$ [isentropic process]
$n = \infty$ [constant volume process]

Processes with exponents other than 0, 1, k, and ∞ are typically found where the working fluid is acted upon by mechanical equipment (e.g., a reciprocating compressor). In a rotating device, the working fluid's pressure, temperature, and volume are determined by what the machine does, not what would occur naturally.

Example

0.5 m^3 of superheated steam at 400 kPa and 300°C is expanded behind a piston until the temperature is 210°C. The steam expands polytropically with a polytropic exponent of 1.3. The final volume and pressure of the steam are 0.884 m^3 and 190.7 kPa, respectively. What is most nearly the total work done during the expansion process?

(A) 24 kJ

(B) 100 kJ

(C) 330 kJ

(D) 420 kJ

Solution

The steam expands polytropically, so use Eq. 18.19 to find the work done during expansion. Use uppercase letters to distinguish the total work and total volume from their per unit mass counterparts.

$$W = (p_2V_2 - p_1V_1)/(1-n)$$

$$= \frac{(190.7 \text{ kPa})(0.884 \text{ m}^3) - (400 \text{ kPa})(0.5 \text{ m}^3)}{1-1.3}$$

$$= 104.7 \text{ kJ} \quad (100 \text{ kJ})$$

The answer is (B).

7. OPEN THERMODYNAMIC SYSTEMS

In an *open system*, mass (the working fluid, substance, matter, etc.) crosses the system boundary. Water flowing through a pump and steam flowing through a turbine represent open systems. In a *steady-state open system*, the mass flow rates into and out of the system are the same.

The first law of thermodynamics can also be written for open systems, but more terms are required to account for the many energy forms. The first law formulation is essentially the *Bernoulli energy conservation equation* extended to nonadiabatic processes.

$$Q = \Delta U + \Delta \text{PE} + \Delta \text{KE} + W_{\text{rev}} + W_{\text{shaft}}$$

Q is the heat flow into the system, inclusive of any losses. It can be supplied from furnace flame, electrical heating, nuclear reaction, or other sources. If the system is adiabatic, Q is zero.

Equation 18.20: Reversible Flow Work

$$w_{rev} = -\int v \, dp + \Delta KE + \Delta PE \qquad 18.20$$

Description

At the boundary of an open system, there is pressure opposing fluid from entering the system. The work required to cause the flow into the system against the exit pressure is called reversible flow work, w_{rev} (also p-V work, *flow energy*, etc.), and is given by Eq. 18.20.[7] Since reversible flow work is work being done on the system, it is always negative.

Example

A boiler feedwater pump receives a steady flow of saturated liquid water at a temperature of 50°C and a pressure of 12.349 kPa. The pressure of the water increases isentropically inside the boiler to 1000 kPa. At 50°C, the specific volume of the water is 0.001012 m³/kg. Changes in kinetic and potential energies are negligible. What is most nearly the work done on the water by the boiler?

(A) 0.64 kJ/kg

(B) 0.87 kJ/kg

(C) 1.0 kJ/kg

(D) 2.3 kJ/kg

Solution

From Eq. 18.20, the reversible flow work is

$$w_{rev} = -\int v \, dp + \Delta KE + \Delta PE = \int v \, dp = -v(p_2 - p_1)$$

$$= \left(-0.001012 \; \frac{m^3}{kg}\right)(1000 \text{ kPa} - 12.349 \text{ kPa})$$

$$= -1.0 \text{ kJ/kg} \quad (1.0 \text{ kJ/kg})$$

The answer is (C).

---

[7](1) The *NCEES Handbook* definition of Eq. 18.20 as "reversible flow work" implies that the equation is valid only for reversible processes. Equation 18.20 can be used with irreversible processes, and the flow work is understood as being the useful work performed after other energy losses. (2) Usually, the thermodynamic interpretation of the term "flow work" is the work done on all of the substance in the system. The form of Eq. 18.20 calculates the "specific" boundary work (i.e., the work done per unit mass of the system). This is not explicitly stated in the *NCEES Handbook*. (3) The inclusion of the kinetic and potential energy terms in the flow work is incorrect. Flow work is strictly $p_2 v_2 - p_1 v_1$. Flow work (power) is the work (power) needed to inject/eject mass into/out of the system: $\dot{m}(p_2 v_2 - p_1 v_1)$.

Equation 18.21: First Law of Thermodynamics for Open Systems

$$\sum \dot{m}_i[h_i + v_i^2/2 + gz_i] - \sum \dot{m}_e[h_e + v_e^2/2 + gz_e]$$
$$+ \dot{Q}_{in} - \dot{W}_{net} = d(m_s u_s)/dt$$

$$18.21$$

Description

The first law of thermodynamics applied to open systems has a simple interpretation: Work done on the substance flowing through the system results in pressure, kinetic, or potential energy changes, a heat transfer, an energy storage, or a combination thereof.

Equation 18.21 is the first law of thermodynamics for open systems with multiple inlets and multiple outlets.[8] In the real world, there are seldom more than two inlets and/or two outlets. In Eq. 18.21, the entering and exiting mass flow rates may be different. If more mass enters the system than exits it, stationary mass will be stored within the system. A stationary stored mass also stores energy by virtue of its temperature (internal energy), pressure (flow energy), and elevation (potential energy). The rate of change in the energy storage within the control volume is represented by the $d(m_s u_s)/dt$ term in which m_s is the mass of fluid in the system, and u_s is the specific energy storage of the system.[9]

Most practical real-world processes do not have an accumulation of substance within them. If the mass flow rates are the same, Eq. 18.21 is known as the *steady-flow energy equation*. Specifically, there are several ways that work done on a steady-flow system can be manifested.

Heat loss:

$$\dot{W}_{in} = \dot{Q}_{in}$$

---

[8](1) Equation 18.21 is technically incomplete. By labeling the heat energy and work terms "in" and "net," respectively, the *NCEES Handbook* partially abandons the standard thermodynamic sign convention that was implicit in Eq. 18.3. In Eq. 18.3, the heat term, Q, implicitly represents $Q_{in} - Q_{out}$, and the work term, W, implicitly represents $W_{out} - W_{in}$. Equation 18.21 defines $\dot{Q}$ and $\dot{W}$ explicitly, but the equation is incomplete ($\dot{Q}_{out}$ has been omitted) and inconsistent ($\dot{W}_{net}$ appears, but not $\dot{Q}_{net}$; $\dot{W}_{net}$ in Eq. 18.21 is the same as W in Eq. 18.3). The standard thermodynamic sign convention applies only to the work term, $\dot{W}_{net}$. In order to properly use Eq. 18.21, the sign of the $\dot{Q}$ term must be changed based on the definitions (subscripts), not based on the standard thermodynamic sign convention. (2) The *NCEES Handbook* is inconsistent in the variable it uses to represent elevation above the datum in energy equations. For fluid subjects, the *NCEES Handbook* uses z in the Bernoulli equation; for thermodynamics subjects, the *NCEES Handbook* uses Z. Since lowercase z is the most common symbol used in engineering practice, this book uses that convention. The equations in this book involving elevation above a datum will differ slightly in appearance from those in the *NCEES Handbook*.

[9]For systems that accumulate mass, the right-hand side of Eq. 18.21 seems to imply that the energy storage within the control volume results in a change in internal energy (temperature) only. This is incorrect, as the energy can be stored in any of the forms represented by the terms in the equation. The correct representation for a rate of change in stored system energy would be dE_{CV}/dt or similar.

Change in pressure, volume, and/or temperature:

$$\frac{\dot{W}_{\text{in}}}{\dot{m}} = h_i - h_e = p_i v_i + u_i - p_e v_e - u_e$$

Change in velocity (kinetic energy):

$$\frac{\dot{W}_{\text{in}}}{\dot{m}} = \frac{v_i^2}{2} - \frac{v_e^2}{2}$$

Change in elevation (potential energy):

$$\frac{\dot{W}_{\text{in}}}{\dot{m}} = g z_i - g z_e$$

Example

The absolute pressure in a rigid, insulated tank is initially zero. The tank volume is 0.040 m³. The tank and steam inlet are at the same elevation. A valve is opened allowing 250°C and 600 kPa steam with negligible velocity to slowly fill the tank. Most nearly, what is the temperature of the steam in the tank after the tank is full?

(A) 250°C

(B) 300°C

(C) 350°C

(D) 400°C

Solution

Use Eq. 18.21. The control volume is the tank. Subscript s refers to the substance in the tank. Subscript i refers to the tank inlet, at the valve. The tank does not have an exit, so all subscript e variables are zero. Subscript 1 refers to what is initially in the tank. Since the tank is initially evacuated, it has no contents, so all subscript 1 variables are zero. Subscript 2 refers to the steam in the tank after filling. Since the tank is insulated, Q_{in} is zero. There is no work done on or by the steam, so W_{net} is zero. Although pressure does not appear explicitly in Eq. 18.21, the final pressure in the tank is $p_2 = p_i$.

$$\sum \dot{m}_i[h_i + v_i^2/2 + g z_i]$$
$$-\sum \dot{m}_e[h_e + v_e^2/2 + g z_e]$$
$$+ \dot{Q}_{\text{in}} - \dot{W}_{\text{net}} = d(m_s u_s)/dt$$
$$m_i h_i = m_2 u_2 - m_1 u_1$$
$$= m_2 u_2$$

Use a mass balance to relate m_i to m_2.

$$\sum m_i - \sum m_e = (m_2 - m_1)_{\text{system}}$$
$$m_i = m_2$$

Combine the two equations and solve for the internal energy of the system.

$$m_i h_i = m_2 u_2$$
$$u = h_i = 2957.2 \text{ kJ/kg}$$

250°C, 600 kPa (0.6 MPa) steam is superheated. From the superheated steam table, its enthalpy is $h = 2957.2$ kJ/kg. Since this is the final internal energy of the steam in the tank, interpolate between 2881.2 kJ/kg and 2962.1 kJ/kg, the internal energies of 350°C and 400°C steam, respectively.

$$\frac{u - u_1}{u_2 - u_1} = \frac{T - T_1}{T_2 - T_1}$$

$$T = \left(\frac{u - u_1}{u_2 - u_1}\right)(T_2 - T_1) + T_1$$

$$= \left(\frac{2957.2 \dfrac{\text{kJ}}{\text{kg}} - 2881.2 \dfrac{\text{kJ}}{\text{kg}}}{2962.1 \dfrac{\text{kJ}}{\text{kg}} - 2881.2 \dfrac{\text{kJ}}{\text{kg}}}\right)$$

$$\times (400°C - 350°C) + 350°C$$

$$= 397.0°C \quad (400°C)$$

The answer is (D).

8. SPECIAL CASES OF OPEN SYSTEMS[10]

Equation 18.22: Constant Volume Process

$$w_{\text{rev}} = -v(p_2 - p_1) \qquad \textbf{18.22}$$

Description

For an adiabatic, steady flow, constant volume (i.e., *isochoric*) process with negligible changes in potential and kinetic energy, Eq. 18.20 reduces to Eq. 18.22.

Example

Water flows steadily at a rate of 60 kg/min through a pump. The water pressure is increased from 50 kPa to 5000 kPa. The average specific volume of water is 0.001 m³/kg. Most nearly, what is the hydraulic power delivered to the water by the pump?

(A) 5 kW

(B) 60 kW

(C) 300 kW

(D) 500 kW

---

[10]The *NCEES Handbook* qualifies Eq. 18.22 through Eq. 18.31 with the title "Special Cases of Open Systems (with no change in kinetic or potential energy)." However, there are additional limitations. Since the heat term has been omitted, these equations are limited to adiabatic systems. Since the mass flow terms have been omitted, there is no possibility of accumulation within the system, so these equations are limited to steady-flow systems.

Solution

The pumping power is

$$\dot{W}_{rev} = -\dot{m}v(p_2 - p_1)$$

$$= \frac{-\left(60 \ \frac{kg}{min}\right)\left(0.001 \ \frac{m^3}{kg}\right)(5000 \ kPa - 50 \ kPa)}{60 \ \frac{s}{min}}$$

$$= -4.95 \ kW \quad (-5 \ kW)$$

The hydraulic power delivered to the water by the pump is 5 KW.

The answer is (A).

Equation 18.23: Constant Pressure Process

$$w_{rev} = 0 \qquad 18.23$$

Description

For an adiabatic, steady flow, constant pressure (i.e., *isobaric*) process with negligible changes in potential and kinetic energy, Eq. 18.20 reduces to Eq. 18.23.

Equation 18.24 and Eq. 18.25: Constant Temperature Process

$$pv = \text{constant} \qquad 18.24$$

$$w_{rev} = RT\ln(v_2/v_1) = RT\ln(p_1/p_2) \qquad 18.25$$

Description

For an adiabatic, steady flow, constant temperature (i.e., *isothermal*) process with negligible changes in potential and kinetic energy, Eq. 18.20 reduces to Eq. 18.25.

Equation 18.26 Through Eq. 18.29: Isentropic Systems

$$pv^k = \text{constant} \qquad 18.26$$

$$w_{rev} = k(p_2v_2 - p_1v_1)/(1-k) \qquad 18.27$$

$$w_{rev} = kR(T_2 - T_1)/(1-k) \qquad 18.28$$

$$w_{rev} = \frac{k}{k-1}RT_1\left[1 - \left(\frac{p_2}{p_1}\right)^{(k-1)/k}\right] \qquad 18.29$$

Values

k is the ratio of specific heats, equal to 1.4 for air.

Description

Equation 18.26 through Eq. 18.29 are for isentropic systems.

Equation 18.30 and Eq. 18.31: Polytropic Systems

$$pv^n = \text{constant} \qquad 18.30$$

$$w_{rev} = n(p_2v_2 - p_1v_1)/(1-n) \qquad 18.31$$

Description

Equation 18.30 and Eq. 18.31 are for polytropic systems. n is the polytropic exponent described in Sec. 18.6. The work term given by Eq. 18.31 for steady flow polytropic systems is different from the corresponding work term given by Eq. 18.19 for closed system polytropic processes. This is because shaft work is calculated from $v \, dp$ (i.e., no change in volume), and boundary work is calculated from $p \, dv$ (i.e., with a change in volume). The derivations are different.

Example

The state of an ideal gas is changed in a steady-state open polytropic process from 400 kPa and 1.2 m³ to 300 kPa and 1.5 m³. The polytropic exponent is 1.3. Most nearly, what is the work performed?

(A) 130 kJ

(B) 930 kJ

(C) 1200 kJ

(D) 9000 kJ

Solution

The work for an open, polytropic process is

$$w_{rev} = n(p_2v_2 - p_1v_1)/(1-n)$$

$$= \frac{(1.3)\left(\begin{array}{c}(300 \ kPa)(1.5 \ m^3) \\ - (400 \ kPa)(1.2 \ m^3)\end{array}\right)}{1-1.3}$$

$$= 130 \ kJ$$

The answer is (A).

9. STEADY-STATE SYSTEMS

Equation 18.32 and Eq. 18.33: Steady-Flow Energy Equation

$$\sum \dot{m}_i(h_i + v_i^2/2 + gz_i) - \sum \dot{m}_e(h_e + v_e^2/2 + gz_e)$$
$$+ \dot{Q}_{in} - \dot{W}_{out} = 0 \qquad 18.32$$

$$\sum \dot{m}_i = \sum \dot{m}_e \qquad 18.33$$

Description

If the mass flow rate is constant, the system is a *steady-flow system*, and the first law is known as the *steady-flow energy equation*, SFEE, given by Eq. 18.32.[11]

The subscripts i and e denote conditions at the in-point and exit of the control volume, respectively. $v^2/2 + gz$ represents the sum of the fluid's kinetic and potential energies. Generally, these terms are insignificant compared with the thermal energy terms.

$\dot{W}_{out}$ is the rate of *shaft work* (i.e., *shaft power*)—work that the steady-flow device does on the surroundings. Its name is derived from the output shaft that serves to transmit energy out of the system. For example, turbines and internal combustion engines have output shafts. $\dot{W}_{out}$ can be negative, as in the case of a pump or compressor.[12]

The enthalpy, h, represents a combination of internal energy and reversible (flow) work ($pv + u$).

Example

An insulated reservoir has three inputs and one output. The first input is saturated water at 80°C with a mass flow rate of 1 kg/s. The second input is saturated water at 30°C with a mass flow rate of 2 kg/s. The output is water with a mass flow rate of 4 kg/s and a temperature of 60°C. The water level of the reservoir remains constant. Velocity and elevation changes are insignificant. What is most nearly the temperature of the third input?

- (A) 60°C
- (B) 70°C
- (C) 80°C
- (D) 100°C

Solution

Since the system is a steady-state system, the sum of the mass flow rates at the entrance is equal to the mass flow rate at the exit. Calculate the mass flow rate of the third input.

---

[11]Equation 18.32 is technically incomplete. By labeling the heat energy and work terms "in" and "out," respectively, the *NCEES Handbook* abandons the standard thermodynamic sign convention that was implicit in Eq. 18.3. In Eq. 18.3, the heat term, Q, implicitly represents $Q_{in} - Q_{out}$, and the work term, W, implicitly represents $W_{out} - W_{in}$. Equation 18.32 defines $\dot{Q}$ and $\dot{W}$ explicitly, but the $\dot{Q}_{out}$ and $\dot{W}_{in}$ terms are omitted. In order to properly use Eq. 18.32, the signs of the $\dot{Q}$ and $\dot{W}$ terms must be changed based on the definitions (subscripts), not based on the standard thermodynamic sign convention.

[12]As already explained, Eq. 18.32 does not use the standard thermodynamic sign convention that was implicit in Eq. 18.3. In order to properly use Eq. 18.32, the signs of the Q and W terms must be changed based on the definitions (subscripts), not based on the standard thermodynamic sign convention.

Use Eq. 18.33.

$$\sum \dot{m}_i = \sum \dot{m}_e$$
$$\dot{m}_1 + \dot{m}_2 + \dot{m}_3 = \dot{m}_e$$
$$\dot{m}_3 = \dot{m}_e - \dot{m}_1 - \dot{m}_2 = 4\ \frac{kg}{s} - 2\ \frac{kg}{s} - 1\ \frac{kg}{s}$$
$$= 1\ kg/s$$

Use Eq. 18.32. Since the reservoir is insulated, the system is adiabatic, and $Q_{in} = 0$. Since the system does no work on the surroundings, $W_{out} = 0$. Since velocity and elevation changes are insignificant, the $v^2/2$ and gz terms can be omitted.

$$\sum \dot{m}(h_i + v_i^2/2 + gz_i)$$
$$- \sum \dot{m}_e(h_e + v_e^2/2 + gz_e)$$
$$+ \dot{Q}_{in} - \dot{W}_{out} = 0$$
$$= \dot{m}_1 h_1 + \dot{m}_2 h_2 + \dot{m}_3 h_3 - \dot{m}_e h_e$$
$$h_3 = \frac{\dot{m}_e h_e - \dot{m}_1 h_1 - \dot{m}_2 h_2}{\dot{m}_3}$$

Although the enthalpies of saturated liquid water could be determined from the steam tables, it is not necessary to do so. All of the inputs and the output are liquid water, so the enthalpies, h, can be represented by $c_p T$. And, since c_p is a common term and is essentially constant over the small temperature range involved, it can be omitted.

$$T_3 \approx \frac{\dot{m}_e T_e - \dot{m}_1 T_1 - \dot{m}_2 T_2}{\dot{m}_3}$$

$$= \frac{\left(4\ \frac{kg}{s}\right)(60°C) - \left(1\ \frac{kg}{s}\right)(80°C)}{ - \left(2\ \frac{kg}{s}\right)(30°C)}{1\ \frac{kg}{s}}$$

$$= 100°C$$

The answer is (D).

10. EQUIPMENT AND COMPONENTS

Equation 18.34 and Eq. 18.35: Nozzles and Diffusers

$h_i + v_i^2/2 = h_e + v_e^2/2$	*18.34*
isentropic efficiency $= \dfrac{v_e^2 - v_i^2}{2(h_i - h_{es})}$ [nozzle]	*18.35*

Description

Since a flowing fluid is in contact with nozzle, orifice, and valve walls for only a very short period of time, flow through them is essentially adiabatic. No work is done on the fluid as it passes through. Since most of the fluid does not contact the nozzle walls, friction is minimal. A lossless adiabatic process is an isentropic process. If the potential energy changes are neglected, Eq. 18.32 reduces to Eq. 18.34.

The *nozzle efficiency* is defined as Eq. 18.35.[13] The subscript "*es*" refers to the exit condition for an isentropic (ideal) expansion.

Example

Steam enters an adiabatic nozzle at 1 MPa, 250°C, and 30 m/s. At one point in the nozzle, the enthalpy drops 40 kJ/kg from its inlet value. What is most nearly the velocity at that point?

(A) 31 m/s

(B) 110 m/s

(C) 250 m/s

(D) 280 m/s

Solution

Use Eq. 18.34.

$$h_i + v_i^2/2 = h_e + v_e^2/2$$

$$v_e = \sqrt{2(h_i - h_e) + v_i^2}$$

$$= \sqrt{(2)\left(40\ \frac{kJ}{kg}\right)\left(1000\ \frac{J}{kJ}\right) + \left(30\ \frac{m}{s}\right)^2}$$

$$= 284\ \text{m/s} \quad (280\ \text{m/s})$$

The answer is (D).

....

Equation 18.36 Through Eq. 18.38: Turbines, Pumps, and Compressors

$$h_i = h_e + w \qquad \textbf{18.36}$$

$$\text{isentropic efficiency} = \frac{h_i - h_e}{h_i - h_{es}} \quad \text{[turbine]} \qquad \textbf{18.37}$$

$$\text{isentropic efficiency} = \frac{h_{es} - h_i}{h_e - h_i} \quad \text{[compressor, pump]}$$

$$\textbf{18.38}$$

Description

A *pump* or *compressor* converts mechanical energy into fluid energy, increasing the total energy content of the fluid flowing through it. *Turbines* can generally be thought of as pumps operating in reverse. A turbine extracts energy from the fluid, converting fluid energy into mechanical energy.

These devices can be considered to be adiabatic because the fluid gains (or loses) very little heat during the short time it passes through them.

Equation 18.36 assumes that the pump or turbine is capable of isentropic compression or expansion.[14] However, due to inefficiencies, the actual exit enthalpy will deviate from the ideal isentropic enthalpy, h_{es}. The isentropic efficiencies are given by Eq. 18.37 and Eq. 18.38.

Equation 18.37 for turbines is equivalent to the ratio $w_{out,actual}/w_{out,ideal}$, while Eq. 18.38 for pumps and compressors is equivalent to the ratio $w_{in,ideal}/w_{in,actual}$. For turbines, the ideal power generated (i.e., the denominator) is larger than the actual power generated. For pumps, the actual work input (the denominator) is larger than the ideal work input. Since these two expressions are essentially reciprocals, it is important to recognize that the larger quantity always is in the denominator.

Equation 18.37 and Eq. 18.38 both imply that it is necessary to know the ideal conditions when determining isentropic efficiencies. This often results in having to solve a problem twice: once with the ideal properties, and once with the actual properties. Actual properties are usually measured and are known. Ideal properties assume a process that has neither friction nor heat loss (i.e., is reversible and adiabatic). By definition, a reversible adiabatic process is isentropic. Two characteristics of isentropic processes are used to determine the ideal state properties: (1) Entropy does not change in an isentropic process (i.e., $\Delta s = 0$). (2) The final pressure is not changed by the inefficiency. For both pumps and turbines, the exit properties can be evaluated at the entrance entropy and the actual exit pressure.[15]

----

[14]The *NCEES Handbook* is inconsistent in its subscripting for work terms. The standard thermodynamic sign convention is required to determine the sign of the work term, w, in Eq. 18.36. Work, w, can be either positive or negative. For turbines (when the system does work on the environment), the w term is positive. For pumps (where the environment does work on the system), the w term is negative. Algebraically, Eq. 18.36 is equivalent to $h_i - w = h_e$. Since, for pumps, w is negative, this is equivalent to $h_i + w_{in} = h_e$. For pump problems asking for the work per unit mass of fluid, answers derived rigorously from Eq. 18.36 would all need to be negative.

[15]The *NCEES Handbook* does not provide a "pressures" saturated water properties table (i.e., a table of properties presented with constant pressure increments). This makes solving pump and turbine problems (and, all steam power cycle problems) extremely inconvenient, as interpolation is almost always required. The inconvenience is particularly extreme when evaluating condenser performance, since the primary descriptive operating condition of a condenser is its pressure. Since interpolation takes more time than is normally available for a single problem on the FE exam, the logical interpretation is that accurate results are not required for these problems. It is probably sufficient to use the values that correspond to the closest pressure in the "temperatures" saturated water properties table.

----

[13]In its section on heat cycles, the *NCEES Handbook* uses the symbol η for (thermal) efficiency. However, it does not use any symbol for (isentropic) efficiency in Eq. 18.35 or in the equations for other components. Isentropic efficiency for nozzles and other components is commonly written as η_s.

Thermodynamics

Example

A turbine produces 3 MW of power by expanding 500°C steam at 1 MPa to a saturated 30 kPa vapor. What is most nearly the isentropic efficiency of the turbine?

- (A) 96.5%
- (B) 98.2%
- (C) 99.1%
- (D) 99.7%

Solution

From the superheated water table, at 1 MPa and 500°C, the entering enthalpy is $h_i = 3478.5$ kJ/kg.

At the exit, the steam is saturated at 30 kPa. (This roughly corresponds to a saturation temperature of 70°C.) Interpolating from the saturated water table, the actual exit enthalpy is

$$h_e = h_{g,p_{\text{sat}}=p_e} = 2625.2 \text{ kJ/kg}$$

The actual enthalpy change (work) is

$$
\begin{aligned}
w_{\text{actual}} &= h_i - h_e \\
&= 3478.5 \frac{\text{kJ}}{\text{kg}} - 2625.2 \frac{\text{kJ}}{\text{kg}} \\
&= 853.3 \text{ kJ/kg}
\end{aligned}
$$

In order to calculate the turbine's isentropic efficiency, it is necessary to calculate the ideal exit enthalpy, h_{es}, that corresponds to isentropic expansion (i.e., the ideal case). The ideal exit enthalpy is calculated from the ideal quality, which in turn, is found from the ideal entropy. If the expansion had been isentropic, the entropy at the exit would have corresponded to the entropy at the entrance of the turbine. The exit pressure is unaffected by the isentropic efficiency.

At the initial state of 1 MPa and 500°C, the entropy is $s_i = 7.7622$ kJ/kg·K. At the final state, the pressure is 30 kPa, which roughly corresponds to a saturation temperature of 70°C. Interpolating from the saturated water table, the saturated liquid and vaporization entropies are $s_f = 0.9430$ kJ/kg·K and $s_{fg} = 6.827$ kJ/kg·K. For an isentropic (ideal) expansion, the ideal quality would have been

$$
\begin{aligned}
x_{es} &= \frac{s_{es} - s_{f,p_{\text{sat}}=p_{es}}}{s_{fg,p_{\text{sat}}=p_{es}}} \\
&= \frac{7.7622 \frac{\text{kJ}}{\text{kg·K}} - 0.9430 \frac{\text{kJ}}{\text{kg·K}}}{6.827 \frac{\text{kJ}}{\text{kg·K}}} \\
&= 0.9989
\end{aligned}
$$

Interpolating from the saturated water table, the saturated liquid and vaporization entropies are $h_f = 288.94$ kJ/kg and $h_{fg} = 2336.2$ kJ/kg. If the expansion had been isentropic, the enthalpy would have been

$$
\begin{aligned}
h_{es} &= h_{f,p_{es}=p_{\text{sat}}} + x_{es}h_{fg,p_{es}=p_{\text{sat}}} \\
&= 288.94 \frac{\text{kJ}}{\text{kg}} + (0.9989)\left(2336.2 \frac{\text{kJ}}{\text{kg}}\right) \\
&= 2622.5 \text{ kJ/kg}
\end{aligned}
$$

The ideal work would have been

$$
\begin{aligned}
w_{\text{ideal}} &= h_i - h_{es} = 3478.5 \frac{\text{kJ}}{\text{kg}} - 2622.5 \frac{\text{kJ}}{\text{kg}} \\
&= 856.0 \text{ kJ/kg}
\end{aligned}
$$

From Eq. 18.37, the isentropic efficiency of the turbine is

$$
\begin{aligned}
\text{isentropic efficiency} &= \frac{h_i - h_e}{h_i - h_{es}} = \frac{w_{\text{actual}}}{w_{\text{ideal}}} \\
&= \frac{853.3 \frac{\text{kJ}}{\text{kg}}}{856.0 \frac{\text{kJ}}{\text{kg}}} \\
&= 0.9968 \quad (99.7\%)
\end{aligned}
$$

The answer is (D).

Equation 18.39: Throttling Valves and Throttling Processes

$$h_i = h_e \qquad \text{18.39}$$

Description

In a *throttling process*, there is no change in system enthalpy, but there is a significant pressure drop. The process is adiabatic, and Eq. 18.32 reduces to Eq. 18.39.

Equation 18.40: Boilers, Condensers, and Evaporators

$$h_i + q = h_e \qquad \text{18.40}$$

Description

An *evaporator* is a device that adds heat, q, to water at low (near atmospheric) pressure and produces saturated steam. A *boiler* adds heat to water (known as *feedwater*) and produces superheated steam. The superheating may occur in the boiler, or there may be a separate unit known as a *superheater*. Both evaporators and boilers add most, if not all, of the heat of vaporization, h_{fg}, to the water. Superheaters add additional energy known as the *superheat* or *superheat energy*.

A *condenser* is a device that removes heat, q, from saturated or superheated steam. Usually, all of the superheat energy is removed, leaving the steam as a saturated liquid. Some of the heat of vaporization may also be removed, resulting in a *subcooled liquid*. Heat removed from the steam is released (usually to the environment) after being carried away by cooling water or air.

Disregarding changes to kinetic and potential energies, Eq. 18.40 describes the operation of evaporators, boilers, and condensers.[16] For both adiabatic and non-adiabatic devices, q represents the change in water or steam energy.

Example

A simple Rankine cycle operates between 20°C and 100°C and uses water as the working fluid. The turbine has an isentropic efficiency of 100%, and the pump has an isentropic efficiency of 65%. The steam leaving the boiler and the water leaving the condenser are both saturated.

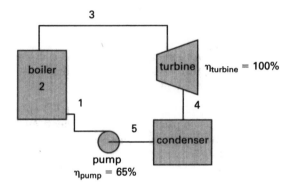

The properties of steam at the two temperatures are as follows.

For a saturation temperature of 20°C,

pressure	2.339 kPa
specific volume	
fluid	0.001002 m³/kg
gas	57.79 m³/kg
enthalpy	
fluid	83.95 kJ/kg
fluid-to-gas	2454.1 kJ/kg
gas	2538.1 kJ/kg
entropy	
fluid	0.2966 kJ/kg·K
gas	8.6672 kJ/kg·K

---

[16]The standard thermodynamic sign convention is required to determine the sign of the work term, q, in Eq. 18.40. Work, q, can be either positive or negative. For evaporators and boilers (where energy enters the system), the q term is positive. For condensers (where energy leaves the system), the q term is negative. Algebraically, Eq. 18.40 is equivalent to, $h_i = h_e - q$. Since, for condensers, q is negative, this is equivalent to $h_i = h_e + q_{out}$. For condenser problems asking for the energy removal per unit mass of steam, answers derived rigorously from Eq. 18.40 would all need to be negative.

For a saturation temperature of 100°C,

pressure	101.35 kPa
specific volume	
fluid	0.001044 m³/kg
gas	1.6729 m³/kg
enthalpy	
fluid	419.04 kJ/kg
fluid-to-gas	2257.0 kJ/kg
gas	2676.1 kJ/kg
entropy	
fluid	1.3069 kJ/kg·K
gas	7.3549 kJ/kg·K

The energy removed from each kilogram of steam in the condenser is most nearly

(A) 84 kJ/kg

(B) 420 kJ/kg

(C) 1500 kJ/kg

(D) 2100 kJ/kg

Solution

At state 3,

$$T_3 = 100°C \quad \text{[saturated]}$$
$$h_3 = h_g = 2676.1 \text{ kJ/kg}$$
$$s_3 = s_g = 7.3549 \text{ kJ/kg·K}$$

At state 4,

$$s_4 = s_3 = 7.3549 \text{ kJ/kg·K}$$
$$x = \frac{s - s_f}{s_{fg}} = \frac{s - s_f}{s_g - s_f}$$
$$= \frac{7.3549 \dfrac{\text{kJ}}{\text{kg·K}} - 0.2966 \dfrac{\text{kJ}}{\text{kg·K}}}{8.6672 \dfrac{\text{kJ}}{\text{kg·K}} - 0.2966 \dfrac{\text{kJ}}{\text{kg·K}}}$$
$$= 0.843$$
$$h_4 = h_f + x h_{fg}$$
$$= 83.95 \frac{\text{kJ}}{\text{kg}} + (0.843)\left(2454.1 \frac{\text{kJ}}{\text{kg}}\right)$$
$$= 2153 \text{ kJ/kg}$$

Use Eq. 18.40. The energy removed is

$$h_i + q = h_e$$
$$q_L = h_{e4} - h_{i5}$$
$$= 2153 \frac{\text{kJ}}{\text{kg}} - 83.95 \frac{\text{kJ}}{\text{kg}}$$
$$= 2069 \text{ kJ/kg} \quad (2100 \text{ kJ/kg})$$

The answer is (D).

Equation 18.41: Heat Exchangers

$$\dot{m}_1(h_{1i} - h_{1e}) = \dot{m}_2(h_{2e} - h_{2i}) \qquad 18.41$$

Description

A *heat exchanger* transfers heat energy from one fluid to another through a wall separating them. Heat exchangers have two working fluids, two entrances, and two exits. In Eq. 18.41, the hot and cold substances are identified by the numerical subscripts.[17]

No work is done within a heat exchanger, and the potential and kinetic energies of the fluids can be ignored. If the heat exchanger is well insulated (i.e., is adiabatic), Eq. 18.32 reduces to Eq. 18.41.

Equation 18.42 and Eq. 18.43: Mixers, Separators, and Open or Closed Feedwater Heaters

$$\sum \dot{m}_i h_i = \sum \dot{m}_e h_e \qquad 18.42$$

$$\sum \dot{m}_i = \sum \dot{m}_e \qquad 18.43$$

Description

A *feedwater heater* uses steam to increase the temperature of water entering the steam generator. The steam can come from any waste steam source but is usually bled off from a turbine. In this latter case, the heater is known as an *extraction heater*. The water that is heated usually comes from the condenser.

Open heaters (also known as *direct contact heaters* and *mixing heaters*) physically mix the steam and water. A *closed feedwater heater* is a traditional closed heat exchanger that can operate at either high or low pressures. There is no mixing of the water and steam in the closed feedwater heater. The cooled stream leaves the feedwater heater as a liquid.

For adiabatic operation, Eq. 18.32 reduces to Eq. 18.42. Since some feedwater heaters have two or more input streams (e.g., a bleed steam input and a condenser liquid input), Eq. 18.42 and Eq. 18.43 are presented as summations.

11. SECOND LAW OF THERMODYNAMICS

The *second law of thermodynamics* can be stated in several ways. The second law can be described in terms of the environment ("The entropy of the environment always increases in real processes."), working fluid ("A substance can be brought back to its original state without increasing the entropy of the environment only in a reversible system."), or equipment ("A machine that

returns the working fluid to its original state requires a heat sink." Or, "A machine rejects more energy than the useful work it performs."). The second law can also be used to define what kinds of processes can occur naturally (spontaneously).

A natural process that starts in one equilibrium state and ends in another will go in the direction that causes the entropy of the system and the environment to increase.

Equation 18.44 and Eq. 18.45: Change of Entropy

$$ds = (1/T)\delta q_{\text{rev}} \qquad 18.44$$

$$s_2 - s_1 = \int_1^2 (1/T)\delta q_{\text{rev}} \qquad 18.45$$

Description

Entropy is a measure of the energy that is no longer available to perform useful work within the current environment. An increase in entropy is known as *entropy production*. The total entropy in a system is equal to the integral of all entropy productions that have occurred over the life of the system.

Equation 18.44 defines the *exact differential* of entropy, ds, in terms of the *inexact differential* of heat, δq.[18] An inexact differential is path-dependent, while the exact differential is not.[19]

Since Eq. 18.44 is stated as an equality, the heat transfer process is implicitly reversible. (If the process is irreversible, an inequality is required.) Equation 18.44 and Eq. 18.45 are restatements of the *inequality of Clausius* (see Eq. 18.56) for reversible processes. Equation 18.45 purports to be the solution to Eq. 18.44.[20]

---

[17]Although there is no single unified convention on how to differentiate between the two working fluids, a common convention is to use the subscripts H and C to designate the hot and cold fluids, respectively. (The *NCEES Handbook* adopts that convention in equations involving the logarithmic mean temperature difference, LMTD.) This helps to avoid errors when assigning temperatures and other properties in equations.

[18]The *NCEES Handbook* uses inexact differentials with Eq. 18.44, Eq. 18.45, Eq. 18.56, and Eq. 18.57, but probably nowhere else. Most notably, they are not used in first-law formulations such as Eq. 18.3, where a presentation consistent with the concept of inexact differentials would be $dU = \delta Q - \delta W$.

[19]*Inexact differentials* are also known as *imperfect differentials* and *incomplete differentials*. The distinction between exact and inexact differentials is unique to the field of thermodynamics. Rudolf Clausius, who formulated the concept of entropy (ca. 1850), recognized that entropy production was path-dependent, although the notational difference between the two types of differentials was apparently introduced only later by Carl Neuman (ca. 1875).

[20](1) The integral notation used by the *NCEES Handbook* in Eq. 18.45 is incorrect. There is no mathematical operation that "maps" an inexact entropy differential to an entropy change because the entropy production is path-dependent. From the (second part of the) fundamental theorem of calculus, a definite integral is defined by its endpoints, not its path. For example,

$$\int_a^b f(x)\,dx = F(b) - F(a)$$

That is not possible with an inexact differential. (2) The temperature, T, used in Eq. 18.44 and Eq. 18.45 is the temperature of the heat sink or heat source. It is the same as $T_{\text{reservoir}}$ used elsewhere in the *NCEES Handbook*.

Equation 18.46 and Eq. 18.47: Entropy Change for Solids and Liquids

$$ds = c(dT/T) \qquad \textbf{18.46}$$

$$s_2 - s_1 = \int c(dT/T) = c_{\text{mean}} \ln(T_2/T_1) \qquad \textbf{18.47}$$

Description

Since entropy production is related to heat transfer, it is logical to incorporate the calculation of thermal energy transfers into the calculation of entropy production. The general relationship between thermal energy and temperature difference for a unit mass is $q = c\Delta T$. The specific heat, c, of liquids and solids is essentially constant over fairly large temperature and pressure ranges. Equation 18.46 substitutes $c\,dT$ for δq in Eq. 18.44. Equation 18.47 substitutes $c\,dT$ for δq in Eq. 18.45.[21]

Equation 18.48: Entropy Production at Constant Temperature

$$\Delta S_{\text{reservoir}} = Q/T_{\text{reservoir}} \qquad \textbf{18.48}$$

Description

Discussions of entropy and the second law invariably lead to use of the terms "heat source" and "heat sink." A *thermal reservoir* is an infinite mass with a constant temperature. When the thermal reservoir supplies energy, it is known as a *heat source*. When the reservoir absorbs energy, it is known as a *heat sink*. Since the reservoir has an infinite mass, its temperature, $T_{\text{reservoir}}$, does not change when energy is supplied or absorbed.[22]

Equation 18.48 defines the total change in reservoir entropy when heat is transferred to or from it.[23] This

change is known as *entropy production*. Equation 18.48 is an equality because Q is explicitly the heat that enters the reservoir, regardless of how much heat was lost in transit.

Example

A large concrete bridge anchorage at 15°C receives 50 GJ of thermal energy from exposure to sunlight. The temperature of the anchorage does not change significantly. What is most nearly the change in entropy within the concrete?

(A) 97 MJ/K

(B) 170 MJ/K

(C) 1900 MJ/K

(D) 3300 MJ/K

Solution

The anchorage is essentially an infinite thermal sink. From Eq. 18.48, the entropy production is

$$\Delta S_{\text{reservoir}} = Q/T_{\text{reservoir}}$$

$$= \frac{(50 \text{ GJ})\left(1000 \dfrac{\text{MJ}}{\text{GJ}}\right)}{15°\text{C} + 273°}$$

$$= 173.6 \text{ MJ/K} \quad (170 \text{ MJ/K})$$

The answer is (B).

Equation 18.49: Isothermal, Reversible Process

$$\Delta s = s_2 - s_1 = q/T \qquad \textbf{18.49}$$

Variation

$$\Delta S = m\Delta s = S_2 - S_1 = \frac{Q}{T_{\text{reservoir}}}$$

Description

For a process taking place at a constant temperature, T (i.e., discharging energy to a constant-temperature reservoir at T), the entropy production in the reservoir depends on the amount of energy transfer.[24] The standard thermodynamic sign convention must be followed when using Eq. 18.49. If heat enters the system, q will be positive, and s_2 will be greater than s_1. Equation 18.49 can be written in terms of total properties, as the variation equation shows.

---

[21](1) The *NCEES Handbook* presents Eq. 18.46 and Eq. 18.47 as equalities. The implication of this is the substance has been assumed to be completely incompressible (i.e., unaffected by pressure). (2) The *NCEES Handbook* presents the last form of Eq. 18.47 as an equality (i.e., as an engineering fact). However, there is no theoretical basis to using the mean specific heat. Using the mean specific heat is merely a computational convenience. If the last form of Eq. 18.47 is presented at all, it should be as an approximation and limited to small temperature changes. (3) The *NCEES Handbook* does not define c_{mean}. Three common definitions are used as simplifications of more rigorous methods: (a) the average of the specific heats at the beginning and ending temperatures, (b) the specific heat at the average temperature, and (c) the specific heat at the midpoint of a long pipe or other linear process.

[22]Although the *NCEES Handbook* identifies the reservoir temperature as $T_{\text{reservoir}}$, it is normal and customary in engineering thermodynamics to designate the environment temperature as T_0.

[23](1) It is important to recognize that Eq. 18.48 is written from the standpoint of the reservoir, not the process, machine, or substance that the reservoir is attached to. (2) The standard thermodynamic sign convention is required to determine the sign of the heat energy term, Q, in Eq. 18.48. Heat energy can be either positive or negative. For a heat sink, where energy enters the reservoir, Q is positive. For a heat source, where energy leaves the reservoir, Q is negative. (3) With a strict sign convention, it is important to recognize that ΔS is strictly defined as $S_2 - S_1$. If heat enters the reservoir (a positive Q), the final entropy will be greater than the initial entropy. (4) Since uppercase letters are used in Eq. 18.48, the intent is to calculate the entropy production in the entire reservoir from the entire heat transfer, not on a per unit mass basis. This is probably just as well, since the reservoir has an infinite mass.

[24](1) The variable T used by the *NCEES Handbook* in Eq. 18.49 is the same as $T_{\text{reservoir}}$ used elsewhere. This variable is commonly shown as T_0 to indicate the temperature of the environment. (2) It is important to recognize that, while Δs is defined as $s_2 - s_1$ (as it would be for any variable), $s_2 - s_1$ is not derived from a definite integral.

Example

An ideal gas undergoes a reversible isothermal expansion at 50°C. 200 kJ of heat enters the process. What is most nearly the change in air entropy at the end of the process?

(A) 0.62 kJ/K

(B) 0.72 kJ/K

(C) 0.90 kJ/K

(D) 1.0 kJ/K

Solution

The change in total entropy after the heat addition is

$$\Delta S = Q/T = \frac{200 \text{ kJ}}{50°C + 273°}$$
$$= 0.619 \text{ kJ/K} \quad (0.62 \text{ kJ/K})$$

The answer is (A).

Equation 18.50 and Eq. 18.51: Isentropic Process

$ds = 0$	*18.50*
$\Delta s = 0$	*18.51*

Description

Isentropic means constant entropy. Entropy does not change in an isentropic process, as shown by Eq. 18.50 and Eq. 18.51.

Equation 18.52 and Eq. 18.53: Adiabatic Process

$\delta q = 0$	*18.52*
$\Delta s \geq 0$	*18.53*

Description

Equation 18.52 and Eq. 18.53 are for adiabatic processes. There is no heat gain or loss in an *adiabatic process*. Processes that are well insulated (or that occur quickly enough to preclude significant heat loss (e.g., supersonic flow through a nozzle)) are often assumed to be adiabatic. An adiabatic process is not necessarily reversible, however. So, even though $q = 0$, it is still possible that other losses may cause entropy to increase. For example, pumps, compressors, and turbines lose or gain negligible heat, but they are affected by fluid and bearing friction.

Example

Air at 80°C originally occupies 0.5 m³ at 200 kPa. The gas is compressed reversibly and adiabatically to 1.20 MPa. What is most nearly the heat flow?

(A) 0 J

(B) 0.69 J

(C) 0.82 J

(D) 1.5 J

Solution

The gas is compressed adiabatically. So, by Eq. 18.52,

$$\delta q = 0 \quad (0 \text{ J})$$

The answer is (A).

Equation 18.54 and Eq. 18.55: Increase of Entropy Principle

$\Delta s_{\text{total}} = \Delta s_{\text{system}} + \Delta s_{\text{surroundings}} \geq 0$	*18.54*
$\Delta \dot{s}_{\text{total}} = \sum \dot{m}_{\text{out}} s_{\text{out}} - \sum \dot{m}_{\text{in}} s_{\text{in}}$ $- \sum (\dot{q}_{\text{external}} / T_{\text{external}}) \geq 0$	*18.55*

Description

The increase of entropy principle states that the change in total entropy (i.e., the entropy of the system and its surroundings) will be greater than or equal to zero.[25]

For all reversible processes, the equalities in Eq. 18.54 and Eq. 18.55 are valid.[26] For irreversible processes, the inequalities are valid.

Example

For an irreversible process, the total change in entropy of the system and surroundings is

(A) ∞

(B) 0

(C) greater than 0

(D) less than 0

Solution

For an irreversible process,

$$\Delta s_{\text{total}} = \Delta s_{\text{system}} + \Delta s_{\text{surroundings}} > 0$$

The answer is (C).

---

[25]This is the principle that leads to the conclusion that the universe is winding down, and that its eventual demise will be as a chaotic collection of random, cold particles.

[26](1) Equation 18.54 is not a formula, per se. It is a symbolic formulation of the statement, "The total entropy change includes the entropy productions of the system and the surroundings." (2) The *NCEES Handbook* presents Eq. 18.54 with lowercase letters, representing entropy changes per unit mass. The sum of the per unit system entropy increase and the per unit surroundings entropy increase would be a meaningless quantity, because the masses are different. The implication that Δs_{total} could be multiplied by any total mass value is incorrect. (3) It is not obvious why the *NCEES Handbook* presents Eq. 18.54 in terms of entropy, while Eq. 18.55 uses entropy per unit time. While both equations are correct, Eq. 18.55 does not follow directly from Eq. 18.54, which appears to be the intention. (4) The subscript "external" has not been defined by the *NCEES Handbook*, although "environment" could be an intended interpretation. For the finite summation to be valid, however, all of the T_{external} terms would have to represent constant heat sink or heat source temperatures. In that case, T_{external} would be the same as the T and $T_{\text{reservoir}}$ variables used elsewhere in the *NCEES Handbook*.

Kelvin-Planck Statement of Second Law (Power Cycles)

The *Kelvin-Planck statement of the second law* effectively says that it is impossible to build a cyclical engine that will have a thermal efficiency of 100%:

> No heat engine can operate in a cycle while transferring heat with a single thermal reservoir.

Figure 18.1(a) shows a violation of the second law in which a heat engine extracts heat from a high-temperature reservoir and converts that heat into an equivalent amount of work. There is no heat transfer to the low-temperature reservoir. (This is signified by the absence of an arrow going to the low-temperature reservoir.) In the real world, there are always losses (e.g., frictional heating) in the conversion of energy to work, and those losses cannot go back to a high-temperature reservoir. They must go to a low-temperature reservoir.

Figure 18.1 Second Law Violations

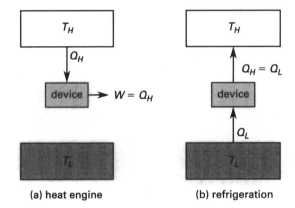

(a) heat engine   (b) refrigeration

Another statement that is equivalent is

> It is impossible to operate an engine operating in a cycle that will have no other effect than to extract heat from a reservoir and turn it into an equivalent amount of work.

This formulation is not a contradiction of the first law of thermodynamics. The first law does not preclude the possibility of converting heat entirely into work—it only denies the possibility of creating or destroying energy. The second law says that if some heat is converted entirely into work, some other energy must be rejected to a low-temperature reservoir (i.e., lost to the surroundings).

A corollary to this formulation of the second law is that the maximum possible efficiency of a heat engine operating between two thermal reservoirs is the Carnot cycle efficiency. It is possible to have a heat engine with a thermal efficiency higher than a particular Carnot cycle engine, but the temperatures of the heat engine would have to be different from the Carnot temperatures.

Equation 18.56 and Eq. 18.57: Inequality of Clausius

$$\oint (1/T)\delta q_{\text{rev}} \leq 0 \qquad \textbf{18.56}$$

$$\int_1^2 (1/T)\delta q \leq s_2 - s_1 \qquad \textbf{18.57}$$

Description

The *inequality of Clausius* is a mathematical statement of the second law of thermodynamics.

Equation 18.56 is the mathematical formulation of the inequality of Clausius, formulated as a *cyclic integral*, indicated by the symbol $\oint$. (A cyclic integral is a line integral evaluated over a closed path.) The left-hand side of Eq. 18.56 is a conceptual operation, not a mathematical operation or function. Equation 18.56 is interpreted as: "When a closed process is returned to its original condition in a cycle, after following the path taken by the substance in the process, the entropy production is, at best, zero, but otherwise, is less than zero."[27] Equation 18.57 purports to be the definition of entropy production.[28]

12. FINDING WORK AND HEAT GRAPHICALLY

Equation 18.58: Temperature-Entropy Diagrams

$$q_{\text{rev}} = \int_1^2 T \, ds \qquad \textbf{18.58}$$

---

[27]The *NCEES Handbook* states the inequality of Clausius incorrectly. The equality holds for the reversible case; the inequality holds for the irreversible case. If the inexact differential of heat were reversible, as indicated by the subscript, then the result would always be "=0," not "<0." The subscript "rev" should be omitted.

[28](1) The integral notation used by the *NCEES Handbook* in Eq. 18.56 and Eq. 18.57 is incorrect. There is no mathematical operation that "maps" an inexact entropy differential to an entropy change because the entropy production is path-dependent. From the (second part of the) fundamental theorem of calculus, a definite integral is defined by its endpoints, not its path. For example,

$$\int_a^b f(x) \, dx = F(b) - F(a)$$

That is not possible with an inexact differential. (2) Additionally, including the limits of integration strengthens the implication that an actual integration of the inexact differential is the intent. (3) Although the *NCEES Handbook* uses a cyclic integral for Eq. 18.56, it does not for Eq. 18.57. (4) The temperature, T, used in Eq. 18.56 and Eq. 18.57 is the temperature of the heat sink or heat source. It is the same as $T_{\text{reservoir}}$ used elsewhere in the *NCEES Handbook*.

Description

A process between two thermodynamic states can be represented graphically. The line representing the locus of quasiequilibrium states between the initial and final states is known as the *path* of the process.

It is convenient to plot the path on a *T-s* diagram. (See Fig. 18.2.) The amount of heat absorbed or released from a system can be determined as the area under the path on the *T-s* diagram, as given by Eq. 18.58 and shown in Fig. 18.2.

Pressure-Volume Diagrams

Similarly, the pressure and volume of a system can be plotted on a *p-V* diagram. Since $w = p\Delta v$, the work done by or on the system can be determined from that graph. (See Fig. 18.3.)

The variables p, V, T, and s are *point functions* because their values are independent of the path taken to arrive at the thermodynamic state. Work and heat (W and Q), however, are *path functions* because they depend on the path taken. For that reason, care should be taken when calculating work and heat by direct integration.[29]

Figure 18.2 *Heat from T-s Diagram*

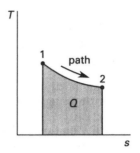

Figure 18.3 *Work from p-V Diagram*

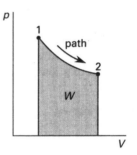

---

[29]As Fig. 18.2 and Fig. 18.3 show, the area under the curve depends on the path taken between points 1 and 2. The area does not merely depend on the values of T and p at points 1 and 2. This is the *de facto* definition of a path function. The calculation cannot involve a definite integral; it must involve a *line integral* (also known as a *path integral, contour integral,* or *curve integral*).

19 Power Cycles and Entropy

Nomenclature

B	width	m
c	specific heat	kJ/kg·K
COP	coefficient of performance	–
F	force	N
h	enthalpy	kJ/kg
HV	heating value	J/kg
I	process irreversibility	kJ/kg
k	ratio of specific heats	–
m	mass	kg
$\dot{m}$	mass flow rate	kg/s
mep	mean effective pressure	Pa
n	number	–
N	number	–
p	pressure	Pa
q	heat energy	kJ
$\dot{q}$	heat energy flow rate	kJ/s
Q	total heat energy	kJ
$\dot{Q}$	rate of heat transfer	kW
r	volumetric compression ratio	–
R	radius	m
s	specific entropy	kJ/kg·K
S	length of stroke	m
S	total entropy	kJ/K
sfc	specific fuel consumption	kg/J
T	absolute temperature	K
T	torque	N·m
u	specific internal energy	kJ/kg
v	velocity	m/s
V	volume	m³
w	specific work	kJ/kg
W	total work	kJ
$\dot{W}$	rate of work (power)	kW
z	elevation	m

Symbols

η	efficiency	–
v	specific volume	m³/s
ϕ	closed-system availability	kJ/kg
Ψ	open-system availability	kJ/kg

Subscripts

a	actual
b	brake
c	Carnot cycle, clearance, compression, or cylinders
d	displacement
f	fuel
H	high
HP	heat pump
L	low
ref	refrigerator
rev	reversible
s	stroke
t	total
T	turbine

1. BASIC CYCLES

It is convenient to show a source of energy as an infinite constant-temperature reservoir. Figure 19.1 illustrates a source of energy known as a *high-temperature reservoir* or *source* reservoir. By convention, the reservoir temperature is designated T_H (or T_{in}), and the heat transfer from it is Q_H (or Q_{in}). The energy derived from such a theoretical source might actually be supplied by combustion, electrical heating, or nuclear reaction.

Figure 19.1 Energy Flow in Basic Cycles

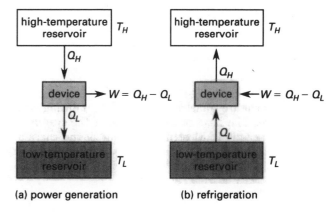

(a) power generation   (b) refrigeration

Similarly, energy is released (i.e., is "rejected") to a low-temperature reservoir known as a *sink reservoir* or *energy sink*. The most common practical sink is the local environment. T_L and Q_L (or T_{out} and Q_{out}) are used to represent the reservoir temperature and energy absorbed. It is common to refer to Q_L as the "rejected energy" or "energy rejected to the environment."

Although heat can be extracted and work can be performed in a single process, a *cycle* is necessary to obtain work in a useful quantity and duration. A cycle is a series of processes that eventually brings the system

Thermodynamics

back to its original condition. Most cycles are continuously repeated.

A cycle is completely defined by the working substance, the high- and low-temperature reservoirs, the means of doing work on the system, and the means of removing energy from the system. (The Carnot cycle depends only on the source and sink temperatures, not on the working fluid. However, most practical cycles depend on the working fluid.)

A cycle will appear as a closed curve when plotted on p-V and T-s diagrams. The area within the p-V and T-s curves represents both the net work and net heat.

2. POWER CYCLES

A *power cycle* is a cycle that takes heat and uses it to do work on the surroundings. The *heat engine* is the equipment needed to perform the cycle.

Equation 19.1: Thermal Efficiency

$$\eta = W/Q_H = (Q_H - Q_L)/Q_H \qquad \textit{19.1}$$

Description

The *thermal efficiency* of a power cycle is defined as the ratio of useful work output to the supplied input energy. W in Eq. 19.1 is the net work, since some of the gross output work may be used to run certain parts of the cycle.[1] For example, a small amount of turbine output power may run boiler feed pumps.

Equation 19.1 shows that obtaining the maximum efficiency requires minimizing the Q_L term. Equation 19.1 also shows that $W_{net} = Q_{net}$. This follows directly from the first law of thermodynamics.

Example

350 MJ of heat are transferred into a system during each power cycle. The heat transferred out of the system is 297.5 MJ per cycle. Most nearly, what is the thermal efficiency of the cycle?

(A) 1.0%

(B) 5.0%

(C) 7.5%

(D) 15%

Solution

From Eq. 19.1, the thermal efficiency is

$$
\begin{aligned}
\eta &= (Q_H - Q_L)/Q_H \\
&= \frac{350 \text{ MJ} - 297.5 \text{ MJ}}{350 \text{ MJ}} \\
&= 0.15 \quad (15\%)
\end{aligned}
$$

The answer is (D).

Equation 19.2 and Eq. 19.3: Carnot Cycle

$$\eta_c = (T_H - T_L)/T_H \qquad \textit{19.2}$$

$$\eta = 1 - \frac{T_L}{T_H} \qquad \textit{19.3}$$

Description

The most efficient power cycle possible is the Carnot cycle. The thermal efficiency of the entire cycle is given by Eq. 19.2.[2] Temperature must be expressed in the absolute scale. Equation 19.2 is easily derived from Eq. 19.1 since $Q = T_{reservoir}\Delta S$. Figure 19.2 shows that the entropy change, ΔS, is the same for the two heat transfer processes.

The *Carnot cycle* is an ideal power cycle that is impractical to implement. However, its theoretical work output sets the maximum attainable from any heat engine, as evidenced by the isentropic (reversible) processes between states (4 and 1) and (2 and 3) in Fig. 19.2. The working fluid in a Carnot cycle is irrelevant.

Example

A Carnot engine operates on steam between 65°C and 425°C. What is most nearly the efficiency?

(A) 19%

(B) 48%

(C) 52%

(D) 81%

Solution

Although the denominator of Eq. 19.2 must be expressed as an absolute temperature, the numerator is

---

[1] (1) The NCEES *FE Reference Handbook* (*NCEES Handbook*) presents the net work differently from the net heat. *Net heat* is the difference between the heat added in the boiler and the heat lost in the condenser. This is shown as $Q_H - Q_L$. In a steam power cycle, for example, the *net work* is the difference between the work done by the turbine and the work used by the feed pumps. Although the pump work is small, it is not zero, and the symmetry of Eq. 19.1 is lost by writing W instead of $W_H - W_L$ or similar. (2) From Eq. 19.1, it is obvious that $Q_H - Q_L = W_H - W_L$ and $Q_{net} = W_{net}$.

[2] (1) Although "diesel cycle" is rarely capitalized in writing, "Carnot cycle" and "Otto cycle" usually are. The *NCEES Handbook* uses a subscripted lowercase c to designate "Carnot." (2) η in Eq. 19.3 is the same as η_c in Eq. 19.2, although these two equations are separated by several pages in the *NCEES Handbook*.

Figure 19.2 Carnot Power Cycle

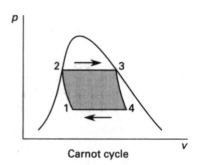

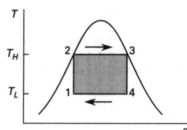

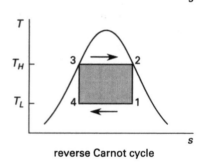

reverse Carnot cycle

a temperature difference. The offset temperatures (e.g., 273° for SI temperatures) cancel. The efficiency is

$$\eta_c = (T_H - T_L)/T_H$$
$$= \frac{425°C - 65°C}{425°C + 273°}$$
$$= 0.516 \quad (52\%)$$

The answer is (C).

Equation 19.4: Rankine Cycle

$$\eta = \frac{(h_3 - h_4) - (h_2 - h_1)}{h_3 - h_2} \qquad 19.4$$

Variation

$$\eta = \frac{W_{\text{out}} - W_{\text{in}}}{Q_{\text{in}}}$$
$$= \frac{Q_{\text{in}} - Q_{\text{out}}}{Q_{\text{in}}}$$
$$= \frac{(h_3 - h_2) - (h_4 - h_1)}{h_3 - h_2}$$

Description

The *Rankine cycle* is similar to the Carnot cycle except that the compression process occurs in the liquid region. (See Fig. 19.3.) The Rankine cycle is closely approximated in steam turbine plants. The thermal efficiency of the Rankine cycle is lower than that of a Carnot cycle operating between the same temperature limits because the mean temperature at which heat is added to the system is lower than T_H.

Figure 19.3 Rankine Cycle

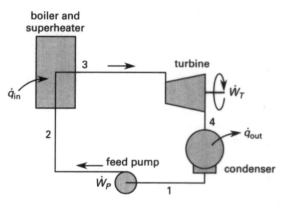

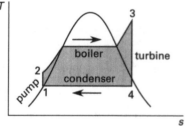

The thermal efficiency of the entire cycle is given by Eq. 19.4 and the variation equations. The enthalpy differences in the numerator of Eq. 19.4 represent work terms between the locations identified by the subscripts. Since $W_{\text{net}} = Q_{\text{net}}$, the equation could also be stated in terms of heat transfers, as is done in the variation equation. The two formulations are rearrangements of the same terms.

Superheating occurs when heat in excess of that required to produce saturated vapor is added to the water. Superheat is used to raise the vapor above the critical temperature, to raise the mean effective temperature at which heat is added, and to keep the expansion primarily in the vapor region to reduce wear on the turbine blades.

Example

A Rankine steam cycle operates between the pressure limits of 600 kPa and 10 kPa. The turbine inlet temperature is 300°C. The liquid water leaving the condenser is saturated. (The enthalpies of the steam at the various points on the cycle are shown in the illustration.) Most nearly, what is the thermal efficiency of the cycle?

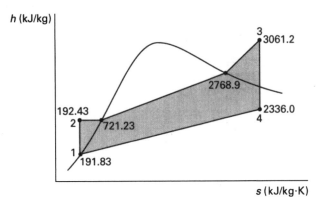

(A) 19%

(B) 25%

(C) 32%

(D) 48%

Solution

Using Eq. 19.4, the thermal efficiency of the Rankine cycle is

$$\eta = \frac{(h_3 - h_4) - (h_2 - h_1)}{h_3 - h_2}$$

$$= \frac{\left(3061.2 \; \frac{kJ}{kg} - 2336.0 \; \frac{kJ}{kg}\right) - \left(192.43 \; \frac{kJ}{kg} - 191.83 \; \frac{kJ}{kg}\right)}{3061.2 \; \frac{kJ}{kg} - 192.43 \; \frac{kJ}{kg}}$$

$$= 0.2526 \quad (25\%)$$

The answer is (B).

Equation 19.5 and Eq. 19.6: Otto Cycle

$$\eta = 1 - r^{1-k} \qquad \text{19.5}$$
$$r = v_1/v_2 \qquad \text{19.6}$$

Description

Combustion power cycles differ from vapor power cycles in that the combustion products cannot be returned to their initial conditions for reuse. Due to the computational difficulties of working with mixtures of fuel vapor and air, combustion power cycles are often analyzed as air-standard cycles.

An *air-standard cycle* is a hypothetical closed system using a fixed amount of ideal air as the working fluid. In contrast to a combustion process, the heat of combustion is included in the calculations without consideration of the heat source or delivery mechanism (i.e., the combustion process is replaced by a process of instantaneous heat transfer from high-temperature surroundings). Similarly, the cycle ends with an instantaneous transfer of waste heat to the surroundings. All processes are considered to be internally reversible. Because the air is assumed to be ideal, it has a constant specific heat.

Actual engine efficiencies for internal combustion engine cycles may be as much as 50% lower than the efficiencies calculated from air-standard analyses. Empirical corrections must be applied to theoretical calculations based on the characteristics of the engine. However, the large amount of excess air used in turbine combustion cycles results in better agreement (in comparison to reciprocating cycles) between actual and ideal performance.

The *air-standard Otto cycle* consists of the following processes and is illustrated in Fig. 19.4.

1 to 2:   isentropic compression ($q = 0$, $\Delta s = 0$)

2 to 3:   constant volume heat addition

3 to 4:   isentropic expansion ($q = 0$, $\Delta s = 0$)

4 to 1:   constant volume heat rejection

Figure 19.4 *Air-Standard Otto Cycle*

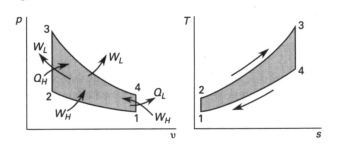

The Otto cycle is a four-stroke cycle because four separate piston movements (strokes) are required to accomplish all of the processes: the intake, compression, power, and exhaust strokes. Two complete crank revolutions are required for these four strokes. Therefore, each cylinder contributes one power stroke every other revolution.

The ideal thermal efficiency for the Otto cycle, Eq. 19.5, can be calculated from the *compression ratio, r*, Eq. 19.6.

Example

What is the ideal efficiency of an air-standard Otto cycle with a compression ratio of 6:1?

(A) 17%

(B) 19%

(C) 49%

(D) 51%

Solution

The ratio of specific heats for air is $k = 1.4$. From Eq. 19.5,

$$\eta = 1 - r^{1-k}$$
$$= 1 - (6)^{1-1.4}$$
$$= 0.512 \quad (51\%)$$

The answer is (D).

3. INTERNAL COMBUSTION ENGINES

The performance characteristics (e.g., horsepower) of internal combustion engines can be reported with or without the effect of power-reducing friction and other losses. A value of a property that includes the effect of friction is known as a *brake value*. If the effect of friction is removed, the property is known as an *indicated value*.[3]

Engine power can be measured by a *Prony brake* (also known as a *de Prony brake* and *absorption dynamometer*), which is basically a device that provides rotational resistance that the engine has to overcome. (See Fig. 19.5.) The engine's output shaft is connected to a rotating hub (wheel, disk, roller, etc.) in such a way that the engine (or, resisting) torque can be adjusted and measured. The resistance torque can be calculated as the product of the applied force and the moment arm, $T = FR$. The term "brake power" is derived from the Prony brake apparatus, since the only power available to turn the rotating hub is what is left over after frictional and windage losses.

Figure 19.5 *Brake Power*

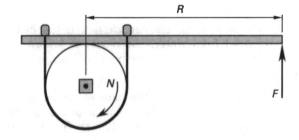

Equation 19.7 Through Eq. 19.13: Brake and Indicated Properties

$$\dot{W}_b = 2\pi T N = 2\pi F R N \qquad 19.7$$

$$\dot{W}_i = \dot{W}_b + \dot{W}_f \qquad 19.8$$

$$\text{mep} = \frac{\dot{W} n_s}{V_d n_c N} \qquad 19.9$$

---

[3]It may be helpful to think of the *i* in "indicated" as meaning "ideal."

$$V_d = \frac{\pi B^2 S}{4} \qquad 19.10$$

$$V_t = V_d + V_c \qquad 19.11$$

$$r_c = V_t / V_c \qquad 19.12$$

$$\text{sfc} = \frac{\dot{m}_f}{\dot{W}} = \frac{1}{\eta \text{HV}} \qquad 19.13$$

Description

Common brake properties are *brake power*, $\dot{W}_b$ (see Eq. 19.7 and Fig. 19.5); *fuel consumption rate*, $\dot{m}_b$; and *brake mean effective pressure*, mep (see Eq. 19.9). *Displacement volume*, V_d, is needed to calculate mep, and is found from Eq. 19.10. B is the diameter of the *cylinder bore*, and S is the length of the *stroke*. Total volume, V_t (see Eq. 19.11), is the sum of displacement volume and *clearance volume*, V_c. Dividing the total volume by the clearance volume yields the *compression ratio*, r_c (see Eq. 19.12).

Common indicated properties are *indicated power*, $\dot{W}_i$ (see Eq. 19.8); *indicated specific fuel consumption*, isfc; and *indicated mean effective pressure*, imep.

Specific fuel consumption, sfc (see Eq. 19.13), is the fuel usage rate divided by the power generated.

The brake and indicated powers differ by the *friction power*, W_f.

Example

An electric motor is tested in a brake that uses a Kevlar belt to apply frictional resistance to a 22 cm radius hub. The tight side tension is measured by spring scale A as 30 N. The slack side tension is measured by spring scale B as 10 N. The motor turns at 1725 rpm.

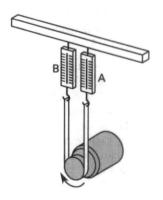

Most nearly, what is the motor horsepower?

(A) 0.17 hp

(B) 0.50 hp

(C) 0.80 hp

(D) 1.1 hp

Solution

The power is

$$\dot{W} = 2\pi TN = 2\pi FRN$$

$$= \frac{2\pi(30 \text{ N} - 10 \text{ N})(22 \text{ cm})\left(1725 \dfrac{\text{rev}}{\text{min}}\right)\left(1.341 \dfrac{\text{hp}}{\text{kW}}\right)}{\left(100 \dfrac{\text{cm}}{\text{m}}\right)\left(60 \dfrac{\text{s}}{\text{min}}\right)\left(1000 \dfrac{\text{W}}{\text{kW}}\right)}$$

$$= 1.07 \text{ hp} \quad (1.1 \text{ hp})$$

The answer is (D).

Equation 19.14 Through Eq. 19.17: Engine Efficiencies

$$\eta_b = \frac{\dot{W}_b}{\dot{m}_f(HV)} \tag{19.14}$$

$$\eta_i = \frac{\dot{W}_i}{\dot{m}_f(HV)} \tag{19.15}$$

$$\eta_i = \frac{\dot{W}_b}{\dot{W}_i} = \frac{\eta_b}{\eta_i} \tag{19.16}$$

$$\eta_v = \frac{2\dot{m}_a}{\rho_a V_d n_c N} \quad \text{[four-stroke cycles only]} \tag{19.17}$$

Description

The *brake thermal efficiency*, η_b (see Eq. 19.14), is found from the brake power, fuel consumption rate, and the *heating value* of the fuel, HV. The *indicated thermal efficiency* (see Eq. 19.15) is found from the indicated power, fuel consumption rate, and the heating value of the fuel. *Mechanical efficiency* (see Eq. 19.16) is the ratio of brake thermal efficiency to indicated thermal efficiency.

Because of friction in the intake system, the presence of expanding exhaust gases, valve timing, and air inertia, a reciprocating engine will not take in as much air as is calculated from the displacement. The *volumetric efficiency* (see Eq. 19.17) is the ratio of the actual amount of air taken in during each intake stroke to the *displacement volume*. This can be calculated from per-stroke characteristics or (as in the case of Eq. 19.17) from per unit time characteristics. In Eq. 19.17, $\dot{m}_a$ is the actual mass of air taken in per unit time for all cylinders; and, the denominator represents the theoretical mass of air based on the atmospheric density outside the engine and the displacement volume. n_c is the number of cylinders in the engine. In a four-stroke engine, where there is one intake stroke for every two revolutions, the rotational speed is divided by 2 to get the number of intake strokes per unit time. That is the source of the "2" in the numerator of Eq. 19.17.

4. REFRIGERATION CYCLES

In contrast to heat engines, in refrigeration cycles, heat is transferred from a low-temperature area to a high-temperature area. Since heat flows spontaneously only from high- to low-temperature areas, refrigeration needs an external energy source to force the heat transfer to occur. This energy source is a pump or compressor that does work in compressing the refrigerant. It is necessary to perform this work on the refrigerant in order to get it to discharge energy to the high-temperature area.

In a power (heat engine) cycle, heat from combustion is the input and work is the desired effect. Refrigeration cycles, though, are power cycles in reverse, and work is the input, with cooling the desired effect. (For every power cycle, there is a corresponding refrigeration cycle.) In a refrigerator, the heat is absorbed from a low-temperature area and is rejected to a high-temperature area. The pump work is also rejected to the high-temperature area.

General refrigeration devices consist of a coil (the evaporator) that absorbs heat, a condenser that rejects heat, a compressor, and a pressure-reduction device (the expansion valve or throttling valve).

In operation, liquid refrigerant passes through the evaporator where it picks up heat from the low-temperature area and vaporizes, becoming slightly superheated. The vaporized refrigerant is compressed by the compressor and in so doing, increases even more in temperature. The high-pressure, high-temperature refrigerant passes through the condenser coils, and because it is hotter than the high-temperature environment, it loses energy. Finally, the pressure is reduced in a throttling process in the expansion valve, where some of the liquid refrigerant also flashes into a vapor.

If the low-temperature area from which the heat is being removed is occupied space (i.e., air is being cooled), the device is known as an *air conditioner*. If the heat is being removed from water, the device is known as a *chiller*. An air conditioner produces cold air; a chiller produces cold water.

Rate of refrigeration (i.e., the rate at which heat is removed) is measured in *tons*. A ton of refrigeration corresponds to 200 Btu/min (12,000 Btu/hr) and 3516 W. The ton is derived from the heat flow required to melt a ton of ice in 24 hours.

Heat pumps also operate on refrigeration cycles. Like standard refrigerators, they transfer heat from low-temperature areas to high-temperature areas. The device shown in Fig. 19.1(b) could represent either a heat pump or a refrigerator. There is no significant difference in the mechanisms or construction of heat pumps and refrigerators; the only difference is the purpose of each.

The main function of a refrigerator is to cool the low-temperature area. The useful energy transfer of a refrigerator is the heat removed from the cold area. A heat pump's main function is to warm the high-temperature area. The useful energy transfer is the heat rejected to the high-temperature area.

Equation 19.18 and Eq. 19.19: Coefficient of Performance

$$COP = Q_H/W \quad \text{[heat pumps]} \qquad 19.18$$

$$COP = Q_L/W \quad \begin{bmatrix} \text{refrigerators and} \\ \text{air conditioners} \end{bmatrix} \qquad 19.19$$

Variation

$$COP_{\text{heat pump}} = COP_{\text{refrigerator}} + 1$$

Values

multiply	by	to obtain
tons of refrigeration	200	Btu/min
tons of refrigeration	12,000	Btu/hr
tons of refrigeration	3516	W

Description

The concept of thermal efficiency is not used with devices operating on refrigeration cycles. Rather, the *coefficient of performance* (COP) is defined as the ratio of useful energy transfer to the work input. The higher the coefficient of performance, the greater the effect for a given work input will be. Since the useful energy transfer is different for refrigerators and heat pumps, the coefficients of performance will also be different.

Example

A refrigeration cycle has a coefficient of performance of 2.2. The cycle exchanges heat between two infinite thermal reservoirs. For each 6 kW of cooling, what is most nearly the power input required?

(A) 1.4 kW

(B) 2.7 kW

(C) 5.5 kW

(D) 8.1 kW

Solution

From Eq. 19.19, the coefficient of performance of a refrigeration cycle is

$$COP = Q_L/W$$

$$W = \frac{Q_L}{COP} = \frac{6 \text{ kW}}{2.2}$$

$$= 2.73 \text{ kW} \quad (2.7 \text{ kW})$$

The answer is (B).

Equation 19.20 and Eq. 19.21: Carnot Refrigeration Cycle

$$COP_c = T_H/(T_H - T_L) \quad \text{[heat pumps]} \qquad 19.20$$

$$COP_c = T_L/(T_H - T_L) \quad \text{[refrigeration]} \qquad 19.21$$

Description

The *Carnot refrigeration cycle* is a Carnot power cycle running in reverse. Because it is reversible, the Carnot refrigeration cycle has the highest coefficient of performance for any given temperature limits of all the refrigeration cycles. As shown in Fig. 19.6, all processes occur within the vapor dome. The coefficients of performance for a Carnot refrigeration cycle, as given by Eq. 19.20 and Eq. 19.21, establish the upper limit of the COP.

As with Carnot power cycles, Eq. 19.20 and Eq. 19.21 are easily derived from Eq. 19.1 since $Q = T_{\text{reservoir}} \Delta S$ and ΔS are the same for the two heat transfer processes.

Figure 19.6 Carnot Refrigeration Cycle

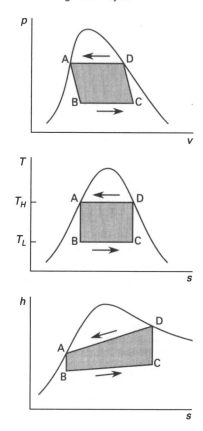

Example

A heat pump takes heat from groundwater at 7°C and maintains a room at 21°C. What is most nearly the maximum coefficient of performance possible for this heat pump?

(A) 1.4

(B) 2.8

(C) 5.6

(D) 21

Solution

The upper limit for the COP of a heat pump is set by the COP of a Carnot heat pump, as given by Eq. 19.20.

$$COP_c = T_H/(T_H - T_L)$$
$$= \frac{21°C + 273°}{(21°C + 273°) - (7°C + 273°)}$$
$$= 21$$

The answer is (D).

Equation 19.22 and Eq. 19.23: Vapor Refrigeration

$$COP_{ref} = \frac{h_1 - h_4}{h_2 - h_1} \qquad 19.22$$

$$COP_{HP} = \frac{h_2 - h_3}{h_2 - h_1} \qquad 19.23$$

Description

The components and processes of a *vapor refrigeration cycle*, also known as a *vapor compression cycle*, are shown in Fig. 19.7.[4] The coefficient of performance for the cycle used as refrigeration and as a heat pump are given by Eq. 19.22 and Eq. 19.23, respectively.[5]

Referring to Fig. 19.7, in the vapor compression cycle, cold liquid refrigerant in a saturated state passes through the *evaporator* and is vaporized while absorbing heat from the environment. In a refrigerator, this is referred to as the *cooling effect*, and it represents the

---

[4]The *NCEES Handbook* refers to the vapor refrigeration cycle as a "reversed rankine" cycle. Although any power cycle can theoretically be reversed to create a refrigeration cycle, a reversed Rankine steam refrigeration cycle is not even a theoretical likelihood. The Rankine steam cycle and the vapor refrigeration cycle both involve vaporization and condensation of the working fluid. Beyond that, however, the working fluids, equipment, magnitudes of heat transfers, and physical sizes are very different.

[5](1) The *NCEES Handbook* is inconsistent in its capitalization of subscripts, which can lead to confusion. (2) A high-pressure (or topping) turbine is usually referred to as an *HP turbine*. Although clear from context, the subscripts "ref" and "HP" in Eq. 19.22 and Eq. 19.23, respectively, are abbreviations for "refrigerator" and "heat pump," not "reference" and "high pressure."

Figure 19.7 *Vapor Refrigeration Cycle*

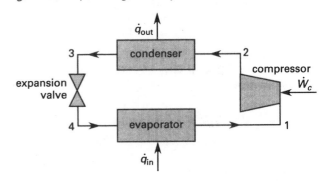

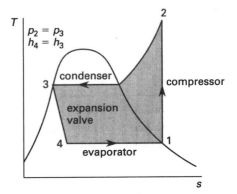

useful energy transfer for a refrigerator. The vaporized refrigerant is then compressed, usually in a reciprocating compressor. This is the process in which work, w_c, is performed on the refrigerant.[6] The refrigerant is heated by the compression, and this heat is removed in a *condenser*. The heat removed comes from two sources: (1) the heat absorbed in the low-temperature region, and (2) the work of compression.

For a heat pump, the heat rejected is the useful energy transfer. This is referred to as the *heating effect*. Finally, the pressure of the cooled vapor is reduced by throttling through an *expansion valve* (*throttling valve*). In the throttling process, pressure decreases, but enthalpy remains constant. Of course, the entropy increases in the throttling process.

Equation 19.24 and Eq. 19.25: Two-Stage Refrigeration Cycle

$$COP_{ref} = \frac{\dot{Q}_{in}}{\dot{W}_{in,1} + \dot{W}_{in,2}} = \frac{h_5 - h_8}{h_2 - h_1 + h_6 - h_5} \qquad 19.24$$

$$COP_{HP} = \frac{\dot{Q}_{out}}{\dot{W}_{in,1} + \dot{W}_{in,2}} = \frac{h_2 - h_3}{h_2 - h_1 + h_6 - h_5} \qquad 19.25$$

---

[6]The *NCEES Handbook* designates the compression power as $\dot{w}_c$. In this case, the subscript c refers to "compression," not to "Carnot" as it did in Eq. 19.20 and Eq. 19.21.

Description

In a two-stage refrigeration cycle, the condenser heat from the low-temperature cycle evaporates a (usually different) refrigerant in the evaporator of the high-temperature cycle. Among other advantages, a two-cycle refrigerator can operate with a greater temperature difference between the hot and cold reservoirs.

A plot of a *two-stage refrigeration cycle* on a *T-s* diagram is shown in Fig. 19.8.[7] The coefficient of performance of a two-stage refrigeration cycle and the heat pump shown are given by Eq. 19.24 and Eq. 19.25, respectively.[8]

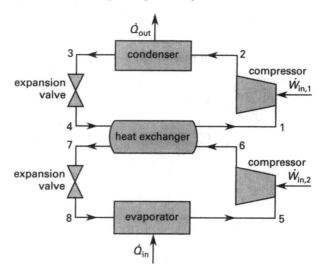

Figure 19.8 *Two-Stage Refrigeration Cycle*

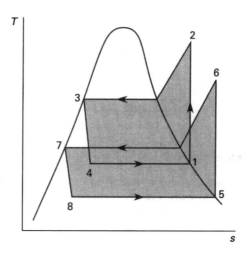

Equation 19.26 and Eq. 19.27: Air-Refrigeration Cycle

$$\text{COP}_{\text{ref}} = \frac{h_1 - h_4}{(h_2 - h_1) - (h_3 - h_4)} \qquad 19.26$$

$$\text{COP}_{\text{HP}} = \frac{h_2 - h_3}{(h_2 - h_1) - (h_3 - h_4)} \qquad 19.27$$

Description

An *air-refrigeration cycle* is essentially a *Brayton gas turbine cycle* operating in reverse.

An air-refrigeration cycle is shown in Fig. 19.9. The coefficient of performance of an air-refrigeration cycle and the heat pump shown are given by Eq. 19.26 and Eq. 19.27, respectively.

5. AVAILABILITY AND IRREVERSIBILITY

Consider some quantity of an "energized" substance. The potential for the substance to release its energy depends not only on the properties of the environment. For example, a hot billet of steel can give off its heat energy, but it can only cool to the temperature of the environment. Similarly, compressed air in a tire can only expand to the pressure of the atmosphere. A heated billet will release energy that is equal to its change in

internal energy, and pressurized air will perform boundary work. If the processes that release energy are reversible (that is, are without friction and heat losses), all of the energy released will all be available for useful work. This will be the maximum possible usefulness for the substance in the local environment.

Exergy is the term that describes energy release all the way down to the local environmental conditions. "Exergy" is synonymous with "availability." An equation used to calculate exergy is known as an *availability function*.

Equation 19.28 and Eq. 19.29: Closed-System Exergy (Availability) Function

$$\phi = (u - u_L) - T_L(s - s_L) + p_L(v - v_L) \qquad 19.28$$

$$w_{\text{max}} = w_{\text{rev}} = \phi_1 - \phi_2 \qquad 19.29$$

---

[7]The *NCEES Handbook* is not consistent in its representation of work and heat terms in power cycle and refrigeration cycle diagrams. For power cycles, the work and heat transfer terms are shown as lowercase w and q, while in refrigeration cycles, the same quantities are represented by W and Q. The same concepts are intended.

[8](1) The two forms of *NCEES Handbook* Eq. 19.24 and Eq. 19.25 are not entirely parallel. While the ratios yield the same numerical result, the numerators and denominators of each form do not refer to the same parameters. Since the numerator and denominator of the first term represent the total heat and work of a cycle, they are total properties with units of kJ (per second). The numerator and denominator of the second term represent the heat and work per unit mass of refrigerant. Based on the *NCEES Handbook*'s convention to use uppercase letters as variables for total properties and lowercase letters as variables for specific properties, it would be incorrect to assume that $\dot{Q}_{\text{in}} = h_5 - h_8$ as Eq. 19.24 suggests. (2) Furthermore, the first form of each equation is a ratio of energy per unit time, while the second form of each equation is a ratio of energy. Again, the ratios are numerically the same, but the numerators and denominators represent different parameters.

Figure 19.9 Air-Refrigeration Cycle

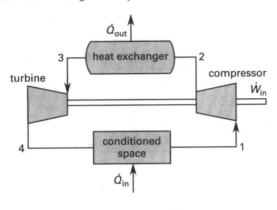

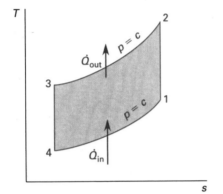

Description

The maximum possible work that can be obtained from a substance is known as the *availability*, ϕ. Availability is independent of the device but is dependent on the temperature of the local environment. Both the first and second law of thermodynamics must be applied to determine availability.

The availability function of a closed system per unit mass is defined by Eq. 19.28.[9] The subscript L refers to the temperature (and pressure) of the low-temperature reservoir (i.e., often, but not necessarily, the local environment), which define the limits of cooling and expansion.[10] Equation 19.29 calculates the maximum useful work from the starting and ending availability functions when the starting and ending conditions are not the local environment. For a process that reduces properties to the local environment (i.e., the final state is condition L), the reversible (i.e., maximum) work is

---

[9] The *NCEES Handbook* sometimes uses extraneous parentheses in its equations. As presented in the *NCEES Handbook*, there is no significance to the parentheses around the first two terms of Eq. 19.28, nor around the first two terms of Eq. 19.30.

[10] The *NCEES Handbook* says that the subscript L "designates environmental conditions...," which is misleading. The properties of the substance, not the environment, are to be used. In Eq. 19.28, u_L represents the internal energy of the substance at the temperature of the low temperature reservoir. It does not mean the internal energy of the low temperature reservoir substance. For steam exhausting to atmospheric pressure, u_L would not represent the internal energy of the atmosphere.

simply the availability calculated in Eq. 19.28, and Eq. 19.29 is not needed.

The availability of a closed system is the same as the change in Helmholtz function, without the effect of boundary work.

Equation 19.30 and Eq. 19.31: Open-System Exergy (Availability) Function

$$\Psi = (h - h_L) - T_L(s - s_L) + \mathrm{v}^2/2 + gz \qquad \textbf{19.30}$$

$$w_{\max} = w_{\mathrm{rev}} = \Psi_1 - \Psi_2 \qquad \textbf{19.31}$$

Description

For an open system, the steady-state availability function, Ψ, is given by Eq. 19.30. Equation 19.30 incorporates terms for kinetic and potential energy which cannot be extracted in a closed system. Equation 19.30 also combines the internal energy and flow work into an enthalpy term, since $h = u + pv$. The subscript L refers to the temperature (and pressure) of the low-temperature reservoir (i.e., often the local environment), which define the limits of cooling and expansion. Equation 19.31 calculates the maximum useful work from the starting and ending availability functions when the starting and ending conditions are not the local environment. For a device (e.g., a turbine) that is discharging directly to the local environment (i.e., the final state is condition L), the reversible (i.e., maximum) work is simply the availability calculated in Eq. 19.30, and Eq. 19.31 is not needed. The availability of a closed system is the same as the change in Helmholtz function, without the effect of boundary work.

The availability of an open system is the same as the change in Gibbs function, without the effects of kinetic and potential energies.

Equation 19.32: Irreversibility

$$I = w_{\mathrm{rev}} - w_{\mathrm{actual}} = T_L \Delta s_{\mathrm{total}} \qquad \textbf{19.32}$$

Description

To achieve the maximum work output, both the process within the control volume and the energy transfers between the system and environment must be reversible. The difference between the maximum and the actual work output is known as the *process irreversibility*, I. $\Delta s_{\mathrm{total}}$ in Eq. 19.32 represents the net entropy production, considering both the substance and the environment. For example, if heat energy were lost from a high-temperature substance, the entropy of that substance would decrease, but the entropy of the environment would increase by an even greater amount. The total would be a net increase in total entropy.

20 Mixtures of Gases, Vapors, and Liquids

Nomenclature

a	Helmholtz function	kJ/kg
$\hat{a}$	activity	–
A	molar Helmholtz function	kJ/kmol
c	specific heat	kJ/kg·K
C	number of components	–
C	molar heat	kJ/kg·kmol
f^L	fugacity of pure liquid	–
$\hat{f}^L$	fugacity in liquid phase	–
f^V	fugacity of pure vapor	–
$\hat{f}^V$	fugacity in vapor phase	–
F	degrees of freedom	–
g	Gibbs function	kJ/kg
G	molar Gibbs function	kJ/kmol
h	specific enthalpy	kJ/kg
h	Henry's law constant	atm
H	molar enthalpy	kJ/kmol
k	constant	–
K	chemical equilibrium constant	–
m	mass	kg
M	molecular weight	kg/kmol
N	number of moles	–
p	absolute pressure	Pa
P	number of phases	–
R	specific gas constant	kJ/kg·K
$\overline{R}$	universal gas constant, 8314	J/kmol·K
s	specific entropy	kJ/kg·K
S	molar entropy	kJ/kmol·K
T	absolute temperature	K
u	specific internal energy	kJ/kg
U	molar internal energy	kJ/kmol
v	stoichiometric coefficient	kmol
V	volume	m^3
x	mole fraction	–
x	volumetric fraction	–
y	mass fraction	–

Symbols

γ	activity coefficient	–
ξ	extent	mol
v	specific volume	m^3/kg
ϕ	relative humidity	%
Φ	fugacity coefficient	–
ω	humidity ratio	–

Subscripts

a	dry air
db	dry bulb
dp	dew point
fg	liquid-to-gas (vaporization)
g	saturation
i	component i
p	constant pressure
sat	saturation
v	constant volume
v	water vapor
wb	wet bulb
*	pure component

1. IDEAL GAS MIXTURES

Equation 20.1 Through Eq. 20.3: Mass Fraction

$$y_i = m_i/m \qquad \text{20.1}$$
$$m = \sum m_i \qquad \text{20.2}$$
$$\sum y_i = 1 \qquad \text{20.3}$$

Description

An ideal gas mixture consists of a mixture of ideal gases, each behaving as if it alone occupied the space.

The *mass fraction*, y_i (also known as the *gravimetric fraction* and *weight fraction*), of a component i in a mixture of components $i = 1, 2, \ldots, n$ is the ratio of the component's mass to the total mixture mass.[1]

Example

A 2 L container holds a mixture of three inert gases. The pressure inside the container is measured at 1.5 atm at a temperature of 293K (room temperature). Two of the three inert gases are known. The first gas, krypton, has a mass fraction of 0.352. The second gas, argon, has a mass fraction of 0.2799. The combined mass of argon

---

[1]In its thermodynamics section, the NCEES *FE Reference Handbook* (*NCEES Handbook*) uses the variable y to mean both mass fraction and mole fraction. It is common in engineering practice to designate mass fraction as w, g, G, M, and sometimes even x. In chemical engineering unit operations (mass transfer), it is common to use x and y both as mole fractions. In fact, the *NCEES Handbook* does this in its coverage of the thermodynamics of vapor-liquid equilibrium. Since the *NCEES Handbook* uses the same variable for two similar fractions, care must be observed when using equations (e.g., Henry's law) that use mass, volume, and mole fractions.

and krypton in the container is 5.630 g. What is most nearly the mass of the third gas?

(A) 2.5 g

(B) 3.1 g

(C) 3.3 g

(D) 5.6 g

Solution

The mass fractions of the three gases are related by $\sum y_i = 1$. This can be used to find the mass fraction of the third gas.

$$y_{\text{Ar}} + y_{\text{Kr}} + y_3 = 1$$
$$0.2799 + 0.352 + y_3 = 1$$
$$y_3 = 0.3681$$

The total mass can be found from the definition of mass fraction ($y_i = m_i/m$) and the combined masses of argon and krypton.

$$\frac{m_{\text{Ar}}}{m} + \frac{m_{\text{Kr}}}{m} = y_{\text{Ar}} + y_{\text{Kr}}$$
$$m = \frac{m_{\text{Ar}} + m_{\text{Kr}}}{y_{\text{Ar}} + y_{\text{Kr}}}$$
$$= \frac{5.630 \text{ g}}{0.2799 + 0.352}$$
$$= 8.910 \text{ g}$$

The mass of the third gas is

$$m_3 = y_3 m$$
$$= (0.3681)(8.910 \text{ g})$$
$$= 3.280 \text{ g} \quad (3.3 \text{ g})$$

The answer is (C).

Equation 20.4 Through Eq. 20.6: Mole Fractions

$$x_i = N_i/N \qquad 20.4$$
$$N = \sum N_i \qquad 20.5$$
$$\sum x_i = 1 \qquad 20.6$$

Description

The *mole fraction*, x_i, of a liquid component i is the ratio of the number of moles of substance i to the total number of moles of all substances in the mixture.

Since equal numbers of moles of any ideal gas occupy the same volume (i.e., approximately 22.4 L/mol at standard conditions), the mole fraction of a gas in a mixture of ideal gases is equal to its volumetric fraction. For chemical reactions that involve all gaseous components, the coefficients of the molecular species represent the number of molecules taking place in the reaction. Since each coefficient is some definite proportion of Avogadro's number, the coefficients also represent the numbers of moles (and, accordingly), the number of volumes of that gas taking part in the reaction.

Example

A 10 mole mixture of three gases is stored in a container. There are 2 moles of helium and 3 moles of nitrogen in the mixture. What is most nearly the mole fraction of the third gas in the mixture?

(A) 0.20

(B) 0.30

(C) 0.50

(D) 0.75

Solution

Use Eq. 20.5 to find the number of moles of the unknown gas in the mixture.

$$N_{\text{mixture}} = \sum N_i$$
$$= N_{\text{He}} + N_{\text{N}_2} + N_i$$
$$N_i = N_{\text{mixture}} - N_{\text{He}} - N_{\text{N}_2}$$
$$= 10 \text{ mol} - 2 \text{ mol} - 3 \text{ mol}$$
$$= 5 \text{ mol}$$

Use Eq. 20.4 to solve for the mole fraction of the unknown gas in the mixture.

$$x_i = \frac{N_i}{N_{\text{mixture}}} = \frac{5 \text{ mol}}{10 \text{ mol}}$$
$$= 0.50$$

The answer is (C).

Equation 20.7 Through Eq. 20.9: Converting Between Mass and Mole Fractions

$$y_i = \frac{x_i M_i}{\sum x_i M_i} \qquad 20.7$$
$$x_i = \frac{y_i/M_i}{\sum(y_i/M_i)} \qquad 20.8$$
$$M = m/N = \sum x_i M_i \qquad 20.9$$

Description

It is possible to convert from mole fraction to mass fraction (see Eq. 20.7) through the molecular weight of the component, M_i (see Eq. 20.9). Similarly, it is possible to convert from mass fraction to mole fraction (see Eq. 20.8).

Example

A gas mixture consisting of 4 moles of N_2, 2.5 moles of CO_2, and an unknown amount of CO is held at two times atmospheric pressure in a container with a volume of 0.13 m^3. The total number of moles in the mixture is 8.5. The temperature of the mixture is 100°C. What is most nearly the mass fraction of CO in the mixture?

(A) 0.10

(B) 0.20

(C) 0.48

(D) 0.53

Solution

Rearrange the equation for total moles in a mixture to find the number of moles of CO in the mixture.

$$N = \sum N_i = N_{N_2} + N_{CO_2} + N_{CO}$$
$$N_{CO} = N - N_{N_2} - N_{CO_2} = 8.5 \text{ mol} - 4 \text{ mol} - 2.5 \text{ mol}$$
$$= 2 \text{ mol}$$

Find the mole fraction of each compound in the mixture.

$$x_i = N_i/N$$
$$x_{N_2} = \frac{4 \text{ mol}}{8.5 \text{ mol}} = 0.471$$
$$x_{CO_2} = \frac{2.5 \text{ mol}}{8.5 \text{ mol}} = 0.294$$
$$x_{CO} = \frac{2 \text{ mol}}{8.5 \text{ mol}} = 0.235$$

Use Eq. 20.7 to find the mass fraction of CO in the mixture.

$$y_i = \frac{x_i M_i}{\sum x_i M_i}$$

$$y_{CO} = \frac{x_{CO} M_{CO}}{x_{N_2} M_{N_2} + x_{CO_2} M_{CO_2} + x_{CO} M_{CO}}$$

$$= \frac{(0.235)\left(28 \ \frac{g}{mol}\right)}{(0.471)\left(28 \ \frac{g}{mol}\right) + (0.294)\left(44 \ \frac{g}{mol}\right)}$$
$$+ (0.235)\left(28 \ \frac{g}{mol}\right)$$

$$= 0.201 \quad (0.20)$$

The answer is (B).

....

Equation 20.10 and Eq. 20.11: Partial Pressure and Dalton's Law

$$p_i = \frac{m_i R_i T}{V} \qquad 20.10$$

$$p = \sum p_i \qquad 20.11$$

Description

The *partial pressure*, p_i, of gas component i in a mixture of nonreacting gases $i = 1, 2, \ldots, n$ is the pressure gas i alone would exert in the total volume at the temperature of the mixture (see Eq. 20.10).

According to *Dalton's law of partial pressures*, the total pressure of a gas mixture is the sum of the partial pressures (see Eq. 20.11).

Example

Three identical rigid containers each store a separate element. The first container holds helium at 90 kPa, the second container holds neon at 120 kPa, and the third container holds xenon at 150 kPa. The contents of the neon and xenon containers are then pumped into the helium container. After the temperature has stabilized, what is most nearly the pressure of the mixture inside the helium container?

(A) 90 kPa

(B) 120 kPa

(C) 150 kPa

(D) 360 kPa

Solution

The partial pressure is the pressure a gas would have if it occupied the container by itself. Using Dalton's law of partial pressures, Eq. 20.11, solve for the total pressure of the mixture inside the container.

$$p = \sum p_i$$
$$= p_{He} + p_{Ne} + p_{Xe}$$
$$= 90 \text{ kPa} + 120 \text{ kPa} + 150 \text{ kPa}$$
$$= 360 \text{ kPa}$$

The answer is (D).

....

Equation 20.12 Through Eq. 20.14: Partial Volume and Amagat's Law

$$V_i = \frac{m_i R_i T}{p} \qquad 20.12$$

$$V = \sum V_i \qquad 20.13$$

$$x_i = p_i/p = V_i/V \qquad 20.14$$

Description

The *partial volume*, V_i, of gas i in a mixture of nonreacting gases is the volume that gas i alone would occupy at the temperature and pressure of the mixture (see Eq. 20.12).

Amagat's law (also known as *Amagat-Leduc's rule*) states that the total volume of a mixture of nonreacting gases is equal to the sum of the partial volumes (see Eq. 20.13).

Thermodynamics

For mixtures of nonreacting ideal gases, the mole fraction, x_i, partial pressure ratio, and volumetric fraction are the same (see Eq. 20.14).

Example

1 g of oxygen and 2 g of helium are mixed in a gas sampling bag. The gases are at 50°C and atmospheric pressure. What is most nearly the volume of the bag?

(A) 0.002 m³

(B) 0.007 m³

(C) 0.012 m³

(D) 0.014 m³

Solution

Find the partial volume of each component.

$$\overline{R} = \frac{8314 \, \dfrac{J}{kmol \cdot K}}{1000 \, \dfrac{J}{kJ}} = 8.314 \, \frac{kJ}{kmol \cdot K}$$

$$V_i = \frac{m_i R_i T}{p} = \frac{m_i \overline{R} T}{p M_i}$$

$$V_{O_2} = \frac{(1 \, g)\left(8.314 \, \dfrac{kJ}{kmol \cdot K}\right)(50°C + 273°)}{(101.3 \, kPa)\left(32 \, \dfrac{kg}{kmol}\right)\left(1000 \, \dfrac{g}{kg}\right)}$$

$$= 8.28 \times 10^{-4} \, m^3$$

$$V_{He} = \frac{(2 \, g)\left(8.314 \, \dfrac{kJ}{kmol \cdot K}\right)(50°C + 273°)}{(101.3 \, kPa)\left(4.00 \, \dfrac{kg}{kmol}\right)\left(1000 \, \dfrac{g}{kg}\right)}$$

$$= 1.33 \times 10^{-2} \, m^3$$

Use Amagat's law to calculate the total volume of the bag.

$$V = \sum V_i = V_{O_2} + V_{He}$$
$$= 8.28 \times 10^{-4} \, m^3 + 1.33 \times 10^{-2} \, m^3$$
$$= 0.0141 \, m^3 \quad (0.014 \, m^3)$$

The answer is (D).

Equation 20.15 Through Eq. 20.17: Gibbs Theorem

$$u = \sum y_i u_i \qquad \text{20.15}$$

$$h = \sum y_i h_i \qquad \text{20.16}$$

$$s = \sum y_i s_i \qquad \text{20.17}$$

Description

Equation 20.15 through Eq. 20.17 are mathematical formulations of *Gibbs theorem* (also known as *Gibbs rule*). This theorem states that the total property (e.g., u, h, or s) of a mixture of ideal gases is the sum of the properties that the individual gases would have if each occupied the total mixture volume alone at the same temperature. Equation 20.17 is stated as an equality, and this requires the mixing of components to be isentropic (i.e., adiabatic and reversible). Each component's entropy must be evaluated at the temperature of the mixture and its partial pressure.[2]

While the *specific* mixture properties (i.e., u, h, s, c_p, c_v, and R) are all gravimetrically weighted, the *molar* properties are not. Molar U, H, S, C_p, and C_v, as well as the molecular weight and mixture density, are all volumetrically weighted.

Example

A mixture contains 50 g of nitrogen and 25 g of carbon dioxide. The gases are stored in a container at 500K. At this temperature, the molar enthalpy of the nitrogen is 5912 J/mol, and the molar enthalpy of the carbon dioxide is 8314 J/mol. What is most nearly the total enthalpy of the mixture?

(A) 5.7 kJ

(B) 8.3 kJ

(C) 11 kJ

(D) 15 kJ

Solution

Although Eq. 20.16 could be used, it would be necessary to calculate the specific enthalpies ($h = H/M$). It is easier to recognize that molar properties of mixtures are volumetrically weighted, and that volumetric fractions of ideal gases are the same as mole fractions. The numbers of moles are

$$N_i = \frac{m_i}{M_i}$$

$$N_{N_2} = \frac{50 \, g}{28 \, \dfrac{g}{mol}} = 1.79 \, mol$$

$$N_{CO_2} = \frac{25 \, g}{44 \, \dfrac{g}{mol}} = 0.57 \, mol$$

---

[2]The *NCEES Handbook* states that u_i and h_i are evaluated "…at T," referring to the temperature of the mixture, while s_i is evaluated "…at T and p_i." Internal energy is indeed a function of only temperature. However, enthalpy depends on the pressure, also, since $h = u + pv$. For a fixed mass of gas occupying a fixed volume, specifying the temperature is sufficient to establish the pressure. So, although the value of pressure is needed to determine enthalpy, it is necessary only to specify temperature if the volume is known. Pressure can be calculated from the equation of state.

Thermodynamics

The mole (volumetric) fractions are

$$x_{N_2} = \frac{N_{N_2}}{N_{N_2} + N_{CO_2}} = \frac{1.79 \text{ mol}}{1.79 \text{ mol } + 0.57 \text{ mol}}$$
$$= 0.759$$

$$x_{CO_2} = 1 - x_{N_2} = 1 - 0.759 = 0.241$$

The molar enthalpy of the mixture is weighted by the mole (volumetric) fractions.

$$H = \sum x_i H_i = (0.759)\left(5912 \frac{J}{mol}\right) + (0.241)\left(8314 \frac{J}{mol}\right)$$
$$= 6492 \text{ J/mol}$$

The total enthalpy of the mixture is

$$H_{total} = NH = \frac{(1.79 \text{ mol} + 0.57 \text{ mol})\left(6492 \frac{J}{mol}\right)}{1000 \frac{J}{kJ}}$$
$$= 15.3 \text{ kJ} \quad (15 \text{ kJ})$$

The answer is (D).

2. VAPOR-LIQUID MIXTURES[3]

Equation 20.18: Henry's Law at Constant Temperature

$$p_i = py_i = hx_i \qquad \textit{20.18}$$

Description

Henry's law states that the partial pressure of a slightly soluble gas above a liquid is proportional to the amount (i.e., mole fraction, x_i) of the gas dissolved in the liquid. This law applies separately to each gas to which the liquid is exposed, as if each gas were present alone. The algebraic form of Henry's law is given by Eq. 20.18, in which h is the *Henry's law constant* with units of pressure.[4]

It is important to recognize that, in Eq. 20.18, x_i is the mole fraction of the gas (solute) in the liquid (solvent), while y_i is the mole fraction of the gas (solute) in the gas mixture above the liquid.

[3]In comparison to the thousands of important and practical thermodynamics facts, principles, laws, and applications, fugacity and the related concept of activity are too esoteric, and the nomenclature sufficiently obtuse, to warrant much attention.
[4](1) *Henry's law* has four different practical formulations, with four different definitions (and values) of Henry's law constant. All of these formulations are in engineering use. It is important to use the Henry's law constant that has units matching the form of the law. (2) The *NCEES Handbook* presents Eq. 20.18 as Henry's law. In fact, only the $p_i = hx_i$ part is Henry's law. The $p_i = y_i p_{total}$ part is not Henry's law, although it is easily derived from the ideal gas laws. If it is associated with anything, $p_i = y_i p_{total}$ is usually presented in conjunction with *Dalton's law of partial pressures* ($p_{total} = \sum p_i = \sum x_i p_{total}$, where x is traditionally used to designate the mole fraction).

Equation 20.19: Raoult's Law for Vapor-Liquid Equilibrium

$$p_i = x_i p_i^* \qquad \textit{20.19}$$

Description

Vapor pressure is the pressure exerted by the solvent's vapor molecules when they are in equilibrium with the liquid. The symbol for the vapor pressure of a pure vapor over a pure solvent at a particular temperature is p_i^*. Vapor pressure increases with increasing temperature.

Raoult's law, given by Eq. 20.19, states that the vapor pressure, p_i, of a solvent is proportional to the mole fraction of that substance in the solution.

According to Raoult's law, the partial pressure of a solution component will increase with increasing temperature (as p_i^* increases) and with increasing mole fraction of that component in the solution. Raoult's law applies to each of the substances in the solution. By *Dalton's law* (see Eq. 20.11) the total vapor pressure above the liquid is equal to the sum of the vapor pressures of each component.

Example

A nonvolatile, nonelectrolytic liquid is combined with a solid to form a solution that just boils at 1 atm pressure. The vapor pressure of the pure liquid is 850 torr. Most nearly, what is the molar percentage of the liquid in the solution?

(A) 64%

(B) 79%

(C) 86%

(D) 89%

Solution

A liquid boils when its vapor pressure equals the pressure of its surroundings. The vapor pressure of the solution is 1 atm or 760 torr. From Raoult's law,

$$p_i = x_i p_i^*$$

$$x_{solvent} = \frac{p_{solution}}{p_{pure\,solvent}} = \frac{760 \text{ torr}}{850 \text{ torr}}$$
$$= 0.894 \quad (89\%)$$

The answer is (D).

Equation 20.20 Through Eq. 20.23: Vapor-Liquid Equilibrium

$$\hat{f}_i^V = \hat{f}_i^L \qquad \textit{20.20}$$
$$\hat{f}_i^L = x_i \gamma_i f_i^L \qquad \textit{20.21}$$
$$\hat{f}_i^L = x_i k_i \qquad \textit{20.22}$$
$$\hat{f}_i^V = y_i \hat{\Phi}_i p \qquad \textit{20.23}$$

Description

Fugacity, sometimes called *actual fugacity*, is an effective pressure which replaces the actual pressure of a real gas in precise chemical equilibrium computations.[5] If real substances (primarily gases) behaved ideally, fugacity would not be required. But, just as the ideal equation of state is replaced with corrected equations for pressure-volume-temperature problems, fugacity replaces actual pressure in vapor-liquid equilibrium calculations. Fugacity can also be used to derive the real gas compressibility factor, although that is not its primary function.

Fugacity, f, is calculated from the actual pressure and the *fugacity coefficient*, Φ. A "hat" (e.g., $\hat{f}$) is used when the component is in a mixture, while the "unhatted" character designates a pure substance.

$$f = \Phi p_{\text{actual}}$$

For any actual pressure, the corresponding fugacity will produce calculated results that match observed results. The specific parameter that fugacity is designed to match (preserve, ensure, obtain, produce, etc.) is the *chemical potential*, μ. Chemical potential is an abstract concept based on Gibbs free energy, another abstract concept. Using the superscript "0" to indicate the standard reference condition of 25°C and either 1 atm (101.325 kPa), 1 bar (100 kPa), or 760 torr, a practical formula that is used to represent chemical potential is

$$\mu = \mu^0 + RT \ln \frac{p}{p^0}$$

Fugacity values, f, of pure substances are experimentally derived such that the parallel formula results in the same chemical potential.

$$\mu^0 + RT \ln \frac{f}{f^0} = \mu^0 + RT \ln \frac{p}{p^0}$$

A multicomponent vapor-liquid system is in equilibrium if the fugacities of each component's liquid and vapor phase are equal (see Eq. 20.20). The fugacity of liquid and vapor components can usually be calculated using Eq. 20.21 and Eq. 20.23, respectively. f represents the fugacity of a pure substance, while $\hat{f}$ represents the fugacity of the substance in a mixture. $\hat{f}^L$ represents the fugacity of the substance in a mixture in the liquid phase, and $\hat{f}^V$ represents the fugacity of the substance in a vapor phase.

For slightly soluble gases, the fugacity of the liquid phase of a component can be calculated using Eq. 20.22. The *activity coefficient* of component i, γ_i, in a multicomponent system is a correction factor for non-ideal behavior of a component in the liquid phase.

Equation 20.24 and Eq. 20.25: Activity Coefficients of a Two-Component System

$$\ln \gamma_1 = A_{12}\left(1 + \frac{A_{12}x_1}{A_{21}x_2}\right)^{-2} \qquad 20.24$$

$$\ln \gamma_2 = A_{21}\left(1 + \frac{A_{21}x_2}{A_{12}x_1}\right)^{-2} \qquad 20.25$$

Description

Equation 20.24 and Eq. 20.25 are derived from the *van Laar model* and can be used to calculate the *activity coefficients* for a two-component system.[6] The constants A_{12} and A_{21} are determined empirically in practice.

Equation 20.26 and Eq. 20.27: Fugacity of a Pure Liquid

$$f_i^L = \Phi_i^{\text{sat}} p_i^{\text{sat}} \exp\{v_i^L(p - p_i^{\text{sat}})/(RT)\} \qquad 20.26$$

$$f_i^L \cong p_i^{\text{sat}} \qquad 20.27$$

Description

The fugacity of a pure liquid component, f_i^L, is given by Eq. 20.26.[7] Similar to the activity coefficient, the *fugacity coefficient* of component i, Φ_i, is an empirically determined value used to correct non-ideal behavior of the vapor-phase component. If pressures are near atmospheric, then the fugacity of a pure liquid component can often be approximated by Eq. 20.27.

3. PSYCHROMETRICS

The study of the properties and behavior of atmospheric air is known as *psychrometrics*. Properties of air are seldom evaluated from theoretical thermodynamic principles, however. Specialized techniques and charts have been developed for that purpose.

Equation 20.28: Total Atmospheric Pressure

$$p = p_a + p_v \qquad 20.28$$

Description

Air in the atmosphere contains small amounts of moisture and can be considered to be a mixture of two ideal gases—dry air and water vapor. All of the thermodynamic rules relating to the behavior of nonreacting gas mixtures apply to atmospheric air. From Dalton's law, for example, the total atmospheric pressure is the sum of the dry air partial pressure and the water vapor pressure. (See Eq. 20.28.)

---

[5]For all of its complexity and elegance, fugacity is just the pressure you have to use in certain kinds of problems in order to get the correct answer. It is an ideal (partial) pressure that has been corrected for real partial pressure behavior. The corrections are empirical, based on experimentation. For that reason, fugacity (like an equations of state for a real gas) is basically a fancy correlation, not a basic engineering principle.

[6]The van Laar model is one of many correlations that are used for fitting observed data into activity coefficients. As with all correlations, the form of the equation has been selected to provide the best data fit. As such, Eq. 20.24 and Eq. 20.25 are useful, but they are correlations, not engineering fundamentals.

[7]The saturation pressure that is represented by p^{sat} in Eq. 20.26 is the same as p_{sat} throughout the rest of the *NCEES Handbook*.

Thermodynamics

Example

An air-water vapor mixture is stored in a fixed-volume container at a temperature of 45°C. The total atmospheric pressure of the mixture is 1.13 atm, and the relative humidity of the mixture is 0.22. The partial pressure of the water vapor is 2.110 kPa. What is most nearly the partial pressure of the dry air in the mixture?

(A) 103 kPa

(B) 105 kPa

(C) 108 kPa

(D) 110 kPa

Solution

Rearrange Eq. 20.28 for total atmospheric pressure to find the partial pressure of the dry air.

$$p = p_a + p_v$$

$$p_a = p - p_v = (1.13 \text{ atm})\left(101.325 \ \frac{\text{kPa}}{\text{atm}}\right) - 2.110 \text{ kPa}$$

$$= 112.4 \text{ kPa} \quad (110 \text{ kPa})$$

The answer is (D).

Equation 20.29: Dew-Point Temperature

$$T_{\text{dp}} = T_{\text{sat}} \quad [p_g = p_v] \qquad \textbf{20.29}$$

Description

Psychrometrics uses three different definitions of temperature. These three terms are *not* interchangeable.

- *dry-bulb temperature, T_{db}:* This is the temperature that a regular thermometer measures if exposed to air.

- *wet-bulb temperature, T_{wb}:* This is the temperature of air that has gone through an adiabatic saturation process. It is measured with a thermometer that is covered with a water-saturated cotton wick.

- *dew-point temperature, T_{dp}:* This is the dry-bulb temperature at which water starts to condense when moist air is cooled in a constant pressure process. The dew-point temperature is equal to the saturation temperature (read from steam tables) for the partial pressure of the vapor.

For every temperature, there is a unique equilibrium vapor pressure of water, p_g, called the *saturation pressure*. If the vapor pressure equals the saturation pressure, the air is said to be saturated. *Saturated air* is a mixture of dry air and water vapor at the saturation pressure. When the air is saturated, all three temperatures are equal.

Unsaturated air is a mixture of dry air and superheated water vapor. When the air is unsaturated, the dew-point temperature will be less than the wet-bulb temperature.

$$T_{\text{dp}} < T_{\text{wb}} < T_{\text{db}} \quad [\text{unsaturated}]$$

Equation 20.30 and Eq. 20.31: Specific Humidity (Humidity Ratio, Absolute Humidity)

$$\omega = m_v/m_a \qquad \textbf{20.30}$$

$$\omega = 0.622 p_v/p_a = 0.622 p_v/(p - p_v) \qquad \textbf{20.31}$$

Description

The amount of water in atmospheric air is specified by the *humidity ratio* (also known as the *specific humidity*), ω. The humidity ratio is the mass ratio of water vapor to dry air. If both masses are expressed in pounds (kilograms), the units of ω are lbm/lbm (kg/kg). However, since there is so little water vapor, the water vapor mass is often reported in *grains* or grams. (There are 7000 grains per pound.) Accordingly, the humidity ratio may have the units of grains per pound or grams per kg. The humidity ratio is expressed as Eq. 20.30 or Eq. 20.32.

The humidity ratio is expressed per pound of dry air, not per pound of the total mixture. Equation 20.31 is derived from the ideal gas law, and 0.622 is the ratio of the specific gas constants for air and water vapor.

Example

One method of removing moisture from air is to cool the air so that the moisture condenses out. What is most nearly the temperature to which air at 100 atm must be cooled at constant pressure in order to obtain a humidity ratio of 0.0001?

(A) −6.0°C

(B) 2.0°C

(C) 8.0°C

(D) 14°C

Solution

Water vapor will condense when it is cooled to its dew-point temperature, which is the same as its saturation temperature. Rearranging Eq. 20.31,

$$\omega = 0.622 p_v/(p - p_v)$$

$$p_v = \left(\frac{\omega}{0.622}\right)(p - p_v) = \frac{\dfrac{\omega p}{0.622}}{1 + \dfrac{\omega}{0.622}}$$

$$= \frac{(0.0001)(100 \text{ atm})\left(101.35 \ \frac{\text{kPa}}{\text{atm}}\right)}{0.622}}{1 + \dfrac{0.0001}{0.622}}$$

$$= 1.629 \text{ kPa}$$

From the steam tables, at 1.629 kPa,

$$T_{\text{sat}} \approx 14°\text{C}$$

The answer is (D).

Thermodynamics

Equation 20.32: Relative Humidity Ratio

$$\phi = p_v/p_g \qquad 20.32$$

Description

The *relative humidity*, ϕ, is a second index of moisture content of air. The relative humidity is the partial pressure of the water vapor divided by the saturation pressure at the dry-bulb temperature.

Example

Atmospheric air at $21°C$ has a relative humidity of 50%. What is most nearly the dew-point temperature?

(A) $7.0°C$

(B) $10°C$

(C) $17°C$

(D) $24°C$

Solution

At $21°C$, the saturation pressure of water is

$$p_g = 2.505 \text{ kPa}$$

From Eq. 20.32, the vapor pressure is

$$\phi = p_v/p_g$$
$$p_v = \phi p_g = (0.5)(2.505 \text{ kPa}) = 1.2525 \text{ kPa}$$

The dew-point temperature is the saturation temperature corresponding to the vapor pressure conditions. From the steam tables at 1.2525 kPa,

$$T_{\text{sat}} = T_{\text{dp}} \approx 10°C$$

The answer is (B).

4. PSYCHROMETRIC CHART

It is possible to develop mathematical relationships for enthalpy and specific volume (the two most useful thermodynamic properties) for atmospheric air. However, these relationships are almost never used. Rather, psychrometric properties are read directly from psychrometric charts, as illustrated in Fig. 20.1 and Fig. 20.2.

A psychrometric chart is easy to use, despite the multiplicity of scales. The thermodynamic state (i.e., the position on the chart) is defined by specifying the values of any two parameters on intersecting scales (e.g., dry-bulb and wet-bulb temperature, or dry-bulb temperature and relative humidity). Once the state has been located on the chart, all other properties can be read directly.

Equation 20.33: Enthalpy of Dry Air

$$h = h_a + \omega h_v \qquad 20.33$$

Variation

$$h_{\text{total}} = m_a h \qquad \begin{bmatrix} \text{for any mass} \\ \text{of dry air} \end{bmatrix}$$

Description

Enthalpy of an air-vapor mixture is the sum of the enthalpies of the air and water vapor. Enthalpy of the mixture per pound of dry air can be read from the psychrometric chart. Equation 20.33 computes enthalpy per pound of dry air. The variation equation illustrates an important point: In psychrometrics, the basis for the total enthalpy of air (an air-water vapor mixture) is the mass of the dry air only (i.e., $h_{\text{total}} = h m_{\text{air}}$), not the total air mass (i.e., not $h_{\text{total}} = h(m_{\text{air}} + m_{\text{vapor}})$).

Example

Atmospheric air has a humidity ratio of 0.008 kg/kg and a dry-bulb temperature of $30°C$. What is most nearly the enthalpy of the air?

(A) 15 kJ/kg

(B) 22 kJ/kg

(C) 30 kJ/kg

(D) 50 kJ/kg

Solution

Using Fig. 20.1, locate the humidity ratio of 0.008 kg/kg along the vertical axis on the right side of the diagram. Move left along the horizontal grid line until the line for the humidity ratio intersects with the vertical grid line for a dry-bulb temperature of $30°C$. Find the diagonal line for enthalpy that intersects with this point. The enthalpy of the air is approximately 50 kJ/kg.

The answer is (D).

Figure 20.1 *ASHRAE Psychrometric Chart No. 1 (SI units)*

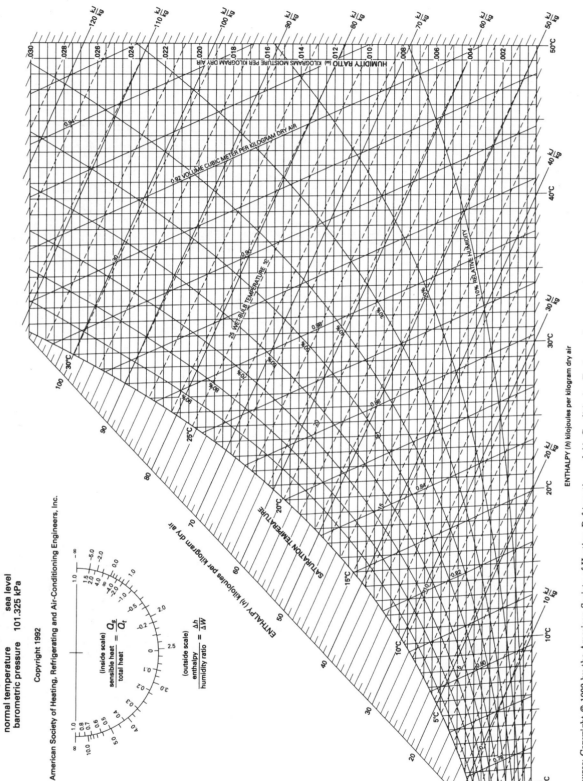

Source: Copyright © 1992 by the American Society of Heating, Refrigeration and Air-Conditioning Engineers, Inc.

Figure 20.2 *ASHRAE Psychrometric Chart No. 1 (customary U.S. units)*

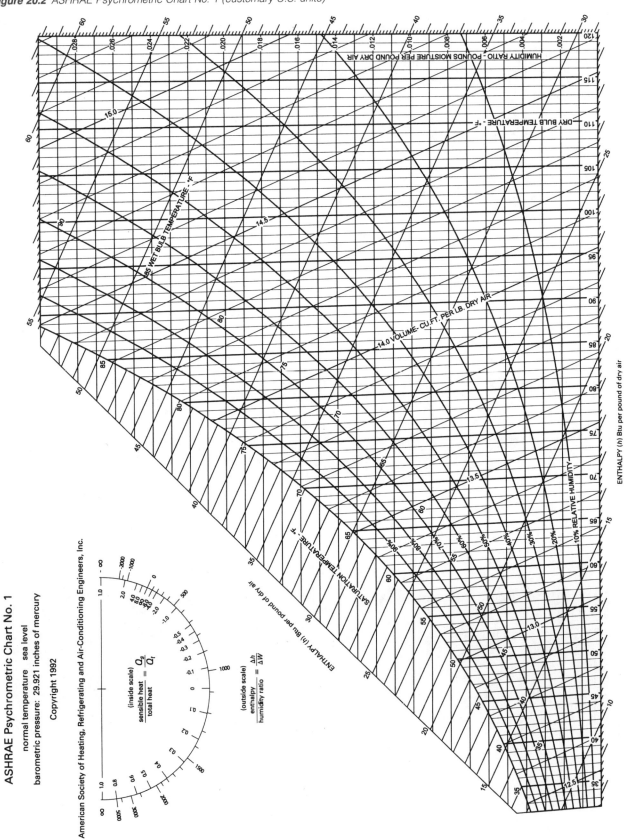

Diagnostic Exam

Topic V: Chemistry

1. A 100 A current passes through a solution of pure water. The electrolysis of water proceeds according to the reaction shown.

$$H_2O \rightarrow H_2 + \left(\tfrac{1}{2}\right)O_2$$
$$\Delta H_f = 285.830 \text{ kJ/mol}$$

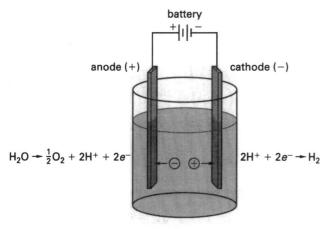

Assuming all of the energy goes into the reaction, what is most nearly the electrical power required to produce oxygen gas at a rate of 50 mg/s?

(A) 0.89 kW

(B) 1.5 kW

(C) 3.1 kW

(D) 9.2 kW

2. 25 mL of a 0.01 mol/L sample of hydrochloric acid (HCl) is slowly titrated with a 0.05 mol/L solution of sodium hydroxide (NaOH). What volume of sodium hydroxide should be used?

(A) 0.50 mL

(B) 2.5 mL

(C) 5.0 mL

(D) 25 mL

3. What type of process does sodium metal experience in the formation of table salt from the reaction shown?

$$2Na(s) + Cl_2(g) \rightarrow 2NaCl(s)$$

(A) reduction

(B) replacement

(C) oxidation

(D) decomposition

4. What is the ideal volume of 49 g of chlorine gas at standard temperature and pressure (STP)?

(A) 11 L

(B) 15 L

(C) 22 L

(D) 31 L

5. A 100 A current passes through a solution of pure water. The electrolysis of water proceeds according to the reaction shown.

$$H_2O \rightarrow H_2 + \tfrac{1}{2}O_2$$
$$\Delta H_f = 285.830 \text{ kJ/mol}$$

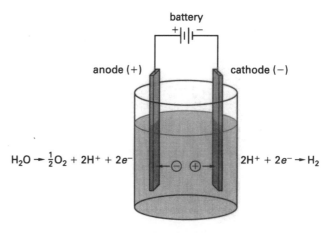

At what rate is oxygen produced?

(A) 8.3 mg/s

(B) 9.3 mg/s

(C) 17 mg/s

(D) 19 mg/s

6. What is a distinguishing characteristic of halogens?

(A) They are phosphorescent.

(B) Next to the noble gases, they are the most chemically inactive group.

(C) They readily accept an electron from another atom to form compounds.

(D) They have a high electrical conductivity.

7. What is the valence (oxidation state) of carbon in sodium carbonate (Na_2CO_3)?

(A) -4

(B) -2

(C) $+2$

(D) $+4$

8. Assuming 100% ionization, most nearly what mass of lead nitrate, $Pb(NO_3)_2$, must be dissolved in 1 L of pure water to produce a solution that contains 20 mg of lead ions?

(A) 26 mg

(B) 32 mg

(C) 43 mg

(D) 52 mg

9. Most nearly, what is the mass of 0.01 gram-moles of Na_2SO_4?

(A) 0.71 g

(B) 1.2 g

(C) 1.4 g

(D) 2.4 g

10. A compound contains 68.94% oxygen and 31.06% of an unknown element by weight. The molecular weight of this compound is 69.7 g/mol. What is this compound?

(A) NO_2

(B) F_2O_2

(C) B_2O_3

(D) SiO_4

SOLUTIONS

1. The number of moles of oxygen gas (O_2) produced is

$$n_{O_2} = \frac{m}{MW} = \frac{50 \times 10^{-3} \frac{g}{s}}{32 \frac{g}{mol}}$$

$$= 0.0015625 \text{ mol/s}$$

The coefficients in the reaction equation can be interpreted as the number of moles. Each mole of oxygen gas produced requires twice as many moles of water. Therefore, the number of moles of water dissociated per second is

$$n_{H_2O} = (2)\left(0.0015625 \frac{mol}{s}\right) = 0.003125 \text{ mol/s}$$

The power required can be calculated from the enthalpy of reaction.

$$P = n_{H_2O} \Delta H_f$$

$$= \left(0.003125 \frac{mol}{s}\right)\left(285.83 \frac{kJ}{mol}\right)$$

$$= 0.893 \text{ kW} \quad (0.89 \text{ kW})$$

The answer is (A).

2. Use the following chemical equations.

$$HCl \rightarrow H^+ + Cl^-$$

$$NaOH \rightarrow Na^+ + OH^-$$

The number of moles of OH^- equals the number of moles of H^+.

Use the following equation to obtain the number of moles of H^+.

$$n_{H^+} = M_{HCl} V_{HCl}$$

$$= \left(0.01 \frac{mol}{L}\right)(0.025 \text{ L})$$

$$= 2.5 \times 10^{-4} \text{ mol}$$

Use the following equation to obtain the number of moles of OH^-.

$$n_{OH^-} = M_{NaOH} V_{NaOH}$$

Rearranging to solve for the volume of NaOH,

$$V_{NaOH} = \frac{n}{M} = \frac{2.5 \times 10^{-4} \text{ mol}}{0.05 \frac{mol}{L}}$$

$$= 0.0050 \text{ L} \quad (5.0 \text{ mL})$$

The answer is (C).

3. Assign oxidation numbers as follows.

$$\overset{0}{2Na} + \overset{0}{Cl_2} \rightarrow \overset{+1}{2Na} + \overset{-1}{2Cl}$$

Sodium goes from a neutral state of charge to a positive state of charge. This is oxidation.

The answer is (C).

4. One mole of any ideal gas at STP occupies 22.4 L. The molecular weight of chlorine gas is

$$MW_{Cl_2} = (2)(AW_{Cl}) = (2)\left(35.453 \ \frac{g}{mol}\right)$$

$$= 70.906 \ g/mol$$

$$V = \left(\frac{m_{Cl_2}}{MW_{Cl_2}}\right)\left(22.4 \ \frac{L}{mol}\right)$$

$$= \left(\frac{49 \ g}{70.906 \ \frac{g}{mol}}\right)\left(22.4 \ \frac{L}{mol}\right)$$

$$= 15.48 \ L \quad (15L)$$

The answer is (B).

5. The reaction that occurs at the cathode is

$$O^{-2} - 2e^- \rightarrow \tfrac{1}{2}O_2$$

The molecular weight of oxygen gas is

$$MW_{O_2} = (2)\left(16 \ \frac{g}{mol}\right) = 32 \ g/mol$$

From Faraday's law, one gram equivalent weight of water is dissociated at the anode for each faraday, or 96 485 C, of electricity passed through the solution. Four electrons are gained for each oxygen (O_2) molecule.

$$m_{grams} = \frac{It(MW)}{(96\,485)(\text{change in oxidation state})}$$

$$\dot{m} = \frac{m_{grams}}{t}$$

$$= \frac{I(MW)}{(96\,485)(\text{change in oxidation state})}$$

$$= \frac{(100 \ A)\left(32 \ \frac{g}{mol}\right)}{\left(96\,485 \ \frac{A\cdot s}{mol}\right)(4)}$$

$$= 8.29 \times 10^{-3} \ g/s \quad (8.3 \ mg/s)$$

The answer is (A).

6. Halogens need one electron to complete an electron shell. They readily accept electrons from other atoms to form compounds.

The answer is (C).

7. Since sodium carbonate is a neutral compound, the sum of the oxidation numbers is zero. Oxygen has an oxidation number of -2, and sodium has an oxidation number of $+1$. The oxidation number of carbon is

$$(2)(+1) + x + (3)(-2) = 0$$

$$x = +4$$

The answer is (D).

8. The combining weights of each element in the molecule are

$$Pb: (1)\left(207.2 \ \frac{g}{mol}\right) = 207.2 \ g/mol$$

$$N: (2)\left(14.007 \ \frac{g}{mol}\right) = 28.01 \ g/mol$$

$$O: (6)\left(15.999 \ \frac{g}{mol}\right) = 95.99 \ g/mol$$

The molecular weight of $Pb(NO_3)_2$ is

$$207.2 \ \frac{g}{mol} + 28.01 \ \frac{g}{mol} + 95.99 \ \frac{g}{mol} = 331.2 \ g/mol$$

Calculate the gravimetric fraction of Pb in $Pb(NO_3)_2$.

$$x_{Pb} = \frac{m_{Pb}}{m_{Pb(NO_3)_2}} = \frac{207.2 \ \frac{g}{mol}}{331.2 \ \frac{g}{mol}}$$

$$= 0.6256$$

The mass of $Pb(NO_3)_2$ required is

$$m_{Pb(NO_3)_2} = \frac{20 \ mg}{0.6256}$$

$$= 31.97 \ mg \quad (32 \ mg)$$

The answer is (B).

9. The combining weights of each element are

$$Na: (2)\left(22.990 \ \frac{g}{mol}\right) = 45.98 \ g/mol$$

$$S: (1)\left(32.066 \ \frac{g}{mol}\right) = 32.07 \ g/mol$$

$$O: (4)\left(15.999 \ \frac{g}{mol}\right) = 64.00 \ g/mol$$

Chemistry

The molecular weight of Na_2SO_4 is

$$45.98 \ \frac{g}{mol} + 32.07 \ \frac{g}{mol} + 64.00 \ \frac{g}{mol} = 142 \ g/mol$$

The mass of Na_2SO_4 is

$$(0.01 \ mol)\left(142 \ \frac{g}{mol}\right) = 1.42 \ g \quad (1.4 \ g)$$

The answer is (C).

10. An analytical procedure results in two equations with three unknowns: the unknown element's atomic weight, the number of unknown atoms, and the number of oxygen atoms. Some deductive reasoning can be used to solve this problem, since (by the law of definite proportions) the number of unknown element atoms per oxygen atom is a (small) whole number (e.g., 1, 2, 3, etc.). However, it is easier to calculate the gravimetric percentage of oxygen in the answer options given.

$$NO_2: \ G_O = \frac{combining \ weight}{total \ weight}$$

$$= \frac{(2)\left(15.999 \ \frac{g}{mol}\right)}{14.007 \ \frac{g}{mol} + (2)\left(15.999 \ \frac{g}{mol}\right)}$$

$$= 0.6955$$

$$F_2O_2: \ G_O = \frac{(2)\left(15.999 \ \frac{g}{mol}\right)}{(2)\left(18.998 \ \frac{g}{mol}\right) + (2)\left(15.999 \ \frac{g}{mol}\right)}$$

$$= 0.4571$$

$$B_2O_3: \ G_O = \frac{(3)\left(15.999 \ \frac{g}{mol}\right)}{(2)\left(10.811 \ \frac{g}{mol}\right) + (3)\left(15.999 \ \frac{g}{mol}\right)}$$

$$= 0.6894$$

$$SiO_4: \ G_O = \frac{(4)\left(15.999 \ \frac{g}{mol}\right)}{28.086 \ \frac{g}{mol} + (4)\left(15.999 \ \frac{g}{mol}\right)}$$

$$= 0.6950$$

B_2O_3 has an oxygen gravimetric fraction of 68.9%, which coincides with the analysis of the unknown compound.

The answer is (C).

21 Inorganic Chemistry

Nomenclature

A	atomic weight	kg/kmol
E	potential	V
EW	equivalent weight	kg/kmol
F	Faraday constant, 96 485	C/mol or A·s/mol
FW	formula weight	kg/kmol
h	Henry's law constant	atm/mole fraction
H	enthalpy	kJ/kmol
k	reaction rate constant	–
K	constant	–
m	mass	kg
m	molality	g/1000 g
M	molarity	mol/L
MW	molecular weight	kg/kmol
n	number of moles	–
N_A	Avogadro's number, 6.022×10^{23}	1/mol
p	pressure	atm
r	rate of reaction	mol/L·s
$\overline{R}$	universal gas constant, 0.08206	atm·L/mol·K
t	time	s
T	absolute temperature	K
V	volume	m^3
x	gravimetric (mass) fraction[1]	–
y	gas volumetric (mole) fraction	–
Z	atomic number	–

Subscripts

0	standard state
a	acid
b	base or boiling
eq	equilibrium
f	formation
p	pressure
r	reaction
SP	solubility product

---

[1]The NCEES *FE Reference Handbook* (*NCEES Handbook*) is inconsistent in the variable it uses for gravimetric fraction (mass fraction). In the Chemistry section (and this chapter), lowercase italic x is used. In the Materials Science section, "wt %" is used. In the Thermodynamics section, x represents mole fraction, and y represents mass fraction. (However, y is then used in the Thermodynamics section to represent a mole fraction for vapor-liquid equilibria.) In the Chemical Engineering section for unit operations (e.g., distillation), x also represents mole fraction.

1. ATOMS AND MOLECULES

Atomic Structure

An *atom* is the smallest subdivision of an element that can take part in a chemical reaction. The atomic nucleus consists of neutrons and protons, which are both also known as *nucleons*. Protons have a positive charge and neutrons have no charge, but the masses of neutrons and protons are essentially the same, one *atomic mass unit* (amu). One amu is exactly $^1/_{12}$ of the mass of an atom of carbon-12, approximately equal to 1.66×10^{-27} kg. The relative atomic weight, or simply *atomic weight*, A, of an atom is approximately equal to the number of protons and neutrons in the nucleus. The *atomic number*, Z, of an atom is equal to the number of protons in the nucleus.

The atomic number determines the way an atom behaves chemically; all atoms with the same atomic number are classified together as the same element. An *element* is a substance that cannot be decomposed into simpler substances during ordinary chemical reactions.

Although an element can have only a single atomic number, atoms of that element can have different atomic weights, and these are known as *isotopes*. The nuclei of isotopes differ from one another only in the number of neutrons.

The atomic number and atomic weight of an element, E, are written in symbolic form as $_ZE^A$, E_Z^A, or $_Z^AE$. For example, carbon is the sixth element; radioactive carbon has an atomic mass of 14. The symbol for carbon-14 is $^{14}_6$C. Because the atomic number and the chemical symbol give the same information, the atomic number can be omitted (e.g., C^{14} or C-14).

A chemical *compound* is a combination of two or more atoms that associate through chemical bonding. A *molecule* is the smallest subdivision of an element or compound that can exist in a natural state.

The Periodic Table

The *periodic table* is organized around the *periodic law:* Properties of the elements are periodic functions of their atomic numbers. (See Table 21.1.) Elements are arranged in order of increasing atomic numbers from left to right. The vertical columns are known as *groups*, numbered in Roman numerals. Each vertical group except 0 and VIII has A and B subgroups (*families*).

Chemistry

Chemistry

Table 21.1 *The Periodic Table of Elements*

The number of electrons in filled shells is shown in the column at the extreme left; the remaining electrons for each element are shown immediately below the symbol for each element. Atomic numbers are enclosed in brackets. Atomic weights (rounded, based on carbon-12) are shown above the symbols. Atomic weight values in parentheses are those of the isotopes of longest half-life for certain radioactive elements whose atomic weights cannot be precisely quoted without knowledge of origin of the element.

METALS   NONMETALS   TRANSITION METALS

periods	I A	II A	III B	IV B	V B	VI B	VII B	VIII B	VIII B	VIII B	I B	II B	III A	IV A	V A	VI A	VII A	O
1 / 0	1.0079 H [1] 1																	4.0026 He [2] 2
2 / 2	6.941 Li [3] 1	9.0122 Be [4] 2											10.811 B [5] 3	12.0115 C [6] 4	14.007 N [7] 5	15.999 O [8] 6	18.998 F [9] 7	20.179 Ne [10] 8
3 / 2,8	22.990 Na [11] 1	24.308 Mg [12] 2											26.981 Al [13] 3	28.086 Si [14] 4	30.974 P [15] 5	32.066 S [16] 6	35.453 Cl [17] 7	39.948 Ar [18] 8
4 / 2,8	39.098 K [19] 8,1	40.078 Ca [20] 8,2	44.956 Sc [21] 9,2	47.88 Ti [22] 10,2	50.941 V [23] 11,2	51.996 Cr [24] 13,1	54.938 Mn [25] 13,2	55.847 Fe [26] 14,2	58.933 Co [27] 15,2	58.68 Ni [28] 16,2	63.546 Cu [29] 18,1	65.39 Zn [30] 18,2	69.723 Ga [31] 18,3	72.61 Ge [32] 18,4	74.921 As [33] 18,5	78.96 Se [34] 18,6	79.904 Br [35] 18,7	83.80 Kr [36] 18,8
5 / 2,8,18	85.468 Rb [37] 8,1	87.62 Sr [38] 8,2	88.906 Y [39] 9,2	91.224 Zr [40] 10,2	92.906 Nb [41] 12,1	95.94 Mo [42] 13,1	(98) Tc [43] 13,2	101.07 Ru [44] 15,1	102.91 Rh [45] 16,1	106.42 Pd [46] 18	107.87 Ag [47] 18,1	112.41 Cd [48] 18,2	114.82 In [49] 18,3	118.71 Sn [50] 18,4	121.75 Sb [51] 18,5	127.60 Te [52] 18,6	126.90 I [53] 18,7	131.29 Xe [54] 18,8
6 / 2,8,18	132.91 Cs [55] 18,8,1	137.33 Ba [56] 18,8,2	* [57–71]	178.49 Hf [72] 32,10,2	180.95 Ta [73] 32,11,2	183.85 W [74] 32,12,2	186.21 Re [75] 32,13,2	190.2 Os [76] 32,14,2	192.22 Ir [77] 32,15,2	195.08 Pt [78] 32,17,1	196.97 Au [79] 32,18,1	200.59 Hg [80] 32,18,2	204.38 Tl [81] 32,18,3	207.2 Pb [82] 32,18,4	208.98 Bi [83] 32,18,5	(209) Po [84] 32,18,6	(210) At [85] 32,18,7	(222) Rn [86] 32,18,8
7 / 2,8,18,32	(223) Fr [87] 18,8,1	226.024 Ra [88] 18,8,2	† [89–103]	(261) Rf [104] 32,10,2	(262) Ha [105] 32,11,2													

* LANTHANIDE SERIES

138.91 La [57] 18,9,2	140.12 Ce [58] 20,8,2	140.91 Pr [59] 21,8,2	144.24 Nd [60] 22,8,2	(145) Pm [61] 23,8,2	150.36 Sm [62] 24,8,2	151.96 Eu [63] 25,8,2	157.25 Gd [64] 25,9,2	158.92 Tb [65] 27,8,2	162.50 Dy [66] 28,8,2	164.93 Ho [67] 29,8,2	167.26 Er [68] 30,8,2	168.93 Tm [69] 31,8,2	173.04 Yb [70] 32,8,2	174.97 Lu [71] 32,9,2

† ACTINIDE SERIES

227.03 Ac [89] 18,9,2	232.04 Th [90] 18,10,2	231.04 Pa [91] 20,9,2	238.03 U [92] 21,9,2	237.05 Np [93] 23,8,2	(244) Pu [94] 24,8,2	(243) Am [95] 25,8,2	(247) Cm [96] 25,9,2	(247) Bk [97] 26,9,2	(251) Cf [98] 28,8,2	(252) Es [99] 29,8,2	(257) Fm [100] 30,8,2	(258) Md [101] 31,8,2	(259) No [102] 32,8,2	(260) Lr [103] 32,9,2

Adjacent elements in horizontal rows (i.e., in different groups) differ in both physical and chemical properties. However, elements in the same column (group) have similar properties. Graduations in properties, both physical and chemical, also occur in the periods (i.e., the horizontal rows).

There are several ways to categorize groups of elements in the periodic table. The broadest categorization of the elements is into metals and nonmetals.[2]

Nonmetals (elements at the right end of the periodic chart) are elements 1, 2, 5–10, 14–18, 33–36, 52–54, and 85–86. The nonmetals include the *halogens* (group VIIA) and the *noble gases* (group 0). Nonmetals are poor electrical conductors and have little or no metallic luster. Most are either gases or brittle solids under normal conditions; only bromine is liquid under ordinary conditions.

Metals are all of the remaining elements. The metals are further subdivided into the *alkali metals* (group IA), the *alkaline earth metals* (group IIA), *transition metals* (all B families and group VIII), the *lanthanides* (also known as *lanthanons*, elements 57–71), and the *actinides* (also known as *actinons*, elements 89–103). Metals have low electron affinities, are reducing agents, form positive ions, and have positive oxidation numbers. They have high electrical conductivities, luster, ductility, and malleability, and generally high melting points.

The electron-attracting power of an atom, which determines much of its chemical behavior, is called its *electronegativity* and is measured on an arbitrary scale of 0 to 4. Fluorine is the most electronegative element and is assigned a value of 4.0. Generally, the most electronegative elements are those at the right end of the periods. Elements with low electronegativities are the metals found at the beginning (i.e., left end) of the periods. Electronegativity generally decreases going down a group. In other words, the trend in any family is toward more metallic properties as the atomic weight increases.

Ions and Electron Affinity

The atomic number, Z, of chlorine is 17, which means there are 17 protons in the nucleus of a chlorine atom. There are also 17 electrons in various shells surrounding the nucleus.

Chlorine has only seven electrons in the outer shell. A stable shell requires eight electrons for chlorine in row three of the periodic table. (Other elements may require 2, 8, 18, or 32 electrons to form a stable outer shell.) In order to achieve this stable configuration, chlorine atoms tend to attract electrons from other atoms, a characteristic known as *electron affinity*. The energy required to remove an electron from a neighboring atom is known as

the *ionization energy*. The electrons attracted by chlorine atoms come from nearby atoms with low ionization energies.

Chlorine, prior to taking a nearby atom's electron, is electrically neutral. The fact that it needs one electron to complete its outer subshell does not mean that chlorine needs an electron to become neutral. On the contrary, the chlorine atom becomes negatively charged when it takes the electron. An atomic nucleus with a charge is known as an *ion*.

Negatively charged ions are known as *anions*. Anions lose electrons at the anode during electrolytic reactions. Anions must lose electrons to become neutral. The loss of electrons is known as *oxidation*.

The charge on an anion is equal to the number of electrons taken from a nearby atom. In the past, this charge has been known as the *valence*. (The term *charge* can usually be substituted for valence.) Valence is equal to the number of electrons that must be gained for charge neutrality. For a chlorine ion, the valence is −1 since it must lose one electron for charge neutrality.

Sodium has one electron in its outer subshell; this electron has a low ionization energy and is very easily removed. If its outer electron is removed, such as when the electron is taken by a chlorine atom, sodium becomes positively charged. (For a sodium ion, the valence is +1.)

Positively charged ions are known as *cations*. Cations gain electrons at the cathode in electrolytic reactions. The gaining of electrons is known as *reduction*. Cations must gain electrons to become neutral.

Ionic and Covalent Bonds

If a chlorine atom becomes an anion by attracting an electron from a sodium atom (which becomes a cation), the two ions will be attracted to each other by electrostatic force. The electrostatic attraction of the positive sodium to the negative chlorine effectively bonds the two ions together. This type of bonding, in which electrostatic attraction is predominant, is known as *ionic bonding*. In an ionic bond, one or more electrons are transferred from the valence shell of one atom to the valence shell of another. There is no sharing of electrons between atoms.

Ionic bonding occurs in compounds containing atoms with high electron affinities and atoms with low ionization energies. Specifically, the difference in electronegativities must be approximately 1.7 or greater for the bond to be classified as predominantly ionic.

Several common gases in their free states exist as diatomic molecules. Examples are hydrogen (H_2), oxygen (O_2), nitrogen (N_2), and chlorine (Cl_2). Since two atoms of the same element will have the same electronegativity and ionization energy, one atom cannot take electrons from the other; the bond formed is not ionic.

---

[2]A third category, that of *metalloids*, defines a group of elements located between the metals and nonmetals on the periodic table and having intermediate properties between metals and nonmetals. For example, silicon ($Z=14$), arsenic ($Z=33$), and tellurium ($Z=52$) have metallic luster but are not conductors in the normal sense.

Chemistry

The electrons in these diatomic molecules are shared equally in order to fill the outer shells. This type of bonding, in which sharing of electrons is the predominant characteristic, is known as *covalent bonding*. Covalent bonds are typical of bonds formed in organic compounds. Specifically, the difference in electronegativities must be less than approximately 1.7 for the bond to be classified as predominantly covalent.

If both atoms forming a covalent bond are the same element, the electrons will be shared equally. This is known as a *nonpolar covalent bond*. If the atoms are not both the same element, the electrons will not be shared equally, resulting in a *polar covalent bond*. For example, the bond between hydrogen and chlorine in HCl is partially covalent and partially ionic in nature. For most compounds, there is no sharp dividing line between ionic and covalent bonds.

Oxidation Number

The *oxidation number* (*oxidation state*) is an electrical charge assigned by a set of prescribed rules. It is actually the charge, assuming all bonding is ionic. In a compound, the sum of the elemental oxidation numbers equals the net charge. For monoatomic ions, the oxidation number is equal to the charge. For neutral molecules, the sum of all oxidation numbers of the elements in the compounds must be zero.

In covalent compounds, all of the bonding electrons are assigned to the ion with the greater electronegativity. For example, nonmetals are more electronegative than metals. Carbon is more electronegative than hydrogen.

For atoms in an elementary free state, the oxidation number is zero. Hydrogen gas is a diatomic molecule, H_2. The oxidation number of the hydrogen molecule, H_2, is zero. The same is true for the atoms in O_2, N_2, Cl_2, and so on. Also, the sum of all the oxidation numbers of atoms in a neutral molecule is zero.

The oxidation number of an atom that forms a covalent bond is equal to the number of shared electron pairs. For example, each hydrogen atom has one electron. There are two electrons (i.e., a single shared electron pair) in each carbon-hydrogen bond in methane (CH_4), so the oxidation number of hydrogen is 1.

Fluorine is the most electronegative element, and it has an oxidation number of -1. Oxygen is second only to fluorine in electronegativity. Usually, the oxidation number of oxygen is -2, except in peroxides, where it is -1, and when combined with fluorine, where it is $+2$. Hydrogen is usually $+1$, except in hydrides, where it is -1.

The oxidation numbers of some common atoms and molecules are listed in Table 21.2.

Table 21.3 gives the standard oxidation potentials for corrosion reactions.

Table 21.2 *Oxidation Numbers of Atoms and Charge Numbers of Radicals**

name	symbol	oxidation or charge number
acetate	$C_2H_3O_2$	-1
aluminum	Al	$+3$
ammonium	NH_4	$+1$
barium	Ba	$+2$
boron	B	$+3$
borate	BO_3	-3
bromine	Br	-1
calcium	Ca	$+2$
carbon	C	$+4, -4$
carbonate	CO_3	-2
chlorate	ClO_3	-1
chlorine	Cl	-1
chlorite	ClO_2	-1
chromate	CrO_4	-2
chromium	Cr	$+2, +3, +6$
copper	Cu	$+1, +2$
cyanide	CN	-1
dichromate	Cr_2O_7	-2
fluorine	F	-1
gold	Au	$+1, +3$
hydrogen	H	$+1$ [-1 in hydrides]
hydroxide	OH	-1
hypochlorite	ClO	-1
iron	Fe	$+2, +3$
lead	Pb	$+2, +4$
lithium	Li	$+1$
magnesium	Mg	$+2$
mercury	Hg	$+1, +2$
nickel	Ni	$+2, +3$
nitrate	NO_3	-1
nitrite	NO_2	-1
nitrogen	N	$-3, +1, +2, +3, +4, +5$
oxygen	O	-2 [-1 in peroxides]
perchlorate	ClO_4	-1
permanganate	MnO_4	-1
phosphate	PO_4	-3
phosphorus	P	$-3, +3, +5$
potassium	K	$+1$
silicon	Si	$+4, -4$
silver	Ag	$+1$
sodium	Na	$+1$
sulfate	SO_4	-2
sulfite	SO_3	-2
sulfur	S	$-2, +4, +6$
tin	Sn	$+2, +4$
zinc	Zn	$+2$

*The information in this table may not be provided in the actual exam.

Compounds

Compounds are combinations of two or more atoms that associate through chemical bonding. *Binary compounds* contain two elements; *ternary (tertiary) compounds* contain three elements. A *chemical formula* is a representation of the relative numbers of each element in the compound. For example, the formula $CaCl_2$ shows that

Table 21.3 *Standard Oxidation Potentials for Corrosion Reactions*[a,b]

corrosion reaction	potential, E_o (volts), versus normal hydrogen electrode
$Au \rightarrow Au^{3+} + 3e^-$	-1.498
$2H_2O \rightarrow O_2 + 4H^+ + 4e^-$	-1.229
$Pt \rightarrow Pt^{2+} + 2e^-$	-1.200
$Pd \rightarrow Pd^{2+} + 2e^-$	-0.987
$Ag \rightarrow Ag^+ + e^-$	-0.799
$2Hg \rightarrow Hg_2^{2+} + 2e^-$	-0.788
$Fe^{2+} \rightarrow Fe^{3+} + e^-$	-0.771
$4(OH)^- \rightarrow O_2 + 2H_2O + 4e^-$	-0.401
$Cu \rightarrow Cu^{2+} + 2e^-$	-0.337
$Sn^{2+} \rightarrow Sn^{4+} + 2e^-$	-0.150
$H_2 \rightarrow 2H^+ + 2e^-$	0.000
$Pb \rightarrow Pb^{2+} + 2e^-$	$+0.126$
$Sn \rightarrow Sn^{2+} + 2e^-$	$+0.136$
$Ni \rightarrow Ni^{2+} + 2e^-$	$+0.250$
$Co \rightarrow Co^{2+} + 2e^-$	$+0.277$
$Cd \rightarrow Cd^{2+} + 2e^-$	$+0.403$
$Fe \rightarrow Fe^{2+} + 2e^-$	$+0.440$
$Cr \rightarrow Cr^{3+} + 3e^-$	$+0.744$
$Zn \rightarrow Zn^{2+} + 2e^-$	$+0.763$
$Al \rightarrow Al^{3+} + 3e^-$	$+1.662$
$Mg \rightarrow Mg^{2+} + 2e^-$	$+2.363$
$Na \rightarrow Na^+ + e^-$	$+2.714$
$K \rightarrow K^+ + e^-$	$+2.925$

[a]Measured at 25°C. Reactions are written as anode half-cells. Arrows are reversed for cathode half-cells.
[b]In some chemistry texts, the reactions and the signs of the values (in this table) are reversed; for example, the half-cell potential of zinc is given as -0.763 V for the reaction $Zn^{2+} + 2e^- \rightarrow Zn$. When the potential E_o is positive, the reaction proceeds spontaneously as written.

there is one calcium atom and two chlorine atoms in one molecule of calcium chloride.

Generally, the numbers of atoms are reduced to their lowest terms. However, there are exceptions. For example, to show the actual number of atoms in each molecule, acetylene is C_2H_2, and hydrogen peroxide is H_2O_2.

For binary compounds with a metallic element, the positive metallic element is listed first. The chemical name ends in the suffix "-ide." For example, NaCl is sodium chloride. If the metal, such as iron, has two oxidation states, the suffix "-ous" is used for the lower state, and "-ic" is used for the higher state. Alternatively, the element name can be used with the oxidation number written in Roman numerals. For example,

$FeCl_2$: ferrous chloride, or iron (II) chloride

$FeCl_3$: ferric chloride, or iron (III) chloride

For binary compounds formed between two nonmetals, the more positive element is listed first. The number of atoms of each element is specified by the prefixes "di-" (2), "tri-" (3), "tetra-" (4), "penta-" (5), and so on. For example,

N_2O_5: dinitrogen pentoxide

Binary acids start with the prefix "hydro-," list the name of the nonmetallic element, and end with the suffix "-ic." For example,

HCl: hydrochloric acid

Ternary compounds generally consist of an element and a radical, with the positive part listed first in the formula. Ternary acids (also known as *oxy-acids*) usually contain hydrogen, a nonmetal, and oxygen, and can be grouped into families with different numbers of oxygen atoms. The most common acid in a family (i.e., the root acid) has the name of the nonmetal and the suffix "-ic." The acid with one more oxygen atom than the root is given the prefix "per-" and the suffix "-ic." The acid containing one less oxygen atom than the root is given the ending "-ous." The acid containing two less oxygen atoms than the root is given the prefix "hypo-" and the suffix "-ous." For example,

$HClO$: hypochlorous acid

$HClO_2$: chlorous acid

$HClO_3$: chloric acid (the root)

$HClO_4$: perchloric acid

Compounds form according to the *law of definite (constant) proportions*: A pure compound is always composed of the same elements combined in a definite proportion by mass. For example, common table salt is always NaCl. It is not sometimes NaCl and other times Na_2Cl or $NaCl_3$ (which do not exist).

Furthermore, compounds form according to the *law of (simple) multiple proportions*: When two elements combine to form more than one compound, the masses of the elements usually combine in ratios of the smallest possible integers.

In order to evaluate whether a compound formula is valid, it is necessary to know the oxidation numbers of the interacting atoms. Although some atoms have more than one possible oxidation number, most do not. The sum of the oxidation numbers must be zero if a neutral compound is to form. For example, H_2O is a valid compound because the two hydrogen atoms have a total positive oxidation number of $2 \times 1 = +2$. The oxygen ion has an oxidation number of -2. These oxidation numbers sum to zero.

On the other hand, $NaCO_3$ is not a valid compound formula. Sodium (Na) has an oxidation number of $+1$. However, the CO_3 ion has a charge number of -2. The correct sodium carbonate molecule is Na_2CO_3.

Table 21.4 gives the common names and molecular formulas of some industrial inorganic chemicals.[3]

---

[3]Table 21.4 contains many archaic terms. "Common names" is misleading, as it is unlikely that some of these names are in common use. Few people would understand the term "wolfram filament" when discussing incandescent light bulbs. Lists of ingredients for paint, fingernail polish, cosmetics, sunscreen, food coloring, and correction fluid do not list titanium dioxide as "anatase" or "rutile," which are two of the naturally occurring ores from which titanium dioxide may be obtained.

Chemistry

Table 21.4 Common Names and Molecular Formulas of Some Industrial Inorganic Chemicals

chemical name	common name	molecular formula
hydrochloric acid	muriatic acid	HCl
hypochlorite ion	–	OCl^{-1}
chlorite ion	–	ClO_2^{-1}
chlorate ion	–	ClO_3^{-1}
perchlorate ion	–	ClO_4^{-1}
calcium sulfate	gypsum	$CaSO_4$
calcium carbonate	limestone	$CaCO_3$
magnesium carbonate	dolomite	$MgCO_3$
aluminum oxide	bauxite	Al_2O_3
titanium dioxide	anatase	TiO_2
titanium dioxide	rutile	TiO_2
ferrous sulfide	pyrite	FeS
magnesium sulfate	epsom salt	$MgSO_4$
sodium carbonate	soda ash	Na_2CO_3
sodium chloride	salt	NaCl
potassium carbonate	potash	K_2CO_3
sodium bicarbonate	baking soda	$NaHCO_3$
sodium hydroxide	lye	NaOH
sodium hydroxide	caustic soda	NaOH
silane	–	SiH_4
ozone	–	O_3
ferrous/ferric oxide	magnetite	Fe_3O_4
mercury	quicksilver	Hg
deuterium oxide[a]	heavy water	$(H^2)_2O$
borane	–	BH_3
boric acid (solution)	eyewash	H_3BO_3
deuterium	–	H^2
tritium	–	H^3
nitrous oxide	laughing gas	N_2O
phosgene[b]	–	$COCl_2$
tungsten	wolfram	W
permanganate ion	–	MnO_4^{-1}
dichromate ion	–	$Cr_2O_7^{-2}$
hydronium ion	–	H_3O^{+1}
sodium chloride (solution)	brine	NaCl
sulfuric acid	battery acid	H_2SO_4

[a](1) "Deuterium oxide" is an obsolete term for heavy water. (2) The *NCEES Handbook* uses "H^2" as the symbol for deuterium. The modern chemical symbol is "D," making D_2O the modern molecular formula for heavy water. (3) When a superscript is used to represent the number of hydrogen atoms, it is almost always written as "^2H," not "H^2." In that case, the molecular symbol would be written as "^2H_2O."

[b]The inclusion of phosgene in this table dates the table to sometime after World War I, when it was common knowledge that poisonous phosgene gas was used as a chemical weapon to clear trenches. Since then, use of phosgene gas during warfare has been prohibited by international convention. While phosgene has some valid industrial and pharmaceutical uses, its inclusion in the *NCEES Handbook* is an anachronism.

Equation 21.1: Moles

$$1 \text{ mol} = 1 \text{ gram mole} \qquad 21.1$$

Description

The *mole* is a measure of the quantity of an element or compound. Specifically, a mole of an element (or compound) will have a mass equal to the element's atomic (or compound's molecular) weight. A mole of any substance will have the same number of atoms as one gram of carbon-12.

The three main types of moles are based on mass measured in grams, kilograms, and pounds. Obviously, a gram-based mole of carbon (12 grams) is not the same quantity as a pound-based mole of carbon (12 pounds). Although "mol" is understood in countries using SI units to mean a gram-mole, the term *mole* is ambiguous, and the units mol (gmol), kmol (kgmol), or lbmol, must be specified, or the type of mole must be spelled out.

"Molar" is used as an adjective when describing properties of a mole. For example, a molar volume is the volume of a mole.

Formula and Molecular Weight; Equivalent Weight

The *formula weight*, FW, of a molecule (compound) is the sum of the atomic weights of all elements in the formula. The *molecular weight*, MW, is the sum of the atomic weights of all atoms in the molecule and is generally the same as the formula weight. The units of molecular weight are g/mol, kg/kmol, or lbm/lbmol.

The *equivalent weight* (i.e., an *equivalent*), EW, is the amount of substance (in grams) that supplies one gram-mole (i.e., 6.02×10^{23}) of reacting units. For acid-base reactions, an acid equivalent supplies one gram-mole of H^+ ions. A base equivalent supplies one gram-mole of OH^- ions. In oxidation-reduction reactions, an equivalent of a substance gains or loses a gram-mole of electrons. Similarly, in electrolysis reactions, an equivalent weight is the weight of substance that either receives or donates one gram-mole of electrons at an electrode.

The equivalent weight can be calculated as the molecular weight divided by the change in oxidation number experienced in a chemical reaction. A substance can have several equivalent weights.

$$EW = \frac{MW}{\Delta \text{oxidation number}}$$

Equation 21.2: Avogadro's Number

$$1 \text{ mol} = 6.02 \times 10^{23} \text{ particles} \qquad 21.2$$

Description

Avogadro's hypothesis states that equal volumes of all gases at the same temperature and pressure contain equal numbers of gas molecules. Specifically, at *standard scientific conditions* (1.0 atm and 0°C), 1 gram-mole of any gas occupies 22.4 L. These characteristics are derived from the kinetic theory of gases under the assumption of an ideal (perfect) gas. Avogadro's hypothesis is approximately valid for real gases at sufficiently low pressures and high temperatures.

Avogadro's law, combined with Charles' and Boyle's laws, can be stated as the *equation of state* for ideal gases.

$$pV = n\overline{R}T$$

$\overline{R}$ is the *universal gas constant*, which has a value of 0.08206 atm·L/mol·K (or 8314 J/kmol·K) and can be used with any gas. The number of moles is n.

One gram-mole of any substance has a number of particles (atoms, molecules, ions, electrons, etc.) equal to 6.02×10^{23}, *Avogadro's number*, N_A. (See Eq. 21.2.) A pound-mole contains approximately 454 times the number of particles in a gram-mole.

Example

The atomic weight of hydrogen is 1.0079 g/mol. What is most nearly the mass of a hydrogen atom?

(A) 1.7×10^{-24} g/atom

(B) 6.0×10^{-23} g/atom

(C) 1.0×10^{-10} g/atom

(D) 1.0 g/atom

Solution

By definition, the mass of an atom is its atomic weight divided by Avogadro's number.

$$m = \frac{1.0079 \; \dfrac{g}{mol}}{6.022 \times 10^{23} \; \dfrac{atoms}{mol}}$$

$$= 1.67 \times 10^{-24} \; \text{g/atom} \quad (1.7 \times 10^{-24} \; \text{g/atom})$$

The answer is (A).

Gravimetric Fraction

The *gravimetric fraction*, x_i, of an element i in a compound is the fraction by weight of that element in the compound. The gravimetric fraction is found from an *ultimate analysis* (also known as a *gravimetric analysis*) of the compound.

$$x_i = \frac{m_i}{m_1 + m_2 + \cdots + m_i + \cdots + m_n} = \frac{m_i}{m_t}$$

The *percentage composition* is the gravimetric fraction converted to percentage.

$$\% \text{ composition} = x_i \times 100\%$$

If the gravimetric fractions are known for all elements in a compound, the *combining weights* of each element can be calculated. (The term *weight* is used even though mass is the traditional unit of measurement.)

$$m_i = x_i m_t$$

Empirical Formula Development

It is relatively simple to determine the *empirical formula* of a compound from the atomic and combining weights of elements in the compound. The empirical formula gives the relative number of atoms (i.e., the formula weight is calculated from the empirical formula).

step 1: Divide the gravimetric fractions (or percentage compositions) by the atomic weight of each respective element.

step 2: Determine the smallest ratio from step 1.

step 3: Divide all of the ratios from step 1 by the smallest ratio.

step 4: Write the chemical formula using the results from step 3 as the numbers of atoms. Multiply through as required to obtain all integer numbers of atoms.

Example

A clear liquid is analyzed, and the following gravimetric percentage compositions are recorded: carbon, 37.5%; hydrogen, 12.5%; oxygen, 50%. What is the chemical formula for the liquid?

(A) CH_4O

(B) $C_3H_{13}O_3$

(C) C_4HO_2

(D) $C_{12}HO_{16}$

Solution

step 1: Divide the percentage compositions by the atomic weights.

$$\text{C:} \quad \frac{37.5 \; g}{12.0115 \; \dfrac{g}{mol}} = 3.122 \; mol$$

$$\text{H:} \quad \frac{12.5 \; g}{1.0079 \; \dfrac{g}{mol}} = 12.4 \; mol$$

$$\text{O:} \quad \frac{50 \; g}{15.999 \; \dfrac{g}{mol}} = 3.125 \; mol$$

step 2: The smallest ratio is 3.122 mol.

Chemistry

step 3: 3.122 and 3.125 are essentially the same. Choose either. Divide all ratios by 3.125 mol. Round to integers.

$$\text{C: } \frac{3.122 \text{ mol}}{3.125 \text{ mol}} = 1$$

$$\text{H: } \frac{12.4 \text{ mol}}{3.125 \text{ mol}} = 4$$

$$\text{O: } \frac{3.125 \text{ mol}}{3.125 \text{ mol}} = 1$$

step 4: The empirical formula is CH_4O.

If it had been known that the liquid behaved as though it contained a hydroxyl (OH) radical, the formula would have been written as CH_3OH. This is recognized as methyl alcohol.

The answer is (A).

2. CHEMICAL REACTIONS

During chemical reactions, bonds between atoms are broken and new bonds are formed. The starting substances are known as *reactants*; the ending substances are known as *products*. In a chemical reaction, reactants are either converted to simpler products or synthesized into more complex compounds.

The coefficients in front of element and compound symbols in chemical reaction equations are the numbers of molecules or moles taking part in the reaction. For gaseous reactants and products, the coefficients also represent the numbers of volumes. This is a direct result of Avogadro's hypothesis, which says that equal numbers of molecules in the gas phase occupy equal volumes at the same conditions.

Because matter cannot be destroyed in a normal chemical reaction (i.e., mass is conserved), the numbers of moles of each element must match on both sides of the equation. When the numbers of each element on both sides match, the equation is said to be *balanced*. The total atomic weights on both sides of the equation will be equal when the equation is balanced.

Balancing simple chemical equations is largely a matter of deductive trial and error. More complex reactions require the use of oxidation numbers.

Types of Chemical Reactions

There are four common types of reactions.

- *direct combination* (or *synthesis*): This is the simplest type of reaction where two elements or compounds combine directly to form a compound.

$$2H_2 + O_2 \rightarrow 2H_2O$$

$$SO_2 + H_2O \rightarrow H_2SO_3$$

- *decomposition* (or *analysis*): Bonds within a compound are disrupted by heat or other energy to produce simpler compounds or elements.

$$2HgO \rightarrow 2Hg + O_2$$

$$H_2CO_3 \rightarrow H_2O + CO_2$$

- *single displacement* (or *replacement*[4]): This type of reaction has one element and one compound as reactants.

$$2Na + 2H_2O \rightarrow 2NaOH + H_2$$

$$2KI + Cl_2 \rightarrow 2KCl + I_2$$

- *double displacement* (or *replacement*): These are reactions with two compounds as reactants and two compounds as products.

$$AgNO_3 + NaCl \rightarrow AgCl + NaNO_3$$

$$H_2SO_4 + ZnS \rightarrow H_2S + ZnSO_4$$

Multiple Reactions

Production of all but the simplest of compounds usually requires multiple reactants and reactions. Within a reaction system (e.g., in nature or in process vessels), there are four ways such multiple reactions can be related.

- *independent reactions*: These are reactions that occur within the same system but have no direct effect on each other.

$$A + B \rightarrow C + D$$

$$E + F \rightarrow G + H$$

- *series reactions* (or *sequential reactions*): These are reactions that must occur in a particular order, with the products of preceding reactions becoming the reactants of subsequent reactions.

$$A + B \rightarrow C + D \rightarrow E + F$$

- *parallel reactions* (or *competing reactions*): These are reactions that begin with the same set of reactants but that result in different sets of products, depending on conditions, catalysts, and relative abundance of some chemical species. Parallel reactions are often responsible for the production of undesirable products from established processes.

$$A + B \rightarrow C + D$$

$$A + B \rightarrow E + F$$

- *complex reactions*: These are combinations of series and parallel reactions.

---

[4]Another name for replacement is *metathesis*.

Homogeneous and Heterogeneous Reactions

Reactions can be classified by the phases (such as gas or liquid) that are present in the reaction system. If the reactants and products in a reaction are all in the same phase, the reaction is said to be *homogeneous*; if not, the reaction is *heterogeneous*. For example, a reaction between hydrogen and oxygen to form water vapor is homogeneous, while a reaction between liquid hydrochloric acid and zinc pellets to form hydrogen gas is heterogeneous.

Catalytic Reactions

A *catalytic reaction* is a reaction that takes place in the presence of a catalyst. A *catalyst* is a substance that increases the rate of a chemical reaction without itself undergoing any permanent chemical change. (The catalyst may change temporarily during the reaction before returning to its original condition.) A catalyst works by lowering the *activation energy*, the minimum energy needed to cause a reaction. In some cases, a catalyst makes possible a reaction that could not occur at all without its presence. Many catalytic reactions are heterogeneous, with a fluid phase passing over a solid catalyst.

Biocatalytic reactions are a special class of catalytic reactions. In a biocatalytic reaction, the catalyst is a biological substance. If the biocatalyst is a protein enzyme, the reaction is referred to as an *enzymatic reaction*. The reactants involved in an enzymatic reaction are referred to as *substrates* to distinguish them from the catalyst. Enzymatic reactions are commonly used to make foodstuff (e.g., cheese) and beverages (e.g., beer and wine).

Balancing Chemical Equations

The coefficients in front of element and compound symbols in chemical reaction equations are the numbers of molecules or moles taking part in the reaction. (For gaseous reactants and products, the coefficients also represent the numbers of volumes. This is a direct result of Avogadro's hypothesis that equal numbers of molecules in the gas phase occupy equal volumes under the same conditions.)[5]

Since atoms cannot be changed in a normal chemical reaction (i.e., mass is conserved), the numbers of moles of each element must match on both sides of the equation. When the numbers of moles of each element match, the equation is said to be "balanced." The total atomic weights on both sides of the equation will be equal when the equation is balanced. The total masses of reactants consumed will be equal to the total masses of products formed.

---

[5]When water is part of the reaction, the interpretation that the coefficients are volumes is valid only if the reaction takes place at a high enough temperature to vaporize the water.

Balancing simple chemical equations is largely a matter of deductive trial and error. More complex reactions require use of oxidation numbers.

Example

Balance the following reaction equation.

$$Al + H_2SO_4 \rightarrow Al_2(SO_4)_3 + H_2$$

(A) $Al + H_2SO_4 \rightarrow Al_2(SO_4)_3 + H_2$

(B) $2Al + H_2SO_4 \rightarrow Al_2(SO_4)_3 + H_2$

(C) $2Al + 3H_2SO_4 \rightarrow Al_2(SO_4)_3 + 3H_2$

(D) $4Al + 3H_2SO_4 \rightarrow 2Al_2(SO_4)_3 + 3H_2$

Solution

As written, the reaction is not balanced. For example, there is one aluminum on the left, but there are two on the right. The starting element in the balancing procedure is chosen somewhat arbitrarily.

step 1: Since there are two aluminums on the right, multiply Al by 2.

$$2Al + H_2SO_4 \rightarrow Al_2(SO_4)_3 + H_2$$

step 2: Since there are three sulfate radicals (SO_4) on the right, multiply H_2SO_4 by 3.

$$2Al + 3H_2SO_4 \rightarrow Al_2(SO_4)_3 + H_2$$

step 3: Now there are six hydrogens on the left, so multiply H_2 by 3 to balance the equation.

$$2Al + 3H_2SO_4 \rightarrow Al_2(SO_4)_3 + 3H_2$$

The answer is (C).

Stoichiometric Reactions

Stoichiometry is the study of the proportions in which elements and compounds react and are formed. A *stoichiometric reaction* (also known as a *perfect reaction* or an *ideal reaction*) is one in which just the right amounts of reactants are present. After the reaction stops, there are no unused reactants.

Stoichiometric problems are known as *weight and proportion problems* because their solutions use simple ratios to determine the masses of reactants required to produce given masses of products, or vice versa. The procedure for solving these problems is essentially the same regardless of the reaction.

step 1: Write and balance the chemical equation.

step 2: Determine the atomic (molecular) weight of each element (compound) in the equation.

Chemistry

step 3: Multiply the atomic (molecular) weights by their respective coefficients and write the products under the formulas.

step 4: Write the given mass data under the weights determined in step 3.

step 5: Fill in the missing information by calculating simple ratios.

Example

Caustic soda (NaOH) is made from sodium carbonate (Na_2CO_3) and slaked lime ($Ca(OH)_2$) according to the given reaction.

$$Na_2CO_3 + Ca(OH)_2 \rightarrow 2NaOH + CaCO_3$$

How many kilograms of caustic soda can be made from 2000 kg of sodium carbonate?

(A) 1200 kg

(B) 1500 kg

(C) 1900 kg

(D) 2700 kg

Solution

Calculate the combining weights.

$$(1 \text{ mol}) \left(\begin{array}{c} (2)\left(22.990 \ \dfrac{g}{mol}\right) + 12.011 \ \dfrac{g}{mol} \\[2mm] + (3)\left(15.999 \ \dfrac{g}{mol}\right) \end{array} \right)$$

$$= 105.988 \text{ g} \quad (106 \text{ g}) \quad [Na_2CO_3]$$

$$(1 \text{ mol})\left(40.078 \ \frac{g}{mol} + (2)\left(15.999 \ \frac{g}{mol}\right) + 1.0079 \ \frac{g}{mol}\right)$$

$$= 74.0918 \text{ g} \quad (74 \text{ g}) \quad [Ca(OH)_2]$$

$$(2 \text{ mol})\left(22.990 \ \frac{g}{mol} + 15.999 \ \frac{g}{mol} + 1.0079 \ \frac{g}{mol}\right)$$

$$= 79.994 \text{ g} \quad (80 \text{ g}) \quad [2NaOH]$$

$$(1 \text{ mol})\left(40.078 \ \frac{g}{mol} + 12.011 \ \frac{g}{mol} + (3)\left(15.999 \ \frac{g}{mol}\right)\right)$$

$$= 100.086 \text{ g} \quad (100 \text{ g}) \quad [CaCo_3]$$

	Na_2CO_3	+	$Ca(OH)_2$	$\rightarrow$	$2NaOH$	+	$CaCO_3$
combining weights	106 g		74 g		80 g		100 g

The simple ratio used is

$$\frac{NaOH}{Na_2CO_3} = \frac{80 \text{ g}}{106 \text{ g}} = \frac{m}{2000 \text{ kg}}$$

$$m_{NaOH} = 1509 \text{ kg} \quad (1500 \text{ kg})$$

The answer is (B).

Nonstoichiometric Reactions

In many cases, it is not realistic to assume a stoichiometric reaction because an excess of one or more reactants is necessary to assure that all of the remaining reactants take part in the reaction. Combustion is an example where the stoichiometric assumption is, more often than not, invalid. Excess air is generally needed to ensure that all of the fuel is burned.

With nonstoichiometric reactions, the reactant that is used up first is called the *limiting reactant*. The amount of product will be dependent on (limited by) the limiting reactant.

The *theoretical yield* or *ideal yield* of a product is the maximum mass of product per unit mass of limiting reactant that can be obtained from a given reaction if the reaction goes to completion. The *percentage yield* is a measure of the efficiency of the actual reaction.

$$\text{percentage yield} = \frac{\text{actual yield} \times 100\%}{\text{theoretical yield}}$$

Oxidation-Reduction Reactions

Oxidation-reduction reactions (also known as *redox reactions*) involve the transfer of electrons from one element or compound to another. Specifically, one reactant is oxidized and the other reactant is reduced.

In *oxidation*, the substance's oxidation state increases, the substance loses electrons, and the substance becomes less negative. Oxidation occurs at the *anode* (positive terminal) in electrolytic reactions.

In *reduction*, the substance's oxidation state decreases, the substance gains electrons, and the substance becomes more negative. Reduction occurs at the *cathode* (negative terminal) in electrolytic reactions.

Whenever oxidation occurs in a chemical reaction, reduction must also occur. For example, consider the formation of sodium chloride from sodium and chlorine. This reaction is a combination of oxidation of sodium and reduction of chlorine. The electron released during oxidation is used up in the reduction reaction.

$$2Na + Cl_2 \rightarrow 2NaCl$$

$$Na \rightarrow Na^+ + e^-$$

$$Cl + e^- \rightarrow Cl^-$$

The substance that causes oxidation to occur (chlorine in the preceding example) is called the *oxidizing agent* and is itself reduced (i.e., becomes more negative) in the process. The substance that causes reduction to occur (sodium in the example) is called the *reducing agent* and is itself oxidized (i.e., becomes less negative) in the process.

The total number of electrons lost during oxidation must equal the total number of electrons gained during reduction. This is the main principle used in balancing redox reactions. Although there are several formal methods of applying this principle, balancing

Chemistry

an oxidation-reduction equation remains somewhat intuitive and iterative.

The oxidation number change method of balancing redox reactions consists of the following steps.

step 1: Write an unbalanced equation that includes all reactants and products.

step 2: Assign oxidation numbers to each atom in the unbalanced equation.

step 3: Note which atoms change oxidation numbers, and calculate the amount of change for each atom. When more than one atom of an element that changes oxidation number is present in a formula, calculate the change in oxidation number for that atom per formula unit.

step 4: Balance the equation so that the number of electrons gained equals the number lost.

step 5: Balance (by inspection) the remainder of the chemical equation as required.

Irreversible and Reversible Reactions

Irreversible reactions proceed in one direction only: if A and B react to form C and D, then C and D usually cannot react to form A and B. In these *irreversible reactions*, the reactants form products that cannot form the reactants again under the same conditions. The combustion of methane is an example of an irreversible reaction.

$$CH_4 + O_2 \rightarrow CO_2 + 2H_2O$$

There are some reactions, however, that can proceed in either direction under identical conditions. In a *reversible reaction,* as the reactants are forming the products, the products are simultaneously breaking down to form the reactants again. As a result, the reactants are never fully consumed. Therefore, a reversible reaction is characterized by the simultaneous presence of all reactants and products.

For example, the chemical equation for the exothermic formation of ammonia from nitrogen and hydrogen is

$$N_2 + 3H_2 \rightleftharpoons 2NH_3 + heat$$

At chemical equilibrium, reactants and products are both present. However, the concentrations of the reactants and products do not continue to change after equilibrium is reached.

Le Chatelier's Principle

Le Châtelier's principle predicts the direction in which a reversible reaction at equilibrium will go when some condition (temperature, pressure, concentration, etc.) is stressed (i.e., changed). This principle states that when an equilibrium state is stressed by a change, a new equilibrium that reduces that stress is reached.

Consider the formation of ammonia from nitrogen and hydrogen. When the reaction proceeds in the forward direction, energy in the form of heat is released and the temperature increases. If the reaction proceeds in the reverse direction, heat is absorbed and the temperature decreases. If the system is stressed by increasing the temperature, the reaction will proceed in the reverse direction because that direction absorbs heat and reduces the temperature.

For reactions that involve gases, the reaction equation coefficients can be interpreted as volumes. In the nitrogen-hydrogen reaction, four volumes combine to form two volumes. If the equilibrium system is stressed by increasing the pressure, then the forward reaction will occur because this direction reduces the volume and pressure.

If the concentration of any substance is increased, the reaction proceeds in a direction away from the substance with the increase in concentration. For example, an increase in the concentration of the reactants shifts the equilibrium to the right, increasing the amount of products formed.

Equation 21.3: Rate and Order of Reactions

$$aA + bB \rightleftharpoons cC + dD \qquad 21.3$$

Description

The time required for a reaction to proceed to equilibrium or completion depends on the rate of reaction. The *rate of reaction*, r, is the change in concentration per unit time, measured in mol/L·s.

$$r = \frac{\text{change in concentration}}{\text{time}}$$

For a reversible reaction such as Eq. 21.3, the *law of mass action* states that the rate of reaction is proportional to the equilibrium molar concentrations, $[X]$ (i.e., the molarities), of the reactants. The constants k_{forward} and k_{reverse} are the reaction rate constants needed to obtain the units of rate.

$$r_{\text{forward}} = k_{\text{forward}}[A]^a[B]^b$$

$$r_{\text{reverse}} = k_{\text{reverse}}[C]^c[D]^d$$

At equilibrium, the forward and reverse speeds of reaction are equal.

The rate of reaction for solutions is generally not greatly affected by pressure, but is affected by the following factors.

- *types of substances in the reaction:* Some substances are more reactive than others.

- *exposed surface area:* For heterogeneous reactions, the rate of reaction is proportional to the amount of contact between the reactants.

- *concentrations:* The rate of reaction increases with increases in concentration.

Chemistry

- *temperature:* For heterogeneous reactions, the rate of reaction increases with increases in temperature.

- *catalysts:* A *catalyst* is a substance that increases the reaction rate without being consumed in the reaction. If a catalyst is introduced, rates of reaction will increase (i.e., equilibrium will be reached more quickly), but the equilibrium will not be changed.

The *order of a reaction* is defined as the total number of reacting molecules in or before the slowest step in the mechanism, as determined experimentally. Consider the reversible reaction given by Eq. 21.3. The order of the forward reaction is $a + b$; the order of the reverse reaction is $c + d$.

Equation 21.4: Equilibrium Constant

$$K_{eq} = \frac{[C]^c [D]^d}{[A]^a [B]^b} \qquad 21.4$$

Variation

$$K_{eq} = \frac{k_{forward}}{k_{reverse}} \quad \text{[reversible reactions]}$$

Description

For reversible reactions, the *equilibrium constant*, K_{eq}, is equal to the ratio of the forward rate of reaction to the reverse rate of reaction. Except for catalysis, the equilibrium constant depends on the same factors affecting the reaction rate. For a reversible reaction, the equilibrium constant is given by the law of mass action.

If any of the reactants or products are in pure solid or pure liquid phases, their concentrations are omitted from the calculation of the equilibrium constant. For example, in weak aqueous solutions, the concentration of water, $[H_2O]$, is very large and essentially constant; therefore, that concentration is omitted.

For gaseous reactants and products, the concentrations (i.e., the numbers of atoms) will be proportional to the partial pressures. An equilibrium constant can be calculated directly from the partial pressures and is given the symbol K_p. For example, for the formation of ammonia gas from nitrogen and hydrogen ($3H_2 + N_2 \rightarrow 2NH_3$), the pressure equilibrium constant is

$$K_p = \frac{[p_{NH_3}]^2}{[p_{N_2}][p_{H_2}]^3}$$

K_{eq} and K_p are not numerically the same, but when the ideal gas law is valid, they are related by

$$K_p = K_{eq}(\overline{R}T)^{\Delta n}$$

Δn is the number of moles of products minus the number of moles of reactants.

Example

What is the equilibrium constant for the following reaction?

$$MgSO_4(s) \rightleftharpoons MgO(s) + SO_3(g)$$

(A) $K_{eq} = \dfrac{[Mg][SO_3]}{2[MgSO_4]}$

(B) $K_{eq} = \dfrac{[MgSO_4]}{[MgO][SO_3]}$

(C) $K_{eq} = [MgO][SO_3]$

(D) $K_{eq} = [SO_3]$

Solution

Solids have a concentration of 1, so $K_{eq} = [SO_3]$.

The answer is (D).

Acids and Bases

An *acid* is any compound that dissociates in water into H^+ ions. (The combination of H^+ and water, H_3O^+, is known as the *hydronium ion*.) This is known as the *Arrhenius theory of acids*. Acids with one, two, and three ionizable hydrogen atoms are called *monoprotic*, *diprotic*, and *triprotic acids*, respectively.

The properties of acids are as follows.

- Acids conduct electricity in aqueous solutions.

- Acids have a sour taste.

- Acids turn blue litmus paper red.

- Acids have a pH between 0 and 7.

- Acids neutralize bases.

- Acids react with active metals to form hydrogen.

$$2H^+ + Zn \rightarrow Zn^{2+} + H_2$$

- Acids react with oxides and hydroxides of metals to form salts and water.

$$2H^+ + 2Cl^- + FeO \rightarrow Fe^{2+} + 2Cl^- + H_2O$$

- Acids react with salts of either weaker or more volatile acids (such as carbonates and sulfides) to form a new salt and a new acid.

$$2H^+ + 2Cl^- + CaCO_3 \rightarrow H_2CO_3 + Ca^{2+} + 2Cl^-$$

A *base* is any compound that dissociates in water into OH^- ions. This is known as the *Arrhenius theory of bases*. Bases with one, two, and three replaceable hydroxide ions are called *monohydroxic*, *dihydroxic*, and *trihydroxic* bases, respectively.

Chemistry

The properties of bases are as follows.

- Bases conduct electricity in aqueous solutions.
- Bases have a bitter taste.
- Bases turn red litmus paper blue.
- Bases have a pH between 7 and 14.
- Bases neutralize acids, forming salts and water.

Equation 21.5: Acids and Bases

$$pH = \log_{10}\left(\frac{1}{[H^+]}\right) \qquad 21.5$$

Variation

$$pOH = -\log_{10}[OH^-] = \log_{10}\frac{1}{[OH^-]}$$

Description

A measure of the strength of an acid or base is the number of hydrogen or hydroxide ions in a liter of solution. Since these are very small numbers, a logarithmic scale is used.

The quantities $[H^+]$ and $[OH^-]$ in square brackets in Eq. 21.5 and the variation equation are the ionic concentrations in moles of ions per liter. The number of moles can be calculated from Avogadro's law by dividing the actual number of ions per liter by 6.02×10^{23}.

A *neutral solution* has a pH of 7. Solutions with pH less than 7 are acidic; the smaller the pH, the more acidic the solution. Solutions with pH more than 7 are basic.

Example

Most nearly, what is the pH if the ionic concentration of H^+ is 5×10^{-6} mol/L?

(A) 5.3

(B) 6.0

(C) 8.0

(D) 8.7

Solution

Use Eq. 21.5.

$$pH = \log_{10}\left(\frac{1}{[H^+]}\right) = -\log_{10}([H^+])$$

$$= -\log_{10}\left(5 \times 10^{-6}\ \frac{mol}{L}\right)$$

$$= 5.3$$

The answer is (A).

Enthalpy of Formation

Enthalpy, H, is a measure of the energy that a substance possesses by virtue of its temperature, pressure, and phase. The *enthalpy of formation (heat of formation)*, ΔH_f, of a compound is the energy absorbed during the formation of one gram-mole of the compound from pure elements. The enthalpy of formation is assigned a value of zero for elements in their free states at 25°C and 1 atm. This is the so-called *standard state*, or *standard temperature and pressure* (STP) for enthalpies of formation. This set of conditions differs from the set of conditions used in industrial hygiene air monitoring, called the *normal conditions* (NTP).

Enthalpy of Reaction

The *enthalpy of reaction (heat of reaction)*, ΔH_r, is the energy absorbed during a chemical reaction under constant volume conditions. It is found by summing the enthalpies of formation of all products and subtracting the sum of enthalpies of formation of all reactants. This is essentially a restatement of the energy conservation principle and is known as *Hess' law of energy summation*.

Equation 21.6: Enthalpy of Reaction

$$\Delta H_r = \sum_{products} \Delta H_f - \sum_{reactants} \Delta H_f \qquad 21.6$$

Description

In Eq. 21.6, ΔH_f is the *enthalpy of formation (heat of formation)*. When tabulated values are reported at 25°C, ΔH_f^0 is known as the *standard enthalpy of formation (standard heat of formation)*. The standard enthalpy of reaction of elements in their free-state configurations is zero.

Reactions that give off energy (i.e., have negative enthalpies of reaction) are known as *exothermic reactions*. Many (but not all) exothermic reactions begin spontaneously. On the other hand, *endothermic reactions* absorb energy and require thermal or electrical energy to begin.

Example

A table of standard 25°C enthalpies of formation contains the following values.

substance	enthalpy, H
CH_4 (g)	−17.90 kcal/mol
H_2O (g)	−57.80 kcal/mol
H_2O (l)	−68.30 kcal/mol
CO_2 (g)	−94.05 kcal/mol

Calculate the lower heat of stoichiometric combustion (standardized to 25°C) of gaseous methane (CH_4) and oxygen assuming all products of combustion remain gaseous.

Chemistry

Solution

The balanced chemical equation for the stoichiometric combustion of methane is

$$CH_4 + 2O_2 \rightarrow 2H_2O + CO_2$$

The enthalpy of formation of oxygen gas (its free-state configuration) is zero. Using Eq. 21.6, the enthalpy of reaction per mole of methane is

$$\Delta H_r = 2\Delta H_{f,H_2O} + \Delta H_{f,CO_2} - \Delta H_{f,CH_4} - 2\Delta H_{f,O_2}$$

$$= (2)\left(-57.80 \ \frac{kcal}{mol}\right) + \left(-94.05 \ \frac{kcal}{mol}\right)$$

$$- \left(-17.90 \ \frac{kcal}{mol}\right) - (2)(0)$$

$$= -191.75 \ kcal/mol \ CH_4 \quad [\text{exothermic}]$$

3. SOLUTIONS

Units of Concentration

There are many units of concentration used to express solution strengths.

- *F—formality:* The number of gram formula weights (i.e., formula weights in grams) per liter of solution.

- *m—molality:* The number of gram-moles of solute per 1000 grams of solvent. A "molal" (i.e., 1 m) solution contains 1 gram-mole per 1000 grams of solvent.

- *M—molarity:* The number of gram-moles of solute per liter of solution. A "molar" (i.e., 1 M) solution contains 1 gram-mole per liter of solution. Molarity is related to normality: $N = M \times \Delta$oxidation number.

- *N—normality:* The number of gram equivalent weights of solute per liter of solution. A solution is "normal" (i.e., 1 N) if there is exactly one gram equivalent weight per liter. Molarity is related to normality: $N = M \times \Delta$oxidation number.

- *x—mole fraction:*[6] The number of moles of solute divided by the number of moles of solvent and all solutes.

- meq/L—*milligram equivalent weights of solute per liter of solution:* calculated by multiplying normality by 1000 or dividing concentration in mg/L by equivalent weight.

- mg/L—*milligrams per liter:* The number of milligrams of solute per liter of solution. Same as ppm for solutions of water.

- ppm—*parts per million:* The number of pounds (or grams) of solute per million pounds (or grams) of solution. Same as mg/L for solutions of water.

- ppb—*parts per billion:* The number of pounds (or grams) of solute per billion (10^9) pounds (or grams) of solution. Same as μg/L for solutions of water.

Solutions of Gases in Liquids

Henry's law states that the amount (i.e., concentration, mass, weight, or mole fraction) of a slightly soluble gas dissolved in a liquid is proportional to the partial pressure of the gas as long as the gas and liquid are nonreacting. This law applies separately to each gas to which the liquid is exposed, as if each gas were present alone. Using Henry's law constant, h, in atmospheres, the algebraic form of Henry's law is given by

$$p_i = y_i p = y_i h$$

Generally, the solubility of gases in liquids decreases with increasing temperature.

Solutions of Solids in Liquids

When a solid is added to a liquid, the solid is known as the *solute* and the liquid is known as the *solvent*. If the dispersion of the solute throughout the solvent is at the molecular level, the mixture is known as a *solution*. If the solute particles are larger than molecules, the mixture is known as a *suspension*.

In some solutions, the solvent and solute molecules bond loosely together. This loose bonding is known as *solvation*. If water is the solvent, the bonding process is also known as *aquation* or *hydration*.

The solubility of most solids in liquids increases with increasing temperature. Pressure has very little effect on the solubility of solids in liquids.

When the solvent has dissolved as much solute as it can, it is known as a *saturated solution*. Adding more solute to an already saturated solution will cause the excess solute to settle to the bottom of the container, a process known as *precipitation*. Other changes (in temperature, concentration, etc.) can be made to cause precipitation from saturated and unsaturated solutions.

Degree of Ionization

The degree to which a material is ionized can be calculated from the following equations.

$$pK_a - pH = \log_{10}\left(\frac{\text{nonionized form}}{\text{ionized form}}\right)$$

$$= \log_{10}\frac{[HA]}{[A^-]} \quad [\text{acids}]$$

$$pK_a - pH = \log_{10}\left(\frac{\text{ionized form}}{\text{nonionized form}}\right)$$

$$= \log_{10}\frac{[HB^+]}{[B]} \quad [\text{bases}]$$

---

[6]The *NCEES Handbook* does not assign a variable to mole fraction in the Chemistry section, which uses the variable x to represent gravimetric fraction. See Ftn. 1.

Example

The liquid form of a medicine has a pK_a of 7.8. Proper dosing requires 3 g of the medicine to be absorbed in an un-ionized form. The blood of a patient has a pH of 7.35. Most nearly, how much of the medicine should be injected in the bloodstream?

(A) 4.7 g

(B) 6.8 g

(C) 11 g

(D) 17 g

Solution

The reaction is

$$HA \rightarrow H^+ + A^-$$

$$pK_a - pH = \log_{10}\left(\frac{HA}{A^-}\right)$$

$$7.8 - 7.35 = \log_{10}\left(\frac{HA}{A^-}\right)$$

$$\frac{HA}{A^-} = 10^{0.45} = 2.818$$

$$A^- = \frac{HA}{2.818} = 0.355HA$$

The degree of ionization is 0.355. The fraction of medicine un-ionized is

$$1 - 0.355 = 0.645$$

The required injection mass is

$$m = \frac{3 \text{ g}}{0.645} = 4.65 \text{ g} \quad (4.7 \text{ g})$$

The answer is (A).

Equation 21.7 and Eq. 21.8: Solubility Product

$$A_m B_n \rightarrow mA^{n+} + nB^{m-} \qquad 21.7$$
$$K_{SP} = [A^+]^m[B^-]^n \qquad 21.8$$

Description

When an ionic solid is dissolved in a solvent, it dissociates, as shown in Eq. 21.7.

If the equilibrium constant is calculated, the terms for pure solids and liquids are omitted. The *solubility product constant*, K_{SP}, consists only of the ionic concentrations (i.e., the molarities).[7] The solubility product for slightly

---

[7](1) The *solubility product constant* is usually referred to as just the *solubility product*. (2) K_{sp} is the common representation of the solubility product. The use of an uppercase subscript, K_{SP}, is essentially unique to the *NCEES Handbook*.

---

soluble solutes is essentially constant at a standard value, as given by Eq. 21.8.

When the product of terms exceeds the standard value of the solubility product, solute will precipitate out until the product of the remaining ion concentrations attains the standard value. If the product is less than the standard value, the solution is not saturated.

The solubility products of nonhydrolyzing compounds are relatively easy to calculate. This encompasses chromates (CrO_4^{2-}), halides (F^-, Cl^-, Br^-, I^-), sulfates (SO_4^{2-}), and iodates (IO_3^-). However, compounds that hydrolyze must be evaluated differently.

Example

Calcium ions (Ca^{2+}) and carbonate ions (CO_3^{2-}) are present in 16°C water at concentrations of 25 mg/L and 15 mg/L, respectively. What is most nearly the solubility product constant for $CaCO_3$?

(A) 1.6×10^{-7} M^2

(B) 5.8×10^{-7} M^2

(C) 1.9×10^{-6} M^2

(D) 9.5×10^{-6} M^2

Solution

The dissociation reaction is

$$CaCO_3 \rightleftharpoons Ca^{2+} + CO_3^{2-}$$

From this reaction, both coefficients are 1 for the solubility product constant equation. Determine the molar concentration of both products.

Use Table 21.1. The atomic weight of calcium, Ca, is 40.078 g/mol. The molecular weight of carbonate, CO_3, is

$$MW_{CO_3} = MW_C + 3MW_O$$
$$= 12.0115 \ \frac{g}{mol} + (3)\left(15.999 \ \frac{g}{mol}\right)$$
$$= 60.0085 \text{ g/mol}$$

The molarities are

$$[Ca^{2+}] = \frac{25 \ \frac{mg}{L}}{\left(40.078 \ \frac{g}{mol}\right)\left(1000 \ \frac{mg}{g}\right)}$$
$$= 6.24 \times 10^{-4} \text{ M}$$

$$[CO_3^{2-}] = \frac{15 \ \frac{mg}{L}}{\left(60.0085 \ \frac{g}{mol}\right)\left(1000 \ \frac{mg}{g}\right)}$$
$$= 2.50 \times 10^{-4} \text{ M}$$

Chemistry

Use Eq. 21.8. The solubility constant is

$$K_{SP} = [A^+]^m[B^-]^n = [Ca^{2+}]^1[CO_3^{2-}]^1$$
$$= (6.24 \times 10^{-4} \text{ M})^1 (2.50 \times 10^{-4} \text{ M})^1$$
$$= 1.56 \times 10^{-7} \text{ M}^2 \quad (1.6 \times 10^{-7} \text{ M}^2)$$

The answer is (A).

Heat of Solution

The *heat of solution*, ΔH, is an amount of energy that is absorbed or released when a substance enters a solution. It can be calculated from the enthalpies of formation of the solution components. For example, the heat of solution associated with the formation of dilute hydrochloric acid from HCl gas and large amounts of water would be represented as follows.

$$HCl(g) \xrightarrow{H_2O} HCl(aq) + \Delta H$$

$$\Delta H = -17.21 \text{ kcal/mol}$$

If a heat of solution is negative (as it is for all aqueous solutions of gases), heat is given off when the solute dissolves in the solvent. This is an *exothermic reaction (process)*. If the heat of solution is positive, heat is absorbed when the solute dissolves in the solvent. This is an *endothermic reaction (process)*.

Boiling and Freezing Points

A liquid boils when its vapor pressure is equal to the surrounding pressure. Because the addition of a nonvolatile solute such as salt to a solvent decreases the vapor pressure (Raoult's law with the assumption of a nonvolatile solute), the temperature of the solution must be increased to maintain the same vapor pressure. The boiling point (temperature), T_b, of a solution is higher than the boiling point of the pure solvent at the same pressure.

The *boiling point elevation* is given by the following equation. K_b is the *molal boiling point constant*, which is a property of the solvent only. The molal boiling point constant for water is $0.512°C/m$.

$$\Delta T_b = mK_b$$
$$= \frac{m_{solute, in\,g}K_b}{(MW)m_{solvent, in\,kg}} \quad \text{[increase]}$$

Similarly, the freezing (melting) point, T_f, will be lower for the solution than for the pure solvent. The freezing point depression, given in the following equation, depends on the *molal freezing point constant*, K_f, a property of the solvent only. The molal freezing point constant for water is $1.86°C/m$.

$$\Delta T_f = -mK_f$$
$$= \frac{-m_{solute, in\,g}K_f}{(MW)m_{solvent, in\,kg}} \quad \text{[decrease]}$$

The given equations are for dilute, nonelectrolytic solutions and nonvolatile solutes.

Example

A solution contains 10 g NaCl in 100 g of aqueous solution. The molal boiling point constant for water is $0.512°C/mol \cdot kg$. The boiling point of the solution is most nearly

   (A) 98°C

   (B) 102°C

   (C) 104°C

   (D) 106°C

Solution

The molecular weight of NaCl is

$$22.990 \frac{g}{mol} + 35.453 \frac{g}{mol} = 58.443 \text{ g/mol}$$

For NaCl, $i = 2$ because it dissociates into two ions. The boiling point rise is

$$\Delta T_b = \frac{im_{solute, in\,g}K_b}{(MW)m_{solute, in\,g}}$$
$$= \frac{(2)(10 \text{ g})\left(0.512 \frac{°C}{mol \cdot kg}\right)\left(1000 \frac{g}{kg}\right)}{\left(58.443 \frac{g}{mol}\right)(100 \text{ g} - 10 \text{ g})}$$
$$= 1.9468°C$$

The boiling point is

$$T_b = 100°C + 1.9468°C = 101.9468°C \quad (102°C)$$

The answer is (B).

Electrochemistry

Some chemical processes, such as plating operations, require electrical energy. Other chemical processes, such as those that occur in batteries, produce electrical energy. *Electrochemistry* relates chemical reactions to electrical phenomena. Meaningful electrochemical reactions are characterized by the movement of electrons from one location to another.

An *electrochemical cell* is an arrangement of components that produces electrical energy. It is composed of two *half-cells*, each of which consists of an electrode in contact with an electrolyte. The electrolytes at the electrodes may be the same or different. A chemical reaction, known as a *half-cell reaction*, involving the transfer of electrons occurs between each electrode and its electrolyte.

For the half-cell reactions to proceed meaningfully, the electrodes must be connected in such a way that electrons can flow from one electrode to the other. This is accomplished, for example, when wires connect the electrodes to terminals of a voltmeter, light bulb, or other electrical device. The electrolytes must also be connected in such a way that allows ions formed at each terminal to interact chemically with one another. This *ionic contact* occurs when both electrodes share the same electrolyte. When each electrode has a different electrolyte, the electrolytes must be separated by a porous membrane or other feature. A common method of maintaining ionic contact between two different electrolytes is a *salt bridge*. A salt bridge consists of a pathway (e.g., the interior of a glass tube or the surface of a strip of filter paper) containing a relatively inert electrolyte (often in the form of a gel to prevent mixing with the electrolytes) through which ions can flow.

The total *potential* of the electrochemical cell is the sum of the potentials of the two half-cell reactions. *Standard potentials* for half-cell reactions are given in Table 21.3. The standard potential is the potential at *standard state*, which is usually defined as a temperature of 25°C and either (for liquids) a concentration of one mole per liter or (for gases) a pressure of 1 atm.

The *Nernst equation* is used to correct standard potentials for nonstandard conditions. For the reaction $A + B \rightarrow C + D$, the Nernst equation is

$$E = E_0 - \frac{\overline{R}T}{nF} \ln \frac{[C][D]}{[A][B]}$$

n is the number of moles of electrons transferred in the reaction. F is the *Faraday constant*, 96 485 C/mol.

Faraday's Laws of Electrolysis

An *electrolyte* is a substance that dissociates in solution to produce positive and negative ions. The solution can be an aqueous solution of a soluble salt, or it can be an ionic substance in molten form.

Electrolysis is the passage of an electric current through an electrolyte driven by an external voltage source. Electrolysis occurs when the positive terminal (the *anode*) and negative terminal (the *cathode*) of a voltage source are placed in an electrolyte. Negative ions (anions) will be attracted to the anode, where they are oxidized. Positive ions (cations) will be attracted to the cathode, where they will be reduced. The passage of ions constitutes the current.

Some reactions that do not proceed spontaneously can be forced to proceed by supplying electrical energy. Such reactions are known as *electrolytic* (*electrochemical*) *reactions*.

Faraday's laws of electrolysis can be used to predict the duration and magnitude of a direct current needed to complete an electrolytic reaction.

law 1: The mass of a substance generated by electrolysis is proportional to the amount of electricity used.

law 2: For any constant amount of electricity, the mass of substance generated is proportional to its equivalent weight.

law 3: One *faraday* of electricity (96 485 C or 96 485 A·s) will produce one gram equivalent weight.

The number of grams of a substance produced at an electrode in an electrolytic reaction can be found from

$$m_{\text{grams}} = \frac{It(\text{MW})}{(96\,485)(\text{change in oxidation state})}$$
$$= (\text{no. of faradays})(\text{GEW})$$

The number of gram-moles produced is

$$n = \frac{m}{\text{MW}}$$
$$= \frac{\text{no. of faradays}}{\text{change in oxidation state}}$$
$$= \frac{It}{(96\,485)(\text{change in oxidation state})}$$

22 Combustion

Contents

Nomenclature

A/F	mass air-fuel ratio	–
$\overline{A/F}$	molar air-fuel ratio	–
c_p	molar heat capacity	J/mol·K
H	molar enthalpy	kcal/mol
m	mass	kg
M	molecular weight	g
N	number of moles	–
T	temperature	°C
v	stoichiometric coefficient	–

Symbols

η	efficiency	–

Subscripts

f	formation or fuel
i	indicated
r	reaction
ref	reference

1. HEATS OF REACTION

The *heating value* of a fuel is the energy that is given off when the fuel is burned (usually in atmospheric air). Heating value can be specified per unit mass, per unit volume, per unit liquid volume (e.g., kJ per liter), and per mole. In engineering practice, the heating value is always stated per measurable unit. The heating value will be specified in kJ/kg for coal, and in kJ/L for oil. Heating values of fuel gases may be given in kJ/m^3 at a specified temperature and pressure that are approximately equal to normal atmospheric conditions (i.e., "room temperature"). However, fuel gases are seldom purchased by volume. In academia, heating values per mole are popular because for conditions close to room temperature, they can be derived from basic chemical and thermodynamic principles, as well as from tables of molar standard *enthalpies of formation* or *heats of formation*, ΔH_f. However, molar basis is rarely encountered outside of academia.

The enthalpy of formation of a compound is the energy absorbed during the formation of one (gram) mole of the compound from pure elements. The enthalpy of formation is defined as zero for elements (including diatomic gases such as H_2 and O_2) in their free states at standard conditions. Any deviation in temperature or phase of an element will change the value from zero. Compounds (combinations of elements) rarely have enthalpies of formation equal to zero.

When a heating value is derived from an enthalpy of formation, it is referred to as an *enthalpy of reaction* or *heat of reaction*, ΔH_r. (The general term, *heat of combustion*, is also used.) When compiled into tables, enthalpies of formation are standardized to some reference condition known as the *thermodynamic superscript*, *standard reference state*, or *reference state*, usually 25°C (298K) and 1 atm (1 bar, 100 torr). Standard enthalpies of reaction that are based on the standard enthalpies of formation are designated as ΔH_r^0.[1]

Enthalpies of reaction can also be determined in a bomb calorimeter. Since the energy given off is determined in a fixed volume, enthalpies of reaction are constant-volume values. This fact is rarely needed, however.

Chemical (including combustion) reactions that give off energy have negative enthalpies of reaction and are known as *exothermic reactions*. Many exothermic reactions begin spontaneously and/or are self-sustaining. *Endothermic reactions* absorb energy and have positive enthalpies for reaction. Endothermic reactions will continue only as long as they have energy sources.

Equation 22.1: Hess' Law

$$\left(\Delta H_r^0\right) = \sum_{\text{products}} v_i\left(\Delta H_f^0\right)_i - \sum_{\text{reactants}} v_i\left(\Delta H_f^0\right)_i \quad 22.1$$

Description

Hess' law (*Hess' law of energy summation*) is used to calculate the enthalpy of reaction from the enthalpies of formation. The enthalpy of reaction is the sum of the enthalpies of formation of the products less the sum of the enthalpies of formation of the reactants. This is illustrated by Eq. 22.1.[2]

---

[1]The NCEES *FE Reference Handbook* (*NCEES Handbook*) uses a degree symbol to designate standard state. In practice, there is considerable variation in designating standard state (e.g., the subscript "std," a stroked lowercase letter o, a superscript zero, or a circle with a horizontal bar that either does extend outside of the circle or does not. Since the intent of the symbol is to designate a zero-energy condition, this book uses a superscripted zero to designate the standard state.

[2](1) The *NCEES Handbook* sometimes uses extraneous parentheses in its equations. There is no mathematical or thermodynamic significance to the parentheses used in Eq. 22.1. As used in the *NCEES Handbook*, the parentheses around the left-hand side are particularly unnecessary. (2) Since the summations are shown to be over all of the "products" and "reactants," the increment variable, i, is not necessary. (3) Since the number of reactants is not the same as the number of products, the *NCEES Handbook*'s use of variable i for both is misleading.

Chemistry

Enthalpy of formation is tabulated on a per mole basis. Since number of moles is proportional to the number of molecules, each molecule of a reactant or product will contribute its enthalpy of formation to the overall reaction. The multiplicity of contributions is accounted for in Eq. 22.1 by the coefficient terms, v_i.[3] These terms are the species coefficients in the stoichiometric chemical reaction equation.

Example

Standard enthalpies of formation for some gases are given.

$C_2H_5OH(l)$	-228 kJ/mol
CO	-111 kJ/mol
CO_2	-394 kJ/mol
$H_2O(g)$	-242 kJ/mol
$H_2O(l)$	-286 kJ/mol
NO	$+30$ kJ/mol

Most nearly, what is the standard enthalpy of reaction for the complete combustion of ethanol, C_2H_5OH?

(A) -1400 kJ/mol

(B) -1300 kJ/mol

(C) -1100 kJ/mol

(D) -910 kJ/mol

Solution

The balanced combustion reaction is

$$C_2H_5OH + 3O_2 \rightarrow 2CO_2 + 3H_2O$$

The standard enthalpy of reaction is evaluated at 25°C, so the water vapor will be in liquid form.

Use Hess' law, Eq. 22.1. The enthalpy of reaction is

$$(\Delta H_r^0) = \sum_{\text{products}} v_i(\Delta H_f^0)_i - \sum_{\text{reactants}} v_i(\Delta H_f^0)_i$$

$$= v_{CO_2}\Delta H_{f,CO_2}^0 + v_{H_2O}\Delta H_{f,H_2O}^0$$
$$- (v_{C_2H_5OH}\Delta H_{f,C_2H_5OH}^0 + v_{O_2}\Delta H_{f,O_2}^0)$$

$$= (2)\left(-394 \; \frac{kJ}{mol}\right) + (3)\left(-286 \; \frac{kJ}{mol}\right)$$
$$- \left((1)\left(-228 \; \frac{kJ}{mol}\right) + (3)\left(0 \; \frac{kJ}{mol}\right)\right)$$

$$= -1418 \; \text{kJ/mol} \quad (-1400 \; \text{kJ/mol})$$

The answer is (A).

---

[3]The *NCEES Handbook* is inconsistent in the variable it uses to represent the stoichiometric coefficients. In Eq. 22.1, the *NCEES Handbook* uses Greek upsilon, v. In the material on the thermodynamics of chemical reaction equilibria, the *NCEES Handbook* uses lowercase italic v. This book uses Greek upsilon, v, for specific volume, and it uses lowercase italic v for stoichiometric coefficients.

Equation 22.2 and Eq. 22.3: Enthalpy of Reaction

$$\Delta H_r^0(T) = \Delta H_r^0(T_{\text{ref}}) + \int_{T_{\text{ref}}}^{T} \Delta c_p \, dT \qquad 22.2$$

$$\Delta c_p = \sum_{\text{products}} v_i c_{p,i} - \sum_{\text{reactants}} v_i c_{p,i} \qquad 22.3$$

Description

Enthalpies of formations are temperature dependent. For example, 286 kJ/mol of energy is released for each (gram) mole of 25°C liquid water produced. However, less energy is released when the water is formed at 50°C. Based on the equation $\Delta h = c_p \Delta T$ (which is good for any substance in any state), the molar enthalpy difference is $\Delta H = Mc_p\Delta T$, where c_p is a constant or mean specific heat, and M is the molecular weight.

Since enthalpies of formation are temperature dependent, enthalpies of reaction are temperature dependent. Equation 22.2, known as *Kirchhoff's law*, illustrates how the enthalpy of reaction at any temperature, T, is modified from the standard enthalpy of reaction.[4] Equation 22.3 calculates Δc_p, the *aggregate molar heat capacity*, from the molar heat capacities of all of the reactants and products.[5]

Example

The standard (25°C) heat of combustion for hydrogen fuel in a bomb calorimeter is -285.8 kJ/mol. The mean specific heat of liquid water between the temperatures of 25°C and 50°C is 4.179 kJ/kg·°C. Most nearly, what is the molar heat of combustion if the combustion products in a 25°C bomb calorimeter are cooled at 50°C instead of the normal 25°C?

(A) -295 kJ/kg

(B) -288 kJ/kg

(C) -284 kJ/kg

(D) -280 kJ/kg

---

[4](1) Although the *NCEES Handbook* frequently uses parentheses to separate multiplicative terms in equations, the parentheses used with (T) and (T_{ref}) in Eq. 22.2 mean "at that temperature." They do not mean multiplication by T and T_{ref}. (2) The use of the delta symbol for the molar aggregate heat capacity, Δc_p, within the integral is traditional to this subject. (3) Standard enthalpies of formation are always per unit mole, so the last term must also be on a molar basis. Although c_p is subsequently defined as a molar heat capacity, the symbol for specific heat capacity (unit mass basis) is used in Eq. 22.2 and Eq. 22.3. (4) The integral in Eq. 22.2 implies that the heat capacity varies with temperature. In practice, it may be approximated by a polynomial correlation that can be integrated. (5) Equation 22.2 is insufficient when a substance experiences a change of phase while cooling or heating.

[5]Some of the comments regarding other equations apply to Eq. 22.3. (1) Although c_p is defined as a molar heat capacity, the symbol for specific heat capacity (unit mass basis) is used. (2) Since the summations are shown to be over all of the "products" and "reactants," the increment variable, i, is not necessary. (3) Since the number of reactants is not the same as the number of products, the *NCEES Handbook*'s use of variable i for both is misleading.

Chemistry

Solution

The stoichiometric chemical reaction equation for the combustion of hydrogen is

$$H_2(g) + \tfrac{1}{2}O_2(g) \rightarrow H_2O(l)$$

One mole of hydrogen produces one mole of water. Roughly based on Eq. 22.2, the molar heat of combustion at 50°C is

$$\Delta H_{r,T} = \Delta H_r^0 + Mc_p(T - T^0)$$
$$= -285.8 \ \frac{kJ}{mol}$$
$$+ \frac{\left(18 \ \frac{g}{mol}\right)\left(4.179 \ \frac{kJ}{kg \cdot °C}\right)(50°C - 25°C)}{1000 \ \frac{g}{kg}}$$
$$= -283.9 \ kJ/mol \quad (-284 \ kJ/mol)$$

The answer is (C).

2. COMBUSTION

Combustion reactions involving organic compounds and oxygen take place according to standard stoichiometric principles. *Stoichiometric air (ideal air)* is the exact quantity of air necessary to provide the oxygen required for complete combustion of the fuel. Stoichiometric oxygen volumes can be determined from the balanced chemical reaction equation. Table 22.1 contains some of the more common chemical reactions.

As Table 22.1 shows, the products of complete combustion of a hydrocarbon fuel are carbon dioxide (CO_2) and water vapor (H_2O).[6] When sulfur is present in the fuel, sulfur dioxide (SO_2) is the normal product. When there is insufficient oxygen for complete combustion, carbon monoxide (CO) will be formed. Atmospheric nitrogen does not, under normal combustion conditions, dissociate and form oxides.

---

[6](1) In a discussion of heat of reaction, the *NCEES Handbook* states about combustion, "The principal products [of combustion] are $CO_2(g)$ and $H_2O(l)$." *g* represents a substance in gaseous form, and *l* represents a substance in liquid form. While this may be correct for some entries in a heat of reaction tabulation that are derived from bomb calorimetry, this is patently in error for industrial and commercial combustion. The heat of combustion ensures that any water formed will, at least initially, appear in the stack (flue) gas as a vapor. In order to avoid the corrosive effects of high temperature liquid water (and sulfuric acid), great efforts are made to ensure that water remains in a vapor state throughout the stack/flue system. (2) From the standpoint of calculating heats of combustion, using the enthalpy of formation of liquid water includes the heat of vaporization in the heat of combustion. This is the definition of *higher heat of combustion*, which is not achievable in boilers, furnaces, and incinerators. Since water vapor is not permitted to condense out, only the lower heat of combustion is available in a combustion (burner/boiler/flue) system. (3) It is standard typographical notation to represent "(*g*)" and "(*l*)" with italic letters to avoid confusion with the chemical species.

Table 22.1 Ideal Combustion Reactions

fuel	formula	reaction equation (excluding nitrogen)
carbon (to CO)	C	$2C + O_2 \rightarrow 2CO$
carbon (to CO_2)	C	$C + O_2 \rightarrow CO_2$
sulfur (to SO_2)	S	$S + O_2 \rightarrow SO_2$
sulfur (to SO_3)	S	$2S + 3O_2 \rightarrow 2SO_3$
carbon monoxide	CO	$2CO + O_2 \rightarrow 2CO_2$
methane	CH_4	$CH_4 + 2O_2$ $\rightarrow CO_2 + 2H_2O$
acetylene	C_2H_2	$2C_2H_2 + 5O_2$ $\rightarrow 4CO_2 + 2H_2O$
ethylene	C_2H_4	$C_2H_4 + 3O_2$ $\rightarrow 2CO_2 + 2H_2O$
ethane	C_2H_6	$2C_2H_6 + 7O_2$ $\rightarrow 4CO_2 + 6H_2O$
hydrogen	H_2	$2H_2 + O_2 \rightarrow 2H_2O$
hydrogen sulfide	H_2S	$2H_2S + 3O_2$ $\rightarrow 2H_2O + 2SO_2$
propane	C_3H_8	$C_3H_8 + 5O_2$ $\rightarrow 3CO_2 + 4H_2O$
n-butane	C_4H_{10}	$2C_4H_{10} + 13O_2$ $\rightarrow 8CO_2 + 10H_2O$
octane	C_8H_{18}	$2C_8H_{18} + 25O_2$ $\rightarrow 16CO_2 + 18H_2O$
olefin series	C_nH_{2n}	$2C_nH_{2n} + 3nO_2$ $\rightarrow 2nCO_2 + 2nH_2O$
paraffin series	C_nH_{2n+2}	$2C_nH_{2n+2} + (3n+1)O_2$ $\rightarrow 2nCO_2$ $+ (2n+2)H_2O$

(Multiply oxygen volumes by 3.773 to get nitrogen volumes.)

Equation 22.4 and Eq. 22.5: Air-Fuel Ratios[7]

$$\overline{A/F} = \frac{\text{no. of moles of air}}{\text{no. of moles of fuel}} \qquad 22.4$$

$$A/F = \frac{\text{mass of air}}{\text{mass of fuel}} = (\overline{A/F})\left(\frac{M_{air}}{M_{fuel}}\right) \qquad 22.5$$

Description

Stoichiometric air requirements are usually stated in units of mass (kilograms) of air for solid and liquid fuels, and in units of volume (cubic meters) of air for gaseous fuels. When stated in terms of mass, the ratio of air to fuel masses is known as the *air-fuel ratio*, A/F, given by Eq. 22.5. The molar air-fuel ratio is of academic interest. Since numbers of moles and numbers of volumes are proportional, the molar air-fuel ratio and volumetric air-fuel ratio are the same.[8]

---

[7]The *NCEES Handbook* presents the definition of the air-fuel ratio under the heading "Incomplete Combustion." The value of the ratio is dependent on the amount of air, but the definition is not. Equation 22.4 and Eq. 22.5 can be used with stoichiometric air and excess air, as well as insufficient air.

[8]As stated, the molar air-fuel ratio is an academic concept. In practice, "air-fuel ratio" always means a ratio of masses. If anything else is intended, the term "air-fuel ratio" should never be used in spoken or written communications without including "molar" or some other qualification.

Chemistry

Atmospheric Air for Combustion

Atmospheric air is a mixture of oxygen, nitrogen, and small amounts of carbon dioxide, water vapor, argon, and other inert gases. If all constituents except oxygen are grouped with the nitrogen, the air composition is as given in Table 22.2. It is helpful in many combustion problems to know the effective molecular weight of air, M_{air}, which is approximately 28.84 g/mol.[9]

Table 22.2 Composition of Dry Air[a]

component	percent by weight	percent by volume
oxygen	23.15	20.95
nitrogen/inerts	76.85	79.05
ratio of nitrogen to oxygen	3.320	3.773[b]
ratio of air to oxygen	4.320	4.773

[a]Inert gases and CO_2 are included as N_2.
[b]The value is also reported by various sources as 3.76, 3.78, and 3.784.

Stoichiometric air includes atmospheric nitrogen. For each volume (or mole) of oxygen, 3.773 volumes (or moles) of nitrogen and other atmospheric gases pass unchanged through the reaction. In combustion reaction equations, it is traditional to use a volumetric (or molar) ratio of 3.76, since that is the ratio based on a rounded air composition of 79% nitrogen and 21% oxygen. For example, the combustion of methane in air would be written as

$$CH_4 + 2O_2 + 2(3.76)N_2 \rightarrow CO_2 + 2H_2O + 7.52N_2$$

Example

In a stoichiometric octane (C_8H_{18}) combustion reaction, what is most nearly the air-fuel ratio?

(A) 12.0
(B) 12.5
(C) 14.7
(D) 15.1

Solution

Find the number of moles of air needed for complete combustion of octane.

$$C_8H_{18} + x(O_2 + 3.76N_2) \rightarrow 8CO_2 + 9H_2O + x(3.76N_2)$$
$$C_8H_{18} + 12.5(O_2 + 3.76N_2)$$
$$\rightarrow 8CO_2 + 9H_2O + 12.5(3.76N_2)$$

[9]In its table of thermodynamic properties of gases, the *NCEES Handbook* rounds this number to 29. Depending on the type of problem, this may or may not be sufficiently precise.

The total number of moles of air to fully combust 1 mole of octane is $(12.5)(1 + 3.76) = 59.5$ mol. Find the mass of air and mass of octane.

$$m_{octane} = NM = (1 \text{ mol})\left(114 \frac{g}{mol}\right)$$
$$= 114 \text{ g}$$
$$m_{air} = (59.5 \text{ mol})\left(29 \frac{g}{mol}\right)$$
$$= 1725.5 \text{ g}$$

Use Eq. 22.5 to find the A/F ratio.

$$A/F = \frac{\text{mass of air}}{\text{mass of fuel}} = \frac{1725.5 \text{ g}}{114 \text{ g}}$$
$$= 15.1$$

The answer is (D).

Equation 22.6: Percent Theoretical Air

$$\text{percent theoretical air} = \frac{(A/F)_{actual}}{(A/F)_{stoichiometric}} \times 100\%$$

$$22.6$$

Description

Complete combustion occurs when all of the fuel is burned. If there is inadequate oxygen, there will be *incomplete combustion*, and some carbon will appear as carbon monoxide in the products of combustion. As shown in Eq. 22.6, the *percent theoretical air* is the actual air-fuel ratio as a percentage of the theoretical air-fuel ratio calculated from the stoichiometric combustion equation.

Example

Assume air is 21% oxygen and 79% nitrogen. What is most nearly the percent theoretical air for the following balanced combustion reaction?

$$CH_4 + (7.14)\text{air} \rightarrow (2)H_2O + CO + (5.64)N_2$$

(A) 50%
(B) 66%
(C) 68%
(D) 75%

Solution

Find the actual air-fuel ratio.

$$(A/F)_{actual} = \frac{\text{mass of air}}{\text{mass of fuel}} = \frac{N_{air}M_{air}}{N_{fuel}M_{fuel}}$$
$$= \frac{(7.14)\left(\begin{array}{c}(0.21 \text{ mol})(32 \text{ g}) \\ + (0.79 \text{ mol})(28 \text{ g})\end{array}\right)}{(1 \text{ mol})(16 \text{ g})}$$
$$= 12.87$$

Balance the following equation to find the number of moles of air for complete combustion.

$$CH_4 + (a)\text{air} \rightarrow (b)H_2O + (d)N_2$$
$$+ (e)CO_2$$

$$CH_4 + (a)((0.21)O_2 + (0.79)N_2) \rightarrow (b)H_2O + (d)N_2$$
$$+ (e)CO_2$$

From a hydrogen balance, 2 moles of water are produced. From a carbon balance, 1 mole of carbon dioxide is produced. Therefore, 2 moles of oxygen are required to react 1 mole of carbon.

$$2 = a(0.21)$$

$$a = 9.52$$

Calculate the stoichiometric air-fuel ratio.

$$
\begin{aligned}
(A/F)_{\text{stoichiometric}} &= \frac{\text{mass of air for complete combustion}}{\text{mass of fuel}} \\
&= \frac{N_{\text{air,combustion}} M_{\text{air,combustion}}}{N_{\text{fuel}} M_{\text{fuel}}} \\
&= \frac{(9.52 \text{ mol})\left(28.84 \dfrac{\text{g}}{\text{mol}}\right)}{(1 \text{ mol})\left(16 \dfrac{\text{g}}{\text{mol}}\right)} \\
&= 17.17
\end{aligned}
$$

Use Eq. 22.6 to find the percent theoretical air.

$$
\begin{aligned}
\text{percent theoretical air} &= \frac{(A/F)_{\text{actual}}}{(A/F)_{\text{stoichiometric}}} \times 100\% \\
&= \frac{12.87}{17.17} \times 100\% \\
&= 75\%
\end{aligned}
$$

The answer is (D).

Equation 22.7: Percent Excess Air

$$\text{percent excess air} = \frac{(A/F)_{\text{actual}} - (A/F)_{\text{stoichiometric}}}{(A/F)_{\text{stoichiometric}}} \times 100\%$$

22.7

Description

Usually 10–50% excess air is required for complete combustion to occur. *Excess air* is expressed as a percentage of the stoichiometric air requirements, as shown in Eq. 22.7. Excess air appears as oxygen and nitrogen along with the products of combustion.

Example

The stoichiometric air requirement for complete combustion of a unit of fuel is 75.2 mol, and the actual air provided is 95.4 mol. What is most nearly the percent excess air?

(A) 4.5%

(B) 6.0%

(C) 21%

(D) 27%

Solution

Use Eq. 22.7 to calculate the percent excess air.

$$
\begin{aligned}
\text{percent excess air} &= \frac{(A/F)_{\text{actual}} - (A/F)_{\text{stoichiometric}}}{(A/F)_{\text{stoichiometric}}} \times 100\% \\
&= \frac{95.4 \text{ mol air} - 75.2 \text{ mol air}}{75.2 \text{ mol air}} \times 100\% \\
&= 26.9\% \quad (27\%)
\end{aligned}
$$

The answer is (D).

Chemistry

23

Organic Chemistry

1. INTRODUCTION

Organic chemistry deals with the formation and reaction of compounds of carbon, many of which are produced by living organisms. Organic compounds typically have one or more of the following characteristics.

- Organic compounds are relatively insoluble in water.[1]

- Organic compounds are soluble in organic solvents.

- Organic compounds are relatively nonionizing.

- Organic compounds are unstable at high temperatures.

The method of naming organic compounds was standardized in 1930 at the International Union Chemistry meeting in Belgium. Names conforming to the established guidelines are known as *IUC names* or *IUPAC names*.[2]

2. FUNCTIONAL GROUPS

Certain combinations of atoms occur repeatedly in organic compounds and remain intact during reactions. Such combinations are called *functional groups* or *moieties*.[3] For example, the radical OH^- is known as a *hydroxyl group*. Table 23.1 contains the most important functional groups. In this table and others similar to it, the symbols R and R' usually denote an attached hydrogen atom or other hydrocarbon chain of any length. They may also denote some other group of atoms. RO designates a hydrocarbon chain with an additional oxygen atom. Aromatic ring structures, usually derivatives of benzene, are designated Ar. Halogen atoms are designated X.

---

[1]This is especially true for hydrocarbons. However, many organic compounds containing oxygen are water soluble. The sugar family is an example of water-soluble compounds.
[2]IUPAC stands for *International Union of Pure and Applied Chemistry*.
[3]Although "moiety" is often used synonymously with "functional group," there is a subtle difference. When a functional group combines into a compound, its moiety may gain/lose an atom from/to its combinant. Therefore, moieties differ from their original functional groups by one or more atoms.

Table 23.1 Functional Groups of Organic Compounds

name	standard symbol	formula	number of single bonding sites
aldehyde		CHO	1
alkyl	[R]	C_nH_{2n+1}	1
alkoxy (alkoxyl)	[RO]	$C_nH_{2n+1}O$	1
amine (amino, $n=2$)		NH_n	$3-n\,[n=0,1,2]$
aryl (aromatic ring)	[Ar]	C_6H_5	1
carbinol		COH	3
carbonyl (keto)	[CO]	CO	2
carboxyl		COOH	1
ester		COO	2
ether		O	2
halogen (halo or halide)	[X]	Cl, Br, I, or F	1
hydroxyl		OH	1
nitrile		CN	1
nitro		NO_2	1

Example

Which functional group is represented by the chemical formula $CH_3CH_2NHCH(CH_3)_2$?

(A) aldehyde

(B) alkyl

(C) amine

(D) ester

Solution

Ethyl, propyl, and possibly methyl groups are present, but those are not listed options. Amine is the only group that contains bonded nitrogen-hydrogen atoms. This is *n*-ethyl-2-propanamine (ethylisopropylamine), a propane (propyl) chain with an amine attached to the second carbon (hence, the "2-") and with an ethyl group attached to the nitrogen. The molecule can be visualized as CH_3CH_2-NH-$CH(CH_3)_2$ or as ethyl-amine-propyl.

The answer is (C).

Table 23.2 *Families (Chemical Classes) of Organic Compounds*

family (chemical class)	structure[a]	example
organic acids		
carboxylic acids	[R]-COOH	acetic acid ((CH_3)COOH)
fatty acids	[Ar]-COOH	benzoic acid (C_6H_5COOH)
alcohols		
aliphatic	[R]-OH	methanol (CH_3OH)
aromatic	[Ar]-[R]-OH	benzyl alcohol ($C_6H_5CH_2OH$)
aldehydes	[R]-CHO	formaldehyde (HCHO)
alkyl halides		
(haloalkanes)	[R]-[X]	chloromethane (CH_3Cl)
amides	[R]-CO-NH_n	β-methylbutyramide ($C_4H_9CONH_2$)
amines	[R]$_{3-n}$-NH_n	methylamine (CH_3NH_2)
	[Ar]$_{3-n}$-NH_n	aniline ($C_6H_5NH_2$)
primary amines	$n = 2$	
secondary amines	$n = 1$	
tertiary amines	$n = 0$	
amino acids	CH-[R]-(NH_2)COOH	glycine ($CH_2(NH_2)COOH$)
anhydrides	[R]-CO-O-CO-[R']	acetic anhydride ($(CH_3CO)_2O$)
arenes (aromatics)	ArH = C_nH_{2n-6}	benzene (C_6H_6)
aryl halides	[Ar]-[X]	fluorobenzene (C_6H_5F)
carbohydrates	$C_x(H_2O)_y$	dextrose ($C_6H_{12}O_6$)
sugars		
polysaccharides		
esters	[R]-COO-[R']	methyl acetate (CH_3COOCH_3)
ethers	[R]-O-[R]	diethyl ether ($C_2H_5OC_2H_5$)
	[Ar]-O-[R]	methyl phenyl ether ($CH_3OC_6H_5$)
	[Ar]-O-[Ar]	diphenyl ether ($C_6H_5OC_6H_5$)
glycols	$C_nH_{2n}(OH)_2$	ethylene glycol ($C_2H_4(OH)_2$)
hydrocarbons		
alkanes (single bonds)[b]	RH = C_nH_{2n+2}	octane (C_8H_{18})
saturated hydrocarbons		
cycloalkanes (cycloparaffins)		
	C_nH_{2n}	cyclohexane (C_6H_{12})
alkenes (double bonds between two carbons)[c]	C_nH_{2n}	ethylene (C_2H_4)
unsaturated hydrocarbons		
cycloalkenes	C_nH_{2n-2}	cyclohexene (C_6H_{10})
alkynes (triple bonds between two carbons)	C_nH_{2n-2}	acetylene (C_2H_2)
unsaturated hydrocarbons		
ketones	[R]-[CO]-[R]	acetone (($CH_3)_2CO$)
nitriles	[R]-CN	acetonitrile (CH_3CN)
phenols	[Ar]-OH	phenol (C_6H_5OH)

[a]See Table 23.1 for definitions of [R], [Ar], [X], and [CO].
[b]Alkanes are also known as the *paraffin series* and *methane series*.
[c]Alkenes are also known as the *olefin series*.

3. FAMILIES OF ORGANIC COMPOUNDS

For convenience, organic compounds are categorized into families, or *chemical classes*. "R" is an abbreviation for the word "radical," though it is unrelated to ionic radicals.[4] Compounds within each family have similar structures based on similar combinations of groups. For example, all alcohols have the structure [R]-OH, where [R] is any alkyl group and OH is the hydroxyl group.

Families of compounds can be further subdivided into subfamilies. For example, the hydrocarbons are classified into *alkanes* (single carbon-carbon bond), *alkenes* (double carbon-carbon bond), and *alkynes* (triple carbon-carbon bond).

Table 23.2 contains the most common organic families, and Table 23.3 gives synthesis routes for various classes of organic compounds.

Table 23.4 gives the common names and molecular formulas for some industrial organic chemicals.

---

[4]"R" may be interpreted as "the *rest* of the molecule."

Table 23.3 *Synthesis Routes for Various Classes of Organic Compounds*

organic acids
  oxidation of primary alcohols
  oxidation of ketones
  oxidation of aldehydes
  hydrolysis of esters
alcohols
  oxidation of hydrocarbons
  reduction of aldehydes
  reduction of organic acids
  hydrolysis of esters
  hydrolysis of alkyl halides
  hydrolysis of alkenes (aromatic hydrocarbons)
aldehydes
  oxidation of primary and tertiary alcohols
  oxidation of esters
  reduction of organic acids
amides
  replacement of hydroxyl group in an acid with an amino
    group
anhydrides
  dehydration of organic acids (withdrawal of one water
    molecule from two acid molecules)
carbohydrates
  oxidation of alcohols
esters
  reaction of acids with alcohols (*ester alcohols*)*
  reaction of acids with phenols (*ester phenols*)
  dehydration of alcohols
  dehydration of organic acids
ethers
  dehydration of alcohol
hydrocarbons
  alkanes: reduction of alcohols and organic acids
            hydrogenation of alkenes
  alkenes: dehydration of alcohols
            dehydrogenation of alkanes
ketones
  oxidation of secondary and tertiary alcohols
  reduction of organic acids
phenols
  hydrolysis of aryl halides

*The reaction of an organic acid with an alcohol is called *esterification.*

Table 23.4 *Common Names and Molecular Formulas of Some Industrial Organic Chemicals*

chemical name	common name	molecular formula
isopropyl benzene	cumene	$C_6H_5CH(CH_3)_2$
vinyl benzene	styrene	$C_6H_5CH{=}CH_2$
vinyl chloride	–	$CH_2{=}CHCl$
ethylene oxide	–	C_2H_4O
p-dihydroxy benzene	hydroquinone	$C_6H_4(OH)_2$
vinyl alcohol	–	$CH_2{=}CHOH$
phenol	carbolic acid	C_6H_5OH
aminobenzene	aniline	$C_6H_5NH_2$
urea	–	$(NH_2)_2CO$
methyl benzene	toluene	$C_6H_5CH_3$
dimethyl benzene	xylene	$C_6H_4(CH_3)_2$
2,2-dimethylpropane	neopentane	$CH_3C(CH_3)_2CH_3$

Example

In what family does the compound C_4H_6 belong?

  (A) alkane

  (B) alkyne

  (C) cycloalkene

  (D) glycol

Solution

The compound formula matches the general formula of C_nH_{2n-2}. This is a cycloalkene. (The compound is cyclobutene.)

The answer is (C).

4. SYMBOLIC REPRESENTATION

The nature and structure of organic groups and families cannot be explained fully without showing the types of bonds between the elements. Figure 23.1 illustrates the symbolic representation of some of the functional groups and families, and Fig. 23.2 illustrates reactions between organic compounds.

Example

Which of the following is an amino acid?

  (A) $CH_3COOCHCH_2$

  (B) CH_3CHNH_2COOH

  (C) CH_3SCH_3

  (D) $C_6H_5CH_2OH$

Solution

An amino acid is a bifunctional molecule consisting of an amine group and a carboxyl group. A carboxyl group has the formula COOH. In option (B), the amine group (ending with NH_2) is visible on the left, and the carboxyl group on the right completes the amino acid.

$$CH_3 - \overset{\overset{\displaystyle NH_2}{|}}{\underset{\underset{\displaystyle H}{|}}{C}} - COOH$$

None of the other options contain either a carboxyl or amine group. This compound is alanine, a popular amino acid used in body building. Option (A) is vinyl acetate. Option (C) is dimethyl sulfide. Option (D) is benzyl alcohol.

The answer is (B).

Figure 23.1 *Representation of Functional Groups and Families*

group	group representation	family	family representation
aldehyde	$-CH$ or $-C$ (with H and O)	aldehyde	$R-C$ (with H and O)
amino	$-NH_2$	amine	$R-NH_2$
aryl (benzene ring)	(benzene ring)	aryl halide	(benzene ring)$-X$
carbonyl (keto)	$-C-$ or $-C$ (with O)	ketone	$R-C$ (with R and O)
carboxyl	$-C-OH$ or $-C$ (with O-H and O)	organic acid	$R-C$ (with O-H and O)
ester	$-C-OR$ or $-C$ (with O-R and O)	ester	$R-C$ (with O-R and O)
hydroxyl	$-OH$	alcohol	$R-OH$

Figure 23.2 *Reactions Between Organic Compounds*

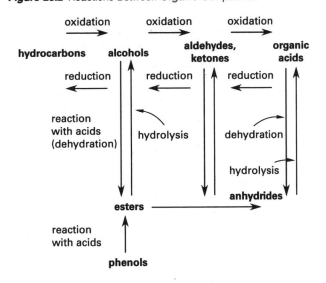

5. ORGANIC COMPOUND NAMES

In the early days of organic chemistry, there were no systematic conventions for naming organic compounds. Accordingly, many *common* or *trivial names* based on sources and/or histories of the compounds were used. These common names did not provide any meaningful information about how the compounds might have been related to one another.

To address the need for systematic, logical naming conventions, the International Union of Pure and Applied Chemistry (IUPAC) developed sets of rules for naming organic compounds (as published in the IUPAC *Blue Book*[5]) and inorganic compounds (as published in the IUPAC *Red Book*[6]). Following the IUPAC rules, the name of each compound communicates an unambiguous structural formula for the compound. Table 23.5 at the end of this chapter contains a selection of IUPAC naming conventions.[7]

Example

What is the most appropriate name for a compound with the structure shown?

(cyclopentane structure with Br)

- (A) bromocyclopentane
- (B) pentabromide
- (C) 2-hexabromide
- (D) bromylpentane

Solution

The structure contains five carbons in a nonaromatic ring, so it is a cyclopentane (C_5H_{10}) compound. With the bromine atom replacing one of the 10 hydrogen atoms, the compound would most likely be designated bromocyclopentane.

The answer is (A).

Example

What is the common name for the compound represented by the shorthand structure shown?

(benzene ring structure with NH_2)

---

[5]*Nomenclature of Organic Chemistry: IUPAC Recommendations and Preferred Names*
[6]*Nomenclature of Inorganic Chemistry: IUPAC Recommendations*
[7]However, unless an IUPAC name is simpler or an unambiguous structural definition is required, the common name of a compound may still be used.

(A) acrylonitrile

(B) aniline

(C) denatured alcohol

(D) hydrocyanide

Solution

The structure shown contains an amine (NH_2) group and a C_6H_5 group. The complete chemical formula is $C_6H_5NH_2$, which is aniline. The full atomic-bond representation is

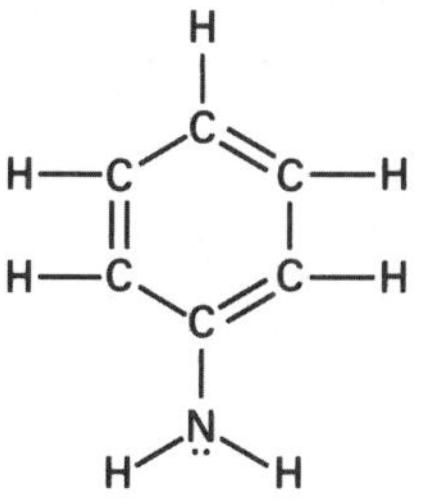

The answer is (B).

Naming Conventions for Alcohols

Alcohols have the general structure of [R]-OH, where the hydroxyl group is attached someplace in the alkyl group chain. An alcohol may be categorized as a primary alcohol, secondary alcohol, or tertiary alcohol, depending on to what the hydroxyl group is attached. In a *primary alcohol* (also known as a *1° alcohol*), the hydroxyl group is attached to a carbon atom (i.e., a *primary carbon atom*) that attaches to only one other carbon atom. In a *secondary alcohol* (also known as a *2° alcohol*), the hydroxyl group is attached to a carbon (i.e., a *secondary carbon atom*) that, itself, also is attached to two other carbon atoms. Similarly, in a *tertiary alcohol* (also known as a *3° alcohol*), the hydroxyl group is attached to a carbon (i.e., a *tertiary carbon atom*) that, itself, also is attached to three other carbon atoms.

Three different naming conventions are in use with alcohols. With the first naming convention, the name of the alkyl group attached to the hydroxyl group is specified as the surname, and the word *alcohol* is added. For example, using this convention, the CH_3OH is named "methyl alcohol," since CH_3 is a methyl group.[8] CH_3CH_2OH (see Fig. 23.3) is the primary alcohol known as "ethyl alcohol," since CH_3CH_2 is an ethyl group. $(CH_3)_2COH$ is a secondary alcohol known as "isopropyl alcohol." $(CH_3)_3COH$ is a tertiary alcohol known as "tert-butyl alcohol."

With the second naming convention (the modified *Geneva system*), the longest carbon chain containing the hydroxyl group determines the surname; the ending *e*

[8]Methyl alcohol is not a primary, secondary, or tertiary alcohol according to the first *naming* convention. It is in a category of its own.

Figure 23.3 Types of Alcohols

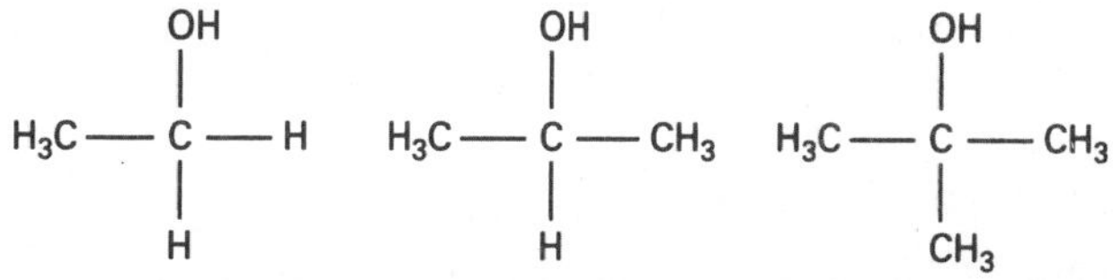

(a) primary alcohol (b) secondary alcohol (c) tertiary alcohol

of the corresponding hydrocarbon is replaced by *ol*; the carbons in the chain containing the hydroxyl group chain are numbered from the end that gives the hydroxyl group the smaller number; and, the side chains are named and numbered by their carbon attachment point numbers. With this system, CH_3OH is named "methanol." The primary alcohol CH_3CH_2OH (see Fig. 23.3) is named "ethanol." The secondary alcohol $(CH_3)_2COH$ is named "1-propanol" or "dimethyl ethanol." The tertiary alcohol $(CH_3)_3COH$ is known as "2-methyl-2-propanol."

With the third (archaic) naming convention, the base alcohol is named *carbinol*, and the higher degree alcohols are considered derivatives of the group. For example, CH_3OH is named "carbinol." The primary alcohol CH_3CH_2OH is named "methyl carbinol." The secondary alcohol $(CH_3)_2COH$ is named "dimethyl carbinol." $(CH_3)_3COH$ is named "trimethyl carbinol."

Example

What name for the alcohol structure shown is NOT correct?

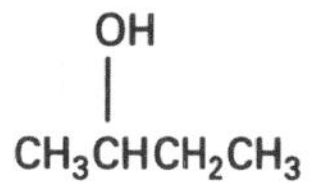

(A) methyl-butyl carbinol

(B) 2-butanol

(C) *sec*-butyl alcohol

(D) *s*-butanol

Solution

The structure is a four-carbon chain, with the hydroxyl on the second carbon. Since the alcoholic carbon is connected to two other carbons, this is a secondary alcohol, which can be indicated with "2-," "sec," or "s." Disregarding the hydroxyl group, the chain has the basic formula of C_4H_9, which is the butyl group. "2-butanol," "s-butanol," "sec-butyl alcohol," and "s-butyl alcohol" are all valid names.

The CH_3 to the left of the alcoholic carbon is a methyl group. The remainder of the chain attached to the alcoholic carbon is CH_3CH_2, which is an ethyl group. Another name for this compound is "methyl-ethyl carbinol."

The answer is (A).

Naming Conventions for Ethers

Ethers are organic compounds in which an oxygen atom is connected to two carbon groups. They have the general structure of [R]-O-[R], where [R] is a carbon chain member of the alkyl group.

Ethers are generally designated by naming according to their alkyl groups in alphabetical order and then adding "ether." In more complex ethers, since the group [R]-O is known as an *alkoxyl group*, ethers may also be referred to by the "alkoxy" derivatives of their alkyl groups. The ether group is named as an alkoxy substituent, in which the "yl" ending of alkyl groups is replaced by "oxy." For example, a methyl group with an attached oxygen, OCH_3, is a *methoxy group*. In cyclic ethers, the prefix *oxa-* is used to indicate replacement of a carbon by an oxygen.

Example

What name for the structure shown is NOT correct?

$$CH_3CH_2 \longrightarrow O \longrightarrow CH_2CH_3$$

(A) diethyl ether

(B) diethylene oxide

(C) ethoxyethane

(D) ethyl-ethyl ether

Solution

The oxygen atom is connected to two identical ethyl groups, so the names "diethyl ether" and "ethyl-ethyl ether" are valid. The combination of the first ethyl group and oxygen, CH_3CH_2O, could be referred to as "ethoxy," making "ethoxyethane" a valid name. The name "diethylene oxide" is attractive, but there are very few organic oxides, and most have relatively simple structures. Rarely do they contain both carbon and hydrogen.

The answer is (B).

Naming Conventions for Carboxylic Acids

A *carboxylic acid* is an acid that contains a carboxyl group (COOH). The general formula of a carboxylic acid is [R]-COOH or [R]-C(O)OH. (The parenthetical "(O)" refers to an oxygen atom that is on its own branch.) The rest of the molecule, [R], may be quite complex and large, as in amino acids. The most simple structures, linear chain carboxylic acids, have specific names which depend on the number of carbon atoms in the molecule.[9] IUPAC names have also been established for carboxylic acids, as shown in Table 23.6.

---

[9]This is a situation requiring rote memorization.

Table 23.6 Carboxylic Acids

carbon atoms	common name	formula
1	formic acid	HCOOH
2	acetic acid	CH_3COOH
3	propionic acid	CH_3CH_2COOH
4	butyric acid	$CH_3(CH_2)_2COOH$
5	valeric acid	$CH_3(CH_2)_3COOH$
6	caproic acid	$CH_3(CH_2)_4COOH$
7	enanthic acid	$CH_3(CH_2)_5COOH$
8	caprylic acid	$CH_3(CH_2)_6COOH$
9	pelargonic acid	$CH_3(CH_2)_7COOH$
10	capric acid	$CH_3(CH_2)_8COOH$

Example

Most likely, what type of molecule is represented by the structure shown?

(A) alcohol

(B) aldehyde

(C) amino acid

(D) ketone

Solution

The molecule contains a carboxyl (COOH) group and two amine groups. This is lysine, an amino acid.

The answer is (C).

Naming Conventions for Aldehydes

The structure of an *aldehyde* is CHO. Aldehydes are commonly formed from the oxidation of alcohols. The common name of an aldehyde is derived from the acid that is formed when aldehyde is oxidized. The acid and aldehyde molecules have the same number of carbon atoms, so the "*ic acid*" in the acid's name is replaced with *aldehyde*.

Example

The alcohol 1-butanol is catalytically oxidized according to the chemical reaction shown in order to form butanoic acid.

What is the name of the intermediate compound formed?

(A) alkoxyanaldehyde

(B) butanaldehyde

(C) methanaldehyde

(D) propanaldehyde

Solution

The intermediate product of oxidizing an alcohol to form an acid gets its name from the acid formed. "*ic acid*" in the acid's name is replaced with *aldehyde*. So, butanoic acid yields "butanoaldehyde," "butanaldehyde," or "butyraldehyde."

The answer is (B).

Naming Conventions for Ketones

Ketones have structures of either [R]-CO-[R] or [R]-CO-[R'] and can be formed when acids are heated in the absence of oxygen, a process known as *pyrolysis*.[10] Accordingly, the ketone's name is derived from the acid's name. Alternatively, the two alkyl groups are named and the compound is identified as a ketone (e.g., "ethyl methyl ketone").

Example

Which process could have formed the compound shown?

$$CH_3CH_2 - \overset{\overset{\displaystyle O}{\displaystyle \|}}{C} - CH_3$$

(A) hydrolysis of a carboxylic acid

(B) pyrolysis of a ketone

(C) pyrolysis of an acid

(D) reduction of a primary alcohol

Solution

The molecule shown (methyl ethyl ketone, MEK, or butanone by common name) has the structure of a ketone, [R]-CO-[R']. Ketones can be formed in the pyrolysis of secondary acids or by oxidizing secondary alcohols.

The answer is (C).

Naming Conventions for Unsaturated Acyclic Hydrocarbons

Unsaturated acyclic hydrocarbons (UAHs) are a class of compounds that (1) contain hydrogen and carbon atoms

only (i.e., the "hydrocarbon" designation), (2) are linear chains (i.e., the "acyclic" designation, as opposed to ring structures) of carbons, and (3) contain at least one double or triple covalent bond between adjacent carbon atoms (i.e., the "unsaturated" designation).

Alkenes[11] are UAHs that contain only a single double covalent bond; *dienes* are UAHs with two double covalent bonds; *trienes* are UAHs with three double covalent bonds. *Alkynes* are UAHs with a single triple covalent bond. (See Table 23.7.)

Example

In what category of hydrocarbons does the structure shown belong?

(A) alkene

(B) alkyne

(C) diene

(D) triene

Solution

The structure is C_5H_8, which follows the C_nH_{2n-2} formula. The compound is either a diene or an alkyne. Since the structure has a carbon triple bond, it is an alkyne. The compound is pentyne.

The answer is (B).

6. FORMATION OF ORGANIC COMPOUNDS

There are usually many ways of producing an organic compound. The types of reactions contained in this section deal only with the interactions between the organic families. The following processes are referred to.

- *oxidation:* replacement of a hydrogen atom with a hydroxyl group

- *reduction:* replacement of a hydroxyl group with a hydrogen atom

- *hydrolysis:* addition of one or more water molecules

- *dehydration:* removal of one or more water molecules

---

[10]Ketone production is relatively rare. The process described in the NCEES *FE Reference Manual* (*NCEES Handbook*), pyrolysis, is rarely used. Ketones are more commonly produced by oxidizing secondary alcohols.

[11]The *NCEES Handbook*'s description of unsaturated acyclic hydrocarbons begins with, "The simplest compounds in this class of hydrocarbon chemicals are olefins or alkenes...." This implies that olefins and alkenes are somehow equivalent. *Olefins* are a class of compounds that, like alkenes, contain only a single double carbon bond. However, olefins include cyclic (ring) structures, while alkenes do not. Thus, alkenes are a subset of olefins.

Table 23.7 Representative Unsaturated Acyclic Hydrocarbons[a]

n	alkenes, C_nH_{2n}	dienes, C_nH_{2n-2}	trienes, C_nH_{2n-4}	alkynes, C_nH_{2n-2}
2	C_2H_4 (ethylene)	–	–	C_2H_2 (ethyne; acetylene)
3	C_3H_6 (propylene)	C_3H_4 (propadiene)	C_3H_2 (propynylidene or propargylene)[b]	C_3H_4 (propyne)
4	C_4H_8 (butene)	C_4H_6 (butadiene)	C_4H_4 (butatriene)	C_4H_6 (butyne)
5	C_5H_{10} (pentene)	C_5H_8 (pentadiene)	C_5H_6 (pentatriene)	C_5H_8 (pentyne)
6	C_6H_{12} (hexene)	C_6H_{10} (hexadiene)	C_6H_8 (hexatriene)	C_6H_{10} (hexyne)
7	C_7H_{14} (heptene)	C_7H_{12} (heptadiene)	C_7H_{10} (heptatriene)	C_7H_{12} (heptyne)

[a]Various other isomers, including cyclic forms, with the same chemical formulas exist.
[b]The two linear forms of C_3H_2 do not follow the naming convention.

Example

What type of reaction is represented by the formation of maltose from two glucose molecules?

$$C_6H_{12}O_6 + C_6H_{12}O_6 \rightarrow C_{12}H_{22}O_{11} + H_2O$$

(A) dehydration

(B) hydration

(C) hydrolysis

(D) reduction

Solution

A water molecule is released (i.e., removed), so this is dehydration.

The answer is (A).

Chemistry

Table 23.5 *Selected IUPAC Organic Compound Naming Conventions*

name component	meaning	example
1-, 2-, 3-, etc.	in a straight carbon chain (e.g., a normal alkyl), the number designates the position (counting from the nearest end) of the carbon atom in the chain to which the named functional group is attached	2-methylpentane, $CH_3C_5H_{11}$ (a methyl group is attached to the second carbon in pentane, replacing the hydrogen at that point)
1,2-	the named functional group is attached to the first and second carbons in the chain	1,2-dichloroethane, $C_2H_4Cl_2$
1-[R]-2-[R'], etc.	one named functional group is attached to the first carbon in the chain, and the second group is attached to the second carbon	2-methy-1-propyl alcohol (2-methyl-1-propanol)
2,2-	the named functional group is attached twice to the second carbon in the chain; same as "di-"	2,2-dichloropentane, $C_5H_{10}Cl_2$
alkyl	contains an alkyl group, formed by removing one hydrogen from an alkane, with a resulting structure of C_nH_{2n+1}; characterized by carbon-carbon single bonds	alkyl bromide (generic), CH_3CH_2Br
aromatic (arene)	contains an aromatic (benzene) ring structure consisting of six carbons: three with single bonds and three with double bonds, each carbon having two hydrogens	benzene
aryl	contains an aromatic (benzene) ring with replacements for one or more hydrogen atoms	iodobenzene, C_6H_5I
butyl	contains a butyl group, generally in the form of C_4H_9 (i.e., $C(CH_3)_3$); has four carbons	butyl nitrite, $C_4H_9NO_2$
cis-	designates the more chemically active stereoisomer of the compound, having two atoms of the same element attached on one side of the molecule	*cis*-1,2-dichloroethane
cyclic (cyclo-)	has a nonaromatic carbon ring structure consisting of three, four, five, six, or seven carbons connected with single bonds; each carbon has two hydrogens or suitable replacements	cyclohexane, C_6H_{12}
di-	contains two of the groups	diethyl sulfate, $(C_2H_5)_2SO_4$
ethyl	contains an ethyl group, C_2H_5, derived from ethane (C_2H_6); has two carbons	ethyl alcohol, C_2H_5OH
halide	contains one or more halogen atoms (e.g., Br, Cl, I, F, etc.)	methyl chloride (chloromethane), CH_3Cl
iso-	is an isomer of an alkyl group (of six carbons or less) having a single one-carbon side-chain (branch) on the next-to-last carbon of the chain (as contrasted with *n*- and neo-)	isopentane, C_5H_{12}

(continued)

name component	meaning	example
m- or meta- (same as 1,3- in ring structures)	in a benzene ring (arene) structure, two attached functional groups skip one positions (i.e., attach to positions 1 and 3)	*m*-xylene
methyl	contains a methyl group, CH_3, derived from methane (CH_4); has one carbon	methyl amine, CH_3NH_2
n-	"normal," meaning that the molecular form is straight-line and does not have side chains; is an isomer of an alkyl group (of six carbons or less) that has a single one-carbon side-chain (branch) on the next-to-last carbon of the chain (as contrasted with iso- and neo-)	*n*-pentane (pentane), C_5H_{12}
neo-	is an isomer of an alkyl group (of six carbons or less) having two one-carbon side-chains (branches) on the next-to-last carbon of the chain, and these side chains are part of a *tert*-butyl group (as contrasted with *n*- and iso-)	neopentane, C_5H_{12}
o- or ortho- (same as 1,2- in ring structures)	in a benzene ring (arene) structure, two attached functional groups occupy adjacent positions (positions 1 and 2)	*o*-xylene
p- or para- (same as 1,4- in ring structures)	in a benzene ring (arene) structure, two attached functional groups occupy opposite positions (positions 1 and 4)	*p*-xylene
phenol(ic)	contains an aromatic (benzene) ring structure in which at least one hydrogen is replaced with a hydroxyl group	phenol, C_6H_6O
propyl	contains a propyl group, C_3H_7, derived from propane (C_3H_7); has three carbons	propyl acetate, $CH_3COOC_3H_7$
sec-	in an alcohol, a secondary alcohol where the carbon attached to the hydroxyl radial is attached to two alkyl groups	sec-propyl alcohol, (isopropryl alcohol) $(CH_3)_2CHOH$
t- or *tert-*	in an alcohol, a tertiary alcohol, where the carbon attached to the hydroxyl radial is attached to three alkyl groups	*tert*-butyl alcohol, $(CH_3)_3COH$
trans-	designates the less chemically active stereoisomer of the compound, having two atoms of the same element attached on opposite sides of the molecule	*trans*-1,2-dichloroethane
vinyl	contains the ethylene molecule ($H_2C{=}CH_2$) minus one hydrogen atom, resulting in the group $-CH{=}CH_2$; characterized by carbon-carbon double bonds	vinyl chloride (chloroethene), C_2H_3Cl

24 Biochemistry

Nomenclature

BOD	biochemical oxygen demand	mg/L
DO	dissolved oxygen	mg/L
F	fraction of influent	–
k	logistic growth rate constant	d^{-1}
K	biodegradation rate constant	d^{-1}
K	reaction rate constant	d^{-1}
K_a	partition coefficient	–
K_{eq}	equilibrium constant	–
L	ultimate BOD	mg/L
MW	molecular weight	g/mol
Ncc	number of carbons in biomass equivalent molecule	–
Ncp	number of carbons in bioproduct equivalent molecule	–
Ncs	number of carbons in substrate equivalent molecule	–
pK_a	ionization constant	–
P	fraction of digester suspended solids	–
Q	volumetric flow rate	m^3/s or m^3/d
Q_o	heat evolved per equivalent of available electrons, 26.95	kcal/mol
RQ	respiratory quotient	–
t	time	d
TWA	time-weighted average	ppm
x	cell or organism number or concentration	–
x_0	initial concentration	mg/L
x_∞	carrying capacity	mg/L
X	suspended solids	–
Y	yield coefficient	–

Symbols

α	energy balance multiplier	–
β	energy balance multiplier	–
γ	degree of reduction	–
δ	energy balance multiplier	–
μ	specific growth rate	d^{-1}

Subscripts

5	5 days
a	acid
b	biomass
d	digester or diluted
I	influent
p	product
s	substrate
v	volatile
x	energy balance multiplier
y	energy balance multiplier
z	energy balance multiplier

1. CELL STRUCTURE

A cell is the fundamental unit of living organisms. Organisms are classified as either prokaryotes or eukaryotes. A *prokaryote* is a cellular organism that does not have a distinct nucleus. Examples of prokaryotes are bacteria and blue-green algae. Figure 24.1 shows the features of a typical prokaryotic cell.

Figure 24.1 *Prokaryotic Cell Features*

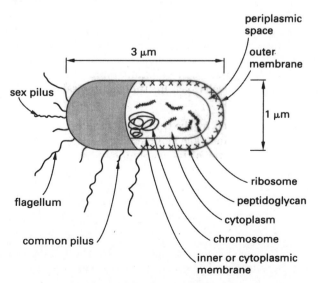

A *eukaryote* is an organism composed of one or more cells containing visibly evident nuclei and *organelles* (structures with specialized functions). Eukaryotic cells are found in protozoa, fungi, plants, and animals. For a long time, it was thought that eukaryotic cells were composed of an outer membrane, an inner nucleus, and a large mass of cytoplasm within the cell. Better experimental techniques revealed that eukaryotic cells also contain many organelles. Figure 24.2 shows the features of a typical animal and a typical plant cell. Note the size scale for the cells in Fig. 24.1 and Fig. 24.2.

Figure 24.2 *Eukaryotic Cell Features*

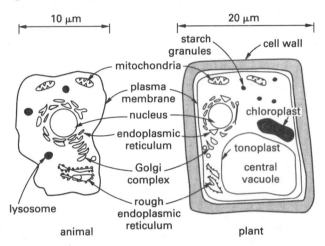

A *cell membrane* consists of a double phospholipid layer in which the polar ends of the molecules point to the outer and inner surfaces of the membrane and the nonpolar ends point to the center. A cell membrane also contains proteins, cholesterol, and glycoproteins. Some organisms, such as plant cells, have a cell wall made of carbohydrates that surrounds the cell membrane. Many single-celled organisms have specialized structures called *flagella* that extend outside the cell and help them to move.

Using the *gram stain process*, some cell membranes can be stained crystal violet for ease of research. These *gram-positive* cells have peptidoglycan walls with surface proteins that accept the crystal violet stain and appear purple. Many pathogenic cells, however, do not readily accept such staining and are referred to as *gram-negative* cells. Gram-negative cells have complex walls with violet stain-resistant lipid outer layers. Gram-negative cells pick up the counterstain safranin and appear pink.

Cell size is limited by the transport of materials through the cell membrane. The volume of a cell is proportional to the cubic power of its average linear dimension; the surface area of a cell is proportional to the square of the average linear dimension. The material (nutrients, etc.) transported through a cell's membrane is proportional to its surface area, while the material composing the cell is proportional to the volume of the cell.

The *cytoplasm* includes all the contents of the cell other than the nucleus. *Cytosol* is the watery solution of the cytoplasm. It contains dissolved glucose and other nutrients, salts, enzymes, carbon dioxide, and oxygen. In eukaryotic cells, the cytoplasm also surrounds the organelles.

The *endoplasmic reticulum* is one example of an organelle that is found in all eukaryotic cells. A rough endoplasmic reticulum has sub-microscopic organelles called ribosomes attached to it. Amino acids are bound together at the surface of ribosomes to form proteins.

The *Golgi apparatus*, or *Golgi complex*, is an organelle found in most eukaryotic cells. It looks like a stack of flattened sacks. The Golgi apparatus is responsible for accepting materials (mostly proteins), making modifications, and packaging the materials for transport to specific areas of the cell.

The *mitochondria* are specialized organelles that are the major site of energy production via aerobic respiration in eukaryotic cells. They convert organic materials into energy.

Lysosomes are other organelles located within eukaryotic cells that serve the specialized function of digestion within the cell. They break lipids, carbohydrates, and proteins into smaller particles that can be used by the rest of the cell.

Chloroplasts are organelles located within plant cells and algae that conduct photosynthesis. Chlorophyll is the green pigment that absorbs energy from sunlight during photosynthesis. During *photosynthesis*, the energy from light is captured and eventually stored as sugar.

Vacuoles are found in some eukaryotic cells. They serve many functions including storage, separation of harmful materials from the remainder of the cell, and maintaining fluid balance or cell size. Vacuoles are separated from the cytoplasm by a single membrane called the tonoplast. Most mature plant cells contain a large central vacuole that occupies the largest volume of any single structure within the cell.

Two main types of nucleic acids are found in cells—*deoxyribonucleic acid* (DNA) and *ribonucleic acid* (RNA). DNA is found within the nucleus of eukaryotic cells and within the cytoplasm in prokaryotic cells. DNA contains the genetic sequence that is passed on during reproduction. This sequence governs the functions of cells by determining the sequence of amino acids that are combined to form proteins. Each species manufactures its own unique proteins.

RNA is found both within the nucleus and in the cytoplasm of prokaryotic and eukaryotic cells. Most RNA molecules are involved in protein synthesis. Messenger RNA (mRNA) serves the function of carrying the DNA sequence from the nucleus to the rest of the cell. Ribosomal RNA (rRNA) makes up part of the ribosomes, where amino acids are bound together to form proteins. Transfer RNA (tRNA) carries amino acids to the ribosomes for protein synthesis.

Example

Destruction of the nuclear membrane of a cell would be LEAST lethal to which of the following cell types?

    (A) eukaryotic animal cell

    (B) eukaryotic plant cell

    (C) prokaryotic animal cell

    (D) prokaryotic bacterium cell

Solution

All cells have a cell membrane that separates cytoplasm from the surrounding environment. However, only eukaryotic cells have membranes surrounding their nuclei. Not all prokaryotic cells have a nuclear membrane. All animal cells are eukaryotic, so prokaryotic animal cells do not exist.

The answer is (D).

Example

What is the difference between a gram-positive cell and a gram-negative cell?

    (A) Gram-negative cell walls have more surface proteins than gram-positive cell walls.

    (B) Gram-negative cell walls have more lipid outer layers than gram-positive cell walls.

    (C) Gram-negative cell walls are less hydrophobic than gram-positive cell walls.

    (D) Gram-negative cell walls are less complex than the gram-positive cell walls.

Solution

Gram-negative cells have more complex walls with hydrophobic, lipid outer layers than gram-positive cells.

The answer is (B).

2. METABOLISM/METABOLIC PROCESSES

Metabolism is a term given to describe all chemical activities performed by a cell. The cell uses *adenosine triphosphate* (ATP) as the principle energy currency in all processes. Those processes that allow the bacterium to synthesize new cells from the energy stored within its body are said to be *anabolic*. All biochemical processes in which cells convert substrate into useful energy and waste products are said to be *catabolic*.

3. CELL TRANSPORT

The transfer of nutrients and other solutes across membrane barriers occurs by means of several mechanisms.

- *Passive diffusion* in cells is similar to the transfer that occurs in nonliving systems. Material moves spontaneously from a region of high concentration to a region of low concentration. The rate of transfer obeys Fick's law, the principle governing passive diffusion in dilute solutions. The rate is proportional to the concentration gradient across the membrane. Since this form of diffusion is powered only by a concentration gradient, it does not require the expenditure of chemical or biological energy (e.g., ATP).

Some cell walls are made up of lipid (fat) layers that are *hydrophobic* (i.e., that are repelled by water). Passive diffusion through lipid cell walls proceeds more easily when a nutrient (or other solute) is small, uncharged or un-ionized (including nonpolar), and lipid-soluble. Lipid-soluble substances include oxygen, carbon dioxide, fatty acids, and some steroidal hormones. Large, charged, polar, and water-soluble nutrients must be carried through cell walls by permeases (i.e., using facilitated diffusion). A *partition coefficient*, K_a, or the solubility of the solute in lipid relative to its solubility in water, can be used to describe the effect of lipid solubility on transport.

$$K_a = \frac{\text{solubility in lipid}}{\text{solubility in water}}$$

It is more common to report $\log K_a$ than K_a because the relationship between pK_a and permeability is fairly linear.

$$pK_a = \log K_a = \log\left(\frac{\text{solubility in lipid}}{\text{solubility in water}}\right)$$

The permeability of a molecule across a membrane of thickness x is

$$\text{permeability}_{\text{cm/s}} = \frac{pK_a\left(\begin{array}{c}\text{diffusion}\\\text{coefficient}\\\text{in water}\end{array}\right)_{\text{cm}^2/\text{s}}}{x_{\text{cm}}}$$

The pH of a solution will have an effect on the partition coefficient. Although the relationship is complex, for weakly acidic and basic solutions, pK_a and pH are approximately related by the *Henderson-Hasselbach equation*. Square brackets designate the molar concentration of component X.

The dissociation of a weak acid may be represented in one of two ways.

$$HA + H_2O \rightleftharpoons H_3O^+ + A^-$$

$$HA \rightleftharpoons H^+ + A^-$$

The Henderson-Hasselbach equation for a weak acid is

$$pK_a - pH = \log\left(\frac{\text{nonionized form}}{\text{ionized form}}\right)$$

$$= \log\frac{[HA]}{[A^-]} \quad \text{[weakly acidic]}$$

The dissociation of a weak base may be represented in one of two ways.

$$B + H_3O^+ \rightleftharpoons H_2O + BH^+$$

$$B + H^+ \rightleftharpoons BH^+$$

The Henderson-Hasselbach equation for a weak base is

$$pK_a - pH = \log\left(\frac{\text{ionized form}}{\text{nonionized form}}\right)$$

$$= \log\frac{[HB^+]}{[B]} \quad \text{[weakly basic]}$$

- Some solutes are carried by *permeases* (also known as *membrane enzymes*, *membrane proteins*, and *protein carriers*) by diffusion. This is known as *facilitated diffusion*. Since facilitated diffusion also occurs naturally from regions of high concentration to regions of low concentration, this process does not require the use of biological energy (e.g., ATP).

- *Active diffusion*, of which there are three categories, is transfer forced by a pressure gradient or "piggyback" function. Some form of chemical or biological energy is used to move a nutrient permease "uphill" ("upstream," etc.), from a region of lower concentration to a region of higher concentration. This type of movement requires the expenditure of chemical or biological energy.

  Membrane pumping, where permeases transport the substance in a direction opposite the direction of passive diffusion.

  Endocytosis, a process where cells absorb a material by surrounding (i.e., engulfing) it with their membrane.

  Exocytosis, during which a secretory vesicle expels material that was within the cell to an area outside the cell.

- Other specialized mechanisms for specific organs.

In the case of many gram-negative bacteria with dual (i.e., inner and outer) wall membranes, including *Escherichia coli* (*E. coli*), sugars and amino acids are transported across the inner plasma membrane by water-soluble proteins located in the periplasmic space between the two membranes.

Example

Which type(s) of diffusion do(es) NOT require the use of biological energy (e.g., ATP)?

I.   facilitated diffusion

II.   active diffusion

III.   passive diffusion

(A) I only

(B) III only

(C) I and III only

(D) I, II, and III

Solution

Neither passive diffusion nor facilitated diffusion require biological energy.

The answer is (C).

Example

In which type(s) of diffusion is the diffusing substance moved from an area of lower concentration to an area of greater concentration?

I.   passive diffusion

II.   facilitated diffusion

III.   active diffusion

(A) I only

(B) III only

(C) I and II only

(D) I, II, and III

Solution

A protein carrier can only move "uphill" (i.e., to an area of greater concentration) if it expends energy in active diffusion.

The answer is (B).

Example

$HC_2H_3O_2$ has an acid dissociation constant of 1.8×10^{-5}. What is the pH of a buffer solution made from 0.25 M $HC_2H_3O_2$ and 0.5 M $C_2H_3O_2^-$?

(A) 0.30

(B) 4.4

(C) 5.0

(D) 5.3

Solution

$HC_2H_3O_2$ is acetic acid, a weak acid. Use the Henderson-Hasselbach equation with the $\log K_a$ equation.

$$pH = pK_a + \log\frac{[A^-]}{[HA]} = pK_a + \log\frac{[C_2H_3O_2^-]}{[HC_2H_3O_2]}$$

$$= -\log(1.8 \times 10^{-5}) + \log\frac{0.50 \text{ M}}{0.25 \text{ M}}$$

$$= 4.7 + 0.30$$

$$= 5.0$$

The answer is (C).

4. ORGANISMAL GROWTH IN A BATCH CULTURE

Bacterial organisms can be grown (cultured) in a nutritive medium. The rate of growth follows the phases depicted in Fig. 24.3.

Figure 24.3 *Organismal Growth in Batch Culture*

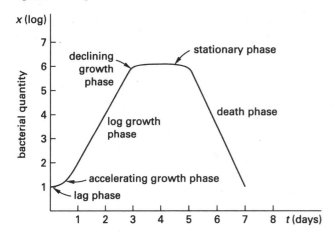

As Fig. 24.3 shows, the *lag phase* begins immediately after inoculation of the microbes into the nutrient medium. In this period, the microbial cells adapt to their new environment. The microbes might have to produce new enzymes to take advantage of new nutrients they are being exposed to, or they may need to adapt to different concentrations of solutes or to different temperatures than they are accustomed to. The length of the lag phase depends on the differences between the conditions the microbes experience before and after inoculation. The cells will start to divide when the requirements for growth are satisfied.

The *exponential,* or *logarithmic, growth phase* follows the lag phase. During this phase the number of cells increases according to the *specific growth rate,* μ.

$$\mu = \frac{1}{x}\left(\frac{dx}{dt}\right)$$

In this equation, x is the cell or organism number or concentration, t is time, and μ is the specific growth rate during the exponential phase.

The *declining growth phase* follows the logarithmic phase. During this phase, one or more essential nutrients are depleted and/or waste products are accumulated at levels that slow cell growth.

The *stationary phase* begins after the decelerating growth phase. During the stationary phase, the net growth rate of the cells is zero. New cells are created at the same rate at which cells are dying.

The *death phase* follows the stationary phase. During this phase, the death rate exceeds the growth rate.

The *logistic equation* is used to represent the population quantity up to (but excluding) the death phase. The curve takes on a *sigmoidal shape,* also known as an *S-curve* or *bounded exponential growth curve.* In the logistic formulation, the specific growth rate is related to the *carrying capacity,* x_∞, which is the maximum population the environment can support. Carrying capacity depends on the specific culture, medium, and conditions. The equation for exponential growth rate is

$$\mu = k\left(1 - \frac{x}{x_\infty}\right)$$

In this equation, k is the *logistic growth rate constant,* and x is the number of organisms at time t. The equation can be written for growth including the initial stationary phase as

$$\frac{dx}{dt} = kx\left(1 - \frac{x}{x_\infty}\right)$$

Integration gives the equation for the number of cells or organisms as a function of time.

$$x = \frac{x_0}{x_\infty}\left(1 - e^{kt}\right)$$

Example

The differential equation that describes the number of bacteria, x, experiencing exponential growth over time, t, is

$$\mu = \frac{1}{x}\left(\frac{dx}{dt}\right)$$

What is a valid expression of the solution to this equation?

(A) $\ln x = C e^{\mu t}$

(B) $\ln x = \mu t + \ln C$

(C) $x = C \ln \mu t$

(D) $x = \mu t + C$

Solution

The solution to the differential equation that represents exponential growth can be expressed in several ways.

$$x = C e^{\mu t}$$
$$\ln x = \mu t + \ln C$$
$$\frac{x}{C} = e^{\mu t}$$
$$\ln \frac{x}{C} = \mu t$$

The answer is (B).

5. MICROORGANISMS

Microorganisms occur in untreated water and represent a potential human health risk if they enter the water supply. Microorganisms include viruses, bacteria, fungi, algae, protozoa, worms, rotifers, and crustaceans.

Microorganisms are organized into three broad groups based upon their structural and functional differences. The groups are called *kingdoms*. The three kingdoms are animals (rotifers and crustaceans), plants (mosses and ferns), and *Protista* (bacteria, algae, fungi, and protozoa). Bacteria and protozoa of the kingdom *Protista* constitute the major groups of microorganisms in the biological systems that are used in secondary treatment of wastewater.

6. PATHOGENS

Many infectious diseases in humans or animals are caused by organisms categorized as *pathogens*. Pathogens are found in fecal wastes that are transmitted and transferred through the handling of wastewater. Pathogens will proliferate in areas where sanitary disposal of feces is not adequately practiced and where contamination of water supply from infected individuals is not properly controlled. The wastes may also be improperly discharged into surface waters, making the water *nonpotable* (unfit for drinking). Certain shellfish can become toxic when they concentrate pathogenic organisms in their tissues, increasing the toxic levels much higher than the levels in the surrounding waters.

Organisms that are considered to be pathogens include bacteria, protozoa, viruses, and *helminths* (worms). Not all microorganisms are considered pathogens. Some microorganisms are exploited for their usefulness in wastewater processing. Most wastewater engineering (and an increasing portion of environmental engineering) involves designing processes and operating facilities that utilize microorganisms to destroy organic and inorganic substances.

7. MICROBE CATEGORIZATION

Carbon is the basic building block for cell synthesis, and it is prevalent in large quantities in wastewater. Wastewater treatment uses microorganisms to treat organic compounds. Therefore, the growth of organisms that use organic material as energy is encouraged.

If a microorganism uses organic material as a carbon supply, it is *heterotrophic*. *Autotrophs* require only carbon dioxide to supply their carbon needs. Organisms that rely only on the sun for energy are called *phototrophs*. *Chemotrophs* extract energy from organic or inorganic oxidation/reduction (redox) reactions. *Organotrophs* use organic materials, while *lithotrophs* oxidize inorganic compounds.

Most microorganisms in wastewater treatment processes are bacteria. Conditions in the treatment plant are readjusted so that chemoheterotrophs predominate.

Each species of bacteria reproduces most efficiently within a limited range of temperatures. Four temperature ranges are used to classify bacteria. Those bacteria that grow best below 68°F (20°C) are called *psychrophiles* (i.e., are *psychrophilic*, also known as *cryophilic*). *Mesophiles* (i.e., *mesophilic bacteria*) grow best at temperatures in a range starting around 68°F (20°C) and ending around 113°F (45°C). Between 113°F (45°C) and 140°F (60°C), the *thermophiles* (*thermophilic bacteria*) grow best. Above 140°F (60°C), *stenothermophiles* grow best.

Because most reactions proceed slowly at these temperatures, cells use *enzymes* to speed up the reactions and control the rate of growth. Enzymes are proteins, ranging from simple structures to complex conjugates, and are specialized for the reactions they catalyze.

The temperature ranges are qualitative and somewhat subjective. The growth range of facultative thermophiles extends from the thermophilic range into the mesophilic range. Bacteria will grow in a wide range of temperatures and will survive at a very large range of temperatures. *E. coli*, for example, is classified as a mesophile. It grows best at temperatures between 68°F (20°C) and 122°F (50°C) but can continue to reproduce at temperatures down to 32°F (0°C).

Nonphotosynthetic bacteria are classified into two heterotrophic and autotrophic groups depending on their sources of nutrients and energy. *Heterotrophs* use organic matter as both an energy source and a carbon source for synthesis. Heterotrophs are further subdivided into groups depending on their behavior toward free oxygen: aerobes, anaerobes, and facultative bacteria. *Obligate aerobes* require free dissolved oxygen while they decompose organic matter to gain energy for growth and reproduction. *Obligate anaerobes* oxidize organics in the complete absence of dissolved oxygen by using the oxygen bound in other compounds, such as nitrate and sulfate. *Facultative bacteria* comprise a group that uses free dissolved oxygen when available but that can also behave anaerobically in the absence of free dissolved oxygen (i.e., *anoxic conditions*). Under anoxic conditions, a group of facultative anaerobes, called *denitrifiers*, utilizes nitrites and nitrates instead of oxygen. Nitrate nitrogen is converted to nitrogen gas in the absence of oxygen. This process is called *anoxic denitrification*.

Autotrophic bacteria (*autotrophs*) oxidize inorganic compounds for energy, use free oxygen, and use carbon dioxide as a carbon source. Significant members of this group are the *Leptothrix* and *Crenothrix* families of *iron bacteria*. These have the ability to oxidize soluble ferrous iron into insoluble ferric iron. Because soluble iron is often found in well waters and iron pipe, these bacteria deserve some attention. They thrive in water pipes where dissolved iron is available as an energy source and bicarbonates are available as a carbon source. As the colonies die and decompose, they release foul tastes and odors and have the potential to cause staining of porcelain or fabrics.

8. VIRUSES

Viruses are parasitic organisms that pass through filters that retain bacteria, can only be seen with an electron microscope, and grow and reproduce only inside living cells, but they can survive outside the host. They are not cells, but particles composed of a protein sheath surrounding a nucleic-acid core. Most viruses of interest in water supply range in size from 10 nm to 25 nm.

The *viron particles* invade living cells, and the viral genetic material redirects cell activities toward production of new viral particles. A large number of viruses are released to infect other cells when the infected cell dies. Viruses are host-specific, attacking only one type of organism.

There are more than 100 types of human enteric viruses. Those of interest in drinking water are *Hepatitis A*, *Norwalk*-type viruses, *Rotaviruses*, *Adenoviruses*, *Enteroviruses*, and *Reoviruses*.

9. BACTERIA

Bacteria are microscopic organisms (*microorganisms*) having round, rodlike, spiral or filamentous single-celled or noncellular bodies. Bacteria are *prokaryotes* (i.e., they lack nuclei structures). They are often aggregated into colonies. Bacteria use soluble food and reproduce through binary fission. Most bacteria are not pathogenic to humans, but they do play a significant role in the decomposition of organic material and can have an impact on the aesthetic quality of water.

10. FUNGI

Fungi are aerobic, multicellular, nonphotosynthetic, heterotrophic, eukaryotic protists. Most fungi are saprophytes that degrade dead organic matter. Fungi grow in low-moisture areas, and they are tolerant of low-pH environments. Fungi release carbon dioxide and nitrogen during the breakdown of organic material.

Fungi are obligate aerobes that reproduce by a variety of methods including fission, budding, and spore formation. They form normal cell material with one-half the nitrogen required by bacteria. In nitrogen-deficient wastewater, they may replace bacteria as the dominant species.

11. ALGAE

Algae are autotrophic, photosynthetic organisms (*photoautotrophs*) and may be either unicellular or multicellular. They take on the color of the pigment that is the catalyst for photosynthesis. In addition to chlorophyll (green), different algae have different pigments, such as carotenes (orange), phycocyanin (blue), phycoerythrin (red), fucoxanthin (brown), and xanthophylls (yellow).

Algae derive carbon from carbon dioxide and bicarbonates in water. The energy required for cell synthesis is obtained through photosynthesis. Algae utilize oxygen for respiration in the absence of light. Algae and bacteria have a symbiotic relationship in aquatic systems, with the algae producing oxygen used by the bacterial population.

In the presence of sunlight, the photosynthetic production of oxygen is greater than the amount used in respiration. At night algae use up oxygen in respiration. If the daylight hours exceed the night hours by a reasonable amount, there is a net production of oxygen. Excessive algal growth (*algal blooms*) can result in supersaturated oxygen conditions in the daytime and anaerobic conditions at night.

Some algae cause tastes and odors in natural water. While they are not generally considered pathogenic to humans, algae do cause turbidity, and turbidity provides a residence for microorganisms that are pathogenic.

12. PROTOZOA

Protozoa are single-celled animals that reproduce by *binary fission* (dividing in two). Most are aerobic chemoheterotrophs (i.e., *facultative heterotrophs*). Protozoa have complex digestive systems and use solid organic matter as food, including algae and bacteria. Therefore, they are desirable in wastewater effluent because they act as polishers in consuming the bacteria.

Flagellated protozoa are the smallest protozoans. The *flagella* (long hairlike strands) provide motility by a whiplike action. Amoeba move and take in food through the action of a mobile protoplasm. Free-swimming protozoa have *cilia*, small hairlike features, used for propulsion and gathering in organic matter.

13. WORMS AND ROTIFERS

Rotifers are aerobic, multicellular chemoheterotrophs. The rotifer derives its name from the apparent rotating motion of two sets of cilia on its head. The cilia provide mobility and a mechanism for catching food. Rotifers consume bacteria and small particles of organic matter.

Many *worms* are aquatic parasites. *Flatworms* of the class *Trematoda* are known as *flukes*, and the *Cestoda* are tapeworms. *Nematodes* of public health concern are *Trichinella*, which causes trichinosis; *Necator*, which causes pneumonia; *Ascaris*, which is the common roundworm that takes up residence in the human intestine (ascariasis); and *Filaria*, which causes filariasis.

14. MOLLUSKS

Mollusks, such as mussels and clams, are characterized by a shell structure. They are aerobic chemoheterotrophs that feed on bacteria and algae. They are a source of food for fish and are not found in wastewater treatment systems to any extent, except in underloaded lagoons. Their presence is indicative of a high level of dissolved oxygen and a very low level of organic matter.

Macrofouling is a term referring to infestation of water inlets and outlets by clams and mussels. *Zebra mussels*, accidentally introduced into the United States in 1986, are particularly troublesome for several reasons. First, young mussels are microscopic, easily passing through intake screens. Second, they attach to anything, even other mussels, producing thick colonies. Third, adult mussels quickly sense some biocides, most notably those that are halogen-based (including chlorine), quickly closing and remaining closed for days or weeks.

The use of biocides in the control of zebra mussels is controversial. Chlorination is a successful treatment that is recommended with some caution since it results in increased toxicity, affecting other species and THM production. Nevertheless, an ongoing biocide program aimed at pre-adult mussels, combined with slippery polymer-based and copper-nickle alloy surface coatings, is probably the best approach at prevention. Once a pipe is colonized, mechanical removal by scraping or water blasting is the only practical option.

15. INDICATOR ORGANISMS

The techniques for comprehensive bacteriological examination for pathogens are complex and time-consuming. Isolating and identifying specific pathogenic microorganisms is a difficult and lengthy task. Many of these organisms require sophisticated tests that take several days to produce results. Because of these difficulties, and also because the number of pathogens relative to other microorganisms in water can be very small, *indicator organisms* are used as a measure of the quality of the water. The primary function of an indicator organism is to provide evidence of recent fecal contamination from warm-blooded animals.

The characteristics of a good indicator organism are: (a) The indicator is always present when the pathogenic organism of concern is present. It is absent in clean, uncontaminated water. (b) The indicator is present in fecal material in large numbers. (c) The indicator responds to natural environmental conditions and to treatment processes in a manner similar to the pathogens of interest. (d) The indicator is easy to isolate, identify, and enumerate. (e) The ratio of indicator to pathogen should be high. (f) The indicator and pathogen should come from the same source (e.g., gastrointestinal tract).

While there are several microorganisms that meet these criteria, *total coliform* and *fecal coliform* are the indicators generally used. *Total coliform* refers to the group of aerobic and facultatively anaerobic, gram-negative, non-spore-forming, rod-shaped bacteria that ferment lactose with gas formation within 48 hr at 95°F (35°C). This encompasses a variety of organisms, mostly of intestinal

origin, including *E. coli*, the most numerous facultative bacterium in the feces of warm-blooded animals. Unfortunately, this group also includes *Enterobacter*, *Klebsiella*, and *Citrobacter*, which are present in wastewater but can be derived from other environmental sources such as soil and plant materials.

Fecal coliforms are a subgroup of the total coliforms that come from the intestines of warm-blooded animals. They are measured by running the standard total coliform fermentation test at an elevated temperature of 112°F (44.5°C), providing a means to distinguish false positives in the total coliform test.

Results of fermentation tests are reported as a *most probable number* (MPN) *index*. This is an index of the number of coliform bacteria that, more probably than any other number, would give the results shown by the laboratory examination; it is not an actual enumeration.

16. DECOMPOSITION OF WASTE

Decomposition of waste involves oxidation/reduction reactions and is classified as aerobic or anaerobic. The type of electron acceptor available for catabolism determines the type of decomposition used by a mixed culture of microorganisms. Each type of decomposition has its own peculiar characteristics that affect its use in waste treatment.

17. AEROBIC DECOMPOSITION

Molecular oxygen (O_2) must be present as the terminal electron acceptor in order for decomposition to proceed by aerobic oxidation. As in natural water bodies, the dissolved oxygen content is measured. When oxygen is present, it is the only terminal electron acceptor used. Hence, the chemical end products of decomposition are primarily carbon dioxide, water, and new cell material as demonstrated by Table 24.1. Odoriferous, gaseous end products are kept to a minimum. In healthy natural water systems, aerobic decomposition is the principal means of self-purification.

A wider spectrum of organic material can be oxidized by aerobic decomposition than by any other type of decomposition. Because of the large amount of energy released in aerobic oxidation, most aerobic organisms are capable of high growth rates. Consequently, there is a relatively large production of new cells in comparison with the other oxidation systems. This means that more biological sludge is generated in aerobic oxidation than in the other oxidation systems.

Aerobic decomposition is the preferred method for large quantities of dilute ($BOD_5 < 500$ mg/L) wastewater because decomposition is rapid and efficient and has a low odor potential. For high-strength wastewater ($BOD_5 > 1000$ mg/L), aerobic decomposition is not suitable because of the difficulty in supplying enough oxygen and because of the large amount of biological sludge that is produced.

Table 24.1 *Waste Decomposition End Products*

substrates	representative end products		
	aerobic decomposition	anoxic decomposition	anaerobic decomposition
proteins and other organic nitrogen compounds	amino acids ammonia → nitrites → nitrates alcohols organic acids $\Big\}$ → CO_2 + H_2O	amino acids nitrates → nitrites → N_2 alcohols organic acids $\Big\}$ → CO_2 + H_2O	amino acids ammonia hydrogen sulfide methane carbon dioxide alcohols organic acids
carbohydrates	alcohols fatty acids $\Big\}$ → CO_2 + H_2O	alcohols fatty acids $\Big\}$ → CO_2 + H_2O	carbon dioxide alcohols fatty acids
fats and related substances	fatty acids + glycerol alcohols lower fatty acids $\Big\}$ → CO_2 + H_2O	fatty acids + glycerol alcohols lower fatty acids $\Big\}$ → CO_2 + H_2O	fatty acids + glycerol carbon dioxide alcohols lower fatty acids

18. ANOXIC DECOMPOSITION

Some microorganisms will use nitrates in the absence of oxygen needed to oxidize carbon. The end products from such *denitrification* are nitrogen gas, carbon dioxide, water, and new cell material. The amount of energy made available to the cell during denitrification is about the same as that made available during aerobic decomposition. The production of cells, although not as high as in aerobic decomposition, is relatively high.

Denitrification is of importance in wastewater treatment when nitrogen must be removed. In such cases, a special treatment step is added to the conventional process for removal of carbonaceous material. One other important aspect of denitrification is in final clarification of the treated wastewater. If the final clarifier becomes anoxic, the formation of nitrogen gas will cause large masses of sludge to float to the surface and escape from the treatment plant into the receiving water. Thus, it is necessary to ensure that anoxic conditions do not develop in the final clarifier.

19. ANAEROBIC DECOMPOSITION

In order to achieve anaerobic decomposition, molecular oxygen and nitrate must not be present as terminal electron acceptors. Sulfate, carbon dioxide, and organic compounds that can be reduced serve as terminal electron acceptors. The reduction of sulfate results in the production of hydrogen sulfide (H_2S) and a group of equally odoriferous organic sulfur compounds called *mercaptans*.

The anaerobic decomposition of organic matter, also known as *fermentation*, generally is considered to be a two-step process. In the first step, complex organic compounds are fermented to low molecular weight *fatty acids* (*volatile acids*). In the second step, the organic acids are converted to methane. Carbon dioxide serves as the electron acceptor.

Anaerobic decomposition yields carbon dioxide, methane, and water as the major end products. Additional end products include ammonia, hydrogen sulfide, and mercaptans. As a consequence of these last three compounds, anaerobic decomposition is characterized by a malodorous stench.

Because only small amounts of energy are released during anaerobic oxidation, the amount of cell production is low. Thus, sludge production is correspondingly low. Wastewater treatment based on anaerobic decomposition is used to stabilize sludges produced during aerobic and anoxic decomposition.

Direct anaerobic decomposition of wastewater generally is not feasible for dilute waste. The optimum growth temperature for the anaerobic bacteria is at the upper end of the mesophilic range. Thus, to get reasonable biodegradation, the temperature of the culture must first be elevated. For dilute wastewater, this is not practical. For concentrated wastes ($BOD_5 >$ 1000 mg/L), anaerobic digestion is quite appropriate.

20. FACTORS AFFECTING DISEASE TRANSMISSION

Waterborne disease transmission is influenced by the latency, persistence, and quantity (dose) of the pathogens. *Latency* is the period of time between excretion of a pathogen and its becoming infectious to a new host. *Persistence* is the length of time that a pathogen remains viable in the environment outside a human host. The *infective dose* is the number of organisms that must be ingested to result in disease.

Chemistry

21. STOICHIOMETRY OF SELECTED BIOLOGICAL SYSTEMS

This section describes four classes of biological reactions involving microorganisms, each with simplified stoichiometric equations.

Stoichiometric problems are known as *weight and proportion problems* because their solutions use simple ratios to determine the masses of reactants required to produce given masses of products, or vice versa. The procedure for solving these problems is essentially the same regardless of the reaction.

step 1: Write and balance the chemical equation. (For convenience in using the degree of reduction equation, reduce the reactant's chemical formula to a single carbon atom.)

step 2: Determine the atomic (molecular) weight of each element (compound) in the equation.

step 3: Multiply the atomic (molecular) weights by their respective coefficients and write the products under the formulas.

step 4: Write the given mass data under the weights determined in step 3.

step 5: Fill in the missing information by calculating simple ratios.

The first biological reaction considered in this section is the production of biomass with a single extracellular product. Water and carbon dioxide are also produced as shown in the reaction equation.

$$C_{Ncs}H_{m \cdot Ncs}O_{n \cdot Ncs} + aO_2 + bNH_3$$
$$\text{[substrate]}$$

$$\rightarrow cC_{Ncc}H_{\alpha \cdot Ncc}O_{\beta \cdot Ncc}N_{\delta \cdot Ncc}$$
$$\text{[biomass]}$$

$$+ dC_{Ncp}H_{x \cdot Ncp}O_{y \cdot Ncp}N_{z \cdot Ncp}$$
$$\text{[bioproduct]}$$

$$+ eH_2O + fCO_2$$

The degrees of reduction (available electrons per unit of carbon) for substrate, s, biomass, b, and product, p, are

$$\gamma_s = 4 + m - 2n$$

$$\gamma_b = 4 + \alpha - 2\beta - 3\delta$$

$$\gamma_p = 4 + x - 2y - 3z$$

Table 24.2 shows typical degrees of reduction, γ. A high *degree of reduction* denotes a low degree of oxidation. Solving for the coefficients requires satisfying the carbon, nitrogen, and electron balances, plus knowing the respiratory coefficient and a yield coefficient. The key biomass production and reduction factors involved

in determining carbon, nitrogen, electron, and energy balances are shown.

$$cNcc + dNcp + f = Ncs \quad \text{[carbon]}$$

$$c\delta Ncc + dzNcp = b \quad \text{[nitrogen]}$$

$$c\gamma_b Ncb + d\gamma_p Ncp = \gamma_s Ncs - 4a \quad \text{[electron]}$$

$$Q_o c\gamma_b Ncc + Q_o d\gamma_p Ncp = Q_o \gamma_s Ncs - Q_o 4a \quad \text{[energy]}$$

Table 24.2 *Composition Data for Biomass and Selected Organic Compounds*

compound	molecular formula	degree of reduction, γ	molecular weight, MW
generic biomass*	$CH_{1.64}N_{0.16}O_{0.52}$	4.17 (NH_3)	24.5
	$P_{0.0054}S_{0.005}$	4.65 (N_3) 5.45 (HNO_3)	
methane	CH_4	8	16.0
n-alkane	C_4H_{32}	6.13	14.1
methanol	CH_4O	6.0	32.0
ethanol	C_2H_6O	6.0	23.0
glycerol	$C_2H_6O_3$	4.67	30.7
mannitol	$C_6H_{14}O_6$	4.33	30.3
acetic acid	$C_2H_4O_2$	4.0	30.0
lactic acid	$C_3H_6O_3$	4.0	30.0
glucose	$C_6H_{12}O_6$	4.0	30.0
formaldehyde	CH_2O	4.0	30.0
gluconic acid	$C_6H_{12}O_7$	3.67	32.7
succinic acid	$C_4H_6O_4$	3.50	29.5
citric acid	$C_6H_8O_7$	3.0	32.0
malic acid	$C_4H_6O_5$	3.0	33.5
formic acid	CH_2O_2	2.0	46.0
oxalic acid	$C_2H_2O_4$	1.0	45.0

*Sulfur is present in proteins.

Adapted from *Biochemical Engineering and Biotechnology Handbook* by B. Atkinson and F. Mavitona, Macmillan, Inc., 1983.

Q_o is the heat evolved per equivalent (gram mole) of available electrons, approximately 26.95 kcal/mol of electrons.

The *respiratory quotient* (RQ) is the CO_2 produced per unit of O_2.

$$RQ = \frac{f}{a}$$

The coefficients c and d are referred to as maximum theoretical *yield coefficients* when expressed per gram of substrate. The yield coefficient can be given either as grams of cells or grams of product per gram of substrate.

$$Y_{ideal,b} = \frac{m_b}{m_s}$$

$$Y_{ideal,p} = \frac{m_p}{m_s}$$

The ideal yield coefficients are related to the actual yield coefficients by the *yield factor*.

$$\text{yield factor} = \frac{Y_{\text{actual}}}{Y_{\text{ideal}}}$$

The second reaction is the aerobic biodegradation of glucose in the presence of oxygen and ammonia. In this reaction, cells are formed, and carbon dioxide and water are the only products. The stoichiometric equation is

$$\underset{\text{[substrate]}}{C_6H_{12}O_6} + aO_2 + bNH_3$$

$$\rightarrow c\underset{\text{[cells]}}{CH_{1.8}O_{0.5}N_{0.2}} + dCO_2 + eH_2O$$

For the stoichiometric equation,

$$a = 1.94$$
$$b = 0.77$$
$$c = 3.88$$
$$d = 2.13$$
$$e = 3.68$$

The coefficient c is the theoretical maximum yield coefficient, which may be reduced by a yield factor.

The third reaction is the anaerobic (no oxygen) biodegradation of organic wastes with incomplete stabilization (i.e., incomplete treatment). Methane, carbon dioxide, ammonia, and water as well as smaller organic waste molecules are the products. The stoichiometric equation is

$$C_aH_bO_cN_d$$

$$\rightarrow nC_wH_xO_yN_z + mCH_4$$

$$+ sCO_2 + rH_2O + (d - nx)NH_3$$

$$s = a - nw - m$$

$$r = c - ny - 2s$$

Knowledge of product composition, yield coefficient, and methane CO_2 ratio is needed.

The fourth reaction is the anaerobic biodegradation of organic wastes with complete stabilization. Besides organic waste, water is consumed in this reaction and the products are methane, carbon dioxide, and ammonia. The stoichiometric equation is

$$C_aH_bO_cN_d + rH_2O$$

$$\rightarrow mCH_4 + sCO_2 + dNH_3$$

$$r = \frac{4a - b - 2c + 3d}{4}$$

$$s = \frac{4a - b + 2c + 3d}{8}$$

$$m = \frac{4a + b - 2c - 3d}{8}$$

Composition data for biomass and selected organic compounds is given in Table 24.2.

Example

An organic compound has an empirical formula of $CH_{1.8}O_{0.5}N_{0.2}$. During combustion in oxygen gas, the compound dissociates and nitrogen is produced. What is the standard heat of combustion for one gram of this compound expressed in kJ/g?

(A) 5.0 kJ/g

(B) 18 kJ/g

(C) 22 kJ/g

(D) 25 kJ/g

Solution

Since this organic compound's chemical formula contains nitrogen, it matches the generic empirical formula for a "biomass." However, the compound is the reactant, not the product, and it should be considered as a substrate.

Write the equation for this compound with $c = \alpha = \beta = \delta = x = y = 0$.

$$CH_{1.8}O_{0.5}N_{0.2} + aO_2 \rightarrow dN_2 + eH_2O + fCO_2$$

Balance this reaction.

carbon, C: $f = 1$
nitrogen, N: $2d = 0.2$; $d = 0.1$
hydrogen, H: $2e = 1.8$; $e = 0.9$
oxygen, O: $2f + e = 0.5 + 2a$; $a = 1.2$

The balanced stoichiometric combustion reaction is

$$CH_{1.8}O_{0.5}N_{0.2} + 1.2O_2 \rightarrow 0.1N_2 + 0.9H_2O + CO_2$$

With $m = 1.8$ and $n = 0.5$, determine the degree of reduction for a substrate with a single carbon atom.

$$\gamma_s = 4 + m - 2n$$
$$= 4 + 1.8 - (2)(0.5)$$
$$= 4.8$$

Use the energy balance equation to determine the molar heat of combustion.

molar heat of combustion

$$= -\gamma_s Q_o$$

$$= -(4.8)\left(-26.95 \ \frac{\text{kcal}}{\text{mol}}\right)\left(4.1868 \ \frac{\text{kJ}}{\text{kcal}}\right)$$

$$= 541.6 \text{ kJ/mol}$$

The approximate molecular weight of this compound is

$$MW = (1)\left(12\ \frac{g}{mol}\right) + (1.8)\left(1\ \frac{g}{mol}\right)$$
$$+ (0.5)\left(16\ \frac{g}{mol}\right) + (0.2)\left(14\ \frac{g}{mol}\right)$$
$$= 24.6\ g/mol$$

The ideal heat of combustion per unit mass is

$$\frac{541.6\ \frac{kJ}{mol}}{24.6\ \frac{g}{mol}} = 22\ kJ/g$$

The answer is (C).

Diagnostic Exam

Topic VI: Heat Transfer

1. Liquid nitrogen at 77K is stored in an uninsulated 1.0 m diameter spherical tank. The tank is exposed to ambient air at 285K. The thermal resistance of the tank material is negligible. The convective heat transfer coefficient of the tank exterior is 30 $W/m^2 \cdot K$. The initial heat transfer from the air to the tank is most nearly

(A) 4.9 kW

(B) 9.8 kW

(C) 16 kW

(D) 20 kW

2. The exterior walls of a house are 3 m high, 0.14 m thick, and 40 m in total length. The thermal conductivity of the walls is 0.038 $W/m \cdot °C$. The interior of the walls is maintained at 20°C when the exterior (outdoor) wall temperature is 0°C. Neglecting corner effects, most nearly, what is the heat transfer through the walls?

(A) 0.023 kW

(B) 0.065 kW

(C) 0.23 kW

(D) 0.65 kW

3. A stainless steel tube (3 cm inside diameter and 5 cm outside diameter) is covered with 4 cm thick insulation. The thermal conductivities of stainless steel and the insulation are 20 $W/m \cdot K$ and 0.06 $W/m \cdot K$, respectively.

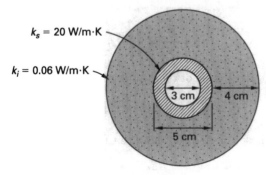

$k_s = 20$ W/m·K

$k_i = 0.06$ W/m·K

3 cm

4 cm

5 cm

If the inside wall temperature of the tube is 500K and the outside temperature of the insulation is 50K, what is most nearly the heat loss per meter of tube length?

(A) 120 W/m

(B) 140 W/m

(C) 160 W/m

(D) 180 W/m

4. A concrete wall is 20 cm thick and has an overall thermal resistance of 0.2 $m^2 \cdot °C/W$. The temperature difference between the two wall surfaces is 5°C. What is most nearly the heat transfer through the wall?

(A) 1.5 W/m^2

(B) 2.5 W/m^2

(C) 13 W/m^2

(D) 25 W/m^2

5. 50°C water flows through a 1.5 m long copper pipe. The thermal conductivity of the copper is 350 $W/m \cdot °C$. The internal diameter of the pipe is 20 mm, and the pipe wall is 5 mm thick. The temperature outside of the pipe is 20°C. What is most nearly the heat transfer through the pipe?

(A) 140 kW

(B) 160 kW

(C) 240 kW

(D) 440 kW

6. A parallel flow tubular heat exchanger cools water from 90°C to 70°C. The coolant increases in temperature from 0°C to 35°C. The log mean temperature difference is most nearly

(A) 28°C

(B) 32°C

(C) 58°C

(D) 62°C

7. A fan moves 25°C air over a 25 W resistive electrical device that has a uniform surface temperature of 100°C. The forced convection heat transfer coefficient is 50 $W/m^2 \cdot °C$. If the cooling fan fails and the device is cooled by natural convection (heat transfer coefficient of 10 $W/m^2 \cdot °C$), the resulting surface temperature of the device will be most nearly

(A) 100°C

(B) 375°C

(C) 400°C

(D) 625°C

8. Two large parallel plates are maintained electrically at uniform temperatures. Plate 1 has a surface temperature of 900K and an emissivity of 0.5. Plate 2 has a surface temperature of 650K and an emissivity of 0.8. The facing surfaces of both plates constitute opaque, diffuse gray radiators. A thin aluminum radiation shield is placed between the plates. The space between the plates is evacuated so that convective effects are negligible. The emissivity of both sides of the radiation shield is 0.15. In steady state, what is most nearly the net heat flux from plate 1 to plate 2?

(A) 940 W/m^2

(B) 1500 W/m^2

(C) 1900 W/m^2

(D) 2600 W/m^2

9. Air at 300K flows at 0.45 m/s over the surface of a 1 m × 1 m flat plate. The average kinematic viscosity of the air is 20.92×10^{-6} m^2/s, the Prandtl number is 0.7, and the thermal conductivity is 30×10^{-3} W/m·K. If the surface temperature of the plate is 400K, the average heat transfer coefficient is most nearly

(A) 2.6 W/m^2·K

(B) 5.0 W/m^2·K

(C) 7.5 W/m^2·K

(D) 8.2 W/m^2·K

10. A metal sphere at 750°C with a diameter of 2 cm is enclosed in a vacuum container. The temperature of the surroundings is −20°C. 12 W of power is needed to maintain the sphere's temperature. What is most nearly the emissivity of the sphere?

(A) 0.15

(B) 0.20

(C) 0.25

(D) 0.28

SOLUTIONS

1. The heat transfer from the tank to the atmosphere is

$$\dot{Q} = hA(T_w - T_\infty) = h\pi D^2(T_w - T_\infty)$$

$$= \frac{\left(30 \ \frac{\text{W}}{\text{m}^2 \cdot \text{K}}\right)\pi(1 \text{ m})^2(77\text{K} - 285\text{K})}{1000 \ \frac{\text{W}}{\text{kW}}}$$

$$= -19.6 \text{ kW}$$

The heat transfer from the air to the tank is 19.6 kW (20 kW).

The answer is (D).

2. Calculate the area.

$$A = Lh = (40 \text{ m})(3 \text{ m}) = 120 \text{ m}^2$$

The heat transfer due to conduction is

$$\dot{Q} = \frac{kA(T_1 - T_2)}{\Delta x}$$

$$= \frac{\left(0.038 \ \frac{\text{W}}{\text{m} \cdot °\text{C}}\right)(120 \text{ m}^2)(20°\text{C} - 0°\text{C})}{(0.14 \text{ m})\left(1000 \ \frac{\text{W}}{\text{kW}}\right)}$$

$$= 0.651 \text{ kW} \quad (0.65 \text{ kW})$$

The answer is (D).

3. The radius of the inside of the tube is

$$r_1 = \frac{D_i}{2} = \frac{3 \text{ cm}}{2} = 1.5 \text{ cm}$$

The radius of the outside of the tube is

$$r_2 = \frac{D_o}{2} = \frac{5 \text{ cm}}{2} = 2.5 \text{ cm}$$

The outer radius of the insulation is

$$r_3 = r_2 + t = 2.5 \text{ cm} + 4 \text{ cm} = 6.5 \text{ cm}$$

The heat transfer through the tube is

$$\dot{Q} = \frac{2\pi k_s L(T_1 - T_2)}{\ln \frac{r_2}{r_1}}$$

The temperature between the tube and insulation is

$$T_2 = T_1 - \frac{\dot{Q} \ln \frac{r_2}{r_1}}{2\pi k_s L}$$

The heat transfer through the insulation is

$$\dot{Q} = \frac{2\pi k_i L(T_2 - T_3)}{\ln \dfrac{r_3}{r_2}}$$

From this, the temperature between the tube and insulation is also

$$T_2 = \frac{\dot{Q} \ln \dfrac{r_3}{r_2}}{2\pi k_i L} + T_3$$

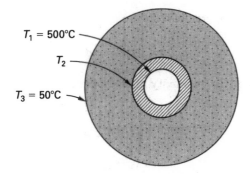

$T_1 = 500°C$
T_2
$T_3 = 50°C$

Equate the two expressions for the temperature between the tube and boundary.

$$T_1 - \frac{\dot{Q} \ln \dfrac{r_2}{r_1}}{2\pi k_s L} = \frac{\dot{Q} \ln \dfrac{r_3}{r_2}}{2\pi k_i L} + T_3$$

Solve for the heat transfer per unit length.

$$\frac{\dot{Q}}{L} = \frac{2\pi (T_1 - T_3)}{\dfrac{\ln \dfrac{r_2}{r_1}}{k_s} + \dfrac{\ln \dfrac{r_3}{r_2}}{k_i}}$$

$$= \frac{2\pi (500\text{K} - 50\text{K})}{\dfrac{\ln \dfrac{2.5 \text{ cm}}{1.5 \text{ cm}}}{20 \ \dfrac{\text{W}}{\text{m}\cdot\text{K}}} + \dfrac{\ln \dfrac{6.5 \text{ cm}}{2.5 \text{ cm}}}{0.06 \ \dfrac{\text{W}}{\text{m}\cdot\text{K}}}}$$

$$= 177.3 \text{ W/m} \quad (180 \text{ W/m})$$

The answer is (D).

4. Find the heat transfer per unit area.

$$\dot{Q} = \frac{A\Delta T}{R}$$

$$\dot{q} = \frac{\dot{Q}}{A} = \frac{\Delta T}{R}$$

$$= \frac{5°C}{0.2 \ \dfrac{\text{m}^2 \cdot °C}{\text{W}}}$$

$$= 25 \text{ W/m}^2$$

The answer is (D).

5. The inner radius of the pipe is 20 mm/2 = 10 mm. The outer radius of the pipe is 10 mm + 5 mm = 15 mm. Calculate the rate of heat transfer through the pipe.

$$\dot{Q} = \frac{2\pi k L(T_1 - T_2)}{\ln \dfrac{r_2}{r_1}}$$

$$= \frac{(2\pi)\left(350 \ \dfrac{\text{W}}{\text{m}\cdot°C}\right)(1.5 \text{ m})(50°C - 20°C)}{\left(\ln \dfrac{15 \text{ mm}}{10 \text{ mm}}\right)\left(1000 \ \dfrac{\text{W}}{\text{kW}}\right)}$$

$$= 244 \text{ kW} \quad (240 \text{ kW})$$

The answer is (C).

6. Calculate the log mean temperature difference for parallel flow in tubular heat exchangers.

$$\Delta T_{\text{lm}} = \frac{(T_{\text{Ho}} - T_{\text{Co}}) - (T_{\text{Hi}} - T_{\text{Ci}})}{\ln \left(\dfrac{T_{\text{Ho}} - T_{\text{Co}}}{T_{\text{Hi}} - T_{\text{Ci}}}\right)}$$

$$= \frac{(70°C - 35°C) - (90°C - 0°C)}{\ln \left(\dfrac{70°C - 35°C}{90°C - 0°C}\right)}$$

$$= 58.23°C \quad (58°C)$$

The answer is (C).

7. Use Newton's law of cooling. At state 1, the fan is operating, and at state 2, the fan has failed.

$$\dot{Q} = hA(T_w - T_\infty)$$

$$\frac{\dot{Q}}{A} = h_1(T_{w,1} - T_\infty) = h_2(T_{w,2} - T_\infty)$$

Solve for the surface temperature of the device after the fan fails.

$$T_{w,2} = \left(\frac{h_1}{h_2}\right)(T_{w,1} - T_\infty) + T_\infty$$

$$= \left(\frac{50 \ \dfrac{\text{W}}{\text{m}^2 \cdot °C}}{10 \ \dfrac{\text{W}}{\text{m}^2 \cdot °C}}\right)(100°C - 25°C) + 25°C$$

$$= 400°C$$

The answer is (C).

Heat Transfer

8. The net heat transfer is

$$\dot{q}_{12} = \frac{\dot{Q}_{12}}{A}$$

$$= \frac{\sigma(T_1^4 - T_2^4)}{A\left(\dfrac{1-\varepsilon_1}{\varepsilon_1 A_1} + \dfrac{1}{A_1 F_{13}} + \dfrac{1-\varepsilon_{3,1}}{\varepsilon_{3,1} A_3}\right.}{\left. + \dfrac{1-\varepsilon_{3,2}}{\varepsilon_{3,2} A_3} + \dfrac{1}{A_3 F_{32}} + \dfrac{1-\varepsilon_2}{\varepsilon_2 A_2}\right)}$$

Since the plates are parallel, $A = A_1 = A_2 = A_3$. Since the plates are large, $F_{13} = F_{32} = 1$. Since the emissivity of the shield is the same on both sides, $\varepsilon_{3,1} = \varepsilon_{3,2}$.

$$\dot{q}_{12} = \frac{\sigma(T_1^4 - T_2^4)}{\dfrac{1-\varepsilon_1}{\varepsilon_1} + \dfrac{1}{1} + \dfrac{1-\varepsilon_3}{\varepsilon_3} + \dfrac{1-\varepsilon_3}{\varepsilon_3} + \dfrac{1}{1} + \dfrac{1-\varepsilon_2}{\varepsilon_2}}$$

$$= \frac{\sigma(T_1^4 - T_2^4)}{\dfrac{1}{\varepsilon_1} + \dfrac{1}{\varepsilon_2} + \dfrac{2}{\varepsilon_3} - 2}$$

$$= \frac{\left(5.67 \times 10^{-8}\ \dfrac{\text{W}}{\text{m}^2 \cdot \text{K}^4}\right)\left((900\text{K})^4 - (650\text{K})^4\right)}{\dfrac{1}{0.5} + \dfrac{1}{0.8} + \dfrac{2}{0.15} - 2}$$

$$= 1856.9\ \text{W/m}^2 \quad (1900\ \text{W/m}^2)$$

The answer is (C).

9. The Reynolds number is

$$\text{Re}_L = \frac{\rho u_\infty L}{\mu} = \frac{u_\infty L}{\nu}$$

$$= \frac{\left(0.45\ \dfrac{\text{m}}{\text{s}}\right)(1\ \text{m})}{20.92 \times 10^{-6}\ \dfrac{\text{m}^2}{\text{s}}}$$

$$= 2.15 \times 10^4$$

Since Re_L is less than 10^5, the flow over the flat plate is laminar. Use a Nusselt correlation.

$$\overline{\text{Nu}}_L = \frac{\overline{h}L}{k} = 0.6640\,\text{Re}_L^{1/2}\text{Pr}^{1/3}$$

$$= (0.6640)(2.15 \times 10^4)^{1/2}(0.7)^{1/3}$$

$$= 86.47$$

$$\overline{h} = \overline{\text{Nu}}_L\left(\frac{k}{L}\right)$$

$$= (86.47)\left(\frac{30 \times 10^{-3}\ \dfrac{\text{W}}{\text{m} \cdot \text{K}}}{1\ \text{m}}\right)$$

$$= 2.59\ \text{W/m}^2 \cdot \text{K} \quad (2.6\ \text{W/m}^2 \cdot \text{K})$$

The answer is (A).

10. Calculate the absolute temperatures of the sphere and the surroundings.

$$T_1 = 750°\text{C} + 273° = 1023\text{K}$$

$$T_2 = -20°\text{C} + 273° = 253\text{K}$$

Use the equation for net energy exchange between two bodies, when the body is small compared to its surroundings.

$$\dot{Q}_{12} = \varepsilon\sigma A(T_1^4 - T_2^4)$$

The surface area of a sphere is

$$A = 4\pi r^2 = \pi d^2$$

Solve for the emissivity of the sphere.

$$\varepsilon = \frac{\dot{Q}_{12}}{\sigma A(T_1^4 - T_2^4)} = \frac{\dot{Q}_{12}}{\sigma\pi d^2(T_1^4 - T_2^4)}$$

$$= \frac{(12\ \text{W})\left(100\ \dfrac{\text{cm}}{\text{m}}\right)^2}{\left(5.67 \times 10^{-8}\ \dfrac{\text{W}}{\text{m}^2 \cdot \text{K}^4}\right)\pi(2\ \text{cm})^2}$$

$$\times \left((1023\text{K})^4 - (253\text{K})^4\right)$$

$$= 0.154 \quad (0.15)$$

The answer is (A).

25 Conduction

Nomenclature

A	area	m^2
Bi	Biot number	–
c_p	specific heat	J/kg·K
G	heat generation rate	W/m^3
h	coefficient of heat transfer	W/m^2·K
h	enthalpy	kJ/kg
k	thermal conductivity	W/m·K
L	thickness	m
m	factor for fins equal to $\sqrt{hP/kA_c}$	1/m
m	mass	kg
P	perimeter	m
$\dot{q}$	heat transfer per unit area	W/m^2
Q	heat energy	J
$\dot{Q}$	rate of heat transfer	W
r	radius	m
R	thermal resistance	K/W
t	thickness	m
t	time	s
T	temperature	K
U	overall coefficient of heat transfer	W/m^2·K
V	volume	m^3
w	width	m
x	distance	m

Symbols

β	decay constant (reciprocal of time constant)	1/s
ρ	mass density	kg/m^3
τ	time constant	s

Subscripts

0	initial
1	inner
2	outer
b	base
c	cross section or corrected
cr	critical
i	initial, inner, or i^{th} layer
m	mean
o	outside
p	pressure
s	solids or surface
th	thermal
∞	bulk fluid

1. INTRODUCTION TO CONDUCTIVE HEAT TRANSFER

Conduction is the flow of heat through solids or stationary fluids. Thermal conductance in metallic solids is due to molecular vibrations within the metallic crystalline lattice and movement of free valence electrons through the lattice. Insulating solids, which have fewer free electrons, conduct heat primarily by the agitation of adjacent atoms vibrating about their equilibrium positions. This vibrational mode of heat transfer is several orders of magnitude less efficient than conduction by free electrons.

In stationary liquids, heat is transmitted by longitudinal vibrations, similar to sound waves. The *net transport theory* explains heat transfer through gases. Hot molecules move faster than cold molecules. Hot molecules travel to cold areas with greater frequency than cold molecules travel to hot areas.

Determining heat transfer by conduction can be an easy task if sufficient simplifying assumptions are made. Major discrepancies can arise, however, when the simplifying assumptions are not met. The following assumptions are commonly made in simple problems.

- The heat transfer is steady-state.

- The heat path is one-dimensional. (Objects are infinite in one or more directions and do not have any end effects.)

- The heat path has a constant area.

- The heat path consists of a homogeneous material with constant conductivity.

- The heat path consists of an isotropic material.[1]

- There is no internal heat generation.

Many real heat transfer cases violate one or more of these assumptions. Unfortunately, problems with closed-form solutions (suitable for working by hand) are in the minority. More complex problems must be solved by appropriate iterative, graphical, or numerical methods.[2]

---

[1]Examples of *anisotropic materials*, materials whose heat transfer properties depend on the direction of heat flow, are crystals, plywood and other laminated sheets, and the core elements of some electrical transformers.
[2]Finite-difference methods are commonly used.

Heat Transfer

Equation 25.1: Fourier's Law of Conduction

$$\dot{Q} = -kA\frac{dT}{dx} \qquad 25.1$$

Values

Table 25.1 Typical Thermal Conductivities at 0°C

substance	k W/m·K
silver	419
copper	388
aluminum	202
brass	97
steel (1% C)	47
lead	35
ice	2.2
glass	1.1
concrete	0.87
water	0.55
fiberglass	0.052
cork	0.043
air	0.024

Description

The steady-state heat transfer by conduction through a flat slab is specified by *Fourier's law*, Eq. 25.1. Fourier's law is written with a minus sign to indicate that the heat flow is opposite the direction of the thermal gradient.

The *thermal conductivity* (also known as the *thermal conductance*), k, is a measure of the rate at which a substance transfers thermal energy through a unit thickness.[3] Units of thermal conductivity are W/m·K or W·cm/m²·K. The conductivity of a substance should not be confused with the *overall conductivity*, U, of an object. Table 25.1 lists representative thermal conductivities for commonly encountered substances.

2. THERMAL RESISTANCE

Equation 25.2: Thermal Resistance, Plane Wall

$$R = \frac{L}{kA} \qquad 25.2$$

Description

For a plane wall, the thermal resistance depends on the thickness (path length), L, and the thermal conductivity, k. Thermal resistance is usually expressed per unit of exposed surface area, as Eq. 25.2 indicates.

---

[3]Another (less-encountered) meaning for *conductivity* is the reciprocal of thermal resistance (conductivity = kA/L).

Example

Concrete has a thermal conductivity of 1.4 W/m·°C. What is most nearly the thermal resistance of a 20 cm thick concrete wall with an exposed face area of 1 m²?

- (A) 0.14 °C/W
- (B) 0.28 °C/W
- (C) 1.4 °C/W
- (D) 7.0 °C/W

Solution

Use Eq. 25.2.

$$R = \frac{L}{kA} = \frac{0.2 \text{ m}}{\left(1.4 \ \dfrac{\text{W}}{\text{m·°C}}\right)(1 \text{ m}^2)}$$

$$= 0.143 \text{ °C/W} \quad (0.14 \text{ °C/W})$$

The answer is (A).

Equation 25.3 and Eq. 25.4: Resistance in Series

$$\dot{Q} = \frac{\Delta T}{R_{\text{total}}} \qquad 25.3$$

$$R_{\text{total}} = \sum R \qquad 25.4$$

Variation

$$\dot{Q} = \frac{T_1 - T_{n+1}}{R_{\text{total}}} = \frac{A(T_1 - T_{n+1})}{\displaystyle\sum_{i=1}^{n} \frac{L_i}{k_i}}$$

Description

For heat transfer through multiple layers (for example, a layered wall or cylinder with several layers of different insulation), each layer is considered a series resistance. The sum of the individual thermal resistances is the total thermal resistance. For example, in Fig. 25.1, the total thermal resistance is given by the sum of the two individual resistances. Typical units are K/W.

Figure 25.1 Composite Slab (Plane) Wall

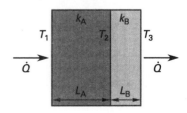

Equation 25.5: Thermal Resistance, Cylindrical Wall

$$R = \frac{\ln\left(\dfrac{r_2}{r_1}\right)}{2\pi kL} \qquad 25.5$$

Description

Equation 25.5 gives the thermal resistance for a cylindrical wall.

Example

A 0.5 m long cast iron pipe has an inner diameter of 6 cm and an outer diameter of 6.5 cm. The thermal conductivity of cast iron is 80 W/m·K. The thermal resistance of the pipe is most nearly

(A) 0.000080 K/W

(B) 0.00032 K/W

(C) 0.00064 K/W

(D) 0.0010 K/W

Solution

The thermal resistance is calculated using Eq. 25.5.

$$R = \frac{\ln\left(\dfrac{r_2}{r_1}\right)}{2\pi kL} = \frac{\ln\dfrac{6.5 \text{ cm}}{6 \text{ cm}}}{(2\pi)\left(80 \dfrac{\text{W}}{\text{m·K}}\right)(0.5 \text{ m})}$$

$$= 0.000318 \text{ K/W} \quad (0.00032 \text{ K/W})$$

The answer is (B).

Equation 25.6: Thermal Resistance, Film

$$R = \frac{1}{hA} \qquad 25.6$$

Description

Heat conducted through solids (walls) to an exposed surface is usually transferred to and removed by a moving fluid. For example, heat transmitted through a heat exchanger wall is removed by a moving coolant. Unless the fluid is extremely turbulent, the fluid molecules immediately adjacent to the exposed surface move much slower than molecules farther away. Molecules immediately adjacent to the wall may be stationary altogether. The fluid molecules that are affected by the exposed surface constitute a layer known as a *film*. The film has a thermal resistance just like any other layer.

Because the film thickness is not easily determined, the thermal resistance of a film is given by a *film coefficient* (*convective heat transfer coefficient* or *unit conductance*), h, with units of W/m²·K.

If the fluid is extremely turbulent, the thermal resistance will be small (that is, h will be very large). In such cases, the wall temperature is essentially the same as the fluid temperature.

Example

A wall has a film coefficient of 3500 W/m²·°C. What is most nearly the thermal resistance of a 1 m² area of the wall?

(A) 0.29 °C/kW

(B) 0.33 °C/kW

(C) 0.68 °C/kW

(D) 3.5 °C/kW

Solution

The thermal resistance is

$$R = \frac{1}{hA} = \frac{(1)\left(1000 \dfrac{\text{W}}{\text{kW}}\right)}{\left(3500 \dfrac{\text{W}}{\text{m}^2\text{·°C}}\right)(1 \text{ m}^2)}$$

$$= 0.286 \text{ °C/kW} \quad (0.29 \text{ °C/kW})$$

The answer is (A).

3. STEADY CONDUCTION THROUGH A PLANE WALL

Equation 25.7: Fourier's Law, Plane Wall

$$\dot{Q} = \frac{-kA(T_2 - T_1)}{L} \qquad 25.7$$

Variations

$$q_{12} = \frac{-k(T_2 - T_1)}{L} = \frac{k(T_1 - T_2)}{L}$$

$$Q_{12} = q_{12}A = \frac{kA(T_1 - T_2)}{L}$$

Description

On its own, heat always flows from a higher temperature to a lower temperature. The heat transfer from high-temperature point 1 to lower-temperature point 2 through an infinite plane of thickness L and homogeneous conductivity, k, is given by Eq. 25.7. (See Fig. 25.2.)

Heat Transfer

Figure 25.2 *Plane Wall*

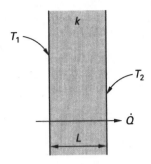

The temperature difference $T_2 - T_1$ is the *temperature gradient* or *thermal gradient*. Heat transfer is always positive. The minus sign in Eq. 25.7 indicates that the heat flow direction is opposite that of the thermal gradient. The direction of heat flow is obvious in most problems, so the minus sign is usually omitted, and the temperature difference is written as $T_1 - T_2$.[4]

Example

A 4 cm thick insulator ($k = 2 \times 10^{-4}$ cal/cm·s·°C) has an area of 1000 cm². If the temperatures on its two sides are 170°C and 50°C, and films are neglected, what is most nearly the heat transfer by conduction?

(A) 0.10 cal/s

(B) 1.2 cal/s

(C) 6.0 cal/s

(D) 30 cal/s

Solution

Use Eq. 25.7.

$$\dot{Q} = \frac{-kA(T_2 - T_1)}{L}$$

$$= \frac{-\left(2 \times 10^{-4}\ \dfrac{\text{cal}}{\text{cm·s·°C}}\right) \times (1000\ \text{cm}^2)(50°\text{C} - 170°\text{C})}{4\ \text{cm}}$$

$$= 6.0\ \text{cal/s}$$

The answer is (C).

Equation 25.8: Temperature at Intermediate Locations

$$\dot{Q} = \frac{T_1 - T_2}{R_A} = \frac{T_2 - T_3}{R_B} \qquad \textbf{25.8}$$

[4]The NCEES *FE Reference Handbook* (*NCEES Handbook*) is inconsistent in its inclusion of the minus sign. After Eq. 25.7, the *NCEES Handbook* abandons the convention for all subsequent heat transfer equations.

Description

The temperature at any point on or within a simple or complex wall can be found if the heat transfer, q or Q, is known. The procedure is to calculate the thermal resistance up to the point of unknown temperature and then to solve for the temperature difference. Since one of the temperatures is known, the unknown temperature is found from the temperature difference.

4. CONDUCTION THROUGH A CYLINDRICAL WALL

The Fourier equation is based on a uniform path length and a constant cross-sectional area. If the heat flow is through an area that is not constant, the *logarithmic mean area*, A_m, should be used in place of the regular area. The log mean area should be used with heat transfer through thick pipe and cylindrical tank walls. (See Fig. 25.3.)

Figure 25.3 *Cylindrical Wall*

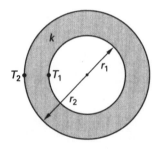

Equation 25.9: Fourier's Law, Cylindrical Wall

$$\dot{Q} = \frac{2\pi kL(T_1 - T_2)}{\ln\left(\dfrac{r_2}{r_1}\right)} \qquad \textbf{25.9}$$

Variation

$$\dot{Q} = qA_m = \frac{kA_m(T_1 - T_2)}{r_2 - r_1}$$

Description

The overall radial heat transfer through an uninsulated hollow cylinder without films is given by Eq. 25.9. This equation disregards heat transfer from the ends and assumes that the length is sufficiently large so that the heat transfer is radial at all locations.

Example

An 8 m long pipe of 15 cm outside diameter is covered with 2 cm of insulation with a thermal conductivity of 0.09 W/m·K.

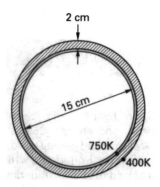

If the inner and outer temperatures of the insulation are 750K and 400K, respectively, what is most nearly the heat loss from the pipe?

(A) 4.5 kW

(B) 6.7 kW

(C) 8.5 kW

(D) 10 kW

Solution

Use Eq. 25.9.

$$\dot{Q} = \frac{2\pi k L (T_1 - T_2)}{\ln\left(\dfrac{r_2}{r_1}\right)}$$

$$= \frac{(2\pi)\left(0.09 \ \dfrac{\text{W}}{\text{m·K}}\right)(8 \text{ m})(750\text{K} - 400\text{K})}{\left(\ln \dfrac{9.5 \text{ cm}}{7.5 \text{ cm}}\right)\left(1000 \ \dfrac{\text{W}}{\text{kW}}\right)}$$

$$= 6.698 \text{ kW} \quad (6.7 \text{ kW})$$

The answer is (B).

Equation 25.10: Critical Insulation Radius

$$r_{\text{cr}} = \frac{k_{\text{insulation}}}{h_\infty} \qquad \textbf{25.10}$$

Description

The addition of insulation to a bare pipe or wire increases the surface area. (See Fig. 25.4.) Adding insulation to a small-diameter pipe may actually increase the heat loss above bare-pipe levels. Adding insulation up to the *critical thickness* is dominated by the increase in surface area. Only adding insulation past the critical thickness will decrease heat loss.[5] The *critical radius* is usually very small (a few millimeters), and it is most relevant in the case of insulating thin wires. The critical radius,

---

[5]There is another, less commonly used, meaning for the term *critical thickness*: the thickest required insulation. In situations where the required insulation thickness is different for energy conservation, condensation control, personnel protection, and process temperature control, the "critical" thickness is the thickness that controls the design.

measured from the center of the pipe or wire, is given by Eq. 25.10. (See Fig. 25.5.) h_∞ is the film coefficient representing resistance to convective heat transfer.

Figure 25.4 Composite Cylinder (insulated pipe)

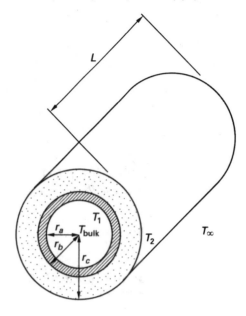

Figure 25.5 Insulation Radius

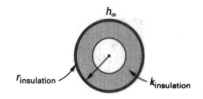

5. TRANSIENT CONDUCTION USING THE LUMPED CAPACITANCE MODEL

If the internal thermal resistance of a body is small in comparison to the external thermal resistance, the *lumped parameter method*, also known as the *lumped capacitance model*, can be used to approximate the transient (time-dependent) heat flow. This method is also referred to as *Newton's method*. Figure 25.6 shows the variables used in the lumped capacitance model.

Figure 25.6 Variables Used in Transient Conduction

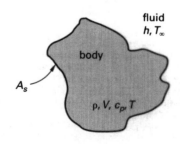

Equation 25.11: Biot Number

$$\mathrm{Bi} = \frac{hV}{kA_s} \ll 1 \qquad 25.11$$

Variation

$$\mathrm{Bi} = \frac{hL_c}{k}$$

Description

The *Biot number*, Bi (also known as the *Biot modulus* and *transient modulus*), is a comparison of the internal thermal resistance to the external resistance of a body. If the Biot number is small (less than 0.1), the internal thermal resistance will be small, and the body temperature will be essentially uniform throughout, during heating or cooling. The length used to calculate the Biot number is the *characteristic length*, L_c, not an external body dimension.

$$L_c = \frac{V}{A_s}$$

Equation 25.12 Through Eq. 25.15: Constant Environment Temperature[6]

$$\dot{Q} = hA_s(T - T_\infty) = -\rho V c_p\left(\frac{dT}{dt}\right) \qquad 25.12$$

$$T - T_\infty = (T_i - T_\infty)e^{-\beta t} \qquad 25.13$$

$$\beta = \frac{hA_s}{\rho V c_p} \qquad 25.14$$

$$\beta = \frac{1}{\tau} \qquad 25.15$$

Description

Equation 25.12 gives the instantaneous heat transfer at a particular moment when the body temperature is known. Equation 25.13 gives the temperature of the body as a function of time. The time variable, t, starts at zero. Equation 25.15 shows that the factor β, known as the *decay constant* or *exponential frequency*, is the reciprocal of the time constant, τ.[7] As with many other engineering subjects, the *time constant* is the length of time required to vary (relax) the property (initial temperature differential in this case) by approximately 63.2%. Stated another way, the instantaneous temperature gradient will be approximately 36.8% of the initial temperature gradient after a duration equal to the time constant.

---

[6]The *NCEES Handbook* refers to the case of a constant environment temperature as the "constant fluid temperature" case. This makes sense in the context of Fig. 25.6. The fluid is the substance surrounding the cooling body. Fluids include gases and are not necessarily liquids.
[7]τ is the time constant. As the reciprocal of the time constant, β has no other interpretation, and its significance is not explained in the *NCEES Handbook*. Equation 25.13 could have been written in terms of the time constant, rather than β. The most common symbol for the decay constant for all subjects is λ. Other symbols (e.g., b, k, r) are used, but the use of β is uncommon.

Equation 25.16: Total Heat Transferred

$$Q_{\text{total}} = \rho V c_p(T_i - T) \qquad 25.16$$

Description

Although the rate of heat transfer varies with time in a transient condition, the energy change can be determined from the starting and ending conditions, independent of duration (time). Equation 25.16 calculates the total energy (heat) change as a function of the initial temperature, T_i, and final temperature, T. Since $T_i - T$ is a temperature difference, temperature can be expressed in any consistent scale. Absolute temperatures are not required.[8]

Example

A hot solid iron sphere (15 cm diameter) is placed in a bath of cold water. The initial temperature of the sphere is 433K. The sphere is cooled to 303K. The density of iron is 7.874 g/cm^3, and the specific heat capacity of iron is 0.45 kJ/kg·K. The total heat transferred from the sphere to the water is most nearly

(A) 190 kJ

(B) 810 kJ

(C) 1000 kJ

(D) 1800 kJ

Solution

The volume of the iron sphere is

$$V = \tfrac{4}{3}\pi r^3$$

$$= \left(\tfrac{4}{3}\pi\right)\left(\frac{15 \text{ cm}}{2}\right)^3$$

$$= 1767.1 \text{ cm}^3$$

Use Eq. 25.16 to calculate the total heat transferred from the iron sphere to the water.

$$Q_{\text{total}} = \rho V c_p(T_i - T)$$

$$= \frac{\left(7.874\,\dfrac{g}{cm^3}\right)(1767.1 \text{ cm}^3)\left(0.45\,\dfrac{kJ}{kg\cdot K}\right)}{1000\,\dfrac{g}{kg}}$$

$$\times (433K - 303K)$$

$$= 814 \text{ kJ} \quad (810 \text{ kJ})$$

The answer is (B).

---

[8]The *NCEES Handbook* specifies that the initial body temperature is in kelvins. However, while use of absolute temperatures is an option, it is not a requirement.

Heat Transfer

6. FINS

Fins (*extended surfaces*) are features that receive and move thermal energy by conduction along their lengths and widths prior to (in most cases) convective and radiative heat removal. They include simple fins, fin tubes, finned channels, and heat pipes. Some simple features (e.g., long wires) can be considered and evaluated as fins even though that is not their intended function.

Equation 25.17 Through Eq. 25.23: Heat Transfer, Fins

$$\dot{Q} = \sqrt{hPkA_c}\,(T_b - T_\infty)\tanh(mL_c) \qquad 25.17$$

$$m = \sqrt{\frac{hP}{kA_c}} \qquad 25.18$$

$$L_c = L + \frac{A_c}{P} \qquad 25.19$$

$$P = 2w + 2t \quad \text{[rectangular]} \qquad 25.20$$

$$A_c = wt \quad \text{[rectangular]} \qquad 25.21$$

$$P = \pi D \quad \text{[pin]} \qquad 25.22$$

$$A_c = \frac{\pi D^2}{4} \quad \text{[pin]} \qquad 25.23$$

Description

An external fin is attached at its base to a source of thermal energy at temperature T_b. The temperature across the face of the fin at any point along its length is assumed to be constant. The far-field temperature of the surrounding environment is T_∞. For *rectangular fins* (also known as *straight fins* or *longitudinal fins*), the cross-sectional area, A_c, is uniform and is given by Eq. 25.21.[9] (See Fig. 25.7.) For a *pin fin* (i.e., a fin with a circular cross section), cross-sectional area, A_c, is given by Eq. 25.23. (See Fig. 25.8.)

Most equations for heat transfer from a fin disregard the small amount of heat transfer from the exposed end. For that reason, the fin is assumed to possess an *adiabatic tip* or *insulated tip*. A simple approximation to the exact solution of a nonadiabatic tip can be obtained by replacing the actual fin length with a corrected length, as given in Eq. 25.19.[10]

---

[9](1) The *NCEES Handbook* refers to the substrate from which the fin extends as the "base," designated by subscript b. In common usage, the area or temperature of the "base of the fin" would refer to the fin. In NCEES usage, "base" refers to the substrate, not the fin at its attachment point. (2) The *NCEES Handbook* uses the subscript c to designate the kind of area (i.e., cross-sectional area, as differentiated from surface area) rather than the object (e.g., f for fin or b for the fin base).

[10]The *NCEES Handbook* is not consistent in its use of subscripts in Eq. 25.19. While c in A_c stands for "cross," c in L_c stands for "corrected."

Figure 25.7 Rectangular Fin

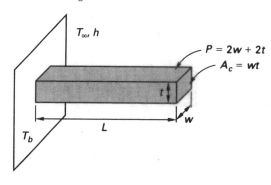

Figure 25.8 Pin Fin

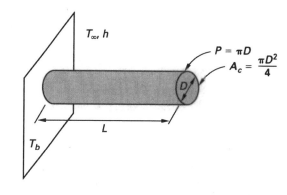

Example

The base of a 1.2 cm × 1.2 cm × 25 cm long rectangular rod is maintained at 150°C by an electrical heating element. The conductivity of the rod is 140 W/m·K. The ambient air temperature is 27°C, and the average film coefficient is 9.4 W/m²·K. The fin has an adiabatic tip. What is most nearly the energy input required to maintain the base temperature?

(A) 0.10 W

(B) 1.7 W

(C) 9.7 W

(D) 100 W

Solution

From Eq. 25.20, the perimeter length is

$$2w + 2t = \frac{(2)(1.2\text{ cm}) + (2)(1.2\text{ cm})}{100\ \frac{\text{cm}}{\text{m}}} = 0.048\text{ m}$$

From Eq. 25.21, the cross-sectional area of the fin is

$$A_c = wt = \frac{(1.2\text{ cm})(1.2\text{ cm})}{\left(100\ \frac{\text{cm}}{\text{m}}\right)^2} = 0.000144\text{ m}^2$$

Use Eq. 25.18.

$$m = \sqrt{\frac{hP}{kA_c}}$$

$$= \sqrt{\frac{\left(9.4 \ \dfrac{\text{W}}{\text{m}^2 \cdot \text{K}}\right)(0.048 \ \text{m})}{\left(140 \ \dfrac{\text{W}}{\text{m} \cdot \text{K}}\right)(0.000144 \ \text{m}^2)}}$$

$$= 4.73 \ 1/\text{m}$$

At steady state, the energy input is equal to the energy loss. Since the tip is adiabatic, it is not necessary to replace the actual length with the corrected length from Eq. 25.19.

From Eq. 25.17, the total heat loss is

$$\dot{Q} = \sqrt{hPkA_c} \ (T_b - T_\infty) \tanh(mL_c)$$

$$= \sqrt{\begin{array}{c} \left(9.4 \ \dfrac{\text{W}}{\text{m}^2 \cdot \text{K}}\right)(0.048 \ \text{m}) \\[2mm] \times \left(140 \ \dfrac{\text{W}}{\text{m} \cdot \text{K}}\right)(0.000144 \ \text{m}^2) \end{array}}$$

$$\times (150^\circ\text{C} - 27^\circ\text{C}) \tanh\left(\left(4.73 \ \dfrac{1}{\text{m}}\right)\left(\dfrac{25 \ \text{cm}}{100 \ \dfrac{\text{cm}}{\text{m}}}\right)\right)$$

$$= 9.72 \ \text{W} \quad (9.7 \ \text{W})$$

The answer is (C).

Heat Transfer

26 Convection

Nomenclature

A	area	m^2
B	separation distance	m
c_p	specific heat	J/kg·K
C	constant	–
C	heat capacity rate	W/K
D	diameter[1]	m
F	correction factor for mean temperature difference	–
g	gravitational acceleration, 9.81[2]	m/s^2
Gr	Grashof number	–
h	film coefficient	W/m^2·K
h_{fg}	latent heat of vaporization	J/kg
k	thermal conductivity	W/m·K
L	characteristic length	m
m	exponent	–
n	exponent	–
NTU	number of transfer units	–
Nu	Nusselt number	–
Pr	Prandtl number	–
$\dot{q}$	heat transfer per unit area	W/m^2
Q	heat transfer	W
$\dot{Q}$	heat transfer rate[3]	W
R	thermal resistance	m^2·K/W

Ra	Rayleigh number	–
Re	Reynolds number	–
T	temperature	K
U	overall coefficient of heat transfer	W/m^2·K
v	velocity	m/s
V	volumetric flow rate	m^3/s
x	distance	m

Symbols

α	thermal diffusivity	m^2/s
β	coefficient of volumetric expansion	1/K
ε	heat exchanger effectiveness	–
θ	angle	deg
μ	absolute viscosity[4]	kg/s·m
ν	kinematic viscosity	m^2/s
ρ	mass density	kg/m^3
σ	surface tension	N/m

Subscripts

b	boiling or bulk
c	cold or condensing
C	cooling
Ci	cold, in
Co	cold, out
D	diameter
e	excess
f	fluid, fowling, or friction
H	hot or hydraulic
Hi	hot, in
Ho	hot, out
i	inside
l	liquid
L	length
lm	log mean
m	mean
o	outside
r	radiation
s	surface
sat	saturated
v	vapor
V	constant volumetric flow rate
w	wall or wire
∞	at infinity or free stream

1. INTRODUCTION TO CONVECTION

Convection is the removal of heat from a surface by a fluid. *Forced convection* is the removal of heat from a surface by a fluid resulting from external surface forces,

---

[1]The *NCEES FE Reference Handbook* (*NCEES Handbook*) uses the symbol D to designate both inside (see Eq. 26.9) and outside (see Eq. 26.2, Eq. 26.6, and Eq. 26.8) diameters. The context must be evaluated to determine the exact meaning.

[2]g also has a value of 1.27×10^8 m/h^2.

[3]The *NCEES Handbook* designates the heat transfer "rate" as $\dot{Q}$, with the top dot used to designate a rate per unit time. (This is known as *Newton's notation.*) While it is not a universal practice, modern usage dispenses with both the "rate" term in the name and the top dot in the symbol.

[4](1) The use of mass units in viscosity values is typical in the subject of convective heat transfer. (2) Most data compilations give fluid viscosity in units of seconds. In the United States, heat transfer is traditionally given on a per hour basis. Therefore, a conversion factor of 3600 is needed when calculating dimensionless numbers from table data. (3) The units kg/s·m are equivalent to $N·s/m^2$ or Pa·s.

Heat Transfer

such as a pump or fan. *Natural convection* (also known as *free convection*) is the removal of heat from a surface by a fluid that moves vertically under the influence of a density gradient.

Equation 26.1: Newton's Law of Cooling

$$\dot{Q} = hA(T_w - T_\infty) \qquad 26.1$$

Variation

$$\dot{Q} = \dot{q}A = hA(T_s - T_\infty)$$

Values

Table 26.1 *Typical Film Coefficients for Natural Convection**

	h (W/m²·K)
no change in phase	
still air	5.0–25.0
condensing	
steam	
horizontal surface	9600–24 400
vertical	4000–11 300
organic solvents	850–2800
ammonia	2800–5700
evaporating	
water	4500–11 300
organic solvents	550–1700
ammonia	1100–2300

*Values outside these ranges have been observed. However, these ranges are typical of those encountered in industrial processes.

Description

Equation 26.1, Newton's law of cooling, is the basic equation used to calculate the steady-state convective heat transfer in both heating and cooling configurations. The *film coefficient (heat transfer coefficient)*, h, is seldom known to great accuracy.[5] (See Table 26.1.) The average film coefficient, $\overline{h}$, is used where there are variations over the heat transfer surface.[6]

In Eq. 26.1, T_∞ is the *bulk temperature* of the environment (air, gas, surrounding liquid, etc.), and T_w is the instantaneous temperature of the cooling body's surface.[7]

---

[5]An error of up to 25% can be expected.
[6]Though $\overline{h}$ has traditionally been used in books on the subject of heat transfer and is used in the *NCEES Handbook*, most modern books use the symbol h. The fact that the film coefficient is an inaccurate, average value is implicit.
[7]The *NCEES Handbook* uses the subscript w to designate "wall." However, Newton's law of cooling does not primarily apply to heat transfer through a wall or even heat transfer from a surface. Newton's law of cooling applies to a cooling (body) object. The "wall" designation might also imply that the temperature is constant. However, the body temperature changes with time. Such is the entire purpose of Newton's law of cooling: to specify the heat transfer and temperature as functions of time. It is common to designate the body temperature at a particular moment in time simply with the variable T. If the temperature of the body is not uniform throughout, the symbol T_s can be used to designate a surface temperature.

Example

An object is cooled by a circulating water bath. At a particular moment, the surface temperature of the body is 373K, while the bulk temperature of the water bath is 353K. The convective film coefficient is 350 W/m²·K. The object has an exposed surface area of 1 m². Most nearly, what is the instantaneous rate of heat transfer from the object?

(A) 3.5 kW

(B) 5.0 kW

(C) 7.0 kW

(D) 8.2 kW

Solution

Using Eq. 26.1, the rate of heat transfer is

$$\dot{Q} = hA(T_w - T_\infty)$$

$$= \frac{\left(350 \ \dfrac{W}{m^2 \cdot K}\right)(1 \ m^2)(373K - 353K)}{1000 \ \dfrac{W}{kW}}$$

$$= 7.0 \ kW$$

The answer is (C).

2. NUSSELT NUMBER

Equation 26.2: Nusselt Number, Cylinder in Crossflow

$$\overline{Nu}_D = \frac{\overline{h}D}{k} = C Re_D^n Pr^{1/3} \qquad 26.2$$

Values

Table 26.2 *Values of C and n for a Known Reynolds Number, Re_D*

Re_D	C	n
1–4	0.989	0.330
4–40	0.911	0.385
40–4000	0.683	0.466
4000–40,000	0.193	0.618
40,000–250,000	0.0266	0.805

Description

The *Nusselt number*, Nu, is defined by Eq. 26.2. The subscript D indicates that the correlation is based on the outside diameter of the cylinder, not some other characteristic dimension. The Nusselt number is sometimes written with a subscript (e.g., Nu_h or Nu_f) to indicate that the fluid properties are evaluated at the film temperature. Since the flow velocity, heat transfer rate, film resistance, and other fluid properties are not the same everywhere around the periphery of the cylinder, the overbars are used to designate average values.

For a cylinder (for example, a tube or wire) in crossflow, Eq. 26.2 can be used with any fluid to calculate the film coefficient.[8] The fluid properties are evaluated at the film temperature. The entire surface area of the tube is used when calculating the heat transfer.

Equation 26.2 can be simplified for air since $Pr^{1/3} \approx 1.00$. This modified equation is known as the *Hilbert-Morgan equation*.

Values of C and n are found from Table 26.2 when Re_D is known.

Example

Water at 25°C flows perpendicularly over a long, hot pipe of 2.9 cm outside diameter. The water has an average convection film coefficient of 4900 W/m²·K and a thermal conductivity of 610×10^{-3} W/m·K. The Nusselt number is most nearly

(A) 230

(B) 290

(C) 350

(D) 410

Solution

The Nusselt number can be calculated with Eq. 26.2.

$$\overline{Nu}_D = \frac{\overline{h}D}{k}$$

$$= \frac{\left(4900 \ \frac{W}{m^2 \cdot K}\right)(2.9 \text{ cm})}{\left(610 \times 10^{-3} \ \frac{W}{m \cdot K}\right)\left(100 \ \frac{cm}{m}\right)}$$

$$= 232.95 \quad (230)$$

The answer is (A).

3. PRANDTL NUMBER

Equation 26.3: Prandtl Number

$$Pr = \frac{c_p \mu}{k} \qquad 26.3$$

Description

The dimensionless *Prandtl number*, Pr, is defined by Eq. 26.3. It represents the ratio of momentum diffusion to thermal diffusion. For gases, the values used in calculating the Prandtl number do not vary significantly with temperature, and so neither does the Prandtl number itself. The values used are for the fluid, not for the surface material.

---

[8]There are more sophisticated correlations.

Example

For air at a temperature of 42°C, the heat capacity is 990 J/kg·K, the viscosity is 1.9×10^{-5} kg/s·m, and the thermal conductivity is 0.028 W/m·K. What is most nearly the Prandtl number?

(A) 0.59

(B) 0.67

(C) 0.75

(D) 0.83

Solution

Using Eq. 26.3, the Prandtl number is

$$Pr = \frac{c_p \mu}{k}$$

$$= \frac{\left(990 \ \frac{J}{kg \cdot K}\right)\left(1.9 \times 10^{-5} \ \frac{kg}{s \cdot m}\right)}{0.028 \ \frac{W}{m \cdot K}}$$

$$= 0.67$$

The answer is (B).

4. RAYLEIGH NUMBER

The dimensionless *Rayleigh number* is the product of the Grashof and Prandtl numbers.[9] It is used as an indicator of the primary heat transfer mechanism. When the Rayleigh number is less than some critical value (specific to the fluid, and usually determined by experimentation), heat transfer is primary by conduction. When the Rayleigh number is more than the critical value, heat transfer is primarily by convection. (The convective flow can be either laminar or turbulent.) The symbol can be written with a subscripted variable (e.g., Ra_L) designating the characteristic dimension with which the critical values have been correlated. For example, the length of a plate (see Eq. 26.4) is generally used, rather than its width. For a cylinder (see Eq. 26.6), the diameter is used.

Equation 26.4 and Eq. 26.5: Rayleigh Number, Flat Plate, Natural Convection

$$Ra_L = \frac{g\beta(T_s - T_\infty)L^3}{\nu^2}Pr \qquad 26.4$$

$$\beta = \frac{2}{T_s + T_\infty} \qquad 26.5$$

---

[9](1) In Eq. 26.4, the *NCEES Handbook* recognizes the Prandtl number but not the *Grashof number*. Rather than define the Grashof number (as it did with the Reynolds, Nusselt, Prandtl, and Rayleigh numbers), the *NCEES Handbook* elects to present a combination of variables, $g\beta(T_s - T_\infty)L^3/\nu^2$. This is equivalent to, and generally presented as, the Grashof number, Gr. (2) The Grashof number relates the buoyancy and viscosity of the fluid, while the Prandtl number relates the momentum diffusivity and thermal diffusivity.

Heat Transfer

Variation

$$\mathrm{Ra} = \frac{L^3 g \beta \rho^2 (T_s - T_\infty) c_p}{k\mu}$$

Description

Equation 26.4 gives the Rayleigh number for a vertical flat plate in a stationary fluid[10] with *characteristic length*, L.[11] The *coefficient of volumetric expansion*, β, for ideal gases is the reciprocal of the absolute film temperature. Gravitational acceleration, g, and viscosity, ν, must have the same unit of time in order to make Eq. 26.4 dimensionless.

For an ideal gas, the *coefficient of thermal expansion*, β (also known as the *volumetric coefficient of expansion*), can be found using Eq. 26.5. The temperatures used in Eq. 26.5 must be absolute temperatures.[12]

Equation 26.6: Rayleigh Number, Cylinder, Natural Convection

$$\mathrm{Ra}_D = \frac{g\beta(T_s - T_\infty)D^3}{\nu^2}\mathrm{Pr} \qquad \textbf{26.6}$$

Description

For a horizontal cylinder in a stationary fluid, the Rayleigh number can be found using Eq. 26.6.

Example

A horizontal pipe has a diameter of 0.21 m and an outer surface temperature of 350K. The pipe is surrounded by stationary water with a bulk temperature of 300K. For water at 300K, the kinematic viscosity is 1.6×10^{-5} m²/s. The coefficient of thermal expansion is 0.0028 K^{-1}, and the Prandtl number is 0.72. What is most nearly the Rayleigh number?

(A) 2.2×10^7

(B) 2.9×10^7

(C) 3.3×10^7

(D) 3.6×10^7

---

[10]The *NCEES Handbook* is not consistent in its designation for "surface." While the subscript w is used in Eq. 26.1 for the surface of an object (wall), Eq. 26.4 and Eq. 26.5 use subscript s.

[11]The length of the side of a square, the mean length of a rectangle, and 90% of the diameter of a circle have historically been used as the *characteristic length*. However, the ratio of surface area to perimeter gives better agreement with experimental data.

[12]The *NCEES Handbook* is not explicit about the temperature at which β, ν, and Pr are evaluated. Since β is the reciprocal of the film temperature, Eq. 26.5 implicitly defines the *film temperature* as the average of the surface and bulk temperatures. This also implicitly identifies the temperature at which other properties are evaluated. Since the *NCEES Handbook* does not provide sufficient tables, it is logical to conclude that the film temperature is irrelevant, because on the exam, all necessary fluid data will be provided within a problem.

Solution

The Rayleigh number can be calculated using Eq. 26.6.

$$\mathrm{Ra}_D = \frac{g\beta(T_s - T_\infty)D^3}{\nu^2}\mathrm{Pr}$$

$$= \left(\frac{\left(9.81 \, \frac{\mathrm{m}}{\mathrm{s}^2}\right)\left(0.0028 \, \frac{1}{\mathrm{K}}\right)}{\left(1.6 \times 10^{-5} \, \frac{\mathrm{m}^2}{\mathrm{s}}\right)^2} \right)(0.72)$$

$$= 3.577 \times 10^7 \quad (3.6 \times 10^7)$$

The answer is (D).

5. REYNOLDS NUMBER

The Reynolds number is used to determine which of the three flow regimes is applicable. Laminar flow over smooth flat plates occurs for Reynolds numbers up to approximately 2×10^5; turbulent flow exists for Reynolds numbers greater than approximately 3×10^6.[13] Transition flow is in between. The distance from the leading edge at which turbulent flow is initially experienced is determined from the *critical Reynolds number*, commonly taken as $\mathrm{Re} = 5 \times 10^5$ for smooth flat plates, though the actual value is highly dependent on surface roughness. Distance, x, is measured from the leading edge.

The free-stream velocity is always zero with natural convection, so the traditional Reynolds number is also always zero. The Grashof and Rayleigh numbers take the place of determining whether flow is laminar or turbulent.[14] The film Reynolds number is used to determine whether condensation is turbulent.

Equation 26.7: Reynolds Number, Flat Plate

$$\mathrm{Re}_L = \frac{\rho \mathrm{v}_\infty L}{\mu} \qquad \textbf{26.7}$$

Description

Use Eq. 26.7 to find the Reynolds number for a flat plate of length L in parallel flow.[15]

---

[13]Turbulent flow can begin at Reynolds numbers less than 3×10^5 if the plate is rough. This discussion assumes the plate is smooth.

[14]Rising air nevertheless has a velocity. The critical Reynolds number for laminar flow of air is approximately 550 (corresponding to a Grashof number of 10^9).

[15]The *NCEES Handbook* is inconsistent in the variable used for velocity in the definition of Reynolds number. The *NCEES Handbook* uses lowercase roman v in the Fluid Mechanics section, uppercase roman V in the Moody diagram, lowercase italic v in the Environmental Engineering section, and uppercase italic V in the Chemical Engineering section. Inconsistent with any of those, the *NCEES Handbook*'s version of Eq. 26.7, adopts the (common heat transfer) convention of using u_∞ as yet another velocity variable for the Reynolds number. This book uses lowercase roman v for velocity.

Example

20°C air flows at 0.75 m/s over and parallel to a wide flat plate 1.7 m long. At 20°C, the density of air is 1.2 kg/m³, and the absolute viscosity is 1.8×10^{-5} kg/m·s. What is most nearly the Reynolds number for the purpose of forced convection?

(A) 12 000
(B) 23 000
(C) 47 000
(D) 85 000

Solution

The Reynolds number can be calculated using Eq. 26.7.

$$\mathrm{Re}_L = \frac{\rho \mathrm{v}_\infty L}{\mu}$$

$$= \frac{\left(1.2 \ \frac{\mathrm{kg}}{\mathrm{m}^3}\right)\left(0.75 \ \frac{\mathrm{m}}{\mathrm{s}}\right)(1.7 \ \mathrm{m})}{1.8 \times 10^{-5} \ \frac{\mathrm{kg}}{\mathrm{m \cdot s}}}$$

$$= 85\,000$$

The answer is (D).

Equation 26.8: Reynolds Number, Cylinder

$$\mathrm{Re}_D = \frac{\rho \mathrm{v}_\infty D}{\mu} \qquad 26.8$$

Description

Use Eq. 26.8 to calculate the Reynolds number for a cylinder of outside diameter D in crossflow.

Example

Water with a bulk temperature of 27°C flows over and across a long, hot pipe of 4 cm diameter at a velocity of 1.2 m/s. At 27°C, the density of water is 1.0×10^3 kg/m³, and the absolute viscosity is 850×10^{-6} kg/s·m. What is most nearly the Reynolds number?

(A) 44 000
(B) 51 000
(C) 56 000
(D) 61 000

Solution

The Reynolds number can be calculated using Eq. 26.8.

$$\mathrm{Re}_D = \frac{\rho \mathrm{v}_\infty D}{\mu}$$

$$= \frac{\left(1.0 \times 10^3 \ \frac{\mathrm{kg}}{\mathrm{m}^3}\right)\left(1.2 \ \frac{\mathrm{m}}{\mathrm{s}}\right)(4 \ \mathrm{cm})}{\left(850 \times 10^{-6} \ \frac{\mathrm{kg}}{\mathrm{s \cdot m}}\right)\left(100 \ \frac{\mathrm{cm}}{\mathrm{m}}\right)}$$

$$= 56\,471 \quad (56\,000)$$

The answer is (C).

Equation 26.9: Reynolds Number, Internal Flow

$$\mathrm{Re}_D = \frac{\rho \mathrm{v}_m D}{\mu} \qquad 26.9$$

Description

Equation 26.9 is the traditional definition of the Reynolds number for internal flow within a circular channel. D is the internal diameter. v_m is defined as the average (mean) fluid velocity. The average velocity is essentially equal to the bulk velocity if the flow is fully turbulent. If the flow is fully developed laminar, the average velocity is 50% of the maximum velocity.

Example

Water at 25°C flows at 3.6 m/s in a pipe with an internal diameter of 0.10 m. The water has a density of 1.0×10^3 kg/m³ and an absolute viscosity of 850×10^{-6} kg/s·m. What is most nearly the Reynolds number?

(A) 35 000
(B) 87 000
(C) 260 000
(D) 420 000

Solution

The Reynolds number for internal flow can be calculated using Eq. 26.9.

$$\mathrm{Re}_D = \frac{\rho \mathrm{v}_m D}{\mu}$$

$$= \frac{\left(1.0 \times 10^3 \ \frac{\mathrm{kg}}{\mathrm{m}^3}\right)\left(3.6 \ \frac{\mathrm{m}}{\mathrm{s}}\right)(0.10 \ \mathrm{m})}{850 \times 10^{-6} \ \frac{\mathrm{kg}}{\mathrm{s \cdot m}}}$$

$$= 423\,529 \quad (420\,000)$$

The answer is (D).

Heat Transfer

6. NATURAL CONVECTION

Natural convection (also known as *free convection*) is the removal of heat from a surface by a fluid that moves vertically under the influence of a density gradient. As a fluid warms, it becomes lighter and rises from the heating surface. The fluid is acted on by buoyant and gravitational forces. The fluid does not have a component of motion parallel to the surface.[16]

Heat transfer by natural convection is attractive from an engineering design standpoint because no motors, fans, pumps, or other equipment with moving parts are required. However, the transfer surface must be much larger than it would be with forced convection.[17]

7. NUSSELT EQUATION

The *Nusselt equation* and correlations of its form are often used to find the film coefficients for convective heating and cooling.

Equation 26.10: Nusselt Equation, Vertical Flat Plate or Cylinder

$$\overline{h} = C\left(\frac{k}{L}\right)\mathrm{Ra}_L^n \qquad 26.10$$

Variation

$$\mathrm{Nu} = \frac{\overline{h}L}{k} = C\mathrm{Ra}^n$$

Values

Table 26.3 Values of C and n for Vertical Plate or Cylinder in Natural Convection

range of Ra_L	C	n
10^4–10^9	0.59	1/4
10^9–10^{13}	0.10	1/3

Description

For a vertical flat plate or a large diameter vertical cylinder in a stationary fluid, the film coefficient can be found using Eq. 26.10.

For laminar convection $(1000 < \mathrm{Ra} < 10^9)$, n has a value of approximately $1/4$. For turbulent convection $(\mathrm{Ra} > 10^9)$, n is approximately $1/3$. For sublaminar convection $(\mathrm{Ra} < 1000)$, n is less than $1/4$ (typically taken as $1/5$), and graphical solutions are commonly used.

---

[16]Rotating spheres and cylinders and vertical plane walls are special categories of convective heat transfer where the fluid has a component of relative motion parallel to the heat transfer surface.

[17]Natural convection requires approximately 2 to 10 times more surface area than does forced convection.

The values of the dimensionless empirical constants C and n can be used with all fluids and any consistent systems of units. Table 26.3 is limited in application to single heat transfer surfaces (for example, a single tube or a single plate).

The thermal conductivity, k, in Eq. 26.10 is for the transfer fluid, not for the surface (wall), and it is evaluated at the film temperature.

Example

The Rayleigh number for a large diameter vertical pipe 1.7 m high is 1.4×10^5. Natural convection occurs in the laminar regime. The thermal conductivity of the pipe is 0.028 W/m·K. What is most nearly the average heat transfer coefficient?

(A) 0.10 W/m²·K

(B) 0.13 W/m²·K

(C) 0.19 W/m²·K

(D) 0.27 W/m²·K

Solution

Natural convection in the laminar regime is characterized by a Rayleigh number less than 10^9. From Table 26.3, $C = 0.59$, and $n = 1/4$ (0.25).

Using Eq. 26.10, the average heat transfer coefficient is

$$\overline{h} = C\left(\frac{k}{L}\right)\mathrm{Ra}_L^n$$

$$= (0.59)\left(\frac{0.028 \ \dfrac{\mathrm{W}}{\mathrm{m \cdot K}}}{1.7 \ \mathrm{m}}\right)(1.4 \times 10^5)^{0.25}$$

$$= 0.1880 \ \mathrm{W/m^2 \cdot K} \quad (0.19 \ \mathrm{W/m^2 \cdot K})$$

The answer is (C).

Equation 26.11: Nusselt Equation, Horizontal Cylinder

$$\overline{h} = C\left(\frac{k}{D}\right)\mathrm{Ra}_D^n \qquad 26.11$$

Values

Table 26.4 Values of C and n for Horizontal Cylinder in Natural Convection

Ra_D	C	n
10^{-3}–10^2	1.02	0.148
10^2–10^4	0.850	0.188
10^4–10^7	0.480	0.250
10^7–10^{12}	0.125	0.333

Heat Transfer

Description

For a horizontal cylinder in a stationary fluid, the film coefficient can be found using Eq. 26.11. Values of C and n are found from Table 26.4.

8. CONDENSING VAPOR

When a vapor condenses on a cooler surface, the condensate forms a thin layer on the surface. This layer insulates the surface and creates a thermal resistance. However, if the condensate falls or flows from the surface (as it would from a horizontal tube), the condensate also removes thermal energy from the surface. Film coefficients are relatively high (on the order of 11.5 kW/m²·K to 22.7 kW/m²·K).[18]

Filmwise condensation occurs when the condensing surface is smooth and free from impurities.[19] A continuous film of condensate covers the entire surface. The film flows smoothly down over the surface under the action of gravity and eventually falls off. However, if the surface contains impurities or irregularities that prevent complete wetting, the film will be discontinuous, a condition known as *dropwise condensation*.[20]

Equation 26.12: Condensation, Outside Horizontal Tubes

$$\overline{\mathrm{Nu}}_D = \frac{\overline{h}D}{k} = 0.729 \left[\frac{\rho_l^2 g h_{fg} D^3}{\mu_l k_l (T_{\mathrm{sat}} - T_s)} \right]^{0.25} \qquad 26.12$$

Description

Equation 26.12, based on Nusselt's theoretical work, predicts film coefficients for the laminar filmwise condensation of a pure saturated vapor on the outside of a horizontal tube with a diameter between 2.5 cm and 7.6 cm. Equation 26.12 is in fair agreement with experimental data, with calculated values generally being

low.[21] Proper units must be observed in order to keep the argument of exponentiation unitless.

Using Nusselt correlations for condensing vapor depends on determining several properties of the resulting condensate. Since the properties of the condensing vapor vary with temperature, these liquid properties are evaluated (by convention) at the average of the saturation and surface temperatures.[22] This implicitly defines the *film temperature*.

$$T_{\mathrm{film}} = \frac{T_s + T_{\mathrm{sat}}}{2}$$

The actual surface temperature is often unknown in initial studies. However, for steam, condensation frequently occurs with a temperature difference of $T_{\mathrm{sat}} - T_s$ between 3°C and 22°C.

Equation 26.13: Condensation, Vertical Surfaces

$$\overline{\mathrm{Nu}}_L = \frac{\overline{h}L}{k_l} = 0.943 \left[\frac{\rho_l^2 g h_{fg} L^3}{\mu_l k_l (T_{\mathrm{sat}} - T_s)} \right]^{0.25} \qquad 26.13$$

Description

Filmwise condensation of pure saturated vapors on vertical surfaces (including the insides and outsides of tubes) is predicted by Eq. 26.13. Equation 26.13 cannot be used for condensation on inclined tubes. The film flow is not parallel with the longitudinal axis of an inclined tube, resulting in an effective inclination angle that varies with location along the tube.[23] As with condensation on horizontal surfaces, the latent heat of condensation is evaluated at the vapor temperature, while the remaining fluid properties are evaluated at the film

---

[18]A film coefficient of 11.5 kW/m²·K is routinely assumed as a first estimate for condensation of steam on the outside of tubes.

[19]Filmwise condensation can always be expected with clean steel and aluminum tubes under ordinary conditions, as well as with heavily contaminated tubes. Dropwise condensation generally requires smooth surfaces with minute amounts of contamination, rather than rough surfaces. Since dropwise condensation can be expected only under carefully controlled conditions, the assumption of filmwise condensation is generally warranted.

[20]Film coefficients for dropwise condensation can be 4 to 8 times larger than for filmwise condensation because the film is thinner and the thermal resistance is smaller.

[21](1) The ρ_l^2 term in the numerator is a simplification. Nusselt's theoretical work correlated heat transfer with the quantity $\rho_l(\rho_l - \rho_v)$. The quantity ρ_l^2 in the *NCEES Handbook* results from assuming the vapor density, ρ_v, is zero. The *NCEES Handbook* does not invoke this simplification for the material presented on boiling and evaporation. (2) The more precise value of the constant 0.729 is often reported as 0.725, which was the value originally derived from a numerical analysis. In practice, highly precise estimates are illusionary, as actual values are found within the range between the two values.

[22]The *NCEES Handbook* gives the guidance, "Evaluate all liquid properties at the average temperature..." However, this guidance does not apply to the latent heat, h_{fg}, which is a property of the vapor. The latent heat should be evaluated at the saturation temperature and pressure.

[23](1) Equation 26.13 was derived by Nusselt with a coefficient of 0.943. However, ripples in the laminar film appear at condensation Reynolds numbers as low as 30 or 40. Experimental data show actual film coefficients are approximately 20% higher than the theoretical. A coefficient of 1.13 in place of 0.943 reflects this increase. Retaining the 0.943 value, however, yields a conservative value. (2) Equation 26.13 can be used to find the condensing film coefficient on a vertical tube when the total condensation is less than 460 kg/h.

Heat Transfer

temperature. The characteristic length, L, in Eq. 26.13 is the surface length.

9. INTRODUCTION TO FORCED CONVECTION

As with natural convection, *forced convection* depends on the movement of a fluid to remove heat from a surface. With forced convection, a fan, a pump, or relative motion causes the fluid motion. If the flow is over a flat surface, the fluid particles near the surface will flow more slowly due to friction with the surface. The *boundary layer* of slow-moving particles comprises the major thermal resistance. The thermal resistance of the tube and other heat exchanger components is often disregarded.

Newton's law of convection, Eq. 26.1, gives the heat transfer for Newtonian fluids in forced convection over exterior surfaces.[24] The film coefficient, h, is also known as the *coefficient of forced convection*. T_∞ is the *free-stream temperature*.

10. FLOW OVER FLAT PLATES

The boundary layer of a fluid flowing over a flat plate is assumed to have a parabolic velocity distribution.[25] The layer has three distinct regions: laminar, transition, and turbulent. From the leading edge, the layer is laminar and the thickness increases gradually until the transition region where the thickness increases dramatically. Thereafter, the boundary layer is turbulent. The laminar region is always present, though its length decreases as velocity increases. Turbulent flow may not develop at all with short plates.

Equation 26.14 and Eq. 26.15: Flat Plate

$$\overline{Nu}_L = \frac{\overline{h}L}{k} = 0.6640 Re_L^{1/2} Pr^{1/3} \quad [Re_L < 10^5] \qquad 26.14$$

$$\overline{Nu}_L = \frac{\overline{h}L}{k} = 0.0366 Re_L^{0.8} Pr^{1/3} \quad [Re_L > 10^5] \qquad 26.15$$

Description

Equation 26.14 and Eq. 26.15 give the Nusselt number for a flat plate in parallel flow. The Reynolds number is used to determine whether the flow regime is laminar ($Re_L < 100\,000$) or turbulent ($Re_L > 100\,000$). The Reynolds number is calculated using the length of the plate as the characteristic length.

---

[24](1) Newton's law of convection is the same for natural and forced convection. Only the methods used to evaluate the film coefficient are different. (2) The results of this chapter do not generally apply to non-Newtonian fluids.

[25]The velocity distribution does not have to be parabolic. In *Couette flow*, there are two closely spaced parallel surfaces, one which is stationary and the other moving with constant velocity. The velocity gradient is assumed to be linear between the plates.

Example

Air at 40°C flows at 0.65 m/s over a square flat plate 1.5 m on each side. The Reynolds number is 56 000. The Prandtl number is 0.71. What is most nearly the average Nusselt number?

(A) 140

(B) 180

(C) 320

(D) 560

Solution

Since $Re_L = 56\,000 < 10^5$, Eq. 26.14 can be used to calculate the average Nusselt number.

$$\begin{aligned}
\overline{Nu}_L &= 0.6640 Re_L^{1/2} Pr^{1/3} \\
&= (0.6640)(56\,000)^{1/2}(0.71)^{1/3} \\
&= 140
\end{aligned}$$

The answer is (A).

11. FLOW INSIDE TUBES

Laminar flow in smooth tubes occurs at Reynolds numbers less than 2300. As with flow over a flat plate, the velocity distribution is parabolic, but the extent of the parabola is limited to the tube radius. In the *entrance region*, the parabola does not extend to the centerline. Further on, a point is reached where the parabolic distribution is complete, and the flow is said to be *fully developed* laminar flow.[26] At that point, the average velocity is one-half of the maximum (centerline) velocity.

Equation 26.16: Laminar Flow Inside Tubes with Uniform Heat Flux

$$Nu_D = 4.36 \quad [\text{uniform heat flux}] \qquad 26.16$$

Variation

$$Nu_D = \frac{\overline{h}D}{k}$$

Description

Equation 26.16 correlates the Nusselt number in the case of laminar flow inside a circular channel with uniform heat flux along the length of flow. Laminar flow is appropriate when $Re_D < 2300$. Equation 26.16 provides an effective means of determining the average film coefficient for this situation. Since the heat flux passing through the tube wall into the fluid is constant along the length of flow, the length of the tube is not relevant.

---

[26]The term *fully developed* is also used when referring to full turbulence. In this section, it is understood that the flow is laminar.

Heat Transfer

Equation 26.17: Laminar Flow Inside Tubes with Constant Surface Temperature

$$\text{Nu}_D = 3.66 \quad \left[\begin{array}{c}\text{constant surface}\\\text{temperature}\end{array}\right] \qquad 26.17$$

Variation

$$\text{Nu}_D = \frac{\overline{h}D}{k}$$

Description

Equation 26.17 correlates the Nusselt number in the case of laminar flow inside a circular channel with constant surface temperature along the length of flow. Laminar flow is appropriate when $\text{Re}_D < 2300$. Equation 26.17 provides an effective means of determining the average film coefficient for this situation. Since the temperature along the tube length is constant along the length of flow, the length of the tube is not relevant.

Equation 26.18: Laminar Flow Inside Tubes with Constant Wall Temperature

$$\text{Nu}_D = 1.86 \left(\frac{\text{Re}_D \text{Pr}}{\frac{L}{D}}\right)^{1/3} \left(\frac{\mu_b}{\mu_s}\right)^{0.14} \qquad 26.18$$

Description

The *Sieder-Tate equation* (also known as *Sieder-Tate correlation*), Eq. 26.18, predicts the average film coefficient along the entire length of laminar flow. In Eq. 26.18, μ_b is the absolute viscosity of the fluid at the bulk temperature, and μ_s is the absolute viscosity of the fluid at the tube's surface (wall) temperature. All of the other fluid properties are evaluated at the bulk temperature.

Example

A fluid flows through a 0.30 m diameter circular tube 2.1 m in length. The following properties have been calculated.

$$\text{Re}_D = 2100$$
$$\text{Pr} = 0.71$$
$$\mu_b = 850 \text{ kg/s·m}$$
$$\mu_s = 860 \text{ kg/s·m}$$

What is most nearly the Nusselt number?

(A) 11
(B) 56
(C) 460
(D) 890

Solution

Since the Reynolds number is less than 2300, flow is laminar.

Use Eq. 26.18 to calculate the Nusselt number.

$$\text{Nu}_D = 1.86 \left(\frac{\text{Re}_D \text{Pr}}{\frac{L}{D}}\right)^{1/3} \left(\frac{\mu_b}{\mu_s}\right)^{0.14}$$

$$= (1.86) \left(\frac{(2100)(0.71)}{\frac{2.1 \text{ m}}{0.30 \text{ m}}}\right)^{1/3} \left(\frac{850 \frac{\text{kg}}{\text{s·m}}}{860 \frac{\text{kg}}{\text{s·m}}}\right)^{0.14}$$

$$= 11$$

The answer is (A).

Equation 26.19: Turbulent Flow Inside Straight Tubes

$$\text{Nu}_D = 0.023 \text{Re}_D^{0.8} \text{Pr}^{1/3} \left(\frac{\mu_b}{\mu_s}\right)^{0.14} \qquad 26.19$$

Description

If there is a large change in viscosity during the heat transfer process, as there would be with oils and other viscous fluids heated in a long tube, the *Sieder-Tate equation* for turbulent flow, Eq. 26.19, should be used instead of the Nusselt equation. Equation 26.19 can be used with both the case of uniform surface temperature and the case of uniform heat flux but is limited to $\text{Re}_D > 10\,000$ and $\text{Pr} > 0.7$.[27]

All fluid properties in Eq. 26.19 are evaluated at the bulk temperature, except for μ_s, which is evaluated at the surface temperature.

Example

A fluid flows through a long circular tube with uniform surface temperature. The following properties have been calculated.

$$\text{Re}_D = 2.2 \times 10^4$$
$$\text{Pr} = 0.75$$
$$\mu_b = 840 \text{ kg/s·m}$$
$$\mu_s = 850 \text{ kg/s·m}$$

---

[27]Although the *NCEES Handbook* specifies a lower limit for the Prandtl number, it does not specify an upper limit. The upper limit is reported by various researchers as 700, 16 700, and 17 000.

Heat Transfer

What is most nearly the Nusselt number?

(A) 53

(B) 62

(C) 320

(D) 1500

Solution

Since the Reynolds number is greater than 10 000, and the Prandtl number is greater than 0.7, Eq. 26.19 may be used. The Nusselt number is

$$\mathrm{Nu}_D = 0.023\mathrm{Re}_D^{0.8}\mathrm{Pr}^{1/3}\left(\frac{\mu_b}{\mu_s}\right)^{0.14}$$

$$= (0.023)(2.2\times10^4)^{0.8}(0.75)^{1/3}\left(\frac{840\ \frac{\mathrm{kg}}{\mathrm{s\cdot m}}}{850\ \frac{\mathrm{kg}}{\mathrm{s\cdot m}}}\right)^{0.14}$$

$$= 62$$

The answer is (B).

Equation 26.20 and Eq. 26.21: Turbulent Liquid Metal Flow in Tubes

$$\mathrm{Nu}_D = 7.0 + 0.025\mathrm{Re}_D^{0.8}\mathrm{Pr}^{0.8} \quad \begin{bmatrix}\text{constant surface} \\ \text{temperature}\end{bmatrix}$$
26.20

$$\mathrm{Nu}_D = 6.3 + 0.0167\mathrm{Re}_D^{0.85}\mathrm{Pr}^{0.93} \quad [\text{uniform heat flux}]$$
26.21

Description

When the surface (wall) temperature is constant, Eq. 26.20 can be used to calculate the average film coefficient for liquid metals (for example, mercury, sodium, and lead-bismuth alloys) experiencing fully developed turbulent flow inside tubes.[28] Fluid properties are evaluated at the mean bulk temperature.

With a constant heat flux, Eq. 26.21 can be used to calculate the average film coefficient for liquid metals with fully developed turbulent flow inside tubes.

Equation 26.20 and Eq. 26.21 are both limited to $0.003 < \mathrm{Pr} < 0.05$.

Example

Liquid metal flows through a tube with constant surface temperature. The Reynolds number is 2700, and the

---

[28]The term *fully developed* is also used when referring to full laminar flow. In this section it is understood that the flow is turbulent.

Prandtl number is 0.04. What is most nearly the Nusselt number?

(A) 2.5

(B) 6.6

(C) 8.1

(D) 9.7

Solution

For liquid metal flow with constant surface temperature, the Nusselt number can be calculated using Eq. 26.20.

$$\mathrm{Nu}_D = 7.0 + 0.025\mathrm{Re}_D^{0.8}\mathrm{Pr}^{0.8}$$
$$= 7.0 + (0.025)(2700)^{0.8}(0.04)^{0.8}$$
$$= 8.06 \quad (8.1)$$

The answer is (C).

12. FLOW THROUGH NONCIRCULAR DUCTS

Equation 26.22: Hydraulic Diameter

$$D_H = \frac{4\times\text{cross-sectional area}}{\text{wetted perimeter}}$$
26.22

Description

A *duct* is any closed channel through which a fluid flows. Tubes and pipes are examples of round ducts. "Ducts" are not limited to air conditioning ducts.

Dimensional analysis shows that a *characteristic length* is required in the Nusselt number, but it does not identify the length to be used. It has been common practice to correlate empirical pressure drop and heat transfer data with the *hydraulic diameter*, D_H, of noncircular (e.g., rectangular, square, elliptical, polygonal) ducts.

Though an approximation, empirical data supports using the hydraulic diameter in most cases. Notable exceptions are flow through ducts with narrow angles (for example, an equilateral triangle with a narrow vertex angle) and flow *parallel* to banks of tubes.

Equation 26.23: Hydraulic Diameter, Annulus

$$D_H = D_o - D_i$$
26.23

Description

Annular flow is the flow of fluid through an annulus. Fluid flow is annular in simple tube-in-tube heat exchangers.[29] For an annulus, Eq. 26.23 gives the hydraulic diameter.

---

[29]Flow is not annular through shell and ube heat exchangers.

13. FLOW OVER SPHERES

Spheres have the smallest surface area-to-volume ratio, so spherical tanks are used where heat transfer is to be minimized. When a sphere experiences motion relative to a surrounding fluid, heat transfer to and from the fluid is predicted by a Nusselt number correlation.

Equation 26.24: Nusselt Number, Flow Over Spheres

$$\overline{Nu}_D = \frac{\overline{h}D}{k} = 2.0 + 0.60Re_D^{1/2}Pr^{1/3} \quad \begin{bmatrix} 1 < Re_D < 70\,000; \\ 0.6 < Pr < 400 \end{bmatrix}$$

$$26.24$$

Description

The Nusselt correlation for fluid flow over a sphere is given by Eq. 26.24. Fluid properties should be evaluated at the film temperature.

Example

The Reynolds number for air flowing over a sphere is 62 000. The Prandtl number is 0.81. What is most nearly the average Nusselt number?

(A) 110

(B) 140

(C) 450

(D) 670

Solution

The average Nusselt number can be calculated using Eq. 26.24.

$$\overline{Nu}_D = 2.0 + 0.6Re_D^{1/2}Pr^{1/3}$$
$$= 2.0 + (0.6)(62\,000)^{1/2}(0.81)^{1/3}$$
$$= 141.27 \quad (140)$$

The answer is (B).

14. BOILING

The temperature at which a liquid boils is known as its *saturation temperature*. Boiling will occur when a liquid is heated to its saturation temperature. A common method of boiling a liquid is to heat it in a vessel (pan, tank, etc.). Boiling will occur when the surface temperature, T_s, is greater than the saturation temperature, T_{sat}.

The behavior of a liquid that is just beginning to boil (i.e., when vapor bubbles start forming on the heated surface) is different from the behavior in a rolling boil. The boiling behavior is affected by the difference in temperature of the surface and the saturation temperature, as well as the

roughness of the heated surface. Boiling behavior is also affected by the presence of external agitation. Boiling behavior is categorized in a number of ways. These boiling regimes are illustrated by a *boiling curve*, such as Fig. 26.1 for water at atmospheric pressure.

Figure 26.1 *Typical Boiling Curve*

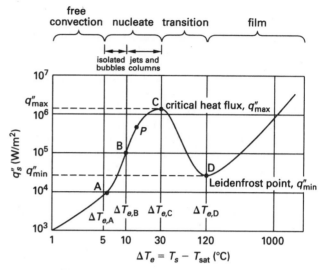

Typical boiling curve for water at one atmosphere: surface heat flux q_s'' as a function of excess temperature, $\Delta T_e = T_s - T_{sat}$

Source: Incropera, Frank P. and David P. DeWitt, *Fundamentals of Heat and Mass Transfer*, 3rd ed., Wiley, 1990.

In *forced convection boiling*, the fluid is moved or agitated by a stirrer or paddle. In *flow boiling*, the fluid is moved through a tube by a pump.

In *pool boiling*, the liquid pool is relatively calm (i.e., quiet, quiescent). Fluid motion near the heated surface is induced by free convection and bubble movement but does not significantly affect the free liquid surface.[30]

In *sub-cooled boiling*, the bulk temperature of the liquid is less than the saturation temperature. Vapor bubbles forming on the heated surface may condense back into liquid form. Fluid movement is induced by the density differences.

In *saturated boiling*, the liquid temperature exceeds the saturation temperature only slightly. Vapor bubbles forming on the heated surface rise into the liquid under the influence of buoyancy.

In *free convection boiling*, the liquid is at the saturation temperature, but there is insufficient heat transfer to cause vapor bubbles to form.[31] With an increasing surface temperature and increasing heat transfer, bubbles

---

[30]In its description of boiling regimes, the *NCEES Handbook* repeatedly refers to "the surface." Since there are two surfaces (the heated surface and the liquid surface), care must be taken in interpreting the *NCEES Handbook*'s descriptions.

[31]The *NCEES Handbook* describes this regime as "Insufficient vapor is in contact with the liquid phase..." While there is insufficient vapor to transfer heat to the liquid, the root cause of free convection boiling is insufficient thermal energy transfer due to a low excess temperature.

Heat Transfer

begin to form. This regime ends with the formation of isolated bubbles, a state known as the *onset of nucleate boiling* (ONB) or *bubble nucleation*.

In *nucleate boiling*, individual bubbles form on the heated surface and rise essentially vertically in succession. The rising vapor progresses from isolated bubbles to streams (jets or slugs) of bubbles. The bubbles rise vertically and follow direct paths until the bubbles become large enough to combine and the liquid begins to move ("roll") and disrupt direct vertical movement. The end of nucleate boiling marks the point where heat transfer (i.e., heat flux) is maximum. This defines the *critical heat flux*, commonly abbreviated as CHF.

In *transition boiling*, the vapor bubbles begin to combine within the liquid, and the liquid movement intermittently interrupts direct vapor bubble movement. The boiling regime oscillates between nucleate boiling and film boiling. The beginning of transition boiling is known as the *departure from nucleate boiling* (DNB) point. The end of the transition boiling regime is known as the *Leidenfrost point*, the minimum heat flux for film boiling.

In *film boiling* (also known as *filmwise boiling*), the heated surface is completely covered by a layer (blanket) of vapor. A significant fraction of the heat flux is via radiation through the vapor layer.

....

Equation 26.25 Through Eq. 26.30: Heat Flux

$$q'' = h(T_s - T_{sat}) = h\Delta T_e \qquad \text{26.25}$$

$$\dot{q}_{nucleate} = \mu_l h_{fg} \left[\frac{g(\rho_l - \rho_v)}{\sigma}\right]^{1/2} \left[\frac{c_{pl}(T_s - T_{sat})}{C_{sf} h_{fg} Pr_l^n}\right]^3$$
$$\text{26.26}$$

$$\dot{q}_{max} = C_{cr} h_{fg} [\sigma g \rho_v^2 (\rho_l - \rho_v)]^{1/4} \qquad \text{26.27}$$

$$\dot{q}_{min} = 0.09 \rho_v h_{fg} \left[\frac{\sigma g(\rho_l - \rho_v)}{(\rho_l + \rho_v)^2}\right]^{1/4} \qquad \text{26.28}$$

$$\dot{q}_{film} = C_{film} \left[\frac{g k_v^3 \rho_v (\rho_l - \rho_v)}{\mu_v D(T_s - T_{sat})} \times [h_{fg} + 0.4 c_{pv}(T_s - T_{sat})]\right]^{1/4}$$
$$\times (T_s - T_{sat})$$
$$\text{26.29}$$

$$C_{film} = \begin{cases} 0.62 \text{ for horizontal cylinders} \\ 0.67 \text{ for spheres} \end{cases} \qquad \text{26.30}$$

Values

Table 26.5 *Values of the Coefficient C_{cr} for Maximum Heat Flux (dimensionless parameter $L^* = L(g(\rho_l - \rho_v/\sigma)^{1/2}))$*

heater geometry	C_{cr}	charac. dimension of heater, L	range of L^*
large horizontal flat heater	0.149	width or diameter	$L^* > 27$
small horizontal flat heater*	$18.9 K_1$	width or diameter	$9 < L^* < 20$
large horizontal cylinder	0.12	radius	$L^* > 1.2$
small horizontal cylinder	$0.12 L^{*-0.25}$	radius	$0.15 < L^* < 1.2$
large sphere	0.11	radius	$L^* > 4.26$
small sphere	$0.227 L^{*-0.5}$	radius	$0.15 < L^* < 4.26$

* $K_1 = \sigma/(g(\rho_l - \rho_v) A_{heater})$

Description

Equation 26.25 is the general calculation of heat flux (i.e., $\dot{Q}/A$, the heat transfer per unit area).[32] In Eq. 26.25, ΔT_e is the *excess temperature*, defined as the difference between the surface temperature and the saturation temperature.

$$\Delta T_e = T_s - T_{sat}$$

Equation 26.26 is the *Rohsenow's correlation* for heat flux in nucleate boiling. C_{sf} is an empirical constant dependent on the fluid and heated surface. The exponent, n, on the Prandtl number is also empirical. The liquid and vapor properties are evaluated at the saturation temperature. Tables of heat-transfer-specific properties are usually used. Since the vapor density, ρ_v, is small, it can be omitted from first approximations.

Equation 26.27 calculates the critical (maximum, peak, etc.) heat flux (CHF) corresponding to the end of nucleate boiling. As Eq. 26.27 shows, the heat flux is proportional to the heat (enthalpy) of vaporization, h_{fg}. When maximum heat flux from a surface is desired, liquid (such as water) with large heats of vaporization should be used. C_{cr} is an empirical constant that depends on the type and orientation of the heating element used. For large, flat heaters (e.g., the bottom of a pan on a stove), C_{cr} is approximately 0.15. (See Table 26.5.) The critical heat flux increases as pressure increases up to approximately one-third of the substance's critical pressure, after which it decreases, reaching zero at the critical pressure.

---

[32]The use of double-primed q (i.e., q'') to designate heat flux is colloquial in the subject of boiling heat transfer. The number of primes indicates the number of length units to be included. Thus, q would have units of W, q' would have units of W/m, and q'' would have units of W/m^2. After introducing this convention, the *NCEES Handbook* subsequently abandons it and uses $\dot{q}$ as the symbol for surface heat flux in every subsequent equation.

Heat Transfer

Equation 26.28, known as the *Zuber equation*, calculates the minimum heat flux corresponding to the beginning of film boiling (i.e., the *Leidenfrost point*).

Many industrial boilers contain tubular resistance (electrical) or hot fluid heaters. Equation 26.29, known as the *Bromley equation*, calculates the heat flux in film boiling for horizontal cylinders and spheres. Equation 26.29 cannot be used directly for heating from flat, horizontal plates.[33] The values of the empirical constant, C_{film}, for film boiling are given by Eq. 26.30.

Example

A horizontal cylindrical heater is used to boil water at a constant temperature of 90°C. The density of the liquid water is 965 kg/m^3, and the density of the water vapor is 0.424 kg/m^3. The surface tension between the liquid and vapor is 0.070 N/m. What is most nearly the peak heat flux with nucleate boiling?

(A) 740 kW/m^2

(B) 860 kW/m^2

(C) 910 kW/m^2

(D) 1100 kW/m^2

Solution

At a temperature of 90°C, from the steam tables, the enthalpy of evaporation is 2283.2 kJ/kg. From Table 26.5, $C_{cr} = 0.12$ for a large horizontal cylinder. Use Eq. 26.27 to calculate the peak heat flux in the nucleate pool.

$$\dot{q}_{max} = C_{cr}h_{fg}[\sigma g\rho_v^2(\rho_l - \rho_v)]^{1/4}$$

$$= (0.12)\left(2283.2\ \frac{kJ}{kg}\right)$$

$$\times \left(\begin{array}{c}\left(0.070\ \frac{N}{m}\right)\left(9.81\ \frac{m}{s^2}\right) \\ \times \left(0.424\ \frac{kg}{m^3}\right)^2 \\ \times \left(965\ \frac{kg}{m^3} - 0.424\ \frac{kg}{m^3}\right)\end{array}\right)^{1/4}$$

$$= 905\ kW/m^2 \quad (910\ kW/m^2)$$

The answer is (C).

15. HEAT EXCHANGERS

In a typical application, two fluids flow through or over a heat exchanger.[34] Heat from the hot fluid passes through the exchanger walls to the cold fluid.[35] The heat transfer mechanism is essentially completely forced convection.

Heat exchangers are categorized into simple *tube-in-tube heat exchangers* (also known as *jacketed pipe heat exchangers*), single-pass shell and tube heat exchangers, multiple-pass shell and tube heat exchangers, and crossflow heat exchangers.[36] *Shell and tube heat exchangers*, also known as *sathes* and *S & T heat exchangers*, consist of a large housing, the *shell*, with many smaller tubes running through it. The *tube fluid* passes through the tubes, while the *shell fluid* passes through the shell and around tubes.[37]

In a *single-pass heat exchanger*, each fluid is exposed to the other fluid only once. Operation is known as *parallel flow* (same as *cocurrent flow*) if both fluids flow in the same direction along the longitudinal axis of the exchanger and *counterflow* (same as *counter current flow*) if the fluids flow in opposite directions.[38] Counterflow is more efficient, and the heat transfer area required is less than that with parallel flow since the temperature gradient is more constant.

For increased efficiency, most exchangers are *multiple-pass heat exchangers*. The tubes pass through the shell more than once, and the shell fluid is routed around baffles.

Equation 26.31: Heat Transfer

$$\dot{Q} = UAF\Delta T_{lm} \qquad \textbf{26.31}$$

Description

Equation 26.31 calculates the steady-state heat transfer (also known as the *heat duty* and *heat load*) in a heat exchanger or feedwater heater.[39] F is a correction factor that depends on the configuration (i.e., crossflow, parallel flow, number of passes, etc.) and corrects the LMTD value for the effectiveness of the heat

---

[33](1) The *NCEES Handbook* gives the heat flux equation for cylinders and spheres, but not for flat horizontal plates, a common configuration (e.g., heating a pan on the stove). (2) Equation 26.29 can also be used for vertical flat plates by using the value of $C_{film} = 0.62$.

[34]These fluids do not have to be liquids. *Air-cooled exchangers* reduce water consumption in traditional cooling applications.

[35]A *recuperative heat exchanger*, typified by the traditional shell and tube exchanger, maintains separate flow channels for each of the fluids. A *regenerative heat exchanger* has only one flow path, to which the two fluids are exposed on an alternating basis.

[36]Fin coil heat exchangers are a special case of crossflow heat exchangers.

[37]Tubular heat exchangers are also known as *shell and tube heat exchangers*.

[38](1) Flow through shell and tube heat exchangers is neither purely parallel nor purely counterflow. Thus, these exchangers are sometimes designated as *parallel counterflow exchangers*. (2) The designation *cocurrent* is not an abbreviation for *counter current*.

[39](1) *Closed feedwater heaters* are heat exchangers whose purpose is to heat water with condensing steam. (2) There are three heat loads referred to in heat exchanger specifications: the *specific heat load*, which is the design heat transfer; the heat released by the hot fluid; and the heat absorbed by the cold fluid. All three would be the same if operation was adiabatic, but due to practical losses, they are not. If they differ by more than 10%, the cause of the discrepancy should be evaluated.

Heat Transfer

exchanger.[40] $F = 1$ when one fluid is condensing or evaporating (i.e., when the temperature of the fluid changing phase does not change in the heat exchanger). Values of F are often read from widely available graphs.[41] Equation 26.31 embodies the *LMTD method*, also known as the *F-factor method*, of accounting for heat exchanger effectiveness. The LMTD method is easily applied to problems when both outlet temperatures are known, such as when calculating the required heat transfer area. When both outlet temperatures are unknown, the NTU method must generally be used.

The *overall heat transfer coefficient, U*, also known as the *overall conductance* and the *overall coefficient of heat transfer*, can be specified for use with either the outside or inside tube areas. The heat transfer is independent of whether the outside or inside area is used. It is more common (and preferred) to use the outside tube area because the outside tube diameter is more easily measured.

In reality, it is very difficult to predict the heat transfer coefficient for most types of commercial heat exchangers. Values can be predicted by comparison with similar units, or "tried-and-true" rules of thumb can be used. One such rule of thumb for baffled shell and tube heat exchangers predicts the clean, average heat transfer coefficient as 60% of the value for the same arrangement of tubes in pure crossflow.

Equation 26.32: Overall Heat Transfer Coefficient

$$\frac{1}{UA} = \frac{1}{h_i A_i} + \frac{R_{fi}}{A_i} + \frac{\ln\left(\frac{D_o}{D_i}\right)}{2\pi k L} + \frac{R_{fo}}{A_o} + \frac{1}{h_o A_o} \qquad 26.32$$

Description

Equation 26.32 calculates the overall heat transfer coefficient, U, for concentric tube and shell and tube heat exchangers from the film coefficients and the tube material conductivities.

The area, A, on the left-hand side of Eq. 26.32 can be either the inside area of the tubing (i.e., based on the inside diameter) or the outside area (i.e., based on the outside diameter). The outside diameter is easier to measure in the field, so that convention is preferred. It is important to specify whether an overall heat transfer coefficient is to be used with the inside or outside area. The *fouling factors, R_{fi} and R_{fo}*, represent the thermal resistance of any accumulations, scale, deposits, and even living biological organisms (e.g., Zebra mussels).

---

[40]The *NCEES Handbook* calls F the *configuration correction factor*. The actual name is the *mean temperature difference correction factor*.
[41]As published by Tubular Exchanger Manufacturers Association, Inc. (TEMA).

Example

A pipe has a thermal conductivity of 0.25 W/m·K and a length of 10 m. The inside diameter of the pipe is 1.2 cm, and the outside diameter of the pipe is 2.0 cm. The outside convective heat transfer coefficient is 10 W/m²·K, and the inside convective heat transfer coefficient is 150 W/m²·K. The fouling factor for the inside of the pipe is 0.0005 m²·K/W, and the fouling factor for the outside of the pipe is negligible. What is most nearly the overall heat transfer coefficient based on the outside surface area?

(A) 2.4 W/m²·K

(B) 3.2 W/m²·K

(C) 7.6 W/m²·K

(D) 9.6 W/m²·K

Solution

Calculate the inside surface area of the pipe.

$$A_i = \pi D_i L = \frac{\pi (1.2 \text{ cm})(10 \text{ m})}{100 \frac{\text{cm}}{\text{m}}} = 0.3770 \text{ m}^2$$

Calculate the outside surface area of the pipe.

$$A_o = \pi D_o L = \frac{\pi (2.0 \text{ cm})(10 \text{ m})}{100 \frac{\text{cm}}{\text{m}}} = 0.6283 \text{ m}^2$$

Use Eq. 26.32.

$$\frac{1}{UA} = \frac{1}{h_i A_i} + \frac{R_{fi}}{A_i} + \frac{\ln\left(\frac{D_o}{D_i}\right)}{2\pi k L} + \frac{R_{fo}}{A_o} + \frac{1}{h_o A_o}$$

$$UA = \left(\frac{1}{h_i A_i} + \frac{R_{fi}}{A_i} + \frac{\ln\left(\frac{D_o}{D_i}\right)}{2\pi k L} + \frac{R_{fo}}{A_o} + \frac{1}{h_o A_o} \right)^{-1}$$

$$= \left(\frac{1}{\left(150 \frac{\text{W}}{\text{m}^2\cdot\text{K}}\right)(0.3770 \text{ m}^2)} + \frac{0.0005 \frac{\text{m}^2\cdot\text{K}}{\text{W}}}{0.3770 \text{ m}^2} + \frac{\ln \frac{2.0 \text{ cm}}{1.2 \text{ cm}}}{(2\pi)\left(0.25 \frac{\text{W}}{\text{m}\cdot\text{K}}\right)(10 \text{ m})} + 0 + \frac{1}{\left(10 \frac{\text{W}}{\text{m}^2\cdot\text{K}}\right)(0.6283 \text{ m}^2)} \right)^{-1}$$

$$= 4.746 \text{ W/K}$$

Based on the outside surface area, the overall heat transfer coefficient is

$$U = \frac{UA}{A_o} = \frac{4.746 \ \dfrac{\text{W}}{\text{K}}}{0.6283 \ \text{m}^2}$$

$$= 7.55 \ \text{W/m}^2\cdot\text{K} \quad (7.6 \ \text{W/m}^2\cdot\text{K})$$

The answer is (C).

16. LOGARITHMIC TEMPERATURE DIFFERENCE

The temperature difference between two fluids is not constant in a heat exchanger. When calculating the heat transfer for a tube whose temperature difference changes along its length, the *logarithmic mean temperature difference*, ΔT_{lm} or LMTD, is used.[42] In the equation shown, ΔT_A and ΔT_B are the temperature differences at ends A and B, respectively, regardless of whether the fluid flow is parallel or counterflow.[43]

$$\Delta T_{\text{lm}} = \frac{\Delta T_A - \Delta T_B}{\ln \dfrac{\Delta T_A}{\Delta T_B}}$$

Equation 26.33: Logarithmic Temperature Difference, Counterflow

$$\Delta T_{\text{lm}} = \frac{(T_{\text{Ho}} - T_{\text{Ci}}) - (T_{\text{Hi}} - T_{\text{Co}})}{\ln\left(\dfrac{T_{\text{Ho}} - T_{\text{Ci}}}{T_{\text{Hi}} - T_{\text{Co}}}\right)} \qquad 26.33$$

Description

For counterflow tubular heat exchangers, the logarithmic temperature difference can be found using Eq. 26.33.

---

[42](1) An exception occurs in HVAC calculations where ΔT at midlength has traditionally been used to calculate the heat transfer in air conditioning ducts. Considering the imprecise nature of HVAC calculations, the added sophistication of using the logarithmic mean temperature difference is probably unwarranted. (2) The symbol ΔT_m is also widely used for the log-mean temperature difference. However, this can also be interpreted as the arithmetic mean temperature. (3) The logarithmic temperature difference is used even with change of phase (e.g., boiling liquid or condensing vapor) and the temperature is constant in one tube.
[43](1) It doesn't make any difference which end is A and which is B. If the numerator is negative, the denominator will also be negative. (2) As ΔT_A and ΔT_B become equal, the equation becomes indeterminate, even though the correct relationship is $\Delta T_{\text{lm}} = \Delta T_A = \Delta T_B$. Also, the first derivative, used in some calculations, is undefined when ΔT_A and ΔT_B are equal, even though the correct value is 0.5. A replacement expression (Underwood, 1933) that avoids these difficulties with (generally) less than a 0.3% error is

$$\Delta T_{\text{lm}} \approx \left(\frac{\Delta T_A^{1/3} + \Delta T_B^{1/3}}{2}\right)^3$$

Equation 26.34: Logarithmic Temperature Difference, Parallel Flow

$$\Delta T_{\text{lm}} = \frac{(T_{\text{Ho}} - T_{\text{Co}}) - (T_{\text{Hi}} - T_{\text{Ci}})}{\ln\left(\dfrac{T_{\text{Ho}} - T_{\text{Co}}}{T_{\text{Hi}} - T_{\text{Ci}}}\right)} \qquad 26.34$$

Description

For parallel flow tubular heat exchangers, the logarithmic temperature difference can be found using Eq. 26.34.

Example

A parallel flow heat exchanger is used to cool oil from 120°C to 60°C. The cooling water enters at 20°C and leaves at 55°C. What is most nearly the log mean temperature difference?

(A) 25°C

(B) 32°C

(C) 140°C

(D) 280°C

Solution

Using Eq. 26.34, the log mean temperature difference for a parallel heat exchanger is

$$\Delta T_{\text{lm}} = \frac{(T_{\text{Ho}} - T_{\text{Co}}) - (T_{\text{Hi}} - T_{\text{Ci}})}{\ln\left(\dfrac{T_{\text{Ho}} - T_{\text{Co}}}{T_{\text{Hi}} - T_{\text{Ci}}}\right)}$$

$$= \frac{(60°\text{C} - 55°\text{C}) - (120°\text{C} - 20°\text{C})}{\ln \dfrac{60°\text{C} - 55°\text{C}}{120°\text{C} - 20°\text{C}}}$$

$$= 31.7°\text{C} \quad (32°\text{C})$$

The answer is (B).

17. NTU METHOD

Some heat exchanger analysis problems, such as where both outlet temperatures are unknown, appear to be unsolvable or solvable only by trial and error using the traditional F-method.[44] However, the *number of transfer units* (NTU) *method* (also known as the *efficiency method* and *effectiveness method*) can be used to handle these problems more easily.[45] The steps in the NTU method depend on whether or not both exit temperatures are known.

---

[44]Actually, any shell and tube heat exchanger with an even number of tube passes has a closed-form analytical solution for the outlet temperature. However, the mathematics are laborious and the form of the solution varies with the type of flow and heat exchanger design.
[45]The names *number of thermal units* (NTU), *heat transfer units* (HTU), and *temperature ratio* (TR) are synonymous with *number of transfer units* (NTU).

Heat Transfer

Equation 26.35 Through Eq. 26.38: Heat Exchanger Effectiveness

$$\varepsilon = \frac{\dot{Q}}{\dot{Q}_{max}} = \frac{\text{actual heat transfer rate}}{\text{maximum possible heat transfer rate}}$$

$$\text{26.35}$$

$$C = \dot{m}c_p \qquad \text{26.36}$$

$$\varepsilon = \frac{C_H(T_{Hi} - T_{Ho})}{C_{min}(T_{Hi} - T_{Ci})} \qquad \text{26.37}$$

$$\varepsilon = \frac{C_C(T_{Co} - T_{Ci})}{C_{min}(T_{Hi} - T_{Ci})} \qquad \text{26.38}$$

Description

The *heat exchanger effectiveness*, ε, is defined as the ratio of the actual heat transfer to the maximum possible heat transfer.[46] This ratio is generally not known in advance.

The first step in the NTU method is to calculate the *thermal capacity rates*, C, for the two fluids, given by Eq. 26.36. It is possible for the two capacity rates to be equal, but usually they are not. The smaller capacity rate is designated C_{min}. The larger is designated C_{max}. The fluid with the smaller capacity rate, C_{min}, will experience the larger temperature change. If the cold fluid has the minimum capacity rate (that is, $C_{min} = C_{cold}$), the effectiveness is given by Eq. 26.38. If the hot fluid has the minimum capacity rate (that is, $C_{min} = C_{hot}$), the effectiveness is given by Eq. 26.37.

Equation 26.39 Through Eq. 26.43: Number of Transfer Units

$$C_r = \frac{C_{min}}{C_{max}} \qquad \text{26.39}$$

$$\text{NTU} = \frac{UA}{C_{min}} \qquad \text{26.40}$$

$$\text{NTU} = \frac{1}{C_r - 1} \ln\left(\frac{\varepsilon - 1}{\varepsilon C_r - 1}\right) \quad \begin{bmatrix} \text{counterflow,} \\ \text{concentric; } C_r < 1 \end{bmatrix}$$

$$\text{26.41}$$

$$\text{NTU} = \frac{\varepsilon}{1 - \varepsilon} \quad \begin{bmatrix} \text{counterflow, concentric;} \\ C_r = 1 \end{bmatrix}$$

$$\text{26.42}$$

$$\text{NTU} = -\frac{\ln[1 - \varepsilon(1 + C_r)]}{1 + C_r} \quad \text{[parallel flow, concentric]}$$

$$\text{26.43}$$

Description

Once the heat capacity ratio is found, Eq. 26.41 through Eq. 26.43 can be used to find the number of transfer units for heat exchangers operating under specific conditions.

[46](1) The maximum possible transfer can occur only if the heat exchanger has an infinite length. (2) Other names used in the literature to define the effectiveness are *efficiency*, *thermodynamic efficiency*, *temperature efficiency*, and *performance parameter*. The symbol P is also used in place of ε.

The number of transfer units for a single-pass counterflow heat exchanger with a heat capacity ratio less than 1 is found from Eq. 26.41. The number of transfer units for a single-pass counterflow heat exchanger with a heat capacity ratio of 1 is found from Eq. 26.42. The number of transfer units for a single-pass parallel flow heat exchanger is found from Eq. 26.43.

Example

A single-pass counterflow heat exchanger is used to cool lubricating oil with cooling water. The exchanger has an effective heat transfer rate of 280 $\text{W/m}^2\cdot\text{K}$ based on an effective heat transfer area of 16 m^2. The thermal capacity rate of lubricating oil is 9.1 kW/K, and the thermal capacity rate of cooling water is 5.2 kW/K. What is most nearly the number of transfer units?

(A) 0.71

(B) 0.86

(C) 0.94

(D) 1.3

Solution

Since 5.2 kW/K < 9.1 kW/K,

$$C_{min} = C_{water} = 5.2 \text{ kW/K} \quad (5.2 \times 10^3 \text{ W/K})$$

Use Eq. 26.40.

$$\text{NTU} = \frac{UA}{C_{min}}$$

$$= \frac{\left(280 \ \dfrac{\text{W}}{\text{m}^2\cdot\text{K}}\right)(16 \text{ m}^2)}{5.2 \times 10^3 \ \dfrac{\text{W}}{\text{K}}}$$

$$= 0.8615 \quad (0.86)$$

The answer is (B).

Equation 26.44 Through Eq. 26.46: Heat Exchanger Effectiveness, Counterflow and Parallel Flow

$$\varepsilon = \frac{1 - \exp[-\text{NTU}(1 - C_r)]}{1 - C_r \exp[-\text{NTU}(1 - C_r)]} \quad \begin{bmatrix} \text{counterflow,} \\ \text{concentric; } C_r < 1 \end{bmatrix}$$

$$\text{26.44}$$

$$\varepsilon = \frac{\text{NTU}}{1 + \text{NTU}} \quad \begin{bmatrix} \text{counterflow, concentric;} \\ C_r = 1 \end{bmatrix} \quad \text{26.45}$$

$$\varepsilon = \frac{1 - \exp[-\text{NTU}(1 + C_r)]}{1 + C_r} \quad \text{[parallel flow, concentric]}$$

$$\text{26.46}$$

Description

If a single-pass counterflow heat exchanger has a heat capacity ratio less than 1, the effectiveness of the exchanger is found from Eq. 26.44.

If a single-pass counterflow heat exchanger has a heat capacity ratio of 1, the effectiveness of the exchanger is found from Eq. 26.45.

For a single-pass parallel flow heat exchanger, the effectiveness of the exchanger is found from Eq. 26.46.

Example

For a counterflow concentric tube heat exchanger, if the number of transfer units is 0.76 and $C_r = 1$, what is most nearly the heat exchanger effectiveness?

(A) 0.27

(B) 0.38

(C) 0.43

(D) 0.67

Solution

Since $C_r = 1$, Eq. 26.45 can be used to calculate the effectiveness.

$$\varepsilon = \frac{\text{NTU}}{1 + \text{NTU}} = \frac{0.76}{1 + 0.76}$$

$$= 0.4318 \quad (0.43)$$

The answer is (C).

18. HEAT TRANSFER IN PACKED BEDS

Equation 26.47: Heat Transfer in Packed Beds

$$Q = h_V V_{\text{bed}}(T_{s,\text{bed particle}} - T_{b,\text{gas}}) \qquad 26.47$$

Description

Catalytic reactors, pebble-bed heat exchangers, and fluidized-bed furnaces are examples of *packed beds*. The heat transfer is based on the bed volume, not on surface area. The film coefficient per unit volume of packed bed, h_V, is a function of the solid particle surface area and the empty cross-sectional area of the bed per unit volume.

Heat Transfer

27 Radiation

Nomenclature

A	area	m^2
F	factor	–
N	number of radiating surfaces	–
P	power	–
$\dot{q}$	unit heat transfer rate	W/m^2
$\dot{Q}$	heat transfer rate	W
T	temperature	K

Symbols

α	absorptivity	–
ε	emissivity	–
ρ	reflectivity	–
σ	Stefan-Boltzmann constant, 5.67×10^{-8}	$W/m^2 \cdot K^4$
τ	transmissivity	–

Subscripts

12	from body 1 to body 2
13	from body 1 to body 3
21	from body 2 to body 1
32	from body 3 to body 2
i	inner or i^{th} surface
j	j^{th} surface
R	reradiating

1. THERMAL RADIATION

Thermal radiation is electromagnetic radiation with wavelengths in the 0.1 μm to 100 μm range. All bodies, even "cold" ones, radiate thermal radiation. Thermal radiation incident to a body can be absorbed, reflected, or transmitted.

Equation 27.1: Radiation Conservation Law

$$\alpha + \rho + \tau = 1 \quad \text{[always]} \qquad 27.1$$

Description

A body in the path of thermal radiation energy can absorb the energy, reflect the energy, or pass the energy through (i.e., transmit the energy). All of the incident energy is accounted for by this requirement. Equation 27.1, written in terms of fractions of the incident energy, is the *radiation conservation law* (*energy conservation law for radiation*).

α is the fraction of energy absorbed (the *absorptivity*), ρ is the fraction reflected (the *reflectivity*), and τ is the fraction transmitted (the *transmissivity*).

Example

Based on the radiation conservation law, if the absorptivity of a body is 48%, and the reflectivity is 36%, what is most nearly the transmissivity?

(A) 16%

(B) 36%

(C) 48%

(D) 84%

Solution

The transmissivity can be found by solving Eq. 27.1.

$$\alpha + \rho + \tau = 1$$
$$\tau = 1 - \alpha - \rho$$
$$= 1 - 0.48 - 0.36$$
$$= 0.16 \quad (16\%)$$

The answer is (A).

Equation 27.2: Radiation Conservation Law for an Opaque Body

$$\alpha + \rho = 1 \quad \text{[opaque]} \qquad 27.2$$

Description

For an opaque body, transmissivity is zero. The radiation conservation law is Eq. 27.2.

Example

For an opaque body, if 62% of the incident energy is reflected, what is most nearly the percentage of energy absorbed?

(A) 19%

(B) 38%

(C) 46%

(D) 62%

Solution

Use Eq. 27.2 to find the absorptivity.

$$\alpha + \rho = 1$$

$$\alpha = 1 - \rho$$

$$= 1 - 0.62$$

$$= 0.38 \quad (38\%)$$

The answer is (B).

2. BLACK, REAL, AND GRAY BODIES

A body will naturally radiate thermal energy into its local environment. The ratio of the actual emitted energy to the ideal emitted power is known as the *emissivity*, ε.

$$\varepsilon = \frac{P_{\text{actual}}}{P_{\text{ideal}}}$$

When a body (or surface) is in thermal equilibrium, any energy it receives by radiation (or other means) is also lost (transmitted). The rates of absorbed and transmitted energies must be the same.[1] Therefore, at thermal equilibrium, the emissivity of a body equals its absorptivity. This statement is known as *Kirchhoff's radiation law*. Kirchhoff's law has a corollary: Emissivity cannot exceed 1.0. Since absorptivity is part of the radiation conservation law, absorptivity cannot exceed 1.0. And, at thermal equilibrium, emissivity equals absorptivity. Therefore, emissivity cannot exceed 1.0.

The emissivity (and, therefore, the emissive power) usually depends on the temperature of the body.[2] A body that emits at constant emissivity, regardless of wavelength, is known as a *gray body*. *Real bodies* are frequently approximated as gray bodies. *Real bodies* do not radiate at the ideal level.

Since absorptivity, α, cannot exceed 1.0, Kirchhoff's radiation law places an upper limit on emissive power. Bodies that radiate at this upper limit ($\alpha = 1$) are known as *black bodies* or *ideal radiators*. A black body emits the maximum possible radiation for its temperature and absorbs all incident energy.[3]

---

[1]In the most general form of Kirchhoff's radiation law, energy must be integrated over all wavelengths and angles.
[2]This is equivalent to saying the emissivity depends on the wavelength of the radiation.
[3]Black-body performance can be approximated but not achieved in practice.

Equation 27.3: Radiant Heat Transfer

$$\dot{Q} = \varepsilon \sigma A T^4 \qquad \text{27.3}$$

Values

$$\sigma = 5.67 \times 10^{-8} \ \text{W/m}^2 \cdot \text{K}^4$$

Description

Radiant heat transfer is the name given to heat transferred by way of thermal radiation. The energy radiated by a hot body at absolute temperature, T, is given by the *Stefan-Boltzmann law*, also known as the *fourth-power law*. In Eq. 27.3, σ is the *Stefan-Boltzmann constant*.

Example

A 17 m² plate with an emissivity of 0.26 is suspended vertically in a large room. If the absolute temperature is 430K, what is most nearly the radiation emitted?

(A) 2.6 kW

(B) 3.5 kW

(C) 6.7 kW

(D) 8.6 kW

Solution

Using Eq. 27.3, the radiation emitted by a body is

$$\dot{Q} = \varepsilon \sigma A T^4$$

$$= \frac{(0.26)\left(5.67 \times 10^{-8} \ \dfrac{\text{W}}{\text{m}^2 \cdot \text{K}^4}\right) \times (17 \ \text{m}^2)(430\text{K})^4}{1000 \ \dfrac{\text{W}}{\text{kW}}}$$

$$= 8.568 \ \text{kW} \quad (8.6 \ \text{kW})$$

The answer is (D).

Equation 27.4: Black Body

$$\alpha = \varepsilon = 1 \quad \text{[black body]} \qquad \text{27.4}$$

Description

For a black body, both emissivity and absorptivity are 1.

Equation 27.5 and Eq. 27.6: Gray Body

$$\alpha = \varepsilon \quad [0 < \alpha < 1; \ 0 < \varepsilon < 1] \qquad 27.5$$

$$\varepsilon + \rho = 1 \quad [\text{opaque gray body}] \qquad 27.6$$

Description

For a gray body, the reflectivity is constant. (See Eq. 27.6.)

Example

If 55% of energy from a gray body is reflected, what is most nearly the emissivity of the body?

(A) 15%

(B) 25%

(C) 45%

(D) 55%

Solution

Solve Eq. 27.6 for emissivity.

$$\varepsilon + \rho = 1$$
$$\varepsilon = 1 - \rho$$
$$= 1 - 0.55$$
$$= 0.45 \quad (45\%)$$

The answer is (C).

3. RECIPROCITY THEOREM FOR RADIATION

Equation 27.7: Reprocity Theorem

$$A_i F_{ij} = A_j F_{ji} \quad [0 \le F_{ij} \le 1] \qquad 27.7$$

Variation

$$A_1 F_{12} = A_2 F_{21}$$

Description

Equation 27.7 is known as the *reciprocity theorem for radiation*. A_i is the surface area (m^2) of surface i. F_{ij} is the *shape factor* (*view factor, configuration factor*), which is the fraction of the radiation leaving surface i that is intercepted by surface j.

Equation 27.8: View Conservation Rule

$$\sum_{j=1}^{N} F_{ij} = 1 \qquad 27.8$$

Description

The thermal energy radiated from a body (surface) must go someplace. For a transmitting body that is completely surrounded by N other distinct receiving bodies (surfaces), it is logical that the total energy intercepted by the receiving bodies will equal the transmitted energy. This logic is repeated for each of the bodies in the collection. Since the shape factor, F_{ij}, is defined as the fraction of the radiation leaving surface i that is intercepted by surface j, Eq. 27.8 represents this logic.[4]

4. NET RADIATION HEAT TRANSFER

When two bodies can "see each other," each will radiate energy to and absorb energy from the other. The net radiant heat transfer between the two bodies is given by

$$\dot{Q}_{12} = \sigma A_1 F_{12}(T_1^4 - T_2^4)$$

F_{12} is the *shape factor*, or *configuration factor*, which depends on the shapes, emissivities, and orientations of the two bodies. If body 1 is small and completely enclosed by body 2, then $F_{12} = \varepsilon_1$.

Equation 27.9: Net Radiation Heat Transfer, Small Body

$$\dot{Q}_{12} = \varepsilon \sigma A(T_1^4 - T_2^4) \qquad 27.9$$

Description

Where body 1 is small and completely enclosed by body 2, the net heat transfer due to radiation is given by Eq. 27.9.

Example

An ideal radiator is maintained at 550K. It is enclosed in a tank with a temperature of 290K. The radiator is small compared to the tank. What is most nearly the net heat transfer between them per unit area of radiator?

(A) 2.9 kW/m²

(B) 3.5 kW/m²

(C) 4.8 kW/m²

(D) 5.4 kW/m²

---

[4](1) This is a *conservation rule*, not a "summation rule" as the NCEES *FE Reference Handbook* (*NCEES Handbook*) refers to it. (2) The summation is shown over N surfaces, but there are $N+1$ surfaces, counting the radiating surface. Only the "other" surfaces (i.e., the intercepting surfaces) are included in the summation. Alternatively, if there were N total surfaces, the summation limit would be $N-1$. (3) Not immediately obvious from Eq. 27.8 as presented in the *NCEES Handbook* is that index variable i is held constant during the summation. The summation is not over all combinations of all surfaces.

Heat Transfer

Solution

Equation 27.9 can be used to determine the heat transfer per unit area. $\varepsilon = 1$ for an ideal radiator.

$$\dot{Q}_{12} = \varepsilon \sigma A (T_1^4 - T_2^4)$$

$$\frac{\dot{Q}_{12}}{A} = \varepsilon \sigma (T_1^4 - T_2^4)$$

$$= \frac{(1.0)\left(5.67 \times 10^{-8} \, \frac{W}{m^2 \cdot K^4}\right)\left((550K)^4 - (290K)^4\right)}{1000 \, \frac{W}{kW}}$$

$$= 4.787 \text{ kW/m}^2 \quad (4.8 \text{ kW/m}^2)$$

The answer is (C).

Equation 27.10: Net Radiation Heat Transfer, Black Bodies

$$\dot{Q}_{12} = A_1 F_{12} \sigma (T_1^4 - T_2^4) \qquad 27.10$$

Description

The net heat transfer due to radiation between two black bodies is given by Eq. 27.10.

Example

A 15 cm thick furnace wall has a 8 cm square inspection port. The interior of the furnace is at 1200°C. The surrounding air temperature is 20°C. The shape factor, F_{12}, is 0.4. The heat loss due to radiation when the inspection port is open is most nearly

(A) 150 W

(B) 440 W

(C) 680 W

(D) 2700 W

Solution

The absolute temperatures are

$$T_{\text{furnace}} = 1200°C + 273° = 1473K$$

$$T_\infty = 20°C + 273° = 293K$$

The radiation heat loss is

$$\dot{Q} = AF_{12}\sigma(T_{\text{furnace}}^4 - T_\infty^4)$$

$$= \frac{\begin{array}{c}(8 \text{ cm})^2(0.4)\left(5.67 \times 10^{-8} \, \frac{W}{m^2 \cdot K^4}\right) \\ \times \left((1473K)^4 - (293K)^4\right)\end{array}}{\left(100 \, \frac{cm}{m}\right)^2}$$

$$= 682.3 \text{ W} \quad (680 \text{ W})$$

The answer is (C).

Equation 27.11: Net Radiation Heat Transfer, Diffuse Gray Surfaces

$$\dot{Q}_{12} = \frac{\sigma(T_1^4 - T_2^4)}{\dfrac{1 - \varepsilon_1}{\varepsilon_1 A_1} + \dfrac{1}{A_1 F_{12}} + \dfrac{1 - \varepsilon_2}{\varepsilon_2 A_2}} \qquad 27.11$$

Description

The net heat transfer due to radiation between two diffuse gray surfaces is given by Eq. 27.11. Equation 27.11 applies to two bodies (surfaces), each of which is completely covered (enclosed) by the other. This configuration is known as a *two-surface enclosure* and is shown in Fig. 27.1.[5] Either body can be designated as body 1, and either can be designated as body 2. Each body radiates to the other; each body receives energy from the other. The emissivities of the two bodies are not necessarily the same. Although the bodies are completely enclosed, this does not necessarily mean that the view factor, F_{12}, is equal to 1.0. The enclosure may not be sufficiently convex to enable all parts of the surfaces to see all other parts of the surfaces. There may be blind spots caused by shadowing.[6] Parallel plates, concentric cylinders, and concentric spheres, however, are *convex hulls* (*convex envelopes*), so $F_{12} = F_{21} = 1.0$.

Figure 27.1 Radiation Between Two Diffuse Gray Surfaces

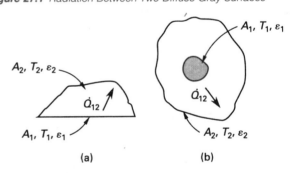

---

[5]The interpretation and application of Fig. 27.1 as presented in the *NCEES Handbook* are not obvious. There are two generalized cases shown. (The *NCEES Handbook* does not include the (a) and (b) designations, and so it appears that something is going on between the left-hand and right-hand sides.) Part (a) shows a radiating flat gray surface that is covered (enclosed) by a dome, which is also a radiating gray surface. Since the dome connects to the surface everywhere along the periphery, the two surfaces "form an enclosure" (as the *NCEES Handbook* phrases it). This constitutes the traditional *two surface enclosure*. In the case of part (b), radiating gray body 1 is completely enclosed by radiating gray body 2. It is not intuitive how the enclosed body 1 of part (b) "forms" an enclosure, as the *NCEES Handbook* phrases it. Nevertheless, Eq. 27.11 applies to both cases.

[6]In a convex hull, every point on the boundary can see every other point on the boundary. A simple visualization of a *convex hull* is the space (envelope) that is obtained by wrapping a rubber band around the vertices or other line intersection points defining the extent of the boundary.

Example

Two plane radiating surfaces separated by a vacuum face each other. Each radiating surface has an emissivity of 0.02. The areas of the two surfaces are large and equal. One surface is at 400K, and the other is at 300K. The view factor, F_{12}, for the 400K surface is 1.0. Most nearly, what is the net heat transfer per unit area from the 400K surface?

(A) 4 W/m²

(B) 6 W/m²

(C) 8 W/m²

(D) 10 W/m²

Solution

Use Eq. 27.11 with $A_1 = A_2$ and $F_{12} = 1.0$.

$$\dot{Q}_{12} = \frac{\sigma(T_1^4 - T_2^4)}{\dfrac{1-\varepsilon_1}{\varepsilon_1 A_1} + \dfrac{1}{A_1 F_{12}} + \dfrac{1-\varepsilon_2}{\varepsilon_2 A_2}}$$

$$\dot{q} = \frac{Q_{12}}{A_1} = \frac{\sigma(T_1^4 - T_2^4)}{\dfrac{1-\varepsilon_1}{\varepsilon_1} + \dfrac{1}{1} + \dfrac{1-\varepsilon_2}{\varepsilon_2}} = \frac{\sigma(T_1^4 - T_2^4)}{\dfrac{1}{\varepsilon_1} + \dfrac{1}{\varepsilon_2} - 1}$$

$$= \frac{\left(5.67 \times 10^{-8} \ \dfrac{\text{W}}{\text{m}^2 \cdot \text{K}^4}\right)\left((400\text{K})^4 - (300\text{K})^4\right)}{\dfrac{1}{0.02} + \dfrac{1}{0.02} - 1}$$

$$= 10.02 \ \text{W/m}^2 \quad (10 \ \text{W/m}^2)$$

The answer is (D).

Equation 27.12: Net Radiation Heat Transfer, Parallel Plates

$$\dot{Q}_{12} = \frac{\sigma(T_1^4 - T_2^4)}{\dfrac{1-\varepsilon_1}{\varepsilon_1 A_1} + \dfrac{1}{A_1 F_{13}} + \dfrac{1-\varepsilon_{3,1}}{\varepsilon_{3,1} A_3} + \dfrac{1-\varepsilon_{3,2}}{\varepsilon_{3,2} A_3} + \dfrac{1}{A_3 F_{32}} + \dfrac{1-\varepsilon_2}{\varepsilon_2 A_2}} \quad 27.12$$

Description

Figure 27.2 illustrates two parallel plates that are separated by a thin internal shield.[7] The significance of the shield thickness is that a thin shield has no mass and cannot store any thermal energy. The significance of being a low-emissivity shield is that the corresponding reflectivity is high. Most of the energy intercepted by the shield is reflected back to the source. The thermal radiation that the shield does receive is re-emitted, some

back to the source plate, and some through to the opposite parallel plate. The lower the emissivity of the shield, the higher its contribution will be to thermal resistance. Equation 27.12 calculates the net heat transfer from plate 1 to plate 2.

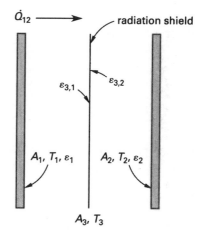

Figure 27.2 *Radiation Shield Between Parallel Plates*

5. RADIATION WITH REFLECTION/ RERADIATION

Surfaces that reradiate absorbed thermal radiation are known as *refractory materials* or *refractories*. (Furnace walls that reradiate almost all of the thermal energy they receive from combustion flames back to boiler tubes are examples of refractories.)

Equation 27.13: Reradiating Surface

$$\dot{Q}_{12} = \frac{\sigma(T_1^4 - T_2^4)}{\dfrac{1-\varepsilon_1}{\varepsilon_1 A_1} + \dfrac{1}{A_1 F_{12} + \left[\left(\dfrac{1}{A_1 F_{1R}}\right) + \left(\dfrac{1}{A_2 F_{2R}}\right)\right]^{-1}} + \dfrac{1-\varepsilon_2}{\varepsilon_2 A_2}} \quad 27.13$$

Description

Equation 27.13 gives the rate of heat transfer between two surfaces with an adjacent third reradiating surface. Reradiating surfaces are considered insulated (adiabatic). (See Fig. 27.3.)

Figure 27.3 *Reradiating Surface*

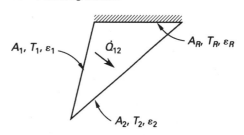

---

[7]This configuration models thermal shielding of cryogenic vessels and temperature sensors.

28 Transport Phenomena

Subscripts

f	friction
H	heat
m	mass
M	mass
w	wall

Nomenclature

a	acceleration	m/s^2
A	area	m^2
c_m	concentration	mol/m^3
c_p	specific heat	kJ/kg·K
D	diameter	m
D_m	diffusion coefficient	m^2/s
f	Darcy friction factor	–
F	force	N
G	mass velocity	kg/m^2·s
h	film coefficient	W/m·K
h	head loss	m
j_A	molar flux	mol/m^2·s
j_H	heat j-number	–
j_M	mass j-number	–
k	thermal conductivity	W/m·K
L	length	m
m	mass	kg
N	mass transfer	mol/s
Nu	Nusselt number	–
p	pressure	Pa
Pr	Prandtl number	–
$\dot{Q}$	heat transfer	W
Re	Reynolds number	–
Sc	Schmidt number	–
Sh	Sherwood number	–
St	Stanton number	–
t	time	s
T	temperature	K
v	velocity	m/s
V	average velocity	m/s
y	distance	m
y	radial distance from inner wall to tube centerline	m

Symbols

μ	absolute viscosity	N·s/m^2
ν	kinematic viscosity	m^2/s
ρ	density	kg/m^3
τ	shear stress	N/m^2

1. INTRODUCTION

Transport phenomena traditionally include the processes of energy transfer, momentum transfer, and mass transfer. Not all engineering students have studied all of these subjects; for instance, mass transfer is typically taught only to chemical engineers. Some students have studied a subject without recognizing it was a subset of transport phenomena; for example, heat transfer is an energy transport phenomenon.

Energy transfer is concerned with the transfer of thermal energy from one location to another. *Mass transfer* is concerned with the transfer of mass from one phase to another. *Momentum transfer* is concerned with the transfer of momentum from one location to another.

All of the transport phenomena have basic concepts in common. The forms of the equations and the methods of solutions are similar. Only the names of the constants and coefficients vary. An analogy can be drawn between the phenomena.

Example

In the investigation of dark energy, the instantaneous motion of a subatomic particle in the three-dimensional space-time continuum is described by the transport phenomena analogous equation shown.

$$\Phi = -\sigma \nabla \phi_A$$

What is the meaning of the subscript "A"?

   (A) cross sectional area

   (B) surface area

   (C) particle density gradient

   (D) the particle, particle type, or some independent property such as temperature or energy level

Solution

The transport phenomena analogy recognizes that the common properties of mass, momentum, and energy (i.e., heat) are described by similar equations. For

Heat Transfer

example, molecular diffusion of substance A in the x-direction is described by Fick's law.

$$\mathbf{j}_{A,x} = -D_A \nabla C_A \equiv j_{A,x} = -D_A \frac{dC_A}{dx}$$

The gradient vector field, ∇, works on the concentration function (variable), C_A, to produce the concentration gradient. In the unified transport phenomena, the flux variable is normalized per unit area or volume. In the case of molecular diffusion, the cross-sectional area has already been incorporated into the flux variable, $j = J/A$. The "A" in the particle motion equation may refer to the particle (analogous to diffusion) or it may refer to some other property, such as temperature (analogous to heat transfer in Fourier's law) or particle energy level.

The answer is (D).

2. UNIT OPERATIONS

There are many applications of transport phenomena. Regardless of the industry or product, transport phenomena can be divided into separate and distinct steps called *unit operations*. The most important unit operations are:

- fluid flow
- heat transfer
- evaporation
- drying
- distillation
- absorption
- membrane separation
- liquid-liquid extraction
- liquid-solid leaching
- crystallization
- mechanical separation processes

3. MOMENTUM TRANSFER

Equation 28.1: Reynolds Number for Flow in a Circular Tube

$$\text{Re} = DV\rho/\mu \qquad \textit{28.1}$$

Description

Momentum transfer is generally studied and referred to as "fluid mechanics." The concepts of flow regime (laminar or turbulent) and viscosity are basic. The flow regime is determined by calculating the *Reynolds number*, Re, which is the ratio of inertia force to viscous force.

Example

What equation is often referred to as "Newton's second law of motion for fluid flow"?

(A) Bernoulli

(B) Colebrook-White

(C) Euler

(D) Navier-Stokes

Solution

Analogous to Newton's second law of motion, $ma = m\,dv/dt$, the *Navier-Stokes equation* of incompressible fluid motion relates the field density, ρ, kinematic viscosity, ν, the velocity, v, and the pressure, p.

$$F - \frac{\nabla p}{\rho} + \nu \nabla^2 \mathbf{v} = \frac{\partial \mathbf{v}}{\partial t} + \mathbf{v} \cdot \nabla \mathbf{v}$$

$$\nu = \frac{\mu}{\rho}$$

The motion equation is always solved with continuity of mass and continuity of energy equations, as well as an equation of state. Together, these equations are referred to as the Navier-Stokes equations, often described as "Newton's second law of motion for fluid flow." Since the most general case results in six equations in six unknowns, only special cases can be solved in closed form.

The answer is (D).

Equation 28.2 and Eq. 28.3: Momentum Transfer

$$\tau_w = -\mu \left(\frac{d\mathbf{v}}{dy}\right)_w = -\frac{f\rho V^2}{8} = \left(\frac{D}{4}\right)\left(-\frac{\Delta p}{L}\right)_f \qquad \textit{28.2}$$

$$\tau_w = \frac{f\rho V^2}{8} \qquad \textit{28.3}$$

Description

When a turbulent fluid flows in a circular tube, the fluid shear stress at the wall is a function of the velocity gradient, dv/dy. The constant of proportionality, μ, is the *absolute viscosity*.

The linear relationship is intrinsic to the definition of a *Newtonian fluid*. However, the connection with momentum transfer may not be obvious. For a constant mass flow rate, *Newton's second law* can be written as

$$F = ma = m\frac{d\mathbf{v}}{dt} = \frac{d(m\mathbf{v})}{dt}$$

$$= \frac{d(\text{momentum})}{dt}$$

The force, F, can be interpreted as the flux of fluid momentum in the radial direction (i.e., away from the wall). The *shear stress*, F/A, can be interpreted as the flux of fluid momentum in the radial direction (i.e., away from the wall) per unit area. The negative sign in Eq. 28.2 indicates that momentum transfer is high when the velocity is low (i.e., at the wall), and vice versa.

The *Fanning friction factor* is defined as the shear stress divided by the kinetic energy per unit volume. The *Darcy friction factor*, f, is four times the Fanning friction factor, hence the factor "4."

$$f = \frac{4\tau_w}{\frac{\frac{1}{2}mV^2}{\text{volume}}} = \frac{8\tau_w}{\rho V^2}$$

The Darcy friction factor is used to calculate friction loss. The traditional *Darcy equation* for friction head loss along a length, L, of pipe is

$$h_f = \frac{-\Delta p}{\rho g} = \frac{fLV^2}{2Dg}$$

Since $h_f = -\Delta p/\rho g$ (where $\Delta p = p_2 - p_1$), the Darcy friction factor can also be written as

$$f = \frac{-2D\Delta p}{\rho L V^2}$$

The momentum transfer in the radial direction per unit area is[1]

$$\frac{F}{A} = \tau_w = -\mu\frac{dv}{dy}$$

$$= -\frac{f\rho V^2}{8} = \left(\frac{D}{4}\right)\left(-\frac{\Delta p}{L}\right)$$

Example

A fluid has a viscosity of 10^{-3} N·s/m^2. Within a lamella of flow near the stationary pipe wall, the fluid velocity in the positive x direction increases by 0.1 m/s over a radial distance in the positive y direction (away from the wall) of 0.001 m. Most nearly, what is the diffusion of momentum within this lamella?

(A) 0.1 N/m^2 in the positive x direction

(B) 0.1 N/m^2 in the negative x direction

(C) 0.1 N/m^2 in the positive y direction

(D) 0.1 N/m^2 in the negative y direction

---

[1]The NCEES *FE Reference Handbook* (*NCEES Handbook*) presents Eq. 28.2 and Eq. 28.3 within inches of each other, both titled "Rate of Transfer," but with different signs. The minus sign at the beginning of Eq. 28.2 is incorrect.

Solution

Using Eq. 28.2 and the sign convention described, the shear stress on the fluid, τ_w, is

$$\tau_w = -\mu\left(\frac{dv}{dy}\right)_w$$

$$= -\left(10^{-3}\frac{\text{N·s}}{\text{m}^2}\right)\left(\frac{0.1\frac{\text{m}}{\text{s}}}{0.001\text{ m}}\right)$$

$$= -0.1 \text{ N/m}^2$$

The negative sign indicates that the shear force opposes the direction of flow. This is because the x momentum is diffusing in the negative y direction. A force in the positive x direction (supplied by a pressure gradient along the direction of fluid flow) maintains the flow field against the shear stress.[2]

The answer is (D).

4. ENERGY TRANSFER

Energy transfer is generally referred to as "heat transfer." The concepts of transfer regime (laminar or turbulent) and thermal conductivity are basic.

Equation 28.4: Prandtl Number

$$\text{Pr} = c_p\mu/k \qquad \qquad 28.4$$

Description

The *Prandtl number*, Pr, is the ratio of the shear component of momentum diffusivity to the heat diffusivity (i.e., the ratio of diffusion of momentum to the diffusion of heat).

Equation 28.5: Heat Flux on Thermal Gradient

$$\left(\frac{\dot{Q}}{A}\right)_w = -k\left(\frac{dT}{dy}\right)_w \qquad \qquad 28.5$$

Description

The *heat flux* depends on the *thermal gradient* (i.e., a *temperature gradient*), dT/dy, in a material. The constant of proportionality, k, is the *thermal conductivity* of the substance.

---

[2]The sign conventions for mechanical stress and momentum diffusion are opposite: $\sigma_{xy} = \tau_{w,xy}$. In mechanics, tension is considered a positive stress and compression is a negative stress. Accordingly, a greater x force on top of a fluid layer than on the bottom would be a positive shear stress: $\sigma_{yx} > 0$. However, a greater x force on the top than on the bottom describes diffusion of x momentum in the negative y direction, and hence, a negative $\tau_{w,yx}$

Heat Transfer

Equation 28.6: Heat Transfer Film Coefficient

$$\left(\frac{\dot{Q}}{A}\right)_w = h\Delta T \qquad 28.6$$

Variation

$$h = \frac{\dfrac{\dot{Q}}{A}}{\Delta T}$$

Description

While Eq. 28.5 is valid specifically for conduction in any substance, the relationship is somewhat different for heat transfer to or from a fluid moving past a solid surface (i.e., a turbulent fluid moving through a tube). The *heat transfer film coefficient*, h, usually just referred to as the *film coefficient*, is defined as the rate of heat transfer per unit area per degree temperature of difference. Unlike with Eq. 28.5, it is not necessary to know the thickness of the film in order to use the film coefficient. In Eq. 28.5 and Eq. 28.6, ΔT is the difference in tube wall and bulk fluid temperatures.

Equation 28.7: Nusselt Number

$$Nu = \frac{hD}{k} \qquad 28.7$$

Description

The dimensionless *Nusselt number* provides a convenient way to calculate the heat transfer coefficient, h, also known as the *film coefficient*. The Nusselt number is the ratio of convective heat transfer, h, to conductive heat transfer, k, in a direction perpendicular to the wall (surface, boundary, etc.). Higher Nusselt numbers, typically in the 100–1000 range, correspond to more active (i.e., turbulent) convection. Nusselt numbers may be for (or up to) a specific location, Nu_x, or they may be an average, $\overline{Nu}$, for use with the entire surface area.

Generally, k is known, but h is not. To find h, Eq. 28.7 is equated to an empirical correlation consisting of various fluid properties combined into other dimensionless numbers, typically the Raleigh and Prandtl numbers (e.g., $Nu = f(Ra, Pr)$) for free convection, or Reynolds and Prandtl (e.g., $Nu = f(Re, Pr)$) for forced convection. Each correlation is limited to one or more specific configurations (e.g., flow inside circular tubes or flow over flat plates) and has limitations which determine whether or not it can be used in a particular situation. For example, the *Dittus-Boelter equation* is only for use with turbulent flow in smooth, round tubes at high Reynolds numbers and with a fluid whose viscosity does not depend significantly on temperature. The Dittus-Boelter equation is

$$Nu_D = 0.023 Re_D^{0.8} Pr^n$$

$$[0.6 \leq Pr \leq 160 \; Re \geq 10{,}000; \; L/D \geq 10;$$

$$n = 0.3 \text{ for liquid cooling;}$$

$$n = 0.4 \text{ for liquid heating}]$$

In Eq. 28.7, D is the *characteristic length*. For flow in a tube, the inside diameter is the characteristic length; for flow over a tube, outside diameter is the characteristic length; for flow parallel to the surface of a flat plate, the characteristic length is some distance in the direction of flow, never exceeding the plate dimension in that direction. For complex shapes, the characteristic length may be defined as volume divided by surface area. Nu is frequently given a subscript indicating what dimension is used as the characteristic length (e.g., "Nu_D").

The thermal conductivity, k, is that of the fluid, not of the pipe or plate. Most properties of the fluid (film) should be evaluated at the *film temperature*, T_{film}, which is an average of the wall surface temperature and the build fluid temperature.

$$T_{\text{film}} = \tfrac{1}{2}(T_w + T_\infty)$$

5. MASS TRANSFER

Mass transfer is generally studied by chemical engineers as background for designing and analyzing such processes as distillation, absorption, drying, and extraction. Mass transfer deals with the migration of a single substance from one location or phase to another.

Quantities of mass transfer are usually calculated per unit area. Although the units of molecules per unit volume or area (per unit time) could be used, it is more common to work with units of moles per unit volume or area (per unit time). The symbol G and the names *mass velocity* or *mass flow rate per unit area* are generally used to represent mass transfer in units of $kg/m^2 \cdot s$.

Equation 28.8: Schmidt Number

$$Sc = \mu/(\rho D_m) \qquad 28.8$$

Description

The flow regime for the *diffusion* of a small amount of substance 1 through substance 2 (i.e., a dilute "mixture" or *dilute solution*) is determined by calculating the *Schmidt number*, Sc. This is the dimensionless ratio of the *molecular momentum diffusivity* (μ/ρ) to the *molecular mass diffusivity*, quantified by the *diffusion coefficient* for substance 1 moving through substance 2, D_m. Typical values for gases range from 0.5 to 2.0. For liquids, the range is about 100 to more than 10,000 for viscous liquids. The Schmidt number is essentially independent of temperature and pressure within "normal" operating conditions.

Equation 28.9: Fick's Law

$$\left(\frac{N}{A}\right)_w = -D_m \left(\frac{dc_m}{dy}\right)_w \qquad 28.9$$

Description

Fick's law describes molecular diffusion of mass for dilute solutions. The number of moles of substance 1 moving through substance 2 (i.e., the *molecular diffusion*) per unit area in the y-direction is given by Eq. 28.9. c_m is the concentration of substance 1 at any particular point. dc_m/dy is the *concentration gradient* of substance 1 in the y-direction. When moles are the units used, N/A is referred to as the *molar flux*.

Equation 28.10: Mass Transfer in Terms of Coefficients

$$\left(\frac{N}{A}\right)_w = k_m \Delta c_m \qquad 28.10$$

Description

For a fluid flowing past a solid surface (as a turbulent fluid flowing in a pipe), the relationship for the inward mass transfer flux, N/A, at the wall is very similar to Eq. 28.6. k_m is the *mass transfer coefficient* for the process, and Δc_m is the difference in the concentrations between the wall and the bulk fluid.

Equation 28.11: Sherwood Number

$$Sh = k_m D/D_m \qquad 28.11$$

Description

The dimensionless *Sherwood number*, Sh, is the ratio of *mass diffusivity* to molecular diffusivity and is given by Eq. 28.11. For turbulent flow through a pipe, D is the pipe inside diameter.

Equation 28.12: Stanton Number

$$St = h/(c_p G) \qquad 28.12$$

Variation

$$St = \frac{Nu}{RePr}$$

Description

The *Stanton number*, St, is the ratio of heat transfer at the wall to the energy transported by the mass flow.

Example

Which dimensionless number is DIFFERENT in common application from the other three?

(A) Prandtl, Pr

(B) Schmidt, Sc

(C) Nusselt, Nu

(D) Stanton, St

Solution

The Prantl number is the ratio of momentum diffusivity (viscosity) to thermal diffusivity. The Schmidt number is the ratio of momentum diffusivity (viscosity) to mass diffusivity. The Nusselt number is the ratio of convective to conductive heat transfer. The Stanton number is the ratio of heat transferred into a fluid to the thermal capacity of fluid. The Prandtl, Nusselt, and Stanton numbers are used in energy transfer application, while the Schmidt number is used in mass transfer applications.

The answer is (B).

6. TRANSPORT PHENOMENA ANALOGIES

It is clear that the structures of the equations that describe the three transport processes are similar. In that sense, momentum, energy, and mass transport processes are *analogous* processes. However, the phrase "transport phenomena analogy" actually has a slightly different meaning.

In some fluid processes, two or three transport processes occur simultaneously. For example, a turbulent fluid flowing through a cooled pipe will experience both frictional and thermal energy losses. It is convenient to use known data from one of the processes to predict the performance of the other process(es). For example, since fluid friction factors (a momentum transport property) are easily calculated, it is desirable to use them to predict the (more difficult) heat and mass transfer rates. Such correlations between the performance data are known as "analogies."

Equation 28.13 and Eq. 28.14: Use of Friction Factors to Predict Transfer Coefficients

$$j_H = \left(\frac{Nu}{RePr}\right) Pr^{2/3} = \frac{f}{8} \qquad 28.13$$

$$j_M = \left(\frac{Sh}{ReSc}\right) Sc^{2/3} = \frac{f}{8} \qquad 28.14$$

Heat Transfer

Description

In the popular Chilton and Colburn analogy, several dimensionless numbers are correlated with dimensionless "*j*-factors" and friction factors.

The Chilton-Colburn *j* factors are accurate analogies between momentum, heat, and mass transfer.[3] The factors can be used to determine an unknown transfer coefficient when one of the other coefficients is known. The entire analogy, representing momentum with friction factor, heat transfer (subscript H), and diffusive mass transfer (subscript M) is

$$j_H = \frac{f}{8} = \frac{h}{c_p G} \mathrm{Pr}^{2/3} = j_M = \frac{k'_c}{V} \mathrm{Sc}^{2/3}$$

The analogy is valid for fully developed turbulent flow in round tubes with $\mathrm{Re} \geq 10\,000$, $0.7 \leq \mathrm{Pr} \leq 160$, and $L/d \geq 60$ (the same constraints as the Titus-Boelter and Sieder-Tate correlations). The friction factor, f, is the *Darcy friction factor* used with the Moody friction factor chart. If the *Fanning friction factor* is used, $f/8$ should be replaced with $f/2$.

Equation 28.13 and Eq. 28.14 represent the relationship between heat and mass only and can be reduced to

$$j_M = \frac{f}{8} = \frac{\mathrm{Sh}}{\mathrm{Re}\,\mathrm{Sc}^{1/3}} = j_H = \frac{\mathrm{Nu}}{\mathrm{Re}\,\mathrm{Pr}^{1/3}}$$

Equation 28.13 and Eq. 28.14 are supported by theoretical derivations for laminar flow over flat plates as well as experimental data. For turbulent flow, the analogies are supported by experimental data for liquids and gases.

Example

Consider a perfectly smooth, flat metal plate of finite length with a gas flowing at velocity, v, parallel to the plate's flat surface. Which statement about a theoretical boundary layer is most INCORRECT?

- (A) A boundary layer exists along the entire plate length.
- (B) The thickness of the boundary layer is less than the characteristic length of the plate.
- (C) Outside of the boundary layer, in the direction perpendicular to the plate's surface, fluid velocity increases parabolically up to v.
- (D) Outside of the boundary layer, fluid flow is fully turbulent.

Solution

As long as the plate is smooth, the boundary layer should persist for the length of the plate. The thickness of the boundary layer is less than the plate length by orders of magnitude. Within the boundary layer, the velocity increases parabolically; outside of the boundary layer, the velocity is constant. Although flow within the boundary layer is considered laminar, flow outside of the boundary layer will be turbulent.

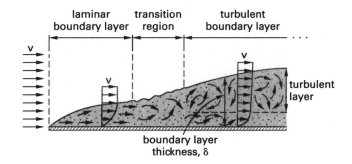

The answer is (C).

---

[3]The terms in the Chilton-Colburn analogy are analogies in the sense that they are analogous to one another. A better term would be "analogs," but "analogies" is commonly used.

Diagnostic Exam

Topic VII: Engineering Sciences

1. A 100 kg block rests on an incline. The coefficient of static friction between the block and the ramp is 0.2. The mass of the cable is negligible, and the pulley at point C is frictionless.

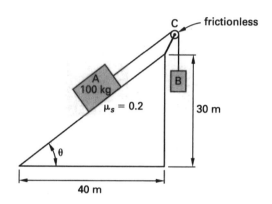

What is the smallest block B mass that will start the 100 kg block moving up the incline?

(A) 44 kg

(B) 65 kg

(C) 76 kg

(D) 92 kg

2. An area is given as shown.

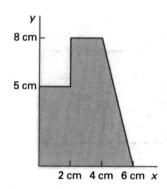

Most nearly, what are the x- and y-coordinates of the area's centroid?

(A) (2.5 cm, 3.4 cm)

(B) (2.8 cm, 3.3 cm)

(C) (3.2 cm, 4.2 cm)

(D) (3.4 cm, 3.7 cm)

3. The cantilever truss shown supports a vertical force of 600 000 N applied at point G.

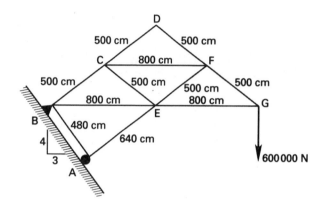

What is most nearly the force in member CF?

(A) 0.96 MN

(B) 1.2 MN

(C) 1.6 MN

(D) 2.3 MN

4. A projectile is fired from a cannon with an initial velocity of 1000 m/s and at an angle of 30° from horizontal. The point of impact is 1500 m below the point of release.

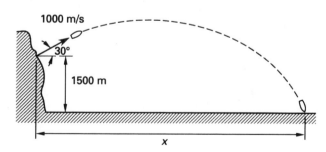

Most nearly, at what distance from the cannon will the projectile strike the ground?

(A) 8200 m

(B) 67 000 m

(C) 78 000 m

(D) 91 000 m

5. A fisherman cuts his boat's engine as it is entering a harbor. The boat comes to a dead stop with its front end touching the dock. The fisherman's mass is 80 kg. He moves 5 m from his seat in the back to the front of the boat in 5 s, expecting to be able to reach the dock. The empty boat has a mass of 300 kg. Disregard all friction. Approximately how far will the fisherman have to jump to reach the dock?

(A) 1.1 m

(B) 1.3 m

(C) 1.9 m

(D) 5.0 m

6. A wheel with a radius of 0.8 m rolls without slipping along a flat surface at 3 m/s. Arc AB on the wheel's perimeter has an interior angle of 90°.

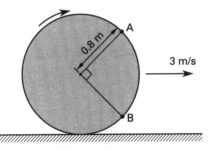

What is most nearly the velocity of point A when point B contacts the ground?

(A) 3.0 m/s

(B) 3.4 m/s

(C) 3.8 m/s

(D) 4.2 m/s

7. A ball is dropped from rest at a point 12 m above the ground into a smooth, frictionless chute. The ball exits the chute 2 m above the ground and at an angle 45° from the horizontal. Air resistance is negligible.

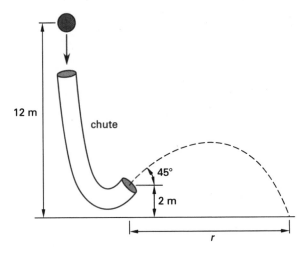

Most nearly, how far will the ball travel in the horizontal direction before hitting the ground?

(A) 12 m

(B) 20 m

(C) 22 m

(D) 24 m

8. Induction motors on a 13.2 kV circuit draw 10 MVA with a power factor of 0.85 lagging. Most nearly, how much capacitive reactive power is needed to correct the power factor to 0.97 lagging?

(A) 2.5 MVAR

(B) 3.1 MVAR

(C) 4.8 MVAR

(D) 5.2 MVAR

9. A transformer circuit is shown.

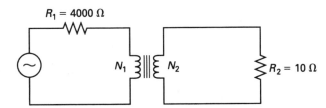

What should be the turns ratio (N_1:N_2) for maximum power transfer in the circuit?

(A) 1:40

(B) 1:20

(C) 20:1

(D) 40:1

10. A circuit consisting of a battery and resistors is shown.

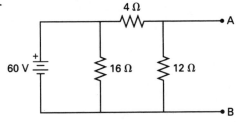

What are most nearly the Thevenin equivalent resistance and voltage, respectively, between terminals A and B?

(A) $R_{Th} = 3.0$ Ω, $V_{Th} = 45$ V

(B) $R_{Th} = 7.5$ Ω, $V_{Th} = 7.5$ V

(C) $R_{Th} = 7.5$ Ω, $V_{Th} = 60$ V

(D) $R_{Th} = 12$ Ω, $V_{Th} = 5.0$ V

SOLUTIONS

1. Choose coordinate axes parallel and perpendicular to the incline.

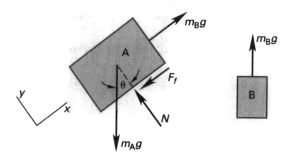

If the block is at rest, all forces are in equilibrium, so the sum of the forces must be zero.

$$\sum F_x = 0$$
$$= m_B g - m_A g(\sin\theta + \mu_s \cos\theta)$$

The smallest mass of block B that will start the block moving is

$$m_B = m_A(\sin\theta + \mu_s \cos\theta) = (100 \text{ kg})\left(\frac{3}{5} + (0.2)\left(\frac{4}{5}\right)\right)$$
$$= 76 \text{ kg}$$

(Note that the frictional force serves to increase, not decrease, the force required to accelerate the block.)

The answer is (C).

2. Divide the shape into regions. Calculate the area and locate the centroid for each region.

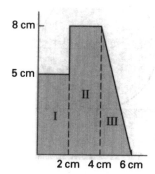

For region I (rectangular),

$$A = (2 \text{ cm})(5 \text{ cm}) = 10 \text{ cm}^2$$
$$x_c = \frac{2 \text{ cm}}{2} = 1 \text{ cm}$$
$$y_c = \frac{5 \text{ cm}}{2} = 2.5 \text{ cm}$$

For region II (rectangular),

$$A = (2 \text{ cm})(8 \text{ cm}) = 16 \text{ cm}^2$$
$$x_c = \frac{2 \text{ cm} + 4 \text{ cm}}{2} = 3 \text{ cm}$$
$$y_c = \frac{8 \text{ cm}}{2} = 4 \text{ cm}$$

For region III (triangular),

$$A = \left(\frac{1}{2}\right)(2 \text{ cm})(8 \text{ cm}) = 8 \text{ cm}^2$$
$$x_c = 4 \text{ cm} + \left(\frac{1}{3}\right)(2 \text{ cm}) = 4.67 \text{ cm}$$
$$y_c = \frac{8 \text{ cm}}{3} = 2.67 \text{ cm}$$

Calculate the centroidal x- and y-coordinates.

$$x_c = \frac{M_{ay}}{A_i} = \frac{\sum x_{c,i} A_i}{\sum A_i}$$

$$= \frac{\begin{array}{c}(1 \text{ cm})(10 \text{ cm}^2) + (3 \text{ cm})(16 \text{ cm}^2)\\ + (4.67 \text{ cm})(8 \text{ cm}^2)\end{array}}{10 \text{ cm}^2 + 16 \text{ cm}^2 + 8 \text{ cm}^2}$$

$$= 2.80 \text{ cm} \quad (2.8 \text{ cm})$$

$$y_c = \frac{M_{ax}}{A_i} = \frac{\sum y_{c,i} A_i}{\sum A_i}$$

$$= \frac{\begin{array}{c}(2.5 \text{ cm})(10 \text{ cm}^2) + (4 \text{ cm})(16 \text{ cm}^2)\\ + (2.67 \text{ cm})(8 \text{ cm}^2)\end{array}}{10 \text{ cm}^2 + 16 \text{ cm}^2 + 8 \text{ cm}^2}$$

$$= 3.25 \text{ cm} \quad (3.3 \text{ cm})$$

The answer is (B).

3. Draw the free-body diagram for the truss.

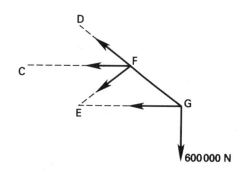

Equilibrium of pin D requires that the forces in members CD and DF are zero.

Use the method of sections. Sum moments about joint E. Clockwise moments are positive.

$$\sum M_E = 0$$
$$= (600\,000\text{ N})(800\text{ cm}) - CF(300\text{ cm})$$
$$CF = \frac{(600\,000\text{ N})(800\text{ cm})}{300\text{ cm}}$$
$$= 1\,600\,000\text{ N}\quad(1.6\text{ MN})$$

The answer is (C).

4. $y = -1500$ m since it is below the launch plane.

$$y = v_0 t \sin\theta - \frac{gt^2}{2}$$
$$\frac{g}{2}t^2 - v_0 t \sin\theta + y = 0$$
$$\left(\frac{9.81\ \frac{\text{m}}{\text{s}^2}}{2}\right)t^2 - \left(1000\ \frac{\text{m}}{\text{s}}\right)t\sin 30° + (-1500\text{ m}) = 0$$
$$\left(4.905\ \frac{\text{m}}{\text{s}^2}\right)t^2 - \left(500\ \frac{\text{m}}{\text{s}}\right)t + (-1500\text{ m}) = 0$$

The time to impact is

$$t = \frac{-b \pm \sqrt{b^2 - 4ac}}{2a}\quad\text{[quadratic formula]}$$

$$= \frac{500\ \frac{\text{m}}{\text{s}} \pm \sqrt{\begin{array}{c}\left(-500\ \frac{\text{m}}{\text{s}}\right)^2 - (4)\left(4.905\ \frac{\text{m}}{\text{s}^2}\right)\\ \times(-1500\text{ m})\end{array}}}{(2)\left(4.905\ \frac{\text{m}}{\text{s}^2}\right)}$$

$$= +104.85\text{ s}, -2.9166\text{ s}$$

The horizontal distance is

$$x = v_0 t \cos\theta$$
$$= \left(1000\ \frac{\text{m}}{\text{s}}\right)(104.85\text{ s})\cos 30°$$
$$= 90\,803\text{ m}\quad(91\,000\text{ m})$$

The answer is (D).

5. The velocity of the fisherman relative to the boat is

$$v = \frac{s}{t} = \frac{5\text{ m}}{5\text{ s}}$$
$$= 1\text{ m/s}$$

If the boat moves as the fisherman moves, the velocity of the fisherman relative to the dock is

$$v'_{\text{fisherman}} = 1\ \frac{\text{m}}{\text{s}} + v'_{\text{boat}}$$

Use the conservation of momentum.

$$\sum m_i v_i = \sum m_i v'_i$$

$$\begin{array}{c}m_{\text{fisherman}}v_{\text{fisherman}}\\ + m_{\text{boat}}v_{\text{boat}}\end{array} = \begin{array}{c}m_{\text{fisherman}}v'_{\text{fisherman}}\\ + m_{\text{boat}}v'_{\text{boat}}\end{array}$$

However, $v_{\text{fisherman}} = v_{\text{boat}} = 0$ initially, so

$$0 = m_{\text{fisherman}}\left(1\ \frac{\text{m}}{\text{s}} + v'_{\text{boat}}\right) + m_{\text{boat}}v'_{\text{boat}}$$

$$v'_{\text{boat}} = \frac{-m_{\text{fisherman}}\left(1\ \frac{\text{m}}{\text{s}}\right)}{m_{\text{fisherman}} + m_{\text{boat}}}$$

$$= \frac{-(80\text{ kg})\left(1\ \frac{\text{m}}{\text{s}}\right)}{80\text{ kg} + 300\text{ kg}}$$

$$= -0.211\text{ m/s}\quad\text{[backward]}$$

The direction the fisherman jumps is positive. Therefore, the distance the fisherman will have to jump is

$$s = v'_{\text{boat}}t = \left(0.211\ \frac{\text{m}}{\text{s}}\right)(5\text{ s})$$
$$= 1.05\text{ m}\quad(1.1\text{ m})$$

The answer is (A).

6. The wheel's radius is 0.8 m. Point B becomes the instantaneous center of rotation when it is in contact with the ground.

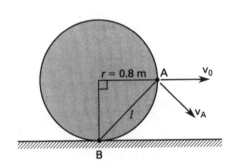

$$l^2 = r^2 + r^2\quad\text{[Pythagorean theorem]}$$
$$= (0.8\text{ m})^2 + (0.8\text{ m})^2$$
$$= 1.28\text{ m}^2$$
$$l = \sqrt{1.28\text{ m}^2} = 1.13\text{ m}$$

The velocity of point A is

$$v_A = \frac{lv_0}{r} = \frac{(1.13\text{ m})\left(3\ \frac{\text{m}}{\text{s}}\right)}{0.8\text{ m}}$$
$$= 4.24\text{ m/s}\quad(4.2\text{ m/s})$$

The answer is (D).

7. The change in elevation between points A and B represents a decrease in the ball's potential energy. The kinetic energy increases correspondingly.

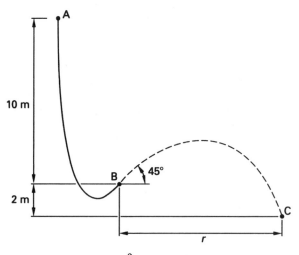

$$mg\Delta h = \frac{mv^2}{2}$$

$$v^2 = 2g\Delta h$$

$$v = \sqrt{2g\Delta h}$$

$$= \sqrt{(2)\left(9.81 \ \frac{m}{s^2}\right)(10 \ m)}$$

$$= 14.0 \ m/s$$

The ball follows the path of a projectile between points B and C.

$$y = \frac{-gt^2}{2} + v_0 \sin(\theta)t$$

$$\left(\frac{g}{2}\right)t^2 - (v_0 \sin \theta)t + y = 0$$

Because the landing point is below the chute exit, $y = -2 \ m$.

$$\left(\frac{9.81 \ \frac{m}{s^2}}{2}\right)t^2 - \left(14.0 \ \frac{m}{s}\right)(\sin 45°)t + (-2 \ m) = 0$$

$$\left(4.91 \ \frac{m}{s^2}\right)t^2 - \left(9.9 \ \frac{m}{s}\right)t + (-2 \ m) = 0$$

Solve for t using the quadratic formula.

$$t = \frac{-b \pm \sqrt{b^2 - 4ac}}{2a}$$

$$= \frac{9.9 \ \frac{m}{s} \pm \sqrt{\left(9.9 \ \frac{m}{s}\right)^2 - (4)\left(4.91 \ \frac{m}{s^2}\right)(-2 \ m)}}{(2)\left(4.91 \ \frac{m}{s^2}\right)}$$

$$= 2.2 \ s, -0.2 \ s$$

Calculate the distance traveled from the x-component of the velocity.

$$x = v_0 t \cos \theta$$

$$= \left(14.0 \ \frac{m}{s}\right)(2.2 \ s)\cos 45°$$

$$= 21.8 \ m \quad (22 \ m)$$

The answer is (C).

8. The complex power in polar (phasor) form is $S = 10 \ MVA \angle \arccos 0.85 = 10 \ MVA \angle 31.79°$. The original real and reactive components are

$$P = S \cos \theta_1 = (10 \ MVA)\cos 31.79° = 8.5 \ MW$$

$$Q = S \sin \theta_1 = (10 \ MVA)\sin 31.79° = 5.268 \ MVAR$$

$$S = P + jQ = 8.5 + j5.268$$

To change the power factor, add capacitive kVAR, which affects the reactive component only. The desired power factor, pf, is 0.97.

$$pf = \cos \theta_2 = 0.97$$

$$\theta_2 = \arccos 0.97$$

$$= 14.07°$$

The power triangles for the current and desired circuits are shown.

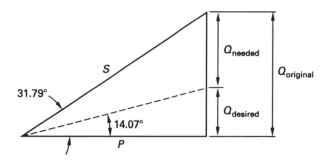

The new reactive power will be

$$Q_{desired} = P\left(\frac{\sin \theta_2}{\cos \theta_2}\right) = P \tan \theta_2$$

$$= (8.5 \ MW)\tan 14.07°$$

$$= 2.13 \ MVAR$$

The capacitive reactive power needed is the difference between the original and desired reactive powers.

$$Q_{needed} = Q_{original} - Q_{desired}$$

$$= 5.268 \ MVAR - 2.13 \ MVAR$$

$$= 3.138 \ MVAR \quad (3.1 \ MVAR)$$

The answer is (B).

9. Maximum power transfer occurs in an ideal transformer in which primary and secondary powers are equal.

$$P_P = P_S$$

$$I_P^2 R_P = I_S^2 R_S$$

$$\left(\frac{I_S}{I_P}\right)^2 = \frac{R_P}{R_S}$$

$$a^2 = \frac{R_P}{R_S}$$

$$a = \sqrt{\frac{R_P}{R_S}} = \sqrt{\frac{4000\ \Omega}{10\ \Omega}}$$

$$= 20$$

$$N_1 : N_2 = 20:1$$

The answer is (C).

10. The Thevenin resistance is the resistance between the two terminals with the voltage source short-circuited. With the circuit shorted, the 16 Ω resistor is effectively bypassed, leaving the other two resistors in a simple parallel arrangement.

$$R_{\mathrm{Th}} = \frac{(4\ \Omega)(12\ \Omega)}{4\ \Omega + 12\ \Omega} = 3.0\ \Omega$$

The Thevenin voltage is the voltage difference between the two terminals, which in this case is equal to the voltage drop across the 12 Ω resistor. Use Ohm's law to find the current $I_{12\,\Omega}$ and to find the voltage across the resistor.

$$V = IR$$

$$I_{12\,\Omega} = \frac{V}{R} = \frac{60\ \mathrm{V}}{4\ \Omega + 12\ \Omega} = 3.75\ \mathrm{A}$$

$$V_{\mathrm{Th}} = I_{12\,\Omega} R = (3.75\ \mathrm{A})(12\ \Omega) = 45\ \mathrm{V}$$

The answer is (A).

29 Trusses

Nomenclature

F	force	N
M	moment	N·m

1. STATICALLY DETERMINATE TRUSSES

A *truss* or *frame* is a set of pin-connected *axial members* (i.e., *two-force members*). The connection points are known as *joints*. Member weights are disregarded, and truss loads are applied only at joints. A *structural cell* consists of all members in a closed loop of members. For the truss to be stable (i.e., to be a *rigid truss*), all of the structural cells must be triangles. Figure 29.1 identifies *chords*, *end posts*, *panels*, and other elements of a typical *bridge truss*.

Figure 29.1 *Parts of a Bridge Truss*

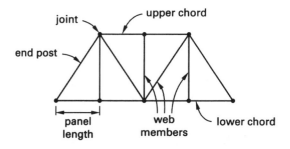

Several types of trusses have been given specific names. Some of the more common named trusses are shown in Fig. 29.2.

Truss loads are considered to act only in the plane of a truss, so trusses are analyzed as two-dimensional structures. Forces in truss members hold the various truss parts together and are known as *internal forces*. The internal forces are found by drawing free-body diagrams.

Although free-body diagrams of truss members can be drawn, this is not usually done. Instead, free-body diagrams of the pins (i.e., the joints) are drawn. A pin in compression will be shown with force arrows pointing toward the pin, away from the member. Similarly, a pin in tension will be shown with force arrows pointing away from the pin, toward the member.

Figure 29.2 *Special Types of Trusses*

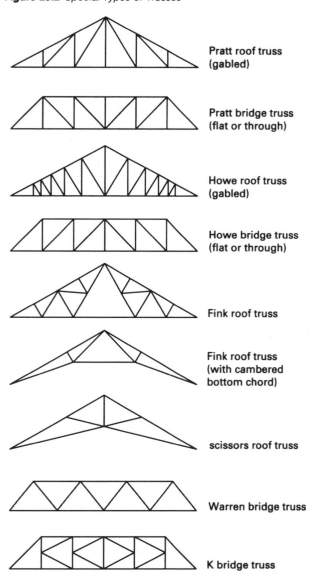

With typical bridge trusses supported at the ends and loaded downward at the joints, the upper chords are almost always in compression, and the end panels and lower chords are almost always in tension.

Since truss members are axial members, the forces on the truss joints are concurrent forces. Only force equilibrium needs to be enforced at each pin; the sum of the forces in each of the coordinate directions equals zero.

Engineering Sciences

Forces in truss members can sometimes be determined by inspection. One of these cases is *zero-force members*. A third member framing into a joint already connecting two collinear members carries no internal force unless there is a load applied at that joint. Similarly, both members forming an apex of the truss are zero-force members unless there is a load applied at the apex. (See Fig. 29.3.)

Figure 29.3 *Zero-Force Members*

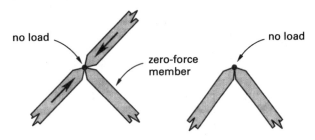

A truss will be *statically determinate* if

$$\text{no. of members} = (2)(\text{no. of joints}) - 3$$

If the left-hand side of the equation is greater than the right-hand side (i.e., there are *redundant members*), the truss is statically indeterminate. If the left-hand side is less than the right-hand side, the truss is unstable and will collapse under certain types of loading.

2. PLANE TRUSS

A *plane truss* (*planar truss*) is a rigid framework where all truss members are within the same plane and are connected at their ends by frictionless pins. External loads are in the same plane as the truss and are applied at the joints only.

Equation 29.1 and Eq. 29.2: Equations of Equilibrium[1]

$$\sum F_n = 0 \qquad \qquad 29.1$$

$$\sum M_n = 0 \qquad \qquad 29.2$$

Description

A plane truss is statically determinate if the truss reactions and member forces can be determined using the equations of equilibrium. If not, the truss is considered statically indeterminate.

---

[1]The NCEES *FE Reference Handbook* (*NCEES Handbook*) uses both n and i for summation variables. Though i is traditionally used to indicate summation, n appears to be used as the summation variable in order to indicate that the summation is over all n of the forces and all n of the position vectors that make up the system (i.e., $i = 1$ to n).

Example

Most nearly, what are reactions F_1 and F_2 for the truss shown?

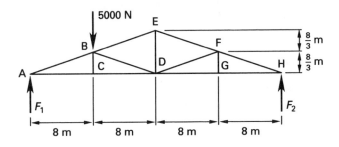

(A) $F_1 = 1000$ N; $F_2 = 4000$ N

(B) $F_1 = 1300$ N; $F_2 = 3800$ N

(C) $F_1 = 2500$ N; $F_2 = 2500$ N

(D) $F_1 = 3800$ N; $F_2 = 1300$ N

Solution

Calculate the reactions from Eq. 29.1 and Eq. 29.2. Let clockwise moments be positive.

$$\sum M_A = 0 = (5000 \text{ N})(8 \text{ m}) - F_2(32 \text{ m})$$

$$F_2 = 1250 \text{ N} \quad (1300 \text{ N}) \quad [\text{upward}]$$

$$\sum F_y = 0 = F_1 + 1250 \text{ N} - 5000 \text{ N}$$

$$F_1 = 3750 \text{ N} \quad (3800 \text{ N}) \quad [\text{upward}]$$

The answer is (D).

3. METHOD OF JOINTS

The *method of joints* is one of the methods that can be used to find the internal forces in each truss member. This method is useful when most or all of the truss member forces are to be calculated. Because this method advances from joint to adjacent joint, it is inconvenient when a single isolated member force is to be calculated.

The method of joints is a direct application of the equations of equilibrium in the x- and y-directions. Traditionally, the method begins by finding the reactions supporting the truss. Next the joint at one of the reactions is evaluated, which determines all the member forces framing into the joint. Then, knowing one or more of the member forces from the previous step, an adjacent joint is analyzed. The process is repeated until all the unknown quantities are determined.

At a joint, there may be up to two unknown member forces, each of which can have dependent x- and y-components. Since there are two equilibrium equations, the two unknown forces can be determined. Even though determinate, however, the sense of a force will often be unknown. If the sense cannot be determined by

Engineering Sciences

logic, an arbitrary decision can be made. If the incorrect direction is chosen, the force will be negative.

Occasionally, there will be three unknown member forces. In that case, an additional equation must be derived from an adjacent joint.

4. METHOD OF SECTIONS

The *method of sections* is a direct approach to finding forces in any truss member. This method is convenient when only a few truss member forces are unknown.

As with the previous method, the first step is to find the support reactions. Then, a cut is made through the truss, passing through the unknown member. (Knowing where to cut the truss is the key part of this method. Such knowledge is developed only by repeated practice.) Finally, all three conditions of equilibrium are applied as needed to the remaining truss portion. Since there are three equilibrium equations, the cut cannot pass through more than three members in which the forces are unknown.

Example

A truss is loaded as shown. The support reactions have already been determined.

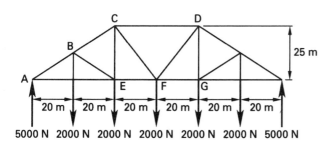

Most nearly, what is the force in member CE?

(A) 1000 N

(B) 2000 N

(C) 3000 N

(D) 4000 N

Solution

Cut the truss as shown.

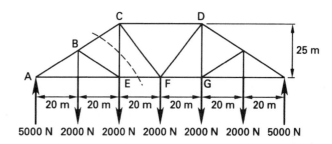

Draw the free body.

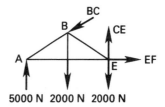

Taking moments about point A will eliminate all of the unknown forces except CE. Let clockwise moments be positive.

$$\sum M_A = 0$$

$$(2000 \text{ N})(20 \text{ m}) + (2000 \text{ N})(40 \text{ m})$$

$$-CE(40 \text{ m}) = 0$$

$$CE = 3000 \text{ N}$$

The answer is (C).

Engineering
Sciences

30 Pulleys, Cables, and Friction

Nomenclature

d	inside diameter	m
D	outside diameter	m
F	force	N
g	gravitational acceleration, 9.81	m/s^2
m	mass	kg
n	number	–
N	normal force	N
P	power	W
r	radius	m
T	torque	N·m
v	velocity	m/s
W	weight	N

Symbols

θ	angle of wrap	radians
μ	coefficient of friction	–
ϕ	angle	deg

Subscripts

k	kinetic
s	static
t	tangential

1. PULLEYS

A *pulley* (also known as a *sheave*) is used to change the direction of an applied tensile force. A series of pulleys working together (known as a *block and tackle*) can also provide *pulley advantage* (i.e., *mechanical advantage*). (See Fig. 30.1.)

If the pulley is attached by a bracket to a fixed location, it is said to be a *fixed pulley*. If the pulley is attached to a load, or if the pulley is free to move, it is known as a *free pulley*.

Most simple problems disregard friction and assume that all ropes (fiber ropes, wire ropes, chains, belts, etc.) are parallel. In such cases, the pulley advantage is equal to the number of ropes coming to and going from the load-carrying pulley. The diameters of the pulleys are not factors in calculating the pulley advantage.

Figure 30.1 *Mechanical Advantage of Rope-Operated Machines*

	fixed sheave	free sheave	ordinary pulley block (*n* sheaves)	differential pulley block
F_{ideal}	W	$\dfrac{W}{2}$	$\dfrac{W}{n}$	$\dfrac{W}{2}\left(1 - \dfrac{d}{D}\right)$

2. CABLES

An *ideal cable* is assumed to be completely flexible, massless, and incapable of elongation; it acts as an axial tension member between points of concentrated loading. The term *tension* or *tensile force* is commonly used in place of member force when dealing with cables.

The methods of joints and sections used in truss analysis can be used to determine the tensions in cables carrying concentrated loads. (See Fig. 30.2.) After separating the reactions into *x*- and *y*-components, it is particularly useful to sum moments about one of the reaction points. All cables will be found to be in tension, and (with vertical loads only) the horizontal tension component

Figure 30.2 *Cable with Concentrated Load*

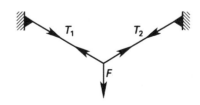

will be the same in all cable segments. Unlike the case of a rope passing over a series of pulleys, however, the total tension in the cable will not be the same in every cable segment.

3. FRICTION

Friction is a force that always resists motion or impending motion. It always acts parallel to the contacting surfaces. The frictional force, F, exerted on a stationary body is known as *static friction, Coulomb friction,* and *fluid friction.* If the body is moving, the friction is known as *dynamic friction* and is less than the static friction.

Equation 30.1 Through Eq. 30.4: Frictional Force[1]

$$F \leq \mu_s N \qquad 30.1$$

$$\boldsymbol{F} < \mu_s \boldsymbol{N} \quad \text{[no slip occurring]} \qquad 30.2$$

$$\boldsymbol{F} = \mu_s \boldsymbol{N} \quad \text{[point of impending slip]} \qquad 30.3$$

$$\boldsymbol{F} = \mu_k \boldsymbol{N} \quad \text{[slip occurring]} \qquad 30.4$$

Values

$$\mu_k \approx 0.75 \mu_s$$

Description

The actual magnitude of the frictional force depends on the *normal force*, N, and the *coefficient of friction*, μ, between the body and the surface. The coefficient of kinetic friction, μ_k, is approximately 75% of the coefficient of static friction, μ_s.

Equation 30.1 is a general expression of the laws of friction. Several specific cases exist depending on whether slip is occurring or impending. Use Eq. 30.2 when no slip is occurring. Equation 30.3 is valid at the point of impending slip (or slippage), and Eq. 30.4 is valid when slip is occurring.

For a body resting on a horizontal surface, the normal force is the weight of the body.

$$N = mg$$

If a body rests on a plane inclined at an angle ϕ from the horizontal, the normal force is

$$N = mg \cos \phi$$

Example

A 35 kg block resting on the 30° incline is shown.

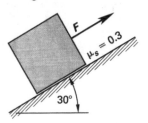

What is most nearly the frictional force at the point of impending slippage?

(A) 37 N

(B) 52 N

(C) 89 N

(D) 100 N

Solution

From Eq. 30.3, the frictional force is

$$\begin{aligned} \boldsymbol{F} = \mu_s \boldsymbol{N} &= \mu_s (mg \cos \phi) \\ &= (0.3)\left((35 \text{ kg})\left(9.81 \frac{\text{m}}{\text{s}^2}\right)\cos 30°\right) \\ &= 89.2 \text{ N} \quad (89 \text{ N}) \end{aligned}$$

The answer is (C).

4. BELT FRICTION

Friction between a belt, rope, or band wrapped around a pulley or sheave is responsible for the transfer of torque. Except when stationary, one side of the belt (the tight side) will have a higher tension than the other (the slack side). (See Fig. 30.3.)

Figure 30.3 *Belt Friction*

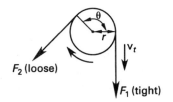

Equation 30.5: Belt Tension Relationship

$$F_1 = F_2 e^{\mu\theta} \qquad 30.5$$

Description

The basic relationship between the belt tensions and the coefficient of friction neglects centrifugal effects and is given by Eq. 30.5. F_1 is the tension on the tight side

---

[1]Although the NCEES *FE Reference Handbook* (*NCEES Handbook*) uses bold, Eq. 30.2 through Eq. 30.4 are not vector equations.

(direction of movement); F_2 is the tension on the other side. The *angle of wrap*, θ, must be expressed in radians.

The net transmitted torque is

$$T = (F_1 - F_2)r$$

The power transmitted to a belt running at tangential velocity v_t is

$$P = (F_1 - F_2)v_t$$

Example

A rope passes over a fixed sheave, as shown. The two rope ends are parallel. A fixed load on one end of the rope is supported by a constant force on the other end. The coefficient of friction between the rope and the sheave is 0.30.

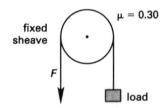

What is most nearly the maximum ratio of tensile forces in the two rope ends?

(A) 1.1

(B) 1.2

(C) 1.6

(D) 2.6

Solution

The angle of wrap, θ, is 180°, but it must be expressed in radians.

$$\theta = (180°)\left(\frac{2\pi \text{ rad}}{360°}\right) = \pi \text{ rad}$$

$$F_1 = F_2 e^{\mu\theta}$$

$$\frac{F_1}{F_2} = e^{(0.30)(\pi \text{ rad})}$$

$$= 2.57 \quad (2.6)$$

Either side could be the tight side. Therefore, the restraining force could be 2.6 times smaller or larger than the load tension.

The answer is (D).

Engineering
Sciences

31 Centroids and Moments of Inertia

Nomenclature

a	subarea	m^2
a	length or radius	m
A	area	m^2
b	base	m
d	distance	m
h	height	m
I	moment of inertia	m^4
I_{xy}	product of inertia	m^4
J	polar moment of inertia	m^4
l	length	m
L	total length	m
m	mass	kg
M	statical moment	m^3
r	radius or radius of gyration	m
v	volume	m^3
V	volume	m^3

Symbols

θ	angle	deg

Subscripts

a	area
c	centroidal
deg	degrees
l	line
o	origin
p	polar
rad	radians
v	volume

1. CENTROIDS

The *centroid* of an area is often described as the point at which a thin, homogeneous plate would balance. This definition, however, combines the definitions of centroid and center of gravity, and implies gravity is required to identify the centroid, which is not true. Nonetheless, this definition provides some intuitive understanding of the centroid.

Centroids of continuous functions can be found by the methods of integral calculus. For most engineering applications, though, the functions to be integrated are regular shapes such as the rectangular, circular, or composite rectangular shapes of beams. For these shapes, simple formulas are readily available and should be used.

Equation 31.1 and Eq. 31.2: First Moment of an Area in the *x-y* Plane[1]

$$M_{ay} = \sum x_n a_n \qquad \textbf{31.1}$$

$$M_{ax} = \sum y_n a_n \qquad \textbf{31.2}$$

Variations

$$M_y = \int x \, dA = \sum x_i A_i$$

$$M_x = \int y \, dA = \sum y_i A_i$$

Description

The quantity $\sum x_n a_n$ is known as the *first moment of the area* or *first area moment* with respect to the *y*-axis. Similarly, $\sum y_n a_n$ is known as the first moment of the area with respect to the *x*-axis. Equation 31.1 and Eq. 31.2 apply to regular shapes with subareas a_n.

The two primary applications of the first moment are determining centroidal locations and shear stress distributions. In the latter application, the first moment of the area is known as the *statical moment*.

Centroid of Line Segments in the *x-y* Plane

For a composite line of total length L, the location of the centroid of a line is defined by the following equations.

$$x_c = \frac{\int x \, dL}{L} = \frac{\sum x_i L_i}{\sum L_i}$$

$$y_c = \frac{\int y \, dL}{L} = \frac{\sum y_i L_i}{\sum L_i}$$

---

[1]The NCEES *FE Reference Handbook* (*NCEES Handbook*) deviates from conventional notation in several ways. Q is the most common symbol for the first area moment (then referred to as the *statical moment*), although symbols S and M are also encountered. To avoid confusion with the moment of a force, the subscript a is used to designate the moment of an area. The *NCEES Handbook* uses a lowercase a to designate the area of a subarea (instead of A_i). The *NCEES Handbook* uses n as a summation variable (instead of i), probably to indicate that the moment has to be calculated from all n of the subareas that make up the total area.

Using the *NCEES Handbook* notation, the equations would be written as

$$L = \sum l_n$$

$$x_{lc} = \frac{\sum x_n l_n}{L}$$

$$y_{lc} = \frac{\sum y_n l_n}{L}$$

Example

Find the approximate x- and y-coordinates of the centroid of wire ABC.

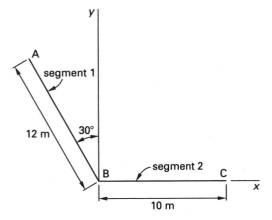

(A) 0.43 m; 1.3 m

(B) 0.64 m; 2.8 m

(C) 2.7 m; 1.5 m

(D) 3.3 m; 2.7 m

Solution

The total length of the line is

$$\sum L_i = 12 \text{ m} + 10 \text{ m} = 22 \text{ m}$$

The coordinates of the centroid of the line are

$$x_c = \frac{\sum x_i L_i}{\sum L_i}$$

$$= \frac{\left(\frac{(-12 \text{ m})\sin 30°}{2}\right)(12 \text{ m}) + \left(\frac{10 \text{ m}}{2}\right)(10 \text{ m})}{22 \text{ m}}$$

$$= 0.64 \text{ m}$$

$$y_c = \frac{\sum y_i L_i}{\sum L_i}$$

$$= \frac{\left(\frac{(12 \text{ m})\cos 30°}{2}\right)(12 \text{ m}) + (0 \text{ m})(10 \text{ m})}{22 \text{ m}}$$

$$= 2.83 \text{ m} \qquad (2.8 \text{ m})$$

The answer is (B).

Equation 31.3 Through Eq. 31.5: Centroid of an Area in the *x-y* Plane[2]

$$A = \sum a_n \qquad \textit{31.3}$$

$$x_{ac} = M_{ay}/A = \sum x_n a_n / A \qquad \textit{31.4}$$

$$y_{ac} = M_{ax}/A = \sum y_n a_n / A \qquad \textit{31.5}$$

Variations

$$x_c = \frac{\int x \, dA}{A} = \frac{\sum x_{c,i} A_i}{A}$$

$$y_c = \frac{\int y \, dA}{A} = \frac{\sum y_{c,i} A_i}{A}$$

Description

The centroid of an area A composed of subareas a_n (see Eq. 31.3) is located using Eq. 31.4 and Eq. 31.5. The location of the centroid of an area depends only on the geometry of the area, and it is identified by the coordinates (x_{ac}, y_{ac}), or, more commonly, (x_c, y_c).

Example

What are the approximate x- and y-coordinates of the centroid of the area shown?

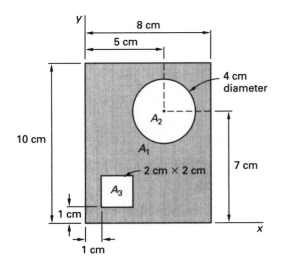

(A) 3.4 cm; 5.6 cm

(B) 3.5 cm; 5.5 cm

(C) 3.9 cm; 4.4 cm

(D) 3.9 cm; 4.8 cm

---

[2]In Eq. 31.4 and Eq. 31.5, the subscript a is used to designate the centroid of an area, but this convention is largely omitted throughout the rest of the *NCEES Handbook*.

Solution

Calculate the total area.

$$A = \sum a_n$$

$$= (8 \text{ cm})(10 \text{ cm}) - \frac{\pi(4 \text{ cm})^2}{4}$$

$$- (2 \text{ cm})(2 \text{ cm})$$

$$= 63.43 \text{ cm}^2$$

Find the first moments about the x-axis and y-axis.

$$M_{ax} = \sum y_n a_n$$

$$= (5 \text{ cm})(80 \text{ cm}^2) + \left(-\pi\left(\frac{4 \text{ cm}}{2}\right)^2 (7 \text{ cm})\right)$$

$$+ \left(-(2 \text{ cm})(4 \text{ cm}^2)\right)$$

$$= 304.03 \text{ cm}^3$$

$$M_{ay} = \sum x_n a_n$$

$$= (4 \text{ cm})(80 \text{ cm}^2) + \left(-(5 \text{ cm})\left(\frac{\pi}{4}\right)(4 \text{ cm})^2\right)$$

$$+ \left(-(2 \text{ cm})(4 \text{ cm}^2)\right)$$

$$= 249.17 \text{ cm}^3$$

The x-coordinate of the centroid is

$$x_c = M_{ay}/A = \frac{249.17 \text{ cm}^3}{63.43 \text{ cm}^2} = 3.93 \text{ cm} \quad (3.9 \text{ cm})$$

The y-coordinate of the centroid is

$$y_c = M_{ax}/A = \frac{304.03 \text{ cm}^3}{63.43 \text{ cm}^2} = 4.79 \text{ cm} \quad (4.8 \text{ cm})$$

The answer is (D).

....

Equation 31.6 Through Eq. 31.9: Centroid of a Volume[3]

$$V = \sum v_n \qquad \text{31.6}$$

$$x_{vc} = \left(\sum x_n v_n\right)/V \qquad \text{31.7}$$

$$y_{vc} = \left(\sum y_n v_n\right)/V \qquad \text{31.8}$$

$$z_{vc} = \left(\sum z_n v_n\right)/V \qquad \text{31.9}$$

[3]The *NCEES Handbook* uses lowercase v to designate the area of a subvolume (instead of V_i). In Eq. 31.7 through Eq. 31.9, the subscript v is used to designate the centroid of a volume, but this convention is largely omitted throughout the rest of the *NCEES Handbook*.

Description

The centroid of a volume V composed of subvolumes v_n (see Eq. 31.6) is located using Eq. 31.7 through Eq. 31.9, which are analogous to the equations used for centroids of areas and lines.

A solid body will have both a center of gravity and a centroid, but the locations of these two points will not necessarily coincide. The earth's attractive force, which is called *weight*, can be assumed to act through the *center of gravity* (also known as the *center of mass*). Only when the body is homogeneous will the *centroid of a volume* coincide with the center of gravity.

Example

The structure shown is formed of three separate solid aluminum cylindrical rods, each with a 1 cm diameter.

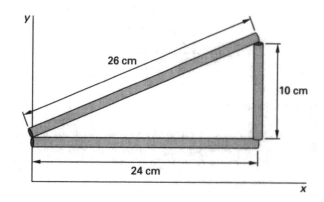

What is the approximate x-coordinate of the centroid of the structure?

(A) 14.0 cm

(B) 15.2 cm

(C) 15.9 cm

(D) 16.0 cm

Solution

Use Eq. 31.7.

$$V_1 = \left(\frac{\pi}{4}\right)(1 \text{ cm})^2 (24 \text{ cm}) = 18.85 \text{ cm}^3$$

$$V_2 = \left(\frac{\pi}{4}\right)(1 \text{ cm})^2 (10 \text{ cm}) = 7.85 \text{ cm}^3$$

$$V_3 = \left(\frac{\pi}{4}\right)(1 \text{ cm})^2 (26 \text{ cm}) = 20.42 \text{ cm}^3$$

$$V = 18.85 \text{ cm}^3 + 7.85 \text{ cm}^3 + 20.42 \text{ cm}^3$$

$$= 47.12 \text{ cm}^3$$

Engineering Sciences

$x_n = \left(\sum x_{vc} v_n\right)/V$

$$= \frac{\left(\dfrac{24\ \text{cm}}{2}\right)(18.85\ \text{cm}^3) + (24\ \text{cm})(7.85\ \text{cm}^3) + \left(\dfrac{24\ \text{cm}}{2}\right)(20.42\ \text{cm}^3)}{47.12\ \text{cm}^3}$$

$$= 14.0\ \text{cm}$$

(The $\pi/4$ and area terms all cancel and could have been omitted.)

The answer is (A).

Equation 31.10 Through Eq. 31.50: Centroid and Area Moments of Inertia for Right Triangles

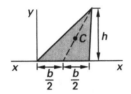

area and centroid

$A = bh/2$	31.10
$x_c = 2b/3$	31.11
$y_c = h/3$	31.12

area moment of inertia

$I_{x_c} = bh^3/36$	31.13
$I_{y_c} = b^3h/36$	31.14
$I_x = bh^3/12$	31.15
$I_y = b^3h/4$	31.16

(radius of gyration)^2

$r_{x_c}^2 = h^2/18$	31.17
$r_{y_c}^2 = b^2/18$	31.18
$r_x^2 = h^2/6$	31.19
$r_y^2 = b^2/2$	31.20

product of inertia

$I_{x_c y_c} = Abh/36 = b^2h^2/72$	31.21
$I_{xy} = Abh/4 = b^2h^2/8$	31.22

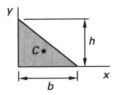

area and centroid

$A = bh/2$	31.23
$x_c = b/3$	31.24
$y_c = h/3$	31.25

area moment of inertia

$I_{x_c} = bh^3/36$	31.26
$I_{y_c} = b^3h/36$	31.27
$I_x = bh^3/12$	31.28
$I_y = b^3h/12$	31.29

(radius of gyration)^2

$r_{x_c}^2 = h^2/18$	31.30
$r_{y_c}^2 = b^2/18$	31.31
$r_x^2 = h^2/6$	31.32
$r_y^2 = b^2/6$	31.33

product of inertia

$I_{x_c y_c} = -Abh/36 = -b^2h^2/72$	31.34
$I_{xy} = Abh/12 = b^2h^2/24$	31.35

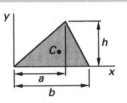

area and centroid

$A = bh/2$	31.36
$x_c = (a+b)/3$	31.37
$y_c = h/3$	31.38

area moment of inertia

$I_{x_c} = bh^3/36$	31.39
$I_{y_c} = [bh(b^2 - ab + a^2)]/36$	31.40
$I_x = bh^3/12$	31.41
$I_y = [bh(b^2 + ab + a^2)]/12$	31.42

Engineering Sciences

(radius of gyration)^2

$$r_{x_c}^2 = h^2/18 \qquad \textbf{31.43}$$

$$r_{y_c}^2 = (b^2 - ab + a^2)/18 \qquad \textbf{31.44}$$

$$r_x^2 = h^2/6 \qquad \textbf{31.45}$$

$$r_y^2 = (b^2 + ab + a^2)/6 \qquad \textbf{31.46}$$

product of inertia

$$I_{x_c y_c} = [Ah(2a - b)]/36 \qquad \textbf{31.47}$$

$$I_{x_c y_c} = [bh^2(2a - b)]/72 \qquad \textbf{31.48}$$

$$I_{xy} = [Ah(2a + b)]/12 \qquad \textbf{31.49}$$

$$I_{xy} = [bh^2(2a + b)]/24 \qquad \textbf{31.50}$$

Description

Equation 31.10 through Eq. 31.50 give the areas, centroids, and moments of inertia for triangles.

The traditional moments of inertia, I_x and I_y (i.e., the second moments of the area), are always positive. However, the *product of inertia*, $I_{x_c y_c}$, listed in Eq. 31.34, is negative. Since the product of inertia is calculated as $I_{xy} = \sum x_i y_i A_i$, where the x_i and y_i are distances from the composite centroid to the subarea A_i, and since these distances can be either positive or negative depending on where the centroid is located, the product of inertia can be either positive or negative.

Example

If a triangle has a base of 13 cm and a height of 8 cm, what is most nearly the vertical distance between the centroid and the radius of gyration about the x-axis?

(A) 0.5 cm

(B) 0.6 cm

(C) 0.7 cm

(D) 0.8 cm

Solution

From Eq. 31.38, the y-component of the centroidal location is

$$y_c = h/3 = \frac{8 \text{ cm}}{3}$$

$$= 2.667 \text{ cm}$$

From Eq. 31.45, the radius of gyration about the x-axis is

$$r_x = \sqrt{\frac{h^2}{6}} = \sqrt{\frac{(8 \text{ cm})^2}{6}}$$

$$= 3.266 \text{ cm}$$

The vertical separation between these two points is

$$r_x - y_c = 3.266 \text{ cm} - 2.667 \text{ cm}$$

$$= 0.599 \text{ cm} \quad (0.6 \text{ cm})$$

The answer is (B).

Equation 31.51 Through Eq. 31.62: Centroid and Area Moments of Inertia for Rectangles

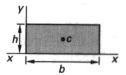

area and centroid

$$A = bh \qquad \textbf{31.51}$$

$$x_c = b/2 \qquad \textbf{31.52}$$

$$y_c = h/2 \qquad \textbf{31.53}$$

area moment of inertia

$$I_x = bh^3/3 \qquad \textbf{31.54}$$

$$I_{x_c} = bh^3/12 \qquad \textbf{31.55}$$

$$J = [bh(b^2 + h^2)]/12 \qquad \textbf{31.56}$$

(radius of gyration)^2

$$r_x^2 = h^2/3 \qquad \textbf{31.57}$$

$$r_{x_c}^2 = h^2/12 \qquad \textbf{31.58}$$

$$r_y^2 = b^2/3 \qquad \textbf{31.59}$$

$$r_p^2 = (b^2 + h^2)/12 \qquad \textbf{31.60}$$

product of inertia

$$I_{x_c y_c} = 0 \qquad \textbf{31.61}$$

$$I_{xy} = Abh/4 = b^2 h^2/4 \qquad \textbf{31.62}$$

Engineering Sciences

Description

Equation 31.51 through Eq. 31.62 give the area, centroids, and moments of inertia for rectangles.

Example

A 12 cm wide × 8 cm high rectangle is placed such that its centroid is located at the origin, $(0,0)$. What is the percentage change in the product of inertia if the rectangle is rotated 90° counterclockwise about the origin?

(A) −32% (decrease)

(B) 0%

(C) 32% (increase)

(D) 64% (increase)

Solution

The product of inertia is zero whenever one or more of the reference axes are lines of symmetry. In this case, both axes are lines of symmetry before and after the rotation. From Eq. 31.61, $I_{x_c y_c} = 0$.

The answer is (B).

Equation 31.63 Through Eq. 31.68: Centroid and Area Moments of Inertia for Trapezoids

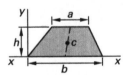

area and centroid

$$A = h(a+b)/2 \qquad 31.63$$

$$y_c = \frac{h(2a+b)}{3(a+b)} \qquad 31.64$$

area moment of inertia

$$I_{x_c} = \frac{h^3(a^2 + 4ab + b^2)}{36(a+b)} \qquad 31.65$$

$$I_x = \frac{h^3(3a+b)}{12} \qquad 31.66$$

$(radius\ of\ gyration)^2$

$$r_{x_c}^2 = \frac{h^2(a^2 + 4ab + b^2)}{18(a+b)} \qquad 31.67$$

$$r_x^2 = \frac{h^2(3a+b)}{6(a+b)} \qquad 31.68$$

Description

Equation 31.63 through Eq. 31.68 give the area, centroids, and moments of inertia for trapezoids.

Example

What are most nearly the area and the y-coordinate, respectively, of the centroid of the trapezoid shown?

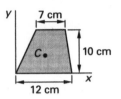

(A) 95 cm^2; 4.6 cm

(B) 110 cm^2; 5.4 cm

(C) 120 cm^2; 6.1 cm

(D) 140 cm^2; 7.2 cm

Solution

The area of the trapezoid is

$$A = h(a+b)/2 = \frac{(10\text{ cm})(7\text{ cm} + 12\text{ cm})}{2}$$

$$= 95\text{ cm}^2$$

From Eq. 31.64, the y-coordinate of the centroid of the trapezoid is

$$y_c = \frac{h(2a+b)}{3(a+b)}$$

$$= \frac{(10\text{ cm})\big((2)(7\text{ cm}) + 12\text{ cm}\big)}{(3)(7\text{ cm} + 12\text{ cm})}$$

$$= 4.56\text{ cm} \quad (4.6\text{ cm})$$

The answer is (A).

Equation 31.69 Through Eq. 31.80: Centroid and Area Moments of Inertia for Rhomboids

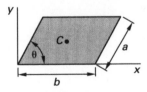

area and centroid

$$A = ab\sin\theta \qquad 31.69$$

$$x_c = (b + a\cos\theta)/2 \qquad 31.70$$

$$y_c = (a\sin\theta)/2 \qquad 31.71$$

area moment of inertia

$$I_{x_c} = (a^3 b \sin^3 \theta)/12 \qquad 31.72$$

$$I_{y_c} = [ab \sin \theta (b^2 + a^2 \cos^2 \theta)]/12 \qquad 31.73$$

$$I_x = (a^3 b \sin^3 \theta)/3 \qquad 31.74$$

$$I_y = [ab \sin \theta (b + a \cos \theta)^2]/3 - (a^2 b^2 \sin \theta \cos \theta)/6 \qquad 31.75$$

(radius of gyration)²

$$r_{x_c}^2 = (a \sin \theta)^2/12 \qquad 31.76$$

$$r_{y_c}^2 = (b^2 + a^2 \cos^2 \theta)/12 \qquad 31.77$$

$$r_x^2 = (a \sin \theta)^2/3 \qquad 31.78$$

$$r_y^2 = (b + a \cos \theta)^2/3 - (ab \cos \theta)/6 \qquad 31.79$$

product of inertia

$$I_{x_c y_c} = (a^3 b \sin^2 \theta \cos \theta)/12 \qquad 31.80$$

Description

Equation 31.69 through Eq. 31.80 give the area, centroids, and moments of inertia for rhomboids.

Equation 31.81 Through Eq. 31.116: Centroid and Area Moments of Inertia for Circles[4]

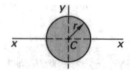

area and centroid

$$A = \pi a^2 \qquad 31.81$$

$$x_c = a \qquad 31.82$$

$$y_c = a \qquad 31.83$$

area moment of inertia

$$I_{x_c} = I_{y_c} = \pi a^4/4 \qquad 31.84$$

$$I_x = I_y = 5\pi a^4/4 \qquad 31.85$$

$$J = \pi r^4/2 \qquad 31.86$$

(radius of gyration)²

$$r_{x_c}^2 = r_{y_c}^2 = a^2/4 \qquad 31.87$$

$$r_x^2 = r_y^2 = 5a^2/4 \qquad 31.88$$

$$r_p^2 = a^2/2 \qquad 31.89$$

product of inertia

$$I_{x_c y_c} = 0 \qquad 31.90$$

$$I_{xy} = Aa^2 \qquad 31.91$$

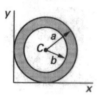

area and centroid

$$A = \pi(a^2 - b^2) \qquad 31.92$$

$$x_c = a \qquad 31.93$$

$$y_c = a \qquad 31.94$$

area moment of inertia

$$I_{x_c} = I_{y_c} = \pi(a^4 - b^4)/4 \qquad 31.95$$

$$I_x = I_y = \frac{5\pi a^4}{4} - \pi a^2 b^2 - \frac{\pi b^4}{4} \qquad 31.96$$

$$J = \pi(r_a^4 - r_b^4)/2 \qquad 31.97$$

(radius of gyration)²

$$r_{x_c}^2 = r_{y_c}^2 = (a^2 + b^2)/4 \qquad 31.98$$

$$r_x^2 = r_y^2 = (5a^2 + b^2)/4 \qquad 31.99$$

$$r_p^2 = (a^2 + b^2)/2 \qquad 31.100$$

product of inertia

$$I_{x_c y_c} = 0 \qquad 31.101$$

$$I_{xy} = Aa^2 \qquad 31.102$$

$$I_{xy} = \pi a^2(a^2 - b^2) \qquad 31.103$$

[4]In Eq. 31.81 through Eq. 31.116, the *NCEES Handbook* designates the radius of a circle or circular segment as a, rather than as the conventional r or R, which are used almost everywhere else in the *NCEES Handbook*.

Engineering Sciences

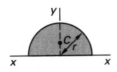

area and centroid

$$A = \pi a^2/2 \qquad \textit{31.104}$$

$$x_c = a \qquad \textit{31.105}$$

$$y_c = 4a/3\pi \qquad \textit{31.106}$$

area moment of inertia

$$I_{x_c} = \frac{a^4(9\pi^2 - 64)}{72\pi} \qquad \textit{31.107}$$

$$I_{y_c} = \pi a^4/8 \qquad \textit{31.108}$$

$$I_x = \pi a^4/8 \qquad \textit{31.109}$$

$$I_y = 5\pi a^4/8 \qquad \textit{31.110}$$

(radius of gyration)^2

$$r_{x_c}^2 = \frac{a^2(9\pi^2 - 64)}{36\pi^2} \qquad \textit{31.111}$$

$$r_{y_c}^2 = a^2/4 \qquad \textit{31.112}$$

$$r_x^2 = a^2/4 \qquad \textit{31.113}$$

$$r_y^2 = 5a^2/4 \qquad \textit{31.114}$$

product of inertia

$$I_{x_c y_c} = 0 \qquad \textit{31.115}$$

$$I_{xy} = 2a^4/3 \qquad \textit{31.116}$$

Description

Equation 31.81 through Eq. 31.116 give the area, centroids, and moments of inertia for circles.

Example

The center of a circle with a radius of 7 cm is located at $(x, y) = (3\text{ cm}, 4\text{ cm})$. Most nearly, what is the minimum distance that the origin of the x- and y-axes would have to be moved in order to reduce the product of inertia to its smallest absolute value?

(A) 3 cm

(B) 4 cm

(C) 5 cm

(D) 7 cm

Solution

The product of inertia of a circle can be a positive value, a negative value, or zero, depending on the location of the axes. The absolute value is zero (i.e., is minimized) when at least one of the axes coincides with a line of symmetry. Although this can be accomplished by moving the origin to the center of the circle (a distance of 5 cm recognizing that this is a 3-4-5 triangle), a shorter move results when the y-axis is moved 3 cm to the right. Then, the y-axis passes through the centroid, which is sufficient to reduce the product of inertia to zero.

The answer is (A).

Equation 31.117 Through Eq. 31.125: Centroid and Area Moments of Inertia for Circular Sectors

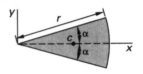

area and centroid

$$A = a^2\theta \qquad \textit{31.117}$$

$$x_c = \frac{2a}{3}\frac{\sin\theta}{\theta} \qquad \textit{31.118}$$

$$y_c = 0 \qquad \textit{31.119}$$

area moment of inertia

$$I_x = a^4(\theta - \sin\theta\cos\theta)/4 \qquad \textit{31.120}$$

$$I_y = a^4(\theta + \sin\theta\cos\theta)/4 \qquad \textit{31.121}$$

(radius of gyration)^2

$$r_x^2 = \frac{a^2}{4}\frac{(\theta - \sin\theta\cos\theta)}{\theta} \qquad \textit{31.122}$$

$$r_y^2 = \frac{a^2}{4}\frac{(\theta + \sin\theta\cos\theta)}{\theta} \qquad \textit{31.123}$$

product of inertia

$$I_{x_c y_c} = 0 \qquad \textit{31.124}$$

$$I_{xy} = 0 \qquad \textit{31.125}$$

Description

Equation 31.117 through Eq. 31.125 give the area, centroids, and moments of inertia for circular sectors. In order to incorporate θ into the calculations, as is done for some of the circular sector equations, the angle must be expressed in radians.

$$\theta_{\text{rad}} = \theta_{\text{deg}}\left(\frac{2\pi}{360°}\right)$$

Example

A grassy parcel of land shaped like a rhomboid has adjacent sides measuring 50 m and 120 m with a 65° included angle. A small, straight creek runs between the opposing acute corners. A goat is humanely tied to the bank of the creek at one of the acute corners by a 40 m long rope. Without crossing the creek, most nearly, on what area of grass can the goat graze?

(A) 450 m²

(B) 710 m²

(C) 910 m²

(D) 26 000 m²

Solution

The creek bisects the 65° angle. The goat sweeps out a circular sector with a 40 m radius.

$$\theta = \frac{\text{swept angle}}{2} = \left(\frac{\frac{65°}{2}}{2}\right)\left(\frac{2\pi}{360°}\right) = 0.2836 \text{ rad}$$

Use Eq. 31.117. The swept area is

$$A = a^2\theta = (40 \text{ m})^2(0.2836 \text{ rad}) = 453.8 \text{ m}^2 \quad (450 \text{ m}^2)$$

The answer is (A).

Equation 31.126 Through Eq. 31.134: Centroid and Area Moments of Inertia for Circular Segments

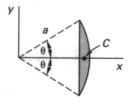

area and centroid

$$A = a^2\left[\theta - \frac{\sin 2\theta}{2}\right] \qquad 31.126$$

$$x_c = \frac{2a}{3}\frac{\sin^3\theta}{\theta - \sin\theta\cos\theta} \qquad 31.127$$

$$y_c = 0 \qquad 31.128$$

area moment of inertia

$$I_x = \frac{Aa^2}{4}\left[1 - \frac{2\sin^3\theta\cos\theta}{3\theta - 3\sin\theta\cos\theta}\right] \qquad 31.129$$

$$I_y = \frac{Aa^2}{4}\left[1 + \frac{2\sin^3\theta\cos\theta}{\theta - \sin\theta\cos\theta}\right] \qquad 31.130$$

(radius of gyration)²

$$r_x^2 = \frac{a^2}{4}\left[1 - \frac{2\sin^3\theta\cos\theta}{3\theta - 3\sin\theta\cos\theta}\right] \qquad 31.131$$

$$r_y^2 = \frac{a^2}{4}\left[1 + \frac{2\sin^3\theta\cos\theta}{\theta - \sin\theta\cos\theta}\right] \qquad 31.132$$

product of inertia

$$I_{x_c y_c} = 0 \qquad 31.133$$

$$I_{xy} = 0 \qquad 31.134$$

Description

Equation 31.126 through Eq. 31.134 give the area, centroids, and moments of inertia for circular segments.

Equation 31.135 Through Eq. 31.145: Centroid and Area Moments of Inertia for Parabolas

area and centroid

$$A = 4ab/3 \qquad 31.135$$

$$x_c = 3a/5 \qquad 31.136$$

$$y_c = 0 \qquad 31.137$$

area moment of inertia

$$I_{x_c} = I_x = 4ab^3/15 \qquad 31.138$$

$$I_{y_c} = 16a^3b/175 \qquad 31.139$$

$$I_y = 4a^3b/7 \qquad 31.140$$

(radius of gyration)²

$$r_{x_c}^2 = r_x^2 = b^2/5 \qquad 31.141$$

$$r_{y_c}^2 = 12a^2/175 \qquad 31.142$$

$$r_y^2 = 3a^2/7 \qquad 31.143$$

product of inertia

$$I_{x_c y_c} = 0 \qquad 31.144$$

$$I_{xy} = 0 \qquad 31.145$$

Description

Equation 31.135 through Eq. 31.145 give the area, centroids, and moments of inertia for parabolas.

Example

The entrance freeway to a city passes under a decorative parabolic arch with a 28 m base and a 200 m height. A famous illusionist contacts the city with a plan to make the city disappear from behind a curtain draped down from the arch. If the drape spans the entire width and height of the opening, and if seams and reinforcement increase the material requirements by 15%, most nearly, how much drapery fabric will be needed?

(A) 3700 m^2

(B) 4300 m^2

(C) 7500 m^2

(D) 8600 m^2

Solution

b is half of the width of the arch.

$$b = \frac{28 \text{ m}}{2} = 14 \text{ m}$$

Use Eq. 31.135. Including the allowance for seams and reinforcement, the required area is

$$A = (1 + \text{allowance})\frac{4ab}{3}$$

$$= (1 + 0.15)\left(\frac{(4)\Big(200 \text{ m}\Big)(14 \text{ m})}{3}\right)$$

$$= 4293 \text{ m}^2 \quad (4300 \text{ m}^2)$$

The answer is (B).

Equation 31.146 Through Eq. 31.153: Centroid and Area Moments of Inertia for Semiparabolas

area and centroid

$A = 2ab/3$		*31.146*
$x_c = 3a/5$		*31.147*
$y_c = 3b/8$		*31.148*

area moment of inertia

$I_x = 2ab^3/15$	*31.149*
$I_y = 2ba^3/7$	*31.150*

$(radius\ of\ gyration)^2$

$r_x^2 = b^2/5$	*31.151*
$r_y^2 = 3a^2/7$	*31.152*

product of inertia

$I_{xy} = Aab/4 = a^2 b^2$	*31.153*

Description

Equation 31.146 through Eq. 31.153 give the area, centroids, and moments of inertia for semiparabolas.

Example

What is most nearly the area of the shaded section above the parabolic curve shown?

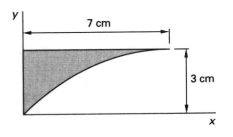

(A) 7 cm^2

(B) 9 cm^2

(C) 11 cm^2

(D) 14 cm^2

Solution

From Eq. 31.146, the semiparabolic area below the curve is

$$A_{\text{below}} = 2ab/3 = \frac{(2)(3 \text{ cm})(7 \text{ cm})}{3}$$

$$= 14 \text{ cm}^2$$

The shaded area above the parabolic curve is

$$A_{\text{above}} = A - A_{\text{below}} = (7 \text{ cm})(3 \text{ cm}) - 14 \text{ cm}^2$$

$$= 7 \text{ cm}^2$$

The answer is (A).

Equation 31.154 Through Eq. 31.160: Centroid and Area Moments of Inertia for General Spandrels (*n*th Degree Parabolas)

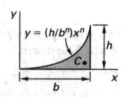

area and centroid

$$A = bh/(n+1) \qquad 31.154$$

$$x_c = \frac{n+1}{n+2} b \qquad 31.155$$

$$y_c = \frac{h}{2}\frac{n+1}{2n+1} \qquad 31.156$$

area moment of inertia

$$I_x = \frac{bh^3}{3(3n+1)} \qquad 31.157$$

$$I_y = \frac{hb^3}{n+3} \qquad 31.158$$

(radius of gyration)^2

$$r_x^2 = \frac{h^2(n+1)}{3(3n+1)} \qquad 31.159$$

$$r_y^2 = \frac{n+1}{n+3} b^2 \qquad 31.160$$

Description

Equation 31.154 through Eq. 31.160 give the area, centroids, and moments of inertia for general spandrels.

Example

For the curve $y = x^3$, what are the approximate coordinates of the centroid of the shaded area between $x = 0$ and $x = 3$ cm?

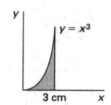

(A) 1.6 cm; 7.8 cm

(B) 1.8 cm; 5.8 cm

(C) 2.0 cm; 18 cm

(D) 2.4 cm; 7.7 cm

Solution

Treat x as the base and y as the height. n is 3 for this spandrel. The height is $h = x^3 = (3)^3 = 27$ cm. The x- and y-coordinates, respectively, are

$$x_c = \left(\frac{n+1}{n+2}\right)b = \left(\frac{3+1}{3+2}\right)(3\text{ cm}) = 2.4\text{ cm}$$

$$y_c = \left(\frac{h}{2}\right)\left(\frac{n+1}{2n+1}\right) = \left(\frac{27\text{ cm}}{2}\right)\left(\frac{3+1}{(2)(3)+1}\right)$$

$$= 7.714\text{ cm} \quad (7.7\text{ cm})$$

The answer is (D).

Equation 31.161 Through Eq. 31.167: Centroids and Area Moments of Inertia for *n*th Degree Parabolas

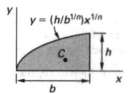

area and centroid

$$A = \frac{n}{n+1} bh \qquad 31.161$$

$$x_c = \frac{n+1}{2n+1} b \qquad 31.162$$

$$y_c = \frac{n+1}{2(n+2)} h \qquad 31.163$$

area moment of inertia

$$I_x = \frac{n}{3(n+3)} bh^3 \qquad 31.164$$

$$I_y = \frac{n}{3n+1} b^3 h \qquad 31.165$$

(radius of gyration)^2

$$r_x^2 = \frac{n+1}{3(n+1)} h^2 \qquad 31.166$$

$$r_y^2 = \frac{n+1}{3n+1} b^2 \qquad 31.167$$

Description

Equation 31.161 through Eq. 31.167 give the area, centroids, and moments of inertia for *n*th degree parabolas.

Equation 31.168: Centroid of a Volume

$$\mathbf{r}_c = \sum m_n \mathbf{r}_n / \sum m_n \qquad \textit{31.168}$$

Description

Equation 31.168 provides a convenient method of locating the centroid of an object that consists of several isolated component masses. The masses do not have to be contiguous and can be distributed throughout space. It is implicit that the vectors that terminate at the submasses' centroids are based at the origin, $(0, 0, 0)$. These vectors have the form of $r_x\mathbf{i} + r_y\mathbf{j} + r_z\mathbf{k}$. The end result is a vector, but since the vector is based at the origin, the vector components can be interpreted as coordinates, (r_{cx}, r_{cy}, r_{cz}).

2. MOMENT OF INERTIA

The *moment of inertia*, I, of an area is needed in mechanics of materials problems. It is convenient to think of the moment of inertia of a beam's cross-sectional area as a measure of the beam's ability to resist bending. Given equal loads, a beam with a small moment of inertia will bend more than a beam with a large moment of inertia.

Since the moment of inertia represents a resistance to bending, it is always positive. Since a beam can be asymmetric in cross section (e.g., a rectangular beam) and be stronger in one direction than another, the moment of inertia depends on orientation. A reference axis or direction must be specified.

The symbol I_x is used to represent a moment of inertia with respect to the x-axis. Similarly, I_y is the moment of inertia with respect to the y-axis. I_x and I_y do not combine and are not components of some resultant moment of inertia.

Any axis can be chosen as the reference axis, and the value of the moment of inertia will depend on the reference selected. The moment of inertia taken with respect to an axis passing through the area's centroid is known as the *centroidal moment of inertia*, I_{x_c} or I_{y_c}. The centroidal moment of inertia is the smallest possible moment of inertia for the area.

Equation 31.169 and Eq. 31.170: Second Moment of the Area

$$I_y = \int x^2 \, dA \qquad \textit{31.169}$$

$$I_x = \int y^2 \, dA \qquad \textit{31.170}$$

Description

Integration can be used to calculate the moment of inertia of a function that is bounded by the x- and y-axes and a curve $y = f(x)$. From Eq. 31.169 and

Eq. 31.170, it is apparent why the moment of inertia is also known as the *second moment of the area* or *second area moment*.

Equation 31.171 and Eq. 31.172: Perpendicular Axis Theorem

$$I_z = J = I_y + I_x = \int (x^2 + y^2) \, dA \qquad \textit{31.171}$$

$$I_z = r_p^2 A \qquad \textit{31.172}$$

Variation

$$J_c = I_{x_c} + I_{y_c}$$

Description

The *polar moment of inertia*, J or I_z, is required in torsional shear stress calculations. It can be thought of as a measure of an area's resistance to torsion (twisting). The definition of a polar moment of inertia of a two-dimensional area requires three dimensions because the reference axis for a polar moment of inertia of a plane area is perpendicular to the plane area.

The polar moment of inertia can be derived from Eq. 31.171.

It is often easier to use the perpendicular axis theorem to quickly calculate the polar moment of inertia.

Perpendicular axis theorem: The moment of inertia of a plane area about an axis normal to the plane is equal to the sum of the moments of inertia about any two mutually perpendicular axes lying in the plane and passing through the given axis.

Since the two perpendicular axes can be chosen arbitrarily, it is most convenient to use the centroidal moments of inertia, as shown in the variation equation.

Example

For the composite plane area made up of two circles as shown, the moment of inertia about the y-axis is 4.7 cm^4, and the moment of inertia about the x-axis is 23.5 cm^4.

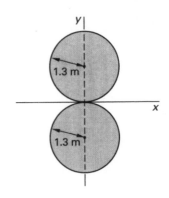

What is the approximate polar moment of inertia of the area taken about the intersection of the x- and y-axes?

(A) 0 cm^4

(B) 14 cm^4

(C) 28 cm^4

(D) 34 cm^4

Solution

Use the perpendicular axis theorem, as given by Eq. 31.171.

$$J = I_y + I_x$$
$$= 4.7 \text{ cm}^4 + 23.5 \text{ cm}^4$$
$$= 28.2 \text{ cm}^4 \quad (28 \text{ cm}^4)$$

The answer is (C).

Equation 31.173 and Eq. 31.174: Parallel Axis Theorem

$$I'_x = I_{x_c} + d_x^2 A \qquad \textit{31.173}$$
$$I'_y = I_{y_c} + d_y^2 A \qquad \textit{31.174}$$

Description

If the moment of inertia is known with respect to one axis, the moment of inertia with respect to another, parallel axis can be calculated from the *parallel axis theorem*, also known as the *transfer axis theorem*. This theorem is used to evaluate the moment of inertia of areas that are composed of two or more basic shapes. d is the distance between the centroidal axis and the second, parallel axis.

The second term in Eq. 31.173 and Eq. 31.174 is often much larger than the first term in each equation, since areas close to the centroidal axis do not affect the moment of inertia considerably. This principle is exploited in the design of structural steel shapes that derive bending resistance from *flanges* located far from the centroidal axis. The *web* does not contribute significantly to the moment of inertia. (See Fig. 31.1.)

Figure 31.1 *Structural Steel Shape*

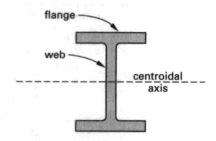

Example

The moment of inertia about the x'-axis of the cross section shown is $334\,000 \text{ cm}^4$. The cross-sectional area is 86 cm^2, and the thicknesses of the web and the flanges are the same.

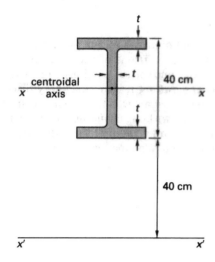

What is most nearly the moment of inertia about the centroidal axis?

(A) $2.4 \times 10^4 \text{ cm}^4$

(B) $7.4 \times 10^4 \text{ cm}^4$

(C) $2.0 \times 10^5 \text{ cm}^4$

(D) $6.4 \times 10^5 \text{ cm}^4$

Solution

Use Eq. 31.173. The moment of inertia around the centroidal axis is

$$I'_x = I_{x_c} + d_x^2 A$$
$$I_{x_c} = I'_x - d_x^2 A$$
$$= 334\,000 \text{ cm}^4 - (86 \text{ cm}^2)\left(40 \text{ cm} + \frac{40 \text{ cm}}{2}\right)^2$$
$$= 24\,400 \text{ cm}^4 \quad (2.4 \times 10^4 \text{ cm}^4)$$

The answer is (A).

Equation 31.175 Through Eq. 31.177: Radius of Gyration

$$r_x = \sqrt{I_x / A} \qquad \textit{31.175}$$
$$r_y = \sqrt{I_y / A} \qquad \textit{31.176}$$
$$r_p = \sqrt{J / A} \qquad \textit{31.177}$$

Engineering Sciences

Variations

$$I = r^2 A$$

$$r_p^2 = r_x^2 + r_y^2$$

Description

Every nontrivial area has a centroidal moment of inertia. Usually, some portions of the area are close to the centroidal axis, and other portions are farther away. The *radius of gyration*, r, is an imaginary distance from the centroidal axis at which the entire area can be assumed to exist without changing the moment of inertia. Despite the name "radius," the radius of gyration is not limited to circular shapes or polar axes. This concept is illustrated in Fig. 31.2.

Figure 31.2 *Radius of Gyration of Two Equivalent Areas*

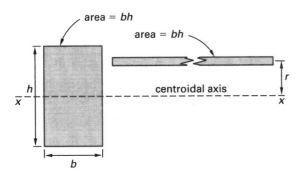

The radius of gyration, r, is given by Eq. 31.175 and Eq. 31.176. The analogous quantity in the polar system is calculated using Eq. 31.177.

Just as the polar moment of inertia, J, can be calculated from the two rectangular moments of inertia, the polar radius of gyration can be calculated from the two rectangular radii of gyration, as shown in the second variation equation.

Example

For the shape shown, the centroidal moment of inertia about an axis parallel to the x-axis is 57.9 cm^4.

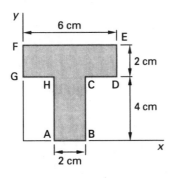

What is the approximate radius of gyration about a horizontal axis passing through the centroid?

(A) 0.86 cm

(B) 1.7 cm

(C) 2.3 cm

(D) 3.7 cm

Solution

The area is

$$A = (2 \text{ cm})(4 \text{ cm}) + (2 \text{ cm})(6 \text{ cm}) = 20 \text{ cm}^2$$

By definition, the radius of gyration is calculated with respect to the centroidal axis. From Eq. 31.175,

$$r_x = \sqrt{I_{x_c}/A} = \sqrt{\frac{57.9 \text{ cm}^4}{20 \text{ cm}^2}} = 1.70 \text{ cm} \quad (1.7 \text{ cm})$$

The answer is (B).

Equation 31.178 and Eq. 31.179: Product of Inertia

$$I_{xy} = \int xy \, dA \qquad \text{31.178}$$

$$I'_{xy} = I_{x_c y_c} + d_x d_y A \qquad \text{31.179}$$

Description

The *product of inertia*, I_{xy}, of a two-dimensional area is found by multiplying each differential element of area by its x- and y-coordinate and then summing over the entire area.

The product of inertia is zero when either axis is an axis of symmetry. Since the axes can be chosen arbitrarily, the area may be in one of the negative quadrants, and the product of inertia may be negative.

The transfer theorem for products of inertia is given by Eq. 31.179. (Both axes are allowed to move to new positions.) d_x and d_y are the distances to the centroid in the new coordinate system, and $I_{x_c y_c}$ is the centroidal product of inertia in the old system.

32 Indeterminate Statics

Nomenclature

I	moment of inertia	m^4
j	number of joints	–
m	number of members (bars)	–
r	number of reactions	–
s	number of special conditions	–

1. INTRODUCTION TO INDETERMINATE STATICS

A structure that is *statically indeterminate* is one for which the equations of statics are not sufficient to determine all reactions, moments, and internal forces. Additional formulas involving deflection are required to completely determine these variables.

Although there are many configurations of statically indeterminate structures, this chapter is primarily concerned with beams on more than two supports and trusses with more members than are required for rigidity.

2. DEGREE OF INDETERMINACY

The *degree of indeterminacy* (*degree of redundancy*) is equal to the number of reactions or members that would have to be removed in order to make the structure statically determinate. For example, a two-span beam on three simple supports is indeterminate (redundant) to the first degree. The degree of indeterminacy of a pin-connected, two-dimensional truss is

$$I = r + m - 2j$$

The degree of indeterminacy of a pin-connected, three-dimensional truss is

$$I = r + m - 3j$$

Rigid frames have joints that transmit moments. The degree of indeterminacy of two-dimensional rigid plane frames is more complex. In the following equation, s is the number of *special conditions* (also known as the number of *equations of conditions*). s is 1 for each internal hinge or a shear release, 2 for each internal roller, and 0 if neither hinges nor rollers are present.

$$I = r + 3m - 3j - s$$

3. INDETERMINATE BEAMS

Three common configurations of beams can easily be recognized as being statically indeterminate. These are the *continuous beam*, *propped cantilever beam*, and *fixed-end beam*, as illustrated in Fig. 32.1.

Figure 32.1 *Types of Indeterminate Beams*

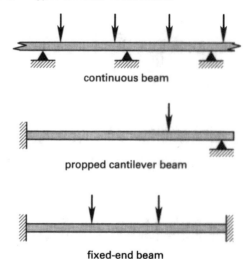

continuous beam

propped cantilever beam

fixed-end beam

Engineering
Sciences

33 Kinematics

Nomenclature

a	acceleration	m/s^2
f	coefficient of friction	–
f	frequency	Hz
g	gravitational acceleration, 9.81	m/s^2
r	position	m
r	radius	m
s	displacement	m
s	distance	m
s	position	m
t	time	s
v	velocity	m/s
x	horizontal distance	m
y	elevation	m

Symbols

α	angular acceleration	rad/s^2
θ	angular position	rad
ρ	radius of curvature	m
ω	angular velocity	rad/s

Subscripts

0	initial
c	constant
f	final
n	normal
r	radial
t	tangential
x	horizontal
y	vertical
θ	transverse

1. INTRODUCTION TO KINEMATICS

Dynamics is the study of moving objects. The subject is divided into kinematics and kinetics. *Kinematics* is the study of a body's motion independent of the forces on the body. It is a study of the geometry of motion without consideration of the causes of motion. Kinematics deals only with relationships among position, velocity, acceleration, and time.

2. PARTICLES AND RIGID BODIES

A body in motion can be considered a *particle* if rotation of the body is absent or insignificant. A particle does not possess rotational kinetic energy. All parts of a particle have the same instantaneous displacement, velocity, and acceleration.

A *rigid body* does not deform when loaded and can be considered a combination of two or more particles that remain at a fixed, finite distance from each other. At any given instant, the parts (particles) of a rigid body can have different displacements, velocities, and accelerations if the body has rotational as well as translational motion.

Equation 33.1 and Eq. 33.2: Instantaneous Velocity and Acceleration

$$\mathbf{v} = d\mathbf{r}/dt \qquad \text{33.1}$$

$$\mathbf{a} = d\mathbf{v}/dt \qquad \text{33.2}$$

Variation

$$\mathbf{a} = \frac{d^2\mathbf{r}}{dt^2}$$

Description

For the position vector of a particle, r, the instantaneous velocity, $\mathbf{v}$, and acceleration, $\mathbf{a}$, are given by Eq. 33.1 and Eq. 33.2, respectively.

Example

The position of a particle moving along the x-axis is given by $r(t) = t^2 - t + 8$, where r is in units of meters

Engineering Sciences

and t is in seconds. What is most nearly the velocity of the particle when $t = 5$ s?

(A) 9.0 m/s

(B) 10 m/s

(C) 11 m/s

(D) 12 m/s

Solution

The velocity equation is the first derivative of the position equation with respect to time.

$$\mathbf{v}(t) = d\mathbf{r}(t)/dt$$
$$= \frac{d}{dt}(t^2 - t + 8)$$
$$= 2t - 1$$
$$\mathbf{v}(5) = (2)(5) - 1$$
$$= 9.0 \text{ m/s}$$

The answer is (A).

3. DISTANCE AND SPEED

The terms "displacement" and "distance" have different meanings in kinematics. *Displacement* (or *linear displacement*) is the net change in a particle's position as determined from the position function, $r(t)$. *Distance traveled* is the accumulated length of the path traveled during all direction reversals, and it can be found by adding the path lengths covered during periods in which the velocity sign does not change. Therefore, distance is always greater than or equal to displacement.

$$s = r(t_2) - r(t_1)$$

Similarly, "velocity" and "speed" have different meanings: *velocity* is a vector, having both magnitude and direction; *speed* is a scalar quantity, equal to the magnitude of velocity. When specifying speed, direction is not considered.

4. RECTANGULAR COORDINATES

The position of a particle is specified with reference to a coordinate system. Three coordinates are necessary to identify the position in three-dimensional space; in two dimensions, two coordinates are necessary. A coordinate can represent a linear position, as in the rectangular coordinate system, or it can represent an angular position, as in the polar system.

Consider the particle shown in Fig. 33.1. Its position, as well as its velocity and acceleration, can be specified in three primary forms: vector form, rectangular coordinate form, and unit vector form.

Figure 33.1 Rectangular Coordinates

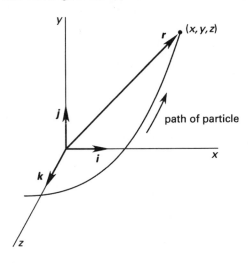

The *vector form* of the particle's position is $\mathbf{r}$, where the vector $\mathbf{r}$ has both magnitude and direction. The *Cartesian coordinate system form* (*rectangular coordinate form*) is (x, y, z).

Equation 33.3: Cartesian Unit Vector Form

$$\mathbf{r} = x\mathbf{i} + y\mathbf{j} + z\mathbf{k} \qquad \text{33.3}$$

Description

The *unit vector form* of a position vector is given by Eq. 33.3.

Example

The position of a particle in Cartesian coordinates over time is $x = 5t$ in the x-direction, $y = 6t$ in the y-direction, and $z = 5t$ in the z-direction. What is the vector form of the particle's position, $\mathbf{r}$?

(A) $\mathbf{r} = 5t\mathbf{i} + 6t\mathbf{j} + 5t\mathbf{k}$

(B) $\mathbf{r} = 5t\mathbf{i} + 6t\mathbf{j} + 6t\mathbf{k}$

(C) $\mathbf{r} = 6t\mathbf{i} + 5t\mathbf{j} + 5t\mathbf{k}$

(D) $\mathbf{r} = 6t\mathbf{i} + 5t\mathbf{j} + 6t\mathbf{k}$

Solution

Using Eq. 33.3, the vector form of the particle's position is

$$\mathbf{r} = x\mathbf{i} + y\mathbf{j} + z\mathbf{k}$$
$$= 5t\mathbf{i} + 6t\mathbf{j} + 5t\mathbf{k}$$

The answer is (A).

5. RECTILINEAR MOTION

Equation 33.4 Through Eq. 33.9: Particle Rectilinear Motion

$$a = \frac{d\mathrm{v}}{dt} \quad \text{[general]} \qquad 33.4$$

$$\mathrm{v} = \frac{ds}{dt} \quad \text{[general]} \qquad 33.5$$

$$a\,ds = \mathrm{v}\,d\mathrm{v} \quad \text{[general]} \qquad 33.6$$

$$\mathrm{v} = \mathrm{v}_0 + a_c t \qquad 33.7$$

$$s = s_0 + \mathrm{v}_0 t + \tfrac{1}{2}a_c t^2 \qquad 33.8$$

$$\mathrm{v}^2 = \mathrm{v}_0^2 + 2a_c(s - s_0) \qquad 33.9$$

Description

A *rectilinear system* is one in which particles move only in straight lines. (Another name is *linear system*.) The relationships among position, velocity, and acceleration for a linear system are given by Eq. 33.4 through Eq. 33.9. Equation 33.4 through Eq. 33.6 show relationships for general (including variable) acceleration of particles.[1] Equation 33.7 through Eq. 33.9 show relationships given constant acceleration, a_c.[2]

When values of time are substituted into these equations, the position, velocity, and acceleration are known as *instantaneous values*.

Equation 33.10 Through Eq. 33.13: Cartesian Velocity and Acceleration

$$\mathbf{v} = \dot{x}\mathbf{i} + \dot{y}\mathbf{j} + \dot{z}\mathbf{k} \qquad 33.10$$

$$\mathbf{a} = \ddot{x}\mathbf{i} + \ddot{y}\mathbf{j} + \ddot{z}\mathbf{k} \qquad 33.11$$

$$\dot{x} = dx/dt = \mathrm{v}_x \qquad 33.12$$

$$\ddot{x} = d^2x/dt^2 = a_x \qquad 33.13$$

Description

The velocity and acceleration are the first two derivatives of the position vector, as shown in Eq. 33.10 and Eq. 33.11.

---

[1]Equation 33.6 can be derived from $a\,dt = d\mathrm{v}$ and $\mathrm{v}\,dt = ds$ by eliminating dt. One scenario where the acceleration depends on position is a particle being accelerated (or decelerated) by a compression spring. The spring force depends on the spring extension, so the acceleration does also.

[2]The NCEES *FE Reference Handbook* (*NCEES Handbook*) is inconsistent in what it uses subscripts to designate. For example, in its Dynamics section, it uses subscripts to designate the location of the accelerating point (e.g., c in a_c for acceleration of the centroid), the direction or related axis (e.g., x in a_x for acceleration in the x-direction), the type of acceleration (e.g., n in a_n for normal acceleration), and the moment in time (e.g., 0 in a_0 for initial acceleration). In Eq. 33.7 through Eq. 33.9, the *NCEES Handbook* uses subscripts to designate the nature of the acceleration (i.e., the subscript c indicates constant acceleration). Elsewhere in the *NCEES Handbook*, the subscript c is used to designate centroid and mass center.

6. CONSTANT ACCELERATION

Equation 33.14 Through Eq. 33.17: Velocity and Displacement with Constant Linear Acceleration

$$a(t) = a_0 \qquad 33.14$$

$$\mathrm{v}(t) = a_0(t - t_0) + \mathrm{v}_0 \qquad 33.15$$

$$s(t) = a_0(t - t_0)^2/2 + \mathrm{v}_0(t - t_0) + s_0 \qquad 33.16$$

$$\mathrm{v}^2 = \mathrm{v}_0^2 + 2a_0(s - s_0) \qquad 33.17$$

Variations

$$\mathrm{v}(t) = a_0 \int dt$$

$$s(t) = a_0 \iint dt^2$$

Description

Acceleration is a constant in many cases, such as a free-falling body with constant acceleration g. If the acceleration is constant, the acceleration term can be taken out of the integrals shown in Sec. 33.5. The initial distance from the origin is s_0; the initial velocity is a constant, v_0; and a constant acceleration is denoted a_0.

Example

In standard gravity, block A exerts a force of $10\,000$ N, and block B exerts a force of 7500 N. Both blocks are initially held stationary. There is no friction, and the pulleys have no mass. Block A has an acceleration of 1.4 m/s^2 once the blocks are released.

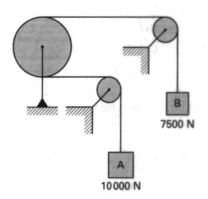

What is most nearly the velocity of block A 2.5 s after the blocks are released?

   (A) 0 m/s

   (B) 3.5 m/s

   (C) 4.4 m/s

   (D) 4.9 m/s

Engineering Sciences

Solution

Use Eq. 33.15 to solve for the velocity of block A.

$$v_A = a_A(t - t_0) + v_0 = \left(1.4 \ \frac{m}{s^2}\right)(2.5 \ s - 0 \ s) + 0 \ \frac{m}{s}$$

$$= 3.5 \ m/s$$

The answer is (B).

Equation 33.18 Through Eq. 33.21: Velocity and Displacement with Constant Angular Acceleration

$$\alpha(t) = \alpha_0 \qquad \text{33.18}$$

$$\omega(t) = \alpha_0(t - t_0) + \omega_0 \qquad \text{33.19}$$

$$\theta(t) = \alpha_0(t - t_0)^2/2 + \omega_0(t - t_0) + \theta_0 \qquad \text{33.20}$$

$$\omega^2 = \omega_0^2 + 2\alpha_0(\theta - \theta_0) \qquad \text{33.21}$$

Description

Equation 33.18 through Eq. 33.21 give the equations for constant angular acceleration.

Example

A flywheel rotates at 7200 rpm when the power is suddenly cut off. The flywheel decelerates at a constant rate of 2.1 rad/s^2 and comes to rest 6 min later. What is most nearly the angular displacement of the flywheel?

(A) 43×10^3 rad

(B) 93×10^3 rad

(C) 140×10^3 rad

(D) 270×10^3 rad

Solution

From Eq. 33.20, the angular displacement is

$$\theta(t) = \alpha_0(t - t_0)^2/2 + \omega_0(t - t_0) + \theta_0$$

$$= \frac{\left(-2.1 \ \frac{rad}{s^2}\right)\left(60 \ \frac{s}{min}\right)^2 (6 \ min - 0 \ min)^2}{2}$$

$$+ \left(7200 \ \frac{rev}{min}\right)\left(2\pi \ \frac{rad}{rev}\right)(6 \ min - 0 \ min)$$

$$+ 0 \ rad$$

$$= 135.4 \times 10^3 \ rad \quad (140 \times 10^3 \ rad)$$

The answer is (C).

7. NON-CONSTANT ACCELERATION

Equation 33.22 and Eq. 33.23: Velocity and Displacement for Non-Constant Acceleration

$$v(t) = \int_{t_0}^{t} a(t) \, dt + v_{t_0} \qquad \text{33.22}$$

$$s(t) = \int_{t_0}^{t} v(t) \, dt + s_{t_0} \qquad \text{33.23}$$

Description

The velocity and displacement, respectively, for non-constant acceleration, $a(t)$, are calculated using Eq. 33.22 and Eq. 33.23.

Example

A particle initially traveling at 10 m/s experiences a linear increase in acceleration in the direction of motion as shown. The particle reaches an acceleration of 20 m/s^2 in 6 seconds.

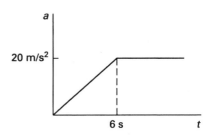

Most nearly, what is the distance traveled by the particle during those 6 seconds?

(A) 60 m

(B) 70 m

(C) 120 m

(D) 180 m

Solution

The expression for the acceleration as a function of time is

$$a(t) = \left(\frac{20 \ \frac{m}{s^2}}{6 \ s}\right) t = \frac{20 \ \frac{m}{s^3}}{6} t$$

From Eq. 33.22, the velocity function is

$$v(t) = \int a(t) \, dt = \int \frac{20}{6} t \, dt = \frac{20}{12} t^2 + C_1$$

Since $v(0) = 10$, $C_1 = 10$.

From Eq. 33.23, the position function is

$$s(t) = \int \mathrm{v}(t)\,dt = \int \left(\frac{20}{12}t^2 + 10\right) dt$$

$$= \frac{20}{36}t^3 + 10t + C_2$$

In a calculation of distance traveled, the initial distance (position) is $s(0) = 0$, so $C_2 = 0$. The distance traveled during the first 6 seconds is

$$s(6) = \int_0^6 \mathrm{v}(t)\,dt = \frac{20}{36}t^3 + 10t \Big|_0^6$$

$$= 180 \text{ m} - 0 \text{ m}$$

$$= 180 \text{ m}$$

The answer is (D).

Equation 33.24 and Eq. 33.25: Variable Angular Acceleration

$$\omega(t) = \int_{t_0}^t \alpha(t)\,dt + \omega_{t_0} \qquad 33.24$$

$$\theta(t) = \int_{t_0}^t \omega(t)\,dt + \theta_{t_0} \qquad 33.25$$

Description

For non-constant angular acceleration, $\alpha(t)$, the angular velocity, ω, and angular displacement, θ, can be calculated from Eq. 33.24 and Eq. 33.25.

8. CURVILINEAR MOTION

Curvilinear motion describes the motion of a particle along a path that is not a straight line. Special examples of curvilinear motion include plane circular motion and projectile motion. For particles traveling along curvilinear paths, the position, velocity, and acceleration may be specified in rectangular coordinates as they were for rectilinear motion, or it may be more convenient to express the kinematic variables in terms of other coordinate systems (e.g., polar coordinates).

9. CURVILINEAR MOTION: PLANE CIRCULAR MOTION

Plane circular motion (also known as *rotational particle motion*, *angular motion*, or *circular motion*) is motion of a particle around a fixed circular path. (See Fig. 33.2.)

Figure 33.2 Plane Circular Motion

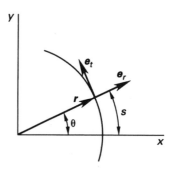

Equation 33.26 Through Eq. 33.31: *x, y, z* Coordinates

$$\mathrm{v}_x = \dot{x} \qquad 33.26$$

$$\mathrm{v}_y = \dot{y} \qquad 33.27$$

$$\mathrm{v}_z = \dot{z} \qquad 33.28$$

$$a_x = \ddot{x} \qquad 33.29$$

$$a_y = \ddot{y} \qquad 33.30$$

$$a_z = \ddot{z} \qquad 33.31$$

Description

Equation 33.26 through Eq. 33.31 give the relationships between acceleration, velocity, and the Cartesian coordinates of a particle in plane circular motion.

Equation 33.32 Through Eq. 33.37: Polar Coordinates

$$\mathrm{v}_r = \dot{r} \qquad 33.32$$

$$\mathrm{v}_\theta = r\dot{\theta} \qquad 33.33$$

$$\mathrm{v}_z = \dot{z} \qquad 33.34$$

$$a_r = \ddot{r} - r\dot{\theta}^2 \qquad 33.35$$

$$a_\theta = r\ddot{\theta} + 2\dot{r}\dot{\theta} \qquad 33.36$$

$$a_z = \ddot{z} \qquad 33.37$$

Description

In *polar coordinates*, the position of a particle is described by a radius, r, and an angle, θ. Equation 33.32 through Eq. 33.37 give the relationships between velocity and acceleration for particles in plane circular motion in a polar coordinate system.

Equation 33.38 Through Eq. 33.41: Rectilinear Forms of Curvilinear Motion

$$v = \dot{s} \qquad \textit{33.38}$$

$$a_t = \dot{v} = \frac{dv}{ds} \qquad \textit{33.39}$$

$$a_n = \frac{v^2}{\rho} \qquad \textit{33.40}$$

$$\rho = \frac{[1 + (dy/dx)^2]^{3/2}}{\left|\dfrac{d^2y}{dx^2}\right|} \qquad \textit{33.41}$$

Description

The relationship between acceleration, velocity, and position in an *ntb coordinate system* is given by Eq. 33.38 through Eq. 33.41.

Equation 33.42 Through Eq. 33.44: Particle Angular Motion

$$\omega = d\theta/dt \qquad \textit{33.42}$$

$$\alpha = d\omega/dt \qquad \textit{33.43}$$

$$\alpha d\theta = \omega d\omega \qquad \textit{33.44}$$

Variation

$$\alpha = \frac{d^2\theta}{dt^2}$$

Description

The behavior of a rotating particle is defined by its angular position, θ, angular velocity, ω, and angular acceleration, α. These variables are analogous to the s, v, and a variables for linear systems. Angular variables can be substituted one-for-one for linear variables in most equations.

Example

The position of a car traveling around a curve is described by the following function of time (in seconds).

$$\theta(t) = t^3 - 2t^2 - 4t + 10$$

What is most nearly the angular velocity after 3 s of travel?

(A) −16 rad/s

(B) −4.0 rad/s

(C) 11 rad/s

(D) 15 rad/s

Solution

The angular velocity is

$$\omega(t) = \frac{d\theta}{dt} = 3t^2 - 4t - 4$$

$$\omega(3) = (3)(3)^2 - (4)(3) - 4$$

$$= 11 \text{ rad/s}$$

The answer is (C).

10. CURVILINEAR MOTION: TRANSVERSE AND RADIAL COMPONENTS FOR PLANAR MOTION

Equation 33.45 Through Eq. 33.49: Polar Coordinate Forms of Curvilinear Motion

$$r = re_r \qquad \textit{33.45}$$

$$v = \dot{r}e_r + r\dot{\theta}e_\theta \qquad \textit{33.46}$$

$$a = (\ddot{r} - r\dot{\theta}^2)e_r + (r\ddot{\theta} + 2\dot{r}\dot{\theta})e_\theta \qquad \textit{33.47}$$

$$\dot{r} = dr/dt \qquad \textit{33.48}$$

$$\ddot{r} = d^2r/dt^2 \qquad \textit{33.49}$$

Variations

$$v = v_r e_r + v_\theta e_\theta$$

$$a = a_r e_r + a_\theta e_\theta$$

Description

The position of a particle in a polar coordinate system may also be expressed as a vector of magnitude r and direction specified by unit vector e_r. Since the velocity of a particle is not usually directed radially out from the center of the coordinate system, it can be divided into two components, called *radial* and *transverse*, which are parallel and perpendicular, respectively, to the unit radial vector. Figure 33.3 illustrates the radial and transverse components of velocity in a polar coordinate system, and the unit radial and unit transverse vectors, e_r and e_θ, used in the vector forms of the motion equations.

Figure 33.3 Radial and Transverse Coordinates

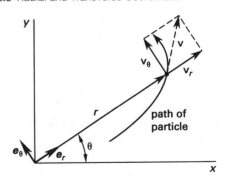

11. CURVILINEAR MOTION: NORMAL AND TANGENTIAL COMPONENTS

Equation 33.50 and Eq. 33.51: Velocity and Resultant Acceleration

$$\mathbf{v} = v(t)\mathbf{e}_t \qquad 33.50$$
$$\mathbf{a} = a(t)\mathbf{e}_t + (v_t^2/\rho)\mathbf{e}_n \qquad 33.51$$

Variation

$$\mathbf{a} = \frac{dv_t}{dt}\mathbf{e}_t + \frac{v_t^2}{\rho}\mathbf{e}_n$$

Description

A particle moving in a curvilinear path will have instantaneous linear velocity and linear acceleration. These linear variables will be directed tangentially to the path, and are known as *tangential velocity*, v_t, and *tangential acceleration*, a_t, respectively. The force that constrains the particle to the curved path will generally be directed toward the center of rotation, and the particle will experience an inward acceleration perpendicular to the tangential velocity and acceleration, known as the *normal acceleration*, a_n. The resultant acceleration, **a**, is the vector sum of the tangential and normal accelerations. Normal and tangential components of acceleration are illustrated in Fig. 33.4. The unit vectors $\mathbf{e_n}$ and $\mathbf{e_t}$ are normal and tangential to the path, respectively. ρ is the instantaneous *radius of curvature*.

Figure 33.4 Normal and Tangential Components

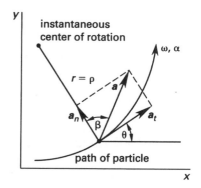

Equation 33.52 Through Eq. 33.56: Vector Quantities for Plane Circular Motion

$$\mathbf{r} = r\mathbf{e}_r \qquad 33.52$$
$$\mathbf{v} = r\omega\mathbf{e}_t \qquad 33.53$$
$$\mathbf{a} = (-r\omega^2)\mathbf{e}_r + r\alpha\mathbf{e}_t \qquad 33.54$$
$$\omega = \dot{\theta} \qquad 33.55$$
$$\alpha = \dot{\omega} = \ddot{\theta} \qquad 33.56$$

Description

For plane circular motion, the vector forms of position, velocity, and acceleration are given by Eq. 33.52, Eq. 33.53, and Eq. 33.54. The magnitudes of the angular velocity and angular acceleration are defined by Eq. 33.55 and Eq. 33.56.

12. RELATIVE MOTION

The term *relative motion* is used when motion of a particle is described with respect to something else in motion. The particle's position, velocity, and acceleration may be specified with respect to another moving particle or with respect to a moving frame of reference, known as a *Newtonian* or *inertial frame of reference*.

Equation 33.57 Through Eq. 33.59: Relative Motion with Translating Axis

$$\mathbf{r}_A = \mathbf{r}_B + \mathbf{r}_{A/B} \qquad 33.57$$
$$\mathbf{v}_A = \mathbf{v}_B + \omega \times \mathbf{r}_{A/B} = \mathbf{v}_B + \mathbf{v}_{A/B} \qquad 33.58$$
$$\mathbf{a}_A = \mathbf{a}_B + \alpha \times \mathbf{r}_{A/B} + \omega \times (\omega \times \mathbf{r}_{A/B}) = \mathbf{a}_B + \mathbf{a}_{A/B} \qquad 33.59$$

Description

The relative position, $\mathbf{r}_A$, velocity, $\mathbf{v}_A$, and acceleration, $\mathbf{a}_A$, with respect to a translating axis can be calculated from Eq. 33.57, Eq. 33.58, and Eq. 33.59, respectively. The angular velocity, ω, and angular acceleration, α, are the magnitudes of the relative position vector, $\mathbf{r}_{A/B}$. (See Fig. 33.5.)

Figure 33.5 Translating Axis

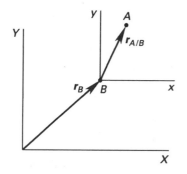

Equation 33.60 Through Eq. 33.62: Relative Motion with Rotating Axis

$$\mathbf{r}_A = \mathbf{r}_B + \mathbf{r}_{A/B} \qquad 33.60$$
$$\mathbf{v}_A = \mathbf{v}_B + \omega \times \mathbf{r}_{A/B} + \mathbf{v}_{A/B} \qquad 33.61$$
$$\mathbf{a}_A = \mathbf{a}_B + \alpha \times \mathbf{r}_{A/B} + \omega \times (\omega \times \mathbf{r}_{A/B})$$
$$+ 2\omega \times \mathbf{v}_{A/B} + \mathbf{a}_{A/B} \qquad 33.62$$

Description

Equation 33.60, Eq. 33.61, and Eq. 33.62 give the relative position, $\mathbf{r}_A$, velocity, $\mathbf{v}_A$, and acceleration, $\mathbf{a}_A$, with respect to a rotating axis, respectively. (See Fig. 33.6.)

Figure 33.6 *Rotating Axis*

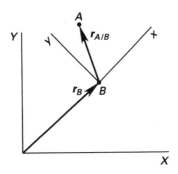

13. LINEAR AND ROTATIONAL VARIABLES

A particle moving in a curvilinear path will also have instantaneous linear velocity and linear acceleration. These linear variables will be directed tangentially to the path and, therefore, are known as *tangential velocity* and *tangential acceleration*, respectively. (See Fig. 33.7.) In general, the linear variables can be obtained by multiplying the rotational variables by the path radius.

Figure 33.7 *Tangential Variables*

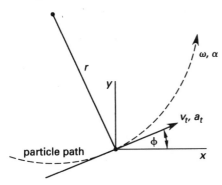

Equation 33.63 Through Eq. 33.66: Relationships Between Linear and Angular Variables

$$v_t = r\omega \qquad \text{33.63}$$

$$a_t = r\alpha \qquad \text{33.64}$$

$$a_n = -r\omega^2 \quad \left[\begin{array}{c}\text{toward the center}\\\text{of the circle}\end{array}\right] \qquad \text{33.65}$$

$$s = r\theta \qquad \text{33.66}$$

Variations

$$v_t = r(2\pi f)$$

$$a_t = \frac{dv_t}{dt}$$

$$a_n = \frac{v_t^2}{r}$$

Description

Equation 33.63 through Eq. 33.65 are used to calculate tangential velocity, v_t, tangential acceleration, a_t, and normal acceleration, a_n, respectively, from their corresponding angular variables. If the path radius, r, is constant, as it would be in rotational motion, the linear distance (i.e., the *arc length*) traveled, s, is calculated from Eq. 33.66.

Example

For the reciprocating pump shown, the radius of the crank is 0.3 m, and the rotational speed is 350 rpm. Two seconds after the pump is activated, the angular position of point A is 35 rad.

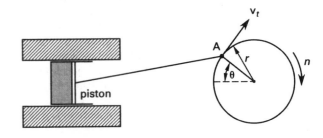

What is most nearly the tangential velocity of point A two seconds after the reciprocating pump is activated?

    (A) 0 m/s

    (B) 1.1 m/s

    (C) 10 m/s

    (D) 11 m/s

Solution

Use the relationship between the tangential and angular variables.

$$\omega = \text{angular velocity of the crank in rad/s}$$

$$= \frac{\left(350\ \frac{\text{rev}}{\text{min}}\right)\left(2\pi\ \frac{\text{rad}}{\text{rev}}\right)}{60\ \frac{\text{s}}{\text{min}}}$$

$$= 36.65\ \text{rad/s}$$

Use Eq. 33.63.

$$v_t = r\omega = (0.3\ \text{m})\left(36.65\ \frac{\text{rad}}{\text{s}}\right) = 11\ \text{m/s}$$

The tangential velocity is the same for any point on the crank at $r = 0.3$ m.

The answer is (D).

14. PROJECTILE MOTION

A projectile is placed into motion by an initial impulse. (Kinematics deals only with dynamics during the flight. The force acting on the projectile during the launch phase is covered in kinetics.) Neglecting air drag, once the projectile is in motion, it is acted upon only by the downward gravitational acceleration (i.e., its own weight). Projectile motion is a special case of motion under constant acceleration.

Consider a general projectile set into motion at an angle θ from the horizontal plane and initial velocity, v_0, as shown in Fig. 33.8. The *apex* is the point where the projectile is at its maximum elevation. In the absence of air drag, the following rules apply to the case of travel over a horizontal plane.

- The trajectory is parabolic.

- The impact velocity is equal to initial velocity, v_0.

- The range is maximum when $\theta = 45°$.

- The time for the projectile to travel from the launch point to the apex is equal to the time to travel from apex to impact point.

- The time for the projectile to travel from the apex of its flight path to impact is the same time an initially stationary object would take to fall straight down from that height.

Figure 33.8 *Projectile Motion*

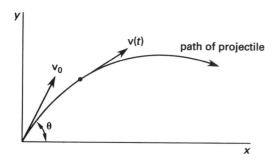

Equation 33.67 Through Eq. 33.72: Equations of Projectile Motion[3]

$a_x = 0$	33.67
$a_y = -g$	33.68
$v_x = v_0 \cos(\theta)$	33.69
$v_y = -gt + v_0 \sin(\theta)$	33.70
$x = v_0 \cos(\theta)t + x_0$	33.71
$y = -gt^2/2 + v_0 \sin(\theta)t + y_0$	33.72

---

[3]As formulated in the *NCEES Handbook*, it is easy to misinterpret the second term in Eq. 33.72. $v_0 \sin(\theta)t$ is the product of three terms: v_0, $\sin\theta$, and t. $\sin(\theta)t$ is not the sine of a function of t. A clearer formulation would have been $v_0 t \sin\theta$.

Variations

$$v_y(t) = v_{y,0} - gt$$

$$y(t) = v_{y,0}t - \tfrac{1}{2}gt^2$$

Description

The equations of projectile motion are derived from the laws of uniform acceleration and conservation of energy.

Example

A golfer on level ground at the edge of a 50 m wide pond attempts to drive a golf ball across the pond, hitting the ball so that it travels initially at 25 m/s. The ball travels at an initial angle of 45° to the horizontal plane. Approximately how far will the golf ball travel?

(A) 32 m

(B) 45 m

(C) 58 m

(D) 64 m

Solution

To determine the distance traveled by the golf ball, the time of impact must be found. At a time of 0 s and the time of impact, the elevation of the ball is known to be 0 m. Rearrange Eq. 33.72 to solve for time, substituting a value of 0 m for the elevation at a time of 0 s and the time of impact.

$$y = -gt^2/2 + v_0 \sin(\theta)t + y_0$$

$$0 \text{ m} = \frac{-gt^2}{2} + v_0 \sin(\theta)t + 0 \text{ m}$$

$$t = \frac{2v_0 \sin\theta}{g}$$

Substitute the expression for the time of impact into Eq. 33.71 and solve. The starting position is 0 m.

$$x = v_0 \cos(\theta)t + x_0$$

$$= v_0 \cos\theta\left(\frac{2v_0 \sin\theta}{g}\right) + x_0$$

$$= \left(25 \ \frac{\text{m}}{\text{s}}\right)\cos 45°\left(\frac{(2)\left(25 \ \frac{\text{m}}{\text{s}}\right)\sin 45°}{9.81 \ \frac{\text{m}}{\text{s}^2}}\right) + 0 \text{ m}$$

$$= 63.7 \text{ m} \quad (64 \text{ m})$$

The answer is (D).

34 Kinetics

Nomenclature

a	acceleration	m/s²
F	force	N
g	gravitational acceleration, 9.81	m/s²
I	mass moment of inertia	kg·m²
m	mass	kg
M	moment	N·m
N	normal force	N
p	momentum	N·s
R	resultant	N
t	time	s
v	velocity	m/s
W	weight	N
x	displacement or position	m

Symbols

α	angular acceleration	rad/s²
μ	coefficient of friction	–
ρ	radius of curvature	m
ϕ	angle	deg

Subscripts

0	initial
c	centroidal
f	final or frictional
i	initial
k	dynamic
n	normal
pc	from point p to point c
r	radial
R	resultant
s	static
t	tangential
θ	transverse

1. INTRODUCTION

Kinetics is the study of motion and the forces that cause motion. Kinetics includes an analysis of the relationship between force and mass for translational motion and between torque and moment of inertia for rotational motion. Newton's laws form the basis of the governing theory in the subject of kinetics.

2. MOMENTUM

The vector *linear momentum (momentum)*, $\mathbf{p}$, is defined by the following equation. It has the same direction as the velocity vector from which it is calculated. Momentum has units of force × time (e.g., N·s).

$$\mathbf{p} = m\mathbf{v}$$

Momentum is conserved when no external forces act on a particle. If no forces act on a particle, the velocity and direction of the particle are unchanged. The *law of conservation of momentum* states that the linear momentum is unchanged if no unbalanced forces act on the particle. This does not prohibit the mass and velocity from changing, however. Only the product of mass and velocity is constant.

3. LINEAR IMPULSE

Impulse is a vector quantity equal to the change in vector momentum. Units of linear impulse are the same as those for linear momentum: N·s. Figure 34.1 illustrates that impulse is represented by the area under the force-time curve.

$$\mathbf{Imp} = \int_{t_1}^{t_2} \mathbf{F}\,dt$$

Figure 34.1 Impulse

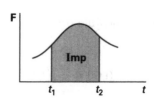

If the applied force is constant, impulse is easily calculated.

$$\mathbf{Imp} = \mathbf{F}(t_2 - t_1)$$

The change in momentum is equal to the impulse. This is known as the *impulse-momentum principle*. For a linear system with constant force and mass,

$$\mathbf{Imp} = \Delta \mathbf{p}$$

Equation 34.1 and Eq. 34.2: Impulse-Momentum Principle for a Particle

$$m\, d\mathbf{v}/dt = \mathbf{F} \qquad 34.1$$
$$m\, d\mathbf{v} = \mathbf{F}\, dt \qquad 34.2$$

Variation

$$\mathbf{F}(t_2 - t_1) = \Delta(m\mathbf{v})$$

Description

The impulse-momentum principle for a constant force and mass demonstrates that the impulse-momentum principle follows directly from Newton's second law.

Example

A 60 000 kg railcar moving at 1 km/h is coupled to a second, stationary railcar. If the velocity of the two cars after coupling is 0.2 m/s (in the original direction of motion) and the coupling is completed in 0.5 s, what is most nearly the average impulsive force on the railcar?

(A) 520 N

(B) 990 N

(C) 3100 N

(D) 9300 N

Solution

The original velocity of the 60 000 kg railcar is

$$\mathrm{v} = \frac{\left(1\ \frac{km}{h}\right)\left(1000\ \frac{m}{km}\right)}{\left(60\ \frac{s}{min}\right)\left(60\ \frac{min}{h}\right)}$$
$$= 0.2777\ \mathrm{m/s}$$

Use the impulse-momentum principle.

$$F\Delta t = m\Delta \mathrm{v}$$
$$F = \frac{m(\mathrm{v}_2 - \mathrm{v}_1)}{t_2 - t_1} = \frac{(60\,000\ \mathrm{kg})\left(0.2\ \frac{m}{s} - 0.2777\ \frac{m}{s}\right)}{0.5\ \mathrm{s} - 0\ \mathrm{s}}$$
$$= -9324\ \mathrm{N} \quad (9300\ \mathrm{N}) \quad \text{[opposite original direction]}$$

The answer is (D).

Equation 34.3: Impulse-Momentum Principle for a System of Particles

$$\sum m_i(\mathbf{v}_i)_{t_2} = \sum m_i(\mathbf{v}_i)_{t_1} + \sum \int_{t_1}^{t_2} \mathbf{F}_i\, dt \qquad 34.3$$

Description

$\sum m_i(\mathbf{v}_i)_{t_2}$ and $\sum m_i(\mathbf{v}_i)_{t_2}$ are the linear momentum at time t_1 and time t_2, respectively, for a system (i.e., collection) of particles. The impulse of the forces $\mathbf{F}$ from time t_1 to time t_2 is

$$\sum \int_{t_1}^{t_2} \mathbf{F}_i\, dt$$

4. IMPACTS

According to Newton's second law, momentum is conserved unless a body is acted upon by an external force such as gravity or friction. In an impact or collision contact is very brief, and the effect of external forces is insignificant. Therefore, momentum is conserved, even though energy may be lost through heat generation and deforming the bodies.

Consider two particles, initially moving with velocities v_1 and v_2 on a collision path, as shown in Fig. 34.2. The conservation of momentum equation can be used to find the velocities after impact, v'_1 and v'_2.

Figure 34.2 *Direct Central Impact*

The impact is said to be an *inelastic impact* if kinetic energy is lost. The impact is said to be *perfectly inelastic* or *perfectly plastic* if the two particles stick together and move on with the same final velocity. The impact is said to be an *elastic impact* only if kinetic energy is conserved.

$$m_1\mathrm{v}_1^2 + m_2\mathrm{v}_2^2 = m_1\mathrm{v}_1'^2 + m_2\mathrm{v}_2'^2 \quad \text{[elastic only]}$$

Equation 34.4: Conservation of Momentum

$$m_1\mathbf{v}_1 + m_2\mathbf{v}_2 = m_1\mathbf{v}'_1 + m_2\mathbf{v}'_2 \qquad 34.4$$

Description

The *conservation of momentum* equation is used to find the velocity of two particles after collision. $\mathbf{v}_1$ and $\mathbf{v}_2$ are the initial velocities of the particles, and $\mathbf{v}'_1$ and $\mathbf{v}'_2$ are the velocities after impact.

Example

A 60 000 kg railcar moving at 1 km/h is instantaneously coupled to a stationary 40 000 kg railcar. What is most nearly the speed of the coupled cars?

(A) 0.40 km/h

(B) 0.60 km/h

(C) 0.88 km/h

(D) 1.0 km/h

Solution

Use the conservation of momentum principle.

$$m_1\mathbf{v}_1 + m_2\mathbf{v}_2 = (m_1 + m_2)\mathbf{v}'$$

$$(60\,000 \text{ kg})\left(1\ \frac{\text{km}}{\text{h}}\right)$$
$$+ (40\,000 \text{ kg})(0) = (60\,000 \text{ kg} + 40\,000 \text{ kg})\mathbf{v}'$$
$$\mathbf{v}' = 0.60 \text{ km/h}$$

The answer is (B).

Equation 34.5: Coefficient of Restitution

$$e = \frac{(v_2')_n - (v_1')_n}{(v_1)_n - (v_2)_n} \qquad \textbf{34.5}$$

Values

inelastic	$e < 1.0$
perfectly inelastic (plastic)	$e = 0$
perfectly elastic	$e = 1.0$

Description

The *coefficient of restitution*, e, is the ratio of relative velocity differences along a mutual straight line. When both impact velocities are not directed along the same straight line, the coefficient of restitution should be calculated separately for each velocity component.

In Eq. 34.5, the subscript n indicates that the velocity to be used in calculating the coefficient of restitution should be the velocity component normal to the plane of impact.

When an object rebounds from a stationary object (an infinitely massive plane), the stationary object's initial and final velocities are zero. In that case, the *rebound velocity* can be calculated from only the object's velocities.

$$e = \left| \frac{v_1'}{v_1} \right|$$

The value of the coefficient of restitution can be used to categorize the collision as elastic or inelastic. For a perfectly inelastic collision (i.e., a plastic collision), as when two particles stick together, the coefficient of restitution is zero. For a perfectly elastic collision, the coefficient of restitution is 1.0. For most collisions, the coefficient of restitution will be between zero and 1.0, indicating a (partially) inelastic collision.

Example

A 2 kg clay ball moving at a rate of 40 m/s collides with a 5 kg ball of clay moving in the same direction at a rate of 10 m/s. What is most nearly the final velocity of both balls if they stick together after colliding?

(A) 10 m/s

(B) 12 m/s

(C) 15 m/s

(D) 19 m/s

Solution

From the coefficient of restitution definition, Eq. 34.5,

$$e = \frac{(v_2')_n - (v_1')_n}{(v_1)_n - (v_2)_n} = 0$$
$$v_2' = v_1' = v'$$

From the conservation of momentum, Eq. 34.4,

$$m_1\mathbf{v}_1 + m_2\mathbf{v}_2 = (m_1 + m_2)\mathbf{v}'$$
$$\mathbf{v}' = \frac{m_1\mathbf{v}_1 + m_2\mathbf{v}_2}{m_1 + m_2}$$
$$= \frac{(2 \text{ kg})\left(40\ \frac{\text{m}}{\text{s}}\right) + (5 \text{ kg})\left(10\ \frac{\text{m}}{\text{s}}\right)}{2 \text{ kg} + 5 \text{ kg}}$$
$$= 18.6 \text{ m/s} \quad (19 \text{ m/s})$$

The answer is (D).

Equation 34.6 and Eq. 34.7: Velocity After Impact

$$(v_1')_n = \frac{m_2(v_2)_n(1 + e) + (m_1 - em_2)(v_1)_n}{m_1 + m_2} \qquad \textbf{34.6}$$

$$(v_2')_n = \frac{m_1(v_1)_n(1 + e) - (em_1 - m_2)(v_2)_n}{m_1 + m_2} \qquad \textbf{34.7}$$

Description

If the coefficient of restitution is known, Eq. 34.6 and Eq. 34.7 may be used to calculate the velocities after impact.

5. NEWTON'S FIRST AND SECOND LAWS OF MOTION

Newton's first law of motion states that a particle will remain in a state of rest or will continue to move with constant velocity unless an unbalanced external force acts on it.

This law can also be stated in terms of conservation of momentum: If the resultant external force acting on a particle is zero, then the linear momentum of the particle is constant.

Newton's second law of motion (conservation of momentum) states that the acceleration of a particle is directly proportional to the force acting on it and is inversely proportional to the particle mass. The direction of acceleration is the same as the direction of force.

Equation 34.8 and Eq. 34.9: Newton's Second Law for a Particle

$$\sum F = d(m\mathbf{v})/dt \qquad \qquad 34.8$$

$$\sum F = m\,d\mathbf{v}/dt = m\mathbf{a} \quad \text{[constant mass]} \qquad 34.9$$

Variation

$$F = \frac{d\mathbf{p}}{dt}$$

Description

Newton's second law can be stated in terms of the force vector required to cause a change in momentum. The resultant force is equal to the rate of change of linear momentum. For a constant mass, Eq. 34.9 applies.

Example

A 3 kg block is moving at a speed of 5 m/s. The force required to bring the block to a stop in 8×10^{-4} s is most nearly

(A) 10 kN

(B) 13 kN

(C) 15 kN

(D) 19 kN

Solution

From Newton's second law, Eq. 34.9, the force required to stop a constant mass of 3 kg moving at a speed of 5 m/s is

$$\sum F = m\,d\mathbf{v}/dt = m(\Delta \mathbf{v}/\Delta t)$$

$$= (3 \text{ kg})\left(\frac{5\,\frac{\text{m}}{\text{s}} - 0\,\frac{\text{m}}{\text{s}}}{(8 \times 10^{-4}\text{ s})\left(1000\,\frac{\text{N}}{\text{kN}}\right)} \right)$$

$$= 18.75 \text{ kN} \quad (19 \text{ kN})$$

The answer is (D).

Equation 34.10 Through Eq. 34.12: Newton's Second Law for a Rigid Body[1]

$$\sum F = m\mathbf{a}_c \qquad \qquad 34.10$$

$$\sum M_c = I_c \boldsymbol{\alpha} \qquad \qquad 34.11$$

$$\sum M_p = I_c \boldsymbol{\alpha} + \boldsymbol{\rho}_{pc} \times m\mathbf{a}_c \qquad 34.12$$

Description

A *rigid body* is a complex shape that cannot be described as a particle. Generally, a rigid body is nonhomogeneous (i.e., the center of mass does not coincide with the volumetric center) or is constructed of subcomponents. In those cases, applying an unbalanced force will cause rotation as well as translation. Newton's second law of motion (conservation of momentum) can be applied to a rigid body, but the law must be applied twice: once for linear momentum and once for angular momentum. Equation 34.10 pertains to linear momentum and relates the net (resultant) force, F, on an object in any direction to the acceleration, $\mathbf{a}_c$, of the object's centroid in that direction.[2] The acceleration is "resisted" by the object's inertial mass, m. Equation 34.11 pertains to angular momentum and relates the net (resultant) moment or torque, M_c, on an object about a centroidal axis to the angular rotational acceleration, $\boldsymbol{\alpha}$, around the centroidal axis.[3] The angular acceleration is resisted by the object's centroidal mass moment of inertia, I_c.

In pure rotation, the object rotates about a centroidal axis. The centroid remains stationary as elements of the rigid body. Equation 34.12 pertains to rotation about any particular axis, p, where $\boldsymbol{\rho}_{pc}$ is the perpendicular vector from axis p to the object's centroidal axis.

Example

A net unbalanced torque acts on a 50 kg cylinder that is allowed to rotate around its longitudinal centroidal axis on frictionless bearings. The cylinder has a radius of 40 cm and a mass moment of inertia of 4 kg·m². The cylinder accelerates from a standstill with an angular acceleration of 5 rad/s².

---

[1]In Eq. 34.10 through Eq. 34.12, the NCEES *FE Reference Handbook* (*NCEES Handbook*) uses bold characters to designate vector quantities (i.e., F, M, a, α, and ρ). Rectilinear components of vectors may be added; and, cross-products are used for multiplication. In most calculations, however, the vector nature of these quantities is disregarded, and only the magnitudes of the quantities are used.

[2]The *NCEES Handbook* is inconsistent in its meaning of a_c. In Eq. 34.10, a_c refers to the acceleration of the centroid, which the *NCEES Handbook* calls "mass center." a_c does not mean constant acceleration as it did earlier in the *NCEES Handbook* Dynamics section.

[3]The *NCEES Handbook* is inconsistent in designating the centroidal parameters. Whereas a_c represents the acceleration of the centroid in Eq. 34.10, and I_c represents the centroidal moment of inertia in Eq. 34.11, the subscript c has been omitted on α, the angular acceleration about the centroidal axis, in Eq. 34.12.

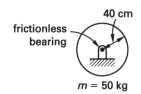

40 cm

frictionless
bearing

m = 50 kg

What is most nearly the unbalanced torque on the cylinder?

(A) 20 N·m

(B) 40 N·m

(C) 200 N·m

(D) 2000 N·m

Solution

Using Eq. 34.11, the magnitude of the moment acting on the cylinder is

$$\sum M_c = I_c \alpha = (4 \text{ kg·m}^2)\left(5 \ \frac{\text{rad}}{\text{s}^2}\right)$$
$$= 20 \text{ N·m}$$

The answer is (A).

Equation 34.13 Through Eq. 34.19: Rectilinear Equations for Rigid Bodies

$$\sum F_x = ma_{xc} \qquad \textit{34.13}$$

$$\sum F_y = ma_{yc} \qquad \textit{34.14}$$

$$\sum M_{zc} = I_{zc}\alpha \qquad \textit{34.15}$$

$$\sum F_x = m(a_G)_x \qquad \textit{34.16}$$

$$\sum F_y = m(a_G)_y \qquad \textit{34.17}$$

$$\sum M_G = I_G\alpha \qquad \textit{34.18}$$

$$\sum M_P = \sum (M_k)_P \qquad \textit{34.19}$$

Description

These equations are the scalar forms of Newton's second law equations, assuming the rigid body is constrained to move in an x-y plane. The subscript zc describes the z-axis passing through the body's centroid. Placing the origin at the body's centroid, the acceleration of the body in the x- and y-directions is a_{xc} and a_{yc}, respectively. α is the angular acceleration of the body about the z-axis. Equation 34.13 through Eq. 34.19 are limited to motion in the x-y plane (i.e., two dimensions). Equation 34.18 calculates the sum of moments about a rigid body's center of gravity (mass center, etc.), G.

Equation 34.19 calculates the sum of moments about any point, P.[4]

6. WEIGHT

Equation 34.20: Weight of an Object

$$W = mg \qquad \textit{34.20}$$

Description

The *weight*, W, of an object is the force the object exerts due to its position in a gravitational field.[5]

Example

A man weighs himself twice in an elevator. When the elevator is at rest, he weighs 713 N; when the elevator starts moving upward, he weighs 816 N. What is most nearly the man's actual mass?

(A) 70 kg

(B) 73 kg

(C) 78 kg

(D) 83 kg

---

[4](1) Equation 34.13 through Eq. 34.15 are prefaced in the *NCEES Handbook* with, "Without loss of generality, the body may be assumed to be in the x-y plane." This statement sounds as though all bodies can be simplified to planar motion, which is not true. The more general three-dimensional case is not specifically presented, so there is no generality to lose. In fact, Eq. 34.15 represents the sum of moments about any point, so this equation *is* the more general case, not the less general case. (2) Equation 34.16, Eq. 34.17, and Eq. 34.18 are functionally the same as Eq. 34.13, Eq. 34.14, and Eq. 34.15 and are redundant. Both sets of equations are limited to the x-y plane. (3) The subscripts c (centroidal or center of mass) and G (center of gravity) refer to the same thing. The change in notation is unnecessary. (4) The subscripts G and P are not defined. (5) The subscript k is not defined, but probably represents an uncommon choice for the first summation variable, normally i. Since k does not appear in the summation symbol, the meaning of M_k must be inferred. (6) Equation 34.18 specifies the point through which the rotational axis passes, but it does not specify an axis, as does Eq. 34.15. Since the equations are limited to the x-y plane, the rotational axis can only be parallel to the z-axis, as in Eq. 34.15. (7) The subscript c has been omitted on α, the angular acceleration about the center of mass, in Eq. 34.15 and Eq. 34.18.

[5](1) The *NCEES Handbook* introduces Eq. 34.20 with the section heading, "Concept of Weight." Units of weight are specified as newtons. In fact, the concept of weight is entirely absent in the SI system. Only the concepts of mass and force are used. The SI system does not support the concept of "body weight" in newtons. It only supports the concept of the force needed to accelerate a body. In presenting Eq. 34.20, the *NCEES Handbook* perpetuates the incorrect ideas that mass and weight are synonyms, and that weight is a fixed property of a body. (2) The *NCEES Handbook* includes a parenthetical "(lbf)" as the unit of weight for U.S. equations. However, Eq. 34.20 cannot be used with customary and normal U.S. units (i.e., mass in pounds) without including the gravitational constant, g_c. In order to make Eq. 34.20 consistent, the *NCEES Handbook* is forced to specify the unit of mass for U.S. equations as lbf-sec²/ft. This (essentially now obsolete) unit of mass is known as a *slug*, something that is not called out in the *NCEES Handbook*. Since a slug is 32.2 times larger than a pound, an examinee using Eq. 34.20 with customary and normal U.S. units could easily be misdirected by the lbf label.

Engineering
Sciences

Solution

The mass of the man can be determined from his weight at rest.

$$W = mg$$
$$m = \frac{W}{g} = \frac{713 \text{ N}}{9.81 \frac{\text{m}}{\text{s}^2}}$$
$$= 72.7 \text{ kg} \quad (73 \text{ kg})$$

The answer is (B).

7. FRICTION

Friction is a force that always resists motion or impending motion. It always acts parallel to the contacting surfaces. If the body is moving, the friction is known as *dynamic friction*. If the body is stationary, friction is known as *static friction*.

The magnitude of the frictional force depends on the normal force, N, and the *coefficient of friction*, μ, between the body and the contacting surface.

$$F_f = \mu N$$

The static coefficient of friction is usually denoted with the subscript s, while the dynamic (i.e., kinetic) coefficient of friction is denoted with the subscript k. μ_k is often assumed to be 75% of the value of μ_s. These coefficients are complex functions of surface properties. Experimentally determined values for various contacting conditions can be found in handbooks.

For a body resting on a horizontal surface, the *normal force*, N, is the weight, W, of the body. If the body rests on an inclined surface, the normal force is calculated as the component of weight normal to that surface, as illustrated in Fig. 34.3. Axes in Fig. 34.3 are defined as parallel and perpendicular to the inclined plane.

$$N = mg \cos \phi = W \cos \phi$$

Figure 34.3 Frictional and Normal Forces

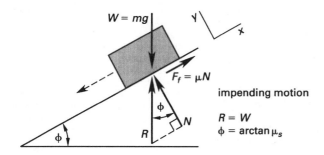

The frictional force acts only in response to a disturbing force, and it increases as the disturbing force increases. The motion of a stationary body is impending when the disturbing force reaches the maximum frictional force, $\mu_s N$. Figure 34.3 shows the condition of impending motion for a block on a plane. Just before motion starts, the resultant, R, of the frictional force and normal force equals the weight of the block. The angle at which motion is just impending can be calculated from the coefficient of static friction.

$$\phi = \arctan \mu_s$$

Once motion begins, the coefficient of friction drops slightly, and a lower frictional force opposes movement. This is illustrated in Fig. 34.4.

Figure 34.4 Frictional Force Versus Disturbing Force

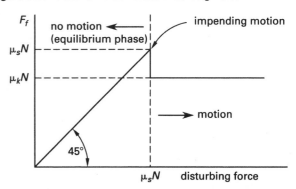

Equation 34.21 Through Eq. 34.24: Laws of Friction

$F \le \mu_s N$	**34.21**
$F < \mu_s N$ [no slip occurring]	**34.22**
$F = \mu_s N$ [point of impending slip]	**34.23**
$F = \mu_k N$ [slip occurring]	**34.24**

Values

$$\mu_k \approx 0.75 \mu_s$$

Description

The laws of friction state that the maximum value of the total friction force, F, is independent of the magnitude of the area of contact. The maximum total friction force is proportional to the normal force, N. For low velocities of sliding, the maximum total frictional force is nearly independent of the velocity. However, experiments show that the force necessary to initiate slip is greater than that necessary to maintain the motion.

Example

A boy pulls a sled with a mass of 35 kg horizontally over a surface with a dynamic coefficient of friction of 0.15.

What is most nearly the force required for the boy to pull the sled?

(A) 49 N

(B) 52 N

(C) 55 N

(D) 58 N

Solution

N is the normal force, and μ_k is the dynamic coefficient of friction. The force that the boy must pull with, F_b, must be large enough to overcome the frictional force. From Eq. 34.24,

$$F_b = F_f = \mu_k N = \mu_k mg$$

$$= (0.15)(35 \text{ kg})\left(9.81 \ \frac{\text{m}}{\text{s}^2}\right)$$

$$= 51.5 \text{ kg·m/s}^2 \quad (52 \text{ N})$$

The answer is (B).

8. KINETICS OF A PARTICLE

Newton's second law can be applied separately to any direction in which forces are resolved into components. The law can be expressed in rectangular coordinate form (i.e., in terms of x- and y-component forces), in polar coordinate form (i.e., in tangential and normal components), or in radial and transverse component form.

Equation 34.25: Newton's Second Law

$$a_x = F_x/m \qquad 34.25$$

Variation

$$F_x = ma_x$$

Description

Equation 34.25 is Newton's second law in rectangular coordinate form and refers to motion in the x-direction. Similar equations can be written for the y-direction or any other coordinate direction. In general, F_x may be a function of time, displacement, and/or velocity.

Example

A car moving at 70 km/h has a mass of 1700 kg. The force necessary to decelerate it at a rate of 40 cm/s² is most nearly

(A) 0.68 N

(B) 42 N

(C) 680 N

(D) 4200 N

Solution

Use Newton's second law.

$$a_x = F_x/m$$

$$F_x = ma_x$$

$$= \frac{(1700 \text{ kg})\left(40 \ \dfrac{\text{cm}}{\text{s}^2}\right)}{100 \ \dfrac{\text{cm}}{\text{m}}}$$

$$= 680 \text{ kg·m/s}^2 \quad (680 \text{ N})$$

The answer is (C).

Equation 34.26 Through Eq. 34.28: Equations of Motion with Constant Mass and Force as a Function of Time

$$a_x(t) = F_x(t)/m \qquad 34.26$$

$$v_x(t) = \int_{t_0}^{t} a_x(t)\,dt + v_{xt_0} \qquad 34.27$$

$$x(t) = \int_{t_0}^{t} v_x(t)\,dt + x_{t_0} \qquad 34.28$$

Variation

$$v_x(t) = \int_{t_i}^{t_f} \frac{F_x(t)}{m}\,dt + v_{x,0}$$

Description

If F_x is a function of time only, then the equations of motion are given by Eq. 34.26, Eq. 34.27, and Eq. 34.28.

Equation 34.29 Through Eq. 34.31: Equations of Motion with Constant Mass and Force

$$a_x = F_x/m \qquad 34.29$$

$$v_x = a_x(t - t_0) + v_{xt_0} \qquad 34.30$$

$$x = a_x(t - t_0)^2/2 + v_{xt_0}(t - t_0) + x_{t_0} \qquad 34.31$$

Variations

$$F_x = ma_x$$

$$v_x(t) = v_{x,0} + \left(\frac{F_x}{m}\right)(t - t_0)$$

$$x(t) = x_0 + v_{x,0}(t - t_0) + \frac{F_x(t - t_0)^2}{2m}$$

Description

If F_x is constant (i.e., is independent of time, displacement, or velocity) and mass is constant, then the equations of motion are given by Eq. 34.29, Eq. 34.30, and Eq. 34.31.

Engineering Sciences

Example

A force of 15 N acts on a 16 kg body for 2 s. If the body is initially at rest, approximately how far is it displaced by the force?

(A) 1.1 m

(B) 1.5 m

(C) 1.9 m

(D) 2.1 m

Solution

The acceleration is found using Newton's second law, Eq. 34.29.

$$a_x = F_x/m = \frac{15 \text{ N}}{16 \text{ kg}} = 0.94 \text{ m/s}^2$$

For a body undergoing constant acceleration, with an initial velocity of 0 m/s, an initial time of 0 s, and a total elapsed time of 2 s, the horizontal displacement is found from Eq. 34.31.

$$x = a_x(t - t_0)^2/2 + v_{xt_0}(t - t_0) + x_{t_0}$$
$$= \frac{\left(0.94 \frac{\text{m}}{\text{s}^2}\right)(2 \text{ s} - 0 \text{ s})^2}{2} + \left(0 \frac{\text{m}}{\text{s}}\right)(2 \text{ s} - 0 \text{ s}) + 0 \text{ m}$$
$$= 1.88 \text{ m} \quad (1.9 \text{ m})$$

The answer is (C).

Equation 34.32 and Eq. 34.33: Tangential and Normal Components

$$\sum F_t = ma_t = m \, dv_t/dt \qquad \textit{34.32}$$
$$\sum F_n = ma_n = m(v_t^2/\rho) \qquad \textit{34.33}$$

Description

For a particle moving along a circular path, the tangential and normal components of force, acceleration, and velocity are related.

Radial and Transverse Components

For a particle moving along a circular path, the radial and transverse components of force are

$$\sum F_r = ma_r$$

$$\sum F_\theta = ma_\theta$$

35 Kinetics of Rotational Motion

Nomenclature

a	acceleration	m/s^2
A	area	m^2
c	number of instantaneous centers	–
d	length	m
F	force	N
g	gravitational acceleration, 9.81	m/s^2
h	height	m
H	angular momentum	N·m·s
I	mass moment of inertia	$kg·m^2$
l	length	m
L	length	m
m	mass[1]	kg
M	mass[1]	kg
M	moment	N·m
n	quantity	–
r	radius of gyration	m
R	mean radius	m
t	time	s
v	velocity	m/s
W	weight	N

Symbols

α	angular acceleration	rad/s^2
θ	angular position	rad
μ	coefficient of friction	–
ρ	density	kg/m^3
ω	angular velocity	rad/s

Subscripts

0	initial
c	centrifugal or centroidal
f	frictional
G	center of gravity
m	mass
n	normal
O	origin or center
s	static
t	tangential

[1]The NCEES *FE Reference Handbook* (*NCEES Handbook*) is inconsistent in its nomenclature usage. It uses both m and M to designate the mass of a object. It generally uses uppercase M to designate the total mass of non-particles (i.e., cylinders). Care must be taken when solving problems involving both mass and moment, as equations for both quantities use the same symbol.

1. MASS MOMENT OF INERTIA

Equation 35.1 Through Eq. 35.4: Mass Moment of Inertia

$$I = \int r^2 \, dm \qquad \textbf{35.1}$$

$$I_x = \int (y^2 + z^2) \, dm \qquad \textbf{35.2}$$

$$I_y = \int (x^2 + z^2) \, dm \qquad \textbf{35.3}$$

$$I_z = \int (x^2 + y^2) \, dm \qquad \textbf{35.4}$$

Description

The *mass moment of inertia* measures a solid object's resistance to changes in rotational speed about a specific axis. Equation 35.1 shows that the mass moment of inertia is calculated as the second moment of the mass.[2] When the origin of a coordinate system is located at the object's center of mass, the radius, r, to the differential element can be calculated from the components of position as

$$r = \sqrt{x^2 + y^2 + z^2}$$

For a homogeneous body with density ρ, Eq. 35.1 can be written as

$$I = \rho \int_V r^2 \, dV$$

I_x, I_y, and I_z are the mass moments of inertia with respect to the x-, y-, and z-axes, respectively. They are not components of a resultant value.

[2](1) There are two closely adjacent sections in the *NCEES Handbook* labeled "Mass Moment of Inertia," each covering the same topic. (2) The integral shown in Eq. 35.1 is implicitly a triple integral (volume integral), more properly shown as $\int_V$ or $\iiint$.

Equation 35.5 and Eq. 35.6: Parallel Axis Theorem

$$I_{new} = I_c + md^2 \qquad \text{35.5}$$
$$I = I_G + md^2 \qquad \text{35.6}$$

Variation

$$I = I_{c,1} + m_1 d_1^2 + I_{c,2} + m_2 d_2^2 + \cdots$$

Description

The *centroidal mass moment of inertia*, I_c, is obtained when the origin of the axes coincides with the object's center of gravity.[3] The *parallel axis theorem*, also known as the *transfer axis theorem*, is used to find the mass moment of inertia about any axis. In Eq. 35.5, d is the distance from the center of mass to the new axis.

For a composite object, the parallel axis theorem must be applied for each of the constituent objects, as shown in the variation equation.

Example

The 5 cm long uniform slender rod shown has a mass of 20 g. The origin of the y-axis corresponds with the rod's center of gravity. The centroidal mass moment of inertia is 42 g·cm².

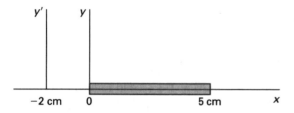

What is most nearly the mass moment of inertia of the rod about the y' axis 2 cm to the left?

(A) 0.12 kg·cm²

(B) 0.33 kg·cm²

(C) 0.45 kg·cm²

(D) 0.91 kg·cm²

Solution

The y' axis is 2 cm from the y-axis. The center of gravity of the rod is located halfway along its length. Use Eq. 35.5.

$$I_{y'} = I_c + md^2 = \frac{42 \text{ g·cm}^2 + (20 \text{ g})(2.5 \text{ cm} + 2 \text{ cm})^2}{1000 \frac{\text{g}}{\text{kg}}}$$

$$= 0.45 \text{ kg·cm}^2$$

The answer is (C).

---

[3]Equation 35.5 and Eq. 35.6 both appear on the same page in the *NCEES Handbook* using different notation. The inconsistent subscripts c and G both refer to the same concept: centroidal (center of gravity, center of mass, etc.).

Equation 35.7: Mass Radius of Gyration

$$r_m = \sqrt{I/m} \qquad \text{35.7}$$

Variation

$$I = r^2 m$$

Description

The *mass radius of gyration*, r_m, of a solid object represents the distance from the rotational axis at which the object's entire mass could be located without changing the mass moment of inertia.

Equation 35.8 Through Eq. 35.19: Properties of Uniform Slender Rods

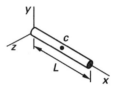

mass and centroid

$$M = \rho LA \qquad \text{35.8}$$
$$x_c = L/2 \qquad \text{35.9}$$
$$y_c = 0 \qquad \text{35.10}$$
$$z_c = 0 \qquad \text{35.11}$$

mass moment of inertia

$$I_x = I_{x_c} = 0 \qquad \text{35.12}$$
$$I_{y_c} = I_{z_c} = ML^2/12 \qquad \text{35.13}$$
$$I_y = I_z = ML^2/3 \qquad \text{35.14}$$

(radius of gyration)²

$$r_x^2 = r_{x_c}^2 = 0 \qquad \text{35.15}$$
$$r_{y_c}^2 = r_{z_c}^2 = L^2/12 \qquad \text{35.16}$$
$$r_y^2 = r_z^2 = L^2/3 \qquad \text{35.17}$$

product of inertia

$$I_{x_c y_c} = 0 \qquad \text{35.18}$$
$$I_{xy} = 0 \qquad \text{35.19}$$

Description

Equation 35.8 through Eq. 35.19 give the properties of slender rods. The center of mass (center of gravity) is located at (x_c, y_c, z_c), designated point c. M is the total mass; A is the cross-sectional area perpendicular to the longitudinal axis; ρ is the mass density, equal to the mass divided by the volume; I is the mass moment of inertia about the subscripted axis, used in calculating rotational acceleration and moments about that axis; and r is the radius of gyration, a distance from the designated axis from the centroid where all of the mass can be assumed to be concentrated. I_{xy} is the *product of inertia*, a measure of symmetry, with respect to a plane containing the subscripted axes. The product of inertia is zero if the object is symmetrical about an axis perpendicular to the plane defined by the subscripted axes.

Example

A uniform rod is 2.0 m long and has a mass of 15 kg. What is most nearly the rod's mass moment of inertia?

(A) 5.0 kg·m^2

(B) 20 kg·m^2

(C) 27 kg·m^2

(D) 31 kg·m^2

Solution

From Eq. 35.14, the mass moment of inertia of the rod is

$$I_{\text{rod}} = ML^2/3 = \frac{(15 \text{ kg})(2.0 \text{ m})^2}{3}$$
$$= 20 \text{ kg·m}^2$$

The answer is (B).

Equation 35.20 Through Eq. 35.34: Properties of Slender Rings

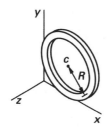

mass and centroid

$M = 2\pi R \rho A$	*35.20*
$x_c = R$	*35.21*
$y_c = R$	*35.22*
$z_c = 0$	*35.23*

mass moment of inertia

$I_{x_c} = I_{y_c} = MR^2/2$	*35.24*
$I_{z_c} = MR^2$	*35.25*
$I_x = I_y = 3MR^2/2$	*35.26*
$I_z = 3MR^2$	*35.27*

(radius of gyration)^2

$r_{x_c}^2 = r_{y_c}^2 = R^2/2$	*35.28*
$r_{z_c}^2 = R^2$	*35.29*
$r_x^2 = r_y^2 = 3R^2/2$	*35.30*
$r_z^2 = 3R^2$	*35.31*

product of inertia

$I_{x_c y_c} = 0$	*35.32*
$I_{y_c z_c} = MR^2$	*35.33*
$I_{xz} = I_{yz} = 0$	*35.34*

Description

Equation 35.20 through Eq. 35.34 give the properties of slender rings. The center of mass (center of gravity) is located at (x_c, y_c, z_c), designated point c, and measured from the mean radius of the ring. M is the total mass; A is the cross-sectional area of the ring; ρ is the mass density, equal to the mass divided by the volume; I is the mass moment of inertia about the subscripted axis, used in calculating rotational acceleration and moments about that axis; and r is the radius of gyration, a distance from the designated axis from the centroid where all of the mass can be assumed to be concentrated. r_{z_c} is the radius of gyration of the ring about an axis parallel to the z-axis and passing through the centroid. I_{xy} is the product of inertia, a measure of symmetry, with respect to a plane containing the subscripted axes. The product of inertia is zero if the object is symmetrical about an axis perpendicular to the plane defined by the subscripted axes.

Example

The period of oscillation of a clock balance wheel is 0.3 s. The wheel is constructed as a slender ring with its 30 g mass uniformly distributed around a 0.6 cm radius. What is most nearly the wheel's moment of inertia?

(A) 1.1×10^{-6} kg·m^2

(B) 1.6×10^{-6} kg·m^2

(C) 2.1×10^{-6} kg·m^2

(D) 2.6×10^{-6} kg·m^2

Solution

From Eq. 35.25, the wheel's moment of inertia is

$$I = MR^2$$

$$= \left(\frac{30\ \text{g}}{10^3\ \frac{\text{g}}{\text{kg}}}\right)\left(\frac{0.6\ \text{cm}}{100\ \frac{\text{cm}}{\text{m}}}\right)^2$$

$$= 1.08 \times 10^{-6}\ \text{kg·m}^2 \quad (1.1 \times 10^{-6}\ \text{kg·m}^2)$$

The answer is (A).

Equation 35.35 Through Eq. 35.46: Properties of Cylinders

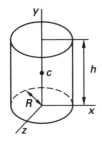

mass and centroid

$M = \pi R^2 \rho h$	*35.35*
$x_c = 0$	*35.36*
$y_c = h/2$	*35.37*
$z_c = 0$	*35.38*

mass moment of inertia

$I_{x_c} = I_{z_c} = M(3R^2 + h^2)/12$	*35.39*
$I_{y_c} = I_y = MR^2/2$	*35.40*
$I_x = I_z = M(3R^2 + 4h^2)/12$	*35.41*

(radius of gyration)²

$r_{x_c}^2 = r_{z_c}^2 = (3R^2 + h^2)/12$	*35.42*
$r_{y_c}^2 = r_y^2 = R^2/2$	*35.43*
$r_x^2 = r_z^2 = (3R^2 + 4h^2)/12$	*35.44*

product of inertia

$I_{x_c y_c} = 0$	*35.45*
$I_{xy} = 0$	*35.46*

Description

Equation 35.35 through Eq. 35.46 give the properties of solid (right) cylinders. The center of mass (center of gravity) is located at (x_c, y_c, z_c), designated point c. M is the total mass; ρ is the mass density, equal to the mass divided by the volume; I is the mass moment of

inertia about the subscripted axis, used in calculating rotational acceleration and moments about that axis; and r is the radius of gyration, a distance from the designated axis from the centroid where all of the mass can be assumed to be concentrated. r_{y_c} is the radius of gyration of the cylinder about an axis parallel to the y-axis and passing through the centroid. I_{xy} is the product of inertia, a measure of symmetry, with respect to a plane containing the subscripted axes. The product of inertia is zero if the object is symmetrical about an axis perpendicular to the plane defined by the subscripted axes.

Example

A 50 kg solid cylinder has a height of 3 m and a radius of 0.5 m. The cylinder sits on the x-axis and is oriented with its longitudinal axis parallel to the y-axis. What is most nearly the mass moment of inertia about the x-axis?

(A) 4.1 kg·m²
(B) 16 kg·m²
(C) 41 kg·m²
(D) 150 kg·m²

Solution

Find the mass moment of inertia using Eq. 35.41.

$$I_x = M(3R^2 + 4h^2)/12$$

$$= \frac{(50\ \text{kg})\left((3)(0.5\ \text{m})^2 + (4)(3\ \text{m})^2\right)}{12}$$

$$= 153.1\ \text{kg·m}^2 \quad (150\ \text{kg·m}^2)$$

The answer is (D).

Equation 35.47 Through Eq. 35.58: Properties of Hollow Cylinders

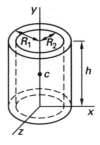

mass and centroid

$M = \pi(R_1^2 - R_2^2)\rho h$	*35.47*
$x_c = 0$	*35.48*
$y_c = h/2$	*35.49*
$z_c = 0$	*35.50*

mass moment of inertia

$$I_{x_c} = I_{z_c} = M(3R_1^2 + 3R_2^2 + h^2)/12 \qquad 35.51$$

$$I_{y_c} = I_y = M(R_1^2 + R_2^2)/2 \qquad 35.52$$

$$I_x = I_z = M(3R_1^2 + 3R_2^2 + 4h^2)/12 \qquad 35.53$$

(radius of gyration)^2

$$r_{x_c}^2 = r_{z_c}^2 = (3R_1^2 + 3R_2^2 + h^2)/12 \qquad 35.54$$

$$r_{y_c}^2 = r_y^2 = (R_1^2 + R_2^2)/2 \qquad 35.55$$

$$r_x^2 = r_z^2 = (3R_1^2 + 3R_2^2 + 4h^2)/12 \qquad 35.56$$

product of inertia

$$I_{x_c y_c} = 0 \qquad 35.57$$

$$I_{xy} = 0 \qquad 35.58$$

Description

Equation 35.47 through Eq. 35.58 give the properties of hollow (right) cylinders. Due to symmetry, the properties are the same for all axes. R_1 is the outer radius, and R_2 is the inner radius. The center of mass (center of gravity) is located at (x_c, y_c, z_c), designated point c. M is the total mass; ρ is the mass density, equal to the mass divided by the volume; I is the mass moment of inertia about the subscripted axis, used in calculating rotational acceleration and moments about that axis; and r is the radius of gyration, a distance from the designated axis from the centroid where all of the mass can be assumed to be concentrated. r_{y_c} is the radius of gyration of the hollow cylinder about an axis parallel to the y-axis and passing through the centroid. I_{xy} is the product of inertia, a measure of symmetry, with respect to a plane containing the subscripted axes. The product of inertia is zero if the object is symmetrical about an axis perpendicular to the plane defined by the subscripted axes.

Example

A hollow cylinder has a mass of 2 kg, a height of 1 m, an outer diameter of 1 m, and an inner diameter of 0.8 m.

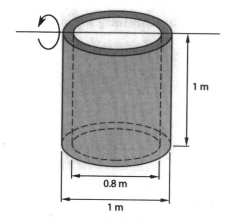

What is most nearly the cylinder's mass moment of inertia about an axis perpendicular to the cylinder's longitudinal axis and located at the cylinder's end?

(A) 0.41 kg·m^2

(B) 0.79 kg·m^2

(C) 0.87 kg·m^2

(D) 1.5 kg·m^2

Solution

The outer radius, R_1, and inner radius, R_2, are

$$R_1 = \frac{1 \text{ m}}{2} = 0.5 \text{ m}$$

$$R_2 = \frac{0.8 \text{ m}}{2} = 0.4 \text{ m}$$

Use Eq. 35.53.

$$I = M(3R_1^2 + 3R_2^2 + 4h^2)/12$$

$$= \frac{(2 \text{ kg})\big((3)(0.5 \text{ m})^2 + (3)(0.4 \text{ m})^2 + (4)(1 \text{ m})^2\big)}{12}$$

$$= 0.87 \text{ kg·m}^2$$

The answer is (C).

Equation 35.59 Through Eq. 35.69: Properties of Spheres

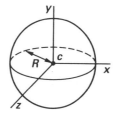

mass and centroid

$$M = \frac{4}{3}\pi R^3 \rho \qquad 35.59$$

$$x_c = 0 \qquad 35.60$$

$$y_c = 0 \qquad 35.61$$

$$z_c = 0 \qquad 35.62$$

mass moment of inertia

$$I_{x_c} = I_x = 2MR^2/5 \qquad 35.63$$

$$I_{y_c} = I_y = 2MR^2/5 \qquad 35.64$$

$$I_{z_c} = I_z = 2MR^2/5 \qquad 35.65$$

(radius of gyration)²

$$r_{x_c}^2 = r_x^2 = 2R^2/5 \qquad \textbf{35.66}$$

$$r_{y_c}^2 = r_y^2 = 2R^2/5 \qquad \textbf{35.67}$$

$$r_{z_c}^2 = r_z^2 = 2R^2/5 \qquad \textbf{35.68}$$

product of inertia

$$I_{x_c y_c} = 0 \qquad \textbf{35.69}$$

Description

Equation 35.59 through Eq. 35.69 give the properties of spheres. The center of mass (center of gravity) is located at (x_c, y_c, z_c), designated point c. M is the total mass; ρ is the mass density, equal to the mass divided by the volume; I is the mass moment of inertia about the subscripted axis, used in calculating rotational acceleration and moments about that axis; and r is the radius of gyration, a distance from the designated axis from the centroid where all of the mass can be assumed to be concentrated. The product of inertia for any plane passing through the centroid is zero because the object is symmetrical about an axis perpendicular to that plane.

2. PLANE MOTION OF A RIGID BODY

General rigid body plane motion, such as rolling wheels, gear sets, and linkages, can be represented in two dimensions (i.e., the plane of motion). Plane motion can be considered as the sum of a translational component and a rotation about a fixed axis, as illustrated in Fig. 35.1.

Figure 35.1 *Components of Plane Motion*

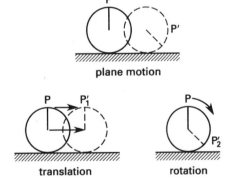

3. ROTATION ABOUT A FIXED AXIS

Instantaneous Center of Rotation

Analysis of the rotational component of a rigid body's plane motion can sometimes be simplified if the location of the body's *instantaneous center* is known. Using the instantaneous center reduces many relative motion problems to simple geometry. The instantaneous center (also known as the *instant center* and IC) is a point at which the body could be fixed (pinned) without

changing the instantaneous angular velocities of any point on the body. For angular velocities, the body seems to rotate about a fixed, instantaneous center.

The instantaneous center is located by finding two points for which the absolute velocity directions are known. Lines drawn perpendicular to these two velocities will intersect at the instantaneous center. (This graphic procedure is slightly different if the two velocities are parallel, as Fig. 35.2 shows.) For a rolling wheel, the instantaneous center is the point of contact with the supporting surface.

Figure 35.2 *Graphic Method of Finding the Instantaneous Center*

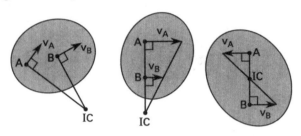

The absolute velocity of any point, P, on a wheel rolling (see Fig. 35.3) with translational velocity, v_O, can be found by geometry. Assume that the wheel is pinned at point C and rotates with its actual angular velocity, $\dot{\theta} = \omega = v_O/r$. The direction of the point's velocity will be perpendicular to the line of length, l, between the instantaneous center and the point.

$$v = l\omega = \frac{l v_O}{r}$$

Figure 35.3 *Instantaneous Center of a Rolling Wheel*

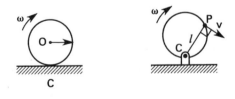

Equation 35.70: Kennedy's Rule

$$c = \frac{n(n-1)}{2} \qquad \textbf{35.70}$$

Description

The location of the instantaneous center can be found by inspection for many mechanisms, such as simple pinned pulleys and rolling/rotating objects. *Kennedy's rule* (law, theorem, etc.) can be used to help find the instantaneous centers when they are not obvious, such as with slider-crank and bar linkage mechanisms. Kennedy's rule states that any three links (bodies), designated as 1, 2, and 3, of a mechanism (that may have more than three links), and undergoing motion relative to one another, will have exactly three associated instantaneous centers, IC_{12}, IC_{13}, and IC_{23}, and those three instant centers will lie on a straight line. Equation 35.70 calculates the number of instantaneous centers for any number of links. c is the number of instantaneous centers, and n is the number of links.

Example

How many instantaneous centers does the linkage shown have?

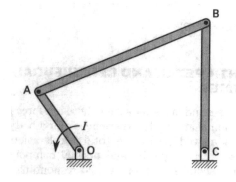

(A) 3

(B) 4

(C) 5

(D) 6

Solution

This is a four-bar linkage. The fourth bar consists of the fixed link between points O and C. Use Eq. 35.70. The number of instantaneous centers is

$$c = \frac{n(n-1)}{2} = \frac{(4)(4-1)}{2} = 6$$

The answer is (D).

Equation 35.71 Through Eq. 35.73: Angular Momentum

$$H_0 = r \times mv \qquad \text{35.71}$$
$$H_0 = I_0\omega \qquad \text{35.72}$$
$$\sum(\text{syst. } \mathbf{H})_1 = \sum(\text{syst. } \mathbf{H})_2 \qquad \text{35.73}$$

Description

The *angular momentum* taken about a point 0 is the moment of the linear momentum vector. Angular momentum has units of distance × force × time (e.g., N·m·s). It has the same direction as the rotation vector

and can be determined from the vectors by use of the right-hand rule (cross product). (See Eq. 35.71.)

For a rigid body rotating about an axis passing through its center of gravity located at point 0, the scalar value of angular momentum is given by Eq. 35.72.[4]

The *law of conservation of angular momentum* states that if no external torque acts upon an object, the angular momentum cannot change. The angular momentum before and after an internal torque is applied is the same. Equation 35.73 expresses the angular momentum conservation law for a system consisting of multiple masses.[5]

Equation 35.74 and Eq. 35.75: Change in Angular Momentum

$$\dot{H}_0 = d(I_0\omega)/dt = M \qquad \text{35.74}$$

$$\sum(\mathbf{H}_{0i})_{t_2} = \sum(\mathbf{H}_{0i})_{t_1} + \sum\int_{t_1}^{t_2} \mathbf{M}_{0i}\,dt \qquad \text{35.75}$$

Variations

$$M = \frac{d\mathbf{H}_O}{dt}$$

$$M = I\frac{d\omega}{dt} = I\alpha$$

Description

Although Newton's laws do not specifically deal with rotation, there is an analogous relationship between applied moment (torque) and change in angular momentum. For a rotating body, the moment (torque), **M**, required to change the angular momentum is given by Eq. 35.74.

The rotation of a rigid body will be about the center of gravity unless the body is constrained otherwise. The scalar form of Eq. 35.74 for a constant moment of inertia is shown in the second variation.

For a collection of particles, Eq. 35.74 may be expanded as shown in Eq. 35.75. Equation 35.75 determines the angular momentum at time t_2 from the angular momentum at time t_1, $\sum(\mathbf{H}_{0i})_{t_1}$, and the angular impulse of the moment between t_1 and t_2, $\sum\int_{t_1}^{t_2}\mathbf{M}_{0i}\,dt$.

---

[4]The *NCEES Handbook* is inconsistent in its representation of the centroidal mass moment of inertia. I_0 and I_c are both used in the Dynamics section for the same concept.

[5]In Eq. 35.73, the nonstandard notation "syst" should be interpreted as the limits of summation (i.e., summation over all masses in the system). This would normally be written as $\sum_{\text{system}}$ or something similar.

Engineering Sciences

Equation 35.76 Through Eq. 35.86: Rotation About an Arbitrary Fixed Axis

$$\sum M_q = I_q \alpha \qquad \textbf{35.76}$$

$$\alpha = \frac{d\omega}{dt} \quad \text{[general]} \qquad \textbf{35.77}$$

$$\omega = \frac{d\theta}{dt} \quad \text{[general]} \qquad \textbf{35.78}$$

$$\omega \, d\omega = \alpha \, d\theta \quad \text{[general]} \qquad \textbf{35.79}$$

$$\omega = \omega_0 + \alpha_c t \qquad \textbf{35.80}$$

$$\theta = \theta_0 + \omega_0 t + \frac{1}{2}\alpha_c t^2 \qquad \textbf{35.81}$$

$$\omega^2 = \omega_0^2 + 2\alpha_c(\theta - \theta_0) \qquad \textbf{35.82}$$

$$\alpha = M_q/I_q \qquad \textbf{35.83}$$

$$\omega = \omega_0 + \alpha t \qquad \textbf{35.84}$$

$$\theta = \theta_0 + \omega_0 t + \alpha t^2/2 \qquad \textbf{35.85}$$

$$I_q \omega^2/2 = I_q \omega_0^2/2 + \int_{\theta_0}^{\theta} M_q \, d\theta \qquad \textbf{35.86}$$

Variations

$$\omega = \int \alpha \, dt = \omega_0 + \left(\frac{M}{I}\right) t$$

$$\theta = \iint \alpha \, dt^2 = \theta_0 + \omega_0 t + \left(\frac{M}{2I}\right) t^2$$

Description

The rotation about an arbitrary fixed axis q is found from Eq. 35.76. Equation 35.77 through Eq. 35.79 apply when the angular acceleration of the rotating body is variable. Equation 35.80 through Eq. 35.82 apply when the angular acceleration of the rotating body is constant.[6] Equation 35.83 through Eq. 35.85 apply when the moment applied to the fixed axis is constant. The change in kinetic energy (i.e., the work done to accelerate from ω_0 to ω) is calculated using Eq. 35.86.

Example

A 50 N wheel has a mass moment of inertia of 2 kg·m^2. The wheel is subjected to a constant 1 N·m torque. What is most nearly the angular velocity of the wheel 5 s after the torque is applied?

(A) 0.5 rad/s

(B) 3 rad/s

(C) 5 rad/s

(D) 10 rad/s

---

[6]The use of subscript c in the *NCEES Handbook* Dynamics section to designate a constant angular acceleration is not a normal and customary engineering usage. Since subscript c is routinely used in dynamics to designate centroidal (mass center), the subscript is easily misinterpreted.

Solution

Use Eq. 35.83 to find the angular acceleration of the wheel when subjected to a 1 N·m moment.

$$\alpha = M_q/I_q = \frac{1\ \text{N·m}}{2\ \text{kg·m}^2}$$

$$= 0.5\ \text{rad/s}^2$$

From Eq. 35.84, the angular velocity after 5 s is

$$\omega = \omega_0 + \alpha t = 0\ \frac{\text{rad}}{\text{s}} + \left(0.5\ \frac{\text{rad}}{\text{s}^2}\right)(5\ \text{s})$$

$$= 2.5\ \text{rad/s} \quad (3\ \text{rad/s})$$

The answer is (B).

4. CENTRIPETAL AND CENTRIFUGAL FORCES

Newton's second law states that there is a force for every acceleration that a body experiences. For a body moving around a curved path, the total acceleration can be separated into tangential and normal components. By Newton's second law, there are corresponding forces in the tangential and normal directions. The force associated with the normal acceleration is known as the *centripetal force*. The centripetal force is a real force on the body toward the center of rotation. The so-called *centrifugal force* is an apparent force on the body directed away from the center of rotation. The centripetal and centrifugal forces are equal in magnitude but opposite in sign.

The centrifugal force on a body of mass m with distance r from the center of rotation to the center of mass is

$$F_c = ma_n = \frac{mv_t^2}{r} = mr\omega^2$$

5. BANKING OF CURVES

If a vehicle travels in a circular path on a flat plane with instantaneous radius r and tangential velocity v_t, it will experience an apparent centrifugal force. The centrifugal force is resisted by a combination of roadway banking (superelevation) and sideways friction. The vehicle weight, W, corresponds to the normal force. For small banking angles, the maximum frictional force is

$$F_f = \mu_s N = \mu_s W$$

For large banking angles, the centrifugal force contributes to the normal force. If the roadway is banked so that friction is not required to resist the centrifugal force, the superelevation angle, θ, can be calculated from

$$\tan \theta = \frac{v_t^2}{gr}$$

36

Energy and Work

Nomenclature

e	coefficient of restitution	–
E	energy	J
F	force	N
g	gravitational acceleration, 9.81	m/s^2
h	height	m
k	spring constant	N/m
m	mass	kg
M	moment	N·m
p	linear momentum	kg·m/s
P	power	W
r	distance	m
s	position	m
t	time	s
T	kinetic energy	J
U	potential energy	J
v	velocity	m/s
W	work	J
x	displacement	m
y	horizontal displacement	m

Symbols

ε	efficiency	–
θ	angle	deg
ω	angular velocity	rad/s

Subscripts

1→2	moving from state 1 to state 2
c	centroidal or constant
e	elastic
f	final or frictional
F	force
g	gravity
IC	instantaneous center
M	moment
n	normal
s	spring
W	weight

1. INTRODUCTION

The *energy* of a mass represents the capacity of the mass to do work. Such energy can be stored and released. There are many forms that the stored energy can take, including mechanical, thermal, electrical, and magnetic energies. Energy is a positive, scalar quantity, although the change in energy can be either positive or negative. *Work*, W, is the act of changing the energy of a mass. Work is a signed, scalar quantity. Work is positive when a force acts in the direction of motion and moves a mass from one location to another. Work is negative when a force acts to oppose motion. (Friction, for example, always opposes the direction of motion and can only do negative work.) The net work done on a mass by more than one force can be found by superposition.

Equation 36.1 Through Eq. 36.6: Work[1]

$$W = \int \mathbf{F} \cdot d\mathbf{r} \qquad \text{36.1}$$

$$U_F = \int F \cos\theta\, ds \quad \text{[variable force]} \qquad \text{36.2}$$

$$U_F = (F_c \cos\theta)\Delta s \quad \text{[constant force]} \qquad \text{36.3}$$

$$U_W = -W\Delta y \quad \text{[weight]} \qquad \text{36.4}$$

$$U_s = -\tfrac{1}{2}k(s_2^2 - s_1^2) \quad \text{[spring]} \qquad \text{36.5}$$

$$U_M = M\Delta\theta \quad \text{[couple moment]} \qquad \text{36.6}$$

Description

The work performed by a force is calculated as a dot product of the force vector acting through a displacement vector, as shown in Eq. 36.1. Since the dot product of two vectors is a scalar, work is a scalar quantity. The integral in Eq. 36.1 is essentially a summation over all forces acting at all distances.[2] Only the component of force in the direction of motion does work. In Eq. 36.2 and Eq. 36.3, the component of force in the direction of motion is $F\cos\theta$, where θ represents the acute angle between the force and the direction vectors. For a single constant force (or, a force resultant), the integral can be

---

[1]In Eq. 36.2 through Eq. 36.6, the variable for work is given as U for consistency with the NCEES *FE Reference Handbook* (*NCEES Handbook*). Normally, it is given as W.

[2](1) The *NCEES Handbook* attempts to distinguish between work and stored energy. For example, the work-energy principle (called the principle of work and energy in the *NCEES Handbook*) is essentially presented as $U_2 - U_1 = W$. However, there is no energy storage associated with a force, say, moving a box across a frictionless surface, which is one of the possible applications of Eq. 36.3. (2) When denoting work associated with a translating body, the *NCEES Handbook* uses both W and U. (3) The *NCEES Handbook* is inconsistent in its use of the variable U, which has three meanings: work, stored energy, and change in stored energy. (4) The *NCEES Handbook* is inconsistent in the variable used to indicate position or distance. Both r and s are used in this section.

dropped, and the differential ds replaced with Δs, as in Eq. 36.3.[3] Equation 36.4 represents the work done in a moving weight, W, a vertical distance, Δy, against earth's gravitational field.[4] Equation 36.5 represents the work associated with an extension or compression of a spring with a spring constant k.[5] Equation 36.6 is the work performed by a couple (i.e., a moment), M, rotating through an angle θ.[6]

2. KINETIC ENERGY

Kinetic energy is a form of mechanical energy associated with a moving or rotating body.

Equation 36.7: Linear Kinetic Energy[7]

$$T = mv^2/2 \qquad 36.7$$

Description

The *linear kinetic energy* of a body moving with instantaneous linear velocity v is calculated from Eq. 36.7.

---

[3]The *NCEES Handbook* is inconsistent in its use of the subscript c. In Eq. 36.3, F_c means a constant force. F_c is not a force directed through the centroid.

[4](1) The subscript W is not associated with work, but rather, is associated with the object (i.e., weight) that is moved. (2) The meaning of the negative sign is ambiguous. If Δy is assumed to mean y_2-y_1, then the negative sign would support a thermodynamic first law interpretation (i.e., work is negative when the surroundings do work on the system). However, Eq. 36.1, Eq. 36.2, Eq. 36.3, and Eq. 36.6 do not have negative signs, so these equations do not seem to be written to be consistent with a thermodynamic sign convention. Δy could mean y_1-y_2, and the negative sign may represent mere algebraic convenience.

[5](1) Equation 36.5 is incorrectly presented in the *NCEES Handbook*. The 1/2 multiplier is incorrectly shown inside the parentheses. (2) There is no mathematical reason why the spring constant, k, cannot be brought outside of the parentheses. (3) Whereas the subscripts F, W, and M in Eq. 36.3, Eq. 36.4, and Eq. 36.6 are uppercase, the subscript s in Eq. 36.5 is lowercase. (4) Whereas the subscripts F, W, and M in Eq. 36.3, Eq. 36.4, and Eq. 36.6 are derived from the source of the energy change (i.e., from the item that moves), the subscript s in Eq. 36.5 is derived from the independent variable that changes. If a similar convention had been followed with Eq. 36.3, Eq. 36.4, and Eq. 36.6, the variables in those equations would have been U_r, U_s, and U_θ. (5) For a compression spring acted upon by an increasing force, $s_2 < s_1$, so the negative sign is incorrect for this application from a thermodynamic system standpoint. (6) This equation is shown in a subsequent column of this section of the *NCEES Handbook* as $U_2 - U_1 = k(x_2^2 - x_1^2)/2$, which is not only a different format, but uses x instead of s, and changes the meaning of U from change in energy to stored energy.

[6]The *NCEES Handbook* associates Eq. 36.6 with a "couple moment," an uncommon term. A property of a couple is the moment it imparts, so it is appropriate to speak of the moment of a couple. Similarly, a property of a hurricane is its wind speed, but referring to the hurricane itself as a hurricane speed would be improper. If Eq. 36.6 is meant to describe a pure moment causing rotation without translation, the terms *couple*, *pure moment*, or *torque* would all be appropriate.

[7](1) The *NCEES Handbook* uses different variables to represent kinetic energy. In its section on Units, KE is used. In its Dynamics section, T is used. (2) In its description of Eq. 36.7, the *NCEES Handbook* uses bold **v**, indicating a vector quantity, but subsequently, does not indicate a vector quantity in the equation. Kinetic energy is not a vector quantity, and a vector velocity is not required to calculate kinetic energy.

Example

A 3500 kg car traveling at 65 km/h skids. The car hits a wall 3 s later. The coefficient of friction between the tires and the road is 0.60, and the speed of the car when it hits the wall is 0.20 m/s. What is most nearly the energy that the bumper must absorb in order to prevent damage to the rest of the car?

(A) 70 J

(B) 140 J

(C) 220 J

(D) 360 kJ

Solution

Using Eq. 36.7, the kinetic energy of the car is

$$T = mv^2/2 = \frac{(3500 \text{ kg})\left(0.20 \, \frac{\text{m}}{\text{s}}\right)^2}{2} = 70 \text{ J}$$

The answer is (A).

Equation 36.8: Rotational Kinetic Energy

$$T = I_{IC}\omega^2/2 \qquad 36.8$$

Description

The *rotational kinetic energy* of a body moving with instantaneous angular velocity ω about an instantaneous center is described by Eq. 36.8. I_{IC} is the mass moment of inertia about the *instantaneous center* of rotation.[8]

Example

A 10 kg homogeneous disk of 5 cm radius rotates on a massless axle AB of length 0.5 m and rotates about a fixed point A. The disk is constrained to roll on a horizontal floor.

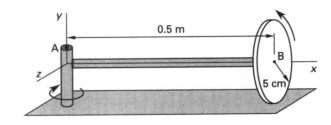

---

[8]Depending on the configuration, in addition to I, the *NCEES Handbook* uses different variables for the mass moment of inertia of bodies with rotational components. For rotation about an arbitrary axis, q, the *NCEES Handbook* uses I_q. For bodies rotating around axes other than their own centroidal axes, it uses I_{IC}. For unconstrained bodies rotating around axes passing through their own centroidal axes, I_{IC} in Eq. 36.8 is the same as I_c in Eq. 36.9 and Eq. 36.10.

Given an angular velocity of 30 rad/s about the x-axis and -3 rad/s about the y-axis, the kinetic energy of the disk is most nearly

(A) 5.6 J

(B) 17 J

(C) 23 J

(D) 34 J

Solution

Assuming the axle is part of the disk, the disk has a fixed point at A. Since the x-, y-, and z-axes are principal axes of inertia for the disk, the kinetic energy is most nearly

$$T = I_{IC}\omega^2/2 = \frac{I_x\omega_x^2}{2} + \frac{I_y\omega_y^2}{2} + \frac{I_z\omega_z^2}{2}$$

$$= \frac{\frac{1}{2}mr^2\omega_x^2}{2} + \frac{mL^2\omega_y^2}{2} + 0$$

$$= \frac{(\frac{1}{2})(10\text{ kg})\left(\frac{5\text{ cm}}{100\frac{\text{cm}}{\text{m}}}\right)^2\left(30\frac{\text{rad}}{\text{s}}\right)^2}{2}$$

$$+ \frac{(10\text{ kg})(0.5\text{ m})^2\left(-3\frac{\text{rad}}{\text{s}}\right)^2}{2} + 0$$

$$= 16.9\text{ J} \quad (17\text{ J})$$

The answer is (B).

Equation 36.9 and Eq. 36.10: Kinetic Energy of Rigid Bodies

$$T = mv^2/2 + I_c\omega^2/2 \qquad 36.9$$
$$T = m(v_{cx}^2 + v_{cy}^2)/2 + I_c\omega_z^2/2 \qquad 36.10$$

Description

Equation 36.9 gives the kinetic energy of a rigid body. For general plane motion in which there are translational and rotational components, the kinetic energy is the sum of the translational and rotational forms. Equation 36.10 gives the kinetic energy for motion in the x-y plane.

Example

A uniform disk with a mass of 10 kg and a diameter of 0.5 m rolls without slipping on a flat horizontal surface, as shown.

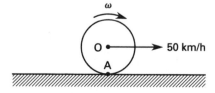

When its horizontal velocity is 50 km/h, the total kinetic energy of the disk is most nearly

(A) 1000 J

(B) 1200 J

(C) 1400 J

(D) 1600 J

Solution

The linear velocity is

$$v_0 = \frac{\left(50\frac{\text{km}}{\text{h}}\right)\left(1000\frac{\text{m}}{\text{km}}\right)}{\left(60\frac{\text{s}}{\text{min}}\right)\left(60\frac{\text{min}}{\text{h}}\right)}$$

$$= 13.89\text{ m/s}$$

The angular velocity is

$$\omega = \frac{v_0}{r} = \frac{13.89\frac{\text{m}}{\text{s}}}{\frac{0.5\text{ m}}{2}}$$

$$= 55.56\text{ rad/s}$$

Using Eq. 36.9, the total kinetic energy is

$$T = mv_0^2/2 + I_c\omega^2/2$$

$$= \frac{mv_0^2}{2} + \frac{\left(\frac{1}{2}mR^2\right)\omega^2}{2}$$

$$= \frac{(10\text{ kg})\left(13.89\frac{\text{m}}{\text{s}}\right)^2}{2}$$

$$+ \frac{\left((\frac{1}{2})(10\text{ kg})\left(\frac{0.5\text{ m}}{2}\right)^2\right)\left(55.56\frac{\text{rad}}{\text{s}}\right)^2}{2}$$

$$= 1447\text{ J} \quad (1400\text{ J})$$

The answer is (C).

Equation 36.11: Change in Kinetic Energy

$$T_2 - T_1 = m(v_2^2 - v_1^2)/2 \qquad 36.11$$

Description

The change in kinetic energy is calculated from the difference of squares of velocity, not from the square of the velocity difference (i.e., $m(v_2^2 - v_1^2)/2 \neq m(v_2 - v_1)^2/2$).

Engineering Sciences

3. POTENTIAL ENERGY

Equation 36.12: Potential Energy in Gravity Field

$$U = mgh \qquad 36.12$$

Description

Potential energy (also known as *gravitational potential energy*), U, is a form of mechanical energy possessed by a mass due to its relative position in a gravitational field. Potential energy is lost when the elevation of a mass decreases. The lost potential energy usually is converted to kinetic energy or heat.

Equation 36.13: Force in a Spring (Hooke's Law)

$$F_s = kx \qquad 36.13$$

Description

A spring is an energy storage device because a compressed spring has the ability to perform work. In a perfect spring, the amount of energy stored is equal to the work required to compress the spring initially. The stored spring energy does not depend on the mass of the spring.

Equation 36.13 gives the force in a spring, which is the product of the *spring constant (stiffness)*, k, and the displacement of the spring from its original position, x.

Example

A spring has a constant of 50 N/m. The spring is hung vertically, and a mass is attached to its end. The spring end displaces 30 cm from its equilibrium position. The same mass is removed from the first spring and attached to the end of a second (different) spring, and the displacement is 25 cm. What is most nearly the spring constant of the second spring?

(A) 46 N/m

(B) 56 N/m

(C) 60 N/m

(D) 63 N/m

Solution

The gravitational force on the mass is the same for both springs. From Hooke's law,

$$F_s = k_1 x_1 = k_2 x_2$$

$$k_2 = \frac{k_1 x_1}{x_2} = \frac{\left(50 \ \frac{N}{m}\right)(30 \ cm)}{25 \ cm} = 60 \ N/m$$

The answer is (C).

Equation 36.14: Elastic Potential Energy

$$U = kx^2/2 \qquad 36.14$$

Description

Given a linear spring with spring constant (stiffness), k, the spring's *elastic potential energy* is calculated from Eq. 36.14.

Example

The 40 kg mass, m, shown is acted upon by a spring and guided by a frictionless rail. When the compressed spring is released, the mass stops slightly short of point A. The spring constant, k, is 3000 N/m, and the spring is compressed 0.5 m.

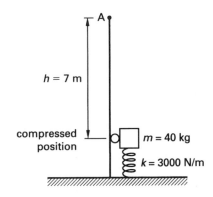

What is most nearly the energy stored in the spring?

(A) 380 J

(B) 750 J

(C) 1500 J

(D) 2100 J

Solution

Using Eq. 36.14, the potential energy is

$$U = kx^2/2 = \frac{\left(3000 \ \frac{N}{m}\right)(0.5 \ m)^2}{2}$$

$$= 375 \ J \quad (380 \ J)$$

The answer is (A).

Equation 36.15: Change in Potential Energy

$$U_2 - U_1 = k(x_2^2 - x_1^2)/2 \qquad 36.15$$

Description

The change in potential energy stored in the spring when the deformation in the spring changes from position x_1 to position x_2 is found from Eq. 36.15.[9]

Equivalent Spring Constant

The entire applied load is felt by each spring in a series of springs linked end-to-end. The *equivalent* (*composite*) *spring constant* for springs in series is

$$\frac{1}{k_{eq}} = \frac{1}{k_1} + \frac{1}{k_2} + \frac{1}{k_3} + \cdots \quad \begin{bmatrix} \text{series} \\ \text{springs} \end{bmatrix}$$

Springs in parallel (e.g., concentric springs) share the applied load. The equivalent spring constant for springs in parallel is

$$k_{eq} = k_1 + k_2 + k_3 + \cdots \quad \begin{bmatrix} \text{parallel} \\ \text{springs} \end{bmatrix}$$

Equation 36.16 Through Eq. 36.18: Combined Potential Energy

$$V = V_g + V_e \qquad 36.16$$
$$V_g = \pm Wy \qquad 36.17$$
$$V_e = +1/2ks^2 \qquad 36.18$$

Description

In mechanical systems, there are two common components of what is normally referred to as potential energy: gravitational potential energy and strain energy.[10] For a system containing a linear, elastic spring that is located at some elevation in a gravitational field, Eq. 36.16 gives the total of these two components.[11] Equation 36.17 gives the potential energy of a weight in a gravitational field.[12] Equation 36.18 gives the strain energy in a linear, elastic spring.[13]

---

[9]The *NCEES Handbook* uses the notation x to denote position, as in Eq. 36.14, and to denote change in length (i.e., a change in position Δx), as in Eq. 36.15. In Eq. 36.5 and Eq. 36.18, the *NCEES Handbook* uses s instead of x as in Eq. 36.15, but the meaning is the same.

[10]Equation 36.16 is not limited to mechanical systems. Potential energy storage exists in electrical, magnetic, fluid, pneumatic, and thermal systems also.

[11](1) Although PE is used in the Units section of the *NCEES Handbook* to identify potential energy, and U is defined as energy in the Dynamics section, here the *NCEES Handbook* introduces a new variable, V, for potential energy. Outside of the conservation of energy equation, this new variable does not seem to be used anywhere else in the *NCEES Handbook*. (2) V_g and V_e have previously (in the *NCEES Handbook*) been represented by U_W and U_s, among others. (3) The subscripts g and e are undefined, but gravitational acceleration is implied for g. The meaning of e is unclear but almost certainly refers to an elastic strain energy.

[12](1) The *NCEES Handbook* is inconsistent in the variable used to represent deflection. Equation 36.17 uses y while other equations in the Dynamics section of the *NCEES Handbook* use Δy and h. y is implicitly the distance from some arbitrary elevation for which $y = 0$ is assigned. (2) This use of $\pm$ is inconsistent with a thermodynamic interpretation of energy, as was apparently used in Eq. 36.4. The sign of the energy would normally be derived from the position, which can be positive or negative. By using $\pm$, the implication is that y is always a positive quantity, regardless of whether the mass is above or below the reference datum.

4. ENERGY CONSERVATION PRINCIPLE

According to the *energy conservation principle*, energy cannot be created or destroyed. However, energy can be transformed into different forms. Therefore, the sum of all energy forms of a system is constant.

$$\sum E = \text{constant}$$

Because energy can neither be created nor destroyed, external work performed on a conservative system must go into changing the system's total energy. This is known as the *work-energy principle*.

$$W = E_2 - E_1$$

Generally, the principle of conservation of energy is applied to mechanical energy problems (i.e., conversion of work into kinetic or potential energy).

Conversion of one form of energy into another does not violate the conservation of energy law. Most problems involving conversion of energy are really special cases. For example, consider a falling body that is acted upon by a gravitational force. The conversion of potential energy into kinetic energy can be interpreted as equating the work done by the constant gravitational force to the change in kinetic energy.

Equation 36.19: Law of Conservation of Energy (Conservative Systems)

$$T_2 + U_2 = T_1 + U_1 \qquad 36.19$$

Description

For *conservative systems* where there is no energy dissipation or gain, the total energy of the mass is equal to the sum of the kinetic and potential (gravitational and elastic) energies.

Example

A projectile with a mass of 10 kg is fired directly upward from ground level with an initial velocity of 1000 m/s. Neglecting the effects of air resistance, what will be the speed of the projectile when it impacts the ground?

(A) 710 m/s

(B) 980 m/s

(C) 1000 m/s

(D) 1400 m/s

---

[13](1) Equation 36.18 has previously been presented with different variables in this section as Eq. 36.14. (2) Equation 36.18 uses s while other equations in the Dynamics section of the *NCEES Handbook* use x. (3) The + symbol is ambiguous, but probably should be interpreted as meaning kinetic energy is always positive. The + is redundant, because the s^2 term is always positive. (4) Equation 36.18 should not be interpreted as one over two times ks^2.

Solution

Use the law of conservation of energy.

$$T_2 + U_2 = T_1 + U_1$$

$$\frac{mv_2^2}{2} + mgh_2 = \frac{mv_1^2}{2} + mgh_1$$

$$(mgh_1 - mgh_2) + \frac{mv_1^2 - mv_2^2}{2} = 0$$

$$0 + \frac{m(v_1^2 - v_2^2)}{2} = 0$$

$$v_2^2 = v_1^2$$

$$v_2 = v_1$$

$$= 1000 \text{ m/s}$$

If air resistance is neglected, the impact velocity will be the same as the initial velocity.

The answer is (C).

Equation 36.20: Law of Conservation of Energy (Nonconservative Systems)

$$T_2 + U_2 = T_1 + U_1 + W_{1\rightarrow 2} \qquad 36.20$$

Description

Nonconservative forces (e.g., friction) are accounted for by the work done by the nonconservative forces in moving between state 1 and state 2, $W_{1\rightarrow 2}$. If the nonconservative forces increase the energy of the system, $W_{1\rightarrow 2}$ is positive. If the nonconservative forces decrease the energy of the system, $W_{1\rightarrow 2}$ is negative.

37 Electrostatics

Nomenclature

A	area	m^2
B	magnetic flux density	T
d	distance	m
E	electric field intensity	N/C or V/m
F	force	N
H	magnetic field strength	A/m
$i(t)$	time-varying current	A
I	constant current	A
J	current density	A/m^2
l	distance moved	m
L	length of a conductor	m
N	number	–
$q(t)$	time-varying charge	C
Q	constant charge	C
r	radius	m
S	surface area	m^2
t	time	s
v	velocity	m/s
v	voltage	V
V	constant voltage or potential difference	V
W	work	J

Symbols

ε	permittivity	F/m or $C^2/N{\cdot}m^2$
μ	permeability	H/m
ρ	flux density	C/m, C/m^2, or C/m^3
ϕ	magnetic flux	Wb
Φ	electric flux	C

Subscripts

0	free space (vacuum)
encl	enclosed
H	magnetic field
L	line or per unit length
S	per unit area, sheet, or surface
V	volume

1. INTRODUCTION

Electric charge is a fundamental property of subatomic particles. The charge on an electron is negative one *electrostatic unit* (esu). The charge on a proton is positive one esu. A neutron has no charge. Charge is measured in the SI system in *coulombs* (C). One coulomb is approximately 6.24×10^{18} esu; the charge of one electron is -1.6×10^{-19} C.

Conservation of charge is a fundamental principle or law of physics. Electric charge can be distributed from one place to another under the influence of an electric field, but the algebraic sum of positive and negative charges in a system cannot change.

2. ELECTROSTATIC FIELDS

An *electric field, E*, with units of newtons per coulomb or volts per meter (N/C, same as V/m) is generated in the vicinity of an electric charge. The imaginary lines of force, as illustrated in Fig. 37.1, are called the *electric flux*, Φ. The direction of the electric flux is the same as the force applied by the electric field to a positive charge introduced into the field. If the field is produced by a positive charge, the force on another positive charge placed nearby will try to separate the two charges, and therefore, the lines of force will leave the first positive charge.

Figure 37.1 *Electric Field Around a Positive Charge*

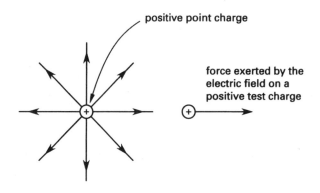

The electric field is a vector quantity having both magnitude and direction. The orientations of the field and flux lines always coincide (i.e., the direction of the electric field vector is always tangent to the flux lines). The total electric flux generated by a point charge is proportional to the charge.

$$\Phi = \frac{Q}{\varepsilon}$$

Equation 37.1 and Eq. 37.2: Forces on Charges

$$\mathbf{F} = Q\mathbf{E} \qquad 37.1$$

$$\mathbf{F}_2 = \frac{Q_1 Q_2}{4\pi\varepsilon r^2}\mathbf{a}_{r12} \qquad 37.2$$

Values

Electric flux does not pass equally well through all materials. It cannot pass through conductive metals at all and is canceled to various degrees by insulating media. The *permittivity* of a medium, ε, determines the flux that passes through the medium. For free space or air, $\varepsilon = \varepsilon_0 = 8.85 \times 10^{-12}$ F/m $= 8.85 \times 10^{-12}$ C^2/N·m^2.

Description

In general, the *force on a test charge Q* in an electric field E is given by Eq. 37.1.

The force experienced by point charge 2, Q_2, in an electric field E created by point charge 1, Q_1, is given by *Coulomb's law*, Eq. 37.2. Because charges with opposite signs attract, Eq. 37.2 is positive for repulsion and negative for attraction. The unit vector $\mathbf{a}_{r12}$ is defined pointing from point charge 1 toward point charge 2. Although the unit vector $\mathbf{a}$ gives the direction explicitly, the direction of force can usually be found by inspection as the direction the object would move when released. Vector addition (i.e., superposition) can be used with systems of multiple point charges.

Example

Two point charges, Q_1 and Q_2, are shown. Q_1 is a charge of 5×10^{-6} C, and Q_2 is a charge of -10×10^{-6} C. The permittivity of the medium is 8.85×10^{-12} F/m, and the distance between the charges is 10 cm.

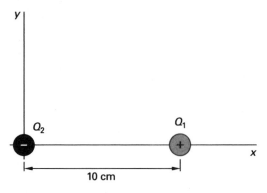

What are most nearly the magnitude and direction of the electrostatic force that acts on Q_2 due to Q_1?

(A) 45 N, from Q_1 to Q_2

(B) 45 N, from Q_2 to Q_1

(C) 89 N, from Q_2 to Q_1

(D) 110 N, from Q_1 to Q_2

Solution

The force is the direction Q_2 would move if unrestrained. Opposite charges attract, so the force is toward Q_1.

Find the magnitude of the force between the point charges using Coulomb's law, Eq. 37.2.

$$
\begin{aligned}
|\mathbf{F}_{21}| &= \frac{Q_1 Q_2}{4\pi\varepsilon r_{12}^2} \\
&= \frac{(5 \times 10^{-6}\ \text{C})(-10 \times 10^{-6}\ \text{C})}{4\pi\left(8.85 \times 10^{-12}\ \dfrac{\text{F}}{\text{m}}\right)\left(\dfrac{10\ \text{cm}}{100\ \dfrac{\text{cm}}{\text{m}}}\right)^2} \\
&= 44.96\ \text{N} \quad (45\ \text{N})
\end{aligned}
$$

The answer is (B).

Equation 37.3: Electric Field Intensity Due to a Point Charge

$$\mathbf{E} = \frac{Q_1}{4\pi\varepsilon r^2}\mathbf{a}_{r12} \qquad 37.3$$

Description

Equation 37.3 is the *electric field intensity* in a medium with permittivity ε at a distance r from a point charge Q_1. The direction of the electric field is represented by the unit vector $\mathbf{a}_{r12}$.

Equation 37.4: Gauss' Law

$$Q_{\text{encl}} = \oiint_S \varepsilon \mathbf{E} \cdot d\mathbf{S} \qquad 37.4$$

Variations

$$\Phi = \frac{\sum q_{\text{encl}}}{\varepsilon} = \frac{Q_{\text{encl}}}{\varepsilon}$$

$$\Phi = \oiint_S \mathbf{E} \cdot d\mathbf{S}$$

Description

Gauss' law states that the total electric flux passing out of an enclosing (closed) surface (i.e., the *Gaussian surface*) is proportional to the total charge within the surface, as shown in the variation equation.

The mathematical formulation of Gauss' law (see Eq. 37.4) states that the total enclosed charge can be determined by summing all of the electric fields on the Gaussian surface, $\mathbf{S}$. The variable $d\mathbf{S}$ is a vector that represents an infinitesimal part of the closed surface, the direction of which is perpendicular to the surface.

Equation 37.5 and Eq. 37.6: Work and Energy in Electric Fields

$$W = -Q \int_{p_1}^{p_2} \mathbf{E} \cdot d\mathbf{l} \qquad 37.5$$

$$W_E = (1/2) \iiint_V \varepsilon |\mathbf{E}|^2 \, dV \qquad 37.6$$

Description

The *work*, W, performed by moving a charge Q_1 radially from point p_1 to point p_2 in an electric field is given by Eq. 37.5.[1] (The dot product of two vectors is a scalar.) The energy stored in an electric field is given by Eq. 37.6.

The work, W, performed in moving a point charge Q_B in the radial direction from distance r_1 to r_2 within a field created by a point charge Q_A is given by

$$W = -\int_{r_1}^{r_2} \mathbf{F} \cdot d\mathbf{r} = -\int_{r_1}^{r_2} \frac{Q_A Q_B}{4\pi\varepsilon r^2} \, dr$$

$$= \left(\frac{Q_A Q_B}{4\pi\varepsilon} \right) \left(\frac{1}{r_2} - \frac{1}{r_1} \right)$$

Work is positive if an external force is required to move the charges (e.g., to bring two repulsive charges together or move a charge against an electric field). Work is negative if the field does the work (allowing attracting charges to approach each other, or allowing repulsive charges to separate).

Work is performed only in moving the charges closer or further apart. Moving one point charge around the other in a constant-radius circle performs no work. In general, no work is performed in moving a charged object perpendicular to an electric field.

For a uniform field (as exists between two charged plates separated by a distance d), the work done in moving an object of charge Q a distance l parallel to the uniform field is given by

$$W = -\mathbf{F} \cdot \mathbf{l} = -EQl = \frac{-V_{\text{plates}} Ql}{d}$$

$$= -Q\Delta V$$

---

[1] In Eq. 37.5, the NCEES *FE Reference Handbook* (*NCEES Handbook*) shows the limits of integration as p_1 and p_2. However, the variable p is not in the equation. Equation 37.5 should not be interpreted too literally, since p is not a variable.

Example

What is most nearly the work required to move a positive charge of 10 C for a distance of 5 m in the same direction as a uniform field of 50 V/m?

(A) $-13\,000$ J

(B) -2500 J

(C) -100 J

(D) -20 J

Solution

From Eq. 37.5, the work required to move a charge in a uniform field is

$$W = -Q \int_{p_1}^{p_2} \mathbf{E} \cdot d\mathbf{l} = -QEl$$

$$= -(10 \text{ C}) \left(50 \; \frac{\text{V}}{\text{m}} \right) (5 \text{ m})$$

$$= -2500 \text{ C} \cdot \text{V} \quad (-2500 \text{ J})$$

The positive charge moves in the direction of the field. Therefore, no external work is required, and the charge returns potential energy to the field.

The answer is (B).

3. VOLTAGE

Voltage is another way to describe the strength of an electric field, using a scalar quantity rather than a vector quantity. The *potential difference*, V, is the difference in electric potential between two points. Potential difference is equal to the work required to move one unit charge from one point to the other. This difference in potential is one volt if one joule of work is expended in moving one coulomb of charge from one point to the other.

Equation 37.7: Electric Field Strength Between Two Parallel Plates

$$E = \frac{V}{d} \qquad 37.7$$

Description

The electric field strength between two parallel plates with potential difference V and separated by a distance d is given by Eq. 37.7.

By convention, the field is directed from the positive plate to the negative plate.

Example

What is most nearly the electric field strength between two plates separated by 0.005 m that are connected across 100 V?

- (A) 0.5 kV/m
- (B) 2 kV/m
- (C) 5 kV/m
- (D) 20 kV/m

Solution

Calculate the electric field strength using Eq. 37.7.

$$E = \frac{V}{d} = \frac{100 \text{ V}}{(0.005 \text{ m})\left(1000 \dfrac{\text{V}}{\text{kV}}\right)} = 20 \text{ kV/m}$$

The answer is (D).

4. CURRENT

Equation 37.8: Current in Electric Fields

$$i(t) = dq(t)/dt \qquad \textbf{37.8}$$

Description

Current, $i(t)$, is the movement of charges. By convention, the current moves in a direction opposite to the flow of electrons (i.e., the current flows from the positive terminal to the negative terminal). Current is measured in amperes (A) and is the time rate change of charge (i.e., the current is equal to the number of coulombs of charge passing a point each second). If $q(t)$ is the instantaneous charge, then the current is given by Eq. 37.8.

If the rate of change in charge is constant, the current is denoted as I.

$$I = \frac{dq}{dt}$$

The *areal current density*, J (usually referred to simply as *current density*), is the current per unit surface area. Since current density is charge per unit time, the current can be written as the rate of change of charge surface density.

$$J = \frac{I}{S} = \frac{\dot{\rho}_S S}{S} = \dot{\rho}_S$$

The *volume current density*, J_V, is the current per unit volume. For a conductor with cross-sectional area S and length L, the volume current density can be expressed as the product of the charge surface density and charge velocity, v.

$$J_V = \frac{I}{V} = \frac{\dot{\rho}_S S}{SL} = \dot{\rho}_S \text{v}$$

The current in the direction perpendicular to a surface, S, can be determined by integrating the current density moving through the surface.

$$i = \int_S \mathbf{J} \cdot d\mathbf{S}$$

Example

The current in a particular DC circuit is numerically equal to one-quarter of the time the current has been running through the circuit. Assuming the circuit starts out carrying no current, what is most nearly the electrical charge delivered by the circuit in 5 s?

- (A) 0.2 C
- (B) 3 C
- (C) 6 C
- (D) 30 C

Solution

Rearrange Eq. 37.8 to solve for charge. Because this calculation assumes that $i(t) = t/4$, the calculation is dimensionally inconsistent.

$$i(t) = dq(t)/dt$$
$$\frac{t}{4} = \frac{dq(t)}{dt}$$
$$\int dq(t) = \int_{0 \text{ s}}^{5 \text{ s}} \frac{t}{4} \, dt$$
$$= \frac{t^2}{8} \bigg|_{0 \text{ s}}^{5 \text{ s}}$$
$$= \frac{(5 \text{ s})^2}{8} - \frac{(0 \text{ s})^2}{8}$$
$$= 3.125 \text{ C} \quad (3 \text{ C})$$

The answer is (B).

5. MAGNETIC FIELDS

A magnetic field can exist only with two opposite, equal poles called the *north pole* and *south pole*. This is unlike an electric field, which can be produced by a single charged object. Figure 37.2 illustrates two common permanent magnetic field configurations. It also illustrates the convention that the lines of magnetic flux are directed from the north pole (i.e., the *magnetic source*) to the south pole (i.e., the *magnetic sink*).

The total amount of magnetic flux in a magnetic field is ϕ, measured in webers (Wb). The flux is given by *Gauss' law* for a magnetic field.

$$\phi = \oint \mathbf{B} \cdot d\mathbf{A} = 0$$

Figure 37.2 Magnetic Fields from Permanent Magnets

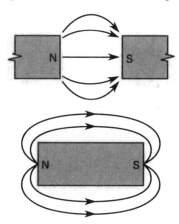

The magnetic flux density, **B**, in teslas (T), equivalent to Wb/m^2, is one of two measures of the strength of a magnetic field. For this reason, it can be referred to as the *strength of the B-field*. (**B** should never be called the magnetic field strength, as that name is reserved for the variable **H**.) **B** is also known as the *magnetic induction*. The magnetic flux density is found by dividing the magnetic flux by an area perpendicular to it. Magnetic flux density is a vector quantity, calculated from

$$\mathbf{B} = \frac{\phi}{A}\mathbf{a}$$

The direction of the magnetic field, illustrated in Fig. 37.3, is given by the *right-hand rule*. In the case of a straight wire, the thumb indicates the current direction, and the fingers curl in the field direction; for a coil, the fingers indicate the current flow, and the thumb indicates the field direction.

Figure 37.3 Right-Hand Rule for the Magnetic Flux Direction in a Coil

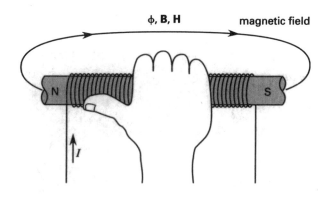

Equation 37.9: Magnetic Field Strength

$$\mathbf{H} = \frac{\mathbf{B}}{\mu} = \frac{I\mathbf{a}_\phi}{2\pi r} \qquad 37.9$$

Values

The *permeability of free space* (air or vacuum) is $\mu_0 = 4\pi \times 10^{-7}$ H/m.

Description

The *magnetic field strength*, **H**, with units of A/m, is derived from the magnetic flux density as in Eq. 37.9.

The magnetic flux density, **B**, is dependent on the *permeability* of the medium much like the electric flux density is dependent on permittivity.

Example

A magnetic field in air has a magnetic flux density of 1×10^{-8} T. What is most nearly the strength of the magnetic field in a vacuum?

(A) 3×10^{-4} A/m

(B) 8×10^{-3} A/m

(C) 3×10^{-2} A/m

(D) 8×10^{-2} A/m

Solution

The permeability of free space is $4\pi \times 10^{-7}$ H/m. The strength of the magnetic field is

$$\mathbf{H} = \frac{\mathbf{B}}{\mu_0} = \frac{1 \times 10^{-8}\ \text{T}}{4\pi \times 10^{-7}\ \frac{\text{H}}{\text{m}}}$$
$$= 7.96 \times 10^{-3}\ \text{A/m} \quad (8 \times 10^{-3}\ \text{A/m})$$

The answer is (B).

Equation 37.10: Force on a Current-Carrying Conductor in a Uniform Magnetic Field

$$\mathbf{F} = I\mathbf{L} \times \mathbf{B} \qquad 37.10$$

Description

The analogy to Coulomb's law, where a force is imposed on a stationary charge in an electric field, is that a magnetic field imposes a force on a moving charge. The force on a wire carrying a current I in a uniform magnetic field **B** is given by Eq. 37.10. **L** is the length vector of the conductor and points in the direction of the current. The force acts at right angles to both the current and magnetic flux density directions.

Equation 37.11: Energy Stored in a Magnetic Field

$$W_H = (1/2)\iiint_V \mu|\mathbf{H}|^2\,dv \qquad 37.11$$

Engineering Sciences

Variation

$$W_H = \tfrac{1}{2} \iiint_V \mathbf{B} \cdot \mathbf{H} \, dV$$

Description

The energy stored in volume V within a magnetic field, $\mathbf{H}$, can be calculated from Eq. 37.11.[2] Equation 37.11 assumes that the $\mathbf{B}$ and $\mathbf{H}$ fields are in the same direction. Assuming the magnetic field is constant throughout the volume V, Eq. 37.11 reduces further to

$$W_H = \frac{\mu H^2 V}{2} = \frac{B^2 V}{2\mu}$$

The integral over any closed surface of magnetic flux density must be zero. Stated another way, magnetic flux lines must follow a closed path, and no matter how large or small the enclosing surface is, the path must be either entirely inside the surface or it must go out and back in. This law is referred to as the "no isolated magnetic charge" or "no magnetic monopoles" law.

6. INDUCED VOLTAGE

Equation 37.12 and Eq. 37.13: Faraday's Law of Induction

$$v = -N \, d\phi/dt \qquad \textbf{37.12}$$

$$\phi = \int_S \mathbf{B} \cdot d\mathbf{S} \qquad \textbf{37.13}$$

Variation

$$v = -NBL\frac{ds}{dt}$$

Description

Faraday's law of induction states that an induced voltage, v, also called the *electromotive force* or emf, will be generated in a circuit when there is a change in the magnetic flux. Figure 37.4 illustrates one of N series-connected conductors cutting across magnetic flux ϕ, calculated from Eq. 37.13.

Figure 37.4 *Conductor Moving in a Magnetic Field*

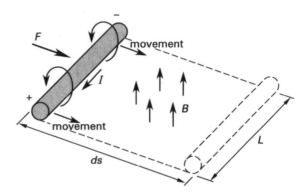

The magnitude of the electromagnetic induction is given by *Faraday's law*, Eq. 37.12. The minus sign indicates the direction of the induced voltage, which is specified by *Lenz's law* to be opposite to the direction of the magnetic field.

Example

A coil has 100 turns. The magnetic flux in the coil decreases from 20 Wb to 10 Wb in 5 s. What is most nearly the magnitude of the voltage induced in the coil?

(A) 50 V

(B) 100 V

(C) 200 V

(D) 300 V

Solution

Use Faraday's law, Eq. 37.12, to find the voltage induced in the coil.

$$v = -N \, d\phi/dt$$
$$= \frac{-(100)(20 \text{ Wb} - 10 \text{ Wb})}{5 \text{ s} - 0 \text{ s}}$$
$$= -200 \text{ V} \quad (200 \text{ V})$$

The answer is (C).

---

[2]The *NCEES Handbook* presents the differential volume in Eq. 37.11 as dv, which is inconsistent with the parallel equation, Eq. 37.6, which uses dV. Since v represents voltage and V is used for volume throughout, the *NCEES Handbook*'s use of dv is considered an error. This section presents Eq. 37.11 with V (not v) and so differs slightly from the *NCEES Handbook*'s equation.

38 Direct-Current Circuits

Nomenclature

A	area	m^2
C	capacitance	F
d	diameter	m
d	distance	m
$i(t)$	time-varying current	A
I	constant current	A
l	length	m
L	inductance	H
L	length	m
N	number of turns	–
P	power	W
$q(t)$	time-varying charge	C
Q	constant charge	C
R	resistance	Ω
t	time	s
v	voltage	V
$v(t)$	time-varying voltage	V
V	constant voltage	V

Symbols

ε	permittivity	F/m or $C^2/N{\cdot}m^2$
μ	permeability	H/m
ρ	resistivity	$\Omega{\cdot}m$
τ	time constant	s

Subscripts

C	capacitive
eq	equivalent
ext	external
fs	full scale
L	inductive or inductor
N	Norton
oc	open circuit
P	parallel
R	reactive
sc	short circuit
S	series
Th	Thevenin

1. INTRODUCTION

Electrical circuits contain active and passive elements. *Active elements* are elements that can generate electric energy, such as voltage and current sources. *Passive elements*, such as capacitors and inductors, absorb or store electric energy; other passive elements, such as resistors, dissipate electric energy.

An *ideal voltage source* supplies power at a constant voltage, regardless of the current drawn. An *ideal current source* supplies power at a constant current independent of the voltage across its terminals. However, real sources have internal resistances that, at higher currents, decrease the available voltage. Therefore, a real voltage source cannot maintain a constant voltage when currents become large. *Independent sources* deliver voltage and current at their rated values regardless of circuit parameters. *Dependent sources* deliver voltage and current at levels determined by voltages or currents elsewhere in the circuit. The symbols for electrical circuit elements and sources are given in Table 38.1.

Table 38.1 *Circuit Element Symbols*

symbol	circuit element
R	resistor
C	capacitor
L	inductor
V or V	independent voltage source
I	independent current source
gV	dependent voltage source
gI	dependent current source

DC Voltage

Voltage, measured in volts (a combined unit equivalent to W/A, C/F, J/C, A/S, and Wb/s), is used to measure the *potential difference* across terminals of circuit elements. Any device that provides electrical energy is called a *seat of an electromotive force* (emf), and the electromotive force is also measured in volts.

2. RESISTORS

Equation 38.1: Resistance

$$R = \frac{\rho L}{A} \qquad 38.1$$

Description

Resistance, R (measured in ohms, Ω), is the property of a circuit or circuit element to oppose current flow. A circuit with zero resistance is a *short circuit*, whereas an *open circuit* has infinite resistance.

Resistors are usually constructed from carbon compounds, ceramics, oxides, or coiled wire. *Resistance* depends on the *resistivity*, ρ (in $\Omega \cdot$m), which is a material property, and the length and cross-sectional area of the resistor. Resistors with larger cross-sectional areas have more free electrons available to carry charge and have less resistance. Each of the free electrons has a limited ability to move, so the electromotive force must overcome the limited mobility for the entire length of the resistor. The resistance increases with the length of the resistor.

Example

A power line is made of copper (resistivity of 1.83×10^{-6} $\Omega \cdot$cm). The wire diameter is 2 cm. What is most nearly the resistance of 5 km of power line?

(A) 0.0032 Ω

(B) 0.29 Ω

(C) 0.67 Ω

(D) 1.8 Ω

Solution

The wire's resistance is directly proportional to its length and inversely proportional to its cross-sectional area. From Eq. 38.1,

$$R = \frac{\rho L}{A} = \frac{\rho L}{\frac{\pi}{4} d^2}$$

$$= \frac{(1.83 \times 10^{-6} \ \Omega \cdot \text{cm})(5 \ \text{km})\left(10^5 \ \frac{\text{cm}}{\text{km}}\right)}{\left(\frac{\pi}{4}\right)(2 \ \text{cm})^2}$$

$$= 0.29 \ \Omega$$

The answer is (B).

Equation 38.2 Through Eq. 38.4: Resistors in Series and Parallel

$$R_S = R_1 + R_2 + \cdots + R_n \qquad 38.2$$

$$R_P = 1/(1/R_1 + 1/R_2 + \cdots + 1/R_n) \qquad 38.3$$

$$R_P = \frac{R_1 R_2}{R_1 + R_2} \qquad 38.4$$

Description

Resistors connected in series share the same current and may be represented by an *equivalent resistance* equal to the sum of the individual resistances. For n resistors in series, the *total resistance*, R_S, is given by Eq. 38.2.

Resistors connected in parallel share the same voltage drop and may be represented by an equivalent resistance equal to the reciprocal of the sum of the reciprocals of the individual resistances. For n resistors in parallel, the total resistance, R_P, is given by Eq. 38.3. The equivalent resistance of two resistors in parallel is given by Eq. 38.4.

Example

For the circuit shown, what is most nearly the equivalent resistance seen by the battery?

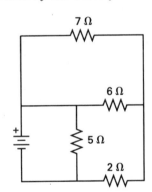

(A) 0.38 Ω

(B) 2.2 Ω

(C) 2.6 Ω

(D) 6.2 Ω

Solution

Redraw the circuit.

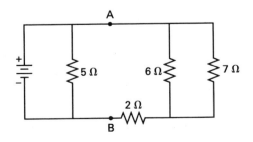

The 7 Ω and 6 Ω resistors are in parallel. The equivalent resistance of terminals A and B is

$$R_{AB} = \frac{(7\ \Omega)(6\ \Omega)}{7\ \Omega + 6\ \Omega} + 2\ \Omega$$
$$= 5.231\ \Omega$$

Using Eq. 38.4, the total resistance seen by the battery is

$$R_P = \frac{R_1 R_2}{R_1 + R_2} = \frac{(5.231\ \Omega)(5\ \Omega)}{5.231\ \Omega + 5\ \Omega}$$
$$= 2.556\ \Omega \quad (2.6\ \Omega)$$

The answer is (C).

Equation 38.5: Joule's Law

$$P = VI = \frac{V^2}{R} = I^2 R \qquad \textbf{38.5}$$

Description

Power is the time rate of energy delivery, usually manifested as a time rate of useful work performed or heat dissipated. In electric circuits, the energy is provided by voltage and/or current sources. In purely resistive circuits, the energy is dissipated in the resistance elements as heat.

In a *direct current* (DC) *circuit*, steady-state voltage and current, V and I respectively, are constant over time.[1] The power dissipated in an individual component with resistance R, or in a circuit with an equivalent resistance R, is given by *Joule's law* (see Eq. 38.5).

Example

A 10 kV power line has a total resistance of 1000 Ω. The current in the line is 10 A. What is most nearly the power lost due to resistive heating?

(A) 1 kW

(B) 10 kW

(C) 100 kW

(D) 1000 kW

Solution

Use Joule's law, Eq. 38.5.

$$P = I^2 R = \frac{(10\ \text{A})^2 (1000\ \Omega)}{1000\ \dfrac{\text{W}}{\text{kW}}} = 100\ \text{kW}$$

---

[1]When energy sources are first connected, *unsteady conditions* (*transient conditions*) may develop.

Check.

$$P = VI = \frac{(10\ \text{kV})\left(1000\ \dfrac{\text{V}}{\text{kV}}\right)(10\ \text{A})}{1000\ \dfrac{\text{W}}{\text{kW}}} = 100\ \text{kW}$$

The answer is (C).

3. CAPACITORS

Equation 38.6 Through Eq. 38.11: Capacitors

$$q_C(t) = C v_C(t) \qquad \textbf{38.6}$$

$$C = q_C(t)/v_C(t) \qquad \textbf{38.7}$$

$$C = \frac{\varepsilon A}{d} \qquad \textbf{38.8}$$

$$i_C(t) = C(dv_C/dt) \qquad \textbf{38.9}$$

$$v_C(t) = v_C(0) + \frac{1}{C}\int_0^t i_C(\tau)\,d\tau \qquad \textbf{38.10}$$

$$\text{energy} = C v_C^2/2 = q_C^2/2C = q_C v_C/2 \qquad \textbf{38.11}$$

Variation

$$Q = CV \qquad [\text{constant } V]$$

Description

A *capacitor* is a device that stores electric charge. Figure 38.1 shows the symbol for a capacitor.[2] A capacitor is constructed as two conducting surfaces separated by an insulator, such as oiled paper, mica, or air. A simple type of capacitor (i.e., the *parallel plate capacitor*) is constructed as two parallel plates. If the plates are connected across a voltage potential, charges of opposite polarity will build up on the plates and create an electric field between the plates. The amount of charge, Q, built up is proportional to the applied voltage. The constant of proportionality, C, is the *capacitance* in farads (F) and depends on the capacitor construction. Capacitance represents the ability to store charge; the greater the capacitance, the greater the charge stored. (See Eq. 38.6 and Eq. 38.7.)

Equation 38.8 gives the capacitance of two parallel plates of equal area A separated by distance d. ε is the *permittivity* of the medium separating the plates.

---

[2]The most generic symbol for a capacitor represents the "plates" by two parallel straight lines. This is the symbol used consistently in this book. A flat-plate capacitor has no polarity, and each plate can hold either positive or negative charges. Some capacitors (e.g., electrolytic capacitors) are polarized, requiring the positive capacitor lead to be connected to the more positive part of the circuit. In Fig. 38.1, the NCEES *FE Reference Handbook* (*NCEES Handbook*) shows a *polarized capacitor*.

Figure 38.1 *Capacitor Symbol*

The current passed by a capacitor is the derivative of the voltage times the capacitance. (See Eq. 38.9.) The voltage depends on the amount of charge (number of charged particles) on the capacitor. Since charged particles are matter, they cannot instantaneously change in quantity. Therefore, the voltage cannot instantaneously change. However, the number of charges leaving the capacitor can instantaneously change, so the current can change instantaneously. Any change in charge on a capacitor is manifested as current in the circuit. Equation 38.10 shows that the voltage across a capacitor changes as the current changes.[3]

The total energy (in joules) stored in a capacitor is given by Eq. 38.11.[4]

Unless voltage changes with time, the amount of charge on a capacitor will not change, and accordingly, there will be no current flow. At steady state, the voltage across all circuit elements in a DC circuit is constant, so no current flows through capacitors in the circuit. At steady state, capacitors in DC circuits behave as open circuits and pass no current.

Example

The capacitor shown has a capacitance of 2.5×10^{-10} F.

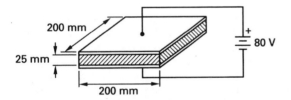

What is most nearly the energy stored in the capacitor?

(A) 0.2 μJ

(B) 0.8 μJ

(C) 1.0 μJ

(D) 20 μJ

---

[3]Not only does the *NCEES Handbook* use the variable reserved for a capacitive time constant, but Eq. 38.10 makes a confusing and unnecessary change of variables in an attempt to differentiate between "real time," t, and "capacitor time," τ. Since the first term, $v_C(0)$, shows that real time starts from $t = 0$ (the same as the lower integration limit of capacitor time), and since real time ends at t (the same as the upper integration limit of capacitor time), t and τ are the same. Equation 38.10 could have been presented as

$$v_C(t) = v_C(0) + \frac{1}{C}\int_0^t i_C(t)\, dt$$

[4]The *NCEES Handbook* is inconsistent in how it represents electrical energy and work terms. The "energy" in Eq. 38.11 and Eq. 38.17 is the same as W_E in Eq. 37.6.

Solution

Using Eq. 38.11, the energy stored in the capacitor is

$$\text{energy} = C v_C^2 / 2$$

$$= \frac{(2.5 \times 10^{-10}\ \text{F})(80\ \text{V})^2 \left(10^6\ \frac{\mu\text{J}}{\text{J}}\right)}{2}$$

$$= 0.8\ \mu\text{J}$$

The answer is (B).

Equation 38.12 and Eq. 38.13: Capacitors in Series and Parallel

$$C_S = \frac{1}{1/C_1 + 1/C_2 + \cdots + 1/C_n} \qquad \textbf{38.12}$$

$$C_P = C_1 + C_2 + \cdots + C_n \qquad \textbf{38.13}$$

Description

The *total capacitance* of capacitors connected in series, C_S, is given by Eq. 38.12. The total capacitance of capacitors connected in parallel, C_P, is given by Eq. 38.13.

Example

What is most nearly the equivalent capacitance between terminals A and B?

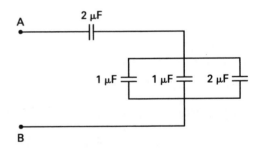

(A) 1.1 μF

(B) 1.3 μF

(C) 2.4 μF

(D) 4.0 μF

Solution

The three capacitors in parallel combine according to Eq. 38.13.

$$C_P = C_1 + C_2 + C_3$$

$$= 1\ \mu\text{F} + 1\ \mu\text{F} + 2\ \mu\text{F}$$

$$= 4\ \mu\text{F}$$

This equivalent capacitance is in series with the 2 μF capacitor. From Eq. 38.12,

$$C_S = \frac{1}{1/C_1 + 1/C_2} = \frac{1}{\dfrac{1}{4\ \mu\text{F}} + \dfrac{1}{2\ \mu\text{F}}}$$

$$= 1.33\ \mu\text{F} \quad (1.3\ \mu\text{F})$$

The answer is (B).

4. INDUCTORS

Equation 38.14 Through Eq. 38.19: Inductors

$L = N^2 \mu A / l$	*38.14*
$v_L(t) = L(di_L/dt)$	*38.15*
$i_L(t) = i_L(0) + \dfrac{1}{L}\displaystyle\int_0^t v_L(\tau)\,d\tau$	*38.16*
$\text{energy} = L i_L^2 / 2$	*38.17*
$L_P = \dfrac{1}{1/L_1 + 1/L_2 + \cdots + 1/L_n}$	*38.18*
$L_S = L_1 + L_2 + \cdots + L_n$	*38.19*

Description

An *inductor* is basically a coil of wire. When connected across a voltage source, current begins to flow in the coil, establishing a magnetic field that opposes current changes. Figure 38.2 shows the symbol for an inductor. Usually, the wire is coiled around a core of magnetic material (high permeability) to increase the inductance. (See Fig. 38.3.) From Faraday's law, the induced voltage across the ends of the inductor is proportional to the change in flux linkage, which in turn is proportional to the current change. The constant of proportionality is the *inductance*, L, expressed in henries (H). (See Eq. 38.14.)[5] Equation 38.19 gives the general definition of an inductor. Energy is stored within the magnetic field and the inductor does not dissipate energy.

Figure 38.2 *Inductor*

$i_L(t) \downarrow$

$L \quad \substack{+ \\ v_L(t) \\ -}$

---

[5]L is the traditional symbol for inductance. However, the *NCEES Handbook* previously used L to designate length (see Eq. 38.12 and Eq. 38.1), which is likely the reason for using l to represent length in Eq. 38.14.

Figure 38.3 *Inductor*

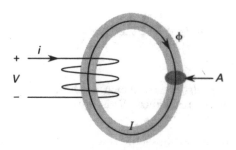

The voltage across an inductor is the derivative of the current times the inductance. (See Eq. 38.15.) The current of inductors cannot change instantaneously, but the voltage can change instantaneously. The current will change as the inductor integrates the voltage to produce a current that opposes the change. (See Eq. 38.16.)

The current depends on the number of charged particles moving through the inductor. Since charged particles are physical matter, they cannot instantaneously change in quantity. Therefore, the current moving in an inductor cannot instantaneously change. However, the voltage across an inductor can instantaneously change. Any change in voltage across an inductor is manifested as a current change in the circuit. Equation 38.16 shows that the current through an inductor changes as the voltage changes.[6]

The total energy (in joules) stored in the electric field of an inductor carrying instantaneous current i_L is given by Eq. 38.17.

The total inductance of inductors connected in parallel, L_P, is given by Eq. 38.18. The total inductance of inductors connected in series, L_S, is given by Eq. 38.19.

5. DC CIRCUIT ANALYSIS

Most circuit problems involve solving for unknown parameters, such as the voltage or current across some element in the circuit. The methods that are used to find these parameters rely on combining elements in series and parallel, and applying Ohm's law or Kirchhoff's laws in a systematic manner.

---

[6]Not only does the *NCEES Handbook* use the variable reserved for an inductive time constant, but *NCEES Handbook* Eq. 38.16 makes a confusing and unnecessary change of variables in an attempt to differentiate between "real time," t, and "inductor time," τ. Since the first term, $i_L(0)$, shows that real time starts from $t=0$ (the same as the lower integration limit of inductor time), and since real time ends at t (the same as the upper integration limit of inductor time), t and τ are the same. Equation 38.16 could have been presented as

$$i_L(t) = i_L(0) + \frac{1}{L}\int_0^t v_L(t)\,dt$$

Equation 38.20: Ohm's Law

$$V = IR \qquad 38.20$$

Description

The *voltage drop*, also known as the *IR drop*, across a circuit with resistance R is given by *Ohm's law*, Eq. 38.20.

Using Ohm's law implicitly assumes a *linear circuit* (i.e., one consisting of linear elements and linear sources). A *linear element* is a passive element whose performance can be represented by a linear voltage-current relationship. The output of a linear source is proportional to the first power of a voltage or current in the circuit. Many components used in electronic devices do not obey Ohm's law over their entire operating ranges.

Example

The voltage between nodes B and E in the circuit shown is 300 V.

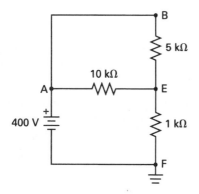

What is most nearly the current flowing between nodes B and E?

(A) 30 mA

(B) 60 mA

(C) 70 mA

(D) 80 mA

Solution

The voltage difference between nodes B and E is given. From Ohm's law,

$$V = IR$$

$$I_{BE} = \frac{V_{BE}}{R} = \frac{(300 \text{ V})\left(1000 \ \frac{\text{mA}}{\text{A}}\right)}{(5 \text{ k}\Omega)\left(1000 \ \frac{\Omega}{\text{k}\Omega}\right)}$$

$$= 60 \text{ mA}$$

The answer is (B).

Equation 38.21 and Eq. 38.22: Kirchhoff's Laws

$$\sum I_{in} = \sum I_{out} \qquad 38.21$$

$$\sum V_{rises} = \sum V_{drops} \qquad 38.22$$

Description

Kirchhoff's current law (KCL) (see Eq. 38.21) states that as much current flows out of a node (connection) as flows into it. Electrons must be conserved at any node in an electrical circuit.

Kirchhoff's voltage law (KVL) (see Eq. 38.22) states that the algebraic sum of voltage drops around any closed path within a circuit is equal to the sum of the voltage rises.

Example

In the steady-state circuit shown, all components are ideal.

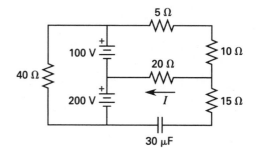

What is most nearly the magnitude of the current through the 20 Ω resistor?

(A) 2.0 A

(B) 2.9 A

(C) 5.0 A

(D) 5.7 A

Solution

Since this is a DC circuit, the capacitor blocks current flow. No current will flow through it.

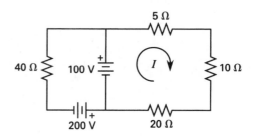

Engineering
Sciences

The only voltage source in the loop containing the 5 Ω, 10 Ω, and 20 Ω resistors is the 100 V battery. Write Kirchhoff's voltage law for the loop.

$$\sum V = \sum IR = I \sum R$$
$$100 \text{ V} = I(5 \text{ }\Omega + 10 \text{ }\Omega + 20 \text{ }\Omega)$$
$$I = 2.86 \text{ A} \quad (2.9 \text{ A})$$

The answer is (B).

Rules for Simple Resistive Circuits

In a simple series (single-loop) circuit, such as the circuit shown in Fig. 38.4, the following rules apply.

Figure 38.4 Simple Series Circuit

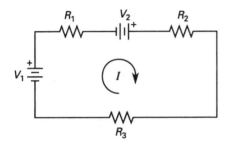

- The current is the same through all circuit elements.

$$I = I_{R_1} = I_{R_2} = I_{R_3}$$

- The *equivalent resistance* is the sum of the individual resistances.

$$R_{\text{eq}} = R_1 + R_2 + R_3$$

- The equivalent applied voltage is the algebraic sum of all voltage sources (polarity considered).

$$V_{\text{eq}} = V_1 + V_2$$

- The sum of the voltage drops across all components is equal to the equivalent applied voltage (KVL).

$$V_{\text{eq}} = I R_{\text{eq}}$$

In a series circuit, the voltage across a resistor is the total circuit voltage times the resistance of that particular resistor divided by the total equivalent resistance. This describes the operation of a *voltage divider* circuit. For example, the voltage across resistor R_1 of Fig. 38.4 is given in

$$V_{R_1} = \left(\frac{R_1}{R_{\text{eq}}}\right) V_{\text{eq}}$$
$$= \left(\frac{R_1}{R_1 + R_2 + R_3}\right)(V_1 + V_2)$$

In a simple parallel circuit with only one active source, such as the circuit shown in Fig. 38.5, the following rules apply.

Figure 38.5 Simple Parallel Circuit

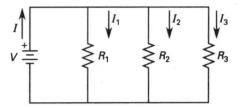

- The voltage drop is the same across all legs.

$$V = V_{R_1} = V_{R_2} = V_{R_3}$$
$$= I_1 R_1 = I_2 R_2 = I_3 R_3$$

- The reciprocal of the equivalent resistance is the sum of the reciprocals of the individual resistances.

$$\frac{1}{R_{\text{eq}}} = \frac{1}{R_1} + \frac{1}{R_2} + \frac{1}{R_3}$$

- The total current is the sum of the leg currents (KCL).

$$I = I_1 + I_2 + I_3$$
$$= \frac{V}{R_1} + \frac{V}{R_2} + \frac{V}{R_3}$$

In a parallel circuit, the current through a resistor is the total circuit current times the total circuit resistance divided by the resistor's resistance. This describes the operation of a *current divider circuit*. For example, the current through resistor R_1 of Fig. 38.5 would be given as

$$I_1 = \left(\frac{R_{\text{eq}}}{R_1}\right) I = \left(\frac{\dfrac{1}{\dfrac{1}{R_1} + \dfrac{1}{R_2} + \dfrac{1}{R_3}}}{R_1}\right) I$$

6. COMMON DC CIRCUIT ANALYSIS METHODS

The circuit analysis techniques in the following sections show how complicated linear circuits are simplified using circuit reduction and how they are analyzed as a system of n simultaneous equations and n unknowns. Circuit analysis can often be used to directly obtain the current or voltage at a component of interest or to reduce the number of simultaneous equations needed.

Engineering Sciences

Use the following procedure to establish the current and voltage drops in a complicated resistive network. The circuit should be viewed from the perspective of the component of interest. Each step in the reduction should result in a circuit that is simpler.

step 1: Combine series voltage and parallel current sources.

step 2: Combine series resistances to make combinations that more closely resemble a component in parallel with the component of interest.

step 3: Combine parallel resistances to make combinations that more closely resemble a component in series with the component of interest. Lines in the circuit represent zero resistance, and components connected by lines are connected to the same node. The lines can be moved to make parallel combinations more recognizable as long as the components remain connected to the node.

step 4: Repeat steps 2 through 4 as many times as needed.

This principle is only valid for linear circuits or non-linear circuits that are operating in a linear range. The superposition theorem can be used to reduce a complicated circuit to multiple less-complicated circuits.

Superposition Method

The *superposition method* can be used to reduce a complicated circuit into several simpler circuits. The *superposition theorem* states that the response of (i.e., the voltage across or current through) a linear circuit element fed by two or more independent sources is equal to the combined responses to each source taken individually, with all other sources set to zero (i.e., voltage sources shorted and current sources opened).

The superposition method determines the response of a component to each of the energy sources in a linear circuit separately and then combines the responses. This requires a circuit analysis for each of the energy sources in the circuit. Superposition works equally well for finding unknown currents and unknown voltages. Superposition tends to be more efficient for less-complicated circuits and when there are more loops and nodes than power sources. The superposition method is inefficient for analyzing complicated circuits.

step 1: Choose one of the voltage or current sources, short all other voltage sources, and open all other current sources.

step 2: Make circuit reductions to simplify the circuit and isolate the component of interest.

step 3: Find the voltage or current for the component of interest.

step 4: Repeat steps 1, 2, and 3 for the other voltage and current sources.

step 5: Sum the voltages or currents using the same conventions for voltage polarity and current direction.

Loop-Current Method

The *loop-current method* (also known as the *mesh current method*) is a direct extension of Kirchhoff's voltage law and is particularly valuable in determining unknown currents in circuits with several loops and energy sources. It requires writing $n-1$ simultaneous equations for an n-loop system.

step 1: Select $n-1$ loops (i.e., one less than the total number of loops).

step 2: Assume current directions for the chosen loops. (The choice of current direction is arbitrary, but some currents may end up being negative in step 4.) Show the direction with an arrow.

step 3: Write Kirchhoff's voltage law for each of the $n-1$ chosen loops. A voltage source is positive when the assumed current direction is from the negative to the positive battery terminal. Voltage drops are always positive.

step 4: Solve the $n-1$ equations (from step 3) for the unknown currents.

Node-Voltage Method

The *node-voltage method* is an extension of Kirchhoff's current law. Although currents can be determined with it, it is primarily used to find voltage potentials at various points (nodes) in the circuit. (A *node* is a point where three or more wires connect.)

step 1: (Optional) Convert all current sources to voltage sources.

step 2: Choose one node as the voltage reference (i.e., 0 V) node. Usually, this will be the circuit ground—a node to which at least one negative battery terminal is connected.

step 3: Identify the unknown voltage potentials at all other nodes referred to the reference node.

step 4: Write Kirchhoff's current law for all unknown nodes. (This excludes the reference node.)

step 5: Write all currents in terms of voltage drops.

step 6: Write all voltage drops in terms of the node voltages.

Equation 38.23 and Eq. 38.24: Source Equivalents

$$R_{eq} = \frac{V_{oc}}{I_{sc}} \quad \text{38.23}$$

$$V_{oc} = V_a - V_b \quad \text{38.24}$$

Description

Source equivalents are simplified models of two-terminal networks. They are used to represent a circuit when it is connected to a second circuit. Source equivalents simplify the analysis because the equivalent circuits are much simpler than the originals.

Thevenin's theorem states that a linear, two-terminal network with dependent and independent sources can be represented by a *Thevenin equivalent* circuit consisting of a voltage source in series with a resistor, as illustrated in Fig. 38.6.[7] The *Thevenin equivalent voltage*, or *open-circuit voltage*, V_{oc}, is the open-circuit voltage across terminals A and B. The *Thevenin equivalent resistance*, R_{eq}, is the resistance across terminals A and B when all independent sources are set to zero (i.e., short-circuiting voltage sources and open-circuiting current sources). The equivalent resistance can also be determined by measuring V_{oc} and the current with terminals A and B shorted together, I_{sc}, and using Eq. 38.23.

Figure 38.6 *Thevenin Equivalent Circuit*

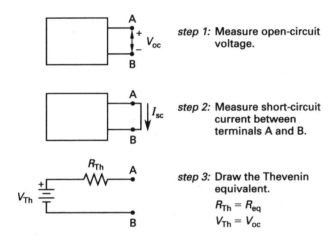

Norton's theorem states that a linear, two-terminal network with dependent or independent sources can be represented by an equivalent circuit consisting of a single current source and resistor in parallel, as shown in Fig. 38.7. The *Norton equivalent current*, I_{sc}, is the *short-circuit current* that flows through a shunt across terminals A and B. The *Norton equivalent resistance*, R_{eq}, is the resistance across terminals A and B when all independent sources are set to zero (i.e., short-circuiting voltage sources and open-circuiting current sources). The *Norton equivalent voltage*, V_{oc}, is measured with terminals open.

Norton's equivalent resistance is equal to Thevenin's equivalent resistance.

The conversions from Norton to Thevenin or from Thevenin to Norton can aid in circuit analysis. The Norton

Figure 38.7 *Norton Equivalent Circuit*

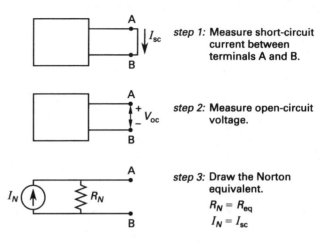

equivalent can be easily converted to a Thevenin equivalent and vice versa with the following equations.

$$R_N = R_{Th}$$

$$V_{Th} = I_N R_N$$

$$I_N = \frac{V_{Th}}{R_{Th}}$$

Maximum Power Transfer

Electric circuits are often designed to transfer power from a source (e.g., generator, transmitter) to a load (e.g., motor, light, receiver). There are two basic types of power transfer circuits. In one type of system, the emphasis is on transmitting power with high efficiency. In this power system, large amounts of power must be transmitted in the most efficient way to the loads. In communication and instrumentation systems, small amounts of power are involved. The power at the transmitting end is small, and the main concern is that the maximum power reaches the load.

The *maximum power transfer* from a circuit will occur when the load resistance equals the Norton or Thevenin equivalent resistance of the source.

7. *RC* AND *RL* TRANSIENTS

When a charged capacitor is connected across a resistor, the voltage across the capacitor will gradually decrease and approach zero as energy is dissipated in the resistor. Similarly, when an inductor through which a steady current is flowing is suddenly connected across a resistor, the current will gradually decrease and approach zero. Both of these cases assume that any energy sources are disconnected at the time the resistor is connected. These gradual decreases represent *transient behavior*. Transient behavior is also observed when a voltage or a current source is initially connected to a circuit with capacitors or inductors.

The *time constant*, τ, for a circuit is the time in seconds it takes for the current or voltage to reach $(1 - 1/e$, where e is the base of natural logarithms, approximately 2.718)

---

[7]The *NCEES Handbook* uses lower-case italic subscripts, *a* and *b*, to designate the terminals in Eq. 38.24. Since this style convention is not used again in the *NCEES Handbook*, and since that convention is inconsistent with this book's style to designate locations, this book uses uppercase Roman subscripts, A and B, in Fig. 38.6 and in this section.

times the difference between the steady-state value and the original value, or approximately 63.3% of its steady-state value. For a series-RL circuit, the time constant is L/R. For a series-RC circuit, the time constant is RC. In general, transient variables will have essentially reached their steady-state values after five time constants (99.3% of the steady-state value).

Equation 38.25 through Eq. 38.30 describe RC and RL transient response for source-free and energizing circuits. Time is assumed to begin when a switch is closed. Decay is a special case of the charging equations where $V = 0$ and either $v_C(0) \neq 0$ or $i_L(0) \neq 0$.

Equation 38.25 Through Eq. 38.27: *RC* Transients

$$v_C(t) = v_C(0)e^{-t/RC} + V\left(1 - e^{-t/RC}\right) \quad [t \geq 0] \qquad \textbf{38.25}$$

$$i(t) = \{[V - v_C(0)]/R\}e^{-t/RC} \quad [t \geq 0] \qquad \textbf{38.26}$$

$$v_R(t) = i(t)R = [V - v_C(0)]e^{-t/RC} \quad [t \geq 0] \qquad \textbf{38.27}$$

Description

Equation 38.25 through Eq. 38.27 describe transient behavior in RC circuits.[8] (See Fig. 38.8.) $v_C(0)$ is the voltage across the terminals of the capacitor when the switch is closed.

Example

The initial voltage across the capacitor is 5 V. At $t = 0$, the switch is closed.

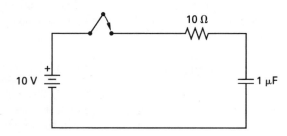

What is most nearly the voltage across the capacitor 10 μs after the switch is closed?

(A) 1.0 V

(B) 5.1 V

(C) 5.4 V

(D) 8.2 V

---

[8]Generally, parentheses, square brackets, and curly brackets are not combined in presenting mathematical equations. Other than designating a multiplicative combination, there is no significance to the curly brackets used in the *NCEES Handbook* Eq. 38.26.

Figure 38.8 *RC Transient Circuit*

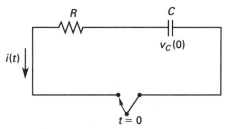

(a) series-*RC*, discharging
(energy source(s) disconnected)

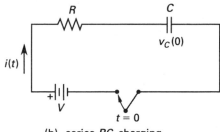

(b) series-*RC*, charging
(energy source(s) connected)

Solution

When the switch closes, the charge on the capacitor begins to increase. From Eq. 38.25,

$$\frac{t}{RC} = \frac{(10 \ \mu s)\left(10^{-6} \ \dfrac{s}{\mu s}\right)}{(10 \ \Omega)(1 \ \mu F)\left(10^{-6} \ \dfrac{F}{\mu F}\right)} = 1$$

$$\begin{aligned} v_C(t) &= v_C(0)e^{-t/RC} + V\left(1 - e^{-t/RC}\right) \\ &= (5 \ \text{V})e^{-1} + (10 \ \text{V})(1 - e^{-1}) \\ &= 8.16 \ \text{V} \quad (8.2 \ \text{V}) \end{aligned}$$

The answer is (D).

Equation 38.28 Through Eq. 38.30: *RL* Transients

$$i(t) = i(0)e^{-Rt/L} + \frac{V}{R}\left(1 - e^{-Rt/L}\right) \quad [t \geq 0] \qquad \textbf{38.28}$$

$$v_R(t) = i(t)R = i(0)Re^{-Rt/L} + V\left(1 - e^{-Rt/L}\right) \qquad [t \geq 0] \qquad \textbf{38.29}$$

$$v_L(t) = L(di/dt) = -i(0)Re^{-Rt/L} + Ve^{-Rt/L} \qquad [t \geq 0] \qquad \textbf{38.30}$$

Description

Equation 38.28 through Eq. 38.30 describe transient behavior in RL circuits. (See Fig. 38.9.) $i(0)$ is the current through the inductor when the switch is closed.

Figure 38.9 *RL Transient Circuit*

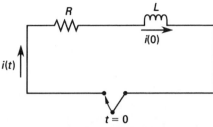

(a) series-*RL*, discharging
(energy source(s) disconnected)

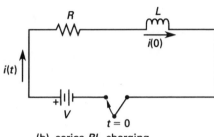

(b) series-*RL*, charging
(energy source(s) connected)

Example

The switch in the circuit shown is closed at $t = 0$.

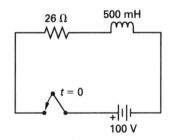

What is the approximate voltage across the inductor at $t = 30$ ms?

(A) 1.0 V

(B) 19 V

(C) 21 V

(D) 48 V

Solution

Use Eq. 38.30.

$$\frac{Rt}{L} = \frac{(26 \ \Omega)(30 \text{ ms})\left(1000 \ \frac{\text{mH}}{\text{H}}\right)}{(500 \text{ mH})\left(1000 \ \frac{\text{ms}}{\text{s}}\right)} = 1.56$$

$$v_L(t) = -i(0)Re^{-Rt/L} + Ve^{-Rt/L}$$

$$v_L(30 \text{ ms}) = 0 + (100 \text{ V})e^{-1.56}$$

$$= 21 \text{ V}$$

The answer is (C).

8. DC VOLTMETERS

A *d'Arsonval meter* movement configured to perform as a *DC voltmeter* is shown in Fig. 38.10.

Figure 38.10 *DC Voltmeter*

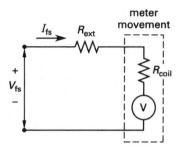

The external resistance is used to limit the current to the full-scale value, I_{fs}, at the desired full-scale voltage, V_{fs}. The electrical relationships in the voltmeter are given by

$$\frac{1}{I_{fs}} = \frac{R_{ext} + R_{coil}}{V_{fs}}$$

The quantity $1/I_{fs}$ is fixed for a given instrument and is called the *sensitivity*. The sensitivity is measured in ohms per volt (Ω/V).

9. DC AMMETERS

A d'Arsonval meter movement configured to perform as a *DC ammeter* is shown in Fig. 38.11.

Figure 38.11 *DC Ammeter*

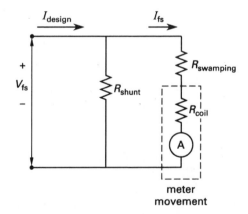

The *shunt resistance* (*swamping resistance*) is used to limit the current to the full-scale value, I_{fs}, at the desired full-scale voltage, V_{fs}. For ammeters, the standard full-scale voltage is 50 mV.[9] The electrical relationships in the ammeter are given by

$$I_{design} = \frac{V_{fs}}{R_{shunt}} + I_{fs}$$

---

[9]The standard ammeter is designed to withstand a 50 mV voltage across the shunt with I_{fs} flowing in the movement at the desired design current flow.

Engineering Sciences

39 Alternating-Current Circuits

Nomenclature

a	turns ratio	–
B	susceptance	S
BW	bandwidth	Hz or rad/s
C	capacitance	H
f	frequency	Hz
G	conductance	S
$i(t)$	time-varying current	A
I	constant current	A
L	inductance	H
N	number of turns	–
pf	power factor	–
P	real power	W
Q	reactive power	VAR
Q	quality factor	–
R	resistance	Ω
S	complex power	VA
t	time	s
T	period	s
$v(t)$	time-varying voltage	V
V	constant voltage	V
x	time-varying general variable	–
X	constant general variable	–
X	reactance	Ω
Y	admittance	S
Z	impedance	Ω

Symbols

θ	phase angle difference	rad
ϕ	offset angle	rad
ω	angular frequency	rad/s

Subscripts

0	at resonance
ave	average
C	capacitive or ideal capacitor
dc	direct current
eff	effective
eq	equivalent
i	imaginary
L	ideal inductor or inductive
max	maximum
P	primary
r	real
rms	effective or root-mean-square
R	ideal resistor
S	secondary

1. ALTERNATING WAVEFORMS

The term *alternating waveform* describes any symmetrical waveform, including square, sawtooth, triangular, and sinusoidal waves, whose polarity varies regularly with time. However, the term *alternating current* (AC) almost always means that the current is produced from the application of a sinusoidal voltage.

Sinusoidal variables can be specified without loss of generality as either sines or cosines. If a sine waveform is used, the instantaneous voltage as a function of time is given by

$$v(t) = V_{max}\sin(\omega t + \phi)$$

V_{max} is the maximum value (also known as the *amplitude*) of the sinusoid. If $v(t)$ is not zero at $t=0$ as in Fig. 39.1, an *offset angle*, ϕ (also known as a *relative phase angle*), must be used.

2. SINE-COSINE RELATIONSHIPS

The choice of sine or cosine to represent AC waveforms is arbitrary, with the only distinction being that the relative phase angle, ϕ, differs by $\pi/2$ radians. The phasor form of complex values is shown relative to the cosine form in the trigonometric form, so it may be necessary to convert a sine representation into a cosine representation or vice versa.

Equation 39.1 and Eq. 39.2: Sine-Cosine Relationships

$$\cos(\omega t) = \sin(\omega t + \pi/2) = -\sin(\omega t - \pi/2) \quad \text{39.1}$$
$$\sin(\omega t) = \cos(\omega t - \pi/2) = -\cos(\omega t + \pi/2) \quad \text{39.2}$$

Description

The trigonometric relationships in Eq. 39.1 and Eq. 39.2 are used to solve problems with alternating currents.

Equation 39.3: Frequency

$$f = 1/T = \omega/2\pi \qquad \textit{39.3}$$

Description

Figure 39.1 illustrates the form of an AC voltage given by $v(t) = V_{\max} \sin(\omega t + \phi)$. The *period* of the waveform is T. (Because the horizontal axis corresponds to time and not to distance, the waveform does not have a wavelength.) The *frequency*, f, of the sinusoid is the reciprocal of the period in hertz (Hz), as shown in Eq. 39.3. *Angular frequency*, ω, in radians per second (rad/s) can also be used.

Figure 39.1 *Sinusoidal Waveform with Phase Angle*

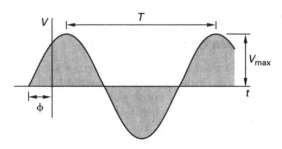

Example

The electric field (in V/m) of a particular plane wave propagating in a dielectric medium is given by

$$\mathbf{E}(t, z) = \mathbf{a_x} \cos\left(10^8 t - \frac{z}{3}\right) - \mathbf{a_y} \sin\left(10^8 t - \frac{z}{3}\right)$$

Time, t, has units of seconds. What is most nearly the field's oscillation frequency?

(A) 6.3 MHz

(B) 7.0 MHz

(C) 16 MHz

(D) 160 MHz

Solution

From the field equation, $\omega = 10^8$ rad/s. From Eq. 39.3,

$$f = \omega/2\pi$$
$$= \frac{10^8 \dfrac{\text{rad}}{\text{s}}}{2\pi\left(10^6 \dfrac{\text{Hz}}{\text{MHz}}\right)}$$
$$= 15.9 \text{ MHz} \quad (16 \text{ MHz})$$

The answer is (C).

3. REPRESENTATION OF SINUSOIDS

There are several equivalent methods of representing a sinusoidal waveform.

- *trigonometric*

$$V_{\max} \cos(\omega t + \phi)$$

- *polar* (or *phasor*)

$$V_{\text{eff}} \angle \phi$$

- *rectangular*

$$V_r + jV_i = V_{\max}(\cos\phi + j\sin\phi)$$

- *exponential*

$$V_{\max} e^{j\phi}$$

In polar, rectangular, or exponential form, the frequency must be specified separately.

When given in polar form, the voltage is usually given as the effective (rms) value (see Sec. 39.5) and not the peak value.[1]

In trigonometric form, ω may be given in either rad/s or deg/s, but rad/s is more common. ϕ is usually given in degrees. This can result in a mismatch in units. (Unfortunately, this is common practice in electrical engineering.) In exponential form, ϕ should always be in radians, but some references use degrees. In polar form and rectangular form, ϕ is usually in degrees.

Equation 39.4 and Eq. 39.5: Trigonometric and Polar (Phasor) Forms

$$P[V_{\max}\cos(\omega t + \phi)] = V_{\text{rms}}\angle\phi = \mathbf{V} \qquad \textit{39.4}$$

$$P[I_{\max}\cos(\omega t + \theta)] = I_{\text{rms}}\angle\theta = \mathbf{I} \qquad \textit{39.5}$$

Description

Equation 39.4 and Eq. 39.5 represent different ways to represent sinusoidal voltages and currents.[2]

---

[1]The convention of using rms values in phasor expressions of sinusoids, as is adopted in the NCEES *FE Reference Handbook* (*NCEES Handbook*), is common but arbitrary. If power is to be calculated from the common $\mathbf{P} = \mathbf{IV}$ (as opposed to from $\mathbf{P} = \frac{1}{2}\mathbf{IV}$), the rms values must be used.

[2](1) The semantics of using the expression "$P[X]$" to designate "the polar form of X" is unique to the *NCEES Handbook*. (2) Although the *NCEES Handbook* uses the equals symbol, =, Eq. 39.4 and Eq. 39.5 are not equations. The equivalence symbol, $\equiv$, should have been used to indicate that these are definitions, not mathematical expressions.

4. AVERAGE VALUE

Equation 39.6 and Eq. 39.7: Average Value

$$X_{\text{ave}} = (1/T)\int_0^T x(t)\,dt \qquad \textbf{39.6}$$

$$X_{\text{ave}} = 2X_{\text{max}}/\pi \quad \begin{bmatrix} \text{full-wave} \\ \text{rectified sinusoid} \end{bmatrix} \qquad \textbf{39.7}$$

Description

Equation 39.6 calculates the *average value* of any periodic variable (e.g., voltage or current).

Waveforms that are symmetrical with respect to the horizontal time axis have an average value of zero, as is shown in Fig. 39.2(a). A full-wave rectified sinusoid is shown in Fig. 39.2(b); the average value of Eq. 39.6 for this waveform is given by Eq. 39.7.

Figure 39.2 *Average and Effective Values*

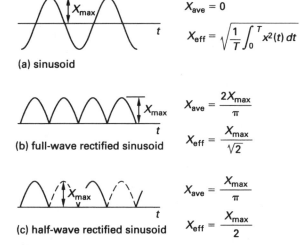

(a) sinusoid

$X_{\text{ave}} = 0$

$X_{\text{eff}} = \sqrt{\dfrac{1}{T}\int_0^T x^2(t)\,dt}$

(b) full-wave rectified sinusoid

$X_{\text{ave}} = \dfrac{2X_{\text{max}}}{\pi}$

$X_{\text{eff}} = \dfrac{X_{\text{max}}}{\sqrt{2}}$

(c) half-wave rectified sinusoid

$X_{\text{ave}} = \dfrac{X_{\text{max}}}{\pi}$

$X_{\text{eff}} = \dfrac{X_{\text{max}}}{2}$

The average value of the half-wave rectified sinusoid as shown in Fig. 39.2(c) is

$$X_{\text{ave}} = \frac{X_{\text{max}}}{\pi} \quad \begin{bmatrix} \text{half-wave} \\ \text{rectified sinusoid} \end{bmatrix}$$

Example

The waveform shown repeats every 10 ms.

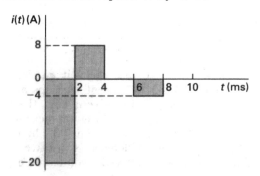

What is most nearly the average value of the waveform?

(A) -20 A

(B) -4.0 A

(C) -3.0 A

(D) 8.0 A

Solution

From Eq. 39.6, the average value of a periodic waveform is

$$I_{\text{ave}} = (1/T)\int_0^T i(t)\,dt$$

$$= \left(\frac{1}{10\text{ ms}}\right)\begin{pmatrix} (-20\text{ A})(2\text{ ms}) + (8\text{ A})(2\text{ ms}) \\ + (0\text{ A})(2\text{ ms}) + (-4\text{ A})(2\text{ ms}) \\ + (0\text{ A})(2\text{ ms}) \end{pmatrix}$$

$$= -3.2\text{ A} \quad (-3.0\text{ A})$$

The answer is (C).

5. EFFECTIVE (rms) VALUES

Equation 39.8 Through Eq. 39.11: Effective Value of Waveforms

$$X_{\text{eff}} = X_{\text{rms}} = \left[(1/T)\int_0^T x^2(t)\,dt\right]^{1/2} \qquad \textbf{39.8}$$

$$X_{\text{eff}} = X_{\text{rms}} = X_{\text{max}}/\sqrt{2} \quad \begin{bmatrix} \text{full-wave} \\ \text{rectified sinusoid} \end{bmatrix} \qquad \textbf{39.9}$$

$$X_{\text{eff}} = X_{\text{rms}} = X_{\text{max}}/2 \quad \begin{bmatrix} \text{half-wave} \\ \text{rectified sinusoid} \end{bmatrix} \qquad \textbf{39.10}$$

$$X_{\text{rms}} = \sqrt{X_{\text{dc}}^2 + \sum_{n=1}^{\infty} X_n^2} \qquad \textbf{39.11}$$

Description

The voltage level of an alternating waveform changes continuously. For power calculations, an alternating waveform is usually characterized by a single voltage value. This value is equivalent to the DC voltage that would have the same heating effect. This is called the *effective value*, also known as the *root-mean-square*, or *rms* value. A DC current of I produces the same heating effect as an AC current of I_{rms}.

Engineering Sciences

The effective value of a general alternating waveform is given by Eq. 39.8. Use Eq. 39.9 for a full-wave rectified sinusoidal waveform, and use Eq. 39.10 for a half-wave rectified sinusoidal waveform.

Equation 39.11 illustrates how the rms value of a combination of waveforms is calculated. X_{dc} is the *DC biasing voltage* across the entire circuit, while the rms values of the component waveforms are designated as X_n.[3]

In the United States, the effective value of the standard voltage used in households is 115–120 V. The polar form of the voltage is commonly depicted as

$$\mathbf{V} \equiv V_{eff} \angle \phi \equiv \left(\frac{V_{max}}{\sqrt{2}} \right) \angle \phi$$

Household voltages and currents can be considered to be effective values unless otherwise specified.

Example

What is most nearly the effective value of the repeating waveform shown?

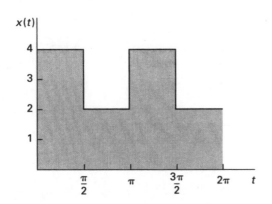

(A) 2.5

(B) 2.8

(C) 3.0

(D) 3.2

---

[3](1) Equation 39.8 through Eq. 39.10 use the same symbol as reactance, although X is intended by the *NCEES Handbook* to indicate some generic variable, such as current or voltage. It is not reactance. (2) Unlike Eq. 39.8, Eq. 39.9, and Eq. 39.10, *NCEES Handbook* Eq. 39.11 does not show $X_{eff} = X_{rms}$, although the two terms are still equivalent. (3) The subscript "dc" in Eq. 39.11 is the same as "DC" that the *NCEES Handbook* uses elsewhere to designate direct current. (4) X_{dc} is defined as the "dc component of $x(t)$." $x(t)$ was used in Eq. 39.8 to designate the native waveform. Since there are numerous waveforms being combined in Eq. 39.11, it is taken on faith that $x(t)$ is intended to represent the combined waveform. (5) The *NCEES Handbook* consistently uses n as a summation index (instead of, for instance, the more common i). Usually, n designates the last term in the summation.

Solution

From Eq. 39.8, for an alternating waveform,

$$
\begin{aligned}
X_{rms} &= \left[(1/T) \int_0^T x^2(t)\, dt \right]^{1/2} \\
&= \sqrt{ \left(\frac{1}{T} \right) \left(\int_0^{T/2} (4)^2\, dt + \int_{T/2}^T (2)^2\, dt \right) } \\
&= \sqrt{ \left(\frac{1}{T} \right) \left(16t \Big|_0^{T/2} + 4t \Big|_{T/2}^T \right) } \\
&= \sqrt{ \left(\frac{1}{T} \right) \left(\frac{16T}{2} + 4T - \frac{4T}{2} \right) } \\
&= 3.16 \quad (3.2)
\end{aligned}
$$

The answer is (D).

6. PHASE ANGLES

Ordinarily, the current and voltage sinusoids in an AC circuit do not peak at the same time. A *phase shift* exists between voltage and current, as illustrated in Fig. 39.3.

Figure 39.3 *Leading Phase Angle Difference*

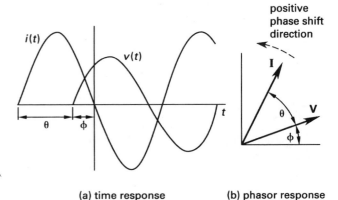

(a) time response   (b) phasor response

This phase shift is caused by the inductors and capacitors in the circuit. Capacitors and inductors have different effects on the phase angle. In a purely resistive circuit, no phase shift exists between voltage and current, and the current is in phase with the voltage.

It is common practice to use the voltage signal as a reference. In Fig. 39.3, the current *leads* (is ahead of) the voltage. In a purely capacitive circuit, the current leads the voltage by 90°; in a purely inductive circuit, the current *lags* behind the voltage by 90°. In a leading circuit, the phase angle difference is positive and the current reaches its peak before the voltage. In a lagging circuit, the phase angle difference is negative and the current reaches its peak after the voltage.

$$v(t) = V_{\max} \sin(\omega t + \phi) \quad \text{[reference]}$$

$$i(t) = I_{\max} \sin(\omega t + \phi + \theta)$$

Each AC *passive circuit element* (resistor, capacitor, or inductor) is assigned an angle, θ, known as its *impedance angle*, that corresponds to the phase angle shift produced when a sinusoidal voltage is applied across that element alone.

7. IMPEDANCE

The term *impedance*, Z (with units of ohms), describes the combined effect circuit elements have on current magnitude and phase. Impedance is a complex quantity with a magnitude and an associated angle, and it is usually written in polar form. However, it can also be written in rectangular form as the complex sum of its *resistive* (*real part*, R) and *reactive* (*imaginary part*, X) *components*, both having units of ohms. The resistive and reactive components combine trigonometrically in the *impedance triangle*, shown in Fig. 39.4. Resistance is always positive, while reactance may be either positive or negative.

Figure 39.4 *Lagging Impedance Triangle*

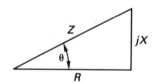

In Fig. 39.4, the impedance is drawn in the complex plane with the real (resistive) part on the horizontal axis and the imaginary (reactive) part on the vertical axis. The impedance in Fig. 39.4 is lagging because the reactive part is positive imaginary. This designation derives from $I = V/Z$ where a positive angle for Z subtracts from the voltage phase angle.

$$\mathbf{Z} \equiv R \pm jX$$

$$R = Z \cos \theta$$

$$X = Z \sin \theta$$

The total resistance is

$$X = X_L - X_C$$

Equation 39.12: Impedance and Ohm's Law

$$\mathbf{Z} = \mathbf{V}/\mathbf{I} \qquad \textit{39.12}$$

Description

Ohm's law for AC circuits with linear circuit elements is similar to Ohm's law for DC circuits. The impedance in an AC circuit is derived from Ohm's law, as shown in Eq. 39.12.

V and I can both be either maximum values or effective values, but never a combination of the two. If the voltage source is specified by its effective value, then the current calculated from $I = V/Z$ will be an effective value.

Equation 39.13: Ideal Resistor

$$\mathbf{Z}_R = R \qquad \textit{39.13}$$

Variation

$$\mathbf{Z}_R \equiv R\angle 0° \equiv R + j0$$

Description

Equation 39.13 defines the impedance of an ideal resistor. An *ideal resistor* has neither inductance nor capacitance. The magnitude of the impedance is the resistance, R, and the phase angle difference is zero. Current and voltage are in phase in an ideal resistor or in a purely resistive circuit.

Equation 39.14 and Eq. 39.15: Ideal Capacitor

$$\mathbf{Z}_C = \frac{1}{j\omega C} = jX_C \qquad \textit{39.14}$$

$$X_C = -\frac{1}{\omega C} \qquad \textit{39.15}$$

Variations

$$\mathbf{Z}_C \equiv X_C \angle -90°$$

$$\mathbf{Z}_C = \frac{-j}{\omega C}$$

Description

Equation 39.14 gives the impedance of an *ideal capacitor* with capacitance, C, in farads (F). An ideal capacitor has neither resistance nor inductance. The magnitude of the impedance is the *capacitive reactance*, X_C, with units of ohms, and the phase angle difference is $-\pi/2$ ($-90°$).[4] Current leads the voltage by 90° in a purely capacitive circuit. Some authors casually define X_C as a positive quantity such that Eq. 39.15 does not have a negative sign. The important thing to know is that the impedance of a capacitor is negative imaginary, regardless of how the reactance sign is defined.

---

[4]Equation 39.15 is derived from Eq. 39.14 and the definition $j^2 = -1$. However, the expression for capacitive reactance is often shown without the negative sign, and capacitive reactance values are normally stated as positive values (e.g., "$X_C = 4\ \Omega$"). In such cases, the negative impedance angle is understood.

Equation 39.16 and Eq. 39.17: Ideal Inductor

$$\mathbf{Z}_L = j\omega L = jX_L \qquad \textit{39.16}$$
$$X_L = \omega L \qquad \textit{39.17}$$

Variation

$$\mathbf{Z}_L = X_L \angle 90°$$

Description

Equation 39.16 gives the impedance of an ideal inductor with inductance, L, in henries (H). An *ideal inductor* has no resistance or capacitance. The magnitude of the impedance is the *inductive reactance*, X_L, with units of ohms (Ω), and the phase angle difference is $\pi/2$ ($90°$). Current lags the voltage by $90°$ in a purely inductive circuit.

Some circuits are shown with the capacitor and inductor impedance, rather than the capacitance or inductance, given in ohms. The reactances are at the circuit's operating frequency and can be used for circuit analysis for current dividers and voltage dividers, as with the DC circuits, although analysis must be done with complex algebra.

Example

A simple circuit consists of an inductor in series with a sinusoidal voltage.

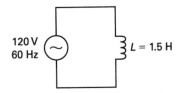

Using the applied voltage as the reference, what is most nearly the current through the inductor?

(A) $0.21 \text{ A} \angle -90°$

(B) $0.21 \text{ A} \angle 90°$

(C) $1.3 \text{ A} \angle -90°$

(D) $1.3 \text{ A} \angle 90°$

Solution

From Eq. 39.17, the reactance is

$$X_L = \omega L = 2\pi f L = 2\pi (60 \text{ Hz})(1.5 \text{ H})$$
$$= 565.5 \ \Omega$$

From Eq. 39.16, the impedance is

$$Z_L = jX_L = j565.5 \ \Omega$$
$$= 565.5 \ \Omega \angle 90°$$

The voltage is $V = 120 \text{ V} \angle 0°$. Rearranging Ohm's law, Eq. 39.12, the current through the inductor is

$$Z = V/I$$
$$I = \frac{V}{Z_{\text{total}}}$$
$$= \frac{120 \text{ V} \angle 0°}{565.5 \ \Omega \angle 90°}$$
$$= 0.21 \text{ A} \angle -90°$$

The answer is (A).

8. ADMITTANCE AND SUSCEPTANCE

The reciprocal of impedance is the complex quantity *admittance*, $\mathbf{Y}$, with units of *siemens* (S). Admittance is particularly useful in analyzing parallel circuits, since admittances of parallel circuit elements add together.

$$\mathbf{Y} = \frac{1}{\mathbf{Z}} = \frac{1}{Z} \angle -\theta$$

The reciprocal of the resistive part of impedance is *conductance*, G. The reciprocal of the reactive part of impedance is *susceptance*, B.

$$G = \frac{1}{R}$$

$$B = \frac{1}{X}$$

By multiplying the numerator and denominator by the *complex conjugate*, admittance can be written in terms of resistance and reactance, and vice versa.

$$\mathbf{Y} = G + jB = \left(\frac{1}{R+jX}\right)\left(\frac{R-jX}{R-jX}\right)$$
$$= \frac{R}{R^2+X^2} - j\left(\frac{X}{R^2+X^2}\right)$$

$$\mathbf{Z} = R + jX = \left(\frac{1}{G+jB}\right)\left(\frac{G-jB}{G-jB}\right)$$
$$= \frac{G}{G^2+B^2} - j\left(\frac{B}{G^2+B^2}\right)$$

Impedances are combined in the same way as resistances: impedances in series are added, while the reciprocals of impedances in parallel are added. For series circuits, the resistive and reactive parts of each impedance element are calculated separately and summed. For parallel circuits, the conductance and susceptance of each element are summed. The total impedance is found by a complex addition of the resistive (conductive) and reactive (susceptive) parts. It is convenient to perform the addition in rectangular form. The given equations represent the magnitude of the combined impedances for series and parallel circuits.

$$Z_{eq} = \sqrt{\left(\sum R\right)^2 + \left(\sum X_L - \sum X_C\right)^2}$$
[series]

$$Z_{eq} = \cfrac{1}{\sqrt{\left(\sum \frac{1}{R}\right)^2 + \left(\sum \frac{1}{X_L} - \sum \frac{1}{X_C}\right)^2}}$$
[parallel]

9. COMPLEX POWER

Equation 39.18: Complex Power Vector

$$\mathbf{S} = \mathbf{VI}^* = P + jQ \qquad \textit{39.18}$$

Description

The *complex power vector*, **S** (also called the *apparent power*), is the vector sum of the real (true, active) power vector, **P**, and the imaginary reactive power vector, **Q**. The complex power vector's units are volt-amps (VA). The components of power combine as vectors in the *complex power triangle*, shown in Fig. 39.5.

Figure 39.5 Lagging (Inductive) Complex Power Triangle

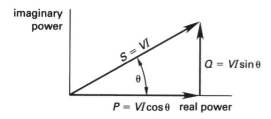

The complex conjugate of the current, **I***, is used in the apparent power, resulting in a positive imaginary part and a positive power angle for a lagging current (which has a negative phase angle compared to the voltage), as shown in Fig. 39.5. For a leading current (which has a positive phase angle compared to the voltage), the power triangle has a negative imaginary part and a negative power angle.

Equation 39.19: Power Factor

$$\text{pf} = \cos\theta \qquad \textit{39.19}$$

Description

The *power factor*, pf (usually given in percent), is $\cos\theta$. The angle θ is called the *power angle* and is the same as the overall impedance angle, or the angle between input

voltage and current in the circuit. These are the voltage and current at the source (usually voltage source), which supplies the electric power to the circuit.

The cosine is positive for both positive and negative angles. The descriptions *lagging* (for an inductive circuit) and *leading* (for a capacitive circuit) must be used with the power factor.

For a purely resistive load, pf = 1, and the average real power is given by Eq. 39.24.

The power factor of a circuit, and therefore, the phase angle difference, can be changed by adding either inductance or capacitance. This is known as *power factor correction*.

Example

An industrial complex is fed by a 13 kV (rms) transmission line. The average current delivered during a month is measured as 140 A. The real power consumed within the industrial complex is 1.7 MW. What power factor should be used in determining the month's electrical services invoice?

(A) 0.72

(B) 0.81

(C) 0.86

(D) 0.93

Solution

The apparent power (complex power) is calculated from the voltage and the measured current. The power factor is

$$\text{pf} = \cos\theta = \frac{P}{S} = \frac{P}{VI}$$

$$= \frac{(1.7 \text{ MW})\left(10^6 \, \frac{\text{W}}{\text{MW}}\right)}{(13 \text{ kV})\left(1000 \, \frac{\text{V}}{\text{kV}}\right)(140 \text{ A})}$$

$$= 0.93$$

The answer is (D).

Example

An ore processing plant with both electrical equipment and coal-fired steam generators has an overall lagging current. In order to reduce electrical costs, what is the best option for the plant to install?

(A) induction motors in parallel with the existing loads

(B) induction motors in series with the existing loads

(C) resistive heater steam generators in place of the coal-fired steam generators

(D) synchronous motors (without anything attached to the output shafts) in parallel with the existing loads

Solution

A lagging power factor indicates that the plant has a predominantly inductive load. Any additional inductive loading, including induction motors, will increase the lagging power factor. Resistive heaters would reduce the lagging power factor but increase electrical costs overall. Unloaded synchronous motors (commonly referred to as *synchronous capacitors*, *synchronous condensers*, and *synchronous compensators*) have leading power factors and will reduce the power factor, pf, reactive power, Q, and the electrical cost.

The answer is (D).

Equation 39.20 Through Eq. 39.24: Real and Reactive Power

$$P = \left(\tfrac{1}{2}\right) V_{max} I_{max} \cos\theta \quad \text{[sinusoids]} \qquad \textbf{39.20}$$

$$P = V_{rms} I_{rms} \cos\theta \qquad \textbf{39.21}$$

$$Q = \left(\tfrac{1}{2}\right) V_{max} I_{max} \sin\theta \quad \text{[sinusoids]} \qquad \textbf{39.22}$$

$$Q = V_{rms} I_{rms} \sin\theta \qquad \textbf{39.23}$$

$$P = V_{rms} I_{rms} = V_{rms}^2/R = I_{rms}^2 R \quad \begin{bmatrix} \text{purely} \\ \text{resistive load;} \\ \text{pf} = 1 \end{bmatrix}$$
$$\textbf{39.24}$$

Variations

$$Q = \frac{V_{rms}^2}{X}$$

$$P_{ave} = V_{rms} I_{rms} \cos 90° = 0$$

$$= V_{rms} I_{rms} \cos(-90°) = 0 \quad \begin{bmatrix} \text{purely} \\ \text{reactive load;} \\ \text{pf} = 0 \end{bmatrix}$$

Description

The *real power*, P, with units of watts (W), is given by Eq. 39.20 and Eq. 39.21. Equation 39.20 is only valid for sinusoids, while Eq. 39.21 is valid for all waveforms.

The *reactive power*, Q, in units of volt-amps reactive (VAR), is the imaginary part of **S**. The reactive power is given by Eq. 39.22 and Eq. 39.23. Equation 39.22 is only valid for sinusoids, while Eq. 39.23 is valid for all waveforms.

For a purely resistive load, pf = 1, and the average real power is given by Eq. 39.24.

For a purely reactive load, pf = 0, and the average real power is given by the second variation equation.

Electric energy is stored in a capacitor or inductor during a fourth of a cycle and is returned to the circuit during the next fourth of the cycle. Only a resistance will actually dissipate energy.

The power factor of a circuit, and, therefore, the phase angle difference, can be changed by adding either inductance or capacitance. This is known as *power factor correction*.

Example

A small batch reactor is heated by a two-phase, 240 V (line-to-line) AC resistance heater for five minutes. 3.2 kW are transferred to the reactor contents in an 84% efficient process. Most nearly, what is the resistance of the heater?

(A) 15 Ω

(B) 18 Ω

(C) 21 Ω

(D) 75 Ω

Solution

A 240 V voltage source is obtained by connecting two 120 V phases in series. Since 120 V is an rms (i.e., effective) value, 240 V is the effective value. Use Eq. 39.24.

$$P = V_{rms}^2/R$$

$$R = \frac{\eta V_{rms}^2}{P}$$

$$= \frac{(0.84)(240 \text{ V})^2}{(3.2 \text{ kW})\left(1000 \; \dfrac{\text{W}}{\text{kW}}\right)}$$

$$= 15.12 \; \Omega \quad (15 \; \Omega)$$

The answer is (A).

Example

The frequency of the current source in the parallel circuit is adjusted until the circuit is purely resistive (i.e., until the power factor is equal to 1.0).

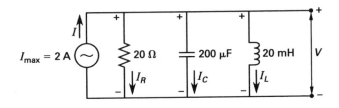

The power dissipated at the adjusted frequency is most nearly

(A) 0 W

(B) 10 W

(C) 20 W

(D) 40 W

Solution

When the circuit is purely resistive, the power factor is equal to 1. Use Eq. 39.24.

$$P = I_{rms}^2 R = \tfrac{1}{2}I_{max}^2 R = \left(\tfrac{1}{2}\right)(2 \text{ A})^2 (20 \text{ }\Omega)$$

$$= 40 \text{ W}$$

The answer is (D).

10. RESONANCE

In a *resonant circuit*, input voltage and current are in phase, and therefore, the phase angle is zero. This is equivalent to saying that the circuit is purely resistive in its response to an AC voltage, although inductive and capacitive elements must be present for resonance to occur. At resonance, the power factor is equal to one, and the reactance, X, is equal to zero, or $X_L + X_C = 0$. The frequency at which the circuit becomes purely resistive, ω_0 or f_0, is the *resonant frequency*.

Equation 39.25 Through Eq. 39.30: Parallel and Series Circuits at Resonant Frequency

$$\omega_0 = \frac{1}{\sqrt{LC}} = 2\pi f_0 \quad \text{[at resonance]} \qquad 39.25$$

$$Z = R \quad \text{[at resonance]} \qquad 39.26$$

$$\omega_0 L = \frac{1}{\omega_0 C} \quad \text{[at resonance]} \qquad 39.27$$

$$BW = \frac{\omega_0}{Q} \quad \text{[in rad/s]} \qquad 39.28$$

$$Q = \frac{\omega_0 L}{R} = \frac{1}{\omega_0 C R} \quad \text{[series-}RLC\text{ circuit]} \qquad 39.29$$

$$Q = \omega_0 R C = \frac{R}{\omega_0 L} \quad \text{[parallel-}RLC\text{ circuit]} \qquad 39.30$$

Variations

$$BW = f_2 - f_1 = \frac{f_0}{Q} \quad \text{[in Hz]}$$

$$BW = \omega_2 - \omega_1 \quad \text{[in rad/s]}$$

Description

Equation 39.25 through Eq. 39.27 apply to both parallel and series circuits at the resonant frequency, where $X_L + X_C = 0$ and pf $= 1$.

In a resonant *series-RLC circuit*, impedance is minimized, and the current and power dissipation are maximized. In a resonant *parallel-RLC circuit*, impedance is maximized, and the current and power dissipation are minimized.

For frequencies below the resonant frequency, a series-*RLC* circuit will be capacitive (leading) in nature.

Above the resonant frequency, the circuit will be inductive (lagging) in nature.

For frequencies below the resonant frequency, a parallel-*RLC* circuit will be inductive (lagging) in nature. Above the resonant frequency, the circuit will be capacitive (leading) in nature.

Circuits can become resonant in two ways. If the frequency of the applied voltage is fixed, the elements must be adjusted so that the capacitive reactance cancels the inductive reactance (i.e., $X_L + X_C = 0$). If the circuit elements are fixed, the frequency must be adjusted.

The behavior of a circuit at frequencies near the resonant frequency is illustrated in Fig. 39.6.

Figure 39.6 *Circuit Characteristics at Resonance*

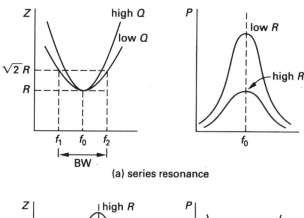

(a) series resonance

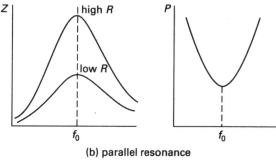

(b) parallel resonance

For both parallel and series resonance, the bandwidth is given by Eq. 39.28.

The half-power points are so named because at those frequencies, the power dissipated in the resistor is half of the power dissipated at the resonant frequency.

$$Z_{f_1} = Z_{f_2} = \sqrt{2}R$$

$$I_{f_1} = I_{f_2} = \frac{V}{Z_{f_1}} = \frac{V}{\sqrt{2}R} = \frac{I_0}{\sqrt{2}}$$

$$P_{f_1} = P_{f_2} = I^2 R = \left(\frac{I_0}{\sqrt{2}}\right)^2 R = \tfrac{1}{2}P_0$$

The *quality factor*, Q, for a circuit is a dimensionless ratio that compares, for each cycle, the reactive energy stored in an inductor to the resistive energy dissipated. The quality factor indicates the shape of the resonance curve. A circuit with a low Q has a broad and flat curve, while one with a high Q has a narrow and peaked curve. The quality factor for a series-RLC circuit is given by Eq. 39.29, and the quality factor for a parallel-RLC circuit is given by Eq. 39.30.

Assuming a fixed primary impedance, maximum power transfer in an AC circuit occurs when the source and load impedances are complex conjugates (resistances are equal and their reactances are opposite). This is equivalent to having a resonant circuit as shown in Fig. 39.7.

Figure 39.7 *Maximum Power Transfer at Resonance*

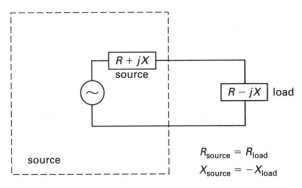

$R_{source} = R_{load}$

$X_{source} = -X_{load}$

Example

For the circuit shown, what are most nearly the resonant frequency and the quality factor?

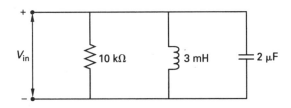

(A) 1.3×10^3 rad/s; 26

(B) 2.0×10^3 rad/s; 0.026

(C) 1.3×10^4 rad/s; 260

(D) 1.5×10^4 rad/s; 2600

Solution

From Eq. 39.25, the resonant frequency is

$$\omega_0 = \frac{1}{\sqrt{LC}}$$

$$= \frac{1}{\sqrt{(3 \times 10^{-3} \text{ H})(2 \times 10^{-6} \text{ F})}}$$

$$= 1.29 \times 10^4 \text{ rad/s} \quad (1.3 \times 10^4 \text{ rad/s})$$

From Eq. 39.30, for a parallel-RLC circuit,

$$Q = \frac{R}{\omega_0 L}$$

$$= \frac{10 \times 10^3 \; \Omega}{\left(1.29 \times 10^4 \; \frac{\text{rad}}{\text{s}}\right)(3 \times 10^{-3} \text{ H})}$$

$$= 258.2 \quad (260)$$

The answer is (C).

11. IDEAL TRANSFORMERS

Transformers are used to change voltages, match impedances, and isolate circuits. They consist of coils of wire wound on a magnetically permeable core. The coils are grouped into primary and secondary windings. The winding connected to the source of electric energy is called the *primary*. The primary current produces a magnetic flux in the core, which induces a current in the *secondary coil*. Core and shell transformer designs are shown in Fig. 39.8.

Figure 39.8 *Core and Shell Transformers*

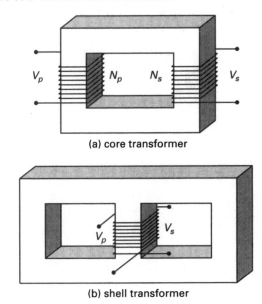

(a) core transformer

(b) shell transformer

Equation 39.31 Through Eq. 39.33: Turns Ratio

$$a = N_1/N_2 \qquad \textit{39.31}$$

$$a = \left|\frac{\mathbf{V}_P}{\mathbf{V}_S}\right| = \left|\frac{\mathbf{I}_S}{\mathbf{I}_P}\right| \qquad \textit{39.32}$$

$$\mathbf{Z}_P = a^2 \mathbf{Z}_S \qquad \textit{39.33}$$

Variations

$$a = \frac{N_P}{N_S}$$

$$\mathbf{Z}_P = \frac{\mathbf{V}_P}{\mathbf{I}_P} = \mathbf{Z}_1 + a^2 \mathbf{Z}_S \quad \text{[phasor/vector addition]}$$

Description

The ratio of numbers of primary to *secondary windings* is the *turns ratio* (*ratio of transformation*), a. If the turns ratio is greater than 1, the transformer decreases voltage and is a *step-down transformer*. If the turns ratio is less than 1, the transformer increases voltage and is a *step-up transformer*.

In a lossless (i.e., 100% efficient) transformer, the power absorbed by the primary winding equals the power generated by the secondary winding, so $I_P V_P = I_S V_S$, and the turns ratio is given by Eq. 39.32.

A lossless transformer is called an *ideal transformer*; its windings are considered to have neither resistance nor reactance.

The impedance seen by the source changes when an impedance is connected to the secondary, as shown in the second variation equation. An impedance of Z_S connected to the secondary of an ideal transformer is equivalent to an impedance of $a^2 Z_S$ connected to the source, as illustrated in Fig. 39.9. It is said that "a secondary impedance of Z_S reflects as $a^2 Z_S$ on the primary side." Real primary circuits have input impedance. Anything attached to the transformer, such as a microphone, will contribute to the input impedance. At a minimum, the transformer's own windings will contribute resistance. The input impedence, $\mathbf{Z}_1$, contributes to the impedance seen by the source (i.e., the primary impedance), as the variation shows.[5]

Figure 39.9 *Equivalent Circuit with Secondary Impedance*

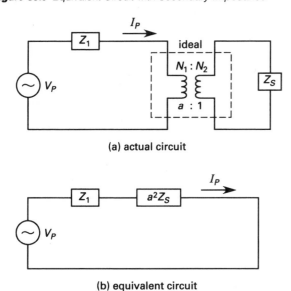

(a) actual circuit

(b) equivalent circuit

---

[5]*NCEES Handbook* Eq. 39.33 assumes the input impedance is zero.

Example

In the ideal transformer shown, coil 1 has 500 turns, V_1 is 1200 V, and V_2 is 240 V.

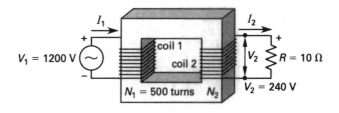

How many turns does coil 2 have?

    (A) 100 turns

    (B) 200 turns

    (C) 500 turns

    (D) 1000 turns

Solution

From Eq. 39.31 and Eq. 39.32, the turns ratio is

$$a = N_1/N_2 = \frac{V_1}{V_2}$$

$$N_2 = \frac{N_1 V_2}{V_1} = \frac{(500 \text{ turns})(240 \text{ V})}{1200 \text{ V}}$$

$$= 100 \text{ turns}$$

The answer is (A).

40 Rotating Machines

Nomenclature

B	magnetic flux density	T
E	induced voltage	V
f	frequency	Hz
I	current	A
K_a	armature constant (constant of the machine)	V·min/Wb
K_f	field constant	H
L	inductance	H
n	speed	rev/min
p	number of poles	–
P	power	W
R	resistance	Ω
t	time	s
T	torque	N·m
V	voltage	V

Symbols

ϕ	magnetic flux	Wb
ω	angular velocity (electrical)	rad/s
Ω	rotational velocity (mechanical)	rad/s

Subscripts

a	armature
e	electrical
f	field
g	generated
h	heat
m	mechanical or motor
s	synchronous

1. AC MACHINES

Rotating machines are broadly categorized as alternating-current (AC) and direct-current (DC) machines. Both categories include machines that use power (i.e., motors) and those that generate power (alternators and generators). Most AC machines can be constructed in either single-phase or polyphase (usually three-phase) configurations. The rotating part of the machine is called the *rotor*. The stationary part of the machine is called the *stator*. (In AC machines, the rotor is the field and the stator is the armature if the machine is a generator, and vice versa if the machine is a motor. The terms "armature" and "field" are not commonly used with AC machines.)

For simplicity, Fig. 40.1 shows only one of the poles for each phase. But there is actually a winding on the opposite side from each winding depicted with the voltage in the opposite polarity, such that the magnetic fields produced by the windings have opposite polarities (north and south). When V_a is a maximum value, the field in the a-phase windings will be in the up direction, while the b-phase and c-phase fields will be equal and negative so their net field will be up also. When V_c reaches the maximum negative voltage, the field in the c-phase windings will be in the negative c-direction (60° counterclockwise), while the a-phase and b-phase fields will be equal and positive so their net field will be in the negative c-direction. When V_b is a maximum value, the field in the b-phase windings will be in the positive b-direction (120° counterclockwise), while the a-phase and c-phase fields will be equal and negative so their net field will be in the positive b-direction. The field makes a complete rotation around the stator for one cycle of three-phase voltage.

Figure 40.1 *Two-Pole AC Machine*

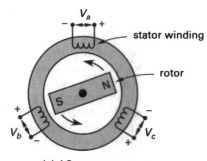

(a) AC generator

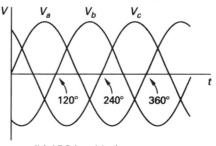

(b) ABC (positive) sequence

In AC machines, the term "pole" refers to a winding in the stationary stator. Each stator winding and its opposite, located directly across the stator, are counted as two poles and constitute a *pole pair*. AC machine stators rarely contain only two poles; a stator may contain

Engineering Sciences

12 or more poles.[1] In an AC generator, each independent pole pair produces a single voltage waveform (i.e., a single phase). The number of independent voltage waveforms produced from an AC generator (one or three, in almost all cases) is equal to the number of independent pole pair sets.

Regardless of the number of phases involved, an AC machine is named according to the number of poles associated with each phase. For example, a single-phase, four-pole AC machine would have four poles arranged around the stator; and a three-phase, two-pole AC machine would have six poles arranged around the stator. The three-phase AC generator shown in Fig. 40.1 is a *two-pole machine* because it has two poles for each phase.

In an AC generator, the number of electrical cycles produced with one mechanical rotation of the rotor depends on the number of poles. With a four-pole machine, one mechanical rotation (360°) of the rotor would produce two electrical cycles (720°) per phase.

Equation 40.1: Synchronous Speed

$$n_s = 120f/p \qquad \text{40.1}$$

Variation

$$n_s = \frac{60\Omega}{2\pi} = \frac{60\omega}{\pi p} \quad \text{[synchronous speed]}$$

Description

In *synchronous motors* and *induction motors*, the stator magnetic field rotates at a speed known as the *synchronous speed*, n_s, given by Eq. 40.1. f is the electrical frequency of the generated potential. p is the number of poles per phase. The actual rotational speed, n, is known as the *mechanical frequency*. Care must be taken to distinguish between the mechanical rotor speed, n (in rpm), the angular mechanical rotor speed, Ω (in rad/s), and the linear and angular electrical frequencies, f and ω (in Hz and rad/s, respectively).[2]

Example

A two-pole AC motor operates on a three-phase, 60 Hz, line-to-line supply with an rms voltage of 240 V. What is most nearly its synchronous speed?

(A) 1000 rpm

(B) 1800 rpm

(C) 2400 rpm

(D) 3600 rpm

---

[1]Four-pole AC machines are the most common.
[2]The NCEES *FE Reference Handbook* (*NCEES Handbook*) does not distinguish between angular armature and electrical speeds, Ω and ω. Consistent with almost all elementary discussions of two-pole motor construction and performance, the *NCEES Handbook* uses ω to represent the angular armature speed. Since $\Omega = 2\omega/p$, angular armature and electrical speeds are the same only for two-pole devices.

Solution

Since this is a two-pole machine, $p = 2$. From Eq. 40.1,

$$n_s = 120f/p$$

$$= \frac{\left(120 \, \dfrac{\dfrac{\text{rev}}{\text{min}}}{\text{Hz}}\right)(60 \text{ Hz})}{2}$$

$$= 3600 \text{ rpm}$$

The answer is (D).

Synchronous Machines

A *synchronous machine* has a permanent magnet or, more typically, a DC electromagnet that produces a constant magnetic field in the rotor. For the electromagnet, a DC current is provided to windings on the rotor and is transferred by stationary brushes making contact with *slip rings* or *collector rings* on the rotating shaft. The term "slip ring" implies that the conductor is a continuous type and slips under the brush. The term "collector ring" implies that the ring collects current from more than one winding (which is typical). A synchronous machine derives its name from the fact that the rotor turns at the synchronous speed, as determined from Eq. 40.1.

The rotor of a synchronous generator is rotated by a mechanical force. As the rotor rotates, both the current and the magnetic field remain constant, but the angle the magnetic field makes with each of the windings changes as the rotor rotates. For each of the stator windings, the flux lines from the rotor will be perpendicular to the loops at some angles and parallel to them at other angles. The current induced in the loops is greatest when the flux lines are perpendicular to the loops, and is zero when the flux lines are parallel to the loops. In this way, a rotating constant magnetic field induces alternating electrical currents in the stator.

The induced voltage, E, is commonly called the *electromotive force* (emf). In an elementary generator, emf is the desired end result to the load. In a motor, emf is also produced but is referred to as *back emf* (*counter emf*), since it opposes the input voltage.

A *synchronous motor* is essentially a synchronous generator operating in reverse. There is no difference in the construction of the machine. Alternating current is supplied to the stationary stator windings. A rotating magnetic field is produced when three-phase power is applied to the windings. Direct current is applied to the windings in the rotor through brushes and slip rings, as in the synchronous generators. The rotor current interacts with the stator field, causing the rotor to turn. Since the stator field frequency is fixed, the motor runs only at a single synchronous speed.

Induction Machines

An *induction motor* is essentially a constant-speed device that receives power through induction, without using brushes or slip rings. A motor can be considered to be a rotating transformer secondary (the rotor) with a stationary primary (the stator). Stator construction in an induction motor is the same as that in a synchronous motor. An emf is induced as the stator field moves past the rotor conductors. The stator field rotates at the synchronous speed given by Eq. 40.1. Since the rotor windings have inductive reactance, the rotor field lags the stator field.

Induction motors are commonly used for most industrial applications. They are more common than synchronous motors since they are more rugged, require less maintenance, and are less expensive. Induction motors consume the majority of all electric power generated and come in sizes from a fraction of a horsepower to many thousand horsepower. They are particularly useful in applications that require smooth starting under a load, such as hoists, mixers, or conveyors. Synchronous motors are much better than induction motors in power-generation applications.

Induction generators have few significant commercial uses, although some applications, such as windmills that generate power via induction motors, are gaining popularity. Induction generators are similar in concept to induction motors and are not discussed further in this chapter.

Equation 40.2: Slip

$$\text{slip} = (n_s - n)/n_s \qquad 40.2$$

Description

For the rotor of an induction motor to have a magnetic field, there must always be a variation in the magnetic flux as a function of time. To have a variation in magnetic flux, the rotor must turn slower than the synchronous speed so that the magnetic fields of the stator and rotor are never in the same direction. The difference in speed is small, but essential. *Percent slip* is the percentage difference in speed between the rotor and stator field. Slip can be expressed as either a decimal or a fraction (e.g., 0.05 slip or 5% slip). A related concept, slip in rpm, is the difference between actual and synchronous speeds.

Induction motor slip is normally 2–5% except for motors designed with a variable rotor resistance that can be increased for speed control.

Example

A three-phase induction motor runs on a 60 Hz power line at 1150 rpm at full load. The motor's stator contains a total of 18 poles. What is most nearly the motor's full-load slip?

(A) 0

(B) 0.042

(C) 0.075

(D) 0.13

Solution

The number of poles per phase is

$$p = \frac{18 \text{ poles}}{3 \text{ phases}} = 6$$

Calculate the synchronous speed of the AC motor from Eq. 40.1.

$$n_s = 120f/p$$
$$= \frac{\left(120 \, \dfrac{\frac{\text{rev}}{\text{min}}}{\text{Hz}}\right)(60 \text{ Hz})}{6}$$
$$= 1200 \text{ rpm}$$

Calculate the slip using Eq. 40.2.

$$\text{slip} = (n_s - n)/n_s$$
$$= \frac{1200 \, \dfrac{\text{rev}}{\text{min}} - 1150 \, \dfrac{\text{rev}}{\text{min}}}{1200 \, \dfrac{\text{rev}}{\text{min}}}$$
$$= 0.042$$

The answer is (B).

2. DC MACHINES

DC machines have a constant magnetic field in the stator (called the *field*). The magnetic field of the rotor (called the *armature*) responds to the stator field. The armature will respond by inducing current if the machine is a generator or by developing torque if the machine is a motor.

Equation 40.3: Magnetic Flux

$$\phi = K_f I_f \quad \text{[in webers]} \qquad 40.3$$

Description

In most DC machines, the magnetic field is provided by electromagnets. (In very small DC machines, it is supplied by permanent magnets.) The stator in a DC machine is composed of windings that produce the machine's field. Field poles are located on the stator and project inward. The poles alternate north and south around the machine and are separated by $360°/p$ around the machine. Each pole has a narrow ferromagnetic (e.g., iron) core around which the winding is wrapped. Each coil may consist of two or more separate windings. The *magnetic flux* produced by the field, Eq. 40.3, is linearly related to the field current, I_f, by a *field constant*, K_f, that depends on the construction of the machine, as long as the current is low enough that saturation does not occur.

The current in the coils creates a magnetic flux in the core that causes currents to circulate in the tiny magnetic domains in the core and reinforces the core's magnetic flux. As the current in the coil increases, the currents in the magnetic domains increase approximately linearly until the currents in the magnetic domains cannot respond fully, and part of the energy in the flux is lost as heat. At this stage, the magnetic material is said to be *saturating*, and the response is no longer linear. As the current in the coil continues to increase, it reaches a point where the currents in the magnetic domains are going as fast as they can, and any additional energy in the flux from the coil is dissipated as heat. At this stage, the magnetic material is said to be *completely saturated*.

The armature circuit and field circuit are actually loops of wire around magnetic materials, but they are each modeled as a resistor in series with an inductor. The variables R_a and L_a are used for the armature resistance and inductance. The variables R_f and L_f are used for the field resistance and inductance. The inductance is usually ignored in DC machines, except for transient conditions.

DC Generators

A *DC generator* is a device that produces DC potential. The actual voltage induced is sinusoidal (i.e., AC). However, brushes on *split-ring commutators* make the connection to the rotating armature and rectify AC potential. The DC generator is, in fact, not DC. Rather, it is rectified AC.

The armature in a simple DC generator consists of a single coil with several turns (loops) of wire. The two ends of the coil terminate at a *commutator*. The commutator consists of a single ring split into two halves known as *segments*. (This arrangement is shown in Fig. 40.2. The field windings and rotor core are omitted for clarity.) The brushes slide on the commutator and make contact with the adjacent segment every half-rotation of the coil. As shown in Fig. 40.2, the coil is nearly perpendicular to the magnetic field, **B**. The magnetic flux through the coil will be at its maximum when the coil is perpendicular to the magnetic field. The magnetic flux through the coil will be zero when the coil is parallel to the magnetic field. The rate of change in the magnetic flux depends on the speed of rotation.

As shown in Fig. 40.2, the current flows from the positive brush to the negative brush. As the coil continues to

Figure 40.2 Commutator Action

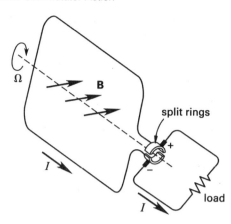

rotate to be parallel to the magnetic field, the commutator rotates such that the brushes are at the gaps between the commutator segments. As the coil rotates further, the positive brush will touch the commutator segment that the negative brush was previously touching, and vice versa. The current will continue in the same direction. Figure 40.3 shows the full-wave rectified voltage that results from this simple arrangement, which is obviously not a constant voltage.

Figure 40.3 Rectified DC Voltage Induced in a Single Coil

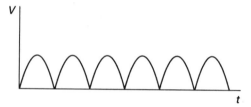

Figure 40.4 shows a simplified DC armature with two coils oriented at $90°$ to each other. (The field windings and rotor core are omitted for clarity.) Each of the coils produces a full-wave rectified emf, as in Fig. 40.3. Since the coils are connected in series (in the modern closed-coil winding arrangement), the emf induced is the sum of the emfs induced in the individual coils, as shown in Fig. 40.5. The voltage induced in each coil of a DC generator with multiple coils is still sinusoidal, but the terminal output is nearly constant, not a (rectified) sinusoid. The slight variations in the voltage are known as *ripple*. The more coils there are, the smoother the DC voltage. The output may be filtered or passed through a DC voltage regulator to reduce or eliminate the ripple.

As shown in Fig. 40.6, DC generators can be modeled as an emf, E_g, which represents the part of the voltage induced in the armature circuit by the magnetic flux as the armature rotates in the field, and a resistor, R_a, which represents the resistance in the armature windings.

Figure 40.4 *Two-Coil, Four-Segment Closed-Coil Armature*

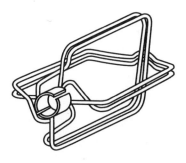

Figure 40.5 *Rectified DC Voltage from a Two-Coil, Four-Segment Generator*

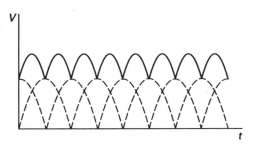

Figure 40.6 *DC Machine Equivalent Circuit*

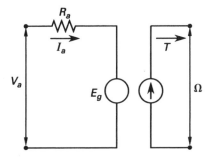

Since the rate of change of the magnetic flux in the armature depends on the speed of rotation, and because the armature is essentially an inductor, the current depends on the magnetic flux. The voltage of an inductor is directly dependent upon the rate of change of the current. Therefore, the induced voltage is directly dependent upon the rotational speed. Also, the greater the magnetic flux, the greater the current induced. Therefore, the induced voltage is directly dependent upon the magnetic flux. The proportionality constant that relates the speed and magnetic flux to the generated voltage is K_a and depends upon the construction of the machine. For the generator model in Fig. 40.6, this results in

$$E_g = K_a n \phi$$

Equation 40.4: Armature Voltage

$$V_a = K_a n\phi \quad \text{[in volts]} \qquad 40.4$$

Description

The *terminal voltage* of the generator model shown in Fig. 40.6 is

$$V_a = E_g + I_a R_a = K_a n\phi + I_a R_a$$

If the armature resistance is disregarded (as in the case of a permanent magnet armature that has no windings), the $I_a R_a$ drop is zero. Only in that case, the terminal and armature voltages are the same, and Eq. 40.4 accurately describes the terminal voltage. K_a is known as the *armature constant* (*armature winding constant* or *machine constant*).[3]

Example

A DC generator armature turns at 1200 rpm. The field's magnetic flux is 0.02 Wb. The armature constant is $K_a = 1$ V·min/Wb. Disregarding armature resistance, what is most nearly the output voltage?

- (A) 12 V
- (B) 20 V
- (C) 24 V
- (D) 48 V

Solution

From Eq. 40.4,

$$\begin{aligned}
V_a &= K_a n\phi \\
&= \left(1 \ \frac{\text{V·min}}{\text{Wb}}\right)\left(1200 \ \frac{\text{rev}}{\text{min}}\right)(0.02 \ \text{Wb}) \\
&= 24 \ \text{V}
\end{aligned}$$

The answer is (C).

DC Motors

DC motors are similar in construction and analysis to DC generators. Electrical power is delivered to a motor by applying a voltage to its terminals, resulting in a flow of armature current, I_a, through the armature. Some of that power is dissipated as heat in the armature winding's resistance, and some is converted into mechanical power to turn the armature. The power for the motor model shown in Fig. 40.6 is determined from

$$P_m = P_h + P_e = I_a^2 R_a + I_a E_g$$

---

[3]Although not incorrect, it is not common to specify the armature voltage (or the armature constant) in terms of n in rpm (as opposed to ω in rad/s).

Equation 40.5 and Eq. 40.6: Mechanical Power and Torque

$$P_m = V_a I_a \quad \text{[in watts]} \qquad 40.5$$

$$T_m = (60/2\pi) K_a \phi I_a \quad \text{[in newton-meters]} \qquad 40.6$$

Variation

$$P_m = T\Omega$$

Description

Disregarding the power dissipated in armature resistance heating, the motor model power equation can be simplified to Eq. 40.5.

The *torque of the DC motor* is linearly related to the strength of the field and the magnetic field of the armature. The stronger each of these fields is, the stronger the force would need to be to bring them into line (which never happens, due to the commutator). The magnetic field of the armature is linearly related to the current of the armature (ignoring magnetization). The proportionality constant that relates the armature current and magnetic flux to the torque voltage is the same K_a as used in Eq. 40.4 and depends upon the construction of the machine.

The mechanical power can also be expressed in terms of torque, T, and the mechanical rotational velocity, Ω, as shown in the variation equation. This can be derived directly from force-velocity-power relationships.

Example

A DC motor armature draws 100 A while generating a magnetic flux of 0.02 Wb. The armature constant is $K_a = 1$ V·min/Wb. What is most nearly the motor's developed torque?

(A) 19 N·m

(B) 24 N·m

(C) 32 N·m

(D) 51 N·m

Solution

From Eq. 40.6,

$$
\begin{aligned}
T_m &= (60/2\pi) K_a \phi I_a \\
&= \left(\frac{60}{2\pi}\right)\left(1\ \frac{\text{V·min}}{\text{Wb}}\right)(0.02\ \text{Wb})(100\ \text{A}) \\
&= 19\ \text{N·m}
\end{aligned}
$$

The answer is (A).

Engineering Sciences

Diagnostic Exam

Topic VIII: Materials Science

1. The results of a tensile test on a round specimen of a given material are shown.

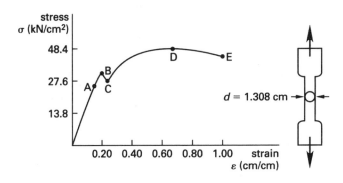

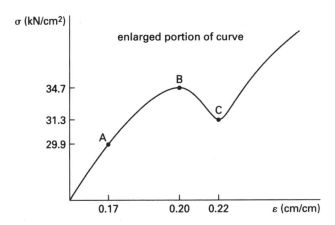

What is most nearly the yield stress?

(A) 14 kN/cm^2

(B) 26 kN/cm^2

(C) 31 kN/cm^2

(D) 48 kN/cm^2

2. The activation energy for creep is 161 kJ/mol for a given alloy. If the applied stress is fixed and the stress sensitivity remains the same, by approximately what factor does the creep rate change when the temperature increases from 350°C to 450°C?

(A) 2.2

(B) 3.0

(C) 74

(D) 220

3. Using the phase diagram given, what is most nearly the percentage of liquid remaining at 600°C that results from the equilibrium cooling of an alloy containing 5% silicon and 95% aluminum?

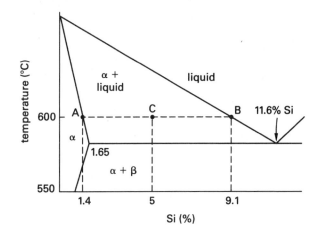

(A) 0.0%

(B) 47%

(C) 53%

(D) 67%

4. The stress-strain curve for a nonlinear, perfectly elastic material is shown. A sample of the material is loaded until the stress reaches the value at point B. Then, the material is unloaded to zero stress.

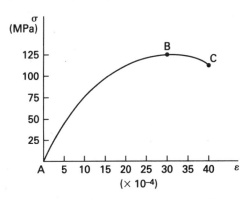

What is most nearly the permanent set in the material?

(A) 0

(B) 0.001

(C) 0.002

(D) 0.003

5. Which statement is FALSE?

(A) The amount or percentage of cold work cannot be obtained from information about change in the area or thickness of a metal.

(B) The process of applying force to a metal at temperatures below the temperature of crystallization in order to plastically deform the metal is called cold working.

(C) Annealing eliminates most of the defects caused by the cold working of a metal.

(D) Annealing reduces the hardness of the metal.

6. Which statement is most accurate regarding the two materials represented in the stress-strain diagrams?

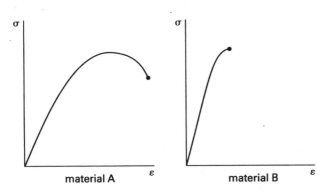

(A) Material B is more ductile and has a lower modulus of elasticity than material A.

(B) Material B would require more total energy to fracture than material A.

(C) Material A will withstand more stress before plastically deforming than material B.

(D) Material B will withstand a higher load than material A but is more likely to fracture suddenly.

7. Which statement is true?

(A) Low-alloy steels are a minor group and are rarely used.

(B) There are three basic types of stainless steels: martensitic, austenitic, and ferritic.

(C) The addition of small amounts of silicon to steel can cause a marked decrease in the yield strength of steel.

(D) The addition of small amounts of molybdenum to low-alloy steels makes it possible to harden and strengthen thick pieces of the metal by heat treatment.

8. The simplified phase diagram of an alloy of components A and B is shown.

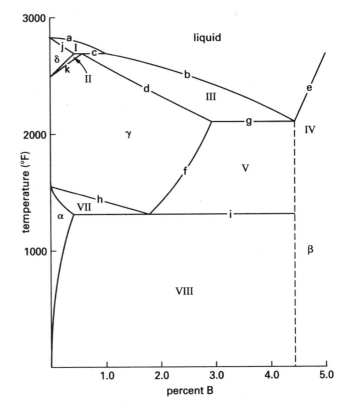

What is most nearly the percentage of solid alloy that will be present at 2300°F if the mixture is 3.0% B and 97% A?

(A) 32%

(B) 40%

(C) 51%

(D) 63%

9. All of the following metals will corrode if immersed in fresh water EXCEPT

(A) copper

(B) nickel

(C) chromium

(D) aluminum

10. The results of a tensile test on a round specimen of a given material are shown.

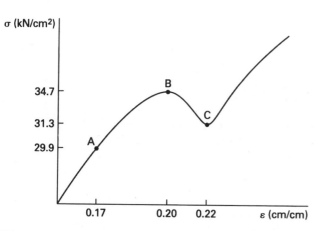

What is most nearly the elastic limit of the material?

- (A) 30 kN/cm^2
- (B) 31 kN/cm^2
- (C) 35 kN/cm^2
- (D) 52 kN/cm^2

SOLUTIONS

1. For this material, the stress required to continue the deformation drops markedly at yield.

$$\sigma_y = 31 \text{ kN/cm}^2$$

The answer is (C).

2. The absolute temperatures are

$$T_1 = 350°C + 273° = 623K$$
$$T_2 = 450°C + 273° = 723K$$

If the applied stress, σ, and the stress sensitivity, n, are fixed, the creep rate increases by a factor of

$$\frac{e^{(-Q/RT_2)}}{e^{(-Q/RT_1)}} = e^{\left(\left(\frac{-Q}{R}\right)\left(\frac{1}{T_2} - \frac{1}{T_1}\right)\right)}$$

$$= e^{\left(\left(\frac{-\left(161\frac{\text{kJ}}{\text{mol}}\right)\left(1000\frac{\text{J}}{\text{kJ}}\right)}{8.314\frac{\text{J}}{\text{mol·K}}}\right) \times \left(\frac{1}{723K} - \frac{1}{623K}\right)\right)}$$

$$= 73.6 \quad (74)$$

The answer is (C).

3. Use the lever rule. At point A, there is 1.4% Si and no liquid, while at point B there is 9.1% Si and all liquid.

$$\% \text{ liquid} = \frac{\text{Si}_C - \text{Si}_A}{\text{Si}_B - \text{Si}_A} \times 100\% = \frac{5\% - 1.4\%}{9.1\% - 1.4\%} \times 100\%$$
$$= 46.75\% \quad (47\%)$$

The answer is (B).

4. A perfectly elastic material exhibits no permanent deformation upon unloading.

The answer is (A).

5. The percentage of cold work can be calculated directly from the reduction in thickness or area of the metal.

The answer is (A).

6. The slope of material B's curve is steeper, so it has the higher modulus of elasticity.

Material A's ultimate strain is greater, so material A is more ductile. Option A is incorrect.

The area under the curve represents the energy (work) required to deform the material, which is the definition of toughness. The area under material A's curve is greater, so material A is tougher. Option B is incorrect.

Material B has no clearly defined yield strength, so compare the stresses reached for any arbitrary amount (e.g., 0.002) of strain. For any strain, material B has a higher stress. Option C is incorrect.

Material B follows the classic stress-strain curve of a brittle material which can fracture suddenly. Option D is correct.

The answer is (D).

7. Low-alloy steels are one of the most commonly used classes of structural steels, so option A is false. There are only two basic types of stainless steels: magnetic (martensitic) and non-magnetic (austenitic). Option B is false. The addition of small amounts of silicon to steel increases both the yield strength and tensile strength. Option C is false. The addition of small amounts of molybdenum to low-alloy steels makes it possible to harden and strengthen thick pieces of the metal by heat treatment. Option D is true.

The answer is (D).

8. Draw the horizontal tie line at 2300°F between line d and line b. Use the horizontal boron (B) scale for convenience. The tie line intersects line d (100% solid γ phase) at approximately $B_{solid} = 2.1\%$ B. The tie line intersects line b (100% liquid phase) at approximately $B_{liquid} = 3.6\%$ B. The percentage of solid with $B_{actual} = 3.0\%$ B is

$$\% \text{ solid} = \frac{B_{liquid} - B_{actual}}{B_{liquid} - B_{solid}} \times 100\%$$

$$= \frac{3.6\% - 3.0\%}{3.6\% - 2.1\%} \times 100\%$$

$$= 0.40 \quad (40\%)$$

The answer is (B).

9. Copper pipes are used extensively in residential water service. In a table of oxidation potentials, copper is higher (i.e., more anodic) than the standard hydrogen reaction, while nickel, chromium, and iron are below the standard hydrogen reaction. There is no galvanic driving potential for the copper ions to go into solution, so copper will not corrode in fresh water.

The answer is (A).

10. Point A is the proportionality limit. The elastic limit is very close to the yield point, point B.

$$\sigma_{\text{elastic limit}} = 34.7 \text{ kN/cm}^2 \quad (35 \text{ kN/cm}^2)$$

The answer is (C).

41 Material Properties and Testing

Nomenclature

a	crack length	m	in
A	area	m^2	ft^2
A	constant	–	–
BHN	Brinell hardness number	–	–
c	specific heat	J/kg·°C	Btu/lbm-°F
C	capacitance	F	F
C	molar specific heat	J/mol·°C	Btu/lbmol-°F
C_V	impact energy	J	ft-lbf
d	diameter	m	ft
d	distance	m	ft
D	diameter	m	ft
E	energy	eV	eV
E	modulus of elasticity	MPa	ksi
F	force	N	lbf
F	load	N	lbf
FS	factor of safety	–	–
g	gravitational acceleration, 9.81 (32.2)	m/s^2	ft/sec^2
G	modulus of rigidity	GPa	ksi
J	flux	$1/m^2{\cdot}s$	$1/ft^2$-sec
k	reduction factor	–	–
K_{IC}	stress intensity factor or fracture toughness	$MPa{\cdot}\sqrt{m}$	$ksi{\cdot}\sqrt{in}$
L	length	m	ft
m	mass	kg	lbm
M	moment	N·m	ft-lbf
MW	molecular weight	kg/kmol	lbm/lbmol

n	stress sensitivity exponent	–	–
N	number of cycles	–	–
P	load	N	lbf
q	charge	C	C
q	reduction in area	–	–
Q	activation energy	J/mol	ft-lbf/lbmol
Q	heat	J	Btu
R	resistance	Ω	Ω
$\overline{R}$	universal gas constant, 8314 (1545)	J/kmol·K	ft-lbf/lbmol-°R
S	strength	MPa	ksi
S'_e	endurance limit	MPa	ksi
t	depth or thickness	m	ft
t	time	s	sec
T	temperature	K	°R
TS	tensile strength	MPa	ksi
V	voltage	V	V
V	volume	m^3	ft^3
w	width	m	ft
Y	geometrical factor	–	–

Symbols

α	thermal expansion coefficient	1/°C	1/°F
β	Andrade's constant	$s^{-1/3}$	$sec^{-1/3}$
γ	electrical conductivity	W/m·K	W/m·K
δ	deformation	m	ft
ε	engineering strain	–	–
ε	permittivity	F/m or $C^2/N{\cdot}m^2$	F/m or $C^2/N{\cdot}m^2$
ε_0	permittivity of a vacuum, 8.85×10^{-12}	F/m or $C^2/N{\cdot}m^2$	F/m or $C^2/N{\cdot}m^2$
κ	dielectric constant	–	–
λ	thermal conductivity	W/m·K	Btu/ft-°R
μ	ductility	–	–
ν	Poisson's ratio	–	–
ρ	density	kg/m^3	lbm/ft^3
ρ	resistivity	$\Omega{\cdot}m$	Ω-ft
σ	engineering stress	MPa	ksi
ϕ	capacity reduction factor	–	–
ϕ	work function	eV	eV

Subscripts

0	initial
a	activation, allowable, or surface
b	size
c	conduction, critical, or load
d	diffusion or temperature
e	effects or endurance
eff	effective
f	failure, final, or fracture

Materials Science

g	gap or glass transition
i	intrinsic
I	intensity
n	nominal
o	original
p	constant pressure or particular
t	tensile or total
T	total or true
u	ultimate
ut	ultimate tensile
v	constant volume
v	valence
y	yield

1. MATERIALS SCIENCE

Materials science is the study of materials to understand their properties, limits, and uses. *Material properties* are key characteristics of a material commonly classified into five main categories: chemical, electrical, mechanical, physical, and thermal.

Mechanical properties describe the relationship between properties and mechanical (i.e., physical) forces, such as stresses, strains, and applied force. Examples of mechanical properties include strength, toughness, ductility, hardness, fatigue, and creep.

Thermal properties are properties that are observed when heat energy is applied to a material. Examples of thermal properties include thermal conductivity, thermal diffusivity, the heat of fusion, and the glass transition temperature.

Electrical properties define the reaction of a material to an electric field. Typical electrical properties are dielectric strength, conductivity, permeability, permittivity, and electrical resistance.

Chemical properties are properties that are evident only when a substance is changed chemically. Common chemical properties include corrosivity, toxicity, and flammability.

Unlike chemical properties, *physical properties* can be observed without altering the material or its structure. Common physical properties include density, melting point, and specific heat.

Some common properties of various materials are given in Table 41.1 and Table 41.2. Additional mechanical properties are given in Table 41.3.

2. MATERIAL SELECTION

Material selection is the process of selecting materials used to design and manufacture a part or product. Material selection is an important component in the design process, as materials must be carefully selected with product performance and manufacturing processes in mind. The goal of materials selection is to meet product performance goals (e.g., strength, ductility, safety) while minimizing costs and waste.

Materials selection typically begins by considering the ideal properties the material would exhibit based on the product's specifications. Then, materials that best exemplify those needs are selected, and a comparison of the selected materials, including costs, is performed. Because many kinds of materials are available, the process often starts by considering broad categories of materials before zeroing in on a specific choice. These general types of materials include

- *ceramics:* glass ceramics, glasses, graphite, and diamond
- *composites:* reinforced plastics, metal-matrix composites, and honeycomb structures
- *ferrous metals:* carbon, alloy, stainless steel, and tool and die steels
- *nonferrous metals and alloys:* aluminum, magnesium, copper, nickel, titanium, superalloys, refractory metals, beryllium, zirconium, low-melting alloys, and precious metals
- *plastics:* thermoplastics, thermosets, and elastomers

The materials selection process is often iterative; selections and comparisons may be done multiple times before finding the optimal material for a given use.

3. CLASSIFICATION OF MATERIALS

When used to describe engineering materials, the terms "strong" and "tough" are not synonymous. Similarly, "weak," "soft," and "brittle" have different engineering meanings. A *strong material* has a high ultimate strength, whereas a *weak material* has a low ultimate strength. A *tough material* will yield greatly before breaking, whereas a *brittle material* will not. (A brittle material is one whose strain at fracture is less than approximately 0.5%.) A *hard material* has a high modulus of elasticity, whereas a *soft material* does not. Figure 41.1 illustrates some of the possible combinations of these classifications, comparing the material's stress, σ, and strain, ε.

4. MECHANICAL PROPERTIES

Mechanical properties are those that describe how a material will react to external forces. Materials are commonly classified by their mechanical properties, including strength, hardness, and roughness. Typical design values of various mechanical properties are given in Table 41.3. Various mechanical properties are covered in the following sections.

5. ENGINEERING STRESS AND STRAIN

Figure 41.2 shows a *load-elongation curve* of *tensile test* data for a ductile ferrous material (e.g., low-carbon steel

Table 41.1 Typical Material Properties*

material	modulus of elasticity, E (Mpsi (GPa))	modulus of rigidity, G (Mpsi (GPa))	Poisson's ratio, ν	coefficient of thermal expansion, α (10^{-6}/°F (10^{-6}/°C))	density, ρ (lbm/in^3 (Mg/m^3))
steel	29.0 (200.0)	11.5 (80.0)	0.30	6.5 (11.7)	0.282 (7.8)
aluminum	10.0 (69.0)	3.8 (26.0)	0.33	13.1 (23.6)	0.098 (2.7)
cast iron	14.5 (100.0)	6.0 (41.4)	0.21	6.7 (12.1)	0.246–0.282 (6.8–7.8)
wood (fir)	1.6 (11.0)	0.6 (4.1)	0.33	1.7 (3.0)	–
brass	14.8–18.1 (102–125)	5.8 (40)	0.33	10.4 (18.7)	0.303–0.313 (8.4–8.7)
copper	17 (117)	6.5 (45)	0.36	9.3 (16.6)	0.322 (8.9)
bronze	13.9–17.4 (96–120)	6.5 (45)	0.34	10.0 (18.0)	0.278–0.314 (7.7–8.7)
magnesium	6.5 (45)	2.4 (16.5)	0.35	14 (25)	0.061 (1.7)
glass	10.2 (70)	–	0.22	5.0 (9.0)	0.090 (2.5)
polystyrene	0.3 (2)	–	0.34	38.9 (70.0)	0.038 (1.05)
polyvinyl chloride (PVC)	<0.6 (<4)	–	–	28.0 (50.4)	0.047 (1.3)
alumina fiber	58 (400)	–	–	–	0.141 (3.9)
aramide fiber	18.1 (125)	–	–	–	0.047 (1.3)
boron fiber	58 (400)	–	–	–	0.083 (2.3)
beryllium fiber	43.5 (300)	–	–	–	0.069 (1.9)
BeO fiber	58 (400)	–	–	–	0.108 (3.0)
carbon fiber	101.5 (700)	–	–	–	0.083 (2.3)
silicon carbide fiber	58 (400)	–	–	–	0.116 (3.2)

*Use these values if the specific alloy and temper are not listed in Table 41.3.

Figure 41.1 Types of Engineering Materials

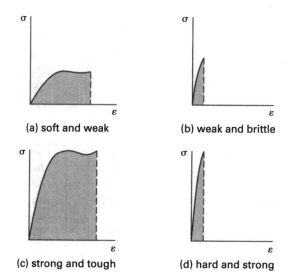

(a) soft and weak

(b) weak and brittle

(c) strong and tough

(d) hard and strong

or other body-centered cubic (BCC) transition metal). In this test, a prepared material sample (i.e., a *specimen*) is axially loaded in tension, and the resulting elongation, ΔL, is measured as the load, F, increases.

When elongation is plotted against the applied load, the graph is applicable only to an object with the same length and area as the test specimen. To generalize the test results, the data are converted to stresses and strains.

Equation 41.1 and Eq. 41.2: Engineering Stress and Strain

$$\sigma = \frac{F}{A_0} \qquad \textit{41.1}$$

$$\varepsilon = \frac{\Delta L}{L_0} \qquad \textit{41.2}$$

Description

Equation 41.1 describes *engineering stress*, σ (usually called *stress*), which is the load per unit original area. Typical units of engineering stress are MPa.

Equation 41.2 describes *engineering strain*, ε (usually called *strain*), which is the elongation of the test specimen expressed as a percentage or decimal fraction of the original length. The units m/m are also sometimes used for strain.[1]

---

[1]In the NCEES *FE Reference Handbook* (*NCEES Handbook*), strain, ε, is the same as creep but is unrelated to permittivity. All three share the same symbol in this section.

Table 41.2 *Properties of Metals*

metal	symbol	atomic weight	density, ρ (kg/m^3) water = 1000	melting point (°C)	melting point (°F)	specific heat (J/kg·K)	electrical resistivity (10^{-8} Ω·m) at 0°C (273.2K)[a]	heat conductivity,[b] λ (W/m·K) at 0°C (273.2K)
aluminum	Al	26.98	2698	660	1220	895.9	2.5	236
antimony	Sb	121.75	6692	630	1166	209.3	39	25.5
arsenic	As	74.92	5776	subl. 613	subl. 1135	347.5	26	–
barium	Ba	137.33	3594	710	1310	284.7	36	–
beryllium	Be	9.012	1846	1285	2345	2051.5	2.8	218
bismuth	Bi	208.98	9803	271	519	125.6	107	8.2
cadmium	Cd	112.41	8647	321	609	234.5	6.8	97
caesium	Cs	132.91	1900	29	84	217.7	18.8	36
calcium	Ca	40.08	1530	840	1544	636.4	3.2	–
cerium	Ce	140.12	6711	800	1472	188.4	7.3	11
chromium	Cr	52	7194	1860	3380	406.5	12.7	96.5
cobalt	Co	58.93	8800	1494	2721	431.2	5.6	105
copper	Cu	63.54	8933	1084	1983	389.4	1.55	403
gallium	Ga	69.72	5905	30	86	330.7	13.6	41
gold	Au	196.97	19 281	1064	1947	129.8	2.05	319
indium	In	114.82	7290	156	312	238.6	8	84
iridium	Ir	192.22	22 550	2447	4436	138.2	4.7	147
iron	Fe	55.85	7873	1540	2804	456.4	8.9	83.5
lead	Pb	207.2	11 343	327	620	129.8	19.2	36
lithium	Li	6.94	533	180	356	4576.2	8.55	86
magnesium	Mg	24.31	1738	650	1202	1046.7	3.94	157
manganese	Mn	54.94	7473	1250	2282	502.4	138	8
mercury	Hg	200.59	13 547	−39	−38	142.3	94.1	7.8
molybdenum	Mo	95.94	10 222	2620	4748	272.1	5	139
nickel	Ni	58.69	8907	1455	2651	439.6	6.2	94
niobium	Nb	92.91	8578	2425	4397	267.9	15.2	53
osmium	Os	190.2	22 580	3030	5486	129.8	8.1	88
palladium	Pd	106.4	11 995	1554	2829	230.3	10	72
platinum	Pt	195.08	21 450	1772	3221	134	9.81	72
potassium	K	39.09	862	63	145	753.6	6.1	104
rhodium	Rh	102.91	12 420	1963	3565	242.8	4.3	151
rubidium	Rb	85.47	1533	38.8	102	330.7	11	58
ruthenium	Ru	101.07	12 360	2310	4190	255.4	7.1	117
silver	Ag	107.87	10 500	961	1760	234.5	1.47	428
sodium	Na	22.989	966	97.8	208	1235.1	4.2	142
strontium	Sr	87.62	2583	770	1418	–	20	–
tantalum	Ta	180.95	16 670	3000	5432	150.7	12.3	57
thallium	Tl	204.38	11 871	304	579	138.2	10	10
thorium	Th	232.04	11 725	1700	3092	117.2	14.7	54
tin	Sn	118.69	7285	232	449	230.3	11.5	68
titanium	Ti	47.88	4508	1670	3038	527.5	39	22
tungsten	W	183.85	19 254	3387	6128	142.8	4.9	177
uranium	U	238.03	19 050	1135	2075	117.2	28	27
vanadium	V	50.94	6090	1920	3488	481.5	18.2	31
zinc	Zn	65.38	7135	419	786	393.5	5.5	117
zirconium	Zr	91.22	6507	1850	3362	284.7	40	23

[a]This is a rounded value. In its *Units* and *Environmental Engineering* sections, the NCEES *FE Reference Handbook* (*NCEES Handbook*) correctly lists the offset between degrees Celsius and Kelvins as 273.15°.
[b]In this table, the *NCEES Handbook* uses lambda, λ, as the symbol for thermal (heat) conductivity. While this usage is not unheard of, it is less common than the use of k, and it is inconsistent with the symbol k used elsewhere in the *NCEES Handbook*.

Table 41.3 *Average Mechanical Properties of Typical Engineering Materials (customary U.S. units)*[a,b]

materials		specific weight, γ (lbf/in³)	modulus of elasticity, E (10³ ksi)	modulus of rigidity, G (10³ ksi)	yield strength, σ_y (ksi)[c] tens.	comp.	shear	ultimate strength, σ_u (ksi)[c] tens.	comp.	shear	% elongation in 2 in specimen	Poisson's ratio, ν	coefficient of thermal expansion, α (10⁻⁶)/°F
metallic													
aluminum wrought alloys	2014-T6	0.101	10.6	3.9	60	60	25	68	68	42	10	0.35	12.8
	6061-T6	0.098	10.0	3.7	37	37	19	42	42	27	12	0.35	13.1
cast iron alloys	gray ASTM 20	0.260	10.0	3.9	–	–	–	26	97	–	0.6	0.28	6.70
	malleable ASTM A197	0.263	25.0	9.8	–	–	–	40	83	–	5	0.28	6.60
copper alloys	red brass C83400	0.316	14.6	5.4	11.4	11.4	–	35	35	–	35	0.35	9.80
	bronze C86100	0.319	15.0	5.6	50	50	–	95	95	–	20	0.34	9.60
magnesium alloy	Am 1004-T61	0.066	6.48	2.5	22	22	–	40	40	22	1	0.30	14.3
steel alloys	structural A36	0.284	29.0	11.0	36	36	–	58	58	–	30	0.32	6.60
	stainless 304	0.284	28.0	11.0	30	30	–	75	75	–	40	0.27	9.60
	tool L2	0.295	29.0	11.0	102	102	–	116	116	–	22	0.32	6.50
titanium alloy	Ti-6Al-4V	0.160	17.4	6.4	134	134	–	145	145	–	16	0.36	5.20
nonmetallic													
concrete	low strength	0.086	3.20	–	–	–	1.8	–	–	–	–	0.15	6.0
	high strength	0.086	4.20	–	–	–	5.5	–	–	–	–	0.15	6.0
plastic reinforced	Kevlar 49	0.0524	19.0	–	–	–	–	104	70	10.2	2.8	0.34	–
	30% glass	0.0524	10.5	–	–	–	–	13	19	–	–	0.34	–
wood select structural grade	Douglas Fir	0.017	1.90	–	–	–	–	0.30[d]	3.78[e]	0.90[e]	–	0.29[f]	–
	White Spruce	0.130	1.40	–	–	–	–	0.36[d]	5.18[e]	0.97[e]	–	0.31[f]	–

[a]Use these values for the specific alloys and temper listed. For all other materials, refer to Table 41.1.
[b]Specific values may vary for a particular material due to alloy or mineral composition, mechanical working of the specimen, or heat treatment. For a more exact value reference books for the material should be consulted.
[c]The yield and ultimate strengths for ductile materials can be assumed to be equal for both tension and compression.
[d]Measured perpendicular to the grain.
[e]Measured parallel to the grain.
[f]Deformation measured perpendicular to the grain when the load is applied along the grain.

Source: Hibbeler, R. C., *Mechanics of Materials*, 4th ed., Prentice Hall, 2000.

Figure 41.2 *Typical Tensile Test of a Ductile Material*

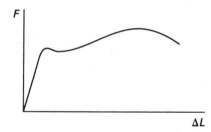

If the stress-strain data are plotted, the shape of the resulting line will be essentially the same as the force-elongation curve, although the scales will differ.

Example

A 100 mm gage length is marked on an aluminum rod. The rod is strained so that the gage marks are 109 mm apart.

The strain is most nearly

(A) 0.001

(B) 0.01

(C) 0.1

(D) 1.0

Solution

From Eq. 41.2, the strain is

$$\varepsilon = \frac{\Delta L}{L_0} = \frac{109 \text{ mm} - 100 \text{ mm}}{100 \text{ mm}} = 0.09 \quad (0.1)$$

The answer is (C).

Equation 41.3 Through Eq. 41.5: True Stress and Strain

$$\sigma_T = \frac{F}{A} \qquad 41.3$$

$$\varepsilon_T = \frac{dL}{L} \qquad \text{41.4}$$

$$\varepsilon_T = \ln(1 + \varepsilon) \qquad \text{41.5}$$

Description

As the stress increases during a tensile test, the length of a specimen increases, and the area decreases. The engineering stress and strain are not *true stress and strain parameters*, σ_T and ε_T, which must be calculated from instantaneous values of length, L, and area, A.[2] Figure 41.3 illustrates engineering and true stresses and strains for a ferrous alloy. Although true stress and strain are more accurate, most engineering work has traditionally been based on engineering stress and strain, which is justifiable for two reasons: (1) design using ductile materials is limited to the elastic region where engineering and true values differ little, and (2) the reduction in area of most parts at their service stresses is not known; only the original area is known.

Figure 41.3 *True and Engineering Stresses and Strains for a Ferrous Alloy*

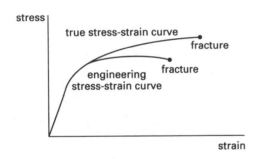

6. STRESS-STRAIN CURVES

Equation 41.6: Hooke's Law

$$\sigma = E\varepsilon \qquad \text{41.6}$$

Variation

$$E = \frac{F/A_0}{\Delta L/L_0} = \frac{FL_0}{A_0 \Delta L}$$

Description

Segment OA in Fig. 41.4 is a straight line. The relationship between the stress and the strain in this linear region is given by *Hooke's law*, Eq. 41.6.

The slope of the line segment OA is the *modulus of elasticity*, E, also known as *Young's modulus* or the *elastic modulus*. Table 41.3 lists approximate values of the modulus of elasticity for materials at room temperature. The modulus of elasticity will be lower at higher temperatures.

[2]The *NCEES Handbook* is inconsistent in representing change in length. ΔL in Eq. 41.2 is the same as dL in Eq. 41.4.

Figure 41.4 *Typical Stress-Strain Curve for Steel*

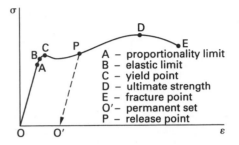

A – proportionality limit
B – elastic limit
C – yield point
D – ultimate strength
E – fracture point
O' – permanent set
P – release point

Example

A test specimen with a circular cross section has an initial gage length of 500 mm and an initial diameter of 60 mm. The specimen is placed in a tensile test apparatus. When the instantaneous tensile force in the specimen is 50 kN, the specimen has a longitudinal elongation of 0.16 mm and a lateral decrease in diameter of 0.01505 mm. What is most nearly the modulus of elasticity?

(A) 30×10^9 Pa

(B) 46×10^9 Pa

(C) 55×10^9 Pa

(D) 70×10^9 Pa

Solution

The area of the 60 mm bar is

$$A_0 = \frac{\pi d_0^2}{4} = \frac{\pi \left(\dfrac{60 \text{ mm}}{1000 \, \frac{\text{mm}}{\text{m}}} \right)^2}{4} = 2.827 \times 10^{-3} \text{ m}^2$$

Using Eq. 41.6 and its variation, the modulus of elasticity is

$$\sigma = E\varepsilon$$

$$E = \frac{\sigma}{\varepsilon} = \frac{FL_0}{A_0 \Delta L} = \frac{(50 \text{ kN})\left(1000 \, \frac{\text{N}}{\text{kN}}\right)(500 \text{ mm})}{(2.827 \times 10^{-3} \text{ m}^2)(0.16 \text{ mm})}$$

$$= 55.26 \times 10^9 \text{ N/m}^2 \quad (55 \times 10^9 \text{ Pa})$$

The answer is (C).

7. POINTS ALONG THE STRESS-STRAIN CURVE

The stress at point A in Fig. 41.4 is known as the *proportionality limit* (i.e., the maximum stress for which the linear relationship is valid). Strain in the *proportional region* is called *proportional* (or *linear*) *strain*.

The *elastic limit*, point B in Fig. 41.4, is slightly higher than the proportionality limit. As long as the stress is kept below the elastic limit, there will be no *permanent set* (*permanent deformation*) when the stress is removed. Strain that disappears when the stress is removed is known as *elastic strain*, and the stress is said to be in the *elastic region*. When the applied stress is removed, the *recovery* is 100%, and the material follows the original curve back to the origin.

If the applied stress exceeds the elastic limit, the recovery will be along a line parallel to the straight line portion of the curve, as shown in the line segment PO'. The strain that results (line OO') is permanent set (i.e., a permanent deformation). The terms *plastic strain* and *inelastic strain* are used to distinguish this behavior from the elastic strain.

For steel, the *yield point*, point C, is very close to the elastic limit. For all practical purposes, the *yield strength* or *yield stress*, S_y, can be taken as the stress that accompanies the beginning of plastic strain. Yield strengths are reported in MPa.

Most nonferrous materials, such as aluminum, magnesium, copper, and other face-centered cubic (FCC) and hexagonal close-packed (HCP) metals, do not have well-defined yield points. In such cases, the yield point is usually taken as the stress that will cause a 0.2% *parallel offset* (i.e., a plastic strain of 0.002), shown in Fig. 41.5. However, the yield strength can also be defined by other offset values or by total strain characteristics.

Figure 41.5 *Yield Strength of a Nonferrous Metal*

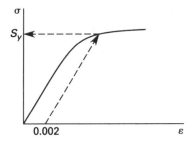

The *ultimate strength* or *tensile strength*, S_u, point D in Fig. 41.4, is the maximum stress the material can support without failure. This property is seldom used in the design of ductile material, since stresses near the ultimate strength are accompanied by large plastic strains.

The *breaking strength* or *fracture strength*, S_f, is the stress at which the material actually fails (point E in Fig. 41.4). For ductile materials, the breaking strength is less than the ultimate strength, due to the necking down in cross-sectional area that accompanies high plastic strains.

8. ALLOWABLE STRESS DESIGN

Once an actual stress has been determined, it can be compared to the *allowable stress*. In engineering design, the term "allowable" always means that a factor of safety has been applied to the governing material strength.

$$\text{allowable stress} = \frac{\text{material strength}}{\text{factor of safety}}$$

For ductile materials, the material strength used is the yield strength. For steel, the factor of safety, FS, ranges from 1.5 to 2.5, depending on the type of steel and the application. Higher factors of safety are seldom necessary in normal, noncritical applications, due to steel's predictable and reliable performance.

$$\sigma_a = \frac{S_y}{\text{FS}} \quad \text{[ductile]}$$

For brittle materials, the material strength used is the ultimate strength. Since brittle failure is sudden and unpredictable, the factor of safety is high (e.g., in the 6 to 10 range).

$$\sigma_a = \frac{S_u}{\text{FS}} \quad \text{[brittle]}$$

If an actual stress is less than the allowable stress, the design is considered acceptable. This is the principle of the *allowable stress design method*, also known as the *working stress design method*.

$$\sigma_{\text{actual}} \leq \sigma_a$$

9. ULTIMATE STRENGTH DESIGN

The allowable stress method has been replaced in most structural work by the *ultimate strength design method*, also known as the *load factor design method*, *plastic design method*, or just *strength design method*. This design method does not use allowable stresses at all. Rather, the member is designed so that its actual *nominal strength* exceeds the required ultimate strength.[3]

The *ultimate strength* (i.e., the required strength) of a member is calculated from the actual *service loads* and multiplicative factors known as *overload factors* or *load factors*. Usually, a distinction is made between dead loads and live loads.[4] For example, the required ultimate moment-carrying capacity in a concrete beam designed according to the American Concrete Institute's *Building Code Requirements for Structural Concrete* (ACI 318) would be[5]

$$M_u = 1.2 M_{\text{dead load}} + 1.6 M_{\text{live load}}$$

The *nominal strength* (i.e., the actual ultimate strength) of a member is calculated from the dimensions and materials. A *capacity reduction factor*, ϕ, of 0.70 to 0.90 is included in the calculation to account for typical

---

[3]It is a characteristic of the ultimate strength design method that the term "strength" actually means load, shear, or moment. Strength seldom, if ever, refers to stress. Therefore, the nominal strength of a member might be the load (in newtons) or moment (in N·m) that the member supports at plastic failure.

[4]*Dead load* is an inert, inactive load, primarily due to the structure's own weight. *Live load* is the weight of all nonpermanent objects, including people and furniture, in the structure.

[5]ACI 318 has been adopted as the source of concrete design rules in the United States.

Materials Science

workmanship and increase required strength. The moment criteria for an acceptable design is

$$M_n \geq \frac{M_u}{\phi}$$

10. DUCTILE AND BRITTLE BEHAVIOR

Ductility is the ability of a material to yield and deform prior to failure.[6] Not all materials are ductile. *Brittle materials*, such as glass, cast iron, and ceramics, can support only small strains before they fail catastrophically, without warning. As the stress is increased, the elongation is linear, and Hooke's law can be used to predict the strain. Failure occurs within the linear region, and there is very little, if any, *necking down* (i.e., a localized decrease in cross-sectional area). Since the failure occurs at a low strain, brittle materials are not ductile.

Figure 41.6 illustrates typical stress-strain curves for ductile and brittle materials.

Figure 41.6 Stress-Strain Curves for Ductile and Brittle Materials

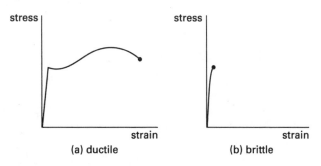

(a) ductile   (b) brittle

Ductility, μ, is a ratio of two quantities—one of which is related to catastrophic failure (i.e., collapse), and the other related to the loss of serviceability (i.e., yielding). For example, a building's ductility might be the ratio of earthquake energy it takes to collapse the building to the earthquake energy that just causes the beams and columns to buckle or doorframes to warp.

$$\mu = \frac{\text{energy at collapse}}{\text{energy at loss of serviceability}}$$

In contrast to the ductility of an entire building, various measures of ductility are calculated from test specimens for engineering materials. Definitions based on length, area, and the volume of the test specimen are in use. If ductility is to be based on test specimen length, the following definition might be used.

$$\mu = \frac{L_u}{L_y} = \frac{\varepsilon_u}{\varepsilon_y}$$

---

[6]The *NCEES Handbook* gives an incorrect and misleading statement when it says "Ductility (also called percent elongation) [is the] permanent engineering strain after failure." Ductility is a ratio of two quantities, not a percentage. Although there are many measures of ductility, none of them involve the permanent (snapped-back) set of a failed member. Percent elongation at failure might be used to categorize a ductile material, but it is not the same as ductility.

Equation 41.7: Percent Elongation

$$\% \text{ elongation} = \left(\frac{\Delta L}{L_o}\right) \times 100\% \qquad 41.7$$

Variation

$$\% \text{ elongation} = \frac{L_f - L_0}{L_0} \times 100\% = \varepsilon_f \times 100\%$$

Description

In contrast to ductility, the *percent elongation at failure* is based on the fracture length, as in Eq. 41.7, or fracture strain, ε_f, shown in Fig. 41.7 and the variation of Eq. 41.7. Percent elongation at failure might be used to categorize a ductile material, but it is not ductility. Since ductility is a ratio of energy absorbed at two points on the loading curve (as represented by the area under the curve), the ultimate length (area, volume, etc.) is measured just prior to fracture, not after. The ultimate length is not the same as what is commonly referred to as "fracture length." The *fracture length* is the length obtained after failure by measuring the two pieces of the failed specimen placed together end-to-end. This length includes the permanent, plastic strain but does not include the recovered elastic strain. The work (energy) required to elastically strain the failed member is not considered with this measure of fracture length.

Figure 41.7 Fracture and Ultimate Strain

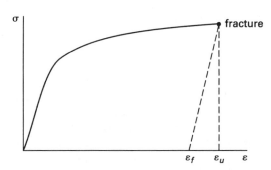

If area is the measured parameter, the term *reduction in area*, q, is used. At failure, the reduction in area due to necking down will be 50% or greater for ductile materials and less than 10% for brittle materials. Reduction in area can be used to categorize a ductile material, but it is not ductility.

$$q_f = \frac{A_0 - A_f}{A_0}$$

11. CRACK PROPAGATION IN BRITTLE MATERIALS

If a material contains a crack, stress is concentrated at the tip or tips of the crack. A crack in the surface of the material will have one tip (i.e., a stress concentration point); an internal crack in the material will have two tips. This increase in stress can cause the crack to propagate (grow) and can significantly reduce the material's ability to bear loads. Other things being equal, a crack in the surface of a material has a more damaging effect.

There are three modes of *crack propagation*, as illustrated by Fig. 41.8.

- *opening or tensile:* forces act perpendicular to the crack, which pulls the crack open, as shown in Fig. 41.8(a). This is known as mode I.

- *in-plane shear or sliding:* forces act parallel to the crack, which causes the crack to slide along itself, as shown in Fig. 41.8(b). This is known as mode II.

- *out-of-plane shear or pushing (pulling):* forces act perpendicular to the crack, tearing the crack apart, as shown in Fig. 41.8(c). This is known as mode III.

Figure 41.8 *Crack Propagation Modes*

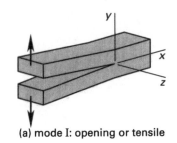

(a) mode I: opening or tensile

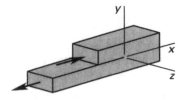

(b) mode II: in-plane shear or sliding

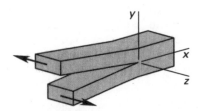

(c) mode III: out-of-plane shear or pushing (pulling)

Equation 41.8: Fracture Toughness

$$K_{IC} = Y\sigma\sqrt{\pi a} \qquad 41.8$$

Values

crack location	geometrical factor, Y
internal	1.0
surface (exterior)	1.1

Description

Fracture toughness is the amount of energy required to propagate a preexisting flaw. Fracture toughness is quantified by a *stress intensity factor*, K, a measure of energy required to grow a thin crack. For a mode I crack (see Fig. 41.8(a)), the stress intensity factor for a crack is designated K_{IC} or K_{Ic}. The stress intensity factor can be used to predict whether an existing crack will propagate through the material. When K_{IC} reaches a critical value, *fast fracture* occurs. The crack suddenly begins to propagate through the material at the speed of sound, leading to catastrophic failure. This critical value at which fast fracture occurs is called *fracture toughness*, and it is a property of the material.

The stress intensity factor is calculated from Eq. 41.8. σ is the nominal stress. a is the crack length. For a surface crack, a is measured from the crack tip to the surface of the material, as shown in Fig. 41.9(a). For an internal crack, a is half the distance from one tip to the other, as shown in Fig. 41.9(b). Y is a dimensionless factor that is dependent on the location of the crack, as shown in the values section.

Figure 41.9 *Crack Length*

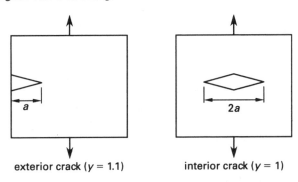

exterior crack ($y = 1.1$)   interior crack ($y = 1$)

When a is measured in meters and σ is measured in MPa, typical units for both the stress intensity factor and fracture toughness are MPa·$\sqrt{m}$, equivalent to MN/m$^{3/2}$. Typical values of fracture toughness for various materials are given in Table 41.4.

Table 41.4 *Representative Values of Fracture Toughness*

material	K_{IC} (MPa·$\sqrt{m}$)	K_{IC} (ksi-$\sqrt{in}$)
Al 2014-T651	24.2	22
Al 2024-T3	44	40
52100 steel	14.3	13
4340 steel	46	42
alumina	4.5	4.1
silicon carbide	3.5	3.2

Example

An aluminum alloy plate containing a 2 cm long crack is 10 cm wide and 0.5 cm thick. The plate is pulled with a uniform tensile force of 10 000 N.

Materials Science

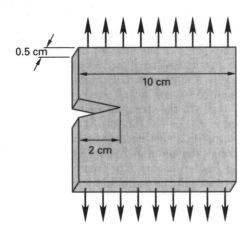

What is most nearly the stress intensity factor at the end of the crack?

(A) 2.1 MPa·$\sqrt{m}$

(B) 5.5 MPa·$\sqrt{m}$

(C) 12 MPa·$\sqrt{m}$

(D) 21 MPa·$\sqrt{m}$

Solution

From Eq. 41.1, the nominal stress is

$$\sigma = \frac{F}{A_0} = \frac{(10\,000 \text{ N})\left(100\,\frac{\text{cm}}{\text{m}}\right)^2}{(10 \text{ cm})(0.5 \text{ cm})}$$

$$= 20 \times 10^6 \text{ N/m}^2 \quad (20 \text{ MPa})$$

Since this is an exterior crack, $Y = 1.1$. Using Eq. 41.8, the stress intensity factor is

$$K_{IC} = Y\sigma\sqrt{\pi a}$$

$$= (1.1)(20 \text{ MPa})\sqrt{\pi\left(\frac{2 \text{ cm}}{100\,\frac{\text{cm}}{\text{m}}}\right)}$$

$$= 5.51 \text{ MPa·}\sqrt{m} \quad (5.5 \text{ MPa·}\sqrt{m})$$

The answer is (B).

12. FATIGUE

A material can fail after repeated stress loadings even if the stress level never exceeds the ultimate strength, a condition known as *fatigue failure*.

The behavior of a material under repeated loadings is evaluated by an *endurance test* (or *fatigue test*). A specimen is loaded repeatedly to a specific stress amplitude, S, and the number of applications of that stress required to cause failure, N, is counted. *Rotating beam tests* that load the specimen in bending, as shown in Fig. 41.10, are more common than alternating deflection and push-pull tests, but are limited to round specimens. The *mean stress* is zero in rotating beam tests.

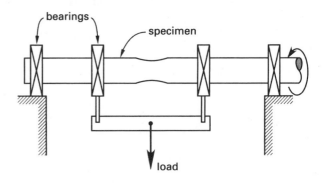

Figure 41.10 *Rotating Beam Test*

This procedure is repeated for different stresses using different specimens. The results of these tests are graphed on a semi-log plot, resulting in the *S-N curve* shown in Fig. 41.11.

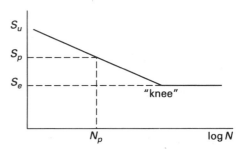

Figure 41.11 *Typical S-N Curve for Steel*

For a particular stress level, such as S_p in Fig. 41.11, the number of cycles required to cause failure, N_p, is the *fatigue life*. S_p is the *fatigue strength* corresponding to N_p.

For steel that is subjected to fewer than approximately 10^3 loadings, the fatigue strength approximately equals the ultimate strength. (Although *low-cycle fatigue theory* has its own peculiarities, a part experiencing a small number of cycles can usually be designed or analyzed as for static loading.) The curve is linear between 10^3 and approximately 10^6 cycles if a logarithmic N-scale is used. Above 10^6 cycles, there is no further decrease in strength.

Below a certain stress level, called the *endurance limit*, *endurance stress*, or *fatigue limit*, S'_e, the material will withstand an almost infinite number of loadings without experiencing failure. This is characteristic of steel and titanium. If a dynamically loaded part is to have an infinite life, the stress must be kept below the endurance limit.

The yield strength is an irrelevant factor in cyclic loading. Fatigue failures are fracture failures, not yielding failures. They start with microscopic cracks at the material surface. Some of the cracks are present initially; others form when repeated cold working reduces the ductility in strain-hardened areas. These cracks grow minutely with each loading. Since cracks start at the location of surface defects, the endurance limit is increased by proper treatment of the surface. Such treatments include polishing, surface hardening, shot peening, and filleting joints.

Equation 41.9 Through Eq. 41.11: Endurance Limit Modifying Factors

$$S_e = k_a k_b k_c k_d k_e S'_e \qquad 41.9$$

$$k_a = a S_{ut}^b \qquad 41.10$$

$$k_b = 1.189 d_{eff}^{-0.097} \quad [8 \text{ mm} \le d \le 250 \text{ mm}] \qquad 41.11$$

Description

The endurance limit is not a true property of the material since the other significant influences, particularly surface finish, are never eliminated. However, representative values of S'_e obtained from ground and polished specimens provide a baseline to which other factors can be applied to account for the effects of surface finish, temperature, stress concentration, notch sensitivity, size, environment, and desired reliability. These other influences are accounted for by *endurance limit modifying factors* that are used to calculate a working endurance strength, S_e, for the material.

The *surface factor*, k_a, is calculated from Eq. 41.10 using values of the factors a and b found from Table 41.5.

Table 41.5 Factors for Calculating k_a

surface finish	a		b
	(kpsi)	(MPa)	
ground	1.34	1.58	−0.085
machined or cold-drawn (CD)	2.70	4.51	−0.265
hot rolled	14.4	57.7	−0.718
as forged	39.9	272.0	−0.995

The *size factor*, k_b, and *load factor*, k_c, are determined for axial loadings from Table 41.6 and for bending and torsion from Table 41.7. For bending and torsion where the diameter, d, is between 8 mm and 250 mm, k_b is calculated from Eq. 41.11.

As the size gets larger, the endurance limit decreases due to the increased number of defects in a larger volume. Since the endurance strength, S'_e, is derived from a circular specimen with a diameter of 7.6 mm, the size modification factor is 1.0 for bars of that size.[7] d_{eff} is the effective dimension.[8] Simplistically, for noncircular cross-sections, the smallest cross-sectional

---

[7]In Table 41.7, the *NCEES Handbook* gives the limits for the use of Eq. 41.11 as "8 mm < d ≤ 250 mm." This should be "8 mm ≤ d ≤ 250 mm" so as to be unambiguous at $d = 8$ mm.

[8]The *NCEES Handbook* does not give any explanation or guidance in determining the effective dimension.

Table 41.6 Endurance Limit Modifying Factors for Axial Loading

size factor, k_b	1
load factor, k_c	
$S_{ut} \le 1520$ MPa	0.923
$S_{ut} > 1520$ MPa	1

Table 41.7 Endurance Limit Modifying Factors for Bending and Torsion

size factor, k_b	
$d \le 8$ mm	1
8 mm $\le d \le$ 250 mm	use Eq. 41.11
$d > 250$ mm	between 0.6 and 0.75
load factor, k_c	
bending	1
torsion	0.577

dimension should be used, and for a solid circular specimen in rotating bending, $d_{eff} = d$. For a nonrotating or noncircular cross section, d_{eff} is obtained by equating the area of material stressed above 95% of the maximum stress to the same area in the rotating-beam specimen of the same length. That area is designated $A_{0.95\sigma}$. For a nonrotating solid rectangular section with width w and thickness t, the effective dimension is

$$d_{eff} = 0.808\sqrt{wt}$$

Values of the *temperature factor*, k_d, and the *miscellaneous effects factor*, k_e, are found from Table 41.8. The miscellaneous effects factor is used to account for various factors that reduce strength, such as corrosion, plating, and residual stress.

Table 41.8 Additional Endurance Limit Modifying Factors

temperature factor, k_d	1 $[T \le 450°C]$
miscellaneous effects factor, k_e	1, unless otherwise specified

Example

A 25 mm diameter machined bar is exposed to a fluctuating bending load in a 200°C environment. The bar is made from ASTM A36 steel, which has a yield strength of 250 MPa, an ultimate tensile strength of 400 MPa, and a density of 7.8 g/cm³. The endurance limit is determined to be 200 MPa. What is most nearly the fatigue strength of the steel?

(A) 95 MPa

(B) 130 MPa

(C) 160 MPa

(D) 200 MPa

Solution

Determine the endurance limit modifying factors.

From Table 41.5, since the surface is machined, $a = 4.51$ MPa, and $b = -0.265$. From Eq. 41.10, the surface factor is

$$k_a = aS_{ut}^b = (4.51 \text{ MPa})(400 \text{ MPa})^{-0.265} = 0.9218$$

Since the diameter is between 8 mm and 250 mm, the size factor is calculated from Eq. 41.11.

$$k_b = 1.189 d_{eff}^{-0.097}$$
$$= (1.189)(25 \text{ mm})^{-0.097}$$
$$= 0.8701$$

From Table 41.7, $k_c = 1$ for bending stress. From Table 41.8, the temperature is less than 450°C, so $k_d = 1$. $k_e = 1$ since it was not specified otherwise.

Using Eq. 41.9, the approximate fatigue strength is

$$S_e = k_a k_b k_c k_d k_e S_e'$$
$$= (0.9218)(0.8701)(1)(1)(1)(200 \text{ MPa})$$
$$= 160.4 \text{ MPa} \quad (160 \text{ MPa})$$

The answer is (C).

13. TOUGHNESS

Toughness is a measure of a material's ability to yield and absorb highly localized and rapidly applied stress. A tough material will be able to withstand occasional high stresses without fracturing. Products subjected to sudden loading, such as chains, crane hooks, railroad couplings, and so on, should be tough. One measure of a material's toughness is the *modulus of toughness*, which is the *strain energy* or work per unit volume required to cause fracture. This is the total area under the stress-strain curve. Another measure is the *notch toughness*, which is evaluated by measuring the *impact energy* that causes a notched sample to fail. At 21°C, the energy required to cause failure ranges from 60 J for carbon steels to approximately 150 J for chromium-manganese steels.

14. CHARPY TEST

In the *Charpy test* (*Charpy V-notch test*), which is popular in the United States, a standardized beam specimen is given a 45° notch. The specimen is then centered on simple supports with the notch down. (See Fig. 41.12.) A falling pendulum striker hits the center of the specimen. This test is performed several times with different heights and different specimens until a sample fractures.

The kinetic energy expended at impact, equal to the initial potential energy, is calculated from the height. It is designated C_V and is expressed in joules. The energy required to cause failure is a measure of

toughness. Without a notch, the specimen would experience uniaxial stress (tension and compression) at impact. The notch allows triaxial stresses to develop. Most materials become more brittle under triaxial stresses than under uniaxial stresses.

Figure 41.12 *Charpy Test*

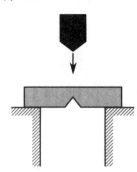

15. DUCTILE-BRITTLE TRANSITION

As temperature is reduced, the toughness of a material decreases. In BCC metals, such as steel, at a low enough temperature the toughness will decrease sharply. The transition from high-energy ductile failures to low-energy brittle failures begins at the *fracture transition plastic* (FTP) *temperature.*

Since the transition occurs over a wide temperature range, the *transition temperature* (also known as the *ductility transition temperature*) is taken as the temperature at which an impact of 20 J will cause failure. (See Table 41.9.) This occurs at approximately −1°C for low-carbon steel.

Table 41.9 *Approximate Ductile Transition Temperatures*

type of steel	ductile transition temperature (°C)
carbon steel	−1
high-strength, low-alloy steel	−18 to −1
heat-treated, high-strength, carbon steel	−32
heat-treated, construction alloy steel	−40 to −62

The appearance of the fractured surface is also used to evaluate the transition temperature. The fracture can be fibrous (from shear fracture) or granular (from cleavage fracture), or a mixture of both. The fracture planes are studied, and the percentages of ductile failure are plotted against temperature. The temperature at which the failure is 50% fibrous and 50% granular is known as *fracture appearance transition temperature* (FATT).

Not all materials have a ductile-brittle transition. Aluminum, copper, other face-centered cubic (FCC) metals,

Materials Science

and most hexagonal close-packed (HCP) metals do not lose their toughness abruptly. Figure 41.13 illustrates the failure energy curves for several materials.

Figure 41.13 Failure Energy versus Temperature

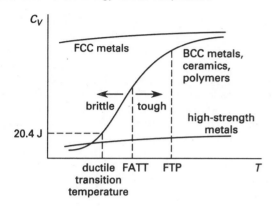

16. CREEP TEST

Creep or *creep strain* is the continuous yielding of a material under constant stress. For metals, creep is negligible at low temperatures (i.e., less than half of the absolute melting temperature), although the usefulness of nonreinforced plastics as structural materials is seriously limited by creep at room temperature.

During a *creep test*, a low tensile load of constant magnitude is applied to a specimen, and the strain is measured as a function of time. The *creep strength* is the stress that results in a specific creep rate, usually 0.001% or 0.0001% per hour. The *rupture strength*, determined from a *stress-rupture test*, is the stress that results in a failure after a given amount of time, usually 100, 1000, or 10,000 hours.

If strain is plotted as a function of time, three different curvatures will be apparent following the initial elastic extension.[9] (See Fig. 41.14.) During the first stage, the *creep rate* ($d\varepsilon/dt$) decreases since strain hardening (dislocation generation and interaction with grain boundaries and other barriers) is occurring at a greater rate than annealing (annihilation of dislocations, climb, cross-slip, and some recrystallization). This is known as *primary creep*.

During the second stage, the creep rate is constant, with strain hardening and annealing occurring at the same rate. This is known as *secondary creep* or *cold flow*. During the third stage, the specimen begins to neck down, and rupture eventually occurs. This region is known as *tertiary creep*.

The secondary creep rate is lower than the primary and tertiary creep rates. The secondary creep rate, represented by the slope (on a log-log scale) of the line during the second stage, is temperature and stress dependent. This slope increases at higher temperatures and stresses.

---

[9]In Great Britain, the initial elastic elongation, ε_0, is considered the first stage. Therefore, creep has four stages in British nomenclature.

Figure 41.14 Stages of Creep

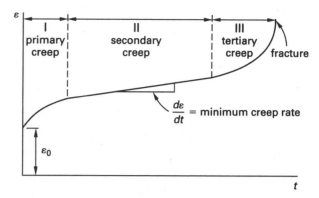

The creep rate curve can be represented by the following empirical equation, known as *Andrade's equation*.

$$\varepsilon = \varepsilon_0(1 + \beta t^{1/3})e^{kt}$$

Dislocation climb (glide and creep) is the primary creep mechanism, although diffusion creep and grain boundary sliding also contribute to creep on a microscopic level. On a larger scale, the mechanisms of creep involve slip, subgrain formation, and grain-boundary sliding.

Equation 41.12: Creep

$$\frac{d\varepsilon}{dt} = A\sigma^n e^{-Q/RT} \qquad \textbf{41.12}$$

Value

$\overline{R} = 8314$ J/kmol·K

Description

Equation 41.12 calculates creep from the strain, ε, time, t, a constant, A, universal gas constant, $\overline{R}$, absolute temperature, T, applied stress, σ, activation energy, Q, and stress sensitivity, n.[10] The activation energy and stress sensitivity are dependent on the material type and glass transition temperature, T_g, as shown in Table 41.10. The exponent $-Q/\overline{R}T$ is unitless.

Table 41.10 Creep Parameters

material	n	Q
polymer		
$< T_g$	2–4	≥ 100 kJ/mol
$> T_g$	6–10	approx. 30 kJ/mol
metals and ceramics	3–10	80–200 kJ/mol

---

[10]The *NCEES Handbook* is inconsistent in the variable it uses for the universal gas constant. R in Eq. 41.12 is the same as $\overline{R}$ in its *Units* section and used almost everywhere in the *NCEES Handbook*.

17. HARDNESS TESTING

Hardness tests measure the capacity of a surface to resist deformation. The main use of hardness testing is to verify heat treatments, an important factor in product service life. Through empirical correlations, it is also possible to predict the ultimate strength and toughness of some materials.

Equation 41.13 and Eq. 41.14: Brinell Hardness Test

$$TS_{MPa} \approx 3.5(BHN) \qquad \textit{41.13}$$

$$TS_{psi} \approx 500(BHN) \qquad \textit{41.14}$$

Description

The *Brinell hardness test* is used primarily with iron and steel castings, although it can be used with softer materials. The *Brinell hardness number*, BHN (or HB or H_B), is determined by pressing a hardened steel ball into the surface of a specimen. The diameter of the resulting depression is correlated to the hardness. The standard ball is 10 mm in diameter and loads are 500 kg and 3000 kg for soft and hard materials, respectively.

The Brinell hardness number is the load per unit contact area. If a load, P (in kilograms), is applied through a steel ball of diameter, D (in millimeters), and produces a depression of diameter, d (in millimeters), and depth, t (in millimeters), the Brinell hardness number can be calculated from

$$\begin{aligned} BHN &= \frac{P}{A_{contact}} = \frac{P}{\pi D t} \\ &= \frac{2P}{\pi D(D - \sqrt{D^2 - d^2})} \end{aligned}$$

For heat-treated plain-carbon and medium-alloy steels, the ultimate tensile strength, TS, can be approximately calculated from the steel's Brinell hardness number, as shown in Eq. 41.13 and Eq. 41.14.

Other Hardness Tests

The *scratch hardness test*, also known as the *Mohs test*, compares the hardness of the material to that of minerals. Minerals of increasing hardness are used to scratch the sample. The resulting *Mohs scale* hardness can be used or correlated to other hardness scales.

The *file hardness test* is a combination of the cutting and scratch tests. Files of known hardness are drawn across the sample. The file ceases to cut the material when the material and file hardnesses are the same.

The *Rockwell hardness test* is similar to the Brinell test. A steel ball or diamond speroconical penetrator (known as a *brale indenter*) is pressed into the material. The machine applies an initial load (60 kgf, 100 kgf, or 150 kgf) that sets the penetrator below surface imperfections.[11] Then, a significant load is applied. The Rockwell hardness, R (or HR or H_R), is determined from the depth of penetration and is read directly from a dial.

Although a number of Rockwell scales (A through G) exist, the B and C scales are commonly used for steel. The *Rockwell B scale* is used with a steel ball for mild steel and high-strength aluminum. The *Rockwell C scale* is used with the brale indentor for hard steels having ultimate tensile strengths up to 2 GPa. The *Rockwell A scale* has a wide range and can be used with both soft materials (such as annealed brass) and hard materials (such as cemented carbides).

Other penetration hardness tests include the *Meyer*, *Vickers*, *Meyer-Vickers*, and *Knoop* tests.

18. THERMAL PROPERTIES

Equation 41.15 and Eq. 41.16: Specific Heat

$$Q = C_p \Delta T \quad \text{[constant pressure]} \qquad \textit{41.15}$$

$$Q = C_v \Delta T \quad \text{[constant volume]} \qquad \textit{41.16}$$

Description

An increase in internal energy is needed to cause a rise in temperature. Different substances differ in the quantity of heat needed to produce a given temperature increase.

The *specific heat* (known as the *specific heat capacity*), c, of a substance is the heat energy, q, required to change the temperature of one unit mass of the substance by one degree. The *molar specific heat*, conventionally designated by C, is the heat energy, Q, required to change the temperature of one mole of the substance by one degree. Specific heat capacity can be presented on a volume basis (e.g., $J/m^3 \cdot °C$), but a *volumetric heat capacity* is rarely encountered in practice outside of composite materials.[12] Even then, values of the volumetric heat capacity must usually be calculated from specific heats (by mass) and densities. The total heat energy required, Q_t, depends on the total mass or total number of moles.[13] Because specific heats of solids and

---

[11]Other Rockwell tests use 15 kgf, 30 kgf, and 45 kgf. The use of kgf units is traditional, and even modern test equipment is calibrated in kgf. Multiply kgf by 9.80665 to get newtons.

[12]The *NCEES Handbook* introduces C_v in reference to "constant volume," then follows it closely with a statement that "the heat capacity of a material can be reported as energy per degree per unit mass or per unit volume." The volumetric heat capacity is not related to C_v and is so rarely encountered that it doesn't have a common differentiating symbol other than "VHC."

[13]The *NCEES Handbook* describes Eq. 41.15 and Eq. 41.16 as the "...amount of heat required to raise the temperature of something..." "Something" here means "an entire object" as opposed to "some material." Without a definition of "something," the equations are ambiguous and misleading, as they are implicitly valid otherwise only for one unit mass or one mole.

liquids are slightly temperature dependent, the mean specific heats are used for processes covering large temperature ranges.

$$Q_t = mc\Delta T$$

$$c = \frac{Q_t}{m\Delta T}$$

The lowercase c implies that the units are J/kg·K. The molar specific heat, designated by the symbol C, has units of J/kmol·K.

$$C = \text{MW} \times c$$

For gases, the specific heat depends on the type of process during which the heat exchange occurs. Molar specific heats for constant-volume and constant-pressure processes are designated by C_v and C_p, respectively.

There are more thermal properties than those listed.

Equation 41.17: Coefficient of Thermal Expansion

$$\alpha = \frac{\varepsilon}{\Delta T} \qquad 41.17$$

Variation

$$\alpha = \frac{\Delta L}{L_0 \Delta T}$$

Description

If the temperature of an object is changed, the object will experience length, area, and volume changes. The magnitude of these changes will depend on the *thermal expansion coefficient* (*coefficient of linear thermal expansion*), α, calculated from the engineering strain, ε, and the change in temperature, ΔT.

19. ELECTRICAL PROPERTIES

Equation 41.18 Through Eq. 41.20: Capacitance

$$q = CV \qquad 41.18$$

$$C = \frac{\varepsilon A}{d} \qquad 41.19$$

$$\varepsilon = \kappa \varepsilon_0 \qquad 41.20$$

Value

$$\varepsilon_0 = 8.85 \times 10^{-12} \text{ F/m (same as C}^2/\text{N·m}^2)$$

Description

A *capacitor* is a device that stores electric charge. A capacitor is constructed as two conducting surfaces separated by an insulator, such as oiled paper, mica, or air. A *parallel plate capacitor* is a simple type of capacitor constructed as two parallel plates. If the plates are connected across a voltage potential, charges of opposite polarity will build up on the plates and create an electric field between the plates. The amount of charge, q, built up is proportional to the applied voltage, V, as shown in Eq. 41.18. The constant of proportionality, C, is the *capacitance* in farads (F) and depends on the capacitor construction. Capacitance represents the ability to store charge; the greater the capacitance, the greater the charge stored.

Equation 41.19 gives the capacitance of two parallel plates of equal area A separated by distance d. ε is the permittivity of the medium separating the plates. The permittivity may also be expressed as the product of the *dielectric constant* (*relative permittivity*), κ, and the *permittivity of a vacuum* (also known as the *permittivity of free space*), ε_0, as shown in Eq. 41.20.

Example

Two square parallel plates (0.04 m × 0.04 m) are separated by a 0.1 cm thick insulator with a dielectric constant of 3.4. What is most nearly the capacitance?

(A) 1.2×10^{-12} F

(B) 1.4×10^{-11} F

(C) 4.8×10^{-11} F

(D) 1.1×10^{-10} F

Solution

From Eq. 41.19 and Eq. 41.20, the capacitance is

$$C = \frac{\varepsilon A}{d} = \frac{\kappa \varepsilon_0 A}{d}$$

$$= \frac{(3.4)\left(8.85 \times 10^{-12} \; \frac{\text{F}}{\text{m}}\right)(0.04 \text{ m})^2 \left(100 \; \frac{\text{cm}}{\text{m}}\right)}{0.1 \text{ cm}}$$

$$= 4.814 \times 10^{-11} \text{ F} \quad (4.8 \times 10^{-11} \text{ F})$$

The answer is (C).

Equation 41.21: Resistivity and Resistance

$$R = \frac{\rho L}{A} \qquad 41.21$$

Description

Resistance, R (measured in ohms, Ω), is the property of a circuit or circuit element to oppose current flow. A circuit with zero resistance is a *short circuit*, whereas an *open circuit* has infinite resistance.

Materials Science

Resistors are usually constructed from carbon compounds, ceramics, oxides, or coiled wire. Resistance depends on the *resistivity*, ρ (in Ω·m), which is a material property, and the length and cross-sectional area of the resistor. (The resistivities of metals at 0°C are given in Table 41.2.) Resistors with larger cross-sectional areas have more free electrons available to carry charge and have less resistance. Each of the free electrons has a limited ability to move, so the electromotive force must overcome the limited mobility for the entire length of the resistor. The resistance increases with the length of the resistor.

Resistivity depends on temperature. For most conductors, it increases with temperature. For most semiconductors, resistivity decreases with temperature.

Example

A standard copper wire has a diameter of 1.6 mm. What is most nearly the resistance of 150 m of wire at 0°C?

(A) 0.91 Ω

(B) 1.2 Ω

(C) 1.5 Ω

(D) 1.7 Ω

Solution

From Table 41.2, the resistivity of copper at 0°C is 1.55×10^{-8} Ω·m. From Eq. 41.21, the resistance is

$$R = \frac{\rho L}{A} = \frac{(1.55 \times 10^{-8}\ \Omega \cdot \text{m})(150\ \text{m})\left(1000\ \dfrac{\text{mm}}{\text{m}}\right)^2}{\left(\dfrac{\pi}{4}\right)(1.6\ \text{mm})^2}$$

$$= 1.156\ \Omega \quad (1.2\ \Omega)$$

The answer is (B).

Semiconductors

Conductors or *semiconductors* are materials through which charges flow more or less easily. When a semiconductor is pure, it is called an *intrinsic semiconductor*. When minor amounts of impurities called *dopants* are added, the materials are termed *extrinsic semiconductors*. The solubility of a dopant determines how well the dopant can *diffuse* (move into areas with low dopant concentration) within the material.

The electrical conductivity of semiconductor materials is affected by temperature, light, electromagnetic field, and the concentration of dopants (impurities). The *solubility of dopant atoms* (i.e., the concentration, typically given in atoms/cm^3) increases very slightly with increasing temperature, reaching a relatively constant maximum in the 1000°C to 1200°C range. Higher concentrations result in precipitation of the doping element into a solid phase. Table 41.11 lists maximum values of dopant solubility. However, there may be limited value in achieving the

maximum values, since some of the dopant atoms may not be electrically active. For example, arsenic (As) has a maximum solubility in *p*-type silicon of approximately 5×10^{-20} atoms/cm^3, but the maximum useful electrical solubility is approximately 2×10^{-20} atoms/cm^3. As calculated from *Fick's first law of diffusion*, the concentration gradient, dC/dx, is a major factor in determining the *electrical flux* (i.e., current), J.

$$J = -D\frac{dC}{dx}$$

Table 41.11 *Some Extrinsic, Elemental Semiconductors*

element	dopant	periodic table group of dopant	maximum solid solubility of dopant (atoms/m^3)
Si	B	III A	600×10^{24}
	Al	III A	20×10^{24}
	Ga	III A	40×10^{24}
	P	V A	1000×10^{24}
	As	V A	2000×10^{24}
	Sb	V A	70×10^{24}
Ge	Al	III A	400×10^{24}
	Ga	III A	500×10^{24}
	In	III A	4×10^{24}
	As	V A	80×10^{24}
	Sb	V A	10×10^{24}

Reprinted with permission from Charles A. Harper, ed., *Handbook of Materials and Processes for Electronics*, copyright © 1970, by The McGraw-Hill Companies, Inc.

Electrons in semiconductors may be bonded or free. Bonding electrons occupy states in the atoms' *valence bands*. Free electrons occupy states in the *conduction bands*. *Holes* are empty states in the valence band. Both holes and electrons can move around, so both are known as *carriers* or *charge carriers*.

Often, a small amount of energy (usually available thermally or provided electrically) is required to fill an *energy gap*, E_g, in order to initiate carrier movement through a semiconductor. The energy gap, often referred to as an *ionization energy*, is the difference in energy between the highest point in the valence band, E_v, and the lowest point in the conduction band, E_c. (The valence band energy may be referred to as the *intrinsic band* energy, and be given the symbol E_i.) The energy gap in insulators is relatively large compared to conductor or semiconductor materials.

$$E_g = E_v - E_c$$

Intrinsic semiconductors are those that occur naturally. When an electron in an intrinsic semiconductor receives enough energy, it can jump to the conduction band and leave behind a hole, a process known as *electron-hole pair production*. For an intrinsic material, electrons and holes are always created in pairs. Therefore, the *activation energy* is half of the energy gap.[14]

$$E_a = \tfrac{1}{2}E_g \quad \text{[intrinsic]}$$

An *extrinsic semiconductor* is created by artificially introducing dopants into otherwise "perfect" crystals. The analysis of energy levels is similar, except that the dopant energies are within the energy band gap, effectively reducing the energy required to overcome the gap. The valence band energy of the dopants may be referred to as a *donor level* (for *n*-type semiconductors) or an *acceptor level* (for *p*-type semiconductors).

At high temperatures, the carrier density approaches the intrinsic carrier concentration. Therefore, for extrinsic semiconductors at high temperatures, the activation energy (ionization energy) is the same as for intrinsic semiconductors, half of the difference in ionization energies. At low temperatures, including normal room temperatures, the carrier density is dominated by the ionization of the donors. At lower temperatures, the activation energy is equal to the difference in ionization energies.

$$E_a = \tfrac{1}{2}(E_g - E_d)$$
$$\approx \tfrac{1}{2}(E_v - E_c) \quad \text{[extrinsic, high temperatures]}$$

$$E_a = E_g - E_d \quad \text{[extrinsic, low temperatures]}$$

Table 41.12 lists ionization energy differences, $E_g - E_d$, and activation energies, E_a, for various extrinsic semiconductors.

Photoelectric Effect

The *work function*, ϕ, is a measure of the energy required to remove an electron from the surface of a metal. It is usually given in terms of electron volts, eV. It is specifically the minimum energy necessary to move an electron from the *Fermi level* of a metal (an energy level below which all available energy levels are filled, and above which all are empty, at 0K) to infinity, that is, the vacuum level. This energy level must be reached in order to move electrons from semiconductor devices into the metal conductors that constitute the remainder of an electrical circuit. It is also the energy level of importance in the design of optical electronic devices.

In photosensitive electronic devices, an incoming photon provides the energy to release an electron and make it available to the circuit, that is, free it so that it may move under the influence of an electric field. This phenomenon whereby a short wavelength photon interacts

[14]The *NCEES Handbook* is inconsistent in the symbols used for activation energy. E_a in Table 41.12 is the same as Q in Eq. 41.12. A common symbol used in practice for diffusion activation energy is Q_d, where the subscript clarifies that the activation energy is for diffusion.

Table 41.12 Impurity Energy Levels for Extrinsic Semiconductors

semiconductor	dopant	$E_g - E_d$ (eV)	E_a (eV)
Si	P	0.044	—
	As	0.049	—
	Sb	0.039	—
	Bi	0.069	—
	B	—	0.045
	Al	—	0.057
	Ga	—	0.065
	In	—	0.160
	Tl	—	0.260
Ge	P	0.012	—
	As	0.013	—
	Sb	0.096	—
	B	—	0.010
	Al	—	0.010
	Ga	—	0.010
	In	—	0.011
	Tl	—	0.010
GaAs	Se	0.005	—
	Te	0.003	—
	Zn	—	0.024
	Cd	—	0.021

with an atom and releases an electron is called the *photoelectric effect*.

Transduction Principles

The *transduction* principle of a given *transducer* (a device that converts a signal to a different energy form) determines nearly all its other characteristics. There are three *self-generating* transduction types: photovoltaic, piezoelectric, and electromagnetic. All other types require the use of an external excitation power source.

In *photovoltaic transduction* (*photoelectric transduction*), light is directed onto the junction of two dissimilar metals, generating a voltage. This type of transduction is used primarily in optical sensors. It can also be used with the measured quantity (known as the *measurand*) controlling a mechanical-displacement shutter that varies the intensity of the built-in light source. *Piezoelectric transduction* occurs because certain crystals generate an electrostatic charge or potential when mechanical forces are applied to the material (i.e., the material is placed in compression or tension or bending forces are applied to it). In *electromagnetic transduction*, the measured quantity is converted into a voltage by a change in magnetic flux that occurs when magnetic material moves relative to a coil with a ferrous core. These self-generating types of transduction are illustrated in Fig. 41.15.

Materials Science

Figure 41.15 *Self-Generating Transducers*

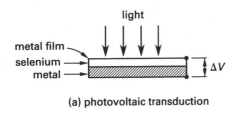

(a) photovoltaic transduction

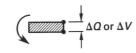

(b) piezoelectric transduction

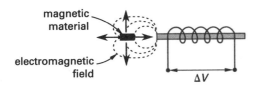

(c) electromagnetic transduction

20. CHEMICAL PROPERTIES

The chemical properties of a substance manifest themselves only when there is a reaction, causing a change in the chemical composition of the substance. For this reason, chemical properties are *reactive properties*. Chemical properties differ from physical properties, which manifest themselves without involving changes in the substance's composition. Since a chemical reaction is not associated with the physical properties of a substance, physical properties are *non-reactive properties*. Physical properties are intrinsic to a substance, whereas chemical properties are associated with a change in the substance.

Physical properties are exemplified by the boiling point of a substance. The boiling point is the temperature at which the substance changes phase between liquid and vapor. For example, the boiling point of water at atmospheric pressure is 100°C. Although there is a phase change, the composition of the substance does not change. Liquid water and steam (vapor) have the same chemical composition: H_2O. Density, thermal conductivity, electrical conductivity, melting point, and enthalpy of vaporization are other examples of physical properties.

An example of a chemical property is the enthalpy (heat) of combustion of a substance. *Enthalpy of combustion* is the energy released when the substance reacts completely with oxygen under standard conditions and with everything in their standard states. When methane reacts with oxygen, 890 kJ of heat are produced per mole of methane. Because the methane (CH_4) changes completely during combustion to carbon dioxide (CO_2) and water (H_2O), enthalpy of combustion is a chemical property. Enthalpy of formation, flammability, and reactivity are other examples of chemical properties.

21. CORROSION

Corrosion is an undesirable degradation of a material resulting from a chemical or physical reaction with the environment. *Galvanic action* results from a difference in oxidation potentials of metallic ions. The greater the difference in oxidation potentials, the greater the galvanic corrosion will be. If two metals with different oxidation potentials are placed in an *electrolytic medium* (e.g., seawater), a *galvanic cell* (*voltaic cell*) will be created. The more electropositive metal will act as an anode and will corrode. The metal with the lower potential, being the cathode, will be unchanged.

A galvanic cell is a device that produces electrical current by way of an oxidation-reduction reaction—that is, chemical energy is converted into electrical energy. Galvanic cells typically have the following characteristics.

- The oxidizing agent is separate from the reducing agent.

- Each agent has its own electrolyte and metallic electrode, and the combination is known as a *half-cell*.

- Each agent can be in solid, liquid, or gaseous form, or can consist simply of the electrode.

- The ions can pass between the electrolytes of the two half-cells. The connection can be through a porous substance, salt bridge, another electrolyte, or other method.

The amount of current generated by a half-cell depends on the electrode material and the oxidation-reduction reaction taking place in the cell. The current-producing ability is known as the *oxidation potential, reduction potential,* or *half-cell potential. Standard oxidation potentials* have a zero reference voltage corresponding to the potential of a *standard hydrogen electrode.*

To specify their tendency to corrode, metals are often classified according to their position in the galvanic series listed in Table 41.13. As expected, the metals in this series are in approximately the same order as their half-cell potentials. However, alloys and proprietary metals are also included in the series.

It is not necessary that two dissimilar metals be in contact for corrosion by galvanic action to occur. Different regions within a metal may have different half-cell potentials. The difference in potential can be due to different phases within the metal (creating very small galvanic cells), heat treatment, cold working, and so on.

In addition to corrosion caused by galvanic action, there is also *stress corrosion, fretting corrosion,* and *cavitation.* Conditions within the crystalline structure can accentuate or retard corrosion. In one extreme type of intergranular corrosion, *exfoliation,* open endgrains separate into layers.

Table 41.14 gives the oxidation potentials for common corrosion reactions.

Table 41.13 *Galvanic Series in Seawater (top to bottom anodic (sacrificial, active) to cathodic (noble, passive))*

magnesium
zinc
Alclad 3S
cadmium
2024 aluminum alloy
low-carbon steel
cast iron
stainless steels (active)
    no. 410
    no. 430
    no. 404
    no. 316
Hastelloy A
lead
lead-tin alloys
tin
nickel
brass (copper-zinc)
copper
bronze (copper-tin)
90/10 copper-nickel
70/30 copper-nickel
Inconel
silver solder
silver
stainless steels (passive)
Monel metal
Hastelloy C
titanium
graphite
gold

Table 41.14 *Standard Oxidation Potentials for Corrosion Reactions[a,b]*

corrosion reaction	potential, E_o (volts), versus normal hydrogen electrode
$Au \rightarrow Au^{3+} + 3e^-$	−1.498
$2H_2O \rightarrow O_2 + 4H^+ + 4e^-$	−1.229
$Pt \rightarrow Pt^{2+} + 2e^-$	−1.200
$Pd \rightarrow Pd^{2+} + 2e^-$	−0.987
$Ag \rightarrow Ag^+ + e^-$	−0.799
$2Hg \rightarrow Hg_2^{2+} + 2e^-$	−0.788
$Fe^{2+} \rightarrow Fe^{3+} + e^-$	−0.771
$4(OH)^- \rightarrow O_2 + 2H_2O + 4e^-$	−0.401
$Cu \rightarrow Cu^{2+} + 2e^-$	−0.337
$Sn^{2+} \rightarrow Sn^{4+} + 2e^-$	−0.150
$H_2 \rightarrow 2H^+ + 2e^-$	0.000
$Pb \rightarrow Pb^{2+} + 2e^-$	+0.126
$Sn \rightarrow Sn^{2+} + 2e^-$	+0.136
$Ni \rightarrow Ni^{2+} + 2e^-$	+0.250
$Co \rightarrow Co^{2+} + 2e^-$	+0.277
$Cd \rightarrow Cd^{2+} + 2e^-$	+0.403
$Fe \rightarrow Fe^{2+} + 2e^-$	+0.440
$Cr \rightarrow Cr^{3+} + 3e^-$	+0.744
$Zn \rightarrow Zn^{2+} + 2e^-$	+0.763
$Al \rightarrow Al^{3+} + 3e^-$	+1.662
$Mg \rightarrow Mg^{2+} + 2e^-$	+2.363
$Na \rightarrow Na^+ + e^-$	+2.714
$K \rightarrow K^+ + e^-$	+2.925

[a]Measured at 25°C. Reactions are written as anode half-cells. Arrows are reversed for cathode half-cells.
[b]NOTE: In some chemistry texts, the reactions and the signs of the values (in this table) are reversed; for example, the half-cell potential of zinc is given as −0.763 V for the reaction $Zn^{2+} + 2e^- \rightarrow Zn$. When the potential E_o is positive, the reaction proceeds spontaneously as written.

Equation 41.22 Through Eq. 41.25: Oxidation-Reduction Corrosion Reactions

$$M^o \rightarrow M^{n+} + ne^- \quad \text{[at the anode]} \quad 41.22$$

$$\tfrac{1}{2}O_2 + 2e^- + H_2O \rightarrow 2OH^- \quad \text{[at the cathode]} \quad 41.23$$

$$\tfrac{1}{2}O_2 + 2e^- + 2H_3O^+ \rightarrow 3H_2O \quad \text{[at the cathode]} \quad 41.24$$

$$2e^- + 2H_3O^+ \rightarrow 2H_2O + H_2 \quad \text{[at the cathode]} \quad 41.25$$

Description

In an oxidation-reduction reaction, such as corrosion, one substance is oxidized and the other is reduced. The oxidized substance loses electrons and becomes less negative; the reduced substance gains electrons and becomes more negative.

Oxidation occurs at the *anode* (positive terminal) in an electrolytic reaction. Equation 41.22 shows the oxidation reaction (or *anode reaction*) of a typical metal, M. The superscript "o" is used to designate the standard, natural state of the atom.

Reduction occurs at the *cathode* (negative terminal) in an electrolytic reaction. Equation 41.23 through Eq. 41.25 list some reduction reactions (or *cathode reactions*) involving hydrogen and oxygen.

22. CORROSION CONTROL METHODS

Corrosion control plays a significant role in minimizing damage due to corrosion. There are several different common approaches to corrosion control. The most appropriate approach depends on the specific situation.

Use of *corrosion-resistant materials* is a common strategy for controlling corrosion. Corrosion-resistant materials are generally much more expensive than common materials, so cost is considered when using considering this strategy. The selection process involves evaluating the corrosion characteristics of the corrosive environment (i.e., the "medium") and the corrosion resistant characteristics of the material in that medium. Stainless steel and special alloys of steel are examples of corrosion-resistant materials.

Materials Science

During design and installation, the predominant approach to corrosion control is avoiding the creation of *galvanic cells*. Galvanic cells are created when combinations of metals with different oxidation potentials come into close proximity with one another. For example, steel screws should not be used in the manufacture or attaching of bronze marine hardware because iron is anodic to copper and will corrode easily. In general, galvanic corrosion of dissimilar metals can be minimized by using metals that are close neighbors in the galvanic series.

Cathodic protection (CP) is widely used in the protection of steel equipment and structures. It involves the use of a *sacrificial anode*, a more active (anodic) metal that protects the more passive (cathodic) object to which it is attached. For example, magnesium strips (serving as the anode) are used to protect steel structures in off-shore platforms. Table 41.14 shows that the magnesium (Mg) oxidation potential is higher, and so it is anodic to steel (which is essentially iron, Fe). The magnesium serves as the anode and corrodes, while the steel serves as the cathode and is protected. The electrochemical reactions are

$$Mg \rightarrow Mg^{2+} + 2e^- \quad \text{[anode (magnesium)]}$$

$$O_2 + 2H_2O + 4e^- \rightarrow 4(OH)^- \quad \text{[cathode (iron)]}$$

Corrosion occurs when electrons are lost from a substance (i.e., the substance oxidizes). With cathodic protection, the direction of the flow of electrons changes from out of the protected material to into the protected material. This can be accomplished as simply as attaching or connecting the sacrificial anode to the protected material, as shown in Fig. 41.16(a). The protection can be enhanced and made even more reliable by forcing a flow of electrons using an active electrical source, as shown in Fig. 41.16(b). This is known as *impressed current cathodic protection*. In cases where the corrosive medium is conductive (i.e., contains ions in a liquid or moist environment), the direct connection between the sacrificial anode and protected part can be omitted.

Sacrificial anodes are also used in the protection of pipelines, subsea structures, wind turbine foundations, the hulls of ships and boats, dock gates, power station intake screens, and water storage tanks. In addition to magnesium, zinc and aluminum are also commonly used as sacrificial anodes, since both aluminum and zinc are anodic compared to steel or iron. The advantages of using sacrificial anodes include lower overall life-cycle cost, little or no required maintenance, high reliability, little or no modifications to the protected object, and long installed life.

Figure 41.16 Sacrificial Anode Corrosion Protection

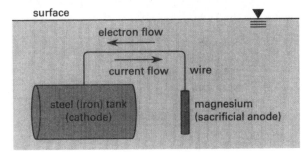

(a) passive protection

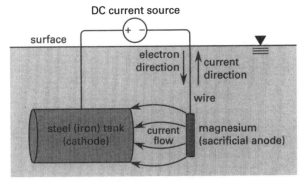

(b) impressed (active) protection

Coatings and linings are generally used to protect large surface areas, such as pipelines. Because they can be used in a variety of corrosive environments, their primary advantage is versatility. Coatings and linings can be used separately to provide cost-effective protection of equipment and structures, but they are often used in conjunction with cathodic protection. For example, galvanizing steel adds a surface coating of zinc, which provides two-pronged protection. The zinc coating is the primary protection. It insulates the steel from the corrosive environment. Second, if the coating is damaged or is imperfect, the zinc becomes the sacrificial anode and protects the steel from the corrosive environment.

Corrosion inhibitors are substances added in small quantities to liquid and gaseous corrosive media to reduce the rate of corrosion. Inhibitors slow down the electrochemical reactions on the protected surface. They typically work by forming a passive oxide film on an anodic surface. Inhibitors are used where large quantities of corrosive fluids are involved. For example, chromate compounds such as $Na_2Cr_2O_7$ are used as corrosion inhibitors in boilers. Use of a corrosion inhibitor is a simple and cost-effective method to control corrosion.

42 Engineering Materials

Symbols

ρ	density	kg/m^3
σ	stress	MPa

Subscripts

a	activation
ave	average
c	composite
d	diffusion
f	finish
g	glass
i	individual
m	melting
o	original or oxidation
qe	quenched end
s	start

Nomenclature

c	specific heat	kJ/kg·K
C	number of components	–
D	diffusion coefficient	m^2/s
D	distance	m
D_o	proportionality constant	m^2/s
DP	degree of polymerization	–
E	energy	kJ
E	modulus of elasticity	GPa
E_o	oxidation potential	V
f	volumetric fraction	–
f_r	modulus of rupture	MPa
f'_c	compressive strength	MPa
F	degrees of freedom	–
L	length	m
m	mass	kg
M	Martensite transformation temperature	°C
MC	moisture content	–
MW	molecular weight	kg/kmol
n	grain size	–
n	number	–
N	number of grains per unit area	$1/m^2$
P	number of phases	–
P_L	points per unit length	$1/m$
Q	activation energy	kJ/kmol
R	universal gas constant, 8314	J/kmol·K
R_C	Rockwell hardness (C-scale)	–
S_V	surface area per unit volume	$1/m$
T	absolute temperature	K
W	water content	%
x	gravimetric fraction	–

1. CHARACTERISTICS OF METALS

Metals are the most frequently used materials in engineering design. Steel is the most prevalent engineering metal because of the abundance of iron ore, simplicity of production, low cost, and predictable performance. However, other metals play equally important parts in specific products.

Metallurgy is the subject that encompasses the procurement and production of metals. *Extractive metallurgy* is the subject that covers the refinement of pure metals from their ores.

Most metals are characterized by the properties in Table 42.1.

Table 42.1 *Properties of Most Metals and Alloys*

high thermal conductivity (low thermal resistance)
high electrical conductivity (low electrical resistance)
high chemical reactivity[a]
high strength
high ductility[b]
high density
high radiation resistance
highly magnetic (ferrous alloys)
optically opaque
electromagnetically opaque

[a]Some alloys, such as stainless steel, are more resistant to chemical attack than pure metals.
[b]Brittle metals, such as some cast irons, are not ductile.

Materials Science

2. BINARY PHASE DIAGRAMS

Most engineering materials are not pure elements but are alloys of two or more elements. Alloys of two elements are known as *binary alloys*. Steel, for example, is an alloy of primarily iron and carbon. Usually one of the elements is present in a much smaller amount, and this element is known as the *alloying ingredient*. The primary ingredient is known as the *host ingredient, base metal*, or *parent ingredient*.

Sometimes, such as with alloys of copper and nickel, the alloying ingredient is 100% soluble in the parent ingredient. Nickel-copper alloy is said to be a *completely miscible alloy* or a *solid-solution alloy*.

The presence of the alloying ingredient changes the thermodynamic properties, notably the freezing (or melting) temperatures of both elements. Usually the freezing temperatures decrease as the percentage of alloying ingredient is increased. Because the freezing points of the two elements are not the same, one of them will start to solidify at a higher temperature than the other. For any given composition, the alloy might consist of all liquid, all solid, or a combination of solid and liquid, depending on the temperature.

A *phase* of a material at a specific temperature will have a specific composition and crystalline structure and distinct physical, electrical, and thermodynamic properties. (In metallurgy, the word "phase" refers to more than just solid, liquid, and gas phases.)

The regions of an *equilibrium diagram*, also known as a *phase diagram*, illustrate the various alloy phases. The phases are plotted against temperature and composition. The composition is usually a gravimetric fraction of the alloying ingredient. Only one ingredient's gravimetric fraction needs to be plotted for a binary alloy.

It is important to recognize that the equilibrium conditions do not occur instantaneously and that an equilibrium diagram is applicable only to the case of slow cooling.

Figure 42.1 is an equilibrium diagram for a copper-nickel alloy. (Most equilibrium diagrams are much more complex.) The *liquidus line* is the boundary above which no solid can exist. The *solidus line* is the boundary below which no liquid can exist. The area between these two lines represents a mixture of solid and liquid phase materials.

Just as only a limited amount of salt can be dissolved in water, there are many instances where a limited amount of the alloying ingredient can be absorbed by the solid mixture. The elements of a binary alloy may be completely soluble in the liquid state but only partially soluble in the solid state.

When the alloying ingredient is present in an amount above the maximum solubility percentage, the alloying

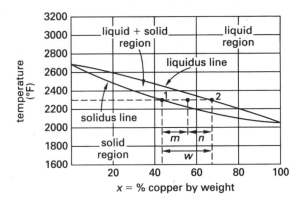

Figure 42.1 *Copper-Nickel Phase Diagram*

ingredient precipitates out. In aqueous solutions, the precipitate falls to the bottom of the container. In metallic alloys, the precipitate remains suspended as pure crystals dispersed throughout the primary metal.

In chemistry, a *mixture* is different from a *solution*. Salt in water forms a solution. Sugar crystals mixed with salt crystals form a mixture.

Figure 42.2 is typical of an equilibrium diagram for ingredients displaying a limited solubility.

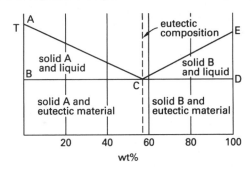

Figure 42.2 *Equilibrium Diagram of a Limited Solubility Alloy*

In Fig. 42.2, the components are perfectly miscible at point C only. This point is known as the *eutectic composition*. A *eutectic alloy* is an alloy having the composition of its eutectic point. The material in the region ABC consists of a mixture of solid component A crystals in a liquid of components A and B. This liquid is known as the *eutectic material*, and it will not solidify until the line BD (the *eutectic line, eutectic point*, or *eutectic temperature*)—the lowest point at which the eutectic material can exist in liquid form—is reached.

Since the two ingredients do not mix, reducing the temperature below the eutectic line results in crystals (layers or plates) of both pure ingredients forming. This is the microstructure of a solid eutectic alloy: alternating pure crystals of the two ingredients. Since two solid substances are produced from a single liquid substance, the process could be written in chemical reaction format as liquid → solid α + solid β. (Alternatively, upon heating, the reaction would be solid α + solid β → liquid.) For this reason, the phase change is called a *eutectic reaction*.

There are similar reactions involving other phases and states. Table 42.2 and Fig. 42.3 illustrate these.

Table 42.2 Types of Equilibrium Reactions

reaction name	type of reaction upon cooling
eutectic	liquid → solid α + solid β
peritectic	liquid + solid α → solid β
eutectoid	solid γ → solid α + solid β
peritectoid	solid α + solid γ → solid β

Figure 42.3 Typical Appearance of Equilibrium Diagram at Reaction Points

reaction name	phase reaction	phase diagram
eutectic	$L \to \alpha(s) + \beta(s)$ cooling	
peritectic	$L + \alpha(s) \to \beta(s)$ cooling	
eutectoid	$\gamma(s) \to \alpha(s) + \beta(s)$ cooling	
peritectoid	$\alpha(s) + \gamma(s) \to \beta(s)$ cooling	

Example

A binary solution with the characteristics and composition of point P is cooled.

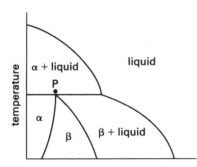

What type of alloy will result?

(A) eutectic

(B) eutectoid

(C) peritectic

(D) peritectoid

Solution

Dropping straight down from point P on the illustration, corresponding to cooling, the α solid and liquid transform into the β solid. This corresponds to the definition of a peritectic reaction. A peritectic alloy will form.

The answer is (C).

3. LEVER RULE

Within a liquid-solid region, the percentage of solid and liquid phases is a function of temperature and composition. Near the liquidus line, there is very little solid phase. Near the solidus line, there is very little liquid phase. The *lever rule* is an interpolation technique used to find the relative amounts of solid and liquid phase at any composition. These percentages are given in fraction (or percent) by weight.

Figure 42.1 shows an alloy with an average composition of 55% copper at 2300°F. (A horizontal line representing different conditions at a single temperature is known as a *tie line*.) The liquid composition is defined by point 2, and the solid composition is defined by point 1. Referring to Fig. 42.1, the gravimetric fraction of solid and liquid can be determined from the lever rule using the line segment lengths m, n, and $w = m + n$ in a method that is analogous to determining steam quality.

$$\text{fraction solid} = 1 - \text{fraction liquid} = \frac{n}{m+n} = \frac{n}{w}$$

$$\text{fraction liquid} = 1 - \text{fraction solid} = \frac{m}{m+n} = \frac{m}{w}$$

Equation 42.1 and Eq. 42.2: Gravimetric Component Fraction

$$\text{wt\% } \alpha = \frac{x_\beta - x}{x_\beta - x_\alpha} \times 100\% \qquad \text{42.1}$$

$$\text{wt\% } \beta = \frac{x - x_\alpha}{x_\beta - x_\alpha} \times 100\% \qquad \text{42.2}$$

Description

Referring to Fig. 42.4, from the lever rule, the gravimetric fractions of solid and liquid phases depend on the lengths of the lines $x - x_\alpha$ and $x_\beta - x$, along with the separation, $x_\beta - x_\alpha$, which may be measured using any convenient scale. (Although the distances can be measured in millimeters or tenths of an inch, it is more convenient to use the percentage alloying ingredient scale.) Then, the fractions of solid and liquid can be calculated from Eq. 42.1 and Eq. 42.2.

The lever rule and method of determining the composition of the two components are applicable to any solution or mixture, liquid or solid, in which two phases are present.

Materials Science

Figure 42.4 *Two-Phase System Phase Diagram*

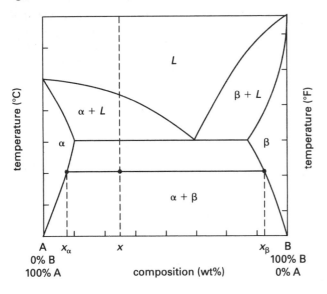

lines, use the horizontal Cu percentage scale. Point A is located at $x_L = 48\%$ Cu, and point B is located at $x_\beta = 91\%$ Cu. Use Eq. 42.2.

$$\text{wt\% } \beta = \frac{x - x_L}{x_\beta - x_L} \times 100\%$$

$$= \frac{70\% - 48\%}{91\% - 48\%} \times 100\%$$

$$= 51.2\% \quad (51\%)$$

The answer is (C).

4. EQUILIBRIUM MIXTURES

Equation 42.3: Gibbs' Phase Rule

$$P + F = C + 2 \qquad 42.3$$

Variation

$$P + F = C + 1 \Big|_{\substack{\text{constant pressure,} \\ \text{constant temperature,} \\ \text{or constant composition}}}$$

Description

Gibbs' phase rule defines the relationship between the number of phases and elements in an equilibrium mixture. For such an equilibrium mixture to exist, the alloy must have been slowly cooled and thermodynamic equilibrium must have been achieved along the way.

At equilibrium, and considering both temperature and pressure to be independent variables, Gibbs' phase rule is Eq. 42.3.

P is the number of phases existing simultaneously; F is the number of independent variables, known as *degrees of freedom*; and C is the number of elements in the alloy. Composition, temperature, and pressure are examples of degrees of freedom that can be varied.

For example, if water is to be stored in a condition where three phases (solid, liquid, gas) are present simultaneously, then $P = 3$, $C = 1$, and $F = 0$. That is, neither pressure nor temperature can be varied. This state corresponds to the *triple point* of water.

If pressure is constant, then the number of degrees of freedom is reduced by one, and Gibbs' phase rule can be rewritten as shown in the variation.

If Gibbs' rule predicts $F = 0$, then an alloy can exist with only one composition.

Example

Consider the Ag-Cu phase diagram given.

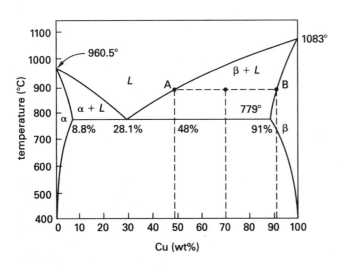

What is most nearly the equilibrium percentage of β in an alloy of 30% Ag, 70% Cu at 900°C?

(A) 0.0%

(B) 22%

(C) 51%

(D) 59%

Solution

Draw the horizontal tie line at 900°C between the liquid, L, phase at point A and the solid, β, at point B. Draw the vertical line that identifies the alloy at $x = 70\%$ Cu. Rather than measure the lengths of the

Example

A system consisting of an open bucket containing a mixture of ice and water is to be warmed from 0°C to 20°C. How many degrees of freedom does the system have?

    (A) 0

    (B) 1

    (C) 2

    (D) 3

Solution

Degrees of freedom and the system properties are instantaneous values. What is happening and what is going to happen to the system are not relevant. Only the current, instantaneous, equilibrium properties are relevant. In this case, the system consists of an open bucket containing ice and water, so the number of phases is $P = 2$. The only substance in the system is water, so the number of components is $C = 1$.

$$P + F = C + 2$$
$$F = C + 2 - P = 1 + 2 - 2$$
$$= 1$$

The answer is (B).

5. UNIFIED NUMBERING SYSTEM

The *Unified Numbering System* (UNS) was introduced in the mid-1970s to provide a consistent identification of metals and alloys for use throughout the world. The UNS designation consists of one of seventeen single uppercase letter prefixes followed by five digits. Many of the letters are suggestive of the family of metals, as Table 42.3 indicates.

Table 42.3 UNS Alloy Prefixes

A	aluminum
C	copper
E	rare-earth metals
F	cast irons
G	AISI and SAE carbon and alloy steels
H	AISI and SAE H-steels
J	cast steels (except tool steels)
K	miscellaneous steels and ferrous alloys
L	low-melting metals
M	miscellaneous nonferrous metals
N	nickel
P	precious metals
R	reactive and refractory metals
S	heat- and corrosion-resistant steels (stainless and valve steels and superalloys)
T	tool steels (wrought and cast)
W	welding filler metals
Z	zinc

6. FERROUS METALS

Steel and Alloy Steel Grades

The properties of steel can be adjusted by the addition of *alloying ingredients*. Some steels are basically mixtures of iron and carbon. Other steels are produced with a variety of ingredients.

The simplest and most common grades of steel belong to the group of *carbon steels*. Carbon is the primary noniron element, although sulfur, phosphorus, and manganese can also be present. Carbon steel can be subcategorized into *plain carbon steel (nonsulfurized carbon steel)*, *free-machining steel (resulfurized carbon steel)*, and *resulfurized and rephosphorized carbon steel*. Plain carbon steel is subcategorized into *low-carbon steel* (less than 0.30% carbon), *medium-carbon steel* (0.30% to 0.70% carbon), and *high-carbon steel* (0.70% to 1.40% carbon).

Low-carbon steels are used for wire, structural shapes, and screw machine parts. Medium-carbon steels are used for axles, gears, and similar parts requiring medium to high hardness and high strength. High-carbon steels are used for drills, cutting tools, and knives.

Low-alloy steels (containing less than 8.0% total alloying ingredients) include the majority of steel alloys but exclude the high-chromium content *corrosion-resistant (stainless) steels*. Generally, low-alloy steels will have higher strength (e.g., double the yield strength) of plain carbon steel. *Structural steel*, *high-strength steel*, and *ultrahigh-strength steel* are general types of low-alloy steel.[1]

High-alloy steels contain more than 8.0% total alloying ingredients.

Table 42.4 lists typical alloying ingredients and their effects on steel properties. The percentages represent typical values, not maximum solubilities.

Tool Steel

Each grade of *tool steel* is designed for a specific purpose. As such, there are few generalizations that can be made about tool steel. Each tool steel exhibits its own blend of the three main performance criteria: toughness, wear resistance, and hot hardness.[2]

Some of the generalizations possible are listed as follows.

---

[1]The ultrahigh-strength steels, also known as *maraging steels*, are very low-carbon (less than 0.03%) steels, with 15–25% nickel and small amounts of cobalt, molybdenum, titanium, and aluminum. With precipitation hardening, ultimate tensile strengths up to 2.8 GPa, yield strengths up to 1.7 GPa, and elongations in excess of 10% are achieved. Maraging steels are used for rocket motor cases, aircraft and missile turbine housings, aircraft landing gear, and other applications requiring high strength, low weight, and toughness.

[2]The ability of a steel to resist softening at high temperatures is known as *hot hardness* and *red hardness*.

Materials Science

Table 42.4 *Steel Alloying Ingredients*

ingredient	range (%)	purpose
aluminum	–	deoxidation
boron	0.001–0.003	increase hardness
carbon	0.1–4.0	increase hardness and strength
chromium	0.5–2	increase hardness and strength
	4–18	increase corrosion resistance
copper	0.1–0.4	increase atmospheric corrosion resistance
iron sulfide	–	increase brittleness
manganese	0.23–0.4	reduce brittleness, combine with sulfur
	> 1.0	increase hardness
manganese sulfide	0.8–0.15	increase machinability
molybdenum	0.2–5	increase dynamic and high-temperature strength and hardness
nickel	2–5	increase toughness, increase hardness
	12–20	increase corrosion resistance
	> 30	reduce thermal expansion
phosphorus	0.04–0.15	increase hardness and corrosion resistance
silicon	0.2–0.7	increase strength
	2	increase spring steel strength
	1–5	improve magnetic properties
sulfur	–	(see *iron sulfide* and *manganese sulfide*)
titanium	–	fix carbon in inert particles; reduce martensitic hardness
tungsten	–	increase high-temperature hardness
vanadium	0.15	increase strength

- An increase in carbon content increases wear resistance and reduces toughness.
- An increase in wear resistance reduces toughness.
- Hot hardness is independent of toughness.
- Hot hardness is independent of carbon content.

Group A steels are air-hardened, medium-alloy cold-work tool steels. Air-hardening allows the tool to develop a homogeneous hardness throughout, without distortion. This hardness is achieved by large amounts of alloying elements and comes at the expense of wear resistance.

Group D steels are high-carbon, high-chromium tool steels suitable for cold-working applications. These steels are high in abrasion resistance but low in machinability and ductility. Some steels in this group are air hardened, while others are oil quenched. Typical uses are blanking and cold-forming punches.

Group H steels are hot-work tool steels, capable of being used in the 600–1100°C range. They possess good wear resistance, hot hardness, shock resistance, and resistance to surface cracking. Carbon content is low, between 0.35% and 0.65%. This group is subdivided according to the three primary alloying ingredients: chromium, tungsten, or molybdenum. For example, a particular steel might be designated as a "chromium hot-work tool steel."

Group M steels are molybdenum high-speed steels. Properties are very similar to the group T steels, but group M steels are less expensive since one part molybdenum can replace two parts tungsten. For that reason, most high-speed steel in common use is produced from group M. Cobalt is added in large percentages (5–12%) to increase high-temperature cutting efficiency in heavy-cutting (high-pressure cutting) applications.

Group O steels are oil-hardened, cold-work tool steels. These high-carbon steels use alloying elements to permit oil quenching of large tools and are sometimes referred to as *nondeforming steels*. Chromium, tungsten, and silicon are typical alloying elements.

Group S steels are shock-resistant tool steels. Toughness (not hardness) is the main characteristic, and either water or oil may be used for quenching. Group S steels contain chromium and tungsten as alloying ingredients. Typical uses are hot header dies, shear blades, and chipping chisels.

Group T steels are tungsten high-speed tool steels that maintain a sharp hard cutting edge at temperatures in excess of 550°C. The ubiquitous 18-4-1 grade T1 (named after the percentages of tungsten, chromium, and vanadium, respectively) is part of this group. Increases in hot hardness are achieved by simultaneous increases in carbon and vanadium (the key ingredient in these tool steels) and special, multiple-step heat treatments.[3]

Group W steels are water-hardened tool steels. These are plain high-carbon steels (modified with small amounts of vanadium or chromium, resulting in high surface hardness but low hardenability). The combination of high surface hardness and ductile core makes group W steels ideal for rock drills, pneumatic tools, and cold header dies. The limitation on this tool steel group is the loss of

---

[3]For example, the 18-4-1 grade is heated to approximately 550°C for two hours, air cooled, and then heated again to the same temperature. The term *double-tempered steel* is used in reference to this process. Most heat treatments are more complex.

hardness that begins at temperatures above 150°C and is complete at 300°C.

Stainless Steel

Adding chromium improves steel's corrosion resistance. Moderate corrosion resistance is obtained by adding 4–6% chromium to low-carbon steel. (Other elements, specifically less than 1% each of silicon and molybdenum, are also usually added.)

For superior corrosion resistance, larger amounts of chromium are needed. At a minimum level of 12% chromium, steel is *passivated* (i.e., an inert film of chromic oxide forms over the metal and inhibits further oxidation). The formation of this protective coating is the basis of the corrosion resistance of *stainless steel*.[4]

Passivity is enhanced by oxidizers and aeration but is reduced by abrasion that wears off the protective oxide coating. An increase in temperature may increase or decrease the passivity, depending on the abundance of oxygen.

Stainless steels are generally categorized into ferritic, martensitic (heat-treatable), austenitic, duplex, and high-alloy stainless steels.[5]

Ferritic stainless steels contain more than 10% to 27% chromium. The body-centered cubic (BCC) ferrite structure is stable (i.e., does not transform to austenite, a face-centered cubic (FCC) structure) at all temperatures. For this reason, ferritic steels cannot be hardened significantly. Since ferritic stainless steels contain no nickel, they are less expensive than austenitic steels. Turbine blades are typical of the heat-resisting products manufactured from ferritic stainless steels.

The so-called *superferritics* are highly resistant to chloride pitting and crevice corrosion. Superferritics have been incorporated into marine tubing and heat exchangers for power plant condensers. Like all ferritics, however, superferritics experience embrittlement above 475°C.

The *martensitic (heat-treatable) stainless steels* contain no nickel and differ from ferritic stainless steels primarily in higher carbon contents. Cutlery and surgical instruments are typical applications requiring both corrosion resistance and hardness.

The *austenitic stainless steels* are commonly used for general corrosive applications. The stability of the austenite (a face-centered cubic structure) depends primarily on

4–22% nickel as an alloying ingredient. The basic composition is approximately 18% chromium and 8% nickel.

The so-called *superaustenitics* achieve superior corrosion resistance by adding molybdenum (typically up to about 7%) or nitrogen (typically up to about 14%).

Cast Iron and Wrought Iron

Cast iron is a general name given to a wide range of alloys containing iron, carbon, and silicon, and to a lesser extent, manganese, phosphorus, and sulfur. Generally, the carbon content will exceed 2%. The properties of cast iron depend on the amount of carbon present, as well as the form (i.e., graphite or carbide) of the carbon.

The most common type of cast iron is *gray cast iron*. The carbon in gray cast iron is in the form of graphite flakes. Graphite flakes are very soft and constitute points of weakness in the metal, which simultaneously improve machinability and decrease ductility. Compressive strength of gray cast iron is three to five times the tensile strength.

Magnesium and cerium can be added to improve the ductility of gray cast iron. The resulting *nodular cast iron* (also known as *ductile cast iron*) has the best tensile and yield strengths of all the cast irons. It also has good ductility (typically 5%) and machinability. Because of these properties, it is often used for automobile crankshafts.

White cast iron has been cooled quickly from a molten state. No graphite is produced from the cementite, and the carbon remains in the form of a carbide, Fe_3C.[6] The carbide is hard and is the reason that white cast iron is difficult to machine. White cast iron is used primarily in the production of malleable cast iron.

Malleable cast iron is produced by reheating white cast iron to between 800°C and 1000°C for several days, followed by slow cooling. During this treatment, the carbide is partially converted to nodules of graphitic carbon known as *temper carbon*. The tensile strength is increased to approximately 380 MPa, and the elongation at fracture increases to approximately 18%.

Mottled cast iron contains both cementite and graphite and is between white and gray cast irons in composition and performance.

Compacted graphitic iron (CGI) is a unique form of cast iron with worm-shaped graphite particles. The shape of the graphite particles gives CGI the best properties of both gray and ductile cast iron: twice the strength of gray cast iron and half the cost of aluminum. The higher strength permits thinner sections. (Some engine blocks are 25% lighter than gray iron castings.) Using computer-controlled refining, volume production of CGI with the consistency needed for commercial applications is possible.

---

[4]Stainless steels are corrosion resistant in oxidizing environments. In reducing environments (such as with exposure to hydrochloric and other halide acids and salts), the steel will corrode.

[5](1) There is a fifth category, that of *precipitation-hardened stainless steels*, widely used in the aircraft industry. (Precipitation hardening is also known as *age hardening*.) (2) The *sigma phase* structure that appears at very high chromium levels (e.g., 24–50%) is usually undesirable in stainless steels because it reduces corrosion resistance and impact strength. A notable exception is in the manufacture of automobile engine valves.

[6]White and gray cast irons get their names from the coloration at a fracture.

Wrought iron is low-carbon (less than 0.1%) iron with small amounts (approximately 3%) of slag and gangue in the form of fibrous inclusions. It has good ductility and corrosion resistance. Prior to the use of steel, wrought iron was the most important structural metal.

7. IRON-CARBON PHASE DIAGRAM

The iron-carbon phase diagram (see Fig. 42.5) is much more complex than idealized equilibrium diagrams due to the existence of many different phases. Each of these phases has a different microstructure and different mechanical properties. By treating the steel in such a manner as to force the occurrence of particular phases, steel with desired wear and endurance properties can be produced.

Allotropes have the same composition but different atomic structures (microstructures), volumes, electrical resistances, and magnetic properties. *Allotropic changes* are reversible changes that occur at the *critical points* (i.e., *critical temperatures*).

Iron exists in three primary allotropic forms: alpha-iron, delta-iron, and gamma-iron. The changes are brought about by varying the temperature of the iron. Heating pure iron from room temperature changes its structure from body-centered cubic (BCC) *alpha-iron* (−273–970°C), also known as *ferrite*, to face-centered cubic (FCC) *gamma-iron* (910–1400°C), to BCC *delta-iron* (above 1400°C).

Iron-carbon mixtures are categorized into *steel* (less than 2% carbon) and *cast iron* (more than 2% carbon) according to the amounts of carbon in the mixtures.

The most important eutectic reaction in the iron-carbon system is the formation of a solid mixture of austenite and cementite at approximately 1129°C. *Austenite* is a solid solution of carbon in gamma-iron. It is nonmagnetic, decomposes on slow cooling, and does not normally exist below 723°C, although it can be partially preserved by extremely rapid cooling.

Figure 42.5 Iron-Carbon Diagram

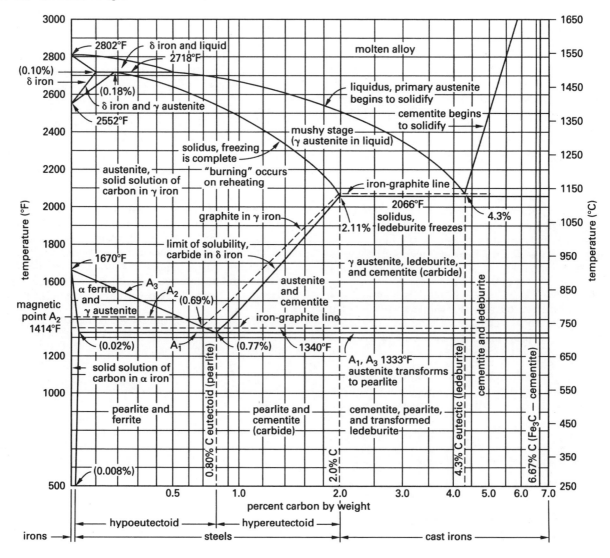

Cementite (Fe$_3$C), also known as *carbide* or *iron carbide*, has approximately 6.67% carbon. Cementite is the hardest of all forms of iron, has low tensile strength, and is quite brittle.

The most important eutectoid reaction in the iron-carbon system is the formation of *pearlite* from the decomposition of austentite at approximately 723°C. Pearlite is actually a mixture of two solid components, ferrite and cementite, with the common *lamellar (layered) appearance*.

Ferrite is essentially pure iron (less than 0.025% carbon) in BCC alpha-iron structure. It is magnetic and has properties complementary to cementite, since it has low hardness, high tensile strength, and high ductility.

8. NONFERROUS METALS

Aluminum and Its Alloys

Aluminum satisfies applications requiring low weight, corrosion resistance, and good electrical and thermal conductivities. Its corrosion resistance derives from the oxide film that forms over the raw metal, inhibiting further oxidation. The primary disadvantages of aluminum are its cost and low strength.

In pure form, aluminum is soft, ductile, and not very strong. Except for use in electrical work, most aluminum is alloyed with other elements. Copper, manganese, magnesium, and silicon can be added to increase its strength, at the expense of other properties, primarily corrosion resistance.[7] Aluminum is hardened by the *precipitation hardening (age hardening)* process.

Silicon occurs as a normal impurity in aluminum, and in natural amounts (less than 0.4%), it has little effect on properties. If moderate quantities (above 3%) of silicon are added, the molten aluminum will have high fluidity, making it ideal for castings. Above 12%, silicon improves the hardness and wear resistance of the alloy. When combined with copper and magnesium (as Mg$_2$Si and AlCuMgSi) in the alloy, silicon improves age hardenability. Silicon has negligible effect on the corrosion resistance of aluminum.

Copper improves the age hardenability of aluminum, particularly in conjunction with silicon and magnesium. Therefore, copper is a primary element in achieving high mechanical strength in aluminum alloys at elevated temperatures. Copper also increases the conductivity of aluminum, but decreases its corrosion resistance.

Magnesium is highly soluble in aluminum and is used to increase strength by improving age hardenability.

Magnesium improves corrosion resistance and may be added when exposure to saltwater is anticipated.

Copper and Its Alloys

Zinc is the most common alloying ingredient in copper. It constitutes a significant part (up to 40% zinc) in brass.[8] (Brazing rod contains even more, approximately 45% to 50%, zinc.) Zinc increases copper's hardness and tensile strength. Up to approximately 30%, it increases the percent elongation at fracture. It decreases electrical conductivity considerably. *Dezincification*, a loss of zinc in the presence of certain corrosive media or at high temperatures, is a special problem that occurs in brasses containing more than 15% zinc.

Tin constitutes a major (up to 20%) component in most bronzes. Tin increases fluidity, which improves casting performance. In moderate amounts, corrosion resistance in saltwater is improved. (*Admiralty metal* has approximately 1%; *government bronze* and *phosphorus bronze* have approximately 10% tin.) In moderate amounts (less than 10%), tin increases the alloy's strength without sacrificing ductility. Above 15%, however, the alloy becomes brittle. For this reason, most bronzes contain less than 12% tin. Tin is more expensive than zinc as an alloying ingredient.

Lead is practically insoluble in solid copper. When present in small to moderate amounts, it forms minute soft particles that greatly improve machinability (2–3% lead) and wearing (bearing) properties (10% lead).

Silicon increases the mechanical properties of copper by a considerable amount. On a per unit basis, silicon is the most effective alloying ingredient in increasing hardness. *Silicon bronze* (96% copper, 3% silicon, 1% zinc) is used where high strength combined with corrosion resistance is needed (e.g., in boilers).

If aluminum is added in amounts of 9–10%, copper becomes extremely hard. Therefore, *aluminum bronze* (as an example) trades an increase in brittleness for increased wearing qualities. Aluminum in solution with the copper makes it possible to precipitation harden the alloy.

Beryllium in small amounts (less than 2%) improves the strength and fatigue properties of copper. These properties make precipitation-hardened *copper-beryllium* (*beryllium-copper*, *beryllium bronze*, etc.) ideal for small springs. These alloys are also used for producing non-sparking tools.

Nickel and Its Alloys

Like aluminum, nickel is largely hardened by precipitation hardening. Nickel is similar to iron in many of its properties, except that it has higher corrosion resistance

---

[7]One ingenious method of having both corrosion resistance and strength is to produce a composite material. *Alclad* is the name given to aluminum alloy that has a layer of pure aluminum bonded to the surface. The alloy provides the strength, and the pure aluminum provides the corrosion resistance.

[8]*Brass* is an alloy of copper and zinc. *Bronze* is an alloy of copper and tin. Unfortunately, brasses are often named for the color of the alloys, leading to some very misleading names. For example, *nickel silver*, *commercial bronze*, and *manganese bronze* are all brasses.

Materials Science

and a higher cost. Also, nickel alloys have special electrical and magnetic properties.

Copper and iron are completely miscible with nickel. Copper increases formability. Iron improves electrical and magnetic properties markedly.

Some of the better-known nickel alloys are *monel metal* (30% copper, used hot-rolled where saltwater corrosion resistance is needed), *K-monel metal*[9] (29% copper, 3% aluminum, precipitation-hardened for use in valve stems), *inconel* (14% chromium, 6% iron, used hot-rolled in gas turbine parts), and *inconel-X* (15% chromium, 7% iron, 2.5% titanium, aged after hot rolling for springs and bolts subjected to corrosion). *Hastelloy* (22% chromium) is another well-known nickel alloy.

Nichrome (15–20% chromium) has high electrical resistance, high corrosion resistance, and high strength at red heat temperatures, making it useful in resistance heating. *Constantan* (40% to 60% copper, the rest nickel) also has high electrical resistance and is used in thermocouples.

Alnico (14% nickel, 8% aluminum, 24% cobalt, 3% copper, the rest iron) and *cunife* (20% nickel, 60% copper, the rest iron) are two well-known nickel alloys with magnetic properties ideal for permanent magnets. Other magnetic nickel alloys are *permalloy* and *permivar*.

Invar, *Nilvar*, and *Elinvar* are nickel alloys with low or zero thermal expansion and are used in thermostats, instruments, and surveyors' measuring tapes.

Refractory Metals

Reactive and *refractory metals* include alloys based on titanium, tantalum, zirconium, molybdenum, niobium (also known as columbium), and tungsten. These metals are used when superior properties (i.e., corrosion resistance) are needed. They are most often used where high-strength acids are used or manufactured.

9. THERMAL PROCESSING

Thermal processing, including hot working, heat treating, and quenching, is used to obtain a part with desirable mechanical properties.

Cold and hot working are both forming processes (rolling, bending, forging, extruding, etc.). The term *hot working* implies that the forming process occurs above the *recrystallization temperature*. (The actual temperature depends on the rate of strain and the cooling period, if any.) *Cold working* (also known as *work hardening* and *strain hardening*) occurs below the recrystallization temperature.

Above the recrystallization temperature, almost all of the internal defects and imperfections caused by hot working are eliminated. In effect, hot working is a "self-healing" operation. A hot-worked part remains

softer and has a greater ductility than a cold-worked part. Large strains are possible without strain hardening. Hot working is preferred when the part must go through a series of forming operations (passes or steps), or when large changes in size and shape are needed.

The hardness and toughness of a cold-worked part will be higher than that of a hot-worked part. Because the part's temperature during the cold working is uniform, the final microstructure will also be uniform. There are many times when these characteristics are desirable, and hot working is not always the preferred forming method. In many cases, cold working will be the final operation after several steps of hot working.

Once a part has been worked, its temperature can be raised to slightly above the recrystallization temperature. This *heat treatment* operation is known as *annealing* and is used to relieve stresses, increase grain size, and recrystallize the grains. Stress relief is also known as *recovery.*

Quenching is used to control the microstructure of steel by preventing the formation of equilibrium phases with undesirable characteristics. The usual desired result is hard steel, which resists plastic deformation. The quenching can be performed with gases (usually air), oil, water, or brine. Agitation or spraying of these fluids during the quenching process increases the severity of the quenching.

Time-temperature-transformation (TTT) *curves* are used to determine how fast an alloy should be cooled to obtain a desired microstructure. Although these curves show different phases, they are not equilibrium diagrams. On the contrary, they show the microstructures that are produced with controlled temperatures or when quenching interrupts the equilibrium process.

TTT curves are determined under ideal, isothermal conditions. However, the curves are more readily available than experimentally determined *controlled-cooling-transformation* (CCT) *curves*. Both curves are similar in shape, although the CCT curves are displaced downward and to the right from TTT curves.

Figure 42.6 shows a TTT diagram for a high-carbon (0.80% carbon) steel. Curve 1 represents extremely rapid quenching. The transformation begins at 216°C, and continues for 8–30 seconds, changing all of the austenite to martensite. The martensitic transformation does not depend on diffusion. Since martensite has almost no ductility, martensitic microstructures are used in applications such as springs and hardened tools where a high elastic modulus and low ductility are needed.

Curve 2 represents a slower quench that converts all of the austenite to fine pearlite.

A horizontal line below the critical temperature is a *tempering* process. If the temperature is decreased rapidly along curve 1 to 270°C and is then held constant along cooling curve 3, bainite is produced. This is the

---

[9]K-monel is one of four special forms of monel metal. There are also H-monel, S-monel, and R-monel forms.

Figure 42.6 *TTT Diagram for High-Carbon Steel*

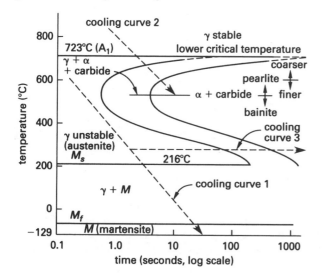

Figure 42.7 *Jominy Hardenability Curves for Six Steel Alloys*

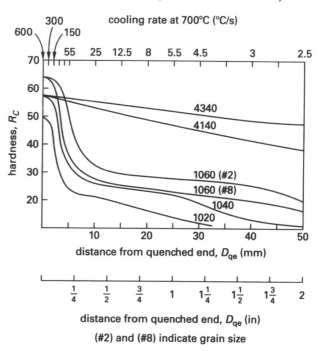

(#2) and (#8) indicate grain size

Van Vlack, L., *Elements of Materials Science and Engineering*, Addison-Wesley, 1989.

principle of *austempering*. Bainite is not as hard as martensite, but it does have good impact strength and fairly high hardness. Performing the same procedure at 180–200°C is *martempering*, which produces *tempered martensite*, a soft and tough steel.

10. HARDNESS AND HARDENABILITY

Hardness is the measure of resistance a material has to plastic deformation. Various *hardness tests* (e.g., Brinell, Rockwell, Meyer, Vickers, and Knoop) are used to determine hardness. These tests generally measure the depth of an impression made by a hardened penetrator set into the surface by a standard force.

Hardenability is a relative measure of the ease by which a material can be hardened. Some materials are easier to harden than others. See Fig. 42.7 for sample hardenability curves for steel.

The hardness obtained also depends on the hardening method (e.g., cooling in air, other gases, water, or oil) and rate of cooling. For example, see Fig. 42.8 and Fig. 42.9. Since the hardness obtained depends on the material, hardening data is often presented graphically. There are a variety of curve types used for this purpose.

Hardenability is not the same as hardness. *Hardness* refers to the ability to resist deformation when a load is applied, whereas *hardenability* refers to the ability to be hardened to a particular depth.

In the *Jominy end quench test*, a cylindrical steel specimen is heated long enough to obtain a uniform grain structure. Then, one end of the specimen is cooled ("quenched") in a water spray, while the remainder of the specimen is allowed to cool by conduction. The cooling rate decreases with increasing distance from the quenched end. When the specimen has entirely cooled, the hardness is determined at various distances from the quenched end. The hardness at the quenched end corresponds to water cooling, while the hardness at

Figure 42.8 *Cooling Rates for Bars Quenched in Agitated Water*

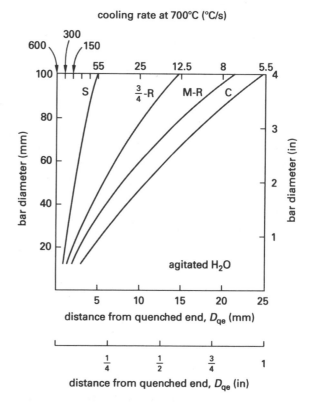

Van Vlack, L., *Elements of Materials Science and Engineering*, Addison-Wesley, 1989.

Materials Science

Figure 42.9 *Cooling Rates for Bars Quenched in Agitated Oil*

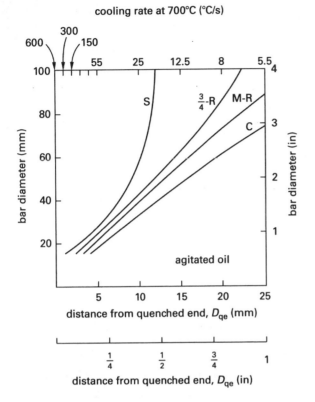

Van Vlack, L., *Elements of Materials Science and Engineering*, Addison-Wesley, 1989.

the opposite end corresponds to air cooling. The same test can be performed with different alloys. *Rockwell hardness* (C-scale) (also known as *Rockwell C hardness*), R_C, is plotted on the vertical axis, while distance from the quenched end, D_{qe}, is plotted on the horizontal scale.[10] The horizontal scale can also be correlated to and calibrated as *cooling rate*. The *Jominy hardenability curves* are used to select an alloy and heat treatment that minimizes distortion during manufacturing. Figure 42.7 illustrates the results of Jominy hardenability tests of six steel alloys.

The position within the bar (see Fig. 42.10) is indicated by the following nomenclature.

- C = center of the bar

- M-R = halfway between the center of the bar and the surface of the bar

- $\frac{3}{4}$-R = three-quarters (75%) of the distance between C and S

- S = surface of the bar

---

[10]In ASTM A255, "Standard Test Methods for Determining Hardenability of Steel," the distance from the quenched end is given the symbol J and is referred to as the *Jominy distance*. Rockwell C hardness is given the standard symbol "HRC." The *initial hardness* at the $J = \frac{1}{16}$ in position is given the symbol "IH."

Figure 42.10 *Bar Positions*

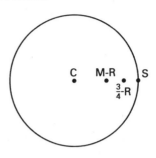

11. METAL GRAIN SIZE

One of the factors affecting hardness and hardenability is the average metal *grain size*. Grain size refers to the diameter of a three-dimensional spherical grain as determined from a two-dimensional micrograph of the metal.

The size of the grains formed depends on the number of nuclei formed during the solidification process. If there are many nuclei, as would occur when cooling is rapid, there will be many small grains. However, if cooling is slow, a smaller number of larger grains will be produced. Fast cooling produces fine-grained material, and slow cooling produces coarse-grained material.

At moderate temperatures and strain rates, fine-grained materials deform less (i.e., are harder and tougher) while coarse-grained materials deform more. For ease of cold-formed manufacturing, coarse-grained materials may be preferred. However, appearance and strength may suffer.

It is difficult to measure grain size because the grains are varied in size and shape and because a two-dimensional image does not reveal volume. Semi-empirical methods have been developed to automatically or semi-automatically correlate grain size with the number of intersections observed in samples. ASTM E112, "Determining Average Grain Size" (and the related ASTM E1382), describes a planimetric procedure for metallic and some nonmetallic materials that exist primarily in a single phase.

Equation 42.4 Through Eq. 42.6: ASTM Grain Size, *n*

$$\frac{N_{actual}}{actual\ area} = \frac{N}{0.0645\ mm^2} \qquad 42.4$$

$$N_{(0.0645\ mm^2)} = 2^{(n-1)} \qquad 42.5$$

$$S_V = 2P_L \qquad 42.6$$

Description

Data on grain size is obtained by counting the number of grains in any small two-dimensional area. The number of grains, N, in a standard area ($0.0645\ mm^2$) can be extrapolated from the observations using Eq. 42.4.

The standard number of grains, N, is the number of grains per square inch in an image of a polished specimen magnified 100 times. Equation 42.5 calculates the ASTM *grain size number* (also known as the ASTM *grain size*), n, from the standard number of grains, N, as the nearest integer greater than 1.

$$n = \frac{\log N + \log 2}{\log 2}$$

The grain-boundary surface area per unit volume, S_V, is taken as twice the number of points of intersection per unit length between the line and boundaries, P_L, as shown by Eq. 42.6. If a random line (any length, any orientation) is drawn across a $100\times$ magnified image (i.e., a *photomicrograph*) of grains, the line will cross some number of grains, say N. For each grain, the line will cross over two grain boundaries, so for a line of length L, the average grain diameter will be

$$D_{ave} = \frac{L}{2N} = \frac{1}{S_V}$$

Example

Eight grains are observed in a 0.0645 mm^2 area of a polished metal surface. What is most nearly the number of grains in an area of 710 mm^2?

   (A) 16 grains

   (B) 120 grains

   (C) 1100 grains

   (D) 88 000 grains

Solution

From Eq. 42.4, and rearranging to solve for N_{actual}, the number of grains in an area of 710 mm^2 is

$$\frac{N_{actual}}{\text{actual area}} = \frac{N}{0.0645 \text{ mm}^2}$$

$$N_{actual} = (\text{actual area})\left(\frac{N}{0.0645 \text{ mm}^2}\right)$$

$$= (710 \text{ mm}^2)\left(\frac{8 \text{ grains}}{0.0645 \text{ mm}^2}\right)$$

$$= 88\,062 \text{ grains} \quad (88\,000 \text{ grains})$$

The answer is (D).

12. AMORPHOUS MATERIALS

Amorphous materials are materials that lack a crystalline structure like liquids, but are rigid and maintain their shape like solids. Glass is a distinct kind of amorphous material that exhibits a glass transition (see Sec. 42.14). Figure 42.11 illustrates the difference between amorphous and crystalline solids over the glass transition temperature, T_g, and melting temperature, T_m.

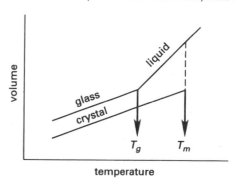

Figure 42.11 Volume-Temperature Curve for Amorphous Materials

13. POLYMERS

Natural Polymers

A *polymer* is a large molecule in the form of a long chain of repeating units. The basic repeating unit is called a *monomer* or just *mer*. (A large molecule with two alternating mers is known as a *copolymer* or *interpolymer*. Vinyl chloride and vinyl acetate form one important family of copolymer plastics.)

Many of the natural organic materials (e.g., rubber and asphalt) are polymers. (Polymers with elastic properties similar to rubber are known as *elastomers*.) Natural rubber is a polymer of the *isoprene latex* mer (formula $[C_5H_8]_n$, repeating unit of $CH_2=CCH_3-CH=CH_2$, systematic name of 2-methyl-1,3-butadiene). The strength of natural polymers can be increased by causing the polymer chains to cross-link, restricting the motion of the polymers within the solid.

Cross-linking of natural rubber is known as *vulcanization*. Vulcanization is accomplished by heating raw rubber with small amounts of sulfur. The process raises the tensile strength of the material from approximately 2.1 MPa to approximately 21 MPa. The addition of carbon black as a reinforcing *filler* raises this value to approximately 31 MPa and provides tear resistance and toughness.

The amount of cross-linking between the mers determines the properties of the solid. Figure 42.12 shows how sulfur joins two adjacent isoprene (natural rubber) mers in *complete cross-linking*.[11] If sulfur does not replace both of the double carbon bonds, *partial cross-linking* is said to have occurred.

Degree of Polymerization

The *degree of polymerization*, DP, is the average number of mers in the molecule, typically several hundred to several thousand.[12] (In general, compounds with degrees of less than ten are called *telenomers* or *oligomers*.) The

---

[11]A tire tread may contain 3–4% sulfur. Hard rubber products, which do not require flexibility, may contain as much as 40–50% sulfur.
[12]Degrees of polymerization for commercial plastics are usually less than 1000.

Figure 42.12 *Vulcanization of Natural Rubber*

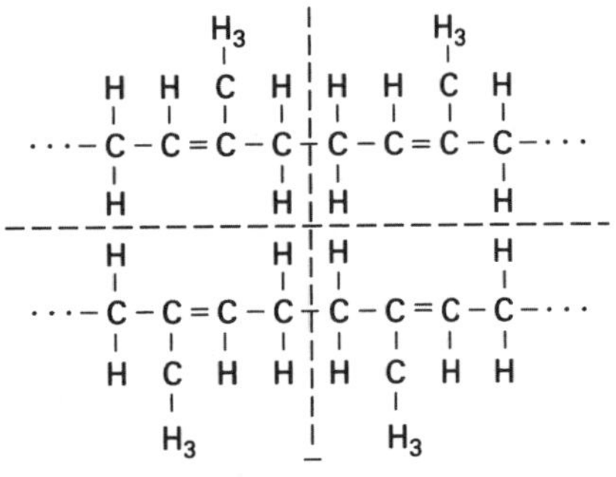

(natural) – 4 mers

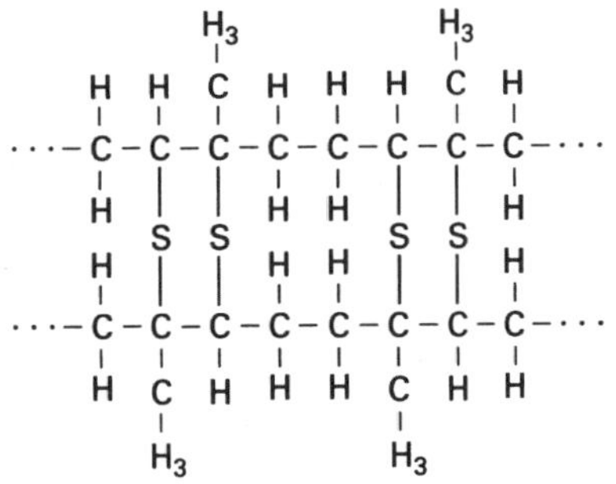

cross-linked

degree of polymerization can be calculated from the mer and polymer molecular weights, MW.

$$DP = \frac{MW_{polymer}}{MW_{mer}}$$

A polymer batch usually will contain molecules with different length chains. Therefore, the degree of polymerization will vary from molecule to molecule, and an average degree of polymerization is reported.

The stiffness and hardness of polymers vary with their degrees. Polymers with low degrees are liquids or oils. With increasing degree, they go through waxy to hard resin stages. High-degree polymers have hardness and strength qualities that make them useful for engineering applications. Tensile strength and melting (softening) point also increase with increasing degree of polymerization.

Synthetic Polymers

Table 42.5 lists some of the common mers. Polymers are named by adding the prefix "poly" to the name of the basic mer. For example, C_2H_4 is the chemical formula for ethylene. Chains of C_2H_4 are called polyethylene.

Polymers are able to form when double (covalent) bonds break and produce reaction sites. The number of bonds in the mer that can be broken open for attachment to other mers is known as the *functionality* of the mer.

Table 42.5 *Names of Common Mers*

name	repeating unit	combined formula
ethylene	CH_2CH_2	C_2H_4
propylene	$CH_2(HCCH_3)$	C_3H_6
styrene	$CH_2CH(C_6H_5)$	C_8H_8
vinyl acetate	$CH_2CH(C_2H_3O_2)$	$C_4H_6O_2$
vinyl chloride	CH_2CHCl	C_2H_3Cl
isobutylene	$CH_2C(CH_3)_2$	C_4H_8
methyl methacrylate	$CH_2C(CH_3)(COOCH_3)$	$C_5H_8O_2$
acrylonitrile	CH_2CHCN	C_3H_3N
epoxide (ethoxylene)	CH_2CH_2O	C_2H_4O
amide (nylon)	$CONH_2$ or $CONH$	$CONH_2$ or $CONH$

Vinyl polymers are a class of polymers where a different element or molecule, R, has been substituted for one of the hydrogen atoms in the repeating units (i.e., $-[CH_2-CHR]-$). For example, the repeating unit of polyethylene is $-[CH_2]-$ or $-[CH_2-CH_2]-$, but the repeating unit of polyvinyl chloride is $-[CH_2-CHCl]-$. *Copolymers* contain two repeating units (i.e., *comonomers*). One of the units may alternate with the other, or it may appear on a regular basis (i.e., periodically) according to some specific statistic or structure. In an *addition polymer*, all of the atoms in the combining molecules appear in the resulting polymer molecule. In a *condensation polymer*, the resulting structure has fewer atoms than the monomers from which it was created. (The missing atoms or fragments "condense out.") Usually, some energy source (e.g., heat) is required to form condensation polymers.

Example

Polyester repeating unit $-[CO(CH_2)_4CO-OCH_2CH_2O]-$ is formed by combining the two repeating units $-[HO_2C-(CH_2)_4-CO_2H]-$ and $-[HO-CH_2CH_2-OH]-$. Most accurately, what type of polymer is the polyester?

(A) addition polymer

(B) comonomer polymer

(C) condensation polymer

(D) vinyl polymer

Solution

Count the number of each atom in the combining and polyester molecules.

element	atoms in combining units	atoms in polyester repeating unit	discrepancy
carbon (C)	$6+2=8$	8	0
hydrogen (H)	$10+6=16$	12	4
oxygen (O)	$4+2=6$	4	2

Four hydrogen atoms and two oxygen atoms (the equivalent of two water molecules) are missing from the resulting polymer unit. Polyester is a condensation polymer.

The answer is (C).

Thermosetting and Thermoplastic Polymers

Most polymers can be softened and formed by applying heat and pressure. These are known by various terms including *thermoplastics, thermoplastic resins,* and *thermoplastic polymers*. Thermoplastics may be either semi-crystalline or amorphous. Polymers that are resistant to heat (and that actually harden or "kick over" through the formation of permanent cross-linking upon heating) are known as *thermosetting plastics*. Table 42.6 lists the common polymers in each category. Thermoplastic polymers retain their chain structures and do not experience any chemical change (i.e., bonding) upon repeated heating and subsequent cooling. Thermoplastics can be formed in a cavity mold, but the mold must be cooled before the product is removed. Thermoplastics are particularly suitable for injection molding. The mold is kept relatively cool, and the polymer solidifies almost instantly.

Figure 42.13 illustrates the relationship between temperature and the strength, σ, or modulus of elasticity, E, of thermoplastics.

Table 42.6 Thermosetting and Thermoplastic Polymers

thermosetting
  epoxy
  melamine
  natural rubber (polyisoprene)
  phenolic (phenol formaldehyde, Bakelite®)
  polyester (DAP)
  silicone
  urea formaldehyde
thermoplastic
  acetal
  acrylic
  acrylonitrile-butadiene-styrene (ABS)
  cellulosics (e.g., cellophane)
  polyamide (nylon)
  polyarylate
  polycarbonate
  polyester (PBT and PET)
  polyethylene
  polymethyl-methacrylate (Plexiglas®, Lucite®)
  polypropylene
  polystyrene
  polytetrafluoroethylene (Teflon®)
  polyurethane
  polyvinyl chloride (PVC)
  synthetic rubber (Neoprene®)
  vinyl

Bakelite® is a trademark of Momentive Specialty Chemicals Inc. Plexiglas® is a trademark of Altuglas International. Lucite® is a trademark of Lucite International. Teflon® and Neoprene® are trademarks of DuPont.

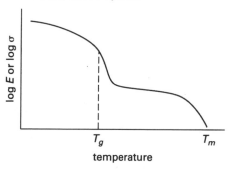

Figure 42.13 Temperature Dependent Strength or Modulus for Thermoplastic Polymers

Thermosetting polymers form complex, three-dimensional networks. Thus, the complexity of the polymer increases dramatically, and a product manufactured from a thermosetting polymer may be thought of as one big molecule. Thermosetting plastics are rarely used with injection molding processes.

Thermosetting compounds are purchased in liquid form, which makes them easy to combine with additives. Thermoplastic materials are commonly purchased in granular form. They are mixed with additives in a *muller* (i.e., a bulk mixer) before transfer to the feed hoppers. Thermoplastic materials can also be molded into small pellets called *preforms* for easier handling in subsequent melting operations. Common additives for plastics include:

- *Plasticizers:* vegetable oils, low molecular weight polymers or monomers

- *Fillers:* talc, chopped glass fibers

- *Flame retardants:* halogenated paraffins, zinc borate, chlorinated phosphates

- *Ultraviolet or visible light resistance:* carbon black

- *Oxidation resistance:* phenols, aldehydes

Fluoropolymers

Fluoropolymers (fluoroplastics) are a class of paraffinic, thermoplastic polymers in which some or all of the hydrogens have been replaced by fluorine.[13] There are seven major types of fluoropolymers, with overlapping characteristics and applications. They include the fully fluorinated fluorocarbon polymers of Teflon® polytetrafluoroethylene (PTFE), fluorinated ethylene propylene (FEP), and perfluoroalkoxy (PFA), as well as the partially fluorinated polymers of polychlorotrifluoroethylene (PCTFE), ethylene tetrafluoroethylene (ETFE), ethylene chlorotrifluoroethylene (ECTFE), and polyvinylidene fluoride (PVDF).

Fluoropolymers compete with metals, glass, and other polymers in providing corrosion resistance. Choosing the right fluoropolymer depends on the operating

---

[13]*Fluoroelastomers* are uniquely different from fluoropolymers. They have their own areas of application.

environment, including temperature, chemical exposure, and mechanical stress.

PTFE, the first available fluoropolymer, is probably the most inert compound known. It has been used extensively for pipe and tank linings, fittings, gaskets, valves, and pump parts. It has the highest operating temperature—approximately 260°C. Unlike the other fluoropolymers, however, it is not a melt-processed polymer. Like a powdered metallurgy product, PTFE is processed by compression and isostatic molding, followed by sintering. PTFE is also the weakest of all the fluoropolymers.

14. GLASS

Glass is a term used to designate any material that has a volumetric expansion characteristic similar to Fig. 42.14. Glasses are sometimes considered to be *supercooled liquids* because their crystalline structures solidify in random orientation when cooled below their melting points. It is a direct result of the high liquid viscosities of oxides, silicates, borates, and phosphates that the molecules cannot move sufficiently to form large crystals with cleavage planes. Glass is considered an amorphous material.

Figure 42.14 *Behavior of a Glass*

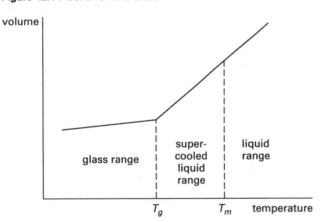

As a liquid glass is cooled, its atoms develop more efficient packing arrangements. This leads to a rapid decrease in volume (i.e., a steep slope on the temperature-volume curve). Since no crystallization occurs, the liquid glass simply solidifies without molecular change when cooled below the melting point. (This is known as *vitrification.*) The more efficient packing continues past the point of solidification.

At the *glass transition temperature* (*fictive temperature*), T_g, the glass viscosity increases suddenly by several orders of magnitude. Since the molecules are more restrained in movement, efficient atomic rearrangement is curtailed, and the volume-temperature curve changes slope. This temperature also divides the region into flexible and brittle regions. At the glass transition temperature, there is a 100-fold to 1000-fold increase in stiffness (modulus of elasticity).

Both organic and inorganic compounds may behave as glasses. *Common glasses* are mixtures of SiO_2, B_2O_3, and various other compounds to modify performance.[14]

15. CERAMICS

Ceramics are compounds of metallic and nonmetallic elements. Ceramics form crystalline structures but have no free valence electrons. All electrons are shared ionically or in covalent bonds. Common examples include brick, portland cement, refractories, and abrasives. (Glass is also considered a ceramic even though it does not crystallize.)

Although perfect ceramic crystals have extremely high tensile strengths (e.g., some glass fibers have ultimate strengths of 700 MPa), the multiplicity of cracks and other defects in natural crystals reduces their tensile strengths to near-zero levels.

Due to the absence of free electrons, ceramics are typically poor conductors of electrical current, although some (e.g., magnetite, Fe_3O_4) possess semiconductor properties. Other ceramics, such as $BaTiO_3$, SiO_2, and $PbZrO_3$, have *piezoelectric* (*ferroelectric*) *qualities* (i.e., generate a voltage when compressed).

Polymorphs are compounds that have the same chemical formula but have different physical structures. Some ceramics, of which *silica* (SiO_2) is a common example, exhibit *polymorphism*. At room temperature, silica is in the form of *quartz*. At 875°C, the structure changes to *tridymite*. A change to a third structure, that of *cristobalite*, occurs at 1470°C.

Ferrimagnetic materials (*ferrites*, *spinels*, or *ferrispinels*) are ceramics with valuable magnetic qualities. Advances in near-room-temperature superconductivity have been based on *lanthanum barium copper oxide* ($La_{2-x}Ba_xCuO_4$), a ceramic oxide, as well as compounds based on yttrium (Y-Ba-Cu-O), bismuth, thallium, and others.

---

[14]This excludes lead-alkali glasses that contain 30–60% PbO.

Diagnostic Exam

Topic IX: Chemical Reaction Engineering

1. A container holds 1000 kg of benzene (C_6H_6). A second container holds 2100 kg of a mixture comprising 40% nitric acid (HNO_3), 52% sulfuric acid (H_2SO_4), and 8% water (H_2O) by mass. The contents of both containers are combined to produce nitrobenzene ($C_6H_5NO_2$) in the reaction shown.

$$C_6H_6 + HNO_3 + H_2O + H_2SO_4 \rightarrow$$
$$C_6H_5NO_2 + 2H_2O + H_2SO_4$$

Most nearly, how much nitric acid, HNO_3, remains when the conversion of benzene is 97% complete?

(A) 21 kg

(B) 45 kg

(C) 58 kg

(D) 70 kg

2. A mixture comprising 50% benzene and 50% toluene by mass is fed at a rate of 20 000 kg/h into a distillation column. The stream of product from the column enters the condenser at a rate of 18 000 kg/h. The distillation column and condenser are configured as shown.

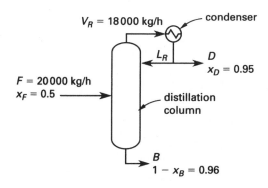

The distillate product from the condenser comprises 95% benzene by mass. The bottoms product comprises 96% toluene by mass. A portion of the product leaving the condenser returns to the column as reflux. Most nearly, what is the reflux rate, R?

(A) 7900 kg/h

(B) 9900 kg/h

(C) 10 000 kg/h

(D) 12 000 kg/h

3. 16 moles of substance B and an equal, stoichiometric amount of substance A are fed into a batch reactor. 4.5 moles of product E are produced according to the reaction shown.

$$A + B \rightarrow E$$

Most nearly, what is the fractional conversion of A?

(A) 0.23

(B) 0.28

(C) 0.37

(D) 0.51

4. A homogenous, liquid phase, elementary thermal decomposition reaction takes place in a batch reactor according to the reaction shown.

$$2A \rightarrow B + C$$

The reactor volume is 350 L, and the operating temperature is 400K. The reaction rate constant is 0.047 L/mol·min. The initial concentration of A is 3 mol/L. Most nearly, what is the conversion of A after the reaction has run for five hours?

(A) 82%

(B) 87%

(C) 95%

(D) 98%

5. Two dilute streams of aqueous potassium nitrate (KNO_3) with the mole percentages shown are combined in a mixer.

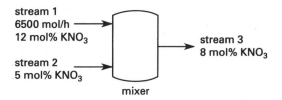

Most nearly, what is the molar flow rate of potassium nitrate in stream 2?

(A) 430 mol/h

(B) 950 mol/h

(C) 2400 mol/h

(D) 8600 mol/h

6. The cylinder of an adiabatic cylinder-frictionless piston assembly is filled with nitrogen. The nitrogen behaves as an ideal gas. 125 J of heat are added to the nitrogen as the piston is allowed to expand the occupied space. The pressure on the piston head remains at 2 atm throughout the process. The gas occupies a volume of 625 cm^3 and is at 25°C before heating. The constant pressure heat capacity of the nitrogen gas at this temperature is 29.1 J/kmol·K. Most nearly, what is the final volume of the gas inside the cylinder?

(A) 690 cm^3

(B) 740 cm^3

(C) 800 cm^3

(D) 950 cm^3

7. 0.85 mol/s of a gas mixture flow into a reactor system at 170°C. The following table gives the rate of reaction of the gas mixture at that temperature.

conversion fraction, X_A	rate of reaction, $-r_A$ (mol/L·s)
0	0.0064
0.1	0.0062
0.2	0.006
0.3	0.0057
0.4	0.0054
0.5	0.005
0.6	0.0046
0.7	0.0041
0.8	0.0036
0.85	0.003

The mixture is fed into a continuous stirred tank reactor (CSTR) where decomposition proceeds to 40% completion. Then, the product of the first CSTR is fed into a second CSTR, where decomposition progresses to 80% completion. Most nearly, the total volume of the two reactors is

(A) 72 L

(B) 96 L

(C) 130 L

(D) 160 L

8. A chemical reaction describing the decomposition of nitric anhydride, N_2O_5, and its reaction rate equation are shown.

$$2N_2O_5 \rightarrow 4NO_2 + O_2$$

$$r = -k[N_2O_5]$$

The nitric anhydride has a concentration of 0.01 mol/L. At 300K, the activation energy is 1.03×10^5 J/mol, and the frequency factor is 4.3×10^{13} s^{-1}. The conversion rate of nitric anhydride at 300K is most nearly

(A) -20×10^{-6} mol/L·s

(B) -9.5×10^{-6} mol/L·s

(C) -2.0×10^{-6} mol/L·s

(D) -0.50×10^{-6} mol/L·s

9. The equation for a reaction is shown.

$$2A + 3B \rightarrow C + 2D$$

Which of these reaction rate relationships is true?

(A) $r_A = -\frac{3}{2}r_B$

(B) $r_B = r_C - r_D$

(C) $r_B = -3r_C + \frac{3}{2}r_D + \frac{3}{2}r_A$

(D) $r_D = -r_A - \frac{2}{3}r_B$

10. Substance A is fed into a continuous stirred tank reactor (CSTR) at a rate of 350 L/min. The initial concentration of A is 0.2 mol/L. The reaction that occurs inside the reactor is

$$A \rightarrow 2B$$

A reciprocal rate plot for the reaction at the reactor conditions is shown.

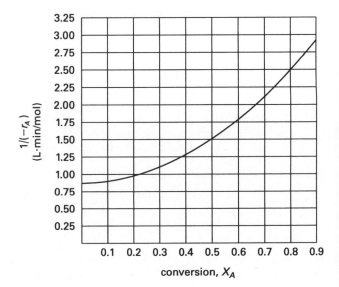

Most nearly, what is the reactor volume required for 75% conversion?

(A) 120 L

(B) 350 L

(C) 480 L

(D) 640 L

SOLUTIONS

1. The initial mass, m, of HNO_3 is

$$m_{HNO_3} = C_{HNO_3} m_{acid} = (0.40)(2100 \text{ kg})$$
$$= 840.0 \text{ kg}$$

The mass of C_6H_6 consumed, m_c, in the reaction is

$$m_{c,C_6H_6} = C_{C_6H_6} m_{C_6H_6} = (0.97)(1000 \text{ kg}) = 970 \text{ kg}$$

The mass of HNO_3 consumed is

$$m_{C_{HNO_3}} = m_{c,C_6H_6}\left(\frac{\text{molecular weight of } HNO_3}{\text{molecular weight of } C_6H_6}\right)$$
$$= (970 \text{ kg})\left(\frac{1.0079 + 14.007 + (3)(15.999)}{(6)(12.011) + (6)(1.0079)}\right)$$
$$= 782.5 \text{ kg}$$

The excess HNO_3 is

$$\text{excess}_{HNO_3} = m_{HNO_3} - m_{c,HNO_3}$$
$$= 840.0 \text{ kg} - 782.5 \text{ kg}$$
$$= 57.5 \text{ kg} \quad (58 \text{ kg})$$

The answer is (C).

2. A more detailed flowsheet is shown.

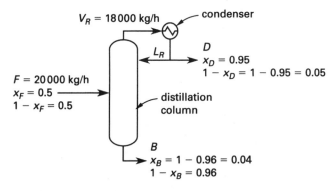

The total mass balance around the column is

$$F = D + B$$

The benzene mass balance around the column is

$$Fx_F = Dx_D + Bx_B$$

Solve these two balances for the distillate flow by substituting the bottoms flow from the total balance into the benzene balance,

$$B = F - D$$

$$Fx_F = Dx_D + (F - D)x_B$$

The bottoms composition of benzene is

$$x_B = 1 - 0.96 = 0.04$$

The distillate flow, D, is

$$D = F\left(\frac{x_F - x_B}{x_D - x_B}\right)$$
$$= \left(20\,000 \ \frac{\text{kg}}{\text{h}}\right)\left(\frac{0.5 - 0.04}{0.95 - 0.04}\right)$$
$$= 10\,110 \text{ kg/h}$$

The overall balance around the condenser is

$$V_R = L_R + D$$

Solve the condenser overall balance for the reflux rate.

$$L_R = V_R - D$$
$$= 18\,000 \ \frac{\text{kg}}{\text{h}} - 10\,110 \ \frac{\text{kg}}{\text{h}}$$
$$= 7890 \text{ kg/h} \quad (7900 \text{ kg/h})$$

The answer is (A).

3. From the reaction stoichiometry, since equal amounts of A and B are fed to the reactor,

$$\text{mol } A \text{ fed} = \text{mol } B \text{ fed} = 16 \text{ mol}$$

$$\text{mol } A \text{ reacted} = \text{mol } E \text{ produced} = 4.5 \text{ mol}$$

$$\text{fractional conversion of } A = \frac{\text{mol } A \text{ reacted}}{\text{mol } A \text{ fed}}$$
$$= \frac{4.5 \text{ mol}}{16 \text{ mol}}$$
$$= 0.28$$

The answer is (B).

4. Since the reaction is elementary, the reaction order depends on the coefficients of the reactants. The reaction is a second-order reaction. The rate of reaction, r_A, is

$$-r_A = kC_A^2$$

The conversion, X_A, for a second-order irreversible reaction is

$$\frac{X_A}{C_{A0}(1 - X_A)} = kt$$

$$\frac{X_A}{1 - X_A} = C_{A0}kt$$

$$X_A = C_{A0}kt - X_A C_{A0}kt$$

$$X_A + X_A C_{A0}kt = C_{A0}kt$$

$$X_A(1 + C_{A0}kt) = C_{A0}kt$$

$$X_A = \frac{C_{A0}kt}{1 + C_{A0}kt}$$

$$= \frac{\left(3 \frac{\text{mol}}{\text{L}}\right)\left(0.047 \frac{\text{L}}{\text{mol·min}}\right) \times (5 \text{ h})\left(60 \frac{\text{min}}{\text{h}}\right)}{1 + \left(3 \frac{\text{mol}}{\text{L}}\right)\left(0.047 \frac{\text{L}}{\text{mol·min}}\right) \times (5 \text{ h})\left(60 \frac{\text{min}}{\text{h}}\right)}$$

$$= 0.9769 \quad (98\%)$$

The answer is (D).

5. The total mass balance around the system is

$$F_1 + F_2 = F_3$$

The KNO_3 balance around the system is

$$x_1 F_1 + x_2 F_2 = x_3 F_3$$

Substitute F_3 from the total mass balance into the component balance equation.

$$x_1 F_1 + x_2 F_2 = x_3(F_1 + F_2)$$

$$= x_3 F_1 + x_3 F_2$$

$$x_2 F_2 - x_3 F_2 = x_3 F_1 - x_1 F_1$$

$$F_2 = \frac{(x_3 - x_1)F_1}{x_2 - x_3}$$

$$= \frac{(0.08 - 0.12)\left(6500 \frac{\text{mol}}{\text{h}}\right)}{0.05 - 0.08}$$

$$= 8667 \text{ mol/h}$$

The molar flow rate of KNO_3 is

$$F_{KNO_3} = x_2 F_2$$

$$= (0.05)\left(8667 \frac{\text{mol}}{\text{h}}\right)$$

$$= 433 \text{ mol/h} \quad (430 \text{ mol/h})$$

The answer is (A).

6. The closed system energy balance is

$$Q - W = \Delta U + \Delta KE + \Delta PE$$

Neglecting kinetic and potential energy changes, the energy balance is

$$Q - W = \Delta U$$

The work, W, done by the gas on the surroundings for a constant pressure process is

$$W = p(V_2 - V_1)$$

The energy balance, inserting the work, and the definition of enthalpy, becomes

$$Q = \Delta U + W$$

$$= U_2 - U_1 + p(V_2 - V_1)$$

$$= (U_2 + pV_2) - (U_1 + pV_1)$$

$$= H_2 - H_1$$

$$= \Delta H$$

For an ideal gas, the number of moles, N, in the cylinder is always

$$N = \frac{pV_1}{\overline{R}T_1}$$

The change in enthalpy, Q, can be calculated using heat capacity at constant pressure and moles from the ideal gas law.

$$Q = \Delta H = Nc_p(T_2 - T_1) = \frac{pV_1}{\overline{R}T_1}c_p(T_2 - T_1)$$

Solving for the final temperature, T_2,

$$T_2 = T_1 + \frac{\overline{R}T_1 Q}{pV_1 c_p}$$

$$= (25°C + 273°) + \frac{\left(8.314 \frac{\text{J}}{\text{mol·K}}\right)(25°C + 273°) \times (125 \text{ J})\left(100 \frac{\text{cm}}{\text{m}}\right)^3}{(2 \text{ atm})\left(101\,325 \frac{\text{Pa}}{\text{atm}}\right) \times (625 \text{ cm}^3)\left(29.1 \frac{\text{J}}{\text{mol·K}}\right)}$$

$$= 382.03\text{K}$$

The ideal gas at the initial and final conditions is

$$\frac{pV_2}{pV_1} = \frac{N\overline{R}T_2}{N\overline{R}T_1}$$

The volume ratio at constant pressure and moles is proportional to the temperature ratio. The final volume, V_2, is

$$\frac{V_2}{V_1} = \frac{T_2}{T_1}$$

$$V_2 = \frac{V_1 T_2}{T_1}$$

$$= \frac{(625 \text{ cm}^3)(382.03\text{K})}{25°\text{C} + 273°}$$

$$= 801.2 \text{ cm}^3 \quad (800 \text{ cm}^3)$$

The answer is (C).

7. At 40% decomposition in the first reactor, the conversion fraction, $X_{A,1}$, is 0.4, and the rate constant, $-r_A$, is 0.0054 mol/L·s. The volume of the first reactor is

$$\frac{V_{\text{CSTR}}}{F_{A,0}} = \frac{X_A}{-r_A}$$

$$V_{\text{CSTR},1} = \frac{X_{A,1} F_{A,0}}{-r_{A,1}}$$

$$= \frac{(0.4)\left(0.85 \frac{\text{mol}}{\text{s}}\right)}{0.0054 \frac{\text{mol}}{\text{L·s}}}$$

$$= 62.96 \text{ L}$$

In the second reactor, at 80% decomposition the conversion fraction is 0.8, and the rate of reaction is 0.0036 mol/L·s. To calculate the volume of the second reactor, the change in mixture conversion is used.

$$\frac{V_{\text{CSTR}}}{F_{A,0}} = \frac{X_A}{-r_A}$$

$$V_{\text{CSTR},2} = \frac{(X_{A,2} - X_{A,1}) F_{A,0}}{-r_{A,2}}$$

$$= \frac{(0.8 - 0.4)\left(0.85 \frac{\text{mol}}{\text{s}}\right)}{0.0036 \frac{\text{mol}}{\text{L·s}}}$$

$$= 94.44 \text{ L}$$

The total volume of both reactors is

$$V_{\text{total}} = V_{\text{CSTR},1} + V_{\text{CSTR},2} = 62.96 \text{ L} + 94.44 \text{ L}$$

$$= 157.40 \text{ L} \quad (160 \text{ L})$$

The answer is (D).

8. Use the Arrhenius equation to find the reaction rate constant.

$$k = A e^{-E_a/\overline{R}T}$$

$$= (4.3 \times 10^{13} \text{ s}^{-1}) e^{-(1.03 \times 10^5 \text{ J/mol})/((8.314 \text{ J/mol·K})(300\text{K}))}$$

$$= 50 \times 10^{-6} \text{ s}^{-1}$$

Use the reaction rate equation. The conversion rate of nitric anhydride is

$$r = -k[\text{N}_2\text{O}_5] = -(50 \times 10^{-6} \text{ s}^{-1})\left(0.01 \frac{\text{mol}}{\text{L}}\right)$$

$$= -0.5 \times 10^{-6} \text{ mol/L·s}$$

The answer is (D).

9. From the reaction equation, the reaction rates are related by

$$-\tfrac{1}{2}r_A = -\tfrac{1}{3}r_B = r_C = \tfrac{1}{2}r_D$$

Substituting equivalent values into option (A) gives

$$r_A = -\tfrac{3}{2}r_B = -\tfrac{3}{2}\left(\tfrac{3}{2}r_A\right)$$

$$= -\tfrac{9}{4}r_A$$

This cannot be true, so option (A) is false. Substituting equivalent values into option (B) gives

$$r_B = r_C - r_D = -\tfrac{1}{3}r_B - \left(-\tfrac{2}{3}r_B\right)$$

$$= \tfrac{1}{3}r_B$$

This cannot be true, so option (B) is false. Substituting equivalent values into option (D) gives

$$r_D = -r_A - \tfrac{2}{3}r_B = r_D - \left(-\tfrac{3}{2}r_D\right)$$

$$= \tfrac{5}{2}r_D$$

This cannot be true, so option (D) is false. Substituting equivalent values into option (C) gives

$$r_B = -3r_C + \tfrac{3}{2}r_D + \tfrac{3}{2}r_A$$

$$= (-3)\left(-\tfrac{1}{3}r_B\right) + \tfrac{3}{2}\left(-\tfrac{2}{3}r_B\right) + \left(\tfrac{3}{2}\right)\left(\tfrac{2}{3}r_B\right)$$

$$= r_B$$

This is true, so option (C) is true.

The answer is (C).

10. From the reciprocal rate plot, at $X_A = 0.75$, $1/(-r_A) = 2.30$ L·min/mol.

The reactor volume, V, at a given percent conversion can be expressed in terms of volumetric flow rate, r, and concentration, C.

$$\frac{V_{\text{CSTR}}}{F_{A,0}} = \frac{X_A}{-r_A}$$

$$V_{\text{CSTR}} = \frac{F_{A,0} X_A}{-r_A}$$

$$= V_0 C_{A,0} X_A \frac{1}{-r_A}$$

$$= \left(350 \ \frac{\text{L}}{\text{min}}\right)\left(0.2 \ \frac{\text{mol}}{\text{L}}\right)\left(\frac{75\%}{100\%}\right)\left(2.30 \ \frac{\text{L·min}}{\text{mol}}\right)$$

$$= 120.8 \text{ L} \quad (120 \text{ L})$$

The answer is (A).

43 Reaction Kinetics[1]

Nomenclature

$\hat{a}$	activity	–
A	pre-exponential or frequency factor	–
A	reactor surface area (for heat transfer to/from surroundings)	m^2
c_p	specific heat capacity	J/kg·K
C	instantaneous molar concentration	mol/L
$\hat{C}$	equilibrium concentration	mol/L
$\hat{f}$	fugacity	Pa or bar
f^0	standard state fugacity	Pa or bar
$[E]$	molar concentration of enzyme	mol/L
E_a	activation energy	J/mol
F	molar flow rate	mol/s
G	Gibbs (function) energy	J/mol
H	enthalpy or energy	J/mol
ΔH_r	heat of reaction	J/mol
k	reaction rate constant	–
K_a	chemical equilibrium constant for use with activities	–
K_c	chemical equilibrium constant for use with concentrations	–
K_M	aggregate reaction rate constant	mol/L
K_p	chemical equilibrium constant for use with pressures	–
m	mass	kg
n	order of the reaction	–
M	ratio of initial product to initial reactant concentrations	–
N	number of CSTRs	–
N	number of moles or molecules	mol
P	pressure	Pa or bar
q	heat flux	W/m^2
r	rate of reaction	various[2]
$\overline{R}$	universal gas constant,[3] 8.314	J/mol·K
$[S]$	molar concentration of substrate	mol/L
$\overline{S}_{DU}$	overall selectivity to desirable product, D	–
SV	space velocity	s^{-1}
t	time or residence time	s
T	absolute temperature	K
$\overline{T}$	mean residence time	s
U	overall heat transfer coefficient	W/m^2·K
v	stoichiometric coefficient[4]	kmol
v	reaction velocity (same as rate, r)	mol/s
V	volume	m^3 or L
X	fractional conversion	–
y	mass fraction	–
Y	instantaneous fractional yield	–
$\overline{Y}$	overall fractional yield	–

Symbols

γ	activity coefficient	–
ε	fractional volume change	–
ξ	extent of reaction	mol
τ	space-time	s

Subscripts

0	initial or zero order
ad	adiabatic

---

[1](1) Although most of the concepts in this chapter are from the *Chemical Engineering* section of the NCEES *FE Reference Handbook* (*NCEES Handbook*), some concepts come from the *Chemistry* section, while others come from the *Chemical Reaction Equilibria* section within *Thermodynamics*. (2) While this chapter tries to present material in a logical and progressive manner, due to the interdependencies within the material and use of some terms, variables, and equations before they are formally introduced, it will probably be necessary to read through this chapter more than once.

---

[2]Units vary depending on the order of the reaction. By convention, the units of activity, fugacity, rate constant, and reaction rate constant are generally disregarded.

[3]The *NCEES Handbook* is inconsistent in the variable used for the universal gas constant. In addition to $\overline{R}$ being used, R is also used.

[4]Although the *NCEES Handbook* uses an italic "vee" for the stoichiometric coefficient, it is more common to use the Greek nu, ν.

c	gaseous reversible reaction
cat	catalyst
D	desired
f	final
p	constant pressure
P	partial pressure
r	heating or cooling medium
t	total
U	undesired

1. CATEGORIES OF REACTIONS

Reaction kinetics, also known as *chemical kinetics*, describes the reaction rates of chemical processes and how the rates are influenced by such factors as reactant concentration, temperature, pressure, activation energy, and catalysts.

By definition, a *chemical reaction* occurs when the chemical composition of a substance changes.[5] For example, when hydrogen peroxide decomposes to water and oxygen, a chemical reaction occurs because the chemical composition of the hydrogen peroxide changes.

$$H_2O_2(l) + \text{light} \rightarrow H_2O(l) + \tfrac{1}{2}O_2(g)$$

There are many ways to categorize chemical reactions. Reactions can be classified according to *mechanism* (e.g., irreversible, reversible, parallel, or in series), *operating mode* (e.g., batch, continuous flow, semi-batch, or semi-flow), *molecularity* (unimolecular, bimolecular, or higher), *order* (integral, nonintegral, hyperbolic), *thermal conditions*[6] (isothermal, adiabatic, or regulated by heat transfer), *phase* (homogeneous or heterogeneous), the presence or absence of a *catalyst*, and *equipment* (stirred tank, packed bed, etc.). Although a reaction cannot be both homogeneous and heterogeneous, a chemical reaction can fall into more than one category.

Some categories of reactions are listed as follows.

- *decomposition reaction*: One or more (typically, one) components decompose in the presence of heat or a catalyst. A decomposition reaction is usually *endothermic* (i.e., requires or absorbs energy).

$$C_2H_6(g) + \text{energy} \rightarrow C_2H_4(g) + H_2(g)$$

[ethane dehydrogenation]

---

[5]Although the condensation of steam can be written as a chemical equation ($H_2O(g) \rightarrow H_2O(l) + \text{energy}$), the chemical structure of the substance does not change. The condensation of steam is a physical change, not a chemical reaction.

[6]All of the reaction kinetics equations in the *NCEES Handbook* invoke the isothermal assumption. This is done purely for the sake of simplifying the analysis. While most reactors can be cooled or heated to provide an isothermal environment, reactions are seldom intrinsically isothermal. In fact, many reactions proceed better if constant temperatures are not imposed on them. There is nothing inherently bad about a nonisothermal environment, other than the mathematics gets more complex. In practice, various methods are used to adjust performance expectations for nonisothermal operation.

- *synthesis reaction*: Multiple components combine to form a different compound. A synthesis reaction is usually exothermic.

$$C_6H_6(l) + 3H_2(g) \rightarrow C_6H_{12}(l) + \text{energy}$$

$$\begin{bmatrix} \text{benzene hydrogenation} \\ \text{to cyclohexane} \end{bmatrix}$$

- *equilibrium reaction*: The main feed component is not completely converted and typically leaves the reactor along with the main product(s). The unreacted feed is usually recovered and recycled.

$$HCCCH_3(g) \rightleftharpoons H_2CCCH_2(g)$$

$$\begin{bmatrix} \text{methyl acetylene in} \\ \text{equilibrium with} \\ \text{propadiene} \end{bmatrix}$$

- *combustion reaction*: A fuel, usually a hydrocarbon, and oxygen combine in an irreversible reaction to produce carbon dioxide and water.

$$C_6H_6 + 7.5O_2 \rightarrow 6CO_2 + 3H_2O + \text{energy}$$

$$\begin{bmatrix} \text{benzene plus oxygen} \\ \text{producing carbon dioxide} \\ \text{and water} \end{bmatrix}$$

- *displacement*, *double displacement*, *disproportionation*, and *metathesis reactions*: Portions of a molecule, either organic or inorganic, change positions with another molecule or portions of another molecule.

$$Mg(s) + 2HCl(aq) \rightarrow MgCl_2(s) + H_2(g)$$

$$\begin{bmatrix} \text{magnesium replacing an} \\ \text{atom of hydrogen in} \\ \text{hydrochloric acid} \\ \text{to form magnesium chloride} \\ \text{and hydrogen gas} \end{bmatrix}$$

Categorization by Phase

Homogeneous reactions occur in a single phase (e.g., two gases combining). Gaseous nitrogen and oxygen combining to form nitrogen oxide ($N_2(g) + O_2(g) \rightarrow 2NO(g)$) is an example of a homogeneous reaction.

Heterogeneous reactions occur in multiple phases (i.e., a gas and a solid combining). For example, the reaction of solid carbon and gaseous oxygen to form carbon dioxide ($C(s) + O_2(g) \rightarrow CO_2(g)$) is a heterogeneous reaction. Reactions in packed fluidized bed reactors, stirred reactors with suspended solid reactants or catalysts, and fixed-bed catalytic reactors are often heterogeneous.

Categorization by Mechanism

Reactions to produce a specific reactant can occur in *series* or in *parallel*.

Reactions in series occur when products are produced by a series of sequential reactions.

$$A \rightarrow B \rightarrow D$$

For example, molecules of propylene assemble to make oligomers of propylene such as hexane, nonene, and dodecene.

$$12C_3H_6 \rightarrow 6C_6H_{12} \rightarrow 4C_9H_{18} \rightarrow 3C_{12}H_{24}$$

Parallel reactions create products from the same reactants simultaneously.

$$A \rightarrow D$$
$$A \rightarrow U$$

For example, cracking propane will form ethylene and propylene simultaneously (as well as hydrogen and carbon).[7]

$$C_3H_8 \rightarrow C_2H_4 + 2H_2 + C$$
$$C_3H_8 \rightarrow C_3H_6 + H_2$$

Categorization by Catalysis

In a *catalytic reaction*, a *catalyst* increases the rate of reaction. Use of a catalyst is often necessary in order to commercialize a process that would otherwise proceed slowly. Catalysts decrease activation energy which increases the rate at which the reaction proceeds. (See Fig. 43.1.) Catalysts can be either homogeneous or heterogeneous. From a processing standpoint, heterogeneous catalysts (e.g., solid catalysts in the presence of liquid or gaseous reactants and products) are more desirable than homogenous reactions (e.g., liquid catalysts with liquid reactants and products) since the catalyst does not leave with the products (i.e., is not lost).

Examples of catalytic reactions include

- hydrogen and acetylene gases passing over a packed bed of supported platinum or palladium at a temperature of about 100°C and a pressure of 30 bars to form ethane ($C_2H_2 + 2H_2 \rightarrow C_2H_6$)

- production of ethylene dichloride by adding (via sparging[8]) chlorine and ethylene gases to a solution of ferric chloride, a soluble catalyst ($Cl_2 + C_2H_4 \rightarrow C_2H_4Cl_2$). The ethylene dichloride is recovered overhead as a vapor and condensed. (Although this catalyst is homogeneous, it does not leave the reactor.)

---

[7]This reaction occurs at high temperature and low pressure in the absence of a catalyst.

[8]*Sparging* is the act of bubbling a gas through a liquid in order to produce some form of mass transfer.

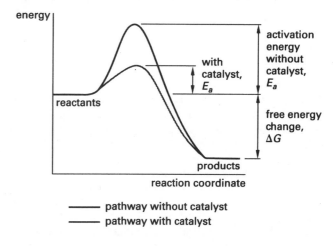

Figure 43.1 *Effect of Catalyst on Activation Energy (E$_a$)*

Categorization by Biocatalysis

Biocatalytic reactions are similar to catalytic reactions, except that the catalysts are enzymes or living organisms. *Bioreactors* (*chemostats*) convert a *substrate*, S, under highly controlled conditions of temperature and media concentration.[9] The combination of the substrate biocatalyst, inert carrier fluids (e.g., water) and, perhaps, the containment vessel is known as a *biosystem*. The performance of most biosystems follow some form of *Monod kinetics* and are often characterized by the *Michaelis-Menten equation*.

Enzymes are proteins that catalyze chemical reactions by lowering the activation energies. Enzymes work by having an *active site* that is carefully designed by nature to bind a particular reactant molecule (known as the *substrate*). Once reaction is complete, the product molecules are released from the enzyme.

Virtually every biological reaction requires an enzyme in order to occur at a significant rate. Each enzyme is specific to a particular reaction. Enzyme kinetics is a complex field, but the basic kinetics of simple enzyme catalysis may be modeled quite simply. The kinetic model for enzyme catalysis explains the fact that the rate depends on the enzyme concentration, [E], even though there is no net change in its concentration over the course of the reaction. Although catalysts are not normally considered to be reactants, enzymes are considered to be *pseudoreactants* because they form an enzyme-substrate complex. The simplest mechanism involves formation of a bound enzyme-substrate complex *ES*, followed by conversion of the complex into the products plus free enzyme.

$$E + S \underset{k_{-1}}{\overset{k_1}{\rightleftharpoons}} ES \overset{k_2}{\rightarrow} E + P$$

---

[9]When a substance is acted upon by a biological organism, or when a substance serves as the "food" for a biological organism, the convention is to refer to the substance as a *substrate*.

With several assumptions, the rate of reaction equation can be developed.[10] The rate of enzyme-catalyzed reaction depends linearly on the enzyme concentration, but in a more complicated way on the substrate concentration, $[S]$. The dependence on the substrate concentration simplifies under certain conditions.

There are three distinct phases of the reaction, between which the order of the reaction can change. (1) During the transient start-up time (prior to steady state), there is a rapid burst of the ES complex being formed. During this phase the empty enzyme sites are being filled. There is no product formation, since a finite amount of time is required for the ES complexes to form. (2) During the steady state phase, the substrate concentration is greater than the enzyme concentration, and the ES complex is formed at the same rate it is decomposed. During this time, the rate of product formation is at v_{max}. (3) When the substrate is depleted to the point where the substrate concentration approaches the enzyme concentration, the production of ES complex and product begins to decrease.

The fermentation of sucrose to carbon dioxide and ethanol $(C_{12}H_{22}O_{11} + H_2O \rightarrow 4CO_2 + 4C_2H_5OH)$ in a bioreactor describes the manufacture of ethanol fuel from biomass. The fermentation enzyme (bacteria, yeast, etc.) is the biocatalyst.

When a reaction involves an enzyme catalyst and the concentration of the enzyme is much smaller than the concentration of the reactant (i.e., the substrate), the concentration of the enzyme will be the controlling factor. Because enzyme-limited reactions do not depend on the substrate concentration, reactions with $[E] << [S]$ are zero-order reactions, and $r = k$. This is because the enzyme is saturated and rate limits the reaction. The rate constant is referred to as the *maximum velocity*, or *v-max*: $r = k = v_{max}$.

When a reaction involves an enzyme catalyst that is available in large quantities compared to the substrate (i.e., $[E] >> [S]$), the concentration of the substrate will be the controlling factor, and the rate will be first-order in $[S]$.

Between the limiting cases of $[E] << [S]$ and $[E] >> [S]$, reaction kinetics are complex. The Michaelis-Menten equation defines the rate of the reaction, r. When only the initial rate is determined by the equation, before the formation of any product, there is no need to consider reverse reactions. K_M is aggregate rate constant that takes into consideration k_1, k_{-1}, and k_2, and that helps

---

[10](1) The reaction that forms the product is assumed to be irreversible, as there is no k_{-2} rate constant. This assumption is made purely to simplify the analysis. (2) The rate equation requires a special application of the steady state approximation. In order for the steady state approximation to be valid, the concentration of the reactive intermediate $([ES])$ must be much less than the concentration of the reactants. In reality, $[ES]$ is not much less than the free enzyme concentration $[E]$. However, because $[E]$ is regenerated in the second step of the mechanism, both $[E]$ and $[ES]$ change much more slowly than $[S]$ and $[P]$, so the steady state approximation is valid.

define the steepness of the Michaelis-Menten curve. (See Fig. 43.2.)

$$r = v = \frac{v_{max}[S]}{K_M + [S]}$$

$$v_0 = \frac{v_{max}[S]}{K_M}$$

$$K_M \equiv \frac{k_2 + k_{-1}}{k_1}$$

$$v_{max} = k_2[E]_0$$

Figure 43.2 *Michaelis-Menten Velocity Curve**

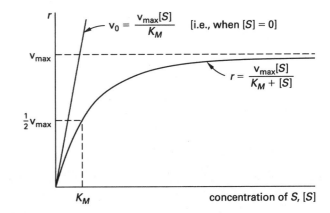

*Despite its similarity to Fig. 43.3, the Michaelis-Menten curve is not a graph of concentration versus time.

Example

At very high concentrations of substrate, S, in an enzyme-catalyzed reaction, what will be the initial rate of reaction?

   (A) $v_0 = v_{max}$

   (B) $v_0 = K_M[S]$

   (C) $v_0 = v_{max}[S]$

   (D) $v_0 = v_{max}[S]/K_M$

Solution

Use the Michaelis-Menten equation. Let $[S]$ become large.

$$\lim_{[S] \to \infty} \left(\frac{v_{max}[S]}{K_M + [S]} \right) = v_{max}$$

When $[S]$ is large, $[S]$ and $K_M + [S]$ cancel. The reaction becomes zero-order, and $r = v = v_{max}$.

The answer is (A).

Categorization by Reaction Direction

Reversible reactions are actually two reactions in one: a *forward reaction* $(A + B \rightarrow C + D)$ and a *reverse reaction* $(C + D \rightarrow A + B)$. As the reactants are forming products, the products are simultaneously forming reactants. This is called a *metathesis reaction*,[11] *double replacement reaction*, or *double displacement reaction*.

Forward reactions favor the forward route. Such reactions are characterized by low free energy and large equilibrium constants. For example, ethylene and butylene react to form two molecules of propylene.

$$C_2H_4 + C_4H_8 \rightarrow 2C_3H_6$$

Reverse reactions favor the reverse route. Such reactions are characterized by high free energy and small equilibrium constants. For example, the two molecules of propylene react to form ethylene and butylene.

$$2C_3H_6 \rightarrow C_2H_4 + C_4H_8$$

Example

How should the reaction shown be described?

$$BaCl_2 + 2AgNO_3 \rightarrow 2AgCl + Ba(NO_3)_2$$

(A) single displacement

(B) double displacement

(C) synthesis

(D) decomposition

Solution

Since the chlorine (Cl) and nitrate (NO_3) anions both switch cations (i.e., switch their positive ions), this is a double displacement reaction. If only one ion of an element moved from one compound to another, the reaction would be a single displacement reaction.

The answer is (B).

2. CHEMICAL REACTION EQUILIBRIA

Reversible reactions are capable of going in either direction and do so to varying degrees (depending on the concentrations and temperature) simultaneously. These reactions are characterized by the simultaneous presence of all reactants and all products. For example, the

chemical equation for the exothermic formation of ammonia from nitrogen and hydrogen is

$$N_2 + 3H_2 \rightleftharpoons 2NH_3 \quad (\Delta H = -92.4 \text{ kJ})$$

At *chemical equilibrium*, reactants and products are both present. Concentrations of the reactants and products do not change after equilibrium is reached.

Not all of a reactant may actually participate in a chemical reaction. There may be thermodynamic reasons, or too much of a reactant may simply be introduced. The *conversion* (*conversion ratio*, *conversion fraction*, etc.) of a reaction is the molar ratio of a reacted component to the component in the feed.

$$\text{conversion fraction} = X = \frac{N_{\text{reacted}}}{N_{\text{feed}}}$$

If the amount of products is limited by the amount of one of the reactants, that reactant is known as the *limiting reactant*.

The reaction may not produce as much product as the stoichiometric reaction equation predicts. There may be thermodynamic reasons, or limiting reactants, or other unexpected chemical losses. The amount of a specific product produced in a reaction is described by its yield. *Yield* (*yield ratio*, *yield fraction*, etc.) is the molar ratio of what is actually produced to the expected (ideal) production.

$$\text{yield fraction} = Y = \frac{N_{\text{obtained}}}{N_{\text{expected}}}$$

The products are not always what are expected, and the reaction may produce undesirable products. There may be limiting reactants (e.g., limited oxygen in a combustion reaction, resulting in the formation of carbon monoxide), or there may be impurities in the feed. The *selectivity* (*selectivity ratio*, *selectivity fraction*) is the molar ratio of desired and undesired products. A selectivity of 1.0 means that there are as many moles of undesired product as there are desired product.

$$\text{selectivity fraction} = S_{DU} = \frac{N_{\text{desired}}}{N_{\text{undesired}}}$$

Equation 43.1: Reversible Reactions

$$\text{moles}_{i,\text{out}} = \text{moles}_{i,\text{in}} + v_i \xi \qquad \textbf{43.1}$$

Description

Equation 43.1 has a logical development. From the conservation of mass, all of a substance that goes into a reaction must either come out or be used. Consider the reaction of hydrogen and oxygen to form water vapor: $2H_2 + O_2 \rightarrow 2H_2O$. The stoichiometric coefficients 2, 1,

---

[11]Older literature may also use the term *double decomposition*, which is a specific form of metathesis reaction in which at least one of the substances does not dissolve in the solvent. In double decomposition, the undissolved substance reacts in its solid form.

and 2 represent the stoichiometric numbers of molecules taking part in the reaction. The coefficients also represent the number of moles and, if the components are gaseous, number of volumes. Since (for reactions with gaseous products and reactants) they are volumes, the stoichiometric coefficients can be given the symbol v. The molar balance is

$$N_{H_2,in} = N_{H_2,out} + v_{H_2}$$

Suppose 3 moles of hydrogen gas and 1 mole of oxygen gas are introduced and the reaction proceeds to completion. Oxygen will be the limiting reactant. 3 moles of hydrogen gas go in, 2 moles will be used up, and 1 mole will come out. The molar balance could be written as

$$N_{H_2,out} = N_{H_2,in} - v_{H_2}$$
$$1 = 3 - 2$$

For reversible reactions, this simple concept is complicated by the fact that the reactions do not necessarily use up all of the limiting reactant. Compared to the stoichiometric quantity, the actual molar fraction utilized or produced is known as the *extent*, ξ. By definition, extent has units of moles, while the stoichiometric constants have no units. Also by definition, the *stoichiometric coefficients*, v_i, are negative for reactants and positive for products.

$$\xi = \frac{\Delta N_{actual}}{\Delta N_{stoichiomeric}} = \frac{\Delta N_{actual}}{v}$$

Suppose 3 moles of hydrogen gas and 1 mole of oxygen gas are combined in a reactor. Further, suppose that some of the hydrogen does not react, and after equilibrium has been established, there are 1.5 moles of hydrogen gas remaining. The hydrogen's extent is

$$\xi_{H_2} = \frac{\Delta N_{H_2,actual}}{v_{H_2}} = \frac{N_{H_2,ending} - N_{H_2,starting}}{v_{H_2}}$$
$$= \frac{1.5 \text{ mol} - 3 \text{ mol}}{-2}$$
$$= 0.75 \text{ mol}$$

Example

A fuel gas is 60% hydrogen (H_2) and 40% carbon monoxide (CO) by volume. 2.5 volumes of fuel gas and 1 volume of oxygen gas (O_2) are combined in a chemical reactor. The reaction goes to completion. For every mole of oxygen in the feed, the effluent consists of 1 mole of water vapor (H_2O), 1 mole of carbon dioxide (CO_2), and $1/2$ mole of hydrogen. Most nearly, what is the hydrogen's extent of reaction?

(A) 0.33

(B) 0.50

(C) 0.67

(D) 0.75

Solution

Volumetric ratios and mole ratios are numerically the same (for reactions with gaseous products and reactants). The actual reaction occurring is

$$2.5(0.6H_2 + 0.4CO) + O_2 \rightarrow H_2O + CO_2 + 0.5H_2$$
$$1.5H_2 + CO + O_2 \rightarrow H_2O + CO_2 + 0.5H_2$$

The limiting component is oxygen (since there is no oxygen in the products). With stoichiometric oxygen, the reaction would have been

$$1.5H_2 + CO + 1.25O_2 \rightarrow 1.5H_2O + CO_2$$

Based on the actual reaction, the hydrogen conversion fraction is

$$X_{H_2} = \frac{N_{H_2,reacted}}{N_{H_2,feed}}$$
$$= \frac{1.5 \text{ mol} - 0.5 \text{ mol}}{1.5 \text{ mol}}$$
$$= 0.667$$

Based on the actual and stoichiometric reactions, the water vapor yield fraction is

$$Y_{H_2O} = \frac{N_{H_2O,obtained}}{N_{H_2O,expected}}$$
$$= \frac{1 \text{ mol}}{1.5 \text{ mol}}$$
$$= 0.667$$

The hydrogen's extent is

$$\xi_{H_2} = \frac{\Delta N_{actual}}{v} = \frac{0.5 - 1.5}{-1.5}$$
$$= 0.667 \quad (0.67)$$

The answer is (C).

3. EQUILIBRIUM CONSTANT

Equation 43.2 Through Eq. 43.6: Equilibrium Constant for Liquids and Solids

$$\hat{a}_i = \frac{\hat{f}_i}{f_i^0} \qquad \textbf{43.2}$$

$$\Delta G^0 = -RT \ln K_a \qquad \textbf{43.3}$$

$$K_a = \frac{(\hat{a}_C^c)(\hat{a}_D^d)}{(\hat{a}_A^a)(\hat{a}_B^b)} = \prod_i (\hat{a}_i)^{v_i} \qquad \textbf{43.4}$$

$$\hat{a}_i = 1 \quad [\text{solids}] \qquad \textbf{43.5}$$

$$\hat{a}_i = x_i \gamma_i \quad [\text{liquids}] \qquad \textbf{43.6}$$

Description

In a reversible reaction, the products and reactants will all be present simultaneously at equilibrium. (See Eq. 43.7.) The chemical *equilibrium constant*, K_a, is derived from an equation of state and is the value that minimizes[12] the Gibbs standard energy change, ΔG^0, in Eq. 43.3.

The equilibrium constant is, essentially, a molar ratio of concentrations of products and reactants. The more the reaction goes towards completion, the larger the equilibrium constant and the greater the concentrations of products. Referring to Eq. 43.7 and denoting the molar concentration of species S as $[S]$, the traditional formula for calculating the equilibrium constant is

$$K_c = \frac{[C]^c[D]^d}{[A]^a[B]^b}$$

The effective concentration of a component (i.e., $[S]$ for component S) in a mixture is known as the *activity*, $\hat{a}$. Then, the equilibrium constant can be written as Eq. 43.4.[13] The equilibrium constant is the ratio of activities of products to reactants.

The concept of activity is related to *fugacity*. (Fugacity is not limited to substances in a gaseous state.) Equation 43.2 shows the relationship of the activity to the component's actual fugacity, $\hat{f}$, and standard state fugacity, f^0, in a mixture.[14]

Special rules are used to define the *activity* (i.e., the concentration) of solid and liquid components. Equation 43.5 ($\hat{a} = 1$) removes the concentrations of any pure solid and/or liquid components from the calculation of the equilibrium constant. Equation 43.6 calculates the activity of a liquid in a mixture as the product of the mole fraction and the experimental (empirical) *activity coefficient*, γ.

Equation 43.7 Through Eq. 43.9: Equilibrium Constant for Mixtures of Ideal Gases

$$aA + bB \rightleftharpoons cC + dD \qquad \text{43.7}$$

$$K_a = K_p = \frac{(p_C^c)(p_D^d)}{(p_A^a)(p_B^b)} = p^{c+d-a-b}\frac{(y_C^c)(y_D^d)}{(y_A^a)(y_B^b)} \qquad \text{43.8}$$

$$\hat{f}_i = y_i p = p_i \qquad \text{43.9}$$

Description

For a reversible reaction with gaseous reactants, an equilibrium can be maintained only if the reactants remain gaseous. (See Eq. 43.7.) If the concentrations are molar, the equilibrium constant for gaseous reversible reactions is designated as K_c. A different equilibrium constant can also be calculated as a ratio of the components' partial pressures, as in Eq. 43.8, in which case, the equilibrium constant is designated as K_p.[15]

The second form of Eq. 43.8 relates the equilibrium constant to the total pressure, $p = \sum p_i$. Equation 43.9 relates the partial pressure to the actual fugacity.[16]

Equation 43.10: Variation of Equilibrium Constant with Temperature

$$\frac{d\ln K}{dT} = \frac{\Delta H^0}{RT^2} \qquad \text{43.10}$$

As Eq. 43.3 shows, the equilibrium constant is related to the equation of state. Since the equation of state contains a temperature term, anything derived from the equation of state depends on temperature. The equilibrium constant's dependency on temperature is given by Eq. 43.10, where ΔH^0 is the standard molar enthalpy of reaction.[17] Although the equilibrium constant varies considerably with temperature, ΔH^0 is reasonably insensitive to the temperature. When integrated, Eq. 43.10 is essentially the same as the Clausius-Clapeyron equation (and the Arrhenius equation for

[12]The *NCEES Handbook* is inconsistent in its nomenclature for the universal gas constant. R in Eq. 43.3 is the same as $\overline{R}$ elsewhere, such as Eq. 43.14.
[13](1) There is no mathematical significance to the parentheses used in Eq. 43.4. It is normal and customary in engineering to designate liquid molecular concentrations and gaseous molar concentrations used in the calculation of (liquid) equilibrium constants with square brackets (e.g., [A]). Parentheses have no such specific meaning. (2) The second form of Eq. 43.4 is an attempt to reduce the first form into the fewest characters. The exponent, v, is the coefficient of the chemical species in the chemical reaction, just as it is in Eq. 43.7. In order to use the second form, the same sign convention for v as is used with Eq. 43.7 must be imposed: v is positive for products and negative for reactants. Although the intent may have been an elegant simplification, nothing except complexity is gained by adding the second form of Eq. 43.4.
[14]The *NCEES Handbook* defines f^0 as the "unit pressure, often 1 bar." This is incorrect. f^0 is the fugacity at the standard state pressure. It is not the standard state pressure itself.

[15](1)The *NCEES Handbook* presents Eq. 43.8 as "$K_a = K_p$," which is incorrect. The numerical values of K_a and K_p are not the same, although they are related. (2) There is no mathematical significance to the parentheses used in Eq. 43.8. To avoid the appearance of using molecular or molar concentrations, it is normal and customary in engineering not to use either square brackets or parentheses with partial pressures in the formulas for gaseous equilibrium constants.
[16]The *NCEES Handbook* is incorrect in writing "$\hat{f}_i = \cdots = p_i$." Just as fugacity for real gases is not the same as the actual pressure, the fugacity of a gaseous component is not the same as its partial pressure. The fugacity is used in place of the partial pressure. If the numerical values were the same, there would be no need for fugacities.
[17]The *NCEES Handbook* refers to ΔH^0 as the "standard enthalpy change of reaction." This, and the similar "standard enthalpy change of combustion," should be avoided.

the temperature dependence of reaction rate constants), and is known as the *van't Hoff equation*.

$$\ln \frac{K_2}{K_1} = \left(\frac{\Delta H^0}{\overline{R}}\right)\left(\frac{T_2 - T_1}{T_1 T_2}\right)$$

4. REACTION RATE

The *rate of reaction* (also known as the *reaction velocity*), r_A, of component A in a chemical reaction is the rate at which the concentration of component A changes. A reactant that decreases in concentration with time has a negative reaction rate (i.e., $r_A < 0$). The rate of a reaction depends primarily on temperature and reactant concentration, although pressure and other factors may also affect the reaction rate. The units of rate are moles per second (i.e., mol/s).

Equation 43.11 through Eq. 43.13: Isothermal, Constant Volume Rate of Reaction

$$-r_A = -\frac{1}{V}\frac{dN_A}{dt} \quad \begin{bmatrix} \text{negative because} \\ A \text{ disappears} \end{bmatrix} \qquad \textbf{43.11}$$

$$-r_A = \frac{-dC_A}{dt} \quad [V \text{ is constant}] \qquad \textbf{43.12}$$

$$-r_A = kf_r(C_A, C_B, \ldots) \qquad \textbf{43.13}$$

Variation

$$-r_A = kf_r(C_A^x, C_B^y, C_C^z, \ldots)$$

Description

Equation 43.11 through Eq. 43.13 describe the rate of consumption of reactant A with time.[18] Equation 43.12, along with $C_A = N_A/V$, is used when volume, V, is constant.[19] Equation 43.13 and its variation are generalized rate equations in which the exponents of each individual component, C_x, give the reaction order for that component. The *reaction rate constant* (or just *rate constant*), k, accounts for temperature effects and is described by the Arrhenius equation. (See Sec. 43.5.) The variation equation is applicable to reactions involving reactants that combine in multiple molecules. No exponents are needed when one reactant molecule forms one product molecule in the absence of other molecules.

---

[18](1) In the *NCEES Handbook*, Eq. 43.11 is accompanied by the explanation, "negative because A disappears." However, there is no need for the two negative signs in Eq. 43.11. For a species that is decreasing, the reaction rate is inherently negative because the amount of reactant A is decreasing, and dN_A/dt is mathematically negative. Clearly, the two negative signs cancel and do nothing mathematically. (2) The volume, V, in Eq. 43.11 is intended to be a variable of time, as opposed to the implicit volume in Eq. 43.12, which is intended to be constant. The representation of volume in Eq. 43.11 should be $V(t)$.

[19]Almost all of the equations in the *NCEES Handbook* relate to constant volume reactors. The exception is covered in Sec. 43.12.

Figure 43.3 shows the reaction rate influence factors such as temperature, concentration, and size of the pieces.

Figure 43.3 *Reaction Rate Influence Factors*

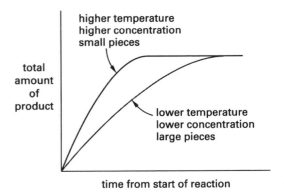

Example

A reaction occurs as shown.

$$S \to P$$

What is the most logical form of the rate equation?

(A) $r = k$

(B) $r = k[S]$

(C) $r = k[P]$

(D) $r = k[S][P]$

Solution

Rate equations do not usually include the concentrations of the products. A logical first assumption would be that the reaction is first order, in which case, the reaction rate would be proportional to the reactant concentration.

The answer is (B).

5. ARRHENIUS EQUATION

Equation 43.14 and Eq. 43.15: Arrhenius Equation[20]

$$k = Ae^{-E_a/\overline{R}T} \qquad \textbf{43.14}$$

$$E_a = \frac{RT_1T_2}{(T_1 - T_2)}\ln\left(\frac{k_1}{k_2}\right) \qquad \textbf{43.15}$$

---

[20](1) The *NCEES Handbook* is inconsistent in its nomenclature for the universal gas constant. Equation 43.14 uses $\overline{R}$ for the universal gas constant, but R is used for the same constant in Eq. 43.15. Both equations should use $\overline{R}$. (2) There is no mathematical significance to the parentheses in the denominator of Eq. 43.15.

Description

The *Arrhenius equation* (see Eq. 43.14) provides a method of calculating the rate constant, k, from relates the *activation energy*, E_a, and the temperature, T. If the *pre-exponential factor*[21] (*collision frequency factor*), A, is unknown, the activation energy can be found from the rate constants at two different temperatures, as shown in Eq. 43.15.

The rate constant usually increases exponentially with temperature. A plot of $\ln k$ versus $1/T$ is a straight line with slope $-E_a/\overline{R}$ and intercept $\ln A$. Figure 43.4 shows the Arrhenius equation plotted as a straight line.

Example

A chemical reaction has rate constants of 50 L/mol·s at 0°C and 100 L/mol·s at 10°C. Most nearly, what is the activation energy?

(A) −45 kJ/mol

(B) −7.7 kJ/mol

(C) 7.7 kJ/mol

(D) 45 kJ/mol

Solution

Use Eq. 43.15.

$$E_A = \frac{RT_1 T_2}{(T_1 - T_2)} \ln\left(\frac{k_1}{k_2}\right)$$

$$= \frac{\left(8.314 \, \dfrac{\text{J}}{\text{mol·K}}\right)(0°C + 273°)(10°C + 273°)}{\left((0°C + 273°) - (10°C + 273°)\right)\left(1000 \, \dfrac{\text{J}}{\text{kJ}}\right)}$$

$$\times \ln\left(\frac{50 \, \dfrac{\text{L}}{\text{mol·s}}}{100 \, \dfrac{\text{L}}{\text{mol·s}}}\right)$$

$$= 44.52 \text{ kJ/mol} \quad (45 \text{ kJ/mol})$$

The answer is (D).

6. REACTION ORDER

Equation 43.16 and Eq. 43.17: Reaction Order

$$-r_A = kC_A^x C_B^y \qquad 43.16$$

$$n = x + y \qquad 43.17$$

---

[21](1) It is unfortunate that the *NCEES Handbook* uses the symbol A for the pre-exponential factor, since A is also used throughout to represent a reactant. (2) The "pre-exponential" factor is just that—a coefficient that precedes the exponential. There is nothing magical about it. As with all exponentials, their coefficients represent the value of the function when the independent variable is zero.

Figure 43.4 *Arrhenius Equation*

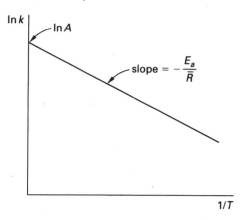

Variation

$$-r_A = kC_A^x$$

Description

For *elementary reactions*, the reaction order is directly related to the *stoichiometry* of the reaction: the overall reaction order is equal to the stoichiometric ratio of reactants to products, and the reaction order with respect to each reactant is equal to the *stoichiometric coefficient* of each individual reactant. For ideal gases, the ratio can also be calculated from ratios of volumes and ratios of partial pressures.

Equation 43.16 gives the reaction rate in terms of the number of moles of A consumed. The reaction is x order with respect to reactant A and y order with respect to reactant B. The overall order, n, is the sum of the exponents and is given by Eq. 43.17. The variation is applicable to reactions in which multiple reactant molecules combine into the product molecules.

Reaction order is empirical and does not need to be a whole number; it is based on the observed dependence of the reaction rate on reactant concentration. In a *first-order reaction*, the reaction rate varies proportionally with the concentration of reactant A, as shown in the rate equation $-r_A = kC_A$. If the reaction is *second order*, the rate equation (i.e., $-r_A = kC_A^2$) shows that the reaction rate varies proportionally with the *square* of the concentration of reactant A. A *zero-order reaction* (i.e., $-r_A = k_0$) is not dependent on reactant concentration.

7. PROCESS CONCEPTS

Reactors often form the central part of a chemical processing plant or refinery. In general terms, many plants consist of a reactor at the front end followed by separation and recovery processes. As such, the reactors often represent about 25–50% of the capital cost for each plant.

The reactor's *selectivity* and *conversion* are key variables in the selection of downstream processing. *Conversion*, X, is the amount of limiting reactant converted to product divided by its stoichiometric limit; this value is expressed as a decimal between zero and one. The definition of *selectivity*, *S*, can vary depending on the reference being consulted. It is either defined as the ratio of desired products to undesired products (and can vary between zero and infinity) or the ratio of desired products to *all* products (and can vary between zero and one).

Figure 43.5 is a general flowsheet detailing the key components of a chemical plant. *Fresh feed* is sent to a reactor, where it is converted according to the stoichiometry of the reaction. From the reactor, the effluent is sent to separations and recovery. In *separations* (the first tower), light ends and light by-products are stripped from the main product stream. In *recovery* (the second tower), a purified product exits overhead, and unreacted feed is recycled to the reactor. A *purge stream* is often necessary to prevent pressure and/or material build-up in the system.

Figure 43.5 *Generalized Chemical Processing Plant*

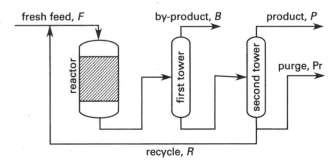

Figure 43.5 defines the fresh feed, *F*, the pure by-product, *B*, the desired pure product, *P*, the recycled (unreacted) feed, *R*, and the purge, Pr, all of which are flow rates typically measured in moles per unit time (e.g., mol/h). The process conversion, selectivity, and purge ratio are

$$\text{conversion} = X = 1 - \frac{\text{Pr} + R}{F + R}$$

$$\text{selectivity} = S = \frac{P}{B} \text{ or } \frac{P}{B + P}$$

$$\text{purge ratio} = \frac{\text{Pr}}{\text{Pr} + R}$$

Equation 43.18: Fractional Conversion of Component A in a Constant Volume Batch Reactor

$$X_A = (C_{A0} - C_A)/C_{A0} \quad [V \text{ is constant}] \qquad 43.18$$

Variation

$$X_A = \frac{N_{A0} - N_A}{N_{A0}} = 1 - \frac{C_A}{C_{A0}}$$

Description

Equation 43.18 defines the fractional *conversion*, X_A, of component *A* in terms of the initial and final concentrations, C_{A0} and C_A, respectively, for a constant volume reactor.[22]

Example

A chemical reaction occurs in a batch reactor and achieves an 80% conversion of component *A* with no change in volume. The initial concentration of component *A* is 0.25 mol/L. Most nearly, what is the final concentration of component *A*?

(A) 0.05 mol/L

(B) 0.10 mol/L

(C) 0.20 mol/L

(D) 1.25 mol/L

Solution

Using Eq. 43.18, the final concentration of component A is

$$X_A = \frac{C_{A0} - C_A}{C_{A0}}$$

$$C_A = C_{A0}(1 - X_A)$$

$$= \left(0.25 \ \frac{\text{mol}}{\text{L}}\right)(1 - 0.80)$$

$$= 0.05 \ \text{mol/L}$$

The answer is (A).

8. REACTOR TYPES

Batch Reactors

The *batch reactor* is the most basic reactor. Batch reactors are heavily utilized in the food and pharmaceutical industries. Batch reactor equipment is versatile and may be used repeatedly for a variety of different reactions. Batch reactor cycles usually involve loading, blending, reacting, unloading, cleaning, preparing, and reloading. Figure 43.6 shows a simplified schematic of a stirred batch reactor.

By definition, a batch reactor has no continuous material flow either into or out of the reactor.[23] In a batch process, the reactor is loaded with ingredients, charged with a catalyst (if necessary), heated (if necessary), and stirred. The reaction mixture remains in the reactor as the reaction progresses for a fixed period of time (i.e., the *residence time*. With a batch reactor, the reactant concentration exponentially decreases with time.

---

[22]The *NCEES Handbook* refers to a "well-mixed, constant-volume batch reactor." The reacting volume that must be constant in order to implement Eq. 43.18 is the volume of the material within the batch reactor. The assumption of constant volume assumes the reacting mixture has a constant density.

[23]However, a batch reactor may have material in it when the reaction starts or ends.

Chem. Reaction
Engineering

Figure 43.6 *Stirred Batch Reactor*

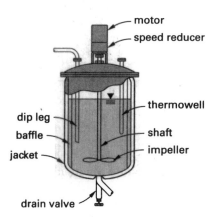

Conversion approaches 100% asymptotically. Conversion will be total if the reactor is allowed to run long enough and the rate is fast enough. However, usually, the reaction will be allowed to proceed for a duration equal to the *optimum cycle time*. Due to the *law of diminishing returns*, the optimum cycle time is usually determined economically.[24]

In a batch reactor, depending on the conditions, the amount of reactant A can be written in terms of number of molecules, N_A, instantaneous volume, $V(t)$, number of moles, n_A, concentration, C_A, and/or conversion, X_A. When the reactant volume is constant, the number of moles of reactant can be replaced with concentration, which is more easily measured than mole quantity.

$$C_A = \frac{N_A}{V} = \frac{N_{A0}(1 - X_A)}{V} = C_{A0}(1 - X_A)$$

[constant volume]

Equation 43.19 and Eq. 43.20: Well-Mixed, Isothermal, Constant-Volume Batch Reactor

$$-r_A = -dC_A/dt \qquad \textbf{43.19}$$

$$t = -C_{A0} \int_0^{X_A} dX_A/(-r_A) \qquad \textbf{43.20}$$

Variations

$$-r_A = \frac{-dN_A}{V(t)\,dt} = \frac{-N_{A0}\,d(1 - X_A)}{V(t)\,dt}$$

$$t = N_{A0} \int_{X_{A_0}}^{X_{A_f}} \frac{dX_A}{V(-r_A)}$$

---

[24]*Law of diminishing returns:* Each consecutive incremental addition of some resource (e.g., reactant) will result in an output improvement (product) that progressively decreases. Basically, as you continue to put in, you obtain less and less improvement.

Description

Equation 43.19 is the basic definition of the reaction rate for component A. Equation 43.20 determines the time necessary to achieve a given conversion, X_A, in a well mixed, constant-volume reactor.[25] C_{A0} is the initial concentration of reactant A.[26]

Differentiating the second variation equation with respect to reaction time produces an equation for the conversion rate.

$$\frac{dX_A}{dt} = \left(\frac{V}{N_{A0}}\right)(-r_A)$$

This equation can be integrated to solve for residence time, as shown in the variation of Eq. 43.20. The results of this integration are shown in Table 43.1 for several reaction orders.[27]

Table 43.1 *Performance Equations for Various Reaction Orders*

reaction order	reaction	rate equation	integrated form
0	$A \to R$	$-r_A = k$	$kt = C_{A_0} - C_A$ $\left[t < \dfrac{C_{A_0}}{k} \right]$
1	$A \to R$	$-r_A = kC_A$	$kt = \ln \dfrac{C_{A_0}}{C_A}$
2	$2A \to R$	$-r_A = kC_A^2$	$kt = \dfrac{1}{C_A} - \dfrac{1}{C_{A_0}}$
3	$3A \to R$	$-r_A = kC_A^3$	$kt = \dfrac{1}{2}\left(\dfrac{1}{C_A^2} - \dfrac{1}{C_{A_0}^2} \right)$

Example

A zero-order chemical reaction, $A \to R$, occurs in a batch reactor. The initial concentration of the reactant A is 1.2 mol/L, and the rate of reaction as a function of fractional conversion is 0.25 mol/L·min. Most nearly, how much time is required to achieve 75% conversion?

(A) 0.51 min

(B) 3.6 min

(C) 56 min

(D) 92 min

---

[25]The *NCEES Handbook* incorrectly places a minus sign before C_{A0} and r_A in Eq. 43.20. If the reaction rate is positive, then the time will be positive.

[26]The *NCEES Handbook* is inconsistent in how it represents "initial time," using such forms as "C_{A0}," "C_{A_0}," "$C_{A=0}$," and "C_{A_o}."

[27]Table 43.1 is not provided in the *NCEES Handbook*.

Solution

Using Eq. 43.20, the time, t, required to achieve 75% conversion is

$$t = -C_{A0} \int_0^{X_A} \frac{dX_A}{-r_A}$$

$$= -1.2 \, \frac{\text{mol}}{\text{L}} \int_0^{0.75} \frac{dX_A}{-0.25 \, \frac{\text{mol}}{\text{L·min}}}$$

$$= (4.8 \text{ min}) X_A \big|_0^{0.75}$$

$$= (4.8 \text{ min})(0.75 - 0)$$

$$= 3.6 \text{ min}$$

The answer is (B).

Example

The initial rate data obtained for the chemical reaction $B \rightarrow D$ are

C_B (mol/L)	$-r_B$ (mol/s)
0.125	1.88×10^{-4}
0.250	3.76×10^{-4}

What is the order exponent, x, of the rate equation?

(A) −0.50

(B) 0.50

(C) 1.0

(D) 2.0

Solution

The rate equation has the form of $-r_B = kC_B^x$. Since doubling the reactant concentration results in doubling the reaction rate, the reaction rate is linearly proportional to the reactant concentration. The order exponent is 1. Prove this analytically.

Write the rate equation twice, and take the ratio of the reaction rates and the concentrations.

$$-r_{B_1} = kC_{B_1}^x$$

$$-r_{B_2} = kC_{B_2}^x$$

$$\frac{r_{B_1}}{r_{B_2}} = \left(\frac{C_{B_1}}{C_{B_2}} \right)^x$$

Solve for x.

$$x = \frac{\log \dfrac{r_{B_1}}{r_{B_2}}}{\log \dfrac{C_{B_1}}{C_{B_2}}} = \frac{\log \dfrac{1.88 \times 10^{-4} \, \frac{\text{mol}}{\text{s}}}{3.76 \times 10^{-4} \, \frac{\text{mol}}{\text{s}}}}{\log \dfrac{0.125 \, \frac{\text{mol}}{\text{L}}}{0.250 \, \frac{\text{mol}}{\text{L}}}} = 1.0$$

The answer is (C).

Example

The initial rate data obtained for the chemical reaction $A + B \rightarrow C$ are

C_A (mol/L)	C_B (mol/L)	$-r_C$ (mol/L·s)
0.1	0.2	0.300×10^{-4}
0.2	0.2	0.424×10^{-4}
0.2	0.3	0.636×10^{-4}

The rate equation is $r_C = kC_A^x C_B^y$. What are the order exponents (x, y) of the rate equation?

(A) (0.5, 1.0)

(B) (1.0, 0.5)

(C) (1.0, 1.0)

(D) (0.5, 0.5)

Solution

Use the first and second data points, where $(C_B)_1 = (C_B)_2$, to determine the order exponent x.

$$\frac{(r_C)_1}{(r_C)_2} = \frac{k(C_A^x)_1 (C_B^y)_1}{k(C_A^x)_2 (C_B^y)_2}$$

$$= \left(\frac{(C_A)_1}{(C_A)_2} \right)^x \left(\frac{(C_B)_1}{(C_B)_2} \right)^y$$

$$= \left(\frac{(C_A)_1}{(C_A)_2} \right)^x$$

Take logarithms of both sides.

$$\log \left(\frac{(r_C)_1}{(r_C)_2} \right) = \log \left(\frac{(C_A)_1}{(C_A)_2} \right)^x$$

$$\log \frac{(r_C)_1}{(r_C)_2} = x \log \frac{(C_A)_1}{(C_A)_2}$$

Solve for x.

$$x = \frac{\log \dfrac{(r_C)_1}{(r_C)_2}}{\log \dfrac{(C_A)_1}{(C_A)_2}}$$

$$= \frac{\log \dfrac{0.300 \times 10^{-4} \, \frac{\text{mol}}{\text{L·s}}}{0.424 \times 10^{-4} \, \frac{\text{mol}}{\text{L·s}}}}{\log \dfrac{0.1 \, \frac{\text{mol}}{\text{L}}}{0.2 \, \frac{\text{mol}}{\text{L}}}}$$

$$= 0.5$$

Use the second and third data points, where $(C_A)_2 = (C_A)_3$, to determine the order exponent y.

$$\frac{(r_C)_2}{(r_C)_3} = \frac{k(C_A^x)_2 (C_B^y)_2}{k(C_A^x)_3 (C_B^y)_3}$$

$$= \left(\frac{(C_A)_2}{(C_A)_3}\right)^x \left(\frac{(C_B)_2}{(C_B)_3}\right)^y$$

$$= \left(\frac{(C_A)_2}{(C_A)_3}\right)^y$$

Take logarithms of both sides and solve for y.

$$y = \frac{\log \dfrac{(r_C)_2}{(r_C)_3}}{\log \dfrac{(C_B)_2}{(C_B)_3}}$$

$$= \frac{\log \dfrac{0.424 \times 10^{-4} \; \frac{\text{mol}}{\text{L·s}}}{0.636 \times 10^{-4} \; \frac{\text{mol}}{\text{L·s}}}}{\log \dfrac{0.2 \; \frac{\text{mol}}{\text{L}}}{0.3 \; \frac{\text{mol}}{\text{L}}}}$$

$$= 1.0$$

The answer is (A).

Stirred Tank Reactors (STRs)

Semibatch (semiflow) reactors are typically called *stirred tank reactors* (STRs) and operate much like batch reactors. These reactors are characterized by a single stirred tank with equipment similar to that of a batch reactor, depending on the specifics of the chemistry (e.g., required catalyst concentration, mixing, refluxing, heat transfer, pressure control). Unlike batch reactors, reactants can be added to and/or product removed from STRs during operation. If run long enough and without process modification, these reactors can become steady state and are described as *continuously stirred tank reactors* (CSTRs).

Semibatch reactors have better selectivity and reactor control as well as higher conversion rates than batch reactors. Control is very important (and apparent) in reactions that have significant heat effects, either exothermic or endothermic. Exothermic reactions release heat and can result in safety issues; however, if the flows are managed properly, along with the required heating or cooling, control of the reactor becomes more viable.

Historically, batch and semibatch reactors have been used in small scale production and are good choices in biotechnology, pharmaceutical, and specialty chemical production. They offer flexibility in operation and low capital costs. Batch systems have operating costs that do not occur in continuous production, such as filling, draining, emptying, and cleaning (which can include steaming or sterilizing). There is also the possibility of contamination when different batches are processed. Consequently, economic analyses must be performed to determine which type of reactor is better suited.

The transient (i.e., nonsteady) performance of a generalized, first-order decomposition reaction occurring in a semibatch reactor is described by the following differential equation.

$$\text{input} - \text{output} - \text{disappearance by reaction}$$
$$= \text{accumulation}$$

$$F_{\text{in}} C_{A,\text{in}} - F_{\text{out}} C_A - k_1 C_A V = V \frac{dC_A}{dt}$$

F_{in} is the molar flow rate into the reactor. $C_{A,\text{in}}$ is the concentration of species A in the inlet stream. F_{out} is the molar flow rate out of the reactor. C_A is the concentration of component A in the reactor or in the outlet stream. k_1 is the first-order reaction rate constant. V is the volume of the reactor. t is time. All of the terms in the mass balance are potentially functions of time.

Some reaction products can inhibit the reaction; this is especially true in biological systems. To reduce the inhibition and favor the main, forward reaction, the products should be removed. Otherwise, either the reverse reaction or a secondary reaction may occur in a semibatch reactor. Removal is achieved by managing the exit flow from the reactor and reducing the concentration of products and by-products.

Continuously Stirred Tank Reactors (CSTRs)

A *continuously stirred tank reactor* (CSTR) consists of a stirred tank with (at least) one continuous material input stream and one continuous material output stream. CSTRs are common in the chemical processing industry, since they are capable of maintaining continuous output without the loading and unloading of material required by batch reactors. CSTRs are rarely used with gases, with the exception of dense phase fluidized bed reactors that are used in some refinery cracking operations.

If the reaction rates are sufficiently fast, only one CSTR may be needed to meet production demands. However, if the reaction rates are slow, several reactors in series may be required to speed production. If there are a large number of CSTRs in series, reaction performance begins to resemble that of a plug flow reactor.

A first-order decomposition reaction occurring in a CSTR has no accumulation, and so, has the steady-state mass balance shown.

$$F_{\text{in}} C_{A,\text{in}} - F_{\text{out}} C_A - k_1 C_A V = 0$$

F_{in} is the molar flow rate into the reactor. $C_{A,in}$ is the concentration of reactant A in the inlet stream. F_{out} is the molar flow rate out of the reactor. C_A is the concentration of reactant A in the reactor or the outlet stream. k_1 is the first-order reaction rate constant. V is the volume of the CSTR.

The modeling of a CSTR assumes total mixing and uniform conditions and properties throughout the vessel. As a result, uniform concentration and temperature can be assumed throughout the reactor. Another intrinsic assumption of the CSTR is steady-state operation. One basic definition of *steady state* is that conditions do not vary with time. In a steady-state operation, molar flow rates into and out of a CSTR are the same (i.e., $F_{in} = F_{out} = F$). The steady-state mass balance can then be solved algebraically to determine the concentration of reactant A in the CSTR.

$$FC_{A,in} - FC_A - k_1 C_A V = 0$$
$$FC_{A,in} = C_A(F + k_1 V)$$
$$C_A = \frac{C_{A,in}}{1 + k_1\left(\frac{V}{F}\right)}$$
$$= \frac{C_{A,in}}{1 + k_1 \tau}$$

Given the almost perfect mixing of a CSTR, it is important that the *residence time* be at least one order of magnitude higher than the *mixing time* (i.e., the time required to homogenize a material—a value typically measured experimentally or calculated from known correlations).

Plug Flow Reactors (PFRs)

Plug flow is a simplified flow model that assumes no axial mixing of components, yielding a uniform concentration at any cross section. *Tubular-flow reactors*, or *plug flow reactors*, (PFR) with highly turbulent flow are often assumed to operate in plug-flow mode. Fluid moves essentially in single-file batches through the reactor. The differential mass balance is expressed in terms of molar flow rate, F_A, as

$$F_A - (F_A + dF_A) = (-r_A)dV$$

From the definition of conversion for a flow reactor, the mass balance is

$$F_A = F_{A0}(1 - X_A)$$

Therefore, the mass balance can also be written as

$$\frac{dX_A}{d\left(\dfrac{V}{F_{A0}}\right)} = -r_A$$

Integration yields

$$\frac{V}{F_{A0}} = \int_{A0}^{X_{Af}} \frac{dX_A}{-r_A}$$

Figure 43.7 shows a serpentine PFR with vertical flow.

Figure 43.7 *Serpentine Plug Flow Reactor (vertical flow)*

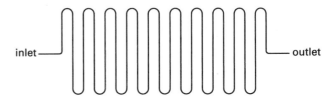

inlet — outlet

Equation 43.21: Space-Time and Space-Velocity

$$SV = 1/\tau \qquad \text{43.21}$$

Variation

$$\tau = \frac{1}{SV} = \frac{C_{A0} V}{F_{A0}} = \frac{V}{\dot{V}_0}$$

Description

Unlike batch and semibatch reactors, which are characterized by residence time, flow reactors are characterized by space-time, τ, and space-velocity, SV. *Space-time* is the time required to process one reactor volume of feed (i.e., how long, on average, a small volume of fluid will remain in the reactor). Space-time is the reactor volume divided by the inlet volumetric feed rate, $V/\dot{V}_0$. *Space-velocity* is the number of reactor volumes of feed that can be processed in a unit of time. Space-velocity is the reciprocal of space-time.

Mean residence time in a plug flow reactor is

$$\overline{T} = \frac{V}{\dot{V}_0} = \tau$$

If the PFR is operating at constant density, and if the inlet and outlet flow rates are the same, then the mean residence time and the space time will be the same.

The performance equations for PFRs with constant density are the same as those for constant volume batch and semibatch reactors if residence time is replaced with space-time. Compare Eq. 43.20 (for batch reactors) with Eq. 43.22.

Equation 43.22: Isothermal Plug Flow Reactor

$$\tau = \frac{C_{A0} V_{\text{PFR}}}{F_{A0}} = C_{A0} \int_0^{X_A} \frac{dX_A}{(-r_A)} \qquad \text{43.22}$$

Description

Equation 43.22 defines the space-time, τ, for a plug flow reactor. F_{A0} is the feed rate of reactant A per unit time. Equation 43.22 can be used to determine the PFR reactor volume needed to achieve any specified conversion fraction.

Example

A plug flow reactor is being designed to process a feed volumetric flow rate of a 2.5 L/s with a component A molar concentration of 0.2 mol/L. The reaction, $A \rightarrow B$, is first-order with a rate constant of 0.05 s^{-1}. What is the required PFR volume to reach 80% conversion?

(A) 20 L

(B) 81 L

(C) 150 L

(D) 250 L

Solution

Using Eq. 43.22, the reactor volume is

$$\tau = C_{A0} \int_0^{X_A} \frac{dX_A}{(-r_A)}$$

$$\frac{C_{A0} V}{F_{A0}} = C_{A0} \int_0^{X_A} \frac{dX_A}{(-r_A)}$$

Since $-r_A = kC_A = kC_{A0}(1 - X_A)$,

$$V = F_{A0} \int_0^{0.8} \frac{dX_A}{kC_{A0}(1 - X_A)}$$

$$= \frac{\dot{V}_0}{k} \int_0^{0.8} \frac{dX_A}{1 - X_A}$$

$$= \left(\frac{2.5 \frac{\text{L}}{\text{s}}}{0.05 \text{ s}^{-1}} \right) \ln \left(\frac{1 - 0.0}{1 - 0.8} \right)$$

$$= 80.5 \text{ L} \quad (81 \text{ L})$$

The answer is (B).

9. BATCH REACTOR OPERATION— IRREVERSIBLE REACTIONS

A batch reactor can be used with reactions of various orders: 0, 1, 2, 3, etc., including reactions that shift orders. The reactions can be irreversible or reversible.[28]

Equation 43.23 Through Eq. 43.27: Zero-Order, Irreversible, Isothermal, Constant Volume Batch Reactor

$$-r_A = kC_A^0 = k(1) \qquad \text{43.23}$$

$$-dC_A/dt = k \qquad \text{43.24}$$

$$C_A = C_{A0} - kt \qquad \text{43.25}$$

$$dX_A/dt = k/C_{A0} \qquad \text{43.26}$$

$$C_{A0} X_A = kt \qquad \text{43.27}$$

Variation

$$X_A = \frac{kt}{C_{A0}}$$

Description

A *zero-order reaction* is one in which the rate is independent of the reactants' concentrations. The rate of the reaction (representing the rate of reactant disappearance and conversion) is equal to zeroth power of the concentration (hence, the zero-order categorization) times the rate constant, k, of that reaction. Equation 43.23 through Eq. 43.27 are applicable to zero-order irreversible reactions occurring in batch reactors that are operating at constant temperature and volume. Equation 43.23 is the rate equation for a zero-order reaction.[29] Equation 43.24 and Eq. 43.25 are the design equations in terms of concentration, while Eq. 43.26 and Eq. 43.27 are the design equations in terms of the fractional conversion. As Eq. 43.24 indicates, the rate is constant. As Eq. 43.25 indicates, the concentration decreases linearly. Figure 43.8 illustrates the rate of a zero-order reaction as a function of concentration. Note that $C_A = 0$ for $t \geq C_{A0}/k$.

---

[28]In thermodynamics, an irreversible process is often one that loses heat or other energy. Some irreversibilities can be prevented by insulating the processes, and thermodynamic irreversibilities can be locally reversed simply by replacing the lost energy. Chemical irreversibilities are more complex. They generally result in chemical changes that require more than the addition of the lost energy to reverse.

[29]In the *NCEES Handbook* Eq. 43.23, the ambiguous term "$k(1)$" is meant to signify k times 1, since concentration to the zeroth power is 1. It does not represent "k of one."

Chem. Reaction
Engineering

Figure 43.8 *Zero-Order Reaction Kinetics*

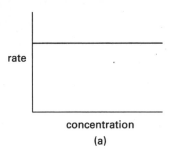

(a)

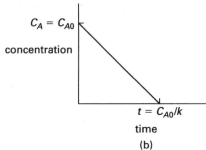

(b)

Example

A zero-order chemical reaction is conducted in an isothermal batch reactor at constant volume. The initial concentration of the reactant is 5 mol/L, and the rate constant is 1.25 mol/L·h. Most nearly, what is the fractional conversion after three hours?

(A) 0.27

(B) 0.49

(C) 0.75

(D) 0.98

Solution

Using Eq. 43.26 and Eq. 43.27, the fractional conversion, X_A, after three hours is

$$kt = C_{A0}X_A$$

$$X_A = \frac{kt}{C_{A0}}$$

$$= \frac{\left(1.25 \ \frac{mol}{L \cdot h}\right)(3 \ h)}{5 \ \frac{mol}{L}}$$

$$= 0.75$$

The answer is (C).

Equation 43.28 Through Eq. 43.32: First-Order, Irreversible, Isothermal, Constant Volume Batch Reactor

$-r_A = kC_A$	*43.28*
$-dC_A/dt = kC_A$	*43.29*
$\ln(C_A/C_{A0}) = -kt$	*43.30*
$dX_A/dt = k(1 - X_A)$	*43.31*
$\ln(1 - X_A) = -kt$	*43.32*

Variation

$$X_A = 1 - e^{-kt}$$

Description

A *first-order reaction* is one in which the rate increases linearly with concentration. The rate of the reaction (representing the rate of reactant disappearance and conversion) is a scalar multiple of the first power of the concentration (hence, the first order categorization), with the scalar being equal to the rate constant, k, of that reaction. Equation 43.28 through Eq. 43.32 are applicable to first-order irreversible reactions occurring in batch reactors that are operating at constant temperature and volume. Equation 43.28 is the rate law for a first-order reaction. Figure 43.9 illustrates the rate of a first-order reaction as a function of concentration as well as the concentration as a function of time. Equation 43.29 and Eq. 43.30 are the design equations in terms of concentration, while Eq. 43.31 and Eq. 43.32 are the design equations in terms of the fractional conversion. As Eq. 43.30 implies and Fig. 43.10 shows, the concentration decreases exponentially.

Figure 43.9 *First-Order Reaction Kinetics*

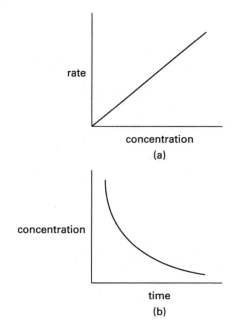

Figure 43.10 Second-Order Reaction Kinetics

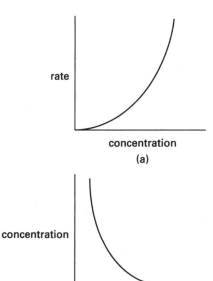

$$-dC_A/dt = kC_A^2 \qquad \textbf{43.34}$$

$$1/C_A - 1/C_{A0} = kt \qquad \textbf{43.35}$$

$$dX_A/dt = kC_{A0}(1 - X_A)^2 \qquad \textbf{43.36}$$

$$X_A/[C_{A0}(1 - X_A)] = kt \qquad \textbf{43.37}$$

Description

A *second-order reaction* is one in which the rate increases with concentration squared. The rate of the reaction (representing the rate of reactant disappearance and conversion) is a scalar multiple of the square (hence, the second order categorization) of the concentration, with the scalar being equal to the rate constant, k, of that reaction. Equation 43.33 through Eq. 43.37 are applicable to second-order irreversible reactions occurring in batch reactors that are operating at constant temperature and volume. Equation 43.33 is the rate law for a second-order reaction. Equation 43.34 and Eq. 43.35 are the design equations in terms of concentration, while Eq. 43.36 and Eq. 43.37 are the design equations in terms of the fractional conversion. As Eq. 43.35 implies and Fig. 43.10 shows, the concentration decreases exponentially, but at a rate greater than a first-order reaction. Figure 43.10 illustrates the rate of a second-order reaction as a function of concentration.

Example

A first-order chemical reaction occurs in an isothermal batch reactor at constant volume. The initial concentration of the reactant is 5 mol/L, and the rate constant is 1.25 h^{-1}. Most nearly, what is the fractional conversion after three hours?

(A) 0.29

(B) 0.52

(C) 0.75

(D) 0.98

Example

A second-order chemical reaction occurs in an isothermal batch reactor at constant volume. The initial concentration of the reactant is 5 mol/L, and the rate constant is 1.25 L/mol·h. Most nearly, what is the final concentration after three hours?

(A) 0.25 mol/L

(B) 0.50 mol/L

(C) 0.75 mol/L

(D) 0.98 mol/L

Solution

Using Eq. 43.32, the fractional conversion, X_A, after three hours is

$$-kt = \ln(1 - X_A)$$
$$X_A = 1 - e^{-kt}$$
$$= 1 - e^{(-1.25\ 1/h)(3\ h)}$$
$$= 0.977 \quad (0.98)$$

The answer is (D).

Solution

Using Eq. 43.35, the reactant concentration after three hours is

$$kt = \frac{1}{C_A} - \frac{1}{C_{A0}}$$

$$C_A = \frac{1}{kt + \dfrac{1}{C_{A0}}}$$

$$= \frac{1}{\left(1.25\ \dfrac{L}{mol \cdot h}\right)(3\ h) + \dfrac{1}{5\ \dfrac{mol}{L}}}$$

$$= 0.253\ mol/L \quad (0.25\ mol/L)$$

The answer is (A).

Equation 43.33 Through Eq. 43.37: Second-Order, Irreversible, Isothermal, Constant Volume Batch Reactor

$$-r_A = kC_A^2 \qquad \textbf{43.33}$$

10. BATCH REACTOR OPERATION— REVERSIBLE REACTIONS

Equation 43.38 Through Eq. 43.44: First-Order, Reversible, Isothermal, Constant Volume Batch Reactor

$$A \underset{k_2}{\overset{k_1}{\rightleftharpoons}} R \qquad \textit{43.38}$$

$$-r_A = -\frac{dC_A}{dt} = k_1 C_A - k_2 C_R \qquad \textit{43.39}$$

$$K_c = k_1/k_2 = \hat{C}_R / \hat{C}_A \qquad \textit{43.40}$$

$$M = C_{R_0} / C_{A_0} \qquad \textit{43.41}$$

$$\frac{d\hat{X}_A}{dt} = \frac{k_1(M+1)}{M + \hat{X}_A}(\hat{X}_A - X_A) \qquad \textit{43.42}$$

$$-\ln\left(1 - \frac{X_A}{\hat{X}_A}\right) = -\ln \frac{C_A - \hat{C}_A}{C_{A_0} - \hat{C}_A} \qquad \textit{43.43}$$

$$-\ln\left(1 - \frac{X_A}{\hat{X}_A}\right) = \frac{(M+1)}{(M + \hat{X}_A)} k_1 t \qquad \textit{43.44}$$

Description

Equation 43.38 through Eq. 43.44 are applicable to reversible first-order reactions occurring in batch reactors that are operating at constant temperature and volume. Equation 43.38 is the general chemical reaction equation for a reversible reaction, with the double arrows and two rate constants—a forward and a reverse constant. Equation 43.39 is the reaction rate equation in terms of the rate of disappearance of component A. The forward reaction reduces the amount of component A while producing product R at a rate that is dependent on the contraction of A, while the reverse reaction reduces the amount of R while producing A at a rate that is dependent on the concentration of the product R.

Equation 43.40 is the definition of the equilibrium constant for a first-order reversible reaction.[30] Equilibrium is achieved when the rate of the forward reaction is the same as the reverse reaction. Equation 43.41 defines the *simplification parameter*, M, as the ratio of the two initial concentrations. M is the ratio of the initial product and initial reactant concentrations, and it is used to simplify the expressions. Equation 43.42 identifies the rate of change in the fractional conversion of A, while Eq. 43.43 and Eq. 43.44 are the integrated forms of Eq. 43.42.[31] Figure 43.11 shows the kinetics for a reversible first-order batch reaction.

---

[30](1) The "hat" above the concentration variable designates the equilibrium value. (2) The *NCEES Handbook* is inconsistent in how it designates equilibrium conditions. Although the hat is used in Eq. 43.40, an asterisk is used in packed columns, and a degree symbol is used with fugacity.

[31]The *NCEES Handbook* is in error in placing a "hat" on the variable X_A on the left-hand side of Eq. 43.42. This X_A is not an equilibrium value; it is an instantaneous value.

Figure 43.11 *Reversible First-Order Batch Reaction Kinetics*

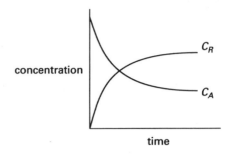

Example

A first-order liquid phase reaction is conducted in an isothermal batch reactor at constant volume. The conversion of A into R is one-for-one. The initial charge is pure reactant A at 1 mol/L, and the forward rate constant is 1.15 h^{-1}. The equilibrium constant for the reaction is 3. Most nearly, what time is required for the concentrations to reach 90% of their equilibrium values?

(A) 0.5 h

(B) 1.0 h

(C) 1.5 h

(D) 2.0 h

Solution

The equilibrium constant has been specified. Use Eq. 43.40.

$$K_c = \frac{\hat{C}_R}{\hat{C}_A} = 3$$

Since the conversion is one-for-one, and since the initial charge of reactant A is 1 mol/L, the molar concentrations of the liquid phase components at any time add up to 1.0 mol/L.

$$\hat{C}_R + \hat{C}_A = 1.0 \text{ mol/L}$$

Rearranging and combining equations, the equilibrium concentration of component A is

$$\hat{C}_R = 3\hat{C}_A = 1.0 \frac{\text{mol}}{\text{L}} - \hat{C}_A$$

$$\hat{C}_A = 0.25 \text{ mol/L}$$

Using Eq. 43.18, the fractional conversion of component A at equilibrium is

$$\hat{X}_A = \frac{C_{A0} - \hat{C}_A}{C_{A0}} = \frac{1.0 \dfrac{\text{mol}}{\text{L}} - 0.25 \dfrac{\text{mol}}{\text{L}}}{1.0 \dfrac{\text{mol}}{\text{L}}} = 0.75$$

Since the initial charge is pure reactant, the initial product $C_{R0} = 0$. Using Eq. 43.41, the ratio of initial concentrations is

$$M = \frac{C_{R0}}{C_{A0}} = 0$$

Using Eq. 43.43 and Eq. 43.44, the time required for the reaction to achieve 90% equilibrium is

$$-\ln\left(1 - \frac{X_A}{\hat{X}_A}\right) = \frac{M+1}{M + \hat{X}_A}k_1 t$$

$$
\begin{aligned}
t &= -\frac{\ln\left(1 - \dfrac{X_A}{\hat{X}_A}\right)}{\dfrac{k_1(M+1)}{M + \hat{X}_A}} \\[2em]
&= -\frac{\ln\left(1 - \dfrac{(0.90)(0.75)}{0.75}\right)}{\dfrac{\left(1.15\,\dfrac{1}{\text{h}}\right)(0+1)}{0 + 0.75}} \\[2em]
&= 1.5 \text{ h}
\end{aligned}
$$

The answer is (C).

11. BATCH REACTOR OPERATION— SHIFTING ORDER[32]

The reaction order can effectively change as some reactions progress. This usually happens in enzyme and surface catalyzed reactions (*enzymolysis*), including fermentation, as a transport condition is incorporated into the chemical kinetics. Depending on concentrations and rate constant values, the reaction can be zero- or first-order, and the order can change with reaction time as reactants are depleted.

In its simplest conceptualization, the complex enzymolysis can be considered as an irreversible reaction.[33]

$$A \underset{k_2}{\overset{k_1}{\rightleftharpoons}} R$$

---

[32]These are, indeed, reactions of changing order. However, since Michaelis-Menten theory and nomenclature are not in the *NCEES Handbook*, and since the role of enzymes in this set of equations is totally transparent, the description, "Shifting Order" is probably less accurate than "Catalyzed Unimolecular Surface Reactions."

[33]Enzyme-catalyzed bioreactions haven't been thought of in this manner since early 1913 when Michaelis and Menten published their model of enzyme catalyzed reactions that is the standard used today. The $E + S$ model described by Michaelis-Menton theory is now dominant.

Equation 43.45 Through Eq. 43.47: Shifting Order, Isothermal, Constant Volume Batch Reactor

$$-r_A = \frac{k_1 C_A}{1 + k_2 C_A} \qquad \textbf{43.45}$$

$$\ln\left(\frac{C_{A_0}}{C_A}\right) + k_2(C_{A_0} - C_A) = k_1 t \qquad \textbf{43.46}$$

$$\frac{\ln(C_{A_0}/C_A)}{C_{A_0} - C_A} = -k_2 + \frac{k_1 t}{C_{A_0} - C_A} \qquad \textbf{43.47}$$

Description

Equation 43.45 through Eq. 43.47 are applicable when the order of a reaction changes, such as with the cases of enzyme and surface catalyzed reactions and fermentation.[34] These equations may not be applicable in other situations of shifting order. Equation 43.45 is the generalized rate equation, and Eq. 43.46 is the solution to the differential form of Eq. 43.45, solved with separation of variables. Equation 43.47 is a simple rearrangement of Eq. 43.46.

Example

A surface-catalyzed reaction is conducted on a gas containing nitrogen dioxide. The rate equation for the conversion of NO_2 in the presence of NH_3 on a catalyst surface is

$$-r_{NO_2} = \frac{(0.002 \text{ s}^{-1}) C_{NO_2}}{1 + \left(15.8 \,\dfrac{\text{L}}{\text{mol}}\right) C_{NO_2}}$$

The initial and final concentrations of NO_2 in the gas are 100×10^{-6} mol/L and 15×10^{-6} mol/L, respectively. There is no change in the total number of moles of the gas. Most nearly, how much time is required to reduce the NO_2 concentration from 100×10^{-6} mol/L to 15×10^{-6} mol/L?

(A) 16 min

(B) 28 min

(C) 63 min

(D) 97 min

Solution

The rate constant equation given in the problem statement defines the rate constants, k_1 and k_2, since it is in the form of Eq. 43.45.

---

[34](1)This topic is poorly and incompletely represented in the *NCEES Handbook*. Except for rote substitution without comprehension, these three equations are difficult to apply. (2) The reaction progression (e.g., $A \rightarrow R$) is not defined. (3) k_1 and k_2 are not defined. They apparently do not represent the two forward rate constants used in Michaelis-Menten, or the rate constants at different temperatures, as in the Arrhenius equation (which uses the same variables). (4) Without this information, it is difficult to make a tie-in to the Michaelis-Menten equation, which also is not mentioned in the *NCEES Handbook*.

Using Eq. 43.46, the time required to reduce the concentration from 100×10^{-6} mol/L to 15×10^{-6} mol/L is

$$\ln\left(\frac{C_{A_0}}{C_A}\right) + k_2(C_{A_0} - C_A) = k_1 t$$

$$t = \frac{\ln\left(\dfrac{C_{A_0}}{C_A}\right) + k_2(C_{A_0} - C_A)}{k_1}$$

$$= \frac{\ln\left(\dfrac{100 \times 10^{-6} \dfrac{\text{mol}}{\text{L}}}{15 \times 10^{-6} \dfrac{\text{mol}}{\text{L}}}\right) + \left(15.8 \dfrac{\text{L}}{\text{mol}}\right)\left(\begin{array}{c}100 \times 10^{-6} \dfrac{\text{mol}}{\text{L}} \\ -15 \times 10^{-6} \dfrac{\text{mol}}{\text{L}}\end{array}\right)}{(0.002 \text{ s}^{-1})\left(60 \dfrac{\text{s}}{\text{min}}\right)}$$

$$= 15.8 \text{ min} \quad (16 \text{ min})$$

The answer is (A).

Alternative Solution

At such low concentrations, the rate law can be approximated as $-r_A \approx k_1 C_A$. For a standard first-order reaction,

$$t = \ln\left(\frac{\dfrac{C_{A_0}}{C_A}}{k_1}\right)$$

$$= \ln\left(\frac{\dfrac{100 \times 10^{-6} \dfrac{\text{mol}}{\text{L}}}{15 \times 10^{-6} \dfrac{\text{mol}}{\text{L}}}}{(0.002 \text{ s}^{-1})\left(60 \dfrac{\text{s}}{\text{min}}\right)}\right)$$

$$= 15.8 \text{ min} \quad (16 \text{ min})$$

The answer is (A).

12. BATCH REACTION OPERATION— VARIABLE VOLUME

In some reactions, the volume of the reacting mass varies with the conversion. This is generally due to the gaseous reactants and products having significantly different densities. (When reactants and products are in the liquid and solid phases, the variations in densities are seldom significant enough to consider.) A variable volume reactor can be implemented as a cylindrical shell with a floating roof. Such reactions generally occur at constant (atmospheric) pressure. A variable volume can

also be implemented in a fixed-volume reactor if the reactor has constant cross-sectional area and a differential pressure measurement is used to control the flow rate out of the reactor. In this instance, at steady state, the variable volume reaction is a constant mass reaction.

Equation 43.48 Through Eq. 43.52: Variable Volume, Isothermal, Constant Pressure Batch Reactor

$$V = V_{X_{A=0}}(1 + \varepsilon_A X_A) \qquad 43.48$$

$$\varepsilon_A = \frac{V_{X_{A=1}} - V_{X_{A=0}}}{V_{X_{A=0}}} = \frac{\Delta V}{V_{X_{A=0}}} \qquad 43.49$$

$$C_A = C_{A0}\left[\frac{1 - X_A}{1 + \varepsilon_A X_A}\right] \qquad 43.50$$

$$t = -C_{A0}\int_0^{X_A} dX_A / [(1 + \varepsilon_A X_A)(-r_A)] \qquad 43.51$$

$$kt = -\ln(1 - X_A) = -\ln\left(1 - \frac{\Delta V}{\varepsilon_A V_{X_{A=0}}}\right) \qquad 43.52$$

[first-order irreversible reaction]

Description

Equation 43.48 through Eq. 43.52 are applicable to variable volume systems. Equation 43.48 gives the reacting mass volume as a function of conversion. This equation defines the volume in terms of the fractional conversion. The parameter ε_A is the *expansion factor* (i.e., the fractional reduction in volume when the reaction is complete), as defined in Eq. 43.49.[35] Equation 43.50 defines the concentration of reactant A as a function of the conversion at that time. Equation 43.50 is the same as Eq. 43.18 with the exception that the revised volume associated with the conversion at that time is used. Equation 43.51 defines the generic reaction time to achieve a particular conversion. Equation 43.52 is the integrated form of Eq. 43.51 for a first-order irreversible reaction.

Example

A first-order, irreversible vapor phase reaction takes place in a floating roof vessel. The first run proceeds normally; the reaction starts with a volume of 2.0 m^3 and ends at completion with a volume of 1.25 m^3. On a subsequent run, the process is interrupted before completion is reached. The initial vessel volume was 200 m^3, and the final volume was 175 m^3. Most nearly, what is the final conversion fraction?

(A) 0.25

(B) 0.33

(C) 0.67

(D) 0.75

---

[35]The variable δ is also used.

Solution

Using Eq. 43.49, the fractional volume change, ε_A, is

$$\varepsilon_A = \frac{\Delta V}{V_{X_{A=0}}} = \frac{125 \text{ m}^3 - 200 \text{ m}^3}{200 \text{ m}^3} = -0.375$$

Using Eq. 43.52, the final conversion is

$$-\ln(1 - X_A) = -\ln\left(1 - \frac{\Delta V}{\varepsilon_A V_{X_{A=0}}}\right)$$

$$1 - X_A = 1 - \frac{\Delta V}{\varepsilon_A V_{X_{A=0}}}$$

$$X_A = \frac{\Delta V}{\varepsilon_A V_{X_{A=0}}}$$

$$= \frac{175 \text{ m}^3 - 200 \text{ m}^3}{(-0.375)(200 \text{ m}^3)}$$

$$= 0.333 \quad (0.33)$$

The answer is (B).

13. CONTINUOUSLY STIRRED TANK REACTOR OPERATION

Perfect mixing is assumed in a continuously stirred tank reactor (CSTR). Therefore, uniform concentration and temperature can be assumed throughout the reactor. The effluent stream has the same properties as the reactor contents.

Write a differential mass balance on the reactor contents. This equation is valid for both steady- and nonsteady-state use.

$$N_{A_0} \frac{dX_A}{dt} = F_{A_0}(X_{A_0} - X_A) + V(-r_A)$$

Setting the derivative equal to zero for steady-state operation produces an equation involving the conversion fraction.

$$X_A - X_{A_0} = \left(\frac{V}{F_{A0}}\right)(-r_A)$$

As previously stated, CSTR reaction performance can resemble that of a plug flow reactor. Accordingly, the space-time concept, previously applied to plug flow reactors, is extended to CSTRs. Using Eq. 43.23 and substituting space-time, τ, for time, t, in Eq. 43.25 produces a simple equation for the change in reactant concentration in constant-density CSTR systems.

$$C_{A0} - C_A = \tau(-r_A)$$

From Eq. 43.28 (or Table 43.1), the rate expression for a first-order reaction is

$$-r_A = kC_A$$

Substituting into the equation for constant-density CSTR systems yields

$$k\tau = \frac{C_{A_0} - C_A}{C_A} = \frac{X_A}{1 - X_A}$$

This can be solved for the conversion fraction as a function of space-time.

$$X_A = 1 - \frac{1}{1 + k\tau}$$

Because the reaction depends on the effluent concentration rather than the feed concentration, conversion is usually lower in a CSTR than in a PFR or batch reactor. However, a CSTR combines the advantages of a PFR (continuous operation) with the advantages of a batch reactor (cost-effective reaction tanks and heat transfer capabilities with either internal coils or external jackets). Additionally, CSTRs may offer advantages in selectivity for complex reactions.

Placing multiple CSTRs in series will increase the ultimate conversion. When analyzing a series of CSTRs, the feed concentration of each reactor is equal to the effluent concentration of the previous reactor in the series.

The mass balance for the second reactor in a series is

$$X_{A_2} - X_{A_1} = \left(\frac{V}{F_{A_0}}\right)(-r_A)$$

For a first-order reaction,

$$X_{A_2} - X_{A_1} = \left(\frac{V C_{A_0}}{F_{A_0}}\right)k(1 - X_{A_2})$$

$$= k\tau(1 - X_{A_2})$$

Substituting the value of X_{A_1} and solving for the overall conversion from the initial inlet to the outlet of the second reactor yields

$$X_{A_2} = 1 - \frac{1}{(1 + k\tau)^2}$$

For N equal-volume CSTRs in series,

$$X_{A_N} = 1 - \frac{1}{(1 + k\tau)^N}$$

Equation 43.53: Isothermal Continuously Stirred Tank Reactor (CSTR)

$$\frac{\tau}{C_{A0}} = \frac{V_{\text{CSTR}}}{F_{A0}} = \frac{X_A}{-r_A} \qquad 43.53$$

Description

Equation 43.53 is the design equation for a constant volume CSTR containing a first-order reaction of the type $A \to R$. The concentration is evaluated at the outlet conditions because the design equation assumes perfect mixing.

Equation 43.54: *N* Isothermal, Constant Volume CSTRs in Series

$$\tau_{N\text{-reactors}} = N\tau_{\text{individual}}$$
$$= \frac{N}{k}\left[\left(\frac{C_{A0}}{C_{AN}}\right)^{1/N} - 1\right] \qquad 43.54$$

Description

Equation 43.54 calculates the space-time for a series of N equal volume CSTRs with a first-order reaction. C_{AN} is the concentration of reactant A leaving the Nth CSTR. The parameters are evaluated at the outlet conditions because the design equation assumes perfect mixing.

Example

A chemical process requiring 99% conversion is designed to utilize a series of equal volume, 0.5 m³ CSTRs. Each reactor has a space-time of 100 min. The reaction rate constant is 0.1 min⁻¹. The initial reactant concentration is 50 mol/m³. How many reactors in series are required?

    (A) 1

    (B) 2

    (C) 3

    (D) 4

Solution

Using Eq. 43.54, solve for N.

$$N\tau = \frac{N}{k}\left[\left(\frac{C_{A0}}{C_{AN}}\right)^{1/N} - 1\right]$$
$$k\tau = \left(\frac{C_{A0}}{C_{AN}}\right)^{1/N} - 1$$

$$1 + k\tau = \left(\frac{C_{A0}}{C_{AN}}\right)^{1/N}$$
$$\log(1 + k\tau) = \frac{1}{N}\log\left(\frac{C_{A0}}{C_{AN}}\right)$$
$$N = \frac{\log\left(\dfrac{C_{A0}}{C_{AN}}\right)}{\log(1 + k\tau)}$$

Since $X_A = 1 - C_A/C_{A0}$,

$$N = \frac{-\log(1 - X_{A,N})}{\log(1 + k\tau)}$$
$$= \frac{-\log(1 - 0.99)}{\log\left(1 + \left(0.1 \text{ min}^{-1}\right)(100 \text{ min})\right)}$$
$$= 1.92 \quad (2)$$

The answer is (B).

14. IRREVERSIBLE REACTIONS IN PARALLEL

In irreversible reactions, the reactants are converted into products, and no reactants are left behind.[36] Such reactions are sometimes called *kinetic reactions*. Yield is determined by the extent of conversion, which depends on the amount of reaction, the size of the reactor, and the operating conditions, especially temperature and pressure. Irreversible reactions (1) proceed only in one direction (forward), (2) can proceed to completion if the reactor is designed appropriately, and (3) have a free energy less than 0 and an equilibrium constant greater than 1. (See Fig. 43.1.)

Some examples of irreversible reactions include

- thermal cracking processes
- combustion and oxidation reactions
- fermentation reactions
- neutralization between strong acids and strong bases
- double decomposition reactions
- most precipitation reactions

As with reversible reactions, there may be additional pathways for irreversible reactions. That is, the reaction may produce undesirable by-products, U, as well as desirable products, D. Undesirable products are created frequently in steam cracking, hydrocracking, and fluidized catalytic reactions. Simultaneous production of

---

[36]In systems of two reactants, no reactants are left behind up to the extent of the limiting reactant.

desirable and undesirable products is an example of having two irreversible reactions in parallel.

$$A \overset{k_D}{\rightarrow} D \quad \text{[desired]}$$

$$A \overset{k_U}{\rightarrow} U \quad \text{[undesired]}$$

Since undesirable products are produced, it is appropriate to define a *yield* variable, Y, which is essentially the instantaneous fraction of reactant, A, that is being converted to the desired product, D, although other definitions exist.

Equation 43.55 Through Eq. 43.60: Two Irreversible, Isothermal Reactions in Parallel

$$-r_A = -dc_A/dt = k_D C_A^x + k_U C_A^y \quad \text{43.55}$$

$$r_D = dc_D/dt = k_D C_A^x \quad \text{43.56}$$

$$r_U = dc_U/dt = k_U C_A^y \quad \text{43.57}$$

$$Y_D = dC_D/(-dC_A) \quad \text{43.58}$$

$$\overline{Y}_D = N_{D_f}/(N_{A_0} - N_{A_f}) \quad \text{43.59}$$

$$\overline{S}_{DU} = N_{D_f}/N_{U_f} \quad \text{43.60}$$

Variation

$$\overline{Y}_U = \frac{N_{U_f}}{N_{A_0} - N_{A_f}}$$

Description

Equation 43.55 is the rate equation for two parallel irreversible reactions. x and y are the stoichiometric coefficients of the products (i.e., are the *order exponents*); x and y are needed only when multiple reactant molecules combine during processing. Equation 43.56 and Eq. 43.57 are the rate equations for the desirable and undesirable reactions, respectively. Equation 43.58, Eq. 43.59, and Eq. 43.60 are definitions for *instantaneous fractional yield*, Y_D, *overall fractional yield*, $\overline{Y}_D$, and *overall selectivity* to the desired product, $\overline{S}_{DU}$, respectively.[37] The subscripts 0 and f represent the initial and final states, respectively.

Example

Two parallel reaction pathways produce desirable and undesirable products simultaneously, as shown.

$$A \overset{k_D}{\rightarrow} D \quad \text{[desired]}$$

$$A \overset{k_U}{\rightarrow} U \quad \text{[undesired]}$$

---

[37] *NCEES Handbook* Eq. 43.60 uses the ambiguous script "*DU*" to represent the selectivity of product D to product U (i.e., the ratio of D to U).

300 mol of A react to completion. The yield of U is 75%. Most nearly, what is the overall selectivity to D?

(A) 1/3

(B) 1/2

(C) 2

(D) 3

Solution

From the variation equation for the overall fractional yield of U, the number of moles of U is

$$\overline{Y}_U = \frac{N_{U_f}}{N_{A_0} - N_{A_f}}$$

$$N_{U_f} = \overline{Y}_U(N_{A_0} - N_{A_f})$$

$$= (0.75)(300 \text{ mol} - 0 \text{ mol})$$

$$= 225 \text{ mol}$$

The overall selectivity to D to U is

$$\overline{S}_{DU} = \frac{N_{D_f}}{N_{U_f}} = \frac{N_{A_0} - N_{U_f}}{N_{U_f}}$$

$$= \frac{300 \text{ mol} - 225 \text{ mol}}{225 \text{ mol}}$$

$$= 0.333 \quad (1/3)$$

The answer is (A).

15. IRREVERSIBLE REACTIONS IN SERIES

In the case of two irreversible, first-order reactions in series resulting in desirable and undesirable products, the rate constants are defined by the following chemical reaction.

$$A \overset{k_D}{\rightarrow} D \overset{k_U}{\rightarrow} U$$

Equation 43.61 Through Eq. 43.63: Two Isothermal, Irreversible, First-Order Reactions in Series

$$r_A = -dC_A/dt = k_D C_A \quad \text{43.61}$$

$$r_D = dC_D/dt = k_D C_A - k_U C_D \quad \text{43.62}$$

$$r_U = dC_U/dt = k_U C_D \quad \text{43.63}$$

Variation

$$\overline{Y}_U = \text{overall fractional yield of } U$$

$$= \frac{N_{U_f}}{N_{A_0} - N_{A_f}}$$

Description

Equation 43.61 is the rate equation for two irreversible reactions in series. Equation 43.62 and Eq. 43.63 are the rate equations for the desirable and undesirable reactions, respectively. Yield and selectivity definitions are the same as for parallel reactions.

Example

A series chemical reaction system is described by the equation shown.

$$A \xrightarrow{k_D} D \xrightarrow{k_U} U$$

300 mol of A are reacted to completion. The yield of U is 25%. Most nearly, what is the overall selectivity to D?

(A) 1/3

(B) 1/2

(C) 2

(D) 3

Solution

From the variation equation for the overall fractional yield of U, the number of moles of U is

$$\overline{Y}_U = \frac{N_{U_f}}{N_{A_0} - N_{A_f}}$$

$$\begin{aligned} N_{U_f} &= \overline{Y}_U(N_{A_0} - N_{A_f}) \\ &= (0.25)(300 \text{ mol} - 0 \text{ mol}) \\ &= 75 \text{ mol} \end{aligned}$$

The overall selectivity to D to U is

$$\begin{aligned} \overline{S}_{DU} &= \frac{N_{D_f}}{N_{U_f}} = \frac{N_{A_0} - N_{U_f}}{N_{U_f}} \\ &= \frac{300 \text{ mol} - 75 \text{ mol}}{75 \text{ mol}} \\ &= 3 \end{aligned}$$

The answer is (D).

16. OPTIMUM YIELD[38]

There is no reason to continue a reaction once the amount of the desired component has peaked. Any continuation will reduce the yield of the desired product from the maximum concentration. *Optimum yield* is achieved when the desired product concentration is maximum.

---

[38]The *NCEES Handbook* refers to this as "optimal" yield.

Equation 43.64 and Eq. 43.65: Optimum Yield for Isothermal, Irreversible Reactions in Series in a PFR

$$\frac{C_{D,\max}}{C_{A_0}} = \left(\frac{k_D}{k_U}\right)^{k_U/(k_U - k_D)} \qquad \textbf{43.64}$$

$$\tau_{\max} = \frac{1}{k_{\log \text{mean}}} = \frac{\ln(k_U/k_D)}{(k_U - k_D)} \qquad \textbf{43.65}$$

Description

For first-order irreversible reactions in series (i.e., $A \rightarrow D \rightarrow U$) occurring in PFRs, the optimum product yield is determined by Eq. 43.64 and Eq. 43.65. Equation 43.64 gives the maximum concentration of the desired product, and Eq. 43.65 gives the corresponding space-time for plug flow reactors.[39]

Example

A series chemical reaction in a PFR system is described by

$$A \xrightarrow{k_D} D \xrightarrow{k_U} U$$

The reaction rate constants for the products D and U are 0.25 L/s and 0.50 L/s, respectively. The feed to the reactor is pure A at a concentration of 2 mol/L. Most nearly, what is the maximum achievable concentration of D?

(A) 0.10 mol/L

(B) 0.50 mol/L

(C) 1.0 mol/L

(D) 1.5 mol/L

Solution

Using Eq. 43.64, the maximum concentration of product D is

$$\frac{C_{D,\max}}{C_{A0}} = \left(\frac{k_D}{k_U}\right)^{k_U/(k_U - k_D)}$$

$$C_{D,\max} = \left(2 \ \frac{\text{mol}}{\text{L}}\right)\left(\frac{0.25 \ \frac{\text{L}}{\text{s}}}{0.50 \ \frac{\text{L}}{\text{s}}}\right)^{(0.5 \text{ L/s})/(0.5 \text{ L/s} - 0.25 \text{ L/s})}$$

$$= 0.50 \text{ mol/L}$$

The answer is (B).

---

[39]In the *NCEES Handbook*, there is no mathematical significance to the parentheses in the denominator of Eq. 43.65.

Equation 43.66 and Eq. 43.67: Optimum Yield for Isothermal, Irreversible Reactions in Series in a CSTR

$$\frac{C_{D,\max}}{C_{A_0}} = \frac{1}{[(k_U/k_D)^{1/2} + 1]^2} \qquad \textbf{43.66}$$

$$\tau_{\max} = 1/\sqrt{k_D k_U} \qquad \textbf{43.67}$$

Description

For first-order irreversible reactions in series (i.e., $A \to D \to U$) occurring in CSTRs, the optimum product yield is determined by Eq. 43.66 and Eq. 43.67. Equation 43.66 gives the maximum concentration of the desired product, and Eq. 43.67 gives the corresponding space-time for plug flow reactors.

Example

A series chemical reaction in a CSTR system is described by

$$A \xrightarrow{k_D} D \xrightarrow{k_U} U$$

The reaction rate constants for the products D and U are 0.25 L/s and 0.50 L/s, respectively. The feed to the reactor is pure A at a concentration of 2 mol/L. Most nearly, what is the maximum achievable concentration of D?

(A) 0.3 mol/L

(B) 0.5 mol/L

(C) 1.0 mol/L

(D) 1.5 mol/L

Solution

Using Eq. 43.66, the maximum concentration of product D is

$$\frac{C_{D,\max}}{C_{A_0}} = \frac{1}{[(k_U/k_D)^{1/2} + 1]^2}$$

$$C_{D,\max} = \frac{C_{A_0}}{\left(\sqrt{\dfrac{k_U}{k_D}} + 1\right)^2}$$

$$= \frac{2 \, \dfrac{\text{mol}}{\text{L}}}{\left(\sqrt{\dfrac{0.5 \, \dfrac{\text{L}}{\text{s}}}{0.25 \, \dfrac{\text{L}}{\text{s}}}} + 1\right)^2}$$

$$= 0.343 \text{ mol/L} \quad (0.3 \text{ mol/L})$$

The answer is (A).

17. MAXIMUM TEMPERATURE CHANGE: ADIABATIC BATCH REACTORS[40]

In a thermodynamic energy balance, the *heat of reaction* (also known as *energy of reaction* and *enthalpy of reaction*), ΔH_r. Since from a thermodynamic standpoint, heat entering a system is positive, heat of reaction is positive for endothermic reactions and negative for exothermic reactions. If the reactor is well insulated such that no heat enters or leaves the reactor, the process and reactor are said to be adiabatic. If the reactor is not well insulated, the nonadiabatic heat transfer (i.e., the heat flux) is represented by the variable, q.[41] The nonadiabatic heat transfer is positive when heat is added to the reactor.

$$m_t c_p \frac{dT}{dt} = V(-\Delta H_r)(-r_A) + qA$$

Reaction temperature, and thereby reaction rate, can be controlled by adding or removing heat through a heat-exchange coil. The heat flux, q, is related to the *overall heat-transfer coefficient*, U, and the temperature gradient, according to

$$q = U(T_r - T)$$

In adiabatic operation, no heat is added or removed from the reactor. To achieve isothermal operation with endothermic reactions, the heat absorbed must be replaced. With exothermic reactions, the heat of reaction must be removed. The maximum possible temperature change is calculated by assuming that the process is adiabatic and, therefore, not isothermal.

In an adiabatic reactor, $q = 0$, so the first equation that can be used to predict the maximum temperature change is

$$m_t c_p \frac{dT}{dt} = V(-\Delta H_r)(-r_A)$$

This can be combined with Eq. 43.20 and $C = N/V$ to produce

$$\frac{dT}{dt} = \left(\frac{-\Delta H_r N_{A_0}}{m_t c_p}\right) \frac{dX_A}{dt}$$

Integrating,

$$T - T_0 = \left(\frac{-\Delta H_r N_{A_0}}{m_t c_p}\right)(X_A - X_{A_0})$$

---

[40]This material is not in the *NCEES Handbook.*
[41]q is a function of time and is per-unit reactor vessel surface area.

The maximum adiabatic temperature change will occur at complete conversion. Using $X_{A_0} = 0$ and $X_A = 1$, the maximum adiabatic temperature change will be

$$\Delta T_{\mathrm{ad}} = \frac{-\Delta H_r N_{A_0}}{m_t c_p}$$

Diagnostic Exam

Topic X: Mass Transfer

1. A distillation column separates 50 000 kg/h of a benzene/toluene (C_6H_6/CH_3) mixture composed of 40% benzene by mass. The mixture is separated into an overhead product composed of 97% benzene by mass and a bottom product of 98% toluene by mass.

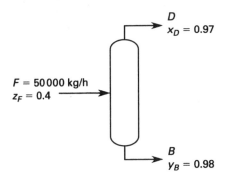

Most nearly, what is the overhead product mass flow rate?

(A) 18 Mg/h

(B) 20 Mg/h

(C) 30 Mg/h

(D) 34 Mg/h

2. A gas bottle 1.2 m in length contains a gaseous mixture of ammonia (NH_3) and hydrogen (H_2). The mixture is at 25°C and 1 atm. At one end of the bottle, the partial pressure of ammonia is 0.8 atm. At the other end of the bottle, the partial pressure of ammonia is 0.4 atm. The diffusion coefficient of ammonia at 25°C is 7.84×10^{-5} m^2/s. There is no convective flow (bulk flow). Most nearly, what is the molar flux of ammonia along the bottle?

(A) 5.4×10^{-4} mol/m$^2 \cdot$s

(B) 1.1×10^{-3} mol/m$^2 \cdot$s

(C) 8.6×10^{-3} mol/m$^2 \cdot$s

(D) 1.6×10^{-2} mol/m$^2 \cdot$s

3. Air is injected into a reaction chamber operating at a constant temperature of 1145K and pressure of 101.3 kPa. Oxygen in the air diffuses through a 1.5 mm thick stagnant gas film to a flat carbon surface, where the oxygen is adsorbed, resulting in an oxygen partial pressure that is effectively zero. The diffusivity of oxygen in the gas film is 1.3×10^{-4} m^2/s. Air is 21 mole % oxygen. Most nearly, what is the gas phase mass transfer coefficient, k_G, for oxygen in the gas film?

(A) 2.2×10^{-8} mol/m$^2 \cdot$s$\cdot$Pa

(B) 4.1×10^{-7} mol/m$^2 \cdot$s$\cdot$Pa

(C) 5.7×10^{-6} mol/m$^2 \cdot$s$\cdot$Pa

(D) 1.0×10^{-5} mol/m$^2 \cdot$s$\cdot$Pa

4. Air is injected into a combustion chamber with a constant temperature of 1145K. Oxygen in the air diffuses through a 2.0 mm thick stagnant gas film to a flat carbon surface, where the oxygen reacts instantaneously to form carbon dioxide (CO_2). The carbon dioxide diffuses away from the carbon. No reaction occurs in the gas film. At the surface of the carbon, the partial pressure of CO_2 is 50.5 kPa. Most nearly, what is the partial pressure of CO_2 1.5 mm from the carbon surface?

(A) 13 kPa

(B) 17 kPa

(C) 25 kPa

(D) 38 kPa

5. Air is injected into a reaction chamber which operates at a constant temperature of 1145K and a pressure of 101.3 kPa. Oxygen in the air diffuses through a 1.5 mm thick stagnant gas film to a flat carbon surface, where the oxygen reacts instantaneously to form carbon monoxide (CO) which diffuses away from the carbon. No reaction occurs in the gas film. The diffusivity of oxygen through the gas film is 1.3×10^{-4} m^2/s. Air is 21 mole % oxygen. Most nearly, what is the rate of oxygen diffusion to the carbon surface?

(A) 0.18 mol/m$^2 \cdot$s

(B) 0.19 mol/m$^2 \cdot$s

(C) 0.21 mol/m$^2 \cdot$s

(D) 0.92 mol/m$^2 \cdot$s

6. A single-stage extraction column is used to extract benzene (C_6H_6) from a cyclohexane (C_6H_{12}) stream as shown. Two streams enter the extraction column: a furfural ($C_5H_4O_2$) solvent stream and a benzene-rich cyclohexane feed stream. The feed stream flows at a rate of 100 kg/h and contains 10% benzene by weight. The solvent stream flows at a rate of 150 kg/h and contains 0.01% benzene by weight. The benzene content in the raffinate product stream is to be reduced to 5% by weight. Cyclohexane and furfural are immiscible.

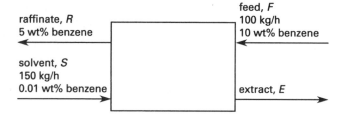

Most nearly, what is the percentage of benzene by weight in the extract stream?

(A) 3.4%

(B) 5.0%

(C) 6.8%

(D) 9.0%

7. Acetic acid (CH_3COOH) is separated from water in a continuous distillation column system with a total condenser and a partial reboiler. The liquid feed stream flows at a rate of 35 380 kg/h, and it contains 60% CH_3COOH by weight and 40% water by weight. Two product streams exit the distillation column. Stream 1 contains 95% (by mole) CH_3COOH and 5% water, and stream 2 contains 10% CH_3COOH and 90% water. The molecular weight of CH_3COOH is 60.05 kg/kmol, and the molecular weight of water is 18.02 kg/kmol. Most nearly, what is the production rate of stream 1?

(A) 280 kg/h

(B) 11 000 kg/h

(C) 15 000 kg/h

(D) 16 000 kg/h

8. A wetted-wall column is used to absorb gaseous ammonia (NH_3) from an ammonia-air stream at room temperature. The partial pressure of NH_3 in the stream is 0.0197 atm. The overall mass transfer coefficient in the vapor phase is 0.582 kmol/m^2·s·atm, and the individual mass transfer coefficient of NH_3 in the vapor phase is 2.27 kmol/m^2·s·atm. The Henry's law constant for the system is 1.26 m^3·atm/kmol. NH_3 behaves as an ideal gas. The partial pressure of NH_3 in both bulk phases is constant. Most nearly, what is the partial pressure of NH_3 in the gas phase at the interface at the top of the column?

(A) 0.015 atm

(B) 0.045 atm

(C) 0.057 atm

(D) 0.24 atm

9. A wetted-wall column is used to absorb gaseous ammonia (NH_3) from an ammonia-air stream. The overall gas phase mass transfer coefficient is 9.20×10^{-3} kmol/m^2·s·atm, and the gas phase specific mass transfer coefficient for NH_3 is 1.31×10^{-2} kmol/m^2·s·atm. The Henry's law constant for the system is 2.70 m^3·atm/kmol. Most nearly, what is the liquid phase mass transfer coefficient for NH_3?

(A) 0.010 m/s

(B) 0.015 m/s

(C) 0.083 m/s

(D) 87 m/s

10. A wetted-wall column absorbs ammonia (NH_3) from a vapor stream. The overall gas phase mass transfer coefficient is 5.6×10^{-2} kmol/m^2·s·atm. 70% of the total resistance to mass transfer is found in the gas phase. Most nearly, what is the gas phase mass transfer coefficient for ammonia?

(A) 0.003 kmol/m^2·s·atm

(B) 0.03 kmol/m^2·s·atm

(C) 0.08 kmol/m^2·s·atm

(D) 13 kmol/m^2·s·atm

SOLUTIONS

1. The total mass balance around the column is

$$F = D + B$$

Substitute the total balance into the benzene balance.

$$Fz_F = Dx_D + Bx_B$$
$$= Dx_D + (F - D)\,x_B$$

The benzine bottoms product, x_B, is

$$x_B = 1 - y_B$$
$$= 1 - 0.98$$
$$= 0.02$$

From the given information, the overhead distillate flow rate, D, is

$$D = F\left(\frac{z_F - x_B}{x_D - x_B}\right) \quad [x_B - x_D \neq 0]$$

$$= \left(\frac{50\,000\ \dfrac{\text{kg}}{\text{h}}}{1000\ \dfrac{\text{kg}}{\text{Mg}}}\right)\left(\frac{0.4 - 0.02}{0.97 - 0.02}\right)$$

$$= 20\ \text{Mg/h}$$

The answer is (B).

2. The molar flux of ammonia, N_A, along the bottle is

$$N_A = \frac{p_A}{P}(N_A + N_B) - \frac{D_m\,\Delta p_A}{\overline{R}\,T\,\Delta z}$$

Since there is no convective flow, the equation simplifies to

$$N_A = -\frac{D_m\,\Delta p_A}{\overline{R}\,T\,\Delta z}$$
$$= -\frac{D_m(p_{A2} - p_{A1})}{\overline{R}\,T(z_2 - z_1)}$$

$$= -\frac{\left(7.84 \times 10^{-5}\ \dfrac{\text{m}^2}{\text{s}}\right)(0.4\ \text{atm} - 0.8\ \text{atm})}{\left(82.06\ \dfrac{\text{m}^3\cdot\text{atm}}{\text{mol}\cdot\text{K}}\right)(25°\text{C} + 273°)(1.2\ \text{m} - 0\ \text{m})}$$

$$= 1.07 \times 10^{-3}\ \text{mol/m}^2\cdot\text{s} \quad (1.1 \times 10^{-3}\ \text{mol/m}^2\cdot\text{s})$$

The answer is (B).

3. Flux for unidirectional diffusion of a gas A through a second stagnant gas B ($N_B = 0$) is

$$N_A = \left(\frac{D_m P}{\overline{R}T(p_B)_{lm}}\right)\left(\frac{p_{A1} - p_{A2}}{z_2 - z_1}\right)$$

The gas mass transfer coefficient definition for gas A through a second stagnant gas B is

$$N_A = k_G(p_{A1} - p_{A2})$$

Solving these two equations for the mass transfer coefficient,

$$k_G = \frac{D_m P}{\overline{R}T(p_B)_{lm}(z_2 - z_1)}$$

Oxygen (gas A) reacts instantaneously at location 2, so the mole fraction there is zero. Since air is 21 mole % oxygen, the oxygen partial pressures are

$$p_{A2} = y_{A2}P$$
$$= 0\ \text{kPa}$$
$$p_{A1} = y_{A1}P = \left(0.21\ \frac{\text{mol}}{\text{mol}}\right)(101.3\ \text{kPa})$$
$$= 21.273\ \text{kPa}$$

The nitrogen (gas B) partial pressures are

$$p_{B2} = P - p_{A2}$$
$$= 101.3\ \text{kPa} - 0\ \text{kPa}$$
$$= 101.3\ \text{kPa}$$

$$p_{B1} = P - p_{A1}$$
$$= 101.3\ \text{kPa} - 21.273\ \text{kPa}$$
$$= 80.027\ \text{kPa}$$

The log mean partial pressure of B is

$$(p_B)_{lm} = \frac{p_{B2} - p_{B1}}{\ln\left(\dfrac{p_{B2}}{p_{B1}}\right)}$$
$$= \frac{101.3\ \text{kPa} - 80.027\ \text{kPa}}{\ln\left(\dfrac{101.3\ \text{kPa}}{80.027\ \text{kPa}}\right)}$$
$$= 90.246\ \text{kPa}$$

Mass Transfer

The mass transfer coefficient is

$$k_G = \frac{D_m P}{\overline{R}T(p_B)_{lm}(z_2 - z_1)}$$

$$= \frac{\left(1.30 \times 10^{-4} \ \frac{m^2}{s}\right)(101.3 \ \text{kPa})\left(1000 \ \frac{mm}{m}\right)}{\left(8.314 \ \frac{Pa \cdot m^2}{mol \cdot K}\right)(1145K)(90.246 \ \text{kPa})(1.5 \ mm)}$$

$$= 1.0219 \times 10^{-5} \ \text{mol/m}^2 \cdot \text{s} \cdot \text{Pa}$$

$$(1.0 \times 10^{-5} \ \text{mol/m}^2 \cdot \text{s} \cdot \text{Pa})$$

The answer is (D).

4. The reaction at the carbon surface involving CO_2 and O_2 is

$$O_2 + C \rightarrow CO_2$$

The steady-state flux of A is

$$N_A = (N_A + N_B)\frac{p_A}{P} - \frac{D_m}{\overline{R}T}\frac{dp_A}{dz}$$

From the stoichiometry of the reaction for CO_2 (gas A) and O_2 (gas B), $N_A = -N_B$, the equimolar steady-state counter diffusion flux is

$$N_A = -\frac{D_m}{\overline{R}T}\frac{dp_A}{dz}$$

Integrating yields a linear partial pressure distribution with a film thickness of

$$\int_{p_{A1}}^{p_A} dp_A = \frac{N_A \overline{R}T}{D_m}\int_{z_1}^{z} dz$$

$$p_A = p_{A1} - \frac{N_A \overline{R}T}{D_m}(z - z_1)$$

$$= p_{A1} - C(z - z_1)$$

C is a constant. Solve for C using the carbon surface (position 1), and the incoming gas (position 2), which is devoid of CO_2.

$$C = -\frac{p_{A2} - p_{A1}}{z_2 - z_1}$$

$$= \frac{0 \ \text{kPa} - 50.5 \ \text{kPa}}{2.0 \ mm - 0 \ mm}$$

$$= 25.25 \ \text{kPa/mm}$$

At a position of 1.5 mm from the carbon surface, the partial pressure is

$$p_A = p_{A1} - C(z - z_1)$$

$$= 50.5 \ \text{kPa} - \left(25.25 \ \frac{\text{kPa}}{mm}\right)(1.5 \ mm - 0 \ mm)$$

$$= 12.625 \ \text{kPa} \quad (13 \ \text{kPa})$$

The answer is (A).

5. Let z be the distance from the surface. The steady-state flux of A is

$$N_A = (N_A + N_B)\frac{p_A}{P} - \left(\frac{D_m}{\overline{R}T}\frac{dp_A}{dz}\right)$$

The reaction at the carbon surface involving CO_2 and O_2 is

$$O_2 + 2C \rightarrow 2CO$$

The O_2 (gas A) and CO_2 (gas B) are in steady-state diffusion through the gas film. From the stoichiometry of the reaction, $N_B = -2N_A$, so the convective terms do not vanish, and the steady-state flux equation must be integrated.

$$\int_{z_1}^{z_2} dz = -\frac{D_m P}{\overline{R}T}\int_{p_{A1}}^{p_{A2}} \frac{dp_A}{PN_A - (N_A + N_B)p_A}$$

The resulting flux of A is[1]

$$N_A = \left(\frac{N_A}{N_A + N_B}\right)\frac{D_m P}{\overline{R}T(z_2 - z_1)}$$

$$\times \ln\left(\frac{\left(\frac{N_A}{N_A + N_B}\right)P - p_{A2}}{\left(\frac{N_A}{N_A + N_B}\right)P - p_{A1}}\right)$$

From the stoichiometry of the reaction, $N_B = -2N_A$, so the flux ratio is

$$\frac{N_A}{N_A + N_B} = \frac{N_A}{N_A - 2N_A} = -1$$

---

[1]Treybal, R. E., *Mass Transfer Operations*, McGraw-Hill, p. 27 (1980).

At position 1, air is 21 mole % oxygen. Oxygen reacts instantaneously at position 2, so the mole fraction there is zero. Then, the oxygen partial pressures are

$$p_{A2} = y_{A2}P = 0 \text{ kPa}$$

$$p_{A1} = y_{A1}P$$

$$= \left(0.21 \, \frac{\text{mol}}{\text{mol}}\right)(101.3 \text{ kPa})$$

$$= 21.273 \text{ kPa}$$

The flux of A is

$$N_A = \left(\frac{N_A}{N_A + N_B}\right)\frac{D_m P}{\overline{R}T(z_2 - z_1)}$$

$$\times \ln\left(\frac{\left(\dfrac{N_A}{N_A + N_B}\right)P - p_{A2}}{\left(\dfrac{N_A}{N_A + N_B}\right)P - p_{A1}}\right)$$

$$= (-1)\left(\frac{\begin{array}{c}\left(1.30 \times 10^{-4} \, \dfrac{\text{m}^2}{\text{s}}\right)(101.3 \text{ kPa}) \\[2mm] \times \left(1000 \, \dfrac{\text{Pa}}{\text{kPa}}\right)\left(1000 \, \dfrac{\text{mm}}{\text{m}}\right) \\ \hline \left(8.314 \, \dfrac{\text{Pa·m}^2}{\text{mol·K}}\right)(1145\text{K}) \\[2mm] \times (1.5 \text{ mm} - 0 \text{ mm}) \end{array}}{}\right.$$

$$\left. \times \ln\left(\frac{\begin{array}{c}(-1)(101.3 \text{ kPa}) \\ - 0 \text{ kPa}\end{array}}{\begin{array}{c}(-1)(101.3 \text{ kPa}) \\ - 21.273 \text{ kPa}\end{array}}\right)\right)$$

$$= 0.175799 \text{ mol/m}^2\text{·s} \quad (0.18 \text{ mol/m}^2\text{·s})$$

The answer is (A).

6. Since cyclohexane and furfural are immiscible, the mass balance for cyclohexane around the stage is

$$F(1 - x_F) = R(1 - x_R)$$

Solve for raffinate flow, R.

$$R = F\left(\frac{1 - x_F}{1 - x_R}\right)$$

$$= \left(100 \, \frac{\text{kg}}{\text{h}}\right)\left(\frac{1 - 0.10}{1 - 0.05}\right)$$

$$= 94.737 \text{ kg/h}$$

The overall mass balance around the stage is

$$F + S = R + E$$

Solve for the extract flow, E.

$$E = F + S - R$$

$$= 100 \, \frac{\text{kg}}{\text{h}} + 150 \, \frac{\text{kg}}{\text{h}} - 94.737 \, \frac{\text{kg}}{\text{h}}$$

$$= 155.26 \text{ kg/h}$$

Since cyclohexane and furfural are immiscible, a mass balance for furfural around the stage is

$$S(1 - y_S) = E(1 - y_E)$$

Solve for the benzene mass fraction in the extract.

$$y_E = 1 - \left(\frac{S}{E}\right)(1 - y_S)$$

$$= 1 - \left(\frac{150 \, \dfrac{\text{kg}}{\text{h}}}{155.26 \, \dfrac{\text{kg}}{\text{h}}}\right)(1 - 0.0001)$$

$$= 0.033975 \quad (3.4\%)$$

The answer is (A).

7. Convert the feed flow rate and composition to a molar basis.

$$F = \left(35\,380 \, \frac{\text{kg}}{\text{h}}\right)\left(\frac{0.6 \, \dfrac{\text{kg AA}}{\text{kg feed}}}{60.05 \, \dfrac{\text{kg AA}}{\text{kmol}}} + \frac{0.4 \, \dfrac{\text{kg water}}{\text{kg feed}}}{18.02 \, \dfrac{\text{kg water}}{\text{kmol}}}\right)$$

$$= 1139 \text{ kmol/h}$$

The mole fraction of the feed is

$$x_F = \frac{\left(35\,380 \, \dfrac{\text{kg feed}}{\text{h}}\right)\left(0.4 \, \dfrac{\text{kg water}}{\text{kg feed}}\right)}{\left(18.02 \, \dfrac{\text{kg water}}{\text{kmol}}\right)\left(1139 \, \dfrac{\text{kmol}}{\text{h}}\right)}$$

$$= 0.6895$$

Write a mass balance equation around the column for water.

$$x_F F = x_D D + x_B B$$

$$= x_D(F - B) + x_B B$$

Rearrange to solve for the bottoms flow rate.

$$B = \frac{F(x_D - x_F)}{x_D - x_B}$$

$$= \frac{\left(1139 \ \frac{\text{kmol}}{\text{h}}\right)(0.9 - 0.6895)}{0.9 - 0.05}$$

$$= 282.1 \ \text{kmol/h}$$

95% of CH_3COOH means that 5% is water by mole. Converting to mass units,

$$\dot{m}_B = B(\text{MW})$$

$$= \left(282.1 \ \frac{\text{kmol}}{\text{h}}\right)$$

$$\times \left(\begin{array}{c} (0.95)\left(60.05 \ \dfrac{\text{kg } CH_3OOH}{\text{kmol}}\right) \\[2mm] + (0.05)\left(18.02 \ \dfrac{\text{kg water}}{\text{kmol}}\right) \end{array} \right)$$

$$= 16\,347 \ \text{kg/h} \quad (16\,000 \ \text{kg/h})$$

The answer is (D).

8. The relationship between the overall coefficient, K_G, and the individual coefficients, k_G and k_L, is

$$\frac{1}{K_G} = \frac{1}{k_G} + \frac{H}{k_L}$$

Solve for the liquid phase individual mass transfer coefficient, k_L.

$$k_L = \frac{H K_G k_G}{k_G - K_G}$$

$$= \frac{\left(1.26 \ \frac{\text{m}^3 \cdot \text{atm}}{\text{kmol}}\right)\left(0.582 \ \frac{\text{kmol}}{\text{m}^2 \cdot \text{s} \cdot \text{atm}}\right)}{2.27 \ \frac{\text{kmol}}{\text{m}^2 \cdot \text{s} \cdot \text{atm}} - 0.582 \ \frac{\text{kmol}}{\text{m}^2 \cdot \text{s} \cdot \text{atm}}}$$

$$= 0.986159 \ \text{kmol} \cdot \text{m}^3 / \text{m}^2 \cdot \text{s} \cdot \text{kmol}$$

The flux of ammonia (gas A) through stagnant air (gas B) is

$$N_A = k_G(p_{AG} - p_{Ai}) = k_L(C_{Ai} - C_{AL})$$

Eliminate flux and obtain an operating line slope.

$$N_A = \frac{p_{Ai} - p_{AG}}{C_{Ai} - C_{AL}} = -\frac{k_L}{k_G}$$

On (C_A, p_A) coordinates, the operating line is a straight line with slope $-k_L/k_G$ going through points (C_{Ai}, p_{Ai}) and (C_{AL}, p_{AG}).

Henry's law for the interface is

$$p_{Ai} = H C_{Ai}$$

The liquid phase concentration, C_{AL}, is zero since the problem statement calls for an analysis for the top of the tower where the water has not had a chance to pick up ammonia. The liquid phase concentration of ammonia will increase with distance down the tower. Solving the operating line and Henry's law simultaneously gives the interfacial concentration.

$$C_{Ai} = \frac{\dfrac{p_{AG}}{H} + \left(\dfrac{k_L}{H k_G}\right) C_{AL}}{1 + \dfrac{k_L}{H k_G}}$$

$$= \frac{\begin{array}{c} \dfrac{0.0197 \ \text{atm}}{1.26 \ \frac{\text{m}^3 \cdot \text{atm}}{\text{kmol}}} \\[4mm] + \left(\dfrac{0.986159 \ \frac{\text{kmol} \cdot \text{m}^3}{\text{m}^2 \cdot \text{s} \cdot \text{kmol}}}{\left(1.26 \ \frac{\text{m}^3 \cdot \text{atm}}{\text{kmol}}\right)\left(2.27 \ \frac{\text{kmol}}{\text{m}^2 \cdot \text{s} \cdot \text{atm}}\right)} \right) \\[4mm] \times \left(0 \ \frac{\text{kmol}}{\text{m}^3}\right) \end{array}}{1 + \dfrac{0.986159 \ \frac{\text{kmol} \cdot \text{m}^3}{\text{m}^2 \cdot \text{s} \cdot \text{kmol}}}{\left(1.26 \ \frac{\text{m}^3 \cdot \text{atm}}{\text{kmol}}\right)\left(2.27 \ \frac{\text{kmol}}{\text{m}^2 \cdot \text{s} \cdot \text{atm}}\right)}}$$

$$= 0.0116263 \ \text{kmol/m}^3$$

From Henry's law, the interfacial partial pressure is

$$p_{Ai} = H C_{Ai}$$

$$= \left(1.26 \ \frac{\text{m}^3 \cdot \text{atm}}{\text{kmol}}\right)\left(0.0116263 \ \frac{\text{kmol}}{\text{m}^3}\right)$$

$$= 0.0146491 \ \text{atm} \quad (0.015 \ \text{atm})$$

The answer is (A).

9. The liquid phase mass transfer coefficient, k'_L, can be calculated from the gas phase and overall mass transfer coefficients and Henry's law constant.

$$\frac{1}{K'_G} = \frac{1}{k'_G} + \frac{H}{k'_L}$$

$$k'_L = H\left(\frac{1}{K'_G} - \frac{1}{k_G}\right)^{-1}$$

$$= \left(2.70 \ \frac{m^3 \cdot atm}{kmol}\right)$$

$$\times \left(\frac{1}{9.20 \times 10^{-3} \ \dfrac{kmol}{m^2 \cdot s \cdot atm}} - \frac{1}{1.31 \times 10^{-2} \ \dfrac{kmol}{m^2 \cdot s \cdot atm}}\right)^{-1}$$

$$= 0.083 \ m/s$$

The answer is (C).

10. The relationship between the overall coefficient and the individual coefficient is

$$\frac{\text{resistance in gas phase}}{\text{resistance in both phases}} = \frac{\dfrac{1}{k_G}}{\dfrac{1}{K_G}} = 0.70$$

The individual gas phase mass transfer coefficient, k_G, is

$$k_G = \frac{K_G}{0.70}$$

$$= \frac{5.6 \times 10^{-2} \ \dfrac{kmol}{m^2 \cdot s \cdot atm}}{0.70}$$

$$= 0.08 \ kmol/m^2 \cdot s \cdot atm$$

The answer is (C).

Mass Transfer

44 Diffusion

Nomenclature

a	specific interfacial surface area	m^2/m^3 or $1/m$
A	area	m^2
c	molar concentration	$kmol/m^3$
c_m	molar concentration of solute in fluid (or generic gradient variable)	$kmol/m^3$ (various)
C	total molar concentration (all components)	$kmol/m^3$
C_{Ai}	molar concentration of component A in the liquid phase at the gas-liquid interface	$kmol/m^3$
C_{AL}	molar concentration of component A in the bulk liquid phase	$kmol/m^3$
C_A^*	molar concentration of component A in equilibrium with the bulk gas vapor composition of A	$kmol/m^3$
D	inside diameter	m
D_{AB}	molecular diffusivity of substance A through substance B	m^2/s
D_m	mass diffusivity (diffusion coefficient)	m^2/s
E_D	eddy diffusivity	m^2/s
f	Fanning friction factor	–
H	Henry's law constant[1]	$m^3 \cdot atm/kmol$
J	molar diffusive flux	$kmol/m^2 \cdot s$
k_m	mass transfer coefficient	m/s
$k_m a$	volumetric mass transfer coefficient	$1/s$
k_G'	gas phase mass transfer coefficient	$kmol/m^2 \cdot s \cdot atm$
k_L'	liquid phase mass transfer coefficient	m/s
K_G'	overall gas phase mass transfer coefficient	$kmol/m^2 \cdot s \cdot atm$
K_L'	overall liquid phase mass transfer coefficient	m/s
n	number of moles	$kmol$
N	total molar flux	$kmol/s$ or $kmol/m^2 \cdot s$
p	total pressure	Pa
p_{Ai}	partial pressure at component A at the gas-liquid interface	Pa
p_{AG}	partial pressure in component A in the bulk gas phase	Pa
p_i	partial pressure of component i	Pa
p_A^*	partial pressure of component A in equilibrium with C_{AL}	Pa
$\overline{R}$	universal gas constant, 8.314 (0.08206)	$kPa \cdot m^3/ kmol \cdot K$ ($m^3 \cdot atm/ kmol \cdot K$)
Re	Reynolds number	–
Sc	Schmidt number	–
Sh	Sherwood number	–
t	time	s
T	absolute temperature	K
v	velocity	m/s
V	average velocity	m/s
V	volume	m^3
x	mole fraction	–
y	distance from inner wall of pipe	m
z	distance in z-direction	m
z	distance from pipe centerline	m
z_+	dimensionless distance from inner wall of pipe	–

Symbols

ε	specific roughness	m
μ	absolute viscosity	$Pa \cdot s$
ν	kinematic viscosity	m^2/s
ρ	mass density	kg/m^3

1. MASS TRANSFER[2]

Mass transfer is the movement of a substance from one location to another location. Mass transfer occurs in many chemical and physical processes such as absorption, adsorption, drying, distillation, and extraction. These processes are usually referred to as *unit operations*.

The time rate of mass transfer per unit area is called *flux*. Flux can be the result of many mechanisms, including convection, thermal processes, mixing, and diffusion. When mass transfer is the result of diffusion, the term *diffusive flux* is used. When flux is the result of convection, the term *convective flux* is used. The rate of mass transfer

---

[1]There are four different formulations of Henry's law, each having different units for Henry's law constant.

[2]Most of the relationships in this chapter apply to dilute solutions only.

is proportional to transfer area, so flux is calculated per unit area. The total flux due to all mechanisms is expressed as the number of moles, N. The total mass transfer per unit area is N/A. *Molar flux* has units of moles of substance per unit area per unit time, such as kmol/m^2·s. *Mass flux* (not used in this chapter) has units of mass per unit area per unit time, such as kg/m^2·s.

Flux can involve a single substance (e.g., osmosis through a membrane), and it can result from two substances being physically combined, which is known as a *binary system.*

With most practical mass transfer operations, the flux of one substance into another results from two mechanisms: diffusion and bulk fluid movement. Depending on the process, one mechanism may be dominant. The bulk fluid movement mass transfer is referred to as *momentum diffusivity*. Momentum diffusivity may be due to pressure or shear stress, or both.

$$N = \text{diffusion mass transfer}$$
$$+ \text{bulk movement mass transfer}$$

Mass transfer rates are limited (i.e., are "rate limited") by conditions in the boundary layer. Real-world effects, such as eddies within the boundary layer, voids, separations from the surface, path restrictions, irregular catalyst presence, and obstructions, can greatly increase actual (as opposed to ideal) *diffusional resistance* within a boundary layer.

2. UNSTEADY-STATE MASS TRANSFER

Prior to steady-state mass transfer, there is a short period of *unsteady-state mass transfer* (USS mass transfer). USS mass transfer of component A in solids (and, in limited situations, in fluids) is governed by *Fick's second law of diffusion.*

$$\frac{\partial C_A}{\partial t} = D_m \frac{\partial^2 C_A}{\partial z^2}$$

This second-order differential equation can be solved analytically for the most basic configurations, but finite-element modeling (FEM) is needed for more complex situations. There are also various graphical means, including *concentration-distance* (C_A-z) *charts* and *time-concentration* (t-C_A) *charts* for common configurations. (See Fig. 44.1.)

Example

Which statement is UNTRUE as it relates to unsteady-state diffusion?

(A) The concentration at a particular location increases with time.

(B) The concentration gradient at a particular location decreases with time.

(C) The concentration at a particular moment depends on the location.

(D) The concentration gradient at a particular location decreases with distance.

Solution

In unsteady-state diffusion, the diffusing species concentration increases with time at a particular location. This translates into the concentration gradient decreasing with time at a particular location.

The answer is (D).

Figure 44.1 *Concentration-Distance Chart for Unsteady-State Diffusion*

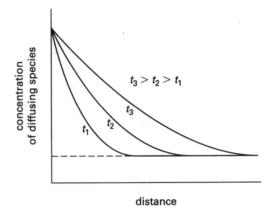

3. CONCENTRATION

In a chemical *solution*, one substance, the *solute*, is dissolved in another substance, the *solvent*. Concentration, c or C, is the measure of how much solute or solvent is in the total volume of the solution.[3] In mass transfer calculations, it is convenient to express concentration in units of moles per unit volume, such as kmol/m^3. In a *dilute solution*, the concentration of the solute is relatively low.

For gases, concentrations can easily be converted to partial pressures by using the *ideal gas law*, $pV = n\overline{R}T$. The concentration, C, is essentially the number of moles, n, per unit volume, V. There are multiple values associated with the universal gas constant, depending on the units used in an analysis. In order to obtain the correct units, the units of pressure for $\overline{R}$ must match the units of pressure for p. When p is given in atmospheres, $\overline{R} = 0.08206$ L·atm/mol·K (equivalent to 0.08206 m^3·atm/kmol·K) should be used.

$$\frac{n_A}{V} = C_A = \frac{p_A}{\overline{R}T}$$

If the total concentration of all molecular species, C, is known, the concentration can be derived from the mole fraction. In gases, the mole fraction is the same as the partial pressure fraction.

$$C_A = x_A C$$
$$C_A = \frac{p_A C}{p} \quad \text{[gases]}$$

---

[3]The NCEES *FE Reference Handbook* (*NCEES Handbook*) is inconsistent and uses both c and C for concentration in the *Chemical Engineering* section.

4. MASS TRANSFER COEFFICIENT

The *mass transfer coefficient*, k_m, is used to calculate the rate of mass transfer from all mechanisms.[4] It is almost always intended to be used in per unit area calculations. To calculate the total flux (as opposed to flux per unit area) in this manner, the mass transfer coefficient and the transfer area, A, must both be known separately.

$$\begin{bmatrix} \text{total rate of} \\ \text{mass transfer} \end{bmatrix} = k_m \begin{bmatrix} \text{interfacial} \\ \text{area} \end{bmatrix} [\text{gradient}]$$

Theoretically, it should be possible to calculate mass transfer coefficients, particularly with well-defined geometrics and areas. However, as is often the case in gas and liquid mixing or with incomplete wetting, the actual interfacial area will be indistinct and difficult to measure. In situations where the area is indistinct, such as when a liquid cascades over packing in a tower, the product of the mass transfer coefficient and a hypothetical area per unit volume of packing, a, is used. a is known as the *specific interfacial surface* or *specific exchange surface*. The product, $k_m a = k_L A$ (for liquid flow), is known as the *volumetric mass transfer coefficient* and the *overall mass transfer coefficient*, even though it is really a liquid transfer coefficient.

$$\begin{bmatrix} \text{total rate of} \\ \text{mass transfer} \\ \text{per unit volume} \end{bmatrix} = k_m a [\text{gradient}]$$

With two-phase systems (e.g., a gas diffusing into a liquid), four different mass transfer coefficients are used. For mass transfer from the gas phase into the liquid phase, the mass transfer coefficient is identified with the subscript "G," as k_G or K_G, depending on whether interfacial or bulk equilibrium properties are used, respectively. For mass transfer from the liquid phase into the gas phase, the mass transfer coefficient is identified with the subscript "L," as k_L or K_L, also depending on whether interfacial or bulk equilibrium properties are used, respectively.

5. GRADIENTS

The type of gradient used to calculate the rate of mass transfer depends on the mass transfer process. For molecular diffusion in liquids, the concentration gradient of a chemical species is commonly used. For thermal processes, a temperature gradient may be used. For gas phase processes, partial pressures may be used. Gradients are assumed to vary linearly over short distances. When the mass transfer occurs over a large distance (e.g., within a long pipe), it may be necessary to use a *logarithmic mean gradient* calculated from the properties at two distant points. Mass transfer coefficients for

use with linear gradients are not interchangeable with mass transfer coefficients for use with logarithmic mean gradients.

Most gradients are negative, since diffusion occurs from a high concentration (pressure, temperature, etc.) to a lower concentration.

$$\text{gradient} = \frac{\Delta C}{\Delta z} = \frac{C_2 - C_1}{z_2 - z_1} \quad \begin{bmatrix} \text{mass transfer in} \\ \text{the } z\text{-direction} \end{bmatrix}$$

6. MOLECULAR DIFFUSION

Molecular diffusion, often referred to as just *diffusion*, is the movement of the molecules of one substance through the molecules of another substance due to thermally induced molecular motion alone. Molecular diffusion occurs because molecules migrate in the direction of a decreasing concentration (i.e., from a region of higher concentration to a region of lower concentration), as shown in Fig. 44.2.

Figure 44.2 Diffusion

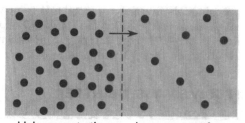

high concentration   low concentration

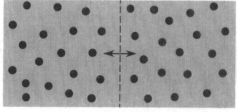

balanced

7. FICK'S FIRST LAW OF DIFFUSION

The rate of molecular diffusion is proportional to the *mass diffusivity*, D_m, also called *molecular diffusivity* or the *diffusion coefficient*. Mass diffusivity is a measure of the ability of the molecules of one substance to move through the molecules of another. Mass diffusivity is affected by temperature, viscosity (in the case of liquids), pressure (in the case of gases), and other factors.

Fick's First Law of Diffusion, often referred to simply as "Fick's Law," describes the molar flux of molecular diffusion through a dilute solution as a result of a *concentration gradient* alone. (The effect of any bulk fluid movement is not included.) When the concentration of substance A in one solution is higher than the concentration in another solution, and if a migration path exists,

---

[4]The mass transfer coefficient is analogous to the heat transfer coefficient in heat transfer.

substance A will diffuse molecularly into the solution with the lower concentration. The amount of substance A per unit area, J_A, that moves in the z-direction is given by Fick's law.

$$J_A = \frac{N_A}{A} = -D_m \frac{\partial c_A}{\partial z} \qquad \left[\begin{array}{l} \text{dilute solutions;} \\ \text{no bulk transfer} \end{array} \right]$$

D_m is the mass diffusivity, c_A is the concentration of substance A at a given point, and z is location in the z-direction. The partial derivative $\partial c_A/\partial z$ is the per unit distance change in concentration gradient of substance A in the z-direction. (The negative sign in the dilute solutions equation indicates the concentration gradient is negative for movement toward a decreasing concentration.)

Diffusion through gases is much higher (on the order of 10^4 to 10^5 times) than through liquids. For example, the diffusivity of air through liquid water is approximately 2×10^{-5} cm^2/s while the diffusivity of water molecules through air is approximately 0.28 cm^2/s.

The mass diffusivity, D_m, is influenced by local pressure and temperature. For liquids, the pressure dependency is given by the *Wilke-Chang equation*, $D_m \sim T^{1.75}/p$ (i.e., approximately inversely proportional to pressure). For ideal gases, as predicted by the Maxwell-Boltzmann distribution, $D_m \sim T^{1.50}/p$, although molecular size influences diffusion through real gases.

8. DIFFUSIVE MASS TRANSFER

Equation 44.1 and Eq. 44.2: Diffusive Mass Transfer at a Wall

$$\left(\frac{N}{A} \right)_w = -D_m \left(\frac{dc_m}{dy} \right)_w \qquad 44.1$$

$$\left(\frac{N}{A} \right)_w = k_m \Delta c_m \qquad 44.2$$

Description

Equation 44.1 and Eq. 44.2 apply to diffusive mass transfer into a dilute fluid at a stationary wall (e.g., as through a membrane or inside a tube). The subscript w indicates that $(N/A)_w$ is the inward mass transfer flux away from the wall of the tube. Equation 44.1 is derived from Fick's law and is appropriate only for dilute solutions. Equation 44.1 is intended to calculate the molar mass transfer in situations dominated by pure diffusion (e.g., osmosis through a membrane into a stationary fluid). y is the perpendicular distance from the wetted wall surface. When used with circular tubes, y is measured toward the centerline and cannot exceed the radius. The derivative $(dc_m/dy)_w$ is the concentration gradient at the wall measured in the y-direction of the solute.

Equation 44.2 is a general definition applicable to mass transfers at surfaces. It can be used to calculate the total mass transfer from all mechanisms. Although Δc_m uses the variable for species concentration, the gradient

should be interpreted more generally, including pressure and temperature differentials, according to the correlation used to develop the mass transfer coefficient, k_m. Equation 44.2 is compatible with processes dominated by convective mass transfer.

Equation 44.3: Diffusive Flux in a Liquid

$$N_A = x_A(N_A + N_B) - CD_m \frac{\partial x_A}{\partial z} \qquad 44.3$$

Description

Equation 44.3 is the general equation for steady-state diffusion in the z-direction of component A through liquid component B. Since there are two substances, this is a *binary system*. This equation shows that the net diffusion per unit area, N_A, is the sum of two mechanisms: bulk fluid transfer and molecular diffusion.[5] The negative sign counteracts the negative gradient. The molecular diffusion is not subtracted from the bulk fluid transfer component.

In Eq. 44.3, x_A is the mole fraction of solute A. The term $x_A(N_A + N_B)$ represents the mass transfer due to bulk fluid flow and other processes. This component is also known as *convective flux* when the bulk fluid flow is due to convection; the term $-CD_m(\partial x_A/\partial z)$ is the diffusive flux. C is the total molar concentration of all components in the system. The total concentration is constant in both liquid systems and nonreactive, constant pressure, constant temperature gaseous systems. The product Cx_A is equivalent to C_A.

In many common processes, the molar diffusive flux of B is equal to that of A but is in the opposite direction. This is not always the case, however, such as in a reactive environment where one mole of A produces two moles of B, $N_B = 2N_A$. In addition, the counterdiffusion is unequal where average velocities of A and B differ significantly, and where molecules of A and B differ significantly in size.

Example

In a binary liquid system, the mass diffusivity of substance A is 0.783×10^{-9} m^2/s, and the total molar concentration of all substances is 50.6 kmol/m^3. The mole fraction differential of substance A is -0.5 per meter. The molar diffusive flux of B is equal to that of A but is in the opposite direction. What is the molar diffusive flux of substance A in the z-direction?

(A) -4.1×10^{-9} mol/m$^2 \cdot$s

(B) -2.0×10^{-5} mol/m$^2 \cdot$s

(C) 1.9×10^{-8} mol/m$^2 \cdot$s

(D) 2.0×10^{-5} mol/m$^2 \cdot$s

---

[5](1) The *NCEES Handbook* is inconsistent in the variables used to represent flux per unit area. In Eq. 44.1 and Eq. 44.2, N/A is used. In Eq. 44.3, N is used. (2) The subscript A refers to component A, not area.

Solution

The diffusive fluxes of A and B due to bulk transfer are equal but in opposite directions, so $N_A + N_B = 0$.

$$N_A = x_A(N_A + N_B) - CD_m \frac{\partial x_A}{\partial z}$$

$$= -CD_m \frac{\partial x_A}{\partial z}$$

$$= -\left(50.6 \ \frac{\text{kmol}}{\text{m}^3}\right)\left(1000 \ \frac{\text{mol}}{\text{kmol}}\right)\left(0.783 \times 10^{-9} \ \frac{\text{m}^2}{\text{s}}\right)$$

$$\quad \times \left(-0.5 \ \frac{1}{\text{m}}\right)$$

$$= 1.981 \times 10^{-5} \ \text{mol/m}^2\text{·s} \quad (2.0 \times 10^{-5} \ \text{mol/m}^2\text{·s})$$

The answer is (D).

Equation 44.4 Diffusive Flux in a Gas

$$N_A = \frac{p_A}{P}(N_A + N_B) - \frac{D_m}{\overline{R}T} \frac{\partial p_A}{\partial z} \qquad \textit{44.4}$$

Description

Diffusive flux of a substance through a gas (i.e., in a binary system) results from the partial pressure gradient (equivalent to the mole fraction gradient) as well as the diffusivity. Equation 44.4 determines the overall molar diffusive flux per unit area of substance A, N_A. In Eq. 44.4, p_A is the partial pressure of gas A, and P is the total pressure of the system. $\overline{R}$ is the universal gas constant, and T is the absolute temperature. Since both the mole fraction and partial pressure decrease with distance, the partial pressure gradient is negative. The negative sign counteracts the negative gradient. The molecular diffusion is not subtracted from the bulk fluid transfer component.

Example

In a gaseous binary system with substances A and B at 100°C, the partial pressure differential of A in the z-direction is -1 kPa/m, and the mass diffusivity is $0.783 \times 10^{-4} \ \text{m}^2/\text{s}$. The molar diffusive flux of B due to bulk fluid transfer is equal to that of A but is in the opposite direction. Most nearly, what is the molar diffusive flux of A through B in the z-direction?

(A) $-2.5 \times 10^{-8} \ \text{mol/m}^2\text{·s}$

(B) $-2.5 \times 10^{-5} \ \text{mol/m}^2\text{·s}$

(C) $2.5 \times 10^{-8} \ \text{mol/m}^2\text{·s}$

(D) $2.5 \times 10^{-5} \ \text{mol/m}^2\text{·s}$

Solution

The diffusive fluxes of substance A and substance B due to bulk transfer are equal but in opposite directions, so $N_A + N_B = 0$.

$$N_A = \frac{p_A}{P}(N_A + N_B) - \frac{D_m}{\overline{R}T} \frac{\partial p_A}{\partial z}$$

$$= -\frac{D_m}{\overline{R}T} \frac{\partial p_A}{\partial z}$$

$$= -\left(\frac{0.783 \times 10^{-4} \ \frac{\text{m}^2}{\text{s}}}{\left(8.314 \ \frac{\text{kPa·m}^3}{\text{kmol·K}}\right)(100°\text{C} + 273°)}\right)\left(-1 \ \frac{\text{kPa}}{\text{m}}\right)$$

$$= 2.525 \times 10^{-5} \ \text{mol/m}^2\text{·s} \quad (2.5 \times 10^{-5} \ \text{mol/m}^2\text{·s})$$

The answer is (D).

Equation 44.5 and Eq. 44.6: Unidirectional Diffusion of Gas A into Stagnant Gas B

$$N_A = \frac{D_m P}{\overline{R}T(p_B)_{lm}} \times \frac{(p_{A2} - p_{A1})}{z_2 - z_1} \qquad \textit{44.5}$$

$$(p_{BM})_{lm} = \frac{p_{B2} - p_{B1}}{\ln\left(\dfrac{p_{B2}}{p_{B1}}\right)} \qquad \textit{44.6}$$

Description

The diffusion of gas A into gas B is generally accompanied by diffusion of gas B into gas A. However, when gas B is stagnant, its molar diffusive flux, N_B, is zero in Eq. 44.4. In this case, the diffusive flux of gas A in the z-direction is given by Eq. 44.5.[6] Subscripts 1 and 2 represent two locations along the mass transfer path.

Equation 44.5 is suitable when the diffusion path is long. $(p_B)_{lm}$ in Eq. 44.5 is the *logarithmic mean* of the partial pressures for gas B, p_{B1} and p_{B2}. The logarithmic mean is calculated from Eq. 44.6, except that when the two pressures are equal, that pressure is the logarithmic mean.[7] The logarithmic mean is always positive.

---

[6](1) The *NCEES Handbook* presents Eq. 44.5 as a positive number. This is inconsistent with Eq. 44.4 and will give the wrong sign to N_A. If Eq. 44.5 is to be a positive number for mass transfer in the z-direction, the numerator of Eq. 44.5 must be $p_{A1} - p_{A2}$. (2) There is no mathematical significance in the use of parentheses in the numerator of Eq. 44.5. The original intent may have been to have absolute value bars around this quantity.

[7]The *NCEES Handbook* gives the left side of Eq. 44.6 as $(p_{BM})_{lm}$. It should be $(p_B)_{lm}$ to match the variable in Eq. 44.5.

The case of stagnation for gas B does not necessarily mean that gas B is motionless. For example, when gas B flows perpendicularly past a tube containing gas A, any molecules of A that diffuse down the tube are swept away. In this situation, the pressure and concentration gradients are constant, which is equivalent to a stagnated gas B at the end of the tube.

Example

One end of the vertical tube shown is placed in 293K water. The tube extends upward for 0.2 m above the water's surface. The total atmospheric pressure is 1.0 atm. The tube is filled with still atmospheric air that is initially completely dry. The vapor pressure of water at 293K is 0.0231 atm, and the diffusivity of water in air is $0.250 \times 10^{-4} \text{ m}^2/\text{s}$.

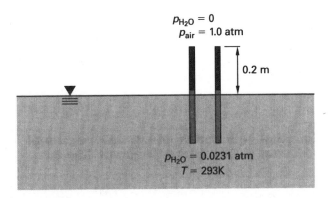

Most nearly, what is the molar diffusive flux of the water vapor through the tube?

(A) $-8.2 \times 10^{-9} \text{ mol/m}^2 \cdot \text{s}$

(B) $-1.2 \times 10^{-4} \text{ mol/m}^2 \cdot \text{s}$

(C) $2.2 \times 10^{-9} \text{ mol/m}^2 \cdot \text{s}$

(D) $1.2 \times 10^{-4} \text{ mol/m}^2 \cdot \text{s}$

Solution

At the bottom of the tube, point 1, the partial pressure of the water in the air is the vapor pressure of water. At the top of the tube, point 2, the air is completely dry, so the partial pressure of the water is zero. Gas A is the water vapor, and gas B is the dry atmospheric air.

$$p_{A1} = 0.0231 \text{ atm}$$

$$p_{B1} = p - p_{A1} = 1.0 \text{ atm} - 0.0231 \text{ atm} = 0.9769 \text{ atm}$$

$$p_{A2} = 0 \text{ atm}$$

$$p_{B2} = p - p_{A2} = 1.0 \text{ atm} - 0 \text{ atm} = 1.0 \text{ atm}$$

The properties are known at the two ends of the tube. 0.2 m (20 cm) is a long distance from the standpoint of diffusion, so using the logarithmic pressure gradient is

appropriate. The logarithmic mean of the two partial pressures for the air is

$$(p_B)_{lm} = \frac{p_{B2} - p_{B1}}{\ln\left(\dfrac{p_{B2}}{p_{B1}}\right)} = \frac{1.0 \text{ atm} - 0.9769 \text{ atm}}{\ln\dfrac{1.0 \text{ atm}}{0.9769 \text{ atm}}}$$

$$= 0.9884 \text{ atm}$$

The molar diffusive flux of the water vapor is given by Eq. 44.5. (In accordance with footnote 6, a negative sign is added to Eq. 44.5 to obtain a positive flux value in the z-direction.)

$$N_A = -\left(\frac{D_m p}{\overline{R}T(p_B)_{lm}}\right)\left(\frac{p_{A2} - p_{A1}}{z_2 - z_1}\right)$$

$$= -\frac{\left(0.250 \times 10^{-4} \dfrac{\text{m}^2}{\text{s}}\right)(1.0 \text{ atm})\left(1.013 \times 10^5 \dfrac{\text{Pa}}{\text{atm}}\right)}{\left(8.314 \dfrac{\text{Pa} \cdot \text{m}^3}{\text{mol} \cdot \text{K}}\right)(293\text{K})(0.9884 \text{ atm})}$$

$$\times \left(\frac{0 \text{ atm} - 0.0231 \text{ atm}}{0.2 \text{ m} - 0 \text{ m}}\right)$$

$$= 1.215 \times 10^{-4} \text{ mol/m}^2 \cdot \text{s} \quad (1.2 \times 10^{-4} \text{ mol/m}^2 \cdot \text{s})$$

The answer is (D).

Equation 44.7 through Eq. 44.9: Equimolar Counterdiffusion in Gases and Liquids

$N_B = -N_A$	**44.7**
$N_A = D_m/(RT) \times [(p_{A1} - p_{A2})/(\Delta z)]$	**44.8**
$N_A = D_m(C_{A1} - C_{A2})/\Delta z$	**44.9**

Description

When gas B has a molar flux that is equal to gas A but in the opposite direction, the $N_A + N_B$ term in Eq. 44.4 is zero. This pattern of diffusion is called *equimolar counterdiffusion* and is described by Eq. 44.7 through Eq. 44.9.[8] Equimolar counterdiffusion commonly occurs with diffusion in nonreacting, constant pressure liquid and constant pressure, constant temperature gas processes, such as between two gas vessels communicating through a tube. Since the pressure on the tube wall is constant, the number of molecules hitting the wall is constant. When a molecule of substance A diffuses into substance B, it must be replaced by a molecule of substance B in order to maintain a constant pressure on the wall. This is a common scenario for most instances of equimolar counterdiffusion.

Equation 44.7 states that the two fluxes are equal but in opposite directions. Equation 44.8 is derived from

---

[8]Δz in Eq. 44.8 and Eq. 44.9 is ambiguous. Since Eq. 44.8 is positive and the partial pressure difference is $p_{A1} - p_{A2}$, Δz must be interpreted as $z_2 - z_1$ for mass transfer in the positive z-direction.

Eq. 44.4 in terms of partial pressures.[9] Equation 44.9 is derived from Eq. 44.8 in terms of concentration using the ideal gas law.

Example

Ammonia gas at 298K diffuses through a 0.5 m long pipe containing nitrogen at constant pressure. At end 1, the partial pressure of ammonia is 1.013×10^5 Pa, and at end 2, the partial pressure is 0.507×10^5 Pa. The diffusivity of ammonia in nitrogen is 0.230×10^{-4} m^2/s. Most nearly, what is the molar diffusive flux of the ammonia gas?

(A) 4.9×10^{-4} mol/m^2·s

(B) 9.4×10^{-4} mol/m^2·s

(C) 7.6×10^{-3} mol/m^2·s

(D) 4.8×10^{-3} mol/m^2·s

Solution

Since the pressure in the pipe is constant along its length, this is a case of equimolar counterdiffusion. Use Eq. 44.8. Ammonia is gas A. The molar diffusive flux of the ammonia gas in the z-direction is

$$N_A = D_m/(RT) \times [(p_{A1} - p_{A2})/(\Delta z)]$$

$$= \left(\frac{0.230 \times 10^{-4} \frac{\text{m}^2}{\text{s}}}{\left(8.314 \frac{\text{Pa·m}^3}{\text{mol·K}}\right)(298\text{K})} \right)$$

$$\times \left(\frac{1.013 \times 10^5 \text{ Pa} - 0.507 \times 10^5 \text{ Pa}}{0.5 \text{ m} - 0 \text{ m}} \right)$$

$$= 9.395 \times 10^{-4} \text{ mol/m}^2\text{·s}$$

$$(9.4 \times 10^{-4} \text{ mol/m}^2\text{·s})$$

The answer is (B).

9. CONVECTIVE MASS TRANSFER

Equation 44.10: Schmidt Number

$$Sc = \mu/(\rho D_m) \qquad \textbf{44.10}$$

Variations

$$Sc \equiv \frac{\nu}{D_m} = \frac{\mu}{\rho D_m}$$

$$Sc = \frac{\text{momentum diffusivity}}{\text{mass diffusivity}}$$

Values

For mass transfer through ideal gases, the Schmidt number is 1, since $\nu = D_m$. For mass transfer through real gases, typical values of the Schmidt number range from 0.5 to 5. For liquids, the Schmidt number can range from about 100 to more than 10,000 in the case of viscous liquids. The Schmidt number is essentially independent of temperature and pressure within normal operating conditions.

Description

Mass transfer within a fluid occurs rapidly when the fluid is in motion or is being mixed. Correlations with dimensionless numbers are often used to quantify mass transfer through fluids experiencing convection and other forms of bulk fluid movement.

The flux of substance A through substance B with bulk movement can be characterized by the *Schmidt number*,[10] Sc. The Schmidt number is the dimensionless ratio of momentum (convective) diffusivity and molecular (mass) diffusivity. That is, the Schmidt number is the ratio of bulk mass transfer to mass transfer by diffusion. Momentum diffusivity is represented by the *kinematic viscosity*, ν, equivalent to μ/ρ. The molecular diffusivity is represented by the mass diffusivity, D_m.

Equation 44.11: Sherwood Number for Turbulent Forced Convective Flow Inside a Circular Tube

$$Sh = \left(\frac{k_m D}{D_m}\right) = 0.023 \left(\frac{DV\rho}{\mu}\right)^{0.8} \left(\frac{\mu}{\rho D_m}\right)^{1/3} \qquad \textbf{44.11}$$

Variations

$$Sh \approx 0.023 (\text{Re})^{0.8} (\text{Sc})^{1/3}$$

$$Sh = \frac{kL}{D} = \frac{\text{mass transfer rate}}{\text{diffusion rate}}$$

Description

The *Sherwood number*, Sh, is the dimensionless ratio of the total rate of mass transfer to the mass diffusivity (i.e., rate of mass transfer due to diffusion alone.)[11]

---

[9](1) Equation 44.8 is poorly formed in the *NCEES Handbook*. The quantity in square brackets represents the partial pressure gradient. It is not part of the denominator with RT. (2) There is no mathematical significance to the parentheses around RT. (3) The *NCEES Handbook* is inconsistent in the variable used for universal gas constant. In Eq. 44.8, the *NCEES Handbook* uses R for the universal gas constant, but it uses $\overline{R}$ in Eq. 44.4 and Eq. 44.5 on the same page.

[10]The Schmidt number is the mass transfer analog to the *Prandtl number*, which is used with heat transfer.
[11]The Sherwood number is the mass transfer analog to the *Nusselt number*, which is used with heat transfer.

Mass Transfer

As the variation to Eq. 44.11 shows, the Sherwood number depends on the Schmidt number and the *Reynolds number*.

Equation 44.11 is a correlation intended specifically for the case of turbulent fluid (gas or liquid) flow inside a circular tube or pipe. Turbulent fluid flow is commonly encountered within a circular pipe when a liquid is pumped. The Sherwood number is calculated as kL/D_m, where k is a transfer coefficient and L is some characteristic dimension to which mass transfer has been correlated. The characteristic dimension used is the inside pipe diameter.

In Eq. 44.11, D is the inside diameter of the tube, and D_m is the mass diffusivity. The variables ρ, μ, and V are the mass density, absolute viscosity, and average velocity, respectively, of the fluid in the tube, respectively, not the pipe properties.[12] This correlation is valid only for $4000 < \text{Re} < 6000$, and $0.6 < \text{Sc} < 3000$.

Example

A circular tube has an inside diameter of 0.02 m and a length of 1 m. Air at 318K and an average pressure of 101.3 kPa flows through the tube at an average velocity of 5.4 m/s. The mass diffusivity of the air is 6.92×10^{-6} m^2/s, the density is 1.114 kg/m^3, and the absolute viscosity is 1.932×10^{-5} Pa·s. Most nearly, what is the mass transfer coefficient?

(A) 1.2×10^{-2} m/s

(B) 2.3×10^{-2} m/s

(C) 3.6×10^{-2} m/s

(D) 4.0×10^{-2} m/s

Solution

From Eq. 13.6, the Reynolds number is

$$\text{Re} = \frac{VD}{\nu} = \frac{VD\rho}{\mu}$$

$$= \frac{\left(5.4 \, \frac{\text{m}}{\text{s}}\right)(0.02 \text{ m})\left(1.114 \, \frac{\text{kg}}{\text{m}^3}\right)}{1.932 \times 10^{-5} \text{ Pa·s}}$$

$$= 6227$$

Since Re > 2000, the air flow is turbulent. Equation 44.11 can be used.

---

[12](1) Other than specifying "turbulent," the *NCEES Handbook* presents Eq. 44.11 without limitations on use. Although dimensional analysis can be used to derive the basic form of Eq. 44.11, the values (e.g., 0.023, 0.8, and 1/3) are derived from experiments. In fact, different researchers have derived different values of the constant and exponents. (2) The first part of Eq. 44.11 is an exact definition of the Sherwood number. The second part of Eq. 44.11 is an approximate correlation that is appropriate only for turbulent flow through circular pipes within specific ranges of the Reynolds and Schmidt numbers. Fe second part of Eq. 44.11 is not a universal definition or an absolute primary law of physics or engineering.

Rearrange Eq. 44.11 to find the mass transfer coefficient.

$$\text{Sh} = \left(\frac{k_m D}{D_m}\right) = 0.023\left(\frac{DV\rho}{\mu}\right)^{0.8}\left(\frac{\mu}{\rho D_m}\right)^{1/3}$$

$$k_m = 0.023\left(\frac{D_m}{D}\right)\left(\frac{DV\rho}{\mu}\right)^{0.8}\left(\frac{\mu}{\rho D_m}\right)^{1/3}$$

$$= (0.023)\left(\frac{6.92 \times 10^{-6} \, \frac{\text{m}^2}{\text{s}}}{0.02 \text{ m}}\right)$$

$$\times \left(\frac{(0.02 \text{ m})\left(5.4 \, \frac{\text{m}}{\text{s}}\right)\left(1.114 \, \frac{\text{kg}}{\text{m}^3}\right)}{1.932 \times 10^{-5} \text{ Pa·s}}\right)^{0.8}$$

$$\times \left(\frac{1.932 \times 10^{-5} \text{ Pa·s}}{\left(1.114 \, \frac{\text{kg}}{\text{m}^3}\right)\left(6.92 \times 10^{-6} \, \frac{\text{m}^2}{\text{s}}\right)}\right)^{1/3}$$

$$= 1.173 \times 10^{-2} \text{ m/s} \quad (1.2 \times 10^{-2} \text{ m/s})$$

The answer is (A).

Equation 44.12 Through Eq. 44.15: Two-Film Theory for Convection with Equimolar Counterdiffusion

$N_A = k'_G(p_{AG} - p_{Ai})$	**44.12**
$N_A = k'_L(C_{Ai} - C_{AL})$	**44.13**
$N_A = K'_G(p_{AG} - p_A^*)$	**44.14**
$N_A = K'_L(C_A^* - C_{AL})$	**44.15**

Description

Two-film theory describes the mass transfer between a gas and a liquid, with equimolar counterdiffusion occurring at the gas-liquid interface through a film of liquid and a film of gas. (See Fig. 44.3.) Phase equilibrium is assumed at the gas-liquid interface. In practical unit operations, such as gas stripping, the flows of gas and liquid result from blowers and fans, so such operations are described as forced convection processes. The bulk transport mechanism dominates.

As shown in Eq. 44.12 through Eq. 44.15, mass transfer in a two-film environment follows the model $N_A = k \times$ driving force. (See Fig. 44.4.) In Eq. 44.12 and Eq. 44.14 the subscript G is used for gas-side properties. In Eq. 44.13 and Eq. 44.15 the subscript L is used for liquid-side properties.

The driving force can be a gas partial pressure difference, or it can be a liquid concentration difference. In Eq. 44.12, p_{Ai} is the partial pressure of A in the gas at

Figure 44.3 *Two-Film Diffusion*

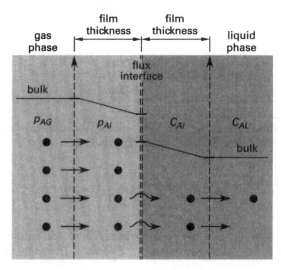

Figure 44.4 *Two-Film Convention Driving Force Variables*

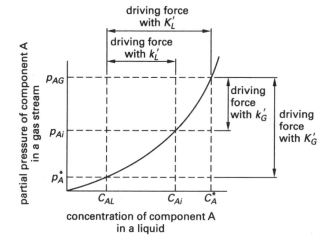

the interface. In Eq. 44.13, C_{Ai} is the concentration of substance A at the interface. Since Eq. 44.12 and Eq. 44.13 each depend on properties within an individual phase (i.e., p_{AG} and p_{Ai} are both gas phase partial pressures), the mass transfer coefficients, k', are known as *individual mass transfer coefficients*. Since the properties at the interface are difficult to measure (making the determination of the corresponding mass transfer coefficients, k', equally difficult), equilibrium properties in the bulk phases are often preferred. The resulting mass transfer coefficients, K', are known as *overall mass transfer coefficients*. In Eq. 44.14, p_A^* is the partial pressure in equilibrium with C_{AL}. C_A^* in Eq. 44.15 is the concentration in equilibrium with p_{AG}. Neither of these conditions actually occur within the substances. They are simply convenient values to use in the correlations that determine the mass transfer coefficients. This simplification allows the mass transfer coefficient to be derived by ignoring the fine details of what is occurring within the system.

Example

What is the liquid-side mass transfer coefficient based on interfacial properties for a liquid film flowing down a vertical wall against an airstream flowing upward?

(A) k'_G

(B) k'_L

(C) K'_G

(D) K'_L

Solution

K'_G and K'_L are both overall mass transfer coefficients that do not depend on interfacial properties. k'_L and k'_G are individual mass transfer coefficients that depend on interfacial properties, but only k'_L is a liquid-side coefficient.

The answer is (B).

Equation 44.16 and Eq. 44.19: Mass Transfer Coefficient Relationships

$$1/K'_G = 1/k'_G + H/k'_L \qquad 44.16$$
$$1/K'_L = 1/Hk'_G + 1/k'_L \qquad 44.17$$
$$p_A^* = HC_{AL} \qquad 44.18$$
$$C_A^* = p_{AG}/H \qquad 44.19$$

Description

It is often necessary to convert between individual and overall mass transfer coefficients. The relationships between the overall mass transfer coefficients, the individual mass transfer coefficients, and *Henry's law constant*, H, are expressed in Eq. 44.16 and Eq. 44.17. Equation 44.18 and Eq. 44.19 illustrate how Henry's law constant can be used to convert bulk properties into bulk equilibrium properties. There are many definitions of Henry's law and its coefficient. H in these equations has units of m³·atm/kmol. Henry's law constant is a *partitioning coefficient* between the gas and liquids.

Example

For gas absorption involving the diffusion of gas A through stagnant gas B, k'_G is 7.11×10^{-3} kmol/m²·s·atm, and k'_L is 4.62 m/s. The total pressure is 1 atm, and the temperature is 298K. The Henry's law constant for the system is 0.746 m³·atm/kmol. For counterdiffusion of gas A and gas B under the same conditions of flow, temperature, and concentration, what is most nearly the overall gas mass transfer coefficient?

(A) 1.5×10^{-4} kmol/m²·s·atm

(B) 7.1×10^{-3} kmol/m²·s·atm

(C) 3.3×10^{-2} kmol/m²·s·atm

(D) 4.6×10^{-2} kmol/m²·s·atm

Solution

Use Eq. 44.16.

$$1/K'_G = 1/k'_G + H/k'_L$$

$$K'_G = \frac{1}{\dfrac{1}{k'_G} + \dfrac{H}{k'_L}}$$

$$= \frac{1}{\dfrac{1}{7.11 \times 10^{-3}\ \dfrac{\text{kmol}}{\text{m}^2 \cdot \text{s} \cdot \text{atm}}} + \dfrac{0.746\ \dfrac{\text{m}^3 \cdot \text{atm}}{\text{kmol}}}{4.62\ \dfrac{\text{m}}{\text{s}}}}$$

$$= 7.102 \times 10^{-3}\ \text{kmol/m}^2 \cdot \text{s} \cdot \text{atm}$$

$$(7.1 \times 10^{-3}\ \text{kmol/m}^2 \cdot \text{s} \cdot \text{atm})$$

The answer is (B).

10. EDDY DIFFUSION[13]

Eddy diffusion theory extends molecular diffusion to situations involving turbulent flow eddies in a fluid. The eddies in the fluid bring about a transfer of dissolved solute. Both molecular and eddy diffusivities are included in the total flux, which depends on the local turbulence intensity.

The molar diffusional flux of solute A, including molecular diffusivity, D_{AB}, and *eddy diffusivity*, E_D, can be found from a modified Fick's law.

$$J_A = -(D_{AB} + E_D)\frac{\partial c_A}{\partial z}$$

An equation that predicts eddy diffusivity in turbulent flow for a circular pipe in the region from $z_+ = 0$ to $z_+ = 45$ with Sc > 1 is

$$\frac{E_D}{\nu} = \frac{0.00090(z_+)^3}{\sqrt{1 + 0.0067(z_+)^2}}$$

In this equation, the eddy diffusivity depends on a dimensionless distance from the wall, z_+.

$$z_+ = \left(\frac{D}{2} - z\right)\left(\frac{\text{v}}{\nu}\right)\sqrt{\frac{f_{\text{Fanning}}}{2}}$$

At the wall, z_+ is zero, so E_D is also zero and molecular diffusivity dominates. The *Fanning friction factor* (which is one-fourth of the *Darcy friction factor*), f_{Fanning}, can be obtained from a friction factor chart or from the *Colebrook equation*.

$$\frac{1}{\sqrt{4f_{\text{Fanning}}}} = -2\log_{10}\left(\frac{2.512}{\text{Re}\sqrt{4f_{\text{Fanning}}}} + \frac{1}{3.7}\left(\frac{\varepsilon}{D}\right)\right)$$

ε is the pipe surface roughness, and D is the pipe diameter. This equation is implicit in the friction factor and must be solved iteratively. The factor of 4 in the equation converts the Fanning friction factor to the Darcy friction factor.

---

[13]No content for this subject is provided in the *NCEES Handbook* even though it is listed in the NCEES outline of exam subjects for the Chemical FE exam.

Diagnostic Exam

Topic XI: Unit Processes

1. A still pot initially contains 55 mole % of solute A and 45 mole % of solute B. The mixture is separated into a distillate and a residue using simple distillation. Solute A is more volatile than solute B, and the relative volatility of solute A to solute B is constant. The initial amount of solute A is 150 moles. At the end of the distillation, 87 moles of distillate are produced, and the residue in the still pot contains 36 mole % of solute A. Most nearly, what is the mole fraction of solute A in the distillate?

(A) 54%

(B) 59%

(C) 72%

(D) 77%

2. In a coal-burning power plant, flue gas containing sulfur dioxide (SO_2) is scrubbed with water. The mole fraction of SO_2 in the discharged gases is 0.004, and the equilibrium constant for SO_2 in water is 40. Assume the system is in equilibrium. Most nearly, what is the equilibrium solute mole fraction in the liquid phase?

(A) 0.00010

(B) 0.016

(C) 0.10

(D) 0.16

3. A packed column is used to clean a gas. The gas phase transfer unit height of the column is 2.0 m. There are 8.2 gas phase transfer units. What is the required height of the packed column?

(A) 4.1 m

(B) 8.3 m

(C) 13 m

(D) 17 m

4. Which of the following statements is NOT a McCabe-Thiele assumption, and therefore, does NOT contribute to identifying the rectifying line in a McCabe-Thiele diagram?

(A) The latent heats of the column components are equal and constant.

(B) The component enthalpy changes and heats of mixing are negligible compared to latent heats.

(C) The column is well insulated so that heat loss is negligible.

(D) The rectifying section pressure is greater than the stripping section pressure.

5. Samples are taken from two adjacent stages of an operating distillation column. The vapor phase mole fraction of the sample from the lower stage is 0.64, and the vapor phase mole fraction of the sample from the upper stage is 0.61. The mole fraction for the vapor in equilibrium with the liquid in the upper stage is 0.60. Most nearly, what is the plate efficiency?

(A) 10%

(B) 25%

(C) 60%

(D) 75%

6. A wetted water column removes solute A through absorption from a mixture of air and solute A. The equilibrium curve of solute A absorption is shown.

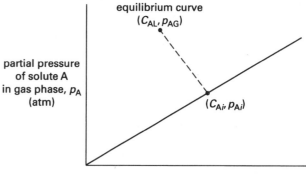

Point (C_{AL}, p_{AG}) in the graph is the bulk-phase concentration and partial pressure, and point (C_{Ai}, p_{Ai}) is the gas-liquid interface concentration and partial pressure. The slope of the dashed line connecting points (C_{AL}, p_{AG}) and (C_{Ai}, p_{Ai}) is -6.67×10^{-2} $m^3 \cdot$atm/kmol. The liquid phase driving force for the process is 4.9×10^{-2} kmol/m^3. Most nearly, what is the partial-pressure driving force in the gas phase?

(A) 2.2×10^{-3} atm

(B) 3.3×10^{-3} atm

(C) 7.3×10^{-3} atm

(D) 5.1×10^{-2} atm

7. A continuous distillation column is used to separate a mixture of cyclohexane (C_6H_{12}) and toluene ($C_6H_5CH_3$). The column feed contains 60% C_6H_{12} and 40% $C_6H_5CH_3$ by mol. The mixture is fed as a saturated vapor. The volatile distillate fraction contains 90% C_6H_{12} by mol. The bottoms (residuum) fraction contains 5% C_6H_{12} by mol. The vapor-liquid equilibrium curve for the distillation of C_6H_{12} and $C_6H_5CH_3$ is shown.

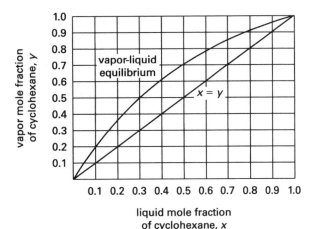

liquid mole fraction
of cyclohexane, x

The reflux ratio is 4. From the top of the column, which stage is the optimum feed location?

(A) stage 3

(B) stage 4

(C) stage 5

(D) stage 6

8. A 1.2 m absorption column is used to remove carbon dioxide (CO_2) from a stream of flue gas. The CO_2 is absorbed using a dilute solution of monoethanolamine (C_2H_7NO), which reacts with the absorbed CO_2. The height of each gas phase transfer unit is 20 cm. The mole fraction of CO_2 in the flue gas entering the column is 0.12. Most nearly, what is the mole fraction of CO_2 in the gas exiting the absorption column?

(A) 0.00030

(B) 0.0016

(C) 0.011

(D) 0.036

9. 57 g of a nonvolatile, nonelectrolytic compound is dissolved in 500 g of 20°C water. The vapor pressure of pure 20°C water is approximately 2.338 kPa. The vapor pressure of the water mixture is approximately 2.32573 kPa. Most nearly, what is the molecular weight of the compound?

(A) 2.1 g/mol

(B) 68 g/mol

(C) 390 g/mol

(D) 3400 g/mol

10. Ethylidene bromide ($C_2H_4Br_2$, 1, 1-dibromoethane) and propylene dibromide ($C_3H_6Br_2$, 1-2-dibromopropane) form an ideal solution when the two are mixed. At 85°C, the vapor pressures of the two pure liquids are 173 mm Hg and 127 mm Hg, respectively. 10 g of ethylidene bromide is dissolved in 80 g of propylene dibromide. Most nearly, what is the partial pressure of propylene dibromide?

(A) 15 mm Hg

(B) 20 mm Hg

(C) 110 mm Hg

(D) 150 mm Hg

SOLUTIONS

1. Use a mole balance equation to determine the amount of residue left in the still pot at the end of the distillation. The initial amount of substance in the still pot, W_0, is the sum of the amount remaining in the still pot, W, and the amount of distilled product, P.

$$W_0 = W + P$$
$$W = W_0 - P = 150 \text{ mol} - 87 \text{ mol}$$
$$= 63 \text{ mol}$$

Rearrange the Rayleigh equation to solve for the relative volatility, α.

$$\ln\left(\frac{W}{W_0}\right) = \frac{1}{\alpha - 1}\ln\left(\frac{x(1 - x_0)}{x_0(1 - x)}\right) + \ln\left(\frac{1 - x_0}{1 - x}\right)$$

$$\alpha = 1 + \frac{\ln\left(\dfrac{x(1 - x_0)}{x_0(1 - x)}\right)}{\ln\left(\dfrac{W}{W_0}\right) - \ln\left(\dfrac{1 - x_0}{1 - x}\right)}$$

$$= 1 + \frac{\ln\left(\dfrac{(0.36)(1 - 0.55)}{(0.55)(1 - 0.36)}\right)}{\ln\left(\dfrac{63 \text{ mol}}{150 \text{ mol}}\right) - \ln\left(\dfrac{1 - 0.55}{1 - 0.36}\right)}$$

$$= 2.5061$$

The mole fraction of solute A in the distilled product is

$$y = \frac{\alpha x}{1 + (\alpha - 1)x}$$
$$= \frac{(2.5061)(0.36)}{1 + (2.5061 - 1)(0.36)}$$
$$= 0.585 \quad (59\%)$$

The answer is (B).

2. The mole fraction of SO_2, y, is the vapor (gas) phase mole fraction. The equilibrium solute mole fraction in the liquid phase, x^*, is

$$x^* = \frac{y}{K} = \frac{0.004}{40}$$
$$= 0.00010$$

The answer is (A).

3. The required height, Z, of the packed column is

$$Z = \text{NTU}_G \cdot \text{HTU}_G = (8.2 \text{ units})\left(2.0 \; \frac{\text{m}}{\text{unit}}\right)$$
$$= 16.4 \text{ m}$$

Since 16.4 m is the minimum height, the required height is 17 m.

The answer is (D).

4. The McCabe-Thiele assumptions are

I.   The latent heats of the column components are equal and constant.

II. The component enthalpy changes and heats of mixing are negligible compared to latent heats.

III. The column is well insulated column so that heat loss is negligible.

IV. There is no pressure drop within the column.

These assumptions lead to the condition of constant molar overflow in the rectifying section. This means that the molar liquid flow rate remains constant as the liquid overflows from one stage to the next. As a result, the rectifying section pressure is NOT assumed to be greater than the stripping section pressure.

The answer is (D).

5. The Murphree plate efficiency, E_{ME}, is

$$E_{\text{ME}} = \frac{y_n - y_{n+1}}{y_n^* - y_{n+1}}$$
$$= \frac{0.61 - 0.64}{0.60 - 0.64}$$
$$= 0.75 \quad (75\%)$$

The answer is (D).

6. The slope, m, of the line joining the two points is

$$m = \frac{p_{Ai} - p_{AG}}{C_{Ai} - C_{AL}}$$

The liquid-phase driving force is the difference in concentrations of solute A at the gas-liquid phase interface and in the bulk liquid phase. Similarly, the partial-pressure driving force in the gas phase is the difference in the partial pressures of solute A in the bulk-gas phase and at the gas-liquid phase interface.

Rearranging the slope equation, the partial-pressure driving force in the gas phase is

$$p_{AG} - p_{Ai} = -m(C_{Ai} - C_{AL})$$
$$= -\left(-6.67 \times 10^{-2} \; \frac{\text{m}^3 \cdot \text{atm}}{\text{kmol}}\right)$$
$$\times \left(4.9 \times 10^{-2} \; \frac{\text{kmol}}{\text{m}^3}\right)$$
$$= 3.27 \times 10^{-3} \text{ atm} \quad (3.3 \times 10^{-3} \text{ atm})$$

The answer is (B).

Unit Processes

7. The rectifying section operating line starts from $x_D = 0.90$ at the $x = y$ line. Determine the y-intercept for the rectifying section operating line by comparing the equation for a straight line, $y = mx + c$, to the equation for the operating line with reflux ratio.

$$y = \frac{R}{R+1}x + \frac{x_D}{R+1}$$

Calculate the value of c for the y-intercept of the rectifying line.

$$c = \frac{x_D}{R+1} = \frac{0.90}{4+1}$$
$$= 0.18$$

Plot the lines as shown.

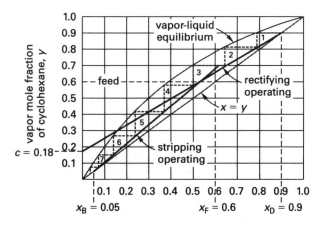

Since the mixture is fed as a saturated vapor, the feed condition factor (i.e., the slope of feed line), q, is 0 (i.e., horizontal) and intersects the rectifying section operating line. The stripping section operating line starts at $x_B = 0.05$ and passes through the intersection point of the feed line and the rectifying section operating line.

Determine the number of theoretical stages by stepping down from the starting point of the rectifying section operating line to the starting point of the stripping section operating line. The number of stages is 8.

The optimum feed location from the top of the column is the stage where the feed line, the rectifying section operating line, and the stripping section operating line intersect. Starting from the top of the column, the optimum feed location is in stage 3.

The answer is (A).

8. The number of gas phase transfer units in a continuous contact column, NTU_G, is

$$Z = \text{NTU}_G \cdot \text{HTU}_G$$

$$\text{NTU}_G = \frac{Z}{\text{HTU}_G}$$

$$= \frac{(1.2 \text{ m})\left(100 \ \frac{\text{cm}}{\text{m}}\right)}{20 \text{ cm}}$$

$$= 6$$

Since C_2H_7NO reacts with CO_2, the mole fraction of CO_2 in the exiting gas stream is determined using the NTU equation for a chemically reacting system.

$$\text{NTU}_G = \ln\left(\frac{y_1}{y_2}\right)$$

$$e^{\text{NTU}_G} = \frac{y_1}{y_2}$$

$$y_2 = \frac{y_1}{e^{\text{NTU}_G}} = \frac{\frac{12\%}{100\%}}{e^6}$$

$$= 0.0002975 \quad (0.00030)$$

The answer is (A).

9. The pressure change, Δp, when the compound is mixed into the water is

$$\Delta p = p_{H_2O} - p_{mix}$$
$$\approx 2.338 \text{ kPa} - 2.32573 \text{ kPa}$$
$$= 0.01227 \text{ kPa}$$

Determine the number of moles of the compound.

$$\Delta p = x_{compound}p_{H_2O}^0 = \left(\frac{n_{compound}}{n_{compound} + n_{H_2O}}\right)p_{H_2O}^0$$

$$n_{compound} = \frac{n_{H_2O}\Delta p}{p_{H_2O}^0 - \Delta p} = \frac{\left(\frac{m_{H_2O}}{MW_{H_2O}}\right)\Delta p}{p_{H_2O}^0 - \Delta p}$$

$$= \frac{\left(\frac{500 \text{ g}}{18 \ \frac{\text{g}}{\text{mol}}}\right)(0.01227 \text{ kPa})}{2.338 \text{ kPa} - 0.01227 \text{ kPa}}$$

$$= 0.1465 \text{ mol}$$

The molecular weight of the compound, $MW_{compound}$, is

$$MW_{compound} = \frac{m_{compound}}{n_{compound}}$$

$$= \frac{57 \text{ g}}{0.1465 \text{ mol}}$$

$$= 388.9 \text{ g/mol} \quad (390 \text{ g/mol})$$

The answer is (C).

10. Determine the molecular weight of each substance.

$$MW_{C_2H_4Br_2} = 2MW_C + 4MW_H + 2MW_{Br}$$

$$= (2)(12.011) + (4)(1.0080) + (2)(79.904)$$

$$= 187.862 \quad (187.9)$$

$$MW_{C_3H_6Br_2} = 3MW_C + 6MW_H + 2MW_{Br}$$

$$= (3)(12.011) + (6)(1.0080) + (2)(79.904)$$

$$= 201.889 \quad (201.9)$$

Determine the number of moles of each substance in the mixture.

$$n_{C_2H_4Br_2} = \frac{m_{C_2H_4Br_2}}{MW_{C_2H_4Br_2}} = \frac{10 \text{ g}}{187.9 \dfrac{\text{g}}{\text{mol}}}$$

$$= 0.0532 \text{ mol}$$

$$n_{C_3H_6Br_2} = \frac{m_{C_3H_6Br_2}}{MW_{C_3H_6Br_2}} = \frac{80 \text{ g}}{201.9 \dfrac{\text{g}}{\text{mol}}}$$

$$= 0.3962 \text{ mol}$$

Applying Raoult's law, the partial pressure of propylene dibromide, $p_{C_3H_6Br_2}$, is

$$p_{C_3H_6Br_2} = x_{C_3H_6Br_2} p^0_{C_3H_6Br_2}$$

$$= \left(\frac{n_{C_3H_6Br_2}}{n_{C_3H_6Br_2} + n_{C_2H_4Br_2}} \right) p^0_{C_3H_6Br_2}$$

$$= \left(\frac{0.3962 \text{ mol}}{0.0532 \text{ mol} + 0.3962 \text{ mol}} \right)(127 \text{ mm Hg})$$

$$= 111.97 \text{ mm Hg} \quad (110 \text{ mm Hg})$$

The answer is (C).

Unit Processes

45 Vapor-Liquid and Gas-Liquid Separation Processes

Nomenclature

a	interfacial area (mass transfer area per unit volume of column)	m^2/m^3
A	absorption factor	–
A	area of column	m^2
B	bottoms product flow rate	kmol/s
C_f	packing factor	–
D	diameter	m
D	overhead (distillate) flow rate	kmol/s
E	efficiency	–
f	fraction of feed vaporized	–
F	feed flow rate	kmol/s
G	gas phase mass velocity	$kmol/m^2 \cdot s$
G	gas phase upflow rate	kmol/s
G'	superficial molar gas mass velocity	$kmol/m^2 \cdot s$
h	specific enthalpy	kJ/kg
H	molar enthalpy	kJ/kmol
H'	Henry's law constant	atm

HETP	height equivalent to a theoretical plate	m
HTU	height of transfer unit	m
K	molar equilibrium ratio	–
K'_G	overall gas phase mass transfer coefficient	$kmol/m^2 \cdot s$
K'_L	overall liquid phase mass transfer coefficient	$kmol/m^2 \cdot s$
L	liquid phase downflow rate	kmol/s
L	liquid phase mass velocity	$kmol/m^2 \cdot s$
L'	superficial molar liquid mass velocity	$kmol/m^2 \cdot s$
N	total number of stages	–
N_{OG}	number of overall gas phase transfer units	–
NTU	number of transfer units	–
p	pressure, total pressure, or vapor pressure	atm
p_i	partial pressure of component i	atm
q	feed quality	–
q	molar heat transfer rate	kW/kmol
Q	total heat transfer rate	kW
R_D	ratio of reflux to overhead product rates	–
S	surface area	m^2
T	temperature	°C
v_0	superficial mass velocity	m/s
V	vapor-phase flow rate	kmol/s
V	volume	m^3
W	total number of moles in still pot (vessel)	–
x	mole fraction of the more volatile component in the liquid phase	–
x^*	equilibrium solute mole fraction in liquid phase	–
X	mole ratio of solute in liquid phase	–
y	mole fraction of the more volatile component in the vapor phase	–
y^*	equilibrium solute mole fraction in gas phase	–
y^*	equilibrium mole fraction vapor in Murphree efficiency	–
Y	mole ratio of solute in gas phase	–
z	feed mole fraction	–
Z	column height	m

Symbols

α	relative volatility	–
ε	porosity (void fraction)	–
μ	absolute viscosity	kg/m·s (Pa·s) (cP in Eckert's correlation)

Unit Processes

ρ	density	kg/m^3
Φ	sphericity	–

Subscripts

a	component a
b	component b
B	bottoms product
C	condenser
D	overhead product (distillate)
EQ	equilibrium
f	saturated fluid
fg	vaporization
F	feed
g	saturated gas
G	gas
i	interstitial
L	saturated liquid
LM	logarithmic mean
m	tray (plate or stage) m in the stripping section
ME	Murphree efficiency
n	tray (plate or stage) n in the rectifying section
o	original condition in still pot, or superficial (overall)
p	particle (packing)
R	reboiler, rectifying section, or reflux
s	solute
S	stripping section
sat	saturated
V	saturated vapor

1. UNIT OPERATIONS

A *unit operation* is a basic single-function process that can be used by itself or combined with other unit operations as part of a larger, multi-step process. Unit operations can be classified by the basic principles used in their analysis and design. These operations include

- fluid flow (momentum transfer) processes (such as filtration and fluidization)

- heat transfer processes (such as evaporation and heat exchange)

- mass transfer processes (such as distillation and extraction separations)

- thermodynamic processes (such as refrigeration and gas liquefaction)

- mechanical processes (such as pulverizing and sieving)

2. SEPARATION PROCESSES

A *separation process* is a process for separating a mixture (known as the *input, feed,* or *charge*) into two or more distinct products, which are sometimes referred to as *fractions*. The purpose of a separation process is to obtain one or more fractions that are more pure or desirable than the original mixture. For example, petroleum can be separated into gasoline, diesel oil, lubricating oil, asphalt, and so on, each of which is better suited for particular uses than the original petroleum.

The most common separation processes are flash distillation, continuous column distillation, batch distillation, absorption, stripping, and extraction.[1]

In the analysis and design of a separation process, it is commonly assumed that the process incorporates a series of *equilibrium stages*. Each equilibrium stage is a discrete portion of the process in which two phases (such as liquid and vapor) reach an equilibrium condition with each other. With this assumption, concentration profiles can be estimated throughout the process without having detailed information about momentum, heat, and mass transfer. The output of one equilibrium stage becomes the input into the next one, and so on, until the desired degree of separation is achieved.

In distillation, the two components in the feed, a and b, have different volatilities. The component with the lower boiling point is known as the "more volatile component." The component with the higher boiling point is known as the "less volatile component." The conditions of the feed with respect to the flash chamber are supercritical for one or both of the components, so the more volatile part of the liquid vaporizes. The mole fraction of the target substance in the liquid phase is designated x. The mole fraction of the target substance in the vapor phase is designated y.

Equation 45.1 and Eq. 45.2: Relative Volatility

$$\alpha = (y/x)_a/(y/x)_b = p_a/p_b \qquad \text{45.1}$$

$$y = \alpha x/[1 + (\alpha - 1)x] \qquad \text{45.2}$$

Description

Equation 45.1 gives the volatility of component a relative to the volatility of component b (i.e., the *relative volatility*) for a binary system.[2] *Raoult's law* $(p_i = x_i p_{i,\text{sat}})$ is used to substitute saturation pressures for mole fractions.[3] p_a is the saturated vapor pressure of component a; p_b is the saturated vapor pressure of component b. Relative volatility is generally assumed to be constant, in which case mole fractions y and x are related by Eq. 45.2.

3. BINARY FLASH DISTILLATION

The simplest separation process is *binary flash distillation*, also known as *equilibrium distillation*. (See Fig. 45.1.) A liquid stream of mixture components a and b (with total feed flow rate F, molar composition

---

[1]Extraction, including liquid-liquid and solid-phase extraction, is not part of the NCEES FE exam outline and is not covered in this book.
[2]The NCEES *FE Reference Handbook* (*NCEES Handbook*) is incorrect in defining p_a and p_b as partial pressures in Eq. 45.1, as these quantities are saturated vapor pressures.
[3](1) Although not specified in the *NCEES Handbook*, p_a and p_b are the saturated vapor pressures. (2) To derive Eq. 45.1, write the partial pressures as $p_i = y_i p = x_i p_{i,\text{sat}}$, where p is the total pressure, and calculate the ratio of the saturation pressures.

$z_a + z_b = 1$, feed temperature T_F, feed pressure p_F, and feed enthalpy H_F) is fed through a valve and into a *flash chamber* (*flash vessel*, *flash drum*, etc.) at temperature T_{vessel} and pressure p_{vessel} (where $p_{vessel} < p_F$).

The overall energy balance is

$$FH_F + Q_{flash} = VH_V + LH_L$$

When the valve and flash chamber are thermally insulated, the flash process can be assumed to be adiabatic, with the heat transfer rate of the flash process, Q_{flash}, equal to zero. In Fig. 45.1, Q_{flash} is the heat entering the vessel. Q_{flash} is negative for heat losses from the vessel, the usual case. Flash distillation is an equilibrium process, as it essentially happens instantaneously within the flash chamber.

The vapor phase (with flow rate V, compositions y_a and y_b, and enthalpy H_V) will be enriched in the more volatile component, while the liquid phase (flow rate L, compositions x_a and x_b, and enthalpy H_L) will be enriched in the less volatile component. This *enrichment* is the basis of the separation process.

Figure 45.1 *Binary Flash Distillation*

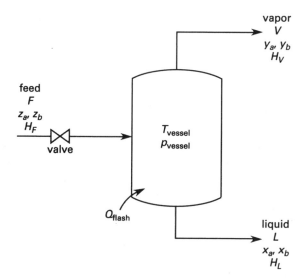

The energy and mass relationship for a binary, nonadiabatic flash distillation has six variables (i.e., six degrees of freedom). The conditions of the feed are usually known, specifying four of these variables: feed flow rate, composition, pressure, and temperature. The pressure in the flash chamber is usually unknown. The last variable is some other related property in the flash chamber (e.g., temperature, fraction of liquid vaporized, or heat transfer rate). The unknown design parameters can be determined by solving the design equations in a systematic manner, either sequentially or simultaneously.

Equation 45.3 and Eq. 45.4: Flash (or Equilibrium) Distillation Mass Balance

$$F = V + L \qquad \text{45.3}$$

$$Fz_F = yV + xL \qquad \text{45.4}$$

Description

Equation 45.3 is the overall material balance for a flash distillation system. The sum of the vapor product flow rate, V, and the liquid product flow rate, L, is equal to the feed flow rate, F. The material flow rates are usually stated in terms of moles per unit time (e.g., kmol/s), although mass flow rates (e.g., kg/s) can also be used with appropriate modifications to the analysis method.[4]

The individual component flow balance is given in Eq. 45.4. In this equation, z_F is the fraction of the target component in the feed, y is the fraction of the target component in the vapor product, and x is the fraction of the target component in the liquid product.

Example

1 mol/h of a mixture of 47 mole % benzene and 53 mole % toluene is separated by flash distillation. At the process exit, the ratio of the mole fraction of benzene in the vapor phase to the mole fraction of benzene in the liquid phase is 1.748. The ratio of the molar vapor flow rate to the molar liquid flow rate is 1.2. Most nearly, what is the mole fraction of benzene in the exit vapor phase?

(A) 0.6

(B) 0.7

(C) 0.8

(D) 0.9

Solution

The ratio of the vapor flow rate to the liquid flow rate is

$$\frac{V}{L} = 1.2$$
$$V = 1.2L$$

Use Eq. 45.3 to find the vapor and liquid flow rates.

$$F = V + L$$
$$1 \ \frac{mol}{h} = 1.2L + L$$
$$L = \frac{1 \ \dfrac{mol}{h}}{2.2} = 0.455 \ mol/h$$

---

[4]The *NCEES Handbook* uses molar rates exclusively.

The ratio of mole fractions of benzene in vapor and liquid phases is defined as the *equilibrium ratio* for benzene, K.

$$K = \frac{y}{x} = 1.748$$

$$y = 1.748x$$

Use Eq. 45.4 to find the mole fraction of benzene in the liquid phase.

$$Fz_F = yV + xL = (1.748x)\,V + xL$$

$$x = \frac{Fz_F}{1.748\,V + L}$$

$$= \frac{\left(1\,\frac{\text{mol}}{\text{h}}\right)(0.47)}{(1.748)\left(0.545\,\frac{\text{mol}}{\text{h}}\right) + 0.455\,\frac{\text{mol}}{\text{h}}}$$

$$= 0.334$$

The mole fraction of benzene in the vapor phase is

$$y = 1.748x = (1.748)(0.334)$$

$$= 0.584 \quad (0.6)$$

The answer is (A).

4. OPERATING LINE

Equation 45.5: Operating Line for Binary Flash Distillation

$$y = \frac{R_{\min}}{R_{\min} + 1}x + \frac{x_D}{R_{\min} + 1} \qquad \textbf{45.5}$$

Variations

$$y = -\left(\frac{L}{V}\right)x + \left(\frac{F}{V}\right)z$$

$$y = -\left(\frac{1-f}{f}\right)x + \frac{z}{f}$$

$$y = -\left(\frac{q}{1-q}\right)x + \left(\frac{1}{1-q}\right)z$$

The *operating line*, represented by the *operating equation*, for a binary system is obtained by solving Eq. 45.3 and Eq. 45.4 for y. The result is shown as the first variation equation. The operating equation can also be written in terms of the *fraction of feed vaporized, f*, which is equal to V/F, the *feed quality, q*, the fraction of liquid remaining, L/F, and the *minimum reflux ratio*, $R_{\min}$, all of which are introduced in subsequent sections.

5. VAPOR-LIQUID EQUILIBRIUM DIAGRAMS

A *vapor-liquid equilibrium* (*VLE*) *diagram* (*equilibrium curve*), as shown in Fig. 45.2, is a graph of the corresponding mole fractional compositions when a liquid and vapor are in equilibrium. Since the entire interior of a distillation or separation column is assumed to be approximately at the same pressure, a VLE diagram represents isobaric (i.e., constant pressure) conditions. The assumption of constant pressure on the column trays is valid except for very low column pressures. Being "in equilibrium" means that the rate of evaporation (liquid changing to vapor) equals the rate of condensation (vapor changing to liquid). A substance at vapor-liquid equilibrium is in a saturated liquid state.

Figure 45.2 VLE Diagram

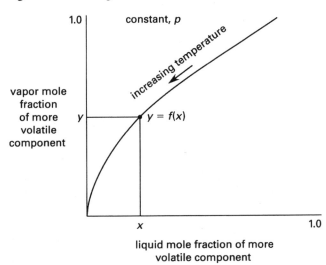

Although for most practical purposes, equilibrium is achieved when the liquid particles are small and surrounded by their vapor, creating large liquid surface areas and open access to the vapor component, equilibrium is not fully achieved at any stage. This nonideal behavior may be accounted for by modifying the equilibrium curve (or the equilibrium data). Since the extent of nonideal performance is usually unknown before a column is built, the nonideal performance is usually determined from plate efficiencies after construction. If known, the resulting *pseudo-equilibrium curve* (or *data*) should be used instead of the ideal curve (data) when graphically solving separation process problems.

6. GRAPHICAL SOLUTION TO FLASH DISTILLATION

Flash distillation problems can be solved with a relatively simple graphical method. With this method, it is assumed that the flash distillation occurs in a single equilibrium stage, an assumption that is valid in practice in large flash chambers. This graphical method plots the operating line on a VLE diagram.

On a VLE diagram, the *operating line*, represented by Eq. 45.5, is a straight line with a slope of $-L/V$ and a y-intercept of $(F/V)z$. Because L and V are both positive, the slope of the operating line is negative. The diagonal straight line starting at the origin is the $y = x$ line. The deviation of the equilibrium line from the $y = x$ line is an indication of how much separation will take place. The resulting diagram, shown in Fig. 45.3, is known as a *McCabe-Thiele diagram*.

Figure 45.3 *McCabe-Thiele Diagram for Binary Flash Distillation*

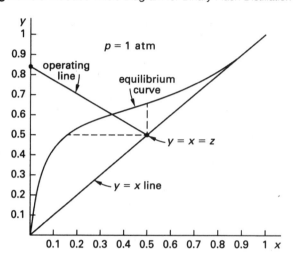

The McCabe-Thiele method is considered to be the simplest and most instructional method for analyzing binary distillation. The method results directly from the fact that the composition at each theoretical tray (or equilibrium stage) is completely determined by the mole fractions of one of the two components. The method relies on the assumption of *constant molar overflow*. Constant molar overflow requires that (1) the molar heats of vaporization of the feed components are equal, (2) for every mole of liquid vaporized, a mole of vapor is condensed, and (3) heat and energy effects (such as heat of solution) are negligible.

Specifying the four feed properties (F, z, T_F, and p_F) and the flash chamber pressure, p, leaves one degree of freedom. Selecting a variable that specifies a condition within the flash vessel, such as a liquid or vapor mole fraction, allows a simple graphical technique to be used to obtain the solution. This also allows the mass and energy balances to be uncoupled and solved sequentially. The intersection of the equilibrium curve and the operating line identifies the compositions of the liquid and vapor leaving the flash chamber. At the point where the operating line intersects the $y = x$ line, $y = x = z$. (See Fig. 45.3.) The maximum vapor composition is found by drawing a vertical line from the $y = x = z$ point to the vapor-liquid equilibrium curve. A horizontal line from the $y = x = z$ point determines the minimum liquid composition. The intersection of the operating line with the equilibrium curve gives the actual composition of the vapor and liquid phases.

7. DIFFERENTIAL DISTILLATION

In *batch distillation*, a fixed quantity of a binary liquid mixture (i.e., the "batch") is placed in the vessel (also called a *still or still pot*) and heated. When the distillation process is complete, the vessel is emptied and charged with a new batch.

Differential distillation (also called *simple distillation* or *Rayleigh distillation*) is the simplest form of batch distillation. A quantity of a binary mixture (the *initial charge*) is placed in the vessel and heated. As the mixture is heated, some of the mixture vaporizes. This vapor is channeled into a condenser and removed from the vessel, so that the amount of charge in the vessel decreases as the process continues.

The liquid remaining in the vessel becomes increasingly concentrated in the less volatile component as the distillation process continues. Similarly, the vapor that is drawn from the vessel will be decreasingly rich in the more volatile component as the distillation process continues.

Equation 45.6 and Eq. 45.7: Differential Distillation Material Balance

$$\ln\left(\frac{W}{W_o}\right) = \int_{x_o}^{x} \frac{dx}{y - x} \qquad 45.6$$

$$\ln\left(\frac{W}{W_o}\right) = \frac{1}{(\alpha - 1)} \ln\left[\frac{x(1 - x_o)}{x_o(1 - x)}\right] + \ln\left[\frac{1 - x_o}{1 - x}\right] \qquad 45.7$$

Description

Equation 45.6 is the differential equation describing the total amount (in moles) of the target component remaining in the distillation vessel given an initial charge, W_o, and vapor products that are removed over time. The concentration of the more volatile target component in the vessel will decrease as the process continues. Equation 45.6 can be written as Eq. 45.7 if there is a constant relative volatility and if Eq. 45.2 is substituted for y. x and y are the liquid and vapor phase fractions, respectively, of the more volatile target component.

Example

A 100 kmol charge of a 28 mole % organic compound in hexane is adiabatically flashed in a still pot. The average relative volatility of the hexane is 3.2 and is assumed to be constant. The final mole fraction of the organic compound in the liquid phase is 0.017. Most nearly, how much organic compound is left in the still pot?

(A) 18 kmol

(B) 47 kmol

(C) 53 kmol

(D) 88 kmol

Solution

The amount of organic compound remaining in the still pot can be determined using Eq. 45.7.

$$\ln\left(\frac{W}{W_o}\right) = \frac{1}{(\alpha - 1)} \ln\left[\frac{x(1 - x_o)}{x_o(1 - x)}\right] + \ln\left[\frac{1 - x_o}{1 - x}\right]$$

$$\ln\left(\frac{W}{100 \text{ kmol}}\right) = \left(\frac{1}{3.2 - 1}\right)\ln\frac{(0.017)(1 - 0.28)}{(0.28)(1 - 0.017)}$$
$$+ \ln\frac{1 - 0.28}{1 - 0.017}$$
$$= -1.726$$
$$W = (100 \text{ kmol})e^{-1.726}$$
$$= 17.8 \text{ kmol} \quad (18 \text{ kmol})$$

The answer is (A).

8. CONTINUOUS DISTILLATION

Continuous distillation is a process in which a stream of the liquid mixture is fed continuously into the vessel and the products of the distillation are continuously removed. *Column distillation* is the practical implementation of continuous distillation. Column distillation occurs in *staged columns* where liquid and vapor intermix in discrete stages on *trays* (*plates*).

Column distillation of binary systems can be considered as a countercurrent cascade of flash distillation units referred to as *stages*. As with flash distillation units, the analysis and design of distillation columns is based on the assumption that the liquid and vapor leaving each stage are in equilibrium, hence the term *equilibrium stage*. The pressure drop between stages is negligible, so the pressure at every stage is the same as the overall column pressure. The stages, however, operate at different temperatures.

Figure 45.4 shows an *N*-stage distillation column, with equilibrium stage 1 at the top and final stage *N* at the bottom. Subscript numbers refer to the stages being exited. Variables x and y represent the liquid and vapor mole fractions, respectively, of the more volatile component. The stream to be distilled, called the *feed*, is introduced into one of the stages in the column. The molar feed flow rate is F.

The portion of the column above the feed point is the *rectifying section*, where the rising vapor becomes enriched with the more volatile component. The flow rates of the vapor and liquid leaving a stage n in the rectifying section are V_n and L_n, respectively. The portion of the column below the feed is the *stripping section*, where the falling liquid is stripped of the more volatile component and becomes enriched with the less volatile component. The flow rates of the vapor and liquid leaving a stage m in the stripping section are V_m and L_m, respectively.

The *overhead stream*, called the *distillate*, contains a greater concentration of the more volatile component.

Figure 45.4 N-Stage Distillation Column[*]

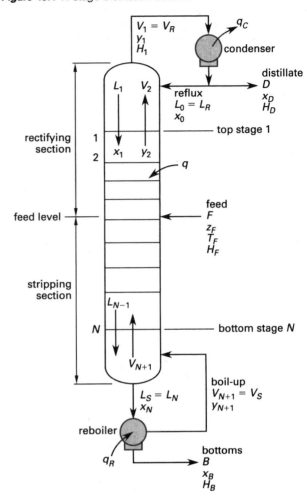

[*]In the *NCEES Handbook*, the corresponding figure to Fig. 45.4 uses uppercase X, not lowercase x, to represent the mole fraction of the more volatile component in the liquid phase.

The stream removed from the bottom of the column is called the *bottoms*, and it consists of the concentrated less volatile component. The molar flow rates of the distillate and bottoms streams are D and B, respectively.

A portion of the condensed distillate, the *liquid reflux*, is fed back into the column. A *partial condenser* condenses only part of the vapor stream leaving the top of a distillation column. The rectifying section operating equation for a column with a partial condenser is the same as that for a column with a total condenser. A partial condenser functions as an equilibrium stage at the top of the column. A partial condenser may be counted as an equilibrium stage. A total condenser does not act as an equilibrium stage.

Reflux is beneficial because it allows heating for a longer time without losing volatile reactants and products. Refluxing helps speed up chemical reactions without having the components evaporate. By cooling the vapor before it escapes from the system, a constant volume of solution is maintained.

The *boil-up*, a portion of the bottoms vaporized by the column's *reboiler*, may be fed back into the column as well. When only a portion of the bottoms is returned to the column, the term *partial reboiler* is used. A *partial reboiler* receives a bottoms stream in liquid form and discharges liquid and vapor streams in equilibrium. Alternatively, the liquid bottoms from a partial reboiler may be split into a bottoms product with the remainder fed to another reboiler to be partially or totally vaporized. A partial reboiler may be counted as an equilibrium stage.

A hypothetical condition known as *total reflux* occurs when all vapor is condensed and returned from the condenser back into the column, and all bottoms are returned from the reboiler back into the column. When the reboiler vaporizes the entire liquid stream, the reboiler is referred to as a *total reboiler*.

The stripping section operating equation for a column with a total reboiler is the same as that for a column with a partial reboiler. The total reboiler does not act as an equilibrium stage since the liquid is simply vaporized without separation occurring.

The column pressure, feed flow rate, feed composition, feed temperature, enthalpy (or quality or vapor fraction), and reflux temperature or enthalpy are usually known. The unknown variables depend on whether a new column is being designed or an existing column is being analyzed. In design situations, the variables associated with a desired separation are specified and the column details are unknown.

Equation 45.8: External Reflux Ratio

$$R_D = L_R/D = (V_R - D)/D \qquad \textbf{45.8}$$

Variation

$$R_D = \frac{L_0}{D}$$

Description

Equation 45.8 defines the *external reflux ratio* (generally referred to as just the *reflux ratio*) in terms of the liquid reflux flow rate entering the first stage of the distillation column, L_R, and the distillate flow rate (*overhead product* flow rate), D. The *minimum reflux ratio*, R_{min}, is the reflux ratio that will result in an infinite number of stages.

Example

An ethanol-water mixture is distilled in a multistage distillation column. The condenser loading is 1000 kmol/h. The flow rate of the distillate is 850 kmol/h. Most nearly, what is the reflux ratio for the ethanol-water distillation column?

(A) 0.18

(B) 0.47

(C) 0.85

(D) 0.93

Solution

From Eq. 45.8, the reflux ratio is

$$R_D = \frac{V_R - D}{D} = \frac{1000 \, \frac{\text{kmol}}{\text{h}} - 850 \, \frac{\text{kmol}}{\text{h}}}{850 \, \frac{\text{kmol}}{\text{h}}}$$

$$= 0.176 \quad (0.18)$$

The answer is (A).

Equation 45.9 and Eq. 45.10: Continuous Distillation Overall Material Balances

$$F = D + B \qquad \textbf{45.9}$$
$$Fz_F = Dx_D + Bx_B \qquad \textbf{45.10}$$

Description

Equation 45.9 gives the overall (molar) material balance for continuous distillation. The sum of the flow rates for the distillate, D, and the bottoms, B, is equal to the feed flow rate, F. Equation 45.10 gives the target component balance for the system.[5] The feed mole fraction is z. Since the vapor product is usually condensed (i.e., is in a liquid phase), the mole fractions in both the distillate and bottoms streams can be represented by x, with a distinction made by subscripts.

Example

A distillation column with a total condenser and a partial reboiler is used to separate an aqueous ethanol mixture containing 20 mole % ethanol. The feed to the column is 1000 kmol/h. The distillate composition is 80 mole % ethanol, and the desired bottoms concentration is 0.02 mole % ethanol. Most nearly, what is the distillate flow?

(A) 100 kmol/h

(B) 150 kmol/h

(C) 250 kmol/h

(D) 500 kmol/h

---

[5]As a vestige of the original reference used to produce it, the *NCEES Handbook* presents Eq. 45.10, Eq. 45.12, Eq. 45.13, Eq. 45.15, and Eq. 45.16 as pertaining to "component A." Although a common practice in some textbooks, this is inconsistent with the labels "a" and "b" used in Eq. 45.1. Component A is understood to represent the more volatile target component (i.e., the desired product).

Solution

Solve Eq. 45.9 for the bottoms flow rate.

$$F = D + B$$

$$B = F - D$$

Substitute the expression for the bottoms flow rate into Eq. 45.8 to find the distillate flow rate.

$$Fz_F = Dx_D + Bx_B$$

$$= Dx_D + (F - D)x_B$$

$$D = \frac{F(z_F - x_B)}{x_D - x_B}$$

$$= \frac{\left(1000 \frac{\text{kmol}}{\text{h}}\right)(0.20 - 0.0002)}{0.80 - 0.0002}$$

$$= 249.8 \text{ kmol/h} \quad (250 \text{ kmol/h})$$

The answer is (C).

9. STAGE MATERIAL BALANCES

The rectifying section of a distillation column is located above the feed stage. In this section, the vapor phase becomes enriched with the more volatile component. The stripping section of a distillation column is located below the feed stage. This is the section where the vapor phase is stripped of the more volatile component.

The subscript n refers to any tray number in the rectifying section. The subscript m refers to any tray number in the stripping section. Trays are numbered from the top down, so the subscript $n+1$ refers to the tray below tray n, and the subscript $m+1$ refers to the tray below tray m.

Equation 45.11 Through Eq. 45.13: Rectifying Section Material Balances

$$V_{n+1} = L_n + D \qquad \text{45.11}$$
$$V_{n+1}y_{n+1} = L_nx_n + Dx_D \qquad \text{45.12}$$
$$y_{n+1} = [L_n/(L_n + D)]x_n + Dx_D/(L_n + D) \qquad \text{45.13}$$

Variation

$$y_{n+1} = \left(\frac{L_n}{V_{n+1}}\right)x_n + \left(\frac{D}{V_{n+1}}\right)x_D$$

Description

Equation 45.11 is the overall material balance around stage n in the rectifying section. Equation 45.12 is the material balance around stage n for the more volatile component alone.[6]

---

[6]The *NCEES Handbook* calls the more volatile component "Component *A*."

The operating equation, Eq. 45.13, for each stage of the rectifying section is derived by combining Eq. 45.11 and Eq. 45.12. The operating equation relates the flow rate, L_n, and concentrations x_n and y_{n+1} of the passing streams. Figure 45.5 shows a rectifying section with its corresponding nomenclature.

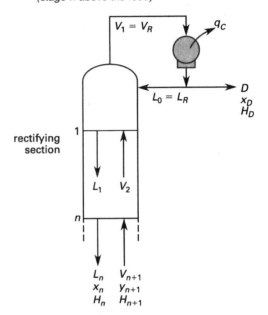

Figure 45.5 *Nomenclature for Rectifying Section (stage n above the feed)*

Example

A binary chemical mixture is processed in a multistage distillation column. The liquid flow rate leaving tray 1 in the rectifying section is 1000 kmol/h; the distillate flow rate is 800 kmol/h with 85 mole % volatile composition. The liquid phase composition leaving tray 1 is 83 mole %. Most nearly, what is the vapor phase mole fraction leaving tray 2?

(A) 82%

(B) 84%

(C) 86%

(D) 88%

Solution

From Eq. 45.13, the mole fraction of the component in the vapor phase leaving tray 2 is

$$y_{n+1} = \frac{L_n}{(L_n + D)x_n} + \frac{Dx_D}{L_n + D}$$

$$y_2 = \left(\frac{L_1}{L_1 + D}\right)x_1 + \frac{Dx_D}{L_1 + D}$$

$$= \left(\frac{1000 \frac{kmol}{h}}{1000 \frac{kmol}{h} + 800 \frac{kmol}{h}}\right)(0.83)$$

$$+ \frac{\left(800 \frac{kmol}{h}\right)(0.85)}{1000 \frac{kmol}{h} + 800 \frac{kmol}{h}}$$

$$= 0.839 \quad (84\%)$$

The answer is (B).

Equation 45.14 Through Eq. 45.16: Stripping Section Material Balances

$$L_m = V_{m+1} + B \qquad 45.14$$
$$L_m x_m = V_{m+1} y_{m+1} + B x_B \qquad 45.15$$
$$y = [L_m/(L_m - B)]x_m - B x_B/(L_m - B) \qquad 45.16$$

Variation

$$y_{m+1} = \left(\frac{L_m}{V_{m+1}}\right)x_m - \left(\frac{B}{V_{m+1}}\right)x_B$$

Description

The material balances in the stripping section are similar to those for the rectifying section. Equation 45.14 gives the overall material balance around stage $m+1$ in the stripping section. Equation 45.15 gives the material balance around stage $m+1$ for the more volatile component alone.[7]

The operating equation, Eq. 45.16, for each stage of the stripping section is derived by combining Eq. 45.14 and Eq. 45.15. The operating equation relates the flow rate, L_m, and concentrations x_m and y_{m+1} of the passing streams. Figure 45.6 shows a stripping section with its corresponding nomenclature.

Example.

In a 14-tray distillation column, the liquid flow leaving tray 12 in the stripping section is 1000 kmol/h, the bottoms flow is 200 kmol/h with 1 mole % volatile composition, and the liquid phase composition leaving tray 12 is 5 mole %. What is the vapor phase mole fraction leaving tray 13?

(A) 1.5%

(B) 3.5%

(C) 5.5%

(D) 6.0%

---

[7]The *NCEES Handbook* calls the more volatile component "Component *A*."

Solution

From Eq. 45.16,

$$y_{m+1} = \frac{L_m x_m}{L_m - B} - \frac{B x_B}{L_m - B}$$

$$y_{13} = \left(\frac{L_{12}}{L_{12} - B}\right)x_{12} - \frac{B x_B}{L_{12} - B}$$

$$= \left(\frac{1000 \frac{kmol}{h}}{1000 \frac{kmol}{h} - 200 \frac{kmol}{h}}\right)(0.05)$$

$$- \frac{\left(200 \frac{kmol}{h}\right)(0.01)}{1000 \frac{kmol}{h} - 200 \frac{kmol}{h}}$$

$$= 0.060 \quad (6.0\%)$$

The answer is (D).

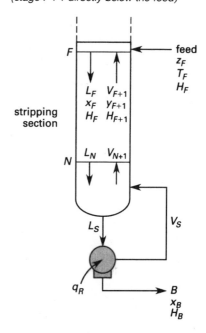

Figure 45.6 *Nomenclature for Stripping Section (stage F + 1 directly below the feed)*

10. DISTILLATION COLUMN OVERALL ENERGY BALANCES[8]

If there is negligible heat loss from the distillation column (i.e., adiabatic operation), the overall energy balance is

$$FH_F + Q_R = Q_C + DH_D + BH_B$$

---

[8]This material is not present in the *NCEES Handbook*.

A total condenser changes the phase of the vapor stream leaving the top of the column. At the top of the column,

$$y_1 = x_D = x_0$$

The material and energy balances for the condenser, respectively, are

$$V_1 = L_R + D$$
$$V_1 H_1 = Q_C + D H_D + L_R H_D$$

An energy balance around the condenser yields the *condenser duty*.

$$Q_C = V_1 H_1 - D H_D - L_R H_D$$
$$= (1 + R) D (H_1 - H_D)$$

The total energy balance around the column yields the *reboiler duty*.

$$Q_R = Q_C + D H_D + B H_B - F H_F$$

11. RECTIFYING SECTION STAGE ENERGY BALANCES[9]

For an adiabatic distillation column, the energy balance for the rectifying section is

$$V_{n+1} H_{n+1} = Q_C + L_n H_n + D H_D$$

At constant pressure, the enthalpies of the saturated liquid and vapor are

$$H_{L,n} = H_n x_n \quad \text{[liquid phase]}$$
$$H_{V,n+1} = H_{n+1} y_{n+1} \quad \text{[vapor phase]}$$

The liquid and vapor mole fractions leaving each stage are related by the *equilibrium ratio*, K_n, which is a function of composition, temperature, and pressure.

$$K_n = \frac{y_n}{x_n}$$

12. STRIPPING SECTION STAGE ENERGY BALANCES[10]

The energy balance for the stripping section of an adiabatic column is

$$V_{m+1} H_{m+1} = L_m H_m - B H_B + Q_R$$

Three more relationships are

$$H_{L,m} = H_m x_m \quad \text{[liquid phase]}$$
$$H_{V,m+1} = H_{m+1} y_{m+1} \quad \text{[vapor phase]}$$
$$K_m = \frac{y_m}{x_m}$$

13. GRAPHICAL SOLUTION TO CONTINUOUS DISTILLATION

The analytical solution of continuous distillation involves solving six equations and six unknown variables; V_{n+1}, L_n, x_n, y_{n+1}, H_n, and H_{n+1} for the rectifying section; and V_{m+1}, L_m, x_m, y_{m+1}, H_m, and H_{m+1} for the stripping section. Although the solution process is cumbersome, it is possible to solve for all six variables sequentially along the entire height of the column. However, graphical techniques similar to that used for flash distillation can be used to simplify column design and analysis considerably.

Graphical solutions for binary system distillation columns are based on the assumption of *constant molar overflow*. This means that the molar flow rate of liquid between stages is constant within each section of the column between the feed and withdrawal points; the molar flow rate of vapor between stages is similarly constant. The constant molar overflow assumption can be expressed as

$$L_n = L_{n+1} = \cdots = L_R$$
$$V_n = V_{n+1} = \cdots = V_R$$
$$L_m = L_{m+1} = \cdots = L_S$$
$$V_m = V_{m+1} = \cdots = V_S$$

Constant molar overflow derives from and is dependent on four conditions: (1) the processes are adiabatic (i.e., column heat losses and sensible heating are negligible); (2) the changes in specific heat are negligible compared to the changes in latent heat; (3) the molar latent heats of vaporization of both components are equal and constant; and (4) heats of mixing are negligible.

With constant molar liquid and vapor flow rates occurring within each section of the column, one operating equation is sufficient to describe the entire rectifying section. The operating equation for the rectifying section (often referred to as the *top operating line*) is a straight line with slope L_R/V_R and y-intercept of $(1 - (L_R/V_R))x_D$. The operating line intersects the $y = x$ line at $y = x = x_D$.

$$y_{n+1} = \left(\frac{L_R}{V_R}\right) x_n + \left(1 - \frac{L_R}{V_R}\right) x_D$$
$$= \left(\frac{L_R}{V_R}\right) x_n + \left(\frac{D}{V_R}\right) x_D \quad \text{[rectifying section]}$$

---

[9]This material is not present in the *NCEES Handbook*.
[10]This material is not present in the *NCEES Handbook*.

The slope, L_R / V_R, is called the *internal reflux ratio*. The slope can be written in terms of the external reflux ratio, L_R / D.

$$\frac{L_R}{V_R} = \frac{\dfrac{L_R}{D}}{1 + \dfrac{L_R}{D}} = \frac{R_D}{1 + R_D}$$

In addition, the y-intercept can be expressed in terms of the external reflux ratio.

$$\frac{Dx_D}{V_R} = \frac{Dx_D}{D + L_R} = \frac{x_D}{1 + R}$$

Then, the operating line in the rectifying section becomes

$$y_{n+1} = \frac{R_D x_n}{1 + R_D} + \frac{x_D}{1 + R}$$

Similarly, one operating equation is sufficient to describe the entire stripping section. The operating equation for the stripping section (often referred to as the *bottom operating line*) is a straight line with slope L_S / V_S and a y-intercept of $\left(1 - (L_S / V_S)\right) x_B$. The operating line intersects the $y = x$ line at $y = x = x_B$.

$$y_{m+1} = \left(\frac{L_S}{V_S}\right) x_m + \left(1 - \frac{L_S}{V_S}\right) x_B$$

$$= \left(\frac{L_S}{V_S}\right) x_m + \left(\frac{B}{V_S}\right) x_B \quad \text{[stripping section]}$$

The graphical solution method uses the equilibrium data (i.e., *VLE diagram*) to relate the mole fractions of liquid and of vapor leaving each stage. The method starts by plotting the operating equation for the rectifying section on an x-y equilibrium (VLE) diagram. The operating equation is used to relate mole fractions between successive stages. Figure 45.7 shows the diagram for a rectifying section in a distillation column with a *total condenser*, so $x_D = y_1$. The operating line represents the operating equation for the rectifying section, which relates x_n to y_{n+1}. y_1 and x_1 are in equilibrium in the first stage. The value of x_1 can be read from the equilibrium data on the diagram. The operating line relates x_1 to y_2 in the next equilibrium stage. x_2 is read directly from y_2 and the equilibrium data. Drawing lines between the equilibrium curve and the operating line produces steps on the diagram, each step representing a theoretical stage in the column. This procedure is called *stepping off stages*.

A similar graphical procedure is used with the stripping section, using the operating equation for the stripping section.

The distillation column's *partial reboiler* is considered to be an equilibrium stage. The vapor leaving the reboiler,

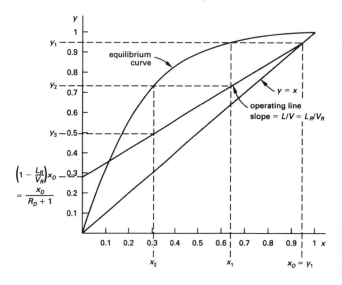

Figure 45.7 *Stepping Off Rectifying Stages*

$y_B = y_{N+1}$, is in equilibrium with the liquid, x_B. The value of y_B is given by the equilibrium data. The operating line relates y_B to x_N in the next equilibrium stage. Starting with x_N, from the operating line, the corresponding value of y_N can be read from the equilibrium data. The same procedure of stepping off stages in the stripping section can be continued up to the feed point.

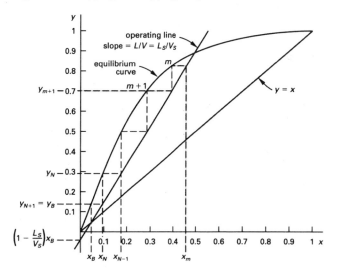

Figure 45.8 *Stepping Off Stripping Stages*

Combining Fig. 45.7 and Fig. 45.8 produces a *McCabe-Thiele diagram* for the entire column, as shown in Fig. 45.9, where the numbers next to the equilibrium curve indicate stages. Stage 6 is a partial reboiler, which is also considered to be an equilibrium stage. The operating line for the rectifying section is used when stepping off stages in the rectifying section. Similarly, the operating line for the stripping section is used to step off stages in the stripping section.

When designing a distillation column, the location of the feed stage is determined as the point at which the stage stepping switches from one operating line to the other. The location of this point can be chosen arbitrarily, but the optimal position of the feed stage is where the two operating lines intersect. Locating the feed stage at the intersection point minimizes the number of stages needed to achieve separation. In Fig. 45.9, the optimal feed position is at stage 3, near the intersection of the operating lines.

Figure 45.9 *McCabe-Thiele Diagram for Entire Column*

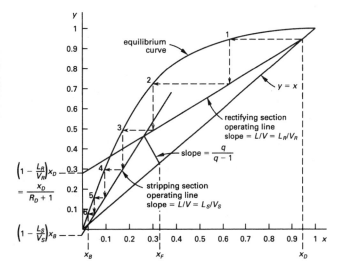

Example

A McCabe-Thiele diagram is to be used to design a distillation column. The ideal and pseudo-equilibrium curves and operating lines for the distillate and rectifying sections are known. The desired distillate and bottoms mole fractions are 0.66 and 0.07, respectively. Based on the pseudo-equilibrium curve, how many stages (including the partial reboiler as a stage) are required?

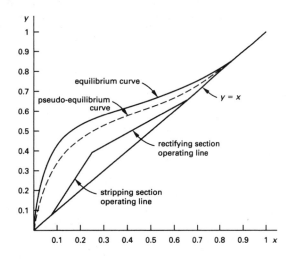

(A) 3

(B) 4

(C) 5

(D) 6

Solution

Draw the McCabe-Thiele diagram between the pseudo-equilibrium curve and the operating lines, starting at $x_D = 0.66$ and ending at $x_B = 0.07$.

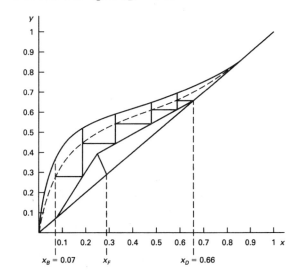

The number of steps (stages) is five.

The answer is (C).

14. OPERATION AT MINIMUM REFLUX

Equation 45.17: Rectifying Operating Line at Minimum Reflux

$$y = \frac{R_{\min}}{R_{\min} + 1} x + \frac{x_D}{R_{\min} + 1} \qquad 45.17$$

Description

Equation 45.17 is the operating line of the rectifying section when the column is operating at the minimum reflux ratio. At the minimum reflux ratio, a column will have an infinite number of trays. Graphically, the minimum reflux ratio occurs where the top operating line touches the equilibrium curve, a point known as the *pinch point*. (See Fig. 45.10.)

Adding more stages to the column after the pinch point produces no further separation. By fixing the point of the top operating line at x_D and pivoting the operating line until it touches the equilibrium curve, the slope of the line and the y-intercept can be found. This information can be used to determine the internal reflux ratio from the operating equation for the rectifying section.

Figure 45.10 Pinch Point in a Binary System

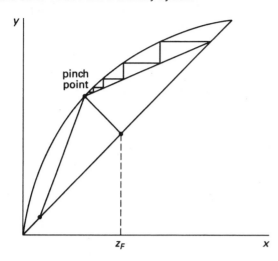

Example

A rectifying line with minimum reflux ratio has a y-intercept of 0.456. The distillate composition is 0.995. Most nearly, what is the minimum reflux ratio?

(A) 0.42

(B) 0.85

(C) 1.2

(D) 1.6

Solution

Use Eq. 45.17 and solve for the minimum reflux ratio.

$$y = \frac{R_{min}}{R_{min}+1}x + \frac{x_D}{R_{min}+1}$$

$$0.456 = \left(\frac{R_{min}}{R_{min}+1}\right)(0) + \frac{x_D}{R_{min}+1}$$

$$= \frac{x_D}{R_{min}+1}$$

$$R_{min} = \frac{x_D}{y} - 1$$

$$= \frac{0.995}{0.456} - 1$$

$$= 1.18 \quad (1.2)$$

The answer is (C).

15. FEED EQUATIONS

The phase and temperature of the feed affect the liquid and vapor flow rates in the column. The *feed quality*, q, is the fraction of the feed that remains liquid. Assuming constant molar overflow, neither the vapor nor the liquid enthalpies change from stage to stage, and the feed quality for any stage is

$$q = \frac{L_S - L_R}{F} = \frac{L_F}{F} = \frac{H_V - H_F}{H_V - H_L}$$

$$= \frac{h_V - h_F}{h_V - h_L}$$

The *fraction of the feed vaporized*, f, is the complement of the feed quality.

$$f = 1 - q = \frac{V_R - V_S}{F} = \frac{V_F}{F}$$

The liquid and the vapor flow rates can be calculated from the feed quality.

$$L_S = L_R + qF$$

$$V_S = V_R - (1-q)F$$

The *feed equation* is a straight line. It can be written in terms of the feed quality, q, and fraction of feed vaporized, f. In all three equations, the first parenthetical quantity represents the slope of the corresponding line.

$$y = \left(\frac{L_R - L_S}{V_R - V_S}\right)x + \left(\frac{F}{V_R - V_S}\right)z_F$$

$$y = \left(\frac{f-1}{f}\right)x + \left(\frac{1}{f}\right)z_F$$

$$y = \left(\frac{q}{q-1}\right)x + \left(\frac{1}{1-q}\right)z_F$$

Equation 45.18 and Eq. 45.19: Feed Operating Line Slope

$$\text{slope} = q/(q-1) \qquad \textbf{45.18}$$

$$q = \frac{\left(\begin{array}{c}\text{heat to convert one mol of feed}\\\text{to saturated vapor}\end{array}\right)}{\text{molar heat of vaporization}} \qquad \textbf{45.19}$$

Variations

$$q = \frac{\left(\begin{array}{c}\text{heat to convert one unit mass of feed}\\\text{to saturated vapor}\end{array}\right)}{\text{specific heat of vaporization}}$$

$$q = \frac{h_g - h}{h_{fg}} = \frac{h_{fg} + h_f - h}{h_{fg}}$$

h_f, h_g, and h_{fg} are the heat (enthalpy) of a saturated liquid, heat (enthalpy) of a saturated vapor (gas), and the latent heat (enthalpy) of vaporization, respectively, and h is the enthalpy of the feed.

Description

Equation 45.18 gives the slope of the feed equation as a function of the feed quality, q, which is a measure of the portion of the feed that remains liquid. (See Eq. 45.19.) Equation 45.19 is written as a ratio of energies (heats or enthalpies).[11] Although Eq. 45.19 uses molar energies, it could just as easily have been written in terms of energies per unit mass. The second variation equation is valid for subcooled liquid feeds. Figure 45.11 shows q and the q-line.

Figure 45.11 *q and the q-Line*

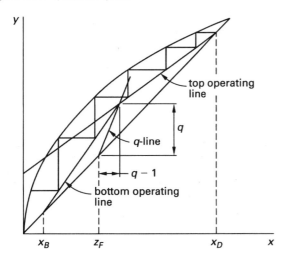

Figure 45.12 shows feed lines, sometimes referred to as *q-lines*, for several possible feed qualities. All feed lines intersect the $y = x$ line at the feed composition, z_F.

Figure 45.12 *Feed Lines for Various Feed Qualities*

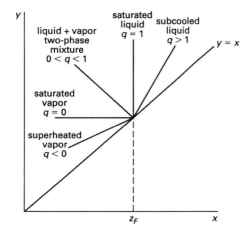

*The *NCEES Handbook* labels this figure as "q-Line Slopes." The diagram shows q-lines with different slopes. However, the slopes of the line are given by Eq. 45.18, not q. These are feed lines.

---

[11]Using the nomenclature of thermodynamics, where x represents the quality of a vapor-liquid mixture, Eq. 45.19 is written as $x = xh_{fg}/h_{fg}$.

With the theoretical condition of total reflux (i.e., when all condenser and reboiler flows are returned to the column), the number of stages to achieve the separation is at its minimum. Total reflux is equivalent to a *maximum reflux ratio*. Total reflux requires that there be no feed stream. With total reflux, $L_R = V_R$, $L_S = V_S$, and $L_R/V_R = L_S/V_S = 1$. The slopes of both the top and the bottom operating lines are 1.0, which is equivalent to the $y = x$ line on the McCabe-Thiele diagram. The minimum number of stages is determined by stepping off stages using the $y = x$ line as the operating line.

Example

A liquid feed has an enthalpy of 58.3 kJ/kg when the enthalpy of saturated liquid is 291.3 kJ/kg and the heat of vaporization is 1759.2 kJ/kg. Most nearly, what is the slope of the feed line?

(A) 0.13

(B) 0.88

(C) 1.1

(D) 8.6

Solution

Because the enthalpy of the liquid is less than the enthalpy of the saturated liquid, the feed is subcooled and $q > 1$.

From Eq. 45.19, the feed quality is

$$
\begin{aligned}
q &= \dfrac{\text{heat to convert one mol of}}{\text{molar heat of vaporization}} \\[6pt]
&= \dfrac{\text{heat to convert one unit mass}}{\text{specific heat of vaporization}} \\[6pt]
&= \dfrac{h_{fg} + h_f - h}{h_{fg}} \\[6pt]
&= \dfrac{1759.2\,\frac{\text{kJ}}{\text{kg}} + 291.3\,\frac{\text{kJ}}{\text{kg}} - 58.3\,\frac{\text{kJ}}{\text{kg}}}{1759.2\,\frac{\text{kJ}}{\text{kg}}} \\[6pt]
&= 1.132
\end{aligned}
$$

From Eq. 45.18, the slope of the feed line is

$$
\begin{aligned}
\text{slope} &= \frac{q}{q-1} \\[6pt]
&= \frac{1.132}{1.132 - 1} \\[6pt]
&= 8.58 \quad (8.6)
\end{aligned}
$$

The answer is (D).

Example

A column with a total condenser and a partial reboiler used to distill an ethyl water-alcohol (C_2H_5OH) mixture is designed according to a McCabe-Thiele diagram and the equilibrium line shown.

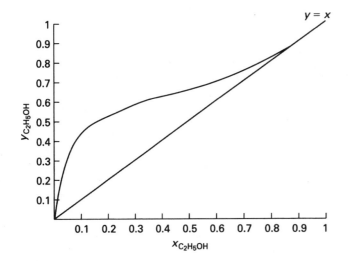

The distillate mole fraction is 0.8. A very low bottoms mole fraction is desired. What is the theoretical minimum number of stages, including the partial reboiler, needed to achieve the separation?

(A) 6

(B) 7

(C) 8

(D) 9

Solution

The $y = x$ line becomes the top and bottom operating lines when determining the minimum number of stages.

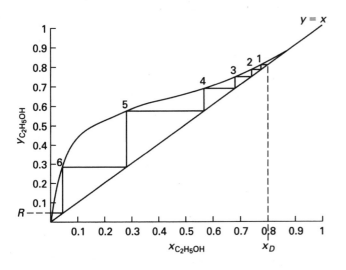

The minimum number of stages required is six.

The answer is (A).

16. STAGE EFFICIENCY

One of the assumptions made when analyzing and designing distillation columns is that the liquid and vapor will come into equilibrium in each distillation stage. However, the actual liquid and vapor compositions achieved usually deviate from the ideal. The *Murphree plate efficiency*, E_{ME}, is a measure of this deviation.

Equation 45.20: Murphree Plate Efficiency

$$E_{ME} = (y_n - y_{n+1})/(y_n^* - y_{n+1}) \qquad 45.20$$

Variations[12]

$$E_{ME} = \frac{y_m - y_{m+1}}{y_m^* - y_{m+1}}$$

$$\text{actual number of plates} = \frac{\text{theoretical number of plates}}{E_{\text{overall}}}$$

In general, the overall efficiency is not the same as the Murphree efficiency.

Description

The Murphree plate efficiency is calculated as the actual change in vapor mole fraction divided by the theoretical change in vapor mole fraction on a specific plate.[13] In Eq. 45.20, y_n is the mole fraction of vapor above equilibrium stage n, y_{n+1} is the mole fraction of vapor entering from the equilibrium stage below n, and y_n^* is the mole fraction of vapor in equilibrium with liquid leaving equilibrium stage n.[14]

The denominator is the vertical distance from the operating line to the equilibrium line. The numerator of Eq. 45.20 is the vertical distance from the operating line to the actual outlet composition of the stage (i.e., to the pseudo-equilibrium line). Figure 45.13 shows that stepping off stages using the Murphree efficiency produces a *pseudo-equilibrium curve*.

The theoretical number of stages (plates) is based on achieving equilibrium on each plate. Since plates are rarely as efficient as theoretical plates, the actual number of plates needed is usually more than the theoretical number. The *overall plate efficiency* (*average plate efficiency*) is used to adjust this number.

---

[12]The *NCEES Handbook*'s use of the subscript n implies that Eq. 45.19 is for the rectifying section only. Equation 45.19, however, can be used for the stripping section, also.

[13]When describing Eq. 45.20, the *NCEES Handbook* ambiguously refers to the vapor mole fraction, y, as "concentration."

[14]The *NCEES Handbook* version of Eq. 45.20 is for the vapor phase only. A similar equation can be written for the liquid phase.

$$E_{ME} = \frac{x_n - x_{n-1}}{x_n^* - x_{n-1}}$$

Figure 45.13 *Developing a Pseudo-Equilibrium Curve from Murphree Efficiencies*

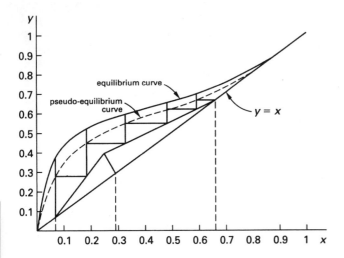

Though considered equilibrium stages in a McCabe-Thiele analysis, partial reboilers and partial condensers are treated separately. Calculation of the overall efficiency excludes partial condensers and partial reboilers, since such equipment operates at efficiencies different from true equilibrium stages in a column.

Example

Samples are taken from two adjacent stages of an operating distillation column. One sample is found to have a vapor phase mole fraction of 0.350, and the other is found to have a vapor phase mole fraction of 0.325. The equilibrium mole fraction for the liquid on the upper stage is 0.375. Most nearly, what is the plate efficiency?

(A) 10%

(B) 25%

(C) 50%

(D) 75%

Solution

Use Eq. 45.20.

$$E_{ME} = \frac{y_n - y_{n+1}}{y_n^* - y_{n+1}}$$

$$= \frac{0.350 - 0.325}{0.375 - 0.325}$$

$$= 0.5 \quad (50\%)$$

The answer is (C).

17. ABSORPTION AND STRIPPING

In *absorption*, a component in a gas stream (the solute) is transferred into a stream of a nonvolatile liquid (the solvent). The solvent is the separating agent; it removes the solute from the feed gas stream and produces a treated gas stream. There are two types of absorption: physical and chemical. In *physical absorption*, a gas component is removed from a process stream because it has greater solubility in the solvent than in its original stream. In *chemical absorption* (reactive absorption), the gas component to be removed reacts with and stays with the solvent.

In *stripping* (also known as *desorption*), a solute from a liquid stream is transferred to an insoluble gas stream. The stripping gas removes the solute from the feed stream and produces a treated liquid. In both cases, absorption drives the mass transfer. Alternatively, a gaseous component can be cleaned from and transferred into the stripping liquid solvent. In either case, the cleaned flow is known as the *raffinate*, while the dirtied flow is known as the *extract*. Figure 45.14 shows a countercurrent packed column used for absorption. The subscripts 1 and 2 refer to the inlet compositions at the bottom (*rich end*) and top (*lean end*), respectively, of the packed column.

Figure 45.14 *Countercurrent Packed Column Absorber*

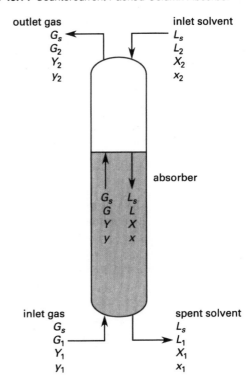

18. ABSORPTION AND STRIPPING IN TRAYED COLUMNS

Trayed columns (*staged columns or trayed towers*) contact liquid and vapor in discrete stages on trays. Absorption and stripping can both be treated as equilibrium-stage operations and analyzed using McCabe-Thiele diagrams. Analysis is simplified by three assumptions: (1) the carrier gas is insoluble in the liquid phase; (2) the solvent is nonvolatile; and (3) the system is isothermal and isobaric (i.e., operates at constant temperature and pressure).

Trayed columns have certain advantages over packed columns. They can handle a wider range of liquid and gas flows than packed columns. (Packed columns are not suitable for very low liquid rates.) Their performance and efficiencies can be predicted better. That is, they can be designed with more assurance than packed columns. It is easier to install cooling in the plates than within packing. Withdrawal of side streams is easier than with packed columns. Trayed columns are easier to clean (e.g., through *manways*), and accordingly, trayed columns can be used with liquids that cause fouling or contain solids.

Since there are at least three components in stripping and absorption (liquid, solute, and solvent), analysis requires a different representation of the equilibrium data from that used for distillation. (See Fig. 45.15.) Usually, a plot is made of solute composition in the gas, y, versus solute composition in the liquid, x. Under the assumptions commonly made with absorption, the equilibrium line is straight.

Figure 45.15 *Absorption Equilibrium Diagram (tray concept)*

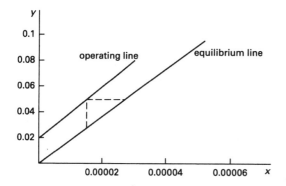

Henry's law relates the concentration of a target solute, A, in the liquid to the partial pressure of solute A in the vapor. Henry's law is valid at low solute concentrations and for solute partial pressures below 1 atm. H' is the *Henry's law constant* with units consistent with pressure, p.[15]

$$p_A = H' x_A$$

Henry's law can be combined with Dalton's law of partial pressures to produce an equation for the mole fraction of the target substance, A. If Henry's law constant, H', is constant over the operating range in the column,

the resulting operating equation represents a straight line (i.e., the operating line).

$$y_A = \left(\frac{H'}{p}\right) x_A$$

Figure 45.16 defines the nomenclature for a *trayed column gas absorber*. If the liquid solvent is nonvolatile and the carrier gas is insoluble in the liquid, and if it is assumed that the gas and the liquid are sufficiently dilute in A, then the liquid flow rates will be approximately constant, as will the gas flow rates.

$$L_N = L_0 = L = \text{constant}$$

$$G_{N+1} = G_1 = G = \text{constant}$$

With these assumptions, compositions in the absorber are specified in terms of mole fractions. The mole balance on A for the dotted envelope of Fig. 45.16 is

$$y_{n+1} G + x_0 L = y_1 G + x_n L$$

Figure 45.16 *Trayed Gas Absorber*

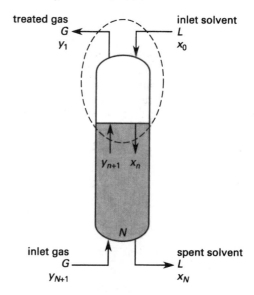

This can be rearranged to the *operating line* shown in Fig. 45.15 with slope (L/G) and intercept $(y_1 - x_0(L/G))$.

$$y_{n+1} = \left(\frac{L}{G}\right) x_n + \left(y_1 - x_0\left(\frac{L}{G}\right)\right)$$

If the liquid and gas phases are not dilute, mole ratios and pure liquid and carrier gas flows must be used for the operating line and the equilibrium curve instead of mole fractions. The usual McCabe-Thiele stepping procedures can then be used in that case.

---

[15]Some authorities represent Henry's law constant variously as k_h, k_j, H, and H_A. In the equation shown, the units of Henry's law constant must be units of pressure. Henry's law constant is usually given in units of atmospheres.

19. ABSORPTION AND STRIPPING IN PACKED COLUMNS

Packed columns (also known as *continuous contact columns* and *packed towers*) can be used in distillation, stripping, absorption, and other processes. While staged columns contact liquid and vapor in discrete stages, packed columns contact the liquid and vapor continuously. Packed columns, therefore, contain no distinct equilibrium stages. Although the equilibrium stage concept is still useful, the design of packed columns is primarily concerned with the height of packing.

Packed columns are generally not suitable for very low liquid flows. However, packed columns have their own advantages over trayed columns. The liquid residual is lower than in a trayed column, which makes packed columns desirable when processing toxic and flammable liquids. Packed columns work better with foaming fluids. Packed columns are less expensive than trayed columns for corrosive liquids. The pressure drop can be lower for packed columns than trayed columns. Packed columns are better suited for vacuum operations. Packed columns offer manufacturing advantages with column diameters smaller than 0.6 m.

Figure 45.17 shows a countercurrent packed column used for absorption. The column is filled with packing. As the fluids pass through the packing, their flows are disrupted and they take long, winding routes (known as *tortuous* routes). The liquid spreads out over the many surfaces of the packing material, and the gas passes over the wetted surfaces. Since there are no equilibrium stages, the nomenclature for a packed column is slightly different than for a staged column. The solvent flow rate, L_s, and the pure gas flow rate, G_s, are essentially constant along the length of the column. However, the

total liquid flow, L, and the total gas flow, G, vary as solute is transferred along the column.

Commercial *packing* is available in various types and sizes. (See Fig. 45.18.) Ceramic packing offers good wettability and corrosion resistance but it has poor strength and is fragile. Metal packing has good wettability and has high strength, but may have low corrosion resistance, particularly at high temperatures. Plastic packing has good strength at low temperatures, good corrosion resistance, and a lower cost, but it may have low wettability at low liquid rates and reduced strength at high temperatures.

Figure 45.18 Commercial Packing Types

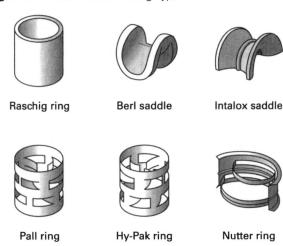

Raschig ring   Berl saddle   Intalox saddle

Pall ring   Hy-Pak ring   Nutter ring

Packing is characterized by its surface-to-volume ratio, a, and intrinsic pressure drop. Pressure drop is characterized by a packing factor, C_f, which describes the packing's flow capacity.[16] It is inversely proportional to packing size. Packing factors for commercial packing can be obtained from manufacturers' or experimental data.

Random packing is poured into a column during installation and allowed to assume random locations and orientations. Small random packing is the most common packed column design. When very low pressure drops and very high flow rates are involved, *stacked packing* (*oriented packing*) can be used; however, only those packing types with symmetrical (e.g., cylindrical) shapes and with sizes (e.g., diameters) larger than 75 mm are practical to stack. When stacked, triangular (diamond) pitch or square pitch can be used within the column. Packing is held in place by special *support trays*.

A variety of shapes have been developed for effective random packing, including *Raschig rings* (short pieces of solid-walled tube), *Pall rings* (similar, but with "fingers" cut from the wall and bent into the middle of the tube), and various saddle shapes such as *Berl saddles* and

Figure 45.17 Components of a Packed Column

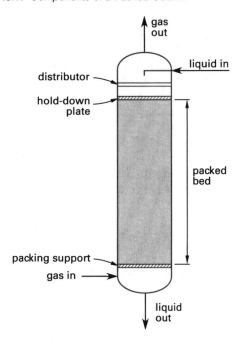

---

[16]In industry literature, the packing factor is also represented by the variable F_p.

Intalox saddles. Table 45.1 gives the surface area per unit volume and the specific roughness of some common kinds and sizes of packing.

Table 45.1 Typical Random Packing Parameters*

packing	nominal size (in (mm))	a (m²/m³)	ε	C_f
Berl saddles, ceramic	½ (13)	210	0.60	240
	1 (25)	260	0.69	110
	1½ (38)	170	0.70	65
Intalox saddles, ceramic	½ (13)	625	0.78	200
	1 (25)	256	0.78	92
	1½ (38)	197	0.81	52
Pall rings, metal	1 (25)	207	0.94	48
	1½ (38)	128	0.95	33
	2 (50)	102	0.96	20
Raschig rings, ceramic	½ (13)	400	0.64	580
	1 (25)	190	0.73	155
	1½ (38)	115	0.78	95

*Exact values vary with manufacturer.

In comparison to numerous small pieces distributed randomly or stacked in a pattern inside the column, *structured packing* consists of manufactured trays (circular disks) or blocks with numerous vertical core passageways in various patterns (e.g., honeycomb, zigzag, perforated plates, etc.) and wire gauze or thin perforated sheets of corrugated metal.

ε in Table 45.1 is the dimensionless *porosity* (*void fraction*) of the packed bed.

$$\varepsilon = \frac{\text{volume of voids in bed}}{\text{total volume of bed}}$$

The *interstitial velocity*, v_i, is the average velocity of fluid through the pores of the column. It is calculated from the *superficial velocity*, v_o.

$$\mathrm{v}_i = \frac{\mathrm{v}_o}{\varepsilon}$$

$$\mathrm{v}_o = \frac{\text{fluid flow rate}}{\text{internal cross-sectional column area}}$$

20. PACKING HEIGHT

To achieve a desired separation, the necessary *height of packing*, Z, of a column can be calculated as the product of the *number of mass transfer units* (NTU) and the *height of a transfer unit* (HTU). (See Fig. 45.19.) The NTU concept is similar to the required number of theoretical stages in a staged column. The NTU relates the contact area required to achieve a stated separation and the driving force in that phase. The HTU concept

Figure 45.19 Transfer Units and Packing Height

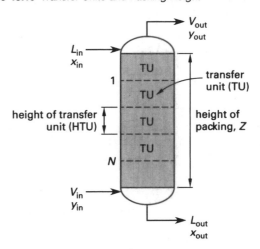

relates flow rate as well as hydraulic and physical factors. The HTU indicates the mass transfer capability of equipment. The smaller the HTU, the more efficient the unit.

Equations for NTUs and HTUs are phase dependent. The height of packing depends on whether the mass transfer is limited by the (amount of) gas phase, liquid phase, or both. In gas absorption, solute concentrations in the gas phase differ widely, resulting in the number of overall gas phase transfer units, NTU_G, and the overall height of a gas phase transfer unit, HTU_G, being used to simplify designs. The liquid phase resistance generally dominates when stripping a solute from a liquid. For this reason, it is a common use of the number of overall liquid phase transfer units, NTU_L, and the overall height of a liquid phase transfer unit, HTU_L, in such situations.

Although a packed column does not have discrete equilibrium stages, the number of equivalent theoretical plates, N_{EQ}, and the *height equivalent of a theoretical plate*, HETP, enable the distillation design methods to be used with packed columns. HETP represents the height of packing that would produce an equivalent separation to a theoretical tray. HETP is a common measure of efficiency given to various types of column packing by manufacturers.

21. THEORETICAL PLATE EQUIVALENTS IN PACKED COLUMNS[17]

Although packed columns do not have discrete equilibrium stages, it is convenient to calculate theoretical numbers for the purpose of comparing column designs and packing efficiencies. The number of theoretical plates (i.e., the number of equilibrium stages) is the same as for staged columns. This equation assumes a

---

[17]This material is not present in the *NCEES Handbook*.

straight operating line and a dilute solution (where Henry's law is applicable).

$$N_{EQ} = \frac{\ln\left(\dfrac{\left(1 - \dfrac{H'G}{pL}\right)\left(y_1 - \dfrac{H'x_2}{p}\right)}{y_2 - \dfrac{H'}{p}} + \dfrac{H'G}{pL}\right)}{\ln\left(\dfrac{pL}{H'G}\right)} = \frac{Z}{\text{HETP}}$$

N_{EQ} is not the same as NTU_G, although the two numbers are related.

$$NTU_G = N_{EQ}\left(\frac{\ln\dfrac{H'G}{pL}}{\dfrac{H'G}{pL} - 1}\right)$$

Equation 45.21: Column Height for Continuous Contact Columns

$$Z = NTU_G \cdot HTU_G = NTU_L \cdot HTU_L \qquad 45.21$$
$$= N_{EQ} \cdot \text{HETP}$$

Description

Equation 45.21 calculates the height of packing in a packed column. The number and height of transfer units based on gas-side factors, the number and height of transfer units based on liquid-side factors, and the number of equilibrium stages and height equivalent of theoretical plates can all be used.[18]

Example

A packing column has four gas transfer units that are each 6 m high. The height of an equivalent theoretical plate is 2 m. How many equilibrium stages are represented?

(A) 4

(B) 8

(C) 12

(D) 24

Solution

Use Eq. 45.21.

$$Z = NTU_G \cdot HTU_G = NTU_L \cdot HTU_L = N_{EQ} \cdot \text{HETP}$$

$$N_{EQ} = \frac{NTU_G HTU_G}{\text{HETP}} = \frac{(4)(6 \text{ m})}{2 \text{ m}} = 12$$

The answer is (C).

---

[18]The *NCEES Handbook* gives no guidance on how N_{EQ} and HETP are determined or used other than Eq. 45.21.

22. HEIGHT OF A TRANSFER UNIT

The expression for the overall *height of a transfer unit*, HTU_G or HTU_L, is given in terms of the *overall mass transfer coefficient*, K'_G or K'_L, and the interfacial area (mass transfer area per unit volume), a. The subscripts G and L indicate that the mass transfer is limited by the gas phase or by the liquid phase, respectively. Values for the height of a transfer unit and mass transfer coefficients are usually obtained from experimental data or correlations. Values of a for manufactured packing are usually provided by the manufacturer or obtained from data tables of standard packing types.

Equation 45.22 and Eq. 45.23: Height of a Transfer Unit[19]

$$HTU_G = \frac{G}{K'_G a} \qquad 45.22$$

$$HTU_L = \frac{L}{K'_L a} \qquad 45.23$$

Description

Equation 45.22 and Eq. 45.23 calculate the height of a transfer unit based on the gas phase and liquid phase properties, respectively. G and L represent the mass velocities of the gas and the liquid, respectively. a is the interfacial area, with units of m^2/m^3.

Example

The height of a gas phase transfer unit is 0.548 m, and there is 0.186 m^2 of mass transfer surface area per cubic meter of packing material. The gas phase mass velocity is 3.852×10^{-3} kmol/$m^2 \cdot$s. Most nearly, what is the mass transfer coefficient?

(A) 3.8×10^{-2} kmol/$m^2 \cdot$s

(B) 4.3×10^{-2} kmol/$m^2 \cdot$s

(C) 4.8×10^{-2} kmol/$m^2 \cdot$s

(D) 5.3×10^{-2} kmol/$m^2 \cdot$s

Solution

Solve Eq. 45.22 for the gas phase mass transfer coefficient.

---

[19]The *NCEES Handbook* uses primed symbols K'_G and K'_L in the context of gas absorption. These are the overall mass transfer coefficients for the case of equimolar counterdiffusion. Gas absorption is rather an example of transfer of A through stagnant B, so the overall mass transfer coefficients should be unprimed, that is K_G and K_L.

$$\mathrm{HTU}_G = \frac{G}{K'_G a}$$

$$K'_G = \frac{G}{\mathrm{HTU}_G a} = \frac{3.852 \times 10^{-3} \; \frac{\mathrm{kmol}}{\mathrm{m^2 \cdot s}}}{(0.548 \; \mathrm{m})\left(0.186 \; \frac{\mathrm{m^2}}{\mathrm{m^3}}\right)}$$

$$= 3.78 \times 10^{-2} \; \mathrm{kmol/m^2 \cdot s}$$

$$(3.8 \times 10^{-2} \; \mathrm{kmol/m^2 \cdot s})$$

The answer is (A).

23. NUMBER OF TRANSFER UNITS

Equation 45.24 and Eq. 45.25: Number of Liquid Phase Transfer Units

$$\mathrm{NTU}_L = \int_{x_1}^{x_2} \frac{dx}{(x^* - x)} \qquad \textit{45.24}$$

$$x^* = y/K \qquad \textit{45.25}$$

Description

In a countercurrent packed column absorber at any liquid concentration (i.e., concentrated or dilute), the general expression for the number of overall liquid phase mass transfer units, NTU_L, is found from Eq. 45.24 and Eq. 45.25. Equation 45.24 is used to calculate the number of transfer units needed based on the liquid-side properties.[20] Equation 45.25 defines x^*, which is the equilibrium solute mole fraction in the liquid phase. K is the equilibrium constant. The subscripts 1 and 2 refer to the compositions at the bottom (rich end) and top (lean end), respectively, of the packed column.[21]

Example

A packed column is used to clean a gas flow. The liquid solvent is introduced at the top with a liquid side mole fraction of $x = 0.125$ and is removed when the solvent reaches a mole fraction of $x = 0.500$. The mass fraction gradient, $x^* - x$, is constant at 0.150 throughout the range of operation. How many theoretical transfer units are needed?

(A) 2.0

(B) 2.5

(C) 3.0

(D) 4.0

Solution

Use Eq. 45.24.

$$\begin{aligned}
\mathrm{NTU}_L &= \int_{x_1}^{x_2} \frac{dx}{(x^* - x)} \\
&= \int_{0.125 \text{ at bottom}}^{0.500 \text{ at top}} \frac{dx}{0.150} \\
&= \frac{x}{0.150}\bigg|_{0.125}^{0.500} \\
&= \frac{0.500 - 0.125}{0.150} \\
&= 2.5
\end{aligned}$$

The answer is (B).

Equation 45.26 and Eq. 45.27: Number of Gas Phase Transfer Units

$$\mathrm{NTU}_G = \int_{y_1}^{y_2} \frac{dy}{(y - y^*)} \qquad \textit{45.26}$$

$$y^* = K \cdot x \qquad \textit{45.27}$$

Description

The number of gas phase mass transfer units, NTU_G, in a countercurrent absorber, regardless of gas concentration (i.e., whether concentrated or dilute), is given by Eq. 45.26.[22] Equation 45.27 defines y^*, which is the equilibrium solute mole fraction in the gas phase. K is the equilibrium constant. The subscripts 1 and 2 refer to the inlet compositions at the bottom (rich end) and top (lean end), respectively, of the packed column.

Example

A packed column is used to clean a gas flow. The gas is introduced at the bottom with a gas side mole fraction of $x = 0.750$ and is removed when the gas reaches a mole fraction of $x = 0.125$. The mass fraction gradient, $y - y^*$, is constant at 0.250 throughout the range of operation. Most nearly, how many theoretical transfer units are needed?

(A) 1.5

(B) 2.5

(C) 3.0

(D) 4.0

---

[20](1) In the *NCEES Handbook*, the integral in Eq. 45.24 is evaluated from x_1 to x_2. In fact, x_1 is at the rich end of the column, and x_2 is at the lean end of the column. The integral should be evaluated from x_2 to x_1, with $x_1 > x_2$. (2) As displayed in the *NCEES Handbook*, there is no mathematical significance to the parentheses around the denominator of Eq. 45.24.

[21]The subscripts do not refer to "in" and "out" or "initial" and "final."

[22](1) In the *NCEES Handbook*, the integral in Eq. 45.26 is evaluated from y_1 to y_2. In fact, y_1 is at the rich end of the column, and y_2 is at the lean end of the column. The integral should be evaluated from y_2 to y_1, with $y_1 > y_2$. (2) As displayed in the *NCEES Handbook*, there is no mathematical significance to the parentheses around the denominator of Eq. 45.26.

Unit Processes

Solution

Use Eq. 45.26.

$$\begin{aligned}
\text{NTU}_G &= \int_{y_1}^{y_2} \frac{dy}{(y - y^*)} \\
&= \int_{0.125 \text{ at bottom}}^{0.750 \text{ at top}} \frac{dy}{0.250} \\
&= \frac{y}{0.25}\Big|_{0.125}^{0.750} \\
&= \frac{0.750 - 0.125}{0.250} \\
&= 2.5
\end{aligned}$$

The answer is (B).

Equation 45.28 and Eq. 45.29: Number of Gas Phase Transfer Units for Dilute Solutions

$$\text{NTU}_G = \frac{y_1 - y_2}{(y - y^*)_{LM}} \qquad \textbf{45.28}$$

$$(y - y^*)_{LM} = \frac{(y_1 - y_1^*) - (y_2 - y_2^*)}{\ln\left(\dfrac{y_1 - y_1^*}{y_2 - y_2^*}\right)} \qquad \textbf{45.29}$$

Description

In a countercurrent absorber with dilute solute concentrations, the operating and equilibrium lines are essentially straight lines. A solution is dilute when the ratio of gas mass velocity to the liquid mass velocity, G/L, and the equilibrium constant, K, are constant throughout the column. Equation 45.28 is a simplified version of Eq. 45.26 for use in the case of *dilute solutions*. The subscripts 1 and 2 refer to the inlet compositions at the bottom (rich end) and top (lean end), respectively, of the packed column.

As the paths of liquids and gas are tortuous and the driving mole fraction gradient varies along the path,[23] the *logarithmic mean mole fraction gradient*, $(y - y^*)_{LM}$, defined by Eq. 45.29 is used.

Example

A packed absorption column is used to remove a solute from a liquid solvent. Samples of a dilute gas phase are taken at the rich and lean ends. The actual and equilibrium mole fractions are tabulated as shown.

location	actual value	equilibrium value
1	0.825	0.750
2	0.300	0.250

---

[23]This is parallel to the use of the logarithmic mean temperature difference with heat transfer along a long-path heat exchanger where the driving thermal gradient varies along the path.

Most nearly, how many transfer units are represented?

(A) 8

(B) 9

(C) 14

(D) 21

Solution

Use Eq. 45.29 to calculate the logarithmic mean mole fraction gradient. The subscripts 1 and 2 refer to the inlet compositions at the bottom (rich end) and top (lean end).

$$\begin{aligned}
(y - y^*)_{LM} &= \frac{(y_1 - y_1^*) - (y_2 - y_2^*)}{\ln\left(\dfrac{y_1 - y_1^*}{y_2 - y_2^*}\right)} \\
&= \frac{(0.825 - 0.750) - (0.300 - 0.250)}{\ln\dfrac{0.825 - 0.750}{0.300 - 0.250}} \\
&= 0.06166
\end{aligned}$$

Use Eq. 45.28 to calculate the number of transfer units.

$$\begin{aligned}
\text{NTU}_G &= \frac{y_1 - y_2}{(y - y^*)_{LM}} \\
&= \frac{0.825 - 0.300}{0.06166} \\
&= 8.1 \quad (8)
\end{aligned}$$

The answer is (A).

24. REACTIVE ABSORPTION

In *reactive absorption*, where an absorption column is used to remove solute from a gas, the transferred solute reacts chemically with the solvent in the liquid phase. As a result, the solute is absent from the treated gas phase and appears as a reaction product in the solvent. If the reaction is considered irreversible, the equilibrium partial pressure of the solute is also zero.

Equation 45.30: Number of Transport Units in a Chemically Reacting Dilute Solution

$$\text{NTU}_G = \ln\left(\frac{y_1}{y_2}\right) \qquad \textbf{45.30}$$

Description

The number of gas phase transfer units, NTU_G, is determined from Eq. 45.30. Equation 45.30 is a simplification of Eq. 45.28 for the case of dilute solutions where the solute reacts with the liquid phase. The subscripts 1 and 2 refer to the inlet compositions at the bottom (rich end) and top (lean end), respectively, of the packed column.

Example

A packed absorption column is used to remove a solute from a gas. The gas mole fraction is dilute throughout the process. The solute reacts chemically in the liquid phase. Samples of the gas phase are taken at the rich and lean ends. The actual and equilibrium mole fractions are tabulated as shown.

location	actual value	equilibrium value
1	0.825	0.300
2	0.750	0.250

Most nearly, how many transfer units are represented?

(A) 0.01

(B) 0.03

(C) 0.05

(D) 0.10

Solution

Use Eq. 45.30.

$$NTU_G = \ln\left(\frac{y_1}{y_2}\right)$$

$$= \ln\frac{0.825}{0.750}$$

$$= 0.0953 \quad (0.10)$$

The answer is (D).

25. PRESSURE DROP IN PACKED BEDS

The liquid and gas flow rates, the type of packing, and the column diameter all contribute to pressure drop through the column packing (a *packed bed*). The expected pressure drop can be calculated from the Ergun formulas or the generalized pressure-drop correlation for packed columns developed by *Sherwood* et al. and modified by *Eckert*. (See Sec. 45.26 and Sec. 45.27.) For a packed column to operate effectively, the pressure drop must be within some acceptable range, typically 200–1200 Pa/m (i.e., per meter of packing height) for columns operating at 1 atm and higher. A design pressure drop of approximately 300–400 Pa/m is recommended for absorbers and strippers.

At pressure drops above 1200 Pa/m, *flooding* becomes problematic. Flooding is characterized by sudden large surges in pressure drop through the column caused by liquid buildup in the column. At pressure drops below 200 Pa/m, areas of poor liquid-gas contact and even dry areas can be expected.

Pressure drops in absorption columns can be changed by using different packing, by reducing the liquid and gas flow rates, and by increasing the column diameter.

26. ERGUN PRESSURE DROP

Equation 45.31: Ergun Equation

$$\frac{\Delta P}{L} = \frac{150 v_o \mu (1 - \varepsilon)^2}{\Phi_s^2 D_p^2 \varepsilon^3} + \frac{175 \rho v_o^2 (1 - \varepsilon)}{\Phi_s D_p \varepsilon^3} \qquad 45.31$$

Description

The pressure drop through a layer of packing can be estimated using the *Ergun equation*, Eq. 45.31. This equation is valid for liquids and gases under both laminar and turbulent flow conditions. The Ergun equation is a semi-empirical equation, obtained partly from theory and partly from experimental data for spherical, cylindrical, and crushed solids packing. Various terms in the equation account for deviations from ideal particles.

The first term in the Ergun equation represents how pressure is affected by viscous forces (accommodating the viscous flow regime), and the second term represents the effect of inertial forces (accommodating the turbulent flow regime). μ and ρ are the absolute viscosity and mass density, respectively. v_o is the *superficial velocity* through the packed bed, the velocity in length per unit time through an empty column. ε is the dimensionless *porosity* (*void fraction*) of the packed bed. (See Sec. 45.19.) The *equivalent diameter* of the packing material, D_p, is the diameter of the sphere that would have the same surface-to-volume ratio.[24]

$$D_p = \frac{6(1 - \varepsilon)}{a}$$

The dimensionless *sphericity*, Φ, is included in the Ergun equation to account for the deviations from spherical form found in commercial packing materials with high surface areas and void fractions. Sphericity is a measure of the roundness of the packing particles. It is the ratio of the surface area of a sphere with the same volume as the particle to the surface area of the particle itself. For a perfectly spherical particle, $\Phi = 1$. Sphericity can be calculated from a particle's volume, V_p, and its surface area, S_p.[25]

$$\Phi = \left(\frac{\pi D_p^2}{\frac{\pi D_p^3}{6}}\right)\left(\frac{V_p}{S_p}\right) = \frac{6V_p}{D_p S_p}$$

The Ergun equation assumes a constant gas or vapor density throughout the packed column. In calculations, this assumption can be approximated by using the gas

---

[24]The *NCEES Handbook* defines D_p as an *average diameter* which works with solid particles of roughly spherical shapes, but not with commercial packing.
[25]This equation is not included in the *NCEES Handbook*, so NCEES will need to provide the equivalent particle diameter, D_p, in problems.

or vapor properties that correspond to the average of the inlet and outlet conditions. Process software can get more accurate results by calculating the pressure drop in very small stages and recalculating the physical properties after each stage.

Example

A column with a packing height of 2 m is filled with 13 mm Berl saddles that have a sphericity of 0.37. The surface area per unit of packed tower volume is 510 m^2/m^3. The porosity of the packing is 0.60. Air flows through the tower with a superficial velocity of 0.56 m/s. The average absolute viscosity of air is 1.84×10^{-5} kg/m·s, and the average density is 1.21 kg/m^3. Most nearly, what is the pressure drop through the packing?

(A) 320 Pa

(B) 640 Pa

(C) 1100 Pa

(D) 2200 Pa

Solution

The average particle diameter is

$$D_p = \frac{6(1-\varepsilon)}{a}$$

$$= \frac{(6)(1-0.60)}{510 \ \frac{m^2}{m^3}}$$

$$= 4.71 \times 10^{-3} \text{ m}$$

Use Eq. 45.31. The pressure drop through the entire column is

$$\frac{\Delta p}{L} = \frac{150 v_o \mu (1-\varepsilon)^2}{\Phi^2 D_p^2 \varepsilon^3} + \frac{1.75 \rho v_o^2 (1-\varepsilon)}{\Phi D_p \varepsilon^3}$$

$$\Delta p = L\left(\frac{150 v_o \mu (1-\varepsilon)^2}{\Phi^2 D_p^2 \varepsilon^3} + \frac{1.75 \rho v_o^2 (1-\varepsilon)}{\Phi D_p \varepsilon^3}\right)$$

$$= (2 \text{ m})\left(\frac{(150)\left(0.56 \ \frac{m}{s}\right)(1.84 \times 10^{-5} \text{ Pa·s})}{(0.37)^2 (4.71 \times 10^{-3} \text{ m})^2 (0.6)^3} + \frac{(1.75)\left(1.21 \ \frac{kg}{m^3}\right)\left(0.56 \ \frac{m}{s}\right)^2 (1-0.6)}{(0.37)(4.71 \times 10^{-3} \text{ m})(0.6)^3}\right)$$

$$= 2168 \text{ Pa} \quad (2200 \text{ Pa})$$

The answer is (D).

27. ECKERT PRESSURE DROP IN PACKED COLUMNS[26]

The *Eckert generalized pressure drop correlation* provides a graphical way of estimating the pressure drop.[27] (See Fig. 45.20.) Both axes are correlated to dimensionless numbers that are dependent on the physical properties and flow rates of the liquid and gas, as well as on the packing.[28] L' and G' represent the superficial liquid velocity and superficial gas velocity, respectively. The superficial velocity is the molar flow rate per unit cross sectional area of column, with units of $kmol/m^2$·s. It can be calculated from the actual flow rates and the column diameter.

$$x\text{-coordinate} = \frac{L'}{G'}\sqrt{\frac{\rho_G}{\rho_L - \rho_G}}$$

$$y\text{-coordinate} = \frac{C_f G'^2 \mu_{cP}^{0.1}}{\rho_G(\rho_L - \rho_G)}$$

$$G' = \frac{G}{A_{column}} = \frac{4G}{\pi D^2}$$

$$L' = \frac{L}{A_{column}} = \frac{4L}{\pi D^2}$$

[26]This material is not present in the *NCEES Handbook*.
[27]The correlation was originally developed by Sherwood, then modified by Eckert, and then modified again by Strigle. It is generally attributed to Eckert.
[28](1) In the equations for the x- and y-coordinates, the quantity $\rho_L - \rho_G$ is approximately equal to ρ_L, so the equation for the x-coordinate may appear simplified in some references. (2) In the equation for the y-coordinate, viscosity must be given in centipoise (cP).

Figure 45.20 *Eckert Generalized Pressure Drop Correlation for Packed Columns*

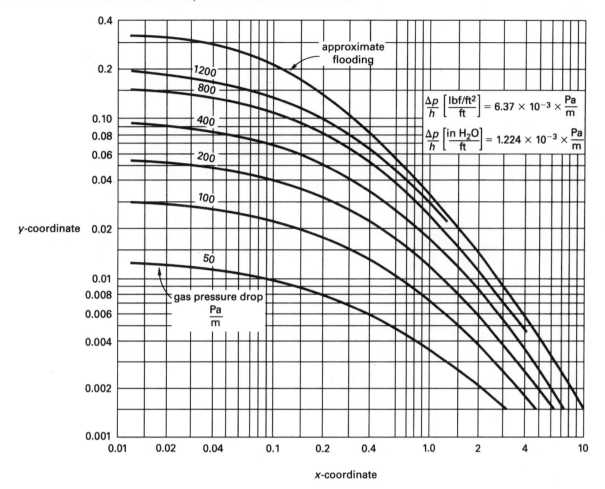

46

Solid-Liquid Processes[1]

Nomenclature

A	surface area	m^2
F_{solid}	mass flow rate of insoluble solids	kg/s
$F_{solvent}$	mass flow rate of solvent	kg/s
k_Y	gas-phase mass transfer coefficient	$kg/s \cdot m^2$
L_s	mass of dry solids	kg
$\dot{m}$	mass flow rate	kg/s
n	number of equilibrium stages	–
N	drying rate	$kg/s \cdot m^2$
N_c	constant period drying rate	$kg/s \cdot m^2$
O_j	overflow liquid mass flow rate leaving stage j	kg/s
t	time	s
t_c	drying time in constant rate drying period	s
T_s	surface temperature	°C
U_j	underflow liquid mass flow rate leaving stage j	kg/s
$\dot{V}$	volumetric flow rate	m^3/s
x	mass fraction of solute in under flow liquid	kg solute/ kg liquid
X	mass ratio of solute to insoluble solids	kg solute/ kg insoluble solids
X	free moisture content	kg liquid/ kg dry solids
X^*	equilibrium moisture content of solid	kg liquid/ kg dry solids
x_i	mass fraction of solute in liquid in stage i	kg solute/ kg liquid
X_c	critical moisture content	kg liquid/ kg dry solids
X_i	moisture content at time i	kg liquid/ kg dry solids
X_T	total moisture content in a solid	kg liquid/ kg dry solids
y	mass fraction of solute in overflow liquid	kg solute/ kg liquid
y_i	mass fraction of solute in liquid in stage i	kg solute/ kg liquid
Y	mass ratio of solute to solvent	kg solute/ kg solvent

Symbols

ε	porosity	kg solute/ kg liquid
ρ	density	kg/m^3
ω	humidity	kg liquid/ kg dry gas

Subscripts

c	critical or constant
D	drying, total
f	falling rate period or fluid
s	solids
sat	saturation
T	total

1. DRYING

Drying is a mass transfer process that involves liquid, such as water or a chemical solvent, from a solid. In some cases, drying can be done mechanically with compression or centrifugation. This chapter describes only drying by thermal evaporation of the liquid.

A material that contains no liquid is called *bone dry*. However, most materials contain some liquid even after drying. For example, dried table salt usually contains approximately 0.5% water. In drying, the liquid is usually removed to some acceptable level.

Drying equipment is specific to the material form (e.g., slab, powder, or continuous sheet). Drying can be performed either in batches or through a continuous process. *Direct dryers* expose the material to hot gas directly, whereas *indirect dryers* transfer heat indirectly by exposing the material to thermal radiation from a hot surface.[2] Since the drying mechanism depends on the form of the material, there is no single, universal analytical treatment for dryers. However, there are some general principles for predicting drying rates.

In addition to hot air drying, other types of drying include dielectric drying, freeze drying (lyophilization), and supercritical drying. This is called *dry basis*.

2. BASIS OF LIQUID CONTENT

Humidity, ω, and *concentrations* (e.g., moisture or solvent content) of liquids in solids, X, are reported as a ratio of liquid mass to dry solid mass. The original or final

---

[1]The topic of "solid-liquid processes" is not included in the *NCEES FE Reference Handbook* (*NCEES Handbook*). However, this chapter is included for your reference because of its relevance to chemical engineering.

[2]The term "adiabatic dryer" is sometimes used in reference to a direct dryer, and "nonadiabatic dryer" is used in reference to an indirect dryer. In both cases, the reference is to a type of dryer, not necessarily to a thermodynamic process or condition.

liquid mass is not included in the basis. Consider 10 kg of dry powder mixed with 1 kg of water. The total mass is 11 kg. The water content of the moist powder on a dry basis is $X = X_T/L_s = 1 \text{ kg}/10 \text{ kg} = 0.10$.

Equilibrium moisture, X^*, is the water content in a solid at equilibrium. *Bound water* (moisture in a substance) is the water content in a substance which exerts vapor pressure less than that of the pure liquid at the same temperature. Unbound moisture is the moisture held by a substance in excess of the equilibrium moisture content corresponding to saturation in the surrounding atmosphere. It is primarily held in the voids of the substance. The equilibrium vapor pressure of bound water is lower than the vapor pressure of pure (unbound) water at the same temperature. The water content in a solid in excess of the equilibrium amount is the *free moisture*, X.

$$X = X_T - X^*$$

3. AIR-WATER SYSTEMS

For water to transfer from a wet solid to a passing air stream, the humidity of the air must be lower than the saturation humidity corresponding to the temperature of the wet solid. (The air near the surface of the solid will be at or near to the temperature of the solid.) If the humidity of the gas is higher than the saturation humidity corresponding to the temperature of the solid, water will condense onto the surface of the solid. Either process will continue until the system reaches equilibrium.

Figure 46.1 shows a drying curve (rate of drying curve) for a solid. The two main parts of the curve are the *constant rate period*, where the rate of drying is constant, and the *falling rate periods* I and II, where the rate of drying decreases. The drying conditions and the particular solid determine the falling rate period curves.

Figure 46.1 *Typical Drying Curve for a Solid*

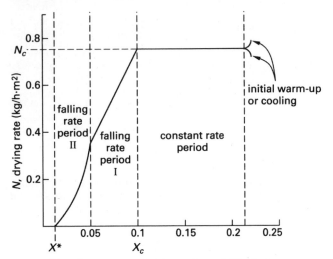

X, moisture content of bone-dry solid (kg/kg)

The drying rate, N, is reported in units of liquid mass removed per unit area per unit time.

In the constant rate period, the drying rate, N_c, is a function of the *gas-phase mass transfer coefficient, k_Y*, the saturation humidity at the liquid-surface temperature, ω_{sat}, and the humidity of the drying air, ω.

$$N_c = k_Y(\omega_{\text{sat}} - \omega)$$

During the constant rate period, the bulk surface temperature of the solid (referred to as the *ultimate surface temperature*), T_s, remains constant due to reboiling of the fluid. This temperature generally equals the saturation (i.e., "boiling") temperature of the liquid for the partial pressure it is exerting. At the critical moisture content, X_c, the drying rate begins to decrease because the solid surface is no longer completely saturated with liquid. The drying process moves into falling rate period I and continues until diffusion from inside the solid begins to govern the drying rate (falling rate period II). Depending on the physical characteristics of the solid, falling rate period II may not occur at all. Drying stops when the solid reaches the *equilibrium moisture content*, X^*.

At any moment, the instantaneous rate of drying, N, of the solids is

$$N = -\left(\frac{L_s}{A}\right)\frac{dX}{dt}$$

The negative sign indicates that the drying rate decreases with time. Rearranging and integrating this equation between two moisture contents, X_1 and X_2, defines the drying time.

$$t = \left(-\frac{L_s}{A}\right)\int_{X_1}^{X_2}\frac{dX}{N} = \left(\frac{L_s}{A}\right)\int_{X_2}^{X_1}\frac{dX}{N}$$

If moisture contents X_1 and X_2 are both greater than the critical moisture content, X_c, then the drying time in the constant rate period is

$$t_c = \frac{L_s(X_1 - X_2)}{AN_c} \quad \text{[constant rate]}$$

If moisture contents X_1 and X_2 are both less than the critical moisture content, X_c, as in the falling rate periods, determining the drying time requires integration. When the curve in a falling rate period is a straight line of the form $N = mX + b$, with m as the slope and b as the y-intercept in this linear region, the corresponding drying time in the falling rate period, t_f, is

$$t_f = \left(-\frac{L_s}{A}\right)\int_{X_1}^{X_2}\frac{dX}{mX + b}$$

$$t_f = \left(\frac{L_s}{mA}\right)\ln\frac{mX_1 + b}{mX_2 + b}$$

$$= \left(\frac{L_s(X_1 - X_2)}{A(N_1 - N_2)}\right)\ln\frac{N_1}{N_2} \quad \text{[falling rate]}$$

Due to a lack of specific drying rate information, it is common to assume that both falling rate curves can be approximated by a single straight line. In this case, the drying time is

$$t_f = \left(\frac{L_s(X_c - X^*)}{N_cA}\right)\ln\frac{X_1 - X^*}{X_2 - X^*} \quad \text{[falling rate]}$$

Example

Material with a 15% moisture content is dried to 2.5% moisture content. The drying curve is shown. The initial moist solids mass is 500 kg, with a drying surface of 1 m²/10 kg (dry basis).

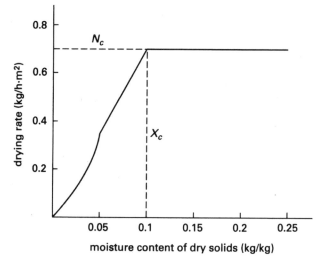

Most nearly, what is the drying time during the constant rate period?

(A) 0.99 h

(B) 1.1 hr

(C) 1.7 h

(D) 2.4 h

Solution

The initial material has 15% moisture, so by definition the initial moisture content, X_1, is

$$X_1 = \frac{X_T}{L_s}$$

$$= \frac{\left(0.15\ \dfrac{\text{kg moisture}}{\text{kg total}}\right)(500\ \text{kg total})}{\left((1 - 0.15)\ \dfrac{\text{kg dry solids}}{\text{kg total}}\right)(500\ \text{kg total})}$$

$$= 0.176\ \text{kg moisture/kg dry solids}$$

In the constant rate period, the moisture content is reduced from 15% to the critical moisture content, X_c. From the drying curve, the critical moisture content, X_c, is

$$X_c = 0.1\ \text{kg moisture/kg dry solids}$$

The corresponding drying rate, N_c, in the constant rate period is

$$N_c = 0.70\ \text{kg moisture/h·m}^2$$

In the constant rate drying period, the drying time, t_c, is

$$t_c = \left(\frac{L_s}{A}\right)\left(\frac{X_1 - X_c}{N_c}\right) = \left(\frac{10\ \text{kg dry solids}}{1\ \text{m}^2}\right)$$

$$\times \left(\frac{0.176\ \dfrac{\text{kg moisture}}{\text{kg dry solids}} - 0.1\ \dfrac{\text{kg moisture}}{\text{kg dry solids}}}{0.70\ \dfrac{\text{kg moisture}}{\text{h·m}^2}}\right)$$

$$= 1.09\ \text{h} \quad (1.1\ \text{h})$$

The answer is (B).

4. LEACHING

Leaching is the process of transferring a soluble solute from a solid to a liquid. A common example of leaching is brewing coffee from ground coffee beans using hot water. An industrial example is the use of water to recover copper sulfate, $CuSO_4$, from insoluble calcined ore. In leaching, some of the solids dissolve in the liquid. Thus, the solids mass is not constant.

Usually, there are five rate steps in a leaching process. (1) The solvent is transferred from the bulk solution to the surface of the solid. (2) The solvent penetrates or diffuses into the solid by intraparticle diffusion. (3) The solute dissolves from the solid into the solvent. (4) The solute diffuses through the mixture to the surface of solid (again by intraparticle diffusion). (5) The solute is transferred to the bulk solution.

Since a portion of the solids dissolves in the liquid, the solid and liquid flow rates vary. Therefore, the analysis of leaching requires accounting for these varying liquid and solid flow rates. Disregarding the entrapment of liquid in the solid underflow and making the following assumptions can simplify leaching into an idealized process.

- Leaching is isothermal.

- Leaching is isobaric.

- No solvent dissolves into the solid.

- No solvent is entrained with the solid.

- The solid matrix is insoluble.

- The heat of mixing is negligible.

- The stages are equilibrium stages.

The variables X and Y used in the operating equation (i.e., the equation of the operating line) for leaching, are defined as follows.

$$Y \equiv \frac{\text{mass of solute}}{\text{mass of solvent}}$$

$$X \equiv \frac{\text{mass of solute}}{\text{mass of insoluble solids}}$$

The operating equation for leaching is

$$Y_{j+1} = \left(\frac{F_{\text{solid}}}{F_{\text{solvent}}}\right)X_j + \left(Y_1 - \left(\frac{F_{\text{solid}}}{F_{\text{solvent}}}\right)X_0\right)$$

This is the equation for a straight line; the ratio of solid feed rate to solvent feed rate, $F_{\text{solid}}/F_{\text{solvent}}$, is the slope. The Y versus X equilibrium data must be obtained experimentally. Using the McCabe-Thiele method, the equilibrium stages can be stepped off between the operating line and the equilibrium line. (See Fig. 46.2.)

Figure 46.2 McCabe-Thiele Leaching Diagram

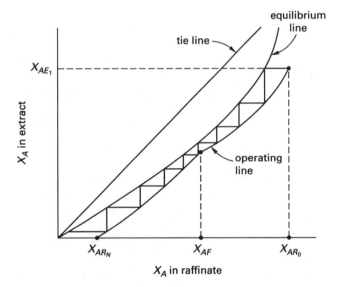

5. WASHING

Washing processes are similar to leaching. However, unlike leaching, washing does not reduce the insoluble solid mass, making washing simpler to analyze.

An example of washing occurs during the process of mining sand from the ocean. Wet sand contains salt, which can be removed by washing the sand with pure water. In washing, the solid (sand) is mixed with the wash liquor (pure water) and then fed to a settler. The solids are removed from the bottom of the settler as the underflow liquid. The clear liquid containing the solute (salt) is removed from the top of the settler as the overflow liquid. Washing can be performed in a single stage or in a countercurrent cascade. (See Fig. 46.3.)

Figure 46.3 Two-Stage Mixer Settler for Countercurrent Washing

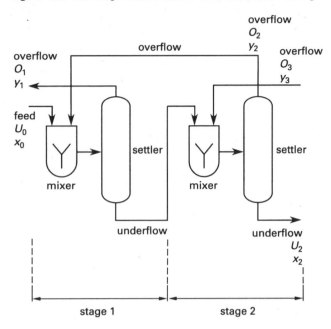

Equilibrium exists when the mass fraction (i.e., concentration) of solute in the overflow liquid, y, and the mass fraction of solute in the underflow liquid, x, are equal.

$$y = x \quad \text{[equilibrium]}$$

For a countercurrent cascade, the operating equation for washing is derived from the steady-state mass balance around the mass-balance envelope in Fig. 46.4.

Figure 46.4 Block Diagram of Countercurrent Washing

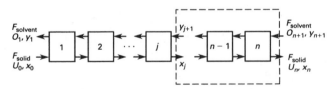

For the operating equation to be a straight line, the overflow and underflow flow rates must be constant throughout the cascade washer. O_i and U_i are the overflow and the underflow liquid flow rates, respectively, leaving stage i.

$$y_{j+1} = \left(\frac{U_j}{O_{j+1}}\right)x_j + \left(\frac{O_{n+1}}{O_{j+1}}\right)y_{n+1} - \left(\frac{U_n}{O_{j+1}}\right)x_n$$

The porosity, ε, of underflow slurry is

$$\varepsilon = \frac{\text{volume voids}}{\text{total volume}} = \frac{\text{volume liquid}}{\text{total volume}}$$

$$1 - \varepsilon = \frac{\text{volume solids}}{\text{total volume}}$$

The underflow mass flow rate leaving stage j is

$$U_j = \dot{V}_{\text{wet solids}}\varepsilon\rho_f$$

The flow rate of solids and/or wet solids is usually determined by process requirements. If the volumetric flow rate of dry solids processed is fixed, then the underflow mass flow rate, U_j, can be expressed as a function of the mass flow rate of dry solids, $\dot{m}_{\text{dry solids}}$; the fluid density, ρ_f; and the dry solids density, ρ_s.

$$U_j = \dot{m}_{\text{dry solids}}\left(\frac{\varepsilon}{1-\varepsilon}\right)\left(\frac{\rho_f}{\rho_s}\right)$$

If the dry solids mass flow rate, the porosity of the underflow, ε, and the densities of the fluid and the dry solids are all constant, then all the underflow mass flow rates from each stage, U_j, are constant.[3]

$$U_j = U = \text{constant}$$

If all the underflow mass flow rates are constant, then the overall mass balance shows that all the overflow mass flow rates from each stage, O_j, must also be constant. If the following conditions exist, a constant flow rate can be assumed.

- There are no solids in the overflow.

- The fluid density, ρ_f, and the dry solids density, ρ_s, are constant.

- The porosity of the underflow, ε, is constant.

The constant flow rate assumption results in a simplified operating equation.

$$y_{j+1} = \left(\frac{U}{O}\right)x_j + \left(y_{n+1} - \left(\frac{U}{O}\right)x_n\right)$$

Sometimes the feed consists only of dry solids, so the feed flow, U_0, is markedly different from the underflow leaving stage 1, U_1. This condition makes it necessary to solve for the first stage separately. Subsequent stages may then be solved, starting from stage 1, to develop the operating equation for the cascade.

Using the operating equation and treating each stage as an equilibrium stage allows a McCabe-Thiele diagram (see Fig. 46.2) to be used for analyses of washing processes. The equilibrium line is the $y=x$ line. Stages are stepped off between the equilibrium line and the operating line, between the inlet and outlet mass fraction of solute in the underflow, x_i.

[3]The constant underflow assumption relies on several assumptions, the most restrictive of which is constant fluid density, ρ_f, which implies constant fluid composition. Since solute transfers from the solids to the fluid, this assumption is inexact. However, it is valid for dilute solutions.

Example

Raw sand ore mined from a beach contains 5.5 wt% salt. The salt content of the sand must be reduced to 0.2 wt% by washing with 2.0 kg pure water per 1000 cm³ wet sand. The raw sand ore enters the washer at a rate of 1000 cm³/h and contains 25% seawater by volume.

The density of seawater is 1.0 g/cm³; the density of the dry sand including air voids is 2.1 g/cm³; and the density of dry sand excluding air voids is 3.4 g/cm³. Most nearly, how many equilibrium washing stages would this separation require?

(A) 1

(B) 2

(C) 3

(D) 4

Solution

Assuming the porosity of solids in the underflow, ε, the densities, ρ_i, and all the underflow and overflow mass flow rates, U_i and O_i, are constant between stages. The applicable operating equation is

$$y_{j+1} = \left(\frac{U}{O}\right)x_j + \left(y_{n+1} - \left(\frac{U}{O}\right)x_n\right)$$

2.0 kg pure water are used per 1000 cm³ of wet sand. Using a basis of 1000 cm³/h of wet sand processed,

$$O = \left(\frac{2.0\ \text{kg water}}{1000\ \text{cm}^3\ \text{wet sand}}\right)\left(1000\ \frac{\text{cm}^3\ \text{wet sand}}{\text{h}}\right)$$
$$= 2.0\ \text{kg water/h}$$

Since the problem statement specified the amount of wet solids processed, determine U, the underflow rate.

The sand contains 25% water by volume. The porosity is

$$\varepsilon = 0.25$$

Using a basis of 1000 cm³/h, U is

$$U = \dot{V}_{\text{wet solids}}\varepsilon\rho_f$$
$$= \frac{\left(1000\ \frac{\text{cm}^3}{\text{h}}\right)(0.25)\left(1.0\ \frac{\text{g}}{\text{cm}^3}\right)}{1000\ \frac{\text{g}}{\text{kg}}}$$
$$= 0.25\ \text{kg/h}$$

The slope of the operating line is

$$\frac{U}{O} = \frac{0.25 \ \dfrac{\text{kg}}{\text{h}}}{2 \ \dfrac{\text{kg}}{\text{h}}} = 0.125$$

The y-intercept for the operating line is

$$y\text{-intercept} = y_{n+1} - \left(\frac{U}{O}\right)x_n = 0 - (0.125)(0.002)$$

$$= -0.00025$$

The equilibrium relationship for washing processes is $y = x$. Plotting the operating line on a y versus x equilibrium diagram shows that this separation requires 1.8 stages. Since the number of stages must be an integer, use 2 equilibrium stages.

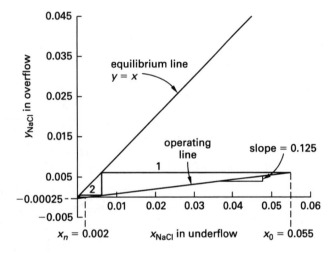

The answer is (B).

Diagnostic Exam

Topic XII: Process Control

1. A block flow diagram for a chemical plant includes two processes, $G_1(s)$ and $G_2(s)$.

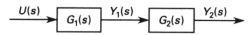

What is the overall transfer function, $Y_2(s)$?

(A) $Y_2(s) = G_1(s) + G_2(s)$

(B) $Y_2(s) = G_1(s) G_2(s) U(s)$

(C) $Y_2(s) = G_1(s) + G_2(s) - Y_1(s)$

(D) $Y_2(s) = G_1(s)/G_2(s)$

2. In the reactor system shown, the air transmitter measures the air level of the tank and sends the data to the analysis controller, which sends instructions to the air compressor. The agitator mixes the air with the fluid in the reactor.

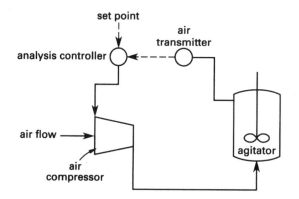

Which component is the actuator?

(A) agitator

(B) air compressor

(C) air transmitter

(D) analysis controller

3. A proportional-integral-derivative (PID) controller is used instead of a proportional-integral (PI) controller when

(A) there is only one energy storage device in the system

(B) large disturbances and noise are present in the system

(C) fast system response is required

(D) there are large transport delays in the system

4. The circuit shown has two resistors with known resistances, a variable resistor, and an unknown resistor.

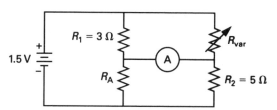

When the variable resistor has a resistance of 2.5 Ω, no current flows through the ammeter. What is the resistance of the unknown resistor?

(A) 1.5 Ω

(B) 1.7 Ω

(C) 4.2 Ω

(D) 6.0 Ω

5. A unit impulse function at $t = 0$ is the input signal for the transfer function shown.

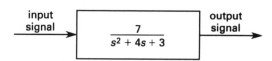

What is the output signal as a function of time?

(A) $y(t) = \left(\frac{7}{2}\right)(e^t + e^{-3t})u(t)$

(B) $y(t) = \left(\frac{7}{2}\right)(e^{-t} + e^{-3t})u(t)$

(C) $y(t) = \left(\frac{7}{2}\right)(e^{-3t} - e^{-t})u(t)$

(D) $y(t) = \left(\frac{7}{2}\right)(e^{-t} - e^{-3t})u(t)$

6. The input signal for the transfer function shown is a step function of height 5 at $t=0$.

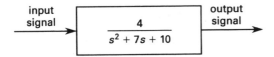

What is the steady-state output?

(A) 0.4

(B) 0.8

(C) 2.0

(D) 2.5

7. Which of the following feedforward transfer function controller equations is realizable?

(A) $G_{ff} = \dfrac{24s(s+1)e^{3s}}{(2s+1)(3s+1)}$

(B) $G_{ff} = \left(-\dfrac{4}{3}\right)\left(\dfrac{3s+1}{2s+1}\right)$

(C) $G_{ff} = \left(+\dfrac{2}{3}\right)\left(\dfrac{(3s+1)e^{2s}}{2s+1}\right)$

(D) $G_{ff} = \left(-\dfrac{1}{9}\right)\left(\dfrac{(2s+1)e^{s}}{s+1}\right)$

8. A closed-loop system is shown.

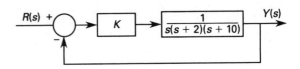

Most nearly, at what gain constant, K, does the system become marginally stable?

(A) 10

(B) 60

(C) 100

(D) 240

9. The characteristic equation for a control system is $s^2 + 4s + K = 0$. What must be the range of K so that all the roots will be real?

(A) $K \le 0$

(B) $K \ge 0$

(C) $K \le 4$

(D) $K \ge 4$

10. What is the purpose of a transducer?

(A) signal detection

(B) amplification

(C) voltage reduction

(D) conversion of one physical quantity to another

Process Control

SOLUTIONS

1. Transfer functions in series multiply together. Solve for the overall transfer function, $Y_2(s)$, by multiplying the individual transfer functions and substituting $Y_1(s)$ into $Y_2(s)$.

$$Y_1(s) = G_1(s)U(s)$$
$$Y_2(s) = G_2(s)Y_1(s)$$
$$= G_1(s)G_2(s)U(s)$$

The answer is (B).

2. In a control loop, the actuator is the final control element that changes the value of the manipulated variable. The manipulated variable in the system is the air flow. The air transmitter communicates with the air compressor through the analysis controller. The actuator for the air flow in the system is the air compressor.

The answer is (B).

3. The purpose of derivative control action is to anticipate the future behavior of the error signal. It allows the controller to quickly take corrective action in response to a sudden change in the error signal.

The answer is (C).

4. The circuit shown in the problem statement is a Wheatstone bridge, which determines an unknown resistance when the bridge is balanced.

$$\frac{R_1}{R_A} = \frac{R_{var}}{R_2}$$

$$R_A = \frac{R_1 R_2}{R_{var}}$$

$$= \frac{(3\ \Omega)(5\ \Omega)}{2.5\ \Omega}$$

$$= 6.0\ \Omega$$

The answer is (D).

5. The transfer function is

$$\frac{Y(s)}{X(s)} = G(s) = \frac{7}{s^2 + 4s + 3}$$

A unit impulse function at $t=0$ is the input signal, which is also called the Dirac delta function.

$$x(t) = \delta(t)$$

The Laplace transform of the unit impulse function is found in a table of Laplace transform pairs.

$$X(s) = \mathcal{L}\big(x(t)\big) = \mathcal{L}\big(\delta(t)\big) = 1$$

The output in the Laplace domain is

$$Y(s) = G(s)X(s)$$
$$= \left(\frac{7}{s^2 + 4s + 3}\right)X(s)$$
$$= \frac{7}{s^2 + 4s + 3}$$
$$= \frac{7}{(s+1)(s+3)}$$

$Y(s)$ as shown is not in the Laplace transform table. Convert it with partial fractions to an entry that is in the table. Assume $Y(s)$ can be transformed into partial fractions.

$$\frac{7}{(s+1)(s+3)} = \frac{A}{s+1} + \frac{B}{s+3}$$

Multiply by the denominator on the left-hand side to cancel the denominators.

$$7 = \left(\frac{A}{s+1} + \frac{B}{s+3}\right)(s+1)(s+3)$$

This gives the polynomial identity

$$7 = (3A + B) + (A + B)s$$

Since two polynomials are equal if and only if their corresponding coefficients are equal, equate the coefficients of like powers of s. This results in the following set of equations.

$$3A + B = 7$$
$$A + B = 0$$

This set of equations has the solutions

$$A = \frac{7}{2}$$
$$B = -\frac{7}{2}$$

The output $Y(s)$ is

$$Y(s) = \frac{A}{s+1} + \frac{B}{s+3}$$
$$= \left(\tfrac{7}{2}\right)\left(\frac{1}{s+1} - \frac{1}{s+3}\right)$$

Take the inverse Laplace transform of $Y(s)$ to get the time domain output $y(t)$.

$$y(t) = \mathcal{L}^{-1}(Y(s))$$

$$= \mathcal{L}^{-1}\left(\left(\tfrac{7}{2}\right)\left(\frac{1}{s+1} - \frac{1}{s+3}\right)\right)$$

$$= \left(\tfrac{7}{2}\right)\left(\mathcal{L}^{-1}\left(\frac{1}{s+1}\right) - \mathcal{L}^{-1}\left(\frac{1}{s+3}\right)\right)$$

$$= \left(\tfrac{7}{2}\right)(e^{-t} - e^{-3t}) \quad \left(\left(\tfrac{7}{2}\right)(e^{-t} - e^{-3t})u(t)\right)$$

The answer is (D).

6. The transfer function is

$$\frac{Y(s)}{X(s)} = G(s) = \frac{4}{s^2 + 7s + 10}$$

The input signal is a unit step function of height 5 at $t = 0$.

$$x(t) = 5u(t)$$

The Laplace transform of the unit step function is found from a table of Laplace transform pairs.

$$X(s) = \mathcal{L}(x(t)) = \mathcal{L}(5u(t)) = \frac{5}{s}$$

The output in the Laplace domain is

$$Y(s) = G(s)X(s)$$

$$= \left(\frac{4}{s^2 + 7s + 10}\right)X(s)$$

$$= \left(\frac{4}{s^2 + 7s + 10}\right)\left(\frac{5}{s}\right)$$

$$= \frac{20}{(s^2 + 7s + 10)s}$$

The steady state output can be obtained from the final value theorem.

$$\lim_{t \to \infty} y(t) = \lim_{s \to \infty} sY(s)$$

$$= \lim_{s \to \infty} s\frac{20}{(s^2 + 7s + 10)s}$$

$$= \lim_{s \to \infty} \frac{20}{s^2 + 7s + 10}$$

$$= \frac{20}{(0)^2 + (7)(0) + 10}$$

$$= 2.0$$

The answer is (C).

7. The e^{3s}, e^{2s}, and e^s factors in options (A), (C), and (D) require the controller to know the input

(disturbance) 3, 2, and 1 in units of time into the future, respectively. This is not possible. The only transfer function that is realizable is option (B). In addition, option (B) has the same order of polynomial in the numerator and denominator, which is a requirement for realization.

The answer is (B).

8. The characteristic equation is given by

$$a_n s^n + a_{n-1} s^{n-1} + a_{n-2} s^{n-2} + \cdots + a_0 = 0$$

$$s^3 + 12s^2 + 20s + K = 0$$

Form the Routh table for the coefficients.

s^3	1	20
s^2	12	K
s	$(240 - K)/12$	0
s^0	K	

The Routh-Hurwitz criterion states that the number of sign changes in the first column of the table equals the number of positive, or unstable, roots. If the value of K is between 0 and 240, there are no sign changes, so the system becomes marginally stable at $K = 240$.

The answer is (D).

9. The characteristic equation is

$$s^2 + 4s + K = 0$$

Use the quadratic formula to find s.

$$s = \frac{-b \pm \sqrt{b^2 - 4ac}}{2a} = \frac{-4 \pm \sqrt{(4)^2 - (4)(1)K}}{(2)(1)}$$

$$= \frac{-4 \pm \sqrt{16 - 4K}}{2}$$

For the roots to be real, the quantity under the radical must be positive.

$$16 - 4K \geq 0$$

Solving for K gives

$$K \leq 4$$

The answer is (C).

10. A transducer measures some type of physical quantity, and converts that quantity into another quantity that is easily measured—often a voltage or gage displacement.

The answer is (D).

47 Control Loops and Control Systems

Nomenclature

a	Routh table parameter	–
A	constant	–
$\mathbf{A}$	system matrix	–
b	Routh table parameter	–
$B(s)$	feedback transfer function	–
$\mathbf{B}$	control vector	–
BW	bandwidth	rad/s or Hz
c	Routh table parameter	–
$\mathbf{C}$	output vector	–
$D(s)$	denominator polynomial	–
$\mathbf{D}$	feed-through vector	–
$e(t)$	error function	–
E	error	–
$E(s)$	transform of error function, $\mathcal{L}(e(t))$	–
$F(s)$	transform of forcing function, $\mathcal{L}(f(t))$	various
$G(s)$	forward transfer function	–
GM	gain margin	–
h	magnitude	–
$H(s)$	reverse transfer function	–
$\mathbf{I}$	identity matrix	–
k	spring constant	N/m

K	error constant	–
K	gain constant	–
K	scale factor	–
L	length	m
$L(s)$	load disturbance	–
m	number of zeros	–
M	magnitude	various
n	degrees of freedom	–
$N(s)$	numerator polynomial	–
OS	overshoot	%
p	pole	–
$p(t)$	arbitrary function	–
$P(s)$	transform of arbitrary function, $\mathcal{L}(p(t))$	–
PM	phase margin	–
Q	quality factor	–
r	root or system degree	–
$r(t)$	time-based response function	–
$R(s)$	transform of response function, $\mathcal{L}(r(t))$	–
s	s-domain variable	–
t	time	s
T	system type	–
$T(s)$	transfer function, $\mathcal{L}(t(t))$	–
$u(t)$	unit step function	–
$\mathbf{u}$	unit vector	–
$\mathbf{U}$	r-dimensional control vector	–
$\mathbf{V}$	input vector	–
x	amplitude of oscillation	various
x	position	m
$\mathbf{x}$	state vector	–
X	state variable	–
$\mathbf{X}$	state vector	–
$X(s)$	input transfer function	–
$\mathbf{y}$	output vector	–
$\mathbf{Y}$	output vector	–
$Y(s)$	output transfer function	–
z	zero	–

Symbols

α	angle on pole-zero diagram	deg
α	termination angle	deg
β	angle on pole-zero diagram	deg
δ	logarithmic decrement	–
ε	a small number	–
ζ	damping ratio	–
σ	asymptote centroid	–
θ	time increment	s
Φ	state transition matrix	–
τ	inverse natural frequency, period of oscillation, or time constant	s
ω	frequency	rad/s

Process Control

Subscripts

a	acceleration
A	asymptote
B	block
C	compensator or controller
d	damped natural
D	derivative
f	feedback or forcing
ff	feed forward
i	incoming
I	integral
m	number of zeros
n	number of poles or undamped natural
o	out or output
p	peak or position
P	proportional
r	damped resonant
s	settling
ss	steady state
v	velocity
z	zero

Figure 47.1 A Feedforward Control System

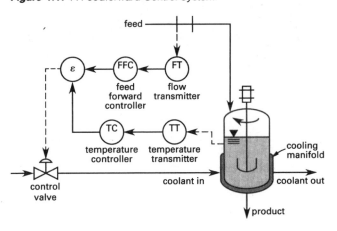

1. CONTROL SYSTEMS AND STRATEGIES

A *control system* monitors a process and makes adjustments to it so that the process's performance is maintained within acceptable limits. For example, a control system may increase or decrease the opening of a fuel valve to keep a furnace performing optimally or switch a heater on and off to keep a mixer at a steady temperature.

The quality of the process that is being monitored and adjusted (for example, temperature) is called the *process variable*. The desired value for the process variable (for example, the ideal temperature for best performance) is called the *set point*. The difference between the set point and the actual value of the process variable is called the *error*.

The two most basic control strategies are feedback control and feedforward control. Both strategies involve making adjustments to what is put into the process, but each uses a different way to determine what these adjustments should be.

A *feedback control system* measures the process variable, compares its actual value against the set point, and uses that information to adjust the input. For example, a feedback control system might measure the rotational speed of a shaft, compare this speed against the speed that is desired, and use this information to decide how much to increase or decrease the voltage provided to the machinery driving the shaft.

A *feedforward control system* measures *disturbances* (outside factors that affect the process) and uses that information to adjust the input. For example, a feedforward control system might measure the temperature of the surrounding environment and use this information to decide whether to turn a heater on or off. Figure 47.1 shows a feedforward control system.

An important disadvantage of a feedforward control system is that it requires a precise understanding of the process being monitored and how it is affected by all possible disturbances. This can make a feedforward control system more time consuming and expensive to implement than a feedback system, which requires only that the process output be monitored.

However, a well-implemented feedforward control system can provide more stability than a feedback control system. A feedback controller reacts only after the process has already been affected, but a feedforward controller can react to disturbances before they affect the process so that the process performance more closely matches its desired level of operation. This can make the process more accurate and reliable and reduce wear and tear on equipment.

In *cascade control*, a feedback control system is improved by combining two feedback loops so one controller (called the primary, outer, or master controller) controls the set point of the other (called the secondary, inner, or slave controller). Cascade control is often used when the process includes a relatively fast-changing variable (such as the flow of a liquid or gas) that affects a relatively slow-changing one (such as system temperature). The primary controller responds to the slower-changing variable by adjusting the set point that the secondary controller uses in responding to the faster-changing variable. When the two variables differ greatly in their behavior, controlling them with separate feedback loops can increase overall stability; on the other hand, both complexity and cost are increased. Figure 47.2 shows a cascade control system diagram.

Ratio control is a special case of feedforward control. The ratio of two process variables is controlled rather than the individual variables. Typical uses of ratio control include fuel-to-air ratios for furnaces, stoichiometric feed ratios for reactors, and reflux ratios for distillation columns.

Figure 47.2 A Cascade Control System Block Diagram

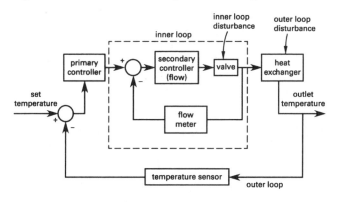

2. INTRODUCTION TO RESPONSE

In addition to the controller, a feedback control system must include a *sensor*, which monitors the process variable, and a *control valve* (also called the *final control element*), which physically affects the process input in some way (such as by increasing or decreasing the fuel supplied to the process).

The sensor measures the process variable and sends a signal to the controller to indicate the amount of error. The controller then sends a signal to the control valve to indicate how the input should be adjusted. This signal from the controller is called the *response*.

When the process is operating at steady state, the controller should provide no response. Real processes always operate under transient conditions, and process variables are never perfectly steady, but the controller should be set so that minor fluctuations are filtered out.

During true transient conditions, such as start-up, shutdown, and changes in load, the controller should be capable of maintaining variable control through the use of changing set points. These may be provided manually, or through a *distributed control system* (DCS). A DCS can be set up to provide a preprogrammed ramp rate of set point changes. Multiple controllers are distributed through the DCS; the feedback or feedforward loops are linked as in cascade control.

After a *step change*—a sudden, discontinuous change from one value to another—in the set point, the system response will cause the process variable to overshoot the set point; further adjustments by the control system will cause the process variable to oscillate around the set point before the process returns to steady state. To reduce or eliminate this effect, a *damping factor* (or *damping ratio*) may be included in the calculations made by the controller.

Overdamping is the use of a damping factor large enough to completely eliminate the overshoot and oscillation. However, overdamping also increases the time the system takes to reach steady state, so it is generally not used except where it is important not to allow the

process variable to overshoot the set point. More common is *underdamping*, which shortens the time to reach steady state but allows some overshoot and oscillation, as shown in Fig. 47.3.

Figure 47.3 Over-, Under-, and Critical Damping

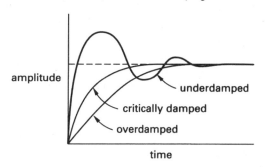

A controller's response time can be characterized by its *time constant*. If a step change is applied to a first-order process (a process that can be modeled by a first-order differential equation), the time constant is the time needed for the process variable to rise or fall by about 63.2% of the total change. The length of time needed for the process to settle at the new set point will be three to five times the controller's time constant.

3. MATCHING MEASURED AND MANIPULATED VARIABLES

The *measured variable* is the variable whose value is sent to the controller by the sensor and transmitter. The *manipulated variable* is the variable that is adjusted by the controller in order to maintain the set point.

In a feedforward control system, the two variables may be for the same stream. For example, a temperature sensor on an inlet stream to a heat exchanger could be used to adjust a control valve that governs the flow of that stream to the heat exchanger.

4. OPEN-LOOP TRANSFER FUNCTIONS

Equation 47.1: Open-Loop, Linear, Time-Invariant Transfer Function

$$\frac{Y(s)}{X(s)} = G(s) = \frac{N(s)}{D(s)} = K \frac{\displaystyle\prod_{m=1}^{M}(s - z_m)}{\displaystyle\prod_{n=1}^{N}(s - p_n)} \qquad 47.1$$

Description

Equation 47.1 is an open-loop, linear, time-invariant transfer function. This is represented by Fig. 47.4.

Figure 47.4 *Open-Loop, Linear, Time-Invariant Transfer Function*

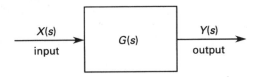

A *time-invariant transfer function* simply affects the input in a manner that does not change with time. In a *linear time-invariant system* (LTI), the transfer function can be expressed as a rational fraction of polynomials. In a *proper control system*, the degree of the denominator polynomial is equal to or larger than the degree of the numerator polynomial. (All realizable control systems are proper. An *improper control system* can be imagined or designed, but it cannot be constructed.) In Eq. 47.1, $G(s)$ is a ratio of two polynomials; $N(s)$ is the numerator polynomial; $D(s)$ is the denominator polynomial; K is the *gain constant* (or, just *gain*), a scalar; z_m are the *zeros*, the roots of the numerator polynomial; and p_n are the *poles*, the roots of the denominator polynomial. In a time-invariant system, the numerator and denominator functions do not depend on time (i.e., they are not $N(s,t)$ or $D(s,t)$).[1]

Example

For an open-loop, linear, time-invariant transfer function,

$$X(s) = \frac{A}{s}$$

$$Y(s) = \frac{A(s+1)}{s(s+2)}$$

What is most nearly the transfer function?

(A) $G(s) = \dfrac{1}{s+2}$

(B) $G(s) = \dfrac{s+1}{s+2}$

(C) $G(s) = \dfrac{s+2}{s+1}$

(D) $G(s) = \dfrac{1}{s+1}$

Solution

From Eq. 47.1,

$$G(s) = \frac{Y(s)}{X(s)} = \frac{\dfrac{A(s+1)}{s(s+2)}}{\dfrac{A}{s}} = \frac{s+1}{s+2}$$

The answer is (B).

---

[1]Circuits that contain capacitors or other energy storage devices are usually not time-invariant. However, a system that shifts or delays the input by a fixed amount of time is considered to be time-invariant.

5. CLOSED-LOOP FEEDBACK SYSTEMS

A basic feedback system consists of two black-box units (a *dynamic unit* and a *feedback unit*), a pick-off point (take-off point), and a *summing point* (*comparator* or *summer*). The output signal is returned as input in a feedback loop (feedback system). (See Fig. 47.5.) The incoming signal, X_i, is combined with the feedback signal, X_f, to give the *error* (*error signal*), E. Whether addition or subtraction is used depends on whether the summing point is additive (i.e., a positive feedback system) or subtractive (i.e., a negative feedback system), respectively. The summing point is assumed to perform positive addition unless a minus sign is present. $E(s)$ is the *error transfer function* (*error gain*).

$$E(s) = \mathcal{L}\big(e(t)\big) = X(s) \pm B(s)$$
$$= X(s) \pm H(s)\,Y(s)$$

Figure 47.5 *Closed-Loop Feedback System*

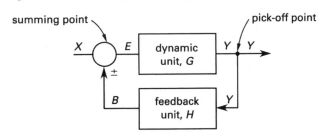

The ratio $E(s)/X(s)$ is the *error ratio* (*actuating signal ratio*).

$$\frac{E(s)}{X(s)} = \frac{1}{1+G(s)H(s)} \quad \text{[negative feedback]}$$

$$\frac{E(s)}{X(s)} = \frac{1}{1-G(s)H(s)} \quad \text{[positive feedback]}$$

Since the dynamic and feedback units are black boxes, each has an associated transfer function. The transfer function of the dynamic unit is known as the *forward transfer function* (*direct transfer function*), $G(s)$. In most feedback systems—amplifier circuits in particular—the magnitude of the forward transfer function is known as the *forward gain* or *direct gain*. $G(s)$ can be a scalar if the dynamic unit merely scales the error. However, $G(s)$ is normally a complex operator that changes both the magnitude and the phase of the error.

The *pick-off point* transmits the output signal, Y, from the dynamic unit back to the feedback element. The output of the dynamic unit is not reduced by the pick-off point. The transfer function of the feedback unit is the *reverse transfer function* (*feedback transfer function*, *feedback gain*, etc.), $H(s)$, which can be a simple magnitude-changing scalar or a phase-shifting function.

In a *unity feedback system* (*unitary feedback system*), $H(s) = 1$, and $B(s) = Y(s)$.[2]

$$B(s) = H(s) Y(s)$$

The ratio $B(s)/X(s)$ is the *feedback ratio* (*primary feedback ratio*).

$$\frac{B(s)}{X(s)} = \frac{G(s)H(s)}{1 + G(s)H(s)} \quad \text{[negative feedback]}$$

$$\frac{B(s)}{X(s)} = \frac{G(s)H(s)}{1 - G(s)H(s)} \quad \text{[positive feedback]}$$

The *loop transfer function* (*loop gain, open-loop gain,* or *open-loop transfer function*), $\pm G(s)H(s)$, is the gain after going around the loop one time.

The *overall transfer function* (*closed-loop transfer function, control ratio, system function, closed-loop gain,* etc.), $Y(s)/X(s)$, is the overall transfer function of the feedback system.

$$T(s) = \frac{Y(s)}{X(s)} = \frac{G(s)}{1 + G(s)H(s)} \quad \text{[negative feedback]}$$

$$T(s) = \frac{Y(s)}{X(s)} = \frac{G(s)}{1 - G(s)H(s)} \quad \text{[positive feedback]}$$

Equation 47.2: Characteristic Equation

$$1 + G_1(s)G_2(s)H(s) = 0 \qquad \textbf{47.2}$$

Description

The quantity $1 + G_1(s)G_2(s)H(s)$ in Eq. 47.2 or $1 + G(s)H(s) = 0$ in the previous sections is the *characteristic equation* and will be a polynomial of the variable s. The *order of the system* is the largest exponent of s in the characteristic equation. (This corresponds to the highest-order derivative in the system equation.) Since the denominator polynomial in a proper control system will always be of a higher degree than the numerator polynomial, the order of the system corresponds to the highest-order term in the denominator polynomial.

Equation 47.2 describes the characteristic equation for a negative feedback system with two forward transfer functions, $G_1(s)$ and $G_2(s)$, in series before the pick-off point. (See Fig. 47.6.)

---

[2]The fact that $B(s)$ can sometimes be the same as $Y(s)$ means that vigilance is required whenever equations containing $Y(s)$ are used. For example, the ratio $Y(s)/X(s)$ refers to the ratio of input to output for all systems, but it can also refer to the primary feedback ratio, $B(s)/X(s)$, in unit feedback systems.

Figure 47.6 *Negative Feedback Control System*

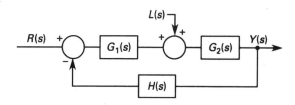

Equation 47.3: Negative Feedback Control System Response

$$Y(s) = \frac{G_1(s)G_2(s)}{1 + G_1(s)G_2(s)H(s)} R(s)$$
$$+ \frac{G_2(s)}{1 + G_1(s)G_2(s)H(s)} L(s) \qquad \textbf{47.3}$$

Description

In a negative feedback system (see Fig. 47.6), the denominator of Eq. 47.3 will be greater than 1.0. Although the closed-loop transfer function, $Y(s)/X(s)$, will be less than $G(s)$, there may be other desirable effects. Generally, a system with negative feedback will be less sensitive to variations in temperature, circuit component values, input signal frequency, and signal noise. Other benefits include distortion reduction, increased stability, and impedance matching. (For circuits to be directly connected in series without affecting their performance, all input impedances must be infinite and all output impedances must be zero.)

Although prone to oscillation, positive feedback systems produce high gain. Some circuits, primarily those incorporating op amps, rely on positive feedback to produce bistable states and hysteresis.

Example

How is the output function $Y(s)$ related to the input functions $R(s)$ and $L(s)$ for the feedback control system shown?

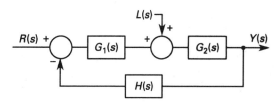

(A) $Y = \dfrac{RG_1G_2}{1 + G_1G_2H} + \dfrac{LG_2}{1 + G_1G_2H}$

(B) $Y = (R + L)\left(\dfrac{G_1G_2}{1 + G_1G_2H} + \dfrac{G_1G_2}{1 + G_1G_2H}\right)$

(C) $Y = \dfrac{RG_1}{1 + G_1G_2H} + \dfrac{LG_2}{1 + G_1G_2H}$

(D) $Y = \dfrac{RG_1G_2}{1 + G_1G_2H} + \dfrac{LG_2}{1 + G_2H}$

Solution

Assume $L(s) = 0$. The feedback is negative.

$$G(s) = G_1(s)\,G_2(s)$$

$$\frac{Y(s)}{R(s)} = \frac{G(s)}{1 + G(s)H(s)}$$

$$Y(s)_1 = \frac{R(s)\,G_1(s)\,G_2(s)}{1 + G_1(s)\,G_2(s)\,H(s)}$$

Assume $R(s) = 0$. The feedback into the $L(s)$ summing point is positive, but the output of $H(s)$ is negated, making this a negative feedback system.

$$G(s) = G_2(s)$$

$$H(s) = G_1(s)H(s)$$

$$\frac{Y(s)}{L(s)} = \frac{G(s)}{1 + G(s)H(s)}$$

$$Y(s)_2 = \frac{L(s)\,G_2(s)}{1 + G_1(s)\,G_2(s)\,H(s)}$$

The transfer function is

$$Y(s) = Y(s)_1 + Y(s)_2$$

$$= \frac{R(s)\,G_1(s)\,G_2(s)}{1 + G_1(s)\,G_2(s)\,H(s)}$$

$$+ \frac{L(s)\,G_2(s)}{1 + G_1(s)\,G_2(s)\,H(s)}$$

This answer is the same as Eq. 47.3.

The answer is (A).

6. FEEDFORWARD SYSTEMS

Like a feedback system, a basic feedforward control system consists of two black-box units (a *dynamic unit* and a *feedforward unit*), a *pick-off point*, and a *summing point*. However, the arrangement of these components is different than in a feedback system. (See Fig. 47.7.)

Figure 47.7 *Feedforward Control System*

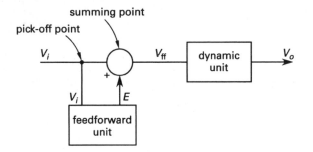

The pick-off point transmits the incoming signal, V_i, to the feedforward unit, which processes it and produces an error signal, E. This error signal is summed with the incoming signal to produce the *feedforward signal*, V_{ff}. The feedforward signal is processed in the dynamic unit, and the resulting output, V_o, should be at the desired set point.

The accuracy of feedforward control depends on the accuracy of the process model that governs how the feedforward unit responds to changes. Feedforward control is often used in conjunction with feedback control, the feedback being used to correct for any inaccuracies in the process model.

The discussion of transfer functions in Sec. 47.5 also applies to feedforward controllers.

7. BLOCK DIAGRAM ALGEBRA

The functions represented by several interconnected black boxes (*cascaded blocks*) can be simplified into a single block operation. Some of the most important simplification rules of block diagram algebra are shown in Fig. 47.8. Case 3 represents the standard feedback model.

8. PREDICTING SYSTEM RESPONSE

The transfer function, $T(s)$, is derived without knowledge of the input and is insufficient to predict the response of the system. The system response, $Y(s)$, will depend on the form of the input function, $X(s)$. Since the transfer function is expressed in the *s*-domain, the forcing and response functions must also be expressed in terms of s.

$$Y(s) = T(s)X(s)$$

The time-based response function, $y(t)$, is found by performing the inverse Laplace transform.

$$y(t) = \mathcal{L}^{-1}\big(Y(s)\big)$$

9. INITIAL AND FINAL VALUES

The initial and final (steady-state) value of any function, $G(s)$, can be found from the *initial* and *final value theorems*, respectively, providing the limits exist. The initial value is found from

$$\lim_{t \to 0^+} g(t) = \lim_{s \to \infty} \big(sG(s)\big) \quad \text{[initial value]}$$

$$\lim_{t \to \infty^-} g(t) = \lim_{s \to 0} \big(sG(s)\big) \quad \text{[final value]}$$

Figure 47.8 Rules for Simplifying Block Diagrams

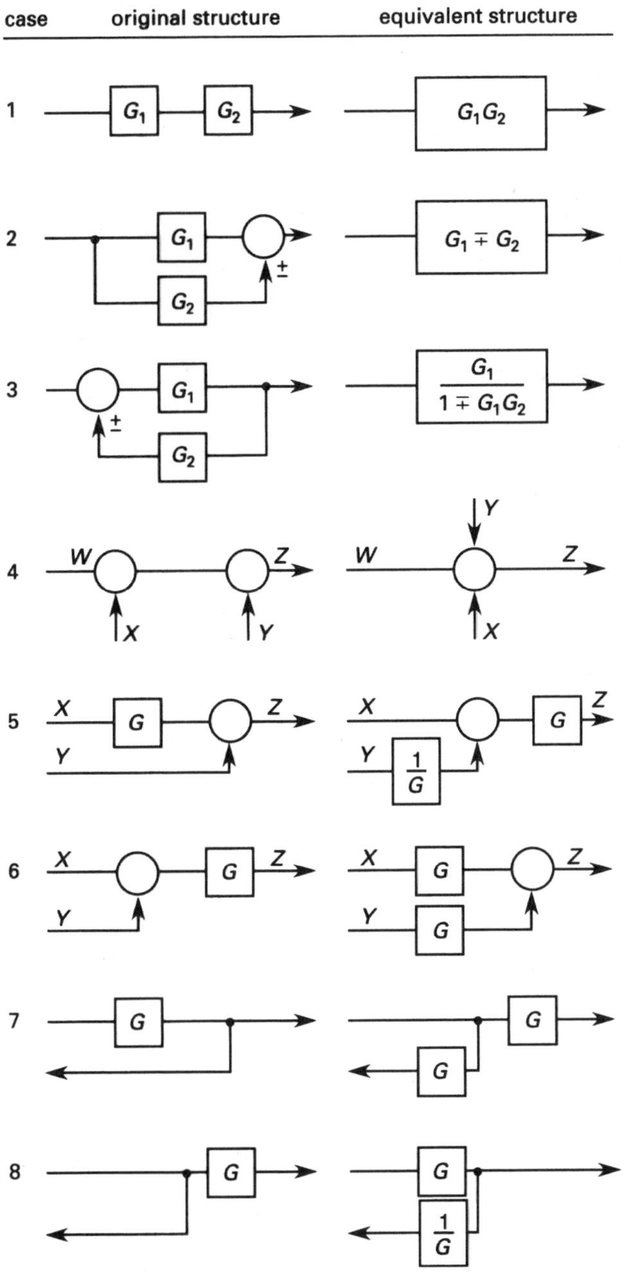

case	original structure	equivalent structure
1	$G_1 \to G_2$	$G_1 G_2$
2	G_1, G_2 with summing junction $\pm$	$G_1 \mp G_2$
3	G_1, G_2 feedback $\pm$	$\dfrac{G_1}{1 \mp G_1 G_2}$
4	$W \to \bigcirc \xrightarrow{X} \bigcirc \xrightarrow{Y} Z$	$W \to \bigcirc \to Z$ with Y, X
5	$X \to G \to \bigcirc \to Z$ with Y	$X \to \bigcirc \to G \to Z$ with $Y \to \frac{1}{G}$
6	$X \to \bigcirc \to G \to Z$ with Y	$X \to G \to \bigcirc \to Z$ with $Y \to G$
7	G with branch	G, G
8	G with branch	G, $\frac{1}{G}$

Equation 47.4: DC Gain with Unit Step Input

$$\text{DC gain} = \lim_{s \to 0} G(s) \qquad \textbf{47.4}$$

Description

Equation 47.4 can be used to determine the DC gain when all poles of the function $G(s)$ have negative real parts. Equation 47.4 is particularly valuable in determining the steady-state response (substitute $Y(s)$ for $G(s)$)

and the steady-state error (substitute $E(s)$ for $G(s)$) for a unit step input. $G(s)$ could be an open-loop or closed-loop transfer function.[3]

10. UNITY FEEDBACK SYSTEM

Equation 47.5: Open-Loop Transfer Function

$$G(s) = \frac{K_B}{s^T} \times \frac{\displaystyle\prod_{m=1}^{M} (1 + s/\omega_m)}{\displaystyle\prod_{n=1}^{N} (1 + s/\omega_n)} \qquad \textbf{47.5}$$

Variation

$$G(s) = \frac{K_B}{s^T} \times \frac{\displaystyle\prod_{m=1}^{M} (s - z_m)}{\displaystyle\prod_{n=1}^{N} (s - p_n)}$$

Description

Equation 47.5 is used with a unity feedback control system model (see Fig. 47.9). A *unity feedback* loop is a feedback loop with a value of 1.

Figure 47.9 Unity Feedback Control System

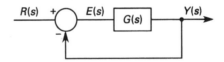

A unity feedback system can be assumed when the dynamics of $H(s)$ are much faster than that of $G(s)$ (such as when $H(s)$ is merely a filter or scalar) so that the feedback function, $B(s)$, is essentially the same as $Y(s)$. In that case, the control system can be replaced by a unity negative feedback system with an open-loop transfer function of $G(s)H(s)$.

---

[3]Equation 47.4 is derived from the final value theorem, but gives the appearance of being incorrect. The apparent error derives from the *NCEES Handbook*'s ambiguous and inconsistent use of $G(s)$. In the *NCEES Handbook*'s version of Fig. 47.4, $G(s)$ is the overall transfer function, which is the ratio of the output to the input, regardless of the nature of the feedback system (positive, negative, unity, open, or closed). In Eq. 47.4, $G(s)$ is the ratio of the output to the input, but specifically only with a unit step input (a condition that is not mentioned in the *NCEES Handbook*). Since the transform of a unit step is $1/s$, from the final value theorem, the output of a system with a unit step after the transients have died out is

$$\text{DC gain}\big|_{\text{unit step}} = \lim_{s \to 0} G(s)$$

Equation 47.4 is a derived equation dependent on environment. It is not an engineering "absolute," and it cannot be used with any other form of input.

Process Control

In Eq. 47.5, the exponent T is known as the *system type*. The system type will be an integer greater than or equal to zero. T is the number of *pure integrators* (also known as *free integrators*) in the open transfer function, equal to the number of s variables that can be factored from $D(s)$, the denominator of $G(s)$.

11. SPECIAL CASES OF STEADY-STATE RESPONSE

In addition to determining the steady-state response from the final value theorem (see Sec. 47.9), the steady-state response to a specific input can be easily derived from the transfer function, $T(s)$, in a few specialized cases. For example, the steady-state response function for a system acted upon by an *impulse* is simply the transfer function. That is, a pulse has no long-term effect on a system. Figure 47.10 lists input functions and their Laplace transforms.

Figure 47.10 *Input Functions and their Laplace Transforms*

name	$f(t)$		$F(s)$
impulse	$f(t) = \begin{cases} 1 & t = 0 \\ 0 & t > 0 \end{cases}$		1
step	$f(t) = 1$		$\dfrac{1}{s}$
ramp	$f(t) = t$		$\dfrac{1}{s^2}$
exponential	$f(t) = e^{at}$		$\dfrac{1}{s - a}$
sine	$f(t) = \sin(\omega t)$		$\dfrac{1}{\omega^2 + s^2}$

The steady-state response for a *step input* (often referred to as a *DC input*) is obtained by substituting zero for s everywhere in the transfer function. (If the step has magnitude h, the steady-state response is multiplied by h.)

The steady-state response for a sinusoidal input is obtained by substituting $j\omega_f$ for s everywhere in the transfer function, $T(s)$. The output will have the same frequency as the input. It is particularly convenient to perform sinusoidal calculations using phasor notation.

12. STEADY-STATE ERROR

Another way of determining the long-term performance of a system is to determine the *steady-state error*, e_{ss}. The steady-state error will always have a value of zero, infinity, or a constant. Ideally, the error $e(t)$ would be zero for both the transient and steady-state cases. Pure gain systems (i.e., ideal linear amplifiers) always have non-zero steady-state errors (i.e., $G(s)$ is a scalar multiplier, K). Integrating systems (i.e., $G(s) = K/s$) can have near-zero steady-state errors.

The steady-state error depends on the type of input, $R(s)$, and the system type. Table 47.1 lists the steady-state errors for unity feedback system types 0, 1, and 2 for unit step, ramp, and parabolic inputs.[4] The system type is the number of pure integrations in the feedforward path, determined as the value of the exponent T (i.e., the "power") of s in the denominator of Eq. 47.5. For a *unit impulse*, $R(s) = 1$; for a *unit step*, $R(s) = 1/s$; for a *unit ramp*, $R(s) = 1/s^2$; and, for a *parabolic input*, $R(s) = 1/s^3$. The values of K_B appearing in Table 47.1 are commonly referred to as *static error constants*. For a unit step, $u(t)$, K_B is known as the *static position error constant*, K_p. For a unit ramp, $tu(t)$, K_B is known as the *static velocity error constant*. For a parabolic input, $\frac{1}{2}t^2 u(t)$, K_B is known as the *static acceleration error constant*, K_a. The steady-state error is found from the final value theorem.

$$e_{ss}(t) = \lim_{t \to \infty} e(t) = \lim_{t \to \infty} \big(r(t) - y(t) \big)$$

$$e_{ss}(s) = \lim_{s \to 0} sE(s) = \lim_{s \to 0} s\big(R(s) - Y(s) \big)$$

$$= \lim_{s \to 0} \frac{sR(s)}{1 - G(s)} \quad \text{[unity feedback]}$$

Table 47.1 *Steady-State Error, e_{ss}, for Unity Feedback*

	type		
input	$T = 0$	$T = 1$	$T = 2$
unit step	$1/(K_B + 1)$	0	0
ramp	∞	$1/K_B$	0
acceleration	∞	∞	$1/K_B$

Example

For the feedback control system shown, what is the steady-state error function, $e_{ss}(t)$, for a ramp input function?

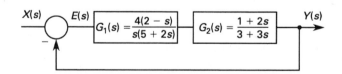

(A) 0

(B) 1/4

(C) 15/8

(D) ∞

---

[4]A parabolic input is also known as an *acceleration input*.

Solution

The open-loop transfer function is

$$G(s) = G_1(s)\,G_2(s)$$
$$= \left(\frac{(4)(2-s)}{s(5+2s)}\right)\left(\frac{1+2s}{3+3s}\right)$$

The DC gain, K_B, for a unit ramp is the static velocity error constant, K_v.

$$K_B = K_v = \lim_{s\to 0} sG(s) = \lim_{s\to 0}\frac{4s(2-s)(1+2s)}{s(5+2s)(3+3s)} \cdot$$
$$= \frac{(4)(2)(1)}{(5)(3)}$$
$$= 8/15$$

Since s^1 appears in the denominator and no additional factoring is possible, this is a type 1 system.

Using the steady-state error analysis table, the steady-state error function, $e_{ss}(t)$, for a ramp input function is

$$e_{ss}(t) = \frac{1}{K_B} = 15/8$$

The answer is (C).

13. DETERMINING THE ERROR CONSTANTS

With a lot of work, the error constant, K_B, in a unity feedback system can be determined by writing the open-loop transfer function, $G(s)$, in *canonical form* and factoring out all constants. Alternatively, the error functions can be found directly from limits on $G(s)$.

$$K_p = \lim_{s\to 0} G(s)$$

$$K_v = \lim_{s\to 0} sG(s)$$

$$K_a = \lim_{s\to 0} s^2 G(s)$$

14. POLES AND ZEROS

A *pole* is a value of s that makes a function, $G(s)$, infinite. Specifically, a pole makes the denominator of $G(s)$ zero. (Pole values are the system *eigenvalues*.) A *zero* of the function makes the numerator of $G(s)$ (and $G(s)$ itself) zero. Poles and zeros need not be real or unique; they can be imaginary and repeated within a function.

A *pole-zero diagram* is a plot of poles and zeros in the *s-plane*—a rectangular coordinate system with real and imaginary axes. A zero is represented by $\bigcirc$; a pole is represented by $\times$. Poles off the real axis always occur in conjugate pairs known as *pole pairs*.

Sometimes it is necessary to derive the function $G(s)$ from its pole-zero diagram. This will be only partially successful since repeating identical poles and zeros are not usually indicated on the diagram. Also, scale factors (scalar constants) are not shown.

15. PREDICTING SYSTEM RESPONSE FROM RESPONSE POLE-ZERO DIAGRAMS

A response pole-zero diagram based on $R(s)$ can be used to predict how the system responds to a specific input. (This pole-zero diagram must be based on the product $T(s)F(s)$ since that is how $R(s)$ is calculated. Plotting the product $T(s)F(s)$ is equivalent to plotting $T(s)$ and $F(s)$ separately on the same diagram.)

The system will experience an *exponential decay* when a single pole falls on the real axis. A pole with a value of $-r$, corresponding to the linear term $(s+r)$, will decay at the rate of e^{-rt}. (See Fig. 47.11.) The quantity $1/r$ is the *decay time constant*, the time for the response to achieve approximately 63% of its steady-state value. The farther left the point is located from the vertical imaginary axis, the faster the motion will die out.

Figure 47.11 *Types of Responses Determined by Pole Location*

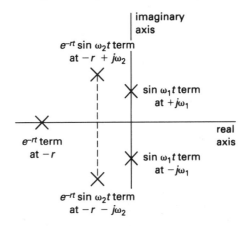

Undamped sinusoidal oscillation will occur if a pole pair falls on the imaginary axis. A conjugate pole pair with the value of $\pm j\omega$ indicates oscillation with a natural frequency of ω rad/s.

Pole pairs to the left of the imaginary axis represent *decaying sinusoidal response*. The closer the poles are to the real (horizontal) axis, the slower will be the oscillations. The closer the poles are to the imaginary (vertical) axis, the slower will be the decay. The *natural frequency*, ω, of undamped oscillation can be determined from a *conjugate pole pair* having values of $r \pm j\omega_f$.

$$\omega = \sqrt{r^2 + \omega_f^2}$$

The magnitude and phase shift can be determined for any input frequency from the pole-zero diagram with the following procedure: Locate the angular frequency,

ω_f, on the imaginary axis. Draw a line from each pole (i.e., a pole-line) and from zero (i.e., a zero-line) of $T(s)$ to this point. (See Fig. 47.12.) The angle of each of these lines is the angle between it and the horizontal real axis. The overall magnitude is the product of the lengths of the zero-lines, L_z, divided by the product of the lengths of the pole-lines, L_p. (The scale factor, K, must also be included because it is not shown on the pole-zero diagram.) The phase is the sum of the pole-angles less the sum of the zero-angles.

$$|T(s)| = K\frac{\prod_z |L_z|}{\prod_p |L_p|} = K\frac{\prod_z \text{length}}{\prod_p \text{length}}$$

$$\angle T(s) = \sum_p \alpha - \sum_z \beta$$

Figure 47.12 *Calculating Magnitude and Phase from a Pole-Zero Diagram*

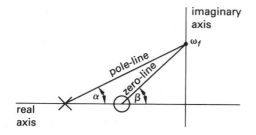

16. FREQUENCY RESPONSE

The gain and phase angle frequency response of a system will change as the forcing frequency is varied. The *frequency response* is the variation in these parameters, always with a sinusoidal input. *Gain* and *phase characteristics* are plots of the steady-state gain and phase angle responses with a sinusoidal input versus frequency. While a linear frequency scale can be used, frequency response is almost always presented against a logarithmic frequency scale.

The steady-state gain response is expressed in decibels, while the steady-state phase angle response is expressed in degrees.

Equation 47.6: Gain Margin

If $\angle G(j\omega_{180}) = -180°$, then
$$\text{GM} = -20\log_{10}(|G(j\omega_{180})|) \qquad \textit{47.6}$$

Variation

$$\text{gain} = 20\log|T(j\omega)| \qquad [\text{in dB}]$$

Description

Gain margin is the additional gain required to produce instability in the unity gain feedback control system. The gain is calculated from Eq. 47.6 where $|G(s)|$ is the absolute value of the steady-state response. A doubling of $|G(j\omega)|$ is referred to as an *octave* and corresponds to a 6.02 dB increase. A tenfold increase in $|G(j\omega)|$ is a *decade* and corresponds to a 20 dB increase.

$$\text{no. of octaves} = \frac{\text{gain}_{2,\text{dB}} - \text{gain}_{1,\text{dB}}}{6.02 \text{ dB}}$$
$$= 3.32 \times \text{no. of decades}$$

$$\text{no. of decades} = \frac{\text{gain}_{2,\text{dB}} - \text{gain}_{1,\text{dB}}}{20 \text{ dB}}$$
$$= 0.301 \times \text{no. of octaves}$$

Equation 47.7: Phase Margin

If $|G(j\omega)| = 1$, then
$$\text{PM} = 180° + \angle G(j\omega_{0\,\text{dB}}) \qquad \textit{47.7}$$

Description

Phase margin is the additional phase required to produce instability in the unity gain feedback control system. The *phase margin* of an amplifier's output signal is the difference between its phase angle and 180°.

Example

A control system has an open-loop gain of 1 (0 dB) at the frequency where its phase angle is $-200°$. What is the phase margin?

(A) $-120°$

(B) $-60°$

(C) $-20°$

(D) $-10°$

Solution

The phase margin is given by

$$\text{PM} = 180° + \angle G(j\omega_{0\,\text{dB}})$$
$$= 180° - 200°$$
$$= -20°$$

The answer is (C).

17. GAIN CHARACTERISTIC

The *gain characteristic* (*M-curve* for magnitude) is a plot of the gain as ω_f is varied. It is possible to make a rough sketch of the gain characteristic by calculating the gain at a few points (pole frequencies, $\omega = 0$, $\omega = \infty$, etc.). The curve will usually be asymptotic to several lines. The frequencies at which these asymptotes intersect are *corner frequencies*. The peak gain, M_p, coincides with the natural (resonant) frequency of the system. The gain characteristic peaks when the forcing frequency equals the natural frequency. It is also said that this peak corresponds to the resonant frequency. Strictly speaking, this is true, although the gain may not actually be resonant (i.e., may not be infinite). Large peak gains indicate lowered stability and large overshoots. The *gain crossover point*, if any, is the frequency at which log(gain) = 0.

The *half-power points* (*cutoff frequencies*) are the frequencies for which the gain is 0.707 (i.e., $\sqrt{2}/2$) times the peak value. This is equivalent to saying the gain is 3 dB less than the peak gain. The *cutoff rate* is the slope of the gain characteristic in dB/octave at a half-power point. The frequency difference between the half-power points is the *bandwidth*, BW. (See Fig. 47.13.) The *closed-loop bandwidth* is the frequency range over which the closed-loop gain falls 3 dB below its value at $\omega = 0$. (The term "bandwidth" often means closed-loop bandwidth.) The *quality factor*, Q, is

$$Q = \frac{\omega_n}{\text{BW}}$$

Figure 47.13 Bandwidth

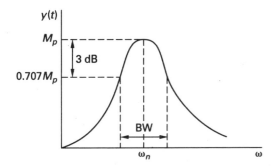

Since a low or negative gain (compared to higher parts of the curve) effectively represents attenuation, the gain characteristic can be used to distinguish between low- and high-pass filters. A *low-pass filter* will have a large gain at low frequencies and a small gain at high frequencies. Conversely, a *high-pass filter* will have a high gain at high frequencies and a low gain at low frequencies.

18. PHASE CHARACTERISTIC

The phase angle response will also change as the forcing frequency is varied. The *phase characteristic* (α *curve*) is a plot of the phase angle as ω_f is varied.

19. STABILITY

A stable system will remain at rest unless disturbed by external influence and will return to a rest position once the disturbance is removed. A pole with a value of $-r$ on

the real axis corresponds to an exponential response of e^{-rt}. Since e^{-rt} is a decaying signal, the system is stable. Similarly, a pole of $+r$ on the real axis corresponds to an exponential response of e^{rt}. Since e^{rt} increases without limit, the system is unstable.

Since any pole to the right of the imaginary axis corresponds to a positive exponential, a *stable system* will have poles only in the left half of the *s*-plane. If there is an isolated pole on the imaginary axis, the response is stable. However, a conjugate pole pair on the imaginary axis corresponds to a sinusoid that does not decay with time. Such a system is considered to be unstable.

Passive systems (i.e., the homogeneous case) are not acted upon by a forcing function and are always stable. In the absence of an energy source, exponential growth cannot occur. *Active systems* contain one or more energy sources and may be stable or unstable.

There are several *frequency response (domain) analysis techniques* for determining the stability of a system, including the Bode plot, root-locus diagram, Routh stability criterion, Hurwitz test, and Nichols chart. The term *frequency response* almost always means the steady-state response to a sinusoidal input.

The value of the denominator of $T(s)$ is the primary factor affecting stability. When the denominator approaches zero, the system increases without bound. In the typical feedback loop, the denominator is $1 \pm G(s)H(s)$, which can be zero only if $|G(s)H(s)| = 1$. It is logical, then, that most of the methods for investigating stability (e.g., Bode plots, root-locus diagrams, Nyquist analysis, and the Nichols chart) investigate the value of the open-loop transfer function, $G(s)H(s)$. Since $\log(1) = 0$, the requirement for stability is that $\log(G(s)H(s))$ must not equal 0 dB.

A negative feedback system will also become unstable if it changes to a positive feedback system, which can occur when the feedback signal is changed in phase more than 180°. Therefore, another requirement for stability is that the phase angle change must not exceed 180°.

20. BODE PLOTS

Bode plots are gain and phase characteristics for the open-loop $G(s)H(s)$ transfer function that are used to determine the *relative stability* of a system. In Fig. 47.14, the gain characteristic is a plot of $20\log(|G(s)H(s)|)$ versus ω for a sinusoidal input. (Bode plots, though similar in appearance to the gain and phase frequency response charts, are used to evaluate stability and do not describe the closed-loop system response.)

The *gain margin* is the number of decibels that the open-loop transfer function, $G(s)H(s)$, is below 0 dB at the *phase crossover frequency* (i.e., where the phase angle is $-180°$). (If the gain happens to be plotted on a linear scale, the gain margin is the reciprocal of the gain at the phase crossover point.) The gain margin must be positive for a stable system, and the larger it is, the more stable the system will be.

Figure 47.14 Gain and Phase Margin Bode Plots

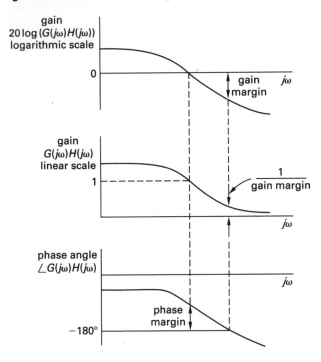

The *phase margin* is the number of degrees the phase angle is above $-180°$ at the *gain crossover point* (i.e., where the logarithmic gain is 0 dB or the actual gain is 1).

In most cases, large positive gain and phase margins will ensure a stable system. However, the margins could have been measured at other than the crossover frequencies. Therefore, a Nyquist stability plot is needed to verify the absolute stability of a system.

21. ROOT-LOCUS DIAGRAMS

A *root-locus diagram* is a pole-zero diagram showing how the poles of $G(s)H(s)$ move when one of the system parameters (e.g., the gain factor) in the transfer function is varied. The diagram gets its name from the need to find the roots of the denominator (i.e., the poles). The locus of points defined by the various poles is a line or curve that can be used to predict *points of instability* or other critical operating points. A point of instability is reached when the line crosses the imaginary axis into the right-hand side of the pole-zero diagram.

A root-locus curve may not be contiguous, and multiple curves will exist for different sets of roots. Sometimes the curve splits into two branches. In other cases, the curve leaves the real axis at *breakaway points* and continues on with constant or varying slopes approaching asymptotes. One branch of the curve will start at each open-loop pole and end at an open-loop zero.

Equation 47.8 Through Eq. 47.10: Poles and Zeros

$$1 + K\frac{(s - z_1)(s - z_2)\cdots(s - z_m)}{(s - p_1)(s - p_2)\cdots(s - p_n)} = 0 \quad [m \le n] \qquad 47.8$$

$$\alpha = \frac{(2k + 1)180°}{n - m} \quad [k = 0, \pm 1, \pm 2, \pm 3, \ldots] \qquad 47.9$$

$$\sigma_A = \frac{\sum\limits_{i=1}^{n} \mathrm{Re}(p_i) - \sum\limits_{i=1}^{m} \mathrm{Re}(z_i)}{n - m} \qquad 47.10$$

Description

In Eq. 47.8, p values are the open-loop poles, and z values are the open-loop zeros. For $m < n$, Eq. 47.9 gives the location at which $n - m$ branches terminate at infinity. This location is the intersection of the real axis with the asymptote angles, or the *asymptote centroid*, and is found by Eq. 47.10, where Re represents the real part.

Example

A control system with negative feedback is shown.

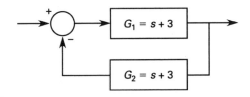

The root-locus diagram of the system has a

(A) pole at $(-3, j0)$ and a zero at $(-3, j0)$

(B) zero at $(-3, j0)$

(C) double zero at $(-3, j0)$

(D) double pole at $(-3, j0)$ and a zero at $(0, 0)$

Solution

The poles and zeros of the root-locus diagram for a control system are the poles and zeros of the open-loop transfer function.

The open-loop transfer function is

$$G(s) = G_1(s)G_2(s) = (s + 3)^2$$

There are no poles and two zeros at $(-3, j0)$.

The answer is (C).

22. PID CONTROLLERS

One widely used control mechanism is the *proportional-integral-derivative controller* (also called a *PID controller* or *three-mode controller*). A PID controller is capable of three modes of control—proportional, integral, and derivative—or a combination of them. These three terms refer to how the controller uses the error (the difference between the set point and the actual value of the process variable) to calculate the adjustment to make to the input.

Proportional control is the simplest method: the adjustment made to the process input is in proportion to the error in the process output. For example, the flow of steam into a system may be increased or decreased in proportion to the difference between the set point and the actual temperature. One disadvantage, however, is that when the set point is changed, proportional control alone cannot bring the process output exactly to the new set point; a small error, called *offset* (also called *proportional-only offset* or *droop*), must persist.

In *integral control* (also called *reset control* or *floating control*), the adjustment made to the process is in proportion to the integral of the error. The adjustment, then, is influenced not only by the magnitude of the error but its duration. Integral control generally moves the process output more quickly toward the set point than proportional control does, and it can also eliminate offset. However, because integral control responds to accumulated past error as well as current error, it does not stop making adjustments at the moment the process variable reaches the set point; this typically causes the process variable to overshoot the set point and reverse direction, making a series of decreasing oscillations around the set point rather than stopping at the set point as soon as it is reached. To reduce this behavior, integral control ($G_c(s) = 1/T_I s$) is typically used in combination with proportional control rather than alone.

In *derivative control* (also known as *rate action control* or *anticipatory control*), the controller calculates the rate at which the process variable is changing and makes adjustments to the process input in proportion to this rate of change. In relatively smooth processes, derivative control can increase stability and hasten recovery from disturbances. If the process is "noisy," however—subject to small, frequent, random fluctuations—derivative control ($G_c(s) = T_D s$) will tend to amplify the fluctuations and make the process less stable, not more.

Equation 47.11: PID Controller Gain

$$G_C(s) = K\left(1 + \frac{1}{T_I s} + T_D s\right) \quad \text{[PID controller]} \quad \textbf{47.11}$$

Variation

$$G_C(s) = K_P + \frac{K_I}{s} + K_D s$$

Description

Equation 47.11 gives the gain for a *proportional-integral-derivative controller*, or *PID controller*. In Eq. 47.11, K is the *proportional gain*; K/T_I is the *integral gain* (usually written as K_I); and KT_D is the *derivative gain* (usually written as K_D). T_I and T_D are the *integral time*

(also known as *reset time*) and *derivative time* (also known as *rate time*), respectively.

Equation 47.12: Lag or Lead Compensator Gain

$$G_C(s) = K\left(\frac{1 + sT_1}{1 + sT_2}\right) \quad \text{[lag or lead compensator]}$$
$$\textbf{47.12}$$

Variation

$$G_C(s) = K\left(\frac{s - z}{s - p}\right) \quad \text{[first-order compensator]}$$

Description

Phase lead-lag (lag-lead) compensators are common control components placed in a feedback circuit (see Sec. 47.24) to improve the frequency response of a control system.[5] They can also be used to reduce steady-state error, reduce resonant peaks, and improve system response by reducing rise time.

Lead compensators shift the output phase to the left on the time line (i.e., the output leads the input), and *lag compensators* shift the output phase to the right on the time line (i.e., the output lags the input). A lead compensator tends to shift the root-locus toward the complex plane, which improves the system stability and response speed. Lead compensators are associated with derivative (s) terms, while lag compensators are associated with integral ($1/s$) terms. Both lead and lag compensators introduce a single pole-zero pair into the transfer function. A lead compensator will have a pole in the complex plane, while a lag compensator will have a pole in the real plane. Figure 47.15 shows the effect of lead compensation on a root-locus diagram.

Figure 47.15 Effect of Lead Compensation on Root-Locus Diagram

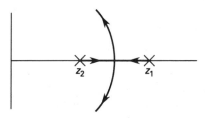

(a) original root-locus diagram

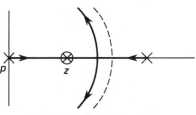

(b) shifted to left by adding pole *p* and zero *z*

---

[5]Compensators may also be referred to as "controllers." The two terms are used interchangeably.

Stated another way, for a lead compensator, the introduced pole is greater than the introduced zero (i.e., $p > z$), and for a lag compensator, $p < z$. Equation 47.12 can be used for a lag or lead compensator, depending on the ratio of constants T_1/T_2. In Eq. 47.12, the zero is located at $1/T_1$, and the pole is located at $1/T_2$. For a lead compensator, $T_2/T_1 < 1$, and for a lag compensator, $T_1/T_2 > 1$.

23. ROUTH CRITERION

Equation 47.13: Linear Characteristic Equation

$$a_n s^n + a_{n-1} s^{n-1} + a_{n-2} s^{n-2} + \cdots + a_0 = 0 \qquad 47.13$$

Description

From Eq. 47.2, the *characteristic equation* of a control system is $1 + G(s)H(s) = 0$. If any of the linear coefficients, a_n, of the characteristic equation are negative, the system will be marginally stable at best, but more likely, unstable. In a linear system (i.e., with the form of Eq. 47.13), the roots (zeros) will be of the form $z = a + jb$ (i.e., will be a real or a complex number), where $a = \mathrm{Re}(z) = \mathrm{Re}(a + jb)$ is the real part, and $jb = \mathrm{Im}(z) = \mathrm{Im}(a + jb)$ is the imaginary part. Stable systems will have roots with negative real parts such as $s = -3$, corresponding to decaying exponential terms such as e^{-3t}.

The *Routh-Hurwitz criterion* uses the coefficients of the polynomial characteristic equation, Eq. 47.13, to determine system stability. A table (the *Routh table* or *Hurwitz matrix*) of these coefficients is formed using the conventions in Eq. 47.14 through Eq. 47.18.

Equation 47.14 Through Eq. 47.18: Routh Table

a_n	a_{n-2}	a_{n-4}	$\cdots$	$\cdots$	$\cdots$	
a_{n-1}	a_{n-3}	a_{n-5}	$\cdots$	$\cdots$	$\cdots$	
b_1	b_2	b_3	$\cdots$	$\cdots$	$\cdots$	47.14
c_1	c_2	c_3	$\cdots$	$\cdots$	$\cdots$	

$$b_1 = \frac{a_{n-1}a_{n-2} - a_n a_{n-3}}{a_{n-1}} \qquad 47.15$$

$$b_2 = \frac{a_{n-1}a_{n-4} - a_n a_{n-5}}{a_{n-1}} \qquad 47.16$$

$$c_1 = \frac{a_{n-3}b_1 - a_{n-1}b_2}{b_1} \qquad 47.17$$

$$c_2 = \frac{a_{n-5}b_1 - a_{n-1}b_3}{b_1} \qquad 47.18$$

Description

The *Routh-Hurwitz criterion* states that the number of sign changes in the first column of the Routh table equals the number of positive (unstable) roots. Therefore, a system will be stable if all entries in the first column have the same sign. The table is organized according to Eq. 47.14. Then, the remaining coefficients are calculated using Eq. 47.15 through Eq. 47.18 until all values are zero.

Special methods are used if there is a zero in the first column but nowhere else in that row. One of the methods is to substitute a small number, represented by ε or δ, for the zero and calculate the remaining coefficients as usual.

The *Routh test* indicates that the necessary conditions for a polynomial to have all its roots in the left-hand plane (i.e., for the system to be stable) are (a) all of the terms must have the same sign; and (b) all of the powers between the highest and the lowest value must have nonzero coefficients, unless all even-power or all odd-power terms are missing. Condition (a) also implies that the coefficient cannot be imaginary.

Example

Which of the following characteristic equations can be stable?

I. $4s^4 + 8s^2 + 3s + 2 = 0$

II. $4s^4 + 2s^3 + 8s^2 + 3s + 2 = 0$

III. $4s^4 + 2s^3 + 8js^2 + 5s + 2 = 0$

IV. $4s^4 + 2s^3 + 8s^2 - 3s + 2 = 0$

(A) I only

(B) II only

(C) I and IV

(D) II and III

Solution

To represent a stable system, according to the Routh test, all consecutive powers of s must be represented. Equation I does not represent a stable system because there is no s^3 term. To be stable, all linear coefficients must be real. Equation III contains the coefficient $8j$, so it does not represent a stable system. Finally, to be stable, all linear coefficients must be positive, so Eq. IV does not represent a stable system. Equation II is the only equation that meets the criteria.

The answer is (B).

Example

A characteristic equation is

$$s^5 + 15s^4 + 185s^3 + 725s^2 - 326s + 120 = 0$$

A Routh table has been constructed as follows.

s^5:	1	185	-326
s^4:	15	725	120
s^3:	136.7	334	
s^2:	761.7	120	
s^1:	-355.5		
s^0:	120		

Which table entry is INCORRECT?

(A) 15

(B) 185

(C) 334

(D) 725

Solution

In this problem,

$$a_n = a_5 = 1$$

$$a_{n-1} = a_4 = 15$$

$$a_{n-2} = a_3 = 185$$

$$a_{n-3} = a_2 = 725$$

$$a_{n-4} = a_1 = -326$$

$$a_{n-5} = a_0 = 120$$

Comparing with Eq. 47.14, values in the s^5 and s^4 rows are all correct. The second s^3 entry should be

$$b_2 = \frac{a_{n-1}a_{n-4} - a_n a_{n-5}}{a_{n-1}} = \frac{a_4 a_1 - a_5 a_0}{a_4}$$

$$= \frac{(15)(-326) - (1)(120)}{15}$$

$$= -334$$

The answer is (C).

24. APPLICATION TO CONTROL SYSTEMS

A control system monitors a process and makes adjustments to maintain performance within certain acceptable limits. Feedback is implicitly a part of all control systems. The *controller* (*control element*) is the part of the control system that establishes the acceptable limits of performance, usually by setting its own reference inputs. The controller transfer function for a proportional controller is a constant: $G_1(s) = K$. The *plant* (*controlled system*) is the part of the system that responds to the controller. Both of these are in the forward loop. The input signal, $R(s)$, in Fig. 47.16 is known in a control system as the *command* or *reference value*. Figure 47.16 is known as a *control logic diagram* or *control logic block diagram*. The controller in Fig. 47.16 can be a PID controller or a compensator (see Sec. 47.22).

Figure 47.16 *Typical Feedback Control System*

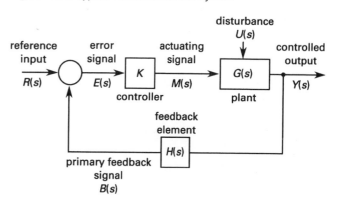

A *servomechanism* is a special type of control system in which the controlled variable is mechanical position, velocity, or acceleration. In many servomechanisms, $H(s) = 1$ (i.e., unity feedback), and it is desired to keep the output equal to the reference input (i.e., maintain a zero-error function). If the input, $R(s)$, is constant, the descriptive terms *regulator* and *regulating system* are used.

25. CONTROL SYSTEM MODELS

Equation 47.19 Through Eq. 47.22: First-Order Control Systems

$$\frac{Y(s)}{R(s)} = \frac{K}{\tau s + 1} \qquad 47.19$$

$$y(t) = y_0 e^{-t/\tau} + KM(1 - e^{-t/\tau}) \qquad 47.20$$

$$\frac{Y(s)}{R(s)} = \frac{K e^{-\theta s}}{\tau s + 1} \qquad 47.21$$

$$y(t) = [y_0 e^{-(t-\theta)/\tau} + KM(1 - e^{-(t-\theta)/\tau})]u(t - \theta) \qquad 47.22$$

Description

Equation 47.19 is the transfer function model for a first-order system, and Eq. 47.20 gives the step response to a step input of magnitude M. Equation 47.21 is used for systems with time delay, such as dead time or transport lag, and Eq. 47.22 gives the step response to a step input of magnitude M, where $u(t)$ is the unit step function.

Equation 47.23 Through Eq. 47.32: Second-Order Control Systems

$$\frac{Y(s)}{R(s)} = \frac{K\omega_n^2}{s^2 + 2\zeta\omega_n s + \omega_n^2} \qquad 47.23$$

$$\omega_d = \omega_n \sqrt{1 - \zeta^2} \qquad \textit{47.24}$$

$$\omega_r = \omega_n \sqrt{1 - 2\zeta^2} \qquad \textit{47.25}$$

$$t_p = \pi / \left(\omega_n \sqrt{1 - \zeta^2} \right) \qquad \textit{47.26}$$

$$M_p = 1 + e^{-\pi\zeta/\sqrt{1-\zeta^2}} \qquad \textit{47.27}$$

$$\%OS = 100 e^{-\pi\zeta/\sqrt{1-\zeta^2}} \qquad \textit{47.28}$$

$$\delta = \frac{1}{m} \ln \left(\frac{x_k}{x_{k+m}} \right) = \frac{2\pi\zeta}{\sqrt{1 - \zeta^2}} \qquad \textit{47.29}$$

$$\omega_d \tau = 2\pi \qquad \textit{47.30}$$

$$T_s = \frac{4}{\zeta \omega_n} \qquad \textit{47.31}$$

$$\frac{Y(s)}{R(s)} = \frac{K}{\tau^2 s^2 + 2\zeta\tau s + 1} \qquad \textit{47.32}$$

Description

Equation 47.23 is the transfer function model for a standard second-order control system, where Eq. 47.24 is the *damped natural frequency*, and Eq. 47.25 is the *damped resonant frequency* or *peak frequency*. Equation 47.26 and Eq. 47.27 are for a normalized, underdamped second-order control system. For one unit step input, Eq. 47.26 gives the time required, t_p, to reach a peak value of M_p, given by Eq. 47.27.

Example

A second-order control system has a control response ratio of

$$\frac{Y(s)}{R(s)} = \frac{1}{s^2 + 0.3s + 1}$$

If the system is acted upon by a unit step, what is most nearly the magnitude of the first oscillatory response peak?

(A) 1.3

(B) 1.6

(C) 2.5

(D) 3.4

Solution

The response transfer function can be written in the form

$$\frac{Y(s)}{R(s)} = \frac{K\omega_n^2}{s^2 + 2\zeta\omega_n s + \omega_n^2}$$

Working with the denominator,

$$\omega_n^2 = 1$$

$$2\zeta = 0.3$$

$$\zeta = 0.15$$

Working with the numerator,

$$K = 1$$

Use Eq. 47.27.

$$M_p = 1 + e^{-\pi\zeta/\sqrt{1-\zeta^2}} = 1 + e^{-\pi(0.15)/\sqrt{1-(0.15)^2}}$$

$$= 1.63 \quad (1.6)$$

The answer is (B).

26. STATE-VARIABLE CONTROL SYSTEM MODELS

While the classical methods of designing and analyzing control systems are adequate for most situations, state model representations are preferred for more complex digital sampling cases, particularly those with multiple inputs and outputs or when behavior is nonlinear or varies with time.

The state variables completely define the dynamic state (position, voltage, pressure, etc.), $x_i(t)$, of the system at time t. (In simple problems, the number of state variables corresponds to the number of *degrees of freedom*, n, of the system.) The n state variables are written in matrix form as a state vector, $\mathbf{X}$.

$$\mathbf{X} = \begin{pmatrix} x_1 \\ x_2 \\ x_3 \\ \vdots \\ x_n \end{pmatrix}$$

It is a characteristic of state models that the state vector is acted upon by a first-degree derivative operator, d/dt, to produce a differential term, $\mathbf{X}'$, of order 1,

$$\mathbf{X}' = \frac{d\mathbf{X}}{dt}$$

The previous equations illustrate the general form of a state model representation: $\mathbf{U}$ is an r-dimensional (i.e., an $r \times 1$ matrix) *control vector*; $\mathbf{Y}$ is an m-dimensional (i.e., an $m \times 1$ matrix) *output vector*; $\mathbf{A}$ is an $n \times n$ *system vector*; $\mathbf{B}$ is an $n \times r$ *control vector*; $\mathbf{C}$ is an $m \times n$ *output vector*; and $\mathbf{D}$ is a *feed-through vector*. The actual unknowns are the x_i state variables. The y_i state variables, which may not be needed in all problems, are only linear combinations of the x_i state variables. (For example, x might represent a spring end position; y might represent stress in the spring. Then, $y = k\Delta x$.)

Equation 47.33 and Eq. 47.34: State and Output Equations

$$\dot{\mathbf{x}}(t) = \mathbf{Ax}(t) + \mathbf{Bu}(t) \quad \text{[state equation]} \quad 47.33$$

$$\mathbf{y}(t) = \mathbf{Cx}(t) + \mathbf{Du}(t) \quad \text{[output equation]} \quad 47.34$$

Variations

$$\mathbf{X}' = \mathbf{AX} + \mathbf{BU} \quad \text{[state equation]}$$

$$\mathbf{Y} = \mathbf{CX} \quad \text{[response equation]}$$

Description

Equation 47.33 is the *state equation*, and Eq. 47.34 is the *output equation* or *response equation*. As is common in transform equations, lowercase letters are used to represent vectors (matrices) containing time-domain values, while uppercase letters are used to represent vectors (matrices) containing *s*-domain values.

A conventional block diagram can be modified to show the multiplicity of monitored properties in a state model, as shown in Fig. 47.17. (The block $\mathbf{I}/s$ is a diagonal identity matrix with elements of $1/s$. This effectively is an integration operator.) The actual physical system does not need to be a feedback system. The form of Eq. 47.33 and Eq. 47.34 is the sole reason that a feedback diagram is appropriate.

Figure 47.17 *State Variable Diagram*

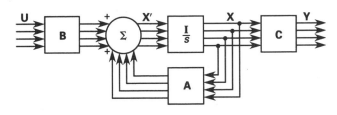

A state variable model permits only first-degree derivatives, so additional x_i state variables are used for higher-order terms (e.g., acceleration).

System controllability exists if all of the system states can be controlled by the inputs, $\mathbf{U}$. In state model language, system controllability means that an arbitrary initial state can be steered to an arbitrary target state in a finite amount of time. *System observability* exists if the initial system states can be predicted from knowing the inputs, $\mathbf{U}$, and observing the outputs, $\mathbf{Y}$. (*Kalman's theorem* based on matrix rank is used to determine system controllability and observability.)

Example

A system is governed by the following differential equations.

$$\dot{x}_2 + a_1\dot{x}_1 + a_0 x_1 = u(t)$$

$$\dot{x}_1 = -x_2$$

The output of the system is

$$y_1(t) = 3x_1$$
$$y_2(t) = 4x_1 - 5\dot{x}_1$$

The output matrix of the state-variable control system model is

(A) $\begin{bmatrix} 3 & 0 \\ 4 & 5 \end{bmatrix}$

(B) $\begin{bmatrix} a_1 \\ a_0 \end{bmatrix}$

(C) $\begin{bmatrix} a_1 & 0 \\ 0 & a_0 \end{bmatrix}$

(D) $\begin{bmatrix} -5 & 3 \\ 4 & 0 \end{bmatrix}$

Solution

The state equation is given by Eq. 47.33 as

$$\dot{\mathbf{x}}(t) = \mathbf{Ax}(t) + \mathbf{Bu}(t)$$

The output equation is given by Eq. 47.34 as

$$\mathbf{y}(t) = \mathbf{Cx}(t) + \mathbf{Du}(t)$$

Representing the given differentials in this format, the vector notation for the system is

$$y_1(t) = 3x_1$$
$$y_2(t) = 4x_1 - 5\dot{x}_1 = 4x_1 + 5x_2$$

$$\begin{bmatrix} \dot{x}_1 \\ \dot{x}_2 \end{bmatrix} = \begin{bmatrix} 0 & 1 \\ -a_0 & -a_1 \end{bmatrix} \begin{bmatrix} x_1 \\ x_2 \end{bmatrix} + \begin{bmatrix} 0 \\ 1 \end{bmatrix} u(t)$$

$$\begin{bmatrix} y_1 \\ y_2 \end{bmatrix} = \begin{bmatrix} 3 & 0 \\ 4 & 5 \end{bmatrix} \begin{bmatrix} x_1 \\ x_2 \end{bmatrix} + \begin{bmatrix} 0 \\ 0 \end{bmatrix} u(t)$$

$$\mathbf{A} = \begin{bmatrix} 0 & 1 \\ -a_0 & -a_1 \end{bmatrix}$$

$$\mathbf{B} = \begin{bmatrix} 0 \\ 1 \end{bmatrix}$$

$$\mathbf{C} = \begin{bmatrix} 3 & 0 \\ 4 & 5 \end{bmatrix}$$

$$\mathbf{D} = \begin{bmatrix} 0 \\ 0 \end{bmatrix}$$

From the variation equation, because $\mathbf{Y} = \mathbf{CX}$, the output matrix is $\mathbf{C}$.

The answer is (A).

Process Control

Equation 47.35 Through Eq. 47.37: Laplace Transform of the State Equation

$$s\mathbf{X}(s) - \mathbf{x}(0) = \mathbf{A}\mathbf{X}(s) + \mathbf{B}\mathbf{U}(s) \qquad 47.35$$

$$\mathbf{X}(s) = \Phi(s)\mathbf{x}(0) + \Phi(s)\mathbf{B}\mathbf{U}(s) \qquad 47.36$$

$$\Phi(s) = [s\mathbf{I} - \mathbf{A}]^{-1} \qquad 47.37$$

Description

Equation 47.35 gives the Laplace transform of the time-invariant state equation, where $\mathbf{X}(s)$ is given by Eq. 47.36, and the Laplace transform of the state transition matrix is found from Eq. 47.37.

Equation 47.38 Through Eq. 47.40: Laplace Transform of the Output Equation

$$\Phi(t) = L^{-1}\{\Phi(s)\} \qquad 47.38$$

$$\mathbf{x}(t) = \Phi(t)\mathbf{x}(0) + \int_0^t \Phi(t-\tau)\,\mathbf{B}\mathbf{u}(\tau)\,d\tau \qquad 47.39$$

$$\mathbf{Y}(s) = \{\mathbf{C}\Phi(s)\,\mathbf{B} + \mathbf{D}\}\mathbf{U}(s) + \mathbf{C}\Phi(s)\,\mathbf{x}(0) \qquad 47.40$$

Description

The state-transition matrix given by Eq. 47.38 can be used in Eq. 47.39. The Laplace transform of the output equation is found from Eq. 47.40. $\{\mathbf{C}\Phi(s)\,\mathbf{B} + \mathbf{D}\}\mathbf{U}(s)$ represents the output or outputs due to the $\mathbf{U}(s)$ inputs, and $\mathbf{C}\Phi(s)\,\mathbf{x}(0)$ represents the output or outputs due to the initial conditions.

Process Control

48 Instrumentation and Control System Hardware

Nomenclature

A	area	m^2	ft^2
C	concentration	various	various
C_v	flow factor	n.a.	$gal/(min\sqrt{psi})$
d	diameter	m	in
E	potential	V	V
I	current	A	A
k	constant	various	various
K_v	flow factor	$m^3/(h\sqrt{bar})$	n.a.
L	length	m	ft
n	quantity	–	–
p	pressure	Pa	lbf/ft^2
P	permeability	C/m^2	C/ft^2
Q	flow quantity	n.a.	gal/min
R	resistance	Ω	Ω
S	proportionality constant	V/pH	V/pH
SG	specific gravity	–	–
t	thickness	m	ft
T	temperature	K	°R
V	voltage	V	V
$\dot{V}$	volumetric flow	m^3/h	gal/min

Symbols

α	temperature coefficient	1/K	1/°R
β	temperature coefficient	$1/K^2$	$1/°R^2$

Subscripts

0	reference
a	measured solution
el	electrode
i	internal buffer (reference solution)
T	at temperature T
v	valve

1. TRANSDUCERS

A *transducer* is any device used to convert one phenomenon (e.g., force) into another phenomenon (e.g., an electrical signal). For example, a microphone is a transducer that transforms sound energy into electrical signals that can be recorded, amplified, or transmitted. Other transducers sense such phenomena as temperature, pressure, physical movement, and light intensity.

The signal from a transducer may be used in a feedback loop to control the physical phenomenon in some way. For example, a thermostat is a transducer that is sensitive to temperature. When the temperature drops below the set-point temperature, the thermostat circuit sends a signal to activate a furnace or other device.

A typical transducer consists of two components, a sensing element and a transmitter. The *transducer sensing element* (*sensor*) used depends on the characteristic to be measured.

- *temperature sensors:* In electrical circuits, these sensors generate voltage, and temperature-dependent resistors modify current and voltage. In pneumatic circuits, fluid-filled pressure elements (e.g., bulbs and coiled tubes) expand and contract against a plate to generate a pneumatic signal.

- *pressure sensors:* In electrical circuits, piezoelectric elements and strain gages are used to generate or modify voltage. In pneumatic circuits, elements such as bellows, bourdon tubes, bulbs, and diaphragms expand and contract in response to pressure changes. Pressure sensors may be connected directly to pressurized fluid streams, or in the case of corrosive fluids, are protected by an inert gas.

- *differential pressure sensors:* These sensors react to differences in two pressures and share features with pressure sensors.

- *fluid flow sensors:* Fluid flow velocity is proportional to the square root of the velocity pressure. Therefore, fluid flow velocity is often measured indirectly by differential pressure devices such as orifice and venturi meters. Turbine flowmeters, vane deflection, and vortex shedding meters can be used to measure velocity directly.

- *liquid level sensors:* Float position, pressure and differential pressure, and electrical (capacitive, inductive, and resistive) sensors can be used to detect liquid level.

Process Control

- *pH sensors:* pH is a logarithmic representation of the concentration of hydrogen ions. Specially designed electrodes and electrode pairs are used to generate an electric potential from the detected hydrogen ion concentration.

- *chemical composition sensors:* When the composition of a substance can be correlated with other characteristics (e.g., pH, conductivity, infrared/ultraviolet absorption), measuring those characteristics will often provide faster and more convenient onsite analysis than traditional offsite laboratory chemical analyses.

- *force and stress sensors:* Force and stress can be measured with pneumatic, piezoelectric, and strain-activated elements.

- *position, presence, proximity, and speed sensors:* In addition to direct-acting electromechanical sensors (e.g., microswitches), capacitive, inductive, and photoelectric sensors can be used to detect and measure the presence of substances without direct contact.

- *illumination sensors:* Photodiodes are used to detect and measure light.

Transducer transmitters are circuits that transform sensor outputs for use by controller inputs. Transmitters can be digital or analog, depending on the nature of their controllers.[1] By necessity, transmitters have very fast response times. Some analog transducers simply pass their sensor outputs on to their controllers (in which case, the sensors are the transducers). Other transmitters may provide little more than isolation and surge protection. However, most analog transmitter circuits provide some *signal conditioning* such as scaling and smoothing. Commercial analog controllers are responsive to voltage or current inputs. Current-sensing controllers often respond to loop currents in the ranges of 4–20 mA or 10–50 mA, so their corresponding transmitters convert sensor output signals into that range.

The output of a control is used to drive an *actuator*, which is a style of motor. Actuators may respond to electrical, pneumatic, or hydraulic controller signals. Actuators provide the pushing, pulling, turning, and moving operations in a process. Actuators, in turn, move some part of the attached mechanism. For example, actuators in a hydraulic loader lift and lower the bucket. Two broad categories of actuators are linear and rotary. A linear actuator is shown in Fig. 48.1. Mechanisms used to develop motion include pistons, hydraulic cylinders, diaphragms, gearing, and stepping motors.

Figure 48.1 Rack and Pinion Linear Actuator

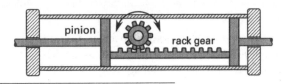

---

[1] Digital controllers are not discussed in this chapter.

Example

Which of the following statements is correct?

   (A) Transducers and actuators are types of sensors.

   (B) Transducers and sensors are types of actuators.

   (C) Actuators and sensors are types of transducers.

   (D) Actuators, sensors, and transducers represent different types of devices.

Solution

Sensors commonly convert monitored characteristics into voltage, current, or resistance. A transducer is anything that converts one characteristic (condition, phenomenon, etc.) into another (e.g., voltage or current). Actuators are linear and rotating devices that convert electrical or pneumatic inputs into kinetic energy (i.e., movement) or vice versa.

The answer is (C).

2. SENSITIVITY

Sensitivity is the ratio of the change in electrical signal magnitude to the change in magnitude of the physical parameter being measured. That is, transducer sensitivity is the ratio of the transmitted signal to the sensor signal. If a transducer exhibits a significant change in an electrical characteristic (voltage, current, resistance, capacitance, or inductance) in response to a change in a parameter, then it is sensitive to that parameter. The greater the change in the electrical signal for the same change in the parameter, the greater the sensitivity of the transducer. It is desirable to use transducers that are sensitive to only one parameter and insensitive to all others. For example, an odometer in a car should respond to only the rotation of the car wheels, and the measurement should not depend on other factors like wind speed, temperature, or humidity.

3. LINEARITY

The *linearity* of a transducer is the degree to which the output (e.g., voltage) is in direct proportion to the parameter being measured. Transducers are usually designed and selected to be linear over the range of measurements. This is not always practical, so a second-order term and even a third-order term may be needed.

4. MEASUREMENT ACCURACY

A measurement is said to be *accurate* if it is substantially unaffected by (i.e., is insensitive to) all variations outside of the measurer's control.

For example, suppose a rifle is aimed at a point on a distant target and several shots are fired. The target point represents the "true value" of a measurement—the value that should be obtained. The impact points represent the measured values—what is actually obtained. The distance from the centroid of the points of impact to the target point is a measure of the alignment accuracy between the barrel and the sights. The difference between the true and measured values is known as the *measurement bias*.

5. MEASUREMENT PRECISION

Precision is not synonymous with accuracy. Precision is a function of the repeatability of the measured results. If an experiment is repeated with identical results, the measurement is said to be precise. In the rifle example from Sec. 48.4, the average distance of each impact from the centroid of the impact group is a measure of precision. It is possible to take highly precise measurements and still have a large bias.

Most measurement techniques that are intended to improve accuracy (e.g., taking multiple measurements and refining the measurement methods or procedures) actually increase the precision.

Sometimes, the term *reliability* is used to describe the precision of a measurement. A *reliable measurement* is the same as a *precise estimate*.

6. MEASUREMENT STABILITY

Stability and *insensitivity* are related terms. (Conversely, *instability* and *sensitivity* are also related.) A stable measurement is insensitive to minor changes in the measurement process.

7. MEASUREMENT RESOLUTION

Resolution represents the smallest change in the measured characteristic that can be detected in the output of the transducer or the transmitter circuit. It is usually expressed as a percentage of *full-scale output* (FSO). Resolution generally is not constant throughout a transducer's range. Published values for commercial devices may be maximum resolution or average resolution. In some cases, resolution can be affected by or be dependent on the electrical characteristics of the transmitter circuit.

Example

Which term BEST describes the maximum expected error associated with a transducer?

(A) range

(B) resolution

(C) accuracy

(D) precision

Solution

Range describes either the range of the monitored condition that a transducer can measure or the range of the transducer output signal. Resolution describes the smallest change in the monitored condition that the

transducer can detect and respond to. Accuracy describes the difference between the maximum possible sensor value and the actual value. Precision describes the consistency of sensor measurements in equilibrium conditions.

The answer is (C).

8. CHEMICAL SENSORS

While the term "transducer" is commonly used for devices that respond to mechanical input (force, pressure, torque, etc.), the term *sensor* is commonly applied to devices that respond to chemical conditions.[2] For example, an electrochemical sensor might respond to a specific gas, compound, or ion (known as a *target substance* or *species*). Two types of electrochemical sensors are in use today: potentiometric and amperometric. Table 48.1 lists types of common chemical sensors.[3]

Potentiometric sensors generate a measurable voltage at their terminals. In electrochemical sensors taking advantage of half-cell reactions at electrodes, the generated voltage is proportional to the absolute temperature, T, and is inversely proportional to the number of electrons, n, (i.e., the valence or oxidation number of the ions) taking part in the chemical reaction at the half-cell. In the following equation (derived from the Nernst equation), p_1 is the partial pressure of the target substance at the measurement electrode and p_2 is the partial pressure of the target substance at the reference electrode.

$$V \propto \left(\frac{T_{\text{absolute}}}{n} \right) \ln \frac{p_1}{p_2}$$

Amperometric sensors (also known as *voltammetric sensors*) generate a measurable current at their terminals. In conventional electrochemical sensors known as *diffusion-controlled cells*, a high-conductivity acid or alkaline liquid electrolyte is used with a gas-permeable membrane that transmits ions from the outside to the inside of the sensor. A reference voltage is applied to two terminals within the electrolyte, and the current generated at a (third) sensing electrode is measured.

[2]The categorization is common but not universal. The terms "transducer," "sensor," "sending unit," and "pickup" are often used loosely.
[3](1) By reproducing Table 48.1 directly, the *NCEES Handbook* perpetuated the errors and inconsistencies present in the original source. (2) The table's title, "common chemical sensors," is inaccurate because many of the devices do not respond to chemical species. Semiconducting oxide devices (bias), piezoelectric devices (force), pyroelectric devices (temperature), and optical devices (radiation) are not chemical sensors. (3) All of the entries in the column marked "principle" should list either "amperometric" or "potentiometric." Instead, the column entries are a hodgepodge of response characteristic, sensor type, construction method, and construction material (belonging in the "materials" column). (4) In the "analyte" column for ion-selective electrode, "Ca²" is an error and should be "Ca²⁺." (5) For example, the resistance of a nonlinear resistance temperature detector (RTD) (see Sec. 48.9) is frequently modeled as

$$R_T = R_0(1 + \alpha T + \beta T^2 + \gamma T^3)$$

Table 48.1 *Examples of Common Chemical Sensors*

sensor type	principle	materials	analyte
semiconducting oxide sensor	conductivity impedance	SnO_2, TiO_2, ZnO_2, WO_3, polymers	O_2, H_2, CO, SO_x, NO_x, combustible hydrocarbons, alcohol, H_2S, NH_3
electrochemical sensor (liquid electrolyte)	amperometric	composite Pt, Au catalyst	H_2, O_2, O_3, CO, H_2S, SO_2, NO_x, NH_3, glucose, hydrazine
ion-selective electrode (ISE)	potentiometric	glass, LaF_3, CaF_2	pH, K^+, Na^+, Cl^-, Ca^2, Mg^{2+}, F^-, Ag^+
solid electrode sensor	amperometric	YSZ, H^+-conductor	O_2, H_2, CO, combustible hydrocarbons
	potentiometric	YSZ, β-alumina, Nasicon	O_2, H_2, CO_2, CO, NO_x, SO_x, H_2S, Cl_2
		Nafion	H_2O, combustible hydrocarbons
piezoelectric sensor	mechanical w/ polymer film	quartz	combustible hydrocarbons, VOCs
catalytic combustion sensor	calorimetric	Pt/Al_2O_3, Pt-wire	H_2, CO, combustible hydrocarbons
pyroelectric sensor	calorimetric	pyroelectric + film	vapors
optical sensors	colorimetric fluorescence	optical fiber/ indicator dye	acids, bases, combustible hydrocarbons, biologicals

Source: *Journal of The Electrochemical Society*, 150(2). ©2003. The Electrochemical Society.

The maximum current generated is known as the *limiting current*. Current is proportional to the concentration, C, of the target substance; the permeability, P; the exposed sensor (membrane) area, A; and the number of electrons transferred per molecule detected, n. The current is inversely proportional to the membrane thickness, t.

$$I \propto \frac{nPCA}{t}$$

Equation 48.1: pH Combined Electrode Potential

$$E_{el} = E^0 - S(\mathrm{pH}_a - \mathrm{pH}_i) \qquad \textbf{48.1}$$

Variations

$$E_{el} = E^0 + S(\log_{10}[H_2^+] - \log_{10}[H_1^+])$$
$$= E^0 + S\left(\log_{10}\frac{[H_2^+]}{[H_1^+]}\right)$$

Description

A pH sensor measures the pH of a solution through a probe connected to an electronic meter. The sensor generates an electrode potential, E_{el}, predicted by Eq. 48.1.

The potential generated by a theoretically perfect pH sensor in a neutral (pH of 7) solution is zero. However, a real sensor has an intrinsic potential known as the *standard electrode potential*, E^0. As it is derived from the *Nernst equation* used for analyzing electrolytic reactions, two electrodes are involved, one for each electrode/solution. In Eq. 48.1, pH_a is the pH of the measured solution as detected by the measurement electrode, and pH_i is the pH of the internal buffer (i.e., reference solution as detected by the reference electrode). (An ideal reference electrode will generate the same reference voltage regardless of the pH.) Simple, uncompensated electrodes that respond directly to solution parameters and that must be used in pairs to obtain the active and reference measurements are known as *single electrodes*. When the two electrodes are combined into a single probe, the term *combination electrode* is used.[4]

Equation 48.1 is the equation of a straight line, so the sensor it describes is implicitly linear in its response over the instrument's pH range, generally between 2 and 11. Since the actual electrical output of a pH sensor is in the millivolt range, the proportionality constant (slope), S, in Eq. 48.1 is in millivolts/pH. The slope is negative because generated potential decreases as pH increases. pH meter scales are usually marked or read directly in pH, not millivolts. Since the Nernst equation and the proportionality constant, S, is temperature dependent, quality pH measuring devices incorporate temperature probes for internal compensation. In order to use Eq. 48.1, it may be necessary to convert molar hydrogen ion concentrations into pH.

$$\mathrm{pH} = -\log_{10}[H^+]$$

Example

The potential of a calomel (Hg_2Cl_2) buffering electrode is 0.241 V. At 25°C, the electrode potential of a silver-AgCl/calomel combined electrode is given by

$$0.558 + 0.059\log[Ag^+] \quad \text{[volts]}$$

---

[4]pH_a can be interpreted as the pH of the *active* sensor, the *acidic* solution, the *alkaline* solution, or (in the case of medical sensors) the *arterial* blood flow. In fact, the subscript is derived from the *activity* of the hydrogen ions in the solution. In general, pH_i is the pH of a reference electrode, so pH_{ref} would be an appropriate variable. However, the *NCEES Handbook* has adopted the subscript specifically for a one-piece, combined sensor having an *inner* buffer reference.

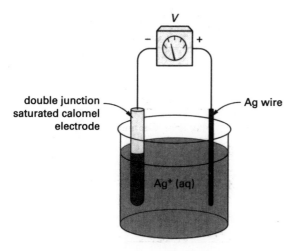

The silver ion concentration is 1 M. Most nearly, what is the standard electrode potential for the silver reaction, $Ag^+ + e^- \rightarrow Ag$?

(A) 0.18 V

(B) 0.30 V

(C) 0.35 V

(D) 0.80 V

Solution

Use Eq. 48.1.

$$E_{el} = E^0 - S(pH_a - pH_i)$$
$$= E^0 + S(pH_i) - S(pH_a)$$
$$= E^0_{Ag^+} - S(\log_{10}[Hg^+]) - S(pH_a)$$

The value of S is not known, but the product of S and $\log_{10}[Hg^+]$ represents the electrode potential of the calomel reference electrode reaction. This is given as 0.241 V.

$$0.558 \text{ V} = E_{Ag^+} - E^0_{Hg^+}$$
$$E_{Ag^+} = 0.558 \text{ V} + E^0_{Hg^+}$$
$$= 0.558 \text{ V} + 0.241 \text{ V}$$
$$= 0.799 \text{ V} \quad (0.80 \text{ V})$$

Since 25°C corresponds to the standardization temperature of electrode potentials, this is the standard electrode potential for the silver reaction.

The answer is (D).

9. RESISTANCE TEMPERATURE DETECTORS

Resistance temperature detectors (RTDs), also known as *resistance thermometers*, change resistance predictably in response to changes in temperature. An RTD is composed of a fine wire wrapped around a form and

protected with glass or a ceramic coating. Nickel and copper are commonly used for industrial RTDs. Platinum is often used in RTDs because it is mechanically and electrically stable, is chemically inert, resists contamination, and can be highly refined to a high and consistent purity. RTDs are connected through resistance bridges to compensate for lead resistance.

Equation 48.2: Resistance Versus Temperature

$$R_T = R_0[1 + \alpha(T - T_0)] \qquad \textbf{48.2}$$

Variations

$$R_T \approx R_0(1 + \alpha\Delta T + \beta\Delta T^2)$$

$$\Delta T = T - T_0$$

Description

With some exceptions, including carbon and silicon, resistance in most conductors increases with temperature. The resistance of RTDs has greater sensitivity and more linear response to temperature than that of standard resistors. The resistance at a given temperature can be calculated from the *coefficients of thermal resistance*, α and β. Higher-order terms—third, fourth, etc.—are used when extreme accuracy is required. The variation of resistance with temperature is nonlinear, though β is small and is often insignificant over short temperature ranges. Therefore, a linear relationship is often assumed and only α is used.[5] In commercial RTDs, α is referred to as the *alpha-value*.

R_0 is the resistance (usually 100 Ω for standard RTDs) at the *reference temperature*, T_0, usually at 0°C (32°F). The first-order approximation in Eq. 48.2 is sufficient in most practical applications.

Similarly to resistors, commercial RTDs are categorized by the precision of their labeled resistance values. RTDs in various classes (e.g., AA, A, B, 1/10 B, etc.) are available, although Class B RTDs are the most common. At 0°C, resistances of Class A RTDs can vary ±0.15% from their nominal values (e.g., 100 Ω), while Class B RTDs can vary ±0.30%. The maximum deviations in the as-delivered resistance and the corresponding temperature response are known as *tolerances*. Identifying RTD tolerance classes by their percentages (e.g., "a 0.12% RTD") should be avoided, since tolerances are actually dependent on temperature and resistance wire (or, manufacturing method, in the case of thin-film RTDs) used, and the percentage tolerances are different for resistance and temperature response. Figure 48.2 shows tolerance values for standard 100 Ω RTDs.

---

[5]Although the *NCEES Handbook* presents Eq. 48.2 as an equality, it is only an approximation. Higher order terms (e.g., the variation equation) may be needed to obtain sufficient accuracy.

Figure 48.2 *Platinum RTD Tolerance Values*

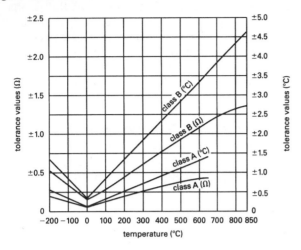

Adapted from tempco.com, 2013, by Tempco Manufactured Products.

10. THERMOCOUPLES

A *thermocouple* is a temperature-measuring device (sensor, circuit, etc.) that generates a voltage difference in response to a temperature difference. A *thermocouple* consists of two wires of dissimilar metals joined at both ends.[6] One set of ends, typically called a *junction*, is kept at a known *reference temperature* while the other junction is exposed to the unknown temperature.[7] (See Fig. 48.3.) In a laboratory, the reference junction is often maintained at the *ice point*, 32°F (0°C), in an ice bath for convenience in later analysis.

Thermocouples are assigned ANSI codes based on their alloy combinations.[8] The ANSI code is often referred to as the thermocouple *type* (i.e., Type J, Type N, etc.). Additionally, each thermocouple has a temperature range and a specific environment in which it can be used. Operating ranges can be inferred from Fig. 48.4.

Thermocouple materials, standard ANSI designations, and approximate useful temperature ranges are given in Table 48.2. The "functional" temperature range can be much larger than the useful range. The most significant factors limiting the useful temperature range, sometimes referred to as the *error limits range*, are linearity, the rate at which the material will erode due to oxidation at higher temperatures, irreversible magnetic effects above magnetic critical points, and longer stabilization periods at higher temperatures.

---

[6]The joint may be made by simply twisting the ends together. However, to achieve a higher mechanical strength and a better electrical connection, the ends should be soldered, brazed, or welded.

[7]The *ice point* is the temperature at which liquid water and ice are in equilibrium. Other standardized temperature references are: *oxygen point*, −297.346°F (90.19K); *steam point*, 212.0°F (373.16K); *sulfur point*, 832.28°F (717.76K); *silver point*, 1761.4°F (1233.96K); and *gold point*, 1945.4°F (1336.16K).

[8]It is not uncommon to list the two thermocouple materials with "vs." (as in *versus*). For example, a copper/constantan thermocouple might be designated as "copper vs. constantan."

Figure 48.3 *Thermocouple Circuits*

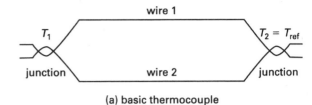

(a) basic thermocouple

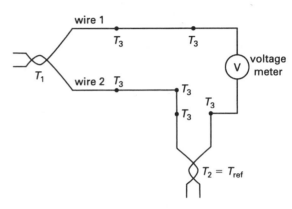

(b) thermocouple in measurement circuit

Figure 48.4 *Thermocouple Hot Junction Voltage (EMF) Versus Temperature (32°F cold junction)*

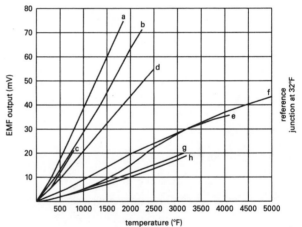

line key
a chromel versus constantan
b iron versus constantan
c copper versus constantan
d chromel versus alumel
e tungsten 5% rhenium versus tungsten 26% rhenium
f tungsten versus tungsten 26% rhenium
g platinum versus platinum 13% rhodium
h platinum versus platinum 10% rhodium

Copyright 2016 by Convectronics as posted on convectronics.com.

For instructional use only. Not for design use. Significant differences can exist between graphed and ideal values. Manufactured devices may exhibit characteristics deviated from theoretical performance.

Table 48.2 Typical Temperature Ranges of Thermocouple Materials[a]

materials	ANSI designation	useful range (°F (°C))
copper-constantan	T	−330 to 660 (−180 to 370)
chromel-constantan	E	−200 to 1650 (0 to 870)
iron-constantan[b]	J	32 to 1380 (0 to 760)
chromel-alumel	K	−330 to 2280 (0 to 1260)
platinum-10% rhodium	S	32 to 2700 (0 to 1480)
platinum-13% rhodium	R	32 to 2700 (0 to 1480)
Pt-6% Rh-Pt-30% Rh	B	1600 to 3100 (870 to 1700)
tungsten-Tu-25% rhenium	–	to 4200[c] (2320)
Tu-5% rhenium-Tu-26% rhenium	–	to 4200[c] (2320)
Tu-3% rhenium-Tu-25% rhenium	–	to 4200[c] (2320)
iridium-rhodium	–	to 3500[c] (1930)
nichrome-constantan	–	to 1600[c] (870)
nichrome-alumel	–	to 2200[c] (1200)

[a]Actual values will depend on wire gauge, atmosphere (oxidizing or reducing), use (continuous or intermittent), and manufacturer.
[b]Nonoxidizing atmospheres only.
[c]Approximate usable temperature range. Error limit range is less.

A voltage is generated when the temperatures of the two junctions are different. This phenomenon is known as the *Seebeck effect*.[9] Referring to the polarities of the voltage generated, one metal is known as the *positive lead* (*positive element*) while the other is the *negative lead* (*negative element*). The generated voltage is small, and thermocouples are calibrated in $\mu V/°F$ or $\mu V/°C$. An amplifier may be required to provide usable signal levels, although thermocouples can be connected in series (a *thermopile*) to increase the value.[10] The accuracy (referred to as the *calibration*) of thermocouples is approximately $1/2$–$3/4$%, though manufacturers produce thermocouples with various guaranteed accuracies.

Equation 48.3: Thermocouple Voltage

$$V = k_T(T - T_{\text{ref}}) \qquad \textbf{48.3}$$

Description

The voltage generated by a thermocouple is given by Eq. 48.3. Since the *thermoelectric constant*, k_T, varies

---

[9]The inverse of the Seebeck effect, that current flowing through a junction of dissimilar metals will cause either heating or cooling, is the *Peltier effect*, though the term is generally used in regard to cooling applications. An extension of the Peltier effect, known as the *Thompson effect*, is that heat will be carried along the conductor. Both the Peltier and Thompson effects occur simultaneously with the Seebeck effect. However, the Peltier and Thompson effects are so minuscule that they can be disregarded.
[10]There is no special name for a combination of thermocouples connected in parallel.

with temperature, thermocouple problems can be solved with published tables of total generated voltage versus temperature.

Generation of thermocouple voltage in a measurement circuit is governed by three laws. The *law of homogeneous circuits* states that the temperature distribution along one or both of the thermocouple leads is irrelevant. Only the junction temperatures contribute to the generated voltage.

The *law of intermediate metals* states that an intermediate length of wire placed within one leg or at the junction of the thermocouple circuit will not affect the voltage generated as long as the two new junctions are at the same temperature. This law permits the use of a measuring device, soldered connections, and extension leads.

The *law of intermediate temperatures* states that if a thermocouple generates voltage V_1 when its junctions are at T_1 and T_2, and it generates voltage V_2 when its junctions are at T_2 and T_3, then it will generate voltage $V_1 + V_2$ when its junctions are at T_1 and T_3.

Example

At a particular temperature, a type-K (chromel-alumel) thermocouple produces a voltage of 10.79 mV. The cold junction is kept at 32°F (0°C) by an ice bath. Most nearly, what is the temperature of the hot junction?

(A) 470°F

(B) 490°F

(C) 500°F

(D) 520°F

Solution

Use Fig. 48.4. Determine the slope of the linear portion of the curve.

$$k_T = \frac{\Delta \text{EMF}}{\Delta T} \approx \frac{55 \text{ mV} - 9 \text{ mV}}{2500°F - 500°F} = 0.023 \text{ mV/°F}$$

Use Eq. 48.3. The temperature of the hot junction is

$$V = k_T(T - T_{\text{ref}})$$
$$T = \frac{V}{k_T} + T_{\text{ref}}$$
$$= \frac{10.79 \text{ mV}}{0.023 \dfrac{\text{mV}}{°F}} + 32°F$$
$$= 501°F \quad (500°F)$$

This temperature is within the linear range of the curve.

The answer is (C).

Process Control

11. STRAIN GAGES

A *bonded strain gage* is a metallic resistance device that is bonded to the surface of the unstressed member. (See Fig. 48.5(a).) The gage consists of a folded metallic conductor (known as the *grid*) on a backing (known as the *substrate*). The grids of strain gages were originally of the folded-wire type. For example, nichrome wire with a total resistance under 1000 Ω was commonly used.

Modern strain gages are generally of the foil type manufactured using printed circuit techniques. Semiconductor gages are used when extreme sensitivity (i.e., a gage factor in excess of 100) is required. However, semiconductor gages are extremely temperature-sensitive.

The substrate and grid experience the same strain as the surface of the member. The resistance of the gage changes as the member is stressed due to changes in conductor cross section and intrinsic changes in resistivity with strain. Temperature effects must be compensated by the circuitry or by using a second unstrained gage as part of the bridge measurement system.

When simultaneous strain measurements in two or more directions are needed, it is convenient to use a commercial *rosette strain gage*. (See Fig. 48.5(b).) A rosette consists of two or more grids properly oriented for application as a single unit.

Table 48.3 at the end of this chapter shows various types of strain gages.

Figure 48.5 *Strain Gage*

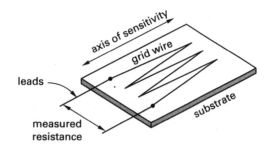

(a) bonded-wire strain gage

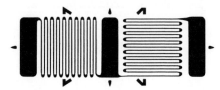

(b) commercial two-element rosette

Equation 48.4: Gage Factor

$$GF = \frac{\Delta R/R}{\Delta L/L} = \frac{\Delta R/R}{\varepsilon} \qquad 48.4$$

Description

The *gage factor* (*strain sensitivity factor*), GF, is the ratio of the fractional change in resistance to the fractional change in length (strain) along the detecting axis of the gage. (See Table 48.4.) The gage factor is a function of the gage material. It can be calculated from the grid material's properties and configuration. The higher the gage factor, the greater the sensitivity of the gage. From a practical standpoint, however, the gage factor and gage resistance are provided by the gage manufacturer. Only the change in resistance is measured.

Values

Table 48.4 *Approximate Gage Factors*

material	GF
constantan	2.0
iron, soft	4.2
isoelastic	3.5
manganin	0.47
monel	1.9
nichrome	2.0
nickel	−12[*]
platinum	4.8
platinum-iridium	5.1

[*]Value depends on amount of preprocessing and cold working.

Example

A strain gage is to be used in measuring the strain on a test specimen. A strain gage with an initial resistance of 120 Ω exhibits a decrease of 0.120 Ω. The gage factor is 2.00. The initial length of the strain gage was 1.000 cm. What is most nearly the final length of the strain gage?

(A) 0.9995 cm

(B) 1.0000 cm

(C) 1.0005 cm

(D) 1.0050 cm

Solution

From Eq. 48.4,

$$GF = \frac{\Delta R/R}{\Delta L/L}$$

Solve for the change in length.

$$\Delta L = \frac{\Delta R/R}{GF/L} = \frac{\dfrac{0.120\ \Omega}{120\ \Omega}}{\dfrac{2.00}{1.000\ \text{cm}}} = 0.0005\ \text{cm}$$

The resistance decreased, so the length of the strain gage also decreased.

$$L_{\text{final}} = L - \Delta L = 1.000\ \text{cm} - 0.0005\ \text{cm}$$
$$= 0.9995\ \text{cm}$$

The answer is (A).

12. WHEATSTONE BRIDGES

The *Wheatstone bridge* shown in Fig. 48.6 is one type of *resistance bridge*. The Wheatstone bridge can be used to determine the unknown resistance of a resistance transducer (e.g., thermistor or resistance-type strain gage), such as R_1 in Fig. 48.6. The potentiometer, R_4, in Fig. 48.6 and Fig. 48.7, is adjusted (i.e., the bridge is "balanced") until no current flows through the meter or until there is no voltage across the meter, hence the name *null indicator*, or alternatively, *zero-indicating bridge* or *null-indicating bridge*. The unknown resistance can also be determined from the voltage imbalance shown by the meter reading, in which case the bridge is known as a *deflection bridge* rather than a null-indicating bridge. When the bridge is balanced and no current flows through the meter leg, the following equations apply.

$$I_1 = I_2$$

$$I_4 = I_3$$

$$V_4 + V_3 = V_1 + V_2$$

$$R_1 R_4 = R_2 R_3 \quad \text{[balanced]}$$

Any one of the four resistances can be the unknown, up to three of the remaining resistances can be fixed or adjustable, and the battery and meter can be connected to either of two diagonal corners. The following bridge law statement can be used to help formulate the proper relationship: *When a series Wheatstone bridge is null-balanced, the ratio of resistance of any two adjacent legs equals the ratio of resistance of the remaining two legs, taken in the same sense.* In this statement, "taken in the same sense" means that both ratios must be formed reading either left to right, right to left, top to bottom, or bottom to top.

Figure 48.6 *Series-Balanced Wheatstone Bridge*

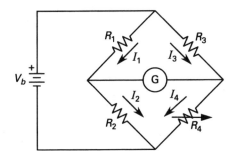

Figure 48.7 *Wheatstone Quarter Bridge*

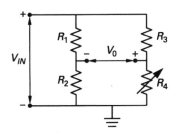

Equation 48.5 and Eq. 48.6: Wheatstone Quarter Bridge Equations

$$\Delta R \ll R \qquad \text{48.5}$$

$$V_0 \approx \frac{\Delta R}{4R} \cdot V_{IN} \qquad \text{48.6}$$

Description

A special case of the Wheatstone bridge circuit is the quarter bridge circuit shown in Fig. 48.7. The quarter bridge circuit has three identical resistors and one resistor with a resistance slightly different from the resistance of the other three. This different resistor is the transducer. The resistance difference, ΔR, can be positive or negative, but it must be relatively small compared to R. (See Eq. 48.5.)

$$R_1 = R_2 = R_3 = R$$

$$R_4 = R + \Delta R$$

Most Wheatstone bridge circuits are difficult to analyze if they are not balanced, but the Wheatstone quarter bridge has a simple approximation, given by Eq. 48.6. Equation 48.6 is useful for many instrumentation applications.[11]

Example

In the circuit shown,

$$V_0 = 0.0500 \text{ V}$$

$$R_1 = R_2 = R_3 = R$$

$$= 10.00 \text{ k}\Omega$$

$$R_4 = 10.25 \text{ k}\Omega$$

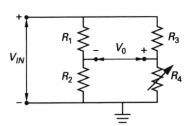

The voltage V_{IN} is most nearly

(A) 2 V

(B) 4 V

(C) 5 V

(D) 8 V

---

[11]The *NCEES Handbook* is not consistent in its use of subscripts. While upper case letters might have been used to designate a constant DC value, there is probably no distinction between "IN" as used in the "*Instrumentation, Measurement, and Controls*" section and "in" as used in the "*Electrical and Computer Engineering*" section.

Solution

The circuit is a Wheatstone bridge. Since R_4 is close to 10 kΩ, the quarter bridge approximation can be used.

$$V_0 \approx \frac{\Delta R}{4R} \cdot V_{IN}$$

$$\begin{aligned}
V_{IN} &\approx \frac{4R}{\Delta R} V_0 \\
&= \left(\frac{(4)(10 \text{ k}\Omega)}{10.25 \text{ k}\Omega - 10 \text{ k}\Omega} \right)(0.0500 \text{ V}) \\
&= 8 \text{ V}
\end{aligned}$$

The answer is (D).

13. VALVE FLOW COEFFICIENTS

Valve flow capacities depend on the geometry of the inside of the valve. The *flow coefficient*, C_v, for a valve (particularly a control valve) relates the flow quantity (in gallons per minute) of a fluid with fluid specific gravity to the pressure drop (in pounds per square inch). (The flow coefficient for a valve is not the same as the coefficient of flow for an orifice or venturi meter.) The flow coefficient is not dimensionally homogeneous.

Metricated countries use a similar concept with a different symbol, K_v, (not the same as the loss coefficient, K) to distinguish the valve flow coefficient from customary U.S. units. K_v is defined[12] as the flow rate in cubic meters per hour of water at a temperature of 16°C with a pressure drop across the valve of 1 bar. To further distinguish it from its U.S. counterpart, K_v may also be referred to as a *flow factor*. C_v and K_v are linearly related by

$$\dot{V}_{\text{m}^3/\text{h}} = K_v \sqrt{\frac{\Delta p_{\text{bars}}}{\text{SG}}} \qquad \text{[SI]}$$

$$Q_{\text{gpm}} = C_v \sqrt{\frac{\Delta p_{\text{psi}}}{\text{SG}}} \qquad \text{[U.S.]}$$

$$K_v = 0.865 C_v$$

When selecting a control valve for a particular application, the value of C_v is first calculated. Depending on the application and installation, C_v may be further modified by dividing by *piping geometry* and *Reynolds number factors*. (These additional procedures are often specified by the valve manufacturer.) Then, a valve with the required value of C_v is selected.

---

[12]Several definitions of both C_v and K_v are in use. A definition of C_v based on Imperial gallons is used in Great Britain. Definitions of K_v based on pressure drops in kilograms-force and volumes in liters per minute are in use. Other differences in the definition include the applicable temperature, which may be given as 5–30°C or 5–40°C instead of 16°C.

Although the flow coefficient concept is generally limited to control valves, its use can be extended to all fittings and valves. The relationship between C_v and the loss coefficient, K, is

$$C_v = \frac{29.9 d_{\text{in}}^2}{\sqrt{K}} \qquad \text{[U.S.]}$$

14. PRESSURE RELIEF AND ISOLATION DEVICES

Nearly all piping systems, from a gas water heater to an oil refinery, require specialty valves and devices, including *pressure relief valves* (PRVs, i.e., *safety relief valves*), *check valves*, *shut-off valves*, and *rupture discs* (RDs, i.e., *burst discs*). These devices protect personnel and facilities from harm and catastrophic failures, respectfully, caused by abnormal operating conditions, power interruptions, or malfunctions in automated processes. Abnormal operating conditions occur when pressure exceeds the system design operating range due to overpressurization or underpressurization. These devices quickly relieve pressure, prevent backflow, or isolate system components. The design and sizing of pressure relief and isolation devices depend on parameters such as flow rate, pressure, temperature, and type of gas or fluid in the system.

A PRV is a normally closed, spring-actuated device installed at strategic locations along a system that automatically opens to relieve pressure when the system's operating pressure reaches a specific pressure. When the overpressure situation abates, the PRV closes, preventing further loss of contents.

An RD is a diaphragm designed to rupture at a predetermined pressure differential and to vent at a specific flow rate in accordance with design specifications. RDs are typically used in tandem with PRVs. Careful consideration should be given to the installation location due to the high noise level occurring during disc rupture. American Society of Mechanical Engineers (ASME) *Boiler and Pressure Vessel Code* (BPVC) Sec. VIII, Div. 1, UG-125 through UG-140 require all pressure vessels to be equipped with overpressure protection devices.

Process Control

Table 48.3 Strain Gages

strain	gage setup	bridge type	sensitivity (mV/V @ 100 $\mu\varepsilon$)*	details
axial		1/4	0.5	Good: Simplest to implement, but must use a dummy gage if compensating for temperature. Also responds to bending strain.
		1/2	0.65	Better: Temperature compensated, but is sensitive to bending strain.
		1/2	1.0	Better: Rejects bending strain, but not temperature. Must use dummy gages if compensating for temperature.
		full	1.3	Best: More sensitive and compensates for both temperature and bending strain.
bending		1/4	0.5	Good: Simplest to implement, but must use a dummy gage if compensating for temperature. Responds equally to axial strain.
		1/2	1.0	Better: Rejects axial strain and is temperature compensated.
		full	2.0	Best: Rejects axial strain and is temperature compensated. Most sensitive to bending strain.
torsional and shear		1/2	1.0	Good: Gages must be mounted at 45° from centerline.
		full	2.0	Best: Most sensitive full-bridge version of previous setup. Rejects both axial and bending strains.

*Although the *NCEES Handbook* defines the variable ε in its *Instrumentation, Measurement, and Controls* section as "strain," the abbreviation "$\mu\varepsilon$" (i.e., "microstrain") is a specific strain value: 10^{-6} (e.g., 10^{-6} mm/mm, etc.).

Adapted from ni.com, 2013, by National Instruments Corporation.

Process Control

Diagnostic Exam

Topic XIII: Process Design and Economics

1. A heat exchanger is used to heat a feed before it reaches a reactor. A piping and instrumentation diagram of the process is shown.

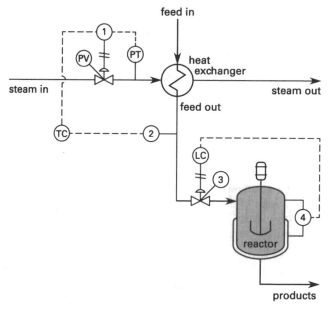

Most likely, how should the instruments, 1, 2, 3, 4, be labeled?

(A) 1 = PC, 2 = TT, 3 = LV, 4 = LT

(B) 1 = PIC, 2 = LT, 3 = LT, 4 = LS

(C) 1 = FC, 2 = TT, 3 = LV, 4 = LT

(D) 1 = FIC, 2 = FT, 3 = LT, 4 = LIC

2. A chemical plant construction project in 1981 cost $8,200,000. The cost included equipment, labor, building, engineering, and supervision. The Chemical Engineering Plant Cost Index (CEPCI) is shown.

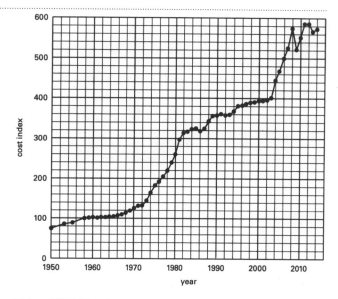

Using CEPCI, what is most nearly the cost of constructing the same chemical plant in 2005?

(A) $10,000,000

(B) $13,000,000

(C) $15,000,000

(D) $20,000,000

3. A company takes out a commercial loan of $1,600,000. The company pays off the loan fully by paying a fixed amount annually for 20 years. The effective annual interest rate on the loan is 3.3%. Most nearly, what is the annual payment?

(A) $90,000

(B) $110,000

(C) $140,000

(D) $180,000

4. The maximum pressure for an adiabatic exothermic reaction is determined to be 80 psig. Given that all of the options are available, which of the following methods represents an inherently safe design for the reaction?

(A) Run the reaction in a reactor designed for 110 psig maximum pressure.

(B) Develop or use a reaction chemistry that is not exothermic, or is only mildly exothermic, so that the pressure buildup is minimal.

(C) Run the reaction in a reactor designed for 80 psig maximum pressure and that has an emergency relief system by gradually adding reactants and by using temperature control.

(D) Run the reaction in a reactor designed for 80 psig maximum pressure by gradually adding reactants and by training an observer-operator to observe and counteract any out-of-control conditions.

5. Cost estimates for a proposed municipal facility are being evaluated. Initial construction cost is anticipated to be $120,000, and annual maintenance expenses are expected to be $6500 for the first 20 years and $2000 for every year thereafter. The facility is to be used and maintained for an indefinite period of time. Using an interest rate of 10% per year, the capitalized cost of this facility is most nearly

(A) $180,000

(B) $190,000

(C) $200,000

(D) $270,000

6. A machine has an initial cost of $18,000 and operating costs of $2500 each year. The salvage value decreases by $3000 each year. The machine is now three years old. Assuming an effective annual interest rate of 12%, the cost of owning and operating the machine for one more year is most nearly

(A) $5500

(B) $6100

(C) $6600

(D) $7100

7. A chemical plant purchased a $140,000 machine that has a useful life of seven years. If the machine has a salvage value of $20,000 at the end of its life, and straight line depreciation is used, the book value at the end of year 4 is most nearly

(A) $20,000

(B) $50,000

(C) $60,000

(D) $70,000

8. $5000 is put into an empty savings account with a nominal interest rate of 5%. No other contributions are made to the account. With monthly compounding, approximately how much interest will have been earned after five years?

(A) $1250

(B) $1380

(C) $1410

(D) $1420

9. An investment currently costs $28,000. If the current inflation rate is 6% and the effective annual return on investment is 10%, approximately how long will it take for the investment's future value to reach $40,000?

(A) 1.8 yr

(B) 2.3 yr

(C) 2.6 yr

(D) 3.4 yr

10. The purchase price of a car is $25,000. Mr. Smith makes a down payment of $5000 and borrows the balance from a bank at 6% nominal annual interest, compounded monthly. Most nearly, what is the required monthly payment to pay off the loan in 5 years?

(A) $350

(B) $400

(C) $450

(D) $500

SOLUTIONS

1. The feed leaving the heat exchanger must be a specific temperature. The temperature transmitter (TT), 2, reads the temperature of the feed exiting the heat exchanger. The TT signals the temperature controller (TC), which in turn signals the pressure controller (PC), 1. The PC also receives signals from the pressure transmitter (PT). If the feed leaving the heat exchanger is not at the required temperature, the PC will increase or decrease the pressure of steam entering the heat exchanger through the pressure valve (PV).

The level transmitter (LT), 4, reads the fluid level inside the reactor. The reading is then sent to the level controller (LC). If the contents of the reactor are not at the desired level, the LC uses the level valve (LV), 3, to adjust the feed.

The answer is (A).

2. The cost, C_2, of constructing the plant in 2005 is determined by using the cost indexes formula, the CE cost index for 1981 ($CEPCI_1$), and the CE cost index for 2005 ($CEPCI_2$) from the CE index curve.

From the CE index curve, $CEPCI_2 = 480$ and $CEPCI_1 = 300$.

$$\begin{aligned}
C_2 &= C_1\left(\frac{CEPCI_2}{CEPCI_1}\right) \\
&= (\$8{,}200{,}000)\left(\frac{480}{300}\right) \\
&= \$13{,}120{,}000 \quad (\$13{,}000{,}000)
\end{aligned}$$

The answer is (B).

3. The loan amount is the present worth, P, and the annual payment is the uniform amount per interest period, A. The annual payment for the loan is

$$\begin{aligned}
A &= P(A/P, i\%, n) \\
&= P\left(\frac{i(1+i)^n}{(1+i)^n - 1}\right) \\
&= (\$1{,}600{,}000)\left(\frac{(0.033)(1+0.033)^{20}}{(1+0.033)^{20} - 1}\right) \\
&= \$110{,}550 \quad (\$110{,}000)
\end{aligned}$$

The answer is (B).

4. Options (A), (C), and (D) do not address the pressure build-up due to an exothermic reactions. Those options accept the build-up and attempt to contain it. Option (B) eliminates the pressure build-up, making the reaction chemistry inherently safe.

The answer is (B).

5. Capitalized costs are present worth values when the analysis period is infinite. In general, capitalized cost $= A/i$, where A is a uniform series of infinite end-of-period cash flows and i is the interest rate per compounding period.

In this problem, there is a one-time cost of \$120,000, an infinite end-of-year series of \$2000, and a uniform series of end-of-year cost of \$6500 − \$2000 = \$4500 in years 1 through 20.

$$\begin{aligned}
\genfrac{}{}{0pt}{}{\text{capitalized}}{\text{cost}} &= \$120{,}000 + A_1(P/A, i, n) + A_2/i \\
&= \$120{,}000 + (\$4500)(P/A, 10\%, 20) \\
&\quad + \frac{\$2000}{0.10} \\
&= \$120{,}000 + (\$4500)(8.5136) \\
&\quad + \frac{\$2000}{0.10} \\
&= \$178{,}311 \quad (\$180{,}000)
\end{aligned}$$

The answer is (A).

6. The cost of owning and operating the machine one more year is equal to the operating costs plus the lost salvage value. The value of having the cash one year earlier must also be considered. After three years, the machine's salvage value is \$9000, and after another year it will be \$6000.

The cost of one more year of ownership and operation is

$$\begin{aligned}
C &= \text{operating cost} + \text{lost salvage value} \\
&\quad + \text{opportunity cost} \\
&= \$2500 + (\$9000 - \$6000) + (0.12)(\$9000) \\
&= \$6580 \quad (\$6600)
\end{aligned}$$

The answer is (C).

7. The straight-line depreciation is

$$\begin{aligned}
D &= \frac{C - S_n}{n} \\
&= \frac{\$140{,}000 - \$20{,}000}{7 \text{ yr}} \\
&= \$17{,}143/\text{yr}
\end{aligned}$$

The book value at the end of year 4 is

$$\begin{aligned}
BV_j &= C - \sum D_j \\
BV_4 &= \$140{,}000 - (4)(\$17{,}143) \\
&= \$71{,}428 \quad (\$70{,}000)
\end{aligned}$$

The answer is (D).

Process Design/ Economics

8. 5% is the nominal rate per year. The effective annual interest rate is

$$i_e = \left(1 + \frac{r}{m}\right)^m - 1$$
$$= \left(1 + \frac{0.05}{12}\right)^{12} - 1$$
$$= 0.05116$$

The total future value after five years is

$$F = P(F/P, i\%, n) = P(1 + i)^n$$
$$= (\$5000)(1 + 0.05116)^5$$
$$= \$6417$$

The interest earned is

$$I_{\text{earned}} = F - P = \$6417 - \$5000$$
$$= \$1417 \quad (\$1420)$$

(This problem can also be solved by calculating the effective interest rate per month and compounding for 60 months.)

The answer is (D).

9. The combined interest rate adjusted for inflation is

$$d = i + f + (i \times f)$$
$$= 0.10 + 0.06 + (0.10)(0.06)$$
$$= 0.166$$

Use the present worth factor to determine the number of periods, n.

$$P = F(1 + d)^{-n}$$
$$\$28,000 = (\$40,000)(1 + 0.166)^{-n}$$
$$0.7 = (1 + 0.166)^{-n}$$

Take the log of both sides.

$$\log 0.7 = \log (1.166)^{-n}$$
$$= -n \log 1.166$$
$$-0.1549 = -n(0.0667)$$
$$n = 2.32 \quad (2.3 \text{ yr})$$

The answer is (B).

10. The present worth, P, of the loan is

$$P = \$25,000 - \$5000 = \$20,000$$

The effective rate per month is

$$i = \frac{6\%}{12} = 0.5\%$$
$$n = (5 \text{ yr})\left(12 \frac{\text{mo}}{\text{yr}}\right) = 60 \text{ mo}$$

Use the capital recovery factor.

$$A = (\$20,000)(A/P, 0.5\%, 60)$$
$$= (\$20,000)(0.0193)$$
$$= \$386 \quad (\$400)$$

The answer is (B).

49 Plant and Process Design

Nomenclature

D	diameter	ft	m
D_m	diffusion coefficient	ft^2/sec	m^2/s
f	frequency	Hz	Hz
L	characteristic length	ft	m
M	year	–	–
n	rotational speed	rpm	rpm
n	scaling exponent	–	–
N	number	–	–
p	number of poles	–	–
pf	power factor	–	–
P	power	hp	kW
R	scale	–	–
Re	Reynolds number	–	–
s	slip	–	–
T	torque	ft-lbf	N·m
v	velocity	ft/sec	m/s

Symbols

η	efficiency	–	–
μ	absolute viscosity	lbf-sec/ft^2	Pa·s
ν	kinematic viscosity	ft^2/sec	m^2/s
ρ	density	lbm/ft^3	kg/m^3

Subscripts

b	blend
f	full-size
p	pumping or pilot

1. EQUIPMENT SPECIFICATIONS

Chemical processing operations and facilities require unique equipment to transport, convert, separate, combine, purify, and condition substances to form new end-product compositions. The starting substances are known as *reactants*; the end-product substances are known as *products*. The reactants are either *converted* to simpler products or *synthesized* into more complex compounds. Chemical processing operations typically involve multiple sequential stages where the product of one stage is the reactant of a subsequent stage.

Transport equipment may incorporate pumps, compressors, turbines, pipes, valves, screw conveyors, and instrumentation. *Conditioning equipment* may incorporate reactors, heat exchangers, condensers, cooling towers, distillation towers, packed towers, evaporators, mixers, dryers, and filters. *Control equipment* may incorporate a programmable logic controller for subsystem operation control or a distributed control system covering an entire chemical processing facility.

Specific equipment requirements are determined during the engineering design phase when vendor equipment specifications, designations, and availability are evaluated to determine if off-the-shelf or custom equipment will be required. Off-the-shelf equipment typically has lower costs, lower risks, and shorter lead times than custom equipment.

Equipment specifications are written documents developed to describe technical requirements, solicit vendor bids, and complement procurement documentation. Equipment specifications are detailed and unambiguous, contain measurable and quantifiable requirements, and commonly include drawings and diagrams. Equipment specifications list the most important requirements, such as

- performance metrics

- operating conditions

- capabilities or capacities

- applicable standards or codes

- size or footprint dimensions

- interfaces (human, electrical, computer, software, hardware, chemical, mechanical, etc.)

- material or material compatibility

- power consumption (or output)

Process Design/ Economics

- requirements for insulation, heat generation rate, and cooling

- maintenance or duty cycle

- test data from start-up and prototypes

- inspection data or reports

2. EQUIPMENT SELECTION

Equipment selection is a function of the relative merits of differing types of equipment, perhaps using different technologies, to perform a given task. Equipment selection is primarily based on the ability of each alternative set of equipment to perform the task. Additional equipment selection factors include economics, safety, fouling, space utilization, lifecycle costs, maintenance, reliability, product quality, waste minimization, and subsequent waste treatment. Selection of equipment strongly depends on the unit operation being considered.

When evaluating and selecting equipment according to specified characteristics, each specified factor (technical, schedule, budget, and risk) should have an associated *importance weight* or *importance rating value*. For example, technical requirements may be twice as important as schedule and cost.

Some equipment specifications may be uncompromising and must be met with full compliance. Some may be stated as a range, allowing for some discretion and latitude in equipment selection. For example, an equipment height of less than 18 ft (6 m) may be essential and given an importance rating of 1.0 to account for ceiling height, while a width specification to match existing equipment may be given an importance rating of only 0.30.

Maintenance, scalability, and reliability are important considerations as well. For example, reactors with tube bundles may be difficult to recharge with catalyst and clean of any fouling that may occur.

Life cycle costs and overall safety must be considered as well. Explosive reactants require adequate relief devices and special barricades that will operate even during runaway and explosion or fire. Reactor size must be taken into account when these systems are selected. In addition, toxic materials require special handling and equipment. Adequate ventilation is also required to minimize worker exposure.

Example

A component in a chemical plant is specified by four main factors, F_1, F_2, F_3, and F_4. F_1 is twice as important as F_3. F_2 is one-third as important as F_1, F_3, and F_4 combined. F_4 is half as important as F_3. What relative importance ratings should be applied to these four factors?

    (A) $F_1 = 2/7$, $F_2 = 3/7$, $F_3 = 1/7$, and $F_4 = 1/14$

    (B) $F_1 = 4/9$, $F_2 = 21/27$, $F_3 = 4/18$, and $F_4 = 1/9$

    (C) $F_1 = 6/21$, $F_2 = 21/63$, $F_3 = 3/21$, and $F_4 = 3/42$

    (D) $F_1 = 3/7$, $F_2 = 1/4$, $F_3 = 3/14$, and $F_4 = 3/28$

Solution

Write the relationships between the importance factors as a set of simultaneous equations in four unknowns.

$$F_1 = 2F_3$$
$$F_2 = \frac{F_1 + F_3 + F_4}{3}$$
$$F_4 = \frac{F_3}{2}$$

Normalize the results by adding the fourth equation.

$$F_1 + F_2 + F_3 + F_4 = 1.0$$

This set of four equations in four unknowns is solved by substitution. It has the solution of $F_1 = 3/7$, $F_2 = 1/4$, $F_3 = 3/14$, and $F_4 = 3/28$.

The answer is (D).

3. REACTOR SPECIFICATIONS

There are several broad categories of *reactors*, many of which are specific to particular processes.[1] Based on mode of operation, three main reactor types are prevalent for reactions confined to one phase: (1) *batch reactors*, (2) *continuous stirred tank reactors* (CSTRs), and (3) *plug flow reactors* (PFRs). Other types of reactors include semi-batch reactors, catalytic reactors, and recycle reactors.

Properly specifying batch reactors, CSTRs, and plug flow reactors requires knowledge of the following.

- comparative performance (against the other reactor types)

- reactor volume for a specific product conversion and production rate

- reaction temperature and pressure, feed temperature and pressure, and feed flow rates

- agitation method and rate (for batch reactors and CSTRs)

- heat transfer rate to heat or cool the reactor contents to the desired temperature[2]

- jacketing, possibly with multiple cooling/heating sections (for plug flow reactor)

- insulation

- number and placement locations of thermocouples throughout the reactor

- pressure transducers, if reaction pressure is monitored

---

[1]See Chap. 43 for more detailed information about the reaction kinetics of specific reactor types.

[2]For a batch reactor or CSTR, this is provided by a jacket, heat transfer coils, an overhead condenser, or (if operated adiabatically) feed flow temperature. Energy balances are used to specify heating and cooling requirements.

- effluent product measurements (temperature, pressure, flow rate, composition, etc.)

- safety relief devices and venting to account for possible runaway reactions, plugging, and loss of heating/ cooling

- safety barricades and fire protection

- filling, draining, and cleaning ports (e.g., flat nozzle handholes and manholes)

- provisions for health and safety of workers, during both operation and maintenance

- fittings—material and ratings

- electrical equipment classification

- controller equipment (e.g., digital controllers or a distributed control system and/or a PLC) to control temperatures, pressures, and feed flows

- controller strategies and logic ladder; state trees

- upstream and downstream separation operations to refine the product and return recycled reactants to the reactor

For reactor selection, the primary criterion is the reactor size or volume necessary to achieve a desired conversion and production rate. (This will vary depending on the type of reactor.) Depending on limitations such as heat transfer surface or product distribution, multiple ideal reactors in series or parallel may be required. A number of cases for different reactor types can be calculated, and the least costly reactor may be considered a candidate for selection.

It may also be desirable to use a separator with a product recycle stream and a recycle stream of unreacted chemicals back to the reactor to improve overall reactant conversion. This would minimize waste from unreacted chemicals and lower waste treatment load. It could also allow for a smaller reactor and a lower single pass conversion. However, additional equipment associated with the separator and recycle system would require more capital. As a result, the trade-off between size of recycle equipment and that of reactor investment would need to be calculated to determine the best economic benefit.

The number and quantity of products are also considered. If only a few high-volume products are required, a CSTR or PFR would be more cost effective than a batch reactor. However, if small quantities of many products will be manufactured in a given time, a batch reactor may be the most cost effective, since it is difficult to transition a high-volume continuous reactor between products without significant transition product losses.

If significant heat transfer to a reactor is required, a PFR may be desirable due to its high surface-to-volume ratio compared to other reactor types. Additional heat transfer equipment may also be required. Heat transfer affects process safety in that a highly exothermic reaction with insufficient heat removal could cause reactor runaway and destruction of plant equipment.

Example

Several types of reactors are being considered for a reaction between two liquid components that takes place at high pressure and high temperature. Large quantities of reactants are involved. What type of reactor is most likely to be able to accommodate this reaction?

(A) large-capacity batch reactor

(B) medium-capacity semi-batch reactor

(C) plug flow reactor

(D) continuous stirred reactor

Solution

Relatively small-diameter vessels are capable of withstanding high pressures with substantial factors of safety. Similarly, insulating to maintain high reaction temperature is simplified by having small diameters. Large flow-throughs are facilitated by having long reaction paths. Altogether, these characteristics are most consistent with plug flow reactors.

The answer is (C).

4. HEAT EXCHANGER SPECIFICATIONS

Heat exchangers are widely utilized in chemical processing operations. A shell and tube heat exchanger with X shell passes and Y tube passes is designated as an X-Y *heat exchanger*. Flow through heat exchangers can be concurrent flow (also known as parallel flow), countercurrent flow, spiral/mixed flow, distributed flow, and vapor/spiral flow.

The Tubular Exchanger Manufacturers Association (TEMA) sets standards for design, fabrication, and material selection. Most manufacturers follow these *TEMA standards.* Heat exchanger types can be described by a three-character TEMA designation. For example, a one-two TEMA E shell and tube heat exchanger would be a shell and tube heat exchanger with one shell pass and two tube passes. Table 49.1 defines some of the conventions that are used in naming tubular heat exchangers. For example, TEMA type AEW refers to a one-pass shell exchanger with a removable head and a floating tubesheet.

Fixed tubesheet and U-tube bundles are the two most common commercial heat exchanger designs. One arrangement is shown in Fig. 49.1. Fixed tubesheet heat exchangers (e.g., TEMA types BEM, AEM, and NEN) use straight tubes and offer the lowest cost heat transfer surface. A series of straight tubes is sealed between flat, perforated metal *tubesheets.* The straight tubes are replaceable and easily cleaned on the inside. Since the tube bundle cannot be removed, the shellside of the tubes can only be cleaned chemically. Therefore, the shellside fluid must be nonfouling. There are no gaskets on the shellside, and the two fluids cannot accidentally mix. The no-gasket, closed-shell design is applicable to

Table 49.1 *TEMA Heat Exchanger Designations*

<u>front-end head[a] *type*</u>

A   channel and removable head
B   bonnet (integral, removable head)
C   channel (integral with tubesheet; removable plate cover)
D   special, high-pressure closure[b]

<u>shell type</u>

E   one-pass shell
F   two-pass shell with longitudinal baffle[b]
F   split flow, one-pass tube[b]
G   split flow, two-pass tube[b]
H   double split flow[b]
J   divided flow, one-pass tube[b]
K   kettle type reboiler[b]

<u>rear-end head type</u>

L   fixed tubesheet (like "A" head)
M   fixed tubesheet (like "B" head)
N   fixed tubesheet (like "C" head)
P   outside packed floating head
S   floating head with backing device (including clamp ring)
T   pull-through floating head
U   U-tube bundle
W   packed floating tubesheet with lantern ring
X   crossflow heat exchanger

[a]The term *head* is synonymous with *cover*.
[b]Specialty exchangers such as reboilers, steam heaters, vapor condensers, and feedwater heaters.

high vacuum and high pressure work as well as to potentially toxic fluids.

Removable-bundle heat exchangers differ in the types of removable heads. *Floating-head exchangers* (TEMA types AEW and BEW) have straight-through tubes with one tubesheet that is fixed to the shell and another that is fixed only to the tubes and "floats" within the shell. Floating-head exchangers are able to compensate for thermal expansion and contraction. Since they are separated by only a gasket, both fluids must be nonvolatile and nontoxic, and operation must be below approximately 300°F (150°C) and 300 psig (2.1 MPa).

With *outside-packed, floating-head exchangers* (TEMA types AEP and BEP), only the shellside fluid is exposed to the packing (i.e., the gasket). Corrosive gases, vapors, and liquids can be circulated in the tubes.

Internal clampring, floating-head exchangers (TEMA types AES and BES) are useful for applications with high-fouling fluids that require frequent inspection and cleaning or with high temperature differentials between the two fluids. Multipass arrangements are possible.

Pull-through floating-head exchangers (TEMA types AET and BET) have a floating head that is bolted directly to the floating tubesheet. The bundle can be removed without removing the shell or floating head covers. The design accommodates fewer tubes for a given shell diameter and offers less surface area than other removable-bundle exchangers. Although expensive, this design permits frequent cleaning.

U-tube heat exchangers (TEMA types AEU and BEU) have a bundle of tubes, each bent in a series of concentrically tighter U-shapes, as shown in Fig. 49.2. They are lower in initial cost since there is only one tubesheet. The tube bundle can be removed for inspection and cleaning. The individual tubes automatically compensate for thermal expansion and contraction. These exchangers are ideal for intermittent service or where thermal shock is expected. However, the insides of the tubes cannot be mechanically cleaned along their entire length, and the tubes cannot be replaced. The design cannot be single-pass on the tube side, and true countercurrent flow is not possible.

Plate-type heat exchangers (often called *panel coil exchangers*, *welded-plate exchangers*, or simply *plate exchangers*) are used as immersion heaters in tanks for the heating or cooling of solutions and as evaporator surfaces where liquids cascade down their sides. They are constructed from two sheets welded and embossed or expanded to form a series of passes through which one of the fluids flows. Figure 49.3 depicts single and double embossed configurations.

Figure 49.1 *Fixed Tubesheet Heat Exchanger (one shell pass; two tube passes)*

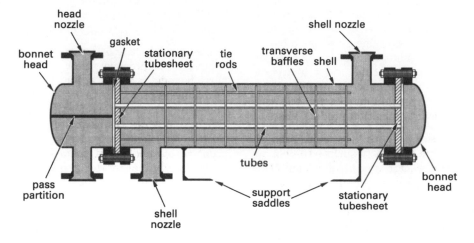

Figure 49.2 *U-Tube Heat Exchanger (one shell pass; two tube passes)*

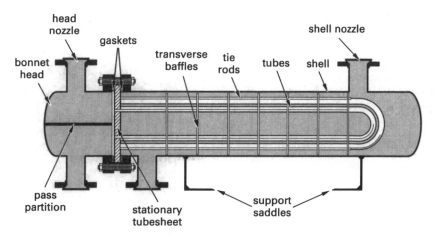

Plate-and-frame heat exchangers combine several thin-gauge plates with embossed flow paths that are separated by an elastomeric or asbestos fiber gasket. Several two-plate units are clamped together into a compact unit that is suitable for large temperature crosses or small temperature approaches.

In addition to a high heat-transfer rate due to thin plate material, other advantages of plate-type and plate-and-frame heat exchangers include compactness, accessibility to both sides of each plate for maintenance and cleaning, and flexibility of design (since the number of plates in the frame can be varied).

For a given heat transfer coefficient, plate heat exchangers have a lower pressure drop. However, the narrow passageways produce high pressure drops in high-volume, low-pressure gas applications. These exchangers are limited to approximately 300 psig (2.1 MPa) and compatible gasket materials.

Air-cooled exchangers include a motor and fan assembly that forces air over a series of (typically coiled) tubes.

Figure 49.3 *Cross Section of Plate-Type Heat Exchanger*

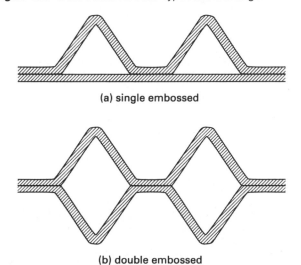

(a) single embossed

(b) double embossed

To increase the heat transfer, the tubes are usually finned (in *fin-tube exchangers*) and use *heavy-duty coil*.

Example

What organization is primarily active in writing and maintaining specifications for cross-flow and plate heat exchangers used with gases and liquids?

(A) TEMA (Tubular Exchanger Manufacturers Association)

(B) ASME (American Society of Mechanical Engineers)

(C) ASHRAE (American Society of Heating, Refrigeration, and Air Conditioning Engineers)

(D) AHRI (Air Conditioning, Heating, and Refrigeration Institute)

Solution

The Air Conditioning, Heating, and Refrigeration Institute (AHRI) publishes numerous standards and guidelines covering cross-flow and plate heat exchangers.

The answer is (D).

5. MOTOR SPECIFICATIONS

Rotating Machines

Rotating machines are categorized as AC and DC machines. Both categories include machines that use electrical power (i.e., motors) and those that generate electrical power (alternators and generators).[3] Machines can be constructed in either single-phase or poly-phase configurations, although single-phase machines may be inferior in terms of economics and efficiency.

Large AC motors are almost always three-phase. However, since the phases are balanced, it is necessary to analyze only one phase of the motor. Torque and power are divided evenly among the three phases.

---

[3]An *alternator* produces AC potential. A *generator* produces DC potential.

Torque and Power

For rotating machines, torque and power are basic operational parameters. It takes mechanical power to turn an alternator or generator. A motor converts electrical power into mechanical power. In SI units, power is given in watts (W) and kilowatts (kW). One horsepower (hp) is equivalent to 0.7457 kW. The relationship between torque and power is

$$T_{\text{ft-lbf}} = \frac{5252 P_{\text{horsepower}}}{n_{\text{rpm}}}$$

$$T_{\text{N·m}} = \frac{9549 P_{\text{kW}}}{n_{\text{rpm}}}$$

There are many important torque parameters for motors. The *starting torque* (also known as *static torque*, *break-away torque*, and *locked-rotor torque*) is the turning effort exerted by the motor when starting from rest. *Pull-up torque* (*acceleration torque*) is the minimum torque developed during the period of acceleration from rest to full speed. *Pull-in torque* (as developed in synchronous motors) is the maximum torque that brings the motor back to synchronous speed. *Nominal pull-in torque* is the torque that is developed at 95% of synchronous speed.

The *full-load torque* (*steady-state torque*) occurs at the rated speed and horsepower. Full-load torque is supplied to the load on a continuous basis. The full-load torque establishes the temperature increase that the motor must be able to withstand without deterioration. The *rated torque* is developed at rated speed and rated horsepower. The maximum torque that the motor can develop at its synchronous speed is the *pull-out torque*. *Breakdown torque* is the maximum torque that the motor can develop without stalling (i.e., without coming rapidly to a complete stop).

Motors are rated according to their output power (*rated power* or *brake power*). A 5 hp motor will normally deliver a full five horsepower when running at its rated speed. While the rated power is not affected by the motor's energy conversion efficiency, η, the electrical power input to the motor is

$$P_{\text{electrical}} = \frac{P_{\text{rated}}}{\eta}$$

Table 49.2 lists standard motor sizes by rated horsepower.[4] The rated horsepower should be greater than the calculated brake power requirements. Since the rated power is the power actually produced, motors are not selected on the basis of their efficiency or electrical power input. The smaller motors listed in Table 49.2 are generally single-phase motors. The larger motors listed are three-phase motors.

[4]For economics, standard motor sizes should be specified.

Table 49.2 *Typical Standard Motor Sizes (horsepower)**

1/8	1/6	1/4	1/3	1/2	3/4
1	1 1/2	2	3	5	7 1/2
10	15	20	25	30	40
50	60	75	100	125	150
200	250				

(Multiply hp by 0.7457 to obtain kW.)
*1/8 hp and 1/6 hp motors are less common.

Nameplate Values

A *nameplate* is permanently affixed to each motor's housing or frame. This nameplate is embossed or engraved with the *rated values* of the motor (*nameplate values* or *full-load values*). Nameplate information may include some or all of the following: voltage,[5] frequency,[6] number of phases, rated power (output power), running speed, duty cycle, locked-rotor and breakdown torques, starting current, current drawn at rated load (in kVA/hp), ambient temperature, temperature rise at rated load, temperature rise at service factor, insulation rating, power factor, efficiency, frame size, and enclosure type.

It is important to recognize that the rated power of the motor is the actual output power (i.e., power delivered to the load), not the input power. Only the electrical power input is affected by the motor efficiency. Figure 49.4 shows an example of a nameplate for a three-phase induction motor.

Figure 49.4 *Nameplate for Three-Phase Induction Motor*

Name of Manufacturer				
ORD. No.	IN4560981324			
TYPE	HIGH EFFICIENCY	FRAME	286T	
H.P.	42	SERVICE FACTOR	1.10	3 PH
AMPS	42	VOLTS	415	Y
R.P.M.	1790	HERTZ	60	4 POLE
DUTY	CONT	DATE	01/15/2016	
CLASS INSUL	F	NEMA DESIGN	B	NEMA NOM. EFF.	95
Address of Manufacturer				

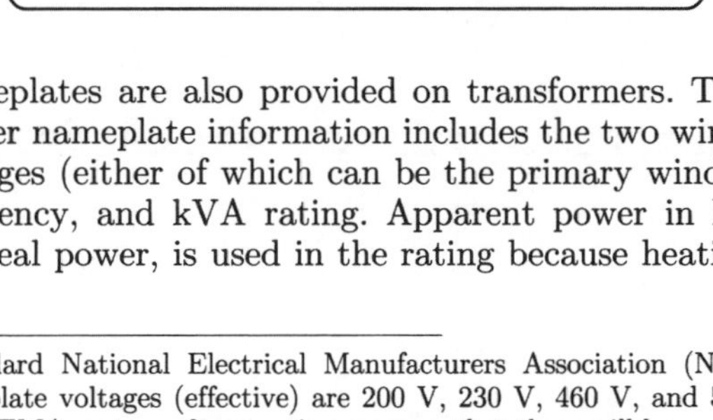

Nameplates are also provided on transformers. Transformer nameplate information includes the two winding voltages (either of which can be the primary winding), frequency, and kVA rating. Apparent power in kVA, not real power, is used in the rating because heating is

[5]Standard National Electrical Manufacturers Association (NEMA) nameplate voltages (effective) are 200 V, 230 V, 460 V, and 575 V. The NEMA motor voltage rating assumes that there will be a voltage drop from the network to the motor terminals. A 200 V motor is appropriate for a 208 V network, a 230 V motor on a 240 V network, and so on. NEMA motors are capable of operating in a range of only ±10% of their rated voltages. Thus, 230 V rated motors should not be used on 208 V systems.
[6]While some 60 Hz motors (notably those intended for 230 V operation) can be used at 50 Hz, most others (e.g., those intended for 200 V operation) are generally not suitable for use at 50 Hz.

proportional to the square of the supply current. As with motors, continuous operation at the rated values will not result in excessive heat buildup.

Duty Cycle

Motors are categorized according to their *duty cycle*: *continuous-duty* (24 hr/day); *intermittent* or *short-time duty* (15 min to 30 min); and *special duty* (application specific). Duty cycle is the amount of time a motor can be operated out of every hour without overheating the winding insulation. The idle time is required to allow the motor to cool.

Induction Motors

The three-phase induction motor is by far the most frequently used motor in industry. In an induction motor, the magnetic field rotates at the synchronous speed. The *synchronous speed* can be calculated from the number of poles and frequency. The frequency, f, is either 60 Hz (in the United States) or 50 Hz (in Europe and other locations). The number of poles, p, must be even.[7] The most common motors have 2, 4, or 6 poles.

$$n_{\text{synchronous}} = \frac{120f}{p} \quad [\text{rpm}]$$

Due to rotor circuit resistance and other factors, rotors (and the motor shafts) in induction motors run slightly slower than their synchronous speeds. The percentage difference is known as the *slip*, s. Slip is seldom greater than 10%, and it is usually much less than that. 4% is a typical value.

$$s = \frac{n_{\text{synchronous}} - n_{\text{actual}}}{n_{\text{synchronous}}}$$

The rotor's actual speed is[8]

$$n_{\text{actual}} = (1 - s)n_{\text{synchronous}}$$

Induction motors are usually specified in terms of the *kVA ratings*. The kVA rating is not the same as the motor power in kilowatts, although one can be calculated from the other if the motor's power factor is

---

[7]There are various forms of the synchronous speed equation. As written, the speed is given in rpm, and the number of poles, p, is twice the number of *pole pairs*. When the synchronous speed is specified as f/p, it is understood that the speed is in revolutions per second (rps) and p is the number of pole *pairs*.

[8]Some motors (i.e., *integral gear motors*) are manufactured with integral speed reducers. Common standard output speeds are 37 rpm, 45 rpm, 56 rpm, 68 rpm, 84 rpm, 100 rpm, 125 rpm, 155 rpm, 180 rpm, 230 rpm, 280 rpm, 350 rpm, 420 rpm, 520 rpm, and 640 rpm. While the integral gear motor is more compact, lower in initial cost, and easier to install than a separate motor with belt drive, coupling, and guard, the separate motor and reducer combination may nevertheless be preferred for its flexibility, especially in replacing and maintaining the motor.

known. The power factor generally varies from 0.8 to 0.9 depending on the motor size.

$$\text{kVA rating} = \frac{P_{\text{kW}}}{\text{pf}}$$

$$P_{\text{kW}} = 0.7457 P_{\text{mechanical,hp}}$$

Induction motors can differ in the manner in which their rotors are constructed. A *wound rotor* is similar to an armature winding in a dynamo. Wound rotors have high-torque and soft-starting capabilities. There are no wire windings at all in a *squirrel-cage* rotor. Most motors use squirrel-cage rotors. Typical torque-speed characteristics of a design B induction motor are shown in Fig. 49.5.

Figure 49.5 *Induction Motor Torque-Speed Characteristic (typical of design B frames)*

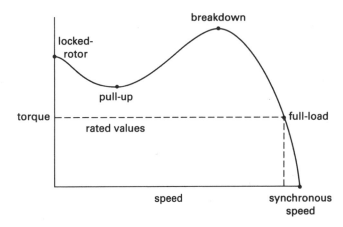

Synchronous Motors

Synchronous motors are essentially dynamo alternators operating in reverse. The stator field frequency is fixed, so regardless of load, the motor runs only at a single speed—the synchronous speed. Stalling occurs when the motor's counter-torque is exceeded. For some equipment that must be driven at constant speed, such as large air or gas compressors, the additional complexity of synchronous motors is justified.

Power factor can be adjusted manually by varying the field current. With *normal excitation* field current, the power factor is 1.0. With *over-excitation*, the power factor is leading, and the field current is greater than normal. With *under-excitation*, the power factor is lagging, and the field current is less than normal.

Since a synchronous motor can be adjusted to draw leading current, it can be used for power factor correction. A synchronous motor used purely for power factor correction is referred to as a *synchronous capacitor* or *synchronous condenser*. A power factor of 80% is often specified or used with synchronous capacitors.

Process Design/ Economics

Example

A screw conveyor carrying plastic pellets to a large plastic injection machine is designed to turn at 100 rpm. A three-phase, 60 Hz, capacitor start, synchronous motor is specified, as is a 12:1 reduction gear set. How many poles (per phase) should the motor have?

(A) 2

(B) 4

(C) 6

(D) 8

Solution

With an output speed from the 12:1 gear set of 100 rpm, the motor must turn at a nominal speed of $(12)(100 \text{ rpm}) = 1200 \text{ rpm}$. This is the synchronous speed.

Solve the synchronous speed equation for the number of poles.

$$p = \frac{120f}{n_{\text{synchronous}}} = \frac{(120)(60 \text{ Hz})}{1200 \frac{\text{rev}}{\text{min}}} = 6$$

The answer is (C).

Example

A paste exhibits dilatant viscosity characteristics. A mixer for the paste requires a direct-drive motor with the following characteristics.

- 720 rpm
- 200 hp
- controllable torque
- variable speed

What type of motor should be specified?

(A) non-exited (self-excited) synchronous

(B) DC-excited synchronous

(C) squirrel cage induction

(D) wound rotor induction

Solution

Commercial induction and synchronous motors both can meet the horsepower and speed requirements. Non-exited synchronous motors are suitable only for fractional horsepower (i.e., small) applications. Excitation is required for high horsepower (i.e., larger than 1 hp) synchronous machines. However, all synchronous motors run at constant speed and are not suitable for variable speed applications. Squirrel cage induction rotors do not have switchable windings, and (in the absence of electronic waveform and frequency modifications) are not candidates for variable speed applications. Wound rotors, although more complex and costly, are suitable for controllable torque and variable speed, as individual windings can be switched into and out of the run circuit.

The answer is (D).

6. CHEMICAL PROCESS REPRESENTATIONS

A commercial chemical process can be diagrammed in several ways. The diagram can vary from a simple block diagram to a full piping diagram with instrumentation details.

Process flow diagrams (PFDs) are block diagrams that are used to present basic processing concepts without details or identifying the specific equipment involved. Each block represents a process (i.e., what must be done), rather than how to achieve it. Block diagrams are useful for preliminary process development, as the various steps are identified before process details are known. Figure 49.6 shows a representative block diagram for a lime kiln operation.

Figure 49.6 *Representative Block Diagram (lime kiln operation)*

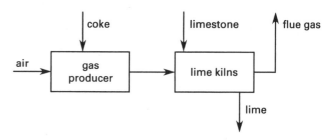

A *process flow sheet* shows the process equipment. The compositions, temperatures, and pressures of each stream may be shown at key locations throughout the process. The heat and the material balances may be included, also. Auxiliary services such as steam, water, air, and circulating oil are usually included. Figure 49.7 shows common and typical symbols used in preparing process flow sheets.[9]

Example

In the diagram shown, A and B are granulated solids conveyed in a water medium. They are transferred to a blender that creates a homogeneous mixture of A and B needed for creating a desired polymer blend. From the blender, a screen is used to separate the largest solids from the water medium and drop them into a screw extruder. The product, the blend of A and B, is pelletized for packaging and sale.

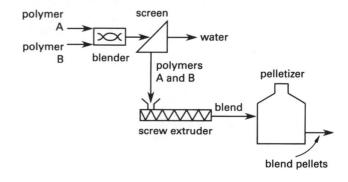

---

[9]There are many variations of these symbols in use. Companies often have established their own unique symbol sets that may differ from these.

What type of process representation is shown?

(A) block diagram

(B) process flow sheet

(C) piping and instrumentation diagram

(D) process tree

Solution

The diagram is more specific than a block diagram. This diagram defines the process and identifies the equipment needed to implement the process. No instrumentation or control lines are shown; and, there are no decision points (common in tree diagrams). This is a process flow sheet.

The answer is (B).

Figure 49.7 *Flow Sheet Symbols*

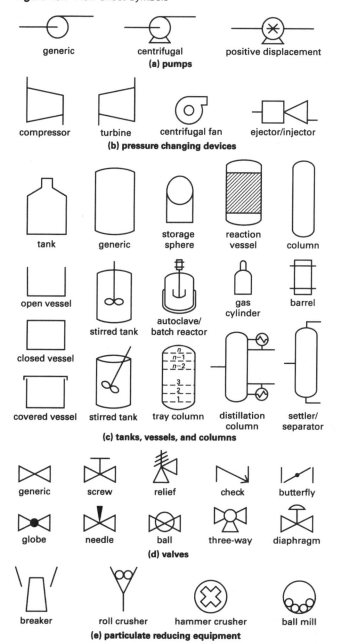

7. PIPING AND INSTRUMENTATION DIAGRAMS (P&IDS)

Piping and instrumentation diagrams (P&IDs), also called process and instrumentation diagrams, are schematic representations of processing operation piping systems including installed equipment and instrumentation. They can range in scope from illustrating a simple oil piping system that provides continuous lubrication to motor bearings with temperature and vibration monitoring and control, to illustrating an oil refinery piping system with complex multi-stage processing, multiple parameter monitoring, and subsystem control loops. Figure 49.8 shows a representative P&ID for a vapor-liquid separator.

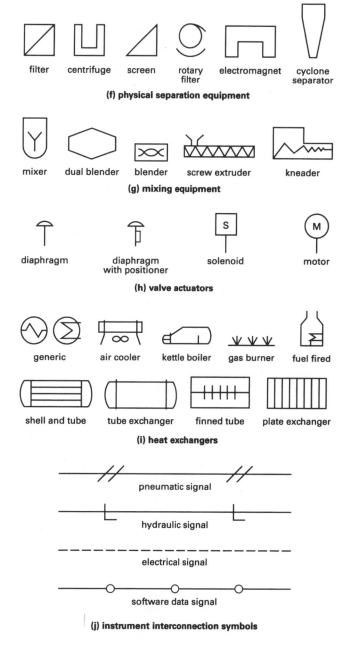

Process Design/ Economics

Figure 49.8 *Representative P&ID for a Vapor-Liquid Separator*

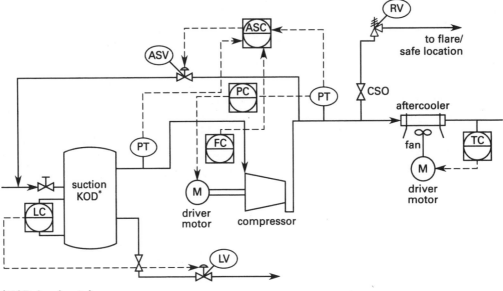

*KOD: knock-out drum

A P&ID shows a piping system layout in a plan, elevation, and/or isometric view, identifying the equipment, piping, instrumentation, and flow paths of the fluids and gases within the system. P&IDs show the interconnection relationships between equipment such as compressors, blowers, pumps, valves, dryers, filters, tanks, heat exchangers, pipes, and instrumentation. P&IDs show instrument locations with respect to equipment and interconnection to the control system. P&IDs show the flow path and conditions of the primary medium being processed, as well as the auxiliary equipment and mediums required for the processing.

P&IDs use symbols and nomenclature with letters and numbers, called *tags*, to identify the type, purpose, and location of equipment and instrumentation in the system. ANSI/ISA-S5.1 provides industry accepted instrumentation symbols and identification. The shape of the tag symbol indicates the type of instrumentation (e.g., a discrete instrument or a PLC) as well as the category of location (e.g., accessible to the operator or outside). The interior of the tag is divided into two lines. (These are illustrated in Fig. 49.9.) The first letter on the top row of a tag identifies what is being measured or monitored. The next letters identify the function of the measurement and communication mode of the measurement, and provide additional information. Table 49.3 lists some tag letters and their definitions. For example, "PT" and "PC" represent a pressure transmitter and a pressure controller, respectively. The number on the bottom line identifies the instrument and equipment item, and similar numbers are used in equipment groups.

Figure 49.9 *General Instrument and Function Symbols*

	primary location normally accessible to operator[c]	field mounted	auxiliary location normally accessible to operator[c]
discrete instruments	IP1[a,b] ⊖	◯	⊖
shared display, shared controls	⊡	⊡	⊡
computer function	⬡	⬡	⬡
programmable logic control	⬓	⬓	⬓

[a]Symbol size may vary according to the user's needs and the type of document. A suggested square and circle size for large diagrams is shown. Consistency is recommended.

[b]Abbreviations of the user's choice such as IP1 (instrument panel no. 1), IC2 (instrument console no. 2), and CC3 (computer console no. 3), etc., may be used when it is necessary to specify instrument or function location.

[c]Normally inaccessible or behind-the-panel devices or functions may be depicted by using the symbol but with dashed horizontal bars, such as

Derived from ANSI/ISA 55.1, Sec. 6.3.

Table 49.3 Tag Letter Format and Nomenclature

ABC 123	first letter A	next letters B C
	measured value	measurement function, communication mode, other
A	analysis	alarm
C		control, controller
D		differential
F	flow, flow rate	
H		hand, high
I	current	indicate, indicator, indicating
L	level	low
M	motor	
P	pressure	
R		relief, recorder
S	surge	safety, switch, solenoid
T	temperature	transmit, transmitter
V	vibration	valve
Z	position	actuator

Derived from ANSI/ISA 55.1, Table 1, "Identification of Letters."

Example

An ethanol-water mixture is pumped into a distillation column through a piping system as shown in the accompanying PI&D.

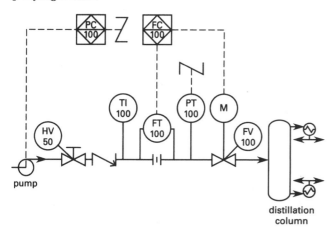

Most likely, how is the flow rate controlled?

(A) An electrical signal from the differential pressure transducer is relayed to a PLC, and an output electrical signal is utilized to control the diaphragm control valve.

(B) Electrical signals from the flow and pressure transmitters are relayed to a PLC, and the electrical signals are utilized to control the flow valve.

(C) An electrical signal from the flow transmitter is relayed to a PLC, and an output electrical signal is utilized to control the flow valve.

(D) An electrical signal from the pressure transmitter is relayed to a PLC, and an output electrical signal is utilized to control the speed of the pump.

Solution

The PI&D depicts a system where fluid being pumped under the control of a pressure controller (PC). The flow first encounters a hand-operating valve (HV) and then a check valve (unlabeled) followed by an orifice meter. The flow transmitter (FT) detects flow through (i.e., pressure drop across) the orifice (unlabeled) and signals the motor (M) to change the state of the flow valve (FV).

The answer is (C).

Example

What does the symbol shown represent?

I. a discrete instrument

II. a programmable logic control

III. primary location normally accessible to the operator

IV. auxiliary location normally accessible to the operator

(A) I and IV

(B) I and III

(C) II and III

(D) II and IV

Solution

From Fig. 49.9, the symbol represents a programmable logic control normally accessible to the operator.

The answer is (C).

8. PLANT SCALE-UP

Many commercial chemical processing plants begin as *pilot studies*. (Pilot studies may also be referred to as *lab bench*, *test bed*, *model*, or *simulation* studies.) In most cases, the pilot study will include a smaller, dimensionally scaled-down version of the equipment and/or process. The pilot study may be referred to as the *model*, and the first implementations of the larger-size versions may be referred to as the *prototypes*.

Scale-up from the model to the prototype is best done in incremental steps, so that the process can be validated with each new scale. When this is not possible, *similitude* concepts can be cautiously applied. The model and full-scale units should possess geometric similarity, kinematic similarity, and dynamic similarity. When a chemical reaction is involved, chemical similitude must also be maintained. Dimensionless numbers (see Table 49.4) various rules of thumb, and experience are used to scale the forces most critical to the process. Dimensionless numbers should be held constant during scale-up.

Geometric similarity requires all corresponding dimensions on the full-scale unit to have the same ratio (i.e., *scale* or *scale factor*, $R = L_f / L_p$) as the successful model. These dimensions include such parameters as vessel and impeller diameters, baffle widths and quantity, and liquid levels.

Kinematic similarity requires corresponding points in each system to have the same velocity ratios (so that fluid streamlines are similar). The characteristic length for agitators is the impeller diameter. When scaling up agitation systems, kinematic similarity may be estimated by keeping a constant tip speed between the pilot model and full-scale system. Dimensionless numbers such as the Reynolds number are also applicable to mixing and agitation by substituting impeller tip speed for the fluid velocity.

Dynamic similarity adds the requirement of equal force ratios at corresponding points in each system. Properties such as gravitation, surface tension, viscosity, pressure, and inertia are included in balancing force ratios. Additional dimensionless numbers can be derived from momentum, mass, and heat conservation equations. The *power number*, Po; *blend number*, N_b; and *pumping number*, N_p, are specifically related to agitated systems.

Chemical similarity relates functional qualities of the chemical reaction to the process. The *Damköhler numbers* are dimensionless numbers for ensuring chemical similitude. These relate the reaction and the mass transfer rates. The mass transfer processes involved may be diffusion or convection, or it may simply be the mass flow rate.

In many cases, true dynamic and kinematic similarity cannot be achieved at the same time between model and full-scale systems. Empirical correlations between dimensionless numbers, derived from lab- or pilot-plant-scale data, are useful in scaling up the desired process parameters.

Example

A pilot study model pipe has an inside diameter one-tenth of the full-scale pipe's diameter. The same fluid is used in the pilot plant and full-scale pipes, and both pipes are smooth. Compared to the velocity in the pilot study, what should be the velocity in the full-scale pipe?

(A) 1/100

(B) 1/10

(C) 10

(D) 100

Table 49.4 Dimensionless Numbers

name	symbol	formula	interpretation
blend	N_b	$n\theta$	related to uniform blending
Damköhler I	Dm_I	$\dfrac{rL}{vC}$	$\dfrac{\text{chemical reaction generation}}{\text{convective mass transfer}}$
Damköhler II	Dm_{II}	$\dfrac{rL^2}{D_m C}$	$\dfrac{\text{chemical reaction generation}}{\text{molecular diffusion mass transfer}}$
Euler	Eu	$\dfrac{\Delta p}{\rho v^2}$	$\dfrac{\text{pressure force}}{\text{convective (inertial) force}}$
Froude	Fr	$\dfrac{v^2}{gD}$	$\dfrac{\text{convective (inertial) force}}{\text{gravitational force}}$
Nusselt	Nu	$\dfrac{hD}{k}$	$\dfrac{\text{total heat transfer}}{\text{molecular conductive heat transfer}}$
Peclet	Pe	$\dfrac{\rho v D c_p}{k}$	(Re)(Pr) $\dfrac{\text{convective heat transfer}}{\text{molecular conductive heat transfer}}$
power	Po	$\dfrac{P}{\rho n^3 D^5}$	$\dfrac{\text{impeller force}}{\text{inertial force}}$
Prandtl	Pr	$\dfrac{\mu c_p}{k} = \dfrac{\nu}{\alpha}$	$\dfrac{\text{momentum diffusivity}}{\text{thermal diffusivity}}$
pumping	N_p	$\dfrac{Q}{nD^3}$	$\dfrac{\text{pumping rate}}{\text{impeller speed}}$
Reynolds	Re	$\dfrac{\rho v D}{\mu} = \dfrac{vD}{\nu}$	$\dfrac{\text{convective (inertial) force}}{\text{molecular (viscous) force}}$
Weber	We	$\dfrac{\rho v^2 D}{\sigma}$	$\dfrac{\text{convective (inertial) force}}{\text{surface tension force}}$

Solution

Given that the pilot-plant-scale pipe is one-tenth the diameter of the full-scale pipe, the scale, R, is

$$D_p = \left(\frac{1}{10}\right) D_f$$

$$R = \frac{D_f}{D_p} = 10$$

Kinematic similarity implies that velocity ratios are the same, necessitating equal Reynolds numbers. Dynamic similarity would be obtained by having equal force ratios, which would necessitate Euler numbers being equal for each case. Since pressure drops are not part of this example, dynamic similarity is not needed.

$$\text{Re}_p = \text{Re}_f$$

$$\left(\frac{Dv\rho}{\mu}\right)_p = \left(\frac{Dv\rho}{\mu}\right)_f$$

Using the same fluid in both the pilot-plant-scale pipe and the full-scale pipe, the fluid properties in the Reynolds numbers will cancel. The full-size fluid velocity must be one-tenth of the velocity in the pilot plant.

$$v_f = \frac{v_p}{R} = \frac{v_p}{10}$$

The answer is (B).

9. COST ESTIMATION

There are several methods and tools available for developing *cost estimates* of plant and process equipment, and there are numerous types and classification systems for cost estimates. Cost estimation requires thorough and in-depth understanding, identification, and evaluation of technical and programmatic requirements, specifications, and scope of work, along with interdependencies to schedule, budget, allocate resources, and manage risks. It is usually difficult to account for all of the factors affecting costs, so simplistic methods are used to reduce cost dependencies to a single variable, such as a cost index of capacity.

Equation 49.1: Scaling of Equipment Cost with Capacity

$$\text{cost of unit A} = \text{cost of unit B} \left(\frac{\text{capacity of unit A}}{\text{capacity of unit B}}\right)^n \quad 49.1$$

Values

Table 49.5 *Typical Exponents (n) for Equipment Cost vs. Capacity*

equipment	size range	exponent
dryer, drum, single vacuum	10 ft^2 to 10^2 ft^2	0.76
dryer, drum, single atmospheric	10 ft^2 to 10^2 ft^2	0.40
fan, centrifugal	$10^3 \text{ ft}^3/\text{min}$ to $10^4 \text{ ft}^3/\text{min}$	0.44
fan, centrifugal	$2 \times 10^4 \text{ ft}^3/\text{min}$ to $7 \times 10^4 \text{ ft}^3/\text{min}$	1.17
heat exchanger, shell and tube, floating head, c.s.	100 ft^2 to 400 ft^2	0.60
heat exchanger, shell and tube, fixed sheet, c.s.	100 ft^2 to 400 ft^2	0.44
motor, squirrel cage, induction, 440 V, explosion proof	5 hp to 20 hp	0.69
motor, squirrel cage, induction, 440 V, explosion proof	20 hp to 200 hp	0.99
tray, bubble cup, c.s.	3 ft to 10 ft diameter	1.20
tray, sieve, c.s.	3 ft to 10 ft diameter	0.86

Description

Equation 49.1 shows how equipment cost can be assumed to vary with capacity.[10] n is the *cost exponent* or *scale exponent*. Typical chemical plant processing equipment and scale factors are shown in Table 49.5.[11] All other important considerations affecting cost estimates (e.g., whether or not the two units are from the same manufacturer or have the same construction details) are disregarded. Often equations such as Eq. 49.1 are applicable only in one direction (i.e., scaling downward or upward).

Example

The purchased cost of a 10 MMBtu/hr thermal oxidizer was $1,125,000.[12] There is now a need for a 6 MMBtu/hr thermal oxidizer. If a scaling exponent of 0.8 is applicable, most nearly, how much will a 6 MMBtu/hr oxidizer cost?

(A) $550,000

(B) $680,000

(C) $750,000

(D) $830,000

---

[10]Although the NCEES *FE Reference Handbook* (*NCESS Handbook*) refers to "capacity," in actuality, size, area, flow rate, and power are the independent variables in the title of Table 49.5.

[11]Although the SI system is used on the FE exam, the items listed in the *NCEES Handbook* and Table 49.5 are indexed to equipment with U.S. dimensions and capacities.

[12]MMBtu is interpreted as "a thousand thousands," or a million. This convention is common in applications involving combustion and other heating processes.

Solution

From Eq. 49.1,

$$\text{cost of unit } A = \text{cost of unit } B \left(\frac{\text{capacity of unit } A}{\text{capacity of unit } B} \right)^n$$

$$= (\$1,125,000) \left(\frac{6 \text{ MMBtu/h}}{10 \text{ MMBtu/h}} \right)^{0.8}$$

$$= \$747,607 \quad (\$750,000)$$

The answer is (C).

10. RULES OF THUMB FOR CHEMICAL PLANT COSTS

For the purpose of early planning, the following rules of thumb can be used to estimate costs and cash required.[13]

- contractors fees (indirect expense): 5% of direct and indirect plant costs

- contingency:[14] 10% of direct and indirect plant costs

- working capital reserves:[15] 15% of the total capital investment

11. CHEMICAL PLANT COST INDEXES

Capital costs rise and fall over time due to inflation and other causes. Cost data from past years can be misleading unless these changes are taken into account. *Cost indexes* have been developed for this purpose.

A cost index is the ratio of an item's actual average cost in a particular time period (such as in a particular year) and the average cost of the same item within the time period that has been defined as the base; this ratio is multiplied by 100. For example, if the year 2000 is chosen as the base, the cost index for that year is 100; if the average cost of the same item is 7% higher in 2003, then the cost index for 2003 is 107.

There are four major indexes commonly used in the chemical industry.

Chemical Engineering Plant Cost Index (*CEPCI*). Primarily an index of process plant construction costs, the CE Index tracks four components: equipment, construction labor, buildings, and engineering and supervision. The CE Index is published in each issue of the monthly magazine *Chemical Engineering*.

Intratec Chemical (IC) Plant Construction Index. The IC Index tracks process plant construction and is

published by Intratec, a consulting company. It has a shorter lag between its index date and its release date than the other indexes.

Marshall and Swift (M&S) Engineering Plant Cost Index. The M&S Index tracks process industry equipment and all industry averages. It is used primarily for calculations involving replacement equipment costs. The M&S Index is published by Marshall and Swift; until 2012 it was published in *Chemical Engineering*.

Nelson-Farrar (NF) Indexes. The NF Indexes are primarily used for petroleum and petrochemical facilities. They are published monthly in the *Oil and Gas Journal*.

Equation 49.2: Indexed Valuations

$$\text{current } \$ = (\text{cost in year } M) \left(\frac{\text{current index}}{\text{index in year } M} \right)$$

49.2

Description

The cost of a project or piece of equipment in the current year is equal to the product of its cost in an earlier year and the ratio of the cost indexes[16] of the current and earlier years.

Example

A heat exchanger was purchased in 1985 for $25,000. The CEPCI cost indexes for 1985 and the current year are 325 and 467, respectively. Most nearly, what is the estimated cost of a similar heat exchanger today?

(A) $27,000

(B) $31,000

(C) $36,000

(D) $41,000

Solution

To find the current cost of the heat exchanger, use Eq. 49.2 with year $M = 1985$.

$$\text{current } \$ = (\text{cost in year } M) \left(\frac{\text{current index}}{\text{index in year } M} \right)$$

$$= (\$25,000) \left(\frac{467}{325} \right)$$

$$= \$35,923 \quad (\$36,000)$$

The answer is (C).

---

[13]These rules of thumb are from the *NCEES Handbook* and should be useful only for taking the FE exam. Other rules of thumb should be validated for real world use.

[14]In the *NCEES Handbook*, the contingency expense is listed under the *indirect costs* heading. Since direct operating costs (on which the contingency is based) are not indirect costs, it is not clear if the contingency is intended to be applied to the plant construction or plant operating costs.

[15]In the *NCEES Handbook*, working capital reserves should not be included under the heading of *indirect costs*.

[16]In Eq. 49.2, the *NCEES Handbook* uses "current $" as a substitute for "current cost."

12. INHERENTLY SAFER DESIGN

Inherently safer technology (IST), also known as *inherently safer design* (ISD), permanently eliminates or reduces hazards to avoid or reduce the consequences of incidents. IST is a philosophy, applied to the design and operation life cycle, including manufacture, transport, storage, use, and disposal. IST is an iterative process that considers such options, including eliminating a hazard, reducing a hazard, substituting a less hazardous material, using less hazardous process conditions, and designing a process to reduce the potential for, or consequences of, human error, equipment failure, or intentional harm. Overall safe design and operation options cover a spectrum from inherent through passive, active and procedural risk management strategies. There is no clear boundary between IST and other strategies.

ISTs are relative: A technology can only be described as inherently safer when compared to a different technology, including a description of the hazard or set of hazards being considered, their location, and the potentially affected population. A technology may be inherently safer than another with respect to some hazards but inherently less safe with respect to others, and may not be safe enough to meet societal expectations.

ISTs are based on an informed decision process: Because an option may be inherently safer with regard to some hazards and inherently less safe with regard to others, decisions about the optimum strategy for managing risks from all hazards are required. The decision process must consider the entire life cycle, the full spectrum of hazards and risks, and the potential for transfer of risk from one impacted population to another. Technical and economic feasibility of options must also be considered.

Four main components of inherently safer design are, as identified by the National Academy of Sciences (NAS):

- *substitution:* replacing one material with another, less hazardous material

- *minimization:* reducing the amount of hazardous material in the process

- *moderation:* reducing hazardous process conditions, such as using lower pressures and/or temperatures

- *simplification:* designing processes to be less complicated, and therefore less prone to failure

Example

From the standpoint of inherently safer design (ISD), every chemical engineer should be familiar with which one of the following industrial disasters?

   (A) Exxon Valdez oil spill (1989)

   (B) Bhopal India spill (1984)

   (C) Three Mile Island nuclear plant incident (1979)

   (D) Pasadena, Texas chemical plant explosion (1989)

Solution

Use of ISD most likely could have avoided history's worst industrial accident, which occurred at the Union Carbide India Ltd. pesticide plant in Bhopal, India in 1984 and killed thousands of people. As documented by the Chemical Safety Board (CSB), water inadvertently entered a storage tank containing more than 80,000 pounds of methyl isocyanate, or MIC, which reacts violently with water. A subsequent runaway reaction overheated the tank and resulted in a massive toxic gas release.

The answer is (B).

13. PLANT ENERGY AND RESOURCE CONSERVATION[17]

Plant and operations design should encompass energy efficiency schemes that reduce energy consumption. Schemes include appropriate material and equipment selection, reusing and recycling processing agents, modeling and optimizing process layout and operations, and effective monitoring and control systems and algorithms. The plant and operations design should also identify and accommodate data acquisition instrumentation, hardware, and software for energy analysis and management, and maintenance operations.

Although much of the energy efficiency of a chemical plant or process is decided by the original architecture and mechanical and process system design, there are additional strategies that can later be used to increase efficiency and reduce energy consumption. These strategies can be grouped into broad categories such as building, equipment and process commissioning, mechanical systems, electricity use, and plumbing. Major impacts on energy efficiency associated with producing chemical products are available through consideration of such nontraditional methods as alternative and renewable energy technologies, transportation methods, building materials, solar panels, biofuels, and reduction of greenhouse gases.

Process commissioning is the process of inspecting, testing, starting up, and adjusting systems and then verifying and documenting that they are operating as intended and meet the design criteria of the contract documents. Commissioning is an expansion of the traditional *testing*, *adjusting*, and *balancing* (TAB) that is commonly performed on mechanical systems, but with greatly broadened scope over a longer time period. Similarly, for chemical processing equipment, the equipment should perform as designed within Standard Operating Conditions (SOCs) at turnover from construction to startup.

---

[17]Many of the suggestions in this section are mandatory requirements for LEED certification or entitle the construction to LEED credits.

The process systems that require commissioning depend on the complexity of the process and the needs of the production management. They may include some or all of the following.

- chemical process equipment, including tanks, pumps, compressors, reactors, separators, columns, piping, insulation, heat exchangers and coolers, and all other unit operations

- mechanical systems, including heating and cooling equipment, air handling equipment, distribution systems, motors, pumps, sensors and controls, dampers, and cooling-tower operation

- electrical systems, including switchgear, controls, emergency generators, fire management, and safety systems

- plumbing systems, including tanks, pumps, water heaters, compressors, and fixtures

- sprinkler systems, including standpipes, alarms, hose cabinets, and controls

- fire management and life safety systems, including alarms and detectors, air handling equipment, smoke dampers, and building communications

- vertical transportation, including elevator controls and escalators

- telecommunication and computer networks

Pinch analysis, also known as *energy integration*, is a technique for designing a process to minimize energy consumption and maximize heat recovery. This technique calculates thermodynamically attainable energy targets for a given process and identifies how to achieve them.

Energy use can be reduced by reducing the reaction time necessary to produce products. For example, oil refineries use fluid catalytic cracking units to reduce the heavy hydrocarbon fractions in oil into smaller molecules such as gasoline and other fuels. This reduces reaction times for conversion on the order of many minutes to seconds, while converting the heaviest fractions to useful products. Modern catalytic cracking units were created through the development and use of better catalysts and more efficient ways to contact the catalyst solids with the oil.

Reduction of electrical and heating energy consumed by mechanical systems should always be considered with buildings that are occupied. The HVAC systems of occupied buildings should conform to ANSI/ASHRAE/IES Standard 90.1, *Energy Standard for Buildings Except Low-Rise Residential Buildings*, or the local energy code, whichever is more stringent.

Example

Which of the following standards or documents should be relied on to design components of occupied buildings that use reduced energy?

    (A) ASHRAE Standard 90.1

    (B) International Building Code

    (C) International Residential Code

    (D) LEED Guide to Building and Construction

Solution

ANSI/ASHRAE/IES Standard 90.1, *Energy Standard for Buildings Except Low-Rise Residential Buildings* is the prevailing standard on designing buildings that use reduced energy.

The answer is (A).

14. SUSTAINABILITY

Many professional societies' codes of ethics stress the importance of incorporating sustainability into engineering design and development. *Sustainability* (also known as *sustainable development* or *sustainable design*) encompasses a wide range of concepts and strategies. However, a general definition of sustainability is any design or development that seeks to minimize negative impacts on the environment so that the present generation's resource needs do not compromise the resource needs of a future generation.

The principles of sustainable design have a wide range of applications; everything from the design of small objects for everyday use to the planning and design of cities can be developed, constructed, and operated in ways that make better use of natural resources. The objective of sustainable design is to create places, products, and services in ways that reduce the use of nonrenewable resources, minimize environmental impact, and relate people to the natural environment. Examples of sustainable design principles include using renewable energy sources, conserving water, and using *sustainable materials* (materials sourced, manufactured, and transported with sustainability in mind).

Many of these factors are of importance to chemical engineers in the design of new plants, which includes new building design traditionally carried out by architects and civil engineers, as well as producing products that are sustainable in production and in use. Factors such as recycling, alternate raw materials and energy sources, water conservation, alternate feedstocks for energy and chemicals as well as reducing environmental footprint and impact. New products such as energy technologies, food, medicine, and quality water are examples where chemical engineers can make an impact. Producing new building materials that are compatible with these objectives is also desirable.

Sustainable building design (also known as *green building*) is an increasingly important part of chemical plant and process design. Sustainable design includes a wide range of concerns, such as the environmental impact of buildings, structures, and equipment; the wise use of materials, energy conservation, use of alternative energy sources, adaptive reuse, indoor air quality, recycling, reuse, and other strategies to achieve a balance between the consumption of environmental resources and the renewal of those resources. Sustainable design considers the full life cycle of a building and the materials that comprise the building. This includes the impact of raw material extraction through its fabrication, installation, operation, maintenance, and disposal. Engineers and architects address many sustainable design issues during the design of a building, and there are several steps that interior designers also take to minimize the environmental impact of interior build-out.

Several organizations have emerged that provide industry-recognized ratings of the relative sustainability of a building or process. These organizations develop objective criteria that designers must follow in order to receive a particular type of rating. Although conforming to the criteria that these organizations establish is not mandatory by any building code, some governmental entities and large corporations may require their designers to follow the organizations' guidelines. The following are some major rating systems.

- The American Institute of Chemical Engineers (AIChE), Institute for Sustainability (IfS) is an umbrella organization encompassing a broad spectrum of activities related to sustainability. It has three sub-entities: the Sustainable Engineering Forum (SEF), the Center for Sustainable Technology Practices (CSTP), and the Youth Council on Sustainability Science and Technology (YCOSST). It also provides oversight to the AIChE Sustainability Index. The AIChE Sustainability Index is used to assess a company's sustainability performance with seven key metrics: strategic commitment to sustainability, sustainability innovation, environmental performance, safety performance, product stewardship, social responsibility, and value-chain management.

- The Leadership in Energy and Environmental Design (LEED) Green Building Rating System is a national, consensus-based building rating system designed to accelerate the development and implementation of green building practices. It was developed by the U.S. Green Building Council (USGBC), which is a national coalition of leaders from all aspects of the building industry working to promote buildings that are environmentally responsible and profitable and that provide healthy places to live and work. In addition to developing the rating system, the full LEED program offers training workshops, professional accreditation, resource support, and third-party certification of building performance. In order for a building to be certified, certain prerequisites must be achieved and enough points must be earned to meet or exceed the program's technical requirements. Points add up to a final score that relates to one of four possible certification levels: certified, silver, gold, and platinum. LEED is one of the primary building rating systems in the United States.

- Energy Star is a program of the Environmental Protection Agency (EPA) and the U.S. Department of Energy, started in 1992. As part of the Energy Star for Buildings and Manufacturing Plants program, Energy Star provides tools and resources to assist architects, business owners, and others involved in the building process to design, build, commission, and manage projects in ways that save energy.

The selection and use of materials in a building, or an entire chemical plant, represents a significant part of the total sustainability of plant design, construction, and operation. As with energy consumption and other sustainability issues, material selection must be made with consideration of the entire life cycle of the plant. However, sustainability issues must be considered along with the traditional concerns of function, cost, appearance, and performance.

Some of the criteria for evaluating the sustainability of a product or process include:

- *Embodied energy:* The material or product should require as little energy as possible for its extraction as a raw material, initial processing, and subsequent manufacture or fabrication into a finished building product. This includes the energy required for transportation of the materials and products during their life cycle. The production of the material should also generate as little waste or pollution as possible.

- *Renewable materials:* A material is sustainable if it comes from sources that can renew themselves within a fairly short time.

- *Recycled content:* The more recycled content a material has, the less raw materials and energy required to process the raw materials into a final product.

- *Energy efficiency:* Materials, products, and assemblies should reduce the energy consumption in a building.

- *Use of local materials:* Using locally produced materials reduces transportation costs and can add to the regional character of a design.

- *Durability:* Durable materials will last longer and generally require less maintenance over the life of a product or building. Even though initial costs may be higher, the life-cycle costs may be less.

- *Low volatile organic compound (VOC) content:* use low-emitting materials, including adhesives and sealants, paints and coatings, flooring systems, composite wood and agrifiber products, and systems furniture and seating.

- *Low toxicity:* Materials should be selected that emit few or no harmful gases, such as chlorofluorocarbons (CFCs), formaldehyde, and others on the EPA's list of hazardous substances.

Process Design/ Economics

- *Moisture problems:* If possible, materials should be selected that prevent or resist the growth of biological contaminants.

- *Maintainability:* Materials and products should be able to be cleaned and otherwise maintained with only nontoxic or low-VOC substances.

- *Potential for reuse and recycling:* Some materials and products are more readily recycled than others. Steel, for example, can usually be separated and melted down to make new steel products. On the other hand, plastics used in construction are difficult to remove and separate.

- *Reusability:* A product should be reusable after it has served its purpose in the original building. This type of product becomes a salvaged material in the life cycle of another building.

All plastics used or produced should be identified for recycling. If possible, compostable plastics should be specified or produced. For example, polyethylene terephthalate (PET) from soft drink containers can be used to manufacture carpet with properties similar to polyesters.

Maintaining health is an important aspect of sustainable design, and one of the basic requirements of health is good indoor air quality (IAQ). In addition to simply maintaining health, the quality of indoor air affects people's sense of well-being and can affect absenteeism, productivity, creativity, and motivation. IAQ is a complex subject because there are hundreds of different contaminants, dozens of causes of poor IAQ, many possible symptoms building occupants may experience, and a wide variety of potential strategies for maintaining good IAQ. These can be classified into four broad categories: eliminate or reduce the sources of pollution, control the ventilation of the building, establish good maintenance procedures, and control occupant activity as it affects IAQ.

Example

Which concept is NOT considered to be an element of sustainable design?

    (A) use of rock obtained from a nearby quarry

    (B) use of hay bales to filter rainwater during construction

    (C) use of textured concrete flooring rather than hardwood

    (D) capture of rainwater for irrigation use

Solution

The use of hay bales to filter rainwater during construction is a pollution prevention methodology, not a sustainable design concept.

The answer is (B).

15. GREEN ENGINEERING

Green building is the implementation of design, construction, and operational strategies that reduce the environmental impact of a building or process during both construction and operation and improve occupants' health, comfort, and productivity throughout the life cycle of the building or process. Green building is closely related to general sustainability (see Sec. 49.15).

Green building guidelines for chemical engineers can be summarized in nine *principles of green engineering*.[18]

1. Engineer processes and products holistically, use systems analysis, and integrate environmental impact assessment tools.

2. Conserve and improve natural ecosystems while protecting human health and well-being.

3. Use life-cycle thinking in all engineering activities.

4. Ensure that all material and energy inputs and outputs are as inherently safe and benign as possible.

5. Minimize depletion of natural resources.

6. Strive to prevent waste.

7. Develop and apply engineering solutions, while being cognizant of local geography, aspirations, and cultures.

8. Create engineering solutions beyond current or dominant technologies; improve, innovate, and invent (technologies) to achieve sustainability.

9. Actively engage communities and stakeholders in development of engineering solutions.

Green buildings and processes come in all shapes and sizes. Some look "green"; they proudly display their vegetated roofs and solar panels and other features that allow them to control energy and water consumption and the like. These sustainable approaches are visible elements of the design.

Just as many green buildings, however, are indistinguishable to the eye from other buildings that are designed and constructed in more traditional ways. There is no "typical" green building or process; each project's mix of sustainability techniques and technologies must be chosen given the opportunities and challenges presented by the project and the site.

During the design phase, project team members must gather and analyze data, develop and evaluate design alternatives, and make decisions regarding the project priorities relative to the opportunities and drawbacks each present. The design phase is led by the design professionals (usually architects or engineers) and considers and integrates the work of the project consultants, such as building systems engineers, specialty

---

[18]Abraham, M.; Nguyen, N. "Green Engineering: Defining principles" – Results from the Sandestin conference. *Environmental Progress*, 2004, 22, 233–236.

systems consultants, landscape architects, and the commissioning agency.

In a Leadership in Energy and Environmental Design (LEED) project, the work of the design team will be guided by the combination of credits determined most appropriate for the project. A variety of issues, technologies, and approaches must be considered to develop a comprehensive building design. Issues under consideration during the design phase include

- sustainable sites (SS)
- water efficiency (WE)
- energy and atmosphere (EA)
- materials and resources (MR)
- indoor environmental quality (IEQ)

The building or chemical process's impact on the environment begins during the construction phase. Before that time, a building or process is just a concept on paper; during construction, that design assumes physical form and begins consuming resources. While the design team will continue to be involved with the project, the day-to-day operations on site are now the responsibility of the general contractor, who must be an integral part of the project team and who will implement the strategy developed in the design phase. A sustainable approach to construction leads to reduced resource use, reduced disturbance of the site, and can also lower costs. Attention to environmental issues during construction also leads to a safer, healthier working environment, first for those who construct the building, and later for those who occupy it.

Environmental guidelines can be established as a part of the construction documents and contract for the project. If contractors are required to follow specific environmental guidelines during the construction process, these requirements must be included in the contract, drawings, and specifications for the project. To develop and implement the guidelines, work with the entire team, including the architect, engineers, and contractors, to consider the best ways to educate contractors about sustainability issues and to get their early commitment to follow sustainability guidance.

The best efforts to reduce negative environmental impacts in the built environment are doomed to failure unless well-crafted operations and maintenance (O&M) procedures are implemented. Furthermore, even the best O&M procedures are of no use unless they are understood and followed by building O&M personnel.

Facility managers play the key role in ensuring that this happens. An integrated team approach can be a big help. In this process, O&M personnel are active participants in the design of a facility and the development of O&M procedures. This integrated team promotes useful procedures that are efficient and—most important—faithfully executed.

16. WASTES

Process wastes are generated during manufacturing. *Intrinsic wastes* are part of the design of the product and manufacturing process. Examples of intrinsic wastes are impurities in the reactants and raw materials, by-products, residues, and spent materials. Reduction of intrinsic wastes usually means redesigning the product or manufacturing process. *Extrinsic wastes*, on the other hand, are usually reduced by administrative controls, maintenance, training, or recycling. Examples of extrinsic wastes are fugitive leaks and discharges during material handling, testing, or process shutdown.

Solid wastes are garbage, refuse, biosolids, and containerized solid, liquid, and gaseous wastes.[19] *Hazardous waste* is defined as solid waste, alone or in combination with other solids, that because of its quantity, concentration, or physical, chemical, or infectious characteristics may either (a) cause an increase in mortality or serious (irreversible or incapacitating) illness, or (b) pose a present or future hazard to health or the environment when improperly treated, stored, transported, disposed of, or managed.

A substance is *reactive* if it reacts violently with water; if its pH is less than 2 or greater than 12.5, it is *corrosive*; if its flash point is less than $140°F$ ($60°C$), it is *ignitable*. The *toxicity characteristic leaching procedure* (TCLP) test is used to determine if the substance is *toxic*. The TCLP tests for the presence of eight metals (e.g., chromium, lead, and mercury) and 25 organic compounds (e.g., benzene and chlorinated hydrocarbons).

The EPA defines hazardous waste as any form of waste that is "dangerous or potentially harmful to our health or the environment." Hazardous wastes be categorized by the nature of their sources. *F-wastes* (e.g., spent solvents and distillation residues) originate from non-specific sources; *K-wastes* (e.g., separator sludge from petroleum refining) are generated from industry-specific sources; *P-wastes* (e.g., hydrogen cyanide) are acutely hazardous discarded commercial products, off-specification products, and spill residues; *U-wastes* (e.g., benzene and hydrogen sulfide) are other discarded commercial products, off-specification products, and spill residues. Other types of hazardous wastes include *characteristic waste* (having characteristics of ignitability, corrosivity, reactivity, and toxicity), *universal waste* (e.g., batteries and pesticides), and *mixed waste* (such as combinations of radioactive and other hazardous waste components).

Two notable "rules" pertain to hazardous wastes. The *mixture rule* states that any solid waste mixed with hazardous waste becomes hazardous. The *derived from rule* states that any waste derived from the treatment of

---

[19]The term *biosolids* is replacing the term *sludge* when it refers to organic waste produced from biological wastewater treatment processes. Sludge from industrial processes and flue gas cleanup (FGC) devices retains its name.

Process Design/ Economics

a hazardous waste (e.g., ash from the incineration of hazardous waste) remains a hazardous waste.[20]

Example

Theoretically, landfilled municipal solid waste (MSW) from households is

(A) combustible

(B) nonhazardous

(C) organic

(D) recyclable

Solution

Many components in municipal solid waste are recyclable, but many are not. And, with the prevalence of recycling, most recyclable materials never make it to the landfill. Similarly, many components are organic and/or combustible, but many are not. Since it is not illegal to dispose of combustible, reactive, corrosive, and toxic (i.e., hazardous) wastes in household garbage, MSW is theoretically nonhazardous. In practice, however, there may be a long way to go before all MSW is actually nonhazardous.

The answer is (B).

17. DISPOSITION OF HAZARDOUS WASTES

When a hazardous waste is disposed of, it must be taken to a registered *treatment, storage, or disposal facility* (TSDF). The EPA's *land ban* specifically prohibits the disposal of hazardous wastes on land prior to treatment. Incineration at sea is also prohibited. Wastes must be treated to specific maximum concentration limits by specific technology prior to disposal in landfills.

Once treated to specific regulated concentrations, hazardous waste residues can be disposed of by incineration or in *land disposal units* (LDUs). All LDUs must meet detailed design and operational standards. The EPA has established design and operating requirements for the following types of LDUs.

- landfills
- surface impoundments
- waste piles
- land treatment units
- injection wells
- salt dome formations
- salt bed formations
- underground mines
- underground caves

---

[20]Hazardous waste should not be incinerated with nonhazardous waste, as all of the ash would be considered hazardous by these rules.

18. GENERAL STORAGE OF HAZARDOUS MATERIALS

Hazardous materials must be properly stored prior to treatment or disposal. Storage of hazardous materials (*hazmats*) is often governed by local building codes in addition to state and federal regulations, including the Resource Conservation and Recovery Act (RCRA). Types of construction, maximum floor areas, and building layout may all be restricted.[21]

Good engineering judgment is called for in areas not specifically governed by the building code. Engineering consideration will need to be given to the following aspects of storage facility design.

- spill containment provisions
- chemical resistance of construction and storage materials
- likelihood of and resistance to explosions
- exiting
- ventilation
- electrical design
- storage method
- personnel emergency equipment
- security
- spill cleanup provisions

19. STORAGE TANKS

Underground Storage Tanks (USTs)

Underground storage tanks (USTs) have traditionally been used to store bulk chemicals and petroleum products. Fire and explosion risks are low with USTs, but subsurface pollution is common since inspection is limited. Since 1988, the U.S. Environmental Protection Agency (EPA) has required USTs to have secondary containment, corrosion protection, and leak detection. UST operators also must carry insurance in an amount sufficient to clean up a tank failure.

Above-Ground Storage Tanks (ASTs)

Because of the cost of complying with UST legislation, *above-ground storage tanks* (ASTs) are becoming more popular. AST strengths and weaknesses are the reverse of USTs: ASTs reduce pollution caused by leaks, but the expected damage due to fire and explosion is greatly increased. Because of this, some local ordinances prohibit all ASTs for petroleum products.[22]

---

[21]For example, flammable materials stored in rack systems are typically limited to heights of 25 ft (8.3 m).

[22]The American Society of Petroleum Operations Engineers (ASPOE) policy statement states, "Above-ground storage of liquid hydrocarbon motor fuels is inherently less safe than underground storage. Above-ground storage of Class 1 liquids (gasoline) should be prohibited at facilities open to the public."

Process Design/ Economics

The following factors should be considered when deciding between USTs and ASTs.

- space available
- zoning ordinances
- secondary containment
- leak-detection equipment
- operating limitations
- economics

Most ASTs are constructed of carbon or stainless steel. These provide better structural integrity and fire resistance than fiberglass-reinforced plastic and other composite tanks. Tanks can be either field-erected or factory-fabricated (capacities greater than approximately 50,000 gal (190 kL)). Factory-fabricated ASTs are usually designed according to UL-142 (Underwriters Laboratories *Standard for Safety*), which dictates steel type, wall thickness, and characteristics of compartments, bulkheads, and fittings. Most ASTs are not pressurized, but those that are must be designed in accordance with the ASME *Boiler and Pressure Vessel Code*, Section VIII.

NFPA 30 (*Flammable and Combustible Liquids Code*, National Fire Protection Association, Quincy, MA) specifies the minimum separation distances between ASTs, other tanks, structures, and public right-of-ways. The separation is a function of tank type, size, and contents. NFPA 30 also specifies installation, spill control, venting, and testing.

ASTs must be double-walled, concrete-encased, or contained in a dike or vault to prevent leaks and spills, and they must meet fire codes. Dikes should have a capacity in excess (e.g., 110% to 125%) of the tank volume. ASTs (as do USTs) must be equipped with overfill prevention systems. Piping should be above-ground wherever possible. Reasonable protection against vandalism and hunters' bullets is also necessary.[23]

Though they are a good idea, leak-detection systems are not typically required for ASTs. Methodology for leak detection is evolving, but currently includes vacuum or pressure monitoring, electronic gauging, and optical and sniffing sensors. Double-walled tanks may also be fitted with sensors within the interstitial space.

Operationally, ASTs present special problems. In hot weather, volatile substances vaporize and represent an additional leak hazard. In cold weather, viscous contents may need to be heated (often by steam tracing).

ASTs are not necessarily less expensive than USTs, but they are generally thought to be so. Additional hidden costs of regulatory compliance, secondary containment, fire protection, and land acquisition must also be considered.

---

[23]Approximately 20% of all spills from ASTs result from vandalism.

Example

Which statement is INCORRECT in regard to labeling containers used to store and transport hazardous wastes?

(A) The hazardous substance in the greatest quantity needs to be listed.

(B) The labeling needs to be provided in English.

(C) Noncorrosive and nonreactive substances may be unlisted.

(D) Concentrations of all substances present must be listed.

Solution

Concentrations of all substances present need not be listed.

The answer is (D).

20. POLLUTANTS

A *pollutant* is a material or substance that is accidentally or intentionally introduced to the environment in a quantity that exceeds what occurs naturally. Not all pollutants are toxic or hazardous, but the issue is moot when regulations limiting permissible concentrations are specific. As defined by regulations in the United States, *hazardous air pollutants* (HAPs), also known as *air toxics*, consist of trace metals (e.g., lead, beryllium, mercury, cadmium, nickel, and arsenic) and other substances for a total of approximately 200 "listed" substances.[24]

Another categorization defined by regulation in the United States separates pollutants into *criteria pollutants* (e.g., sulfur dioxide, nitrogen oxides, carbon monoxide, volatile organic compounds, particulate matter, and lead) and *noncriteria pollutants* (e.g., fluorides, sulfuric acid mist, reduced sulfur compounds, vinyl chloride, asbestos, beryllium, mercury, and other heavy metals).

21. POLLUTANT SOURCES

A *pollution source* is any facility that produces pollution. A *generator* is any facility that generates hazardous waste. The term "major" (i.e., a *major source*) is defined differently for each class of nonattainment areas.[25]

The combustion of fossil fuels (e.g., coal, fuel oil, and natural gas) to produce steam in electrical generating plants is the most significant source of airborne

---

[24]The most dangerous air toxics include asbestos, benzene, cadmium, carbon tetrachloride, chlorinated dioxins and dibenzofurans, chromium, ethylene dibromide, ethylene dichloride, ethylene oxide and methylene chloride, radionuclides, vinyl chloride, and emissions from coke ovens. Most of these substances are carcinogenic.

[25]A *nonattainment area* is classified as marginal, moderate, serious, severe, or extreme based on the average pollution (e.g., ozone) level measured in the area.

Process Design/ Economics

pollution. For this reason, this industry is among the most highly regulated.

Specific regulations pertain to generators of hazardous waste. Generators of more than certain amounts (e.g., 100 kg/month) must be registered (i.e., with the Environmental Protection Agency (EPA)). Restrictions are placed on generators in the areas of storage, personnel training, shipping, treatment, and disposal.

22. POLLUTION PREVENTION AND CONTROL

Pollution prevention (sometimes referred to as "P2") can be achieved in a number of ways. Some examples include, in order of desirability,

- source reduction

- recycling

- waste separation and concentration

- waste exchange

- energy/material recovery

- waste treatment

- disposal

The most desirable method is reduction at the source. Reduction is accomplished through process modifications, raw material quantity reduction, substitution of materials, improvements in material quality, and increased efficiencies. However, traditional *end-of-pipe* treatment and disposal processes after the pollution is generated remain the main focus. *Pollution control* is the limiting of pollutants in a planned and systematic manner.

Pollution control, hazardous waste, and other environmental regulations vary from nation to nation and are constantly changing. Therefore, regulation-specific issues, including timetables, permit application processes, enforcement, and penalties for violations, are either omitted from this chapter or discussed in general terms.[26] Those that are given should be considered merely representative and typical of the general range of values.

23. ENVIRONMENTAL REGULATIONS

Specific regulations often deal with parts of the *environment*, such as the atmosphere (i.e., the "air"), oceans and other surface water, subsurface water, and the soil. *Nonattainment areas* are geographical areas identified by regulation that do not meet national ambient air quality standards (NAAQS). Nonattainment is usually the result of geography, concentrations of industrial facilities, and excessive vehicular travel, and can apply

[26]Another factor complicating the publication of specific regulations is that the maximum permitted concentrations and emissions depend on the size and nature of the source.

to any of the regulated substances (e.g., ozone, oxides of sulfur and nitrogen, and heavy metals).

The Environmental Protection Agency (EPA) is a federal agency formed to develop and enforce regulations implemented by environmental laws enacted by Congress. The agency is responsible for researching and setting national standards for a variety of environmental programs. It delegates to states and (through the American Indian Environmental Office) to tribes the responsibility for issuing permits and for monitoring and enforcing compliance. Where national standards are not met, the EPA can issue sanctions and take other steps to assist the states and tribes in reaching desired levels of environmental quality. The EPA also conducts research to advance environmental education.

Water Pollution Control Act, Clean Water Act (CWA), and Water Quality Act

Federal Water Pollution Control Act was passed by Congress in 1972, with major amendments in the Clean Water Act (CWA) of 1977 and the Water Quality Act of 1987. Water bodies are defined as "seas, lakes, rivers, streams, and tributaries which support or could support fish, recreation, or industrial use, consistent with the terminology of the Clean Water Act." Small human-made water bodies constructed for stormwater retention, fire suppression, or recreation are not included, but human-made water bodies and wetlands are included if they were constructed for the purpose of restoring an area's natural ecology.

The National Pollutant Discharge Elimination System (NPDES) program was authorized by the Clean Water Act in 1972 and regulates wastewater discharge pollutants from industrial, municipal, and other facilities into surface waters, storm sewers, water treatment facilities, and other waterbodies.

Under the CWA, it is unlawful to discharge any pollutant from a point source into navigable waters without a National Pollutant Discharge Elimination System (NPDES) permit. *Point sources* are discrete conveyances such as pipes or man-made ditches. Individual homes are typically connected to a municipal system, a septic system, or otherwise do not have a surface discharge, and thus do not require an NPDES permit. However, industrial, municipal, and any other facilities whose discharges go directly to surface waters must obtain a permit.

Resource Conservation and Recovery Act (RCRA)

The Resource Conservation and Recovery Act (RCRA) is federal law passed by Congress in 1976, with major amendments in the Federal Hazardous and Solid Waste Amendments (HSWA) in 1984, the Federal Facility Compliance Act of 1992 and the Land Disposal Program Flexibility Act of 1996. The RCRA governs the disposal of solid and hazardous waste. The RCRA includes the underground storage tank (UST) program which regulates the underground storage of petroleum and hazardous substances; the hazardous waste program which regulates hazardous waste from generation to disposal; and the solid waste program which regulates nonhazardous industrial and municipal solid waste.

Toxic Substances Control Act

The Toxic Substances Control Act (TSCA) of 1976 authorizes the EPA to set requirements for reporting, record-keeping, and testing, and to set restrictions, on chemical substances. The EPA's authority under the TSCA does NOT include food, drugs, cosmetics, pesticides, and certain other substances that are regulated by other agencies or are otherwise exempt. The EPA also maintains the *TSCA Inventory*, which lists more than 83,000 chemicals.

More specifically, the TSCA grants the EPA the authority to

- require notification for new chemical substances before manufacture

- require manufacturers, importers, and processors to test substances when there is a risk or exposure of concern

- issue *Significant New Use Rules* (SNURs) when a new use of a substance could result in exposures to, or releases of, a substance of concern

- set and enforce reporting and other requirements for chemical imports and exports

- set reporting and record-keeping requirements for the manufacture, import, processing, and distribution of chemical substances

- require all toxic substances to be reported immediately to the EPA

Example

Many curb sewer inlets are now marked with the message shown.

With what federal regulation is this marking associated?

   (A) CWA

   (B) NPDES

   (C) RCRA

   (D) TSCA

Solution

The National Pollution Discharge Elimination System (NPDES) specifically requires this type of marking.

The answer is (B).

Process Design/ Economics

50

Engineering Economics

Nomenclature

A	annual amount or annual value	$
B	present worth of all benefits	$
BV	book value	$
C	initial cost, or present worth of all costs	$
d	interest rate per period adjusted for inflation	decimal or %
D	depreciation	$
f	inflation rate per period	decimal or %
F	future worth (future value)	$
G	uniform gradient amount	$
i	interest rate per period	decimal or %
j	number of compounding periods or years	–
m	number of compounding periods per year	–
n	total number of compounding periods or years	–
P	present worth (present value)	$
r	nominal rate per year (rate per annum)	decimal or %
S	salvage value	$
t	time	yr

Subscripts

0	initial
e	effective
j	jth year or period
n	final year or period

1. INTRODUCTION

In its simplest form, an *engineering economic analysis* is a study of the desirability of making an investment.[1] The decision-making principles in this chapter can be applied by individuals as well as by companies. The nature of the spending opportunity or industry is not important. Farming equipment, personal investments, and multimillion dollar factory improvements can all be evaluated using the same principles.

Similarly, the applicable principles are insensitive to the monetary units. Although *dollars* are used in this chapter, it is equally convenient to use pounds, yen, or euros.

Finally, this chapter may give the impression that investment alternatives must be evaluated on a year-by-year basis. Actually, the *effective period* can be defined as a day, month, century, or any other convenient period of time.

2. YEAR-END AND OTHER CONVENTIONS

Except in short-term transactions, it is simpler to assume that all receipts and disbursements (cash flows) take place at the end of the year in which they occur.[2] This is known as the *year-end convention*. The exceptions to the year-end convention are initial project cost (purchase cost), trade-in allowance, and other cash flows that are associated with the inception of the project at $t = 0$.

On the surface, such a convention appears grossly inappropriate since repair expenses, interest payments, corporate taxes, and so on seldom coincide with the end of a year. However, the convention greatly simplifies engineering economic analysis problems, and it is justifiable on the basis that the increased precision associated with a more rigorous analysis is not warranted (due to the numerous other simplifying assumptions and estimates initially made in the problem).

---

[1]This subject is also known as *engineering economics* and *engineering economy*. There is very little, if any, true economics in this subject.
[2]A *short-term transaction* typically has a lifetime of five years or less and has payments or compounding that are more frequent than once per year.

Process Design/ Economics

There are various established procedures, known as *rules* or *conventions*, imposed by the Internal Revenue Service on U.S. taxpayers. An example is the *half-year rule*, which permits only half of the first-year depreciation to be taken in the first year of an asset's life when certain methods of depreciation are used. These rules are subject to constantly changing legislation and are not covered in this book. The implementation of such rules is outside the scope of engineering practice and is best left to accounting professionals.

3. CASH FLOW

The sums of money recorded as receipts or disbursements in a project's financial records are called *cash flows*. Examples of cash flows are deposits to a bank, dividend interest payments, loan payments, operating and maintenance costs, and trade-in salvage on equipment. Whether the cash flow is considered to be a receipt or disbursement depends on the project under consideration. For example, interest paid on a sum in a bank account will be considered a disbursement to the bank and a receipt to the holder of the account.

Because of the time value of money, the timing of cash flows over the life of a project is an important factor. Although they are not always necessary in simple problems (and they are often unwieldy in very complex problems), *cash flow diagrams* can be drawn to help visualize and simplify problems that have diverse receipts and disbursements.

The following conventions are used to standardize cash flow diagrams.

- The horizontal (time) axis is marked off in equal increments, one per period, up to the duration of the project.

- *Receipts* are represented by arrows directed upward. *Disbursements* are represented by arrows directed downward. The arrow length is approximately proportional to the magnitude of the cash flow.

- Two or more transfers in the same period are placed end to end, and these may be combined.

- Expenses incurred before $t=0$ are called *sunk costs*. Sunk costs are not relevant to the problem unless they have tax consequences in an after-tax analysis.

For example, consider a mechanical device that will cost $20,000 when purchased. Maintenance will cost $1000 each year. The device will generate revenues of $5000 each year for five years, after which the salvage value is expected to be $7000. The cash flow diagram is shown in Fig. 50.1(a), and a simplified version is shown in Fig. 50.1(b).

In order to evaluate a real-world project, it is necessary to present the project's cash flows in terms of standard cash flows that can be handled by engineering economic analysis techniques. The standard cash flows are single

Figure 50.1 *Cash Flow Diagrams*

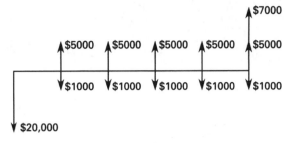

(a) cash flow diagram

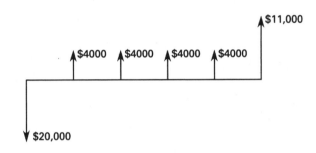

(b) simplified cash flow diagram

payment cash flow, uniform series cash flow, and gradient series cash flow.

A *single payment cash flow* can occur at the beginning of the time line (designated as $t=0$), at the end of the time line (designated as $t=n$), or at any time in between.

The *uniform series cash flow*, illustrated in Fig. 50.2, consists of a series of equal transactions starting at $t=1$ and ending at $t=n$. The symbol A (representing an *annual amount*) is typically given to the magnitude of each individual cash flow.

Figure 50.2 *Uniform Series*

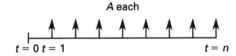

Notice that the cash flows do not begin at the beginning of a year (i.e., the year 1 cash flow is at $t=1$, not $t=0$). This convention has been established to accommodate the timing of annual maintenance and other cash flows for which the *year-end convention* is applicable. The year-end convention assumes that all receipts and disbursements take place at the end of the year in which they occur. The exceptions to the year-end convention are *initial project cost* (purchase cost), *trade-in allowance*, and other cash flows that are associated with the inception of the project at $t=0$.

The *gradient series cash flow*, illustrated in Fig. 50.3, starts with a cash flow (typically given the symbol G) at $t=2$ and increases by G each year until $t=n$, at which

Figure 50.3 Gradient Series

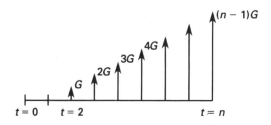

time the final cash flow is $(n - 1)G$. The value of the gradient at $t = 1$ is zero.

4. TIME VALUE OF MONEY

Consider $100 placed in a bank account that pays 5% effective annual interest at the end of each year. After the first year, the account will have grown to $105. After the second year, the account will have grown to $110.25.

The fact that $100 grows to $105 in one year at 5% annual interest is an example of the *time value of money* principle. This principle states that funds placed in a secure investment will increase in value in a way that depends on the elapsed time and the interest rate.

The interest rate that is used in calculations is known as the *effective interest rate*. If compounding is once a year, it is known as the *effective annual interest rate*. However, effective quarterly, monthly, or daily interest rates are also used.

5. DISCOUNT FACTORS

Assume that there will be no need for money during the next two years, and any money received will immediately go into an account and earn a 5% effective annual interest rate. Which of the following options would be more desirable?

option a: receive $100 now

option b: receive $105 in one year

option c: receive $110.25 in two years

None of the options is superior under the assumptions given. If the first option is chosen, $100 will be immediately placed into a 5% account, and in two years the account will have grown to $110.25. In fact, the account will contain $110.25 at the end of two years regardless of which option is chosen. Therefore, these alternatives are said to be *equivalent*.

The three options are equivalent only for money earning a 5% effective annual interest rate. If a higher interest rate can be obtained, then the first option will yield the most money after two years. So, equivalence depends on the interest rate, and an alternative that is acceptable to one decision maker may be unacceptable to another who invests at a higher rate. The procedure for determining the equivalent amount is known as *discounting*.

Table 50.1: Discount Factors

Table 50.1 Discount Factors for Discrete Compounding

factor name	converts	symbol	formula
single payment compound amount	P to F	$(F/P, i\%, n)$	$(1+i)^n$
single payment present worth	F to P	$(P/F, i\%, n)$	$(1+i)^{-n}$
uniform series sinking fund	F to A	$(A/F, i\%, n)$	$\dfrac{i}{(1+i)^n - 1}$
capital recovery	P to A	$(A/P, i\%, n)$	$\dfrac{i(1+i)^n}{(1+i)^n - 1}$
uniform series compound amount	A to F	$(F/A, i\%, n)$	$\dfrac{(1+i)^n - 1}{i}$
uniform series present worth	A to P	$(P/A, i\%, n)$	$\dfrac{(1+i)^n - 1}{i(1+i)^n}$
uniform gradient present worth	G to P	$(P/G, i\%, n)$	$\dfrac{(1+i)^n - 1}{i^2(1+i)^n} - \dfrac{n}{i(1+i)^n}$
uniform gradient future worth*	G to F	$(F/G, i\%, n)$	$\dfrac{(1+i)^n - 1}{i^2} - \dfrac{n}{i}$
uniform gradient uniform series	G to A	$(A/G, i\%, n)$	$\dfrac{1}{i} - \dfrac{n}{(1+i)^n - 1}$

*See Eq. 50.9.

Description

The discounting factors are listed in symbolic and formula form. For more detail on individual factors, see the commentary accompanying Eq. 50.1 through Eq. 50.10. Normally, it will not be necessary to calculate factors from these formulas. Values of these cash flow (discounting) factors are tabulated in Table 50.4 through Table 50.13 (at the end of this chapter) for various combinations of i and n. For intermediate values, computing the factors from the formulas may be necessary, or linear interpolation can be used as an approximation. The interest rate used must be the effective rate per period for all discounting factor formulas. The basis of the rate (annually, monthly, etc.) must agree with the type of period used to count n. It would be incorrect to use an effective annual interest rate if n was the number of compounding periods in months.

6. SINGLE PAYMENT EQUIVALENCE

The equivalent future amount, F, at $t=n$, of any *present amount*, P, at $t=0$ is called the *future worth*. The equivalence of any future amount to any present amount is called the *present worth*. Compound amount factors may be used to convert from a known present amount to a future worth or vice versa.

Equation 50.1: Single Payment Future Worth

$$F = P(1 + i)^n \qquad 50.1$$

Variation

$$F = P(F/P, i\%, n)$$

Description

The factor $(1 + i)^n$ is known as the *single payment compound amount factor*. Rather than actually writing the formula for the compound amount factor (which converts a present amount to a future amount), it is common convention to substitute the standard functional notation of $(F/P, i\%, n)$, as shown in the variation equation. This notation is interpreted as, "Find F, given P, using an interest rate of $i\%$ over n periods."

Example

A 40-year-old consulting engineer wants to set up a retirement fund to be used starting at age 65. $20,000 is invested now at 6% compounded annually. The amount of money that will be in the fund at retirement is most nearly

(A) $84,000

(B) $86,000

(C) $88,000

(D) $92,000

Solution

Determine the future worth of $20,000 in 25 years. From Eq. 50.1,

$$\begin{aligned} F &= P(1 + i)^n \\ &= (\$20{,}000)(1 + 0.06)^{25} \\ &= \$85{,}837 \quad (\$86{,}000) \end{aligned}$$

The answer is (B).

Equation 50.2: Single Payment Present Worth

$$P = F(1 + i)^{-n} \qquad 50.2$$

Variations

$$P = F(P/F, i\%, n)$$

$$P = \frac{F}{(1 + i)^n}$$

Description

The factor $(1 + i)^{-n}$ is known as the *single payment present worth factor*.

Example

$2000 will become available on January 1 in year 8. If interest is 5%, what is most nearly the present worth of this sum on January 1 in year 1?

(A) $1330

(B) $1350

(C) $1400

(D) $1420

Solution

From January 1 in year 1 to January 1 in year 8 is seven years. From Eq. 50.2, the present worth is

$$\begin{aligned} P &= F(1 + i)^{-n} \\ &= (\$2000)(1 + 0.05)^{-7} \\ &= \$1421 \quad (\$1420) \end{aligned}$$

The answer is (D).

7. UNIFORM SERIES EQUIVALENCE

A cash flow that repeats at the end of each year for n years without change in amount is known as an *annual amount* and is given the symbol A. (This is shown in Fig. 50.2.)

Although the equivalent value for each of the n annual amounts could be calculated and then summed, it is more expedient to use one of the uniform series factors.

Equation 50.3: Uniform Series Future Worth

$$F = A\left(\frac{(1 + i)^n - 1}{i}\right) \qquad 50.3$$

Variation

$$F = A(F/A, i\%, n)$$

Process Design/
Economics

Description

Use the *uniform series compound amount factor* to convert from an annual amount to a future amount.

Example

$20,000 is deposited at the end of each year into a fund earning 6% interest. At the end of ten years, the amount accumulated is most nearly

(A) $150,000

(B) $180,000

(C) $260,000

(D) $280,000

Solution

The amount accumulated at the end of ten years is

$$F = A\left(\frac{(1+i)^n - 1}{i}\right)$$

$$= (\$20,000)\left(\frac{(1+0.06)^{10} - 1}{0.06}\right)$$

$$= \$263,616 \quad (\$260,000)$$

The answer is (C).

Equation 50.4: Uniform Series Annual Value of a Sinking Fund

$$A = F\left(\frac{i}{(1+i)^n - 1}\right) \qquad \text{50.4}$$

Variation

$$A = F(A/F, i\%, n)$$

Description

A *sinking fund* is a fund or account into which annual deposits of A are made in order to accumulate F at $t = n$ in the future. Because the annual deposit is calculated as $A = F(A/F, i\%, n)$, the (A/F) factor is known as the *sinking fund factor*.

Example

At the end of each year, an investor deposits some money into a fund earning 7% interest. The same amount is deposited each year, and after six years the account contains $1600. The amount deposited each time is most nearly

(A) $190

(B) $220

(C) $240

(D) $250

Solution

Use the sinking fund factor from Eq. 50.4 to find the annual value.

$$A = F\left(\frac{i}{(1+i)^n - 1}\right) = (\$1600)\left(\frac{0.07}{(1+0.07)^6 - 1}\right)$$

$$= \$224 \quad (\$220)$$

The answer is (B).

Equation 50.5: Uniform Series Present Worth

$$P = A\left(\frac{(1+i)^n - 1}{i(1+i)^n}\right) \qquad \text{50.5}$$

Variation

$$P = A(P/A, i\%, n)$$

Description

An *annuity* is a series of equal payments, A, made over a period of time. Usually, it is necessary to "buy into" an investment (a bond, an insurance policy, etc.) in order to fund the annuity. In the case of an annuity that starts at the end of the first year and continues for n years, the purchase price, P, is calculated using the *uniform series present worth factor*.

Example

A sum of money is deposited into a fund at 5% interest. $400 is withdrawn at the end of each year for nine years, leaving nothing in the fund at the end. The amount originally deposited is most nearly

(A) $2600

(B) $2800

(C) $2900

(D) $3100

Solution

Find the present worth using the present worth factor.

$$P = A\left(\frac{(1+i)^n - 1}{i(1+i)^n}\right) = (\$400)\left(\frac{(1+0.05)^9 - 1}{(0.05)(1+0.05)^9}\right)$$

$$= \$2843 \quad (\$2800)$$

The answer is (B).

Process Design/ Economics

Equation 50.6: Uniform Series Annual Value Using the Capital Recovery Factor

$$A = P\left(\frac{i(1+i)^n}{(1+i)^n - 1}\right) \qquad 50.6$$

Variation

$$A = P(A/P, i\%, n)$$

Description

The *capital recovery factor* is often used when comparing alternatives with different lifespans. A comparison of two possible investments on the simple basis of their present values may be misleading if, for example, one alternative has a lifespan of 11 years and the other has a lifespan of 18 years. The capital recovery factor can be used to convert the present value of each alternative into its equivalent annual value, using the assumption that each alternative will be renewed repeatedly up to the duration of the longest-lived alternative.

8. UNIFORM GRADIENT EQUIVALENCE

A common situation involves a uniformly increasing cash flow. If the cash flow has the proper form (see Fig. 50.3), its present worth can be determined by using the *uniform gradient factor*, also called the *uniform gradient present worth factor*. The uniform gradient factor, $(P/G, i\%, n)$, finds the present worth of a uniformly increasing cash flow. By definition of a uniform gradient, the cash flow starts in year 2, not in year 1. Similar factors can be used to find the cash flow's future worth and annual worth.

There are three common difficulties associated with the form of the uniform gradient. The first difficulty is that the first cash flow starts at $t = 1$. This convention recognizes that annual costs, if they increase uniformly, begin with some value at $t = 1$ (due to the year-end convention), but do not begin to increase until $t = 2$. The tabulated values of (P/G) have been calculated to find the present worth of only the increasing part of the annual expense. The present worth of the base expense incurred at $t = 1$ must be found separately with the (P/A) factor.

The second difficulty is that, even though the $(P/G, i\%, n)$ factor is used, there are only $n - 1$ actual cash flows. n must be interpreted as the *period number* in which the last gradient cash flow occurs, not the number of gradient cash flows.

Finally, the sign convention used with gradient cash flows can be confusing. If an expense increases each year, the gradient will be negative, since it is an expense. If a revenue increases each year, the gradient will be positive. In most cases, the sign of the gradient depends on whether the cash flow is an expense or a revenue.

Equation 50.7: Uniform Gradient Present Worth

$$P = G\left(\frac{(1+i)^n - 1}{i^2(1+i)^n} - \frac{n}{i(1+i)^n}\right) \qquad 50.7$$

Variation

$$P = G(P/G, i\%, n)$$

Description

Equation 50.7 is used to find the present worth, P, of a cash flow that is increasing by a uniform amount, G. This factor finds the value of only the increasing portion of the cash flow; if the value at $t = 1$ is anything other than zero, its present worth must be found separately and added to get the total present worth.

Equation 50.8 and Eq. 50.9: Uniform Gradient Future Worth

$$F = G\left(\frac{(1+i)^n - 1}{i^2} - \frac{n}{i}\right) \qquad 50.8$$

$$F/G = (F/A - n)/i = (F/A) \times (A/G) \qquad 50.9$$

Variation

$$F = G(F/G, i\%, n)$$

Description

Equation 50.8 is used to find the future worth, F, of a cash flow that is increasing by a uniform amount, G. This factor finds the value of only the increasing portion of the cash flow; if the value at $t = 1$ is anything other than zero, its future worth must be found separately and added to get the total future worth.

Equation 50.9 shows how the future worth factor, $(F/G, i\%, n)$, is closely related to (and, therefore, can be calculated quickly from) the factor $(F/A, i\%, n)$, or from the factors $(F/A, i\%, n)$ and $(A/G, i\%, n)$.

Equation 50.10: Uniform Gradient Uniform Series Factor

$$A = G\left(\frac{1}{i} - \frac{n}{(1+i)^n - 1}\right) \qquad 50.10$$

Variation

$$A = G(A/G, i\%, n)$$

Description

Equation 50.10 is used to find the equivalent annual worth, A, of a cash flow that is increasing by a uniform amount, G. This factor finds the value of only the increasing portion of the cash flow; if the value at $t=1$ is anything other than zero, its annual worth must be found separately and added to get the total annual worth.

Example

The maintenance cost on a house is expected to be $1000 the first year and to increase $500 per year after that. Assuming an interest rate of 6% compounded annually, the maintenance cost over 10 years is most nearly equivalent to an annual maintenance cost of

(A) $1900

(B) $3000

(C) $3500

(D) $3800

Solution

Use the uniform gradient uniform series factor to determine the effective annual cost, remembering that the first year's cost, $A_1 = \$1000$, must be added separately. The annual increase is $G = \$500$. The effective annual cost of the increasing portion of the costs alone is

$$A = G\left(\frac{1}{i} - \frac{n}{(1+i)^n - 1}\right)$$

$$= (\$500)\left(\frac{1}{0.06} - \frac{10}{(1+0.06)^{10} - 1}\right)$$

$$= \$2011$$

The total effective annual cost is

$$A_{\text{total}} = A_1 + A$$

$$= \$1000 + \$2011$$

$$= \$3011 \quad (\$3000)$$

The answer is (B).

9. FUNCTIONAL NOTATION

There are several ways of remembering what the functional notation means. One method of remembering which factor should be used is to think of the factors as *conditional probabilities*. The conditional probability of event A given that event B has occurred is written as $P\{A|B\}$, where the given event comes after the vertical bar. In the standard notational form of discounting factors, the given amount is similarly placed after the

slash. The desired factor (i.e., A) comes before the slash. (F/P) would be a factor to find F given P.

Another method of remembering the notation is to interpret the factors algebraically. The (F/P) factor could be thought of as the fraction F/P. The numerical values of the discounting factors are consistent with this algebraic manipulation. The (F/A) factor could be calculated as $(F/P)(P/A)$. This consistent relationship can be used to calculate other factors that might be occasionally needed, such as (F/G) or (G/P).

10. NONANNUAL COMPOUNDING

If $100 is invested at 5%, it will grow to $105 in one year. If only the original principal accrues interest, the interest is known as *simple interest*, and the account will grow to $110 in the second year, $115 in the third year, and so on. Simple interest is rarely encountered in engineering economic analyses.

More often, both the principal and the interest earned accrue interest, and this is known as *compound interest*. If the account is compounded yearly, then during the second year, 5% interest continues to be accrued, but on $105, not $100, so the value at year end will be $110.25. The value after the third year will be $115.76, and so on.

The interest rate used in the discount factor formulas is the *interest rate per period*, i (called the *yield* by banks). If the interest period is one year (i.e., the interest is compounded yearly), then the interest rate per period, i, is equal to the *effective annual interest rate*, i_e. The effective annual interest rate is the rate that would yield the same accrued interest at the end of the year if the account were compounded yearly.

The term *nominal interest rate*, r (*rate per annum*), is encountered when compounding is more than once per year. The nominal rate does not include the effect of compounding and is not the same as the effective annual interest rate.

Equation 50.11: Effective Annual Interest Rate

$$i_e = \left(1 + \frac{r}{m}\right)^m - 1 \qquad \text{50.11}$$

Description

The effective annual interest rate, i_e, can be calculated if the nominal rate, r, and the number of compounding periods per year, m, are known. If there are m compounding periods during the year (two for semiannual compounding, four for quarterly compounding, twelve for monthly compounding, etc.), the *effective interest rate per period*, i, is r/m. The effective annual interest rate, i_e, can be calculated from the effective interest rate per period by using Eq. 50.11.

Example

Money is invested at 5% per annum and compounded quarterly. The effective annual interest rate is most nearly

    (A) 5.1%

    (B) 5.2%

    (C) 5.4%

    (D) 5.5%

Solution

The rate per annum is the nominal interest rate. Use Eq. 50.11 to calculate the effective annual interest rate.

$$i_e = \left(1 + \frac{r}{m}\right)^m - 1$$
$$= \left(1 + \frac{0.05}{4}\right)^4 - 1$$
$$= 0.05095 \quad (5.1\%)$$

The answer is (A).

11. DEPRECIATION

Tax regulations do not generally allow the purchase price of an asset or other property to be treated as a single deductible expense in the year of purchase. Rather, the cost must be divided into portions, and these artificial expenses are spread out over a number of years. The portion of the cost that is allocated to a given year is called the *depreciation*, and the period of years over which these portions are spread out is called the *depreciation period* (also known as the *service life*).

When depreciation is included in an engineering economic analysis problem, it will increase the asset's after-tax present worth (profitability). The larger the depreciation is, the greater the profitability will be. For this reason, it is desirable to make the depreciation in each year as large as possible and to accelerate the process of depreciation as much as possible.

The *depreciation basis* of an asset is that part of the asset's purchase price that is spread over the depreciation period. The depreciation basis may or may not be equal to the purchase price.

A common depreciation basis is the difference between the purchase price and the expected salvage value at the end of the depreciation period (i.e., depreciation basis = $C - S_n$).

In the *sum-of-the-years' digits* (SOYD) method of depreciation, the digits from 1 to n inclusive are added together. An easy way to calculate this sum is the formula

$$\sum_{j=1}^{n} j = \frac{n(n+1)}{2}$$

The depreciation in year j is found from

$$D_j = \frac{n+1-j}{\sum\limits_{j=1}^{n} j}(C - S_n)$$

Using this method, the depreciation from one year to the next decreases by a constant amount.

Equation 50.12: Straight Line Method

$$D_j = \frac{C - S_n}{n} \qquad \text{50.12}$$

Description

With the *straight line method*, depreciation is the same each year. The depreciation basis ($C - S_n$) is divided uniformly among all of the n years in the depreciation period.

Example

A computer will be purchased at $3900. The expected salvage value at the end of its service life of 10 years is $1800. Using the straight line method, the annual depreciation for this computer is most nearly

    (A) $210

    (B) $230

    (C) $260

    (D) $280

Solution

From Eq. 50.12, the annual depreciation using the straight line method is

$$D_j = \frac{C - S_n}{n}$$
$$= \frac{\$3900 - \$1800}{10}$$
$$= \$210$$

The answer is (A).

Equation 50.13: Modified Accelerated Cost Recovery System (MACRS)

$$D_j = (\text{factor})C \qquad 50.13$$

Values

Table 50.2 *Representative MACRS Depreciation Factors*

year j	recovery period (years)			
	3	5	7	10
	recovery rate (percent)			
1	33.33	20.00	14.29	10.00
2	44.45	32.00	24.49	18.00
3	14.81	19.20	17.49	14.40
4	7.41	11.52	12.49	11.52
5		11.52	8.93	9.22
6		5.76	8.92	7.37
7			8.93	6.55
8			4.46	6.55
9				6.56
10				6.55
11				3.28

Description

In the United States, property placed into service in 1981 and thereafter must use the *Accelerated Cost Recovery System* (ACRS), and property placed into service after 1986 must use the *Modified Accelerated Cost Recovery System* (MACRS) or another statutory method. Other methods, including the straight line method, cannot be used except in special cases.

Under ACRS and MACRS, the cost recovery amount in a particular year is calculated by multiplying the initial cost of the asset by a factor. (This initial cost is not reduced by the asset's salvage value.) The factor to be used varies depending on the year and on the total number of years in the asset's cost recovery period. These factors are subject to continuing legislation changes. Representative MACRS depreciation factors are shown in Table 50.2.

Example

A groundwater treatment system costs $2,500,000. It is expected to operate for a total of 130,000 hours over a period of 10 years, and then have a $250,000 salvage value. During the system's first year in service, it is operated for 6500 hours. Using the MACRS method, its depreciation in the third year is most nearly

(A) $160,000

(B) $250,000

(C) $360,000

(D) $830,000

Solution

MACRS depreciation depends only on the original cost, not on the salvage cost or hours of operation. From Table 50.2, the factor for the third year of a 10 year recovery period is 14.40%. From Eq. 50.13,

$$D_j = (\text{factor})C$$
$$D_3 = (0.1440)(\$2,500,000)$$
$$= \$360,000$$

The answer is (C).

12. BOOK VALUE

The difference between the original purchase price and the accumulated depreciation is known as the *book value*, BV. The book value is initially equal to the purchase price, and at the end of each year it is reduced by that year's depreciation.

Figure 50.4 compares how the ratio of book value to initial cost changes over time under the straight line and the MACRS methods.

Figure 50.4 *Book Value with Straight Line and MACRS Methods*

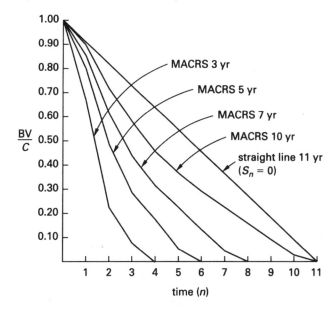

Equation 50.14: Book Value

$$BV = \text{initial cost} - \sum D_j \qquad 50.14$$

Description

In Eq. 50.14, BV is the book value at the end (not the beginning) of the jth year—that is, after j years of depreciation have been subtracted from the original purchase price.

Example

A machine initially costing $25,000 will have a salvage value of $6000 after five years. Using MACRS depreciation, its book value after the third year will be most nearly

(A) $5500

(B) $7200

(C) $10,000

(D) $14,000

Solution

Book value is the initial cost less the accumulated depreciation; the salvage value is disregarded. Use Eq. 50.14 and the MACRS factors for a five-year recovery period.

$$
\begin{aligned}
\text{BV} &= \text{initial cost} - \sum D_j \\
&= \text{initial cost} - (D_1 + D_2 + D_3) \\
&= \text{initial cost} - \begin{pmatrix} (\text{factor}_1)(\text{initial cost}) \\ + (\text{factor}_2)(\text{initial cost}) \\ + (\text{factor}_3)(\text{initial cost}) \end{pmatrix} \\
&= (1 - \text{factor}_1 - \text{factor}_2 - \text{factor}_3)(\text{initial cost}) \\
&= (1 - 0.20 - 0.32 - 0.192)(\$25{,}000) \\
&= \$7200
\end{aligned}
$$

The answer is (B).

13. EQUIVALENT UNIFORM ANNUAL COST

Alternatives with different lifespans will generally be compared by way of *equivalent uniform annual cost*, or EUAC. An EUAC is the annual amount that is equivalent to all of the cash flows in the alternative.

The EUAC differs in sign from all of the other cash flows. Costs and expenses expressed as EUACs, which would normally be considered negative, are considered positive. Conversely, benefits and returns are considered negative. The term *cost* in the designation EUAC serves to make clear the meaning of a positive number.

14. CAPITALIZED COST

The present worth of a project with an infinite life is known as the *capitalized cost*. Capitalized cost is the amount of money at $t = 0$ needed to perpetually support the project on the earned interest only. Capitalized cost is a positive number when expenses exceed income.

Normally, it would be difficult to work with an infinite stream of cash flows since most discount factor tables do not list factors for periods in excess of 100 years. However, the (A/P) discount factor approaches the interest rate as n becomes large. Since the (P/A) and (A/P) factors are reciprocals of each other, it is possible to divide an infinite series of annual cash flows by the interest rate in order to calculate the present worth of the infinite series.

Equation 50.15: Capitalized Costs for an Infinite Series

$$
\text{capitalized costs} = P = \frac{A}{i} \qquad \textit{50.15}
$$

Description

Equation 50.15 can be used when the annual costs are equal in every year. If the operating and maintenance costs occur irregularly instead of annually, or if the costs vary from year to year, it will be necessary to somehow determine a cash flow of equal annual amounts that is equivalent to the stream of original costs (i.e., to determine the EUAC).

Example

The construction of a volleyball court will cost $1200, and annual maintenance cost is expected to be $300. At an effective annual interest rate of 5%, the project's capitalized cost is most nearly

(A) $2000

(B) $3000

(C) $7000

(D) $20,000

Solution

The cost of the project consists of two parts: the construction cost of $1200 and the annual maintenance cost of $300. The maintenance cost is an infinite series of annual amounts, so use Eq. 50.15 to find its present worth.

$$
\begin{aligned}
P_{\text{maintenance}} &= \frac{A}{i} = \frac{\$300}{0.05} \\
&= \$6000
\end{aligned}
$$

Add the present worth of the initial construction cost to get the total present worth (i.e., the capitalized cost) of the project.

$$
\begin{aligned}
P_{\text{total}} &= P_{\text{construction}} + P_{\text{maintenance}} \\
&= \$1200 + \$6000 \\
&= \$7200 \quad (\$7000)
\end{aligned}
$$

The answer is (C).

Process Design/Economics

15. INFLATION

To be meaningful, economic studies must be performed in terms of constant-value dollars. Several common methods are used to allow for *inflation*. One alternative is to replace the effective annual interest rate, i, with a value adjusted for inflation, d.

Equation 50.16: Interest Rate Adjusted for Inflation

$$d = i + f + (i \times f) \qquad 50.16$$

Description

In Eq. 50.16, f is a constant *inflation rate* per year. The inflation-adjusted interest rate, d, can be used to compute present worth.

Example

An investment of \$20,000 earns an effective annual interest of 10%. The value of the investment in five years, adjusted for an annual inflation rate of 6%, is most nearly

   (A) \$27,000

   (B) \$32,000

   (C) \$42,000

   (D) \$43,000

Solution

The interest rate adjusted for inflation is

$$d = i + f + (i \times f)$$
$$= 0.10 + 0.06 + (0.10)(0.06)$$
$$= 0.166$$

To determine the future worth of the investment, adjusted for inflation, use d instead of i in Eq. 50.1.

$$F = P(1 + d)^n$$
$$= (\$20{,}000)(1 + 0.166)^5$$
$$= \$43{,}105 \quad (\$43{,}000)$$

The answer is (D).

16. CAPITAL BUDGETING (ALTERNATIVE COMPARISONS)

In the real world, the majority of engineering economic analysis problems are alternative comparisons. In these problems, two or more mutually exclusive investments compete for limited funds. A variety of methods exists for selecting the superior alternative from a group of proposals. Each method has its own merits and applications.

Present Worth Analysis

When two or more alternatives are capable of performing the same functions, the economically superior alternative will have the largest present worth. The *present worth method* is restricted to evaluating alternatives that are mutually exclusive and that have the same lives. This method is suitable for ranking the desirability of alternatives.

Annual Cost Analysis

Alternatives that accomplish the same purpose but that have unequal lives must be compared by the *annual cost method*. The annual cost method assumes that each alternative will be replaced by an identical twin at the end of its useful life (i.e., infinite renewal). This method, which may also be used to rank alternatives according to their desirability, is also called the *annual return method* or *capital recovery method*.

The alternatives must be mutually exclusive and repeatedly renewed up to the duration of the longest-lived alternative. The calculated annual cost is known as the *equivalent uniform annual cost* (EUAC) or *equivalent annual cost* (EAC). Cost is a positive number when expenses exceed income.

Rate of Return Analysis

An intuitive definition of the *rate of return* (ROR) is the effective annual interest rate at which an investment accrues income. That is, the rate of return of an investment is the interest rate that would yield identical profits if all money was invested at that rate. Although this definition is correct, it does not provide a method of determining the rate of return.

The present worth of a \$100 investment invested at 5% is zero when $i = 5\%$ is used to determine equivalence. Therefore, a working definition of rate of return would be the effective annual interest rate that makes the present worth of the investment zero. Alternatively, rate of return could be defined as the effective annual interest rate that makes the benefits and costs equal.

A company may not know what effective interest rate, i, to use in engineering economic analysis. In such a case, the company can establish a minimum level of economic performance that it would like to realize on all investments. This criterion is known as the *minimum attractive rate of return*, or MARR.

Once a rate of return for an investment is known, it can be compared with the minimum attractive rate of return. If the rate of return is equal to or exceeds the minimum attractive rate of return, the investment is qualified (i.e., the alternative is viable). This is the basis for the rate of return method of alternative viability analysis.

If rate of return is used to select among two or more investments, an *incremental analysis* must be performed. An incremental analysis begins by ranking the alternatives in order of increasing initial investment.

Then, the cash flows for the investment with the lower initial cost are subtracted from the cash flows for the higher-priced alternative on a year-by-year basis. This produces, in effect, a third alternative representing the costs and benefits of the added investment. The added expense of the higher-priced investment is not warranted unless the rate of return of this third alternative exceeds the minimum attractive rate of return as well. The alternative with the higher initial investment is superior if the incremental rate of return exceeds the minimum attractive rate of return.

Finding the rate of return can be a long, iterative process, requiring either interpolation or trial and error. Sometimes, the actual numerical value of rate of return is not needed; it is sufficient to know whether or not the rate of return exceeds the minimum attractive rate of return. This comparative analysis can be accomplished without calculating the rate of return simply by finding the present worth of the investment using the minimum attractive rate of return as the effective interest rate (i.e., $i = $ MARR). If the present worth is zero or positive, the investment is qualified. If the present worth is negative, the rate of return is less than the minimum attractive rate of return and the additional investment is not warranted.

The present worth, annual cost, and rate of return methods of comparing alternatives yield equivalent results, but they are distinctly different approaches. The present worth and annual cost methods may use either effective interest rates or the minimum attractive rate of return to rank alternatives or compare them to the MARR. If the incremental rate of return of pairs of alternatives are compared with the MARR, the analysis is considered a rate of return analysis.

17. BREAK-EVEN ANALYSIS

Break-even analysis is a method of determining when the value of one alternative becomes equal to the value of another. It is commonly used to determine when costs exactly equal revenue. If the manufactured quantity is less than the *break-even quantity*, a loss is incurred. If the manufactured quantity is greater than the break-even quantity, a profit is made.

An alternative form of the break-even problem is to find the number of units per period for which two alternatives have the same total costs. Fixed costs are spread over a period longer than one year using the EUAC concept. One of the alternatives will have a lower cost if production is less than the break-even point. The other will have a lower cost if production is greater than the break-even point.

The *pay-back period*, PBP, is defined as the length of time, n, usually in years, for the cumulative net annual profit to equal the initial investment. It is tempting to introduce equivalence into pay-back period calculations, but the convention is not to.

18. BENEFIT-COST ANALYSIS

The *benefit-cost ratio method* is often used in municipal project evaluations where benefits and costs accrue to different segments of the community. With this method, the present worth of all benefits (irrespective of the beneficiaries), B, is divided by the present worth of all costs, C. If the benefit-cost ratio, B/C, is greater than or equal to 1.0, the project is acceptable. (Equivalent uniform annual costs can be used in place of present worths.)

When the benefit-cost ratio method is used, disbursements by the initiators or sponsors are *costs* and added to C. Disbursements by the users of the project are known as *disbenefits* and subtracted from B. It is often difficult to decide whether a cash flow should be regarded as a cost or a disbenefit. The placement of such cash flows can change the value of B/C, but cannot change whether B/C is greater than or equal to 1.0. For this reason, the benefit-cost ratio alone should not be used to rank competing projects.

If ranking is to be done by the benefit-cost ratio method, an incremental analysis is required, as it is for the rate-of-return method. The incremental analysis is accomplished by calculating the ratio of differences in benefits to differences in costs for each possible pair of alternatives. If the ratio exceeds 1.0, alternative 2 is superior to alternative 1. Otherwise, alternative 1 is superior.

Equation 50.17: Analysis Criterion

$$B - C \geq 0 \text{ or } B/C \geq 1 \qquad 50.17$$

Description

A project is acceptable if its benefit-cost ratio equals or exceeds 1 (i.e., $B/C \geq 1$). This will be true whenever $B - C \geq 0$.

Example

A large sewer system will cost $175,000 annually. There will be favorable consequences to the general public equivalent to $500,000 annually, and adverse consequences to a small segment of the public equivalent to $50,000 annually. The benefit-cost ratio is most nearly

(A) 2.2

(B) 2.4

(C) 2.6

(D) 2.9

Solution

The adverse consequences worth $50,000 affect the users of the project, not its initiators, so this is a disbenefit. The benefit-cost ratio is

$$B/C = \frac{\$500{,}000 - \$50{,}000}{\$175{,}000} = 2.57 \quad (2.6)$$

The answer is (C).

19. SENSITIVITY ANALYSIS, RISK ANALYSIS, AND UNCERTAINTY ANALYSIS

Data analysis and forecasts in economic studies require estimates of costs that will occur in the future. There are always uncertainties about these costs. However, these uncertainties are an insufficient reason not to make the best possible estimates of the costs. Nevertheless, a decision between alternatives often can be made more confidently if it is known whether or not the conclusion is sensitive to moderate changes in data forecasts. *Sensitivity analysis* provides this extra dimension to an economic analysis.

The sensitivity of a decision to various factors is determined by inserting a range of estimates for critical cash flows and other parameters. If radical changes can be made to a cash flow without changing the decision, the decision is said to be *insensitive* to uncertainties regarding that cash flow. However, if a small change in the estimate of a cash flow will alter the decision, that decision is said to be very *sensitive* to changes in the estimate. If the decision is sensitive only for a limited range of cash flow values, the term *variable sensitivity* is used. Figure 50.5 illustrates these terms.

Figure 50.5 Types of Sensitivity

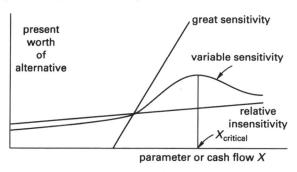

An established semantic tradition distinguishes between risk analysis and uncertainty analysis. *Risk* is the possibility of an unexpected, unplanned, or undesirable event occurring (i.e., an occurrence not planned for or predicted in risk analysis). *Risk analysis* addresses variables that have a known or estimated probability distribution. In this regard, statistics and probability theory can be used to determine the probability of a cash flow varying between given limits. On the other hand, *uncertainty analysis* is concerned with situations in which there is not enough information to determine the probability or frequency distribution for the variables involved.

As a first step, sensitivity analysis should be performed one factor at a time to the dominant factors. Dominant cost factors are those that have the most significant impact on the present value of the alternative.[3] If

warranted, additional investigation can be used to determine the sensitivity to several cash flows varying simultaneously. Significant judgment is needed, however, to successfully determine the proper combinations of cash flows to vary. It is common to plot the dependency of the present value on the cash flow being varied in a two-dimensional graph. Simple linear interpolation is used (within reason) to determine the critical value of the cash flow being varied.

20. ACCOUNTING PRINCIPLES

Basic Bookkeeping

An accounting or *bookkeeping system* is used to record historical financial transactions. The resultant records are used for product costing, satisfaction of statutory requirements, reporting of profit for income tax purposes, and general company management.

Bookkeeping consists of two main steps: recording the transactions, followed by categorization of the transactions.[4] The transactions (receipts and disbursements) are recorded in a *journal* (*book of original entry*) to complete the first step. Such a journal is organized in a simple chronological and sequential manner. The transactions are then categorized (into interest income, advertising expense, etc.) and posted (i.e., entered or written) into the appropriate *ledger account*.[5]

Together, the ledger accounts constitute the *general ledger* or *ledger*. All ledger accounts can be classified into one of three types: *asset accounts*, *liability accounts*, and *owners' equity accounts*. Strictly speaking, income and expense accounts, kept in a separate journal, are included within the classification of owners' equity accounts.

Together, the journal and ledger are known simply as "the books" of the company, regardless of whether bound volumes of pages are actually involved.

Balancing the Books

In a business environment, *balancing the books* means more than reconciling the checkbook and bank statements. All accounting entries must be posted in such a way as to maintain the equality of the *basic accounting equation*,

$$\text{assets} = \text{liability} + \text{owner's equity}$$

In a *double-entry bookkeeping system*, the equality is maintained within the ledger system by entering each transaction into two balancing ledger accounts. For example, paying a utility bill would decrease the cash account (an asset account) and decrease the utility expense account (a liability account) by the same amount.

---

[3]In particular, engineering economic analysis problems are sensitive to the choice of effective interest rate, i, and to accuracy in cash flows at or near the beginning of the horizon. The problems will be less sensitive to accuracy in far-future cash flows, such as subsequent generation replacement costs.

[4]These two steps are not to be confused with the *double-entry bookkeeping method*.
[5]The two-step process is more typical of a *manual bookkeeping system* than a computerized *general ledger system*. However, even most computerized systems produce reports in journal entry order, as well as account summaries.

Transactions are either *debits* or *credits*, depending on their sign. Increases in asset accounts are debits; decreases are credits. For liability and equity accounts, the opposite is true: Increases are credits, and decreases are debits.[6]

Cash and Accrual Systems[7]

The simplest form of bookkeeping is based on the *cash system*. The only transactions that are entered into the journal are those that represent cash receipts and disbursements. In effect, a checkbook register or bank deposit book could serve as the journal.

During a given period (e.g., month or quarter), expense liabilities may be incurred even though the payments for those expenses have not been made. For example, an invoice (bill) may have been received but not paid. Under the *accrual system*, the obligation is posted into the appropriate expense account before it is paid.[8] Analogous to expenses, under the accrual system, income will be claimed before payment is received. Specifically, a sales transaction can be recorded as income when the customer's order is received, when the outgoing invoice is generated, or when the merchandise is shipped.

Financial Statements

Each period, two types of corporate financial statements are typically generated: the *balance sheet* and *profit and loss* (P&L) *statement*.[9] The profit and loss statement, also known as a *statement of income and retained earnings*, is a summary of sources of *income* or *revenue* (interest, sales, fees charged, etc.) and *expenses* (utilities, advertising, repairs, etc.) for the period. The expenses are subtracted from the revenues to give a *net income* (generally, before taxes).[10] Figure 50.6 illustrates a simplified profit and loss statement.

The *balance sheet* presents the *basic accounting equation* in tabular form. The balance sheet lists the major categories of assets and outstanding liabilities. The difference between asset values and liabilities is the *equity*.

[6]There is a difference in sign between asset and liability accounts. An increase in an expense account is actually a decrease. The accounting profession, apparently, is comfortable with the common confusion that exists between debits and credits.

[7]There is also a distinction made between cash flows that are known and those that are expected. It is a *standard accounting principle* to record losses in full, at the time they are recognized, even before their occurrence. In the construction industry, for example, losses are recognized in full and projected to the end of a project as soon as they are foreseeable. Profits, on the other hand, are recognized only as they are realized (typically, as a percentage of project completion). The difference between cash and accrual systems is a matter of *bookkeeping*. The difference between loss and profit recognition is a matter of *accounting convention*. Engineers seldom need to be concerned with the accounting principles and conventions.

[8]The expense for an item or service might be accrued even *before* the invoice is received. It might be recorded when the purchase order for the item or service is generated, or when the item or service is received.

[9]Other types of financial statements (*statements of changes in financial position, cost of sales statements, inventory and asset reports*, etc.) also will be generated, depending on the needs of the company.

[10]Financial statements also can be prepared with percentages (of total assets and net revenue) instead of dollars, in which case they are known as *common size financial statements*.

Figure 50.6 Simplified Profit and Loss Statement

revenue			
interest	2000		
sales	237,000		
returns	(23,000)		
net revenue		216,000	
expenses			
salaries	149,000		
utilities	6000		
advertising	28,000		
insurance	4000		
supplies	1000		
net expenses		188,000	
period net income			28,000
beginning retained earnings			63,000
net year-to-date earnings			91,000

This equity represents what would be left over after satisfying all debts by liquidating the company.

Figure 50.7 is a simplified balance sheet.

Figure 50.7 Simplified Balance Sheet

ASSETS

current assets			
cash	14,000		
accounts receivable	36,000		
notes receivable	20,000		
inventory	89,000		
prepaid expenses	3000		
total current assets		162,000	
plant, property, and equipment			
land and buildings	217,000		
motor vehicles	31,000		
equipment	94,000		
accumulated depreciation	(52,000)		
total fixed assets		290,000	
total assets			452,000

LIABILITIES AND OWNERS' EQUITY

current liabilities			
accounts payable	66,000		
accrued income taxes	17,000		
accrued expenses	8000		
total current liabilities		91,000	
long-term debt			
notes payable	117,000		
mortgage	23,000		
total long-term debt		140,000	
owners' and stockholders' equity			
stock	130,000		
retained earnings	91,000		
total owners' equity		221,000	
total liabilities and owners' equity			452,000

There are several terms that appear regularly on balance sheets.

- *current assets:* cash and other assets that can be converted quickly into cash, such as accounts receivable, notes receivable, and merchandise (inventory). Also known as *liquid assets.*

- *fixed assets:* relatively permanent assets used in the operation of the business and relatively difficult to convert into cash. Examples are land, buildings, and equipment. Also known as *nonliquid assets.*

- *current liabilities:* liabilities due within a short period of time (e.g., within one year) and typically paid out of current assets. Examples are accounts payable, notes payable, and other accrued liabilities.

- *long-term liabilities:* obligations that are not totally payable within a short period of time (e.g., within one year).

Analysis of Financial Statements

Financial statements are evaluated by management, lenders, stockholders, potential investors, and many other groups for the purpose of determining the *health of the company.* The health can be measured in terms of *liquidity* (ability to convert assets to cash quickly), *solvency* (ability to meet debts as they become due), and *relative risk* (of which one measure is *leverage*—the portion of total capital contributed by owners).

The analysis of financial statements involves several common ratios, usually expressed as percentages. The following are some frequently encountered ratios.

- *current ratio:* an index of short-term paying ability.

$$\text{current ratio} = \frac{\text{current assets}}{\text{current liabilities}}$$

- *quick* (or *acid-test*) *ratio:* a more stringent measure of short-term debt-paying ability. The *quick assets* are defined to be current assets minus inventories and prepaid expenses.

$$\text{quick ratio} = \frac{\text{quick assets}}{\text{current liabilities}}$$

- *receivable turnover:* a measure of the average speed with which accounts receivable are collected.

$$\text{receivable turnover} = \frac{\text{net credit sales}}{\text{average net receivables}}$$

- *average age of receivables:* number of days, on the average, in which receivables are collected.

$$\text{average age of receivables} = \frac{365}{\text{receivable turnover}}$$

- *inventory turnover:* a measure of the speed, on the average, with which inventory is sold.

$$\text{inventory turnover} = \frac{\text{cost of goods sold}}{\text{average unit cost of inventory}}$$

- *days supply of inventory on hand:* number of days, on the average, that the current inventory would last.

$$\text{days supply of inventory on hand} = \frac{365}{\text{inventory turnover}}$$

- *book value per share of common stock:* number of dollars represented by the balance sheet owners' equity for each share of common stock outstanding.

$$\text{book value per share of common stock} = \frac{\text{common shareholders' equity}}{\text{number of outstanding shares}}$$

- *gross margin:* gross profit as a percentage of sales. (Gross profit is sales less cost of goods sold.)

$$\text{gross margin} = \frac{\text{gross profit}}{\text{net sales}}$$

- *profit margin ratio:* percentage of each dollar of sales that is net income.

$$\text{profit margin} = \frac{\text{net income before taxes}}{\text{net sales}}$$

- *return on investment ratio:* shows the percent return on owners' investment.

$$\text{return on investment} = \frac{\text{net income}}{\text{owners' equity}}$$

- *price-earnings ratio:* an indication of the relationship between earnings and market price per share of common stock; useful in comparisons between alternative investments.

$$\text{price-earnings} = \frac{\text{market price per share}}{\text{earnings per share}}$$

21. ACCOUNTING COSTS AND EXPENSE TERMS

The accounting profession has developed special terms for certain groups of costs. When annual costs are incurred due to the functioning of a piece of equipment, they are known as *operating and maintenance* (O&M) *costs.* The annual costs associated with operating a business (other than the costs directly attributable to production) are known as *general, selling, and administrative* (GS&A) *expenses.*

Direct labor costs are costs incurred in the factory, such as assembly, machining, and painting labor costs. *Direct material costs* are the costs of all materials that go into production.[11] Typically, both direct labor and direct material costs are given on a per-unit or per-item basis.

---

[11]There may be complications with pricing the material when it is purchased from an outside vendor and the stock on hand derives from several shipments purchased at different prices.

The sum of the direct labor and direct material costs is known as the *prime cost*.

There are certain additional expenses incurred in the factory, such as the costs of factory supervision, stock-picking, quality control, factory utilities, and miscellaneous supplies (cleaning fluids, assembly lubricants, routing tags, etc.) that are not incorporated into the final product. Such costs are known as *indirect manufacturing expenses* (IME) or *indirect material and labor costs*.[12] The sum of the per-unit indirect manufacturing expense and prime cost is known as the *factory cost*.

Research and development (R&D) *costs* and *administrative expenses* are added to the factory cost to give the *manufacturing cost* of the product.

Additional costs are incurred in marketing the product. Such costs are known as *selling expenses* or *marketing expenses*. The sum of the selling expenses and manufacturing cost is the *total cost* of the product. Figure 50.8 illustrates these terms.[13] Typical classifications of expenses are listed in Table 50.3.

Figure 50.8 *Costs and Expenses Combined*

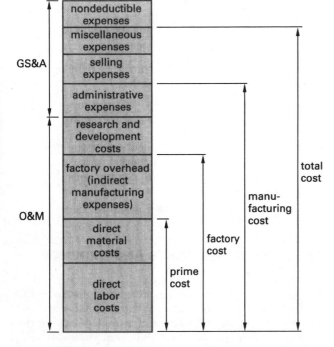

The distinctions among the various forms of cost (particularly with overhead costs) are not standardized. Each company must develop a classification system to deal with the various cost factors in a consistent manner. There are also other terms in use (e.g., *raw materials, operating supplies, general plant overhead*), but these terms must be interpreted within the framework of each company's classification system. Table 50.3 is typical of such classification systems.

[12]The *indirect material and labor costs* usually exclude costs incurred in the office area.
[13]*Total cost* does not include income taxes.

Table 50.3 *Typical Classification of Expenses*

direct labor expenses
  machining and forming
  assembly
  finishing
  inspection
  testing
direct material expenses
  items purchased from other vendors
  manufactured assemblies
factory overhead expenses (*indirect manufacturing expenses*)
  supervision
  benefits
    pension
    medical insurance
    vacations
  wages overhead
    unemployment compensation taxes
    social security taxes
    disability taxes
  stock-picking
  quality control and inspection
  expediting
  rework
  maintenance
  miscellaneous supplies
    routing tags
    assembly lubricants
    cleaning fluids
    wiping cloths
    janitorial supplies
  packaging (materials and labor)
  factory utilities
  laboratory
  depreciation on factory equipment
research and development expenses
  engineering (labor)
  patents
  testing
  prototypes (material and labor)
  drafting
  O&M of R&D facility
administrative expenses
  corporate officers
  accounting
  secretarial/clerical/reception
  security (protection)
  medical (nurse)
  employment (personnel)
  reproduction
  data processing
  production control
  depreciation on nonfactory equipment
  office supplies
  office utilities
  O&M of offices
selling expenses
  marketing (labor)
  advertising
  transportation (if not paid by customer)
  outside sales force (labor and expenses)
  demonstration units
  commissions
  technical service and support
  order processing
  branch office expenses
miscellaneous expenses
  insurance
  property taxes
  interest on loans
nondeductible expenses
  federal income taxes
  fines and penalties

22. COST ACCOUNTING

Cost accounting is the system that determines the cost of manufactured products. Cost accounting is called *job cost accounting* if costs are accumulated by part number or contract. It is called *process cost accounting* if costs are accumulated by departments or manufacturing processes.

Cost accounting is dependent on historical and recorded data. The unit product cost is determined from actual expenses and numbers of units produced. Allowances (i.e., budgets) for future costs are based on these historical figures. Any deviation from historical figures is called a *variance*. Where adequate records are available, variances can be divided into *labor variance* and *material variance*.

When determining a unit product cost, the direct material and direct labor costs are generally clear-cut and easily determined. Furthermore, these costs are 100% variable costs. However, the indirect cost per unit of product is not as easily determined. Indirect costs (*burden, overhead*, etc.) can be fixed or semivariable costs. The amount of indirect cost allocated to a unit will depend on the unknown future overhead expense as well as the unknown future production (*vehicle size*).

A typical method of allocating indirect costs to a product is as follows.

step 1: Estimate the total expected indirect (and overhead) costs for the upcoming year.

step 2: Determine the most appropriate vehicle (basis) for allocating the overhead to production. Usually, this vehicle is either the number of units expected to be produced or the number of direct hours expected to be worked in the upcoming year.

step 3: Estimate the quantity or size of the overhead vehicle.

step 4: Divide expected overhead costs by the expected overhead vehicle to obtain the unit overhead.

step 5: Regardless of the true size of the overhead vehicle during the upcoming year, one unit of overhead cost is allocated per unit of overhead vehicle.

Once the prime cost has been determined and the indirect cost calculated based on projections, the two are combined into a *standard factory cost* or *standard cost*, which remains in effect until the next budgeting period (usually a year).

During the subsequent manufacturing year, the standard cost of a product is not generally changed merely because it is found that an error in projected indirect costs or production quantity (vehicle size) has been made. The allocation of indirect costs to a product is assumed to be independent of errors in forecasts. Rather, the difference between the expected and actual expenses, known as the *burden (overhead) variance*, experienced during the year is posted to one or more *variance accounts*.

Burden (overhead) variance is caused by errors in forecasting both the actual indirect expense for the upcoming year and the overhead vehicle size. In the former case, the variance is called *burden budget variance*; in the latter, it is called *burden capacity variance*.

Process Design/Economics

Table 50.4 *Factor Table i = 0.50%*

n	P/F	P/A	P/G	F/P	F/A	A/P	A/F	A/G
1	0.9950	0.9950	0.0000	1.0050	1.0000	1.0050	1.0000	0.0000
2	0.9901	1.9851	0.9901	1.0100	2.0050	0.5038	0.4988	0.4988
3	0.9851	2.9702	2.9604	1.0151	3.0150	0.3367	0.3317	0.9967
4	0.9802	3.9505	5.9011	1.0202	4.0301	0.2531	0.2481	1.4938
5	0.9754	4.9259	9.8026	1.0253	5.0503	0.2030	0.1980	1.9900
6	0.9705	5.8964	14.6552	1.0304	6.0755	0.1696	0.1646	2.4855
7	0.9657	6.8621	20.4493	1.0355	7.1059	0.1457	0.1407	2.9801
8	0.9609	7.8230	27.1755	1.0407	8.1414	0.1278	0.1228	3.4738
9	0.9561	8.7791	34.8244	1.0459	9.1821	0.1139	0.1089	3.9668
10	0.9513	9.7304	43.3865	1.0511	10.2280	0.1028	0.0978	4.4589
11	0.9466	10.6670	52.8526	1.0564	11.2792	0.0937	0.0887	4.9501
12	0.9419	11.6189	63.2136	1.0617	12.3356	0.0861	0.0811	5.4406
13	0.9372	12.5562	74.4602	1.0670	13.3972	0.0796	0.0746	5.9302
14	0.9326	13.4887	86.5835	1.0723	14.4642	0.0741	0.0691	6.4190
15	0.9279	14.4166	99.5743	1.0777	15.5365	0.0694	0.0644	6.9069
16	0.9233	15.3399	113.4238	1.0831	16.6142	0.0652	0.0602	7.3940
17	0.9187	16.2586	128.1231	1.0885	17.6973	0.0615	0.0565	7.8803
18	0.9141	17.1728	143.6634	1.0939	18.7858	0.0582	0.0532	8.3658
19	0.9096	18.0824	160.0360	1.0994	19.8797	0.0553	0.0503	8.8504
20	0.9051	18.9874	177.2322	1.1049	20.9791	0.0527	0.0477	9.3342
21	0.9006	19.8880	195.2434	1.1104	22.0840	0.0503	0.0453	9.8172
22	0.8961	20.7841	214.0611	1.1160	23.1944	0.0481	0.0431	10.2993
23	0.8916	21.6757	233.6768	1.1216	24.3104	0.0461	0.0411	10.7806
24	0.8872	22.5629	254.0820	1.1272	25.4320	0.0443	0.0393	11.2611
25	0.8828	23.4456	275.2686	1.1328	26.5591	0.0427	0.0377	11.7407
30	0.8610	27.7941	392.6324	1.1614	32.2800	0.0360	0.0310	14.1265
40	0.8191	36.1722	681.3347	1.2208	44.1588	0.0276	0.0226	18.8359
50	0.7793	44.1428	1,035.6966	1.2832	56.6452	0.0227	0.0177	23.4624
60	0.7414	51.7256	1,448.6458	1.3489	69.7700	0.0193	0.0143	28.0064
100	0.6073	78.5426	3,562.7934	1.6467	129.3337	0.0127	0.0077	45.3613

Process Design/
Economics

Table 50.5 *Factor Table i = 1.00%*

n	P/F	P/A	P/G	F/P	F/A	A/P	A/F	A/G
1	0.9901	0.9901	0.0000	1.0100	1.0000	1.0100	1.0000	0.0000
2	0.9803	1.9704	0.9803	1.0201	2.0100	0.5075	0.4975	0.4975
3	0.9706	2.9410	2.9215	1.0303	3.0301	0.3400	0.3300	0.9934
4	0.9610	3.9020	5.8044	1.0406	4.0604	0.2563	0.2463	1.4876
5	0.9515	4.8534	9.6103	1.0510	5.1010	0.2060	0.1960	1.9801
6	0.9420	5.7955	14.3205	1.0615	6.1520	0.1725	0.1625	2.4710
7	0.9327	6.7282	19.9168	1.0721	7.2135	0.1486	0.1386	2.9602
8	0.9235	7.6517	26.3812	1.0829	8.2857	0.1307	0.1207	3.4478
9	0.9143	8.5650	33.6959	1.0937	9.3685	0.1167	0.1067	3.9337
10	0.9053	9.4713	41.8435	1.1046	10.4622	0.1056	0.0956	4.4179
11	0.8963	10.3676	50.8067	1.1157	11.5668	0.0965	0.0865	4.9005
12	0.8874	11.2551	60.5687	1.1268	12.6825	0.0888	0.0788	5.3815
13	0.8787	12.1337	71.1126	1.1381	13.8093	0.0824	0.0724	5.8607
14	0.8700	13.0037	82.4221	1.1495	14.9474	0.0769	0.0669	6.3384
15	0.8613	13.8651	94.4810	1.1610	16.0969	0.0721	0.0621	6.8143
16	0.8528	14.7179	107.2734	1.1726	17.2579	0.0679	0.0579	7.2886
17	0.8444	15.5623	129.7834	1.1843	18.4304	0.0643	0.0543	7.7613
18	0.8360	16.3983	134.9957	1.1961	19.6147	0.0610	0.0510	8.2323
19	0.8277	17.2260	149.8950	1.2081	20.8109	0.0581	0.0481	8.7017
20	0.8195	18.0456	165.4664	1.2202	22.0190	0.0554	0.0454	9.1694
21	0.8114	18.8570	181.6950	1.2324	23.2392	0.0530	0.0430	9.6354
22	0.8034	19.6604	198.5663	1.2447	24.4716	0.0509	0.0409	10.0998
23	0.7954	20.4558	216.0660	1.2572	25.7163	0.0489	0.0389	10.5626
24	0.7876	21.2434	234.1800	1.2697	26.9735	0.0471	0.0371	11.0237
25	0.7798	22.0232	252.8945	1.2824	28.2432	0.0454	0.0354	11.4831
30	0.7419	25.8077	355.0021	1.3478	34.7849	0.0387	0.0277	13.7557
40	0.6717	32.8347	596.8561	1.4889	48.8864	0.0305	0.0205	18.1776
50	0.6080	39.1961	879.4176	1.6446	64.4632	0.0255	0.0155	22.4363
60	0.5504	44.9550	1,192.8061	1.8167	81.6697	0.0222	0.0122	26.5333
100	0.3697	63.0289	2,605.7758	2.7048	170.4814	0.0159	0.0059	41.3426

Table 50.6 *Factor Table i = 1.50%*

n	P/F	P/A	P/G	F/P	F/A	A/P	A/F	A/G
1	0.9852	0.9852	0.0000	1.0150	1.0000	1.0150	1.0000	0.0000
2	0.9707	1.9559	0.9707	1.0302	2.0150	0.5113	0.4963	0.4963
3	0.9563	2.9122	2.8833	1.0457	3.0452	0.3434	0.3284	0.9901
4	0.9422	3.8544	5.7098	1.0614	4.0909	0.2594	0.2444	1.4814
5	0.9283	4.7826	9.4229	1.0773	5.1523	0.2091	0.1941	1.9702
6	0.9145	5.6972	13.9956	1.0934	6.2296	0.1755	0.1605	2.4566
7	0.9010	6.5982	19.4018	1.1098	7.3230	0.1516	0.1366	2.9405
8	0.8877	7.4859	26.6157	1.1265	8.4328	0.1336	0.1186	3.4219
9	0.8746	8.3605	32.6125	1.1434	9.5593	0.1196	0.1046	3.9008
10	0.8617	9.2222	40.3675	1.1605	10.7027	0.1084	0.0934	4.3772
11	0.8489	10.0711	48.8568	1.1779	11.8633	0.0993	0.0843	4.8512
12	0.8364	10.9075	58.0571	1.1956	13.0412	0.0917	0.0767	5.3227
13	0.8240	11.7315	67.9454	1.2136	14.2368	0.0852	0.0702	5.7917
14	0.8118	12.5434	78.4994	1.2318	15.4504	0.0797	0.0647	6.2582
15	0.7999	13.3432	89.6974	1.2502	16.6821	0.0749	0.0599	6.7223
16	0.7880	14.1313	101.5178	1.2690	17.9324	0.0708	0.0558	7.1839
17	0.7764	14.9076	113.9400	1.2880	19.2014	0.0671	0.0521	7.6431
18	0.7649	15.6726	126.9435	1.3073	20.4894	0.0638	0.0488	8.0997
19	0.7536	16.4262	140.5084	1.3270	21.7967	0.0609	0.0459	8.5539
20	0.7425	17.1686	154.6154	1.3469	23.1237	0.0582	0.0432	9.0057
21	0.7315	17.9001	169.2453	1.3671	24.4705	0.0559	0.0409	9.4550
22	0.7207	18.6208	184.3798	1.3876	25.8376	0.0537	0.0387	9.9018
23	0.7100	19.3309	200.0006	1.4084	27.2251	0.0517	0.0367	10.3462
24	0.6995	20.0304	216.0901	1.4295	28.6335	0.0499	0.0349	10.7881
25	0.6892	20.7196	232.6310	1.4509	30.0630	0.0483	0.0333	11.2276
30	0.6398	24.0158	321.5310	1.5631	37.5387	0.0416	0.0266	13.3883
40	0.5513	29.9158	524.3568	1.8140	54.2679	0.0334	0.0184	17.5277
50	0.4750	34.9997	749.9636	2.1052	73.6828	0.0286	0.0136	21.4277
60	0.4093	39.3803	988.1674	2.4432	96.2147	0.0254	0.0104	25.0930
100	0.2256	51.6247	1,937.4506	4.4320	228.8030	0.0194	0.0044	37.5295

Process Design/
Economics

Table 50.7 Factor Table i = 2.00%

n	P/F	P/A	P/G	F/P	F/A	A/P	A/F	A/G
1	0.9804	0.9804	0.0000	1.0200	1.0000	1.0200	1.0000	0.0000
2	0.9612	1.9416	0.9612	1.0404	2.0200	0.5150	0.4950	0.4950
3	0.9423	2.8839	2.8458	1.0612	3.0604	0.3468	0.3268	0.9868
4	0.9238	3.8077	5.6173	1.0824	4.1216	0.2626	0.2426	1.4752
5	0.9057	4.7135	9.2403	1.1041	5.2040	0.2122	0.1922	1.9604
6	0.8880	5.6014	13.6801	1.1262	6.3081	0.1785	0.1585	2.4423
7	0.8706	6.4720	18.9035	1.1487	7.4343	0.1545	0.1345	2.9208
8	0.8535	7.3255	24.8779	1.1717	8.5830	0.1365	0.1165	3.3961
9	0.8368	8.1622	31.5720	1.1951	9.7546	0.1225	0.1025	3.8681
10	0.8203	8.9826	38.9551	1.2190	10.9497	0.1113	0.0913	4.3367
11	0.8043	9.7868	46.9977	1.2434	12.1687	0.1022	0.0822	4.8021
12	0.7885	10.5753	55.6712	1.2682	13.4121	0.0946	0.0746	5.2642
13	0.7730	11.3484	64.9475	1.2936	14.6803	0.0881	0.0681	5.7231
14	0.7579	12.1062	74.7999	1.3195	15.9739	0.0826	0.0626	6.1786
15	0.7430	12.8493	85.2021	1.3459	17.2934	0.0778	0.0578	6.6309
16	0.7284	13.5777	96.1288	1.3728	18.6393	0.0737	0.0537	7.0799
17	0.7142	14.2919	107.5554	1.4002	20.0121	0.0700	0.0500	7.5256
18	0.7002	14.9920	119.4581	1.4282	21.4123	0.0667	0.0467	7.9681
19	0.6864	15.6785	131.8139	1.4568	22.8406	0.0638	0.0438	8.4073
20	0.6730	16.3514	144.6003	1.4859	24.2974	0.0612	0.0412	8.8433
21	0.6598	17.0112	157.7959	1.5157	25.7833	0.0588	0.0388	9.2760
22	0.6468	17.6580	171.3795	1.5460	27.2990	0.0566	0.0366	9.7055
23	0.6342	18.2922	185.3309	1.5769	28.8450	0.0547	0.0347	10.1317
24	0.6217	18.9139	199.6305	1.6084	30.4219	0.0529	0.0329	10.5547
25	0.6095	19.5235	214.2592	1.6406	32.0303	0.0512	0.0312	10.9745
30	0.5521	22.3965	291.7164	1.8114	40.5681	0.0446	0.0246	13.0251
40	0.4529	27.3555	461.9931	2.2080	60.4020	0.0366	0.0166	16.8885
50	0.3715	31.4236	642.3606	2.6916	84.5794	0.0318	0.0118	20.4420
60	0.3048	34.7609	823.6975	3.2810	114.0515	0.0288	0.0088	23.6961
100	0.1380	43.0984	1,464.7527	7.2446	312.2323	0.0232	0.0032	33.9863

Process Design/ Economics

Table 50.8 Factor Table i = 4.00%

n	P/F	P/A	P/G	F/P	F/A	A/P	A/F	A/G
1	0.9615	0.9615	0.0000	1.0400	1.0000	1.0400	1.0000	0.0000
2	0.9246	1.8861	0.9246	1.0816	2.0400	0.5302	0.4902	0.4902
3	0.8890	2.7751	2.7025	1.1249	3.1216	0.3603	0.3203	0.9739
4	0.8548	3.6299	5.2670	1.1699	4.2465	0.2755	0.2355	1.4510
5	0.8219	4.4518	8.5547	1.2167	5.4163	0.2246	0.1846	1.9216
6	0.7903	5.2421	12.5062	1.2653	6.6330	0.1908	0.1508	2.3857
7	0.7599	6.0021	17.0657	1.3159	7.8983	0.1666	0.1266	2.8433
8	0.7307	6.7327	22.1806	1.3686	9.2142	0.1485	0.1085	3.2944
9	0.7026	7.4353	27.8013	1.4233	10.5828	0.1345	0.0945	3.7391
10	0.6756	8.1109	33.8814	1.4802	12.0061	0.1233	0.0833	4.1773
11	0.6496	8.7605	40.3772	1.5395	13.4864	0.1141	0.0741	4.6090
12	0.6246	9.3851	47.2477	1.6010	15.0258	0.1066	0.0666	5.0343
13	0.6006	9.9856	54.4546	1.6651	16.6268	0.1001	0.0601	5.4533
14	0.5775	10.5631	61.9618	1.7317	18.2919	0.0947	0.0547	5.8659
15	0.5553	11.1184	69.7355	1.8009	20.0236	0.0899	0.0499	6.2721
16	0.5339	11.6523	77.7441	1.8730	21.8245	0.0858	0.0458	6.6720
17	0.5134	12.1657	85.9581	1.9479	23.6975	0.0822	0.0422	7.0656
18	0.4936	12.6593	94.3498	2.0258	25.6454	0.0790	0.0390	7.4530
19	0.4746	13.1339	102.8933	2.1068	27.6712	0.0761	0.0361	7.8342
20	0.4564	13.5903	111.5647	2.1911	29.7781	0.0736	0.0336	8.2091
21	0.4388	14.0292	120.3414	2.2788	31.9692	0.0713	0.0313	8.5779
22	0.4220	14.4511	129.2024	2.3699	34.2480	0.0692	0.0292	8.9407
23	0.4057	14.8568	138.1284	2.4647	36.6179	0.0673	0.0273	9.2973
24	0.3901	15.2470	147.1012	2.5633	39.0826	0.0656	0.0256	9.6479
25	0.3751	15.6221	156.1040	2.6658	41.6459	0.0640	0.0240	9.9925
30	0.3083	17.2920	201.0618	3.2434	56.0849	0.0578	0.0178	11.6274
40	0.2083	19.7928	286.5303	4.8010	95.0255	0.0505	0.0105	14.4765
50	0.1407	21.4822	361.1638	7.1067	152.6671	0.0466	0.0066	16.8122
60	0.0951	22.6235	422.9966	10.5196	237.9907	0.0442	0.0042	18.6972
100	0.0198	24.5050	563.1249	50.5049	1,237.6237	0.0408	0.0008	22.9800

Table 50.9 Factor Table i = 6.00%

n	P/F	P/A	P/G	F/P	F/A	A/P	A/F	A/G
1	0.9434	0.9434	0.0000	1.0600	1.0000	1.0600	1.0000	0.0000
2	0.8900	1.8334	0.8900	1.1236	2.0600	0.5454	0.4854	0.4854
3	0.8396	2.6730	2.5692	1.1910	3.1836	0.3741	0.3141	0.9612
4	0.7921	3.4651	4.9455	1.2625	4.3746	0.2886	0.2286	1.4272
5	0.7473	4.2124	7.9345	1.3382	5.6371	0.2374	0.1774	1.8836
6	0.7050	4.9173	11.4594	1.4185	6.9753	0.2034	0.1434	2.3304
7	0.6651	5.5824	15.4497	1.5036	8.3938	0.1791	0.1191	2.7676
8	0.6274	6.2098	19.8416	1.5938	9.8975	0.1610	0.1010	3.1952
9	0.5919	6.8017	24.5768	1.6895	11.4913	0.1470	0.0870	3.6133
10	0.5584	7.3601	29.6023	1.7908	13.1808	0.1359	0.0759	4.0220
11	0.5268	7.8869	34.8702	1.8983	14.9716	0.1268	0.0668	4.4213
12	0.4970	8.3838	40.3369	2.0122	16.8699	0.1193	0.0593	4.8113
13	0.4688	8.8527	45.9629	2.1239	18.8821	0.1130	0.0530	5.1920
14	0.4423	9.2950	51.7128	2.2609	21.0151	0.1076	0.0476	5.5635
15	0.4173	9.7122	57.5546	2.3966	23.2760	0.1030	0.0430	5.9260
16	0.3936	10.1059	63.4592	2.5404	25.6725	0.0990	0.0390	6.2794
17	0.3714	10.4773	69.4011	2.6928	28.2129	0.0954	0.0354	6.6240
18	0.3505	10.8276	75.3569	2.8543	30.9057	0.0924	0.0324	6.9597
19	0.3305	11.1581	81.3062	3.0256	33.7600	0.0896	0.0296	7.2867
20	0.3118	11.4699	87.2304	3.2071	36.7856	0.0872	0.0272	7.6051
21	0.2942	11.7641	93.1136	3.3996	39.9927	0.0850	0.0250	7.9151
22	0.2775	12.0416	98.9412	3.6035	43.3923	0.0830	0.0230	8.2166
23	0.2618	12.3034	104.7007	3.8197	46.9958	0.0813	0.0213	8.5099
24	0.2470	12.5504	110.3812	4.0489	50.8156	0.0797	0.0197	8.7951
25	0.2330	12.7834	115.9732	4.2919	54.8645	0.0782	0.0182	9.0722
30	0.1741	13.7648	142.3588	5.7435	79.0582	0.0726	0.0126	10.3422
40	0.0972	15.0463	185.9568	10.2857	154.7620	0.0665	0.0065	12.3590
50	0.0543	15.7619	217.4574	18.4202	290.3359	0.0634	0.0034	13.7964
60	0.0303	16.1614	239.0428	32.9877	533.1282	0.0619	0.0019	14.7909
100	0.0029	16.6175	272.0471	339.3021	5638.3681	0.0602	0.0002	16.3711

Process Design/
Economics

Table 50.10 Factor Table i = 8.00%

n	P/F	P/A	P/G	F/P	F/A	A/P	A/F	A/G
1	0.9259	0.9259	0.0000	1.0800	1.0000	1.0800	1.0000	0.0000
2	0.8573	1.7833	0.8573	1.1664	2.0800	0.5608	0.4808	0.4808
3	0.7938	2.5771	2.4450	1.2597	3.2464	0.3880	0.3080	0.9487
4	0.7350	3.3121	4.6501	1.3605	4.5061	0.3019	0.2219	1.4040
5	0.6806	3.9927	7.3724	1.4693	5.8666	0.2505	0.1705	1.8465
6	0.6302	4.6229	10.5233	1.5869	7.3359	0.2163	0.1363	2.2763
7	0.5835	5.2064	14.0242	1.7138	8.9228	0.1921	0.1121	2.6937
8	0.5403	5.7466	17.8061	1.8509	10.6366	0.1740	0.0940	3.0985
9	0.5002	6.2469	21.8081	1.9990	12.4876	0.1601	0.0801	3.4910
10	0.4632	6.7101	25.9768	2.1589	14.4866	0.1490	0.0690	3.8713
11	0.4289	7.1390	30.2657	2.3316	16.6455	0.1401	0.0601	4.2395
12	0.3971	7.5361	34.6339	2.5182	18.9771	0.1327	0.0527	4.5957
13	0.3677	7.9038	39.0463	2.7196	21.4953	0.1265	0.0465	4.9402
14	0.3405	8.2442	43.4723	2.9372	24.2149	0.1213	0.0413	5.2731
15	0.3152	8.5595	47.8857	3.1722	27.1521	0.1168	0.0368	5.5945
16	0.2919	8.8514	52.2640	3.4259	30.3243	0.1130	0.0330	5.9046
17	0.2703	9.1216	56.5883	3.7000	33.7502	0.1096	0.0296	6.2037
18	0.2502	9.3719	60.8426	3.9960	37.4502	0.1067	0.0267	6.4920
19	0.2317	9.6036	65.0134	4.3157	41.4463	0.1041	0.0241	6.7697
20	0.2145	9.8181	69.0898	4.6610	45.7620	0.1019	0.0219	7.0369
21	0.1987	10.0168	73.0629	5.0338	50.4229	0.0998	0.0198	7.2940
22	0.1839	10.2007	76.9257	5.4365	55.4568	0.0980	0.0180	7.5412
23	0.1703	10.3711	80.6726	5.8715	60.8933	0.0964	0.0164	7.7786
24	0.1577	10.5288	84.2997	6.3412	66.7648	0.0950	0.0150	8.0066
25	0.1460	10.6748	87.8041	6.8485	73.1059	0.0937	0.0137	8.2254
30	0.0994	11.2578	103.4558	10.0627	113.2832	0.0888	0.0088	9.1897
40	0.0460	11.9246	126.0422	21.7245	259.0565	0.0839	0.0039	10.5699
50	0.0213	12.2335	139.5928	46.9016	573.7702	0.0817	0.0017	11.4107
60	0.0099	12.3766	147.3000	101.2571	1253.2133	0.0808	0.0008	11.9015
100	0.0005	12.4943	155.6107	2199.7613	27,484.5157	0.0800	–	12.4545

Table 50.11 *Factor Table i = 10.00%*

n	P/F	P/A	P/G	F/P	F/A	A/P	A/F	A/G
1	0.9091	0.9091	0.0000	1.1000	1.0000	1.1000	1.0000	0.0000
2	0.8264	1.7355	0.8264	1.2100	2.1000	0.5762	0.4762	0.4762
3	0.7513	2.4869	2.3291	1.3310	3.3100	0.4021	0.3021	0.9366
4	0.6830	3.1699	4.3781	1.4641	4.6410	0.3155	0.2155	1.3812
5	0.6209	3.7908	6.8618	1.6105	6.1051	0.2638	0.1638	1.8101
6	0.5645	4.3553	9.6842	1.7716	7.7156	0.2296	0.1296	2.2236
7	0.5132	4.8684	12.7631	1.9487	9.4872	0.2054	0.1054	2.6216
8	0.4665	5.3349	16.0287	2.1436	11.4359	0.1874	0.0874	3.0045
9	0.4241	5.7590	19.4215	2.3579	13.5735	0.1736	0.0736	3.3724
10	0.3855	6.1446	22.8913	2.5937	15.9374	0.1627	0.0627	3.7255
11	0.3505	6.4951	26.3962	2.8531	18.5312	0.1540	0.0540	4.0641
12	0.3186	6.8137	29.9012	3.1384	21.3843	0.1468	0.0468	4.3884
13	0.2897	7.1034	33.3772	3.4523	24.5227	0.1408	0.0408	4.6988
14	0.2633	7.3667	36.8005	3.7975	27.9750	0.1357	0.0357	4.9955
15	0.2394	7.6061	40.1520	4.1772	31.7725	0.1315	0.0315	5.2789
16	0.2176	7.8237	43.4164	4.5950	35.9497	0.1278	0.0278	5.5493
17	0.1978	8.0216	46.5819	5.5045	40.5447	0.1247	0.0247	5.8071
18	0.1799	8.2014	49.6395	5.5599	45.5992	0.1219	0.0219	6.0526
19	0.1635	8.3649	52.5827	6.1159	51.1591	0.1195	0.0195	6.2861
20	0.1486	8.5136	55.4069	6.7275	57.2750	0.1175	0.0175	6.5081
21	0.1351	8.6487	58.1095	7.4002	64.0025	0.1156	0.0156	6.7189
22	0.1228	8.7715	60.6893	8.1403	71.4027	0.1140	0.0140	6.9189
23	0.1117	8.8832	63.1462	8.9543	79.5430	0.1126	0.0126	7.1085
24	0.1015	8.9847	65.4813	9.8497	88.4973	0.1113	0.0113	7.2881
25	0.0923	9.0770	67.6964	10.8347	98.3471	0.1102	0.0102	7.4580
30	0.0573	9.4269	77.0766	17.4494	164.4940	0.1061	0.0061	8.1762
40	0.0221	9.7791	88.9525	45.2593	442.5926	0.1023	0.0023	9.0962
50	0.0085	9.9148	94.8889	117.3909	1163.9085	0.1009	0.0009	9.5704
60	0.0033	9.9672	97.7010	304.4816	3,034.8164	0.1003	0.0003	9.8023
100	0.0001	9.9993	99.9202	13,780.6123	137,796.1234	0.1000	–	9.9927

Table 50.12 *Factor Table i = 12.00%*

n	P/F	P/A	P/G	F/P	F/A	A/P	A/F	A/G
1	0.8929	0.8929	0.0000	1.1200	1.0000	1.1200	1.0000	0.0000
2	0.7972	1.6901	0.7972	1.2544	2.1200	0.5917	0.4717	0.4717
3	0.7118	2.4018	2.2208	1.4049	3.3744	0.4163	0.2963	0.9246
4	0.6355	3.0373	4.1273	1.5735	4.7793	0.3292	0.2092	1.3589
5	0.5674	3.6048	6.3970	1.7623	6.3528	0.2774	0.1574	1.7746
6	0.5066	4.1114	8.9302	1.9738	8.1152	0.2432	0.1232	2.1720
7	0.4523	4.5638	11.6443	2.2107	10.0890	0.2191	0.0991	2.5515
8	0.4039	4.9676	14.4714	2.4760	12.2997	0.2013	0.0813	2.9131
9	0.3606	5.3282	17.3563	2.7731	14.7757	0.1877	0.0677	3.2574
10	0.3220	5.6502	20.2541	3.1058	17.5487	0.1770	0.0570	3.5847
11	0.2875	5.9377	23.1288	3.4785	20.6546	0.1684	0.0484	3.8953
12	0.2567	6.1944	25.9523	3.8960	24.1331	0.1614	0.0414	4.1897
13	0.2292	6.4235	28.7024	4.3635	28.0291	0.1557	0.0357	4.4683
14	0.2046	6.6282	31.3624	4.8871	32.3926	0.1509	0.0309	4.7317
15	0.1827	6.8109	33.9202	5.4736	37.2797	0.1468	0.0268	4.9803
16	0.1631	6.9740	36.3670	6.1304	42.7533	0.1434	0.0234	5.2147
17	0.1456	7.1196	38.6973	6.8660	48.8837	0.1405	0.0205	5.4353
18	0.1300	7.2497	40.9080	7.6900	55.7497	0.1379	0.0179	5.6427
19	0.1161	7.3658	42.9979	8.6128	63.4397	0.1358	0.0158	5.8375
20	0.1037	7.4694	44.9676	9.6463	72.0524	0.1339	0.0139	6.0202
21	0.0926	7.5620	46.8188	10.8038	81.6987	0.1322	0.0122	6.1913
22	0.0826	7.6446	48.5543	12.1003	92.5026	0.1308	0.0108	6.3514
23	0.0738	7.7184	50.1776	13.5523	104.6029	0.1296	0.0096	6.5010
24	0.0659	7.7843	51.6929	15.1786	118.1552	0.1285	0.0085	6.6406
25	0.0588	7.8431	53.1046	17.001	133.3339	0.1275	0.0075	6.7708
30	0.0334	8.0552	58.7821	29.9599	241.3327	0.1241	0.0041	7.2974
40	0.0107	8.2438	65.1159	93.0510	767.0914	0.1213	0.0013	7.8988
50	0.0035	8.3045	67.7624	289.0022	2,400.0182	0.1204	0.0004	8.1597
60	0.0011	8.3240	68.8100	897.5969	7,471.6411	0.1201	0.0001	8.2664
100	–	8.3332	69.4336	83,522.2657	696,010.5477	0.1200	–	8.3321

Process Design/ Economics

Table 50.13 Factor Table i = 18.00%

n	P/F	P/A	P/G	F/P	F/A	A/P	A/F	A/G
1	0.8475	0.8475	0.0000	1.1800	1.0000	1.1800	1.0000	0.0000
2	0.7182	1.5656	0.7182	1.3924	2.1800	0.6387	0.4587	0.4587
3	0.6086	2.1743	1.9354	1.6430	3.5724	0.4599	0.2799	0.8902
4	0.5158	2.6901	3.4828	1.9388	5.2154	0.3717	0.1917	1.2947
5	0.4371	3.1272	5.2312	2.2878	7.1542	0.3198	0.1398	1.6728
6	0.3704	3.4976	7.0834	2.6996	9.4423	0.2859	0.1059	2.0252
7	0.3139	3.8115	8.9670	3.1855	12.1415	0.2624	0.0824	2.3526
8	0.2660	4.0776	10.8292	3.7589	15.3270	0.2452	0.0652	2.6558
9	0.2255	4.3030	12.6329	4.4355	19.0859	0.2324	0.0524	2.9358
10	0.1911	4.4941	14.3525	5.2338	23.5213	0.2225	0.0425	3.1936
11	0.1619	4.6560	15.9716	6.1759	28.7551	0.2148	0.0348	3.4303
12	0.1372	4.7932	17.4811	7.2876	34.9311	0.2086	0.0286	3.6470
13	0.1163	4.9095	18.8765	8.5994	42.2187	0.2037	0.0237	3.8449
14	0.0985	5.0081	20.1576	10.1472	50.8180	0.1997	0.0197	4.0250
15	0.0835	5.0916	21.3269	11.9737	60.9653	0.1964	0.0164	4.1887
16	0.0708	5.1624	22.3885	14.1290	72.9390	0.1937	0.0137	4.3369
17	0.0600	5.2223	23.3482	16.6722	87.0680	0.1915	0.0115	4.4708
18	0.0508	5.2732	24.2123	19.6731	103.7403	0.1896	0.0096	4.5916
19	0.0431	5.3162	24.9877	23.2144	123.4135	0.1881	0.0081	4.7003
20	0.0365	5.3527	25.6813	27.3930	146.6280	0.1868	0.0068	4.7978
21	0.0309	5.3837	26.3000	32.3238	174.0210	0.1857	0.0057	4.8851
22	0.0262	5.4099	26.8506	38.1421	206.3448	0.1848	0.0048	4.9632
23	0.0222	5.4321	27.3394	45.0076	244.4868	0.1841	0.0041	5.0329
24	0.0188	5.4509	27.7725	53.1090	289.4944	0.1835	0.0035	5.0950
25	0.0159	5.4669	28.1555	62.6686	342.6035	0.1829	0.0029	5.1502
30	0.0070	5.5168	29.4864	143.3706	790.9480	0.1813	0.0013	5.3448
40	0.0013	5.5482	30.5269	750.3783	4,163.2130	0.1802	0.0002	5.5022
50	0.0003	5.5541	30.7856	3,927.3569	21,813.0937	0.1800	–	5.5428
60	0.0001	5.5553	30.8465	20,555.1400	114,189.6665	0.1800	–	5.5526
100	–	5.5556	30.8642	15,424,131.91	85,689,616.17	0.1800	–	5.5555

Process Design/ Economics

Diagnostic Exam

Topic XIV: Safety, Health, and Environment

1. LD_{50} means the dose in

  (A) mg/d at which the test animals died after 50 days of exposure

  (B) mg/d at which 50% of the test animals died

  (C) mg per kg of body mass at which 50% of the test animals died

  (D) mg per 50 kg of body mass at which the test animals died

2. A person consumes 2.0 L/d of water. A chemical in the drinking water is at a concentration of 50 μg/L. The person has a mass of 72 kg. Over a period of 30 days, what is most nearly the total dose of chemical the person consumes?

  (A) 0.046 μg/kg·d

  (B) 0.16 μg/kg·d

  (C) 1.4 μg/kg·d

  (D) 42 μg/kg·d

3. Two machines run simultaneously in a work environment. One machine has a measured sound pressure of 3 Pa, and the other machine has a measured sound pressure of 12 Pa. The standard reference sound pressure is 2×10^{-5} Pa. What is most nearly the total sound level?

  (A) 110 dB

  (B) 120 dB

  (C) 130 dB

  (D) 140 dB

4. Which of the following is NOT found on material safety data sheets?

  (A) chemical abstract series number

  (B) price

  (C) reactivity

  (D) spill procedures

5. The cancer slope factor for trichloroethylene (TCE) by the oral route is 4.6×10^{-2} $(\text{mg/kg·d})^{-1}$. A community is exposed to TCE in drinking water at a level of 5.6 μg/L over a period of 5 yr. The people in the community have an average mass of 73 kg and ingest an average of 1.8 L/d of drinking water. The average lifetime expectancy for people in the community is 72 yr. What is most nearly the increased cancer risk for the population?

  (A) 4.4×10^{-7}

  (B) 2.5×10^{-6}

  (C) 5.6×10^{-6}

  (D) 9.4×10^{-5}

6. According to Occupational Safety and Health Administration (OSHA) permissible noise exposure levels, what is most nearly the time a worker is permitted to be exposed to a sound pressure level of 120 dB?

  (A) 0.13 h

  (B) 0.54 h

  (C) 2.0 h

  (D) 8.0 h

7. A factory worker is exposed to carbon monoxide (CO) for 2 h on working days for 10 yr. The fraction of days in the year the worker is exposed is 0.75. The worker breathes 1.47 m³/h of air during that time, and the concentration of carbon monoxide in the air is 88 μg/m³. The worker has a mass of 72 kg and has a life expectancy of 80 yr. What is most nearly the airborne chronic daily intake?

  (A) 3.4×10^{-4} mg/kg·d

  (B) 5.2×10^{-4} mg/kg·d

  (C) 8.7×10^{-4} mg/kg·d

  (D) 1.3×10^{-3} mg/kg·d

8. The no-observed-adverse-effect level (NOAEL) for copper cyanide (CuCN) is 25 mg/kg·d, with an uncertainty factor of 5000. The chronic daily intake in a community is 10 μg/kg·d. The average person in the community has a mass of 68 kg. What is most nearly the safe human dose?

(A) 0.15 mg/d

(B) 0.34 mg/d

(C) 0.44 mg/d

(D) 0.53 mg/d

9. A village water supply is contaminated with heptachlor (a class B carcinogen) for a period of 8 yr. Heptachlor has a bioconcentration factor of 15 700 L/kg in fish and a concentration of 0.03×10^{-3} mg/L in water. An average villager has a 67 yr life span, a mass of 68 kg, and a fish consumption rate of 8 g/d. The fraction ingested is 1, and the exposure frequency is 1. What is most nearly the chronic daily intake from fish consumption?

(A) 6.6×10^{-6} mg/kg·d

(B) 11×10^{-6} mg/kg·d

(C) 28×10^{-6} mg/kg·d

(D) 51×10^{-6} mg/kg·d

10. Construction workers are exposed to sound pressure levels of 85 dB for 2 hr, 95 dB for 2.5 hr, and 87 dB for 2 hr. What is most nearly the total noise dose that the workers experience?

(A) 67%

(B) 85%

(C) 92%

(D) 97%

SOLUTIONS

1. LD_{50} is the dose at which 50% of the test animals exhibited a fatal response. Dose is measured in milligrams of toxicant per kilogram of body mass. Only option C is true.

The answer is (C).

2. The dose is

$$\text{dose} = \frac{\text{mass of chemical}}{\text{body weight} \cdot \text{exposure time}}$$

$$= \frac{\left(50 \; \frac{\mu g}{L}\right)\left(2.0 \; \frac{L}{d}\right)(30 \text{ d})}{(72 \text{ kg})(30 \text{ d})}$$

$$= 1.4 \; \mu g/kg \cdot d$$

The answer is (C).

3. Calculate the sound pressure level of each machine.

$$\text{SPL}_{3\,\text{Pa}} = 10 \log_{10}\left(\frac{p}{p_0}\right)^2$$

$$= 10 \log_{10}\left(\frac{3 \text{ Pa}}{2 \times 10^{-5} \text{ Pa}}\right)^2$$

$$= 103.5 \text{ dB}$$

$$\text{SPL}_{12\,\text{Pa}} = 10 \log_{10}\left(\frac{p}{p_0}\right)^2$$

$$= 10 \log_{10}\left(\frac{12 \text{ Pa}}{2 \times 10^{-5} \text{ Pa}}\right)^2$$

$$= 115.6 \text{ dB}$$

The total sound pressure is

$$\text{SPL}_{\text{total}} = 10 \log_{10}\sum 10^{\text{SPL}/10}$$

$$= 10 \log_{10}(10^{103.5\,\text{dB}/10} + 10^{115.6\,\text{dB}/10})$$

$$= 115.8 \text{ dB} \quad (120 \text{ dB})$$

The answer is (B).

4. Price is not included on material safety data sheets.

The answer is (B).

Safety/Health/Environment

5. The product of ingestion rate (IR) and the exposure frequency (EF) is given as 1.8 L/d. Calculate the chronic daily intake for water ingestion.

$$CDI = \frac{(CW)(IR)(EF)(ED)}{(BW)(AT)}$$

$$= \frac{\left(5.6 \ \frac{\mu g}{L}\right)\left(1.8 \ \frac{L}{d}\right)(5 \text{ yr})}{(73 \text{ kg})(72 \text{ yr})\left(1000 \ \frac{\mu g}{mg}\right)}$$

$$= 9.59 \times 10^{-6} \text{ mg/kg·d}$$

Calculate the risk.

$$\text{risk} = (CDI)(CSF)$$

$$= \left(9.59 \times 10^{-6} \ \frac{mg}{kg \cdot d}\right)\left(4.6 \times 10^{-2} \ \left(\frac{mg}{kg \cdot d}\right)^{-1}\right)$$

$$= 4.4 \times 10^{-7}$$

The answer is (A).

6. The exposure time permitted is

$$T = (2 \text{ h})^{((105-SPL)/5)} = (2 \text{ h})^{((105 \text{ dB}-120 \text{ dB})/5)}$$

$$= 0.125 \text{ h} \quad (0.13 \text{ h})$$

The answer is (A).

7. The airborne chronic daily intake is

$$CDI = \frac{(CA)(IR)(ET)(EF)(ED)}{(BW)(AT)}$$

$$= \frac{\left(88 \ \frac{\mu g}{m^3}\right)\left(1.47 \ \frac{m^3}{h}\right)\left(2 \ \frac{h}{d}\right)(0.75)(10 \text{ yr})}{(72 \text{ kg})(80 \text{ yr})\left(1000 \ \frac{\mu g}{mg}\right)}$$

$$= 3.4 \times 10^{-4} \text{ mg/kg·d}$$

The answer is (A).

8. The safe human dose (SHD) is

$$SHD = \frac{(NOAEL)W}{UF}$$

$$= \frac{\left(25 \ \frac{mg}{kg \cdot d}\right)(68 \text{ kg})}{5000}$$

$$= 0.34 \text{ mg/d}$$

The answer is (B).

9. The concentration of heptachlor in fish (CF) is

$$BCF = \frac{C_{org}}{C}$$

$$C_{org} = CF = (BCF)C$$

$$= \left(15\,700 \ \frac{L}{kg}\right)\left(0.03 \times 10^{-3} \ \frac{mg}{L}\right)$$

$$= 0.471 \text{ mg/kg}$$

The chronic daily intake from fish consumption is

$$CDI = \frac{(CF)(IR)(FI)(EF)(ED)}{(BW)(AT)}$$

$$= \frac{\left(0.471 \ \frac{mg}{kg}\right)\left(8 \ \frac{g}{d}\right)(1)(1)(8 \text{ yr})}{(68 \text{ kg})(67 \text{ yr})\left(1000 \ \frac{g}{kg}\right)}$$

$$= 6.616 \times 10^{-6} \text{ mg/kg·d}$$

$$(6.6 \times 10^{-6} \text{ mg/kg·d})$$

The answer is (A).

10. Calculate the exposure time permitted at each sound pressure level (SPL).

$$T_1 = (2 \text{ h})^{((105-SPL)/5)} = (2 \text{ h})^{((105 \text{ dB}-85 \text{ dB})/5)}$$

$$= 16 \text{ h}$$

$$T_2 = (2 \text{ h})^{((105-SPL)/5)} = (2 \text{ h})^{((105 \text{ dB}-95 \text{ dB})/5)}$$

$$= 4 \text{ h}$$

$$T_3 = (2 \text{ h})^{((105-SPL)/5)} = (2 \text{ h})^{((105 \text{ dB}-87 \text{ dB})/5)}$$

$$= 12.1 \text{ h}$$

Calculate the mixed noise dose.

$$D = 100\% \times \sum \frac{C_i}{T_i} = (100\%)\left(\frac{C_1}{T_1} + \frac{C_2}{T_2} + \frac{C_3}{T_3}\right)$$

$$= (100\%)\left(\frac{2 \text{ h}}{16 \text{ h}} + \frac{2.5 \text{ h}}{4 \text{ h}} + \frac{2 \text{ h}}{12.1 \text{ h}}\right)$$

$$= 91.5\% \quad (92\%)$$

The answer is (C).

Safety/Health/ Environment

51 Toxicology

1. EXPOSURE PATHWAYS

The human body has three primary exposure pathways: dermal absorption, inhalation, and ingestion. (See Fig. 51.1.) The eyes are an additional exposure pathway because they are particularly vulnerable to damage in the workplace.

2. DERMAL ABSORPTION

The skin is composed of the epidermis, the dermis, and the subcutaneous layer. The *epidermis* is the upper layer, which is composed of several layers of flattened and scale-like cells. These cells do not contain blood vessels; they obtain their nutrients from the underlying dermis. The cells of the epidermis migrate to the surface, die, and leave behind a protein called *keratin*. Keratin is the most insoluble of all proteins, and, together with the scale-like cells, provides extreme resistance to substances and environmental conditions. Beneath the epidermis lies the *dermis*, which contains blood vessels, connective tissue, hair follicles, sweat glands, and other glands. The dermis supplies the nutrients for itself and for the epidermis. The innermost layer is called the *subcutaneous fatty tissue*, which provides a cushion for the skin and connection to the underlying tissue.

The condition of the skin, and the chemical nature of any toxic substance it contacts, affect whether and at what rate the skin absorbs that substance. The epidermis is impermeable to many gases, water, and chemicals. However, if the epidermis is damaged by cuts and abrasions, or is broken down by repeated exposure to soaps, detergents, or organic solvents, toxic substances can readily penetrate and enter the bloodstream.

Chemical burns, such as those from acids, can also destroy the protection afforded by the epidermis, allowing toxicants to enter the bloodstream. Inorganic chemicals (and organic chemicals that are dissolved in water) are not readily absorbed through healthy skin. However, many organic solvents are lipid- (fat-) soluble and can easily penetrate skin cells and enter the body. After a toxicant has penetrated the skin and entered the bloodstream, the blood can transport it to target organs in the body.

Equation 51.1 and Eq. 51.2: Dermal Contact

$$AD = \frac{(CW)(SA)(PC)(ET)(EF)(ED)(CF)}{(BW)(AT)} \quad \begin{bmatrix} \text{water} \\ \text{contact} \end{bmatrix}$$

$$51.1$$

$$AD = \frac{(CS)(CF)(SA)(AF)(ABS)(EF)(ED)}{(BW)(AT)} \quad \begin{bmatrix} \text{soil} \\ \text{contact} \end{bmatrix}$$

$$51.2$$

Description

Equation 51.1 defines the absorbed dose (AD) for dermal contact with water. Equation 51.2 defines the absorbed dose for dermal contact with soil. Variables used in Eq. 51.1 and Eq. 51.2 are the soil contaminant absorption factor (ABS); soil-to-skin-adherence factor (AF); averaging time (AT); body weight (BW); volumetric conversion factor (CF), equal to 0.001 L/cm^3 for water or 10^{-6} kg/mg for soil; chemical concentration in soil (CS); chemical concentration in water (CW); exposure duration (ED); exposure frequency (EF); exposure time (ET); permeability constant (PC); and skin surface area available for contact (SA).

3. INHALATION

The *respiratory tract* consists of the nasal cavity, pharynx, larynx, trachea, primary bronchi, bronchioles, and alveoli. Molecular transfer of oxygen and carbon dioxide between the bloodstream and the lungs occurs in the millions of air sacs, known as *alveoli*. Toxicants that reach the alveoli can be transferred to the blood through the respiratory system.

Toxicants that reach the alveoli are not transferred to the blood at the same rates. Various factors can increase or decrease the transfer rate of one toxicant relative to another. Both the respiration rate and the duration of exposure affect the mass of toxicant transferred to the bloodstream.

Safety/Health/Environment

Figure 51.1 *Exposure Routes for Chemical Agents*

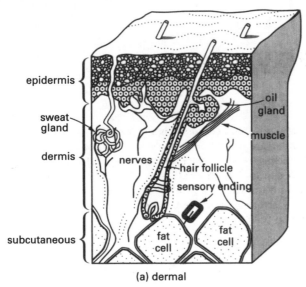

(a) dermal

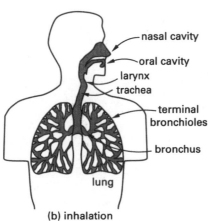

(b) inhalation

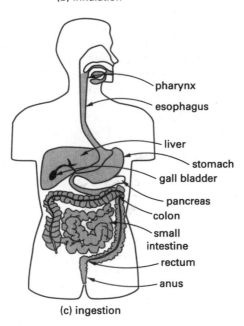

(c) ingestion

Equation 51.3: Inhalation of Airborne (Vapor Phase) Chemicals

$$CDI = \frac{(CA)(IR)(ET)(EF)(ED)}{(BW)(AT)} \quad \begin{bmatrix} \text{vapor} \\ \text{inhalation} \end{bmatrix} \quad 51.3$$

Description

Equation 51.3 defines the chronic daily intake (CDI) for inhalation of airborne (vapor phase) chemicals. Variables used in Eq. 51.3 are averaging time (AT), body weight (BW), contaminant concentration in air (CA), exposure duration (ED), exposure frequency (EF), exposure time (ET), and inhalation rate (IR).

4. INGESTION

The *small intestine* is the principal site in the digestive tract where the beneficial nutrients from food, and toxics from contaminated substances, are absorbed. Within the small intestine, millions of *villi* (projections) provide a large surface area to absorb substances into the bloodstream. The primary function of the *large intestine* is to absorb water and store digestion waste.

Toxic substances are absorbed in the large and small intestines at various rates depending on the specific toxicant, its molecular size, and its degree of lipid (fat) solubility. Small molecular size and high lipid solubility increase absorption of toxicants in the digestive tract.

Equation 51.4 Through Eq. 51.7: Chronic Daily Intake

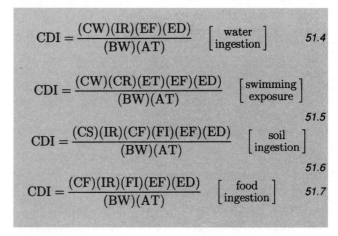

Description

Equation 51.4 defines the chronic daily intake (CDI) for ingestion in drinking water. Equation 51.5 defines the chronic daily intake for ingestion while swimming. Equation 51.6 defines the chronic daily intake for ingestion of chemicals in soil. Equation 51.7 defines the chronic daily intake for ingestion of contaminated fruits, vegetables, fish, and shellfish. Variables used in Eq. 51.4 through Eq. 51.7 are averaging time (AT); body weight (BW); volumetric conversion factor (CF), equal to

Safety/Health/
Environment

0.001 L/cm^3 for soil; concentration in fish (CF);[1] contact rate (CR); chemical concentration in soil (CS); chemical concentration in water (CW); exposure duration (ED); exposure frequency (EF); exposure time (ET); fraction ingested (FI); and ingestion rate (IR).

Example

A village water supply is contaminated with benzene (C_6H_6) from a leaking underground storage tank. The tank has been leaking for 11 years. The average concentration of benzene in the water supply during this period is 50×10^{-3} mg/L. A 65 kg villager has a 70 yr life span, a water consumption of 35 L/d, and an exposure frequency of 1. For this villager, what is most nearly the chronic daily intake for water ingestion?

(A) 3.1×10^{-3} mg/kg·d

(B) 4.2×10^{-3} mg/kg·d

(C) 5.4×10^{-3} mg/kg·d

(D) 6.9×10^{-3} mg/kg·d

Solution

Use Eq. 51.4.

$$CDI = \frac{(CW)(IR)(EF)(ED)}{(BW)(AT)}$$

$$= \frac{\left(50 \times 10^{-3}\ \frac{mg}{L}\right)\left(35\ \frac{L}{d}\right)(1)(11\ yr)}{(65\ kg)(70\ yr)}$$

$$= 4.231 \times 10^{-3}\ mg/kg·d \quad (4.2 \times 10^{-3}\ mg/kg·d)$$

The answer is (B).

5. THE EYE

Transparent tissue on the surface of the front of the eye is known as the *cornea* and is the most likely eye tissue to come in contact with toxic substances. Chemicals (e.g., organic solvents) in liquid, dust, vapor, gas, aerosol, and mist forms can enter the bloodstream through the eye. During the absorption process, chemicals can cause *keratitis*, an inflammation of the outer layer of the eye.

6. LOCAL AND SYSTEMIC EFFECTS

Hazards and toxic substances (*toxicants*) can have local and/or systemic effects on the body. A *local effect* is an adverse health effect that occurs at the point or in the area of exposure. For example, an acid burn will be limited to the skin area contacted by the acid. Points of contact

---

[1]The NCEES *FE Reference Handbook* (*NCEES Handbook*) uses the variable CF for both *conversion factor* (see Eq. 51.6) and *concentration in fish* (see Eq. 51.7). This second usage is not defined, implying that CF in Eq. 51.6 and Eq. 51.7 are the same variable. The two usages are not the same, nor are the units the same. Conversion factor has typical units of L/cm^3, while concentration in fish has units of mg/kg.

include skin, mucous membranes, respiratory tract, gastrointestinal system, and eyes. Absorption is not necessary and often does not occur when the effect remains local.

A *systemic effect* is an adverse health effect that generally occurs away from the initial exposure point. Absorption is usually necessary for the substance to be transported to a different location. In order to exert their effects, substances with systemic effects often accumulate in *target organs*. Some substances, such as lead, that cause systemic effects are *cumulative poisons*. The concentrations of these substances increase in target organs through repeated chronic exposures. The effects are usually not noticed until a critical concentration is reached.

When toxic substances are not eliminated fast enough to keep up with the exposure, the liver, kidneys, and central nervous system are the target organs and are commonly affected systemically.

For systemic effects to occur, the rate of accumulation of a toxicant must exceed the body's ability to excrete (eliminate) it or to biotransform it (transform it to less harmful substances). A toxicant can be eliminated from the body through the *kidneys*, which are the primary organs for eliminating toxicants from the body. The kidneys biotransform a toxicant into a water-soluble form and then eliminate in the urine. The *liver* is also an important organ for eliminating toxicants from the body, first by biotransformation, then by excretion into bile where it is eliminated through the small intestine as feces.

A toxicant may also be stored in tissues for long periods before an effect occurs. Toxic substances stored in tissue (primarily in fat but also in bones, the liver, and the kidneys) may exert no effect for many years, or at all within the affected person's life. For instance, the pesticide DDT can be stored in body fat for many years and not exert any adverse effect on the body.

7. EFFECTS OF EXPOSURE TO TOXICANTS

After toxicants are absorbed into the body through one or more of the pathways, a wide variety of effects on the human body are possible. When the toxic agents concentrate in target tissue or organs, the agents may interfere with the normal functioning of enzymes and cells or may cause genetic mutations.

Pulmonary Toxicity

Pulmonary toxicity refers to adverse effects on the respiratory system from toxic agents. Examples are

- damage to the nasal passages and nerve cells

- *nasal cancer*

- *bronchitis*, excessive mucus secretion

- *pulmonary edema*, the excessive accumulation of fluid in the alveoli of the lungs

- *fibrosis*, an increased amount of connective tissue

- *silicosis*, the deposition of connective tissue around alveoli

- *emphysema*, the inability of lungs to expand and contract

Cardiotoxicity

Cardiotoxicity refers to the effects of toxic agents on the heart.

- The heart rate may be changed, and the strength of contractions may be diminished.

- Certain metals can affect the contractions of the heart and can interfere with cell metabolism.

- Carbon monoxide can result in a decrease in the oxygen supply, causing improper functioning of the nervous system controlling heart rate.

Hematoxicity

Hematoxicity refers to damage to the body's blood supply, which includes red blood cells, white blood cells, platelets, and plasma. The *red blood cells* transport oxygen to the body's cells and carbon dioxide to the lungs. *White blood cells* perform a variety of functions associated with the immune system.

- *Platelets* are important in blood clotting.

- *Plasma* is the noncellular portion of blood and contains proteins, nutrients, gases, and waste products.

- Benzene, lead, methylene chloride, nitrobenzene, naphthalene, and insecticides are capable of red blood cell destruction and can cause a decrease in the oxygen-carrying capacity of the blood. The resulting anemia can affect normal nerve cell functioning and control of the heart rate, and can cause shortness of breath, pale skin, and fatigue.

- Carbon tetrachloride, pesticides, benzene, and ionizing radiation can affect the ability of the bone marrow to produce red blood cells.

- Mercury, cadmium, and other toxicants can affect the ability of the kidneys to stimulate the bone marrow to produce more red blood cells when needed to counteract low oxygen levels in the blood.

- Some chemicals, including carbon monoxide, can interfere with the blood's capacity to carry oxygen, resulting in lowered blood pressure, dizziness, fainting, increased heart rate, muscular weakness, nausea, and, after prolonged exposure, death.

- Hydrogen cyanide and hydrogen sulfide can stimulate cells in the aorta, causing increased heart and respiratory rate. At high concentrations, death can result from respiratory failure.

- Benzene, carbon tetrachloride, and trinitrotoluene can suppress stem cell production and the production of white blood cells. This can affect the clotting mechanism and the immune system.

- Benzene can cause low levels of white blood cells, a condition known as *leukemia*.

Hepatoxicity

Hepatoxicity refers to adverse effects on the liver that impede its ability to function properly. The liver converts carbohydrates, fats, and proteins to maintain the proper levels of glucose in the blood and converts excess protein and carbohydrates to fat. It also converts excess amino acids to ammonia and urea, which are removed in the kidneys. The liver also provides storage of vitamins and beneficial metals, as well as carbohydrates, fats, and proteins. Red blood cells that have degenerated are removed by the liver. Substances needed for other metabolic processes are provided by the liver. Finally, the liver detoxifies metabolically produced substances and toxicants that enter the body.

- Hexavalent chromium and arsenic cause cell damage in the liver.

- Carbon tetrachloride and alcohol can cause damage and death of liver cells, a condition known as *cirrhosis of the liver*.

- Chemicals or viruses can cause inflammation of the liver, known as *hepatitis*. Cell death and enlargement of the liver can occur.

Nephrotoxicity

Nephrotoxicity refers to adverse effects on the kidneys. The kidneys excrete ammonia as urea to rid the body of metabolic wastes. They maintain blood pH by exchanging hydrogen ions for sodium ions, and maintain the ion and water balance by excreting excess ions or water as needed. They also secrete hormones needed to regulate blood pressure. Like the liver, the kidneys function to detoxify substances.

- Heavy metals—primarily lead, mercury, and cadmium—cause impaired cell function and cell death. These metals can be stored in the kidneys, interfering with the functioning of enzymes in the kidneys.

- Chloroform and other organic substances can cause cell dysfunction, cell death, and cancer.

- Ethylene glycol can cause renal failure from obstruction of the normal flow of liquid through the kidneys.

Neurotoxicity

Neurotoxicity refers to toxic effects on the nervous system, which consists of the *central nervous system* (CNS) and the *peripheral nervous system* (PNS). The central nervous system includes the brain and the spinal cord, while the peripheral nervous system includes the remaining nerves, which are distinguished as sensory and motor nerves.

Safety/Health/
Environment

Neurotoxic effects fall into two basic types: *destruction* of nerve cells and *interference* with neurotransmission.

Immunotoxicity

Immunotoxicity refers to toxic effects on the immune system, which includes the lymph system, blood cells, and antibodies in the blood.

Reproductive Toxicity

The effect of toxicants on the reproductive system is known as *reproductive toxicity*. For the male reproductive system, toxicants primarily affect the division of sperm cells and the development of healthy sperm. For the female reproductive system, toxicants can affect the endocrine system, the brain, and the reproductive tract.

Carcinogens

The distinguishing feature of cancer is the uncontrolled growth of cells into masses of tissue called *tumors*. Tumors may be *benign*, in which the mass of cells remains localized, or *malignant*, in which the tumors spread through the bloodstream to other sites within the body. This latter process is known as *metastasis* and determines whether the disease is characterized as cancer. The term *neoplasm* (new and abnormal tissue) is also used to describe tumors.

Cancer occurs in three stages: initiation, promotion, and progression. During the *initiation* stage, a cell mutates and the DNA is not repaired by the body's normal DNA repair mechanisms. During the *promotion* stage, the mutated cells increase in number and undergo differentiation to create new genes. During the *progression* stage, the cancer cells invade adjacent tissue and move through the bloodstream to other sites in the body.

8. DOSE-RESPONSE RELATIONSHIPS

Dose-response relationships are used to relate the response of an organism to dose levels of toxicants. The shapes of the curves and the labeling of the dose-response graphs are slightly different for noncarcingens (see Fig. 51.2 and Fig. 51.4) and carcinogens (see Fig. 51.3), but the basic concepts are similar. Figure 51.2 is a typical dose-response graph for a *threshold toxicant*.[2]

Carcinogens generally do not exhibit thresholds of response at low doses of exposure, and any exposure is assumed to have an associated risk. The EPA method used to evaluate carcinogenic risk assumes no threshold and a linear response to any amount of exposure. *Noncarcinogens* (systemic toxicants) are chemicals that do not produce tumors (or gene mutations) but instead adversely interfere with the functions of enzymes in the body, which thereby causes abnormal metabolic

---

[2]Since Fig. 51.2 clearly shows the percentage of test animals (organisms) that experienced fatalities, the response for this curve is death. The label on the vertical axis, "toxic response," probably should be "fatal response." Not all dose-response curves involve fatal responses.

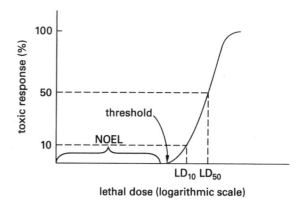

Figure 51.2 *Threshold Toxicant Dose-Response Curve*

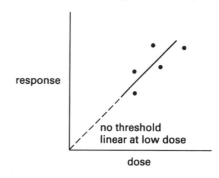

Figure 51.3 *No-Threshold Carcinogenic Dose-Response Curve*

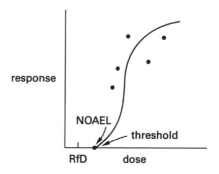

Figure 51.4 *Noncarcinogenic Dose-Response Curve*

responses. Noncarcinogens have a dose threshold below which no adverse health response can be measured.

Equation 51.8: Dose

$$dose = \left(\frac{mass\ of\ chemical}{body\ weight \cdot exposure\ time} \right) \qquad 51.8$$

Safety/Health/ Environment

Description

For both carcinogenic and noncarcinogenic substances, *dose* is calculated from Eq. 51.8.[3]

Dose-Response Curves

An objective of toxicity tests is to establish the *dose-response curve*, as illustrated in Fig. 51.2.

Several important features of a dose-response curve are described as follows.

- *response:* the ordinate

- *dose:* the abscissa

- *no-observed effect:* The range of the curve below which no effect is observed is the range of no-observed effect. The lower end of the range is known as the *threshold*, the *no-effect level* (NEL), or the *no-observed-effect level* (NOEL).

- *lowest-observed effect:* The dose where minor effects first can be measured, but the effects are not directly related to the response being measured, is known as the *lowest-observed-effect level* (LOEL).

- *no-observed-adverse effect:* The dose where effects related to the response being measured first can be measured is known as the *no-observed-adverse-effect level* (NOAEL).

- *lowest-observed-adverse effect:* The dose where effects related to the response being measured first can be measured, and are the same effects as the effects observed at the higher doses, is known as the *lowest-observed-adverse-effect level* (LOAEL).

- *frank effect:* The *frank-effect level* (FEL) dose marks the point where maximum effects are observed with little increase in effect for increasing dose.

The *lethal dose* or *lethal concentration* is the concentration of toxicant at which a specified percentage of test animals die. The lethal dose is expressed as the mass of toxicant per unit mass of test animal. For example, LD_{50} means the dose in milligrams of toxicant per kilogram of body mass at which 50% of the test animals died.

For acute tests involving inhalation as the exposure pathway, the concentration, in parts per million, of the toxicant in air is used. If the toxicant is in particulate form, the concentration in milligrams of toxic particles per cubic meter of air is used. For example, LC_{50} means the concentration of the toxicant in air at which 50% of the test animals died.

For some toxicants, primarily those believed to be carcinogens, there is no recognized threshold. Any dose is considered to have an effect even though such effect may be unmeasurable at low doses. Such toxicants have no

safe exposure level. Lead is an example of a toxicant with no threshold dose. Figure 51.3 is typical of a *no-threshold toxicant*, such as a carcinogen. At low doses, the dose-response curve is considered to be linear.

Figure 51.4 illustrates a noncarcinogenic dose response curve. The threshold and no-observed-adverse-effect level (NOAEL) are coincident. The *reference dose* (RfD) is the dose that is considered to be safe. It is established by reducing the NOAEL according to an accepted method.

9. EPA METHODS

Several approaches exist for establishing a safe human dose from the data obtained from toxicological and epidemiological studies. The most common approach is to use the reference dose recommended by the EPA.

Equation 51.9: Reference Dose

$$RfD = \frac{NOAEL}{UF} \qquad \text{51.9}$$

Description

The dose below which adverse health effects in humans are not measurable or observable is defined by the EPA as the reference dose. The reference dose is the safe daily intake that is believed not to cause adverse health effects. The reference dose is route specific. For gases and vapors (exposure by the inhalation pathway), the threshold may be defined as the *reference concentration* (RfC). Sometimes the term "reference dose" is also used for exposure by the inhalation pathway.

Example

The NOAEL for copper cyanide (CuCN) is 25 mg/kg·d. The total uncertainty factor (UF) is 5000. What is most nearly the reference dose?

(A) 3.0×10^{-3} mg/kg·d

(B) 5.0×10^{-3} mg/kg·d

(C) 7.0×10^{-3} mg/kg·d

(D) 9.0×10^{-3} mg/kg·d

Solution

Use Eq. 51.9 to find the reference dose.

$$RfD = \frac{NOAEL}{UF} = \frac{25 \dfrac{mg}{kg \cdot d}}{5000}$$

$$= 5.0 \times 10^{-3} \text{ mg/kg·d}$$

The answer is (B).

---

[3](1) Although the *NCEES Handbook* has established variables for body weight (*W* and BW) and exposure time (ET and ED), it does not use them in this equation. (2) Although mass is referred to in the numerator, weight is referred to in the denominator. Despite this inconsistency, body "weight" must be expressed in kilograms, not newtons.

Equation 51.10: Hazard Index

$$\text{HI} = \text{chronic daily intake/RfD} \qquad \textbf{51.10}$$

Description

The *chronic daily intake* (CDI), also known as the *estimated exposure dose* (EED), is the amount of a toxicant that the organism absorbs daily. It is the actual daily dose. The *hazard index* (HI) is a fraction calculated by dividing the actual dose (chronic daily intake, CDI) by what is considered to be a safe dose (the reference dose, RfD). The total hazard index is used to assess whether the exposure (i.e., risk) is acceptable. The total hazard index is the sum of the hazard indexes of each toxicant over the sum of all routes of exposure. If the total hazard index exceeds 1.0, the exposure (risk) is unacceptable.

Example

Methanol (CH_3OH) has a chronic daily intake of 29 μg/kg·d and a reference dose of 500 μg/kg·d. What is most nearly the hazard index of methanol?

(A) 0.032

(B) 0.045

(C) 0.058

(D) 0.076

Solution

Use Eq. 51.10.

$$\text{HI} = \text{chronic daily intake/RfD}$$

$$= \frac{29 \ \dfrac{\mu g}{kg \cdot d}}{500 \ \dfrac{\mu g}{kg \cdot d}}$$

$$= 0.058$$

The answer is (C).

Equation 51.11: Safe Human Dose

$$\text{SHD} = \text{RfD} \times W = \frac{\text{NOAEL} \times W}{\text{UF}} \qquad \textbf{51.11}$$

Description

The reference dose is the dose per unit mass of an organism, specifically, a test animal. The *safe human dose* (SHD) is the dose for a "standard" human. In Eq. 51.11, SHD is the safe human dose (mg/d), NOAEL is the threshold dose per unit mass (mg/kg·d) from the dose-response curve, UF is the total *uncertainty factor*

(which defines the reliability of the test data), and W is the mass of the standard test human (typically 70 kg).[4]

Example

The reference dose for copper cyanide (CuCN) is 5.0×10^{-3} mg/kg·d. An average weight adult male has a mass of 70 kg. What is most nearly the safe human dose for an average adult male?

(A) 0.35 mg/d

(B) 0.44 mg/d

(C) 0.67 mg/d

(D) 0.81 mg/d

Solution

Using Eq. 51.11, the safe human dose is

$$\text{SHD} = \text{RfD} \times W$$

$$= \left(5.0 \times 10^{-3} \ \frac{mg}{kg \cdot d}\right)(70 \text{ kg})$$

$$= 0.35 \text{ mg/d}$$

The answer is (A).

Direct Human Exposure

The EPA's classification system for carcinogenicity is based on a consensus of expert opinion called *weight of evidence*.

The EPA maintains a database of toxicological information known as the Integrated Risk Information System (IRIS). The IRIS data include chemical names, chemical abstract service registry numbers (CASRNs), reference doses for systemic toxicants, carcinogen potency factors (CPFs) for carcinogens, and the carcinogenicity group classification, which is shown in Table 51.1.

Table 51.1 EPA Carcinogenicity Classification System

group	description
A	human carcinogen
B1 or B2	probable human carcinogen
	B1 indicates that human data are available.
	B2 indicates sufficient evidence in animals and inadequate or no evidence in humans.
C	possible human carcinogen
D	not classifiable as to human carcinogenicity
E	evidence of noncarcinogenicity for humans

---

[4](1) In Eq. 51.11, the *NCEES Handbook* uses W for weight, while in other equations, it uses BW, "mass," or "weight." (2) Although the *NCEES Handbook* defines W as "the weight of the adult male," numerical values must have units of kilograms, not newtons.

The dose-response for carcinogens differs substantially from that of noncarcinogens. For carcinogens, it is believed that any dose can cause a response (i.e., mutation of DNA).

Since there are no levels (no thresholds) of carcinogens that could be considered safe for continued human exposure, a judgment must be made as to the acceptable level of exposure, which is typically chosen to be an excess lifetime cancer risk of 1×10^{-6} (0.0001%). *Excess lifetime cancer risk* refers to the incidence of cancers developed in the exposed animals minus the incidence in the unexposed control animals. For whole populations exposed to carcinogens, the number of total excess cancers, EC, is the product of the total exposed population, EP, and the probability of excess cancer, R.

$$EC = EP \times R$$

Under the EPA approach, the *carcinogen potency factor* (CPF) is the slope of the dose-response curve at very low exposures. The CPF is the probability of risk produced by lifetime exposure to 1.0 mg/kg·d of the known or potential human carcinogen. The CPF is also called the *potency factor* or *cancer slope factor* (CSF) and has units of $(mg/kg \cdot d)^{-1}$. The CPF is pathway (route) specific. The CPF is obtained by extrapolation from the high doses typically used in toxicological studies. (See Table 51.2.)

Equation 51.12: Added Risk of Cancer

$$\text{risk} = \text{dose} \times \text{toxicity} = CDI \times CSF \qquad 51.12$$

Description

Equation 51.12 multiplies the slope factor by the long-term daily intake (chronic daily intake, CDI) to obtain the lifetime added risk for daily doses other than 1.0 mg/kg·d.

Example

The water supply of a village with a stable population is contaminated with benzene (C_6H_6). The chronic daily intake (CDI) is 1.2×10^{-3} mg/kg·d. The cancer slope factor (CSF) for benzene by the oral route is 2.0×10^{-2} $(mg/kg \cdot d)^{-1}$. What is most nearly the added risk of cancer for villagers who drink the water?

(A) 2.4×10^{-5}

(B) 3.5×10^{-5}

(C) 4.4×10^{-5}

(D) 5.5×10^{-5}

Table 51.2 EPA Recommended Values for Estimating Intake

parameter	standard value
average body weight, adult female	65.4 kg
average body weight, adult male	78 kg
average body weight, child*	
6–11 months	9 kg
1–5 years	16 kg
6–12 years	33 kg
amount of water ingested, adult	2.3 L/d
amount of water ingested, child	1.5 L/d
amount of air breathed, adult female	11.3 m³/d
amount of air breathed, adult male	15.2 m³/d
amount of air breathed, child (3–5 years)	8.3 m³/d
amount of fish consumed, adult	6 g/d
water swallowing rate, while swimming	50 mL/h
inhalation rates	
adult (6 hours/day)	0.98 m³/h
adult (2 hours/day)	1.47 m³/h
child	0.46 m³/h
skin surface available, adult male	1.94 m²
skin surface available, adult female	1.69 m²
skin surface available, child	
3–6 years (male and female)	0.720 m²
6–9 years (male and female)	0.925 m²
9–12 years (male and female)	1.16 m²
12–15 years (male and female)	1.49 m²
15–18 years (female)	1.60 m²
15–18 years (male)	1.75 m²
soil ingestion rate, child 1–6 years	> 100 mg/d
soil ingestion rate, person > 6 years	50 mg/d
skin adherence factor, gardener's hands	0.07 mg/cm²
skin adherence factor, wet soil	0.2 mg/cm²
exposure duration	
lifetime (carcinogens; for noncarcinogens use actual exposure duration)	75 yr
at one residence, 90th percentile	30 yr
national median	5 yr
averaging time	(ED)(365 d/yr)
exposure frequency (EF)	
swimming	7 d/yr
eating fish and shellfish	48 d/yr
exposure time (ET)	
shower, 90th percentile	12 min
shower, 50th percentile	7 min

*Data in this category taken from Copeland, T., A. M. Holbrow, J. M. Otan, et al., "Use of probabilistic methods to understand the conservatism in California's approach to assessing health risks posed by air contaminants," *Journal of the Air and Waste Management Association*, vol. 44, pp. 1399–1413, 1994.

Safety/Health/ Environment

Solution

From Eq. 51.12, the added risk is

$$\text{risk} = \text{CDI} \times \text{CSF}$$

$$= \left(1.2 \times 10^{-3} \ \frac{\text{mg}}{\text{kg·d}}\right)\left(2.0 \times 10^{-2} \cdot \left(\frac{\text{mg}}{\text{kg·d}}\right)^{-1}\right)$$

$$= 2.4 \times 10^{-5}$$

The answer is (A).

Equation 51.13: Bioconcentration Factors

$$\text{BCF} = C_{\text{org}}/C \qquad \textbf{51.13}$$

Description

Besides setting factors for direct human exposure to toxicants through water ingestion, inhalation, and skin contact, the EPA also has developed *bioconcentration factors* (BCFs) (also referred to as *steady-state BCFs*) so that the human intake from consumption of fish and other foods can be determined. Bioconcentration factors have been developed for many toxicants and provide a relationship between a toxicant concentration in the tissue of an organism and the concentration in a medium (e.g., water). The concentration in an organism equals the product of the BCF and the concentration in the medium. Not all chemicals or other substances will bioaccumulate, and the BCF pertains to a specific type of organism, such as fish.

$$C_{\text{organism}} = \text{BCF} \times C_{\text{medium}}$$

The bioconcentration factor can be stated with several different units. For fish and other surface water organisms, it is usually the ratio of concentration in the organism (mg/kg) to the concentration in water (mg/L). In that case, the bioconcentration factor has units of L/kg. The factor also can be unitless when ratios of two concentrations have the same units.

Bioconcentration factors for selected chemicals in fish are given in Table 51.3. The substances are arranged in descending order of BCFs to illustrate the substances that have a high potential to bioaccumulate in fish.

The BCFs can be applied to determine the total dose to humans who ingest fish from water contaminated with toxicants that bioaccumulate. This dose would be added to the dose received from drinking the contaminated water.

Example

A fish with a mercury concentration of of 120 000 ppm is found in water with a mercury concentration of 22 ppm. What is most nearly the bioconcentration factor of mercury in the fish?

(A) 1500

(B) 2500

(C) 5500

(D) 7500

Solution

The bioconcentration factor of mercury in the fish can be calculated using Eq. 51.13.

$$\text{BCF} = C_{\text{org}}/C$$

$$= \frac{120\,000 \text{ ppm}}{22 \text{ ppm}}$$

$$= 5455 \quad (5500)$$

The answer is (C).

Table 51.3 *Typical Bioconcentration Factors for Fish**

substance	BCF (L/kg)
polychlorinated biphenyls	100 000
4,4' DDT	54 000
DDE	51 000
heptachlor	15 700
chlordane	14 000
toxaphene	13 100
mercury	5500
2,3,7,8-tetrachlorodibenzo-*p*-dioxin (TCDD)	5000
dieldrin	4760
copper	200
cadmium	81
lead	49
zinc	47
arsenic	44
tetrachloroethylene	31
aldrin	28
carbon tetrachloride	19
chromium	16
chlorobenzene	10
benzene	5.2
chloroform	3.75
vinyl chloride	1.17
antimony	1

*For illustrative purposes only. Subject to change without notice. Local regulations may be more restrictive than federal.

10. ACGIH METHODS

ACGIH is a professional association of industrial hygienists and practitioners from associated professions. It has developed its own methods for establishing safe exposures (known as *threshold limit values*) to industrial chemicals (liquids and gases in particular). The ACGIH methods are different from the EPA methods but have achieved significant acceptance throughout the world.

Threshold Limit Values

Two statutory limits quantify the concentration of a gas in air that a worker can be safely exposed to. These are *permissible exposure limit* (PEL), established by OSHA, and *threshold limit value* (TLV), established by ACGIH. The PEL is a regulatory exposure limit for workers. The TLV is a concentration in air that nearly all workers can be exposed to daily without any adverse effects.

The ACGIH method of establishing safe exposures uses predetermined TLVs for both noncarcinogens and carcinogens. Table 51.4 lists typical values.

Table 51.4 *Threshold Limit Values*

compound	TLV
ammonia	25
chlorine	0.5
ethyl chloride	1000
ethyl ether	400

ACGIH establishes three types of TLVs: the time-weighted average concentration, the short-term exposure limits, and the ceiling threshold limits. The *maximum time-weighted average concentration* (TLV-TWA) is the concentration that all workers may be exposed to during an 8-hour day and 40-hour week. The TLV-TWA applies only to exposure through inhalation.

Short-term exposure limits (TLV-STELs) are time-weighted average concentrations that have been established by ACGIH for most (but not all) airborne toxicants. TLV-STELs are maximum concentrations to which workers may be exposed for up to 15 minutes (i.e., "short periods"), but may never be exceeded. TLV-STELs are established to specifically avoid such adverse health effects as irritation, chronic or irreversible tissue damage, and *narcosis* (alteration of consciousness, including unconsciousness) of sufficient degree to increase the likelihood of accidental injury, impair self-rescue, or materially reduce work efficiency. In addition to being a maximum 15-minute average concentration, the TLV-STELs exposures are limited to a maximum of four 15-minute periods per day, and they require at least one 60-minute exposure-free period between any consecutive TLV-STEL exposure periods. In all periods, the daily TLV-TWA concentration must be observed.

The *ceiling threshold limit values* (TLV-C) are the values that should not be exceeded at any time during the workday. If instantaneous sampling is infeasible, the sampling period for the TLV-C can be up to 15 minutes in duration. Also, the TLV-TWA should not be exceeded.

For mixtures of substances, the *equivalent exposure* calculated over 8 hours is the one-hour equivalent exposure calculated from the sum of the individual exposures.

$$E = \frac{1}{8}\sum_{i=1}^{n} C_i t_i$$

The *hazard ratio* is the concentration of the contaminant divided by the exposure limit of the contaminant. For mixtures of substances, the total hazard ratio is the sum of the individual hazard ratios and must not exceed unity. This is known as the *law of additive effects*. For this law to apply, the effects from the individual substances in the mixture must act on the same organ. If the effects do not act on the same organ, then each of the individual hazard ratios must not exceed unity. The equivalent exposure of a mixture of gases is

$$E_m = \sum_{i=1}^{n} \frac{C_i}{\text{PEL}_i}$$

11. NIOSH METHODS

NIOSH was established by the Occupational Safety and Health Act of 1970. NIOSH is part of the Centers for Disease Control and Prevention (CDC) and is the only federal institute responsible for conducting research and making recommendations for the prevention of work-related illnesses and injuries. NIOSH's responsibilities include

- investigating hazardous working conditions as requested by employers or workers

- evaluating hazards ranging from chemicals to machinery

- creating and disseminating methods for preventing disease, injury, and disability

- conducting research and providing recommendations for protecting workers

- providing education and training to persons preparing for or actively working in the field of occupational safety and health

As part of the rule-making (i.e., mandatory OSHA requirements), NIOSH establishes RELs. RELs are based on hard science (experiments and studies). In contrast, OSHA PELs are the values that have been written into law after everyone (including opponents of stricter controls) have had their influence. In an ideal world, PELs established by OSHA would be based on NIOSH RELs. However, the connection between the two is complicated by factors of politics, economics, and expediency.

The NIOSH RELs are TWA concentrations for up to a 10-hour workday during a 40-hour workweek. A ceiling REL should not be exceeded at any time. The "skin" designation means there is a potential for dermal absorption, so skin exposure should be prevented as necessary through the use of good work practices and gloves, coveralls, goggles, and other appropriate equipment.

52 Industrial Hygiene

Nomenclature[1]

a	speed of sound	m/s
A	area	m^2
A	asymmetry angle	deg
ABS	absorption factor	–
AD	absorbed dose	mg/kg·d
AF	adherence factor	mg/cm^2
AT	averaging time	yr
BCF	bioconcentration factor	L/kg
BW	body mass	kg
C	concentration	ppm, mg/m^3
C_0	discharge coefficient	–
C_{fi}	volumetric percent of fuel gas	%
C_i	time of noise exposure at specified level	s
C_i	volumetric percent of flammable gas	%
CA	contaminant concentration in air	mg/m^3
CDI	chronic daily intake	mg/kg·d
CF	contaminant concentration in fish	mg/kg
CF	volumetric conversion factor	L/cm^3
CM	coupling multiplier for lifting	–
CPF	carcinogen potency factor (same as cancer slope factor)[2]	(mg/kg·d)$^{-1}$
CR	contact rate	d^{-1}
CS	chemical concentration in soil	mg/kg
CSF	cancer slope factor (same as carcinogen potency factor)	(mg/kg·d)$^{-1}$
CW	chemical concentration in water	mg/L
D	dose	mg/kg·d
D	vertical distance	in
E	time-weighted average exposure	ppm, mg/m^3
E_m	equivalent exposure of mixture	–
EC	excess cancers	–
ED	exposure duration	yr
EED	estimated exposure dose	mg/kg·d
EF	exposure factor	–
EP	exposed population	–
ET	exposure time	h/d
f	frequency	Hz
F	frequency of task	1/min
FI	fraction injested	–
FM	frequency multiplier for lifting	–
H	horizontal hand distance	in
HI	hazard index	–
I	intensity	W/m^2
IR	ingestion rate	L/d
IR	inhalation rate	m^3/h
k	decay constant	s^{-1}
k	nonideal mixing factor	–
k	rate constant	s^{-1}
K	mass transfer coefficient	m/s
LD$_{50}$	median lethal dose	mg/kg
LFL	lower flammability limit	%
LOAEL	lowest-observed-adverse-effect level	mg/kg·d
LOEL	lowest-observed-effect level	mg/kg·d
M	molecular weight	kg/kmol
N	number of atoms	–
NOAEL	no-observed-adverse-effect level	mg/kg·d
NOEL	no-observed-effect level	mg/kg·d
p	pressure	Pa
p^{sat}	saturation vapor pressure	Pa
PC	permeability constant	cm/h
PEL	permissible exposure limit	mg/m^3
Q	heat gain	W
Q	quantity	various
r	distance	m

---

[1]This chapter covers many different safety and health topics. The NCEES *FE Reference Handbook* (*NCEES Handbook*) has adopted the convention of presenting variables and their units exactly as they are given in the various sources used to compile it. The results are overlapping variables and inconsistent units within this chapter and inconsistencies with the rest of the *NCEES Handbook* (and this book). For example, SA and A_S are both used for surface area; units of the permeability constant (as required to be consistent with other unit definitions) are the nonstandard cm/h; g_c with units of m/s^2 is the variable for standard gravity; Q_m is the variable used for a mass generation rate; time has units of percent, seconds, hours, days, and years; and variables for time include t, T, ET, and ED.

---

[2]It is idiosyncratic of the subject that units of the cancer slope factor are stated as (mg/kg·d)$^{-1}$ rather than as kg·d/mg.

Safety/Health/Environment

R	risk; probability of excess cancer	–
R_g	universal gas constant	$kPa·m^3/$
		$kmol·K$
RfD	reference dose	mg/kg·d
RWL	recommended weight limit	lbf
SA	surface area	cm^2
SHD	safe human dose	mg/d
SPL	sound pressure level	dB
SWL	sound power level	dB
t	rest time, percent of period	%
t	time	s
T	temperature	K
T_i	exposure time	h
TLV	threshold limit value	–
UF	uncertainty factor	–
UFL	upper flammability limit	%
V	vertical hand distance	in
V	volume	m^3
VM	vertical multiplier for lifting	–
W	mass	kg
W	power	W

Symbols

α	floor inclination angle	deg
λ	wavelength	m
μ	coefficient of friction	–
ρ	density	kg/m^3
τ	half-life	s

Subscripts

0	initial condition or reference
C	convection
E	evaporation
f	fuel
g	gauge
H	hole
L	liquid
m	mixed or mixture
max	maximum
M	metabolic
n	noise or number
org	organism
rest	resting
rms	root-mean-square
R	radiation
S	storage or surface
t	time or time period
V	volumetric
W	sound power

1. INDUSTRIAL HYGIENE

Industrial hygiene is the art and science of identifying, evaluating, and controlling environmental factors (including stress) that may cause sickness, health impairment, or discomfort among workers or citizens of the community. Industrial hygiene involves the recognition of health hazards associated with work operations and processes, evaluations, measurements of the magnitude of hazards, and determining applicable control methods. Occupational health seeks to reduce hazards leading to illness or impairment for which a worker may be compensated under a worker protection program.

The fundamental law governing worker protection in the United States is the 1970 federal *Occupational Safety and Health Act*. It requires employers to provide a workplace that is free from hazards by complying with specified safety and health standards. Employees must also comply with standards that apply to their own conduct. The federal regulatory agency responsible for administering the Occupational Health and Safety Act is the Occupational Safety and Health Administration (OSHA). OSHA sets standards, investigates violations of the standards, performs inspections of plants and other facilities, investigates complaints, and takes enforcement action against violators. OSHA also funds state programs, which are permitted if they are at least as stringent as the federal program. OSHA standards are given in Title 29 of the *Code of Federal Regulations* (CFR).

The 1970 Occupational Safety and Health Act also established the *National Institute for Occupational Safety and Health* (NIOSH). NIOSH is responsible for safety and health research and makes recommendations for regulations. The recommendations are known as *recommended exposure limits* (RELs). Among other activities, NIOSH also publishes health and safety criteria and notifications of health hazard alerts, and is responsible for testing and certifying respiratory protective equipment.

While the U.S. Environmental Protection Agency (EPA) provides exposure limits and risk factors for environmental cleanup projects, the American Conference of Governmental Industrial Hygienists (ACGIH) provides recommendations to industrial hygienists about workplace exposure.

Table 52.1 lists some of the organizations that regulate worker and environmental safety.

2. OVERVIEW OF HAZARDS

There are four basic types of hazards with which industrial hygiene is concerned: chemical hazards, biological hazards, physical hazards, and ergonomic hazards. *Chemical hazards* result from chemicals such as gases, vapors, or particulates in harmful concentrations. Besides inhalation, chemical hazards may affect workers by absorption through the skin. *Biological hazards* include exposure to biological organisms that may lead to illness. *Physical hazards* include radiation, noise, vibration, electricity, and excessive heat or cold. *Ergonomic hazards* include work procedures and arrangements that require motions that result in biomechanical stress and injury.

Safety/Health/Environment

Table 52.1 Safety/Regulatory Agencies

acronym	name	jurisdiction
CSA	Canadian Standards Association	nonprofit standards organization
FAA	Federal Aviation Administration	federal regulatory agency
IEC	International Electrotechnical Commission	nonprofit standards organization
ITSNA	Intertek Testing Services NA, Inc. (formerly Edison Testing Labs)	nationally recognized testing laboratory
MSHA	Mine Safety and Health Administration	federal regulatory agency
NFPA	National Fire Protection Association	nonprofit trade association
OSHA	Occupational Safety and Health Administration	federal regulatory agency
UL	Underwriters Laboratories	nationally recognized testing laboratory
USCG	United States Coast Guard	federal regulatory agency
USDOT	United States Department of Transportation	federal regulatory agency
USEPA	United States Environmental Protection Agency	federal regulatory agency

3. CHEMICAL HAZARDS

Chemical hazards are one of the types of health hazards employers must manage. Chemical hazards can be characterized in terms of *corrosivity, flammability, toxicity,* and *reactivity*.

Corrosivity

Corrosivity is the ability of a material to damage other materials (e.g., to rust metal) or organisms (e.g., to burn skin). Examples of corrosive substances include strong acids, such as sulfuric acid and hydrochloric acid; caustics or alkalis, such as sodium hydroxide; and halogens, such as chlorine. A metal container that has been used to store a corrosive substance is also a chemical hazard.

Corrosivity indicates a material's rate of corrosion. Usually expressed in terms of mass (weight) loss or loss in thickness of the material being tested, corrosion rates are determined by exposing standardized steel samples

to corrosive materials. The mass (weight) loss is expressed in units of milligrams lost per 100 dm^2 per day (MDD); and the loss in thickness is expressed in units of mils lost per year (MPY). A mil is a unit of thickness equal to one thousandth of an inch.

Flammability

Flammability is the ability of a material to burn or ignite. Flammable solids burn easily once they come in contact with an ignition source. The flammability of a material is determined by fire testing. The National Fire Protection Association (NFPA) and U.S. Department of Transportation (USDOT) define a flammable liquid as a liquid with a *flash point* below 37.8°C (100°F). Section 52.8 presents more details on flammable and combustible liquids.

Toxicity

Toxicity is the degree of damage a toxic material causes to an organism. The damage can be to an entire organism, such as an animal or a plant, or to particular organs, such as the eyes or lungs. The *dosage* or quantity of the toxic material ingested is an important factor of toxicity. A commonly used indicator of toxicity is the *lethal dose* (LD) required to affect a certain percentage of the tested population. For example, LD_{50} represents the median single lethal dose (usually by oral ingestion or skin exposure) that can kill 50% of test subjects as determined by laboratory tests. Most toxic materials are chemical compounds, such as pesticides and poisons. However, toxic materials can also be biological in nature, such as bacteria, which can cause disease, or physical in nature, such as asbestos, which can cause inflammation and scarring of the lungs.

Reactivity

Reactivity is the propensity of a material to chemically react with its surroundings or other materials. Highly reactive materials are chemically unstable and can corrode, decompose, polymerize, burn, or explode under normal environmental conditions (e.g., normal temperature, moisture, and pressure). For example, reactive waste materials such as used lithium-sulfur batteries are chemically unstable under normal conditions. They can emit toxic fumes and can explode. Substances with low reactivity are chemically stable and do not react with their surroundings under normal conditions.

Reactivity is also related to the rate at which a substance reacts (i.e., it refers to the speed of the reaction). Fast reactions can be difficult to control and can be hazardous.

4. RISK ASSESSMENT

Risk assessment is an analytical process that determines the probability that some mishap will occur. It is applied primarily to human health risks and safety hazards or accidents from industrial practices.

Health risk assessment is based on toxicological studies. Data are usually gathered from exposure of laboratory animals to chemicals or processes, although human exposure data can be used if available.[3] Tests are run to determine if a chemical is a *carcinogen* (i.e., causes cancer), a *mutagen* (i.e., causes mutations), a *teratogen* (i.e., causes birth defects), a *nephrotoxin* (i.e., harms the kidneys), a *neurotoxin* (i.e., damages the brain or nervous system), or a *genotoxin* (i.e., harms the genes).

Safety risk assessment of potentially unsafe equipment or practices is traditionally called *hazard analysis*. Attention has traditionally focused on processes and equipment. However, hazard analysis is also applicable to material storage, shipping, transportation, and waste disposal. An analysis of each point in the process is performed using a fault-tree.

Fault-tree analysis (FTA) is a deductive logic modeling method. Unwanted events (i.e., failures or accidents) are first assumed. Then, the conditions are identified that could bring about the events. All possible contributors to an unwanted event are considered. The event is shown graphically at the top of a network. It is linked through various event statements, logic gates, and probabilities to more basic fault events located laterally and below, producing a graphical tree having branches of sequences and system conditions.

Failure modes and effect analysis (FMEA) is essentially the reverse of fault-tree analysis. It starts with the components of the system and, by focusing on the weaknesses and failure susceptibilities, evaluates how the components can contribute to the unwanted event. The basis for FMEA is, essentially, a parts list showing how each assembly is broken down into subassemblies and basic components. The appearance of an FMEA analysis is tabular, with the columns being used for failure modes, causes, symptoms, redundancies, consequences of failure, frequencies, probabilities, and so on.

Life-cycle analysis is used to prevent failures of items that have limited lives or that accumulate damage. Though the probability of failure may ideally be low, the history of the equipment may also be unknown. Life cycle analysis commonly focuses on time, history, and condition to determine the *degree of damage* (i.e., any reduction in useful life) present in a unit. One approach is to use reporting systems (i.e., operating logs and historical data). Another approach uses ongoing *material conditions monitoring* (MCM) and testing of the actual equipment.

Equation 52.1 and Eq. 52.2: Risk

$$\text{risk} = \text{hazard} \times \text{probability} \quad 52.1$$

$$\text{risk} = \text{hazard} \times \text{exposure} \quad [\text{biological hazard}] \quad 52.2$$

[3]The best-known exposure test is the *Ames test*, wherein microbes, cells, or test animals are exposed to a chemical in order to determine its carcinogenicity.

Variations

$$\text{risk} = \text{hazard} \times \text{vulnerability}$$

$$\text{risk} = \text{hazard} \times \text{dose}$$

Description

A *hazard* is anything that has the ability (capacity) to cause damage or harm. Typically, a hazard is quantified in terms of the loss or damage it creates. *Risk* calculated from Eq. 52.1 involves both probability and severity. With biological hazards (e.g., pathogens, chemicals, radiation), *probability* is replaced by the term *exposure*, and risk is calculated using Eq. 52.2. There are other ways of formulating Eq. 52.1. Probability can be replaced with *vulnerability*. Exposure can be replaced with *dose*. Technically, risk itself is dimensionless.[4]

5. HAZARD COMMUNICATION

Two important safety policies required by OSHA are use and distribution of *material safety data sheets* (MSDSs) and labeling of containers of hazardous materials. A third OSHA requirement is that employers must provide hazard information and safety training to workers. The OSHA *Hazard Communication Standard* is given in 29 CFR Part 1910.1200.

MSDS

An MSDS, also known as a *safety data sheet* (SDS) or *product data sheet* (PDS), provides key information about a chemical or substance so that users or emergency responders can determine safe use procedures and necessary emergency response actions. An MSDS provides information about the identification of the material and its manufacturer, identification of hazardous components and their characteristics, physical and chemical characteristics of the ingredients, fire and explosion hazard data, reactivity data, health hazard data, precautions for safe handling and use, and recommended control measures for use of the material. An MSDS must be kept on file by a designated site safety officer for any chemical that is stored, handled, or used on-site.

Container Labeling

Labels are required on hazardous material containers. Labels should provide essential information for the safe use and storage of hazardous materials. Failure to provide adequate labeling of hazardous material containers is a common violation of OSHA standards.

The fire/hazard diamond shown in Fig. 52.1 is a common component of labels on chemical storage/transport containers. It is also found on the MSDS.

[4]For example, "sharp blade" is a hazard, and "sharp blade × probability" is the probability of an injury from the blade. Equation 52.1 and Eq. 52.2, as given in the *NCEES Handbook*, are probably intended to be descriptive (illustrative), rather than used for actual calculations.

Figure 52.1 Hazard Diamond

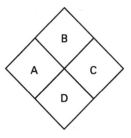

Positions A through C use a numerical scale of 0 (no hazard) to 4 (extreme hazard). Position A (blue) indicates a health hazard, position B (red) indicates flammability, and position C (yellow) indicates reactivity. Position D (white) indicates other hazards, such as whether a material is alkaline, corrosive, or radioactive.

Packaging is required to include *signal words* describing the toxicity of materials. Table 52.2 describes the signal words for pesticides.

Table 52.2 Pesticide Toxicity Categories

signal word on label	toxicity category	acute-oral LD_{50} for rats	amount needed to kill an average size adult	notes
Danger—Poison	highly toxic	≤ 50	≤ 1 tsp	skull and cross-bones; keep out of reach of children
Warning	moderately toxic	50–500	1–6 tsp	keep out of reach of children
Caution	slightly toxic	500–5000	1 oz–1 pint	keep out of reach of children
Caution	relatively nontoxic	> 5000	≥ 1 pint	keep out of reach of children

Globally Harmonized System of Classification and Labeling of Chemicals

The *Globally Harmonized System of Classification and Labeling of Chemicals* (GHS) is a document created by the United Nations. The GHS defines the physical and environmental hazards of chemicals, classifies processes based on defined hazard criteria, and communicates hazardous information and protective measures on labels and on MSDSs. The GHS is not a standard or regulation, but it provides countries with the basic building blocks to develop or modify an existing national program of hazard classification. The GHS closely follows many of the standards set out in U.S. Department of Transportation and OSHA regulations.

Rather than using numbers or codes to identify hazards, the GHS recommends using *pictograms* to address acute toxicity (see Fig. 52.2), physical hazards, and environmental hazards associated with the storage, handling, or transportation of chemicals. Figure 52.3 lists GHS pictograms and hazard classes. Figure 52.4 provides transport pictograms for use in the transportation of hazardous materials or wastes. In addition to hazard pictograms, the GHS also recommends using *signal words* and *hazard statements*, which are already incorporated into MSDSs and other hazard communications.

A Globally Harmonized System (GHS) label includes, from top to bottom, the following items.

- product name, including hazardous components
- hazard pictogram(s)
- signal word(s)
- physical, health, and environmental hazard statements
- supplemental information
- precautionary pictograms
- first aid statements
- name, address, and telephone number of the manufacturer

Figure 52.2 Acute Oral Toxicity Categories

	category 1	category 2	category 3	category 4	category 5
LD_{50}	≤ 5 mg/kg	> 5 < 50 mg/kg	≥ 50 < 300 mg/kg	≥ 300 < 2000 mg/kg	≥ 2000 < 5000 mg/kg
pictogram	☠	☠	☠	!	no symbol
signal word	danger	danger	danger	warning	warning
hazard statement	fatal if swallowed	fatal if swallowed	toxic if swallowed	harmful if swallowed	may be harmful if swallowed

Example

A bulk load of potassium chlorate, $KClO_3$, is transported by rail to a fireworks manufacturer. How should the rail cars be marked?

(A) explosive

(B) flammable

(C) oxidizer

(D) pyrophoric

Solution

Potassium chlorate is a strong oxidizer. By itself, it is not explosive, flammable, or pyrophoric.

The answer is (C).

Figure 52.3 *GHS Pictograms and Hazard Classes*

oxidizers

environmental toxicity

explosives;
self reactives;
organic peroxides

acute toxicity (severe)

corrosives

gases under pressure

carcinogen;
respiratory sensitive;
reproductive toxicity;
target organ toxicity;
mutagenicity;
aspiration toxicity

flammables;
self reactives;
pyrophorics;
self heating;
emits flammable gas;
organic peroxides

irritant;
dermal sensitizer;
acute toxicity (harmful);
narcotic effects;
respiratory tract irritation

Figure 52.4 *Transport Pictograms*

flammable liquid;
flammable gas;
flammable aerosol
(black and red)

flammable solid;
self-reactive substances
(black, white,
and red stripes)

pyrophorics;
(spontaneously combustible);
self-heating substances
(black, white, and red)

substances, which in
contact with water,
emit flammable gases
(dangerous when wet)
(black and purple)

oxidizing gases;
oxidizing liquids;
oxidizing solids
(black and yellow)

explosive divisions
1.1, 1.2, 1.3
(black and red)

explosive division 1.4
(black and red)

explosive division 1.5
(black and red)

explosive division 1.6
(black and red)

gases under pressure
(black and green)

acute toxicity (severe):
oral, dermal, inhalation
(black and white)

corrosive
(black and white)

marine pollutant
(black and white)

organic peroxides
(black, red, and yellow)

Worker Information and Training

OSHA requires that employers provide workers with information about the potential health hazards from exposure to hazardous chemicals that they use in the workplace. It also requires employers to provide adequate training to workers on how to safely handle and use hazardous materials.

6. PERSONAL PROTECTIVE EQUIPMENT

Personal protective equipment (PPE) refers to a variety of equipment worn to protect workers from various chemical, biological, physical, and ergonomic hazards. OSHA standards and regulations require employers to prevent and control workplace hazards through the use of engineering controls, such as building a barrier between the worker and the hazard; administrative controls, such as rotating workers; and safe work practices, such as general workplace and operation-specific rules. If, after implementing these controls and practices, employees may still be exposed to workplace hazards, employers must provide PPE. PPE can include gloves, hard hats, safety glasses, safety shoes, masks and respirators, ear plugs/muffs, coveralls, vests, full body suits, safety harnesses, and fire-retardant clothing. PPE is intended to protect from dermal contact with and

inhalation of hazardous substances, as well as exposure to unsafe noise levels and other physical hazards.

Respirators should be used for emergency and backup protection but not as a primary source of protection from hazardous vapors. To reduce leakage around the edges of the face mask, respirators must be properly fitted for each worker, and workers need to be trained in proper respirator use. In addition, a worker may feel a false sense of security while wearing a respirator. In some circumstances, both engineering controls and PPE must be used to protect against inhalation hazards.

Besides inhalation, dermal contact is an important concern when working with hazardous solvents. Barriers, guards, and other engineering controls should be provided to keep the worker from coming into contact with the solvent. However, since some contact may occur even with such controls in use, PPE should also be provided. Protective clothing includes aprons, face shields, goggles, impact-resistant glasses with side-shields, and gloves. The manufacturer's recommendations should be followed for use of all protective clothing and equipment.

Proper glove selection is critical in protecting against dermal hazards. The composition of the glove and the conditions under which gloves will be worn must be considered. The time it takes for solvents to penetrate gloves that are commonly thought of as "protective" is surprisingly short. For example, methyl chloride will permeate a neoprene glove in less than 15 minutes. Both the permeability and the abrasion resistance of gloves must be considered. The manufacturer should provide the breakthrough time and the permeation rate for gloves. The breakthrough time and permeation rate are specific to the chemical being handled and the composition and thickness of the glove. Gloves are given a degradation rating for a given chemical based on the permeation rate (see Table 52.3).

Table 52.3 Permeation Rates and Degradation Ratings

permeation rate (μg/cm^2·min)	degradation rating
< 0.9	excellent (E)
0.9–9	very good (VG)
9–90	good (G)
90–900	fair (F)
900–9000	poor (P)
> 9000	not recommended (NR)

Protective eyewear should be provided when there is a risk of chemicals splashing into the eyes, or when workers are exposed to particulates that could either cause damage to the eye or be absorbed by the eye. For chemical splash protection, workers should wear unvented chemical goggles, indirect-vented chemical goggles, or indirect-vented eyecup goggles. A face shield may also be needed. Direct-vented goggles and normal eyeglasses should not be used, and contact lenses should not be worn.

In addition to chemical hazards that could result in inhalation or dermal contact, workers may be exposed to physical and environmental hazards such as excess noise (see Sec. 52.15), electricity (see Sec. 52.18), or equipment and conditions that pose physical harm. PPE is also required in these circumstances if mechanical means are not sufficient to fully protect a worker.

When employees must work in areas that exceed the sound pressure levels provided in OSHA regulations, they should wear PPE, such as ear plugs or ear muffs, in addition to following the OSHA regulations for the length of exposure to certain noise levels.

Example

A manufacturer has submitted a sample of its 22 mil nitrile gloves for testing. The permeation rate of the glove material was found to be approximately 7400 μg/cm^2·min for water-based acids, caustics, and noncorrosive inorganic poisons. With what chemicals may this glove be used?

(A) acids

(B) caustics

(C) noncorrosive poisons

(D) none of the above

Solution

Permeation rates of gloves vary from near zero to over 9000 μg/cm^2·min, which is the upper threshold for a "poor" rating. This glove has a permeation rate that is near the "not recommended" range. Although a 22 mil nitrile glove should offer better protection than demonstrated by its poor degradation rating, something in the manufacturing process has degraded the protection offered by this material. This glove should not be used for anything that is potentially harmful.

The answer is (D).

7. EXPOSURE FACTORS FOR GASES AND VAPORS

The most frequently encountered hazard in the workplace is exposure to gases and vapors from solvents and chemicals. Several factors define the exposure potential for gases and vapors. The most important are how a material is used and what engineering or personal protective controls exist. If the inhalation route of entry is controlled, dermal contact may still be a major route of exposure.

Vapor pressure of a substance is related to temperature. Vapor pressure affects the concentration of the substance in vapor form above the liquid and is dependent upon the temperature and the properties of the substance. Processes that operate at lower temperatures are inherently less hazardous than processes that operate at higher temperatures.

Reactivity affects the hazard potential because the products may be volatile or nonvolatile depending on the properties of the combining substances. Table 52.4 illustrates how various hazardous materials interact.

Solvents

Solvents are highly volatile, which means they vaporize readily. The major route of exposure to solvents is

Table 52.4 *Hazardous Waste Compatibility*

no.	name	1	2	3	4	5	6	7	8	9	10	11	12	13	14	15	16	17	18	19	20	21	104	105	106	107
1	acids, minerals, non-oxidizing	1																								
2	acids, minerals, oxidizing		2																							
3	acids, organic	G/H		3																						
4	alcohols & glycols	H	H/F	H/P	4																					
5	aldehydes	H/P	H/F	H/P		5																				
6	amides	H	H/GT				6																			
7	amines, aliphatic & aromatic	H	H/GT	H	H			7																		
8	azo compounds, diazo comp, hydrazines	H/G	H/GT	H/G	H/G	H			8																	
9	carbamates	H/G	H/GT						H/G	9																
10	caustics	H	H	H		H				H/G	10															
11	cyanides	GT/GF	GT/GF	GT/GF						G		11														
12	dithiocarbamates	H/GF/F	H/GF/F	H/GF/GT				GF/GT		U	H/G		12													
13	esters	H	H/F							H/G	H			13												
14	ethers	H	H/F												14											
15	fluorides, inorganic	GT	GT	GT												15										
16	hydrocarbons, aromatic		H/F														16									
17	halogenated organics	H/GT	H/F/GT					H/GT	H/G			H/GF	H					17								
18	isocyanates	H/G	H/F/GT	H/G	H/P			H/P	H/G	H/P/G	H/G	U							18							
19	ketones	H	H/F						H/G	H	H									19						
20	mercaptans & other organic sulfides	GT/GF	H/F/GT						H/G									H	H	H	20					
21	metal, alkali & alkaline earth, elemental	GF/H/F	GF/H/F	GF/H	GF/H	GF/H	GF/H	GF/H	GF/H	GF/H	GF/H	GF/GT/H	GF/H					H/E	GF/H	GF/H	GF/H	21				
104	oxidizing agents, strong	H/GT		H/GT	H/F	H/F	H/F/GT	H/E/GT	H/F	H/F			H/F	H/GT				H/F	H/F/GT	H/F	H/F/GT	H/F/GT/GE	104			
105	reducing agents, strong	H/GF	H/F/GT	H/GF	H/GF/F	GF/H/F	GF/H	H					H/GT	H/F				H/T	GF/H	GF/H	GF/H		H/F/E	105		
106	water & mixtures containing water	H	H					G										H/G			GF/H				GF/GT	106	
107	water reactive substances	EXTREMELY REACTIVE! Do not mix with any chemical or waste material.																								107

Bottom axis columns: 1 2 3 4 5 6 7 8 9 10 11 12 13 14 15 16 17 18 19 20 21 22 23 24 25 26 27 28 29 30 31 32 33 34 101 102 103 104 105 106 107

key

reactivity code	consequences
H	heat generation
F	fire
G	innocuous & non-flammable gas
GT	toxic gas generation
GF	flammable gas generation
E	explosion
P	polymerization
S	solubilization of toxic material
U	may be hazardous but unknown

example:

H	heat generation,
F	fire, and toxic gas
GT	generation

through the respiratory system and dermal contact. Most solvents are very *lipid soluble*, which means they are soluble in fat. Because of this, they readily cross into the bloodstream.

Solvents are widely used throughout industry for many purposes, and their safe use is an important industrial hygiene concern. It is essential that accurate MSDS information be provided to employees on the physical properties and the toxicological effects of exposure to the solvents in their workplaces.

Equation 52.3 Through Eq. 52.6: Vaporized Liquid

$$Q_m = MKA_S p^{\text{sat}}/R_g T_L \qquad 52.3$$

$$Q_m = A_H C_0 (2\rho g_c p_g)^{1/2} \qquad 52.4$$

$$C_{\text{ppm}} = Q_m R_g T \times 10^6 / k Q_V p M \qquad 52.5$$

$$Q_V t = V \ln[(C_1 - C_0)/(C_2 - C_0)] \qquad 52.6$$

Safety/Health/Environment

Description

Equation 52.3 gives the vaporization rate from a liquid surface. Equation 52.4 gives the flow rate of a liquid from a hole in the wall of a process unit. Equation 52.5 gives the concentration of a vaporized liquid in a ventilated space. Equation 52.6 gives the sweep through concentration change in a vessel.

Equation 52.3 is based on the ideal gas law ($pV = mRT$ in conventional units), where the specific gas constant, R, is calculated from the universal gas constant, R_g, and molecular weight, M. In Eq. 52.3, K is the mass transfer coefficient with units of velocity; A_S is the solvent exposed surface area; and p^{sat} is the saturation pressure of the solvent at the temperature of the liquid, T_L. If consistent units are used, the vapor generation rate, Q_m, will have units of mass per unit time.[5]

Equation 52.4 is based on *Torricelli's speed of efflux*, calculated from Bernoulli's equation. The velocity term ($\mathrm{v} = \sqrt{2gh}$) calculated from Bernoulli's equation is multiplied by the area of the hole, the discharge coefficient, and the density. If consistent units are used, the result will have units of mass per unit time.[6]

Equation 52.5 calculates the ratio of the rate volume of generated solvent and the rate volume of ventilation air (or gas), and then converts the volumetric fraction to parts per million by multiplying by 10^6. In Eq. 52.5, k is a factor intended to account for nonideal mixing.[7] The *nonideal mixing factor* is the fraction of gas that mixes perfectly with the purge gas, leaving the remainder of the vessel contents at full strength. The factor is less than 1.0; recommended values depend on both the toxicity of the vessel contents and the degree of turbulence within the vessel. Values are particularly low for vessels not specifically designed for purging, because the vessels are prone to short-circuiting in the purge gas flow path. A value of $1/10$ is recommended for this worst-case scenario.

Equation 52.6 calculates the ideal volume (*sweep through volume*) of *dilution air* required to reduce the

concentration of a solvent from C_1 to C_2.[8] This kind of process is known as *inerting*, *gas purging*, and *gas sweeping*. C_0 is the concentration of the solvent in the purge air, usually zero.

Example

A 4 m^3 tank initially contains atmospheric air. The tank is purged with a nitrogen mixture consisting of 1 mol of oxygen for every 99 mol of nitrogen. The purge gas flow rate is 2.0 m^3/min. The tank is optimized for purging, so the nonideal mixing factor is $1/10$. Most nearly, how long will it take for the volumetric fraction of oxygen in the tank to reach 6%?

(A) 4 min

(B) 9 min

(C) 20 min

(D) 30 min

Solution

The oxygen mole fraction of the purge gas is

$$x = \frac{n_{O_2}}{n_{\text{total}}} = \frac{n_{O_2}}{n_{O_2} + n_{N_2}} = \frac{1 \text{ mol}}{1 \text{ mol} + 99 \text{ mol}}$$
$$= 0.01$$

For ideal gases, mole fractions are volumetric fractions. Air is approximately 21% oxygen and 79% nitrogen by volume. So, the volumetric fraction of oxygen in air is 0.21.

Use Eq. 52.6. Incorporate the nonideal mixing factor, k.

$$kQ_V t = V \ln\left((C_1 - C_0)/(C_2 - C_0)\right)$$

$$t = \frac{V \ln \dfrac{C_1 - C_0}{C_2 - C_0}}{kQ_V}$$

$$= \frac{(4 \text{ m}^3)\ln \dfrac{0.21 - 0.01}{0.06 - 0.01}}{\left(\dfrac{1}{10}\right)\left(2.0 \ \dfrac{\text{m}^3}{\text{min}}\right)}$$

$$= 27.7 \text{ min} \quad (30 \text{ min})$$

The answer is (D).

---

[5]The variables used in Eq. 52.3 differ markedly from those used in the Thermodynamics section and other sections of the *NCEES Handbook*. They are also inconsistent with the variables and methods used by ACGIH and OSHA.

[6](1) Equation 52.4 appears to calculate a volumetric flow rate (analogous to $Q = C_d A\sqrt{2gh}$ or $Q = C_d A\sqrt{2gp/\rho}$ using commonly encountered units). However, the density term normally placed outside of the radical ($\dot{m} = C_d A\rho\sqrt{2gh}$) has been brought under the radical. (2) The *NCEES Handbook* uses g_c and its corresponding definition as the "gravitational constant." However, the acceleration due to gravity ($g = 9.81$ m/s^2) is intended. Equation 52.4 does not use the customary U.S. gravitational constant ($g_c = 32.2$ lbm-ft/lbf-sec^2).

[7](1) The *NCEES Handbook* uses the variable Q twice in Eq. 52.5 to represent *quantity* (i.e., an amount). The quantity is a mass in the numerator but a volume in the denominator, as differentiated through the subscripts. (2) The mass transfer coefficient, K, in Eq. 52.3 is not the same as the nonideal mixing factor, k, in Eq. 52.5. The temperature of the liquid, T_L, in Eq. 52.3 is not the same as the temperature of the vapor, T, in Eq. 52.5. The saturation pressure, p^{sat}, in Eq. 52.3 is not the same as the ambient pressure, p, in Eq. 52.5.

[8](1) The *NCEES Handbook* names this the "sweep through concentration change," but the equation calculates a volumetric flow rate. (2) Inerting with an inert purge gas is used to prevent a combustible mixture from forming from the remaining tails of an emptied fuel container. Although the vapor enclosure is referred to as a "vessel" in Eq. 52.6, the equation can also be used for a room. In fact, this equation is known as the *room purge equation* by industrial hygienists. (3) Although the nonideal mixing factor included in Eq. 52.5 is needed, it has been omitted from Eq. 52.6. (4) A lot of simplifying assumptions are required to get to Eq. 52.6. The equation essentially describes an exponent decay.

8. GASES AND FLAMMABLE OR COMBUSTIBLE LIQUIDS

Hazardous gases fall into four main types: cryogenic liquids, simple asphyxiants, chemical asphyxiants, and all other gases whose hazards depend on their properties.

Cryogenic liquids can vaporize rapidly, producing a cold gas that is more dense than air and displacing oxygen in confined spaces.

Simple asphyxiants, which include helium, neon, nitrogen, hydrogen, and methane, can dilute or displace oxygen. *Chemical asphyxiants*, which include carbon monoxide, hydrogen cyanide, and hydrogen sulfide, can pass into blood cells and tissue and interfere with blood-carrying oxygen.

Equation 52.7 and Eq. 52.8: Flammability

> LFL = lower flammability limit (volume % in air)
>
> *52.7*
>
> UFL = upper flammability limit (volume % in air)
>
> *52.8*

Description

The term *flammable* refers to the ability of an ignition source to propagate a flame throughout the vapor-air mixture and have a closed-cup flash point below 37.8°C (100°F) and a vapor pressure not exceeding 272 atm at 37.8°C (100°F). The phrase *closed-cup flash point* refers to a method of testing for flash points of liquids. The term *combustible* refers to liquids with flash points above 37.8°C (100°F).

For each airborne flammable substance, there are minimum and maximum concentrations in air between which flame propagation will occur. The lower concentration in air is known as the *lower flammability limit* (LFL) or *lower explosive limit* (LEL). The upper limit is known as the *upper flammability limit* (UFL) or *upper explosive limit* (UEL). Below the LEL, there is not enough fuel to propagate a flame. Above the UEL, there is not enough air to propagate a flame. The lower the LEL, the greater the hazard from a flammable liquid. For many common liquids and gases, the LEL is a few percent, and the UEL is 6–12%. If a concentration in air is less than the permissible exposure limit (PEL) or the threshold limit value (TLV), the concentration will be less than the LEL. The occupational safety requirements for handling and using flammable and combustible liquids are given in Subpart H of 29 CFR 1910.106.

See Table 52.5 for a listing of some combustible materials and their LFLs and UFLs.

LFLs and UFLs are both temperature dependent, as Fig. 52.5 illustrates. Figure 52.5 shows the region of flammable mixtures. T_L and T_u represent the lower and upper *flash point temperatures*, while AIT represents the

autoignition temperature. A flash point is the lowest temperature at which the vapor will ignite in the presence of an ignition source. The autoignition temperature is the lowest temperature at which an ignition source is not required to ignite the vapor.

Table 52.5 *Examples of Flammability Limits (volumetric percent in air)*

compound	LFL	UFL
ethyl alcohol	3.3	19
ethyl ether	1.9	36
ethylene	2.7	36
methane	5	15
propane	2.1	9.5

Figure 52.5 *Flammable Region for Gas Mixtures*

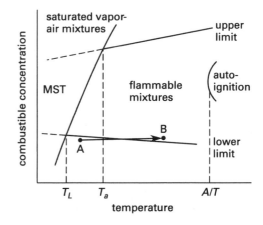

Equation 52.9 and Eq. 52.10: Le Chatelier's Rule

$$\sum_{i=1}^{n} (C_i / \text{LFL}_i) \geq 1 \quad \begin{bmatrix} \text{requirement for} \\ \text{flammability} \end{bmatrix} \qquad 52.9$$

$$\text{LFL}_m = \frac{100\%}{\sum\limits_{i=1}^{n} (C_{fi} / \text{LFL}_i)} \qquad 52.10$$

Description

The LFL of mixtures of multiple flammable gases in air can be estimated using an empirical rule (law, theory, etc.) developed by Le Châtelier (which is presented in simplified form as Eq. 52.9). In Eq. 52.9, C_i is the volumetric percent of flammable gas i in air alone. If the sum of terms is greater than 1.0, the mixture is considered to be above the mixture LFL.

Equation 52.10 calculates an approximation of the lower flammability limit of the fuel mixture, LFL_m, using the same theory. In Eq. 52.10, C_{fi} is the volumetric percent of fuel gas in the mixture.

9. EVALUATION AND CONTROL OF HAZARDOUS SOLVENTS

Vapor-Hazard Ratio

The toxicological effects from aqueous solutions include dermatitis, throat irritation, and bronchitis.

One indicator of hazards from solvent vapors and gases is the *vapor-hazard ratio number*, which is the equilibrium vapor pressure in ppm at 25°C (77°F) divided by the TLV in ppm. The higher the ratio, the greater the hazard. The vapor-hazard ratio accounts for the volatility of a solvent as well as its toxicity. To assess the overall hazard, the vapor-hazard ratio should be evaluated in conjunction with the TLV, ignition temperature, flash point, toxicological information, and degree of exposure.

The best control method is not to use a solvent that is hazardous. Sometimes a process can be redesigned to eliminate the use of a solvent. The following evaluation steps are recommended.

- Use water or an aqueous solution when possible.

- Use a *safety solvent* if it is not possible to use water. Safety solvents have vapor inhibitors and high flash points.

- Use a different process when possible to avoid use of a hazardous solvent.

- Provide a properly designed ventilation system if toxic solvents must be used.

- Never use highly toxic or highly flammable solvents (benzene, carbon tetrachloride, gasoline).

Ventilation

The most effective way to prevent inhalation of vapors from solvents is to provide closed systems or adequate local exhaust ventilation. If limitations exist on the use of closed systems or local exhaust ventilation, then workers should be provided with personal protective equipment.

10. INORGANIC PARTICULATE HAZARDS

Particulates include dusts, fumes, fibers, and mists. *Dusts* have a wide range of sizes and usually result from a mechanical process such as grinding. *Fumes* are extremely small particles, less than 1 μm in diameter, and result from combustion and other processes. *Fibers* are thin and long particulates, with asbestos being a prime example. *Mists* are suspended liquids that float in air, such as from the atomization of cutting oil. All of these types of particulate can pose an inhalation hazard if they reach the lungs.

There are four factors that affect the health risk from exposure to particulates: (1) the types of particulate, (2) the length of exposure, (3) the concentration of particulates in the breathing zone, and (4) the size of particulates in the breathing zone.

The type of particulate can determine the type of health effect that may result from the exposure. Both organic and inorganic dusts can produce allergic effects, dermatitis, and systemic toxic effects. Particulates that contain free silica can produce pneumoconiosis from chronic exposure. *Pneumoconiosis* is lung disease caused by fibrosis from exposure to both organic and inorganic particulates. Other particulates can cause systemic toxicity to the kidneys, blood, and central nervous system. Asbestos fibers can cause lung scarring and cancer.

The critical duration of exposure varies with the type of particulate.

The concentration of particulates in the breathing zone is the primary factor in determining the health risk from particulates.

The fourth exposure factor is the size of the particulates. With one known exception, particulates larger than approximately 5 μm cannot reach the alveoli, or inner recesses, of the lungs before being trapped and expelled from the body through the digestive system or from the mouth and nose. Protection is afforded by the presence of mucus and cilia in the nasal passages, throat, larynx, trachea, and bronchi. The exception is asbestos fibers, which can reach the alveoli even though fibers may be larger than 5 μm. Particles smaller than 5 μm are considered respirable dusts and pose an exposure hazard when present in the breathing zone.

Silica

Silica (SiO_2) has several associated health hazards. The crystalline form of free silica (quartz) deposited in the lungs causes the growth of fibrous tissue around the deposit. The fibrous tissue reduces the amount of normal lung tissue, thereby reducing the ability of the lungs to transfer oxygen. When the heart tries to pump more blood to compensate, heart strain and permanent damage or death may result. This condition is known as *silicosis*. Mycobacterial infection occurs in about 25% of silicosis cases. Smokers exposed to silica dust have a significantly increased chance of developing lung cancer.

Asbestos

Asbestos is generically described as a naturally occurring, fibrous, hydrated mineral silicate. Asbestos mining, construction activities, and working in shipyards are possible exposure activities. Inhalation of short asbestos fibers can cause *asbestosis*, a kind of pneumoconiosis, as a nonmalignant scarring of the lungs. *Bronchogenic carcinoma* is a malignancy (cancer) of the lining of the lung's air passages. *Mesothelioma* is a diffuse malignancy of the lining of the chest cavity or the lining of the abdomen.

The onset of illness seems to be roughly correlated with length and diameter of inhaled asbestos fibers. Fibers 2 μm in length cause asbestosis. Mesothelioma is associated with fibers 5 μm long. Fibers longer than 10 μm produce lung cancer. Fiber diameters greater than 3 μm are more likely to cause asbestosis or lung cancer, while fibers 3 μm or less in diameter are associated with mesothelioma.

The OSHA regulations for protection from exposure to asbestos are extensive. They require an employer to perform an exposure assessment in many cases. Monitoring must be performed by a *competent person*, which is defined by OSHA as one who is capable of identifying asbestos hazards and selecting control strategies and who has the authority to make corrective changes. The regulations also specify when medical surveillance is required, when personal protection must be provided, and the engineering controls and work practices that must be implemented.

Lead

The body does not use lead for any metabolic purpose, so any exposure to lead is undesirable. Lead dust and fumes can pose a severe hazard. Acute large doses of lead can cause systemic poisoning or seizures. Chronic exposure can damage the blood-forming bone marrow and the urinary, reproductive, and nervous systems. Lead is probably a human carcinogen, although whether it is causative or facilitative remains subject to research.

Beryllium

Inhalation of metallic beryllium, beryllium oxide, or soluble beryllium compounds can lead to *chronic beryllium disease* (*berylliosis*). Ingestion and dermal contact do not pose a documented hazard, so maintaining beryllium dusts and fumes below the TLV in the breathing zone is a critical protection measure.

Chronic beryllium disease is characterized by granulomas on the lungs, skin, and other organs. The disease can result in lung and heart dysfunction and enlargement of certain organs. Beryllium has been classified as a suspected human carcinogen.

Coal Dust

Coal dust can cause chronic bronchitis, silicosis, and *coal worker's pneumoconiosis*, also known as *black lung disease*.

Welding Fumes

Exposure to welding fumes can cause a disease known as *metal fume fever*. This disease results from inhalation of extremely fine oxide particles that have been freshly formed as fume. Zinc oxide fume is the most common source, but magnesium oxide, copper oxide, and other metallic oxides can also cause metal fume fever. Metal fume fever is of short duration, with symptoms including fever and shaking chills appearing 4 hours to 12 hours after exposure.

Radioactive Dust

Radioactive dusts can cause toxicity in addition to the effects from ionizing radiation. Inhalation of radioactive dust can result in deposition of radionuclides in the body, which may enter the bloodstream and affect individual organs.

Control measures should be instituted to prevent workers from inhaling radioactive dust, either by restricting access or by providing appropriate personal protection such as respirators. Engineering controls to capture radioactive dust are an absolute necessity to minimize worker exposure.

11. CONTROL OF PARTICULATES

Ventilation is the most effective method for controlling particulates. Enclosed processes should be used wherever possible. Equipment can be enclosed so that only the feed and discharge openings are open. With adequate pressure, enclosed equipment can be nearly as effective as closed processes. Large automated equipment can sometimes be placed in separate enclosures. Workers would have to wear personal protective equipment to enter the enclosures. Local exhaust ventilation with hooded enclosures can be very effective at controlling particulate emissions into general work areas. Where complete enclosure and local exhaust methods are not sufficient, *dilution ventilation* will be necessary to control particulates in the work area. In some instances, the work process can be changed from a dry to a wet process to reduce particulate generation.

Unlike for vapor protection, respirators are an effective means of controlling worker exposure to particulates that remain in the work area after engineering controls have been applied or when access to dusty areas is intermittent. Respirators may also be used to provide additional protection or comfort to workers in areas where local or general ventilation is effective. The NIOSH guidelines for selection of respirators should be followed to ensure that the respirator will be effective at removing the specific particulate to which the workers are exposed.

12. BIOLOGICAL HAZARDS

Biological Agents

Approximately 200 biological agents are known to produce infectious, allergenic, toxic, and carcinogenic reactions in workers. These agents and their reactions are as follows.

- Microorganisms (viruses, bacteria, fungi) and the toxins they produce cause infection and allergic reactions. A wide variety of biological organisms can be inhaled as particulates causing respiratory diseases and allergies. Examples include dust that contains anthrax spores from the wool or bones of infected animals, and fungi spores from grain and other agricultural produce.

- Arthropod (crustaceans), arachnid (spiders, scorpions, mites, and ticks), and insect bites and stings cause skin inflammation, systemic intoxication, transmission of infectious agents, and allergic reactions.

- Allergens and toxins from plants cause dermatitis from skin contact, rhinitis (inflammation of the nasal mucus membranes), and asthma from inhalation.

- Protein allergens (urine, feces, hair, saliva, and dander) from vertebrate animals cause allergic reactions.

Also posing potential biohazards are lower plants other than fungi (e.g., lichens, liverworts, and ferns) and invertebrate animals other than arthropods (e.g., parasites, flatworms, and roundworms).

Microorganisms may be divided into prokaryotes and eukaryotes. *Prokaryotes* are organisms having DNA that is not physically separated from its cytoplasm (cell plasma that does not include the nucleus). They are small, simple, one-celled structures, less than 5 μm in diameter, with a primitive nuclear area consisting of one chromosome. Reproduction is normally by binary fission in which the parent cell divides into two daughter cells. All bacteria, both single-celled and multicellular, are prokaryotes, as are blue-green algae.

Eukaryotes are organisms having a nucleus that is separated from the cytoplasm by a membrane. Eukaryotes are larger cells (greater than 20 μm) than prokaryotes, with a more complex structure, and each cell contains a distinct membrane-bound nucleus with many chromosomes. They may be single-celled or multicellular, reproduction may be asexual or sexual, and complex life cycles may exist. This class of microorganisms includes fungi, algae (except blue-green), and protozoa.

Since prokaryotes and eukaryotes have all of the enzymes and biological elements to produce metabolic energy, they are considered organisms.

In contrast, a *virus* does not contain all of the elements needed to reproduce or sustain itself and must depend on its host for these functions. Viruses are nucleic acid molecules enclosed in a protein coat. A virus is inert outside of a host cell and must invade the host cell and use its enzymes and other elements for the virus's own reproduction. Viruses can infect very small organisms such as bacteria, as well as humans and animals. Viruses are 20 μm to 300 μm in diameter.

Smaller than the viruses by an order of magnitude are *prions*, small proteinaceous infectious particles. Prions have properties similar to viruses and cause degenerative diseases in humans and animals.

Infection

The invasion of the body by pathogenic microorganisms and the reaction of the body to them and to the toxins they produce is called an *infection*. Infection may be *endogenous* where microorganisms that are normally present in the body (*indigenous*) at a particular site (such as *E. coli* in the intestinal tract) reach another site (such as the urinary tract), causing infection there.

Infections from microorganisms not normally found on the body are called *exogenous* infections.

The most common routes of exposure to infectious agents are through cuts, punctures, and bites (insect and animal); abrasions of the skin; inhalation of aerosols generated by accidents or work practices; contact between mucous membranes or contaminated material; and ingestion. In laboratory and medical settings, transmission of blood-borne pathogens can occur through handling of blood products and human tissue.

Biohazardous Workplaces and Activities

Although engineers have long been concerned with waterborne diseases and their prevention in the design and operation of water supply and wastewater systems, pathogens may also be encountered in the workplace through air or direct contact.

Microbiology and Public Health Laboratories

Workers in laboratories handling infectious agents experience a risk of infection.

Health Care Facilities

Health care facilities—such as hospitals, medical offices, blood banks, and outpatient clinics—present numerous opportunities for exposure to a wide variety of hazardous and toxic substances, as well as to infectious agents.

Biotechnology Facilities

Biotechnology involves a much greater scope and complexity than the historical use of microorganisms in the chemical and pharmaceutical industries. This technology deals with DNA manipulation and the development of products for medicine, industry, and agriculture. The microorganisms used by the biotechnology industry often are genetically engineered plant and animal cells. Allergies can be a major health issue.

Animal Facilities

Workers exposed to animals are at risk for animal-related allergies and infectious agents. Occupations include agricultural workers, veterinarians, workers in zoos and museums, taxidermists, and workers in animal-product processing plants.

Zoonotic diseases (diseases that affect both humans and animals) are the most common diseases reported by laboratory workers. Work acquired infections from non-human primates are common.

Some of the diseases of concern in animal facilities include Q fever, hantavirus, Ebola, Marburg viruses, and simian immunodeficiency viruses.

Agriculture

Agricultural workers are exposed to infectious microorganisms through inhalation of aerosols, contact with broken skin or mucus membranes, and inoculation from injuries. Farmers and horticultural workers may be exposed to fungal diseases. Food and grain handlers may be exposed to parasitic diseases. Workers who process animal products may acquire bacterial skin diseases such as anthrax from contaminated hides, tularemia from skinning infected animals, and erysipelas from contaminated fish, shellfish, meat, or poultry. Infected turkeys, geese, and ducks can expose poultry workers to *psittacosis*, a bacterial infection. Workers handling grain may be exposed to *mycotoxins* from fungi and *endotoxins* from bacteria.

Utility Workers

Workers maintaining water systems may be exposed to Legionella pneumophila (Legionnaires' disease). Sewage collection and treatment workers may be exposed to enteric bacteria, hepatitis A virus, infectious bacteria, parasitic protozoa (*Giardia*), and allergenic fungi.

Solid waste handling and disposal facility workers may be exposed to blood-borne pathogens from infectious wastes.

Wood-Processing Facilities

Wood-processing workers may be exposed to bacterial endotoxins and allergenic fungi.

Mining

Miners may be exposed to zoonotic bacteria, mycobacteria, fungi, and untreated runoff water and wastewater.

Forestry

Forestry workers may be exposed to zoonotic diseases (*rabies* virus, *Russian spring fever* virus, *Rocky Mountain spotted fever*, *Lyme disease*, and *tularemia*) transmitted by ticks and fungi.

Blood-Borne Pathogens

The risk from hepatitis B and human immunodeficiency virus (HIV) in health care and laboratory situations led OSHA to publish standards for occupational exposure to blood-borne pathogens. Some blood-borne pathogens are summarized as follows.

Human Immunodeficiency Virus (HIV)

HIV is the blood-borne virus that causes acquired immunodeficiency syndrome (AIDS). Contact with infected blood or other body fluids can transmit HIV. Transmission may occur from unprotected sexual intercourse, sharing of infected needles, accidental puncture wounds from contaminated needles or sharp objects, or transfusion with contaminated blood.

Symptoms of HIV include swelling of lymph nodes, pneumonia, intermittent fever, intestinal infections, weight loss, and tuberculosis. Death typically occurs from severe infection causing respiratory failure due to pneumonia.

Hepatitis

The hepatitis virus affects the liver. Symptoms of infection include jaundice, cirrhosis and liver failure, and liver cancer.

Hepatitis A can be contracted through contaminated food or water or by direct contact with blood or body fluids such as blood or saliva. Hepatitis B, known as *serum hepatitis*, may be transmitted through contact with infected blood or body fluids, or through blood transfusions. Hepatitis B is the most significant occupational infector of health care and laboratory workers. Hepatitis C is similar to hepatitis B, but can also be transmitted by shared needles, accidental puncture wounds, blood transfusions, and unprotected sex.

Hepatitis D occurs when one of the other hepatitis viruses replicates. Individuals with chronic hepatitis D often develop cirrhosis of the liver. Chronic hepatitis may be present in carriers.

Syphilis

The bacterium responsible for the transmission of syphilis is called *treponema pallidum pallidum*. (Treponema pallidum has four subspecies, so the extra "pallidum" indicates the virus that specifically causes syphilis.) Syphilis is almost always transmitted through sexual contact, though it may be transmitted in utero through the placenta from mother to fetus. This is known as *congenital syphilis*.

Toxoplasmosis

Toxoplasmosis is caused by a parasitic organism called *Toxoplasma gondii*, which may be transmitted by ingestion of contaminated meat, across the placenta, and through blood transfusions and organ transplants.

Rocky Mountain Spotted Fever

Ticks infected with the pathogen *Rickettsia rickettsii* pass this disease from pets and other animals to humans. Symptoms and effects include headache, rash, fever, chills, nausea, vomiting, cardiac arrhythmia, and kidney dysfunction. Death may occur from renal failure and shock.

Bacteremia

Bacteremia is the presence of bacteria in the bloodstream, whether associated with active disease or not.

Bacteria- and Virus-Derived Toxins

Some toxins are derived from bacteria and viruses. The effects of these toxins vary from mild illness to debilitating illness or death.

Botulism

The organism *Clostridium botulinum* produces the toxin that is responsible for botulism. There are four types of botulism: food-borne, infant, adult enteric (intestinal), and wound. *Food botulism* is associated with poorly preserved foods and is the most widely recognized form. *Infant botulism* can occur in the second month after birth when the bacteria colonize the intestinal tract and produce the toxin. *Adult enteric botulism* is similar to infant botulism. *Wound botulism* occurs when the spores enter a wound through contaminated soil or needles. The toxin is absorbed in the bloodstream and blocks the release of a neurotransmitter. Severe cases can result in respiratory paralysis and death.

Lyme Disease

Lyme disease is transmitted to humans through bites of ticks infected with *Borrelia burgdorferi*.

Tetanus

Tetanus occurs from infection by the bacterium *Clostridium tetani*, which produces two exotoxins, tetanolysin and tetanospasmin. Routine immunizations prevent the disease.

Toxic Shock Syndrome

Toxic shock syndrome (TSS) is caused by the bacterium *Staphylococcus aureus*, which produces a pyrogenic toxin.

Ebola (African Hemorrhagic Fever)

The Ebola and Marburg viruses produce an acute hemorrhagic fever in humans. Symptoms include headache, progressive fever, sore throat, and diarrhea.

Hantavirus

The hantavirus is found in rodents and shrews of the southwest and is spread by contact with their excreta.

Tuberculosis

Tuberculosis (TB) is a bacterial disease from *Mycobacterium tuberculosis*. Humans are the primary source of infection. TB affects a third of the world's population outside the United States. A drug-resistant strain is a serious problem worldwide, including in the United States. The risk of contracting active TB is increased among HIV-infected individuals.

Legionnaires' Disease

Legionnaires' disease (legionellosis) is a type of pneumonia caused by inhaling the bacteria *Legionella pneumophilia*. Symptoms include fever, cough, headache, muscle aches, and abdominal pain. People usually recover in a few weeks and suffer no long-term consequences. *Legionellae* are common in nature and are associated with heat-transfer systems, warm-temperature water, and stagnant water. Sources of exposure include sprays from cooling towers or evaporative condensers and fine mists from showers and humidifiers. Proper design and operation of ventilation, humidification, and water-cooled heat-transfer equipment and other water systems equipment can reduce the risk. Good system maintenance includes regular cleaning and disinfection.

13. RADIATION

Radiation can be either nonionizing or ionizing. *Nonionizing radiation* includes electric fields, magnetic fields, electromagnetic radiation, radio frequency and microwave radiation, and optical radiation and lasers.

Dealing with *ionizing radiation* from nuclear sources requires special skills and knowledge. A specialist in health physics should be consulted whenever ionizing radiation is encountered.

Nuclear Radiation

Nuclear radiation is a term that applies to all forms of radiation energy that originate in the nuclei of radioactive atoms. Nuclear radiation includes alpha particles, beta particles, neutrons, X-rays, and gamma rays. The common property of all nuclear radiation is an ability to be absorbed by and transfer energy to the absorbing body.

The preferred unit of ionizing radiation, given in the National Council on Radiation Protection and Measurement's (NCRP's) *Recommended Limitations for Exposure to Ionizing Radiation*, is the mSv. Sv is the symbol for sievert, which is the SI unit of absorbed dose times the *quality factor* of the radiation as compared to gamma radiation. The exposure dose is measured in grays (Gy). The gray is equal to 1 J of absorbed energy per kilogram of matter. A summary of ionizing radiation units is given in Table 52.6.

The radiation dose in air (measured in grays) is modified by a *radiation weighting factor*, Q (formerly known as a *quality factor*), in order to determine the effective radiation dose (measured in sieverts) in tissue. The value of the weighting factor depends on the type of radiation, the type of tissue, and the energy spectrum. Generic values are 1.0 for X-ray, gamma, and beta radiation; 10 for high-energy protons (other than recoil protons) and neutrons on unknown energy; and 20 for alpha particles, multiply charged particles, fission fragments, and heavy particles of unknown charge. Values for neutrons range from 5 to 20, depending on neutron energy.

Alpha Particles

Alpha particles consist of two protons and two neutrons, with an atomic mass of four. Alpha particles combine with electrons from the absorbed material and become helium atoms. Alpha particles have a positive charge of two units and react electrically with human tissue. Because of their large mass, they can travel only about 10 cm in air and are stopped by the outer layer of the skin. Alpha-emitters are considered to be only internal radiation hazards, which requires alpha particles to be ingested by eating or breathing. They affect the bones, kidney, liver, lungs, and spleen.

Table 52.6 *Units for Measuring Ionizing Radiation*

property	SI
energy absorbed	gray (Gy)
	1 J/kg
	1 Gy = 100 rad (obsolete)
biological effect	sievert (Sv)
	Gy × quality factor
	1 Sv = 100 rem (obsolete)

Beta Particles

Beta particles are electrically charged particles ejected from the nuclei of radioactive atoms during disintegration. They have a negative charge of one unit and the mass of an electron. High-energy beta particles can penetrate in human tissue to a depth of 20 mm to 130 mm and travel up to 9 m in air. Skin burns can result from an extremely high dose of low-energy beta radiation, and some high-energy beta sources can penetrate deep into the body, but beta-emitters are primarily internal radiation hazards, which would require them to be ingested. Beta particles are more hazardous than alpha particles because they can penetrate deeper into tissue. High-energy beta radiation can produce a secondary radiation called *bremsstrahlung*. These are X-rays produced when electrons (i.e., beta particles) pass near the nuclei of other atoms. Bremsstrahlung radiation is proportional to the energy of the beta particle and the atomic number of the adjacent nucleus. Materials with low atomic numbers (e.g., plexiglass) are preferred shielding materials.

Neutrons

Neutron particles have no electrical charge and are released upon disintegration of certain radioactive materials. Their range in air and in human tissue depends on their kinetic energy, but the average depth of penetration in human tissue is 60 mm. Neutrons lose velocity when they are absorbed or deflected by the nuclei with which they collide. However, the nuclei are left with higher energy that is later released as protons, gamma rays, beta particles, or alpha particles. It is these secondary emissions from neutrons that produce damage in tissue.

X-Rays

X-rays are produced by electron bombardment of target materials and are highly penetrating electromagnetic radiation. X-rays have a valuable scientific and commercial use in producing shadow pictures of objects. The energy of an X-ray is inversely proportional to its wavelength. X-rays of short wavelength are called *hard*, and they can penetrate several centimeters of steel. Long wavelength X-rays are called *soft*, and they are less penetrating. The power of X-rays and gamma rays to penetrate matter is called *quality*. *Intensity* is the energy flux density.

Gamma Rays

Gamma rays, or gamma radiation, are a class of electromagnetic photons (radiation) emitted from the nuclei of radioactive atoms. They are highly penetrating and are an external radiation hazard. Gamma rays are emitted spontaneously from radioactive materials, and the energy emitted is specific to the radionuclide. Gamma rays present an internal exposure problem because of their deep penetrating ability.

Radioactive Decay

Radioactive decay is measured in terms of *half-life*, the time to lose half of the activity of the original material. Decay activity can be calculated from the *decay constant*, k.

$$N = N_o e^{-kt} = N_o e^{-0.693t/\tau}$$

The half-life can be calculated from the decay rate constant.

$$\tau = t_{1/2} = 0.693/k$$

14. RADIATION EFFECTS ON HUMANS

Ionizing radiation transfers energy to human tissue when it passes through the body. *Dose* refers to the amount of radiation that a body absorbs when exposed to ionizing radiation. The effects on the body from external radiation are quite different from the effects from internal radiation. Internal radiation is spread throughout the body to tissues and organs according to the chemical properties of the radiation. The effects of internal radiation depend on the energy and the residence time within the body. The principal effect of radiation on the body is destruction of or damage to cells. Damage may affect reproduction of cells or cause mutation of cells.

The effects of ionizing radiation on individuals include skin, lung, and other cancers; bone damage; cataracts; and a shortening of life. Effects on the population as a whole include possible damage to human reproductive elements, thereby affecting the genes of future generations.

Safety Factors

An environmental engineer should be aware of the basic safety factors for limiting dose. These factors are time, distance, and shielding.

The dose received is directly related to the time exposed, so reducing the time of exposure will reduce the dose. An individual's time of exposure can also be limited by spreading the exposure time among more workers.

Distance is another safety factor that can be changed to reduce the dose. The intensity of external radiation decreases as the inverse of the square of the distance.

By increasing the distance to a source from 2 m to 20 m, for example, the exposure would be reduced to 1% (i.e., $(2 \text{ m}/20 \text{ m})^2$).

Shielding involves placing a mass of material between a source and workers. The objective is to use a high-density material that will act as a barrier to X-ray and gamma-ray radiation. Lead and concrete are often used, with lead being the more effective material because of its greater density. For neutrons, different material is needed than for X-rays and gamma rays because neutrons produce secondary radiation from collisions with nuclei. Neutron shielding requires a light nucleus material. Typically water or graphite is used.

The shielding properties of materials are often compared using the *half-value thickness*, which is the thickness of the material required to reduce the radiation to half of the incident value. The half-value properties vary with the radiation source.

15. SOUND AND NOISE

Characteristics of Sound

Sound is pressure variation in air, water, or some other medium that the human ear can detect. *Noise* is unwanted, unpleasant, or painful sound. The *frequency* of sound is the number of pressure variations per second, measured in cycles per second, or hertz (Hz). The frequency range of human audible sound is approximately 20–20 000 Hz.

Sound passes through a medium at the *speed of sound*, a, equal to the product of the wavelength and the frequency.

$$a = f\lambda$$

The speed of sound is dependent upon the medium, as illustrated in Table 52.7.

Table 52.7 *Speed of Sound in Various Media*

medium	speed of sound (m/s)	condition
air	330	1 atm, 0°C
water	1490	1 atm, 20°C
aluminum	4990	1 atm
steel	5150	1 atm

Sound Pressure

Sound pressure measures the intensity of sound and is the variation in atmospheric pressure. (See Fig. 52.6.) The *root-mean-square* (*rms*) *pressure* is used.

$$p_{\text{rms}} = \sqrt{\frac{\sum p_i^2}{n}}$$

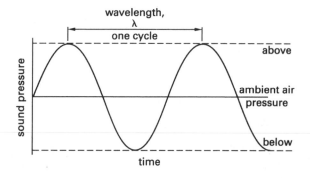

Figure 52.6 *Characteristics of a Sound Wave*

Equation 52.11 Through Eq. 52.14: Sound Pressure Level

$$SPL = 10\log_{10}(p^2/p_0^2) \qquad 52.11$$

$$SPL_{\text{total}} = 10\log_{10}\sum 10^{SPL/10} \qquad 52.12$$

$$\Delta SPL = 10\log_{10}(r_1/r_2)^2 \quad \text{[point source]} \qquad 52.13$$

$$\Delta SPL = 10\log_{10}(r_1/r_2) \quad \text{[line source]} \qquad 52.14$$

Description

Equation 52.11 is used to calculate the *sound pressure level* (SPL). SPL is measured in decibels relative to a reference level, p_0, of 20 μPa, the threshold of hearing at a reference frequency of 1000 Hz. The sound pressure level from multiple sources is calculated from Eq. 52.12.

Equation 52.13 is used to calculate the attenuation due to a point source, and Eq. 52.14 is used to calculate the attenuation due to a line source.

Example

The sound pressure at a construction site is 20 Pa. The standard reference pressure is 2×10^{-5} Pa. What is most nearly the sound pressure level?

(A) 100 dB

(B) 120 dB

(C) 140 dB

(D) 160 dB

Solution

Use Eq. 52.11 to calculate the sound pressure level.

$$
\begin{aligned}
SPL &= 10\log_{10}(p^2/p_0^2) \\
&= 10\log_{10}\left(\frac{(20 \text{ Pa})^2}{(2 \times 10^{-5} \text{ Pa})^2}\right) \\
&= 120 \text{ dB}
\end{aligned}
$$

The answer is (B).

Sound Power

Sound power, W, is the absorbed (or transmitted) sound energy per unit time (in watts). The sound power level, SWL, is measured in decibels, calculated relative to a reference level, W_0, of 10^{-12} W.

$$\text{SWL} = 10 \log\left(\frac{W}{W_0}\right)$$

Sound Intensity

Sound intensity is an areal function of the sound power of a source.

$$I = \frac{W}{4\pi r^2}$$

Loudness

Loudness as perceived by humans is primarily a function of sound pressure but is also affected by frequency because the human ear is more sensitive to high-frequency sounds than low-frequency sounds.

Noise

Noise can cause psychological and physiological damage and can interfere with workers' communication, thereby affecting safety. Exposure to excessive noise for a sufficient time can result in hearing loss. The permissible OSHA noise exposure levels (i.e., sound pressure levels) are given in Table 52.8.

Table 52.8 Typical Permissible Noise Exposure Levels

noise level (dBA)	permissible time (hr)
80	32
85	16
90	8
95	4
100	2
105	1
110	0.5
115	0.25
120	0.125
125	0.063
130	0.031

When the daily noise exposure is composed of two or more periods of noise exposure at different levels, their combined effect should be considered, rather than the individual effect of each. If the sum of the fractions $C_1/T_1 + C_2/T_2 + \cdots + C_n/T_n$ exceeds 1.0, then the mixed exposure should be considered to exceed the limit value. C indicates the total time of exposure at a specified noise level, and T indicates the total time of exposure permitted at that level.

Hearing Loss

Hearing loss can be caused by sudden intense noise over only a few exposures. This type of loss is known as *acoustic trauma*. Hearing loss can also be caused by exposure over a long duration (months or years) to hazardous noise levels. This type is known as *noise-induced hearing loss*. The permanence and nature of the injury depends on the type of hearing loss.

The main risk factors associated with hearing loss are the intensity of the noise (sound pressure level), the type of noise (frequency), daily exposure time (hours per day), and the total work duration (years of exposure). These are known as *noise exposure factors*. Generally, exposure to sound levels above 115 dBA is considered hazardous, and exposure to levels below 70–75 dBA is considered safe from risk of permanent hearing loss. Also, noise with predominant frequencies above 500 Hz is considered to have a greater potential to cause hearing loss than lower-frequency sounds.

Classes of Noise Exposure

Noise exposure can be classified as continuous noise, intermittent noise, and impact noise. *Continuous noise* is broadband noise of a nearly constant sound pressure level and frequency to which a worker is exposed 8 hours daily and 40 hours weekly. *Intermittent noise* involves exposure to a specific broadband sound pressure level several times a day. *Impact noise* is a sharp burst of short duration sound.

OSHA has established permissible noise exposures, known as permissible exposure levels (PELs). The PELs are equivalent to a continuous 8-hour exposure at a sound pressure level of 90 dBA, which is established as a 100% dose. For other exposure durations, OSHA has established relationships between the sound level and the exposure time. Every 5 dBA increase in noise cuts the allowable exposure time in half. Sound pressure levels below 90 dBA are not considered hazardous.

For intermittent noise, the time characteristics of the noise must be determined. Both short-term and long-term exposure must be measured. A dosimeter is typically used to measure exposure to intermittent noises.

For impact noises, workers should not be exposed to peaks of more than 140 dBA under any circumstances. The threshold limit value for impulse noise should not exceed the values provided in Table 52.9.

Table 52.9 Typical Threshold Limit Values for Impact Noise

peak sound level (dB)	maximum number of daily impacts
140	100
130	1000
120	10 000

Safety/Health/Environment

Equation 52.15: Noise Dose

$$D = 100\% \times \sum \frac{C_i}{T_i} \quad \text{52.15}$$

Description

When workers are exposed to different noise levels during the day, the *noise dose*, D, can be calculated from Eq. 52.15. If D equals or exceeds 100%, the mixed dose exceeds the OSHA standard.

In Eq. 52.15, T_i is the exposure time permitted at the corresponding sound pressure level. It can be calculated from a formula or read directly from Table 52.8. At 50%, the noise dose, D, is equivalent to eight hours of exposure to an 85 dBA time-weighted average (TWA) noise source. At 100%, the noise dose is equivalent to eight hours of exposure to a 90 dBA noise source.

$$T_i = (2 \text{ h})^{(105 - \text{SPL})/5} \quad [80 \text{ dBA} \leq \text{SPL} \leq 130 \text{ dBA}]$$

If D is equal to or greater than 50% but less than 100%, or equivalently, if the eight-hour TWA sound level is at or above 85 dBA, OSHA mandates the employer to implement a hearing conservation program. A *hearing conservation program* includes the mandatory elements of (1) providing employee hearing tests, (2) supplying hearing protection equipment when requested by an employee, and (3) providing ongoing noise monitoring.

If D is greater than 100%, some type of noise abatement program will be required to reduce the sound levels.

Employees should never be exposed to *impulse (impact) sound* sources exceeding 140 dBA.

Example

A construction worker leaves a 90 dBA environment after five hours and works in an 85 dBA environment for three more hours. What is most nearly the mixed dose that the worker has received?

(A) 69%

(B) 76%

(C) 81%

(D) 95%

Solution

Determine the time permitted at the sound pressure level (SPL) for each environment using Table 52.8.

$$T_1 = 8 \text{ h at 90 dBA}$$
$$T_2 = 16 \text{ h at 85 dBA}$$

Use Eq. 52.15 to determine the mixed dose.

$$D = 100\% \times \sum \frac{C_i}{T_i}$$
$$= 100\% \times \left(\frac{5 \text{ h}}{8 \text{ h}} + \frac{3 \text{ h}}{16 \text{ h}}\right)$$
$$= 81.25\% \quad (81\%)$$

The answer is (C).

Noise Control

A *hearing conservation program* should include noise measurements, noise control measures, hearing protection, audiometric testing of workers, and information and training programs. Employees are required to properly use the protective equipment provided by employers.

After the noise exposure is compared with acceptable noise levels, the degree of noise reduction needed can be determined. Noise reduction measures can comprise the following three basic methods applied in order.

1. changing the process or equipment

2. limiting the exposure

3. using hearing protection

Administrative controls include changing the exposure of workers to high noise levels by modifying work schedules or locations so as to reduce workers' exposure times. Administrative controls include any administrative decision that limits a worker's exposure to noise.

Personal hearing protection is the final noise control measure, to be implemented only after engineering controls are implemented. Protective devices do not reduce the noise hazard and may not be totally effective, so engineering controls are preferred over hearing protection. Protective devices include helmets, earplugs, canal caps, and earmuffs. Earplugs may be used with helmets to increase the level of noise reduction.

An important characteristic of personal hearing protection is the *noise reduction rating* (NRR). The NRR is established by the EPA and must be printed on the package of a device. The NRR can be used to determine whether a device provides sufficient hearing protection.

Audiometry

Audiometry is the measurement of hearing acuity. It is used to assess a worker's hearing ability by measuring the individual's threshold sound pressure level at various frequencies (250–6000 Hz). The threshold audiogram can be used to create a baseline of hearing ability and to determine changes over time and identify changes resulting from noise control measures. Baseline and annual hearing tests are required where workers are exposed to more than a TWA over 8 hours of 85 dBA. The average change from the baseline is used to measure the degree of hearing impairment.

16. HEAT AND COLD STRESS

Thermal Stress

Heat and cold, or *thermal*, stress involves three zones of consideration relative to industrial hygiene. In the middle is the *comfort zone*, where workers feel comfortable in the work environment. On either side of the comfort zone is a *discomfort zone* where workers feel uncomfortable with the heat or cold, but a health risk is not present. Outside of each discomfort zone is a *health risk zone* where there is a significant risk of health disorders due to heat or cold. Industrial hygiene is primarily concerned with controlling worker exposure in the health risk zone.

The analysis of thermal stress involves taking a *heat balance* of the human body with the objective of determining whether the net heat storage is positive, negative, or zero. A simplified form of the heat balance is

$$Q_S = Q_M + Q_R + Q_C + Q_E$$

Q_S is the body's heat storage (accumulation) rate; Q_M is the metabolic heat generation rate; Q_R is the net heat gain (loss) by radiation; Q_C is the heat gain (loss) by convection; and Q_E is the latent heat gain (loss) by evaporation. If the storage, Q_S, is zero, heat gain is balanced by heat loss, and the body is in equilibrium. If Q_S is positive, the body is gaining heat; and if Q_S is negative, the body is losing heat.

The heat balance is affected by environmental and climatic conditions, work demands, and clothing. The metabolic rate, Q_M, is more significant for heat stress than for cold stress when compared with radiation and convection. The metabolic rate can affect heat gain by one to two orders of magnitude compared to radiation and convection, but it affects heat loss to about the same extent as radiation and convection.

Clothing affects the thermal balance through insulation, permeability, and ventilation. *Insulation* provides resistance to heat flow by radiation, convection, and conduction. *Permeability* affects the movement of water vapor and the amount of evaporative cooling. *Ventilation* influences evaporative and convective cooling.

Heat Stress

Heat stress can increase body temperature, heart rate, and sweating, which together constitute *heat strain*.

The most serious heat disorder is *heatstroke*, because it involves a high risk of death or permanent damage. Fortunately, heatstroke is rare. Of lesser severity, *heat exhaustion* is the most commonly observed heat disorder for which treatment is sought. *Dehydration* is usually not noticed or reported, but without restoration of water loss, dehydration leads to heat exhaustion. The symptoms of these key heat stress disorders are as follows.

- *heatstroke:* chills, restlessness, irritability
- *heat exhaustion:* fatigue, weakness, blurred vision, dizziness, headache
- *dehydration:* no early symptoms, fatigue or weakness, headache, dry mouth

Appropriate first aid and medical attention should be sought when any heat stress disorder is recognized.

Control of Heat Stress

Controls that are applicable to any heat stress situation are known as *general controls*. General controls include worker training, heat stress hygiene, and medical monitoring.

Specific controls are controls that are put in place for a particular job. They include engineering controls, administrative controls, and personal protection. *Engineering controls* include changing the physical work demands to reduce the metabolic heat gain, reducing external heat gain from the air or surfaces, and enhancing external heat loss by increasing sweat evaporation and decreasing air temperature. *Administrative controls* include scheduling the work to allow worker acclimatization to occur, leveling work activity to reduce peak metabolic activity, and sharing or scheduling work so the heat exposure of individual workers is reduced. *Personal protection* includes using systems to circulate air or water through tubes or channels around the body, wearing ice garments, and wearing reflective clothing.

Cold Stress

The body reacts to cold stress by reducing blood circulation to the skin to insulate itself. The body also shivers to increase metabolism. These mechanisms are ineffective against long-term extreme cold stress, so humans react by increasing clothing for more insulation, increasing body activity to increase metabolic heat gain, and finding a warmer location.

There are two main hazards from cold stress: hypothermia and tissue damage. *Hypothermia* depresses the central nervous system, causing sluggishness and slurred speech, and progresses to disorientation and unconsciousness. To avoid hypothermia, the minimum core body temperature must be above 96.8°F (36°C) for prolonged exposure and above 95°F (35°C) for occasional exposure of short duration.

Worker training, cold stress hygiene, and medical surveillance can control cold stress. Engineering controls, administrative controls, and personal protection measures can also be used to control cold stress.

17. ERGONOMICS

Ergonomics is the study of human characteristics to determine how a work environment should be designed to make work activities safe and efficient. It includes both physiological and psychological effects on the worker, as well as health, safety, and productivity aspects.

Work-Rest Cycles

Excessively heavy work should be broken by frequent short rest periods to reduce cumulative fatigue. The percentage of time a worker should rest can be estimated by the following equation. Q_M is the metabolic heat gain rate.

$$t_{\text{rest}} = \frac{Q_{M,\text{max}} - Q_M}{Q_{M,\text{rest}} - Q_M} \times 100\%$$

Equation 52.16: NIOSH Lifting Equation

$$\text{RWL} = 51(10/H)(1 - 0.0075|V - 30|)(0.82 + 1.8/D) \\ \times (1 - 0.0032A)(\text{FM})(\text{CM})$$

$$52.16$$

Description

Improper lifting and handling is the most common cause of injury in the workplace. Heavy loads can strain the body, particularly the lower back. Even light or small objects can cause risk of injury to the body if they are handled in a way that requires strain-inducing stretching, reaching, or lifting. Low-back injuries are common in construction, because lifting conditions are rarely optimal.

In 1993, NIOSH developed an evaluation tool called the *NIOSH lifting equation* that predicts the relative risk of the task. The *recommended weight limit*, RWL, is calculated from Eq. 52.16. H is the horizontal distance (in inches) of the hand from the midpoint of the line joining the inner ankle bones to a point projected on the floor directly below the load center. V is the vertical distance of the hands from the floor. D is the vertical travel distance of the hands between the origin and destination of the lift. A is the asymmetry angle (in degrees), which accounts for torso twisting during lifting. FM is the *frequency multiplier*, which is affected by the lifting frequency. CM is the *coupling multiplier*, which accounts for the ease of holding the load (i.e., *coupling quality*). *Good coupling* occurs when a load has handles. *Fair coupling* occurs when a load has handles that are not easy to hold and lift. *Poor coupling* is where the loads are hard to grab and lift. NIOSH lists the frequency multipliers (see Table 52.10) and the coupling quality and multipliers (see Table 52.11 and Table 52.12, respectively) in its publication 94-110, *Applications Manual for the Revised NIOSH Lifting Equation.*

Table 52.10 Frequency Multipliers

F (min⁻¹)	≤8 hr/day <30 in	≤8 hr/day ≥30 in	≤2 hr/day <30 in	≤2 hr/day ≥30 in	≤1 hr/day <30 in	≤1 hr/day ≥30 in
0.2	0.85	0.85	0.95	0.95	1.00	1.00
0.5	0.81	0.81	0.92	0.92	0.97	0.97
1	0.75	0.75	0.88	0.88	0.94	0.94
2	0.65	0.65	0.84	0.84	0.91	0.91
3	0.55	0.55	0.79	0.79	0.88	0.88
4	0.45	0.45	0.72	0.72	0.84	0.84
5	0.35	0.35	0.60	0.60	0.80	0.80
6	0.27	0.27	0.50	0.50	0.75	0.75
7	0.22	0.22	0.42	0.42	0.70	0.70
8	0.18	0.18	0.35	0.35	0.60	0.60
9	0.00	0.15	0.30	0.30	0.52	0.52
10	0.00	0.13	0.26	0.26	0.45	0.45
11	0.00	0.00	0.00	0.23	0.41	0.41
12	0.00	0.00	0.00	0.21	0.37	0.37
13	0.00	0.00	0.00	0.00	0.00	0.34
14	0.00	0.00	0.00	0.00	0.00	0.31
15	0.00	0.00	0.00	0.00	0.00	0.28

Table 52.11 Coupling Quality

container				loose part/ irregular object	
optimal design		non-optimal design	comfortable grip	uncomfortable grip	
optimal handles or cutouts	non-optimal handles or cutouts			fingers flexed 90°	fingers not flexed 90°
GOOD	fingers flexed 90° → FAIR	POOR	GOOD	FAIR	POOR
	fingers not flexed 90° → POOR				

Table 52.12 Coupling Multipliers

coupling quality	$V < 30$ in (75 cm)	$V \geq 30$ in (75 cm)
GOOD	1.00	1.00
FAIR	0.95	1.00
POOR	0.90	0.90

Equation 52.17 Through Eq. 52.24: Biomechanics of the Human Body

$$H_x + F_x = 0 \qquad 52.17$$
$$H_y + F_y = 0 \qquad 52.18$$
$$H_z + W + F_z = 0 \qquad 52.19$$
$$T_{Hxz} + T_{Wxz} + T_{Fxz} = 0 \qquad 52.20$$
$$T_{Hyz} + T_{Wyz} + T_{Fyz} = 0 \qquad 52.21$$
$$T_{Hxy} + T_{Fxy} = 0 \qquad 52.22$$
$$F_x = \mu F_z \qquad 52.23$$
$$F_x = \alpha F_z \cos\alpha \qquad 52.24$$

Description

Equation 52.17 through Eq. 52.24 relate to *biomechanics* of the human body. They are simply statements of the requirements for static equilibrium. H represents forces on the hand; F represents forces on the feet; and T represents twisting moments (torques). x, y, and z are the directions of the coordinate axes. (See Fig. 52.7.)

Biomechanics treats the human body as a system of jointed linkages capable of certain ranges of motion and torque/strength generating capability. By specifying a person's posture for a given task and knowing population anthropometric dimensions and strength capability, one can evaluate/design the task appropriately. The analysis process is based on static equilibrium.

Figure 52.7 Biomechanics of the Human Body

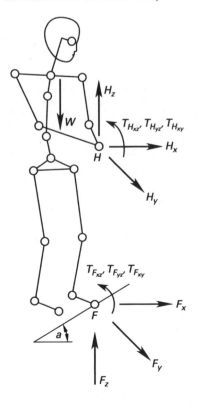

18. ELECTRICAL SAFETY

Electrical safety hazards include shock, arc flash, explosion, and fire. Electrical safety programs must comply with state and federal requirements and are tailored to the specific needs of the workplace. Safety measures include overcurrent protection, such as circuit breakers, grounding, flame-resistant clothing, and insulated tools.

Electrical shock, when current runs through or across the body, is a hazard that can occur in nearly all industries and workplaces. Shock hazard is a function of current, commonly measured in milliamps. Table 52.13 lists levels of current and their effects on a human body.

Table 52.13 Effects of Current on Humans

current level	probable effect on human body
1 mA	Perception level. Slight tingling sensation. Still dangerous under certain conditions.
5 mA	Slight shock felt; not painful, but disturbing. Average individual can let go. However, strong involuntary reactions to shocks in this range may lead to injuries.
6–16 mA	Painful shock, begin to lose muscular control. Commonly referred to as the freezing current or "let-go" range.
17–99 mA	Extreme pain, respiratory arrest, severe muscular contractions. Individual cannot let go. Death is possible.
100–2000 mA	Ventricular fibrillation (uneven, uncoordinated pumping of the heart). Muscular contraction and nerve damage begins to occur. Death is likely.
> 2000 mA	Cardiac arrest, internal organ damage, and severe burns. Death is probable.

19. CUMULATIVE TRAUMA DISORDERS

Cumulative trauma disorders (CTDs) can occur in almost any work situation. CTDs result from repeated stresses that are not excessive individually but, over time, cause disorders, injuries, and the inability to perform a job. High repetitiveness, or continuous use of the same body part results in fatigue followed by cumulative muscle strain. These cumulative injuries are usually incurred by tendons, tendon sheaths, and soft tissue. Moreover, the cumulative injuries can result in damage to nerves and restricted blood flow. CTDs are common in the hand, wrist, forearm, shoulder, neck, and back. Bone and the spinal vertebrae may also be damaged.

The manifestations of CTDs on soft tissues include stretched and strained muscles, rough or torn tendons, inflammation of tendon sheaths, irritation and inflammation of bursa, and stretched (sprained) ligaments. Nerves can be affected by pressure from tendons or other soft tissue, resulting in loss of muscle control, numbness, tingling, or pain, and loss of response of nerves that control automatic functions such as body

Safety/Health/Environment

temperature and sweating. Blood vessels may be compressed, resulting in restricted blood flow and impaired control of tissues (muscles) dependent on that blood supply. Vibration, such as from operating vibrating tools, can cause the arteries in the fingers and hands to close down, resulting in numbness, tingling, and eventually loss of sensation and control.

Industrial hygienists have defined *high repetitiveness* as a cycle time of less than 30 s, or more than 50% of a cycle time spent performing the same fundamental motion. If the work activity requires the muscles to remain contracted at about 15–20% of their maximum capability, circulation can be restricted, which also contributes to CTDs. Also, severe deviation of the wrists, forearms, and other body parts can contribute to CTDs. The following are some examples of CTDs.

Carpal Tunnel Syndrome

Carpal tunnel syndrome (CTS) is the best known CTD. The American Industrial Hygiene Association has described CTS as an occupational illness of the hand and arm system. CTS results from rapid, repetitious finger and wrist movements.

The wrist has a "tunnel" created by the carpal bones on the outer side and ligaments, which are firmly attached to the bones, across the inner side. In the carpal tunnel, which is roughly oval in shape, are tendons and tendon sheaths of the fingers, several nerves, and arteries. If the wrist is bent up or down or flexed from side to side, the space in the carpal tunnel is reduced. Swelling of the tendons or tendon sheaths can place pressure on the nerves, blood vessels, and tendons. Activities that can lead to this disorder include grinding, sanding, hammering, keyboarding, and assembly work.

Cubital Tunnel Syndrome

Cubital tunnel syndrome occurs from compression of the nerve in the forearm below the elbow and results in tingling, numbness, or pain in the fingers. Leaning over a workbench and resting the forearm on a hard surface or edge typically causes this disorder.

Epicondylitis

Epicondylitis is also known as "tennis elbow" and "golfer's elbow." It results from irritation of the tendons of the elbow. It is caused by forceful wrist extensions, repeated straightening and bending of the elbow, and impacting throwing motions.

Ganglionitis

Ganglionitis is a swelling of a tendon sheath in the wrist. Activities that can lead to this disorder include grinding, sanding, sawing, cutting, and using pliers and screwdrivers.

Neck Tension Syndrome

Neck tension syndrome is characterized by an irritation of the muscles of the neck. It commonly occurs after repeated or sustained overhead work.

Pronator Syndrome

Pronator syndrome compresses a nerve in the forearm. It results from rapid and forceful strenuous flexing of the elbow and wrist. Activities that can lead to this disorder include buffing, grinding, polishing, and sanding.

Tendonitis

Tendonitis is an inflammation of a tendon where its surface becomes thickened, bumpy, and irregular. Tendon fibers may become frayed or torn. This disorder can result from repetitious, forceful movements, contact with hard surfaces, and vibrations.

Shoulder tendonitis is irritation and swelling of the tendon or bursa of the shoulder. It is caused by continuous elevation of the arm.

Tenosynovitis

Tenosynovitis is characterized by swelling of tendon sheaths and irritation of the tendon. It is known as *DeQuervain's syndrome* when it affects the thumb. Activities that can lead to this disorder include grinding, polishing, sanding, sawing, cutting, and using screwdrivers.

Trigger finger is a special case of tenosynovitis that results in the tendon of the trigger finger becoming nearly locked so that its forced movement is jerky. It comes from using hand tools with sharp edges pressing into the tissue of the finger or where the tip of the finger is flexed but the middle part is straight.

Thoracic Outlet Syndrome

Thoracic outlet syndrome is characterized by reduced blood flow to and from the arm due to compression of nerves and blood vessels between the collarbone and the ribs. It results in a numbing of the arm and constrains muscular activities.

Ulnar Artery Aneurysm

Ulnar artery aneurysm is characterized by a weakening of an artery in the wrist, causing an expansion that presses on the nerve. This often occurs from pounding or pushing with the heel of the hand, as in assembly work.

Ulnar Nerve Entrapment

Ulnar nerve entrapment involves pressure on a nerve in the wrist. It occurs from prolonged flexing of the wrist and repeated pressure on the palm. Activities that can lead to this disorder include carpentry, brick laying, and using pliers and hammers.

White Finger

White finger is also known as "dead finger," *Raynaud's syndrome*, or *vibration syndrome*. In this disorder, the finger turns cold and numb, tingles, and loses sensation and control. This is a result of insufficient blood supply, which causes the finger to turn white. It results from closure of the arteries due to vibrations. Gripping vibrating tools, especially in the cold, is a common cause.

Safety/Health/ Environment

53 Discharge Water Quality

Nomenclature

C	concentration	mg/L
EW	equivalent weight	g/mol
m	mass	g
meq	milliequivalent concentration	meq
MW	molecular weight	g/mol

1. WATER CHEMISTRY UNITS OF CONCENTRATION

There are many ways that the amount of compounds dissolved in water can be presented. For water supply and wastewater calculations, concentrations can be presented as substance or as one of several equivalents. For example, an analysis might report a sulfate concentration as 38 mg/L as substance, 790 meq/L, or 40 mg/L as $CaCO_3$.

The *as substance concentration* is the gravimetric amount of the substance in the volumetric water basis. If water contains 3.8 mg/L of CO_2, then there are 3.8 mg of CO_2 in a liter of water.

The *milliequivalent concentration* is the number of equivalent weights of the substance in the volumetric water basis times 1000 meq/eq. The *equivalent weight* is the molecular weight divided by the oxidation number (valence, or number of charges on the ion).

$$\text{meq} = \frac{m}{\text{EW}} \times 1000 \ \frac{\text{meq}}{\text{eq}}$$

$$\text{EW} = \frac{\text{MW}}{\text{oxidation number}}$$

The molecular weight of $CaCO_3$ is approximately 100 g/mol and the calcium ion is Ca^{++}, so the equivalent weight is approximately 50 g/mol. If a water sample contains 25 mg/L of $CaCO_3$, concentration might be reported as 0.5 meq/L.

Water chemistry concentrations are often reported in *$CaCO_3$ equivalents*. The *as $CaCO_3$ concentration* is the amount of $CaCO_3$ in mg/L that would contribute the same number of ionic charges as the reported compound. The $CaCO_3$ equivalent is calculated from the ratio of equivalent weights.

$$C_{\text{as } CaCO_3} = C_{\text{as substance}} \left(\frac{\text{EW}_{CaCO_3}}{\text{EW}_{\text{substance}}} \right)$$

$$= C_{\text{as substance}} \left(\frac{50.1 \ \frac{\text{g}}{\text{mol}}}{\text{EW}_{\text{substance}}} \right)$$

Example

The concentration of magnesium in a water sample is 31 mg/L as substance. Most nearly, what is the $CaCO_3$ equivalent concentration?

(A) 12 mg/L as $CaCO_3$

(B) 64 mg/L as $CaCO_3$

(C) 130 mg/L as $CaCO_3$

(D) 250 mg/L as $CaCO_3$

Solution

The molecular weight of magnesium, Mg^{++}, is 24.305 g/mol. The magnesium ion is doubly charged, so the equivalent weight is

$$\text{EW}_{Mg^{++}} = \frac{\text{MW}_{Mg^{++}}}{\text{oxidation number}} = \frac{24.305 \ \frac{\text{g}}{\text{mol}}}{2}$$

$$= 12.153 \ \text{g/mol}$$

The concentration as $CaCO_3$ is

$$C_{Mg^{++}, \text{as } CaCO_3} = C_{Mg^{++}, \text{as substance}} \left(\frac{\text{EW}_{CaCO_3}}{\text{EW}_{Mg^{++}}} \right)$$

$$= \left(31 \ \frac{\text{mg}}{\text{L}} \right) \left(\frac{50.1 \ \frac{\text{g}}{\text{mol}}}{12.153 \ \frac{\text{g}}{\text{mol}}} \right)$$

$$= 127.8 \ \text{mg/L} \quad (130 \ \text{mg/L as } CaCO_3)$$

The answer is (C).

Safety/Health/Environment

2. CATIONS AND ANIONS IN NEUTRAL SOLUTIONS

Equivalency concepts provide a useful check on the accuracy of water analyses. For the water to be electrically neutral, the sum of anion equivalents must equal the sum of cation equivalents.

Concentrations of dissolved compounds in water are usually expressed in mg/L, not equivalents. However, anionic and cationic substances can be converted to their equivalent concentrations in milliequivalents per liter (meq/L) by dividing their concentrations in mg/L by their equivalent weights.

$$C_{\text{meq/L}} = \frac{C_{\text{mg/L}}}{\text{EW}_{\text{g/mol}}}$$

Since water is an excellent solvent, it will contain the ions of the inorganic compounds to which it is exposed. A chemical analysis listing these ions does not explicitly determine the compounds from which the ions originated. Several graphical methods, such as bar graphs and Piper (Hill) trilinear diagrams, can be used for this purpose. The bar graph, also known as a *milliequivalent per liter bar chart*, is constructed by listing the cations (positive ions) in the sequence of calcium (Ca^{++}), magnesium (Mg^{++}), iron (Fe^{+++}), sodium (Na^{+}), and potassium (K^{+}), and pairing them with the anions (negative ions) in the sequence of carbonate (CO_3^{--}), bicarbonate (HCO_3^{-}), sulfate (SO_4^{--}), chloride (Cl^{-}), nitrate (NO_3^{-}), and fluoride (F^{-}). A bar chart can be used to deduce the hypothetical combinations of positive and negative ions that would have resulted in the given water analysis.

Example

A water analysis of lake water has the results shown, with all values reported as $CaCO_3$.

alkalinity	151.5 mg/L
sodium	120.0 mg/L
calcium	127.5 mg/L
iron (III)	0.107 mg/L
magnesium	43.5 mg/L
potassium	8.24 mg/L
chloride	39.5 mg/L
fluoride	1.05 mg/L
nitrate	1.06 mg/L
sulfate	106 mg/L

What is one of the most likely compounds dissolved in the water?

(A) $CaSO_4$

(B) $MgCl_2$

(C) $Fe_2(SO_4)_3$

(D) $NaHCO_3$

Solution

Since this is lake water, alkalinity can be assumed to be the result of dissolved CO_2 and the formation of bicarbonate, HCO_3^{-}.

Calculate the milliequivalent concentrations for each component. Since the $CaCO_3$ equivalent has been given, the molecular and equivalent weights are not needed. The milliequivalent concentration is calculated as

$$C_{\text{meq/L}} = \frac{C_{\text{as CaCO}_3,\text{mg/L}}}{\text{EW}_{\text{CaCO}_3}} = \frac{C_{\text{as CaCO}_3,\text{mg/L}}}{50.1 \ \frac{\text{g}}{\text{mol}}}$$

ion/component	$CaCO_3$ equivalent (mg/L)	milliequivalents (meq/L)
Na^{+}	120.0	2.40
Ca^{++}	127.5	2.54
Fe^{+++}	0.107	0.00214
Mg^{++}	43.5	0.868
K^{+}	8.24	0.164
HCO_3^{-}	151.5	3.02
Cl^{-}	39.5	0.788
F^{-}	1.05	0.0210
NO_3^{-}	1.06	0.0212
SO_4^{--}	106.0	2.12

Draw a milliequivalent bar chart following the Ca-Mg-Fe-Na-K and CO3-HCO3-SO4-Cl-NO3-F sequences.

Most likely, none of the compounds listed except $Fe_2(SO_4)_3$ contributed to the lake water.

The answer is (C).

3. ACIDITY

Acidity is a measure of acids in solution. Acidity in surface water is caused by formation of carbonic acid (H_2CO_3) from carbon dioxide in the air. Carbonic acid is aggressive and must be neutralized to eliminate a cause of water pipe corrosion. If the pH of water is greater than 4.5, carbonic acid ionizes to form bicarbonate. If the pH is greater than 8.3, carbonate ions form.

Measurement of acidity is done by titration with a standard basic measuring solution. Acidity in water is typically given in terms of the $CaCO_3$ equivalent that would neutralize the acid.

4. ALKALINITY

Alkalinity is a measure of the ability of a water to neutralize acids (i.e., to absorb hydrogen ions without significant pH change). The principal alkaline ions are OH^-, CO_3^{--}, and HCO_3^-. Other radicals, such as NO_3^-, also contribute to alkalinity, but their presence is rare. The measure of alkalinity is the sum of concentrations of each of the substances measured as equivalent $CaCO_3$.

5. HARDNESS

Hardness in natural water is caused by the presence of polyvalent (but not singly charged) metallic cations. Principal cations causing hardness in water and the major anions associated with them are presented in Table 53.1. Because the most prevalent of these species are the divalent cations of calcium and magnesium, total hardness is typically defined as the sum of the concentration of these two elements and is expressed in terms of milligrams per liter as $CaCO_3$. (Hardness is occasionally expressed in units of *grains per gallon*, where 7000 grains are equal to a pound.)

Table 53.1 *Principal Cations and Anions Indicating Hardness*

cations	anions
Ca^{++}	HCO_3^-
Mg^{++}	SO_4^{--}
Sr^{++}	Cl^-
Fe^{++}	NO_3^-
Mn^{++}	SiO_3^{--}

Carbonate hardness is caused by cations from the dissolution of calcium or magnesium carbonate and bicarbonate in the water. Carbonate hardness is hardness that is chemically equivalent to alkalinity, where most of the alkalinity in natural water is caused by the bicarbonate and carbonate ions.

Noncarbonate hardness is caused by cations from calcium (i.e., calcium hardness) and magnesium (i.e., magnesium hardness) compounds of sulfate, chloride, or silicate that are dissolved in the water. Noncarbonate hardness is equal to the total hardness minus the carbonate hardness.

Hardness can be classified as shown in Table 53.2. Although high values of hardness do not present a health risk, they have an impact on the aesthetic acceptability of water for domestic use. (Hardness reacts with soap to reduce its cleansing effectiveness and to form scum on the water surface.) Where feasible, carbonate hardness in potable water should be reduced to the 25–40 mg/L range and total hardness reduced to the 50–75 mg/L range.

Water containing bicarbonate (HCO_3^-) can be heated to precipitate carbonate (CO_3^{--}) as a *scale*. Water used in steam-producing equipment (e.g., boilers) must be essentially hardness-free to avoid deposit of scale.

Table 53.2 *Relationship of Hardness Concentration to Classification*

hardness (mg/L as $CaCO_3$)	classification
0 to 60	soft
61 to 120	moderately hard
121 to 180	hard
181 to 350	very hard
> 350	saline; brackish

Noncarbonate hardness, also called *permanent hardness*, cannot be removed by heating. It can be removed by precipitation softening processes (typically the lime-soda ash process) or by ion exchange processes using resins selective for ions causing hardness.

Hardness is measured in the laboratory by titrating the sample using a standardized solution of ethylenediaminetetraacetic acid (EDTA) and an indicator dye such as Eriochrome Black T. The sample is titrated at a pH of approximately 10 until the dye color changes from red to blue. The standardized solution of EDTA is usually prepared such that 1 mL of EDTA is equivalent to 1 mg/L of hardness.

6. HARDNESS AND ALKALINITY

Hardness is caused by multi-positive ions. Alkalinity is caused by negative ions. Both positive and negative ions are present simultaneously. Therefore, an alkaline water can also be hard.

With some assumptions and minimal information about the water composition, it is possible to determine the ions in the water from the hardness and alkalinity. For example, Fe^{++} is an unlikely ion in most water supplies, and it is often neglected.

If hardness and alkalinity (both as $CaCO_3$) are the same and there are no monovalent cations, then there are no SO_4^{--}, Cl^-, or NO_3^- ions present. That is, there is no noncarbonate (permanent) hardness. If hardness is greater than the alkalinity, however, then noncarbonate hardness is present, and the carbonate (temporary) hardness is equal to the alkalinity. If hardness is less than the alkalinity, then all hardness is carbonate hardness, and the extra HCO_3^- comes from other sources (such as $NaHCO_3$).

7. NATIONAL PRIMARY DRINKING WATER REGULATIONS

Following passage of the Safe Drinking Water Act in the United States, the Environmental Protection Agency (EPA) established minimum primary drinking water regulations. These regulations set limits on the amount of various substances in drinking water. Every public water supply serving at least 15 service connections or 25 or more people must ensure that its water meets these minimum standards.

Safety/Health/Environment

Accordingly, the EPA has established the National Primary Drinking Water Regulations and the National Secondary Drinking Water Regulations. The primary standards establish *maximum contaminant levels* (MCL) and *maximum contaminant level goals* (MCLG) for materials that are known or suspected health hazards. The MCL is the enforceable level that the water supplier must not exceed, while the MCLG is an unenforceable health goal equal to the maximum level of a contaminant that is not expected to cause any adverse health effects over a lifetime of exposure.

8. NATIONAL SECONDARY DRINKING WATER REGULATIONS

The national secondary drinking water regulations, outlined in Table 53.3, are not designed to protect public health. Instead, they are intended to protect "public welfare" by providing helpful guidelines regarding the taste, odor, color, and other aesthetic aspects of drinking water.

Table 53.3 National Secondary Drinking Water Regulations (Code of Federal Regulations (CFR) Title 40, Ch. I, Part 143)

contaminant	suggested levels	effects
aluminum	0.05–0.2 mg/L	discoloration of water
chloride	250 mg/L	salty taste and pipe corrosion
color	15 color units	visible tint
copper	1.0 mg/L	metallic taste and staining
corrosivity	noncorrosive	taste, staining, and corrosion
fluoride	2.0 mg/L	dental fluorosis
foaming agents	0.5 mg/L	froth, odor, and bitter taste
iron	0.3 mg/L	taste, staining, and sediment
manganese	0.05 mg/L	taste and staining
odor	3 TON*	"rotten egg," musty, and chemical odor
pH	6.5–8.5	low pH—metallic taste and corrosion; high pH—slippery feel, soda taste, and deposits
silver	0.1 mg/L	discoloration of skin and graying of eyes
sulfate	250 mg/L	salty taste and laxative effect
total dissolved solids (TDS)	500 mg/L	taste, corrosivity, and soap interference
zinc	5 mg/L	metallic taste

*threshold odor number

9. COLOR

Color in water is caused by substances in solution, known as *true color*, and by substances in suspension, mostly organics, known as *apparent* or *organic color*. Iron, copper, manganese, and industrial wastes all can cause color. Color is aesthetically undesirable, and it stains fabrics and porcelain bathroom fixtures.

Water color is determined by comparison with standard platinum/cobalt solutions or by spectrophotometric methods. The standard color scales range from 0 (clear) to 70. Water samples with more intense color can be evaluated using a dilution technique.

10. TURBIDITY

Turbidity is a measure of the light-transmitting properties of water and is comprised of suspended and colloidal material. Turbidity is expressed in *nephelometric turbidity units* (NTU). Viruses and bacteria become attached to these particles, where they can be protected from the bactericidal and viricidal effects of chlorine, ozone, and other disinfecting agents. The organic material included in turbidity has also been identified as a potential precursor to carcinogenic disinfection by-products.

Turbidity in excess of 5 NTU is noticeable by visual observation. Turbidity in a typical clear lake is approximately 25 NTU, and muddy water exceeds 100 NTU. Turbidity is measured using an electronic instrument called a nephelometer, which detects light scattered by the particles when a focused light beam is shown through the sample.

11. SOLIDS

Solids present in a sample of water can be classified in several ways.

- *total solids* (TS): Total solids are the material residue left after the evaporation of the sample. Total solids include total suspended solids and total dissolved solids.

- *total suspended solids* (TSS): The material retained on a standard glass-fiber filter disk is defined as the suspended solids in a sample. The filter is weighed before filtration, dried, and weighed again. The gain in weight is the amount of suspended solids. Suspended solids can also be categorized into *volatile suspended solids* (VSS) and *fixed suspended solids* (FSS).

- *total dissolved solids* (TDS): These solids are in solution and pass through the pores of the standard glass-fiber filter. Dissolved solids are determined by passing the sample through a filter, collecting the filtrate in a weighed drying dish, and evaporating the liquid. The gain in weight represents the dissolved solids. Dissolved solids can be categorized into *volatile dissolved solids* (VDS) and *fixed dissolved solids* (FDS).

- *total volatile solids* (TVS): The residue from one of the previous determinations is ignited to constant weight in an electric muffle furnace. The loss in weight during the ignition process represents the volatile solids.

- *total fixed solids* (TFS): The weight of solids that remain after the ignition used to determine volatile solids represents the fixed solids.

- *settleable solids*: The volume (mL/L) of settleable solids is measured by allowing a sample to settle for one hour in a graduated conical container (*Imhoff cone*).

Waters with high concentrations of suspended solids are classified as *turbid waters*. Waters with high concentrations of dissolved solids often can be tasted. Therefore, a limit of 500 mg/L has been established in the National Secondary Drinking Water Regulations for dissolved solids.

Safety/Health/Environment

54 Air Quality

Nomenclature

d	diameter	m
DRE	destruction and removal efficiency	%
g	gravitational constant, 9.81	m/s^2
P	particulate matter concentration	mg/m^3
Re	Reynolds number	–
SG	specific gravity	–
v_t	terminal settling velocity	m/s
W	mass feed rate	kg/h
Y	oxygen concentration measured in stack gas	%

Symbols

μ	absolute viscosity	lbm/ft-hr
ρ	density	kg/m^3

Subscripts

c	corrected
f	fluid
g	gravitational constant
m	measured
p	particles or pollutant

1. CLEAN AIR ACT

Air quality issues have been addressed by federal legislation since 1955, and many updated statutes and amendments have followed. The statute providing regulatory authority for national air quality issues is the Clean Air Act (CAA) and amendments. Federal air pollution regulations address air quality issues through a variety of enforcement and incentive mechanisms.

The CAA was amended in 1990 to require the establishment of *National Ambient Air Quality Standards* (NAAQS) in the United States. The NAAQS implements a uniform national air pollution control program. For each pollutant included in the NAAQS, primary and secondary standards are defined. *Primary standards* are intended to protect human health, and *secondary standards* are intended to protect public welfare. The NAAQS are applied to a limited number of substances, known as the *criteria pollutants*. The NAAQS and the criteria pollutants to which they apply are presented in Table 54.1.

To ensure application of the NAAQS in each of the 50 states, the United States Environmental Protection Agency (EPA) has required the states to prepare and submit *State Implementation Plans* (SIP). Each SIP must address a control strategy for each of the criteria pollutants included in the NAAQS. If a state is unable or unwilling to submit an SIP acceptable to the EPA, the EPA has the authority to develop an SIP for the state and then require the state to enforce it.

Prevention of significant deterioration (PSD) is a program that is designed to protect air quality in clean-air areas that already meet the NAAQS. To apply the PSD principle, the EPA has developed a regional classification system. Class I regions, generally including national monuments and parks and national wilderness areas, allow very little air quality deterioration and, thereby, limit economic development. Class II regions allow moderate air quality deterioration, and class III regions allow deterioration to the secondary NAAQS.

Because of geographic conditions, population density, climate, and other factors, some areas in the United States are unable to meet the NAAQS. These areas are designated as *nonattainment areas* (NA). Any new or modified sources in nonattainment areas are required to control emissions at the *lowest achievable emission rate* (LAER). Recognizing the economic problems posed by the CAA in nonattainment areas, the EPA has allowed some flexibility, through *emissions trading*, in how industry goes about meeting the compliance requirements. Emissions trading can occur by the following methods.

- *Emission Reduction Credit:* By applying a higher level of treatment than required by regulation, businesses earn credits that can be redeemed through other CAA programs.

- *Offset:* The offset allows industrial expansion by letting businesses compensate for new emissions by offsetting them with credits acquired by other businesses already existing in the region.

- *Bubble:* All activities in a single plant or among a group of proximate industries can emit at various rates as long as the resulting total emission does not exceed the allowable emission for an individual source.

- *Netting:* Businesses can expand without acquiring a new permit as long as the net increase in emissions is not significant. Consequently, by applying improved control technology, businesses can earn credits that can be subsequently applied to plant expansion with no net increase in emissions.

Safety/Health/Environment

Table 54.1 *National Ambient Air Quality Standards*[a]

pollutant	time basis	primary standard[b]	secondary standard
carbon monoxide (CO)	8 h	9 ppm (10 mg/m^3)	none
	1 h	35 ppm (40 mg/m^3)	none
nitrogen dioxide (NO$_2$)	1 h	100 ppb	none
	mean annual	53 ppb (100 μg/m^3)	same
ground-level ozone (O$_3$)	8 h	0.075 ppm (147 μg/m^3)	same
sulfur dioxide (SO$_2$)	3 h	none	0.50 ppm (1300 μg/m^3)
	1 h	0.075 ppm	none
PM10	24 h	150 μg/m^3	same
PM2.5	annual	12 μg/m^3	15 μg/m^3
	24 h	35 μg/m^3	same
lead (Pb)	3 mo	0.15 μg/m^3	same
hydrocarbons[c]	–	–	–

[a] Typical values. The NAAQS are subject to ongoing legislation, and current values may be revised from those presented here.
[b] Equivalents in parentheses may or may not be part of the NAAQS.
[c] Hydrocarbon standards have been revoked by the EPA since hydrocarbons, as precursors of ozone, are adequately regulated by ozone standards.

Source: National Ambient Air Quality Standards (NAAQS), U.S. Environmental Protection Agency, epa.gov.

New source performance standards (NSPS) are applied to specific source categories (e.g., asphalt concrete plants, incinerators) to prevent the emergence of other air pollution problems from new construction. The NSPS are intended to reflect the best emission control measures achievable at reasonable cost, and are to be applied during initial construction of new industrial facilities.

National Emission Standards for Hazardous Air Pollutants (NESHAP) focus on hazardous pollutants not included as criteria pollutants. *Hazardous pollutants* are compounds that pose particular, usually localized, hazards to the exposed population. On a localized level, their impact is more severe than that of the criteria pollutants. Compounds identified as hazardous pollutants are limited to asbestos, inorganic arsenic, benzene, mercury, beryllium, radionuclides, vinyl chloride, and coke oven emissions. The list is short because onerous demands are placed on the EPA by the CAA for including pollutants and because a very high level of abatement, based on risk to the exposed population, is required once a pollutant is listed. The 1990 CAA amendments listed 187 other *hazardous air pollutants* (HAPs) whose abatement is based on the *maximum achievable control technology* (MACT), a less demanding criterion than risk. Many states have adopted their own regulations to designate and control other hazardous air pollutants.

The 1990 CAA amendments placed limits on allowable emissions of sulfur oxides (SO$_x$) and nitrogen oxides (NO$_x$)—two categories of compounds associated with *acid rain*. These limits require acid deposition emission controls, which are implemented through a system of emission allowances applied to fossil fuel-fired boilers in the 48 contiguous states. Control options may include such things as use of low-sulfur fuels, flue-gas desulfurization, and replacing older facilities.

The 1977 CAA amendments authorized the EPA to regulate any substance or practice with reasonable potential to damage the stratosphere, especially with regard to *ozone*. In response to this regulatory authority, the EPA has closely regulated the production and sale of chlorinated fluorocarbons (CFCs) through out-and-out bans on their production and use, taxes on emissions from their use, and a permitting program.

Criteria Pollutants

The EPA uses the following six criteria pollutants as indicators of air quality. It has established a maximum concentration for each pollutant that, when exceeded, may negatively impact human health.

- ozone
- carbon monoxide
- nitrogen dioxide
- sulfur dioxide
- particulate matter
- lead

The threshold concentrations for criteria pollutants are called *national ambient air quality standards* (NAAQS). (See Table 54.1.) If these standards are exceeded, the area may be designated as being in nonattainment. A *nonattainment area* is defined in the Clean Air Act and Amendments of 1990 as a locality where air pollution levels persistently exceed the national ambient air quality standards, or a locality that contributes to ambient air quality in a nearby area that fails to meet standards.

Nitrogen Oxides

Nitrogen oxides (NOx) incorporate the sum total of the oxides of nitrogen, the most significant of which are nitric oxide (NO) and nitrogen dioxide (NO_2). NOx are emitted from motor vehicles, power plants, and other combustion operations. NOx gases play a major role in the formation of ozone and nitrogen-bearing particles, which are both associated with adverse health effects. NOx also contribute to the formation of acid rain. Adverse environmental effects of NOx include visibility impairment, acidification of freshwater bodies, increases in levels of toxins harmful to fish and other aquatic life, and changes in the populations of some species of vegetation in wetland and terrestrial systems. Exposure to NO_2 may lead to changes in airway responsiveness and lung function in individuals with pre-existing respiratory illnesses and increases in respiratory illnesses in children (5–12 yr). Long-term exposures to NO_2 may lead to increased susceptibility to respiratory infection and may cause alterations in the lung.

Ozone

Ground level ozone (O_3) is a primary component in smog and may adversely affect human health. This is in contrast to *stratospheric ozone*, which occurs naturally and provides a protective layer against ultraviolet radiation. O_3 is not emitted directly into the air, but is formed by the reaction of volatile organic compounds (VOCs) and NOx in the presence of heat and sunlight. Ground level ozone forms readily in the atmosphere, usually during hot summer weather. Because O_3 is formed in the atmosphere over time, the highest levels of ozone typically occur 50–100 km away from the location of the highest ozone precursor (NOx and VOCs) emissions. VOCs are emitted from a variety of sources, including motor vehicles, chemical plants, refineries, factories, consumer and commercial products, and other industrial sources. O_3 concentrations are largely affected by climate and weather patterns.

Health effects from O_3 exposure include increased susceptibility to respiratory infection, lung inflammation, aggravation of pre-existing respiratory diseases such as asthma, significant decreases in lung function, and increased respiratory symptoms, such as chest pain and cough. Environmental impacts of ground level ozone include reductions in agricultural and commercial forest yields, reduced growth and survivability of tree seedlings, and increased plant susceptibility to disease, pests, and other environmental stresses. In 1997, the EPA revised the national ambient air quality standards for ozone by replacing the one hour ozone standard of 0.12 parts per million (ppm) with a new eight hour standard of 0.08 ppm.

Acid Rain

Acid rain describes a mixture of wet and dry atmospheric deposition containing higher than normal amounts of nitric and sulfuric acids. The precursors, or chemical forerunners, of acid rain result from both natural sources, such as volcanoes and decaying vegetation, and from manufactured sources, primarily sulfur dioxide (SO_2)

and nitrogen oxide (NOx) emissions from fossil fuel combustion. Acid rain is associated with the acidification of soils, lakes, and streams, accelerated corrosion of buildings and monuments, and reduced visibility. In the United States, about two-thirds of all SO_2 and one-quarter of all NOx come principally from coal burning power plants. Acid rain occurs when SO_2 or NOx gases react in the atmosphere with water, oxygen, and other chemicals to form mild solutions of sulfuric or nitric acid.

Wet deposition is acidic rain, fog, mist, and snow. It affects a variety of plants and animals, with the ultimate impact determined by factors including the acidity of the liquid solutions, the chemistry and buffering capacity of the soils involved, and the types of fish, trees, and other living organisms dependent on the water.

Dry deposition is the incorporation of acids into dust or smoke, which then accumulate on the ground, buildings, homes, cars, and trees. Dry deposited acids and particles can be washed from these surfaces by rainstorms, leading to increased acidity in the runoff. About half of the acidity in the atmosphere falls back to earth through dry deposition.

Carbon Monoxide

Carbon monoxide (CO) is a colorless, odorless, and, at high levels, poisonous gas, formed when the carbon in fuel is not burned completely. It is a component of motor vehicle exhaust, which contributes about 60% of all CO emissions nationwide and 95% of all emissions in cities. High concentrations of CO generally occur in areas with heavy traffic congestion. Other sources of CO emissions include refineries and industrial processes and non-transportation fuel combustion, including wood and refuse burning. Peak atmospheric CO concentrations typically occur in winter months when CO automotive emissions are greater and nighttime atmospheric inversion conditions are more frequent. *Atmospheric inversions* occur when pollutants are trapped in a cold layer of air beneath a warmer one. High concentrations of CO enter the bloodstream through the lungs and reduce oxygen delivery to the body, which can cause visual impairment and reduce work capacity, manual dexterity, learning ability, and performance in complex tasks.

Smog

Smog is essentially a mixture of pollutants and water and is considered air pollution. *Classic smog* results from large amounts of coal burning in a geographical area and is caused by a mixture of smoke and sulfur dioxide (SO_2). *Photochemical smog* is a mixture of air pollutants, such as nitrogen oxides (NOx), tropospheric ozone, volatile organic compounds (VOCs), peroxyacyl nitrates, and aldehydes. Photochemical smog components are generally highly reactive and oxidizing. Photochemical smog is caused by a reaction between sunlight and the various chemical components listed. Smog is of principal concern in urban areas.

Lead

Lead (Pb) is a metallic element that occurs naturally in soil, rocks, water, and food. Lead was largely used as an additive to paints and gasoline until the 1970s when its use was significantly reduced. Current sources of Pb are lead-acid car batteries and industrial coatings. Approximately 70% of Pb pollution comes from smelters, power plants fueled by coal, and lead used in the processing of oil shale. Exposure to Pb can occur through inhalation of air containing Pb and ingestion of Pb in food, water, soil, or dust. Excessive Pb exposure can cause seizures, mental retardation, and/or behavioral disorders.

Radon

Radon is a radioactive noble gas formed by the natural decay of uranium, which is found in nearly all soils. Radon is considered a health hazard because it causes lung cancer. It is colorless, odorless, and tasteless. Radon gas can accumulate in buildings and drinking water. It causes an estimated 20,000 deaths per year in the United States. Radon is a significant contaminant that impacts indoor air quality worldwide, because it enters buildings through cracks and holes in the foundation, or through well water. Radon contamination inside of homes is of particular concern and cannot be remedied once detected. Radon concentrations in the air are typically expressed in units of picocuries per liter (pCi/L) of air. Concentrations can also be expressed in *working levels* (WL) rather than picocuries per liter (1 pCi/L = 0.004 WL).

PCBs

Polychlorinated biphenyls (PCBs) are manufactured mixtures of up to 209 individual chlorinated compounds (congeners). Because of their nonflammability, chemical stability, high boiling point, and electrical insulating properties, PCBs were used in hundreds of industrial and commercial applications, including electrical, heat transfer, and hydraulic equipment. They were also used as plasticizers in paints, plastics, and rubber products, as well as in pigments, dyes, carbonless copy paper, and many other applications. PCBs can be either oily liquids or solids, colorless to light yellow, and have no odor or taste. Some PCBs can exist as air vapor. However, the production of PCBs was stopped in the United States in 1977 because of PCB buildup in the environment, which gave rise to adverse health effects. Products made before 1977 that may still contain PCBs include old fluorescent lighting fixtures, transformers and capacitors, and cutting and hydraulic oils.

PCB waste is defined by the EPA as waste containing PCBs as a result of a spill, release, or other unauthorized disposal. The wastes are categorized according to the concentrations of PCBs that they contain and their date of disposal or spillage. PCB contaminated sites are most often cleaned up using *incineration* processes. Alternative technologies are allowed if the performance of these technologies is equivalent to incineration's *destruction and removal efficiency* (DRE) of 99.9999%. Both stationary and mobile incinerators can be used. The choice is often based on cost. Approximately 10 commercial incinerators have been approved for PCB disposal in the United States, and incineration has been used at approximately 65 hazardous waste sites with PCB contaminated soils. Alternative PCB treatment technologies include disposal in chemical waste landfills, high-efficiency boilers, thermal desorption, chemical dehalogenation or dechlorination, solvent extraction, soil washing, vitrification, and bioremediation.

Particulate Matter

Particulate matter (PM), or particle pollution, is a complex mixture of extremely small particles and liquid droplets and is one of the primary forms of air pollution. PM is emitted from numerous industrial, mobile (e.g., cars), residential, and even natural sources. PM is made up of a number of components, including acids (such as nitrates and sulfates), organic chemicals, metals, and soil or dust particles. Particle size is linked to the potential for causing health problems. The EPA focuses on particles that are 10 μm in diameter or smaller because particles of this size, which generally pass through the throat and nose and enter the lungs, can cause adverse health effects. The EPA groups particle pollution into two categories: *inhalable coarse particles* (between 2.5 μm and 10 μm in diameter) and *fine particles* ($\leq$ 2.5 μm diameter). Inhalable particles are found near roadways and dusty industries, while fine particles are found in smoke and haze. PM is generally designated by putting the maximum size (in numbers of microns) in the subscript. For example, $PM_{2.5}$ designates fine particles.

Sulfur Dioxide

Sulfur dioxide (SO_2) is the main sulfur-containing compound produced in the combustion of coal or petroleum. Oxidation of SO_2 in the presence of nitrogen oxides (NOx) in the atmosphere produces hydrogen sulfide (H_2S), an acid that ultimately produces acid rain. Over 65% of the SO_2 released into the air in the United States comes from electrical utilities, especially those that burn coal. Other sources of SO_2 are industrial facilities that derive their products from raw materials including metallic ore, coal, and crude oil, or that burn coal or oil to produce process heat (petroleum refineries, cement manufacturing, and metal processing facilities). Health and environmental impacts of SO_2 are respiratory illness, visibility impairment, acid rain, plant and water damage, and aesthetic damage to structures.

Dioxins

Dioxins are halogenated organic compounds, most commonly polychlorinated dibenzofurans (PCDFs) and polychlorinated dibenzodioxins (PCDDs). The most toxic dioxin is 2,3,7,8-tetrachlorodibenzo-*p*-dioxin (TCDD). Dioxins are formed as an unintentional by-product of many industrial processes involving chlorine, such as waste incineration, chemical and pesticide manufacturing, and pulp and paper bleaching. Two dioxins, polybrominated dibenzofurans and dibenzo-dioxins, have been discovered as impurities in brominated flame retardants. Dioxins are persistent in the environment and readily bioaccumulate up the food chain because they are fat soluble. Dioxins are of particular concern at sites where indirect exposure pathways (e.g., ingestion of fruits/vegetables, meat, dairy products, and breast milk) are of concern. The toxicity of other dioxins and chemicals, such as PCBs, which act like dioxins, are measured in relation to TCDD using *toxic equivalence factors* (TEFs). *2,3,7,8-TCDD toxic equivalents* (TEQs) are determined by multiplying the compound concentrations by their respective TEFs and summing them.

2. RESOURCE CONSERVATION AND RECOVERY ACT (RCRA)

Regulations promulgated under the Resource Conservation and Recovery Act (RCRA) impose restrictions on emissions from hazardous waste incinerators. The regulations define *principal organic hazardous constituents* (POHCs) present in the waste feed and establish performance standards for their destruction and removal in the incineration process.

Equation 54.1: Destruction and Removal Efficiency

$$DRE = \frac{W_{in} - W_{out}}{W_{in}} \times 100\% \qquad 54.1$$

For most wastes, the performance standard for POHCs is defined by a *destruction and removal efficiency* (*DRE*) of 99.99% (referred to as "four nines") for each POHC in the waste feed. However, for selected wastes, primarily those associated with dioxins and furans, a *DRE* of 99.9999% ("six nines") is required. The *DRE* includes the destruction efficiency of the incinerator as well as the removal efficiency of any air pollution control equipment. The *DRE* is calculated from Eq. 54.1.

The RCRA regulations also apply to hydrochloric acid (HCl) and particulate emissions. The HCl emissions are based on a mass emission rate of 1.8 kg/h, and the particulate emissions must be corrected for oxygen. In the following equation, Y must be expressed in percent.

$$P_c = P_m \left(\frac{14}{21 - Y} \right)$$

Example

The POHC feed rate to an incinerator is 2.73 kg/h. Stack emissions monitoring shows the POHC at 0.011 kg/h, particulate at 176 mg/m^3, and O_2 at 5%. What is the *DRE* for the POHC?

(A) 98.7%

(B) 99.2%

(C) 99.6%

(D) 99.8%

Solution

Use Eq. 54.1.

$$DRE = \frac{W_{in} - W_{out}}{W_{in}} \times 100\%$$

$$= \frac{2.73 \, \frac{kg}{h} - 0.011 \, \frac{kg}{h}}{2.73 \, \frac{kg}{h}} \times 100\%$$

$$= 99.60\% \quad (99.6\%)$$

The answer is (C).

3. POLLUTANT CLASSIFICATION

Introduction

Air pollutants may be classified as primary and secondary. *Primary pollutants* are those that exist in the air in the same form in which they were emitted. An example of a primary pollutant is carbon monoxide emitted directly from gasoline-powered automobiles. *Secondary pollutants* are those that are formed in the air from other emitted compounds. Ground-level ozone formed from the nitrogen dioxide photolytic cycle is an example of a common secondary pollutant. Primary and secondary pollutants are not the same as, and should not be confused with, the primary and secondary NAAQS. The major source of air pollutants in the United States and most other industrialized countries is fossil fuel combustion from both stationary and mobile sources. To a much lesser degree, pollutants may be generated from wind erosion, plant and animal bio-effluents, vapors from architectural coatings and other hydrocarbon-based materials (e.g., new asphalt pavements), industrial manufacturing, and a variety of other sources.

Particulate Matter

Suspended *particulate matter* (PM) is produced in a variety of forms. These include *dust* from manufacturing, agricultural, and construction activities, *fumes* produced by chemical processes, *mists* formed by condensed water vapor, *smoke* generated by incomplete combustion from transportation and industrial activities, and *sprays* formed by atomization of liquids. Natural wind erosion and forest fires also produce dust and smoke. The health risks associated with particulate matter exposure are exacerbated by the adsorption from the air of toxic organics onto particle surfaces.

Particulate matter may be liquid or solid with a broad size range, from 0.005 μm to 100 μm. The *particulate size* is taken as the aerodynamic diameter of an equivalent perfect sphere having the same settling velocity. Particulate matter in the size range from 0.1 μm to 10 μm presents the greatest potential impact on air quality. Particulate matter smaller than 0.1 μm tends to coagulate into larger particles, and particles larger than 10 μm settle relatively rapidly. Of particular concern is particulate matter with aerodynamic diameters less than 10 μm (PM10) and less than 2.5 μm (PM2.5) for which NAAQS are included. Particles making up PM2.5 are called *respirable particulate matter* and present the greatest potential risk to human health because they are able to reach and remain in the lung. The sources, forms, and sizes of particulate matter are summarized in Table 54.2.

Particulate lead pollution is most pronounced in areas where automobile traffic has historically been high. Although the use of unleaded gasoline has greatly reduced airborne lead pollution from automobile

Table 54.2 *Particulate Matter Types*

type	source	form	size
dust	emissions from handling and mechanical activities (e.g., grain elevators, sandblasting, construction)	solid	1 μm to 10 mm
fume	condensed vapors from chemical reactions (e.g., calcination, distillation, sublimation reactions)	solid	0.03 μm to 0.3 μm
mist	condensed vapors from chemical reactions (e.g., sulfuric acid from sulfur trioxide)	liquid	0.5 μm to 3 μm
smoke	particulate emissions from combustion activities (e.g., forest fires, coal-fired power plants)	solid	0.05 μm to 1 μm
spray	atomized liquids (e.g., sprayed coatings, pesticide application)	liquid	10 μm to 10 mm

Adapted from: Vesilind, P.A., J.J. Pierce, and R.F. Weiner, 1994, *Environmental Engineering*, 3rd ed., Butterworth-Heinemann, Newton, MA.

emissions, residual contamination persists in soils along roadways, in parks, and in other areas where soil is exposed. This contaminated soil can act as an ongoing source of airborne lead when disturbed by vehicle traffic, construction activities, natural wind action, and other events.

An important source of airborne lead not associated with leaded gasoline use is chipping and flaking lead-based paint. When lead-based paint, used either indoors or outdoors, is sanded or when it chips or flakes, the lead-containing particles become airborne and can be inhaled or contribute to soil contamination. Lead is no longer used in paints intended for indoor applications in the United States, but it is still used in some exterior paints.

Exposure to lead-based paint presents a significant health hazard to children and is associated with behavioral changes, learning disabilities, and other health problems. Lead is included in the NAAQS.

Asbestos is a fibrous silicate material that is inert, strong, and incombustible. As *asbestos-containing materials* (ACM) age or are damaged, asbestos fibers are released into the air. Although the greatest potential exposure to asbestos occurs indoors where asbestos-containing materials were used in construction, soil containing mineral asbestos has contributed to significant asbestos exposures in some areas of the United States.

The most substantial asbestos exposure has historically been occupational. Asbestos poses health risks in relatively localized areas from industrial and construction-related activities. Asbestos is included in the national emission standards for hazardous air pollutants (NESHAP).

Equation 54.2: Stokes' Law

$$v_t = \frac{g(\rho_p - \rho_f)d^2}{18\mu} = \frac{g\rho_f(SG - 1)d^2}{18\mu} \qquad 54.2$$

Values

Table 54.3 *Air Viscosity and Density*

temperature (°F)	absolute viscosity, μ (lbm/ft-hr)	density, ρ (lbm/ft³)
80	0.045	0.0734
100	0.047	0.0708

Description

The *terminal settling velocity* for a particle of particulate matter can be calculated from Eq. 54.2, *Stokes' law*.

Safety/Health/ Environment

Example

What is most nearly the settling velocity for particles with an aerodynamic diameter of 10 μm and a density of 43.7 lbm/ft^3 in an atmosphere at 80°F and 14.17 psi?

(A) 1.1×10^{-5} ft/sec

(B) 6.6×10^{-4} ft/sec

(C) 6.4×10^{-3} ft/sec

(D) 6.7×10^{-3} ft/sec

Solution

From Table 54.3, air viscosity, μ_g, is 0.045 lbm/ft-hr and air density, ρ_g, is 0.0734 lbm/ft^3 at 80°F.

Use Eq. 54.2.

$$v_t = \frac{g(\rho_p - \rho_f)d^2}{18\mu}$$

$$= \frac{\left(32.2 \frac{\text{ft}}{\text{sec}^2}\right)\left(43.7 \frac{\text{lbm}}{\text{ft}^3} - 0.0734 \frac{\text{lbm}}{\text{ft}^3}\right)}{(18)\left(0.045 \frac{\text{lbm}}{\text{ft-hr}}\right)\left(10^6 \frac{\mu\text{m}}{\text{m}}\right)^2}$$

$$\times (10 \ \mu\text{m})^2\left(3600 \frac{\text{sec}}{\text{hr}}\right)\left(3.28 \frac{\text{ft}}{\text{m}}\right)^2$$

$$= 6.72 \times 10^{-3} \text{ ft/sec} \quad (6.7 \times 10^{-3} \text{ ft/sec})$$

The answer is (D).

Equation 54.3: Settling Reynolds Number

$$\text{Re} = \frac{v_t\rho d}{\mu} \quad \text{54.3}$$

Description

Equation 54.3 can be used to calculate the *settling Reynolds number*. Equation 54.3 is valid for particle diameters between about 0.1 μm and 100 μm and for settling velocities less than about 1.0 m/s, both of which are well within the range of interest in air pollution applications.

Gaseous Pollutants

The principal gaseous pollutants of concern include oxides of nitrogen, sulfur, and carbon; ozone; and hydrocarbons. Most gaseous pollutants are associated with combustion processes and the characteristics of the fuels burned. Potential emissions from combustion increase as combustion efficiency decreases, as available oxygen decreases, and as the presence of compounds other than carbon and hydrogen increases. The sources of gaseous pollutants and their fates in the atmosphere are summarized in Table 54.4.

The two nitrogen oxides (NOx) that are significant air pollutants are nitric oxide (NO) and nitrogen dioxide

Table 54.4 Gaseous Emissions Sources

gas	source	fate
methane and other organic gases and vapors	combustion, volatilization	reaction with ·OH
carbon monoxide and carbon dioxide	combustion	oxidation
nitrous oxide and nitrogen dioxide	combustion	oxidation, precipitation, dry deposition
sulfur dioxide and sulfur trioxide	combustion	oxidation, precipitation, dry deposition

(NO$_2$). Most nitrogen oxide emissions occur from fuel release and fossil fuel combustion as relatively benign NO. However, NO can rapidly oxidize to NO$_2$, which is associated with respiratory ailments and which reacts with hydrocarbons in the presence of sunlight to form photochemical oxidants. Photochemical oxidants are a significant constituent of smog. Nitrogen dioxide also reacts with the hydroxyl radical (·OH) in the atmosphere to form nitric acid (HNO$_3$), a contributor to acid rain.

Most sulfur emissions result from the combustion of fossil fuels, especially coal. When burned, the fuels release the sulfur primarily as sulfur dioxide (SO$_2$) with much smaller amounts of sulfur trioxide (SO$_3$). These are known as *sulfur oxides* (SOx). Sulfur oxide aerosols contribute to particulate matter concentrations in the air and are a constituent of acid rain. Sulfur oxides also can significantly impact long-range visibility.

Combustion of fossil fuels under ideal conditions produces carbon dioxide (CO$_2$) and water. However, where conditions reduce combustion efficiency, *carbon monoxide* (CO) is formed instead of CO$_2$. Although CO$_2$ is associated with global warming, it typically is not included as an air pollutant. Carbon monoxide, however, represents as much as half the total mass of air pollutants emitted in the United States, with most of these emissions coming from motor vehicles. Carbon monoxide does not seem to present any particular environmental hazards, but it does present a potential serious threat to human health by displacing oxygen (O$_2$) in the blood.

Ozone (O$_3$) forms in the stratosphere and at ground level. Stratospheric ozone is maintained through naturally occurring processes and acts as a filter to ultraviolet light radiation from the sun. Stratospheric ozone is not an air pollutant. Ground-level ozone, on the other hand, is a secondary pollutant formed through reactions with nitrogen oxides and hydrocarbons in the presence of sunlight, making it a *photochemical oxidant*. Among the photochemical oxidants, ozone is the most common. As a secondary pollutant, emissions limitations are not applicable to ozone. Instead, emission controls for those compounds, such as nitrogen oxides and hydrocarbons, that are precursors to ozone formation are required. Ground-level ozone is a major contributor to photochemical smog and is an irritant to the mucosa and lungs.

Hydrocarbons are included as air pollutants because of their role in the formation of photochemical oxidants, particularly ground-level ozone. The hydrocarbons involved in air pollution are primarily alkanes, comprising the methane series. The methane series includes methane (CH_4), ethane (C_2H_6), propane (C_3H_8), butane (C_4H_{10}), and so on. Alkanes have the general formula C_nH_{2n+2}. If an alkane loses one hydrogen atom, it becomes an alkyl radical (i.e., $\cdot CH_3$) that is able to readily react with other compounds. When oxidized, alkanes form compounds of the carbonyl group (*aldehydes*), where an oxygen atom is double-bonded to a terminal carbon atom. Formaldehyde (CH_2O) and acrolein (CH_2CHCOH) are common aldehyde constituents of photochemical smog that are associated with eye irritation.

Hydrocarbons include other aliphatic chemicals, in addition to alkanes, and some aromatic hydrocarbons. Chlorinated solvents such as carbon tetrachloride, tetra-chloroethene, 1,1,1-trichloroethane, trichloroethene, and their degradation products; and benzene, toluene, ethylbenzene, and xylene are among the most common of these and comprise those compounds typically called *volatile organic compounds* (VOCs). Occurrence of VOCs in the air can be a major contributor to formation of photochemical smog.

Structural formulas for a few representative aliphatic and aromatic hydrocarbons are shown in Fig. 54.1.

Hazardous Air Pollutants

The hazardous air pollutants identified under NESHAP as HAPs comprise approximately 200 chemicals and related groups of compounds, and have been identified as potentially significant sources of health risk to humans. These pollutants occur from a variety of sources and include asbestos; compounds of antimony, arsenic, beryllium, lead, manganese, mercury, and nickel; benzene and benzene derivatives such as toluene, xylene, and phenols; halogen gases; halogenated alkanes; chloroform and related compounds; and radionuclides.

Allergens and Microorganisms

Allergens and microorganisms comprise *biological air pollutants* and include such things as algae, animal dander and hair, bacteria, excreta from animals and insects, pollens, mold spores, viruses, and other organic structures, living or dead. Exposure to biological air pollutants can be very dramatic, as in the outbreak of Legionnaires' disease at the American Legion convention in Philadelphia in 1976, or, more commonly, in the form of the symptoms of allergies and other common illnesses.

4. AIR QUALITY MONITORING

Introduction

Air quality monitoring involves three different categories of measurements: emissions, ambient air, and meteorological. Emissions measurements apply to both stationary and mobile sources and are taken by sampling or monitoring at the point where the emission leaves the source. Ambient air is measured to provide a "control" against which to compare emissions monitoring results and to assess pollutant levels. Meteorological measurements include wind speed and direction, air temperature profiles, and other parameters necessary to evaluate pollutant dispersion and fate.

Monitoring may collect longer-term, continuous samples or short-duration, single-measurement grab samples. The type of sample is dictated by the pollutant being measured, the pollutant source, and the objective of the sampling event.

Particulate Matter

Particulate monitors are of three general types: filtering devices, impactor devices, and photometric devices.

The most common filtering device is the *high-volume* or *hi-vol sampler*. The sampler, typically operated over a period of 24 h or less, uses a blower to draw a large volume of air across a filter material. For gross measurements of total particulate, a single filter may be used. If it is necessary to classify particulate by particle size, stacked filters with openings selected to provide the desired particle size distribution are used. A hi-vol sampler is illustrated in Fig. 54.2.

Impactor (impingement) samplers rely on velocity, air flow geometry, and individual particle mass to sample and characterize particulate matter. The sampler works in a cascading fashion, with successively larger particles being collected on a membrane or film in successive stages. The air flow through each stage occurs at successively higher velocities so the higher-mass particles are unable to follow the gas flow-path and, therefore, impinge on the membrane. Impactor samplers are used for sampling particulate matter with diameters below 10 μm and are especially effective for particles with diameters less than 2.5 μm. An impactor sampler is illustrated in Fig. 54.3.

Photometric devices provide an indirect measurement of particles by correlating particulate concentration with light intensity. The air stream is passed between a light beam and detector. As concentrations of particulate increase, the light becomes more scattered, and the light intensity measured by the detector is less. Photometric devices offer the advantage of having a continuous real-time read-out, but they do not differentiate among particle sizes.

Safety/Health/ Environment

Figure 54.1 *Representative Organic Chemicals Important to Air Pollution*

Aliphatics – straight or branched chain
Alkanes (methane series)
all C-C bonds are single

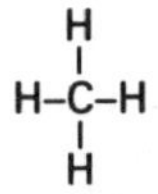

methane (CH$_4$)

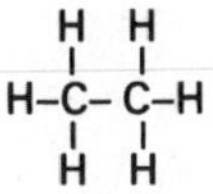

ethane (C$_2$H$_6$)

Alkynes or *acetylenes*
at least one C-C bond is triple

H–C≡C–H

ethyl acetylene (C$_2$H$_2$)

Alkenes or *olefins* (ethene series)
at least one C-C bond is double

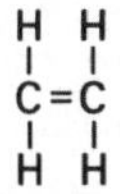

ethene (C$_2$H$_4$)

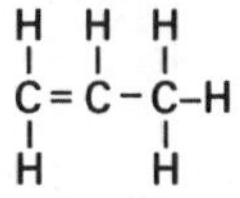

propene (C$_3$H$_6$)

Aldehydes
oxidation products of primary alcohols
contain carbonyl group

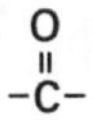

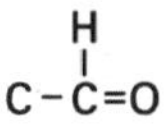

methanol or formaldehyde (CH$_2$O)

Aromatics – contain the 6-carbon
benzene ring

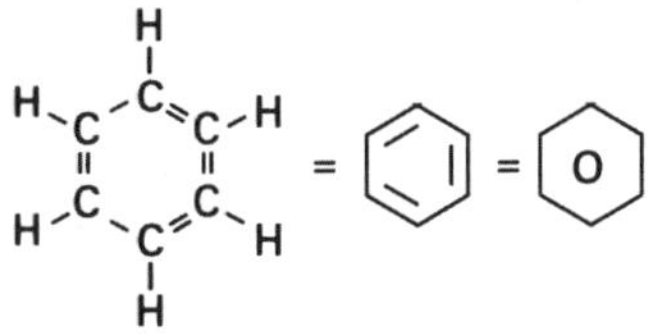

benzene (C$_6$H$_6$)

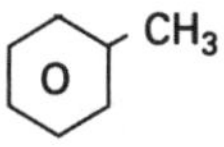

toluene (C$_7$H$_9$)

Halogenated aliphatics and aromatics –
(includes VOCs and CFCs)

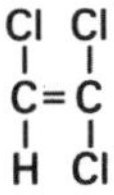

trichloroethene (CHClCCl$_2$)

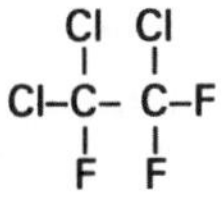

CFC 113 (1,1,2-trichloro-1,2,2-trifluoroethane)

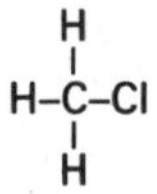

chloroform (CH$_3$Cl)

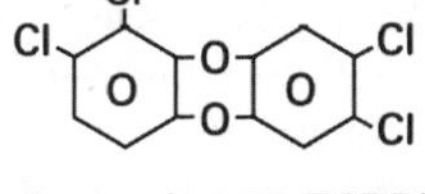

dioxins (2,3,7,8-TCDD)

Gases

Unlike particulate matter, gases do not typically require long collection periods, being well-suited to detection by monitoring instruments that can provide continuous-flow measurements with instantaneous results. Devices employed for gas monitoring are typically specific to only one gas, so multiple instruments may be required to monitor gas mixtures. Monitoring of gaseous emissions employs wet chemistry techniques; visible, infrared, and ultraviolet light analyzers; and electrochemical analyzers. Alternatively, samples can be collected in appropriate containers and analyzed using gas chromatography. Gas chromatography is useful when there are several compounds with similar chemical structures such as N$_2$, O$_2$, CO$_2$, CO, and CH$_4$, or such as VOCs.

Figure 54.2 *Hi-Vol Particulate Sampler*

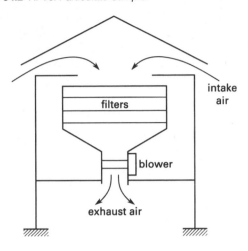

Figure 54.3 *Impactor Particulate Sampler*

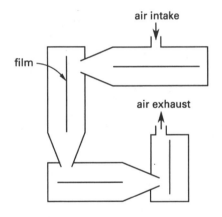

Figure 54.4 *Ringelmann Scale For Smoke Opacity*

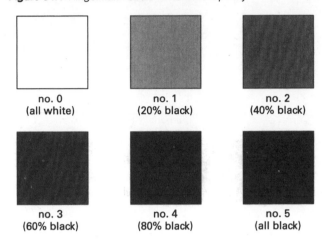

Smoke

Smoke presents an observable, inexpensive, and unobtrusive opportunity to assess emissions from stacks. Regulatory agency staff, plant operations personnel, and other interested persons can become certified to provide quantitative measurements of smoke opacity based solely on observation.

Observations of qualified persons are compared to opacity standards represented by the *Ringelmann scale*. The Ringelmann scale uses a percent opacity from completely transparent (Ringelmann no. 0) to completely opaque (Ringelmann no. 5) and is applied to black or dark smoke. For white smoke, opacity is expressed as "percent opacity" rather than by Ringelmann number. Each Ringelmann number corresponds to 20% opacity, so 5% opacity, the accuracy with which opacity observations are typically reported, would represent $^1/_4$ Ringelmann number. The Ringelmann scale is summarized in Fig. 54.4.

5. AIR QUALITY INDEX

The *Air Quality Index* (AQI)[1] was developed by the EPA to quantify and describe the daily air quality for the general public. The Air Quality Index is divided into six categories of health concern. Each category has a range of AQIs associated with it. A higher AQI indicates lower air quality. The categories are color-coded to simplify identification of hazardous air quality conditions by the public. For example, green corresponds to good air quality, while maroon signalizes hazardous air quality. The EPA requires the AQI to be calculated and reported daily for cities with a population exceeding 350,000 people.

The AQI is determined by five of the six criteria pollutants[2] regulated by the Clean Air Act: ground level ozone (O_3), total suspended particulates (TSP), carbon monoxide (CO), sulfur dioxide (SO_2), and nitrogen dioxide (NO_2). Each AQI category is a separate scale for each criteria pollutant that is based on the concentrations deemed permissible by the NAAQS. AQI values greater than 100 indicate unhealthy air quality, while values less than 100 indicate satisfactory air quality. The AQI of an air sample can be calculated based on the pollutants present.

Each AQI category has a range of pollutant concentrations associated with it. The upper and lower limits of an AQI category's pollutant concentrations are called *breakpoints*. The low and high breakpoint concentrations are the lowest and highest concentrations acceptable in each category, respectively. For example, an air sample is determined to have a pollutant concentration[3] of 8.2 ppm of CO, which falls in Table 54.5 between 4.5 ppm and 9.4 ppm. Therefore, the sample is in the moderate AQI category.

---

[1]The EPA created the Air Quality Index as a replacement for the *Pollutant Standards Index* (PSI) in 1999.

[2]Lead is currently the only criteria pollutant not monitored using the AQI.

[3]It is common practice to round the pollutant concentration up to the nearest whole number.

Table 54.5 *Air Quality Index Categories*

AQI category	color	AQI range	description
good	green	0 to 50	Air quality is considered satisfactory, and air pollution poses little or no risk
moderate	yellow	51 to 100	Air quality is acceptable. However, for some pollutants, there may be a moderate health concern for a very small number of people who are unusually sensitive to air pollution.
unhealthy for sensitive groups	orange	101 to 150	Members of sensitive groups may experience health effects. The general public is not likely to be affected.
unhealthy	red	151 to 200	Everyone may begin to experience health effects. Members of sensitive groups may experience more serious health effects.
very unhealthy	purple	201 to 300	A health alert is issued that everyone may experience more serious health effects.
hazardous	maroon	301 to 500	A health warning of emergency conditions is issued. The entire population is more likely to be affected.

The AQI of an air sample is determined using linear interpolation. If an air sample has several pollutants present, the AQI is calculated for each pollutant, and the greatest AQI is selected (i.e., reported) as the daily AQI.

6. AIR POLLUTION PREVENTION

Pollution prevention, also commonly referred to as *source reduction*, was promulgated in 1990 in the Pollution Prevention Act. *Pollution prevention* is the process of reducing or eliminating waste at the source; the modification of waste-producing processes that enter a waste stream; promoting the uses of nontoxic or less toxic substances; implementing conservation techniques; and reusing materials rather than wasting them. This act also detailed pollution prevention practices (such as recycling, green engineering, and sustainable agriculture) that improve efficiency in the use of energy, water, and other natural resources.

Air pollution generally results from the combustion of hydrocarbon-based fuels, but may also result from dust at construction sites and other disturbed areas. Particle removal is dependent on particle size, but the calculation of collection efficiency is based only on the mass percentage collected. A variety of processes and control devices have been developed to remove particulate matter from gas streams, including cyclones, wet scrubbers, electrostatic precipitators, and baghouses. (See Table 54.6 for removal efficiencies of control devices.) Electrostatic precipitators and baghouses are among the most efficient technologies for removing particulate matter and are thus generally used for air pollution prevention.

Table 54.6 *Typical Removal Efficiencies of Various Particulate Matter Control Devices*

control device	removal efficiency (%)			
	$<1 \mu m$	$1–3 \mu m$	$3–10 \mu m$	$>10 \mu m$
electrostatic precipitator	96.5	98.3	99.1	99.5
fabric filter	100	99.8	>99.9	>99.9
venturi scrubber	>70.0	99.5	>99.8	>99.8
cyclone	11	54	85	95

Safety/Health/ Environment

Diagnostic Exam

Topic XV: Computational Tools

1. The following flowchart represents an algorithm.

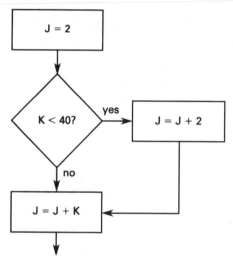

Which of the given structured programming segments correctly translates the algorithm?

- (A) J = 2
  IF K < 40 THEN J = J + 2
  J = J + K
- (B) J = 2
  IF K < 40 THEN J = J + 2
  ELSE J = J + K
- (C) J = 2
  DO WHILE K < 40
  J = J + 2
  ENDWHILE
  J = J + K
- (D) J = 2
  DO UNTIL K < 40
  ENDUNTIL
  J = J + K

2. An operating system is being developed for use in the control unit of an industrial machine. For efficiency, it is decided that any command to the machine will be represented by a combination of four characters, and that only eight distinct characters will be supported. What is the minimum number of bits required to represent one command?

- (A) 5
- (B) 7
- (C) 12
- (D) 16

3. For the program segment shown, the input value of X is 5. What is the output value of T?

```
INPUT X
N = 0
T = 0
    DO WHILE N < X
    T = T + X^N
    N = N + 1
    END DO
OUTPUT T
```

- (A) 156
- (B) 629
- (C) 781
- (D) 3906

4. A spreadsheet has been developed to calculate the cycle time for an assembly line required to meet a given production demand. Cells A1 through A5 contain the task times, cell B1 contains the number of stations, and cell B2 contains the cycle time. The equation for balance delay is

$$\frac{\text{(no. of stations)(cycle time)} - \text{sum of task times}}{\text{(no. of stations)(cycle time)}}$$

If cell C1 is to contain the balance delay for the solution, what formula should be put into it?

(A) (SUM(A1:A5) + B1*B2)/(B1 + B2)

(B) (B1 − (SUM(A1:A5)/B1*B2))/(SUM(B1:B2))

(C) (B1*B2 − SUM(A1:A5))/SUM(A1:A5)

(D) (B1*B2 − SUM(A1:A5))/(B1*B2)

5. The cells in a spreadsheet application are defined as follows.

	A	B	C	D
1	3	12	= SUM(C2:D4)	7
2	5	= AVERAGE(C2:D4)	−1	= C3*2
3	= D2^2	= C1 − B2	= A1 + B4	−10
4	= ABS(B3)	−4	8	= MIN(A1:A3)
5				

What is the value of cell A4?

(A) 0.83

(B) 1.0

(C) 2.5

(D) 8.3

6. What operation is typically represented by the following program flowchart symbol?

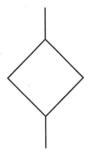

(A) input-output

(B) processing

(C) storage

(D) branching

7. The number 2 is entered into cell A1 in a spreadsheet. The formula A1 + A1 is entered into cell B1. The contents of cell B1 are then copied and pasted into cell C1. The number displayed in cell C1 is

(A) 2

(B) 4

(C) 6

(D) 8

8. Based on the following program segment, which IF, ANSWER statement is true?

I = 1

J = 2

K = 3

L = 4

ANSWER = C

IF (I > J) OR (K < L) THEN ANSWER = A

IF (I < J) OR (K > L) THEN ANSWER = B

IF (I > J) OR (K > L) THEN I = 5

IF (I < L) AND (K < L) THEN ANSWER = D

(A) A

(B) B

(C) C

(D) D

9. How many cells are in the range C5:Z30?

(A) 575

(B) 598

(C) 600

(D) 624

10. In a spreadsheet, the formula A3 + $B3 + D2 is entered into cell C2. The contents of cell C2 are copied and pasted into cell D5. The formula in cell D5 is

(A) A3 + C$2 + C4

(B) B6 + $C4 + C4

(C) A3 + $B6 + E5

(D) A3 + $B2 + B2

SOLUTIONS

1. The decision block indicates an IF statement because both outcomes progress toward the end. If the flowchart looped back for one of the outcomes, the chart might indicate a DO WHILE or DO UNTIL loop. The flowchart indicates that the operation $J = J + K$ will occur regardless of the outcome of the IF statement; therefore, there is no ELSE as shown in option B.

The answer is (A).

2. Since $2^3 = 8$, three bits are sufficient to represent 8 characters. Since there are 4 characters per command, a string with a minimum of $4 \times 3 = 12$ bits is needed.

The answer is (C).

3. The program segment defines the series

$$\sum_{0}^{N=X-1} X^N$$

The value of T at the conclusion of the loop can be computed as

N	T	
0	1	
1	$1+5$	$= 6$
2	$1+5+25$	$= 31$
3	$1+5+25+125$	$= 156$
4	$1+5+25+125+625$	$= 781$

The answer is (C).

4. The spreadsheet formula is

$$(B1*B2 - SUM(A1:A5))/(B1*B2)$$

The answer is (D).

5. Calculating the value of cell A4 requires the calculation of all of those cells that affect its result, directly or indirectly.

$$C3 = A1 + B4 = 3 + (-4) = -1$$

$$D2 = C3*2 = -1*2 = -2$$

$$A3 = D2\char`^2 = (-2)\char`^2 = 4$$

$$D4 = MIN(A1:A3) = MIN(A1,A2,A3) = MIN(3,5,4)$$
$$= 3$$

$$C1 = SUM(C2:D4) = C2 + C3 + C4 + D2 + D3 + D4$$
$$= -1 + (-1) + 8 + (-2) + (-10) + 3$$
$$= -3$$

$$B2 = AVERAGE(C2:D4)$$
$$= (C2 + C3 + C4 + D2 + D3 + D4)/6 = -3/6$$
$$= -0.5$$

$$B3 = C1 - B2 = -3 - (-0.5) = -2.5$$

$$A4 = ABS(B3) = |-2.5| = 2.5$$

The answer is (C).

6. Branching, comparison, and decision operations are typically represented by the diamond symbol.

The answer is (D).

7. When the formula $\$A\$1 + A1$ is copied into cell C1, it becomes $\$A\$1 + B1$. The number displayed in cell B1 is $2 + 2 = 4$. The number displayed in cell C1 is $2 + 4 = 6$.

The answer is (C).

8. The code must be followed all the way from beginning to end. After the first IF, ANSWER is A; after the second, it is B. The third IF evaluates to "false," so the ANSWER remains B, but the fourth statement is true, so the ANSWER is D.

The answer is (D).

9. A range from m to n, inclusive, has $m - n + 1$ elements, not $m - n$.

$$\text{no. of rows} = (30 - 5) + 1 = 26$$
$$\text{no. of columns} = (Z - C) + 1 = (26 - 3) + 1$$
$$= 24$$
$$\text{no. of cells} = (\text{no. of rows})(\text{no. of columns})$$
$$= (26)(24)$$
$$= 624$$

The answer is (D).

10. The first absolute cell reference is unchanged by the paste operation and remains $\$A\$3$.

The second cell reference will have the row reference increased by three and become $\$B6$.

The third cell reference will have the column reference increased by one and the row reference increased by three and will become E5.

The answer is (C).

55 Computer Software

1. CHARACTER CODING

Alphanumeric data refers to characters that can be displayed or printed, including numerals and symbols ($, %, &, etc.) but excluding *control characters* (tab, carriage return, form feed, etc.). Since computers can handle binary numbers only, all symbolic data must be represented by binary codes. *Coding* refers to the manner in which alphanumeric data and control characters are represented by sequences of bits.

The *American Standard Code for Information Interchange*, ASCII, is a seven-bit code permitting 128 (2^7) different combinations. It is commonly used in desktop computers, although use of the high order (eighth) bit is not standardized. ASCII-coded magnetic tape and disk files are used to transfer data and documents between computers of all sizes that would otherwise be unable to share data structures.

The *Extended Binary Coded Decimal Interchange Code*, EBCDIC (pronounced eb'-sih-dik), is in widespread use in IBM mainframe computers. It uses eight bits (one byte) for each character, allowing a maximum of 256 (2^8) different characters.

Since strings of binary digits (bits) are difficult to read, the *hexadecimal* (or "packed") format is used

to simplify working with EBCDIC data. Each byte is converted into two strings of four bits each. The two strings are then converted to hexadecimal. Since $(1111)_2 = (15)_{10} = (F)_{16}$, the largest possible EBCDIC character is coded FF in hexadecimal.

Example

The American Standard Code for Information Interchange (ASCII) control character "LF" stands for

(A) line feed

(B) large format

(C) limit function

(D) large font

Solution

Lines of text displayed on monitors and printers are often followed by invisible carriage return (CR) and line feed (LF) control characters that determine the location of the next character.

The answer is (A).

2. PROGRAM DESIGN

A *program* is a sequence of computer instructions that performs some function. The program is designed to implement an *algorithm*, which is a procedure consisting of a finite set of well-defined steps. Each step in the algorithm usually is implemented by one or more instructions (e.g., READ, GOTO, OPEN, etc.) entered by the programmer. These original "human-readable" instructions are known as *source code statements*.

Except in rare cases, a computer will not understand source code statements. Therefore, the source code is translated into machine-readable object code and absolute memory locations. Eventually, an executable program is produced.

Programs use variables to store values, which may be known or unknown. A *declaration* defines a variable, specifies the type of data a variable can contain (e.g., INTEGER, REAL, etc.), and reserves space for the variable in the program's memory. *Assignments* give values to variables (e.g., $X = 2$). *Commands* instruct the program to take a specific action, such as END, PRINT, or INPUT. *Functions* are specific operations (e.g., calculating the SUM of several values) that are grouped into a unit that can be called within the program.

If the executable program is kept on disk or tape, it is normally referred to as *software*. If the program is placed in ROM (read-only memory) or EPROM (erasable programmable read-only memory), it is referred to as *firmware*. The computer mechanism itself is known as the *hardware*.

3. FLOWCHARTS

A *flowchart* is a step-by-step drawing representing a specific procedure or algorithm. Figure 55.1 illustrates the most common flowcharting symbols. The *terminal symbol* begins and ends a flowchart. The *input/output symbol* defines an I/O operation, including those to and from keyboard, printer, memory, and permanent data storage. The *processing symbol* and *predefined process symbol* refer to calculations or data manipulation. The *decision symbol* indicates a point where a decision must be made or two items are compared. The *connector symbol* indicates that the flowchart continues elsewhere. The *off-page symbol* indicates that the flowchart continues on the following page. Comments can be added in an *annotation symbol*.

Figure 55.1 *Flowcharting Symbols*

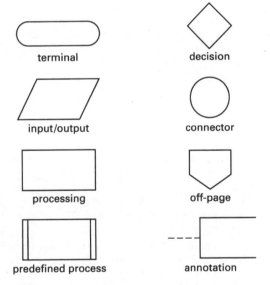

terminal

decision

input/output

connector

processing

off-page

predefined process

annotation

Example

The flowchart shown represents the summer training schedule for a college athlete. What is the regularly scheduled workout on Wednesday morning?

   (A) bike

   (B) rest

   (C) run

   (D) weights

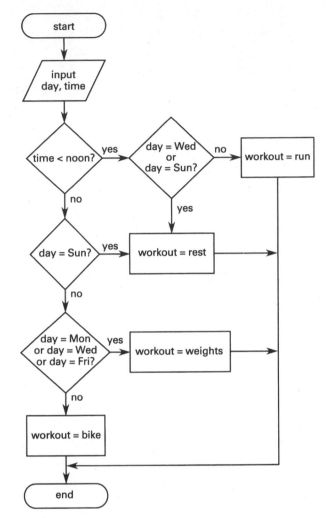

Solution

The day and time being considered are Wednesday morning. At the first decision follow the "yes" branch, which leads to the question "Is the day Wednesday or Sunday?" Again the answer is "yes," so the scheduled workout is actually a resting day.

The answer is (B).

4. LOW-LEVEL LANGUAGES

Programs are written in specific languages, of which there are two general types: low-level and high-level. Low-level languages include machine language and assembly language.

Machine language instructions are intrinsically compatible with and understood by the computer's central processing unit (CPU). They are the CPU's native language. An instruction normally consists of two parts: the operation to be performed (*op-code*) and the operand expressed as a storage location. Each instruction ultimately must be expressed as a series of bits, a form known as *intrinsic machine code*. However, octal and hexadecimal coding are more convenient. In either case, coding a machine language program is tedious and seldom done by hand.

Assembly language is more sophisticated (i.e., is more symbolic) than machine language. Mnemonic codes are used to specify the operations. The operands are referred to by variable names rather than by the addresses. Blocks of code that are to be repeated verbatim at multiple locations in the program are known as *macros* (*macro instructions*). Macros are written only once and are referred to by a symbolic name in the source code.

Assembly language code is translated into machine language by an *assembler* (*macro-assembler* if macros are supported). After assembly, portions of other programs or function libraries may be combined by a *linker*. In order to run, the program must be placed in the computer's memory by a *loader*. Assembly language programs are preferred for highly efficient programs. However, the coding inconvenience outweighs this advantage for most applications.

5. HIGH-LEVEL LANGUAGES

High-level languages are easier to use than low-level languages because the instructions resemble English. High-level statements are translated into machine language by either an interpreter or a compiler. Table 55.1 shows a comparison of a typical ADD command in different computer languages. A *compiler* performs the checking and conversion functions on all instructions only when the compiler is invoked. A true stand-alone executable program is created. An *interpreter*, however, checks the instructions and converts them line by line into machine code during execution but produces no stand-alone program capable of being used without the interpreter. (Some interpreters check syntax as each statement is entered by the programmer. Some languages and implementations of other languages blur the distinction between interpreters and compilers. Terms such as *pseudo-compiler* and *incremental compiler* are used in these cases.)

Table 55.1 Comparison of Typical ADD Commands

language	instruction
intrinsic machine code	1111 0001
machine language	1A
assembly language	AR
high-level language	+

Example

A portion of computer code contains the instructions "LR," "SC," and "JL." Most likely, these are

(A) machine language

(B) assembly language

(C) object code

(D) interpreted code

Solution

Assembly language usually consists of short commands (e.g., "LR" for loading a value into the register, "SC" for storing a result in location C, and "JL" for jumping if the result is less than). Assembly language is translated into binary machine language (also known as "object code") by an interpreter or compiler program.

The answer is (B).

6. RELATIVE COMPUTATIONAL SPEED

Certain languages are more efficient (i.e., execute faster) than others. (Efficiency can also, but seldom does, refer to the size of the program.) While it is impossible to be specific, and exceptions abound, assembly language programs are fastest, followed in order of decreasing speed by compiled, pseudo-compiled, and interpreted programs.

Similarly, certain program structures are more efficient than others. For example, when performing a repetitive operation, the most efficient structure will be a single equation, followed in order of decreasing speed by a stand-alone loop and a loop within a subroutine. Incrementing the loop variables and managing the exit and entry points is known as *overhead* and takes time during execution.

7. STRUCTURE, DATA TYPING, AND PORTABILITY

A language is said to be *structured* if subroutines and other procedures each have one specific entry point and one specific return point. (Contrast this with BASIC, which permits (1) a GOSUB to a specific subroutine with a return from anywhere within the subroutine and (2) unlimited GOTO statements to anywhere in the main program.) A language has *strong data types* if integer and real numbers cannot be combined in arithmetic statements.

A *portable language* can be implemented on different machines. Most portable languages are either sufficiently rigidly defined (as in the cases of ADA and C) to eliminate variants and extensions, or (as in the case of Pascal) are compiled into an intermediate, machine-independent form. This so-called *pseudocode* (*p-code*) is neither source nor object code. The language is said to have been "ported to a new machine" when an interpreter is written that converts p-code to the appropriate machine code and supplies specific drivers for input, output, printers, and disk use. (Some companies have produced Pascal engines that run p-code directly.)

8. STRUCTURED PROGRAMMING

Structured programming (also known as *top-down programming*, *procedure-oriented programming*, and *GOTO-less programming*) divides a procedure or algorithm into parts known as subprograms, subroutines, modules, blocks, or procedures. (The format and readability of the source code—improved by indenting nested structures, for example—do not define structured programming.) Internal subprograms are written by the programmer; external subprograms are supplied in a library by another source. Ideally, the mainline program will consist entirely of a series of calls (references) to these subprograms. Liberal use is made of FOR/NEXT, DO/WHILE, and DO/UNTIL commands. Labels and GOTO commands are avoided as much as possible.

Very efficient programs can be constructed in languages that support *recursive calls* (i.e., permit a subprogram to call itself). Recursion requires less code, but recursive calls use more memory. Some languages permit recursion; others do not.

Variables whose values are accessible strictly within the subprogram are *local variables*. *Global variables* can be referred to by the main program and all other subprograms.

Calculations are performed in a specific order in an instruction, with the contents of parentheses done first. The symbols used for mathematical operations in programming are

+	add
−	subtract
*	multiply
/	divide

Raising one expression to the power of another expression depends on the language used. Examples of how X^B might be expressed are

$$X**B$$

$$X\char`^B$$

Example

Two floating point numbers, A and B, are passed to a recursive subroutine. If number A is greater than or equal to 7, the subroutine returns the value of number A to the main program. If number A is less than the smaller of 7 and number B, number A is divided by 2. The subroutine then passes the quotient and number B to itself. Most likely, what condition might this subroutine encounter?

(A) overflow

(B) miscompare

(C) infinite loop

(D) type mismatch

Solution

Consider the case where A = 6 and B = 9. Since 6 is not greater than or equal to 7, control is not returned to the main program. Since 6 is less 7 (the smaller of 7 and 9), 6

is divided by 2, yielding a quotient of 3. The subroutine then passes A = 3 and B = 9 to itself. Since 3 (the new number A) is not greater than or equal to 7, control is not returned to the main program. Since 3 is less 7 (the smaller of 7 and 9), 3 is divided by 2, yielding a quotient of 1.5. The subroutine then passes A = 1.5 and B = 9 to itself. The process repeats indefinitely, since A will never be greater than 7. In some cases, the quotient might become small enough to terminate with an underflow error. (Nothing is said about the data type (variable class) of the number 7 or the comparison rules, so a type mismatch would require additional assumptions.)

The answer is (C).

Conditions and Statements

Following are brief descriptions of some commonly used structured programming functions.

IF THEN statements: In an IF <condition> THEN <action> statement, the condition must be satisfied, or the action is not executed and the program moves on to the next operation. Sometimes an IF THEN statement will include an ELSE statement in the format of IF <condition> THEN <action 1> ELSE <action 2>. If the condition is satisfied, then action 1 is executed. If the condition is not satisfied, action 2 is executed.

DO/WHILE loops: A set of instructions between the DO/WHILE <condition> and the ENDWHILE lines of code is repeated as long as the condition remains true. The number of times the instructions are executed depends on when the condition is no longer true. The variable or variables that control the condition must eventually be changed by the operations, or the WHILE loop will continue forever.

DO/UNTIL loops: A set of instructions between the DO/UNTIL <condition> and the ENDUNTIL lines of code is repeated as long as the condition remains false. The number of times the instructions are executed depends on when the condition is no longer false. The variable or variables that control the condition must eventually be changed by the operations, or the UNTIL loop will continue forever.

FOR loops: A set of instructions between the FOR <counter range> and the NEXT <counter> lines of code is repeated for a fixed number of loops that depends on the counter range. The counter is a variable that can be used in operations in the loop, but the value of the counter is not changed by anything in the loop besides the NEXT <counter> statement.

GOTO: A GOTO operation moves the program to a number designator elsewhere on the program. The GOTO statement has fallen from favor and is avoided whenever possible in structured programming.

Computational Tools

Example

A computer structured programming segment contains the following program segment.

$$\text{Set } G = 1 \text{ and } X = 0$$
$$\text{DO WHILE } G \le 5$$
$$G = G * X + 1$$
$$X = G$$
$$\text{ENDWHILE}$$

What is the value of G after the segment is executed?

- (A) 5
- (B) 26
- (C) 63
- (D) The loop never ends.

Solution

The first execution of the WHILE loop results in

$$G = (1)(0) + 1 = 1$$
$$X = 1$$

The second execution of the WHILE loop results in

$$G = (1)(1) + 1 = 2$$
$$X = 2$$

The third execution of the WHILE loop results in

$$G = (2)(2) + 1 = 5$$
$$X = 5$$

The WHILE condition is still satisfied, so the instruction is executed a fourth time.

$$G = (5)(5) + 1 = 26$$
$$X = 26$$

The answer is (B).

9. MODULAR AND OBJECT-ORIENTED PROGRAMMING

Modular programming builds efficient larger applications from smaller parts known as *modules*. Ideally, modules deal with unique functions and concerns that are not dealt with by other modules. For example, a module that looks up airline flight numbers could be shared by a reservation-booking module and seat-assigning module.

It would not be efficient for the booking and seating modules to each have their own flight number search routine.

Modular programming is intrinsic to *object-oriented programming* (OOP). Java, Python, C++, some variants of Pascal, Visual Basic .NET, and Ruby are *object-oriented programming languages* (OOPL). OOP defines objects with predefined characteristics and behaviors and also defines all the interactions with, functions of, and operations on the objects. Objects are essentially data structures (i.e., databases) and their associated functionalities. OOP exhibits characteristics such as class, inheritance, data hiding, data abstraction, polymorphism, encapsulation, interface, and package.

For example, consider a program called "Business Trip Reservation System" created in an OOP environment. "Aircraft" would be an *object class* that has *object variables* of aircraft name, passenger capacity, speed, and travel cost per distance. A particular instance (component or record) would have an object variable name value of "Learjet," an associated passenger capacity of 12, a speed of 300 knots/hr, and an operating cost of $100 per mile traveled. Other related object classes in the program might also be established for "Customer" and "Reservation." One of the advantages of OOP is modularity. That is, an existing module for one class can be used without reprogramming for another class. For example, "Short-Haul Aircraft" would be a *subclass* of the "Aircraft" class and would have the same characteristics and functionality as the "Aircraft" class by virtue of *inheritance*. In a *package* or *namespace*, the related classes of "Aircraft," "Customer," and "Reservation" would be integrated in such ways as to prevent duplication of object variable names and to produce good code organization.

As a result of *data abstraction*, an end user could specify the value of a key to be used to retrieve "Learjet" characteristics, but the end user would have no knowledge of the look-up method (e.g., hashing, binary tree, linear search). As a result of *encapsulation*, an end user would not see or have access to the actual data structure (i.e., internal representation) being accessed. Only the object itself is allowed to modify component variable values, and only by using the functionality built into the object. As a result of *data hiding*, if "Learjet" had other characteristics (e.g., fuel capacity) in the database, an object-oriented program working with a subclass of "Short-Haul Aircraft" would not be able to access, change, or corrupt those characteristics in the database, since that data would be invisible to it.

Objects also define and contain the *activities* (i.e., interactions with and operations to be performed on the data types). When designing objects it is necessary to anticipate all ways in which the objects will be used. For example, it might be possible to view characteristics, compare characteristics, reserve, pay for, and cancel a reservation for a Learjet. How the object's functionality is presented to the user is determined by the object's *interface*. As a result of *polymorphism*, a common form of functionality may be used with different objects (e.g., highlighting and clicking on a specific aircraft, a reservation, or a customer will all result in information being displayed, even though the information values and their formats differ).

10. HIERARCHY OF OPERATIONS

Operations in an arithmetic statement are performed in the order of exponentiation first, multiplication and division second, and addition and subtraction third. In the event that there are two consecutive operations with the same hierarchy (e.g., a multiplication followed by a division), the operations are performed in the order encountered, normally left to right (except for exponentiation, which is right to left).[1] Parentheses can modify this order; operations within parentheses are always evaluated before operations outside. If nested parentheses are present in an expression, the expression is evaluated outward starting from the innermost pair.

Example

Based on standard processing and hierarchy of operations, what would most nearly be the result of the following calculation?

$$459 + 306 / (6*5^2 + 3)$$

- (A) 3
- (B) 5
- (C) 459
- (D) 461

Solution

The standard mathematical processing hierarchy is represented by the acronym "BODMAS": Brackets (parentheses), Order (exponentiation or powers), Division and Multiplication (evaluated left to right), and Addition and Subtraction (evaluated left to right).

Evaluate the quantity in the parentheses first. Within the parentheses, evaluate the powered quantity first: $5^2 = 25$. There are no more powered quantities within the parentheses. Evaluate division and multiplication within the parentheses from left to right: $6*25 = 150$. Evaluate addition and subtraction within the parentheses: $150 + 3 = 153$. This is the value of the parenthetical expression. The original expression is now $459 + 306/153$. There are no brackets. There are no powers. Evaluate the division: $306/153 = 2$. Evaluate the addition: $459 + 2 = 461$.

The answer is (D).

---

[1]In most implementations, a statement will be scanned from left to right. Once a left-to-right scan is complete, some implementations then scan from right to left; others return to the equals sign and start a second left-to-right scan. Parentheses should be used to define the intended order of operations.

11. SIMULATORS

A *simulator* is a computer program that replicates a real world system. A *digital model* is used in the simulation and incorporates variables representing the physical characteristics and behavioral properties of the system to be simulated (i.e., is a representation of the system itself). The simulator operates a model over a period of time to gain a detailed understanding of the system's characteristics and dynamics. Simulation makes it possible to test and examine multiple design scenarios before selecting or finalizing a design. Simulators can explore various design alternatives that may not be safe, feasible, or economically possible in real life.

Examples of simulators include the response of buildings to seismic forces, stormwater flowing through a drainage structure during a large storm event, automobiles merging onto a freeway during rush hour, and so on. In addition to testing various designs, simulators are used for training, education, and even entertainment.

12. SPREADSHEETS

Spreadsheet application programs (often referred to as *spreadsheets*) are computer programs that provide a table of values arranged in rows and columns and that permit each value to have a predefined relationship to the other values. If one value is changed, the program will also change other related values.

In a spreadsheet, the items are arranged in rows and columns. The rows are typically assigned with numbers (1, 2, 3, ...) along the vertical axis, and the columns are assigned with letters (A, B, C, ...), as is shown in Fig. 55.2.

Figure 55.2 *Typical Spreadsheet Cell Assignments*

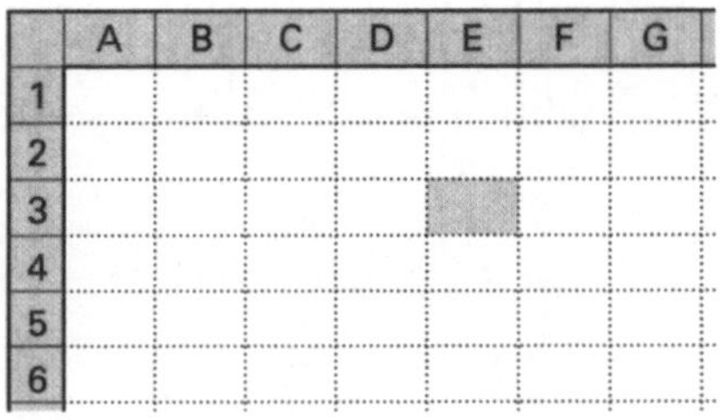

A *cell* is a particular element of the table identified by an address that is dependent on the row and column assignments of the cell. For example, the address of the shaded cell in Fig. 55.2 is E3. A cell may contain a number, a formula relating its value to another cell or cells, or a label (usually descriptive text).

When the contents of one cell are used for a calculation in another cell, the address of the cell being used must be referenced so the program knows what number to use.

When the value of a cell containing a formula is calculated, any cell addresses within the formula are calculated as the values of their respective cells. Cell addressing can be handled one of two ways: relative addressing or absolute addressing. When a cell is copied using *relative addressing*, the row and cell references will be changed automatically. *Absolute addressing* means that when a cell is copied, the row and cell references remain unchanged. Relative addressing is the default. Absolute addressing is indicated by a "$" symbol placed in front of the row or column reference (e.g., $C1 or C$1).

An *absolute cell reference* identifies a particular cell and will have a "$" before both the row and column designators. For example, A1 identifies the cell in the first column and first row, A3 identifies the cell in the first column and third row, and C1 identifies the cell in the third column and first row, regardless of the cell the reference is located in. If the absolute cell reference is copied and pasted into another cell, it continues referring to the exact same cell.

An *absolute column, relative row cell reference* has an absolute column reference (indicated with a "$") and a relative row reference. For example, the cell reference $A1 depends on what row it is entered in; if it is copied into a cell in the final row, it really refers to a cell that is in the first column in the final row. If this reference is copied and pasted into a cell in the third row, the reference will become $A3. Similarly, a reference of $A3 in the second row refers to a cell in the first column one row below the current row. If this reference is copied and pasted into the third row, it becomes $A4.

A *relative column, absolute row cell reference* has a "$" on the row designator, and the column reference depends on the column it is entered in. For example, a cell reference of B$4 in the fourth column (column D) refers to a cell two columns to the left in the fourth row. If this reference is copied and pasted into the sixth column (column F), it becomes D$4.

A cell reference that does not include a "$" is entirely dependent on the cell in which is it located. For example, a cell reference to B4 in the cell C2 refers to a cell that is one column to the left and two rows below. If this reference is copied and pasted into cell D3, it becomes C5.

The syntax for calculations with rows, columns, or blocks of cells can differ from one brand of spreadsheet to another.

Cells can be called out in square or rectangular blocks, usually for a SUM function. The difference between the row and column designations in the call will define the block. For example, SUM(A1:A3) says to sum the cells A1, A2, and A3; SUM(D3:D5) says to sum the cells D3, D4, and D5; and SUM(B2:C4) says to sum cells B2, B3, B4, C2, C3, and C4.

Example

The cells in a spreadsheet are initialized as shown. The formula B1 + A1*A2 is entered into cell B2 and then copied into cells B3 and B4.

	A	B		
1	3	111		
2	4			
3	5			
4	6			
5				

What value will be displayed in cell B4?

(A) 123

(B) 147

(C) 156

(D) 173

Solution

When the formula is copied into cells B3 and B4, the relative references will be updated. The resulting spreadsheet (with formulas displayed) should look like

	A	B
1	3	111
2	4	B1 + A1*A2
3	5	B2 + A1*A3
4	6	B3 + A1*A4
5		

$$B4 = B3 + (A1)(A4)$$
$$= B2 + (A1)(A3) + (A1)(A4)$$
$$= B1 + (A1)(A2) + (A1)(A3) + (A1)(A4)$$
$$= 111 + (3)(4) + (3)(5) + (3)(6)$$
$$= 156$$

The answer is (C).

13. SPREADSHEETS IN ENGINEERING

Spreadsheets are powerful, widely used computational tools in various engineering disciplines. Their popularity is often attributed to widespread availability, ease of use, and robust functionality. A spreadsheet can be used to collect data, identify and analyze trends within a data set, perform statistical calculations, quickly execute a series of interconnected calculations, and determine the effect of one variable change on other related variables, among other uses.

Applications include calculating open channel flow, scheduling construction activities, determining distribution of moments on a beam, calculating design parameters for a batch reactor, calculating the efficiency of a pump or motor, graphing the relationship between voltages and currents in circuit analysis, and determining costs over time in cost estimation.

Some benefits of using spreadsheets for engineering calculations and applications include the ability to reproduce calculations faithfully, to save and later review input and results, to compare results based on different input values, and to plot results graphically. Drawbacks include the difficulty of programming complex calculations (though many spreadsheets are commercially available), the lack of transparency for how results are calculated, and the resulting challenge of debugging inaccurate results. Spreadsheet calculations can be used in iterative design calculations, but results may need to be validated by some other means.

14. FIELDS, RECORDS, AND FILE TYPES

A collection of *fields* is known as a *record*. For example, name, age, and address might be fields in a personnel record. Groups of records are stored in a *file*.

A *sequential file* structure (typical of data on magnetic tape) contains consecutive records and must be read starting at the beginning. An *indexed sequential file* is one for which a separate index file (see Sec. 55.15) is maintained to help locate records.

With a *random (direct access) file structure*, any record can be accessed without starting at the beginning of the file.

15. FILE INDEXING

It is usually inefficient to place the records of an entire file in order. (A good example is a mailing list with thousands of names. It is more efficient to keep the names in the order of entry than to sort the list each time names are added or deleted.) Indexing is a means of specifying the order of the records without actually changing the order of those records.

An *index* (*key* or *keyword*) file is analogous to the index at the end of this book. It is an ordered list of items with references to the complete record. One field in the data record is selected as the *key field* (*record index*). More than one field can be indexed. However, each field will require its own index file. The sorted keys are usually kept in a file separate from the data file. One of the standard search techniques is used to find a specific key.

16. SORTING

Sorting routines place data in ascending or descending numerical or alphabetical order.

With the method of *successive minima*, a list is searched sequentially until the smallest element is found and brought to the top of the list. That element is then skipped, and the remaining elements are searched for the smallest element, which, when found, is placed after the previous minimum, and so on. A total of $n(n-1)/2$ comparisons will be required. When n is large, $n^2/2$ is sometimes given as the number of comparisons.

In a *bubble sort*, each element in the list is compared with the element immediately following it. If the first element is larger, the positions of the two elements are reversed (swapped). In effect, the smaller element "bubbles" to the top of the list. The comparisons continue to be made until the bottom of the list is reached. If no swaps are made in a pass, the list is sorted. A total of approximately $n^2/2$ comparisons are needed, on the average, to sort a list in this manner. This is the same as for the successive minima approach. However, swapping occurs more frequently in the bubble sort, slowing it down.

In an *insertion sort*, the elements are ordered by rewriting them in the proper sequence. After the proper position of an element is found, all elements below that position are bumped down one place in the sequence. The resulting vacancy is filled by the inserted element. At worst, approximately $n^2/2$ comparisons will be required. On average, there will be approximately $n^2/4$ comparisons.

Disregarding the number of swaps, the number of comparisons required by the successive minima, bubble, and insertion sorts is on the order of n^2. When n is large, these methods are too slow. The *quicksort* is more complex but reduces the average number of comparisons (with random data) to approximately $n \times \log n / \log 2$, generally considered as being on the order of $n \log n$. (However, the quicksort falters, in speed, when the elements are in near-perfect order.) The maximum number of comparisons for a *heap sort* is $n \times \log_2 n = n \times \log n / \log 2$, but it is likely that even fewer comparisons will be needed.

17. SEARCHING

If a group of records (i.e., a list) is randomly organized, a particular element in the list can be found only by a *linear search* (*sequential search*). At best, only one comparison and, at worst, n comparisons will be required to find something (an event known as a *hit*) in a list of n elements. The average is $n/2$ comparisons, described as being on the order of n. (The term *probing* is synonomous with *searching*.)

If the records are in ascending or descending order, a binary search will be superior. (A binary search is unrelated to a binary tree. A binary tree structure (see Sec. 55.19) greatly reduces search time but does not use a sorted list.) The search begins by looking at the middle element in the list. If the middle element is the sought-for element, the search is over. If not, half the list can be disregarded in further searching since elements in that portion will be either too large or too small. The middle element in the remaining part of the list is investigated, and the procedure continues until a hit occurs

Computational Tools

or the list is exhausted. The maximum number of required comparisons in a list of n elements will be $\log_2 n = \log n / \log 2$ (i.e., on the order of $\log n$).

Example

A binary search is used to determine if a particular test number is present in a sequence of 101 integers derived from a Fibonacci sequence. The numbers are sorted in decreasing magnitude. The smallest number in the sequence is 144. What is the maximum number of comparisons needed to determine if the test number is present?

    (A) 6

    (B) 7

    (C) 11

    (D) 49

Solution

A binary search looks at half of an active list at a time. Comparing continues until the active list contains only a single value. The index of the middle integer in a sorted list of 101 integers is at index $(1 + 101)/2 = 51$. The first comparison is to integer 51.

The active list now contains 50 integers. The second comparison is to the middle integer at index $(1 + 50)/2 = 25$. The active list now contains a maximum of 25 items. The third comparison is to the integer at index $(26 + 50)/2 = 38$. The active list now contains 12 items. The fourth comparison is to the integer at index $(26 + 37)/2 = 31$. The active list now contains a maximum of 6 items. The fifth comparison is to the integer at index $(32 + 37)/2 = 35$. The active list now contains a maximum of 3 items. The sixth comparison is to the integer at index $(32 + 34)/2 = 33$. The active list now contains 1 item, which is used in the seventh and final comparison.

Since $\log_2 128 = 7$ (i.e., $2^7 = 128$), up to 128 integers can be searched with 7 comparisons.

The answer is (B).

18. HASHING

An index file is not needed if the record number (i.e., the storage location for a read or write operation) can be calculated directly from the key, a technique known as *hashing*. The procedure by which a numeric or nonnumeric key (e.g., a last name) is converted into a record number is called the *hashing function* or *hashing algorithm*. Most hashing algorithms use a remaindering modulus—the remainder after dividing the key by the number of records, n, in the list. Excellent results are obtained if n is a prime number; poor results occur if n is a power of 2. (Finding a record in this manner requires it to have been written in a location determined by the same hashing routine.)

Not all hashed record numbers will be valid. A *collision* occurs when an attempt is made to use a record number that is already in use. Chaining, linear probing, and double hashing are techniques used to resolve such collisions.

19. DATABASE STRUCTURES

Databases can be implemented as indexed files, linked lists, and tree structures; in all three cases, the records are written and remain in the order of entry.

An *indexed file* such as that shown in Fig. 55.3 keeps the data in one file and maintains separate index files (usually in sorted order) for each key field. The index file must be recreated each time records are added to the field. A *flat file* has only one key field by which records can be located. Searching techniques (see Sec. 55.17) are used to locate a particular record. In a *linked list* (*threaded list*), each record has an associated *pointer* (usually a record number or memory address) to the next record in key sequence. Only two pointers are changed when a record is added or deleted. Generally, a linear search following the links through the records is used. Figure 55.4(a) shows an example of a linked list structure.

Figure 55.3 *Key and Data Files*

key file

key	record
ADAMS	3
JONES	2
SMITH	1
THOMAS	4

data file

record	last name	first name	age
1	SMITH	JOHN	27
2	JONES	WANDA	39
3	ADAMS	HENRY	58
4	THOMAS	SUSAN	18

Pointers are also used in *tree structures*. Each record has one or more pointers to other records that meet certain criteria. In a binary tree structure, each record has two pointers—usually to records that are lower and higher, respectively, in key sequence. In general, records in a tree structure are referred to as *nodes*. The first record in a file is called the *root node*. A particular node will have one node above it (the *parent* or *ancestor*) and one or more nodes below it (the *daughters* or *offspring*). Records are found in a tree

Computational Tools

Figure 55.4 Database Structures

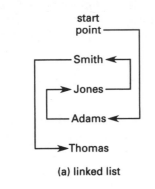

(a) linked list

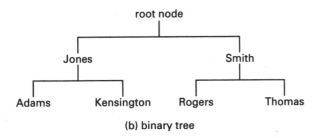

(b) binary tree

by starting at the root node and moving sequentially according to the tree structure. The number of comparisons required to find a particular element is $1 + (\log n/\log 2)$, which is on the order of $\log n$. Figure 55.4(b) shows an example of a binary tree structure.

20. HIERARCHICAL AND RELATIONAL DATA STRUCTURES

A *hierarchical database* contains records in an organized, structured format. Records are organized according to one or more of the indexing schemes. However, each field within a record is not normally accessible. Figure 55.5 shows an example of a hierarchical structure.

Figure 55.5 A Hierarchical Personnel File

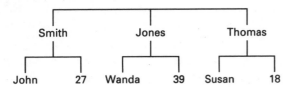

A *relational database* stores all information in the equivalent of a matrix. Nothing else (no index files, pointers, etc.) is needed to find, read, edit, or select information. Any information can be accessed directly by referring to the field name or field value. Figure 55.6 shows an example of relational structure.

Figure 55.6 A Relational Personnel File

rec. no.	last	first	age	
1	Smith	John	27	
2	Jones	Wanda	39	
3	Thomas	Susan	18	

21. ARTIFICIAL INTELLIGENCE

Artificial intelligence (AI) in a machine implies that the machine is capable of absorbing and organizing new data, learning new concepts, reasoning logically, and responding to inquiries. AI is implemented in a category of programs known as *expert systems* that "learn" rules from sets of events that are entered whenever they occur. (The manner in which the entry is made depends on the particular system.) Once the rules are learned, an expert system can participate in a dialogue to give advice, make predictions and diagnoses, or draw conclusions.

Diagnostic Exam

Topic XVI: Ethics and Professional Practice

1. Seventeen years ago, Susan designed a corrugated steel culvert for a rural road. Her work was accepted and paid for by the county engineering department. Last winter, the culvert collapsed as a loaded logging truck passed over. Although there were no injuries, there was damage to the truck and roadway, and the county tried unsuccessfully to collect on Susan's company's bond. The judge denied the claim on the basis that the work was done too long ago. This defense is known as

(A) privity of contract

(B) duplicity of liability

(C) statute of limitations

(D) caveat emptor

2. Ethics requires you to take into consideration the effects of your behavior on which group(s) of people?

I.   your employer

II.   the nonprofessionals in society

III.   other professionals

(A) II only

(B) I and II

(C) II and III

(D) I, II, and III

3. Which of the following terms is NOT related to ethics?

(A) integrity

(B) honesty

(C) morality

(D) profitability

4. What actions can be taken by a state regulating agency against a design professional who violates one or more of its rules of conduct?

I.   the professional's license may be revoked or suspended

II.   notice of the violation may be published in the local newspaper

III.   the professional may be asked to make restitution

IV.   the professional may be required to complete a course in ethics

(A) I and II

(B) I and III

(C) I and IV

(D) I, II, III, and IV

5. Which organizations typically do NOT enforce codes of ethics for engineers?

(A) technical societies (e.g., ASCE, ASME, IEEE)

(B) national professional societies (e.g., the National Society of Professional Engineers)

(C) state professional societies (e.g., the Michigan Society of Professional Engineers)

(D) companies that write, administer, and grade licensing exams

6. An engineer working for a big design firm has decided to start a consulting business, but it will be a few months before she leaves. How should she handle the impending change?

(A) The engineer should discuss her plans with her current employer.

(B) The engineer may approach the firm's other employees while still working for the firm.

(C) The engineer should immediately quit.

(D) The engineer should return all of the pens, pencils, pads of paper, and other equipment she has brought home over the years.

7. During the day, an engineer works for a scientific research laboratory doing government research. During the night, the engineer uses some of the lab's equipment to perform testing services for other consulting engineers. Why is this action probably unethical?

(A) The laboratory has not given its permission for the equipment use.

(B) The government contract prohibits misuse and misappropriation of the equipment.

(C) The equipment may wear out or be broken by the engineer and the replacement cost will be borne by the government contract.

(D) The engineer's fees to the consulting engineers can undercut local testing services' fees because the engineer has a lower overhead.

8. An engineer spends all of his free time (outside of work) gambling illegally. Is this a violation of ethical standards?

(A) No, the engineer is entitled to a life outside of work.

(B) No, the engineer's employer, his clients, and the public are not affected.

(C) No, not as long as the engineer stays debt-free from the gambling activities.

(D) Yes, the engineer should associate only with reputable persons and organizations.

9. During routine inspections, a field engineer discovers that one of the company's pipelines is leaking hazardous chemicals into the environment. The engineer recommends that the line be shut down so that seals can be replaced and the pipe can be inspected more closely. His supervisor commends him on his thoroughness, and says the report will be passed on to the company's maintenance division. The engineer moves on to his next job, assuming things will be taken care of in a timely manner. While working in the area again several months later, the engineer notices that the problem hasn't been corrected and is in fact getting worse. What should the engineer do?

(A) Give the matter some more time. In a large corporate environment, it is understandable that some things take longer than people would like them to.

(B) Ask the supervisor to investigate what action has been taken on the matter.

(C) Personally speak to the director of maintenance and insist that this project be given high priority.

(D) Report the company to the EPA for allowing the situation to worsen without taking any preventative measures.

10. An engineering firm receives much of its revenue from community construction projects. Which of the following activities would it be ethical for the firm to participate in?

(A) Contribute to the campaigns of local politicians.

(B) Donate money to the city council to help finance the building of a new city park.

(C) Encourage employees to volunteer in community organizations.

(D) Rent billboards to increase the company's name recognition.

Ethics/ Prof. Practice

SOLUTIONS

1. Most states have statutes of limitations. Unless a crime or fraudulent act has been committed, defects appearing after a certain amount of time are not actionable.

The answer is (C).

2. Ethical behavior places restrictions on behavior that affect you, your employer, other engineers, your clients, and society as a whole.

The answer is (D).

3. Ethical actions may or may not be profitable.

The answer is (D).

4. All four punishments are commonly used by state engineering licensing boards.

The answer is (D).

5. Companies that write, administer, and grade licensing exams typically do not enforce codes of ethics for engineers.

The answer is (D).

6. There is nothing wrong with wanting to go into business for oneself. The ethical violation occurs when one of the parties does not know what is going on. Even if the engineer acts ethically, takes nothing, and talks to no one about her plans, there will still be the appearance of impropriety if she leaves later. The engineer should discuss her plans with her current employer. That way, there will be minimal disruption to the firm's activities. The engineer shouldn't quit unless her employer demands it. (Those pencils, pens, and pads of paper probably shouldn't have been brought home in the first place.)

The answer is (A).

7. Options (A), (B), (C), and (D) may all be valid. However, the rationale for specific ethical prohibitions on using your employer's equipment for a second job is economic. When you don't have to pay for the equipment, you don't have to recover its purchase price in your fees for services.

The answer is (D).

8. The gambling is illegal. Engineers should do nothing that brings them and their profession into disrepute. It is impossible to separate people from their professions. When engineers participate in disreputable activities, it casts the entire profession in a bad light.

The answer is (D).

9. While it is true that corporate bureaucracy tends to slow things down, several months is too lengthy a period for an environmental issue. On the other hand, it is by no means clear that the company is ignoring the situation. There could have been some action taken that the engineer is unaware of, or extenuating circumstances that are delaying the repair. To go outside the company or even over the head of his supervisor would be premature without more information. The engineer should ask his previous supervisor to look into the issue, and should only take further measures if he is dissatisfied with the response.

The answer is (B).

10. Contributing to local politics, either to individual campaigns or in the form of a gift to the city, would be seen as an attempt to gain political favor. The renting of billboards, while not as well-defined an issue, implies the sort of self-laudatory advertising that ethical professionals prefer to avoid. Encouraging the company's employees to volunteer their own time to the community is acceptable because the company is unlikely to get any specific benefit from it.

The answer is (C).

Ethics/
Prof. Practice

56 Professional Practice

1. AGREEMENTS AND CONTRACTS

General Contracts

A *contract* is a legally binding agreement or promise to exchange goods or services.[1] A written contract is merely a documentation of the agreement. Some agreements must be in writing, but most agreements for engineering services can be verbal, particularly if the parties to the agreement know each other well.[2] Written contract documents do not need to contain intimidating legal language, but all agreements must satisfy three basic requirements to be enforceable (binding).

- There must be a clear, specific, and definite *offer* with no room for ambiguity or misunderstanding.

- There must be some form of conditional future *consideration* (i.e., payment).[3]

- There must be an *acceptance* of the offer.

There are other conditions that the agreement must meet to be enforceable. These conditions are not normally part of the explicit agreement but represent the conditions under which the agreement was made.

- The agreement must be *voluntary* for all parties.

- All parties must have *legal capacity* (i.e., be mentally competent, of legal age, not under coercion, and uninfluenced by drugs).

- The purpose of the agreement must be *legal*.

For small projects, a simple *letter of agreement* on one party's stationery may suffice. For larger, complex projects, a more formal document may be required. Some clients prefer to use a *purchase order*, which can function as a contract if all basic requirements are met.

Regardless of the format of the written document—letter of agreement, purchase order, or standard form—a contract should include the following features.[4]

- introduction, preamble, or preface indicating the purpose of the contract

- name, address, and business forms of both contracting parties

- signature date of the agreement

- effective date of the agreement (if different from the signature date)

- duties and obligations of both parties

- deadlines and required service dates

- fee amount

- fee schedule and payment terms

- agreement expiration date

- standard boilerplate clauses

- signatures of parties or their agents

- declaration of authority of the signatories to bind the contracting parties

- supporting documents

Example

Which feature is NOT a standard feature of a written construction contract?

(A) identification of both parties

(B) specific details of the obligations of both parties

(C) boilerplate clauses

(D) subcontracts

Solution

A written contract should identify both parties, state the purpose of the contract and the obligations of the parties, give specific details of the obligations (including relevant dates and deadlines), specify the consideration,

---

[1]Not all agreements are legally binding (i.e., enforceable). Two parties may agree on something, but unless the agreement meets all of the requirements and conditions of a contract, the parties cannot hold each other to the agreement.

[2]All states have a *statute of frauds* that, among other things, specifies what types of contracts must be in writing to be enforceable. These include contracts for the sale of land, contracts requiring more than one year for performance, contracts for the sale of goods over $500 in value, contracts to satisfy the debts of another, and marriage contracts. Contracts to provide engineering services do not fall under the statute of frauds.

[3]Actions taken or payments made prior to the agreement are irrelevant. Also, it does not matter to the courts whether the exchange is based on equal value or not.

[4]*Construction contracts* are unique unto themselves. Items that might also be included as part of the *contract documents* are the agreement form, the general conditions, drawings, specifications, and addenda.

state the boilerplate clauses to clarify the contract terms, and leave places for signatures. Subcontracts are not required to be included, but may be added when a party to the contract engages a third party to perform the work in the original contract.

The answer is (D).

Agency

In some contracts, decision-making authority and right of action are transferred from one party (the owner, or *principal*) who would normally have that authority to another person (the *agent*). For example, in construction contracts, the engineer may be the agent of the owner for certain transactions. Agents are limited in what they can do by the scope of the agency agreement. Within that scope, however, an agent acts on behalf of the principal, and the principal is liable for the acts of the agent and is bound by contracts made in the principal's name by the agent.

Agents are required to execute their work with care, skill, and diligence. Specifically, agents have *fiduciary responsibility* toward their principal, meaning that the agent must be honest and loyal. Agents are liable for damages resulting from a lack of diligence, loyalty, and/or honesty. If the agents misrepresented their skills when obtaining the agency, they can be liable for breach of contract or fraud.

Standard Boilerplate Clauses

It is common for full-length contract documents to include important *boilerplate clauses*. These clauses have specific wordings that should not normally be changed, hence the name "boilerplate." Some of the most common boilerplate clauses are paraphrased here.

- Delays and inadequate performance due to war, strikes, and acts of God and nature are forgiven (*force majeure*).

- The contract document is the complete agreement, superseding all prior verbal and written agreements.

- The contract can be modified or canceled only in writing.

- Parts of the contract that are determined to be void or unenforceable will not affect the enforceability of the remainder of the contract (*severability*). Alternatively, parts of the contract that are determined to be void or unenforceable will be rewritten to accomplish their intended purpose without affecting the remainder of the contract.

- None (or one, or both) of the parties can (or cannot) assign its (or their) rights and responsibilities under the contract (*assignment*).

- All notices provided for in the agreement must be in writing and sent to the address in the agreement.

- Time is of the essence.[5]

- The subject headings of the agreement paragraphs are for convenience only and do not control the meaning of the paragraphs.

- The laws of the state in which the contract is signed must be used to interpret and govern the contract.

- Disagreements shall be arbitrated according to the rules of the American Arbitration Association.

- Any lawsuits related to the contract must be filed in the county and state in which the contract is signed.

- Obligations under the agreement are unique, and in the event of a breach, the defaulting party waives the defense that the loss can be adequately compensated by monetary damages (*specific performance*).

- In the event of a lawsuit, the prevailing party is entitled to an award of reasonable attorneys' and court fees.[6]

- Consequential damages are not recoverable in a lawsuit.

Example

An engineering consultant has signed a standard owner-engineer contract to build a scale model of a bridge in time for the owner to present the proposal to a financing committee. After the scale model has been built, it is destroyed in a building fire that consumes the consultant's building. The consultant is unable to rebuild the model in time. A breach of contract judgment against the consultant is most likely NOT obtainable due to which legal argument?

(A) caveat emptor

(B) privity of contract

(C) force majeure

(D) strict liability in tort

Solution

Standard contracts have force majeure clauses that excuse nonperformance due to "acts of God" and other unforeseen events such as weather and acts of terrorism.

The answer is (C).

Subcontracts

When a party to a contract engages a third party to perform the work in the original contract, the contract with the third party is known as a *subcontract*. Whether or not responsibilities can be subcontracted under the original contract depends on the content of the *assignment clause* in the original contract.

---
[5]Without this clause in writing, damages for delay cannot be claimed.
[6]Without this clause in writing, attorneys' fees and court costs are rarely recoverable.

Parties to a Construction Contract

A specific set of terms has developed for referring to parties in consulting and construction contracts. The *owner* of a construction project is the person, partnership, or corporation that actually owns the land, assumes the financial risk, and ends up with the completed project. The *developer* contracts with the architect and/or engineer for the design and with the contractors for the construction of the project. In some cases, the owner and developer are the same, in which case the term *owner-developer* can be used.

The *architect* designs the project according to established codes and guidelines but leaves most stress and capacity calculations to the *engineer*.[7] Depending on the construction contract, the engineer may work for the architect, or vice versa, or both may work for the developer.

Once there are approved plans, the developer hires *contractors* to do the construction. Usually, the entire construction project is awarded to a *general contractor*. Due to the nature of the construction industry, separate *subcontracts* are used for different tasks (electrical, plumbing, mechanical, framing, fire sprinkler installation, finishing, etc.). The general contractor who hires all of these different *subcontractors* is known as the *prime contractor* (or *prime*). (The subcontractors can also work directly for the owner-developer, although this is less common.) The prime contractor is responsible for all acts of the subcontractors and is liable for any damage suffered by the owner-developer due to those acts.

Construction is managed by an agent of the owner-developer known as the *construction manager*, who may be the engineer, the architect, or someone else.

Standard Contracts for Design Professionals

Several professional organizations have produced standard agreement forms and other standard documents for design professionals.[8] Among other standard forms,

---

[7]On simple small projects, such as wood-framed residential units, the design may be developed by a *building designer*. The legal capacities of building designers vary from state to state.

[8]There are two main sources of standardized construction and design agreements: the *Engineers Joint Contract Documents Committee* (EJCDC) and *American Institute of Architects* (AIA). Consensus documents, known as *ConsensusDOCS*, for every conceivable situation have been developed by EJCDC. EJCDC includes the American Society of Civil Engineers (ASCE), the American Council of Engineering Companies (ACEC), National Society of Professional Engineers' (NSPE's) Professional Engineers in Private Practice Division, Associated General Contractors of America (AGC), and more than fifteen other participating professional engineering design, construction, owner, legal, and risk management organizations, including the Associated Builders and Contractors; American Subcontractors Association; Construction Users Roundtable; National Roofing Contractors Association; Mechanical Contractors Association of America; and National Plumbing-Heating-Cooling Contractors Association. The AIA has developed its own standardized agreements in a less collaborative manner. Though popular with architects, AIA provisions are considered less favorable to engineers, contractors, and subcontractors who believe the AIA documents assign too much authority to architects, too much risk and liability to contractors, and too little flexibility in how construction disputes are addressed and resolved.

notices, and agreements, the following standard contracts are available.[9]

- standard contract between engineer and client

- standard contract between engineer and architect

- standard contract between engineer and contractor

- standard contract between owner and construction manager

Besides completeness, the major advantage of a standard contract is that the meanings of the clauses are well established, not only among the design professionals and their clients but also in the courts. The clauses in these contracts have already been litigated many times. Where a clause has been found to be unclear or ambiguous, it has been rewritten to accomplish its intended purpose.

Consulting Fee Structure

Compensation for consulting engineering services can incorporate one or more of the following concepts.

- *lump-sum fee:* This is a predetermined fee agreed upon by client and engineer. This payment can be used for small projects where the scope of work is clearly defined.

- *unit price:* Contract fees are based on estimated quantities and unit pricing. This payment method works best when required materials can be accurately identified and estimated before the contract is finalized. This payment method is often used in combination with a lump-sum fee.

- *cost plus fixed fee:* All costs (labor, material, travel, etc.) incurred by the engineer are paid by the client. The client also pays a predetermined fee as profit. This method has an advantage when the scope of services cannot be determined accurately in advance. Detailed records must be kept by the engineer in order to allocate costs among different clients.

- *per diem fee:* The engineer is paid a specific sum for each day spent on the job. Usually, certain direct expenses (e.g., travel and reproduction) are billed in addition to the per diem rate.

- *salary plus:* The client pays for the employees on an engineer's payroll (the salary) plus an additional percentage to cover indirect overhead and profit plus certain direct expenses.

---

[9]The Construction Specifications Institute (CSI) has produced standard specifications for materials. The standards have been organized according to a UNIFORMAT structure consistent with ASTM Standard E1557.

retainer: This is a minimum amount paid by the client, usually in total and in advance, for a normal amount of work expected during an agreed-upon period. Usually, none of the retainer is returned, regardless of how little work the engineer performs. The engineer can be paid for additional work beyond what is normal, however. Some direct costs, such as travel and reproduction expenses, may be billed directly to the client.

incentive: This type of fee structure is based on established target costs and fees and lists minimum and maximum fees and an adjustment formula. The formula may be based on performance criteria such as budget, quality, and schedule. Once the project is complete, payment is calculated based on the formula.

percentage of construction cost: This method, which is widely used in construction design contracts, pays the architect and/or the engineer a percentage of the final total cost of the project. Costs of land, financing, and legal fees are generally not included in the construction cost, and other costs (plan revisions, project management labor, value engineering, etc.) are billed separately.

Example

Which fee structure is a nonreturnable advance paid to a consultant?

(A) per diem fee

(B) retainer

(C) lump-sum fee

(D) cost plus fixed fee

Solution

A *retainer* is a (usually) nonreturnable advance paid by the client to the consultant. While the retainer may be intended to cover the consultant's initial expenses until the first big billing is sent out, there does not need to be any rational basis for the retainer. Often, a small retainer is used by the consultant to qualify the client (i.e., to make sure the client is not just shopping around and getting free initial consultations) and as a security deposit (to make sure the client does not change consultants after work begins).

The answer is (B).

Mechanic's Liens

For various reasons, providers of material, labor, and design services to construction sites may not be promptly paid or even paid at all. Such providers have, of course, the right to file a lawsuit demanding payment, but due to the nature of the construction industry, such relief may be insufficient or untimely. Therefore, such providers have the right to file a *mechanic's lien* (also known as a *construction lien, materialman's lien,*

supplier's lien, or *laborer's lien*) against the property. Although there are strict requirements for deadlines, filing, and notices, the procedure for obtaining (and removing) such a lien is simple. The lien establishes the supplier's security interest in the property. Although the details depend on the state, essentially the property owner is prevented from transferring title of (i.e., selling) the property until the lien has been removed by the supplier. The act of filing a lawsuit to obtain payment is known as "perfecting the lien." Liens are perfected by forcing a judicial foreclosure sale. The court orders the property sold, and the proceeds are used to pay off any lienholders.

Discharge of a Contract

A contract is normally discharged when all parties have satisfied their obligations. However, a contract can also be terminated for the following reasons:

- mutual agreement of all parties to the contract
- impossibility of performance (e.g., death of a party to the contract)
- illegality of the contract
- material breach by one or more parties to the contract
- fraud on the part of one or more parties
- failure (i.e., loss or destruction) of consideration (e.g., the burning of a building one party expected to own or occupy upon satisfaction of the obligations)

Some contracts may be dissolved by actions of the court (e.g., bankruptcy), passage of new laws and public acts, or a declaration of war.

Extreme difficulty (including economic hardship) in satisfying the contract does not discharge it, even if it becomes more costly or less profitable than originally anticipated.

2. PROFESSIONAL LIABILITY

Breach of Contract, Negligence, Misrepresentation, and Fraud

A *breach of contract* occurs when one of the parties fails to satisfy all of its obligations under a contract. The breach can be *willful* (as in a contractor walking off a construction job) or *unintentional* (as in providing less than adequate quality work or materials). A *material breach* is defined as nonperformance that results in the injured party receiving something substantially less than or different from what the contract intended.

Normally, the only redress that an *injured party* has through the courts in the event of a breach of contract is to force the breaching party to provide *specific performance*—that is, to satisfy all remaining contract provisions and to pay for any damage caused. Normally, *punitive damages* (to punish the breaching party) are unavailable.

Negligence is an action, willful or unwillful, taken without proper care or consideration for safety, resulting in damages to property or injury to persons. "Proper care" is a subjective term, but in general it is the diligence that would be exercised by a reasonably prudent person.[10] Damages sustained by a negligent act are recoverable in a tort action. (See "Torts.") If the plaintiff is partially at fault (as in the case of *comparative negligence*), the defendant will be liable only for the portion of the damage caused by the defendant.

Punitive damages are available, however, if the breaching party was fraudulent in obtaining the contract. In addition, the injured party has the right to void (nullify) the contract entirely. A *fraudulent act* is basically a special case of *misrepresentation* (i.e., an intentionally false statement known to be false at the time it is made). Misrepresentation that does not result in a contract is a tort. When a contract is involved, misrepresentation can be a breach of that contract (i.e., *fraud*).

Unfortunately, it is extremely difficult to prove *compensatory fraud* (i.e., fraud for which damages are available). Proving fraud requires showing *beyond a reasonable doubt* (a) a reckless or intentional misstatement of a material fact, (b) an intention to deceive, (c) it resulted in misleading the innocent party to contract, and (d) it was to the innocent party's detriment.

For example, if an engineer claims to have experience in designing steel buildings but actually has none, the court might consider the misrepresentation a fraudulent action. If, however, the engineer has some experience, but an insufficient amount to do an adequate job, the engineer probably will not be considered to have acted fraudulently.

Example

The owner of a construction site is aware that the state driving license of one of its heavy machinery operators has been suspended for multiple driving under the influence (DUI) violations. In order to secure a desirable standing and contract with an abstinence-based commune, the owner misrepresents the non-drinking status of the construction crew. A serious injury occurs when the operator drives a loader over the leg of a member of the commune while intoxicated. Most likely, the commune will be able to obtain a judgment against the operator based on

(A) negligence

(B) breach of contract

(C) misrepresentation

(D) fraud

Solution

All elements necessary to obtain a judgement against the owner based on fraud are present in this scenario. The operator is, most likely, guilty only of negligence.

The answer is (A).

Torts

A *tort* is a civil wrong committed by one person causing damage to another person or person's property, emotional well-being, or reputation.[11] It is a breach of the rights of an individual to be secure in person or property. In order to correct the wrong, a civil lawsuit (*tort action* or *civil complaint*) is brought by the alleged injured party (the *plaintiff*) against the *defendant*. To be a valid *tort action* (i.e., lawsuit), there must have been injury (i.e., damage). Generally, there will be no contract between the two parties, so the tort action cannot claim a breach of contract.[12]

Tort law is concerned with compensation for the injury, not punishment. Therefore, tort awards usually consist of general, compensatory, and special damages and rarely include punitive and exemplary damages. (See "Damages" for definitions of these damages.)

Strict Liability in Tort

Strict liability in tort means that the injured party wins if the injury can be proven. It is not necessary to prove negligence, breach of explicit or implicit warranty, or the existence of a contract (*privity of contract*). Strict liability in tort is most commonly encountered in product liability cases. A defect in a product, regardless of how the defect got there, is sufficient to create strict liability in tort.

Case law surrounding defective products has developed and refined the following requirements for winning a strict liability in tort case. The following points must be proved.

- The product was defective in manufacture, design, labeling, and so on.

- The product was defective when used.

- The defect rendered the product unreasonably dangerous.

- The defect caused the injury.

- The specific use of the product that caused the damage was reasonably foreseeable.

---

[10]Negligence of a design professional (e.g., an engineer or architect) is the absence of a *standard of care* (i.e., customary and normal care and attention) that would have been provided by other engineers. It is highly subjective.

[11]The difference between a *civil tort* (*lawsuit*) and a *criminal lawsuit* is the alleged injured party. A *crime* is a wrong against society. A criminal lawsuit is brought by the state against a defendant.

[12]It is possible for an injury to be both a breach of contract and a tort. Suppose an owner has an agreement with a contractor to construct a building, and the contract requires the contractor to comply with all state and federal safety regulations. If the owner is subsequently injured on a stairway because there was no guardrail, the injury could be recoverable both as a tort and as a breach of contract. If a third party unrelated to the contract was injured, however, that party could recover only through a tort action.

Manufacturing and Design Liability

Case law makes a distinction between *design professionals* (architects, structural engineers, building designers, etc.) and manufacturers of consumer products. Design professionals are generally consultants whose primary product is a design service sold to sophisticated clients. Consumer product manufacturers produce specific product lines sold through wholesalers and retailers to the unsophisticated public.

The law treats design professionals favorably. Such professionals are expected to meet a *standard of care* and skill that can be measured by comparison with the conduct of other professionals. However, professionals are not expected to be infallible. In the absence of a contract provision to the contrary, design professionals are not held to be guarantors of their work in the strict sense of legal liability. Damages incurred due to design errors are recoverable through tort actions, but proving a breach of contract requires showing negligence (i.e., not meeting the standard of care).

On the other hand, the law is much stricter with consumer product manufacturers, and perfection is (essentially) expected of them. They are held to the standard of strict liability in tort without regard to negligence. A manufacturer is held liable for all phases of the design and manufacturing of a product being marketed to the public.[13]

Prior to 1916, the court's position toward product defects was exemplified by the expression *caveat emptor* ("let the buyer beware").[14] Subsequent court rulings have clarified that "... a manufacturer is strictly liable in tort when an article [it] places on the market, knowing that it will be used without inspection, proves to have a defect that causes injury to a human being."[15]

Although all defectively designed products can be traced back to a design engineer or team, only the manufacturing company is usually held liable for injury caused by the product. This is more a matter of economics than justice. The company has liability insurance; the product design engineer (who is merely an employee of the company) probably does not. Unless the product design or manufacturing process is intentionally defective, or

unless the defect is known in advance and covered up, the product design engineer will rarely be punished by the courts.[16]

Example

An engineer designs and self-manufactures a revolutionary racing bicycle. The engineer uses finite element analysis (FEA) software to perfect the design, has the design checked by a reputable authority, and subjects the major components that he manufactures to nondestructive testing. After three years of heavier-than-anticipated usage, one of the engineer's bicycles disintegrates in a race, killing its rider. In his defense, the engineer may claim

(A) privity of contract

(B) standard of care

(C) statute of limitations

(D) contributory negligence

Solution

The scenario does not contain information about initial and ongoing testing, but the engineer has done everything described in a competent manner. While the engineer may still be held responsible in some manner, he has met a normal standard of care in the design and manufacturing of his bicycles.

The answer is (B).

Damages

An injured party can sue for *damages* as well as for specific performance. Damages are the award made by the court for losses incurred by the injured party.

- *General* or *compensatory damages* are awarded to make up for the injury that was sustained.

- *Special damages* are awarded for the direct financial loss due to the breach of contract.

- *Nominal damages* are awarded when responsibility has been established but the injury is so slight as to be inconsequential.

- *Liquidated damages* are amounts that are specified in the contract document itself for nonperformance.

- *Punitive* or *exemplary damages* are awarded, usually in tort and fraud cases, to punish and make an example of the defendant (i.e., to deter others from doing the same thing).

- *Consequential damages* provide compensation for indirect losses incurred by the injured party but not directly related to the contract.

---

[13]The reason for this is that the public is not considered to be as sophisticated as a client who contracts with a design professional for building plans.

[14]1916, *MacPherson v. Buick.* MacPherson bought a Buick from a car dealer. The car had a defective wheel, and there was evidence that reasonable inspection would have uncovered the defect. MacPherson was injured when the wheel broke and the car collapsed, and he sued Buick. Buick defended itself under the ancient *prerequisite of privity* (i.e., the requirement of a face-to-face contractual relationship in order for liability to exist), since the dealer, not Buick, had sold the car to MacPherson, and no contract between Buick and MacPherson existed. The judge disagreed, thus establishing the concept of *third-party liability* (i.e., manufacturers are responsible to consumers even though consumers do not buy directly from manufacturers).

[15]1963, *Greenman v. Yuba Power Products.* Greenman purchased and was injured by an electric power tool.

---

[16]The engineer can expect to be discharged from the company. However, for strategic reasons, this discharge probably will not occur until after the company loses the case.

Insurance

Most design firms and many independent design professionals carry *errors and omissions insurance* to protect them from claims due to their mistakes. Such policies are costly, and for that reason, some professionals choose to "go bare."[17] Policies protect against inadvertent mistakes only, not against willful, knowing, or conscious efforts to defraud or deceive.

3. PROTECTION OF INTELLECTUAL PROPERTY

Creations of the mind—inventions, literary and artistic works, symbols, names, images, and designs used in commerce—represent *intellectual property* (IP) that can (at least, initially) be protected for commercial use and financial gain. IP includes inventions and designs, manufacturing processes, chemical formulas, identifying names and marks, and creative works such as architectural and building designs, listings of computer code, artwork and illustrations, material specifications, screenplays, novels, music, and web pages. Some ownership rights may not require public disclosure, identification, and formal registration processes, but registration must usually occur in order to reserve all financial and punitive rights. The processes required, the duration of protection, and the protections granted depend on the country in which the rights are to be reserved. Protection in one country often affords the IP owner with protection in other developed countries.[18] A few countries do not respect any IP rights.[19] Some countries claim to respect IP rights but do not prosecute violators. Even in developed countries, many individuals routinely disregard IP rights through illegal copying and sharing.

In the United States, IP is protected by trademarks, patents, and copyrights. *Trademarks* protect selected names, words, phrases, symbols, and sounds. Trademarks indicate the unique owner and/or provider of products and services, as well as identifying their origin. A trademark gives a business a particular advantage and is often the source of valuable commercial *goodwill*. A *service mark* is used when a business sells a service rather than a product. The term "trademark" often refers to both trademarks and service marks. In the United States, the United States Patent and Trademark Office (USPTO)

handles registration. Unregistered trademarks are indicated by the symbol "TM." The designation "SM" is used for unregistered service marks. Formal application or registration is not required to use "TM" and "SM". However, the trademark registration symbol, "®," can only be used when a trademark has actually been registered with the USPTO. It cannot be used before the trademark has been registered, or even while the registration is pending.

In the United States, physical inventions, machines, chemical formulas and compositions, some genetic coding, and methods of processing and manufacturing, as well as changes or improvements to those methods, are protected by *patents*. According to the USPTO, there are three types of patents. A *utility patent* protects "... any new and useful process, machine, article of manufacture, or composition of matter, or any new and useful improvement thereof." A *design patent* protects "... a new, original, and ornamental design for an article of manufacture," and a *plant patent* protects an invention or discoveries related to asexual reproduction of "... any distinct and new variety of plant." Application to the U.S. Patent and Trademark Office is required to obtain patent protection. The phrase "Patent Pending" or similar is used to indicate that a patent application has been filed. It is a warning that a patent might be issued; it does not mean that patent protection is assured.

In the United States, *copyrights* protect the ownership of original works of creativity (literary, musical, and dramatic works; photographs; audio and visual recordings; software; and other intellectual works) that are fixed in a tangible medium of expression (e.g., printed on paper, painted on canvas, or recorded magnetically). According to the U.S. Copyright Office, protection of a work begins as soon as "it is created and fixed in a tangible form that is perceptible either directly or with the aid of a machine or device." The copyright symbol informs others of the owner's control over and claim to the production, distribution, display, and/or performance of the work. In the United States, the *U.S. Copyright Office*, a division of the *Library of Congress*, handles copyright registration. While it is not necessary to formally file for copyright protection, doing so will make it much easier to obtain legal enforcement of a copyright and to collect financial damages.

Example

An engineer has developed a process control algorithm that uniquely mixes chemicals through a systematic valve opening and closing sequence. The algorithm has been implemented in a sequence of microprocessor code contained in erasable programmable read-only memory (EPROM) chips distributed to customers. Most likely, how should the microprocessor code be protected against commercial exploitation by third parties?

(A) trade secret

(B) copyright

(C) utility patent

(D) design patent

---

[17]Going bare appears foolish at first glance, but there is a perverted logic behind the strategy. One-person consulting firms (and perhaps, firms that are not profitable) are "judgment-proof." Without insurance or other assets, these firms would be unable to pay any large judgments against them. When damage victims (and their lawyers) find this out in advance, they know that judgments will be uncollectable. So the lawsuit often never makes its way to trial.

[18]Although some aspects of the protection may vary between countries, signatories to the Berne Convention generally respect copyrights registered in all participating countries.

[19]Countries whose enforcement activities are conducive to the routine and blatant disregard of intellectual property ownership rights and that are on the 2014 U.S. Trade Representative's (USTR's) annual *Priority Watch List* ("*Black List*") are Algeria, Argentina, Chile, China, India, Indonesia, Pakistan, Russia, Thailand, and Venezuela. Numerous other countries with some, but inadequate, enforcement are also placed on the USTR *Watch List*.

Ethics/ Prof. Practice

Solution

Although the algorithm logic might qualify for patent protection, the sequence of microprocessor code would be protected by copyright. A trade secret can be protected only by the owner keeping the information secret (or, through a contract with a third party to keep the information secret).

The answer is (B).

57 Ethics

1. CODES OF ETHICS

Creeds, Rules, Statutes, Canons, and Codes

It is generally conceded that an individual acting on his or her own cannot be counted on to always act in a proper and moral manner. Creeds, rules, statutes, canons, and codes all attempt to complete the guidance needed for an engineer to do "...the correct thing."

A *creed* is a statement or oath, often religious in nature, taken or assented to by an individual in ceremonies. For example, the *Engineers' Creed* adopted by the National Society of Professional Engineers (NSPE) is[1]

> As a Professional Engineer, I dedicate my professional knowledge and skill to the advancement and betterment of human welfare.
>
> I pledge...
>
> . . . to give the utmost of performance;
> . . . to participate in none but honest enterprise;
> . . . to live and work according to the laws of man and the highest standards of professional conduct;
> . . . to place service before profit, the honor and standing of the profession before personal advantage, and the public welfare above all other considerations.
>
> In humility and with need for Divine Guidance, I make this pledge.

A *rule* is a guide (principle, standard, or norm) for conduct and action in a certain situation, or a regulation governing procedure. A *statutory rule*, or statute, is enacted by the legislative branch of state or federal government and carries the weight of law. Some U.S. engineering registration boards have statutory *rules of professional conduct*.

---

[1]The *Faith of an Engineer* adopted by the Accreditation Board for Engineering and Technology (ABET) is a similar but more detailed creed.

A *canon* is an individual principle or body of principles, rules, standards, or norms. A *code* is a system of principles or rules. For example, the code of ethics of the American Society of Civil Engineers (ASCE) contains the following seven canons.

1. Engineers shall hold paramount the safety, health, and welfare of the public and shall strive to comply with the principles of sustainable development in the performance of their professional duties.

2. Engineers shall perform services only in areas of their competence.

3. Engineers shall issue public statements only in an objective and truthful manner.

4. Engineers shall act in professional matters for each employer or client as faithful agents or trustees and shall avoid conflicts of interest.

5. Engineers shall build their professional reputation on the merit of their services and shall not compete unfairly with others.

6. Engineers shall act in such a manner as to uphold and enhance the honor, integrity, and dignity of the engineering profession and shall act with zero tolerance for bribery, fraud, and corruption.

7. Engineers shall continue their professional development throughout their careers and shall provide opportunities for the professional development of those engineers under their supervision.

Example

Relative to the practice of engineering, which one of the following best defines "ethics"?

(A) application of United States laws

(B) rules of conduct

(C) personal values

(D) recognition of cultural differences

Solution

Ethics are the rules of conduct recognized in respect to a particular class of human actions or governing a particular group, culture, and so on.

The answer is (B).

Purpose of a Code of Ethics

Many different sets of *codes of ethics* (*canons of ethics, rules of professional conduct*, etc.) have been produced by various engineering societies, registration boards, and other organizations.[2] The purpose of these ethical guidelines is to guide the conduct and decision making of engineers. Most codes are primarily educational. Nevertheless, from time to time they have been used by the societies and regulatory agencies as the basis for disciplinary actions.

Fundamental to ethical codes is the requirement that engineers render faithful, honest, professional service. In providing such service, engineers must represent the interests of their employers or clients and, at the same time, protect public health, safety, and welfare.

There is an important distinction between what is legal and what is ethical. Many legal actions can be violations of codes of ethical or professional behavior. For example, an engineer's contract with a client may give the engineer the right to assign the engineer's responsibilities, but doing so without informing the client would be unethical.

Ethical guidelines can be categorized on the basis of who is affected by the engineer's actions—the client, vendors and suppliers, other engineers, or the public at large. (Some authorities also include ethical guidelines for dealing with the employees of an engineer. However, these guidelines are no different for an engineering employer than they are for a supermarket, automobile assembly line, or airline employer. Ethics is not a unique issue when it comes to employees.)

Example

Complete the sentence: "Guidelines of ethical behavior among engineers are needed because

- (A) engineers are analytical and they don't always think in terms of right or wrong."
- (B) all people, including engineers, are inherently unethical."
- (C) rules of ethics are easily forgotten."
- (D) it is easy for engineers to take advantage of clients."

Solution

Untrained members of society are at the mercy of the professionals (e.g., doctors, lawyers, engineers) they

employ. Even a cab driver can take advantage of a new tourist who doesn't know the shortest route between two points. In many cases, the unsuspecting public needs protection from unscrupulous professionals, engineers included, who act in their own interest.

The answer is (D).

2. NCEES MODEL LAW

Introduction[3]

Engineering is considered to be a "profession" rather than an occupation because of several important characteristics shared with other recognized learned professions, law, medicine, and theology: special knowledge, special privileges, and special responsibilities. Professions are based on a large knowledge base requiring extensive training. Professional skills are important to the well-being of society. Professions are self-regulating, in that they control the training and evaluation processes that admit new persons to the field. Professionals have autonomy in the workplace; they are expected to utilize their independent judgment in carrying out their professional responsibilities. Finally, professions are regulated by ethical standards.

The expertise possessed by engineers is vitally important to public welfare. In order to serve the public effectively, engineers must maintain a high level of technical competence. However, a high level of technical expertise without adherence to ethical guidelines is as much a threat to public welfare as is professional incompetence. Therefore, engineers must also be guided by ethical principles.

The ethical principles governing the engineering profession are embodied in codes of ethics. Such codes have been adopted by state boards of registration, professional engineering societies, and even by some private industries. An example of one such code is the NCEES Rules of Professional Conduct, found in Section 240 of *Model Rules* and presented here. As part of his/her responsibility to the public, an engineer is responsible for knowing and abiding by the code. Additional rules of conduct are also included in *Model Rules*.

The three major sections of Model Rules address (1) Licensee's Obligation to Society, (2) Licensee's Obligation to Employers and Clients, and (3) Licensee's Obligation to Other Licensees. The principles amplified in these sections are important guides to appropriate behavior of professional engineers.

Application of the code in many situations is not controversial. However, there may be situations in which applying the code may raise more difficult issues. In particular, there may be circumstances in which terminology in the code is not clearly defined, or in which two sections of the

---

[2]All of the major engineering technical and professional societies in the United States (ASCE, IEEE, ASME, AIChE, NSPE, etc.) and throughout the world have adopted codes of ethics. Most U.S. societies have endorsed the *Code of Ethics of Engineers* developed by the Accreditation Board for Engineering and Technology (ABET), formerly the Engineers' Council for Professional Development (ECPD). The National Council of Examiners for Engineering and Surveying (NCEES) has developed its *Model Rules* as a guide for state registration boards in developing guidelines for the professional engineers in those states.

[3]Adapted from C. E. Harris, M. S. Pritchard, and M. J. Rabins, *Engineering Ethics: Concepts and Cases*, copyright © 1995 by Wadsworth Publishing Company, pg. 27–28.

code may be in conflict. For example, what constitutes "valuable consideration" or "adequate" knowledge may be interpreted differently by qualified professionals. These types of questions are called *conceptual issues*, in which definitions of terms may be in dispute. In other situations, *factual issues* may also affect ethical dilemmas. Many decisions regarding engineering design may be based upon interpretation of disputed or incomplete information. In addition, *trade-offs* revolving around competing issues of risk vs. benefit, or safety vs. economics may require judgments that are not fully addressed simply by application of the code.

No code can give immediate and mechanical answers to all ethical and professional problems that an engineer may face. Creative problem solving is often called for in ethics, just as it is in other areas of engineering.

Example

Which organizations typically do NOT enforce codes of ethics for engineers?

- (A) technical societies (e.g., ASCE, ASME, IEEE)
- (B) national professional societies (e.g., the National Society of Professional Engineers)
- (C) state professional societies (e.g., the Michigan Society of Professional Engineers)
- (D) companies that write, administer, and grade licensing exams

Solution

Companies that write, administer, and grade licensing exams typically do not enforce codes of ethics for engineers.

The answer is (D).

Licensee's Obligation to Society[4]

1. Licensees, in the performance of their services for clients, employers, and customers, shall be cognizant that their first and foremost responsibility is to the public welfare.

2. Licensees shall approve and seal only those design documents and surveys that conform to accepted engineering and surveying standards and safeguard the life, health, property, and welfare of the public.

3. Licensees shall notify their employer or client and such other authority as may be appropriate when their professional judgment is overruled under circumstances where the life, health, property, or welfare of the public is endangered.

4. Licensees shall be objective and truthful in professional reports, statements, or testimony. They shall include all relevant and pertinent information in such reports, statements, or testimony.

5. Licensees shall express a professional opinion publicly only when it is founded upon an adequate knowledge of the facts and a competent evaluation of the subject matter.

6. Licensees shall issue no statements, criticisms, or arguments on technical matters which are inspired or paid for by interested parties, unless they explicitly identify the interested parties on whose behalf they are speaking and reveal any interest they have in the matters.

7. Licensees shall not permit the use of their name or firm name by, nor associate in the business ventures with, any person or firm which is engaging in fraudulent or dishonest business of professional practices.

8. Licensees having knowledge of possible violations of any of these Rules of Professional Conduct shall provide the board with the information and assistance necessary to make the final determination of such violation.

Example

While working to revise the design of the suspension for a popular car, an engineer discovers a flaw in the design currently being produced. Based on a statistical analysis, the company determines that although this mistake is likely to cause a small increase in the number of fatalities seen each year, it would be prohibitively expensive to do a recall to replace the part. Accordingly, the company decides not to issue a recall notice. What should the engineer do?

- (A) The engineer should go along with the company's decision. The company has researched its options and chosen the most economic alternative.
- (B) The engineer should send an anonymous tip to the media, suggesting that they alert the public and begin an investigation of the company's business practices.
- (C) The engineer should notify the National Transportation Safety Board (NTSB), providing enough details for them to initiate a formal inquiry.
- (D) The engineer should resign from the company. Because of standard nondisclosure agreements, it would be unethical as well as illegal to disclose any information about this situation. In addition, the engineer should not associate with a company that is engaging in such behavior.

Solution

The engineer's highest obligation is to the public's safety. In most instances, it would be unethical to take some public action on a matter without providing the

---

[4]Adapted from *FE Reference Handbook*, 9th Ed., pg. 3–4, © 2013 by the National Council of Examiners for Engineering and Surveying® (ncees.org).

Ethics/ Prof. Practice

company with the opportunity to resolve the situation internally. In this case, however, it appears as though the company's senior officers have already reviewed the case and made a decision. The engineer must alert the proper authorities, the NTSB, and provide them with any assistance necessary to investigate the case. To contact the media, although it might accomplish the same goal, would fail to fulfill the engineer's obligation to notify the authorities.

The answer is (C).

Licensee's Obligation to Employer and Clients

1. Licensees shall undertake assignments only when qualified by education or experience in the specific technical fields of engineering or surveying involved.

2. Licensees shall not affix their signatures or seals to any plans or documents dealing with subject matter in which they lack competence, nor to any such plan or document not prepared under their direct control and personal supervision.

3. Licensees may accept assignments for coordination of an entire project, provided that each design segment is signed and sealed by the licensee responsible for preparation of that design segment.

4. Licensees shall not reveal facts, data, or information obtained in a professional capacity without the prior consent of the client or employer except as authorized or required by law. Licensees shall not solicit or accept gratuities, directly or indirectly, from contractors, their agents, or other parties in connection with work for employers or clients.

5. Licensees shall make full prior disclosures to their employers or clients of potential conflicts of interest or other circumstances which could influence or appear to influence their judgment or the quality of their service.

6. Licensees shall not accept compensation, financial or otherwise, from more than one party for services pertaining to the same project, unless the circumstances are fully disclosed and agreed to by all interested parties.

7. Licensees shall not solicit or accept a professional contract from a governmental body on which a principal or officer of their organization serves as a member. Conversely, licensees serving as members, advisors, or employees of a government body or department, who are the principals or employees of a private concern, shall not participate in decisions with respect to professional services offered or provided by said concern to the governmental body which they serve.

Example

Plan stamping is best defined as the

(A) legal action of signing off on a project you didn't design but are taking full responsibility for

(B) legal action of signing off on a project you didn't design or check but didn't accept money for

(C) illegal action of signing off on a project you didn't design but did check

(D) illegal action of signing off on a project you didn't design or check

Solution

It is legal to stamp (i.e., sign off on) plans that you personally designed and/or checked. It is illegal to stamp plans that you didn't personally design or check, regardless of whether you got paid. It is legal to work as a "plan checker" consultant.

The answer is (D).

Licensee's Obligation to Other Licensees

1. Licensees shall not falsify or permit misrepresentation of their, or their associates', academic or professional qualifications. They shall not misrepresent or exaggerate their degree of responsibility in prior assignments nor the complexity of said assignments. Presentations incident to the solicitation of employment or business shall not misrepresent pertinent facts concerning employers, employees, associates, joint ventures, or past accomplishments.

2. Licensees shall not offer, give, solicit, or receive, either directly or indirectly, any commission, or gift, or other valuable consideration in order to secure work, and shall not make any political contribution with the intent to influence the award of a contract by public authority.

3. Licensees shall not attempt to injure, maliciously or falsely, directly or indirectly, the professional reputation, prospects, practice, or employment of other licensees, nor indiscriminately criticize other licensees' work.

Example

Without your knowledge, an old classmate applies to the company you work for. Knowing that you recently graduated from the same school, the director of engineering shows you the application and resume your friend submitted and asks your opinion. It turns out that your friend has exaggerated his participation in campus organizations, even claiming to have been an officer in an engineering society that you are sure he was never in. On the other hand, you remember him as being a highly intelligent student and believe that he could really help the company. How should you handle the situation?

- (A) You should remove yourself from the ethical dilemma by claiming that you don't remember enough about the applicant to make an informed decision.

- (B) You should follow your instincts and recommend the applicant. Almost everyone stretches the truth a little in their resumes, and the thing you're really being asked to evaluate is his usefulness to the company. If you mention the resume padding, the company is liable to lose a good prospect.

- (C) You should recommend the applicant, but qualify your recommendation by pointing out that you think he may have exaggerated some details on his resume.

- (D) You should point out the inconsistencies in the applicant's resume and recommend against hiring him.

Solution

Engineers are ethically obligated to prevent the misrepresentation of their associates' qualifications. You must make your employer aware of the incorrect facts on the resume. On the other hand, if you really believe that the applicant would make a good employee, you should make that recommendation as well. Unless you are making the hiring decision, ethics requires only that you be truthful. If you believe the applicant has merit, you should state so. It is the company's decision to remove or not remove the applicant from consideration because of this transgression.

The answer is (C).

3. ETHICAL CONSIDERATIONS

Ethical Priorities

There are frequently conflicting demands on engineers. While it is impossible to use a single decision-making process to solve every ethical dilemma, it is clear that ethical considerations will force engineers to subjugate their own self-interests. Specifically, the ethics of engineers dealing with others need to be considered in the following order from highest to lowest priority.

- society and the public
- the law
- the engineering profession
- the engineer's client
- the engineer's firm
- other involved engineers
- the engineer personally

Example

To whom/what is a registered engineer's foremost responsibility?

- (A) client
- (B) employer
- (C) state and federal laws
- (D) public welfare

Solution

The purpose of engineering registration is to protect the public. This includes protection from harm due to conduct as well as competence. No individual or organization may legitimately direct a registered engineer to harm the public.

The answer is (D).

Dealing with Clients and Employers

The most common ethical guidelines affecting engineers' interactions with their employer (the *client*) can be summarized as follows.[5]

- Engineers should not accept assignments for which they do not have the skill, knowledge, or time to complete.

- Engineers must recognize their own limitations. They should use associates and other experts when the design requirements exceed their abilities.

- The client's interests must be protected. The extent of this protection exceeds normal business relationships and transcends the legal requirements of the engineer-client contract.

- Engineers must not be bound by what the client wants in instances where such desires would be unsuccessful, dishonest, unethical, unhealthy, or unsafe.

- Confidential client information remains the property of the client and must be kept confidential.

- Engineers must avoid conflicts of interest and should inform the client of any business connections or interests that might influence their judgment. Engineers should also avoid the *appearance* of a conflict of interest when such an appearance would be detrimental to the profession, their client, or themselves.

---

[5]These general guidelines contain references to contractors, plans, specifications, and contract documents. This language is common, though not unique, to the situation of an engineer supplying design services to an owner-developer or architect. However, most of the ethical guidelines are general enough to apply to engineers in the industry as well.

Ethics/ Prof. Practice

- The engineers' sole source of income for a particular project should be the fee paid by their client. Engineers should not accept compensation in any form from more than one party for the same services.

- If the client rejects the engineer's recommendations, the engineer should fully explain the consequences to the client.

- Engineers must freely and openly admit to the client any errors made.

All courts of law have required an engineer to perform in a manner consistent with normal professional standards. This is not the same as saying an engineer's work must be error-free. If an engineer completes a design, has the design and calculations checked by another competent engineer, and an error is subsequently shown to have been made, the engineer may be held responsible, but will probably not be considered negligent.

Example

You are an engineer in charge of receiving bids for an upcoming project. One of the contractors bidding the job is your former employer. The former employer laid you off in a move to cut costs. Which of the following should you do?

I.   say nothing

II. inform your present employer of the situation

III. remain objective when reviewing the bids

(A) II only

(B) I and II

(C) I and III

(D) II and III

Solution

Registrants should remain objective at all times and should notify their employers of conflicts of interest or situations that could influence the registrants' ability to make objective decisions.

The answer is (D).

Dealing with Suppliers

Engineers routinely deal with manufacturers, contractors, and vendors (*suppliers*). In this regard, engineers have great responsibility and influence. Such a relationship requires that engineers deal justly with both clients and suppliers.

An engineer will often have an interest in maintaining good relationships with suppliers since this often leads to future work. Nevertheless, relationships with suppliers must remain highly ethical. Suppliers should not be encouraged to feel that they have any special favors coming to them because of a long-standing relationship with the engineer.

The ethical responsibilities relating to suppliers are listed as follows.

- The engineer must not accept or solicit gifts or other valuable considerations from a supplier during, prior to, or after any job. An engineer should not accept discounts, allowances, commissions, or any other indirect compensation from suppliers, contractors, or other engineers in connection with any work or recommendations.

- The engineer must enforce the plans and specifications (i.e., the *contract documents*) but must also interpret the contract documents fairly.

- Plans and specifications developed by the engineer on behalf of the client must be complete, definite, and specific.

- Suppliers should not be required to spend time or furnish materials that are not called for in the plans and contract documents.

- The engineer should not unduly delay the performance of suppliers.

Example

In dealing with suppliers, an engineer may

(A) unduly delay vendor performance if the client agrees

(B) spend personal time outside of the contract to ensure adequate performance

(C) prepare plans containing ambiguous design-build references as cost-saving measures

(D) enforce plans and specifications to the letter, without regard to fairness

Solution

An engineer not only may, but is required to, ensure performance consistent with plans and specifications. If a job is intentionally or unintentionally underbid, the engineer will have to use personal time to complete the project.

The answer is (B).

Dealing with Other Engineers

Engineers should try to protect the engineering profession as a whole, to strengthen it, and to enhance its public stature. The following ethical guidelines apply.

- An engineer should not attempt to maliciously injure the professional reputation, business practice, or employment position of another engineer. However, if there is proof that another engineer has acted unethically or illegally, the engineer should advise the proper authority.

- An engineer should not review someone else's work while the other engineer is still employed unless the other engineer is made aware of the review.

- An engineer should not try to replace another engineer once the other engineer has received employment.

- An engineer should not use the advantages of a salaried position to compete unfairly (i.e., moonlight) with other engineers who have to charge more for the same consulting services.

- Subject to legal and proprietary restraints, an engineer should freely report, publish, and distribute information that would be useful to other engineers.

Dealing with (and Affecting) the Public

In regard to the social consequences of engineering, the relationship between an engineer and the public is essentially straightforward. Responsibilities to the public demand that the engineer place service to humankind above personal gain. Furthermore, proper ethical behavior requires that an engineer avoid association with projects that are contrary to public health and welfare or that are of questionable legal character.

- Engineers must consider the safety, health, and welfare of the public in all work performed.

- Engineers must uphold the honor and dignity of their profession by refraining from self-laudatory advertising, by explaining (when required) their work to the public, and by expressing opinions only in areas of their knowledge.

- When engineers issue a public statement, they must clearly indicate if the statement is being made on anyone's behalf (i.e., if anyone is benefitting from their position).

- Engineers must keep their skills at a state-of-the-art level.

- Engineers should develop public knowledge and appreciation of the engineering profession and its achievements.

- Engineers must notify the proper authorities when decisions adversely affecting public safety and welfare are made (a practice known as *whistle-blowing*).

Example

Whistle-blowing is best described as calling public attention to

(A) your own previous unethical behavior

(B) unethical behavior of employees under your control

(C) secret illegal behavior by your employer

(D) unethical or illegal behavior in a government agency you are monitoring as a private individual

Solution

"Whistle-blowing" is calling public attention to illegal actions taken in the past or being taken currently by your employer. Whistle-blowing jeopardizes your own good standing with your employer.

The answer is (C).

Competitive Bidding

The ethical guidelines for dealing with other engineers presented here and in more detailed codes of ethics no longer include a prohibition on *competitive bidding*. Until 1971, most codes of ethics for engineers considered competitive bidding detrimental to public welfare, since cost cutting normally results in a lower quality design. However, in a 1971 case against the National Society of Professional Engineers that went all the way to the U.S. Supreme Court, the prohibition against competitive bidding was determined to be a violation of the Sherman Antitrust Act (i.e., it was an unreasonable restraint of trade).

The opinion of the Supreme Court does not *require* competitive bidding—it merely forbids a prohibition against competitive bidding in NSPE's code of ethics. The following points must be considered.

- Engineers and design firms may individually continue to refuse to bid competitively on engineering services.

- Clients are not required to seek competitive bids for design services.

- Federal, state, and local statutes governing the procedures for procuring engineering design services, even those statutes that prohibit competitive bidding, are not affected.

- Any prohibitions against competitive bidding in individual state engineering registration laws remain unaffected.

- Engineers and their societies may actively and aggressively lobby for legislation that would prohibit competitive bidding for design services by public agencies.

Example

Complete the sentence: "The U.S. Department of Justice's successful action in the 1970s against engineering codes of ethics that formally prohibited competitive bidding was based on the premise that

(A) competitive bidding allowed minority firms to participate."

(B) competitive bidding was required by many government contracts."

(C) the prohibitions violated antitrust statutes."

(D) engineering societies did not have the authority to prohibit competitive bidding."

Solution

The U.S. Department of Justice's successful challenge was based on antitrust statutes. Prohibiting competitive bidding was judged to inhibit free competition among design firms.

The answer is (C).

58 Licensure

1. ABOUT LICENSING

Engineering licensing (also known as *engineering registration*) in the United States is an examination process by which a state's *board of engineering licensing* (typically referred to as the "engineers' board" or "board of registration") determines and certifies that an engineer has achieved a minimum level of competence.[1] This process is intended to protect the public by preventing unqualified individuals from offering engineering services.

Most engineers in the United States do not need to be licensed.[2] In particular, most engineers who work for companies that design and manufacture products are exempt from the licensing requirement. This is known as the *industrial exemption*, something that is built into the laws of most states.[3]

Nevertheless, there are many good reasons to become a licensed engineer. For example, you cannot offer consulting engineering services in any state unless you are licensed in that state. Even within a product-oriented corporation, you may find that employment, advancement, and managerial positions are limited to licensed engineers.

Once you have met the licensing requirements, you will be allowed to use the titles *Professional Engineer* (PE), *Structural Engineer* (SE), *Registered Engineer* (RE), and/or *Consulting Engineer* (CE) as permitted by your state.

Although the licensing process is similar in each of the 50 states, each has its own licensing law. Unless you offer consulting engineering services in more than one state, however, you will not need to be licensed in the other states.

2. THE U.S. LICENSING PROCEDURE

The licensing procedure is similar in all states. You will take two examinations. The full process requires you to complete two applications, one for each of the two examinations. The first examination is the *Fundamentals of Engineering* (FE) *examination*, formerly known (and still commonly referred to) as the *Engineer-In-Training* (EIT) *examination*.[4] This examination is designed for students who are close to finishing or have recently finished an undergraduate engineering degree. Seven versions of the exam are offered: chemical, civil, electrical and computer, environmental, industrial, mechanical, and other disciplines. Examinees are encouraged to take the module that best corresponds to their undergraduate degree. In addition to the discipline-specific topics, each exam covers subjects that are fundamental to the engineering profession, such as mathematics, probability and statistics, ethics, and professional practice.

The second examination is the *Professional Engineering* (PE) *examination*, also known as the *Principles and Practices* (P&P) *examination*. This examination tests your ability to practice competently in a particular engineering discipline. It is designed for engineers who have gained at least four years' post-college work experience in their chosen engineering discipline.

The actual details of licensing qualifications, experience requirements, minimum education levels, fees, and examination schedules vary from state to state. Contact your state's licensing board for more information. You will find contact information (websites, telephone numbers, email addresses, etc.) for all U.S. state and territorial boards of registration at **ppi2pass.com/ stateboards**.

3. NATIONAL COUNCIL OF EXAMINERS FOR ENGINEERING AND SURVEYING

The *National Council of Examiners for Engineering and Surveying* (NCEES) in Seneca, South Carolina, writes, publishes, distributes, and scores the national FE and

---

[1]Licensing of engineers is not unique to the United States. However, the practice of requiring a degreed engineer to take an examination is not common in other countries. Licensing in many countries requires a degree and may also require experience, references, and demonstrated knowledge of ethics and law, but no technical examination.

[2]Less than one-third of the degreed engineers in the United States are licensed.

[3]Only one or two states have abolished the industrial exemption. There has always been a lot of "talk" among engineers about abolishing it, but there has been little success in actually doing so. One of the reasons is that manufacturers' lobbies are very strong.

---

[4]The terms *engineering intern* (EI) and *intern engineer* (IE) have also been used in the past to designate the status of an engineer who has passed the first exam. These uses are rarer but may still be encountered in some states.

Ethics/ Prof. Practice

PE examinations.[5] The individual states administer the exams in a uniform, controlled environment as dictated by NCEES.

4. ROLE OF THE STATE LICENSING BOARDS

The *state engineering licensing boards* are responsible for ensuring that engineering work and related construction are safe and adequately protect the public. Licensing boards accomplish these goals by (1) ensuring the competence of engineering practitioners within the state, and (2) enforcing the rules of proper and ethical engineering practice.

State licensing boards ensure the competence of engineers by requiring that certain types of engineering work be performed by registered PEs (i.e., *licensed engineers* and *registered engineers*). There are rigorous qualifications for becoming a PE. The specifics vary with each state but generally include requiring each candidate to pass the Fundamentals of Engineering (FE) and Principles and Practice of Engineering (PE) exams in their discipline, have a degree from an ABET-accredited engineering program, and have a minimum of four years' work experience. In some states, in order to retain PE status, licensees must regularly update their knowledge and improve their skills.

When an engineer, licensed or not, compromises board rules, particularly with respect to public health and safety, a licensing board will investigate and may take disciplinary action. In addition to revocation of licenses, temporary suspensions, and monetary fines, disciplinary action often include *public censure*. The websites of licensing boards often contain current and historical reports of such *enforcement actions* for the public to read. These reports serve a dual purpose of keeping the engineering community informed about board actions and making it clear that engineers will pay a price for substandard and/or unethical engineering work.

Example

Seven years ago, two neighbors settled a long-standing dispute about the location of the property line between their farms. The settlement resulted in the neighbors "recording" their agreement by driving a 1.2 m length of rebar into the ground next to a stone cairn at each of the agreed-upon plat corners. No money was exchanged, and there was no written record or contract. Since then, one of the farms was sold to a new owner who promptly installed buried pipes to carry water and bulk liquid fertilizer along the previously-disputed property line. Upon inspection, the county engineer discovered that the pipes were, in fact, on the adjacent neighbor's property as specified in an 1892 recorded survey. What might subsequently transpire?

(A) Nothing will happen, because the statute of limitations has lapsed since 1892.

(B) Nothing will happen, because the two original neighbors effectively changed the legal property line when they both agreed to the locations of the marked corners.

(C) The new owner will have to move the buried pipes.

(D) The state licensing board will investigate and determine the true location of the property line.

Solution

The two original neighbors appear to have practiced land surveying without licenses. Recorded deeds and surveyors' maps do not expire, and statutes of limitations do not apply. The good relationship between the two original neighbors aside, future land owners were not protected because the agreement was not properly recorded. In fact, without consideration or a written contract, the agreement was most likely a violation of the *statute of frauds*, which requires the transfer of land to be according to a written agreement. If the state licensing board gets involved, it will be only to investigate the behavior of the two original neighbors. Although the neighbors' actions would generally result in an enforcement action, the state licensing board would not determine the true location of the property line.

The answer is (C).

5. UNIFORM EXAMINATIONS

Although each state has its own licensing law and is, theoretically, free to administer its own exams, none does so for the major disciplines. All states have chosen to use the NCEES exams. The exams from all the states are graded by NCEES. Each state adopts the cut-off passing scores recommended by NCEES. These practices have led to the term *uniform examination*.

6. RECIPROCITY AMONG STATES

With minor exceptions, having a license from one state will not permit you to practice engineering in another state. You must have a professional engineering license from each state in which you work. Most engineers do not work across state lines or in multiple states, but some do. Luckily, it is not too difficult to get a license from every state you work in once you have a license from one of them.

All states use the NCEES examinations. If you take and pass the FE or PE examination in one state, your certificate or license will be honored by all of the other states. Upon proper application, payment of fees, and proof of your license, you will be issued a license by the new state. Although there may be other special requirements

---

[5]National Council of Examiners for Engineering and Surveying, 280 Seneca Creek Road, Seneca, SC 29678, (800) 250-3196, ncees.org.

imposed by a state, it will not be necessary to retake the FE or PE examinations.[6] The issuance of an engineering license based on another state's licensing is known as *reciprocity* or *comity*.

---

[6]For example, California requires all civil engineering applicants to pass special examinations in seismic design and surveying in addition to their regular eight-hour PE exams. Licensed engineers from other states only have to pass these two special exams. They do not need to retake the PE exam.

Index

INDEX - B

Biomechanics of the human body, 52-22
Bioreactor, 43-3
Biosolid, 49-19 (ftn)
Biosystem, 43-3
Biot
    modulus, 25-6
    number, 25-6
Biotransformation, 51-3
Biphenyl, polychlorinated, 54-4
Black
    body, 27-2
    body, net heat transfer, 27-4
    lung disease, 52-12
Blade
    fixed, 13-15
    velocity, tangential, 13-16
Blend number, 49-12
Block
    and tackle, 30-1
    cascaded, 47-6
    diagram algebra, 47-6
    diagram, reduction, 47-6
    diagram, simplification, 47-6
Blood
    -borne pathogen, 52-14
    cell, red, 51-4
    cell, white, 51-4
Blower, 16-1, 16-7
    sparging, 16-7
Board of engineering licensing, 58-1
Bode plot, 47-11
    gain, 47-12 (fig)
    phase margin, 47-12 (fig)
Body
    black, 27-2
    gray, 27-2
    real, 27-2
    rigid, 33-1, 34-4, 36-3
    rigid, kinetic energy, 36-3
    rigid, Newton's second law for, 34-4
Boil-up, 45-7
Boiler, feedwater, 18-12
Boilerplate clause, 56-2
Boiling
    curve, 26-11
    departure from nucleate, 26-12
    film, 26-12
    filmwise, 26-12
    flow, 26-11
    forced convection, 26-11
    free convection, 26-11
    nucleate, 26-12
    nucleate, Rohsenow's equation, 26-12
    onset of nucleate, 26-12
    point, 17-4
    point constant, 21-16
    point elevation, 21-16
    pool, 26-11
    regime, 26-11
    saturated, 26-11
    sub-cooled, 26-11
    transition, 26-12
Bomb, calorimeter, 22-1
Bond
    covalent, 21-4
    ionic, 21-3
    nonpolar covalent, 21-4
    polar covalent, 21-4
Bonded strain gage, 48-8
Bone dry, 46-1
Book
    balancing, 50-13
    of original entry, 50-13
    value, 50-9
    value of stock, per share, 50-15
Bookkeeping, 50-13
    double-entry, 50-13
    system, 50-13
Boom, sonic, 15-2
Bore, 19-5
Borrelia burgdorferi, 52-15
Bottom operating line, 45-11

Bottoms, 45-6
Botulism, 52-14
    adult enteric, 52-15
    food, 52-14, 52-15
    infant, 52-15
    wound, 52-15
Bound water, 46-2
Boundary
    irregular, 9-2
    layer, 26-8
    system, 18-1
    work, 18-3
Bounded exponential curve, 24-5
Boyle's law, 15-2, 18-5
Brake
    fuel consumption rate, 19-5
    mean effective pressure, 19-5
    power, 16-2, 19-5
    Prony, 19-5
    properties, 19-5
    thermal efficiency, 19-6
    value, 19-5
Brale indenter, 41-14
Brass, 42-9 (ftn)
Brayton gas turbine cycle, 19-9
Breach
    material, 56-4
    of contract, 56-4
    unintentional, 56-4
    willful, 56-4
Break-even
    analysis, 50-12
    quantity, 50-12
Breakaway point, 47-12
Breaking strength, 41-7
Breakpoint, 54-10
Breakthrough time, 52-7
Bremsstrahlung, 52-16
Bridge
    deflection, 48-9
    null-indicating, 48-9
    resistance, 48-9
    salt, 21-17
    truss, 29-1
    Wheatstone, 48-9
    zero-indicating, 48-9
Brinell hardness
    number, 41-14
    test, 41-14, 42-11
British thermal unit, 17-5
Brittle material, 41-2, 41-8
    behavior, 41-8
    crack propagation, 41-8
Bromley equation, 26-13
Bronchitis, 51-3
Bronchogenic carcinoma, 52-11
Bronze, 42-9 (ftn)
    admiralty, 42-9
    aluminum, 42-9
    beryllium, 42-9
    commercial, 42-9 (ftn)
    government, 42-9
    manganese, 42-9 (ftn)
    phosphorus, 42-9
    silicon, 42-9
Btu, 17-5
Bubble, 54-1
    nucleation, 26-12
    sort, 55-8
Budgeting, capital, 50-11
Buick, MacPherson versus, 56-6 (ftn)
Building
    designer, 56-3 (ftn)
    green, 49-16, 49-18
Bulk
    temperature, 26-2
    velocity, 13-4
Buoyancy
    center of, 12-6
    theorem, 12-6
Buoyant force, 12-6

Burden, 50-17
    budget variance, 50-17
    capacity variance, 50-17
    variance, 50-17
Burst disc, 48-10

C

Cable, 30-1
    ideal, 30-1
$CaCO_3$
    concentration, 53-1
    equivalent, 53-1
Calcium hardness, 53-3
Calculus, integral, 7-8
    fundamental theorem of, 7-8
Calibration, 48-7
Call, recursive, 55-4
Calorimeter, 22-1
Cancer, 51-5
    initiation stage, 51-5
    nasal, 51-3
    progression stage, 51-5
    promotion stage, 51-5
    risk, excess lifetime, 51-8
Canon, 57-1
    of ethics, 57-2
Canonical form, 47-9
Capacitance, 38-3, 41-15
    total, 38-4
Capacitive reactance, 39-5
Capacitor, 38-3, 41-15
    electrolytic, 38-3 (ftn)
    energy storage, 38-4
    ideal, 39-5
    parallel, 38-4
    parallel plate, 38-3, 41-15
    polarized, 38-3 (ftn)
    series, 38-4
    synchronous, 39-8
Capacity
    aggregate molar heat, 22-2
    carrying, 24-5
    heat, 17-7
    heat, gases, 17-8 (tbl)
    heat, liquids, 17-7 (tbl)
    heat, solids, 17-7 (tbl)
    legal, 56-1
    reduction factor, 41-7
    specific heat, 41-14
    volumetric heat, 41-14
Capillary
    action, 11-6
    depression, 11-6
    rise, 11-6
Capital
    budgeting, 50-11
    recovery, 50-3 (tbl)
    recovery factor, 50-6
    recovery method, 50-11
Capitalized cost, 50-10
    for an infinite series, 50-10
Carbide, 42-9
Carbinol, 23-1 (tbl), 23-5
Carbohydrate, 23-3 (tbl)
Carbon
    atom, primary, 23-5
    atom, secondary, 23-5
    atom, tertiary, 23-5
    black, 42-15
    monoxide, 54-3, 54-7
    steel, 42-5
    steel, nonsulfurized, 42-5
    steel, rephosphorized, 42-5
    steel, resulfurized, 42-5
    temper, 42-7
Carbonate hardness, 53-3
Carbonic acid, 53-2
Carbonyl, 23-1 (tbl)
Carboxylic acid, 23-2 (tbl), 23-6

INDEX - C

INDEX - L

INDEX - M

INDEX - M

INDEX - S